Springer Series in Statistic

Asvisors:
P. Bickel, P. Diggle, S. Fienberg, K. Krickeberg,
I. Olkin, N. Wermuth, S. Zeger

Springer
New York
Berlin
Heidelberg
Barcelona
Budapest
Hong Kong
London
Milan
Paris
Santa Clara
Singapore
Tokyo

Springer Series in Statistics

Andersen/Borgan/Gill/Keiding: Statistical Models Based on Counting Processes.
Andrews/Herzberg: Data: A Collection of Problems from Many Fields for the Student and Research Worker.
Anscombe: Computing in Statistical Science through APL.
Berger: Statistical Decision Theory and Bayesian Analysis, 2nd edition.
Bolfarine/Zacks: Prediction Theory for Finite Populations.
Borg/Groenen: Modern Multidimensional Scaling: Theory and Applications
Brémaud: Point Processes and Queues: Martingale Dynamics.
Brockwell/Davis: Time Series: Theory and Methods, 2nd edition.
Daley/Vere-Jones: An Introduction to the Theory of Point Processes.
Dzhaparidze: Parameter Estimation and Hypothesis Testing in Spectral Analysis of Stationary Time Series.
Fahrmeir/Tutz: Multivariate Statistical Modelling Based on Generalized Linear Models.
Farrell: Multivariate Calculation.
Federer: Statistical Design and Analysis for Intercropping Experiments.
Fienberg/Hoaglin/Kruskal/Tanur (Eds.): A Statistical Model: Frederick Mosteller's Contributions to Statistics, Science and Public Policy.
Fisher/Sen: The Collected Works of Wassily Hoeffding.
Good: Permutation Tests: A Practical Guide to Resampling Methods for Testing Hypotheses.
Goodman/Kruskal: Measures of Association for Cross Classifications.
Gouriéroux: ARCH Models and Financial Applications.
Grandell: Aspects of Risk Theory.
Haberman: Advanced Statistics, Volume I: Description of Populations.
Hall: The Bootstrap and Edgeworth Expansion.
Härdle: Smoothing Techniques: With Implementation in S.
Hartigan: Bayes Theory.
Heyer: Theory of Statistical Experiments.
Huet/Bouvier/Gruet/Jolivet: Statistical Tools for Nonlinear Regression: A Practical Guide with S-PLUS Examples.
Jolliffe: Principal Component Analysis.
Kolen/Brennan: Test Equating: Methods and Practices.
Kotz/Johnson (Eds.): Breakthroughs in Statistics Volume I.
Kotz/Johnson (Eds.): Breakthroughs in Statistics Volume II.
Kres: Statistical Tables for Multivariate Analysis.
Le Cam: Asymptotic Methods in Statistical Decision Theory.
Le Cam/Yang: Asymptotics in Statistics: Some Basic Concepts.
Longford: Models for Uncertainty in Educational Testing.
Manoukian: Modern Concepts and Theorems of Mathematical Statistics.
Miller, Jr.: Simultaneous Statistical Inference, 2nd edition.
Mosteller/Wallace: Applied Bayesian and Classical Inference: The Case of *The Federalist Papers.*

(continued after index)

Per Kragh Andersen
Ørnulf Borgan
Richard D. Gill
Niels Keiding

Statistical Models Based on Counting Processes

With 128 Illustrations

Per Kragh Andersen/Niels Keiding
Statistical Research Unit
University of Copenhagen
Copenhagen
Denmark

Ørnulf Borgan
Institute of Mathematics
University of Oslo
Oslo
Norway

Richard D. Gill
Faculty of Mathematics
University of Utrecht
Utrecht
The Netherlands

Mathematics Subject Classification (1991): 60G07, 60G12, 60J25

Library of Congress Cataloging-in-Publication Data
Statistical models based on counting processes / Per Kragh Andersen
... [et al.].
p. cm.—(Springer series in statistics)
Includes bibliographical references and index.
ISBN 0-387-94519-9
1. Point processes. 2. Mathematical statistics. I. Andersen, Per Kragh. II. Series.
QA274.42.S75 1992
519.2′3—dc20 92-11354

Printed on acid-free paper.

This volume was originally published in hardcover. ISBN: 0-387-97872-0

Production managed by Dimitry L. Loseff; manufacturing supervised by Vincent Scelta.
Typeset by Asco Trade Typesetting Ltd., Hong Kong.
Printed and bound by R.R. Donnelley & Sons, Harrisonburg, Virginia.
Printed in the United States of America.

9 8 7 6 5 4 3 2 (Corrected second printing, 1997)

ISBN 0-387-94519-9 Springer-Verlag New York Berlin Heidelberg SPIN 10552481

Preface

One of the most remarkable examples of fast technology transfer from new developments in mathematical probability theory to applied statistical methodology is the use of counting processes, martingales in continuous time, and stochastic integration in *event history analysis*. By this (or generalized survival analysis), we understand the study of a collection of individuals, each moving among a finite (usually small) number of states. A basic example is moving from alive to dead, which forms the basis of *survival analysis*. Compared to other branches of statistics, this area is characterized by the dynamic temporal aspect, making modelling via the intensities useful, and by the special patterns of incompleteness of observation, of which *right-censoring* in survival analysis is the most important and best known example.

In this book our aim is to present a theory for handling such statistical problems, when observation is assumed to happen in continuous time. In most of the book, a minimum of assumptions are made (non- and semi-parametric theory), although one chapter considers parametric models. The detailed plan of the book is described in Chapter I.

Our desire has been to give an exposition in terms of the necessary mathematics (including previously unfamiliar tools that have proved useful), but at the same time illustrate most methods by concrete practical applications, primarily from our own biostatistical experience. We also wanted to indicate some recent lines of research. As a result, the book moves along at varying levels of mathematical sophistication and also at varying levels of subject-matter discussion. We realize that this will frustrate some readers, in particular those who had expected to find a smooth textbook. In Section I.1, we suggest a reading guide for various categories of readers, and in Section I.2, we give a synopsis of the developments of the field up to now.

The suggestion of writing a monograph on this material came from Klaus Krickeberg in 1982. In assembling a group of authors, we wanted to include our colleague and friend Odd O. Aalen, who started the theory and keeps contributing to it; however, Odd wanted to concentrate on building up medical statistics in Oslo, Norway. The work on the book has become something of a life-style for the four of us, involving drafting sections individually and then meeting, usually in Copenhagen, to discuss in great detail the merits and otherwise of the latest contributions. These meetings always lasted for three full days, with discussions going on at the dinner table to the wonder of our wives and children, none of whom believed we really wanted (were able to?) finish the project. We will miss these opportunities for very positive professional and human interaction, and the concept of "Vol. II" has become our consolation. (Do not worry, this will take a while!)

We have enjoyed extremely positive interest from all colleagues. First, our Nordic colleagues who participated in getting the theory developed have always been ready to follow up with good advice. Besides Odd Aalen, we have interacted particularly with Martin Jacobsen, as well as with Elja Arjas, Nils Lid Hjort, Søren Johansen, and Gert Nielsen. Overseas, we owe particular thanks to Dorota Dabrowska, who read several chapters in great detail and kept us informed of several fascinating new developments. We have been given good advice and been entrusted with prepublication versions of new results by many colleagues, including Chang-Jo F. Chung, Hani Doss, Priscilla Greenwood, Piet Groeneboom, Jean Jacod, John P. Klein, Tze Leung Lai, Ian McKeague, Jens Perch Nielsen, Michael Sørensen, Yangin Sun, Aad van der Vaart, Bert van Es, Mark van Pul, Wolfgang Wefelmeyer, and Jon A. Wellner.

As with any effort of this kind, countless drafts have been produced, some before the age of mathematical word processing. Although later phases have sent (most of) the authors themselves to the TeX keyboards (in Copenhagen with much support from Peter Dalgaard), many versions were done by our secretaries in Copenhagen (in particular, Susanne Kragskov and Kathe Jensen), Oslo (Dina Haraldsson and Inger Jansen), Amsterdam (Carolien Pol-Swagerman), and Utrecht (Karin Berlang and Diana van Doorn). Most calculations and graphs were done (in S) by Lars Sommer Hansen and Klaus Krøier, and Mark van Pul and Kristian Merckoll did some additional calculations.

Our work with the book has been done as part of our appointments at the University of Copenhagen (Per Kragh Andersen, Niels Keiding, the latter for part of the period as Danish Ministry of Education Research Professor), the University of Oslo (Ørnulf Borgan), and Centre for Mathematics and Computer Science, Amsterdam, and the University of Utrecht (Richard D. Gill). Some work was done during visits to the Department of Biostatistics, University of Washington; the Medical Research Council Biostatistics Unit, Cambridge, England; the Department of Mathematics, University of Western Australia; and the Department of Statistics, The Ohio State University.

We acknowledge the hospitality and interest everywhere. Travel costs and secretarial and computing assistance have been financed by our institutions, with supplements from the Danish Medical Research Council (grant no. 12-9229), the Norwegian Council for Research in Science and the Humanities, and Johan and Mimi Wessmann's foundation.

February 1992

København Per Kragh Andersen
Oslo Ørnulf Borgan
Utrecht Richard D. Gill
København Niels Keiding

Contents

CHAPTER I

Introduction

I.1. General Introduction to the Book

I.1.1. Event History Analysis

As explained in the Preface, we study a collection of individuals, each moving among a finite (usually small) number of states. The exact transition times in continuous time form the modelling basis of the phenomena, although often these times are only incompletely observed. The best known example of such incomplete observation is *right-censoring* in classical survival analysis; here, not all of a set of independent lifetimes are observed, so that for some of them it is only known that they are larger than some specific value. We shall mention other kinds of censoring such as interval-censoring and left-censoring as well as *truncation* and *filtering*. Left-truncation in survival analysis means that an individual is included only if its lifetime is larger than some value, whereas by filtering, we mean that the individual is not under observation all the time, but only when a suitable indicator process is switched on.

During the 1960s, an extensive theory of martingales in continuous time and stochastic integration was developed, primarily in France. In the early 1970s, these tools were applied in control theory and electrical engineering.

I.1.2. Counting Processes, Martingales, Stochastic Integrals, Product-Integration

For event history analysis, a major contribution was the 1975 Berkeley Ph.D. dissertation "Statistical Inference for a Family of Counting Processes" by

O.O. Aalen. The basic nonparametric statistical problems for censored data were studied in terms of the conditional *intensity* (or *intensity process*) of the *counting process* that records the (uncensored) events as time proceeds. Formulating many statistical models of interest in terms of a multiplicative decomposition of the intensity process of this counting process into a product of a purely deterministic factor (*hazard*) and an observable factor, Aalen's main contribution was to show that the models may then be analyzed in a unified manner by these tools. The difference between the counting process and the integrated intensity process is a *martingale*. Exact (that is, nonasymptotic) results such as unbiasedness, as well as estimators of variability, are obtained by applying results on *stochastic integration* with respect to this martingale. Asymptotic statistical theory follows using *martingale central limit theory*. Finally, to convert the results to the conventional distribution function (or *survival function*), the general tool of *product-integration* is useful (Aalen and Johansen, 1978). This tool, however, has been very slow in penetrating statistical theory; for a review see Gill and Johansen (1990).

I.1.3 General Aim of This Book

Our aim in this book is, first, to present the above theory, completing some of its mathematical details, as well as surveying a range of topics developed since the first journal publication of the theory (Aalen, 1978b; Aalen and Johansen, 1978). A brief survey of the historical development of the subject is given in Section I.2. Second, we include some very concrete applications, providing detailed empirical discussions of a number of practical examples, mainly from our own experience. We here also consider methods for assessing goodness of fit in detail, displaying results graphically, and adapting the methods of statistical inference to the occasionally complicated observational patterns that one meets in practice. Third, the full mathematical exposition of the theory requires some mathematical tools that we have found useful to summarize.

At the end of this section, we list some of the aims that we do *not* seek to fulfill in this work. Before that, let us first explain the plan of the book, and then suggest how different categories of readers could navigate through the somewhat heterogeneous material presented.

I.1.4. Plan of the Book

To emphasize the practical purpose of the methods developed here, brief presentations are given in Section I.3 of 19 practical examples that will serve as illustrations in the following chapters. The examples are chosen to illustrate various patterns of event history often met in practice, focusing in particular on the various relevant time scales (calendar time, age, duration since onset of disease, etc.).

Chapter II presents the mathematical tools necessary for a full appreciation of the development in later chapters. A textbook-type treatment would have exceeded the limits of the book considerably; instead, we attempt a brief exposition that nevertheless contains precise statements of the definitions and results used later. Section II.1 explains informally but also briefly the key concepts in the basic example of classical survival analysis.

In Chapter III, we provide a rather detailed discussion of the specification of the statistical models in terms of the intensity process. The bulk of this chapter is a mathematical discussion of possible patterns of censoring and other kinds of incomplete observation (truncation, filtering), with extensive discussion of specific examples, mostly motivated by the empirical studies presented in Section I.3.

Chapter IV is an exposition of *nonparametric estimation* in Aalen's multiplicative intensity model. This includes the "Nelson–Aalen" estimator of the cumulative intensity, kernel smoothing methods for the intensity itself, the "Kaplan–Meier" estimator of the survival function and its generalization: the estimator of the transition matrix of a nonhomogeneous Markov process proposed by Aalen and Johansen (1978). The methods are illustrated and further specified in relation to the empirical studies. Chapter V treats k-sample *hypothesis tests* of the relevance of suitable stratifications, including one-sample tests for comparison with a standard. Linear rank tests are emphasized and illustrated with several empirical examples, while there are brief introductions to various specialized rank tests, tests based on the complete intensity process, and sequential tests.

Several basic aspects of *parametric* inference in life testing and epidemiology are usefully studied in the present framework. Chapter VI develops maximum likelihood and M-estimators and illustrates these methods on several examples. Asymptotic statistical theory is given, as well as methods for assessing goodness of fit, including generalization of the "total time on test" methodology from reliability.

Often, stratification is not sufficient and more specific *regression models* are required to obtain the desired insight into the influence of covariates on the intensities of occurrence of the relevant events. In survival analysis, the "semiparametric" regression model and the associated tool of partial likelihood, both due to Cox (1972), were seminal. Cox's model postulates the intensity to be a product of a parametric function of the covariates and an arbitrary function of time, and this model generalizes directly to counting processes. Chapter VII contains an extensive mathematical and applied statistical discussion of this approach. Other regression models are also discussed, particularly a (multivariate) linear intensity model due to Aalen (1980).

The methods developed in Chapters IV–VII are motivated by reasonable though sometimes rather ad hoc mathematical–statistical arguments (moment equations and similar estimating equations, unorthodox "maximum likelihood" constructions). Results on optimality or efficiency of the proposed methods are not directly available from the derivations; in fact, only a few scattered results on these topics are available in the literature. Chapter VIII provides a brief introduction to LeCam's theory of local asymptotic normality and contiguity, as well as the recent general mathematical theory of semiparametric and nonparametric estimation (Bickel et al., 1993; van der Vaart, 1988). It is shown that this provides a sufficiently general mathematical framework to allow efficiency results to be derived, which may sometimes be linked to generalized maximum likelihood properties.

Many further developments are possible within the general framework of statistical models based on counting processes. We restrict ourselves to brief indications of two lines of developments. First, heterogeneity among individuals is not always usefully described by stratification or regression modelling; sometimes a description as random variation may be necessary if not all relevant covariates are measured or to describe correlation among individuals. *Random effect models* are very natural and well known for the normal distribution, but, as shown in Chapter IX, it is still possible to define models and suggest estimation and hypothesis testing methods for event history models with random individual variation (often called random *frailty* or "dependent failure times"). In Chapter X, we report briefly on the fast growing area of generalization of the methods of *multivariate time scales* (calendar time and age; calendar time and duration on trial; duration on trial and duration on a particular, possibly intermittent, treatment). Specific surveys are given of sequential analysis of censored data with staggered entry and of multivariate hazard measures and multivariate Kaplan–Meier estimators for dependent survival times.

I.1.5. Reading Guides

It is obvious from the above synopsis of the book that various categories of readers may be interested in different parts of the material presented.

The *biostatistician* or *reliability engineer* interested primarily in the practical methods should survey our empirical examples in Section I.3 and study the introductory Section II.1, but should be able to follow the specific descriptions in Chapters IV–VII, IX, and X without a detailed study of the mathematical prerequisites in the rest of Chapter II and the detailed mathematical discussion in Chapter III. However, a cursory reading of the first half of Chapter III would be necessary. It is perhaps appropriate to emphasize that this book is *not* intended as a textbook in biostatistics—the mathematical theory has guided our selection of methods, and sometimes other approaches will be preferable in practice.

The *statistician* wanting an introduction to this area should go through Chapter I to see which substantive areas motivated the methods described, become familiar with the underlying mathematics in Chapter II, and understand the general points of censoring, truncation, and filtering in Chapter III. From this viewpoint, one may prefer a more cursory reading of Chapters IV–VII including several illustrative examples.

The *mathematical–statistical specialist* (in this field) looking for "what's new" would, in principle, not need Chapter II, though we hope that at least some from this group will appreciate our summary. The density of unfamiliar material may be highest in Chapters III, IV.4, VI.3, VII (later sections), VIII, IX, and X.3, but there are innovations throughout.

Finally, some readers may be *probabilists*, wanting an explanation of the statistical application of tools (cf. Chapter II) well known to them. They should read Chapter III carefully and study the empirical examples in Chapter I and selected cases about how the empirical problems raised there are handled in the statistical methodology Chapters IV–VII and IX.

I.1.6. Some Areas Not Covered

The book stops short of covering a number of important related theoretical and practical issues.

First, throughout the book we use only continuous-time methods. In most chapters, we study non- and semiparametric methods which are most relevant for small and medium-sized data sets with detailed information. For larger data sets, it will often be preferable to use the "parametric" log–linear intensity models deriving from assumptions of piecewise constant intensities, see Chapter VI and Section VII.6. The data also may be available only in grouped time, in which case generalized linear models based on the binomial distribution may be useful; these are not discussed here. Such methods will often be relevant in *epidemiological*, *demographic*, and *actuarial* applications. Second, our discussion has been influenced by our background in biostatistics; further applications in *reliability and industrial life testing* are also relevant at the theoretical level, as Arjas (1989) showed in his survey. Finally, although we have developed (and/or extended) many *computer algorithms* in connection with the practical examples, we have not carried any of these through to a state where we should like others to run the risk of using them. The graphs were made in S (Becker et al., 1988).

We should finally mention that as we finalized this book, the excellent textbook by Fleming and Harrington (1991) appeared. These authors provided a comprehensive introduction to the basic mathematical theory needed for the counting process approach and surveyed the by now classical statistical theory for censored survival data, using very well-documented examples from their own experience as illustration. It is likely that many readers will appreciate having access to Fleming and Harrington's book before and during the study of the present monograph.

I.2. Brief Survey of the Development of the Subject

In this section, we outline some general tendencies in the historical development of statistical inference for event history data, based on counting processes. Specific details are contained in the relevant chapters and their associated bibliographic remarks.

I.2.1. Aalen's Ph.D. Dissertation and Its Background

The point of departure for this monograh was O.O. Aalen's Ph.D. dissertation (1975). Here, concepts from classical demographical/actuarial life table analysis were studied using continuous-time martingale theory (Meyer, 1966), stochastic integration (Kunita and Watanabe, 1967), and counting process theory [Brémaud, 1972; Jacod, 1973, 1975; surveyed by Brémaud and Jacod (1977)] such as applied in control theory and electrical engineering (Boel, Varaiya, and Wong, 1975a,b). The asymptotic results may then be obtained from martingale central limit theory, in Aalen's (1975, 1977) original version based on the discrete-time results by McLeish (1974).

There were several lines in practical statistics leading up to this work. The classical life table analysis was interpreted statistically by Kaplan and Meier (1958) (Kaplan and Meier "almost simultaneously prepared separate manuscripts that were in many respects similar"; cf. the Notes about Authors on pp. 562–563 of the *JASA* issue). Sverdrup (1965) and Hoem (1969a,b, 1971) embedded well-established techniques from demography and actuarial mathematics into statistical inference for Markov processes. Nelson (1969) had presented a "hazard plot for incomplete failure time data" which had also been studied by Altshuler (1970) in the context of competing risks in animal experiments. Aalen (1972) in his Norwegian cand.real. (master's) thesis independently discussed this estimator (in this monograph we call it the Nelson–Aalen estimator). In survival analysis, nonparametric rank tests for censored data had begun to appear (Gehan, 1965; Mantel, 1966; Efron, 1967; Breslow, 1970; Peto and Peto, 1972).

For the classical competing risks model, Aalen (1976) gave a systematic exact and asymptotic theory for the cumulative hazard estimator, whereas Aalen (1978a), further elaborated by Fleming (1978a,b), gave estimators for various transition probabilities in this model. The asymptotic theory here was primarily based on the approach by Pyke and Shorack (1968) as utilized in the important paper by Breslow and Crowley (1974) on weak convergence of the Nelson–Aalen and Kaplan–Meier estimators.

I.2.2. Early Publications on Statistical Inference for Counting Processes

The journal article by Aalen (1978b) (based on his Ph.D. thesis) presented the general counting process framework, the Nelson–Aalen estimator of the cumulative intensity in the multiplicative intensity model, two-sample test statistics, exact and asymptotic properties of estimators and test statistics based on stochastic integrals and martingale theory, and gave examples.

Aalen, with Johansen, quickly followed up with their important paper from 1978 introducing the product-integral as the canonical transformation from (generalized) hazard or intensity to distribution function, thus enabling estimation of the latter as well as more general transition probabilities in nonhomogeneous continuous-time Markov processes. (A comprehensive survey of product-integration and statistical applications was given by Gill and Johansen, 1990.) Some of the ideas of Aalen and Johansen (1978) were developed independently by Fleming (1978a).

Gill (1980a), in a republication of his Ph.D. thesis from 1979, narrowed the focus to estimation and two-sample tests in the classical censored survival data problem. In this framework, he explored in detail the range of the above-mentioned mathematical tools, showing that it even extended that of the then-available classical techniques. Generalization to k-sample tests for counting process models and remarks on one-sample tests were given by Andersen et al. (1982), who also surveyed the already very extensive literature on two-sample tests for censored survival data. The early development of statistical theory for counting processes was surveyed by Jacobsen (1982), who emphasized a coherent (and self-contained) exposition. The martingale central limit theory soon came to be based on Rebolledo (1978, 1979, 1980a) or Helland (1982).

I.2.3. Cox's Semiparametric Regression Model, Prentice's Linear Rank Tests

The above development was relatively little influenced by the path-breaking semiparametric regression model for survival data of Cox (1972). Cox assumed the hazard (death intensity) to be a product of an unspecified function of time common to all individuals and a known function of a linear combination of covariates with (regression) coefficients to be estimated. Cox's tool in this estimation was an innovative concept that later (Cox, 1975) was to be termed the "partial likelihood." The article by Cox (1972) inspired an impressive research activity, in the early years primarily in understanding the nature of the partial likelihood. Prentice (1978) based his approach to nonparametric rank tests for censored survival data on a marginal rank likelihood which continues to be influential; see Cuzick (1985, 1988) and Clayton and Cuzick (1985a). Much of this development was collected in the monograph by Kalbfleisch and Prentice (1980) which is still the authority on censoring and likelihood and on the hazard function approach to models for several types of failure. There have been few attempts at further mathematical rigorization of Kalbfleisch and Prentice's classification of censoring patterns, the most important being by Arjas and Haara (1984) and Arjas (1989). The former was somewhat further developed and exemplified by Andersen et al. (1988).

The other aspect of the Cox model is asymptotics. Cox's (1972, 1975) own justification was essentially heuristic, with a very clear martingale basis, even if this word was never mentioned. Except for the paper by Næs (1982), the first rigorous proofs (see details in Chapter VII) went along different routes. However, Andersen and Gill (1982) integrated the Cox model fully into the counting process framework (thereby generalizing it beyond the survival data framework) and constructed a transparent proof of consistency and asymptotic normality of the estimators using the martingale central limit theory tools quoted above. Further results were given by Prentice and Self (1983), and the status of the counting process/martingale approach was surveyed by Andersen and Borgan (1985).

I.2.4. Consolidation in the 1980s

The question of nonparametric maximum likelihood interpretation of the empirical distribution for censored survival data [first discussed by Kaplan and Meier, 1958, cf. Johansen, 1978] was revived (Jacobsen 1982, 1984) and cross-fertilized with the Cox partial likelihood discussion (Johansen, 1983; Andersen and Gill, 1982). A sieve approach (Karr, 1987; Pons and Turckheim, 1987) should also be mentioned here.

The early theory only discussed estimation of the cumulative intensity, or cumulative distribution function. Nonparametric estimation of the intensity is a smoothing problem; Ramlau-Hansen's (1983a,b) version was directly tuned to the present framework.

Fully parametric models were studied by Borgan (1984) using tools similar to those of Andersen and Gill (1982), followed by Hjort (1986, 1990a, 1992a).

Completions of the mathematical theory appeared along the way; thus, Gill (1983a) uncovered precise conditions for weak convergence over the whole support of the nonparametric estimator of a distribution function from censored data, and Keiding and Gill (1990) embedded the random truncation model into the framework, thereby obtaining asymptotic theory for this model.

Many specific applications and modifications were worked out; we mention here some lines of particular interest.

Aalen's (1980) *linear regression* or "matrix multiplicative" model generalized his original multiplicative intensity model to allow for covariates with time-varying regression "coefficients." Definitive asymptotic results for this model (which we believe will be powerful in several applications) were only recently provided by Huffer and McKeague (1991).

An alternative approach to *nonparametric regression modelling* (requiring smoothing of the time-dependent regression coefficients) was initiated by Beran (1981) in an unpublished report that was followed up by Dabrowska (1987) and placed in the present framework by McKeague and Utikal (1990a,b).

A number of *graphical goodness-of-fit* procedures for semiparametric (Cox) regression models have been suggested, several building rather directly on the martingale structures (Crowley and Storer, 1983; Arjas, 1988; Barlow and Prentice, 1988; Therneau et al., 1990; Fleming and Harrington, 1991).

There has been a growing desire to describe *heterogeneity* not only by different values of covariates but also by a probability distribution. An important paper here by Clayton and Cuzick (1985a) worked in the "classical" rank test framework of Prentice (1978). Counting process approaches were indicated in the discussion of Clayton and Cuzick's paper by Prentice and Self, cf. Self and Prentice (1986), and, in particular, by Gill, cf. Nielsen et al. (1992) and Chapter IX. Aalen (1987a, 1988b, 1990) gave several empirical illustrations and general suggestions on these models for "frailty," as the random heterogeneity is often termed. The related area of allowing covariate measurement error has developed slowly since the paper by Prentice (1982).

Bayesian modelling in the present framework was given by Hjort (1990b), who specified the class of beta processes to be used as priors, leading to the Nelson–Aalen and Kaplan–Meier estimators for cumulative intensities and survival functions, respectively, in the limit for vague priors. Phelan (1990) generalized this analysis to Markov renewal processes.

Sequential analysis of event history data is an obvious possibility, and the mathematical structure adapts this readily. The most important contributions are by Sellke and Siegmund (1983), Slud (1984), and Gu and Lai (1991).

Among the adaptations of the techniques to *special sampling frames* we mention the case-cohort models of Self and Prentice (1988) and the survey of models for retrospective epidemiological data of Keiding (1991); Pons and Turckheim (1987, 1988a,b) studied a Cox regression model with periodic underlying intensity.

The methods have gained increasing importance in sociology and econometrics; cf. Tuma et al. (1979), Tuma and Hannan (1984), Heckman and Singer (1984a), Blossfeld et al. (1989), and also Manton and Stallard (1988).

I.2.5. Current Developments

At the beginning of the 1990s, our view is that several areas of the theory of statistical models based on counting processes are now maturely developed, although details are still being filled in. Within the original framework, a generally active area is that of the frailty models discussed earlier, for which suggestions for Bayesian estimation via Gibbs sampling were given by Clayton (1991). Also, bootstrap techniques (Hjort, 1985b) are being adapted to this framework; cf. Section IV.1.4.

Two very active areas, however, both somewhat transcend our original framework.

First, the asymptotic statistical viewpoint of contiguity and local asymptotic normality developed originally by LeCam (1960) has cross-fertilized with the considerable mathematical–statistical activity in semiparametric models; see van der Vaart (1988) and Bickel et al. (1993). We show in Chapter VIII that we here have tools to give rather general efficiency results for our counting process models and to link these to generalized maximum likelihood properties. The corresponding asymptotic techniques include a modern version of von Mises' generalized delta method (Gill, 1989) which interprets and generalizes Breslow and Crowley's (1974) proof of asymptotic distribution of the Kaplan–Meier estimator. Greenwood and Wefelmeyer (1990) gave a general local asymptotic normality result for partial likelihoods in this context and derived efficiency results for the estimators in the Cox regression model and Aalen's linear intensity model.

Second, although the estimation of the intensities of a multivariate counting process is basic to the whole theory presented here, the apparently equally basic task of providing a canonical nonparametric estimator of a multivariate (even bivariate) distribution function from censored data has proved surprisingly intricate. Recent work by Dabrowska (1988, 1989b) has helped clarify this issue. In Section X.3, we survey this area and explain why the basic intensity framework needs considerable elaboration in more dimensions, and how the generalized von Mises calculus again becomes fruitful; see also Gill (1990).

I.3. Presentation of Practical Examples

The methodology to be discussed in the present monograph is relevant to a fairly broad class of event history analyses. In this section, we want to present a number of concrete examples, primarily from medicine and biology. A main purpose is to use the many practical problems as motivation for the development of the theory in later chapters, where the examples will be discussed further and used for illustration of the developed methods.

In most cases, we study a collection of individuals, each moving between a finite (usually small) number of states. It is then convenient to illustrate the individual behavior by a diagram with boxes representing states and arrows between boxes indicating possible direct transitions, cf. Hoem (1976).

We shall make a note in each particular example about the relevant *time scale*(*s*) and return to a general discussion in Section X.1.

I.3.1. Survival Analysis

The simplest kind of event history analysis is classical survival analysis, where a collection of individuals are observed from some entry time until a particular event (such as death) happens, cf. Figure I.3.1. Often it is impossible to wait for the event to happen for all individuals; for some, it is only known

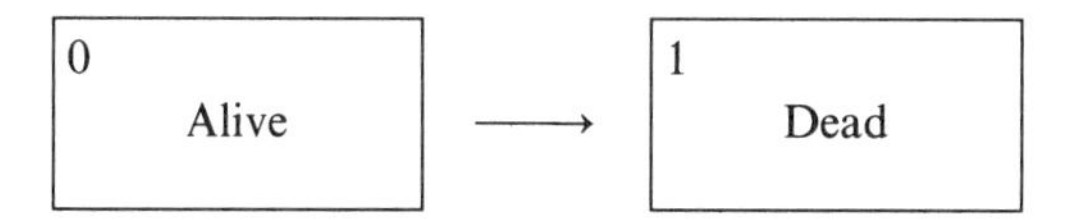

Figure I.3.1. Survival analysis.

that the event had not yet happened at some specified time and in this case the observation of the time to the occurrence of the event is *right-censored.*

We first present the main example in the monograph.

EXAMPLE I.3.1. Survival with Malignant Melanoma

In the period 1962–77, 225 patients with malignant melanoma (cancer of the skin) had a radical operation performed at the Department of Plastic Surgery, University Hospital of Odense, Denmark. That is, the tumor was completely removed together with the skin within a distance of about 2.5 cm around it. All patients were followed until the end of 1977, that is, it was noted if and when any of the patients died.

This is a *historically prospective clinical study* with the object of assessing the effect of risk factors on survival. The time variable viewed as most important is *time since operation.* Among the possible risk factors screened for significance were the *sex* and *age at operation* of the patient. Furthermore, clinical *characteristics of the tumor* such as tumor width and location on the body were considered as well as various histological classifications (that is, obtained by examination of the tissue), including tumor thickness, growth patterns, types of malignant cells, and ulceration. The latter factor is dichotomous and scored as "present" if the surface of the melanoma viewed in a microscope shows signs of ulcers and as "absent" otherwise. The material from 20 patients did not permit a histological evaluation and only the remaining 205 patients are considered here. Note that risk factors (*covariates*) are both qualitative and quantitative but that attention has been restricted to *information known at entry.* In other studies, the main attention may well be toward studying the influence of current values of risk factors on the propensity to die. The latter studies require more advanced statistical methods and are perhaps for that reason much less commonly made. We return to an example of this in Example I.3.4.

Note that the survival time is known only for those patients who died before the end of 1977. The rest of the patients are *censored* at the duration in the study obtained then. Fourteen patients died of causes unrelated to cancer and were in most analyses considered censored at death. The latter point is further discussed in Example I.3.9.

Also note that although the patients enter the study at different calendar times (*staggered entry*), all patients enter at time 0 in the time scale (time since operation) used as the response variable.

For a preliminary presentation of these data, Figure I.3.2 shows empirical

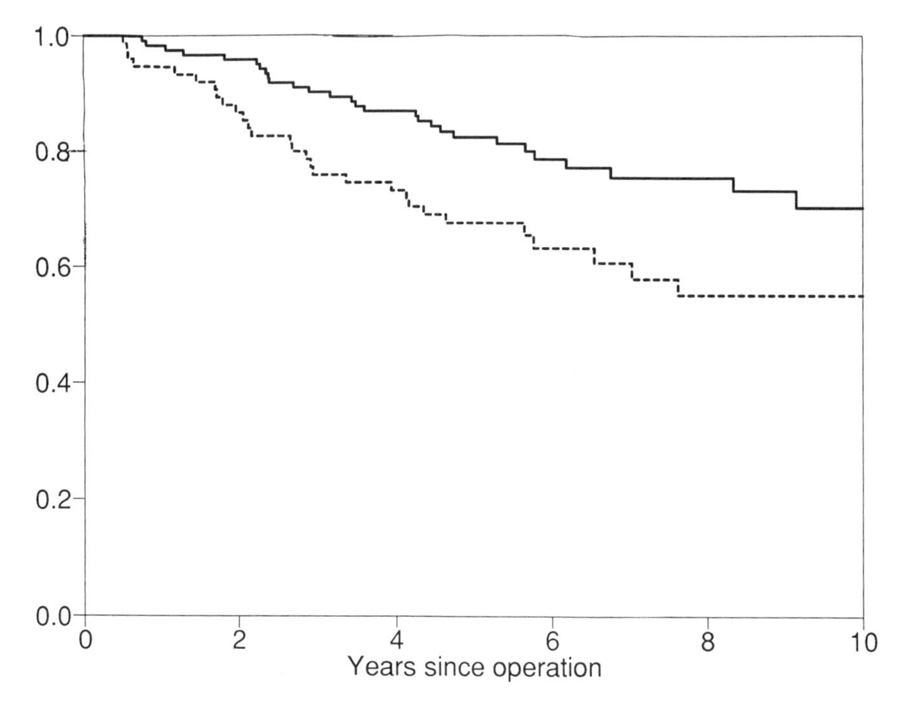

Figure I.3.2. Empirical survival curves for male (---) and female (—) patients with malignant melanoma.

survival curves (Kaplan–Meier plots, to be specified in Section IV.3) for the data stratified by sex. The curves represent, for any number t of years since operation, the estimated probability of surviving past t and it is seen that the estimated survival probability is always higher for women. Further aspects of the data are discussed in Example I.3.9 and the full data set is quoted in the Appendix together with standard life tables for the Danish population 1971–75 to be used in some of the examples mentioned later.

The melanoma data will be used as illustrations throughout the book and are discussed and analyzed in the following examples:

Example III.2.7	Discussion of censoring patterns.
Example IV.1.2	Nelson–Aalen estimate for the integrated intensity of dying from the disease for men and women.
Example IV.1.11	Nonparametric models for excess and relative mortality.
Example IV.1.12	Confidence limits and confidence bands for integrated death intensity for men.
Example IV.3.2	Kaplan–Meier estimate of the survival function for men with confidence limits and bands.

Example IV.3.3	Calculation of "expected" and "corrected" survival functions.
Example IV.3.6	Confidence interval for the lower quartile.
Example V.2.3	Nonparametric tests comparing mortality for men and women.
Example V.2.11	Conservative approximations to nonparametric tests.
Example V.3.1	Estimation of the ratio between the hazards for men and women.
Example V.3.2	Stratified log-rank tests: sex and tumor thickness.
Example V.3.5	Tests for trend: tumor thickness.
Example VI.1.5	Parametric models with constant excess or relative mortality.
Example VI.2.4	M-estimators for constant excess or relative mortality parameters.
Example VI.3.1	Check of model with constant mortality rates for men and women.
Example VI.3.5	Total time on test plots for models with constant relative mortality.
Example VI.3.6	Total time on test plots for model with constant excess mortality.
Example VI.3.10	Check of model with constant excess mortality.
Example VII.2.5	Examples of simple Cox regression models.
Example VII.2.8	Confidence limits for the integrated hazard in a Cox regression model.
Example VII.2.9	Estimation of survival function with confidence limits in a Cox regression model.
Example VII.3.1	Graphical checks of proportional hazards.
Example VII.3.2	Test for log-linearity of the effect of tumor thickness.
Example VII.3.3	Possible threshold effect of tumor thickness.
Example VII.3.4	Tests for proportional hazards using time-dependent covariates.
Example VII.3.5	Goodness-of-fit of a two-sample proportional hazards model.
Example VII.3.6	Residual plots in Cox regression model.
Example VII.3.7	Regression diagnostics in Cox regression model.
Example VII.4.2	Nonparametric additive regression models.
Example VII.4.3	Nonparametric models for excess and relative mortality.

Example VII.6.4	Regression models with piecewise constant baseline hazard.
Example IX.4.3	A frailty model for survival with malignant melanoma.
Example X.1.1	Theoretical note on time scales: age and time since operation.

The data and single-factor survival analyses were presented by Drzewiecki et al. (1980a,b) and Cox regression analyses by Drzewiecki and Andersen (1982). The data have been used later for illustration of survival analysis techniques by Andersen (1982), Andersen and Væth (1984, 1989), Arjas (1989), and Doss and Gill (1992). □

In Example I.3.1, two examples of mechanisms causing censored observation were given: Patients could die from causes unrelated to that under study or patients could survive past the closing date of the study. In general, it is very common in similar studies that patients are lost to follow-up during the study period for various reasons. It is then an issue of considerable importance whether loss to follow-up in some sense may be considered "independent of survival." As we shall see in Section III.2.2, exact statistical modelling of this concept of *independent censoring* is not at all easy.

The staggered entry of the patients in the example had the consequence that the time under study at the closing date was not the same for all patients. Had the decision to close the study been based upon, e.g., how many patients had already died, it would have been possible that the decision to follow one patient for only 5 years depended on what happened to another patient after 7 years of follow-up. Thus, the staggered entry may disturb the attractive property that events in the process only depend on the "past." (As some readers will already have guessed, this is the *martingale* property.) We return to this in Example I.3.6; see also Example III.2.11.

In Example I.3.1, the basic time scale was time since operation, and all patients were followed from the time origin. In the next example, we use age as basic time variable and persons are not observed from birth but rather with *delayed entry*.

EXAMPLE I.3.2. Mortality of Diabetics in the County of Fyn

Fyn county in Denmark contains about 450,000 inhabitants. On 1 July 1973, 1499 of these suffered from insulin-dependent diabetes mellitus (diabetes for short). This was ascertained by recording all insulin prescriptions in the National Health Service files for this county during a 5-month period covering the above date, and subsequently checking each patient's medical record at the general practitioner and, when relevant, the hospital. Survival status for all patients was assessed by 1 January 1982, using the Danish centralized person registry.

It was desired to study the age-specific mortality of diabetics, taking into account the duration of the disease and possibly calendar time. It is seen that here the basic time variable is *age*. Additional time variables are duration of disease and calendar time. A new problem is that patients do not all enter the study at the same age, which is here the basic time variable. First, we are only interested in mortality of diabetics and it, thus, only makes sense to follow patients after disease onset. But second, the cross-sectional sampling induces a *length bias* (longer survivals have a higher probability of being sampled). As will be discussed further in Example III.3.6, one way of avoiding this bias is to let each patient only contribute informaton from the age at the sampling date 1 July 1973 to the age at death or emigration, if before 1 January 1982, and otherwise to the age at 1 January 1982. Such survival data are said to be *left-truncated* and (as in Example I.3.1) *right-censored.*

For a preliminary presentation of the data, see Table I.3.1; the mortality is seen to increase with age, and males have a higher mortality than females.

These data will be discussed and analyzed in the following examples:

Examples III.3.1 and III.3.6 — Theoretical introduction and modelling of age at survey as left-truncating time.

Example III.5.1 — Example of a "defined" time-dependent covariate.

Table I.3.1. The Prevalent Population of Diabetics in the County of Fyn on 1 July 1973 and the Deaths During the Following $8\frac{1}{2}$ Years

Age 1 July 1973 (Years)	No. Alive 1 July 1973	Deaths July 1973–Dec. 1981	Proportion of Deaths
		Males	
0–29	207	9	.043
30–39	115	22	.191
40–49	124	25	.202
50–59	122	43	.352
60–69	126	78	.619
70+	89	77	.865
		Females	
0–29	146	6	.041
30–39	89	10	.112
40–49	83	10	.120
50–59	107	29	.271
60–69	143	69	.483
70+	148	113	.769

Example IV.1.3	Nelson–Aalen estimate of the integrated death intensity from left-truncated and right-censored survival data.
Example IV.1.13	Confidence limits and confidence bands for the integrated death intensity.
Example IV.2.1	First example of kernel-smoothed estimate of the death intensity.
Example IV.2.4	Kernel-smoothed estimate of the death intensity: optimal bandwidth, estimation of local bias, and confidence limits.
Example IV.3.4	Kaplan–Meier estimate of conditional survival functions from left-truncated survival data, with confidence limits and confidence bands.
Example V.1.3	One-sample test of the hypothesis that the age-specific mortality of diabetics equals that of the general population.
Example VI.1.4	Gompertz–Makeham fit of the death intensity: maximum likelihood.
Example VI.2.3	Gompertz–Makeham fit of the death intensity: M-estimates.
Example VI.3.3	Goodness-of-fit of the Gompertz intensity function: Nelson–Aalen plots.
Example VI.3.7	Goodness-of-fit of the Gompertz intensity function: generalized Pearson χ^2 tests.
Example VI.3.9	Goodness-of-fit of the Gompertz–Makeham intensity function: Khmaladze's innovations approach.
Example VII.2.14	A Cox regression model for the relative mortality.
Example X.1.2	Theoretical note on the time scales: age, duration of disease, calendar time.

These data were collected by Green et al. (1981) and the mortality in a 7-year follow-up period was evaluated by Green and Hougaard (1984). The Cox regression model for the relative mortality was developed by Andersen et al. (1985) with further diabetological interpretation by Green et al. (1985). The possibilities of recovering information on diabetes *incidence* (intensity) in the period before the survey date from the retrospective information provided at the survey were explored by Keiding et al. (1989, 1990). □

The next example also concerns left-truncated and right-censored survival data, this time collected on a population level.

EXAMPLE I.3.3. Suicides Among Nonmanual Workers in Denmark 1970–80

Based on data from the Danish census of 9 November 1970 and official mortality statistics from the following 10 years, the mortality of the total Danish labor force aged 20–64 years on the census day 9 November 1970 was studied. The occupation of each person was ascertained from the census form so that the mortalities of different occupations (as given on the census form) could be compared.

We shall be particularly interested in suicides for nonmanual workers. Table I.3.2 displays the number of suicides and the number of person years

Table I.3.2. Number of Suicides and Person-Years Lived 1970–80 for Danish Nonmanual Workers. Variables as Registered 9 November 1970

		Job Status Group (See Text)				
Suicides	Age	1	2	3	4	Total
	20–24	2	14	31	51	98
	25–29	10	48	46	46	150
	30–34	26	51	62	54	193
	35–39	20	40	68	47	175
Males	40–44	28	46	79	42	195
	45–49	20	49	93	44	206
	50–54	21	22	72	37	152
	55–59	19	21	45	34	119
	60–64	6	10	21	17	54
	Total	152	301	517	372	1342
	20–24	0	12	29	61	102
	25–29	5	10	27	65	107
	30–34	6	17	20	59	102
	35–39	7	13	26	66	112
Females	40–44	6	12	22	68	108
	45–49	5	18	24	77	124
	50–54	4	5	27	54	90
	55–59	6	8	17	38	69
	60–64	2	3	7	17	29
	Total	41	98	199	505	843
	20–24	2	26	60	112	200
	25–29	15	58	73	111	257
	30–34	32	68	82	113	295
	35–39	27	53	94	113	287
Males + Females	40–44	34	58	101	110	303
	45–49	25	67	117	121	330
	50–54	25	27	99	91	242
	55–59	25	29	62	72	188
	60–64	8	13	28	34	83
	Total	193	399	716	877	2185

Table I.3.2 (*continued*)

Person-years lived	Age	Job Status Group 1	2	3	4	Total
	20–24	13439	67623	153241	225630	459933
	25–29	52787	190702	235107	173850	652446
	30–34	53274	160758	214221	107147	535400
	35–39	50507	118822	196515	82600	448444
Males	40–44	54022	99817	193692	72675	420206
	45–49	48709	85475	197656	73449	405289
	50–54	41212	61701	170852	62681	336446
	55–59	32668	45934	121908	53577	254087
	60–64	22965	37160	83207	45255	188587
	Total	369583	867992	1566399	896864	3700838
	20–24	2366	79773	199778	610034	891951
	25–29	14047	124452	141463	398147	678109
	30–34	12332	77837	83926	250654	424749
	35–39	9197	53345	70240	218983	351765
Females	40–44	9448	43669	64346	217532	334995
	45–49	10144	39214	67743	209441	326542
	50–54	8729	29833	61724	164024	264310
	55–59	6646	23467	52605	117013	199731
	60–64	4866	18858	32705	69418	125847
	Total	77775	490448	774530	2255246	3597999
	20–24	15805	147396	353019	835664	1351884
	25–29	66834	315154	376570	571997	1330555
	30–34	65606	238595	298147	357801	960149
	35–39	59704	172167	266755	301583	800209
Males + Females	40–44	63470	143486	258038	290207	755201
	45–49	58853	124689	265399	282890	731831
	50–54	49941	91534	232576	226705	600756
	55–59	39314	69401	174513	170590	453818
	60–64	27831	56018	115912	114673	314434
	Total	447358	1358440	2340929	3152110	7298837

lived for nonmanual workers specified by sex, age, and four job status categories: (1) academics, (2) advanced nonacademic training, (3) extensive practical training, (4) other nonmanual worker.

These data will be discussed and analyzed in the following examples.

Example VI.1.1 Estimation of age- and sex-specific suicide intensities.

Example VI.1.2 Analysis of the suicide intensity of female academics using the Standardized Mortality Ratio (SMR).

Example VI.1.3 The dependence of the suicide intensity on the registered background factors analyzed by log-linear intensity models.

The data were extracted from a large population-based study of occupational mortality in Denmark (Andersen, 1985, in Danish, of which an English presentation was given in Statistical Report of the Nordic Countries 49, 1988).

□

Frequently, survival data are collected in a *randomized clinical trial*. Next, we study an example of this.

EXAMPLE I.3.4. Randomized Clinical Trial Concerning Prednisone Treatment of Liver Cirrhosis: CSL 1

From 1962 to 1969, 532 patients with histologically verified liver cirrhosis at several hospitals in Copenhagen were included in a randomized clinical trial with random allocation either to treatment with the hormone *prednisone* or to an inactive *placebo* treatment. We shall here consider only 488 patients in which the initial biopsy (liver tissue sample) could be reevaluated using more restrictive criteria: 251 received prednisone and 237 placebo.

The main purpose of the trial was to ascertain whether prednisone prolonged survival for patients with cirrhosis. Time is again time since entry (i.e., randomization) and patients were followed until September 1974 (where drug administration was discontinued). Thus, patients are censored if alive on 1 October 1974 or if lost to follow-up alive before that date. One hundred and fifty placebo-treated patients were observed to die and 142 prednisone treated.

There was little or no difference in survival between the two treatment groups (see Figure I.3.3), but detailed analyses of the influence of various prognostic factors revealed important interactions with the treatment. Two such regression analyses were carried out; one only taking factors known at the time of entry into the trial into account and one including also variables recorded at follow-up visits during the trial, cf. the discussion in Example I.3.1. The visits were scheduled to take place after 3, 6, and 12 months of treatment and, thereafter, once a year, but the actual follow-up times varied considerably around the scheduled times.

The variables recorded at the time of entry include basic factors like sex and age and several histological classifications of the liver biopsy. Also, a large number of clinical and biochemical variables were recorded both at the time of entry and at the follow-up visits. The clinical variables include information on alcohol consumption, nutritional status, bleeding and degree of ascites (excess fluid in the abdomen), whereas the most important biochemical variables are albumin, bilirubin, alkaline phosphatase, and prothrombin. The latter will be further discussed in Example I.3.12.

The CSL 1 survival data will be discussed and analyzed in the following examples:

Example III.2.10 Discussion of censoring patterns.

Example III.5.3 Note on possible inclusion of time-dependent covariates.

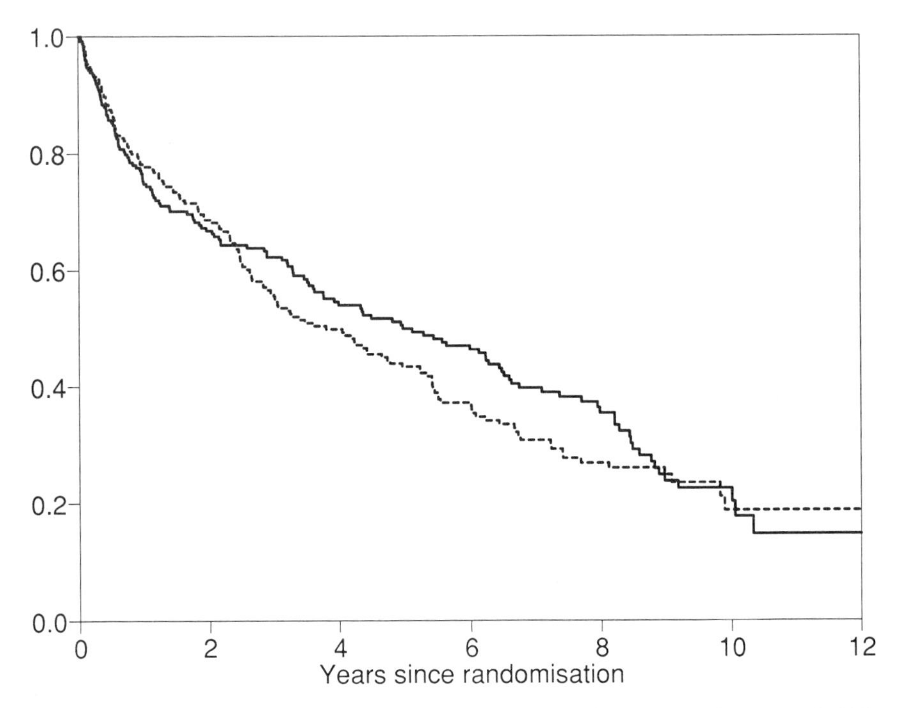

Figure I.3.3. Empirical survival curves for 251 prednisone-treated (—) and 237 placebo-treated (- - -) patients with liver cirrhosis.

Example VII.1.2	Models for the effect of alcohol consumption on the mortality.
Example VII.2.16	Regression models with time-fixed and time-dependent covariates.
Example X.1.3	Theoretical note on time scales: age and time on study.

The trial and the data were first presented by The Copenhagen Study Group for Liver Diseases (1969, 1974). The histological criteria for the evaluation of the diagnosis of cirrhosis as well as preliminary survival analyses were presented by Schlichting et al. (1982a,b,c), whereas the results of more definitive regression analyses with time-fixed covariates were discussed by Schlichting et al. (1983) and Christensen et al. (1985). A regression model with time-dependent covariates was presented by Christensen et al. (1986) and further discussed by Andersen (1986). Finally, Andersen et al. (1983b,c) and Ramlau-Hansen (1983a) used the CSL 1 survival data for illustrating various methodological points in survival analysis. □

The next example also concerns a randomized clinical trial and exemplifies detailed use of left-truncation and right-censoring.

EXAMPLE I.3.5. Prognostic Significance of Residual Cancer Tissue After Diagnostic Biopsy in Breast Cancer

The trials of the Danish Breast Cancer Cooperative Group (DBCG) cover more than 90% of all operable primary breast cancers in Denmark. About 70% of the patients at the participating centers are eligible for the trials. Here we shall consider only premenopausal patients considered to be at high risk based on histological findings. The trials were started during 1978, and based upon the experience until 31 December 1981 (Figure I.3.4, area A), an effect on

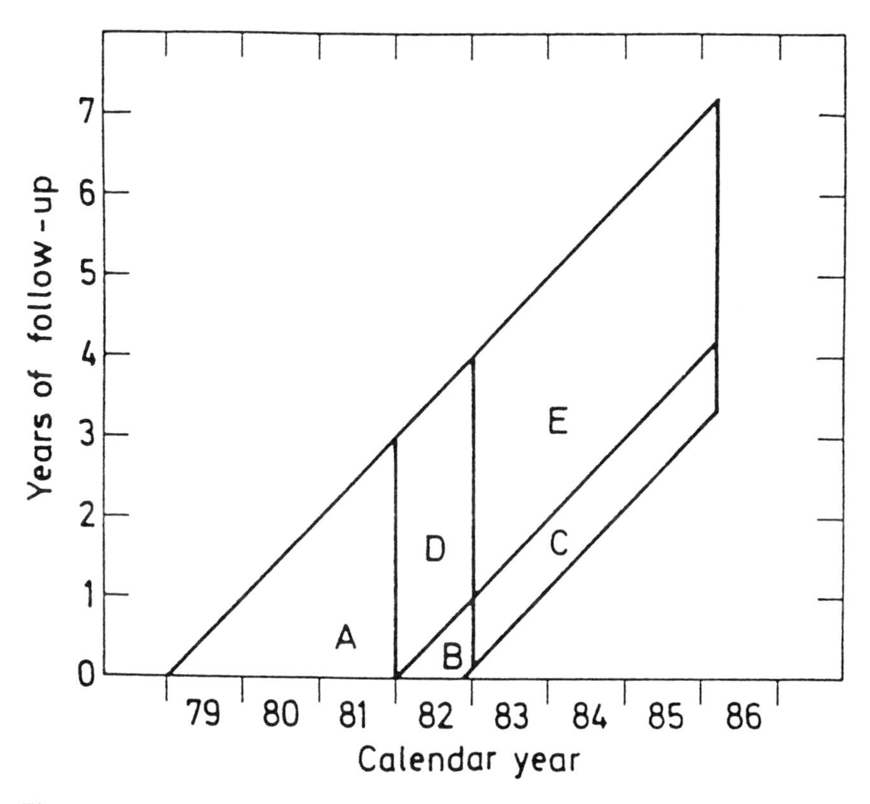

Figure I.3.4. Time × duration (Lexis) diagram of the DBCG-77 trials. A patient with time of operation equal to t and duration of follow-up at recurrence or death equal to X may be represented by the line with unit slope starting at the point $(t, 0)$ and ending at the point $(t + X, X)$ for patients who recurred or died before time τ = 28 February 1986, or $(\tau, \tau - t)$ for patients who were still alive and recurrence free at that date. The preliminary finding was based on the information in area A, and the traditional independent data set for verifying an unexpected finding in A would be based on $B \cup C$. However, much more information is obtained by including also $D \cup E$, and, in fact, already $B \cup D$ would have yielded the independent confirmation not achieved by $B \cup C$. (From Keiding et al., 1987 with kind permission by John Wiley & Sons, Ltd.)

recurrence-free survival of the presence of residual cancer tissue (RCT) after diagnostic biopsy was noted. The diagnostic biopsy is a tissue specimen that is examined by the pathologist for the presence of malignant (i.e., cancerous) cells. If such cells were found, the whole breast was removed shortly thereafter, usually within $\frac{1}{2}$–1 h after the biopsy was taken, and only these patients are included in this study. It was considered unexpected and not easily interpretable to find a connection between the presence of cancer tissue in the biopsy cavity and recurrence-free survival, and a reanalysis on an independent set of data was, therefore, judged necessary before the finding could be considered an established fact within breast cancer prognostics. Ordinarily, one would use patients accrued *after* 1 January 1982. Since accrual to this protocol was closed toward the end of 1982, only a few patients were available, and even an average follow-up time of almost 4 years was not enough to reproduce the early finding (Figure I.3.4, areas B and C). Alternatively, the recurrence-free survivors on 1 January 1982 might be included (Figure I.3.4, areas D and E), counted with *delayed entry* with the duration obtained on that date: These patients are included with *left-truncated* disease durations.

The analysis of these data is given in Example V.2.5. The data were presented and analyzed by Keiding et al. (1987), with full surgical discussion by Watt–Boolsen et al. (1989). □

The next example concerns a randomized clinical trial, this time with a *paired design.*

EXAMPLE I.3.6. Length of Remission of Leukemia Patients

At 11 American hospitals, a total of 62 children with acute leukemia responded to primary treatment with prednisone in the sense that most or all signs of the disease in the bone marrow disappeared (patients entered into *partial* or *complete remission*). These patients were randomized to *remission maintenance therapy* in the following way: At each hospital, patients were paired according to remission status (partial or complete) and one patient in a pair was randomly assigned to 6-mercaptopurine (6-MP), the other to placebo.

The remission lengths, i.e., the interval from time of randomization (~ remission) to time of relapse (time when the disease resumed) for 21 complete pairs of patients are given in Table I.3.3. In addition to these complete pairs, the study included several "half-pairs" (Gehan, private communication), whose remission times are not given. The trial was designed sequentially: When a patient relapsed, this was counted as a preference for the other treatment and the trial was stopped once the difference between the two numbers of preferences had reached significance (cf. Fig. 8 of Freireich et al., 1963). Table I.3.3 shows that there were 18 preferences for 6-MP and only 3 for placebo. Unfortunately, the entry times are not available.

Time from remission to relapse is the basic time variable. Patients were censored at the conclusion of the study; this *stopping rule*, however, depended on the remissions of other patients in such a way that the standard time-

Table I.3.3. Remission Lengths in 21 Pairs of Leukemia Patients Treated with 6-MP or Placebo. Censored observations are indicated by a +. (From Freireich et al., 1963.)

Pair	Remission Status	Placebo	6-MP
1	Partial	1	10
2	Complete	22	7
3	Complete	3	32+
4	Complete	12	23
5	Complete	8	22
6	Partial	17	6
7	Complete	2	16
8	Complete	11	34+
9	Complete	8	32+
10	Complete	12	25+
11	Complete	2	11+
12	Partial	5	20+
13	Complete	4	19+
14	Complete	15	6
15	Complete	8	17+
16	Partial	23	35+
17	Partial	5	6
18	Complete	11	13
19	Complete	4	9+
20	Complete	1	6+
21	Complete	8	10+

ordering was violated; cf. the discussion after Example I.3.1. As an example, the decision to censor the 6-MP-treated patients in pairs 19, 20, and 21 after 9, 6, and 10 weeks depends on what happened after 23 weeks when the first member (the placebo) of pair 16 relapsed.

The data will be discussed and analyzed in the following examples:

Example VII.2.13 Survival analysis of matched pairs.

Example IX.4.2 Survival in matched pairs. A frailty model for remission lengths in leukemia.

The data were first presented by Freireich et al. (1963) and have since then been used very frequently for illustration in the survival data literature, in most cases without consideration of the paired design and sequential stopping rule (e.g., Gehan, 1965; Cox, 1972; Aitkin and Clayton, 1980; Kalbfleisch and Prentice, 1980, p. 246; Cox and Oakes, 1984, pp. 7, 38; Andersen and Væth, 1984; Wei, 1984; and McCullagh and Nelder, 1989, p. 425); exceptions are Oakes (1982a) and Siegmund (1985, Chapter V). □

We next present a classical example from the literature on *left-censoring.*

EXAMPLE I.3.7. Baboon Descent

Baboons in the Amboseli Reserve, Kenya, sleep in the trees and descend for foraging at some time of the day, here defined as the time when half of the members of the study troop had descended (the "median" descent time). Observers often arrive later in the day than this descent and for such days they can only ascertain that descent took place *before* a particular time, so that the descent times are *left-censored.* It is desired to evaluate the distribu-

Table I.3.4. Times of Arrival of the Observer at Site of Baboon Troop and of the Median Descent Time for the Troop, if this Event had Not Already Happened. (Wagner and Altmann, 1973.)

Descent Time Data for Days on Which Arrival Was Before Median Descent

	Date	Arrival Time	Descent Time		Date	Arrival Time	Descent Time
1	25-11-63	0655	0656	30	5-6-64	0825	0844
2	29-10-63	0657	0659	31	17-7-64	0842	0845
3	5-11-63	0658	0720	32	12-6-64	0815	0846
4	12-2-64	0715	0721	33	28-2-64	0730	0848
5	29-3-64	0634	0743	34	14-5-64	0830	0850
6	14-2-64	0738	0747	35	7-7-64	0831	0855
7	18-2-64	0729	0750	36	6-7-64	0822	0858
8	1-4-64	0727	0751	37	2-7-64	0837	0858
9	8-2-64	0732	0754	38	17-3-64	0803	0859
10	26-5-64	0758	0758	39	10-6-64	0848	0859
11	19-2-64	0731	0805	40	11-3-64	0830	0900
12	7-6-64	0758	0808	41	23-7-64	0807	0904
13	22-6-64	0753	0810	42	27-2-64	0723	0905
14	24-5-64	0753	0811	43	31-3-64	0750	0905
15	21-2-64	0750	0815	44	10-4-64	0824	0907
16	13-2-64	0734	0815	45	22-4-64	0833	0908
17	11-6-64	0805	0820	46	7-3-64	0832	0910
18	21-6-64	0756	0820	47	29-2-64	0815	0910
19	13-3-64	0820	0825	48	13-5-64	0758	0915
20	12-7-64	0817	0827	49	20-4-64	0830	0920
21	30-6-64	0758	0828	50	27-4-64	0801	0930
22	5-5-64	0823	0831	51	28-4-64	0835	0930
23	12-5-64	0817	0832	52	23-4-64	0900	0932
24	25-4-64	0831	0832	53	4-3-64	0845	0935
25	26-3-64	0810	0833	54	6-5-64	0840	0935
26	18-3-64	0813	0836	55	26-6-64	0815	0945
27	15-3-64	0711	0840	56	25-3-64	0722	0948
28	6-3-64	0755	0842	57	8-7-64	0821	0952
29	11-5-64	0817	0844	58	21-4-64	0810	1027

Table I.3.4 (*continued*)

Descent Time Data for Days on Which Arrival Was After Median Descent

	Date	Arrival time		Date	Arrival Time		Date	Arrival Time
1	1-12-63	0705	32	13-10-63	0840	63	2-5-64	1012
2	6-11-63	0710	33	4-7-64	0845	64	1-3-64	1018
3	24-10-63	0715	34	3-5-64	0850	65	17-10-63	1020
4	26-11-63	0720	35	25-5-64	0851	66	23-10-63	1020
5	18-10-63	0720	36	24-11-63	0853	67	25-7-64	1020
6	7-5-64	0730	37	15-7-64	0855	68	13-7-64	1031
7	7-11-63	0740	38	16-2-64	0856	69	8-6-64	1050
8	23-11-63	0750	39	10-3-64	0857	70	9-3-64	1050
9	28-11-63	0750	40	28-7-64	0858	71	26-4-64	1100
10	27-11-63	0753	41	18-6-64	0858	72	14-10-63	1205
11	28-5-64	0755	42	20-2-64	0858	73	18-11-63	1245
12	5-7-64	0757	43	2-8-64	0859	74	2-3-64	1250
13	28-3-64	0800	44	27-5-64	0900	75	8-5-64	1405
14	23-3-64	0805	45	28-10-64	0905	76	1-7-64	1407
15	26-10-63	0805	46	15-5-64	0907	77	12-10-63	1500
16	11-7-64	0805	47	10-5-64	0908	78	31-7-64	1531
17	27-7-64	0807	48	27-6-64	0915	79	6-10-63	1535
18	9-6-64	0810	49	11-10-63	0915	80	19-6-64	1556
19	24-6-64	0812	50	17-2-64	0920	81	29-6-64	1603
20	16-10-63	0812	51	22-10-63	0920	82	9-5-64	1605
21	25-2-64	0813	52	10-7-64	0925	83	9-10-63	1625
22	6-6-64	0814	53	14-7-64	0926	84	8-3-64	1625
23	22-11-63	0815	54	11-4-64	0931	85	11-2-64	1653
24	10-10-63	0815	55	23-5-64	0933	86	30-5-64	1705
25	2-11-63	0815	56	30-7-64	0943	87	5-3-64	1708
26	23-6-64	0817	57	18-7-64	0945	88	26-2-64	1722
27	24-4-64	0823	58	29-7-64	0946	89	4-5-64	1728
28	3-7-64	0830	59	16-7-64	0950	90	12-3-64	1730
29	29-4-64	0831	60	22-7-64	0955	91	25-10-63	1730
30	4-8-63	0838	61	15-10-63	0955	92	29-11-63	1750
31	7-10-63	0840	62	19-10-63	1005	93	22-2-64	1801
						94	22-3-64	1829

tion of descent time. Table I.3.4 shows the data: Most observed descents took place between 0800 and 0930, but quite often arrival was within or after this period.

The data will be discussed and analyzed in the following examples:

Example III.4.1 Theoretical note on left-censoring.

Example IV.3.5 Kaplan–Meier estimation of the distribution function in the reverse time scale.

The data and the estimation problem were originally presented by Altmann and Altmann (1970, pp. 81–82), cf. Wagner and Altmann (1973), who developed a complicated ad hoc procedure. Ware and DeMets (1976) remarked that the problem may be transformed into an ordinary survival analysis by *reversing* the time scale: By this procedure, the remaining time (until midnight, say) becomes right-censored. Csörgő and Horváth (1985) fitted a normal distribution to the data and developed versions of Kolmogorov–Smirnov and Cramér–von Mises statistics to assess this fit. Samuelsen (1985), in his unpublished Norwegian M.Sc. thesis, used the data to illustrate his estimator for doubly censored data, published (Samuelsen, 1989), unfortunately, without this illustration. □

I.3.2. Several Types of Failure

In the previous subsection, the very simplest example of event history analysis was studied: One ("absorbing") event may happen for each of a number of individuals. Sometimes the event is classified into one of several categories, typically causes of death or other "failures." We present two examples of this situation: a classical one taken from the literature and the melanoma survival data introduced in Example I.3.1.

Example I.3.8. Causes of Death in Radiation-Exposed Male Mice

In a laboratory experiment designed to study the effect of radiation on life length, two groups of RFM strain male mice were given a radiation dose of 300 rad at an age of 5–6 weeks. The first group of 95 mice lived in a conventional laboratory environment while the second group was in a germ-free environment. After the death of the mice, necropsy was performed by a

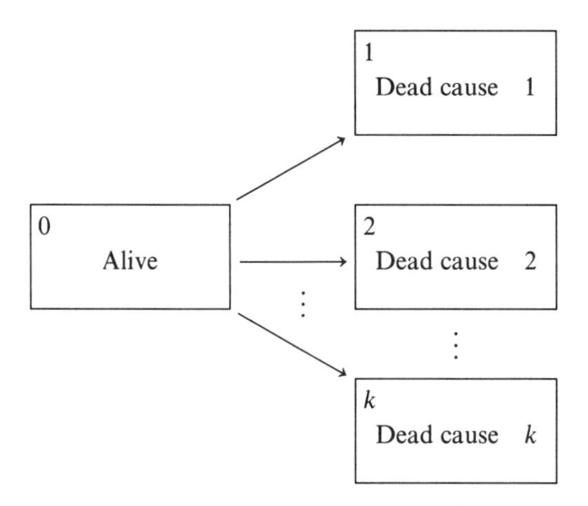

Figure I.3.5. Several types of failure.

Table I.3.5. Necropsy Data for RFM Male Mice Exposed to 300 rads X-Radiation at 5–6 Weeks of Age. (From Hoel and Walburg, 1972.)

Cause of Death	Individual Ages at Death (Days)
	A. Conventional Mice (95)
Thymic Lymphoma (23%)	159, 189, 191, 198, 200, 207, 220, 235, 245, 250, 256, 261, 265, 266, 280, 343, 356, 383, 403, 414, 428, 432
Recticulum cell sarcoma (40%)	317, 318, 399, 495, 525, 536, 549, 552, 554, 557, 558, 571, 586, 594, 596, 605, 612, 621, 628, 631, 636, 643, 647, 648, 649, 661, 663, 666, 670, 695, 697, 700, 705, 712, 713, 738, 748, 753
Other Causes (37%)	163, 179, 206, 222, 228, 249, 252, 282, 324, 333, 341, 366, 385, 407, 420, 431, 441, 461, 462, 482, 517, 517, 524, 564, 567, 586, 619, 620, 621, 622, 647, 651, 686, 761, 763
	B. Germ-Free Mice (82)
Thymic Lymphoma (35%)	158, 192, 193, 194, 195, 202, 212, 215, 229, 230, 237, 240, 244, 247, 259, 300, 301, 321, 337, 415, 434, 444, 485, 496, 529, 537, 624, 707, 800
Reticulum cell sarcoma (18%)	430, 590, 606, 638, 655, 679, 691, 693, 696, 747, 752, 760, 778, 821, 986
Other Causes (47%)	136, 246, 255, 376, 421, 565, 616, 617, 652, 655, 658, 660, 662, 675, 681, 734, 736, 737, 757, 769, 777, 800, 807, 825, 855, 857, 864, 868, 870, 870, 873, 882, 895, 910, 934, 942, 1015, 1019

pathologist, and it was ascertained whether the cause of death was thymic lymphoma, reticulum cell sarcoma (both specific types of cancer), or other causes. Table I.3.5 records the ages at death; the most striking immediate observation being that thymic lymphoma is a dominating cause during the first $1-1\frac{1}{2}$ years, and then loses importance, particularly for conventional mice.

One purpose of the experiment was to assess the temporal patterns of the various causes of death and to compare these over the different laboratory environments. The basic time variable is given as age, but since all the mice were irradiated at almost the same age, it would have been equivalent (and perhaps more intrinsically meaningful) to study time since treatment. A major point is how far one may study one cause of death in isolation.

The data will be further discussed and analyzed in Example IV.4.1.

The basic experiment was described by Walburg and Cosgrove (1969). We quote the data from the expository article by Hoel and Walburg (1972) on statistical analysis of survival experiments; in a companion paper, Hoel (1972), using (almost) the same data, focused on competing risks methodology. (Hoel's article contains, in Table 1, an additional four conventional mice which died at 40, 42, 51, and 62 days, all of "other causes.") The data have been used repeatedly for illustration in the literature on competing risks methodology. Thus, Kalbfleisch and Prentice (1980) quoted the data for use by the readers as exercises; Aalen (1982b) demonstrated the use of his estimate of the integrated cause-specific death intensity, some of whose results were quoted by Andersen and Væth (1984) in their introductory text on survival analysis (in Danish) as well as in Example IV.4.1. Aalen (1982b) also illustrated the calculation of an empirical transition probability matrix for Markov processes using these data. Finally, Matthews (1988) used the data to exemplify a procedure for deriving confidence limits for what is sometimes termed the "cumulative incidence function." The latter may be interpreted as a Markov process transition probability, so that its estimation and confidence limits are special cases of the theory to be developed here; this was noted by Keiding and Andersen (1989) and is also quoted in Example IV.4.1. □

EXAMPLE I.3.9. Survival with Malignant Melanoma

In the data described in Example I.3.1 concerning survival until the end of 1977 among 205 patients with malignant melanoma operated on at Odense University Hospital in the period 1962–77, it may also be of interest to study deaths due to causes other than malignant melanoma. In that case, we have a competing risks model as depicted in Figure I.3.5 with two causes of failure and both cause specific death intensities are then parameters of interest. As mentioned in Example I.3.1, 57 patients died from malignant melanoma, 14 died from other causes, and the remaining 134 patients alive 1 January 1978 are right-censored.

The competing risks aspects of the melanoma data will be illustrated in the following examples:

Example IV.1.11	Nonparametric models for excess and relative mortality.
Example IV.4.1	Theoretical discussion of competing risks.
Example V.1.5	One sample log-rank test comparing intensity of death from causes other than the disease with the general Danish population mortality.
Example VII.2.5	Examples of simple Cox regression models. □

I.3.3. Illness–Death Models

A more detailed event history analysis may be performed when individuals may change status before the absorbing event (death). Figure I.3.6 illustrates the classical illness–death model, where an individual may switch between the states "healthy" and "diseased" before death. The model is also known as a "disability and death" model. Simpler versions of the model excludes the

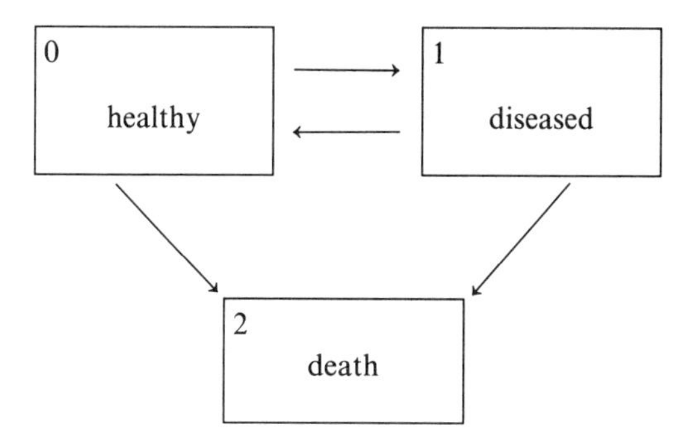

Figure I.3.6. An illness–death model.

Table I.3.6. Psychiatric Admissions and Discharges During the First Year After Having Given Birth for All 71,378 Women Who Gave Birth in Denmark in 1975

Month After Birth:	1	2	3	4	5	6	7	8	9	10	11	12
Admissions	63	34	37	34	31	26	29	17	37	29	25	22
Discharges	22	46	36	39	25	31	32	25	28	28	29	24

possibility of death or does not allow "recovery," that is, the diseased → healthy transition.

Our first example concerns only the healthy ⇄ diseased transitions.

EXAMPLE I.3.10. Psychiatric Admissions for Women Having Just Given Birth

As part of a larger research program to evaluate the introduction of "free" abortion into Denmark by 1 October 1973, the psychiatric career was studied for women who during 1975 had an induced abortion or gave birth. We shall here study only the latter group. Data were obtained by linking the register of births (maintained by the Danish National Board of Health) to the psychiatric register of the Institute of Psychiatric Demography at the University of Aarhus, Denmark. In this way information on dates of admissions to, and discharges from, psychiatric hospitals during the period 1 October 1973–31 December 1976 was obtained for each of the 71,378 women who gave birth in 1975. In addition, age, marital status, number of previous births, and geographical region was known for each woman.

Table I.3.6 shows the number of admissions and discharges for each month during the first year after the birth, the most striking variation being the clustering of admissions right after birth.

We shall be particularly interested in studying transitions in and out of the psychiatric ward *as functions of time since the woman gave birth.* Furthermore, we shall study the influence on the admission rate of the above-mentioned supplementary information as well as of previous psychiatric "career," that is, previous admissions. A difficulty is that the latter are only available insofar as they fall within the sampling window 1 October 1973–31 December 1976.

These data will be discussed and analyzed in the following examples:

Examples III.5.2 and III.5.6	Theoretical notes on using time since last admission as time-dependent covariate, and the possible incomplete observation of this.
Example IV.1.4	Nelson–Aalen estimates of the integrated admission and discharge intensities.
Example IV.2.5	Kernel-smoothed estimates of admission and discharge intensities (with confidence limits), emphasizing estimation in the tails and definition and estimation of the optimal bandwidth.
Example IV.4.2	Aalen–Johansen estimators of transition *probabilities*, with confidence limits.
Example VII.2.17	Cox regression analysis of the admission intensity, using the covariates marital status, age, parity, geographical region, and prior admissions.
Example X.1.4	Theoretical note on the time scales involved: time since birth of the child, duration since last discharge from psychiatric ward, and age of the mother.

A more detailed presentation of the problem and the data was given by Andersen and Rasmussen (1986); see Rasmussen (1983, in Danish) for background on the study. Andersen and Gill (1982) used analyses of these data as an example of their Cox regression model for counting processes. An excerpt of the analyses of Examples IV.1.4, IV.2.5, and IV.4.2 was published by Keiding and Andersen (1989). □

Next, an example with no recovery (diseased → healthy excluded) is given.

EXAMPLE I.3.11. Nephropathy for Diabetics

The Steno Memorial Hospital in Greater Copenhagen was established in 1933 and has since then served as a diabetes specialist hospital for patients from the whole of Denmark. Patients are referred to the Steno from general practitioners and/or other hospitals. About two-thirds of the patients return to the referring physician or local hospital for continued control, while the remainder attend the outpatient clinic at the Steno for longer periods. From the medical records at the Steno, all patients (1503 males and 1224 females) referred between 1933 and 1981 and in whom the diagnosis of insulin-dependent diabetes mellitus was established before age 31 years and between 1933 and 1972 were included in the study (Borch-Johnsen et al., 1985; Ramlau-Hansen et al., 1987). The patients were followed from first contact with the hospital to death, emigration, or 31 December 1984. Thus, the observation of a patient is *left-truncated* (as in Example I.3.2) at the time of first contact to the Steno and *right-censored* if the patient emigrated, was lost to follow-up, or was alive on 1 January 1985.

One of the major complications of insulin-dependent diabetes is *diabetic nephropathy* (DN) which is a sign of kidney failure and defined to be present if at least four samples of 24-h urine at time intervals of at least one month contain more than 0.5 g protein. One hundred and fifteen patients had DN at first admission and 616 developed DN during the observation period. Among these 731 patients, 451 died during the observation period, whereas 267 patients died without DN. Thus, observations include transitions between the states of a model like the one shown in Figure I.3.6 (with no possibility of recovery), where state 0: *alive without DN* corresponds to the *healthy* state, and state 1: *alive with DN* to the *diseased* state.

The seriousness of DN is reflected by the fact that the number of person-years at risk in state 0 is 44,561 and only 5,024 in state 1.

For all 2727 patients, information on sex, date of birth, age at diabetes onset, date of first contact with the hospital, and date last seen was available. In the 731 patients observed in state 1, the year of onset of DN was known in 594, whereas for the remaining 137 the only information available was a year where the patient was last seen without DN and a year where he or she was first seen with DN. This means that for the latter 137 patients the observation of the transition from state 0 to state 1 is *censored on the interval* between the above-mentioned dates. The median interval length was 14 years; the minimum and maximum were 1 year and 34 years, respectively. In the examples in this monograph, the data will be analyzed pretending that the exact time of onset of DN was observed. This is done by calculating for each of the 137 patients (97 males and 40 females) a predicted time of onset of DN measured from onset of diabetes. The prediction was carried out as follows, separately for males and females since males and females are expected to have different intensities of developing DN (Andersen et al., 1983a; Borch–Johnsen et al., 1985; Ramlau–Hansen et al., 1987; Andersen, 1988): Let T_j be the time from diabetes onset to DN for the 594 patients who did get DN and for whom the time of onset of DN was observed. For each of the remaining 137 who did get DN, the time of onset of DN was predicted as

$$\hat{T}_i = \frac{\sum_j T_j I(T_j \in I_i)}{\sum_j I(T_j \in I_i)},$$

where I_i is the time interval in which the $0 \to 1$ transition was known to take place. In Example IV.2.3 where the first analyses of the data are presented, we study the usefulness of this prediction procedure.

It is seen that several time scales are involved here: First, there are *age* and *diabetes duration*, but also *calendar time* and (for death with DN) *duration of DN* may play a role. It is not obvious from the outset which of the former time scales to consider as the basic one, and one purpose of the analyses of the data is to try to make this choice, if necessary.

The Steno Memorial Hospital data on mortality and nephropathy for insulin-dependent diabetics are discussed and analyzed in the following examples:

Example III.3.7	Theoretical discussion of models for the truncation and censoring mechanisms.
Example III.4.2	Discussion of interval censoring.
Example IV.2.3	Kernel-smoothed estimates for the DN intensity for males. Effect of interval censoring.
Example IV.4.3	Nonparametric estimation of transition probabilities in the three-state disability model.
Example V.2.13	Nonparametric comparison of mortality among patients with or without DN.
Example VII.1.1	Discussion of type-specific covariates in regression models for the transition intensities.
Example VII.2.2	Discussion of partial likelihoods for multiplicative regression models.
Example VII.2.11	Presentation of regression models including calculation of life insurance premiums.
Example X.1.5	Note on time scales: age, calendar time, duration of diabetes, and duration of DN.

The data for patients diagnosed between 1933 and 1952 were presented by Andersen et al. (1983a) and further analyzed by Borch–Johnsen et al. (1985) and Andersen (1988). Analyses of these data together with a part of the Fyn County data (Example I.3.2) in a relative mortality regression model were presented by Green et al. (1985) and Andersen et al. (1985). Borch–Johnsen et al. (1986) and Kofoed–Enevoldsen et al. (1987) analyzed the mortality and the DN intensity of all patients diagnosed before 1981, whereas Ramlau-Hansen et al. (1987) used these data for the calculation of a suggestion for new life insurance premiums for Danish insulin-dependent diabetics. These new premiums are now used in Denmark. □

Finally, we present an example using the full model illustrated in Figure I.3.6.

EXAMPLE I.3.12. Abnormal Prothrombin Levels in Liver Cirrhosis

In the randomized clinical trial concerning prednisone treatment of liver cirrhosis (presented in Example I.3.4), various biochemical and clinical characteristics were measured at follow-up visits for the duration of the trial.

Table I.3.7. Transitions Between Low and Normal Prothrombin States and from These to Death for 488 Patients Followed with Liver Cirrhosis

	Prednisone Treatment		Placebo Treatment	
Transition	Number of Transitions	Intensity per Year	Number of Transitions	Intensity per Year
From low to normal	159	0.735	155	0.563
From normal to low	131	0.189	143	0.241
From low to dead	92	0.423	96	0.349
From normal to dead	50	0.0721	54	0.0912

We shall concentrate on the effects of the *prothrombin index*, a measurement based on a blood test of coagulation factors II, VII, and X produced by the liver. The index was scored as abnormal when less than 70% of normal values and normal otherwise.

Note that the state of having abnormal prothrombin values is *reversible*: Patients may return to normal values. A further complication is that the prothrombin status is not, in fact, registered continuously, but rather at irregular and infrequent follow-up visits. In Table I.3.7 and in the analyses of the data to be presented in this monograph, we shall, however, neglect this fact and define changes of values to take place *at* the time points where patients are examined at follow-up visits to the hospital. Consider, for instance, a patient with a prothrombin index of 50% at day 0 (time of start of treatment). If he or she has a prothrombin index of 80% at the next visit to the hospital at day 90, then we assume the patient to be in state 1: "low prothrombin" in the entire interval [0, 90 days], and that he or she makes a transition to state 0: "normal prothrombin" at day 90. If, instead, the patient had died at day 40, then we would assume that the transition to the state 2: "dead" at that day, was a transition from state 1.

Table I.3.7 contains the observed number of the four possible transitions for both treatment groups together with estimated intensity per year which is calculated as the number of transitions divided by the total time (in years) lived by the patients in the "from" state. It appears that prednisone-treated patients are more likely to go to normal prothrombin status (and stay there), but on the other hand, those prednisone patients who have low prothrombin values are more likely to die than placebo patients with low prothrombin. Qualitative, not to speak of quantitative statements regarding prognosis therefore require rather detailed modelling.

The prothrombin data from the CSL 1 trial will be discussed and analyzed in the following examples:

Example IV.4.4 Nelson–Aalen estimates of the integrated transition intensities and nonparametric transition probability estimates in a three-state Markov process.

Example V.2.6 Nonparametric comparison of transition intensities between the two treatment groups.

Example VI.1.7 Estimation in a parametric model with piecewise constant intensities.

Example VII.2.10 Semiparametric proportional transition intensity models. Estimation of transition probabilities.

The follow-up data (including prothrombin) from the CSL 1 trial were presented by Christensen et al. (1985, 1986) and further analyzed by Andersen (1986). Estimation of the transition probabilities in non- and semiparametric models were discussed by Andersen et al. (1991a) and in parametric models by Andersen et al. (1991b), the latter study including a discussion of the effect of assuming the transition times to be observed. □

I.3.4. Other Models

In this section, we collect a number of examples not covered by the above headings. The first two examples concern movements between several states.

Example I.3.13. Pustulosis Palmo-plantaris and Menopause

Pustulosis palmo-plantaris is a chronically recurrent skin disease localized to the palms of the hands and the soles of the feet. In a consecutive series of 100 patients at the Department of Dermatovenerology, Finsen Institute, Copenhagen, 85 of the patients were women and, as seen from Table I.3.8, most first appearances happened between the ages of 35 and 55 years. There are no known causes of the disease and it was, therefore, considered of interest to search for a possible connection between the events of menopause as reported by the women and the first appearance of pustulosis palmo-plantaris.

An event history model for the events of pustulosis and natural as well as artificial menopause (induced surgically or medically) is indicated in Fig. I.3.7; it is seen that the hypothesis of an influence of the event of menopause

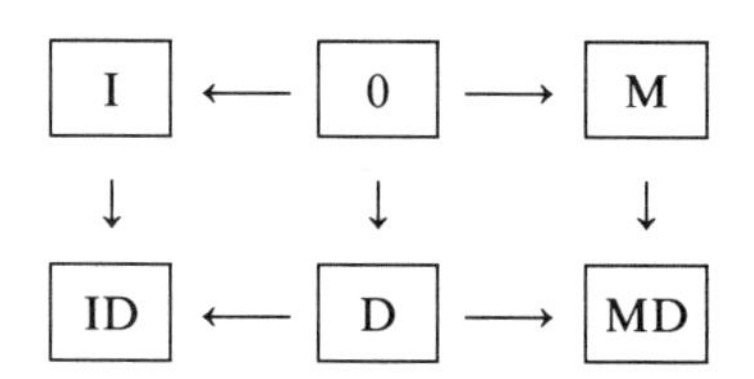

Figure I.3.7. Event history model for the occurrence of natural and induced menopause, and the outbreak of pustulosis palmo-plantaris. 0: no event has occurred; M: (natural) menopause has occurred; D: disease has broken out; I: induced menopause has occurred.

Table I.3.8. Age of First Appearance, Age of Natural or Induced (*) Menopause and Age Last Seen for 85 Female Patients with Pustulosis Palmo-plantaris

First Appear.	Meno-pause	Last Seen	First Appear.	Meno-pause	Last Seen	First Appear.	Meno-pause	Last Seen	First Appear.	Meno pause	Last Seen
17	—	18	40	50	74	49	46	49	54	43*	56
19	—	22	40	47	48	49	49	55	56	50	63
23	—	33	40	50	52	49	45	74	57	47	64
24	—	26	40	46	59	50	52	54	57	37	60
27	—	38	40	51*	55	50	44*	53	58	48	60
27	—	40	41	—	41	50	50*	50	59	50	61
29	—	30	41	40	54	50	45*	50	59	47	64
29	—	32	41	38*	43	51	50	53	60	50	62
30	—	31	43	43	44	51	48*	52	60	51	62
30	—	32	43	47	50	51	49	51	61	55	66
31	—	33	44	—	46	51	48	66	62	54*	67
31	—	35	44	—	44	51	46	62	64	48	66
33	—	37	45	46*	47	52	47	64	64	42	68
34	31*	69	45	45*	47	53	48	56	64	40	66
35	—	36	45	43	47	53	48	54	64	49*	64
35	32*	67	45	—	46	53	52	56	64	47*	65
36	—	37	45	45	55	53	46*	56	66	47	68
36	—	40	47	43	51	53	50*	54	67	52*	71
37	28*	38	48	49	70	53	43	54	67	49	79
37	—	37	48	40	50	54	48	54	70	45	71
38	39	39	48	47	49	54	46	54	73	45	74
39	37*	42									

upon the incidence of pustulosis corresponds to asserting that the transitions $I \rightarrow ID$, $0 \rightarrow D$, and $M \rightarrow MD$ happen at the same rate. However, sampling takes place only from the current (so-called prevalent) population of patients with pustulosis, and it is necessary to specify a model for this sampling procedure before statistical inferences can be drawn.

In Example V.2.8, we specify the model as a Markov process and quote the results of tests for the hypothesis of equal occurrence intensity (incidence) of the disease before and after menopause.

A more detailed discussion was given by Aalen et al. (1980) and Borgan (1980); Borgan and Gill (1982) used the data to illustrate their approach to age-specific case-control theory. □

EXAMPLE I.3.14. Bone Marrow Transplantation: the Development and Prognostic Significance of Cytomegalovirus Infection and Graft-Versus-Host Disease

In five Nordic centers, a total of 190 leukemia patients were bone marrow transplanted (BMT) in 1980–85 and followed up until the end of 1985. For each patient, it was recorded whether (and if so, when) any of the following side effects developed: cytomegalovirus (CMV) infection, acute graft-versus-host disease (GvHD), chronic GvHD, and times of relapse and/or death were also noted. A number of supplementary variables, known at BMT and suspected of influencing prognosis, were also taken into account; these included age and sex of donor and patient, CMV-immunity status (before transplantation) of donor and patient, disease stage, calendar time, and center. The three side effects are all suspected of influencing the risk of developing others as well as of relapse of the leukemia and death, and the object of the analysis is to disentangle these effects. A complete diagram of the possible transitions of the kind illustrated in Fig. I.3.7 would be too complicated, but Fig. I.3.8 indicates the possible transitions when the temporal

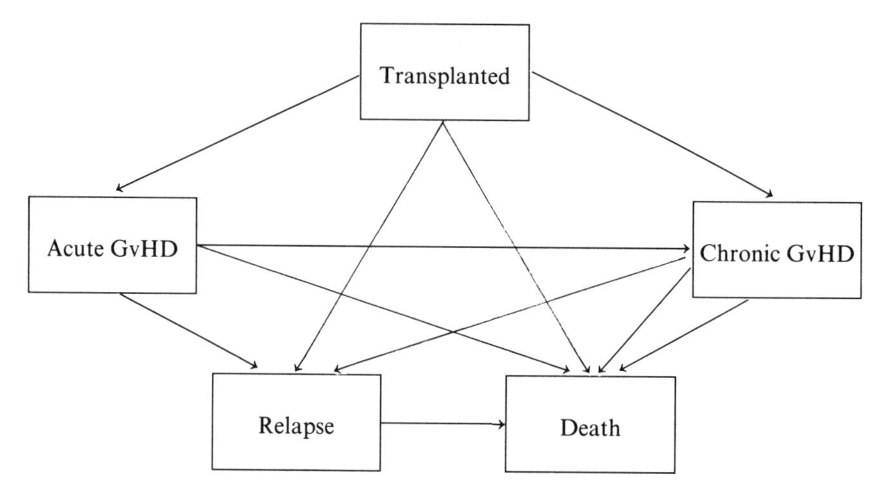

Figure I.3.8. Simplified diagram of post bone marrow transplant event history.

pattern of CMV infection is ignored; in this case, the structure becomes ordered, as it appears from the figure.

The data will be analyzed and discussed in the following examples:

Example V.2.4 Illustration of five-sample test comparing the occurrence of chronic GvHD between the five centers.

Example V.2.12 Illustration of conservative approximation to the above five-sample test.

Example VII.2.18 The "antileukemic effect" of GvHD discussed via proportional intensity models.

Risk factors for the occurrence of GvHD and leukemic relapse were discussed in a preliminary communication by Jacobsen et al. (1987), whereas Jacobsen et al. (1990) gave a rather elaborate discussion of the so-called graft-versus-leukemia activity, in particular whether GvHD decreases the risk of relapse. □

The next example deals with survival data where attention focuses on the possible *dependence* between related individuals.

Example I.3.15. Heritability of Life Length

The Danish adoption registry contains information on all Danish children adopted by unrelated families between 1924 and 1947 (a total of 14,427 persons). A subsample was taken from this registry containing the 1003 children from the registry born between 1924 and 1926. The children were followed from adoption to death, emigration, or 31 December 1987, and information of the life lengths of their biological parents and their adoptive parents was obtained. For all individuals, information on cause of death was also obtained.

The purpose of the study was to relate the life length of the adoptive child to those of the biological parents and to those of the adoptive parents, respectively, to study the possible effects of heritability and environmental factors on longevity. Thus, explicit modelling of the *correlation* between these life lengths is desired and standard models for independent survival times are insufficient.

For the analyses to be reported in this monograph, we follow Sørensen et al. (1988) and further choose those adoptees who (1) are transferred to the adoptive family no later than at age 7 years and (2) survive at least until age 15 years. Table I.3.9 gives some summary statistics for this sample.

The data will be discussed and analyzed in the following example:

Example IX.4.1 Heritability of life length. Association between the life times of adoptees and their parents.

The data were first described and analysed using classical techniques for survival data by Sørensen et al. (1988). The frailty models to be discussed in

Table I.3.9. Number of Parents of the Four Different Types and Number of Deaths in a Study of Heritability of Life Lengths

Parent	No. of Pairs	No. of Dead Adoptees	No. of Dead Parents	No. of Pairs Where Both Members Died
Adoptive father (AF)	924	180	896	175
Adoptive mother (AM)	958	187	867	167
Biological father (BF)	809	158	717	147
Biological mother (BM)	925	181	649	128

Chapter IX were introduced by Nielsen et al. (1992) who also used these data as illustration. □

The next three examples come from biology. The first one deals with the mating pattern of Drosophila flies.

EXAMPLE I.3.16. Mating of Drosophila Flies

In each mating experiment, 30 female virgin flies and 40 male virgin flies of the type Drosophila melanogaster were inserted into an observation arena, the "pornoscope," consisting of a circular plastic bowl, 1 cm high with diameter 17 cm, covered by a transparent lid. The flies were observed continuously and times of initiation and termination of matings were recorded. So, in particular, the number of ongoing matings is known for each time point. The experiment was terminated at the first time instant after 45 min from initiation of the first mating at which no matings were going on—no later, however, than 60 min after initiation of the first mating. The observation times are unpaired in the sense that it is unknown which termination times correspond to which initiation times.

Four experiments are considered here: two "homogametic," where the male and female flies belong to the same race (either a black ebony race or a yellow oregon race), and two "heterogametic," where males and females are of different races. One of the purposes of the experiments was to study the possible sexual selection, i.e., whether, for instance, a male ebony fly is more liable to mate with a female ebony fly than with a female oregon fly. The data are summarised in Table I.3.10.

The data are discussed and analyzed in the following examples:

Example III.1.10 A counting process model for the mating experiment.

Example III.2.12 Discussion of the censoring pattern.

Example III.5.5 A more realistic counting process model for the mating experiment.

Example IV.1.5 Nelson–Aalen plots of the integrated mating intensity estimates.

Table I.3.10. Times of Initiation (T+) and Termination (T−) of Mating in Four Experiments. (Seconds after Insertion into Pornoscope.)

Experiment Type (female, male)							
(ebony, ebony)		(oregon, oregon)		(ebony, oregon)		(oregon, ebony)	
T+	T−	T+	T−	T+	T−	T+	T−
143	807	555	1749	184	539	257	1288
180	1079	742	1890	252	1244	958	2101
184	1167	746	1890	268	1253	1482	2599
303	1216	795	2003	331	1337	1633	2747
380	1260	934	2089	345	1375	2029	2935
431	1317	967	2116	368	1441	2123	2970
455	1370	982	2116	419	1457	2131	3016
475	1370	1043	2140	467	1513	2254	3296
500	1384	1055	2148	502	1533	2330	3364
514	1384	1067	2166	614	1593	2627	3484
521	1434	1081	2166	614	1593	2690	
552	1448	1296	2347	697	1638	3372	
558	1504	1353	2375	735	1784	3423	
606	1520	1361	2451	745	1862	3732	
650	1534	1462	2507	908	2096		
667	1685	1731	2544	1180	2327		
683	1730	1985	2707	1330	2391		
782	1755	2051	2831	1541	2411		
799	1878	2292	3064	1605	2502		
849	2212	2335	3377	1891	2770		
901	2285	2514	3555	2291	3177		
995	2285	2570	3596	2305	3207		
1131	2285	2970		2429	3457		
1216	2411			2499	3545		
1591	2462			3059	3620		
1702	2591			3228			
2212	3130			3394			
				3557			

Example IV.1.10 Theoretical discussion of large sample properties of the Nelson–Aalen estimators.

Example V.2.7 Nonparametric comparison of mating intensities.

Example V.3.3 Nonparametric comparison of mating intensities using stratified tests.

The data came from unpublished experiments conducted by F.B. Christiansen in 1969. They were used for illustration by Aalen (1978b) and Aalen and Hoem (1978). □

The second biological example is about the feeding pattern of rabbits.

EXAMPLE I.3.17. Feeding Habits of Rabbits

To study the feeding pattern of New Zealand rabbits, Jolivet et al. (1983) recorded the feeding times of rabbits during a week: The ending time of each intake has been measured and also the weight of eaten food for each intake. Figure I.3.9 shows the eating times and quantities consumed for one of the rabbits; obviously the rabbit is more active at night than during the day, so that when viewed over a week, a *periodic* pattern is observed.

Jolivet et al. modelled the sequence of feeding times by a Poisson process having a periodic intensity and gave results about the effect of aging on that intensity. Pons and Turckheim (1988a,b) considered how the past feeding behavior may change the intensity of the occurrence of the feeding times: A dynamical model would assume that the intensity is a function of time, of the past feeding times, and possibly of the quantities eaten at earlier feeding times.

The data are discussed and analyzed in the following examples:

Example III.5.4 — A model for the intensity of the point process of feeding times.

Example VII.2.15 — Theoretical properties of a periodic regression model and excerpts of the results obtained by Pons and Turckheim (1988a).

A bibliography for the example is contained in the description above. □

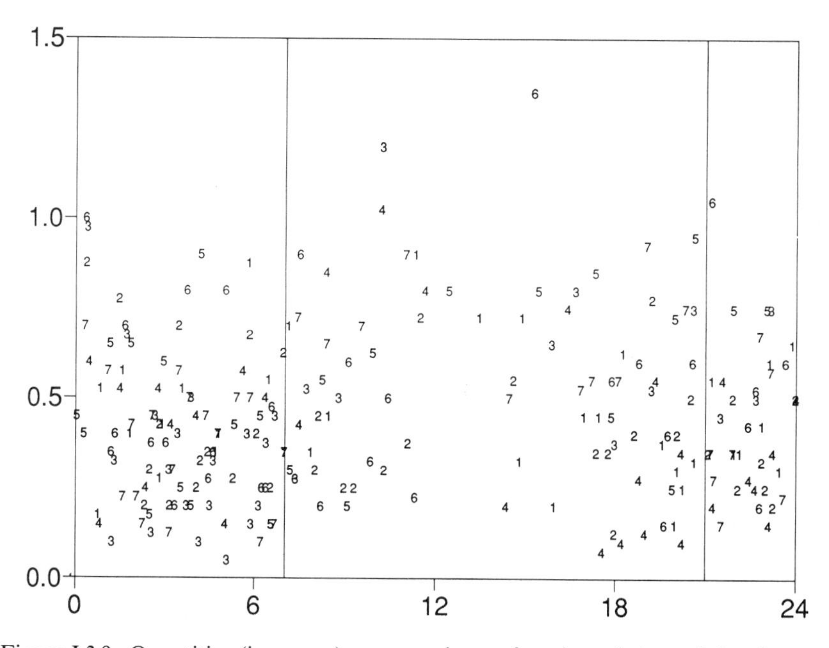

Figure I.3.9. Quantities (in grams) consumed as a function of time of day for one rabbit. The number is the number of the day of the week, whereas the vertical lines indicate the limit between day (7–21) and night. (Redrawn from Pons and Turckheim, 1988b.)

In the third biological example, not only do we follow life events for each individual (here death and emigration) but we also record (and model) births and immigrations.

EXAMPLE I.3.18. Dynamics of a Natural Baboon Troop

Altmann and Altmann (1970) recorded births (B), deaths (D), emigrations (E), and immigrations (I) over a year of a baboon troop in the Amboseli reserve, Kenya; cf. Table I.3.11.

Table I.3.11. Birth (B), Immigration (I), Death (D), and Emigration (E) in Altmann and Altmann's Main Study Troop

After This Many Days	At This Troop size	This Event Occurred
41	40	B
5	41	B
22	42	B
2	43	D
17	42	D
26	41	I
0	42	I
55	43	B
35	44	I
20	45	E
5	44	D
6	43	E
32	42	D
4	41	D
0	40	D
22	39	D
10	38	B
0	39	B
7	40	D
4	39	B
17	40	D
11	39	E
3	38	B
4	39	D
8	38	D
2	37	D
5	36	B
10	37	B
Final	38	—

From Cohen (1969); based upon Altmann and Altmann (1970), Tables III and IV.

As an initial step toward modelling the dynamics of social behavior in a baboon troop like this, it was of interest to investigate a birth–immigration–death–emigration model with a constant intensity of immigration and constant intensities per individual of birth, death, and emigration. In particular, it would be desirable to be able to interpret the group as being in an equilibrium state as governed by the B, D, I, and E rates; comparisons with cross-sectional studies of groups of baboons of this and other species might then be used as direct data on the social behavior of these animals.

However, the model is primarily to be thought of as a baseline against which to compare more refined hypotheses, of which Altmann and Altmann (1970, pp. 207–208) mentioned several, such as (1) dependence between individuals in the group with respect to, e.g., death, (2) birth rate depending on the number of reproductive individuals (or reproductive females only), (3) age-dependent mortality rate, (4) only adult males seem to emigrate, so emigration rates should be related to the number of these, (5) group fission may occur.

The data will be discussed and analyzed in the following examples:

Example III.1.6 Theoretical note specifying the birth–immigration–death–emigration process as a counting process.

Example VI.1.6 Maximum likelihood estimation.

Example VI.3.2 Goodness-of-fit assessed from Nelson–Aalen plots with confidence bands.

Example VI.3.4 Goodness-of-fit assessed by total time on test plots and statistics.

Example VI.3.8 Goodness-of-fit assessed by Khmaladze's adjusted residual approach.

The data were collected and reported by Altmann and Altmann (1970) and further analyzed by Cohen (1969), using, among other things, a jackknife approach for inference in the birth–immigration–death–emigration process. Keiding (1977) added comments to Cohen's analysis from the point of view of counting processes. □

Our final example comes from technology.

Example I.3.19. Software Reliability

Moek (1984) reported a study on an information system for registering aircraft movements. The system can be classified as a transaction-oriented on-line system where relatively small jobs are submitted by several users at arbitrary points of time. The jobs concern information retrieval as well as the update of existing information. The information system consists of three identical subsystems, each comprised of a PDP 11-34 computer with 256 KB direct access memory, working together in such a way that each user has continually at his disposal the most recent information. A distributed data base system is used for this purpose. The software consists of about 600 subroutines with an average length of 70 lines of executable code written in BASIC-PLUS-2 (a compiler-oriented language) and about 40 subroutines with an average length of 200 lines of executable code written in assembler. Each subsystem has 20 MB disc capacity. The number of user terminals connected is 35 and the number of line printers is 4. The entire system was developed in-house and, after some testing, was carried over to the users who continued testing in the operational environment. Failure data collected during the latter stage are given in Table I.3.12. Some faults caused the same failure more than once because immediate correction of the faults was not always possible. Only the first occurrence of such failures was counted in Table I.3.12. A main purpose of the study is to estimate the total number of remaining faults in the software.

These data will be discussed and analyzed in the following examples:

Example III.1.12 Theoretical note introducing two different stochastic models.

Table I.3.12. Failure Data for an Information System (Seconds CPU Time)

Failure No.	Interfailure Time	Failure Time	Failure No.	Interfailure Time	Failure Time
1	880	880	23	4450	133210
2	3430	4310	24	4860	138070
3	2860	7170	25	640	138710
4	11760	18930	26	3990	142700
5	4750	23680	27	26840	169540
6	240	23920	28	2270	171810
7	2300	26220	29	200	172010
8	8570	34790	30	39180	211190
9	4620	39410	31	14910	226100
10	1060	40470	32	14670	240770
11	3820	44290	33	16310	257080
12	14800	59090	34	38410	295490
13	1770	60860	35	1120	296610
14	24270	85130	36	30560	327170
15	4800	89930	37	6210	333380
16	470	90400	38	120	333500
17	40	90440	39	20210	353710
18	10170	100610	40	26400	380110
19	1120	101730	41	37800	417910
20	980	102710	42	74220	492130
21	24300	127010	43	84440	576570
22	1750	128760			

Example VI.1.8	Maximum likelihood estimation in these models.
Example VI.1.12	Theoretical note on the asymptotic properties of maximum likelihood estimators.
Example VI.2.5	M-estimation in the Littlewood model.

The data were first presented by Moek (1984) and were analyzed by Geurts, Hasselaar, and Verhagen (1988) and van Pul (1992a). □

CHAPTER II

The Mathematical Background

Despite the great variety of examples introduced in Chapter I, and the equally great variety of statistical questions which arise from them, we will be able to study both with just a handful of basic tools from the theory of stochastic processes: the theory of *counting processes* and their *intensity processes*, the theory of *stochastic integration*, and *martingale central limit theory*, all centering around the mathematical concepts of *martingale*, *predictable process*, and *filtration*. The present chapter surveys and summarizes the basic theory as we will need it and also gives some basic mathematical material on product-integrals and functional differentiation (the functional delta-method).

The first section gives a heuristic introduction to the basic concepts in the context of one of our basic examples: censored survival data (see Example I.3.1). Key concepts such as filtration, martingale, and predictable process are introduced informally, and their relationships in stochastic integration and martingale central limit theory are illustrated through a statistical example. The aim is to provide intuition for readers not anxious to study the general theory of stochastic processes in great depth. In fact, only quite a small part of this extensive and elaborate theory is needed.

Subsequent sections give a more formal and precise account of the theory. As we will explain later, the core material for the book is the following: basic probability theory in Sections II.2, II.3, and II.4.1, asymptotic theory in Section II.5, and product-integration in II.6. Model building and likelihood constructions need the more advanced material in the rest of Section II.4 and in Section II.7. Section II.8 is not needed at a first reading. The whole chapter is intended to form a compendium of results for later use, and the reader should not feel obliged to study it all in detail before proceeding with the rest of the book.

It would certainly be helpful if the reader already has some familiarity with such notions as *martingale* and *stopping time*, at least, in discrete time. For an introduction to discrete-time martingale theory, see almost any intermediate-level course in probability theory; e.g., Sections 7.7–7.9 of Grimmett and Stirzaker (1982); and for an extensive treatment, including the measure-theoretic approach to conditional expectation and the Radon–Nikodym derivative, see, e.g., Chung (1974) or Breiman (1968). Fleming and Harrington (1991, Chapter 1 and Appendix A) gave a useful summary.

Section II.2 sets the scene, defining stochastic processes, filtrations, and stopping times. Some important notational conventions are also introduced here; these and some others are summarized at the end of the present introductory material.

Section II.3 contains the basic theory of stochastic integration as we shall be using it, tied up with the notions of predictable process and local martingale. The continuous-time theory depends on the fundamental Doob–Meyer decomposition, separating a process into a systematic (predictable) part and a purely random (martingale) part. The elementary discrete-time version of this theorem, together with discrete-time stochastic integration, can be found, for instance, in the book by Chung (1974, Theorem 9.3.2 and Exercises 9.3.9 and 9.3.16). In discrete time, the stochastic integral, an almost trivial object, is known as Doob's "martingale transform."

Now that our basic calculus has been set out, we can start in Section II.4, studying the objects which the book is really about: counting processes. They are related to the general theory of martingales and predictable processes through the key notion of the stochastic (predictable) intensity process belonging to a given counting process. In fact, the integrated intensity process is the systematic part (or compensator) in the Doob–Meyer decomposition of the counting process itself. The basic theory of counting processes is presented in Section II.4.1. Subsequent subsections contain more specialized topics—miscellaneous results on building counting process models which will be much used in Chapter III (though less in other parts of the book).

Section II.5 introduces central limit theorems for (continuous time) martingales. The main theorem, due to Rebolledo (1979), will be continually used to prove the asymptotic normality of statistical estimators, test statistics, etc. The conditions of the main theorem are specialized to the case we will need all the time: martingales which are stochastic integrals of predictable processes with respect to counting process martingales (compensated counting processes). Here, the reader will need some familiarity with the theory of convergence in distribution (weak convergence) of stochastic processes. Fleming and Harrington (1991, Appendix B) gave a clear summary and further references.

Section II.6 contains a nonstochastic interlude on the theory of product-integrals, with its applications (in statistics and probability theory) to hazard rates and Markov processes. The theory is not especially deep: One should consider it as a way to provide an evocative notation for a simple mathematical operation which crops up time and time again in apparently different contexts—likelihood expressions, building transition probabilities from transition rates, distributions from hazard rates. For a first reading, the most important thing is the definition and the applications. Material on special properties of the product-integral can be consulted when needed.

The next two sections are more specialized and could also be omitted at a first reading.

Section II.7 uses product-integrals to represent likelihoods for counting process models. The basic results are given in Section II.7.1; more advanced material (martingale connections, partial likelihood) in the other subsections are needed mainly in Chapter III for our discussion of independent and noninformative censoring. The reader must here be familiar with the notion of the Radon–Nikodym derivative of one measure with respect to another as providing a generalization of both probability density, mass function, and likelihood ratio.

Section II.8 is concerned with a very different topic. It develops the idea of differentiating functions of functions (e.g., the mapping from a distribution function to its quantile function) with a view toward an infinite-dimensional version of the delta-method (variously known as the method of propagation of errors, first-order Taylor expansion, linear approximation, etc.), familiar from elementary statistics (Rao, 1973, Section 6a.2). This supplies a calculus for generating new weak convergence results from old, extending the range of the martingale central limit theorem. This material is not central to the book and the reader could consult it when needed (mainly in Chapters IV and VIII).

Finally, Section II.9 contains historical and bibliographic remarks, together with some technical notes on specialized matters.

Notation

Vectors and matrices are (usually) printed in bold type. This also applies to vector or matrix functions of time t. (In Chapter VIII, bold is also used for other purposes.)

Integrals are Lebesgue–Stieltjes integrals. For a crash course in modern integration theory, see, e.g., Breiman (1968, Appendix) or Rudin (1976, Chapter 11). We use both of the notations $\int \cdots \mathrm{d}F(t)$ and $\int \cdots F(\mathrm{d}t)$ to denote the integral of some function with respect to (the measure generated by) F; no difference in meaning is intended; some contexts just make one or the other notation less ambiguous or more intuitive than the other. Product-integrals ($\prod$) are defined in Section II.6.

$I(\cdots)$ is usually used for the indicator of the set $\{\cdots\}$; $\mathbf{I}$ stands for an identity matrix, $\mathscr{I}$ for an observed or asymptotic information matrix, ι for an identity mapping ($\iota(t) = t$) and $\boldsymbol{\iota}$ for a vector of identity mappings [$\boldsymbol{\iota}(t) = (t, \ldots, t)$].

When vectors really are used as vectors they are to be considered *column* vectors. Often however we just use vector notation to indicate a collection of objects, and write, e.g., $\mathbf{N} = (N_1, \ldots, N_k)$, whereby we do not distinguish between a row or column vector.

The transpose of a vector is indicated by the superscript $^\mathsf{T}$; t usually denotes a time (also s, u, v, τ); $\mathscr{T}$ is a collection of times (the interval $[0, \tau)$ or $[0, \tau]$, where $\tau = \infty$ is also allowed); T is usually a random time (a stopping time, in fact). There is nothing special about the initial time 0, this could also have been replaced by an arbitrary value σ. The indicator function of a time interval $(s, t]$ is denoted $I_{(s,t]}$.

The fixed time interval $\mathscr{T} = [0, \tau)$ or $[0, \tau]$ will usually be in the background at any point in the book. When an integral $\int \cdots$ without limits on the integral sign is mentioned, we mean the *function* of time t, obtained by integrating over the interval $[0, t]$ for each $t \in \mathscr{T}$. The same convention applies to the product-integral $\prod \cdots$.

F and f, A and α, Λ, and λ, usually denote a function and its density or derivative (also in vector or matrix versions). ΔF is the function giving the jumps of F, $\Delta F(t) = F(t) - F(t-)$, supposing F to be right-continuous. In particular therefore, $\int \cdots F(\mathrm{d}t) = \int \cdots f(t)\,\mathrm{d}t$ if F has a density f, whereas $\int \cdots F(\mathrm{d}t) = \sum \cdots \Delta F(t)$ if F is a step-function.

$D(\mathscr{T})$ is the space of right-continuous functions with left-hand limits on $\mathscr{T}$; the so-called *cadlag* functions. The space $D(\mathscr{T})$ itself is often called the Skorohod space, reflecting the usual choice of metric on this space in classical weak convergence theory (Billingsley, 1968; Pollard, 1984).

The notations $\stackrel{\mathscr{D}}{\to}$ and $\stackrel{\mathrm{P}}{\to}$ stand for convergence in distribution and probability, respectively.

The notations $\langle \cdot \rangle$ and $[\cdot]$ for predictable and optional variation process are introduced in Section II.3. The context should always show when a time interval $[s, t]$ and when an optional covariation or "square bracket" process $[M, M']$ is meant.

II.1. An Informal Introduction to the Basic Concepts

In subsequent sections of this chapter, we shall give a formal survey of the theory which we will use to analyze statistical models based on counting processes. By way of introduction, however, we will first discuss the key concepts in a very informal manner in the context of one of the basic examples. At the same time, we will see how the mathematical background lends itself beautifully to the treatment of statistical models formulated in terms of the *hazard rate* and possibly involving *censoring*; thus, also introducing topics treated in depth in Chapters III and IV.

Consider first a sample of n (uncensored) continuously distributed survival times $X_1, \ldots, X_n$ from a survival function S with hazard rate function α; thus,

$\alpha = f/(1 - F)$ where $F = 1 - S$ is the distribution function and f the density of the X_i. The hazard rate α completely determines the distribution through the relations

$$S(t) = \mathrm{P}(X_i > t) = \prod_0^t [1 - \alpha(s)\,\mathrm{d}s] = \exp\left(-\int_0^t \alpha(s)\,\mathrm{d}s\right),$$

where the product-integral $\prod(1 - \alpha)$ is explained later in this section, see (2.1.13), and studied in detail in Section II.6. One can interpret α by the heuristic

$$\mathrm{P}(X_i \in [t, t + \mathrm{d}t) | X_i \geq t) = \alpha(t)\,\mathrm{d}t. \tag{2.1.1}$$

We will consider the nonparametric estimation of the hazard rate or, rather, the cumulative or integrated hazard rate

$$A(t) = \int_0^t \alpha(s)\,\mathrm{d}s. \tag{2.1.2}$$

Typically, in survival analysis problems, complete observation of $X_1, \ldots, X_n$ is not possible. Rather, one only observes $(\tilde{X}_i, D_i)$, $i = 1, \ldots, n$, where D_i is a "censoring indicator," a zero-one valued random variable describing whether X_i or only a lower bound to X_i is observed (really D_i indicates uncensored):

$$X_i = \tilde{X}_i \quad \text{if } D_i = 1,$$
$$X_i > \tilde{X}_i \quad \text{if } D_i = 0.$$

We shall consider $\tilde{X}_1, \ldots, \tilde{X}_n$ as random times; at these times, the value of the corresponding D_i becomes available, and we know whether the corresponding event is a failure or a censoring. Thus, all n survival periods start together at time $t = 0$.

As an example, Figures II.1.1 and II.1.2 depict the observations of 10 randomly selected patients from the data on survival with malignant melanoma introduced in Example I.3.1: First, in the original calendar time scale, and second, in the survival time scale t years since operation. This latter time scale is the one we concentrate on in our illustration of stochastic process concepts. A filled circle corresponds to $D_i = 1$ (a failure), an open circle to $D_i = 0$ (a censoring).

Further analysis is impossible without assumptions on censoring. We will make the most general assumption which still allows progress: the assumption of *independent censoring* (to which we return in Section III.2.2), which means that at any time t (in the survival time scale) the survival experience in the future is not statistically altered (from what it would have been without censoring) by censoring and survival experience in the past. To formalize this notion, we must be able to talk mathematically about past and future. This will be done through the concept of a *filtration* or *history* $(\mathscr{F}_t)_{t \geq 0}$, $\mathscr{F}_t$ representing the available data at time t. Write $\mathscr{F}_{t-}$ correspondingly for the available data just before time t. A specification of $(\mathscr{F}_t)_{t \geq 0}$ can only be done

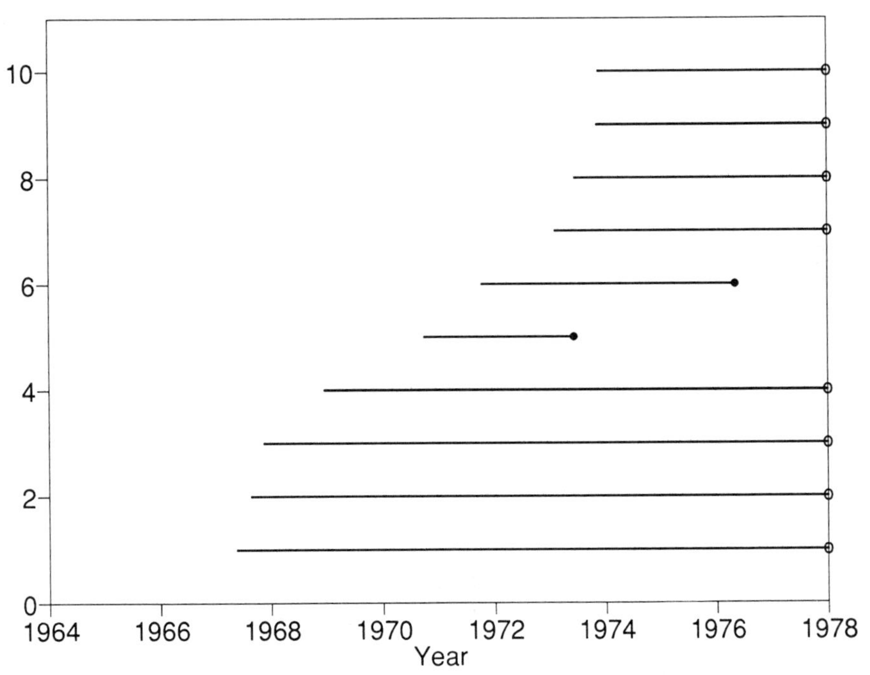

Figure II.1.1. Ten observations from the malignant melanoma study, calendar time (years).

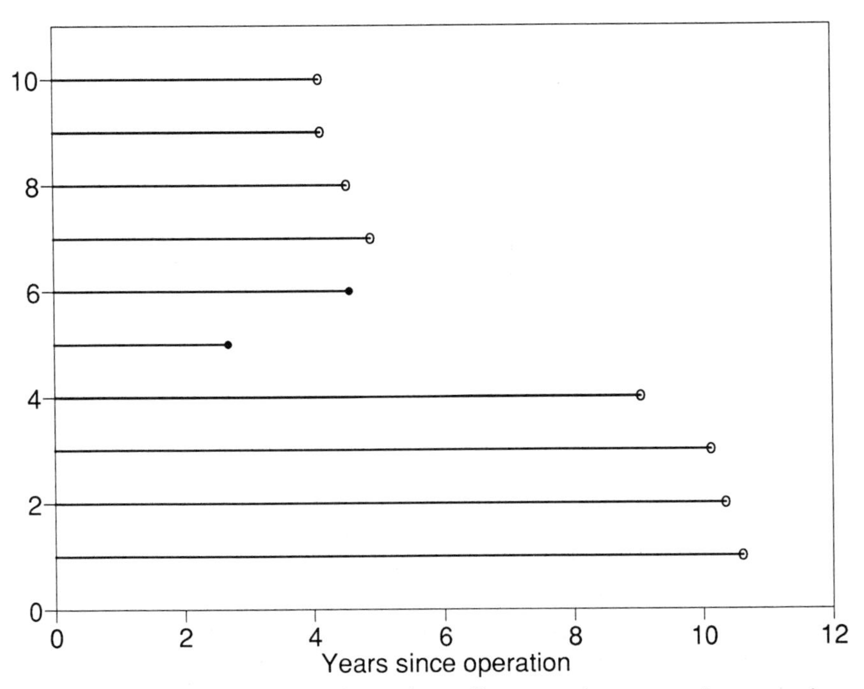

Figure II.1.2. Ten observations from the malignant melanoma study, years since operation (survival time).

relative to some observer, and different observers may collect more or less information—this interplay will be a central theme in model building; see especially Chapters III and IX. But for all obervers, as time proceeds, more information becomes available.

The notion of a filtration is defined formally in Section II.2, as an increasing family of σ-algebras defined on the sample space. In our simple example, we will simply take $\mathscr{F}_t$ to mean the values of $\tilde{X}_i$ and D_i for all i such that $\tilde{X}_i \leq t$, otherwise just the information that $\tilde{X}_i > t$. For $\mathscr{F}_{t-}$ the obvious changes must be made: $\leq$ becomes $<$ and the $>$ becomes $\geq$.

The independent censoring assumption can now be written (still very informally) as

$$\mathrm{P}(\tilde{X}_i \in [t, t + \mathrm{d}t), D_i = 1 | \mathscr{F}_{t-}) = \begin{cases} \alpha(t)\,\mathrm{d}t & \text{if } \tilde{X}_i \geq t \\ 0 & \text{if } \tilde{X}_i < t, \end{cases} \tag{2.1.3}$$

compare this to (2.1.1). Replacing the probablity on the left-hand side by the expectation of an indicator random variable, and summing over i, we get

$$\begin{aligned} \mathrm{E}(\#\{i\colon \tilde{X}_i \in [t, t + \mathrm{d}t), D_i = 1\} | \mathscr{F}_{t-}) &= \#\{i\colon \tilde{X}_i \geq t\} \cdot \alpha(t)\,\mathrm{d}t \\ &= Y(t)\alpha(t)\,\mathrm{d}t \\ &= \lambda(t)\,\mathrm{d}t, \end{aligned} \tag{2.1.4}$$

where we have defined the processes Y and λ by

$$Y(t) = \#\{i\colon \tilde{X}_i \geq t\},$$

the number at risk just before time t for failing in the time interval $[t, t + \mathrm{d}t)$, or the size of the risk set, and

$$\lambda(t) = Y(t)\alpha(t).$$

Now formula (2.1.4) can be interpreted as a *martingale property* involving a certain *counting process*; in this case, the process $N = (N(t))_{t \geq 0}$ counting the observed failures

$$N(t) = \#\{i\colon \tilde{X}_i \leq t, D_i = 1\}$$

and its *intensity process* λ. Let us write $\mathrm{d}N(t)$ or $N(\mathrm{d}t)$ for the increment $N((t + \mathrm{d}t)-) - N(t-)$ of N over the small time interval $[t, t + \mathrm{d}t)$; note that this quantity is precisely the number whose conditional expectation is taken in (2.1.4). With this notation, we can, therefore, rewrite (2.1.4) as

$$\mathrm{E}(\mathrm{d}N(t) | \mathscr{F}_{t-}) = \lambda(t)\,\mathrm{d}t. \tag{2.1.5}$$

Note that the intensity process is random, through dependence on the conditioning random variables in $\mathscr{F}_{t-}$.

To explain the meaning of martingale property, first define the integrated or *cumulative intensity process* Λ by

$$\Lambda(t) = \int_0^t \lambda(s)\,\mathrm{d}s, \quad t \geq 0,$$

and the *compensated counting process* or *counting process martingale* M by

$$M(t) = N(t) - \Lambda(t)$$

or, equivalently,

$$\mathrm{d}N(t) = \mathrm{d}\Lambda(t) + \mathrm{d}M(t) = \lambda(t)\,\mathrm{d}t + \mathrm{d}M(t) = Y(t)\alpha(t)\,\mathrm{d}t + \mathrm{d}M(t). \quad (2.1.6)$$

Consider the conditional expectation, given the strict past $\mathscr{F}_{t-}$, of the increment (or difference) of the process M over the small time interval $[t, t + \mathrm{d}t)$; by (2.1.6), we find

$$\begin{aligned} \mathrm{E}(\mathrm{d}M(t)|\mathscr{F}_{t-}) &= \mathrm{E}(\mathrm{d}N(t) - \mathrm{d}\Lambda(t)|\mathscr{F}_{t-}) \\ &= \mathrm{E}(\mathrm{d}N(t) - \lambda(t)\,\mathrm{d}t|\mathscr{F}_{t-}) \\ &= \mathrm{E}(\mathrm{d}N(t)|\mathscr{F}_{t-}) - \lambda(t)\,\mathrm{d}t \\ &= 0, \end{aligned} \quad (2.1.7)$$

where the last step is precisely the equality (2.1.5), noting that $\lambda(t)\,\mathrm{d}t$ is fixed (nonrandom) given $\mathscr{F}_{t-}$. Now, relation (2.1.7) says that Λ is the *compensator* of N, or that $M = N - \Lambda$ is a *martingale*: Such a process is characterized by the relation $\mathrm{E}(\mathrm{d}M(t)|\mathscr{F}_{t-}) = 0$ for all t.

In wide generality, we have that any counting process N, that is, a process taking the values 0, 1, 2, ... in turn and registering by a jump from the value $k - 1$ to k the time of the kth occurrence of a certain type of event, has an intensity process λ defined by $\lambda(t)\,\mathrm{d}t = \mathrm{E}(\mathrm{d}N(t)|\mathscr{F}_{t-})$. The intensity process is characterized by the fact that $M = N - \Lambda$, where Λ is the corresponding cumulative intensity process, is a martingale. Note that the processes λ and Λ are *predictable* (Section II.3.1); that is to say, $\lambda(t)$ is fixed *just before* time t, and given $\mathscr{F}_{t-}$ we know $\lambda(t)$ already [but not yet $N(t)$, for instance].

The martingale property says that the conditional expectation of increments of M over small time intervals, given the past at the beginning of the interval, is zero. This is (heuristically, at least) equivalent to the more familiar definition of a martingale (see Section II.3.1)

$$\mathrm{E}(M(t)|\mathscr{F}_s) = M(s) \quad (2.1.8)$$

for all $s < t$, which, in fact, just requires the same property for all intervals $(s, t]$: For adding up the increments of M over small subintervals $[u, u + \mathrm{d}u)$ partitioning $[s + \mathrm{d}s, t + \mathrm{d}t) = (s, t]$, we find

$$\begin{aligned} \mathrm{E}(M(t)|\mathscr{F}_s) - M(s) &= \mathrm{E}(M(t) - M(s)|\mathscr{F}_s) \\ &= \mathrm{E}\left(\int_{s<u\le t} \mathrm{d}M(u)|\mathscr{F}_s\right) \\ &= \int_{s<u\le t} \mathrm{E}(\mathrm{d}M(u)|\mathscr{F}_s) \\ &= \int_{s<u\le t} \mathrm{E}(\mathrm{E}(\mathrm{d}M(u)|\mathscr{F}_{u-})|\mathscr{F}_s) \\ &= 0. \end{aligned}$$

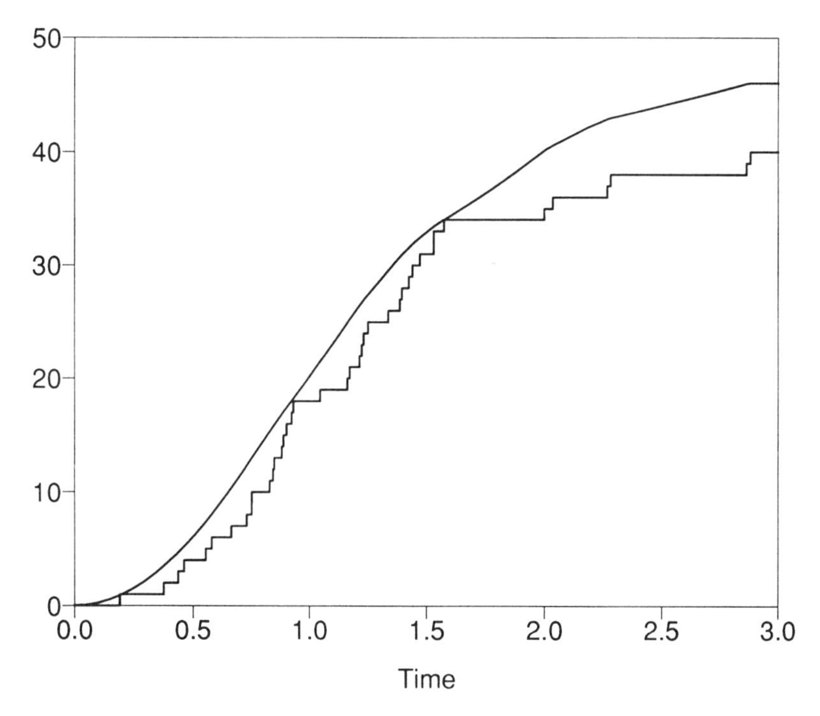

Figure II.1.3. Counting process $N(t)$ and its compensator $\Lambda(t)$, based on $n = 50$ independent randomly censored survival times with hazard $\alpha(t) = t$ (hazard for censoring distribution $0.15t$).

Version (2.1.8) of the martingale property is much easier to make the basis of a mathematical theory.

We can consider a martingale as being a pure noise process. The systematic part of a counting process is its compensator: a smoothly varying and predictable process, which, subtracted from the counting process, leaves unpredictable zero-mean noise. This is illustrated in Figures II.1.3 and II.1.4, which show a Monte Carlo realization of a counting process N and its compensator Λ, and the associated martingale $M = N - \Lambda$. The counting process counts failures in a sample of 50 independent randomly censored failure times; the failure-time distribution has hazard rate $\alpha(t) = t$, the censoring distribution has hazard rate $0.15t$ (both Weibull distibutions). Censoring and failure are independent; thus, in this case $(\tilde{X}_i, D_i) = (\min(X_i, U_i), I(X_i \leq U_i))$ where the failure times X_i and the censoring times U_i are all independent with the specified distributions.

The notions of martingale and compensator will turn up again and again. A special case of this is the concept of the *predictable variation* $\langle M \rangle$ of a martingale (Section II.3.2). Consider the process M^2. Though M was pure noise, M^2 has a tendency to increase over time. We look at its systematic component (its compensator), called M's *predictable variation process* and

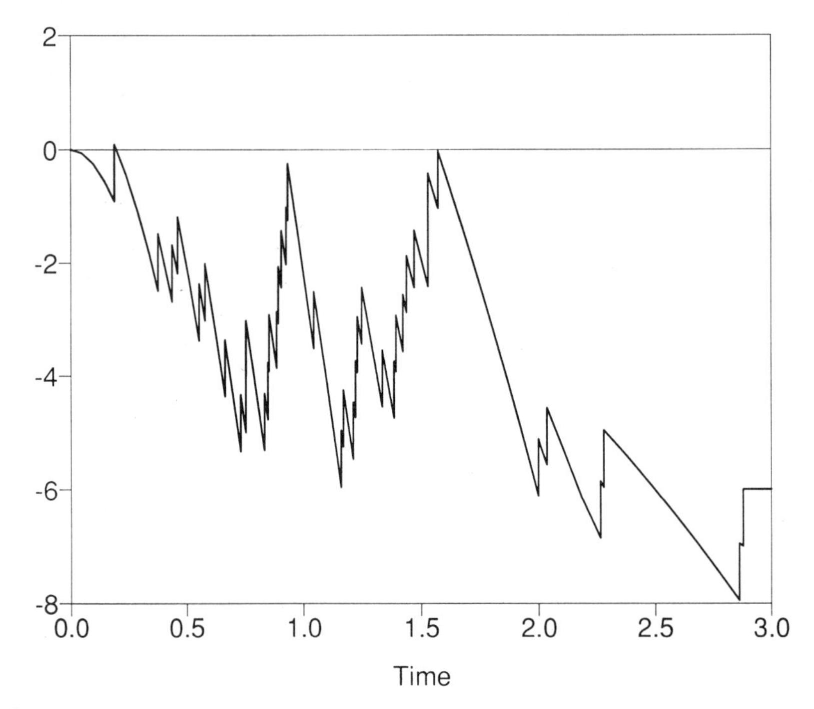

Figure II.1.4. Martingale $M(t) = N(t) - \Lambda(t)$ based on the same simulated data as in Figure II.1.3.

denoted by $\langle M \rangle$. Now

$$\begin{aligned} \mathrm{d}(M^2)(t) &= M((t + \mathrm{d}t)-)^2 - M(t-)^2 \\ &= (M(t-) + \mathrm{d}M(t))^2 - M(t-)^2 \\ &= (\mathrm{d}M(t))^2 + 2\,\mathrm{d}M(t)M(t-). \end{aligned}$$

Because $\mathrm{E}(\mathrm{d}M(t)M(t-)|\mathscr{F}_{t-}) = M(t-)\mathrm{E}(\mathrm{d}M(t)|\mathscr{F}_{t-}) = 0$, we have

$$\mathrm{E}(\mathrm{d}(M^2)(t)|\mathscr{F}_{t-}) = \mathrm{E}((\mathrm{d}M(t))^2|\mathscr{F}_{t-}),$$

that is, the increments of the compensator of M^2 are the conditional *variances* of increments of M (since their conditional means are zero). We summarize this very important fact as

$$\mathrm{var}(\mathrm{d}M(t)|\mathscr{F}_{t-}) = \mathrm{d}\langle M \rangle(t).$$

Now, because no two uncensored survival times fall into the same small interval (the increments of N over small time intervals are zero or one), $\mathrm{d}M(t)$ is just a zero-one random variable minus its conditional expectation $\mathrm{d}\Lambda(t)$ and, therefore, $\mathrm{var}(\mathrm{d}M(t)|\mathscr{F}_{t-}) = \mathrm{d}\Lambda(t)(1 - \mathrm{d}\Lambda(t)) \approx \mathrm{d}\Lambda(t)$. (Note that in a discrete-time situation, the last approximation will not hold and binomial variances will appear.) Thus, if M is a compensated counting process (and the compensator in question, Λ, is continuous), then M's predictable variation process $\langle M \rangle$ is simply Λ itself.

This remarkable fact is related to the fact that for a Poisson random variable, mean and variance coincide. A counting process N behaves locally at time t, and conditional on the past, just like a Poisson process with rate $\lambda(t)$. So, conditional means and variances of increments over small time intervals both coincide with the conditional local rate.

So far, we have only considered probabilistic matters. To introduce the next topic, stochastic integration, we shall turn to the statistical problem of nonparametric estimation of the cumulative hazard rate A, see (2.1.2). Using the version $\mathrm{d}N(t) = Y(t)\alpha(t)\,\mathrm{d}t + \mathrm{d}M(t)$ of (2.1.6), we obtain, on multiplying throughout by $1/Y(t)$ [supposing for the moment $Y(t)$ to be positive],

$$\frac{\mathrm{d}N(t)}{Y(t)} = \alpha(t)\,\mathrm{d}t + \frac{\mathrm{d}M(t)}{Y(t)}. \tag{2.1.9}$$

Now, if $\mathrm{d}M(t)$ is just noise, the same must be true of $\mathrm{d}M(t)$ times $1/Y(t)$: We have, because $Y(t)$ is predictable,

$$\mathrm{E}\left(\frac{\mathrm{d}M(t)}{Y(t)}\,\middle|\,\mathscr{F}_{t-}\right) = \frac{\mathrm{E}(\mathrm{d}M(t)|\mathscr{F}_{t-})}{Y(t)} = 0. \tag{2.1.10}$$

The conditional variance of the noise does change: It is now

$$\operatorname{var}\left(\frac{\mathrm{d}M(t)}{Y(t)}\,\middle|\,\mathscr{F}_{t-}\right) = \frac{\operatorname{var}(\mathrm{d}M(t)|\mathscr{F}_{t-})}{Y(t)^2} = \mathrm{d}\langle M \rangle(t)\left(\frac{1}{Y(t)}\right)^2. \tag{2.1.11}$$

Let us introduce some more notation. To take account of the possibility that $Y(t)$ may be zero, define

$$J(t) = I(Y(t) > 0), \qquad H(t) = J(t)/Y(t)$$

(with $0/0 = 0$); let

$$\hat{A}(t) = \int_{0<s\le t} H(s)\,\mathrm{d}N(s), \qquad A^*(t) = \int_{0<s\le t} J(s)\alpha(s)\,\mathrm{d}s$$

and, finally,

$$Z(t) = \int_{0<s\le t} H(s)\,\mathrm{d}M(s).$$

Then replacing t in (2.1.9) by s and integrating over $0 < s \le t$ gives

$$\hat{A}(t) = A^*(t) + Z(t), \tag{2.1.12}$$

where the left-hand side is indeed an estimator of $A(t)$: $\hat{A}(t)$ (called the Nelson–Aalen estimator and studied in Section IV.1) is just the sum over failure times up to and including time t of the reciprocals of the corresponding risk set sizes; whereas, on the right-hand side, $A^*(t)$ is essentially $A(t)$ itself [we only omit contributions $\alpha(s)\,\mathrm{d}s$ where the risk set is empty, which hopefully hardly ever happens if nonparametric estimation of $A(t)$ is to be meaningful]. Finally, also on the right-hand side, $Z(t)$ is, by (2.1.10), the value at time t of the martingale Z, formed on integrating the predictable process H with respect to the martingale M. Its predictable variation process is, by (2.1.11), the integral of the *square* of H with respect to the predictable variation process of M:

$$Z(t) = \int_{0<s\le t} H(s)\,\mathrm{d}M(s), \qquad \langle Z\rangle(t) = \int_{0<s\le t} H(s)^2\,\mathrm{d}\langle M\rangle(s).$$

This is a general result on *stochastic integration*. We have used it to express statistical estimation error as a martingale, from which a great deal of useful consequences can be derived, both exact or small sample results and asymptotic or large sample results. We discuss here the large sample consequences, via the *martingale central limit theorem* (Section II.5.1).

Typically, with a large sample size n, the random variation in the sample averages $Y(t)/n$ and $N(t)/n$ is small. Suppose, in particular, that $Y(t)/n$ for all t is close to a deterministic function $y(t)$. Then we find by an easy calculation that for the martingale $W^{(n)} = \sqrt{n}(\hat{A} - A^*) = \sqrt{n}Z$ [cf. (2.1.12)], essentially $\sqrt{n}(\hat{A} - A)$, the conditional variances [by (2.1.11)] $n\lambda(t)\,\mathrm{d}t/Y(t)^2$ of its increments $\sqrt{n}\,\mathrm{d}M(t)/Y(t)$ are approximately equal to $\alpha(t)\,\mathrm{d}t/y(t)$ while its many jumps are very small: of a size of the order of $1/\sqrt{n}$. So $W^{(n)}$ has an almost deterministic predictable variation process $\langle W^{(n)}\rangle \approx \int \alpha/y$, whereas its sample paths are almost continuous.

Now these properties actually characterize the (approximate) distribution of $W^{(n)}$: There is precisely one *continuous* martingale $W^{(\infty)}$ with *deterministic* predictable variation $\langle W^{(\infty)}\rangle = \int \alpha/y$, and that is a Gaussian martingale with this *variance function*. Put another way, on making the deterministic time change $t \to \langle W^{(\infty)}\rangle(t)$, $W^{(\infty)}$ becomes a standard Brownian motion or Wiener process. Plotting $\sqrt{n}(\hat{A}(t) - A(t))$ against an estimator of its predictable variation, e.g., $n\int_{0<s\le t}\mathrm{d}N(s)/Y(s)^2$, we see for large n this limiting Brownian motion: a process with independent Gaussian increments, with means zero and variances equal to the lengths of the corresponding time intervals. This result can be used, for instance, to construct confidence bands for the unknown function A, as we will do in Section IV.1.3.

The variance estimator $n\int_{0<s\le t}\mathrm{d}N(s)/Y(s)^2$ is exactly equal to the sum of the squares of the increments of the martingale $W^{(n)}$ over the interval $[0, t]$. Considered as a process, this is called the *optional variation process* of $W^{(n)}$ and denoted $[W^{(n)}]$ (see Section II.3.2); thus

$$[W^{(n)}](t) = \int_{0<s\le t} \{\mathrm{d}W^{(n)}(s)\}^2.$$

Recall that the predictable variation process $\langle W^{(n)}\rangle$ is defined by the same sum but with conditional expectations taken of each squared increment:

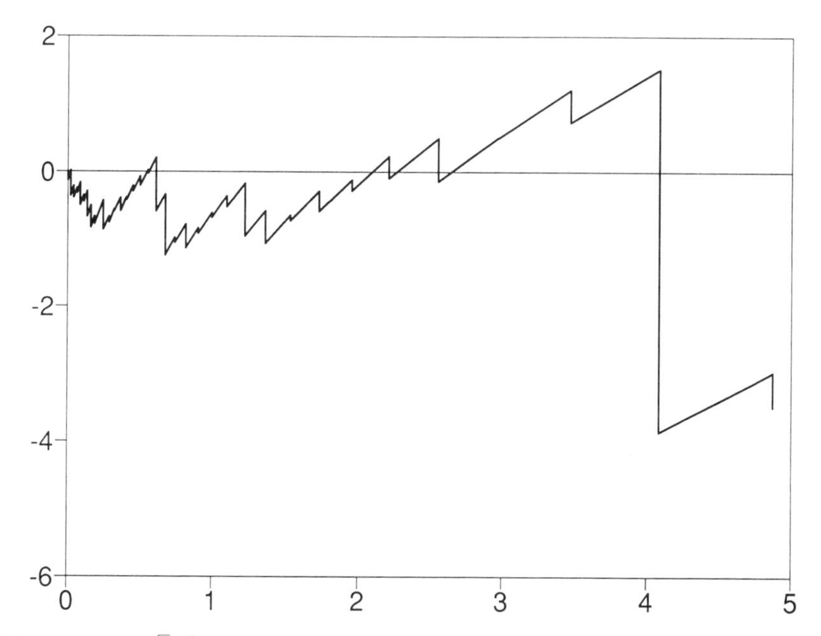

Figure II.1.5. $\sqrt{n}(\hat{A}(t) - A(t))$ plotted against its estimated variance, the optional variation process $n\int_0^t \mathrm{d}N(u)/Y(u)^2$; same data as in Figures II.1.3 and II.1.4.

$$\langle W^{(n)} \rangle (t) = \int_{0<s\leq t} \mathrm{E}(\mathrm{d}W^{(n)}(s)^2 | \mathscr{F}_{s-}).$$

For a Brownian motion, predictable and optional variations coincide and equal the (unconditional) variance function of the process.

These results are illustrated in Figure II.1.5, which shows the realization of $\sqrt{n}(\hat{A}(t) - A(t))$ plotted against its estimated variance, the optional variation process, for the same data which were used in Figures II.1.3 and II.1.4. This is beginning, indeed, to look like a typical realization of a Brownian motion.

From A, we can go on to estimate the survival function S; rewriting the relation between the two as

$$S(t) = \prod_{s=0}^{t} (1 - \mathrm{d}A(s)) \tag{2.1.13}$$

in *product-integral* notation suggests the estimator $\hat{S}(t) = \prod_{s=0}^{t}(1 - \mathrm{d}\hat{A}(s))$, a product-limit estimator; in fact, the famous Kaplan–Meier estimator (see Section IV.3). Note that for continuous A, because $1 - \mathrm{d}A(s) \approx \exp(-\mathrm{d}A(s))$, we can also write $S(t) = \exp(-A(t))$. However, $\hat{A}$ is a jump function and its product-integral is just a finite product and also a step-function. The statistical properties of the Kaplan–Meier estimator can be derived from slightly more elaborate martingale properties; in fact, $\hat{S}/S - 1$ turns out to be a stochastic integral with respect to M and, therefore, also a martingale.

Product-integration will be explained at length in Section II.6. A formal definition of the right-hand side of (2.1.13) is as a limit of approximating finite products. Because $\mathrm{d}A(s) = \alpha(s)\,\mathrm{d}s$ can be interpreted as $\mathrm{P}(X \in [s, s + \mathrm{d}s) | X \geq s)$, see (2.1.1), we have that $1 - \mathrm{d}A(s)$ is $\mathrm{P}(X \geq s + \mathrm{d}s | X \geq s)$ and multiplying such conditional probabilities over small intervals $[s, s + \mathrm{d}s)$ partitioning $[0, t + \mathrm{d}t)$ gives $\mathrm{P}(X \geq t + \mathrm{d}t)$ or just $\mathrm{P}(X > t)$.

These estimators were derived by solving a natural estimating equation, putting the noise equal to zero in the equation $\mathrm{d}N(t) = \mathrm{d}\Lambda(t) + \mathrm{d}M(t)$. In fact, a nonparametric maximum likelihood motivation is also possible, based on the fact that relation (2.1.5) essentially specifies the conditional distribution of increments of N (because they can only take the values zero and one) given the past. Putting these conditional distributions together, we can build up the whole distribution of N.

We are neglecting here the fact that from one small time interval to the next, not just failures can occur but also other events, censorings, and their distribution has to be considered too if one wants to build up the distribution of the whole observed data. We can write, rather informally,

$$\mathrm{P}(\text{data}) = \prod_{0<t<\infty} \mathrm{P}(\mathrm{d}N(t) | \mathscr{F}_{t-}) \mathrm{P}(\text{other events in } \mathrm{d}t | \mathrm{d}N(t), \mathscr{F}_{t-}). \quad (2.1.14)$$

In Section II.7, we show that this likelihood expression can also be given a precise mathematical interpretation, again via the notion of product-integration.

As we mentioned, this can be used to show that the Nelson–Aalen estimator $\hat{A}$, as well as the Kaplan–Meier estimator and its generalization to Markov processes, has an interpretation as nonparametric maximum likelihood estimator; see, in particular, Section IV.1.5. The behavior of the likelihood under censoring is studied in Section III.2, and parametric maximum likelihood estimators in Chapter VI. Efficiency of tests and estimators is analyzed using large sample approximations to the likelihood in Chapter VIII.

Consider now a parametric estimation problem in which the hazard rate α depends on a parameter θ, say. If θ does not enter in the specification of the second factors in (2.1.14), we say that we have *noninformative censoring*, to which we return in Section III.2.3. In this case, using (2.1.4) as the specification of the conditional mean of the zero-one variable $\mathrm{d}N(t)$, we can write

$$\mathrm{P}(\text{data}) \propto \prod_{0<t<\infty} ((\lambda^{\theta}(t)\,\mathrm{d}t)^{\mathrm{d}N(t)}(1 - \lambda^{\theta}(t)\,\mathrm{d}t)^{1-\mathrm{d}N(t)}),$$

where the intensity process, depending on θ, is $\lambda^{\theta}(t) = Y(t)\alpha^{\theta}(t)$. This can be simplified somewhat. First, we can neglect the factors $\mathrm{d}t$ in the first part of the product because these will cancel when we form likelihood ratios. Second, by a Taylor expansion, $1 - \lambda^{\theta}(t)\,\mathrm{d}t \approx \exp(-\lambda^{\theta}(t)\,\mathrm{d}t)$, and a product of exponentials is an exponential of a sum. This means we can write the likelihood as

$$L(\theta) \propto \left(\prod_{0<t<\infty} \lambda^\theta(t)^{\mathrm{d}N(t)} \right) \exp\left(-\int_0^\infty \lambda^\theta(t)\,\mathrm{d}t \right) \tag{2.1.15}$$

and, hence, the log-likelihood

$$\log L(\theta) = \int_0^\infty \log \lambda^\theta(t)\,\mathrm{d}N(t) - \int_0^\infty \lambda^\theta(t)\,\mathrm{d}t + \text{const}$$

and, finally, the score function

$$\begin{aligned}\frac{\partial}{\partial\theta}\log L(\theta) &= \int_0^\infty \frac{\partial}{\partial\theta}\log \lambda^\theta(t)\,\mathrm{d}N(t) - \int_0^\infty \frac{\partial}{\partial\theta}\lambda^\theta(t)\,\mathrm{d}t \\ &= \int_0^\infty \left(\frac{\partial}{\partial\theta}\log\lambda^\theta(t)\right)\mathrm{d}M^\theta(t),\end{aligned}$$

where $\mathrm{d}M^\theta(t) = \mathrm{d}N(t) - \lambda^\theta(t)\,\mathrm{d}t$.

If we had calculated the likelihood based on the data up to time t, we would have obtained the same results but with the integrals over $[0, \infty)$ replaced by integrals over $[0, t]$. This result shows us that the statistically extremely important score function, seen as a process (using the data up to time t for each t), is a martingale: It can be written as a stochastic integral, this time of the derivative of the log intensity process (a predictable process) with respect to the counting process martingale. One can interpret the result as identifying the total score for the data as the sum of the scores for the infinitesimal conditional experiments: At time t observe $\mathrm{d}N(t)$, a zero-one variable with mean $\lambda^\theta(t)\,\mathrm{d}t$ given $\mathscr{F}_{t-}$.

The martingale property of the score based on the data up to each time instant t holds also without the assumption of noninformative censoring [which states that the second factor in (2.1.14) does not depend on θ]. Without that assumption, (2.1.15) is called the *partial likelihood* for θ based on the counting process N (thus ignoring censoring); see Section II.7.3. The fact that the martingale property of the *partial score process* is maintained is part of the explanation that partial-likelihood-based statistical procedures have many of the familiar asymptotic properties of ordinary likelihood methods. This will be especially apparent in Section VI.1 where maximum likelihood estimators are studied.

II.2. Preliminaries: Processes, Filtrations, and Stopping Times

This section and the next present a compact survey of the "general theory of processes" as we need it in this book. Many topics usually central in complete treatments of this theory are ignored because in the applications in our book they do not arise. Many purely technical points (especially concerning measurability questions) are also ignored. At some points, we diverge slightly from the usual definitions for the sake of a streamlined presentation; these divergences are discussed in the bibliographic comments. A complete but also very compact survey of the theory was given by Jacod and Shiryayev (1987) in the first chapters of their book, on which this extract is based. A survey including self-contained derivations of some of the key results was given by Fleming and Harrington (1991), who aimed at statistical applications in survival analysis. Further help in becoming at home in the vast literature surrounding this topic is given in the bibliographic comments.

We are going to model the occurrence in time of random events; in fact, discrete events occurring in continuous time. So we fix a continuous-time interval

$$\mathcal{T} = [0, \tau) \quad \text{or} \quad [0, \tau]$$

for a given terminal time τ, $0 < \tau \leq \infty$. Note that the terminal time point τ may or may not be included; this varies from application to application. We write $\bar{\mathcal{T}} = [0, \tau]$, the time interval augmented with its endpoint if it was not first present.

Let $(\Omega, \mathscr{F}, \mathrm{P})$ be a probability space. A *filtration*

$$(\mathscr{F}_t \colon t \in \mathcal{T}),$$

also called a *history*, is an *increasing right-continuous* family of sub-σ-algebras of $\mathscr{F}$. In the standard theory we use, it is often assumed also to be *complete* in the strong sense that, for every t, the σ-algebra $\mathscr{F}_t$ contains all P-null sets of $\mathscr{F}$. However, the assumption can be safely omitted, subject only to a very minor reformulation of the results of the standard theory; see the bibliographic comments and Jacod and Shiryayev (1987).

When the complete set of assumptions hold, we say that $(\mathscr{F}_t)$ *satisfies the usual conditions* (*les conditions habituelles*):

$$\begin{aligned} &\mathscr{F}_s \subseteq \mathscr{F}_t \subseteq \mathscr{F} \quad \text{for all } s < t && \text{(increasing)} \\ &\mathscr{F}_s = \bigcap_{t > s} \mathscr{F}_t \quad \text{for all } s && \text{(right continuous)} \qquad (2.2.1) \\ &A \subset B \in \mathscr{F}, \mathrm{P}(B) = 0 \Rightarrow A \in \mathscr{F}_0 && \text{(complete).} \end{aligned}$$

The σ-algebra $\mathscr{F}_t$ is interpreted as follows: It contains all events (up to null sets) whose occurrence or not is fixed by time t. There is also a pre-t σ-algebra $\mathscr{F}_{t-}$, the smallest σ-algebra containing all $\mathscr{F}_s$, $s < t$; it contains events fixed strictly before t.

A stochastic process X is just a time-indexed collection of random variables $(X(t) \colon t \in \mathcal{T})$. The process X is called *adapted* (to the filtration) if $X(t)$ is $\mathscr{F}_t$ measurable for each t. We write $X(t, \omega)$ for the realized value of $X(t)$ at the point $\omega \in \Omega$. This way we can think of X not only as a function of ω for fixed t (i.e., as a random variable), but also as a function of t for fixed ω. This function is called a *sample path* of X. The process X is called *cadlag* (*continu à droite, limité à gauche*) if its sample paths $(X(t,\omega): t \in \mathcal{T})$, for almost all ω, are right-continuous with left-hand limits. The set of cadlag functions is often denoted $D(\mathcal{T})$, the Skorohod space of weak convergence theory (Billingsley, 1968),

We often describe a filtration as being the filtration generated by a stochastic process X (perhaps taking values in a quite general measurable space). This means that $\mathcal{F}_t$ is the σ-algebra generated by $X(s)$, $s \leq t$. We also have then that $\mathcal{F}_{t-}$ is generated by $X(s)$, $s < t$. Frequently, we add to $\mathcal{F}_t$ events supposed to be fixed at time zero or, equivalently, add to the random variables generating $\mathcal{F}_t$ certain other random variables whose values are supposed to be fixed at time zero.

An important result of Courrège and Priouret (1965) states that the filtration generated by a right-continuous *jump process* is right-continuous: a jump process X being a process such that for each t and ω, $X(s,\omega)$ is constant in $s \in [t, t+\varepsilon)$ for some $\varepsilon > 0$ (depending on t and ω, in general). It can be shown from this that filtrations generated naturally when considering discrete events occurring in continuous time are automatically right-continuous. Extending the filtration "at time zero" as we have just described leaves it right-continuous. In particular, completion of a filtration by adding null events to every $\mathcal{F}_t$ preserves right-continuity.

A *stopping time* T is a random variable taking values in $\bar{\mathcal{T}}$ such that

$$\{T \leq t\} \in \mathcal{F}_t \quad \text{for all } t \in \mathcal{T}.$$

An important example of a stopping time is the first time a process exceeds a given value: For instance, if X is cadlag and adapted, then $T = \inf\{t: |X(t)| \geq c\}$ is a stopping time. Any fixed time s is also a stopping time. Intuitively, the time T of a random event is a stopping time if at each fixed time t one can observe whether or not the event has already occurred. One can define σ-algebras $\mathcal{F}_T$ and $\mathcal{F}_{T-}$ having the interpretation as being all events which have occurred up to and including the stopping time T or strictly before the stopping time. The former, $\mathcal{F}_T$, can be characterized as the σ-algebra generated by T together with all the random variables $X(T)$, for any cadlag adapted process X. The σ-algebra of events strictly before T has $X(T)$ replaced by $X(T-)$ in this description.

Many operations can be used to build new stopping times from old; for instance, the minimum or maximum of two stopping times is also a stopping time. Adding to a stopping time T a non-negative $\mathcal{F}_T$ measurable random variable gives a new stopping time.

There is an intimate relation between stopping times and the main object we study in this book: counting processes. This is exemplified by considering the case of a single event occurring at a random time instant:

EXAMPLE II.2.1. The One-Jump Counting Process

Let T denote the time of some random event. The indicator process $(I(T \leq t))$ is a cadlag process, equal to zero until time T, then jumping to the value 1 at time T (if the event ever occurs), and then staying at that value. One easily checks that the indicator process $(I(T \leq t))$ is adapted if and only if T is a stopping time. This process is the simplest example of a *counting process*, to be defined formally in Section II.4.1. □

If X is a stochastic process and T a stopping time, it is not self-evident that $X(T)$ is, indeed, a random variable, i.e., that $X(T(\omega), \omega)$ is measurable as a function of $\omega \in \Omega$. Similar measurability questions would arise whenever we carry out operations on the process X which involve its values at more than a countable number of time points t. However, all the processes we shall meet are nice enough that this is never a problem. For instance, all cadlag processes satisfy the measurability requirements. In the future, we will ignore this potential difficulty.

Given a stochastic process X and a stopping time T, one often wants to consider the *stopped process* X^T defined by

$$X^T(t) = X(t \wedge T),$$

where $s \wedge t = \min(s, t)$. If X is cadlag and adapted and T is a stopping time, then X^T is cadlag and adapted too. We sketch a proof of this just to give some flavor of what is involved in a complete treatment of these basic parts of the theory. That X^T is cadlag is obvious. To prove adaptedness of X^T, introduce $T_n = 2^{-n}(\lfloor 2^n T \rfloor + 1)$, where $\lfloor \cdot \rfloor$ denotes *entier*. Then T_n is a stopping time, $T < T_n \leq T + 2^{-n}$, and $T_n \downarrow T$ as $n \to \infty$. Also, T_n takes only countably many values. By an explicit calculation, one can check that X^{T_n} is adapted. But for each t, $X^{T_n}(t) \to X^T(t)$ as $n \to \infty$ by right-continuity of X, showing that X^T is adapted too.

Apart from arising very naturally in applications, stopping times are important in the theory through the notion of *localization*. Here a process is supposed to have some nice property only up to a stopping time, which may, however, be taken to be arbitrarily large. We then derive results for the stopped process, and afterward extend them to the original by letting the stopping times increase. In this way, nice results for a rather stringently defined class of processes may be carried over, only mildly weakened, to a much larger class.

First, we define the notion of a *localizing* sequence of stopping times. This is simply a sequence of stopping times T_n which is nondecreasing and which satisfies

$$\mathrm{P}(T_n \geq t) \to 1 \quad \text{as } n \to \infty \text{ for all } t \in \mathscr{T}.$$

We then say that a process X has a certain property *locally* if a localizing sequence (T_n) exists such that, for each n, the process $I(T_n > 0)X^{T_n}$ has the property. An extremely important example of this is the following. A process X is *locally bounded* if a localizing sequence (T_n) and constants c_n exist such that, for each n,

$$\sup_{t \leq T_n} |X(t)| \leq c_n \quad \text{a.s. on the event } \{T_n > 0\}.$$

Here is one special case which we will refer to often: Any left-continuous adapted process with right-hand limits is locally bounded. To see this, suppose the process X has all its paths left-continuous and define $T_n = \inf\{t: |X(t)| > n\}$. One may check that this is a stopping time by first writing ($\mathbb{Q}$ the rationals)

$$\{T_n < t\} = \bigcup_{u<t, u \in \mathbb{Q}} \{|X(u)| > n\} \in \mathscr{F}_t.$$

Next, writing $\{T_n \leq t\} = \bigcap_{t<s<t+\varepsilon} \{T_n < s\} \in \mathscr{F}_{t+\varepsilon}$ for any $\varepsilon > 0$ (restricting to rational s) and using the right-continuity of the filtration shows that $\{T_n \leq t\} \in \mathscr{F}_t$. By left-continuity, the process $I(T_n > 0)X^{T_n}$ has all its paths bounded in absolute value by n. One may also check that the events $\{T_n \geq t\}$ are nondecreasing in n and $\bigcup_n \{T_n \geq t\} = \Omega$ for each $t \in \mathscr{T}$, showing that the sequence (T_n) is indeed a localizing sequence for X, for the property "boundedness." This example illustrates the use of adding the indicator function $I(T_n > 0)$ to the definition; otherwise, a process could only be locally bounded if its value at time zero were bounded by a constant.

Stochastic integration, i.e., the forming of the integral of one stochastic process with respect to another, will play a very important role in the rest of the book. This will always be for us a *pathwise* operation: for, given $\omega \in \Omega$, one forms an ordinary Lebesgue–Stieltjes integral over a given time interval. When no range of integration is explicitly given, we mean the result obtained by integrating over the interval $[0, t]$ for each $t \in \mathscr{T}$. For example, for two stochastic processes X and Y, $\int X \, \mathrm{d}Y$ denotes the *stochastic process*

$$t \mapsto \int_0^t X(s) \, \mathrm{d}Y(s) = \int_{[0,t]} X(s) \, \mathrm{d}Y(s), \tag{2.2.2}$$

defined for each ω and t such that

$$\int_{[0,t]} |X(s)| \, |\mathrm{d}Y(s)| < \infty. \tag{2.2.3}$$

Here, Y is assumed to be a cadlag process with paths of locally bounded variation; i.e., $\int_{[0,t]} |\mathrm{d}Y(s)|$ is finite for all $t \in \mathscr{T}$, for almost all $\omega \in \Omega$. We call such a process Y a *finite variation process*, and the process $\int |\mathrm{d}Y|$ is called its (total) *variation process*. By convention, $Y(0-) = 0$, so that for $t = 0$, $\int_0^t X \, \mathrm{d}Y = X(0)Y(0)$. [Usually $Y(0) = 0$ in our applications.] If X is nonnegative and Y is nondecreasing, we need not impose the condition (2.2.3) as long as we allow the process $\int X \, \mathrm{d}Y$ to take the value ∞ too. A further useful notational convention is to omit the integrating function or measure $\mathrm{d}Y$ when just Lebesgue measure is meant; so $\int X$ means the process $t \mapsto \int_0^t X(s) \, \mathrm{d}s$. We sometimes use the notation $\int_0^t X(s) Y(\mathrm{d}s)$ instead of $\int_0^t X(s) \, \mathrm{d}Y(s)$, either to distinguish between different possible integration variables, or to emphasize the role of Y as a measure rather than as a (distribution) function.

To a cadlag process X, we associate its left-continuous modification X_-, defined by

$$X_-(t) = X(t-),$$

and its jump process ΔX, defined by

$$\Delta X(t) = X(t) - X(t-).$$

Two stochastic processes whose paths coincide outside of a P-null set are called *indistinguishable*. Later, when we claim uniqueness of certain processes, we mean uniqueness modulo indistinguishability. Also, when we say a process is left-continuous, right-continuous, etc., we mean that it is indistinguishable from a process all of whose paths have these properties.

II.3. Martingale Theory

II.3.1. Martingales, Predictable Processes, Compensators

Two kinds of stochastic processes play important and complementary roles in the general theory of processes and, hence, in the theories of stochastic integration and of counting processes: martingales and predictable processes, especially finite variation predictable processes (compensators). Martingales can be considered as a kind of pure random noise process; see the examples in Section II.1, especially Figures II.1.4 and II.1.5. On the other hand, a predictable process has a kind of regularity or semideterministic behavior. Many stochastic processes can be written as the sum of a (local) martingale and a finite variation predictable process. In such a decomposition (a generalization of the celebrated Doob–Meyer decomposition for submartingales), we are splitting the process into a random and a systematic part. The latter is called the compensator of the process because subtracted off the process, a local martingale (unsystematic noise) is left.

We give the definitions and important properties of these processes, starting with martingales and local martingales.

A *martingale* is a cadlag adapted process M which is *integrable*, i.e.,

$$\mathrm{E}(|M(t)|) < \infty \quad \text{for all } t \in \mathscr{T}$$

and satisfies the *martingale property*

$$\mathrm{E}(M(t)|\mathscr{F}_s) = M(s) \quad \text{for all } s \leq t. \tag{2.3.1}$$

The process is a *submartingale* if this is replaced by the inequality

$$\mathrm{E}(M(t)|\mathscr{F}_s) \geq M(s) \quad \text{for all } s \leq t. \tag{2.3.2}$$

When we have (2.3.2) with the inequality reversed, M is called a *supermartingale*. A martingale is called *square integrable* if

$$\sup_{t \in \mathscr{T}} \mathrm{E}(M(t)^2) < \infty. \tag{2.3.3}$$

A square integrable martingale can be extended to $\overline{\mathscr{T}}$; i.e., even if $\tau \notin \mathscr{T}$, the limit $\lim_{t \to \tau} M(t)$ exists almost surely, and *defining* $M(\tau)$ as this limit, one forms a square integrable martingale on the extended time interval.

A martingale defined on $\overline{\mathcal{T}}$ is automatically uniformly integrable, in the strong sense that the set of random variables $M(T)$, for all stopping times T, is uniformly integrable:

$$\lim_{c \uparrow \infty} \sup_{T} \mathrm{E}(|M(T)| I(|M(T)| > c)) = 0.$$

A uniformly integrable martingale satisfies Doob's optional sampling theorem: that is, it satisfies the martingale property (2.3.1) with the fixed times $s \leq t$ replaced by stopping times $S \leq T$.

For a process to be a martingale or square integrable martingale, by definition certain integrability conditions have to be satisfied. These conditions may be hard to verify in practice or not even true at all. On the other hand, for some of the major applications (limit theory for instance) they can be avoided through the technique of localization introduced in the previous section. Especially important become the notions of a *local martingale* and a *local square integrable martingale.*

Written out in full, a *local martingale* is a process M such that an increasing squence of stopping times (T_n) exists,

$$\mathrm{P}(T_n \geq t) \to 1 \quad \text{as } n \to \infty \text{ for all } t \in \mathcal{T},$$

such that the stopped processes

$$I(T_n > 0)M^{T_n}$$

are martingales for each n. A local martingale is clearly cadlag and adapted. Not so obvious is the fact that any local martingale is actually a local uniformly integrable martingale; in other words, the localizing sequence (T_n) can always be chosen so that the stopped processes are actually martingales on $\overline{\mathcal{T}}$.

Similarly, a *local square integrable martingale* is a process M as above such that the localizing sequence can be chosen making $I(T_n > 0)M^{T_n}$ a square integrable martingale.

Doob's optional sampling theorem guarantees that when we consider later two different processes each having certain properties locally, e.g., a locally bounded process and a local martingale, we can choose a single localizing sequence for both processes simultaneously. This technique is used time and time again in building up the general theory of stochastic integration.

Given a local (square integrable) martingale M and a particular stopping time T, it is often useful to know if the stopped process $I(T > 0)M^T$ is a (square integrable) martingale. Not every stopping time will have this property. We give some useful conditions for this at the end of Section II.3.2.

A class of processes complementary to the martingales is the class of *predictable processes.* A stochastic process H is called predictable if, as a function of $(t, \omega) \in \mathscr{T} \times \Omega$, it is measurable with respect to the σ-algebra on $\mathscr{T} \times \Omega$ generated by the *left-continuous adapted processes* (to be precise, generated by those adapted processes *all* of whose paths are left-continuous). Many equivalent characterizations exist; for instance, a process H is predictable if and only if $H(T)$ is $\mathscr{F}_{T-}$-measurable for all stopping times T. Thus, the value of a predictable process at time T is fixed just before the time itself. For us, it is important to note that at any rate the left-continuous adapted processes themselves are predictable (or indistinguishable from a predictable process). Also, any deterministic measurable function, considered as a stochastic process, is predictable. (This holds true because the left-continuous deterministic step-functions on $\mathscr{T}$ generate its Borel σ-algebra; considered as processes, they are trivially left-continuous and adapted. This illustrates that some right-continuous processes are predictable too. We will see more examples soon.) Occasionally useful is the fact that a cadlag predictable process is locally bounded: By predictability, one can, in effect, stop the process strictly before it exceeds a given value. (This is proved using properties of so-called predictable stopping times, which we do not introduce here.)

There is an important orthogonality between martingales and *finite variation predictable processes*. This is due to the fact that if a process at the same time is both a local martingale and a predictable finite variation process, then it is a trivial or constant process. Because continuous adapted processes are predictable (they are left-continuous and adapted), a continuous local martingale which is not constant has paths of unbounded variation on each bounded interval. For instance, standard Brownian motion (the Wiener process) on a finite time interval is a square integrable martingale with continuous paths of unbounded variation on each bounded interval.

A deep result on which most of the subsequent theory depends (compensators, stochastic integration) is the *Doob–Meyer* decomposition. First, we give the decomposition in its narrow sense (for a certain class of submartingales), and then describe how it can be extended to a much wider class of processes through the idea of localization.

Suppose X is a submartingale such that the class of random variables $X(T)$, T an arbitrary stopping time, is uniformly integrable; this is called a submartingale of class D (Doob). Then the *Doob–Meyer* decomposition theorem states the existence of a cadlag nondecreasing predictable process $\tilde{X}$ such that

$$M = X - \tilde{X}$$

is a uniformly integrable martingale, zero at time zero; moreover, $\tilde{X}$ is integrable [in fact, $\mathrm{E}(\tilde{X}(\tau)) < \infty$, even if $\tau \notin \mathscr{T}$]. The process $\tilde{X}$ can be constructed as a limit in probability of discrete approximations; $\tilde{X}(t)$ is approximated by $\sum \mathrm{E}(X(t_i) - X(t_{i-1}) | \mathscr{F}_{t_{i-1}})$, where $0 = t_0 < t_1 < \cdots < t_n = t$ is a fine partition of $[0, t]$. So, heuristically (cf. Section II.1),

$$\mathrm{d}\tilde{X}(t) = \mathrm{E}(\mathrm{d}X(t) | \mathscr{F}_{t-})$$

and

$$dM(t) = dX(t) - E(dX(t)|\mathscr{F}_{t-}).$$

The predictable part $\tilde{X}$ is the sum of conditional expectations, given the past, of increments of X (nondecreasing because X is a submartingale), whereas M is the sum of increments minus their conditional expectations, hence a martingale.

The orthogonality between finite variation predictable processes and martingales gives the uniqueness of the decomposition: If we had two decompositions $X = \tilde{X} + M = X' + M'$, then $\tilde{X} - \tilde{X}' = M' - M$ would be a finite variation predictable martingale, zero at time zero, and hence constant and equal to zero throughout $\mathscr{T}$.

Obviously, for a supermartingale, one can state the same result but with $\tilde{X}$ now nonincreasing instead of nondecreasing. More generally, the difference of two submartingales of class D has a unique Doob–Meyer decomposition with the predictable part now of finite variation (the difference of two nondecreasing processes).

Next, by localization, we can avoid the integrability conditions altogether. Suppose X is a cadlag adapted process. We say that $\tilde{X}$ is the *compensator* of X if $\tilde{X}$ is a predictable, cadlag, and finite variation process such that $X - \tilde{X}$ is a local martingale, zero at time zero. If a compensator exists it is unique. It turns out that *X has a compensator if and only if X is the difference of two local submartingales.* (Part of the reason for this is that a local submartingale is locally a submartingale of class D).

Here is a simple example: the standard Poisson process.

EXAMPLE II.3.1. Poisson Process

Let $N = (N(t))$ be a standard Poisson process on the line $[0, \infty)$ with the filtration generated by the process itself (we sometimes denote this particular counting process by Π). Since the standard Poisson process has independent increments with means equal to the length of the corresponding interval, it follows that M defined by $M(t) = N(t) - t$ is a martingale. Define the identity function ι by

$$\iota(t) = t.$$

The process equal to the function ι is predictable and nondecreasing, so, by uniqueness, it must be the compensator $\tilde{N}$ of N. In fact, $M = N - \tilde{N} = \Pi - \iota$ is locally square integrable: take as localizing sequence a sequence of constants, or the times of the jumps of the process N. □

Processes having a compensator are called *special semimartingales*. By definition, they can be written as the sum of a local martingale and a finite variation predictable process. A *semimartingale* (a more general object still) is the sum of a local martingale and a cadlag adapted finite variation process (i.e., the finite variation part is not necessary predictable). Semimartingales have become a most important class of processes, but for us their general theory is not so relevant.

All (local) sub- and supermartingales have compensators. In particular, nondecreasing, non-negative locally integrable cadlag processes have compensators (which are also nondecreasing) because such processes are trivially local submartingales. By Jensen's inequality, the square of a square integrable martingale is a submartingale. Therefore, the square of a local square integrable martingale is a local submartingale and also has a nondecreasing compensator. The product of two local square integrable martingales MM' can be written as a difference of two squares of local square integrable martingales $[MM' = \frac{1}{4}(M + M')^2 - \frac{1}{4}(M - M')^2]$ and has a compensator of finite variation. These special cases will turn up in the theory of the next subsection.

II.3.2. Predictable and Optional Variation

Suppose M and M' are local square integrable martingales. Then, as mentioned in the previous subsection, M^2 is a local submartingale and MM' is the difference of two local submartingales. These processes, therefore, have compensators which are denoted by $\langle M, M \rangle$ and $\langle M, M' \rangle$, respectively. As shorthand, we write $\langle M \rangle$ for $\langle M, M \rangle$. Their defining properties are therefore: $\langle M \rangle = \langle M, M \rangle$ and $\langle M, M' \rangle$ are the unique finite variation cadlag predictable processes (in fact, $\langle M \rangle$ is nondecreasing) such that

$$M^2 - \langle M \rangle$$

and

$$MM' - \langle M, M' \rangle$$

are local martingales, zero at time zero.

The process $\langle M \rangle$ is called the *predictable variation process* of M and, similarly, $\langle M, M' \rangle$ the *predictable covariation process* of M and M'. Being compensators, they can be constructed as the limits of discrete approximations allowing a heuristic interpretation (cf. Section II.1) as sums of conditional variances and covariances of increments of M and M':

$$\mathrm{d}\langle M \rangle(t) = \mathrm{var}(\mathrm{d}M(t) | \mathscr{F}_{t-})$$

and

$$\mathrm{d}\langle M, M' \rangle(t) = \mathrm{cov}(\mathrm{d}M(t), \mathrm{d}M'(t) | \mathscr{F}_{t-}).$$

The predictable covariation process is bilinear and symmetric, just like an ordinary covariance:

$$\begin{aligned} \langle aM + bM', M'' \rangle &= a\langle M, M'' \rangle + b\langle M', M'' \rangle, \\ \langle M, M' \rangle &= \langle M', M \rangle. \end{aligned} \tag{2.3.4}$$

The predictable (co)variation process is also called the *pointed brackets process of* M (and M'). If

$$\langle M, M' \rangle = 0,$$

we say that M and M' are *orthogonal*. This means that the product MM' is itself a local martingale.

We often deal with a vector of local square integrable martingales $\mathbf{M} = (M_1, \ldots, M_k)$. We then write $\langle \mathbf{M} \rangle = (\langle M_h, M_j \rangle)$ for the *matrix* of predictable covariation processes of each pair of components of $\mathbf{M}$. The matrix of processes $\langle \mathbf{M}, \mathbf{M}' \rangle$ is defined similarly.

The predictable variation process, at time t, of a local square integrable martingale M is the limit in probability of approximations of the form $\sum \operatorname{var}(M(t_i) - M(t_{i-1}) | \mathscr{F}_{t_{i-1}})$ for ever finer partitions $0 = t_0 < t_1 \cdots < t_n = t$. When we omit the conditional expectation here but just take the limit (in probability) of sums of squares $\sum (M(t_i) - M(t_{i-1}))^2$, we obtain another nondecreasing process called the *optional variation process* of M. This new process exists when M is just a local martingale (not necessarily locally square integrable). It is not predictable, in general. It is denoted by $[M]$ and also called M's *square brackets process*, as well as M's quadratic variation process. Analogously, we have an *optional covariation process* or square bracket process (or quadratic covariation process) $[M, M']$ for two local martingales M and M'; and we can define a matrix optional covariation process for two vector local martingales. When M and M' are continuous, their optional covariation process coincides with the predictable covariation process. On the other hand, when M and M' are finite variation local martingales (as will always be the case in our applications), the optional variation process has an explicit and simple form

$$[M](t) = \sum_{s \le t} \Delta M(s)^2 = M(t)^2 - 2 \int_0^t M(s-) \, \mathrm{d}M(s),$$

$$[M, M'](t) = \sum_{s \le t} \Delta M(s) \Delta M'(s)$$

$$= M(t)M'(t) - \int_0^t M(s-) \, \mathrm{d}M'(s) - \int_0^t M'(s-) \, \mathrm{d}M(s).$$

The process $[M]$ is nondecreasing, cadlag, and

$$M^2 - [M]$$

is a local martingale. If $[M]$ is actually locally integrable, then the local martingale M is actually locally square integrable and $\langle M \rangle$ is the compensator of $[M]$. More generally, $\langle M, M' \rangle$ is the compensator of $[M, M']$. From this, it follows that if M and M' are finite variation local square integrable martingales with no jump times in common, so $\Delta M \Delta M' = 0$ and hence $[M, M'] = 0$, then $\langle M, M' \rangle = 0$ and M and M' are orthogonal (their product is a local martingale).

We exhibit the predictable and optional variation processes for the martingale M of Example II.3.1, the compensated Poisson process, and also for the Wiener process or standard Brownian motion.

EXAMPLE II.3.2. Poisson Process (Continued)

Let $N = \Pi$ be a standard Poisson process as in Example II.3.1 and define $M = \Pi - \iota$, where ι is the identity function $\iota(t) = t$. Using the fact that the variance of a Poisson random variable equals its mean and the independent increments of the Poisson process, one can check that $M^2 - \iota$ is a martingale; so, by uniqueness, the process ι is the compensator of M^2, that is, $\langle M \rangle(t) = t$ for all t, as well as being the compensator of Π itself. On the other hand, because M is a finite variation local martingale with jumps of size $+1$ only at the jump times of Π, we find that $[M] = \Pi$. □

EXAMPLE II.3.3. Standard Brownian Motion (the Wiener Process)

It can be shown that there exists a *continuous* process W which is zero at time zero and has independent normally distributed increments with means zero and variances equal to the lengths of the corresponding intervals. With respect to the natural filtration generated by the process, W is a local square integrable martingale (localized by constant times) and $\langle W \rangle = [W] = \text{var}(W) = \iota$, the identity function. W is called *the Wiener process* or *standard Brownian motion.* More generally, a time-transformed Brownian motion or Wiener process has continuous sample paths, independent, normally distributed mean zero increments. Its predictable and optional variation process are both equal to its variance function, a continuous nondecreasing function. □

The pointed and square bracket processes can be used to verify if a given stopping time localizes a local square integrable martingale. We may omit the multiplicative factor $I(T > 0)$, superfluous anyway if the local martingale is zero at time zero. For a local square integrable martingale M and a stopping time T, we have that M^T is a square integrable martingale if and only if

$$\mathrm{E}\langle M \rangle(T) < \infty$$

and also if and only if

$$\mathrm{E}[M](T) < \infty.$$

If M is a finite variation local martingale and T a stopping time, then M^T is a uniformly integrable martingale of integrable variation if and only if

$$\mathrm{E}\int_0^T |\mathrm{d}M(s)| < \infty.$$

Finally, we will make much use of the following simple way to check that a stopping time localizes a local martingale. If M is the difference of two cadlag increasing processes X and $\tilde{X}$, it suffices to verify that either $\mathrm{E}X(T) < \infty$ or $\mathrm{E}\tilde{X}(T) < \infty$ (by the use of the optional stopping theorem, localizing sequences of stopping times, and monotone convergence). For instance, applying this to $[M] - \langle M \rangle$, it follows that $\langle M \rangle$ has a finite expectation at a given time (or stopping time) if and only if $[M]$ does.

II.3.3. Stochastic Integration

We shall only be concerned with stochastic integrals which have a pathwise interpretation, as we stated in Section II.2: $\int X \,\mathrm{d}Y$ is the ordinary pathwise Lebesgue–Stieltjes integral (over $[0, t]$, for each $t \in \mathscr{T}$) defined for processes X and Y such that $\int_0^t |X(s, \omega)||Y(\mathrm{d}s, \omega)|$ is finite for almost all ω, for each $t \in \mathscr{T}$. However, this integral has special and valuable properties when the integrand X is a *predictable process* H and we integrate with respect to a process Y which is a *local (square integrable) martingale* M. Under the appropriate (local) integrability conditions, the resulting process $\int H \,\mathrm{d}M$ is a local (square integrable) martingale. Moreover, its predictable and optional variation processes can be obtained simply from those of M.

Theorem II.3.1. *Suppose M is a finite variation local square integrable martingale, H is a predictable process, and $\int H^2 \,\mathrm{d}[M]$ is locally integrable or $\int H^2 \,\mathrm{d}\langle M \rangle$ is just locally finite (automatically true if H is locally bounded). Then $\int H \,\mathrm{d}M$ is a local square integrable martingale, and*

$$\left[\int H \,\mathrm{d}M\right] = \int H^2 \,\mathrm{d}[M],$$

$$\left\langle \int H \,\mathrm{d}M \right\rangle = \int H^2 \,\mathrm{d}\langle M \rangle.$$

The predictable process H is locally bounded if it is left-continuous and has right-hand limits.

With weaker assumptions than those of Theorem II.3.1, we get weaker conclusions. Suppose always that M is a finite variation local martingale and H is predictable. If we just assume the process $\int |H||\mathrm{d}M|$ is locally integrable (automatically true if H is locally bounded), then $\int H \,\mathrm{d}M$ is a local martingale and $[\int H \,\mathrm{d}M] = \int H^2 \,\mathrm{d}[M]$. Alternatively, suppose the process $\int H^2 \,\mathrm{d}[M]$ is locally integrable (automatically true if H is locally bounded and M is locally square integrable). Then $\int H \,\mathrm{d}M$ is a local square integrable martingale and $[\int H \,\mathrm{d}M] = \int H^2 \,\mathrm{d}[M]$.

Formulas for predictable and optional covariation processes of stochastic integrals follow the same form: We have

$$\left\langle \int H \,\mathrm{d}M, \int K \,\mathrm{d}M' \right\rangle = \int HK \,\mathrm{d}\langle M, M' \rangle, \tag{2.3.5}$$

$$\left[\int H\,\mathrm{d}M, \int K\,\mathrm{d}M'\right] = \int HK\,\mathrm{d}[M, M']. \tag{2.3.6}$$

Sometimes we use vector and matrix versions of these formulas; for instance, for vectors $\mathbf{M}$ and $\mathbf{M}'$ and matrices $\mathbf{H}$ and $\mathbf{K}$, we have

$$\left\langle \int \mathbf{H}\,\mathrm{d}\mathbf{M}, \int \mathbf{K}\,\mathrm{d}\mathbf{M}' \right\rangle = \int \mathbf{H}\,\mathrm{d}\langle \mathbf{M}, \mathbf{M}'\rangle \mathbf{K}^\top, \tag{2.3.7}$$

$$\left[\int \mathbf{H}\,\mathrm{d}\mathbf{M}, \int \mathbf{K}\,\mathrm{d}\mathbf{M}'\right] = \int \mathbf{H}\,\mathrm{d}[\mathbf{M}, \mathbf{M}']\mathbf{K}^\top. \tag{2.3.8}$$

The following remarks give some indication of how these results can be derived. To begin with, consider as integrand a predictable process H of the form $H(t) = XI_{(u,v]}(t)$, where X is an $\mathscr{F}_u$-measurable random variable and $u < v$ are two fixed time points. This process is adapted and left-continuous, and hence predictable. Suppose M is a martingale and $\mathrm{E}(\int_0^\tau |\mathrm{H}||\mathrm{d}M|) = \mathrm{E}(|X| \int_{(u,v]} |\mathrm{d}M|) < \infty$. Then it is easy to check from the definitions that $\int H\,\mathrm{d}M$ is a martingale and $[\int H\,\mathrm{d}M] = \int H^2\,\mathrm{d}[M]$. If M is square integrable and $\mathrm{E}(\int_0^\tau H^2\,\mathrm{d}\langle M\rangle) = \mathrm{E}(X^2(\langle M\rangle(v) - \langle M\rangle(u))) < \infty$, one can also easily check that $\int H\,\mathrm{d}M$ is a square integrable martingale with $\langle \int H\,\mathrm{d}M\rangle = \int H^2\,\mathrm{d}\langle M\rangle$. Next, the results are extended to predictable processes which are a sum of a finite number of such processes; these are called *simple* predictable processes. Now we note that the class of predictable processes is generated by the simple predictable processes and apply monotone class arguments and localization to get general results.

II.4. Counting Processes

II.4.1. Counting Processes and Their Intensities

A multivariate counting process is a stochastic process which can be thought of as registering the occurrences in time of a number of types of disjoint, discrete events. We suppose that either a filtration is already given, relative to which the process is adapted, or that one constructs the so-called self-exciting filtration generated by the process itself; see the following text.

So, suppose a filtration $(\mathscr{F}_t : t \in \mathscr{T})$ on a probability space $(\Omega, \mathscr{F}, \mathrm{P})$ is given, satisfying the usual conditions (2.2.1), except possibly completeness. A multivariate counting process

$$\mathbf{N} = (N_1, \ldots, N_k)$$

is a vector of k adapted cadlag processes, all zero at time zero, with paths which are piecewise constant and nondecreasing, having jumps of size $+1$ only; no two components jumping simultaneously. We suppose $N_h(t)$ is almost surely finite for each h and all $t \in \mathcal{T}$. If $\tau \notin \mathcal{T}$, then $N_h(\tau) = \lim_{t\uparrow\tau} N_h(t)$ may, however, be infinite. The process is called *self-exciting* if the accompanying filtration is that generated by the process itself; the filtration $(\mathcal{N}_t)$ defined by

$$\mathcal{N}_t = \sigma(\mathbf{N}(s): s \le t) \quad \text{for all } t,$$

possibly augmented by null sets. Often, we deal with a larger filtration $(\mathcal{F}_t)$ having the special form

$$\mathcal{F}_t = \mathcal{F}_0 \vee \mathcal{N}_t;$$

thus, $\mathcal{F}_t$ is generated by $\mathcal{F}_0$ together with $\{\mathbf{N}(s), s \le t\}$. These filtrations are automatically right-continuous and increasing and can be made complete as well if desired (see Section II.2).

To a counting process $\mathbf{N}$ we can associate a sequence of *jump times* and *jump marks*, giving the time and type of each event. Because no two components of $\mathbf{N}$ jump simultaneously, the sum of the components is also a counting process. Defining then

$$N_{\bullet} = \sum_{h=1}^{k} N_h,$$

we can construct stopping times

$$0 < T_1 \le T_2 \le T_3 \le \cdots$$

and random variables

$$J_1, J_2, \ldots,$$

taking values in $\{1, 2, \ldots, k\} \cup \{0\}$ such that for $n \le N_{\bullet}(\tau)$, we have $T_n \in \mathcal{T}$, $J_n \ne 0$, $T_n > T_{n-1}$,

$$N_{\bullet}(T_n) = n \quad \text{and} \quad \Delta N_{J_n}(T_n) = 1.$$

For $n > N_{\bullet}(\tau)$, we set $J_n = 0$ and $T_n = \tau$ just to keep everything well defined. As just said, the variables T_n are stopping times; also J_n is $\mathcal{F}_{T_n}$-measurable. This is essentially the notation for a *marked point process*, to which we return at the end of this subsection.

Because the components of a counting process $\mathbf{N}$ are adapted, cadlag, locally bounded ($0 \le N_h^{T_n} \le n$), and nondecreasing, they are local submartingales and have compensators Λ_h which we collect together in a vector $\mathbf{\Lambda}$. Thus, each Λ_h is a nondecreasing predictable process, zero at time zero, such that

$$M_h = N_h - \Lambda_h \tag{2.4.1}$$

is a local martingale. Because $0 \le \Delta N_h \le 1$, the same can be shown to hold for $\Delta\Lambda_h$ for each h. [Consider, for instance, the predictable process $I(\Delta\Lambda_h > 1)$ and look at its stochastic integral with respect to M. The result should be a local martingale. However, N_h makes jumps of size one only, so

$\int I(\Delta\Lambda_h > 1)(\mathrm{d}N_h - \mathrm{d}\Lambda_h)$ is nonincreasing and zero at time zero, and must, therefore, be identically zero.] The compensator Λ_h can be considered as the *integrated* or *cumulative* stochastic intensity of the counting process N_h, for reasons which will become clear later [cf. (2.4.4)–(2.4.7)]. We define $\Lambda_{\bullet} = \sum_{h=1}^{k} \Lambda_h$; $\Lambda_{\bullet}$ is the compensator of the counting process $N_{\bullet}$, so we also have $0 \leq \Delta\Lambda_{\bullet} \leq 1$.

We next show that M_h is a local square integrable martingale, with a predictable variation process (the compensator of M_h^2) which can be simply described in terms of Λ_h (the compensator of N_h). Now the sequence of stopping times T_n can be considered as a localizing sequence, making $\mathbf{N}$ locally bounded because $N_{\bullet}(T_n) \leq n$ for each n. Also, because Λ_h is cadlag and predictable, it is locally bounded. Combining localizing times for N_h and Λ_h, one finds that, for each h, $M_h = N_h - \Lambda_h$ is a *local square integrable martingale*, in fact, of locally bounded variation. A little further argument shows that M_h may be localized to a square integrable martingale by the sequence T_n [or by any other sequence for which $\mathrm{E}N_{\bullet}(T_n) < \infty$ for each n].

As we mentioned, the predictable variation process of $\mathbf{M}$ can be simply expressed in terms of $\mathbf{\Lambda}$. One finds that

$$\begin{aligned} \langle M_h \rangle &= \Lambda_h - \int \Delta\Lambda_h \,\mathrm{d}\Lambda_h, \\ \langle M_h, M_{h'} \rangle &= -\int \Delta\Lambda_h \,\mathrm{d}\Lambda_{h'} \quad (h \neq h'). \end{aligned} \tag{2.4.2}$$

In particular, when $\mathbf{\Lambda}$ is *continuous*, this becomes

$$\begin{aligned} \langle M_h \rangle &= \Lambda_h, \\ \langle M_h, M_{h'} \rangle &= 0 \quad (h \neq h'). \end{aligned} \tag{2.4.3}$$

This means that the counting process martingales are *orthogonal* (cf. Section II.3.2) when an intensity process exists.

These facts can be proved by first calculating $[\mathbf{M}]$, the matrix of processes $[M_h, M_{h'}]$, and showing that its compensator $\langle \mathbf{M} \rangle$ has the elements just described. The continuous case is especially simple, so we look at that first. In that case, M_h is a local finite variation martingale with jumps coinciding with those of N_h and of the same size ($+1$ only). So the sum of squares of jumps of M_h equals N_h:

$$[M_h] = N_h.$$

It is locally integrable with compensator Λ_h, so M_h is locally square integrable and $\langle M_h \rangle = \Lambda_h$. On the other hand, for $h \neq h'$, M_h and $M_{h'}$ have no jumps in common at all, so $[M_h, M_{h'}] = 0$. The compensator of this process is the zero process, so $\langle M_h, M_{h'} \rangle = 0$ too. The same arguments were used in the special case of the standard Poisson process; see Examples II.3.1 and II.3.2.

In general, when $\mathbf{\Lambda}$ may have jumps, direct calculation gives

$$
\begin{aligned}
[M_h] = \sum (\Delta M_h)^2 &= \int \Delta M_h \, \mathrm{d}M_h \\
&= \int (\Delta N_h - \Delta\Lambda_h)(\mathrm{d}N_h - \mathrm{d}\Lambda_h) \\
&= N_h - 2\int \Delta\Lambda_h \, \mathrm{d}N_h + \int \Delta\Lambda_h \, \mathrm{d}\Lambda_h,
\end{aligned}
$$

where we use the fact that $(\Delta N_h)^2 = \Delta N_h$. We want to calculate the compensator of this process. The compensator of the first of the three terms on the right-hand side is already known to be Λ_h. For the second, note that $\Delta\Lambda_h$ is a bounded predictable process and M_h a local martingale, so $\int \Delta\Lambda_h \, \mathrm{d}M_h = \int \Delta\Lambda_h \, \mathrm{d}N_h - \int \Delta\Lambda_h \, \mathrm{d}\Lambda_h$ is a local martingale too. Thus, the predictable process $\int \Delta\Lambda_h \, \mathrm{d}\Lambda_h$ is the compensator of $\int \Delta\Lambda_h \, \mathrm{d}N_h$. The third term is already predictable and, therefore, its own compensator. Combining, we find that $\Lambda_h - \int \Delta\Lambda_h \, \mathrm{d}\Lambda_h$ is the compensator of $[M_h]$. This gives us the required result for $\langle M_h \rangle$. For the predictable covariation, a similar calculation (for $h \neq h'$) gives

$$
[M_h, M_{h'}] = -\int \Delta\Lambda_h \, \mathrm{d}M_{h'} - \int \Delta\Lambda_{h'} \, \mathrm{d}M_h + \int \Delta\Lambda_h \, \mathrm{d}\Lambda_{h'}.
$$

The compensator of this process is

$$
-\int \Delta\Lambda_h \, \mathrm{d}\Lambda_{h'} - \int \Delta\Lambda_{h'} \, \mathrm{d}\Lambda_h + \int \Delta\Lambda_h \, \mathrm{d}\Lambda_{h'} = -\int \Delta\Lambda_h \, \mathrm{d}\Lambda_{h'}.
$$

In most of this book, we will be interested in the so-called *absolutely continuous case*, the case in which *intensities* exist. We say that N_h has *intensity process* λ_h if λ_h is a predictable process and

$$
\Lambda_h(t) = \int_0^t \lambda_h(s) \, \mathrm{d}s \quad \text{for all } t \tag{2.4.4}
$$

(some authors would not require predictability of λ_h). If N_h is integrable and λ_h is left-continuous with right-hand limits, one can easily show that

$$
\lim_{\Delta t \downarrow 0} \frac{1}{\Delta t} \mathrm{E}(N_h(t + \Delta t) - N_h(t) \mid \mathscr{F}_t) = \lambda_h(t+) \quad \text{a.s.} \tag{2.4.5}
$$

Under further conditions, one can strengthen this to

$$
\lim_{\Delta t \downarrow 0} \frac{1}{\Delta t} \mathrm{P}(N_h(t + \Delta t) - N_h(t) \geq 1 \mid \mathscr{F}_t) = \lambda_h(t+) \quad \text{a.s.} \tag{2.4.6}
$$

or even to

$$
\lim_{\Delta t \downarrow 0} \frac{1}{\Delta t} \mathrm{P}(N_h(t + \Delta t) - N_h(t) = 1 \mid \mathscr{F}_t) = \lambda_h(t+) \quad \text{a.s.,} \tag{2.4.7}
$$

justifying the name we have given to λ_h. Whereas (2.4.5)–(2.4.7) are useful in the interpretation of the intensity process, the regularity conditions needed when taking them as starting point (see Aven, 1985) do not have any further role, and none of (2.4.5)–(2.4.7) have direct operational value, so there is no use in demanding them to hold.

Clearly, an integrated or cumulative intensity process $\boldsymbol{\Lambda}$ exists and is unique, up to indistinguishability, by the existence and uniqueness of compensator. An intensity process $\boldsymbol{\lambda}$, if it exists, is, however, not unique because we have only defined it here by requiring predictability and $\int \boldsymbol{\lambda} = \boldsymbol{\Lambda}$. (If $\boldsymbol{\Lambda}$ is absolutely continuous and its derivative satisfies a modest measurability condition, then a predictable version does exist.) If an intensity process which is left-continuous with right-hand limits *exists*, then it will be unique, among such processes, and up to indistinguishability.

We next give one of the most important examples of a counting process. Though it is extremely simple, it is the basis of a great deal of further developments. This is the case of a one-component one-jump self-exciting counting process (cf. Example II.2.1).

EXAMPLE II.4.1. One-Jump Process

Let T be a non-negative random variable with absolutely continuous distribution function F, survival function $S = 1 - F$, density f, and hazard rate $\alpha = f/S$. Let τ_F be the upper limit of the support of F, but take $\tau = \infty$ and $\mathcal{T} = [0, \infty]$ even if $\tau_F < \infty$. Note that

$$\int_0^t \alpha(u)\,\mathrm{d}u = -\log(1 - F(t)) < \infty$$

for all $t < \tau_F$ though $\int_0^{\tau_F} \alpha = \infty$. Define the univariate counting process N by

$$N(t) = I(T \leq t)$$

and let $(\mathcal{N}_t)$ be the filtration it generates; in fact,

$$\mathcal{N}_t = \sigma\{N(s): s \leq t\} = \sigma\{T \wedge t, I(T \leq t)\}.$$

Define the left-continuous adapted process Y by

$$Y(t) = I(T \geq t).$$

We show that N has compensator Λ defined by

$$\Lambda(t) = \int_0^t Y(s)\alpha(s)\,\mathrm{d}s$$

and, hence, N has intensity process λ defined by

$$\lambda(t) = Y(t)\alpha(t).$$

To do this, we note that Λ is predictable (it is continuous and adapted), so we need only verify that $M = N - \Lambda$ is a local martingale. In fact, it is a square integrable martingale on $\bar{\mathscr{T}}$ by our general theory earlier. To show the martingale property, it suffices to verify that

$$\mathrm{E}(M(\infty)|\mathscr{N}_t) = M(t)$$

because this implies, for $s < t$,

$$\begin{aligned}\mathrm{E}(M(t)|\mathscr{N}_s) &= \mathrm{E}(\mathrm{E}(M(\infty)|\mathscr{N}_t)|\mathscr{N}_s)\\ &= \mathrm{E}(M(\infty)|\mathscr{N}_s)\\ &= M(s).\end{aligned}$$

Consider, first, the case $t = 0$. Because $\mathscr{N}_0$ is trivial, we must show $\mathrm{E}M(\infty) = 0$. Now $N(\infty) = 1$, whereas

$$\begin{aligned}\mathrm{E}\Lambda(\infty) &= \mathrm{E}\int_0^{\tau_F} Y(t)\alpha(t)\,\mathrm{d}t\\ &= \int_0^{\tau_F} \mathrm{P}(T \geq t)\alpha(t)\,\mathrm{d}t\\ &= \int_0^{\tau_F} S(t)\frac{f(t)}{S(t)}\,\mathrm{d}t = \int_0^{\tau_F} f(t)\,\mathrm{d}t = 1.\end{aligned}$$

We next show that the result for general t follows from that for $t = 0$. When conditioning on $\mathscr{N}_t = \sigma(T \wedge t, I(T \leq t))$ we have to consider two separate cases: conditional on $T = s \leq t$ for some $s \leq t$ and conditional on $T > t$. In the first case, $M(\infty) = M(t) = M(s)$ and there is nothing to prove. In the second case, $M(\infty) - M(t) = 1 - \int_t^{\tau_F} I(T \geq s)\alpha(s)\,\mathrm{d}s$ and we must show that the expectation of this random variable, given $T > t$, is zero. But conditional on $T > t$, T now has hazard rate $\alpha I_{(t,\tau_F)}$. So this case is just the same as the case $t = 0$ only with a different hazard rate. □

Various extensions of this example are almost as easy to obtain. For instance, if T does not have a density, then one may still define an *integrated* or *cumulative hazard function* A by $A(t) = \int_0^t (1 - F(s-))^{-1}\,\mathrm{d}F(s)$. The compensator of N is the process $\Lambda = \int Y\,\mathrm{d}A$ with Y defined as above. The calculation above goes through essentially unchanged. We see that, in this example, N has an *intensity process* if and only if the distribution of T has a density.

Another extension to both counting process and filtration is to let, at time T, further information become available in the form of a *mark* J, a random variable taking values in $\{1, \ldots, k\}$, telling us the *type* of the event which has just occurred. In fact, this puts us in the so-called *competing risks models*; see Example III.1.5. Define a k-variate counting process $\mathbf{N}$ by letting $N_h(t) = I(T \leq t, J = h)$, $t \in [0, \infty)$, $h \in \{1, \ldots, k\}$. Let the process be self-exciting by taking $\mathscr{N}_t = \sigma\{T \wedge t, I(T \leq t), J \cdot I(T \leq t)\}$. Suppose T has survival function S and that (T, J) has a joint density $f_h(t)$ with respect to Lebesgue times counting measure. Then with $\alpha_h = f_h/S$, one finds that the compensator of N_h is $\Lambda_h = \int Y\alpha_h$, where $Y(t) = I(T \geq t)$ as before; so N_h has intensity process $\lambda_h(t) = \alpha_h(t)Y(t)$. The calculation of Example II.4.1 needs only minor modifications.

With these extensions, a completely general result for the intensity of a self-exciting counting process can be finally obtained. We will give this result (Jacod's representation) in Section II.7.1.

Earlier we have shown how, for the counting process martingale $\mathbf{M} = \mathbf{N} - \mathbf{\Lambda}$, the predictable and optional covariation process $\langle \mathbf{M} \rangle$ and $[\mathbf{M}]$ can be expressed in terms of $\mathbf{\Lambda}$ and $\mathbf{N}$. These relations have consequences via Theorem II.3.1 for integrals of predictable processes with respect to $\mathbf{M}$. Suppose, for simplicity, that the counting process has an intensity process $\boldsymbol{\lambda}$; so $\mathbf{\Lambda} = \int \boldsymbol{\lambda}$ is absolutely continuous. Let $\mathbf{H}$ be a $p \times k$ matrix of locally bounded predictable processes; for instance, $\mathbf{H}$ might be left-continuous and adapted; or cadlag and predictable; or deterministic and locally bounded. Let $\int \mathbf{H}\,\mathrm{d}\mathbf{M}$ denote the p-vector of rowwise sums of integrals of components of $\mathbf{H}$ with respect to the k components of $\mathbf{M}$. By Theorem II.3.1, the result is a vector of finite variation local square integrable martingales. Using diag of a vector to denote the diagonal matrix with the components of the vector on the diagonal, we obtain the following proposition [cf. (2.3.7) and (2.3.8)]:

Proposition II.4.1. *Let the counting process* $\mathbf{N}$ *have intensity process* $\boldsymbol{\lambda}$, *let* $\mathbf{M} = \mathbf{N} - \int \boldsymbol{\lambda}$, *and let* $\mathbf{H}$ *be locally bounded and predictable. Then* $\mathbf{M}$ *and* $\int \mathbf{H}\,\mathrm{d}\mathbf{M}$ *are local square integrable martingales with*

$$
\begin{aligned}
\langle \mathbf{M} \rangle &= \operatorname{diag} \int \boldsymbol{\lambda}, \\
[\mathbf{M}] &= \operatorname{diag} \mathbf{N}, \\
\left\langle \int \mathbf{H}\,\mathrm{d}\mathbf{M} \right\rangle &= \int \mathbf{H} \operatorname{diag} \boldsymbol{\lambda} \mathbf{H}^{\top}, \\
\left[\int \mathbf{H}\,\mathrm{d}\mathbf{M} \right] &= \int \mathbf{H}\,\mathrm{d}(\operatorname{diag} \mathbf{N}) \mathbf{H}^{\top}.
\end{aligned}
\tag{2.4.8}
$$

Written out componentwise, and with δ_{hl} denoting a Kronecker delta, these important equations become

$$
\begin{aligned}
\langle M_h, M_l \rangle &= \delta_{hl} \int \lambda_h, \\
[M_h, M_l] &= \delta_{hl} N_h, \\
\left\langle \sum_h \int H_{jh}\,\mathrm{d}M_h, \sum_l \int H_{j'l}\,\mathrm{d}M_l \right\rangle &= \sum_h \int H_{jh} H_{j'h} \lambda_h, \\
\left[\sum_h \int H_{jh}\,\mathrm{d}M_h, \sum_l \int H_{j'l}\,\mathrm{d}M_l \right] &= \sum_h \int H_{jh} H_{j'h}\,\mathrm{d}N_h.
\end{aligned}
\tag{2.4.9}
$$

II.4.1.1. *Marked Point Processes*

Sometimes more than a finite number of different types of events are possible though the process counting all of them is still a counting process (there are a finite total number of events in finite time intervals). The different types may even vary continuously; think, for instance, of a continuous numerical measurement becoming available at the time of each possible event. It is now no longer convenient to count each type separately. Rather, one counts over aggregates of types; for instance, measurements in certain intervals or, more generally, in any given Borel set. The processes counting events with type in some sets will each be counting processes, but there can be overlap for overlapping sets.

We will accommodate these possibilities by considering $\mathbf{N}$ and $\mathbf{\Lambda}$ as *measures* on the product space: time cross type (or mark). The counting process $\mathbf{N}$ corresponds to a counting measure on the set of points (T_1, J_1), (T_2, J_2), etc. (jump time, jump type, or mark). Usually one replaces $\mathbf{N}$ and $\mathbf{\Lambda}$ by μ and ν but we will continue to use the old symbols with the new interpretation.

So let $(E, \mathscr{E})$ be some measurable space of marks or types. In the case of an ordinary multivariate counting process, E is just $\{1, \ldots, k\}$ and $\mathscr{E}$ is its power set: all possible subsets of E. We say that $\mathbf{N}$ is a marked point process with respect to a given filtration and a given mark space if $\mathbf{N}$ is a (random) counting measure on the product space $(\mathscr{T} \times E, \mathscr{B}(\mathscr{T}) \otimes \mathscr{E})$, where $\mathscr{B}$ denotes a Borel σ-algebra and where $\mathbf{N}$ satisfies the following conditions: N_A defined by

$$N_A(t) = \mathbf{N}([0, t] \times A)$$

is a counting process for every $A \in \mathscr{E}$ (an adapted cadlag step-function, jumps of size $+1$ only, zero at time zero). It follows from this that, for disjoint sets $A_1, \ldots, A_k \in \mathscr{E}$, the process $(N_{A_1}, \ldots, N_{A_k})$ is a multivariate counting process in the ordinary sense.

The process $(N_{A_1}, \ldots, N_{A_k})$, therefore, has a compensator $(\Lambda_{A_1}, \ldots, \Lambda_{A_k})$. It turns out that one can extract all these compensators from a single so-called *predictable measure* $\mathbf{\Lambda}$ on $(\mathscr{T} \times E, \mathscr{B}(\mathscr{T}) \otimes \mathscr{E})$ in the same way as is done for $\mathbf{N}$:

$$\Lambda_A(t) = \mathbf{\Lambda}([0, t] \times A).$$

Letting (T_n, J_n) be the points of the marked point process, it turns out that T_n is a stopping time and J_n is $\mathscr{F}_{T_n}$-measurable [with values in $(E, \mathscr{E})$] for each $n = 1, 2, \ldots$.

II.4.2. The Innovation Theorem

In this and the next subsections, we present a collection of results showing how intensity processes change under various operations: change of filtration, formation of products, starting, stopping and filtering. This provides the model-builder with a wide range of techniques for deriving new counting process models from old as we do in Chapter III. Another technique (change of probability measure) will be discussed in Section II.7.1, as well as how to start from scratch.

Consider a counting process $\mathbf{N}$, adapted to both of two filtrations $(\mathscr{F}_t)$ and $(\mathscr{G}_t)$ with $\mathscr{F}_t \subseteq \mathscr{G}_t$ for all t; we say the filtrations are *nested.* This corresponds to two different levels of information, or two observers. In both cases, $\mathbf{N}$ is observed. Suppose the counting process has intensity process $\boldsymbol{\lambda}$ with respect to the larger filtration $(\mathscr{G}_t)$. One may ask what its intensity process with respect to the smaller filtration $(\mathscr{F}_t)$ is. This will generally differ from $\boldsymbol{\lambda}$; after all, one is given less information to condition on. The answer is given by the *innovation theorem*, which states that there exists an $(\mathscr{F}_t)$-predictable process $\tilde{\boldsymbol{\lambda}}$ such that

$$\tilde{\boldsymbol{\lambda}}(t) = \mathrm{E}(\boldsymbol{\lambda}(t) | \mathscr{F}_{t-}) \quad \text{for all } t$$

and such that $\tilde{\boldsymbol{\lambda}}$ is the $(\mathscr{F}_t)$-intensity process of $\mathbf{N}$. (More generally, t can be replaced by any $(\mathscr{F}_t)$-*predictable stopping time* T.)

If $\mathbf{N}$ has intensity process $\boldsymbol{\lambda}$ with respect to a filtration $(\mathscr{G}_t)$ and $\mathbf{N}$ is adapted and $\boldsymbol{\lambda}$ is predictable with respect to a smaller filtration $(\mathscr{F}_t)$, then $\boldsymbol{\lambda}$ is also the $(\mathscr{F}_t)$-intensity process of $\mathbf{N}$, by uniqueness of the intensity process.

II.4.3. Product Spaces

Probability models are often built by combining independent components in a product space. This construction is also valuable in building models based on counting processes and proceeds as follows.

Suppose we are given two filtered probability spaces $(\Omega^{(i)}, \mathscr{F}^{(i)}, (\mathscr{F}_t^{(i)}, t \in \mathscr{T}), \mathrm{P}^{(i)})$, $i = 1, 2$, on each of which is defined a counting process $\mathbf{N}^{(i)}$ with intensity process $\boldsymbol{\lambda}^{(i)}$. We can now form the product space on which the processes $\mathbf{N}^{(i)}$ and $\boldsymbol{\lambda}^{(i)}$ are defined in the obvious way, and naturally define a product filtration by

$$\Omega = \Omega^{(1)} \times \Omega^{(2)}, \qquad \mathscr{F} = \mathscr{F}^{(1)} \otimes \mathscr{F}^{(2)},$$

$$\mathscr{F}_t = \mathscr{F}_t^{(1)} \otimes \mathscr{F}_t^{(2)}, \qquad \mathrm{P} = \mathrm{P}^{(1)} \otimes \mathrm{P}^{(2)}.$$

Actually, $(\mathscr{F}_t)$ is not necessarily right-continuous. A sufficient condition for this is that $\mathscr{F}_t^{(i)} = \mathscr{F}_0^{(i)} \vee \sigma\{\mathbf{N}^{(i)}(s): s \leq t\}$ in which case

$$\mathscr{F}_t = \mathscr{F}_0^{(1)} \otimes \mathscr{F}_0^{(2)} \vee \sigma\{\mathbf{N}^{(1)}(s), \mathbf{N}^{(2)}(s): s \leq t\}$$

which is obviously right-continuous. It is easy to check that the processes $\lambda_h^{(i)}(t)$ are also predictable, viewed as defined on Ω and with respect to P and $(\mathscr{F}_t)$. Moreover, the $N_h^{(i)} - \int \lambda_h^{(i)}$ remain local martingales on the product space. Thus, the intensity processes of $\mathbf{N}^{(i)}$ remain the same and, indeed,

$(\mathbf{N}^{(1)}, \mathbf{N}^{(2)})$ forms a multivariate counting process with intensity process $(\boldsymbol{\lambda}^{(1)}, \boldsymbol{\lambda}^{(2)})$ because the probability is zero of having simultaneous jumps of $\mathbf{N}^{(1)}$ and $\mathbf{N}^{(2)}$ (by continuity of their compensators).

Instead of this combination of independent component processes, we will occasionally need to combine *conditionally independent* components. The situation now is that there is one probability space $(\Omega, \mathscr{F}, \mathrm{P})$ on which $\mathbf{N}(t) = (\mathbf{N}^{(1)}(t), \mathbf{N}^{(2)}(t))$ is defined; we consider two filtrations $(\mathscr{F}_t^{(1)})$ and $(\mathscr{F}_t^{(2)})$ and assume that they are *conditionally independent* given some σ-algebra $\mathscr{A} \subseteq \mathscr{F}$, that is, if $A \in \mathscr{F}_t^{(1)}$, $B \in \mathscr{F}_t^{(2)}$, $C \in \mathscr{A}$, $\mathrm{P}(C) > 0$, then

$$\mathrm{P}(A \cap B \mid C) = \mathrm{P}(A \mid C)\mathrm{P}(B \mid C).$$

Often, $\mathscr{F}_t^{(i)} = \sigma\{\mathbf{N}^{(i)}(s) \colon s \le t\}$. The σ-algebra $\mathscr{A}$ models events which are supposed to happen "at time 0."

Suppose that the counting processes $\mathbf{N}^{(1)}$, $\mathbf{N}^{(2)}$ have intensity processes $\boldsymbol{\lambda}^{(1)}$, $\boldsymbol{\lambda}^{(2)}$ with respect to the two filtrations $(\mathscr{F}_t^{(1)})$ and $(\mathscr{F}_t^{(2)})$. Define

$$\mathscr{F}_t = \mathscr{A} \vee \mathscr{F}_t^{(1)} \vee \mathscr{F}_t^{(2)},$$

so that if the $\mathscr{F}_0^{(i)}$ are trivial, we have $\mathscr{F}_0 = \mathscr{A}$. One may check directly that any $(\mathscr{A} \vee \mathscr{F}_t^{(1)})$-martingale is also an $(\mathscr{F}_t)$-martingale, which is the key step in verifying that the $(\mathscr{F}_t)$-compensator of $(\mathbf{N}^{(1)}, \mathbf{N}^{(2)})$ may be obtained by combining the $(\mathscr{A} \vee \mathscr{F}_t^{(i)})$-compensators of $\mathbf{N}^{(i)}(t)$. This leads to the desired result that $(\mathbf{N}^{(1)}, \mathbf{N}^{(2)})$ has intensity process $(\boldsymbol{\lambda}^{(1)}, \boldsymbol{\lambda}^{(2)})$ with respect to the combined filtration $(\mathscr{F}_t)$.

II.4.4. Starting, Stopping, and Filtering

That *stopping* a counting process preserves its intensity (or compensator) is a nice exercise in stochastic integration: Let $\mathbf{N}$ be a multivariate counting process with compensator $\boldsymbol{\Lambda}$, let $\mathbf{M} = \mathbf{N} - \boldsymbol{\Lambda}$, and let T be a stopping time. The indicator function $I_{[0,T]}$ of the stochastic interval $[0, T]$ is left-continuous and adapted, hence predictable and (locally) bounded. Observe that $\mathbf{N}^T = \int I_{[0,T]} \, \mathrm{d}\mathbf{N}$ and $\boldsymbol{\Lambda}^T = \int I_{[0,T]} \, \mathrm{d}\boldsymbol{\Lambda}$ (componentwise integration). Because $\mathbf{M}^T = \mathbf{N}^T - \boldsymbol{\Lambda}^T = \int I_{[0,T]} \, \mathrm{d}\mathbf{M}$, we find that $\mathbf{M}^T$ is a vector of local martingales. Also $\mathbf{N}^T$ is a new multivariate counting process, counting events up to and including time T only. Because the integral of one predictable process with respect to another is again predictable, $\boldsymbol{\Lambda}^T$ is a finite variation predictable process and is, therefore, by uniqueness of compensators, the compensator of $\mathbf{N}^T$. If $\boldsymbol{\Lambda}$ admits of an intensity $\boldsymbol{\lambda}$, we see that $\mathbf{N}^T$ has the intensity process $\boldsymbol{\lambda} I_{[0,T]}$.

One now can proceed to reduce the filtration replacing $(\mathscr{F}_t)$ by $(\mathscr{F}_{t \wedge T})$. Because $\mathbf{N}^T$ and $\boldsymbol{\Lambda}^T$ are adapted with respect to this smaller filtration, by the innovation theorem (Section II.4.2), $\boldsymbol{\lambda} I_{[0,T]}$ remains the intensity process of $\mathbf{N}^T$ with respect to it.

More generally, one could consider $\tilde{\mathbf{N}} = \int C \, d\mathbf{N}$ for any predictable, zero-one valued process C; $\tilde{\mathbf{N}}$ is a counting process with intensity $\boldsymbol{\lambda} C$. We call this *filtering*, which we discuss at length in Section III.4.

We now discuss how *starting* a counting process (at a stopping time), while conditioning on the past at that moment, also preserves its intensity. Let

$$\mathbf{N} = (N_h, h = 1, \ldots, k)$$

be a multivariate counting process on a space $(\Omega, \mathscr{F})$ with compensator $\boldsymbol{\Lambda}$ and intensity process $\boldsymbol{\lambda}$ with respect to a filtration $(\mathscr{F}_t)$. Furthermore, we let V be an $(\mathscr{F}_t)$-stopping time and consider an event $A \in \mathscr{F}_V$. The process $\mathbf{N}$ started at V is defined as

$${}_V\mathbf{N}(t) = \mathbf{N}(t) - \mathbf{N}(t \wedge V).$$

We want to study the process $\mathbf{N}$, starting from the time V, *given* that the event A (prior to V) has actually occurred. We call the process ${}_V\mathbf{N}$, *under this conditional distribution*, a left-truncated process. The proposition below states that left-truncation of $\mathbf{N}$ by the event A (before V) preserves the intensity of $\mathbf{N}$ after time V. For ease of presentation, we suppose $\mathrm{P}(A) > 0$.

Proposition II.4.2. *The left-truncated counting process ${}_V\mathbf{N}$ has intensity process ${}_V\boldsymbol{\lambda} = \boldsymbol{\lambda} I_{(V,\infty)}$, i.e.,*

$${}_V\boldsymbol{\lambda}(t) = \boldsymbol{\lambda}(t) I_{(V,\infty)}(t),$$

with respect to the filtration ${}_V\mathscr{F}_t$ given by

$${}_V\mathscr{F}_t = \mathscr{F}_t \vee \mathscr{F}_V$$

and the conditional probability P^A given by

$$\mathrm{P}^A(F) = \mathrm{P}(F \cap A)/\mathrm{P}(A), \quad F \in \mathscr{F}.$$

A proof of Proposition II.4.2 was given by Andersen et al. (1988). A result similar to Proposition II.4.2 can be obtained for *any* event $A \in \mathscr{F}_V$, not necessarily satisfying $\mathrm{P}(A) > 0$, using the technical apparatus of proper regular conditional probabilities and Blackwell spaces; see Jacobsen (1982, Exercise 8, p. 51, and Appendix 1.)

II.5. Limit Theory

II.5.1. A Martingale Central Limit Theorem

Our main tool for studing the large sample properties of statistical methods for counting process models will be the martingale central limit theorem. There exist many versions of this, as well as generalizations to semimartingales. Here we present a version close to that given by Rebolledo (1980a) for locally square integrable martingales because in our applications we will be applying it to stochastic integrals with respect to the basic counting process martingales, which are locally square integrable. As we suggested in Section II.1, two conditions are required for a local square integrable martingale to be approximately Gaussian: It must have close to continuous paths, and its predictable variation process must be close to deterministic.

For each $n = 1, 2, \ldots$, let $\mathbf{M}^{(n)} = (M_1^{(n)}, \ldots, M_k^{(n)})$ be a vector of k local square integrable martingales, possibly defined on different sample spaces and filtrations for each n. Also, for each $\varepsilon > 0$, let $\mathbf{M}_\varepsilon^{(n)}$ be a vector of k local square integrable martingales, containing all the jumps of components of $\mathbf{M}^{(n)}$ larger in absolute value than ε: So $M_h^{(n)} - M_{\varepsilon,h}^{(n)}$ is also a local square integrable martingale and $|\Delta M_h^{(n)} - \Delta M_{\varepsilon,h}^{(n)}| \leq \varepsilon$. We write $\langle \mathbf{M}^{(n)} \rangle$ for the $k \times k$ matrix of processes $\langle M_h^{(n)}, M_{h'}^{(n)} \rangle$ (Section II.3.2).

Next, let $\mathbf{M}^{(\infty)}$ be a continuous Gaussian vector martingale with $\langle \mathbf{M}^{(\infty)} \rangle = [\mathbf{M}^{(\infty)}] = \mathbf{V}$, a continuous deterministic $k \times k$ positive semidefinite matrix-valued function on $\mathscr{T}$, with positive semidefinite increments, zero at time zero. So $\mathbf{M}^{(\infty)}(t) - \mathbf{M}^{(\infty)}(s) \sim \mathscr{N}(\mathbf{0}, \mathbf{V}(t) - \mathbf{V}(s))$ (a multivariate normal distribution) and is independent of $(\mathbf{M}^{(\infty)}(u); u \leq s)$ for all $0 \leq s \leq t$. Given a function $\mathbf{V}$ with the above-mentioned properties, the process $\mathbf{M}^{(\infty)}$ always exists.

With these preparations made, we can now state the martingale central limit theorem in a convenient form.

Theorem II.5.1 (Rebolledo's theorem). *Let $\mathscr{T}_0 \subseteq \mathscr{T}$ and consider the conditions*

$$\langle \mathbf{M}^{(n)} \rangle (t) \xrightarrow{\mathrm{P}} \mathbf{V}(t) \quad \textit{for all } t \in \mathscr{T}_0 \textit{ as } n \to \infty, \tag{2.5.1}$$

$$[\mathbf{M}^{(n)}](t) \xrightarrow{\mathrm{P}} \mathbf{V}(t) \quad \textit{for all } t \in \mathscr{T}_0 \textit{ as } n \to \infty, \tag{2.5.2}$$

$$\langle M_{\varepsilon h}^{(n)} \rangle (t) \xrightarrow{\mathrm{P}} 0 \quad \textit{for all } t \in \mathscr{T}_0, h \textit{ and } \varepsilon > 0 \textit{ as } n \to \infty. \tag{2.5.3}$$

Then either of (2.5.1) *and* (2.5.2), *together with* (2.5.3), *imply*

$$(\mathbf{M}^{(n)}(t_1), \ldots, \mathbf{M}^{(n)}(t_l)) \xrightarrow{\mathscr{D}} (\mathbf{M}^{(\infty)}(t_1), \ldots, \mathbf{M}^{(\infty)}(t_l)) \quad \textit{as } n \to \infty \tag{2.5.4}$$

for all $t_1, \ldots, t_l \in \mathscr{T}_0$; moreover, both (2.5.1) *and* (2.5.2) *then hold.*

If, furthermore, $\mathscr{T}_0$ is dense in $\mathscr{T}$ and contains τ if $\tau \in \mathscr{T}$, then the same conditions imply

$$\mathbf{M}^{(n)} \xrightarrow{\mathscr{D}} \mathbf{M}^{(\infty)} \quad \textit{in } (D(\mathscr{T}))^k \textit{ as } n \to \infty \tag{2.5.5}$$

and $\langle \mathbf{M}^{(n)} \rangle$ and $[\mathbf{M}^{(n)}]$ converge uniformly on compact subsets of $\mathscr{T}$, in probability, to $\mathbf{V}$.

Here, $(D(\mathscr{T}))^k$ is the space of $\mathbb{R}^k$-valued cadlag functions on $\mathscr{T}$ endowed with the Skorohod topology, and $\xrightarrow{\mathscr{D}}$ denotes weak convergence as described, e.g., by Billingsley (1968). Condition (2.5.1) states that the predictable variation processes converge in probability to a deterministic function; the *Lindeberg condition* (2.5.3) states that the jumps of $\mathbf{M}^{(n)}$ become small as $n \to \infty$.

We reformulate these conditions in the case of most interest to us, stochastic integrals with respect to counting processes having a continuous compensator. For each $n = 1, 2, \ldots$, let $\mathbf{N}^{(n)}$ be a k_n-variate counting process with intensity process $\boldsymbol{\lambda}^{(n)}$. Let $\mathbf{H}^{(n)}$ be a $k \times k_n$ matrix of locally bounded predictable processes. Now define

$$M_j^{(n)}(t) = \sum_{h=1}^{k_n} \int_0^t H_{jh}^{(n)}(s)(\mathrm{d}N_h^{(n)}(s) - \lambda_h^{(n)}(s)\,\mathrm{d}s),$$

$$M_{j\varepsilon}^{(n)}(t) = \sum_{h=1}^{k_n} \int_0^t H_{jh}^{(n)}(s) I(|H_{jh}^{(n)}(s)| > \varepsilon)(\mathrm{d}N_h^{(n)}(s) - \lambda_h^{(n)}(s)\,\mathrm{d}s),$$

for $j = 1, \ldots, k$ and $\varepsilon > 0$; $\mathbf{M}^{(n)} = (M_1^{(n)}, \ldots, M_k^{(n)})$ and $\mathbf{M}_\varepsilon^{(n)} = (M_{1\varepsilon}^{(n)}, \ldots, M_{k\varepsilon}^{(n)})$. Then $\mathbf{M}_\varepsilon^{(n)}$ contains all the jumps of $\mathbf{M}^{(n)}$ larger than ε and both are k-variate local square integrable martingales. The bracket processes in (2.5.1)–(2.5.3) are given (cf. Proposition II.4.1) by

$$\langle M_j^{(n)}, M_{j'}^{(n)} \rangle (t) = \sum_{h=1}^{k_n} \int_0^t H_{jh}^{(n)}(s) H_{j'h}^{(n)}(s) \lambda_h^{(n)}(s)\,\mathrm{d}s, \tag{2.5.6}$$

$$[M_j^{(n)}, M_{j'}^{(n)}](t) = \sum_{h=1}^{k_n} \int_0^t H_{jh}^{(n)}(s) H_{j'h}^{(n)}(s)\,\mathrm{d}N_h^{(n)}(s), \tag{2.5.7}$$

$$\langle M_{j\varepsilon}^{(n)}, M_{j\varepsilon}^{(n)} \rangle (t) = \sum_{h=1}^{k_n} \int_0^t (H_{jh}^{(n)}(s))^2 I(|H_{jh}^{(n)}(s)| > \varepsilon) \lambda_h^{(n)}(s)\,\mathrm{d}s. \tag{2.5.8}$$

We will usually deal with situations in which k_n is fixed and $\lambda_h^{(n)}(s)$ gets larger as $n \to \infty$ for all h and s. To balance this, we will arrange matters so that $H_{jh}^{(n)}(s)$ gets smaller [taking care of (2.5.3)] and so that $H_{jh}^{(n)}(s) H_{j'h}^{(n)}(s) \lambda_h^{(n)}(s)$ converges in probability to a finite, nonzero deterministic limit as $n \to \infty$, for each j, j', and h [taking care of (2.5.1)]; see Section II.1 for a simple example. Thus, the following situation will frequently arise: For some sequence of processes $X^{(n)}$, it is given that

$$X^{(n)}(s) \xrightarrow{\mathrm{P}} f(s) \quad \text{as } n \to \infty \tag{2.5.9}$$

for almost all $s \in \mathcal{T}$, where the deterministic function f satisfies

$$\int_0^\tau |f(s)|\,\mathrm{d}s < \infty. \tag{2.5.10}$$

Under which supplementary conditions may we conclude

$$\int_0^\tau X^{(n)}(s)\,\mathrm{d}s \xrightarrow{\mathrm{P}} \int_0^\tau f(s)\,\mathrm{d}s? \tag{2.5.11}$$

If $\tau < \infty$, then a sufficient condition for (2.5.11) is obviously $\|X^{(n)} - f\|_\infty \xrightarrow{\mathrm{P}} 0$, where $\|\cdot\|_\infty$ denotes the supremum norm on $\mathcal{T}$. However, this is often too strong or too difficult to verify. What we really need is a kind of dominated convergence theorem giving convergence of integrals under assumptions of pointwise convergence of the integrands. Two weaker sets of conditions, one requiring uniform integrability (and leading to convergence in mean) and the other requiring uniform, in probability, bounds (and leading to convergence in probability) are now given:

Proposition II.5.2 (Helland, 1983). *Suppose* (2.5.9) *holds and, furthermore,*

$$\lim_{C\uparrow\infty}\sup_n \mathrm{E}(|X^{(n)}(s)|I(|X^{(n)}(s)|>C))=0 \quad \textit{for all } s \tag{2.5.12}$$

(i.e., $(X^{(n)}(s), n=1,2,\ldots)$ is uniformly integrable) and

$$\mathrm{E}|X^{(n)}(s)|\le k(s) \quad \textit{for all } s, n, \tag{2.5.13}$$

where

$$\int_0^\tau k(s)\,\mathrm{d}s<\infty.$$

Then (2.5.10) *also holds and*

$$\mathrm{E}\left(\sup_t\left|\int_0^t X^{(n)}(s)\,\mathrm{d}s-\int_0^t f(s)\,\mathrm{d}s\right|\right)\to 0 \quad \textit{as } n\to\infty. \tag{2.5.14}$$

Proposition II.5.3 (Gill, 1983b). *Suppose* (2.5.9) *and* (2.5.10) *hold and, furthermore, for all $\delta>0$, there exists k_δ with $\int_0^\tau k_\delta<\infty$ such that*

$$\liminf_{n\to\infty}\mathrm{P}(|X^{(n)}(s)|\le k_\delta(s) \textit{ for all } s)\ge 1-\delta. \tag{2.5.15}$$

Then

$$\sup_t\left|\int_0^t X^{(n)}(s)\,\mathrm{d}s-\int_0^t f(s)\,\mathrm{d}s\right|\xrightarrow{\mathrm{P}}0. \tag{2.5.16}$$

Under both sets of conditions, we get (2.5.11) and more. The two sets are overlapping (neither implies the other) though Helland's conditions (called by him "convergence boundedly in L_1") are typically stronger than Gill's.

Proposition II.5.3 has the following very simple proof. If we replace $X^{(n)}$ by the process $\mathrm{sign}(X^{(n)})(|X^{(n)}|\wedge k_\delta)$, we obtain a sequence of bounded processes $X_\delta^{(n)}$, converging pointwise in probability to the function $f_\delta=\mathrm{sign}(f)(|f|\wedge k_\delta)$. Because k_δ is almost everywhere finite, we have almost everywhere convergence in mean of $X_\delta^{(n)}$ to f_δ. Now we can apply the Lebesgue dominated convergence theorem to the sequence of functions $\mathrm{E}|X_\delta^{(n)}-f_\delta|$ (bounded by $2k_\delta$) and conclude that $\int_0^\tau \mathrm{E}|X_\delta^{(n)}-f_\delta|$ converges to zero. From this, we may conclude that $\int_0^\tau X_\delta^{(n)}$ converges in mean, and hence in probability, to $\int_0^\tau f_\delta$. Finally, we let δ converge to zero to get the required result, for the probability that $\int_0^\tau X_\delta^{(n)}\neq\int_0^\tau X^{(n)}$ is less than or equal to δ, whereas $\int_0^\tau f_\delta\to\int_0^\tau f$ as $\delta\to 0$ by monotone convergence.

II.5.2. Further Results

Here we collect some further results often useful in proving limit theorems, though not all of them specifically asymptotic in nature. The first of them, the inequality of Lenglart (1977), will be used time and time again, especially in verifying the conditions of the martingale central limit theorem. In fact, it serves as an important lemma in the very proof of that theorem (Rebolledo, 1980a). The other results, concerning stochastic time change and Kurtz' (1983) results on convergence of a suitably rescaled counting process to a deterministic function, will only be used for some special cases: the pornoscope example (Examples I.3.16 and IV.1.10), the software reliability example (Examples I.3.19 and VI.1.12), and the Total Time on Test Plot (Section VI.3.2). The pornoscope and software reliability examples have in common that the counting processes in these models are defined directly by specifying their intensity processes rather than being built up by aggregation of simple independent (and often identically distributed) components. In the latter case, the Glivenko–Cantelli theorem and similar empirical process theory will provide the "convergence in probability" assumptions of the martingale central limit theorem. The Total Time on Test Plot is inspired directly by stochastic time change theory.

II.5.2.1. *Lenglart's Inequality*

Suppose $\tilde{X}$ is the (nondecreasing) compensator of a local submartingale X. It turns out that we can bound the probability of a large value of X anywhere in the whole time interval $\mathscr{T}$ in terms just of the probability of a large value of $\tilde{X}$ in the endpoint τ. One says that $\tilde{X}$ dominates X. The inequality of Lenglart (1977) is then, for any $\eta > 0$ and $\delta > 0$,

$$\mathrm{P}\left(\sup_{\mathscr{T}} X > \eta\right) \leq \frac{\delta}{\eta} + \mathrm{P}(\tilde{X}(\tau) > \delta). \tag{2.5.17}$$

Special cases are that X is $\int H\,\mathrm{d}N$, the integral of a non-negative predictable process H with respect to a counting process N (dominated by $\int H\,\mathrm{d}\Lambda$), and that X is the square of a local square integrable martingale M (dominated by $\langle M \rangle$). In the latter case, the inequality can be rewritten as

$$\mathrm{P}\left(\sup_{\mathscr{T}} |M| > \eta\right) \leq \frac{\delta}{\eta^2} + \mathrm{P}(\langle M \rangle(\tau) > \delta). \tag{2.5.18}$$

Thus, if $\langle M \rangle(\tau)$ is small, M is small in absolute value throughout $\mathscr{T}$.

II.5.2.2. *Stochastic Time Change*

Suppose $\mathbf{N}$ is a multivariate counting process with continuous compensator $\mathbf{\Lambda}$ satisfying $\Lambda_h(\tau) = \infty$ almost surely for each h. Let $\mathbf{\Lambda}^{-1}$ be the vector of right-continuous inverses Λ_h^{-1}, defined on the new time interval $[0, \infty)$. We define the time-transformed multivariate counting process

$$\tilde{\mathbf{N}} = \mathbf{N} \circ \mathbf{\Lambda}^{-1} \tag{2.5.19}$$

as the vector of individually transformed processes $N_h \circ \Lambda_h^{-1}$. Graphically, this time transformation corresponds to plotting N_h against its compensator Λ_h instead of against the natural time coordinate. The surprising result is that $\tilde{\mathbf{N}}$ is equal in distribution to a vector $\mathbf{\Pi}$ of *independent* unit rate Poisson processes (see Examples II.3.1 and II.3.2).

If the condition $\Lambda_h(\tau) = \infty$ almost surely does not hold, one can still get a partial result by appending an independent unit rate Poisson process to $\tilde{N}_h$ from the time $\Lambda_h(\tau)$. In this way, one can see that $\tilde{N}_h$ is equal to Π_h, stopped at the random time $\Lambda_h(\tau)$. In the univariate case, $\Lambda(\tau)$ is a stopping time relative to the time-transformed filtration $(\mathscr{F}_{\Lambda^{-1}(u)})_{u \geq 0}$, but, in general, no such simple property holds.

These results are related to results on general random time transformations of a counting process. Consider a multivariate counting process $\mathbf{N}$ on the closed time interval $[0, \tau]$ with intensity process $\boldsymbol{\lambda}$, and let R be a nondecreasing, adapted, continuous process. It is, therefore, predictable. The times $R^{-1}(r) = \inf\{t: R(t) \geq r\}$ are stopping times, and by applying Doob's optional stopping theorem to (a localized version of) the counting process local martingale $\mathbf{M} = \mathbf{N} - \int \boldsymbol{\lambda}$ and these stopping times, one finds that $\mathbf{M} \circ R^{-1}$ is a vector of local martingales with respect to the time-transformed filtration $(\mathscr{F}_{R^{-1}(r)})_{r \geq 0}$. Because $\mathbf{N} \circ R^{-1}$ is a multivariate counting process with respect to this filtration, uniqueness of compensators shows that its compensator is $\int_0^{R^{-1}(\cdot)} \boldsymbol{\lambda}$. If the paths of R are absolutely continous with density R', the intensity process of $\mathbf{N} \circ R^{-1}$ is, therefore, $(\lambda/R') \circ R^{-1}$. The transformed counting process and its intensity are "stopped" at the stopping time $R(\tau)$.

Random time transformation will be used in Section VI.3.2 on the total time on test plot. The results on transformation to a Poisson process lead to the following topic: Kurtz' (1983) limit theorems for self-exciting counting processes.

II.5.2.3. *Kurtz' Theorems*

Consider a multivariate counting process $\mathbf{N}$ with its self-exciting filtration $(\mathscr{N}_t)$. The compensator of $\mathbf{N}$, being predictable with respect to the filtration generated by $\mathbf{N}$, must be a "nonanticipating" functional of $\mathbf{N}$; that is, for each t, $\mathbf{\Lambda}(t)$ is a certain function (depending on t) of the path of $\mathbf{N}$ up to but not including time t. Write $\mathbf{N}^{t-}$ for the process $\mathbf{N}$ stopped "just before time t," i.e., $\mathbf{N}^{t-}(s) = \mathbf{N}(s)$ for $s < t$ and $\mathbf{N}^{t-}(s) = \mathbf{N}(t-)$ for $s \geq t$. With this notation, the nonanticipating nature of $\mathbf{\Lambda}$ means that we can consider $\mathbf{\Lambda}(t)$, not as a function of t and of $\omega \in \Omega$, but as a function of t and of $\mathbf{N}^{t-}(\cdot, \omega)$.

Random time transformation to a Poisson process now implies that in the self-exciting case, one can actually construct $\mathbf{N}$ in distribution by starting with a vector of independent unit rate Poisson processes $\mathbf{\Pi}$ and then defining $\mathbf{N}$ pathwise as the solution of the implicit equation

$$\mathbf{N} \circ \mathbf{\Lambda}^{-1} = \mathbf{\Pi}, \tag{2.5.20}$$

cf. (2.5.19). Intuitively speaking, given a realisation of $\mathbf{\Pi}$, if we have already constructed the path of $\mathbf{N}$ up to time $t-$, then, in the case when $\mathbf{N}$ has an intensity process $\boldsymbol{\lambda}$, we can calculate $\Lambda_h(t)$ and $\lambda_h(t)$ for each h from the path of $\mathbf{N}$ so far, and then define $N_h(t + \mathrm{d}t) = \Pi_h(\Lambda_h(t) + \lambda_h(t)\,\mathrm{d}t)$. Kurtz (1983) gave conditions on $\mathbf{\Lambda}(\cdot\,; \mathbf{N})$ which guarantee that the solution of (2.5.20) is unique.

Furthermore, limit theorems can be obtained using the fact that for a Poisson process, the law of large numbers and the central limit theorem hold. Let $\boldsymbol{\iota}$ be the vector of identity mappings $\iota_h(t) = t$ and let $\mathbf{\Pi}^{(n)}$ be the process $\mathbf{\Pi}^{(n)}(t) = \mathbf{\Pi}(nt)$. Then we have that, almost surely, $\mathbf{\Pi}^{(n)}/n$ converges uniformly on bounded intervals to $\boldsymbol{\iota}$ (the strong law of large numbers) and that $\sqrt{n}(\mathbf{\Pi}^{(n)}/n - \boldsymbol{\iota})$ converges in distribution to a vector of independent standard Wiener processes (see Example II.3.3).

These results are applied by defining a sequence of counting processes $\mathbf{N}^{(n)}$ by the modification of (2.5.20):

$$\mathbf{N}^{(n)} \circ \mathbf{\Lambda}^{-1}(\cdot\,; \mathbf{N}^{(n)}) = \mathbf{\Pi}^{(n)}. \tag{2.5.21}$$

It turns out that under suitable smoothness conditions on the functional dependence of $\mathbf{\Lambda}$ on $\mathbf{N}$, both law of large numbers and central limit theorem carry over from $\mathbf{\Pi}^{(n)}$ to $\mathbf{N}^{(n)}$. In the following version of the law of large numbers (all that we will have use for later), we describe a sequence of counting processes $\mathbf{N}^{(n)}$ corresponding to (2.5.21), as well as the limiting function to which $\mathbf{N}^{(n)}/n$ converges.

Theorem II.5.4 [Kurtz' (1983) law of large numbers]. *Let $\beta_h(t, \mathbf{x}) = \beta_h(t, \mathbf{x}^{t-})$ be nonanticipating non-negative functions of an element $\mathbf{x} \in (D(\mathcal{T}))^k$ and $t \in \mathcal{T}$ such that $\sup_{s \le t} \beta_h(s, \mathbf{x}) \le C_1 + C_2 \sup_{s<t} |\mathbf{x}(s)|$ and $\sup_{s \le t} |\beta_h(s, \mathbf{x}) - \beta_h(s, \mathbf{y})| \le C \sup_{s<t} |\mathbf{x}(s) - \mathbf{y}(s)|$, for all $\mathbf{x}$, $\mathbf{y} \in (D(\mathcal{T}))^k$, and for certain constants C_1, C_2, and C. Let $a_n \to \infty$ be a sequence of positive constants. Let $\mathbf{N}^{(n)}$ be the multivariate counting process with intensity process $\boldsymbol{\lambda}^{(n)} = a_n \boldsymbol{\beta}(\cdot, a_n^{-1}\mathbf{N}^{(n)})$ and let $\mathbf{X}$ be the unique solution of the equation $\mathbf{X}(t) = \int_0^t \boldsymbol{\beta}(s, \mathbf{X})\,\mathrm{d}s$ for all t. Then $\sup_{s \le t} |a_n^{-1}\mathbf{N}^{(n)} - \mathbf{X}| \xrightarrow{\mathrm{P}} 0$ as $n \to \infty$ for all $t \in \mathcal{T}$.*

II.6. Product-Integration and Markov Processes

In Section II.1, we saw the informal use of the *product-integral* both in formulas for likelihood functions and in the representation of a survival functional in terms of its cumulative hazard function. The notation used here, $\prod$, is supposed to suggest a continuous version of the ordinary product $\prod$, just as the integral $\int$ generalizes the sum $\sum$ (see Gill and Johansen, 1990).

In the next section, we will take up the first topic and use product-integration to represent the likelihood function of a process observed on a time interval $[0, \tau]$ as an infinite product of conditional likelihoods for the development of the process in each infinitesimal time interval $[t, t + \mathrm{d}t)$, $t \in [0, \tau]$, given its past history. Exactly the same construction appears in the relationship between a *survival function* and its *hazard rate* or, more generally, its *hazard measure*. In multistate event history analysis, this relationship becomes that between the matrix of infinitesimal intensities or *transition rates* and the *transition probability matrix* of a continuous-time *Markov process*. These relationships will be reviewed below. They are central to our study of the Kaplan–Meier estimator and its generalization to Markov processes in Sections IV.3 and IV.4.

Though statistical models are often specified for continuous-time processes, natural statistical estimators are often discrete in nature. This makes it important to be able to handle discrete and continuous versions of the above relationship within a uniform framework. In fact, the basic and simple analytic tool of product-integration makes the discrete-time multiplication of a series of Markov one-step transition matrices and the continuous-time solution of the Kolmogorov differential equations two special cases of the product-integration of the matrix *intensity measure* of a Markov process.

The following results are taken from Gill and Johansen (1990), where complete proofs are given. Further background comments are given in the bibliographic comments to this chapter. In Section X.3.1, we will look at product-integration from a different point of view, leading to extensions to multivariate time.

Definition II.6.1. Let $\mathbf{X}(t)$, $t \in \mathscr{T}$, be a $p \times p$ matrix of cadlag functions of locally bounded variation. We define

$$\mathbf{Y} = \prod (\mathbf{I} + \mathrm{d}\mathbf{X}),$$

the product-integral of $\mathbf{X}$ over intervals of the form $[0, t]$, $t \in \mathscr{T}$, as the following $p \times p$ matrix function:

$$\mathbf{Y}(t) = \prod_{s \in [0,t]} (\mathbf{I} + \mathbf{X}(\mathrm{d}s)) = \lim_{\max |t_i - t_{i-1}| \to 0} \prod (\mathbf{I} + \mathbf{X}(t_i) - \mathbf{X}(t_{i-1})), \quad (2.6.1)$$

where $0 = t_0 < t_1 < \cdots < t_n = t$ is a partition of $[0, t]$ and the matrix product is taken in its natural order from left to right. In the leftmost term of the product, $\mathbf{X}(0)$ must be replaced by $\mathbf{X}(0-) = \mathbf{0}$ because the left endpoint 0 is included in the interval $[0, t]$.

We similarly define the product-integral over an arbitrary subinterval of $\mathscr{T}$, taking care of the endpoints in the natural way. The first result on product-integration is that the limit, indeed, always exists.

When the function $\mathbf{X}$ is a step-function, i.e., its components are the distribution functions of discrete measures, the product-integral becomes just a finite product over the jump times of $\mathbf{X}$ of the identity matrix plus the jumps of $\mathbf{X}$; thus,

$$\mathbf{Y} = \prod (\mathbf{I} + \Delta \mathbf{X}).$$

In the scalar case, $p = 1$, the order of multiplication does not matter, and one can separate the jumps of X from its continuous part and get

$$\prod (1 + \mathrm{d}X) = \exp(X^c) \prod (1 + \Delta X), \tag{2.6.2}$$

where $X^c = X - \sum \Delta X$ and $\Delta X = X - X_-$. So if X is not only scalar but also continuous (and of bounded variation), the product-integral is just the ordinary exponential $\prod (1 + \mathrm{d}X) = \exp(X)$.

A most important property of product-integration is the *multiplicativity* of product-integrals over disjoint intervals, which follows easily from the definition: for $0 \le s \le t \le u$, we have

$$\prod_{(s,u]} (\mathbf{I} + \mathrm{d}\mathbf{X}) = \prod_{(s,t]} (\mathbf{I} + \mathrm{d}\mathbf{X}) \prod_{(t,u]} (\mathbf{I} + \mathrm{d}\mathbf{X}).$$

Not only does the product-integral exist, but it is also the unique solution of a certain integral equation. This is why it was introduced by Volterra (1887).

Theorem II.6.1. $\prod (\mathbf{I} + \mathrm{d}\mathbf{X})$ *exists and is (componentwise) a cadlag function of locally bounded variation. It is the unique solution to the integral equation*

$$\mathbf{Y}(t) = \mathbf{I} + \int_{s \in [0,t]} \mathbf{Y}(s-)\mathbf{X}(\mathrm{d}s). \tag{2.6.3}$$

This result, like most of the results on product-integration, is actually rather intuitive. The right-hand side of (2.6.1) is the solution of the naturally associated discrete difference scheme for numerically solving (2.6.3) [more formally, it is the result of applying the first-order Euler scheme to the numerical solution of (2.6.3)]. We will see (2.6.3) shortly also as an integral version of the *Kolmogorov forward equation* for a Markov process.

Next we give a few key results on product-integration. The first one, Duhamel's equation, can be thought of as the continuous version of the elementary equality

$$\prod_{i=1}^{n} (\mathbf{I} + \mathbf{A}_i) - \prod_{i=1}^{n} (\mathbf{I} + \mathbf{B}_i) = \sum_{i=1}^{n} \left(\prod_{j=1}^{i-1} (\mathbf{I} + \mathbf{A}_j)(\mathbf{A}_i - \mathbf{B}_i) \prod_{j=i+1}^{n} (\mathbf{I} + \mathbf{B}_j) \right),$$

valid for matrices $\mathbf{A}_i$, $\mathbf{B}_i$, $i = 1, \ldots, n$.

Theorem II.6.2. *Let* $\mathbf{Y} = \prod (\mathbf{I} + \mathrm{d}\mathbf{X})$, $\mathbf{Y}' = \prod (\mathbf{I} + \mathrm{d}\mathbf{X}')$. *Then*

$$\mathbf{Y}(t) - \mathbf{Y}'(t) = \int_{s \in [0,t]} \prod_{[0,s)} (\mathbf{I} + \mathrm{d}\mathbf{X})(\mathbf{X}(\mathrm{d}s) - \mathbf{X}'(\mathrm{d}s)) \prod_{(s,t]} (\mathbf{I} + \mathrm{d}\mathbf{X}'). \tag{2.6.4}$$

When $\mathbf{Y}'(t)$ is nonsingular, one may right-multiply throughout in (2.6.4) by $\mathbf{Y}'(t)^{-1} = (\prod_{(s,t]}(\mathbf{I} + \mathrm{d}\mathbf{X}'))^{-1}\mathbf{Y}'(s)^{-1}$ by multiplicativity. This leads to the version of (2.6.4) we will use most often:

$$\begin{aligned}\mathbf{Y}(t)\mathbf{Y}'(t)^{-1} - \mathbf{I} &= \int_{s\in[0,t]} \prod_{[0,s)} (\mathbf{I} + \mathrm{d}\mathbf{X})(\mathbf{X}(\mathrm{d}s) - \mathbf{X}'(\mathrm{d}s))\left(\prod_{[0,s]} (\mathbf{I} + \mathrm{d}\mathbf{X}')\right)^{-1} \\ &= \int_0^t \mathbf{Y}(s-)(\mathbf{X}(\mathrm{d}s) - \mathbf{X}'(\mathrm{d}s))\mathbf{Y}'(s)^{-1}. \qquad (2.6.5)\end{aligned}$$

In Section II.8, we give functional *continuity* and even *differentiability* results on product-integration derived from the Duhamel equation.

The next result generalizes Theorem II.6.1, showing that the product-integral arises in the solution of a whole class of integral equations.

Theorem II.6.3. *Let* $\mathbf{Z}$, $\mathbf{W}$ *be* $k \times p$ *matrix cadlag functions. For given* $\mathbf{W}$, *the unique solution* $\mathbf{Z}$ *of the Volterra equation*

$$\mathbf{Z}(t) = \mathbf{W}(t) + \int_0^t \mathbf{Z}(s-)\mathbf{X}(\mathrm{d}s) \qquad (2.6.6)$$

is

$$\begin{aligned}\mathbf{Z}(t) &= \mathbf{W}(t) + \int_0^t \mathbf{W}(s-)\mathbf{X}(\mathrm{d}s) \prod_{(s,t]} (\mathbf{I} + \mathrm{d}\mathbf{X}) \\ &= \mathbf{W}(0) \prod_{[0,t]} (\mathbf{I} + \mathrm{d}\mathbf{X}) + \int_0^t \mathbf{W}(\mathrm{d}s) \prod_{(s,t]} (\mathbf{I} + \mathrm{d}\mathbf{X}). \qquad (2.6.7)\end{aligned}$$

Finally, we give two results which are sometimes useful, one giving an explicit series representation of the product-integral, and the other on its determinant in the continuous case:

Theorem II.6.4. [Péano Series, Péano (1888)].

$$\prod_{[0,t]} (\mathbf{I} + \mathrm{d}\mathbf{X}) = \mathbf{I} + \sum_{n=1}^{\infty} \int \cdots \int_{0 \le s_1 < \cdots < s_n \le t} \mathbf{X}(\mathrm{d}s_1) \cdots \mathbf{X}(\mathrm{d}s_n). \qquad (2.6.8)$$

Theorem II.6.5. *Suppose* $\mathrm{d}\mathbf{X}(t) = \mathbf{U}(t)\,\mathrm{d}t$ *and* $\mathbf{X}(0) = \mathbf{0}$. *Then*

$$\det \prod_{[0,t]} (\mathbf{I} + \mathrm{d}\mathbf{X}) = \exp\left(\int_0^t \operatorname{trace} \mathbf{U}(s)\,\mathrm{d}s\right). \qquad (2.6.9)$$

Next we show how product-integration arises in the relation between survival functions and hazard measures.

Theorem II.6.6. *Let* S *be a survival function of a positive random variable (survival time)* T, *i.e.,* $S(t) = \mathrm{P}(T > t)$ *for all* $t \ge 0$ *and* $S(0) = 0$. *Define the cumulative or integrated hazard function*

$$A(t) = -\int_0^t \frac{S(\mathrm{d}s)}{S(s-)}. \tag{2.6.10}$$

Then

$$S(t) = \prod_{[0,t]} (1 - \mathrm{d}A), \tag{2.6.11}$$

for all t such that $A(t) < \infty$.

PROOF. Note that for all t such that $S(t-) > 0$ and, hence, $A(t) < \infty$,

$$S(t) = 1 - \int_0^t S(s-)A(\mathrm{d}s).$$

Now apply Theorem II.6.1 with $X = -A$ and $\mathcal{T} = \{t: S(t-) > 0\}$. If $\mathcal{T} = [0, \tau]$ for some $\tau < \infty$, then $S(\tau-) > 0$ but $S(\tau) = 0$. Consequently, $A(t) = A(\tau) < \infty$ for all $t \geq \tau$ and the relation (2.6.11) holds for all $t \geq \tau$ too. If, however, $\mathcal{T} = [0, \tau)$ for some $\tau \leq \infty$, one can show by some straightforward analysis that $A(t) = A(\tau) = \infty$ for all $t \geq \tau$ and (2.6.11) cannot be extended (except by way of definition) to $t \geq \tau$. □

When S is absolutely continuous so that $F = 1 - S$ has density f, we define the hazard rate $\alpha = f/(1 - F)$. The integrated hazard is then $A(t) = \int_0^t \alpha(s)\,\mathrm{d}s$ and relation (2.6.11) by (2.6.2) becomes

$$S(t) = \prod_0^t (1 - \alpha(s)\,\mathrm{d}s) = \exp\left(-\int_0^t \alpha(s)\,\mathrm{d}s\right). \tag{2.6.12}$$

If, however, S is discrete, one can define the discrete hazard function $\alpha(t) = \mathrm{P}(T = t \mid T \geq t)$. Then $A(t) = \sum_{s \leq t} \alpha(s)$ and (2.6.11) becomes

$$S(t) = \prod_{s \leq t} (1 - \alpha(s)).$$

The cumulative hazard function arose in the compensator of the counting process registering, by a jump at time T, the end of the survival time, cf. Example II.4.1 and the remarks following this example. Let T be a survival time and define $\mathcal{F}_t = \sigma\{T \wedge t, I(T \leq t)\}$; let $N(t) = I(T \leq t)$ and $Y(t) = I(T \geq t)$. Then N has compensator Λ given by

$$\Lambda(t) = \int_0^t Y(s)A(\mathrm{d}s).$$

Next, we extend these results to an almost arbitrary finite state-space, continuous-time Markov process. We explain the implied restriction afterward. First we need a definition. We say that a $p \times p$ matrix function $\mathbf{A}$ corresponds to a locally finite *intensity measure of a Markov process* on a time interval $\mathcal{T}$ if the A_{hj}, $h \neq j$, are nondecreasing cadlag functions, zero at time zero, $A_{hh} = -\sum_{j \neq h} A_{hj}$, and $\Delta A_{hh}(t) \geq -1$ for all t. The function A_{hj} is called the integrated intensity function for transitions from state h to state j, whereas A_{hh} is the negative integrated intensity function for transitions out of state h.

Theorem II.6.7. *Let the matrix function* **A** *correspond to an intensity measure. Define*

$$\mathbf{P}(s,t) = \prod_{(s,t]} (\mathbf{I} + \mathrm{d}\mathbf{A}), \quad s \le t;\ s, t \in \mathscr{T}. \tag{2.6.13}$$

Then **P** *is the transition matrix of a Markov process with state space* $\{1, \ldots, p\}$ *and intensity measure* **A**. *The process can be constructed as follows (starting from any time instant in any state): Given the process is in state h at time t_0, it remains in this state for a length of time with integrated hazard function*

$$-(A_{hh}(t) - A_{hh}(t_0)), \quad t_0 \le t \le \inf\{u \ge t_0 \colon \Delta A_{hh}(u) = -1\}.$$

Given that it jumps out of state h at time t, it jumps into state $j \ne h$ with probability $(\mathrm{d}A_{hj}/(-\mathrm{d}A_{hh}))(t)$.

The theorem is not proved here; one has to show that the construction described in the theorem really does define a process (making only finitely many jumps in finite time intervals), that this process is Markov, and that its transition matrices are given by (2.6.13).

An important special case is when the A_{hj} are absolutely continuous, $A_{hj} = \int \alpha_{hj}$ for certain *intensity functions* or *transition rates* or *transition intensities* α_{hj}. The sojourn times in a given state h are then continuously distributed with hazard rate function $-\alpha_{hh}$ (starting from the time of entry into state h). The conditional jump probabilities (given a jump out of state h at time t) equal $\alpha_{hj}(t)/(-\alpha_{hh}(t))$. We let $\boldsymbol{\alpha}$ be the matrix of these transition intensities. The integral equation (2.6.3) for the product-integral of **A** can be rewritten as an integral form of the Kolmogorov forward differential equations for the transition matrix **P** with (time varying) intensities α_{hj}: $\mathbf{P}(s,t)$ is the unique solution of the equations

$$\mathbf{P}(s,s) = \mathbf{I},$$

$$\frac{\partial}{\partial t}\mathbf{P}(s,t) = \mathbf{P}(s,t)\boldsymbol{\alpha}(t)$$

because (starting at time s instead of 0) Eq. (2.6.3) becomes

$$\mathbf{P}(s,t) = \mathbf{I} + \int_{u \in (s,t]} \mathbf{P}(s,u)\boldsymbol{\alpha}(u)\,\mathrm{d}u.$$

Another special case is when the A_{hj} are step-functions. Now the Markov process is a discrete-time Markov chain and the matrix $\mathbf{I} + \Delta\mathbf{A}(t)$ is the transition matrix for a jump at time t. The product-integral (2.6.13) describes the transition matrix for the time interval $(s, t]$ as the product of the transition matrices for each possible jump time between s and t (including t). The sojourn times have discrete distributions with discrete hazard function $-\Delta A_{hh}$ for leaving state h. Given a jump out of h occurs at time t, the new state is state j with probability $\Delta A_{hj}(t)/(-\Delta A_{hh}(t))$. For a discrete-time Markov process, one can define $\alpha_{hj}(t) = \mathrm{P}(X(t) = j | X(t-) = h)$, $h \neq j$, and $\alpha_{hh}(t) = -\mathrm{P}(X(t) \neq h | X(t-) = h)$. So, $1 + \alpha_{hh}(t) = \mathrm{P}(X(t) = h | X(t-) = h)$. Equation (2.6.13) becomes the product of the transition matrices $\mathbf{I} + \boldsymbol{\alpha}(u) = \mathbf{I} + \Delta\mathbf{A}(u)$ for each time instant u in $(s, t]$ at which a transition is possible.

In some of the examples of Section IV.4.1, we will need a form of the Kolmogorov forward equation for the situation with general integrated intensity functions. Using the definition (2.6.13), the integral equation (2.6.3) can be rewritten, for fixed s and for $t \geq s$, as

$$\begin{aligned} \mathbf{P}(s, s) &= \mathbf{I}, \\ \mathbf{P}(s, \mathrm{d}t) &= \mathbf{P}(s, t-)\mathbf{A}(\mathrm{d}t). \end{aligned} \tag{2.6.14}$$

Finite state-space Markov processes whose transition matrices are *not* product-integrals of finite intensity measures are those such that for some state h and times u and v, given the process is in state h at time u, its future sojourn time in this state has infinite cumulative hazard up to time v. If one artificially kept such a state occupied by letting a new individual start in that state every time the state was vacated, an infinite number of jumps out of the state would occur in finite time. The original process is certain to leave the state in finite time, but does not certainly leave at any specific time instant. It would be all right to include such states in the theory provided care is taken that this possibility does not lead to a positive probability of an infinite number of jumps in a finite time interval. In particular, if the destination state j for an infinite intensity A_{hj} is *absorbing*, thus all intensities for leaving j are zero, this is not a problem because on leaving h for j there is no possibility of reentering h again.

The intensity measure reappears in the compensator of the counting processes registering each type of jump of the process (Jacobsen, 1982, p. 120). This is worth summarizing in a last theorem:

Theorem II.6.8. *Let* $\mathbf{A}$ *correspond to the intensity measure of a Markov process* X. *Let* $\mathscr{F}_t = \sigma\{X(s): s \leq t\}$; *define*

$$\begin{aligned} Y_h(t) &= I(X(t-) = h), \\ N_{hj}(t) &= \#\{s \leq t: X(s-) = h, X(s) = j\}, \quad h \neq j. \end{aligned}$$

Then $\mathbf{N} = (N_{hj}, h \neq j)$ *is a multivariate counting process and its compensator with respect to* $(\mathscr{F}_t) = (\sigma(X(0)) \vee \mathscr{N}_t)$ *has components*

$$\Lambda_{hj}(t) = \int_0^t Y_h(s) A_{hj}(\mathrm{d}s). \tag{2.6.15}$$

Equivalently, the processes M_{hj} *defined by*

$$M_{hj} = N_{hj} - \int Y_h \, \mathrm{d}A_{hj}$$

are martingales.

In the absolutely continuous case with transition intensities $\boldsymbol{\alpha}$, and $\mathbf{A}(t) = \int_0^t \boldsymbol{\alpha}(s)\,\mathrm{d}s$, (2.6.15) gives us that $\mathbf{N}$ has intensity $\boldsymbol{\lambda} = (\lambda_{hj}: h \neq j)$ with

$$\lambda_{hj}(t) = Y_h(t)\alpha_{hj}(t).$$

The result on the intensity of the one-jump counting process following Theorem II.6.6 can be seen as a special case of Theorem II.6.8 by letting T be the time of the only transition of a two-state Markov process from state 0 to state 1, the process starting in state 0 at time 0.

II.7. Likelihoods and Partial Likelihoods for Counting Processes

II.7.1. Jacod's Formulas

In this section, we will consider a multivariate counting process $\mathbf{N} = (N_1, \ldots, N_k)$ under various probability measures and, hence, with a varying (cumulative) intensity process. The counting process will be defined on a fixed probability space and filtration satisfying the usual conditions except possibly completeness (see Section II.2). We will work throughout under the special assumptions

$$\begin{gathered} \mathscr{F}_t = \mathscr{F}_0 \vee \sigma\{\mathbf{N}(s): s \leq t\} = \mathscr{F}_0 \vee \mathscr{N}_t, \\ \mathscr{T} = \overline{\mathscr{T}} = [0, \tau], \qquad \mathscr{F} = \mathscr{F}_\tau. \end{gathered} \tag{2.7.1}$$

This means, in particular, that $\mathbf{N}(\tau)$ and $\boldsymbol{\Lambda}(\tau)$ are both finite.

We can think of a probability measure P on $\mathscr{F}$ as being built up in the following way. First, specify P on $\mathscr{F}_0$. Then, specify the conditional distribution, given $\mathscr{F}_0$, of the time $T_1 \in (0, \tau]$ of the first jump of $\mathbf{N}$. Next, specify the conditional distribution of the *type* of the first jump $J_1 \in \{0, 1, \ldots, k\}$ given $\mathscr{F}_0$ and T_1, where $J_1 = 0$ is only allowed if $T_1 = \tau$ (and means that there is no first jump of $\mathbf{N}$). Go on with a conditional distribution of $T_2 \in (T_1, \tau]$ given $\mathscr{F}_0$, T_1, J_1; and so on. The construction could only fail to define a proper counting process model through an explosion: Infinitely many jumps occur before time τ with positive probability. We would like to be sure that this cannot happen. However, the probability of an explosion is a very complicated functional of all the conditional distributions we have just described. We will typically have to make do with simple sufficient conditions (i.e., conditions which are much stronger than necessary). Theorems II.7.4 and II.7.5 illustrate the problem: The first (the *Girsanov theorem*) makes the weakest possible assumption, but it is quite impossible to verify without making strong boundedness conditions. The second of the theorems makes a much stronger assumption but, at least, it is easy to check.

Before coming to construction theorems, we consider alternative ways to *describe* the distribution of a counting process. The conditional distributions of the T_n, given the past, can also be specified through their cumulative hazard functions. Multiplying the hazard rate at time t by the probability of a jump of type h at time t, given one occurs then, produces type-specific hazards; conversely, from k type-specific hazards for (T_n, J_n) given the past, we can reconstruct the total hazard and the conditional, time-dependent probability of an event of each type, given one occurs.

Now the type-specific hazards are, intuitively speaking, just the intensity processes of the components of $\mathbf{N}$. So this informal argument (see Section II.1 and also the discussion following Example II.4.1) suggests that under (2.7.1), the (cumulative) intensity process of $\mathbf{N}$ is determined by the conditional probability distributions just described and, conversely, from the intensity processes, we can recover the conditional distributions and, hence, together with knowledge of P on $\mathscr{F}_0$, reconstruct P on $\mathscr{F}$.

Instead of going from jump to jump (taking big steps through $\mathscr{T}$), one could also go from infinitesimal time interval to infinitesimal time interval, with the probability of a jump of type h in the interval $[t, t + \mathrm{d}t)$, given the past ($\mathscr{F}_0$ and the path of $\mathbf{N}$ up to time t), being just $\mathrm{d}\Lambda_h(t)$. This gives an alternative way of writing down probability densities or likelihood ratios which may be easier to interpret and leads to the notion of *partial likelihood* in a natural way.

We now make these ideas rigorous by stating some theorems due to Jacod and others. Recall (cf. Section II.4.4) that we write ${}_TX$ for a process X *started* at time T, i.e., ${}_TX(t) = X(t) - X^T(t)$.

Theorem II.7.1 (Jacod's Formula for the Intensity Process). *Under* (2.7.1), *let* P_n *be a regular version of the joint conditional distribution of* (T_n, J_n) *given* $\mathscr{F}_{T_{n-1}} = \mathscr{F}_0 \vee \sigma\{T_1, J_1, \ldots, T_{n-1}, J_{n-1}\}$. *Then, on the time interval* $(T_{n-1}, T_n]$, *we have*

$$ {}_{T_{n-1}}\Lambda_h(t) = \int_{T_{n-1}}^{t} \frac{P_n(\mathrm{d}s, \{h\})}{P_n([s, \tau], \bar{E})}, \quad h = 1, \ldots, k, $$

where $E = \{1, \ldots, k\}$ *and* $\bar{E} = E \cup \{0\}$. *Conversely, on* (T_{n-1}, T_n), *we have*

$$ P_n((t, \tau], \bar{E}) = \prod_{T_{n-1}}^{t} (1 - \mathrm{d}\Lambda_{\cdot}(s)) $$

($\cdot$ *denoting summation over* $1, \ldots, k$) *and, with* $P_n(h|t)$, *the* P_n*-conditional distribution of* J_n *given* $T_n = t$,

$$ P_n(h|t) = \frac{\mathrm{d}\Lambda_h(t)}{\mathrm{d}\Lambda_{\cdot}(t)}, \quad h = 1, \ldots, k. $$

This result is a generalization of Example II.4.1, the "one-jump process." The theorem also holds, with notational modification only, for a marked point process (introduced at the end of Section II.4.1): Replace the discrete "mark"

h by an element $\mathrm{d}x$ of a general mark space $(E, \mathscr{E})$ and interpret ratios of differentials as transition measures. We return to this in Section II.7.3.

To compute likelihood ratios, one could use the "converse" part of the preceding theorem to extract the joint conditional distribution of $T_1, J_1, T_2, J_2, \ldots$ given $\mathscr{F}_0$ from $\mathbf{\Lambda}$. It is more convenient, however, to recast this in terms of the infinitesimal experiments described earlier. We use product-integral notation (Section II.6) together with some even more nonstandard notational conventions which are explained after the statement of the theorem.

Theorem II.7.2 (Jacod's Formula for the Likelihood Ratio). *Suppose* (2.7.1) *holds and* P *and* $\tilde{\mathrm{P}}$ *are two probability measures on the filtered probability space under which* $\mathbf{N}$ *has compensators* $\mathbf{\Lambda}$ *and* $\tilde{\mathbf{\Lambda}}$, *respectively. Suppose* $\tilde{\mathrm{P}}$ *is absolutely continuous with respect to* P, *written* $\tilde{\mathrm{P}} \ll \mathrm{P}$. *Then,*

$$\tilde{\Lambda}_h \ll \Lambda_h \quad \textit{for all } h, \text{ P-}a.s.,$$

$$\Delta\Lambda_.(t) = 1 \quad \textit{for any } t \textit{ implies } \Delta\tilde{\Lambda}_.(t) = 1, \text{ P-}a.s.$$

and

$$\begin{aligned}\frac{\mathrm{d}\tilde{\mathrm{P}}}{\mathrm{dP}} &= \frac{\mathrm{d}\tilde{\mathrm{P}}}{\mathrm{dP}}\bigg|_{\mathscr{F}_0} \frac{\prod_{t\in[0,\tau]}\left(\prod_h \mathrm{d}\tilde{\Lambda}_h(t)^{\Delta N_h(t)}(1-\mathrm{d}\tilde{\Lambda}_.(t))^{1-\Delta N_.(t)}\right)}{\prod_{t\in[0,\tau]}\left(\prod_h \mathrm{d}\Lambda_h(t)^{\Delta N_h(t)}(1-\mathrm{d}\Lambda_.(t))^{1-\Delta N_.(t)}\right)} \\ &= \frac{\mathrm{d}\tilde{\mathrm{P}}}{\mathrm{dP}}\bigg|_{\mathscr{F}_0} \prod_t \prod_h \left(\frac{\mathrm{d}\tilde{\Lambda}_h}{\mathrm{d}\Lambda_h}(t)\right)^{\Delta N_h(t)} \frac{\prod_{t\in[0,\tau]:\Delta N_.(t)\neq 1}(1-\mathrm{d}\tilde{\Lambda}_.(t))}{\prod_{t\in[0,\tau]:\Delta N_.(t)\neq 1}(1-\mathrm{d}\Lambda_.(t))}. \end{aligned} \tag{2.7.2}$$

The first line of (2.7.2) is more intuitively easy to interpret: It describes the likelihood ratio as a ratio of products of conditional (and infinitesimal) multinomial experiments. The second line gives the intended formal mathematical meaning. The following remarks explain how to make the step between the two versions and can be considered as an algorithm for making sense of expressions like (2.7.2) in the future.

Because everything is real-valued, the order of the terms in the first version of (2.7.2) can be changed at will. The "jump parts" [i.e., those with superscript $\Delta N_h(t)$] only occur at a finite number of time points and can be taken as a discrete product outside the product-integrals. The product-integrals which are left are effectively taken over the subset of $t \in [0, \tau]$ for which $\Delta N_.(t) = 0$, i.e., such that $t \neq T_n$ for any n. Because $\Delta\Lambda_.(t) = 1 \Rightarrow \Delta N_.(t) = 1$, P-a.s., this means we are not embarrassed by a term $(1 - \Delta\Lambda_.(t))$ with $\Delta\Lambda_.(t) = 1$ in the denominator. By our assumption (2.7.1), $\Lambda_.(\tau) < \infty$, P-a.s., so the product-integral in the denominator is not zero. The product-integrals can finally, by (2.6.2), be evaluated as the ordinary products of the negative exponentials of the continuous parts of $\tilde{\Lambda}_.$ and $\Lambda_.$, together with the contributions $1 - \Delta\tilde{\Lambda}_.$ and $1 - \Delta\Lambda_.$ over the times in $[0, \tau]$ where $\mathbf{N}$ does not jump.

The jump parts in the first version of (2.7.2) should be intepreted by taking ratios and forming Radon–Nikodym derivatives $((\mathrm{d}\tilde{\Lambda}_h/\mathrm{d}\Lambda_h)(t))^{\Delta N_h(t)}$; according to the first statement of the theorem, this derivative is finite and we will have a finite number of such terms with $t = T_n$ and $h = J_n$ only.

Theorem II.7.2 also holds for marked point processes and we work with such a version in Section II.7.3.

Usually, we will have continuous or even absolutely continuous compensators. In these cases the product-integrals by (2.6.2) simplify to exponentials; and in the case with intensities, the Radon–Nikodym derivative of one *integrated intensity process* with respect to another becomes simply the ratio of the intensities. These cases are covered in the next corollary.

Corollary II.7.3 (Continuous and Absolutely Continuous Case). *If* $\mathbf{\Lambda}$ *and* $\tilde{\mathbf{\Lambda}}$ *are* P-*a.s. continuous, then*

$$\frac{d\tilde{P}}{dP} = \frac{d\tilde{P}}{dP}\bigg|_{\mathscr{F}_0} \frac{\prod_{h,t} d\tilde{\Lambda}_h(t)^{\Delta N_h(t)} \exp(-\tilde{\Lambda}_{\cdot}(\tau))}{\prod_{h,t} d\Lambda_h(t)^{\Delta N_h(t)} \exp(-\Lambda_{\cdot}(\tau))} \tag{2.7.3}$$

and if they are actually absolutely continuous, then

$$\frac{d\tilde{P}}{dP} = \frac{d\tilde{P}}{dP}\bigg|_{\mathscr{F}_0} \frac{\prod_{h,t} \tilde{\lambda}_h(t)^{\Delta N_h(t)} \exp(-\tilde{\Lambda}_{\cdot}(\tau))}{\prod_{h,t} \lambda_h(t)^{\Delta N_h(t)} \exp(-\Lambda_{\cdot}(\tau))}. \tag{2.7.4}$$

The products in (2.7.4) are just $\prod_n \tilde{\lambda}_{J_n}(T_n)$ and $\prod_n \lambda_{J_n}(T_n)$. [A result of Brémaud (1981) states that a predictable intensity λ_h is a.s. unique on the jump times of N_h. So $d\tilde{P}/dP$ *is* well defined by this expression.]

In statistical applications, we study *likelihood ratios* formed by taking Radon–Nikodym derivatives of members of the family of probability measures under consideration with respect to one fixed reference distribution. Likelihoods are only needed up to a proportionality factor. This means that one need only specify the numerators of the likelihood ratios in (2.7.2)–(2.7.4); we have indicated how the resulting formal expressions can be manipulated by formal algebra to give well-defined likelihood functions. So (dropping the tildes), we can rephrase the results of Theorem II.7.2 and Corollary II.7.3 as

$$dP = dP|_{\mathscr{F}_0} \mathop{\pi}_{t \in [0,\tau]} \left(\prod_h d\Lambda_h(t)^{\Delta N_h(t)} (1 - d\Lambda_{\cdot}(t))^{1-\Delta N_{\cdot}(t)} \right), \tag{2.7.2'}$$

$$dP = dP|_{\mathscr{F}_0} \prod_{h,t} d\Lambda_h(t)^{\Delta N_h(t)} \mathop{\pi}_0^{\tau} (1 - d\Lambda_{\cdot})$$

$$= dP|_{\mathscr{F}_0} \prod_{h,t} d\Lambda_h(t)^{\Delta N_h(t)} \exp(-\Lambda_{\cdot}(\tau)), \tag{2.7.3'}$$

$$dP \propto dP|_{\mathscr{F}_0} \prod_{h,t} \lambda_h(t)^{\Delta N_h(t)} \mathop{\pi}_0^{\tau} (1 - d\Lambda_{\cdot})$$

$$= dP|_{\mathscr{F}_0} \prod_{h,t} \lambda_h(t)^{\Delta N_h(t)} \exp(-\Lambda_{\cdot}(\tau)), \tag{2.7.4'}$$

for the general, continuous, and absolutely continuous case, respectively. In going from (2.7.3′) to (2.7.4′), we have substituted $d\Lambda_h(t) = \lambda_h(t)\,dt$ and dropped the factors dt which cancel anyway on forming ratios.

An even more informal way of writing the first of these equations will also be useful. Here, we simply replace $\Delta N_h(t)$, $\Delta N_{\cdot}(t)$ everywhere by $\mathrm{d}N_h(t)$, $\mathrm{d}N_{\cdot}(t)$:

$$\mathrm{dP} = \mathrm{dP}|_{\mathscr{F}_0} \prod_{t \in [0,\tau]} \left(\prod_h \mathrm{d}\Lambda_h(t)^{\mathrm{d}N_h(t)} (1 - \mathrm{d}\Lambda_{\cdot}(t))^{1-\mathrm{d}N_{\cdot}(t)} \right). \tag{2.7.2''}$$

The intended mathematical interpretation is not changed; however, (2.7.2″) especially carries the suggestion that one might be able to calculate a Radon–Nikodym derivative $\mathrm{d}\tilde{\mathrm{P}}/\mathrm{dP}$ by partitioning $[0, \tau]$ into small subintervals, computing increments of $\mathbf{N}$, $\mathbf{\Lambda}$, and $\tilde{\mathbf{\Lambda}}$ over the subintervals and forming the ratio of corresponding finite products; then one takes the limit as the partition becomes finer and finer.

Conditions can be written down under which the Radon–Nikodym derivative $\mathrm{d}\tilde{\Lambda}_h/\mathrm{d}\Lambda_h$ is indeed a limit of approximating discrete ratios, and the product-integral is actually *defined* as a limit of approximating finite products. However, a rigorous derivation (under appropriate conditions on the partitions) is not important for us because we are more concerned with the suggestive nature of the formula. This corresponds to the interpretation of $\mathrm{d}\mathbf{\Lambda}(t)$ as the vector of conditional probabilities that N_h has a jump in the time interval $[t, t + \mathrm{d}t)$, $h = 1, \ldots, k$, given the past $\mathscr{F}_{t-}$. The likelihood is correspondingly written as a product of conditional multinomial probabilities for the infinitesimal subexperiments "in $[t, t + \mathrm{d}t)$ observe $\mathrm{d}\mathbf{N}(t)$."

We next make a technical comment on Theorem II.7.2 and Corollary II.7.3 concerning the condition that $\tilde{P}$ be absolutely continuous with respect to P. A sufficient condition for $\tilde{\mathrm{P}} \ll \mathrm{P}$ in the case when intensity processes exist is that $\tilde{\lambda}_h = 0$ where $\lambda_h = 0$, $\tilde{\mathrm{P}}$-almost surely, and $\tilde{\mathrm{P}}(\Lambda_{\cdot}(\tau) < \infty) = 1$. More complicated necessary and sufficient conditions are also available, due to Kabanov, Liptser, and Shiryayev (1976). However, because all these conditions are formulated in terms of conditions to hold $\tilde{\mathrm{P}}$-almost surely, they are not of much help in *defining* $\tilde{\mathrm{P}}$ from P via an expression for $\mathrm{d}\tilde{\mathrm{P}}/\mathrm{dP}$.

In statistical applications, we will want to consider a whole family of probability measures P, not necessarily mutually absolutely continuous, and, therefore, cannot always apply Theorem II.7.2 to obtain $\mathrm{d}\tilde{\mathrm{P}}/\mathrm{dP}$ for each $\tilde{\mathrm{P}}$, P considered; this should take the value $+\infty$ on the $\tilde{\mathrm{P}}$-singular part of Ω with respect to P. However, for any two probability measures $\tilde{\mathrm{P}}$ and P, there exists a third, say, $\mathrm{Q} = \frac{1}{2}(\tilde{\mathrm{P}} + \mathrm{P})$, which dominates both P and $\tilde{\mathrm{P}}$. We can, therefore, calculate dP/dQ and $\mathrm{d}\tilde{\mathrm{P}}/\mathrm{dQ}$ by Theorem II.7.2 and finally put

$$\frac{\mathrm{d}\tilde{\mathrm{P}}}{\mathrm{dP}} = \frac{\mathrm{d}\tilde{\mathrm{P}}}{\mathrm{dQ}} \bigg/ \frac{\mathrm{dP}}{\mathrm{dQ}} \quad \text{where } \frac{\mathrm{dP}}{\mathrm{dQ}} > 0,$$

$$\frac{\mathrm{d}\tilde{\mathrm{P}}}{\mathrm{dP}} = \infty \quad \text{where } \frac{\mathrm{dP}}{\mathrm{dQ}} = 0.$$

Note that $\mathrm{dP}/\mathrm{dQ} + \mathrm{d}\tilde{\mathrm{P}}/\mathrm{dQ} = 2$, so we never have $\mathrm{dP}/\mathrm{dQ} = \infty$. Because the denominators of $\mathrm{d}\tilde{\mathrm{P}}/\mathrm{dQ}$ and dP/dQ given by (2.7.2) coincide, this comes down to calculating $\mathrm{d}\tilde{\mathrm{P}}/\mathrm{dP}$ by (2.7.2) as it stands; the result is P-a.s. finite but possibly infinite with positive $\tilde{\mathrm{P}}$ probability. Because

$$\mathrm{E_P}\left(\frac{\mathrm{d}\tilde{\mathrm{P}}}{\mathrm{dP}}\right) = \mathrm{E_P}\left(\frac{\mathrm{d}\tilde{\mathrm{P}}}{\mathrm{dP}} I\left(\frac{\mathrm{d}\tilde{\mathrm{P}}}{\mathrm{dP}} < \infty\right)\right) = \tilde{\mathrm{P}}(A),$$

where $A = \{\mathrm{d}\tilde{\mathrm{P}}/\mathrm{dP} < \infty\}$, we have $\tilde{\mathrm{P}} \ll \mathrm{P}$ if and only if

$$\mathrm{E_P}\left(\frac{\mathrm{d}\tilde{\mathrm{P}}}{\mathrm{dP}}\right) = 1.$$

From this last result, we can now derive a result, called a Girsanov-type theorem by analogy with a similar technique for constructing diffusion processes, on the construction of a counting process with a given intensity process by reference to a given counting process model. (We do not use the theorem in this book but the Girsanov theorem is too famous to be omitted.) We state this in the case when intensities exist (in fact, the "reference" probability is most often in applications chosen to correspond to a standard Poisson process Π on a finite time interval $[0, \tau]$):

Theorem II.7.4 (Girsanov). *Suppose* $\mathbf{N}$ *has intensity process* $\boldsymbol{\lambda}$ *under* P *and* (2.2.1) *holds. Let* $\tilde{\boldsymbol{\lambda}}$ *be any predictable process whose components are* P-*a.s. zero where those of* $\boldsymbol{\lambda}$ *are zero. Let* $\tilde{\mathrm{P}}_0$ *be a given probability measure on* $\mathscr{F}_0$, *absolutely continuous with respect to* P *restricted to this sub-*σ*-algebra. Let* Z *denote the right-hand side of* (2.7.4). *Then there exists a probability measure* $\tilde{\mathrm{P}}$ *on* $\mathscr{F}$, *which is absolutely continuous with respect to* P, *agrees with* $\tilde{\mathrm{P}}_0$ *on* $\mathscr{F}_0$, *and is such that* $\mathbf{N}$ *has intensities* $\tilde{\boldsymbol{\lambda}}$ *with respect to* $\tilde{\mathrm{P}}$, *if and only if*

$$\mathrm{E_P}(Z) = 1. \tag{2.7.5}$$

Note that we automatically have $\mathrm{E_P}(Z) \leq 1$. Unfortunately, there is no easy general way to check (2.7.5). Brémaud (1981) gave some sufficient conditions which are sometimes useful.

Another construction result due to Jacobsen (1982) is possible in which a reference probability is not needed, provided we work with a "canonical" choice of Ω. Also, we make some strong boundedness assumptions about the intensities. We will use this construction result for the pornoscope example (Examples I.3.16 and III.1.10) and the software reliability example (Examples I.3.19 and III.1.12), both of which are most conveniently specified through their intensity processes. First, we give this in the case when $\mathscr{F}_0$ is trivial. Let Ω be the collection of all possible sample paths of a multivariate counting process on $[0, \tau]$, i.e., the collection of all functions from $[0, \tau]$ to $\mathbb{N}_0^k$, zero at time zero, piecewise constant, right continuous, with jumps of size $+1$ only, no two components having simultaneous jumps. ($\mathbb{N}_0$ denotes the non-negative integers here.) The multivariate counting process $\mathbf{N}$ is defined on this Ω simply by $\mathbf{N}(\omega) = \omega$. On Ω we impose the filtration $(\mathscr{N}_t)$, $\mathscr{N}_t = \sigma\{\mathbf{N}(s) \colon s \leq t\}$.

The idea of the construction is to specify the intensity process of the counting process as a function of its own past (past jump times and types) as well as the present time instant. So one must specify a large family of functions giving the intensity at each time t, for each number n of previous jumps, and for each precise set of jump times t_i and types j_i, $i = 1, \ldots, n$. Assuming the integrated intensity is finite for any conceivable realization of $\mathbf{N}$ saves us from the possibility of explosion:

Theorem II.7.5 (Jacobsen's Construction). *Suppose* $\lambda_n(t; t_1, \ldots, t_n; j_1, \ldots, j_n)$ *are given measurable non-negative k-variate functions defined for* $0 < t_1 < t_2 < \cdots < t_n \le t \le \tau, j_1, \ldots, j_n \in \{1, \ldots, k\}$. *For a given path* $\mathbf{N}$, *define*

$$\lambda(t; \mathbf{N}) = \lambda_n(t; t_1, \ldots, t_n; j_1, \ldots, j_n)$$

if $\mathbf{N}$ *has* T_i, J_i *satisfying* $T_i = t_i$, $J_i = j_i$; $i = 1, \ldots, n$; $T_{n+1} \ge t$. *Suppose* $\int_0^\tau \lambda(t; \mathbf{N})\,\mathrm{d}t < \infty$ *for all* $\mathbf{N}$. *Then we can define* P *on* Ω *such that* $\mathbf{N}$ *has intensity process* $\boldsymbol{\lambda}(\mathbf{N}) = \lambda(t; \mathbf{N})$ *with respect to* $(\mathscr{N}_t)$.

By including dependence on an element z of a measurable space $(\mathscr{Z}, \mathscr{B})$ in λ_n and adding a probability distribution P_0 on $\mathscr{Z}$, one can extend this in the obvious way to a probability on the filtration $\mathscr{F}_t = \mathscr{B} \otimes \mathscr{N}_t$ on $\mathscr{Z} \times \Omega$.

II.7.2. Martingale Properties of Likelihood Processes

Suppose (2.7.1) holds for each of $\tilde{\mathrm{P}}$ and P, $\tilde{\mathrm{P}} \ll \mathrm{P}$, on a common filtration. Introduce the *likelihood process* L defined by

$$L(t) = \left.\frac{\mathrm{d}\tilde{\mathrm{P}}}{\mathrm{d}\mathrm{P}}\right|_{\mathscr{F}_t}. \tag{2.7.6}$$

Our aim is to show that this process has certain martingale properties, which can be extremely valuable in statistical applications. We shall also see that these properties only depend on the fact that $\mathbf{N}$ has compensator $\boldsymbol{\Lambda}$ or $\tilde{\boldsymbol{\Lambda}}$ under the two relevant probability measures, and that the special structure assumed in (2.7.1) is not needed. This is connected to the theory of *partial likelihood* which we discuss in the next section.

We can generalize and rewrite (2.7.2) as follows:

$$\begin{aligned}
L(t) &= L(0)\frac{\prod_{s\le t}\left(\prod_h \mathrm{d}\tilde{\Lambda}_h(s)^{\Delta N_h(s)}(1-\mathrm{d}\tilde{\Lambda}_{.}(s))^{1-\Delta N_{.}(s)}\right)}{\prod_{s\le t}\left(\prod_h \mathrm{d}\Lambda_h(s)^{\Delta N_h(s)}(1-\mathrm{d}\Lambda_{.}(s))^{1-\Delta N_{.}(s)}\right)} \\
&= L(0)\prod_{s\le t}\left(\left(1-\frac{\mathrm{d}\tilde{\Lambda}_{.}(s)-\mathrm{d}\Lambda_{.}(s)}{1-\Delta\Lambda_{.}(s)}\right)^{1\ \Delta N_{.}(s)}\right. \\
&\qquad \times\left.\left(1-\sum_h\left(\frac{\mathrm{d}\tilde{\Lambda}_h}{\mathrm{d}\Lambda_h}(s)-1\right)\mathrm{d}N_h(s)\right)^{\Delta N_{.}(s)}\right) \\
&= L(0)\prod_{s\le t}\left(1+\sum_h\left(\frac{\mathrm{d}\tilde{\Lambda}_h}{\mathrm{d}\Lambda_h}(s)-1-\frac{\Delta\tilde{\Lambda}_{.}(s)-\Delta\Lambda_{.}(s)}{1-\Delta\Lambda_{.}(s)}\right)\right. \\
&\qquad \times\left.(\mathrm{d}N_h(s)-\mathrm{d}\Lambda_h(s))\vphantom{\sum_h}\right),
\end{aligned} \tag{2.7.7}$$

where the last two steps follow by fomula (2.6.5) for the ratio of two real product-integrals and by some simple but tedious algebra.The derivation and the interpretation of (2.7.7) is not important; what is important is that by the Volterra integral equation characterization for product-integrals (Theorem II.6.1), L is the unique solution of the equation

$$L(t) = L(0) + \sum_h \int_0^t L(s-)\left(\frac{\mathrm{d}\tilde{\Lambda}_h}{\mathrm{d}\Lambda_h}(s) - 1 - \frac{\Delta\tilde{\Lambda}_{\bullet}(s) - \Delta\Lambda_{\bullet}(s)}{1 - \Delta\Lambda_{\bullet}(s)}\right)\mathrm{d}M_h(s), \tag{2.7.8}$$

where $M_h = N_h - \Lambda_h$, $h = 1, \ldots, k$, are $(\mathrm{P}, (\mathscr{F}_t))$-local martingales. Because L is cadlag and adapted, its left-continuous modification L_- is predictable and locally bounded. If the predictable processes

$$\frac{\mathrm{d}\tilde{\Lambda}_h}{\mathrm{d}\Lambda_h} - 1 - \frac{\Delta\tilde{\Lambda}_{\bullet} - \Delta\Lambda_{\bullet}}{1 - \Delta\Lambda_{\bullet}}$$

are locally bounded too, we may conclude that L itself—being the sum of integrals of locally bounded predictable processes with respect to local martingales—is a local martingale too.

In fact, one can show directly from (2.7.6) that L is always a *martingale* when $\tilde{\mathrm{P}} \ll \mathrm{P}$. However, it is extremely useful that local martingale properties of the process

$$L(0)\frac{\prod_{s\le t}\left(\prod_h \mathrm{d}\tilde{\Lambda}_h(s)^{\Delta N_h(s)}(1 - \mathrm{d}\tilde{\Lambda}_{\bullet}(s))^{1-\Delta N_{\bullet}(s)}\right)}{\prod_{s\le t}\left(\prod_h \mathrm{d}\Lambda_h(s)^{\Delta N_h(s)}(1 - \mathrm{d}\Lambda_{\bullet}(s))^{1-\Delta N_{\bullet}(s)}\right)} \tag{2.7.9}$$

[and also with $L(0)$ deleted] can be derived quite independently of the interpretation, under (2.7.1) and $\tilde{\mathrm{P}} \ll \mathrm{P}$, of (2.7.9) as the likelihood ratio process $(L(t)) = ((\mathrm{d}\tilde{\mathrm{P}}/\mathrm{dP})|_{\mathscr{F}_t})$. In fact, local martingale properties of *score processes* and of *information processes* are also preserved, as we shall see later. These results are connected to the fact (which we explore further in the next section) that (2.7.9) can always be interpreted as the *partial likelihood* (ratio) based on observation of $\mathbf{N}$ and relative to some arbitrary filtration $(\mathscr{F}_t)$ not necessarily of the form $\mathscr{F}_t = \mathscr{F}_0 \vee \mathscr{N}_t$.

We finally sketch the related martingale properties of score and information processes. Suppose (2.7.1) holds for all $\{\mathrm{P}_\theta: \theta \in \Theta\}$, for some open set $\Theta \subset \mathbb{R}$. Suppose all P_θ are dominated by a fixed probability measure, Q say. For simplicity, we assume that all P_θ coincide on $\mathscr{F}_0$ and work also in the absolutely continuous case: Under P_θ, $\mathbf{N}$ has compensator $\mathbf{\Lambda}^\theta = \int \boldsymbol{\lambda}^\theta$ for certain intensity processes $\boldsymbol{\lambda}^\theta$. We consider the likelihood function as depending on both $t \in \mathscr{T}$ and $\theta \in \Theta$; dropping the denominator in (2.7.4) (which does not depend on θ), we have the likelihood at time t, as a function of θ, is proportional to

$$L(\theta, t) = \exp(-\Lambda^\theta_{\bullet}(t)) \prod_{T_n \le t} \lambda^\theta_{J_n}(T_n) \tag{2.7.10}$$

and, hence,

$$\log L(\theta; t) = \sum_h \int_0^t \log \lambda_h^\theta(s)\,\mathrm{d}N_h(s) - \Lambda_\cdot^\theta(t)$$
$$= \sum_h \int_0^t (\log \lambda_h^\theta(s)\,\mathrm{d}N_h(s) - \lambda_h^\theta(s)\,\mathrm{d}s).$$

To obtain the *score process*, we differentiate with respect to θ. Supposing the derivative may be taken under the integral sign, we obtain

$$\frac{\partial}{\partial\theta}\log L(\theta; t) = \sum_h \int_0^t \frac{\partial}{\partial\theta}\log \lambda_h^\theta(s)(\mathrm{d}N_h(s) - \lambda_h^\theta(s)\,\mathrm{d}s), \tag{2.7.11}$$

so that again, under local boundedness or integrability conditions, the score process is a $(\mathrm{P}_\theta, (\mathscr{F}_t))$-local martingale. Again, this holds without the assumptions leading to (2.7.10) being the likelihood functon. This result will be especially useful in Section VI.1 on maximum likelihood estimators.

Differentiating again with respect to θ to find the *observed information* at θ, we find (if the differentiation can can again be taken under the integral sign)

$$\frac{\partial^2}{\partial\theta^2}\log L(\theta; t) = \sum_h \int_0^t \frac{\partial^2}{\partial\theta^2}\log \lambda_h^\theta(s)\,\mathrm{d}M_h(s) - \sum_h \int_0^t \left(\frac{\partial}{\partial\theta}\log \lambda_h^\theta(s)\right)^2 \lambda_h^\theta(s)\,\mathrm{d}s. \tag{2.7.12}$$

Note that

$$\left\langle \frac{\partial}{\partial\theta}\log L(\theta; \cdot)\right\rangle = \sum_h \int \left(\frac{\partial}{\partial\theta}\log \lambda_h^\theta\right)^2 \lambda_h^\theta,$$

so that $\langle(\partial/\partial\theta)\log L(\theta; \cdot)\rangle$ is the compensator not only of $((\partial/\partial\theta)\log L(\theta; \cdot))^2$ but also of the process $-(\partial^2/\partial\theta^2)\log L(\theta; \cdot)$; this is a version of the well-known result: The variance of the score coincides with the expected information.

II.7.3. Partial Likelihood

In Section II.7.1, we showed, cf. (2.7.2″), how the likelihood based on observation of a self-exciting multivariate counting process $\mathbf{N}$ with compensator $\mathbf{\Lambda}$ could be suggestively written in the form

$$\mathrm{dP} = \prod_{t\in[0,\tau]} \left(\prod_h \mathrm{d}\Lambda_h(t)^{\mathrm{d}N_h(t)}(1 - \mathrm{d}\Lambda_\cdot(t))^{1-\mathrm{d}N_\cdot(t)}\right). \tag{2.7.13}$$

We also saw in Section II.7.2 that even if the process was not self-exciting—the filtration is larger than that generated by $\mathbf{N}$—then "likelihood ratios" and "score functions" calculated from the right-hand side of (2.7.13) still preserve various martingale properties, important in deriving statistical results for likelihood-based inference.

Here we link these results with the ideas of partial likelihood on the one hand (Cox, 1975; Kalbfleisch and Prentice, 1980) and with likelihoods for marked point processes on the other (Arjas and Haara, 1984; Arjas, 1989). The treatment will be heuristic and will belong more to statistical folklore than to mathematical probability theory. However, it is useful to discuss partial likelihood, even if only informally, in an abstract or general setting where the notation is not encumbered by specific details of particular applications. We will come to such applications in Chapter III; see, in particular, Section III.2.2.

The heuristic interpretation of (2.7.13) as a product of conditional multinomial probabilities ($n = 1$; $k + 1$ cells) connects naturally to the notion of *partial likelihood* (Cox, 1975). Suppose the data X in a statistical problem can be represented as a sequence of smaller pieces of data X_0, X_1, ..., X_n. The density of X can be factored as the product of conditional densities $p(x_i|x_0, \ldots, x_{i-1})$ of each X_i given its predecessors $X_0, \ldots, X_{i-1}$. As a function of unknown parameters θ of the distribution of X, this is the likelihood for θ based on X. Cox (1975) claimed that if one deletes some of the factors, e.g., all the even numbered ($i = 0, 2, \ldots$), what remains (the product over $i = 1, 3, \ldots$) can still be used as a basis for statistical inference. In particular, the standard theory of large sample distributional properties of maximum likelihood estimators, likelihood ratio tests and score tests, will also apply to the *partial likelihood.*

The reason for this is actually martingale based, though Cox (1975) did not explicitly refer to martingales in his paper. The components of the partial score $(\partial/\partial\theta)\log p(X_i|X_0, X_1, \ldots, X_{i-1}; \theta)$, $i = 1, 3, \ldots$, form a so-called discrete-time martingale difference sequence and, from this, it follows that the log partial likelihood satisfies the usual relations between expectations of first and second derivatives. Moreover, if the number of components is large and their individual influence small, the martingale central limit theorem could supply asymptotic normality of the partial score.

This suggests that we could also delete terms from (2.7.13) as desired and use what remains as a partial likelihood. If one deletes the contributions for the time interval $\mathrm{d}t$ according to the value of a predictable indicator process C, the result is called "the partial likelihood for the filtered counting process $\int C\,\mathrm{d}\mathbf{N}$," to which we return in Section III.4.

It is more useful to base partial likelihoods on further factorizations of (2.7.13). Suppose the process $\mathbf{N}$ registers two distinct kinds of events, which may even occur simultaneously. Typically, these will be *failure events* and *censoring* or *covariate* events. We can correspondingly factor each multinomial likelihood in (2.7.13) into a likelihood based on the marginal distribution of the first type of event and the likelihood based on the conditional distribution of the second given the first. Then we could form a partial likelihood by keeping only the first (marginal) pieces of each pair corresponding to the failure events. Such a partial likelihood (based on failure events, disregarding censoring and covariate events) is often used in survival analysis; see especially Sections III.2 and III.5 of this volume. One could also apply the idea to classification of the events into more than two levels, keeping just the contribution from one particular level conditional on the previous levels. The famous Cox partial likelihood (Example VII.2.1) used to derive the usual estimator in the Cox regression model (Cox, 1972) is of this type: One factors according to whether or not there is a failure event; if so, then to which individual it occurs, and then, given the preceding, one looks at possible censorings or changes in covariates. The Cox partial likelihood corresponds to keeping just the middle term of each triple (which individual failed, given there was a failure); one disregards the first term of each triple saying whether or not there is a failure, and the third term, saying what censoring or covariate events took place. In both these examples, the partial likelihood makes inference easier by discarding factors which depend in a complicated or even unknown way on nuisance parameters.

To formalize these ideas, we can use the fact that a classification of the events registered by $\mathbf{N}$ corresponds to aggregation or grouping of some or all of the $k + 1$ multinomial cells in (2.7.13) to a smaller number of cells. Multinomial probabilities will then be written as products of marginal (multinomial) probabilities for the coarse classification of events and conditional probabilities for the fine classification given the first one. There is no difficulty at all in extending this to a hierarchy of classifications. A partial likelihood is formed by collecting just the contributions from one particular level of the hierarchy of classifications.

It is convenient at this point to generalize to marked point process notation (see Section II.4.1); the ideas are not changed though some technical details become more delicate through the added generality. In fact, we will only consider countable mark spaces in the rest of this book, but the restriction does not lead to substantial simplification.

The likelihood in the self-exciting case (2.7.13) is now written as

$$\prod_{t}\left(\prod_{x \in E} \mathbf{\Lambda}(\mathrm{d}t, \mathrm{d}x)^{\mathbf{N}(\mathrm{d}t, \mathrm{d}x)}(1 - \mathbf{\Lambda}(\mathrm{d}t, E))^{1 - \mathbf{N}(\mathrm{d}t, E)}\right), \tag{2.7.14}$$

where $\mathbf{N}$ is a marked point process with marks in $(E, \mathscr{E})$ and $\mathbf{\Lambda}$ is its compensator; $\mathbf{N}$ and $\mathbf{\Lambda}$ are considered as (random) measures on $(\mathscr{T} \times E, \mathscr{B}(\mathscr{T}) \otimes \mathscr{E})$. The interpretation of (2.7.14) remains the same: Conditional on the past at time t, there is an event in the time interval $[t, t + \mathrm{d}t)$, or just $\mathrm{d}t$ for short, with mark in $\mathrm{d}x$ [so $\mathbf{N}(\mathrm{d}t, \mathrm{d}x) = 1$] with probability $\mathbf{\Lambda}(\mathrm{d}t, \mathrm{d}x)$; there is no event [$\mathbf{N}(\mathrm{d}t, E) = 0$] with probability $1 - \mathbf{\Lambda}(\mathrm{d}t, E)$. Likelihood *ratios* are formed, as explained in the discussion following Theorem II.7.2, by collecting the finite number of contributions to (2.7.14) where $\mathbf{N}(\mathrm{d}t, \mathrm{d}x) = 1$ and forming Radon–Nikodym derivatives $\mathbf{\Lambda}(\mathrm{d}t, \mathrm{d}x)/\tilde{\mathbf{\Lambda}}(\mathrm{d}t, \mathrm{d}x) = (\mathrm{d}\mathbf{\Lambda}/\mathrm{d}\tilde{\mathbf{\Lambda}})(t, x)$. The remaining contributions are standard product-integrals over $\mathscr{T}$ less the finite number of time points where an event occurs.

We need a convenient notation for a hierarchy of classifications of type of event, the finest level being provided by $(E, \mathscr{E})$ itself. Arjas (1989) proposed that such classifications be regarded as a function g on the mark space E to a (typically smaller) space G, so g groups or aggregates the marks in E, putting them together into certain classes characterized by the value of g. This is called the *premark approach* with the idea that the occurrence of the premark is some kind of signal that more is going to come. A complication arises because we may want to group some of the events in E with the "nonevent," thus discarding some of the time points at which events occur altogether. Thus, the premark could indicate failures; the full mark adds to the description of what failures occur at a given time also what censorings occur. It is possible for the premark to be empty while the full mark does indicate that an event has taken place (censoring without simultaneous failure). Rather than the term premark, we will use the more neutral and accurate term *reduced mark*. Another common terminology is to talk of *innovative* and *noninnovative* events (or informative and noninformative): The innovative events are the occurrences of the premarks; the noninnovative events are whatever else can happen.

We show how a function g on the mark space reduces or aggregates $\mathbf{N}$ to a "smaller" marked point process $\mathbf{N}^g$ with reduced mark space G. The compensator $\mathbf{\Lambda}^g$ of $\mathbf{N}^g$ will be easily computable from $\mathbf{\Lambda}$, the compensator of $\mathbf{N}$. Then we use the multinomial analogy to suggest how (2.7.14) would factor under this aggregation. A complication comes from having to deal with "no event" (both for $\mathbf{N}$ and for $\mathbf{N}^g$) explicitly. We then write down the analogous factorization of (2.7.14) and show how each of the terms arising can be given a formal (measure-theoretic) interpretation, taking account of the fact that the final expression should make mathematical sense on forming a ratio with a similar expression under a different probability measure (same $\mathbf{N}$, different compensator $\tilde{\mathbf{\Lambda}}$). So the approach is a heuristic one which leads us to a factorization of (2.7.14) whose validity can be established by a direct (tedious) calculation. The approach also leads to interesting conjectures: If the conditional terms which we would like to delete to form a partial likelihood do *not* depend on the parameter of interest anyway, then the result is not just a *partial* likelihood but actually the *full* likelihood for this parameter.

Let $\mathbf{N}$ be a self-exciting marked point process with compensator $\mathbf{\Lambda}$ and mark space E. Let $\varnothing$ be a point *not* in E which we call the "empty mark," and use to represent "no event." Let $\bar{E} = E \cup \{\varnothing\}$. Let G be another mark space, suppose $\varnothing \notin G$, and let $\bar{G} = G \cup \{\varnothing\}$. (Alternatively one can use a different symbol for the empty mark in each space E and G, but this makes things look more complicated than necessary.) Let $g: \bar{E} \to \bar{G}$ be a measurable mapping such that $g(\varnothing) = \varnothing$, "the empty mark in E" is mapped to "the empty mark in G." Typically, g will be a many-to-one mapping. (E and G are equipped with σ-algebras $\mathscr{E}$ and $\mathscr{G}$; $\bar{E}$ and $\bar{G}$ are given the σ-algebras on $\bar{E}$, $\bar{G}$, *generated* by $\mathscr{E}$ and $\mathscr{G}$.)

Let $\mathbf{N}^g$ be the marked point process with mark space G defined by $\mathbf{N}^g((0,t] \times A) = \mathbf{N}((0,t] \times g^{-1}(A))$. Thus, $\mathbf{N}^g$ as a point process has points $(T_n, g(J_n))$ for those points (T_n, J_n) of $\mathbf{N}$ such that $g(J_n) \neq \varnothing$. The compensator of $\mathbf{N}^g$ is $\mathbf{\Lambda}^g$ defined by

$$\mathbf{\Lambda}^g((0,t] \times A) = \mathbf{\Lambda}((0,t] \times g^{-1}(A)).$$

The conditional distribution, given $\mathscr{F}_{t-}$ of the events in dt, can now be built up as follows:

with probability $\mathbf{\Lambda}^g(\mathrm{d}t, \mathrm{d}y)$, the reduced process has an event in $\mathrm{d}t \times \mathrm{d}y$, i.e., $\mathbf{N}^g(\mathrm{d}t, \mathrm{d}y) = 1$; with probability $1 - \mathbf{\Lambda}^g(\mathrm{d}t, G)$, it has no event, i.e., $\mathbf{N}^g(\mathrm{d}t, G) = 0$;

given the reduced process has an event in $\mathrm{d}t \times \mathrm{d}y$, i.e., $\mathbf{N}^g(\mathrm{d}t, \mathrm{d}y) = 1$, the original process has one in $\mathrm{d}t \times \mathrm{d}x$, i.e., $\mathbf{N}(\mathrm{d}t, \mathrm{d}x) = 1$ [for x with $g(x) = y$], with conditional probability $\mathbf{\Lambda}(\mathrm{d}t, \mathrm{d}x)/\mathbf{\Lambda}^g(\mathrm{d}t, \mathrm{d}y)$;

but *given* the reduced process has *no* event in $\mathrm{d}t \times \mathrm{d}y$, i.e., $\mathbf{N}^g(\mathrm{d}t, G) = 0$, the original process has an event in $\mathrm{d}t \times \mathrm{d}x$, i.e., $\mathbf{N}(\mathrm{d}t, \mathrm{d}x) = 1$ [for x with $g(x) = \varnothing$], with conditional probability $\mathbf{\Lambda}(\mathrm{d}t, \mathrm{d}x)/(1 - \mathbf{\Lambda}^g(\mathrm{d}t, G))$; the original process has no event, i.e., $\mathbf{N}(\mathrm{d}t, E) = 0$, with the complementary conditional probability $1 - \mathbf{\Lambda}(\mathrm{d}t, g^{-1}(\varnothing))/(1 - \mathbf{\Lambda}^g(\mathrm{d}t, G))$.

Combining all these possibilities in the same order as we have just described them suggests that (2.7.14) can be rewritten as

$$\begin{aligned}
\mathrm{dP} = \prod_t \Bigg\{ & \left(\prod_{y \in G} \mathbf{\Lambda}^g(\mathrm{d}t, \mathrm{d}y)^{\mathbf{N}^g(\mathrm{d}t, \mathrm{d}y)} (1 - \mathbf{\Lambda}^g(\mathrm{d}t, G))^{1 - \mathbf{N}^g(\mathrm{d}t, G)} \right) \\
& \cdot \prod_{y \in G} \left(\prod_{x: g(x) = y} \left(\frac{\mathbf{\Lambda}(\mathrm{d}t, \mathrm{d}x)}{\mathbf{\Lambda}^g(\mathrm{d}t, \mathrm{d}y)} \right)^{\mathbf{N}(\mathrm{d}t, \mathrm{d}x)} \right)^{\mathbf{N}^g(\mathrm{d}t, \mathrm{d}y)} \\
& \cdot \left(\prod_{x: g(x) = \varnothing} \left(\frac{\mathbf{\Lambda}(\mathrm{d}t, \mathrm{d}x)}{1 - \mathbf{\Lambda}^g(\mathrm{d}t, G)} \right)^{\mathbf{N}(\mathrm{d}t, \mathrm{d}x)} \right. \\
& \cdot \left. \left(1 - \frac{\mathbf{\Lambda}(\mathrm{d}t, g^{-1}(\varnothing))}{1 - \mathbf{\Lambda}^g(\mathrm{d}t, G)} \right)^{1 - \mathbf{N}(\mathrm{d}t, E)} \right)^{1 - \mathbf{N}^g(\mathrm{d}t, G)} \Bigg\}.
\end{aligned} \tag{2.7.15}$$

The intuitive and probabilistic interpretation of (2.7.15) is not really difficult, even if the formula is rather long. A rigorous mathematical interpretation is, however, a little delicate.

The first line of (2.7.15) can be mathematically interpreted as it stands (on taking a ratio with $\mathrm{d}\tilde{\mathrm{P}}$, forming Radon–Nikodym derivatives and product-integrals) and gives us the *partial likelihood* based on $\mathbf{N}^g$ (with the rest of the information in $\mathbf{N}$ ignored). Note that this partial likelihood has exactly the same form as the likelihood based on $\mathbf{N}^g$ alone in the case that this process is self-exciting; however, now $\mathbf{\Lambda}^g$ may depend on the whole past of $\mathbf{N}$, not just on $\mathbf{N}^g$. The second and last two lines provide the other part of the factorization and are harder to interpret mathematically. We look at each part in turn.

The second line gives a contribution $\mathbf{\Lambda}(\mathrm{d}t, \mathrm{d}x)/\mathbf{\Lambda}^g(\mathrm{d}t, \mathrm{d}y)$ for t, x, and $y = g(x)$ which corresponds to events of both $\mathbf{N}$ and $\mathbf{N}^g$, a finite number of terms. Because $\mathbf{\Lambda}^g$ is a "marginalization" of $\mathbf{\Lambda}$, one can "disintegrate" $\mathbf{\Lambda}$ [restricted to $\mathscr{T} \times E \backslash g^{-1}(\varnothing)$] into the product of the image $\mathbf{\Lambda}^g$ of $\mathbf{\Lambda}$ and a *transition probability measure* $\mathbf{\Lambda}(\mathrm{d}x | t, y)$ on $\{x: g(x) = y\}$; thus, one may write

$$\Lambda(\mathrm{d}t, \mathrm{d}x) = \Lambda^g(\mathrm{d}t, \mathrm{d}y)\Lambda(\mathrm{d}x|t, y)$$

in the sense that an integral over t and x on the left-hand side equals a triple integral over t, y, and $x = g(y)$ on the right-hand side. Intuitively, for $y = g(x)$, we write the probability of a mark in $\mathrm{d}x$ and time $\mathrm{d}t$, given the past, as the probability of a reduced mark in $\mathrm{d}y$ and time $\mathrm{d}t$, given the past, times the probability of a mark in $\mathrm{d}x$ given a reduced mark y at time t. So $\Lambda(\mathrm{d}t, \mathrm{d}x)/\Lambda^g(\mathrm{d}t, \mathrm{d}y)$ has a mathematical interpretation as $\Lambda(\mathrm{d}x|t, y)$, and the ratio of two such terms (for compensators Λ and $\tilde{\Lambda}$ under two different probability measures P and $\tilde{\mathrm{P}}$) can be mathematically interpreted as the Radon–Nikodym derivative, for the given point t, y, of the transition probability measure $\Lambda(\cdot\,|t, y)$ with respect to $\tilde{\Lambda}(\cdot\,|t, y)$ on $\{x\colon g(x) = y\}$.

The third line also only occurs in a finite number of terms. This suggests mathematically interpreting $\Lambda(\mathrm{d}t, \mathrm{d}x)/(1 - \Lambda^g(\mathrm{d}t, G))$ as $(1 - \Lambda^g(\{t\} \times G))^{-1} \cdot \Lambda(\mathrm{d}t, \mathrm{d}x)$, ratios of which can be mathematically interpreted as a Radon–Nikodym derivative $(\mathrm{d}\Lambda/\mathrm{d}\tilde{\Lambda})(t, x)$ times the function $(1 - \tilde{\Lambda}^g(\{t\} \times G))/(1 - \Lambda^g(\{t\} \times G))$.

The fourth line is present for all t such that $\mathbf{N}$ has no event in $\mathrm{d}t$. Because

$$1 - \frac{\Lambda(\mathrm{d}t, g^{-1}(\varnothing))}{1 - \Lambda^g(\mathrm{d}t, G)} = \frac{1 - \Lambda(\mathrm{d}t, g^{-1}(\varnothing)) - \Lambda^g(\mathrm{d}t, G)}{1 - \Lambda^g(\mathrm{d}t, G)} = \frac{1 - \Lambda(\mathrm{d}t, E)}{1 - \Lambda^g(\mathrm{d}t, G)},$$

we can mathematically interpret a product over t of terms like this as a ratio of product-integrals.

To sum up, we may rewrite (2.7.15) as

$$\begin{aligned}\mathrm{dP} = \prod_t \Bigg\{ &\left(\prod_{y \in G} \Lambda^g(\mathrm{d}t, \mathrm{d}y)^{\mathbf{N}^g(\mathrm{d}t, \mathrm{d}y)} (1 - \Lambda^g(\mathrm{d}t, G))^{1 - \mathbf{N}^g(\mathrm{d}t, G)} \right) \\ &\cdot \prod_{x: g(x) \neq \varnothing} \Lambda(\mathrm{d}x|t, g(x))^{\mathbf{N}(\mathrm{d}t, \mathrm{d}x)} \prod_{x: g(x) = \varnothing} \left(\frac{\Lambda(\mathrm{d}t, \mathrm{d}x)}{1 - \Lambda^g(\{t\} \times G)} \right)^{\mathbf{N}(\mathrm{d}t, \mathrm{d}x)} \\ &\cdot \left(\frac{1 - \Lambda(\mathrm{d}t, E)}{1 - \Lambda^g(\mathrm{d}t, G)} \right)^{1 - \mathbf{N}(\mathrm{d}t, E)} \Bigg\},\end{aligned} \tag{2.7.16}$$

which may be mathematically interpreted by taking ratios and forming Radon–Nikodym derivatives and product-integrals. The first line is the partial likelihood based on $\mathbf{N}^g$; the next two lines (together) form a partial likelihood based on the rest of $\mathbf{N}$.

A special case is when E and G are countable and Λ is absolutely continuous on $\mathscr{T} \times E$ with respect to Lebesgue measure times counting measure. This means that $\mathbf{N}$ is a counting process in the usual sense (with a countable number of components) and $\mathbf{N}^g$ is some aggregration of $\mathbf{N}$; both have intensities, say λ_x: $x \in E$ and λ_y^g, $y \in G$. Then $\lambda_y^g(t) = \sum_{x: g(x) = y} \lambda_x(t)$; the transition measure $\Lambda(\mathrm{d}x|t, g(x))$ is a probability measure on the finite set $\{x\colon g(x) = y\}$ and its atoms are simply $\lambda_x(t)/\lambda_y^g(t)$. The atomic part $1 - \Lambda^g(\{t\} \times G)$ disappears and the factorization is

$$\begin{aligned}
\mathrm{dP} \propto \prod_t \Bigg\{ &\left(\prod_y \lambda_y^g(t)^{N_y^g(\mathrm{d}t)} (1 - \lambda_\cdot^g(t)\,\mathrm{d}t)^{1-\mathrm{N}^g(\mathrm{d}t)} \right) \\
&\cdot \prod_{x:g(x)\neq\varnothing} \left(\frac{\lambda_x(t)}{\lambda_{g(x)}^g(t)} \right)^{N_x(\mathrm{d}t)} \prod_{x:g(x)=\varnothing} \lambda_x(t)^{N_x(\mathrm{d}t)} \left(\frac{1-\lambda_\cdot(t)\,\mathrm{d}t}{1-\lambda_\cdot^g(t)\,\mathrm{d}t} \right)^{1-N_\cdot(\mathrm{d}t)} \Bigg\} \\
\propto \prod_{t,y} &\lambda_y^g(t)^{N_y^g(\mathrm{d}t)} \exp\left(-\int_0^\tau \lambda_\cdot^g(t)\,\mathrm{d}t \right) \prod_{t,x:g(x)\neq\varnothing} \left(\frac{\lambda_x(t)}{\lambda_{g(x)}^y(t)} \right)^{N_x(\mathrm{d}t)} \\
&\cdot \prod_{t,x:g(x)=\varnothing} \lambda_x(t)^{N_x(\mathrm{d}t)} \exp\left(-\int_0^\tau (\lambda_\cdot(t) - \lambda_\cdot^g(t))\,\mathrm{d}t \right).
\end{aligned}$$

One easily sees that this really is a factorization of the full likelihood

$$\mathrm{dP} \propto \prod_t \left(\prod_x \lambda_x(t)^{N_x(\mathrm{d}t)} (1 - \lambda_\cdot(t)\,\mathrm{d}t)^{1-N_\cdot(\mathrm{d}t)} \right),$$

as must, of course, be the case.

It is possible (see the next paragraph) to give conditions under which the factorization (2.7.16) is correct; probably it is also possible to show that the two factors of (2.7.16) (the first line versus the last three) can be obtained as limits of proper, finite, partial likelihoods in the sense of Cox (1975); see the results of Slud (1991). However, such mathematical results, though supporting the intuition, do not lead to any useful mathematical properties of partial likelihood which cannot be obtained easily and directly. So, it is quite justified to leave the matter in its present informal state: We have described a valuable heuristic tool, not presented a formal mathematical theory.

The following remarks form a technical aside on the regularity conditions on the mark spaces E and G needed to justify all the steps here. Arjas and Haara (1984) assumed that E is the product of a countable set with a Polish space, and g is the coordinate projection onto the first coordinate. An interesting generalization due to Arjas (1989) would be to let the function g be replaced by a predictable process, so that classification of events may change randomly (but predictably) in time. We do not, however, follow up this idea here. Arjas and Haara (1984) used a representation of E as a product space (innovative versus noninnovative marks), with the classification of events seen as a *projection* on the first coordinate (aggregation over the second). Again, the need to consider "no event" leads to an involved notation. The reduced mark approach and the product mark space approach are mathematically equivalent.

II.8 The Functional Delta-Method

The delta-method is a popular and elementary tool of asymptotic statistics. One can summarize it in the following statements: Suppose for some random p-vectors $\mathbf{T}_n$ and a sequence of numbers $a_n \to \infty$,

$$a_n(\mathbf{T}_n - \boldsymbol{\theta}) \xrightarrow{\mathscr{D}} \mathbf{Z} \quad \text{as } n \to \infty,$$

where $\boldsymbol{\theta} \in \mathbb{R}^p$ is fixed. Suppose $\boldsymbol{\phi}\colon \mathbb{R}^p \to \mathbb{R}^q$ is differentiable at $\boldsymbol{\theta}$ with $q \times p$ matrix $\boldsymbol{\phi}'(\boldsymbol{\theta})$ of partial derivatives. Then

$$a_n(\boldsymbol{\phi}(\mathbf{T}_n) - \boldsymbol{\phi}(\boldsymbol{\theta})) \xrightarrow{\mathscr{D}} \boldsymbol{\phi}'(\boldsymbol{\theta})\cdot \mathbf{Z} \quad \text{in } \mathbb{R}^q$$

and, indeed, $a_n(\boldsymbol{\phi}(\mathbf{T}_n) - \boldsymbol{\phi}(\boldsymbol{\theta}))$ is asymptotically equivalent to $\boldsymbol{\phi}'(\boldsymbol{\theta})\cdot a_n(\mathbf{T}_n - \boldsymbol{\theta})$.

Our statistical models will frequently be non- or semiparametric models, and we will estimate cumulative hazard functions, survival functions, and the like, getting their large sample properties by application of the martingale central limit theorem. However, we will often want to transfer asymptotic normality to various functionals of interest (cf. Section IV.3.4). Also, in Section VIII.3 we need to transfer asymptotic efficiency from estimators to functionals of estimators. All this can be done using an infinite-dimensional version of the delta-method.

This indeed exists, but to make its use more elegant, we will need to switch from the usual weak convergence theory for $D[0, \tau]$ based on the Skorohod metric (Billingsley, 1968) to a perhaps less familiar theory due to Dudley (1966) and popularized by Pollard (1984) based on the supremum norm. The σ-algebra on $D[0, \tau]$ used in this theory is exactly the same, but is now interpreted as the σ-algebra generated by the supremum-norm open balls in $D[0, \tau]$, or equivalently as the σ-algebra generated by the *coordinate mappings* $x \mapsto x(t)$ mapping $D[0, \tau]$ to $\mathbb{R}$. (These coincidences perhaps explain the great usefulness of the otherwise rather complicated Skorohod metric.) The open-ball σ-algebra is *strictly smaller* than the (supremum-norm) Borel σ-algebra; this is related to the nonseparability of $D[0, \tau]$ under the supremum norm.

Weak convergence in $D[0, \tau]$, supremum norm, now means convergence of the expectations of all bounded, real, continuous *measurable* functions of the random element concerned. If the limiting process has continuous sample paths, weak convergence in the sense of the Skohorod metric and in the sense of the supremum norm are exactly equivalent. Otherwise, sup-norm convergence is *stronger*.

We will consider random elements of spaces like $B = D[0, \tau]^p \times \mathbb{R}^q$ endowed with, e.g., the max-supremum norm, or another convenient equivalent norm. This makes B a Banach space (not necessarily separable). Weak convergence of random elements of B has already been described; we now need to introduce a concept of *differentiability* suitable for statistical applications, allowing us to generalize the delta-method.

This is provided by the concept of *Hadamard* or *compact* differentiability, intermediate between the more familiar but too strong notion of *Fréchet* or *bounded* differentiability, and the too weak *Gâteaux* or *directional* differentiability. See Gill (1989) for a more extensive introduction to these matters, on which this summary is based.

Many equivalent formulations of Hadamard differentiability exist. Most convenient for us is the following definition:

Definition II.8.1. Let B, B' be spaces of the type just described. $\phi: B \to B'$ is compactly or Hadamard differentiable at a point $\theta \in B$ if and only if a continuous, linear map

$$\mathrm{d}\phi(\theta): B \to B'$$

exists (called the derivative of ϕ at the point θ) such that for all real sequences $a_n \to \infty$ and all convergent sequences $h_n \to h \in B$,

$$a_n(\phi(\theta + a_n^{-1}h_n) - \phi(\theta)) \to \mathrm{d}\phi(\theta)\cdot h \quad \text{as } n \to \infty. \tag{2.8.1}$$

The notation $\mathrm{d}\phi(\theta)\cdot h$ for the linear mapping $\mathrm{d}\phi(\theta)$ acting on the element $h \in B$ is supposed to suggest multiplication. When B and B' are Euclidean and $\boldsymbol{\phi}$ and $\boldsymbol{\theta}$ are real vectors, $\mathrm{d}\phi(\theta)$ as a linear map from one Euclidean space to another can be represented by a matrix multiplication, and $\mathrm{d}\phi(\theta)$ can be identified with the matrix of partial derivatives of the components of $\boldsymbol{\phi}$ with respect to those of $\boldsymbol{\theta}$.

This allows us to give a functional version of the delta-method.

Theorem II.8.1. *Let T_n be a sequence of random elements of B, $a_n \to \infty$ a real sequence, such that*

$$a_n(T_n - \theta) \xrightarrow{\mathscr{D}} Z$$

for some fixed point $\theta \in B$ and a random element Z of B. Suppose $\phi: B \to B'$ is compactly differentiable at θ. Then

$$a_n(\phi(T_n) - \phi(\theta)) \xrightarrow{\mathscr{D}} \mathrm{d}\phi(\theta)\cdot Z$$

and, moreover,

$$a_n(\phi(T_n) - \phi(\theta)) \quad \textit{and} \quad \mathrm{d}\phi(\theta)\cdot a_n(T_n - \theta)$$

are asymptotically equivalent.

The proof of Theorem II.8.1 is extremely simple. One invokes the so-called Skorohod–Dudley almost sure representation theorem (see, e.g., Pollard, 1984 or 1990) to establish [because of the weak convergence of $a_n(T_n - \theta)$ to Z] the existence of versions of T_n and of Z, all defined on a single probability space, and such that $a_n(T_n - \theta)$ converges *almost surely* to Z. Now one applies the definition of compact differentiability to show that, on this new probability space, $a_n(\phi(T_n) - \phi(\theta))$ converges almost surely to $\mathrm{d}\phi(\theta)\cdot Z$. Almost sure convergence implies convergence in distribution, but the new and the old versions of $a_n(\phi(T_n) - \phi(\theta))$ have exactly the same distributions for each n by the Skorohod–Dudley construction. Hence, we get the required result.

Sometimes, (2.8.1) is not satisfied for all sequences, but only for sequences $h_n \in B$ with limit h in a subset $H \subseteq B$. If the limiting process Z has sample paths in H, then this is still enough for the conclusions of the delta-method. We, therefore, make the following definition:

Definition II.8.2. Let B, B', ϕ, θ, and $\mathrm{d}\phi(\theta)$ be as in Definition II.8.1, and suppose $H \subseteq B$ exists such that for all sequences $h_n \in B$ with $h_n \to h \in H$, (2.8.1) holds. Then we say that ϕ is tangentially compactly differentiable at θ (tangentially to H).

Then we have the delta-method again:

Theorem II.8.2. *Let T_n, B, ϕ, θ, and Z be as in Theorem* II.8.1, *except that ϕ is only differentiable tangential to H, but Z has sample paths in H. Then the conclusions of Theorem* II.8.1 *still remain true.*

Sometimes a functional ϕ is only defined on some subset $E \subseteq B$ (maybe, T_n only takes values in E, so the definition of ϕ outside E should be arbitrary). It turns out that one need only verify (2.8.1) for $\theta + a_n^{-1}h_n \in E$ (for either proper or only tangential differentiability): ϕ can then be extended to B so as to be (tangentially) differentiable at θ:

Lemma II.8.3. *Let $E \subseteq B$ and let ϕ: $E \to B'$ be given. Suppose $\theta \in E$ and* (2.8.1) *holds for all sequences a_n, h_n such that $\theta + a_n^{-1}h_n \in E$ for all n and $h_n \to h$ ($\in H$, in the tangential case). Then ϕ can be defined on all of B so as to be (tangentially) differentiable at θ with derivative* $\mathrm{d}\phi(\theta)$.

Before we move to applications, here is a most important remark. Compact differentiability (also tangential differentiability, under the obvious compatibility conditions) satisfies the *chain rule*: The composition of differentiable functions is differentiable, with derivative equal to the composition (product) of the derivatives; or

$$\mathrm{d}(\psi \circ \phi)(\theta) = \mathrm{d}\psi(\phi(\theta)) \cdot \mathrm{d}\phi(\theta).$$

This means that from the differentiability of the few basic functionals in what follows, the differentiability of a large class of composed functionals follows too without any further work. As we saw in the sketch of the proof of Theorem II.8.1, the functional delta-method is really no more than a way to package the Skorohod–Dudley almost sure convergence theorem. The chain rule further allows us to modularize its application: We decompose a complicated functional into a composition of simple and perhaps familiar functionals whose differentiability is known or easy to check.

Now for applications. The most common are quantiles and inverses, integrals and product-integrals, and composition. We mean here a different sort of composition to that referred to just now: Take two elements x and y of $D[0, \tau]$ such that y also takes values in $[0, \tau]$ and form the new element $x(y(\cdot)) \in D[0, \tau]$. All these mappings are (tangentially) differentiable at points satisfying natural restrictions, and so therefore are compositions of these mappings (e.g., the inverse of the product-integral of the ordinary integral of one element of $D[0, \tau]$ with respect to another).

Proposition II.8.4 (The pth Quantile). *Let E consist of all nondecreasing elements of $B = D[0, \tau]$, and let $\phi(x) = x^{-1}(p) = \inf\{t\colon x(t) \geq p\}$ for a fixed $p \in \mathbb{R}$. Suppose $\theta \in E$ is such that $0 < \theta^{-1}(p) < \tau$ and θ is ordinarily differentiable at $\theta^{-1}(p)$. Then ϕ is tangentially differentiable at θ, taking $H = \{h \in B\colon$ h is continuous at $\theta^{-1}(p)\}$, with derivative*

$$\mathrm{d}\phi(\theta)\cdot h = -\frac{h(\theta^{-1}(p))}{\theta'(\theta^{-1}(p))}.$$

Proposition II.8.5 (The Inverse). *Let E be as in Proposition* II.8.4 *and let $\phi(x) = x^{-1}$ restricted to a fixed interval $[p_0, p_1]$. Suppose $\theta \in E$ is such that θ is (ordinarily) continuously differentiable on $[\theta^{-1}(p_0) - \varepsilon, \theta^{-1}(p_1) + \varepsilon]$ with positive derivative there, for some $\varepsilon > 0$. Then ϕ is tangentially differentiable at θ, taking H to be $C[0, \tau]$, the subset of continuous functions. The derivative is given by*

$$\mathrm{d}\phi(\theta)\cdot h = -\left(\frac{h}{\theta'}\right) \circ \theta^{-1}$$

restricted to $[p_0, p_1]$.

In both of the preceding propositions, ϕ was only defined on a subset of $D[0, \tau]$, but Lemma II.8.3 is applicable to extend it in some way to all of $D[0, \tau]$. A similar remark must be made for all our following examples.

Proposition II.8.6 (Integration). *Let $E = D[0, \tau] \times E_M$ where $E_M \subseteq D[0, \tau]$ is the set of functions of total variation bounded by the constant M. Let*

$$\phi\colon E \to D[0, \tau]$$

be defined by $\phi(x, y) = \int x \,\mathrm{d}y$. Then ϕ is compactly differentiable at a point (x, y) of E such that x is of bounded variation too, with

$$\mathrm{d}\phi(x, y)\cdot(h, k) = \int h \,\mathrm{d}y + \int x \,\mathrm{d}k,$$

where the last integral is defined by the integration by parts formula (k may not be of bounded variation)

$$\int x \,\mathrm{d}k = xk - x(0)k(0) - \int k_- \,\mathrm{d}x.$$

A similar *continuity* property of this mapping holds under less stringent conditions: If $(x_n, y_n) \in E$ is a sequence converging (in supremum norm) to $(x, y) \in E$, where x need not be of bounded variation, then $\int x_n \,\mathrm{d}y_n \to \int x \,\mathrm{d}y$ in supremum norm. This result is a version of the classical *Helly–Bray lemma*; see Gill (1989, Lemma 3 and subsequent discussion).

The following result on product-integration (Definition II.6.1) is actually a consequence of the Duhamel equation (2.6.4).

Proposition II.8.7 (Product-Integration). *Let $E_M^{k^2} \subseteq (D[0,\tau])^{k^2}$ be the set of $k \times k$ matrix cadlag functions with components of total variation bounded by the constant M. Let ϕ: $E_M^{k^2} \to (D[0,\tau])^{k^2}$ be defined by*

$$\phi(\mathbf{X}) = \prod (\mathbf{I} + d\mathbf{X}).$$

Then ϕ is compactly differentiable at each point of $E_M^{k^2}$ with derivative $d\phi(\mathbf{X})\cdot\mathbf{H}$ given by

$$(d\phi(\mathbf{X})\cdot\mathbf{H})(t) = \int_{s\in[0,t]} \prod_{[0,s)} (\mathbf{I} + d\mathbf{X})\mathbf{H}(ds) \prod_{(s,t]} (\mathbf{I} + d\mathbf{X}), \tag{2.8.2}$$

where the last integral is defined by application (twice) of the integration by parts formula.

For completeness, the integration by parts formula applied to $\int \prod (\mathbf{I} + d\mathbf{X})\, d\mathbf{Z}$ and to $\mathbf{Z} = \int d\mathbf{H} \prod (\mathbf{I} + d\mathbf{X})$ gives

$$(d\phi(\mathbf{X})\cdot\mathbf{H})(t) = \prod_{(0,t]} (\mathbf{I} + d\mathbf{X})\mathbf{H}(t)$$
$$+ \int_{s\in[0,t]} \prod_{[0,s)} (\mathbf{I} + d\mathbf{X})(\mathbf{X}(ds)\mathbf{H}(s) - \mathbf{H}(s)\mathbf{X}(ds)) \prod_{(s,t]} (\mathbf{I} + d\mathbf{X}),$$

showing that $d\phi(\mathbf{X})$ really is a continuous linear map from $(D[0,\tau])^{k^2}$ to itself. A version of this formula can be found in Fleming (1978b).

One can also obtain a simple continuity result from the Duhamel equation: If $\mathbf{X}_n$, $\mathbf{X}$ in $E_M^{k^2}$ are such that $\mathbf{X}_n \to \mathbf{X}$ in supremum norm, then $\prod (\mathbf{I} + d\mathbf{X}_n) \to \prod (\mathbf{I} + d\mathbf{X})$ in supremum norm too.

Finally, a result on composition:

Proposition II.8.8 (Composition). *Let $E \subseteq (D[0,\tau])^2$ be the set of nondecreasing functions on $[0,\tau]$ and let $\theta = (x, y)$ be a fixed point in E such that $0 < y(0) \leq y(\tau) < \tau$ and such that x is (ordinarily) continuously differentiable on $[0,\tau]$. Then the mapping $(x, y) \mapsto x \circ y$ from E to $D[0,\tau]$ is compactly differentiable tangentially to $C[0,\tau] \times D[0,\tau]$ at θ with derivative*

$$d\phi(x, y)\cdot(h, k) = h \circ y + x' \circ y \cdot k.$$

The composition and inverse mappings occur when plotting one empirical function versus another as in a P-P plot for instance; a plot of $Y(t)$ against $X(t)$ for t in some interval is actually a plot of $Y(X^{-1}(p))$ against p for certain p. Product-integration occurs when computing survival functions from hazard functions. One can imagine elaborate examples such as the quantile residual lifetime function seen as a functional of the cumulative hazard function. (The residual lifetime distribution of a survival time T is actually a family of distributions, namely, of the residual lifetime $T - t$ conditional on $T > t$ for each $t > 0$.)

II.9. Bibliographic Remarks

Martingale theory has a long history and plays a fundamental role in all parts of modern probability theory. A first appearance seems to have been in the paper by Bernstein (1926) under the name *variables enchaînées*. In fact, this paper is devoted to central limit theory and, in particular, its generalization to sums of *dependent* random variables; in particular, Markov chains (*chaînes simples*) were considered. Bernstein's (1926) theorem (§9, Fundamental Lemma) can be considered a first central limit theorem for (discrete-time) semimartingales, deriving asymptotic normality from convergence of the predictable characteristics of the process. Lévy (1935) also studied martingales as part of his path-breaking work on central limit theory, and again proved a central limit theorem for discrete-time martingales (still under the name *variables enchaînées*).

The *word* martingale has a much longer prehistory. Leaving early equestrian, nautical, and Rabelaisian uses aside (see Rabelais, 1542), it was first used in an applied probability context to denote a betting system at Monte Carlo in which one counts on a long run of one color being very likely to be compensated soon by a run of the other color. The essence of the mathematical concept of martingale is, however, that this does not work: In a fair game, whatever betting system one uses (dependent on past observation), the fact is not altered that your expected winnings remain zero; in other words, integration of a predictable process with respect to a martingale produces a martingale. A version of this result was already given by Bernstein (1926, §11, Example).

The same idea was the basis of Richard von Mises' notion of a collective as a foundation for probability. A collective is a sequence such that whatever predictable rule is used to select subsequences, the law of large numbers remains valid for them. The word martingale was introduced to mathematics by Ville (1939) in part of a research program to discredit the theory of collectives. Using martingale theory, Ville showed that collectives exist for which the relative frequency of (say) ones in initial segments of the sequence *always* exceeds its limiting value; in other words, the object that was supposed to form the basis of probability theory did not itself satisfy the theorems of probability. This was just what the mathematical community had been waiting for, and the von Mises theory was abandoned in favor of the Kolmogorov axiomatization, though neither Kolmogorov nor von Mises saw the relevance of the example and something like collectives reappeared later in Kolmogorov's complexity theory. [See van Lambalgen (1987a,b) for an historical and foundational analysis. Kolmogorov and von Mises were interested in understanding randomness, not describing probability.]

Section II.1

Our heuristic description of the theory has developed from Gill (1984) and Andersen and Borgan (1985).

Sections II.2 and II.3

The pioneer of martingale theory, both in discrete and continuous time, was Doob (1940, 1953), who also gave the fundamental optional stopping theorem and the almost sure convergence theorem "for processes having the E-property" as they were first called.

Stochastic integration was pioneered by Itô (1944) [following the earlier work by Wiener (1923)] who gave a meaning to $\int_0^\tau H\,dW$ when W is a Wiener process (or Brownian motion) and H a random function; because the paths of W are extremely erratic, this integral cannot be defined in a pathwise sense (unless H is so nice that integration by parts can be used). The construction of $\int H\,dW$ used some basic properties of W connected to what later became known as its predictable variation process: Not only is W a martingale, but so also is $W^2 - \iota$. This implies that $\langle W \rangle = \iota$, or the compensator of the submartingale W^2 is the identity function ι.

This observation, already made by Doob, motivated Meyer and his co-workers (Dellacherie, Doléans-Dade, and others) in Strasbourg in their search for the right conditions for what became known as the Doob–Meyer decomposition (Meyer, 1962), which they subsequently used to build a theory of stochastic integration. The story is long and complex and our remarks here obviously cannot do it justice.

Dellacherie's (1972) section theorem dealt with the highly delicate and fundamental measurability issues. Kunita and Watanabe (1967) furnished the idea of predictable covariation. The theory of stochastic integration was laid out by Meyer (1967) and put into its final form by Meyer (1976) [complete with the notions of *local martingale* going back to Itô and Watanabe (1965) and *semimartingale*, introduced by Doléans-Dade and Meyer (1970)].

Other celebrated results are Itô's (1951) formula, the theory of Doléans-Dade's (1970) exponential semimartingale, the characterization due independently to Dellacherie (1980) and to Bichteler (1979) of semimartingales as the largest class of processes for which a sensible theory of stochastic integration is possible, the so-called Burkholder–Davis–Gundy inequality, and so on. We do not make use of these results in this volume.

Theories of stochastic integration have also been developed from slightly different points of view by Métivier and Pellaumail (1980) among others.

Nowadays, the standard and complete work on continuous-time stochastic process theory, martingale theory, and stochastic integration is the book by Dellacherie and Meyer (1982), the second in a series of books covering many other topics at great depth. The theory was developed through alternative routes by Protter (1990) and by von Weizsäcker and Winkler (1990). Useful summaries were given by Jacod and Shiryayev (1987) and Daley and Vere-Jones (1988). In particular, for each topic, the book by Jacod and Shiryayev (1987) explicitly shows how the results simplify on specializing to the discrete-time case. The recent book by Liptser and Shiryayev (1989) also gives an excellent more extensive treatment of the theory, including central limit theory.

The hardest parts of the mathematical theory are contained in the Doob–Meyer decomposition and in Dellacherie's section theorems. If one only needs to apply the theory to a restricted class of applications, it may be possible to avoid some of these aspects. Jacobsen (1982), Fleming and Harrington (1991), and others each succeed in giving more elementary derivations of parts of the theory. Protter (1990), by starting from the Dellacherie–Bichteler characterization of semimartingale, was able to give most of the theory without recourse to the very heavy apparatus usually involved at the start of the "standard approach." Brown (1988), on the other hand, gave a complete proof of the Doob–Meyer decomposition using only more elementary tools. This proof was based on the idea of discretization and passage to the continuous limit (in discrete time, the Doob-Meyer decomposition is trivial).

We have made some minor adjustments to the usual definitions to suit our applications. As was explained in detail by Jacod (1979) or more briefly by Jacod and Shiryayev (1987, section I.1, especially Lemma I.1.19), it is not necessary to make the usual assumption of completeness of the filtration, which would otherwise be troublesome for statistical applications in which a whole family of probability measures is involved. Von Weiszäcker and Winkler (1990), in fact, showed how the whole theory can be set up without this assumption.

Our definition of localization is chosen to give the right meaning, both when the time interval $\mathscr{T}$ is open on the right and when it is closed on the right. Some other definitions also deviate from the usual ones to take care of the same possibilities. In our applications, martingales are nearly always local square integrable and of bounded variation (integrals of predictable processes with respect to counting process martingales), and in our survey of the theory, we concentrate on this case. Semimartingales are hardly needed and we have therefore neglected them, though in the modern theory they form the most fundamental object. The expert will also miss predictable stopping times from our survey.

Section II.4

The idea of studying counting processes (or point processes) through their stochastic intensity was taken up seriously by Snyder (1972, 1975) with a view toward engineering applications in communication theory: filtering, prediction, smoothing, and so on (a different kind of filtering than the sort we study in Sections II.4.4 and III.4). A breakthrough came with the Berkeley thesis of Brémaud (1972) who realized that the previously rather heuristic notion of intensity could be made completely rigorous and connected to very powerful mathematical tools through noting that its operative value lies in the fact that *the integrated intensity process of the counting process coincides with its compensator.* The idea was taken up and further extended by Dolivo (1974), Boel, Varaiya, and Wong (1975a,b), and van Schuppen and Wong (1974) among others. About the same time Jacod, spending a year in Princeton, came across Brémaud's thesis and wrote his (1973, 1975) papers connecting compensators and likelihood functions. The statistical implications of the theory was first realized by Aalen whose (1975) Berkeley thesis is the inspiration of this work; see Sections I.1 and I.2. The monograph of Jacobsen (1982) represents the first appearance of such material in book form. The book by Fleming and Harrington (1991) is the first extensive treatment aimed at applications in survival analysis. For surveys of counting process theory, see, for instance, Brémaud and Jacod (1977), Brémaud (1981), Karr (1986), and Daley and Vere-Jones (1988).

The idea of filtering a counting process in the sense used here is due to Aalen (1975). The notion of starting a counting process was introduced by Andersen et al. (1988) with help from M. Jacobsen. Fundamental results on the structure of point process filtrations were given by Courrège and Priouret (1965).

Section II.5

As we indicated above, the history of martingales and the history of the central limit theorem are closely intertwined. After Bernstein's (1926) and Lévy's (1935) early work, the martingale central limit theorem for discrete-time martingales was taken up by Billingsley (1961), Brown (1971), Dvoretsky (1972), and McLeish (1974), among many others. Aalen's (1975) thesis [see also Aalen (1977)] contained a continuous-time martingale central limit theorem for counting process martingale integrals built on McLeish's theorem. This line of work was continued by Helland (1982, 1983) and Fleming and Harrington (1991). Rebolledo (1979) gave the first really general continuous-time martingale central limit theorem; see also Rebolledo (1980a). One of the key tools in his approach was Lenglart's (1977) inequality, which we also use time and time again. Rebolledo (1979) used his own tightness criterion, but a more powerful one was given by Aldous (1978a).

Independently of Rebolledo, authors such as Jacod and Memin (1980), Kłopotowski (1980), and Liptser and Shiryayev (1980), worked on the central limit problem for semimartingales, with the more general ultimate aim to develop theorems with as limiting process *any* semimartingale (not just Gaussian limits) and to develop *necessary* as well as sufficient conditions [for this latter, see also Rebolledo (1980b)]. The developments were surveyed by Shiryayev (1981). This line of work culminated (for the time being, at least), in the book by Jacod and Shiryayev (1987).

An interesting contribution was made by Aldous (1978b), unfortunately never published, who gave a weak convergence theory in which filtration-related properties converge too (e.g., predictable processes converge to predictable processes, martingales to martingales, and so on).

We present the version of the martingale central limit theorem relevant for our applications: the case of a vector of local square integrable martingales.

Time transformations for counting processes were treated by Aalen (1975) and Aalen and Hoem (1978). Independently, Kurtz (1983) used the idea of time transformation of a counting process to a Poisson process to obtain very powerful laws of large numbers and central limit theorems. The conditions for convergence in probability of integrals were proposed by Helland (1983) and Gill (1983b), at the same meeting.

Section II.6

Product-integration was introduced by Volterra (1887) as part of the theory of differential equations. Though its significance in probability and statistics was seen by such authors as Arley (1943) and Dobrushin (1953), it somehow never quite became a well-known theory, and it has been reinvented many times. Cox (1972) gives a side reference to product-integration (which he was aware of through Arley's contribution).

Johansen (1977, reprinted 1986) gave a treatment of product-integration aimed at applications in Markov processes. The monograph on product-integration by Dollard and Friedman (1979), though a mine of information on many aspects of the theory, neglects the continuous–discrete interplay which is so important for our applications. The presentation here is based on Gill and Johansen (1990) who gave complete proofs and many further results. Our notation for product-integral (a script capital letter pi) is due to them.

Readers familiar with stochastic analysis will recognize the product-integral as Doléans–Dade's exponential semimartingale: the solution $Y = \mathscr{E}(X)$ of the stochastic differential equation $\mathrm{d}Y(t) = Y(t-)\,\mathrm{d}X(t)$, with initial condition $Y(0) = 1$.

Section II.7

Likelihood representations for general counting process models were first given by Jacod (1973, 1975). Use of product-integral notation to make the otherwise rather involved formulas more transparent goes back to Johansen (1983). Partial likelihood was invented by Cox (1975) to retrospectively justify the statistical methods he had proposed (Cox, 1972) for analyzing his regression model for censored survival times; see Section VII.2. Kalbfleisch and Prentice (1980) used the idea to informally discuss likelihood constructions for survival data. Our discussion builds on the work of Arjas and Haara (1984) who were the first to formulate the notions of noninformative and independent censoring rigorously, by formulating these notions in terms of likelihoods and intensity processes of general marked point processes. Arjas (1989) and Arjas, Haara, and Norros (1992) developed the ideas further.

Asymptotic statistical theory for partial likelihood was developed by Wong (1986). Slud (1991) derived our "continuous" partial likelihoods rigorously as limits of partial likelihood based on factorizations of an experiment into a finite number of terms. Jacod (1987, 1990a,b), inspired by an early manuscript version of the present material, shows how asymptotic properties of partial likelihoods to some extent have the same striking consequences as those of full likelihood (contiguity and local asymptotic normality; see Chapter VIII).

Section II.8

The material on the functional delta-method given here is largely taken from the paper by Gill (1989), whose own work was inspired by Reeds' (1976) Harvard thesis. The word delta-method is due to C.R. Rao. The ideas of the functional delta-method go back to von Mises (1947), and many other authors built on his work. Reeds (1976) was the first to point out that an elegant von Mises-type theory could be based on the notions of Hadamard or compact differentiability, rather than the Gâteaux- (directional differentiation) or Fréchet- (bounded differentiability) based theories favored by other authors. Incidentally, both Fréchet and Gâteaux were pupils of Hadamard (another famous pupil being P. Lévy who also made some contributions to this area). The idea of compact differentiation can be found in the early work of Hadamard, but it was Fréchet (1937) who recognized its significance and named it after his teacher, after first having developed his own notion of differentiation.

Reeds' (1976) work, not easily available, has been largely neglected or criticized, partly because of his rather polemic style. His main result, a compact differentiability-based proof of the asymptotic normality of M-estimators, was corrected by Heesterman and Gill (1992). Further developments were given by Sheehy and Wellner (1990a,b) using an even further generalized weak convergence theory (van der Vaart and Wellner, 1990, Pollard, 1990).

Section II.9

Historical and bibliographic surveys of stochastic analysis can be found in Dellacherie and Meyer (1982), Liptser and Shiryayev (1989), and Protter (1990). For counting processes, see Brémaud (1981); for product-integration, see Gill and Johansen (1990); for the functional delta-method, see Reeds (1976).

CHAPTER III

Model Specification and Censoring

In Section I.3, several examples of life history data were presented as a concrete motivation for the later development and analysis of statistical models for event history data. Chapter II presented the mathematical background for the study of such models, and concepts like counting processes, marked point processes, filtrations, compensators, likelihoods, and martingales were introduced. In the present chapter, we shall use these concepts repeatedly for the purpose of *specifying* statistical models for life history data.

Thus, one major purpose of this chapter is to provide the reader with a series of concrete models based on counting processes to have in mind when, in later chapters, we discuss statistical inference for general counting processes. Some of the models are specifically designed for the practical examples introduced in Section I.3 and will be used as the basis for analysis in later chapters.

Though, as in these examples, incomplete information of various kinds is inevitable in working with life history data, we shall first in Section III.1 consider models for *completely observed data.* Several such examples are given and the *multiplicative intensity model* of Aalen (1975, 1978b) is introduced. Another major point of this chapter is to give a thorough discussion of patterns of incomplete observation like censoring, truncation, and filtering within the counting process framework, and, in Section III.2, the most common form of incomplete information, *right-censoring*, is introduced. Various models for right-censoring mechanisms and the concepts of *independent* censoring and *noninformative* censoring are discussed. Section III.3 contains discussion of *left-truncation.* In Section III.4, the concept of *filtering* is introduced and more *general patterns of censoring* are discussed, also in connection with left-truncation. Often, the statistician is not interested in specifying a model for the whole system under consideration including the distribution of covariates and the censoring mechanism (or he or she is simply unable to do this), and, therefore, in Section III.5, possibilities of analyzing *partially specified* models are discussed. One of the main points of Sections III.2–III.5 will be to discuss conditions under which important features of the original model are *preserved* under the various schemes of incomplete observation.

In general, we shall consider a multivariate counting process

$$\mathbf{N} = (\mathbf{N}(t), t \in \mathscr{T})$$

defined on a measurable space $(\Omega, \mathscr{F})$ equipped with a filtration $(\mathscr{F}_t)$ to which $\mathbf{N}$ is adapted. Recall from Section II.2 that $\mathscr{T}$ is either $[0, \tau)$ or $[0, \tau]$ with $\tau \leq \infty$. We consider a doubly indexed family

$$\mathscr{P} = \{\mathrm{P}_{\theta\phi}: (\theta, \phi) \in \Theta \times \Phi\}$$

of probability measures on $(\Omega, \mathscr{F})$. The filtration $(\mathscr{F}_t)$, together with each of these probability measures $\mathrm{P}_{\theta\phi}$, satisfies the "usual conditions" (2.2.1) except possibly for completeness.

The interpretation of the parameters is that θ is the parameter of interest parametrizing the transition intensities for the events under study, whereas ϕ is a *nuisance parameter* typically parametrizing the distribution of censoring and covariates. In some cases, there are no nuisance parameters ϕ. The $P_{\theta\phi}$-compensator of $\mathbf{N}$ with respect to $(\mathscr{F}_t)$ is denoted $\mathbf{\Lambda}^\theta(t)$ or $\mathbf{\Lambda}(t, \theta)$, thereby assuming that it does not depend on ϕ. The sets Θ and Φ may be subsets of either finite-dimensional Euclidean spaces, corresponding to parametric models, or of more general function spaces, corresponding to nonparametric models.

III.1. Examples of Counting Process Models for Complete Life History Data. The Multiplicative Intensity Model

In this section, we shall see how the counting process $\mathbf{N}$ and the filtration $(\mathscr{F}_t)$ may be constructed in several examples when there is no censoring. We will either have $\mathscr{F}_t$ as the *self-exciting filtration*

$$\mathscr{N}_t = \sigma\{\mathbf{N}(u), 0 \leq u \leq t\}$$

or

$$\mathscr{F}_t = \mathscr{F}_0 \vee \mathscr{N}_t$$

with $\mathscr{F}_0$ generated by a random variable $\mathbf{X}_0$ which can be thought of as being realized at time 0; see Section II.4.1. Thus, the observations available to the researcher at time t consist of $(\mathbf{N}(u), 0 \leq u \leq t)$ and (when relevant) $\mathbf{X}_0$. We shall often say "the observations at time t are the σ-algebra $\mathscr{F}_t$," thereby meaning that at time t it can be determined whether or not any event $A \in \mathscr{F}_t$

has occurred or equivalently at time t the random variables that generate $\mathscr{F}_t$ can be observed.

III.1.1. Models for Survival Data

We first consider the simplest possible model corresponding to the *one-jump process* already discussed in Example II.4.1.

EXAMPLE III.1.1. A Single Non-negative Random Variable

Let X be a non-negative random variable with absolutely continuous distribution function F, survival function $S = 1 - F$, density f, and hazard rate function $\alpha = f/S$. We assume that the distribution of X depends on a (finite- or infinite-dimensional) parameter θ and write $\alpha^\theta(t)$ or $\alpha(t, \theta)$ for the hazard function, $F^\theta(t)$ or $F(t, \theta)$ for the distribution function, etc. We take $\tau = \infty$ and $\mathscr{T} = [0, \infty]$. Then, by Example II.4.1, the stochastic process N defined by

$$N(t) = I(X \leq t) \tag{3.1.1}$$

is a univariate counting process with compensator

$$\Lambda(t, \theta) = \int_0^t \alpha(u, \theta) Y(u) \, du$$

with respect to the self-exciting filtration $(\mathscr{F}_t) = (\mathscr{N}_t)$ and the probability P_θ corresponding to the distribution F^θ of X. Here,

$$Y(t) = I(X \geq t) = 1 - N(t-). \tag{3.1.2}$$

The likelihood for θ based on observation of N on $\mathscr{T}$ is

$$L(\theta) = \prod_{t \in \mathscr{T}} \{(\alpha(t, \theta) Y(t))^{\Delta N(t)} (1 - \alpha(t, \theta) Y(t) \, dt)^{1 - \Delta N(t)}\},$$

cf. (2.7.4′), which reduces to

$$L(\theta) = \prod_t (\alpha(t, \theta) Y(t))^{\Delta N(t)} \exp\left(-\int_0^\infty \alpha(u, \theta) Y(u) \, du\right)$$

$$= \alpha(X, \theta) \exp\left(-\int_0^X \alpha(t, \theta) \, dt\right) = \alpha(X, \theta) S(X, \theta) = f(X, \theta),$$

the density function $f(\cdot, \theta)$ evaluated at X. □

We next study models for survival data as exemplified in Section I.3.1.

EXAMPLE III.1.2. Uncensored Survival Data

Let $X_1, \ldots, X_n$ be independent non-negative random variables, X_i having hazard function $\alpha_i(t, \theta)$. As in Example III.1.1, we let $\tau = \infty$ and $\mathscr{T} = [0, \infty]$, and define for each $i = 1, \ldots, n$ stochastic processes $N_i(t)$ and $Y_i(t)$ by (3.1.1) and (3.1.2).

Identifying i with an "individual" and X_i with the "survival time" or "failure time" of that individual, then N_i counts 1 only at the time X_i when individual i dies, and $Y_i(t) = 1$ if individual i is still alive or at risk just before time t.

Obviously, $\mathbf{N} = (N_1, \ldots, N_n)$ is a multivariate counting process with respect to the self-exciting filtration $(\mathcal{N}_t)$ generated by $\mathbf{N}$. By the independence of the X_i, it then follows from the product construction given in Section II.4.3 that N_i has intensity process $(\alpha_i(t, \theta) Y_i(t))$ and compensator

$$\Lambda_i(t, \theta) = \int_0^t \alpha_i(u, \theta) Y_i(u) \, du$$

with respect to $(\mathcal{N}_t)$ and P_θ, the joint distribution of the X_i. (Alternatively this result may be derived directly from Jacod's representation Theorem II.7.1.) This shows that N_i has the same compensator as in the previous example but now with respect to a larger filtration. The likelihood for $\mathbf{N}$ is by (2.7.4')

$$L(\theta) = \prod_{t \in \mathcal{T}} \left\{ \prod_{i=1}^n (\alpha_i(t, \theta) Y_i(t))^{\Delta N_i(t)} \left(1 - \sum_{i=1}^n \alpha_i(t, \theta) Y_i(t) \, dt \right)^{1 - \Delta N_\cdot(t)} \right\},$$

where

$$N_\cdot(t) = \sum_{i=1}^n N_i(t).$$

Similar to Example III.1.1, the likelihood

$$\begin{aligned} L(\theta) &= \prod_{i=1}^n \alpha_i(X_i, \theta) \exp\left(- \sum_{i=1}^n \int_0^{X_i} \alpha_i(u, \theta) \, du \right) \\ &= \prod_{i=1}^n \alpha_i(X_i, \theta) S_i(X_i, \theta) \end{aligned}$$

reduces to the product of the density functions $f_i(\cdot, \theta)$ evaluated at the X_i.

When $X_1, \ldots, X_n$ are independent and identically distributed (i.i.d.) with hazard function $\alpha(\cdot, \theta)$, the likelihood reduces to

$$L(\theta) = \prod_{t \in \mathcal{T}} \left\{ \prod_{i=1}^n (\alpha(t, \theta) Y_i(t))^{\Delta N_i(t)} (1 - \alpha(t, \theta) Y_\cdot(t) \, dt)^{1 - \Delta N_\cdot(t)} \right\},$$

where

$$Y_\cdot(t) = \sum_{i=1}^n Y_i(t) = n - N_\cdot(t-).$$

Thus, in this case, because the Y_i are indicator processes, $L(\theta)$ is equal to

$$\prod_{t \in \mathcal{T}} \{\alpha(t, \theta)^{\Delta N_\cdot(t)} (1 - \alpha(t, \theta)(n - N_\cdot(t-)) \, dt)^{1 - \Delta N_\cdot(t)}\},$$

showing that the *aggregated* process $N_\cdot$ is *sufficient* for θ. This corresponds to the fact that the ordered observations $X_{(1)} \le X_{(2)} \le \cdots \le X_{(n)}$ are sufficient. We may also say that the self-exciting filtration after aggregation is sufficient for θ. Note that, in the i.i.d. case, $N_{\bullet}$ is a univariate counting process with intensity process

$$\lambda_{\bullet}(t, \theta) = \alpha(t, \theta) Y_{\bullet}(t)$$

with respect to $(\mathscr{N}_t)$ and P_θ. Thus, the counting process $N_{\bullet}$ obtained by *aggregation* of the *individual* counting processes N_i, each having a multiplicative intensity, is a univariate counting process with the same intensity structure. The intensity process for the aggregated counting process $N_{\bullet}$ is a product of an *individual intensity* $\alpha^\theta(t)$ and a process $Y_{\bullet}(t)$ which can be interpreted as the *number of individuals at risk* for failing just before time t. □

EXAMPLE III.1.3. A Model for Relative Mortality

Let, as in Example III.1.2, $X_1, \ldots, X_n$ be independent non-negative random variables and assume that the distribution of X_i is absolutely continuous with hazard rate function $\mu_i(t)\alpha_0(t, \theta)$. Here $\mu_i(\cdot)$ is assumed to be a *known* hazard rate function, for instance a population-based quantity known from vital statistics, and $\alpha_0(\cdot, \theta)$ is an unknown (time- or age-dependent) *relative mortality* common to all i. Now $\mathbf{N} = (N_1, \ldots, N_n)$ is a multivariate counting process, N_i having intensity process given by

$$\lambda_i(t, \theta) = \alpha_0(t, \theta)\mu_i(t) I(X_i \ge t), \quad i = 1, \ldots, n,$$

with respect to $(\mathscr{N}_t)$ and P_θ. In this case, the likelihood is proportional to

$$\prod_{t \in \mathscr{T}} \{\alpha_0(t, \theta)^{\Delta N_{\bullet}(t)} (1 - \alpha_0(t, \theta) Y_{\bullet}^{\mu}(t)\, dt)^{1 - \Delta N_{\bullet}(t)}\},$$

where

$$Y_{\bullet}^{\mu}(t) = \sum_{i=1}^{n} \mu_i(t) I(X_i \ge t).$$

Thus, the pair $(N_{\bullet}, Y_{\bullet}^{\mu})$ is sufficient for θ, whereas the self-exciting filtration after aggregation is not. By aggregation, a univariate counting process $N_{\bullet} = N_1 + \cdots + N_n$ is obtained with intensity process $\lambda_{\bullet}(t, \theta) = \alpha_0(t, \theta) Y_{\bullet}^{\mu}(t)$. Thus, again, the intensity process for the aggregated counting process $N_{\bullet}$ has a multiplicative form, but in this case $Y_{\bullet}^{\mu}(t)$ is no longer simply the number at risk for failing at t. □

III.1.2. Markov Process Models

We next consider a general Markov process model which is fundamental both for later theoretical examples (e.g., Examples IV.1.9 and IV.2.2 and the examples in Section IV.4) and for our later analysis of the data presented in

Examples I.3.8–I.3.13. We return to the practical examples later where the concrete models are further specified (Examples IV.1.4, IV.4.1–IV.4.4, and V.2.8).

EXAMPLE III.1.4. A Finite-State Markov Process

Let $(X(t), t \in \mathscr{T})$ be a Markov process with finite state space S and right-continuous sample paths and suppose that the initial distribution, i.e., the distribution of $X(0) = X_0$, say, depends on parameters ϕ (and possibly on θ too).

We shall assume the existence of locally integrable *transition intensities* $\alpha_{hj}^{\theta}(t) = \alpha_{hj}(t, \theta)$ from state h to state j, $h \neq j$; see Theorem II.6.7 and the discussion after this theorem. Note that in a given model some $\alpha_{hj}^{\theta}(\cdot)$ may be zero for all values of θ; see, for instance, Example III.1.5. Then

$$\mu_h^{\theta} = \sum_{j \in S} \alpha_{hj}^{\theta}$$

is the force of transition out of the state h, and when $\mu_h^{\theta}(t) = 0$ for all t (and θ), we say that the state h is *absorbing*.

Let $N_{hj}(t)$ be the number of direct transitions for X from h to j, $h \neq j$, in $[0, t]$. Then $\mathbf{N} = (N_{hj}(\cdot), h \neq j)$ and X_0 are equivalent to X in the sense that observation of $X(u)$, $0 \leq u \leq t$ gives the same data as observing X_0 and $\mathbf{N}$ on $[0, t]$. We define $(\mathscr{N}_t)$ to be the self-exciting filtration for $\mathbf{N}$ and let $\mathscr{F}_t = \mathscr{N}_t \vee \mathscr{F}_0$ with $\mathscr{F}_0$ generated by X_0. The $\mathrm{P}_{\theta\phi}$-intensity process for the multivariate counting process $\mathbf{N}$ with respect to $(\mathscr{F}_t)$ is now, by Theorem II.6.8,

$$\lambda_{hj}^{\theta\phi}(t) = \lambda_{hj}^{\theta}(t) = \alpha_{hj}^{\theta}(t) Y_h(t),$$

where $Y_h(t) = I(X(t-) = h)$ is the indicator for X being in the state h just before time t. Thus, the intensity process only depends on θ, and again we have a multiplicative intensity structure.

Next, assume that given X_{i0}, $i = 1, \ldots, n$, independent copies $X_1(\cdot), \ldots, X_n(\cdot)$ of $X(\cdot)$ are constructed with $X_i(0) = X_{i0}$; let $\mathbf{X}_0 = (X_{10}, \ldots, X_{n0})$ and define a multivariate counting process $\mathbf{N} = (N_{hji}, i = 1, \ldots, n;\ h \neq j)$ from $(X_1(\cdot), \ldots, X_n(\cdot))$ as above. Then, by the conditional independence of the $X_i(\cdot)$ and by the product construction (Section II.4.3), it is seen that N_{hji} has $P_{\theta\phi}$-intensity process

$$\lambda_{hji}^{\theta\phi}(t) = \lambda_{hji}^{\theta}(t) = \alpha_{hj}^{\theta}(t) Y_{hi}(t),$$

where $Y_{hi}(t) = I(X_i(t-) = h)$ is the indicator for X_i being in state h at time $t-$.

The multivariate counting process $\mathbf{N} = (N_{hji}, i = 1, \ldots, n;\ h, j \in S, h \neq j)$, therefore, has a multiplicative intensity process with respect to the filtration generated by all n Markov processes which only depends on θ. The likelihood, by (2.7.4′), takes the form

$$L(\theta, \phi) = L_0(\theta, \phi) L_\tau(\theta),$$

where

$$L_0(\theta, \phi) = \mathrm{P}_{\theta\phi}(\mathbf{X}_0)$$

and

$$L_\tau(\theta) = \prod_{t \in \mathcal{T}} \left\{ \prod_i \prod_{h \neq j} (\alpha_{hj}^\theta(t)\, Y_{hi}(t))^{\Delta N_{hji}(t)} \left(1 - \sum_i \sum_{h \neq j} \alpha_{hj}^\theta(t)\, Y_{hi}(t)\, \mathrm{d}t\right)^{1 - \sum_i \sum_{h \neq j} \Delta N_{hji}(t)} \right\}.$$

Then $L_\tau(\theta)$ equals

$$\prod_{t \in \mathcal{T}} \left\{ \prod_{h \neq j} \alpha_{hj}^\theta(t)^{\Delta N_{hj\cdot}(t)} \left(1 - \sum_{h \neq j} \alpha_{hj}^\theta(t)\, Y_{h\cdot}(t)\, \mathrm{d}t\right)^{1 - \sum_{h \neq j} \Delta N_{hj\cdot}(t)} \right\},$$

where

$$N_{hj\cdot}(t) = \sum_{i=1}^{n} N_{hji}(t)$$

and

$$Y_{h\cdot}(t) = \sum_{i=1}^{n} Y_{hi}(t),$$

the latter being a function of $\mathbf{X}_0$ and $(N_{hj\cdot}, h \neq j)$. Thus, if the distribution of $\mathbf{X}_0$ only depends on ϕ, then for each fixed $\phi \in \Phi$, $L_\tau(\theta)$ is the full likelihood for θ, otherwise $L_\tau(\theta)$ is only a *partial* likelihood. At any rate, $L_\tau(\theta)$ is the full *conditional* likelihood given the initial states $\mathbf{X}_0$, and $(\mathbf{X}_0, N_{hj\cdot}; h \neq j))$ is *sufficient* for θ, the second part being a multivariate counting process with intensity process

$$\lambda_{hj\cdot}^\theta(t) = \alpha_{hj}^\theta(t)\, Y_{h\cdot}(t).$$

Again, aggregation leads to a counting process with a multiplicative intensity structure where the first factor is an intensity on the individual level and the second is a process indicating the number at risk just before time t for experiencing events of given types. □

We next specialize to the model relevant for the examples in Section I.3.2.

EXAMPLE III.1.5. Competing Risks

As a very special case of Example III.1.4 one may consider two states 0 (alive) and 1 (dead) and assume $\alpha_{10}(t, \theta) \equiv 0$ (that is, state 1 is absorbing) and the initial distribution degenerate at 0. This yields the independent identically distributed uncensored survival times of Example III.1.2, with hazard function equal to the transition intensity $\alpha_{01}(t, \theta)$; see Figure I.3.1.

Another special case of the Markov process example is the *competing risks* model, obtained by considering one transient state 0 (alive) and absorbing states $h = 1, \ldots, k$ [so that $\alpha_{hj}(t, \theta) \equiv 0$ for $h = 1, \ldots, k$ and all j]; see Figure I.3.5. State h corresponds to "dead by cause h." The initial distribution is degenerate at 0 and the transition intensities $\alpha_{0h}(t, \theta)$, $h = 1, \ldots, k$, are termed "cause specific hazard functions."

It is easily seen that the competing risks model is equivalent to considering independent random variables $X_{i1}, \ldots, X_{ik}$, $i = 1, \ldots, n$, with hazard functions $\alpha_{01}(t, \theta), \ldots, \alpha_{0k}(t, \theta)$ and the multivariate counting process

$$\mathbf{N}(t) = (N_1(t), \ldots, N_k(t))$$

with

$$N_h(t) = \sum_{i=1}^{n} I\left(\min_l X_{il} = X_{ih} \leq t\right).$$

In reliability theory, $\min_l X_{il}$ is interpreted as the lifetime for a series system of k independent components with lifetimes $X_{i1}, \ldots, X_{ik}$. It has been debated extensively how useful this reliability interpretation is in biomedical contexts. In particular, even if the competing risks model may be generated from a set of independent "latent" (or "underlying") lifetimes, these are often hypothetical. References to a discussion of the interpretability and testability of the latent lifetime model include Cox (1959), Tsiatis (1975), Kalbfleisch and Prentice (1980, Chapter 7), and Cox and Oakes (1984, Chapter 9). □

III.1.3. The Multiplicative Intensity Model

In Examples III.1.1–III.1.5, the *individual* counting processes N_{hi} satisfy the *multiplicative intensity model*

$$\lambda_{hi}^{\theta}(t) = \alpha_{hi}^{\theta}(t) Y_{hi}(t), \quad h = 1, \ldots, k; i = 1, \ldots, n,$$

where $\alpha_{hi}^{\theta}(t)$ is deterministic and $Y_{hi}(t)$ is a *predictable* process which is *observable* in the sense that it does not depend on the parameter θ. In these examples, $\alpha_{hi}^{\theta}(t)$ is an *individual* force of transition, relative hazard, or hazard of *type* h and the process $Y_{hi}(t)$ contains information on whether or not individual i is at risk for experiencing an event of type h at time t. For instance (with a slight abuse of notation), h could correspond to a transition from one state to another in a Markov process.

When $\alpha_{hi}^{\theta} = \alpha_h^{\theta}$ for all i, the *aggregated counting process* $N_{h\cdot}$ satisfies the multiplicative intensity model

$$\lambda_h^{\theta}(t) = \alpha_h^{\theta}(t) Y_h(t), \quad h = 1, \ldots, k,$$

with the predictable and observable process $Y_h(t)$ giving the size of the *risk set* for that type of event just before time t [in Examples III.1.2 and III.1.4 but not in Example III.1.3, $Y_h(t)$ was simply the *number* at risk for a type h transition just before time t]. The multiplicative intensity model was introduced by Aalen (1975, 1978b) who discussed the nonparametric case (including its sufficiency and completeness properties) where no further assumptions on the α_h^{θ} are made. Estimation and testing in this model will be discussed in the next two chapters. Borgan (1984) treated models parametrized by $\boldsymbol{\theta} \in \mathbb{R}^q$ and that model (as well as other parametric counting process models) will be discussed in Chapter VI.

Another example of a multiplicative intensity model is the following.

EXAMPLE III.1.6. The Dynamics of a Natural Baboon Troop. A Markov Birth–Immigration–Death–Emigration Process

For the data presented in Example I.3.18, a possible model is the Markov birth–immigration–death–emigration (BIDE) process $\{X(t): 0 \leq t \leq \tau\}$, with $X(t)$ specifying the size of the troop at time t, and with intensities λ (birth), μ (death) ρ (emigration), and ν (immigration), that is,

$$\mathrm{P}(X(t+\varepsilon) = j \mid X(t) = h) = \begin{cases} (\lambda h + \nu)\varepsilon + o(\varepsilon), & j = h+1 \\ 1 - (\lambda h + \mu h + \rho h + \nu)\varepsilon + o(\varepsilon), & j = h \\ (\mu + \rho)h\varepsilon + o(\varepsilon), & j = h-1 \\ o(\varepsilon) & \text{otherwise.} \end{cases}$$

Let $N_1(t)$, $N_2(t)$, $N_3(t)$, and $N_4(t)$ be the number of births, the number of deaths, the number of emigrations, and the number of immigrations, respectively, in $[0, t]$, and let

$$R(t) = \int_0^t X(u)\,du$$

be the "total time at risk before t." Then $\mathbf{N}(t) = (N_1(t), N_2(t), N_3(t), N_4(t))$ is a multivariate counting process with intensity process $(\lambda X(t-), \mu X(t-), \rho X(t-), \nu)$. That is, the multiplicative intensity model is satisfied with the constant intensity functions λ, μ, ρ, and ν. The likelihood based on complete observation over $[0, \tau]$ is

$$\lambda^{N_1(\tau)} \mu^{N_2(\tau)} \rho^{N_3(\tau)} \nu^{N_4(\tau)} e^{-(\lambda+\mu+\rho)R(\tau) - \nu\tau}.$$

This provides an example of a counting process formulation of a Markov process with countable state-space. □

III.1.4. Regression Models

The next example presents a model for completely observed life history data for which aggregation does not lead to Aalen's multiplicative intensity model.

EXAMPLE III.1.7. Relative Risk Regression Models with Time-Independent Covariates

Let $(X_i, \mathbf{Z}_i)$, $i = 1, \ldots, n$, be random variables with X_i non-negative and each $\mathbf{Z}_i$ p-dimensional. We shall assume that $X_1, \ldots, X_n$ are conditionally independent given $\mathbf{Z} = (\mathbf{Z}_1, \ldots, \mathbf{Z}_n)$, that the marginal distribution of $\mathbf{Z}$ depends on parameters ϕ (and possibly θ too), and that the conditional distribution of X_i given $\mathbf{Z} = \mathbf{z} = (\mathbf{z}_1, \ldots, \mathbf{z}_n)$ has hazard function $\alpha_i(t, \theta)$ as in Example III.1.2. We shall consider models of the form

$$\alpha_i(t, \theta) = \alpha_0(t, \gamma) r(\boldsymbol{\beta}^\mathsf{T} \mathbf{z}_i) \tag{3.1.3}$$

with $\theta = (\gamma, \boldsymbol{\beta})$, $\boldsymbol{\beta} \in \mathbb{R}^p$, and the *relative risk* function $r(\cdot)$ being non-negative. The main example is the semiparametric *Cox regression model* (Cox, 1972) where r is the exponential function and γ is infinite dimensional, but parametric models with $\boldsymbol{\gamma} \in \mathbb{R}^q$ may also be considered. In any case, $\mathbf{Z}_1, \ldots, \mathbf{Z}_n$ are *covariates* upon which we want to condition and ϕ is a nuisance parameter. In some cases, it is reasonable to assume $(X_1, \mathbf{Z}_1), \ldots, (X_n, \mathbf{Z}_n)$ to be i.i.d., but when some of the X_i correspond to lifetimes of individuals from the same family or community (e.g., Example I.3.15), the above assumption of conditional independence of the X_i given $\mathbf{Z}$ is more realistic.

From X_i, $i = 1, \ldots, n$, we define stochastic processes $N_i(\cdot)$ and $Y_i(\cdot)$ from (3.1.1) and (3.1.2) and we let $(\mathscr{N}_t)$ be the filtration generated by the multivariate counting process $\mathbf{N} = (N_1, \ldots, N_n)$. Furthermore, we let $\mathscr{F}_0$ be generated by $\mathbf{Z}$, define $\mathscr{F}_t = \mathscr{F}_0 \vee \mathscr{N}_t$, and let $\mathrm{P}_{\theta\phi}$ be the probability measure corresponding to the distribution of $(X_i, \mathbf{Z}_i)$, $i=1, \ldots, n$. The $(\mathrm{P}_{\theta\phi}, (\mathscr{F}_t))$-compensator for $\mathbf{N}$ can now be found directly from Theorem II.7.1 or alternatively from the conditional independence version of the product construction (Section II.4.3).

This shows that the compensator for N_i only depends on θ and that it equals

$$\Lambda_i(t, \theta) = \int_0^t Y_i(u) \alpha_i(u, \theta)\, du.$$

The likelihood, by (2.7.4′), takes the form

$$L(\theta, \phi) = L_0(\theta, \phi) L_\tau(\theta)$$

where

$$L_0(\theta, \phi) = \mathrm{P}_{\theta\phi}(\mathbf{Z})$$

and

$$L_\tau(\theta) = \prod_{t \in \mathscr{T}} \left\{ \prod_{i=1}^n (\alpha_i(t, \theta) Y_i(t))^{\Delta N_i(t)} \left(1 - \sum_i \alpha_i(t, \theta) Y_i(t)\, dt\right)^{1 - \Delta N.(t)} \right\}.$$

If the distribution of $\mathbf{Z}$ does not depend on θ, then for each fixed $\phi \in \Phi$, $L_\tau(\theta)$ is the full likelihood for θ; otherwise, it is a partial likelihood and the full conditional likelihood given $\mathbf{Z}$. No sufficiency reduction of $(N_i(\cdot), \mathbf{Z}_i)$, $i = 1, \ldots, n$, is possible when $\boldsymbol{\beta}$ is unknown. It is easily seen that $L_\tau(\theta)$ is equal to

$$\prod_{i=1}^n \alpha_0(X_i, \gamma) r(\boldsymbol{\beta}^\mathsf{T} \mathbf{z}_i) \exp\left(-r(\boldsymbol{\beta}^\mathsf{T} \mathbf{z}_i) \int_0^{X_i} \alpha_0(u, \gamma)\, du\right),$$

the density for the conditional distribution of $\mathbf{X} = (X_1, \ldots, X_n)$ given $\mathbf{Z} = \mathbf{z}$, evaluated at $\mathbf{X}$.

Combining the present example with Example III.1.3, a regression model for the *relative* mortality is obtained (Andersen et al., 1985). Here the hazard rate function for X_i given $\mathbf{Z} = \mathbf{z}$ is

$$\alpha_i(t, \theta) = \alpha_0(t, \gamma)\mu_i(t)r(\boldsymbol{\beta}^\mathsf{T}\mathbf{z}_i),$$

where $\theta = (\gamma, \boldsymbol{\beta})$ and $\mu_i(\cdot)$ is *known*. Also, Example III.1.4 can be combined with the present one into regression models for the transition intensities in a Markov process of the form

$$\alpha_{hji}(t, \theta) = \alpha_{hj0}(t, \gamma)r(\boldsymbol{\beta}^\mathsf{T}\mathbf{z}_{hi}).$$

Here, *type-specific covariates* $\mathbf{Z}_{hi}$ may be defined from the vector of basic covariates $\mathbf{Z}_i$ for individual i, reflecting the fact that some of these basic covariates may affect the different transition intensities differently (e.g., Andersen and Borgan, 1985). We shall return to a more detailed discussion of multiplicative regression models where $\mathbf{Z}$ is observable in Sections VII.2 and VII.3.

Sometimes there may be unobservable covariates affecting the transition intensities or there may be covariates which one deliberately leaves out. Examples include family studies (e.g., Example I.3.15) where each member of a family shares the value of a common unobserved "frailty." Alternatively, in some studies, one may simply not have been able to record important covariates. We shall return to the analysis of models with unobserved covariates in Chapter IX. □

Next, two examples of *additive* intensity models will be mentioned.

EXAMPLE III.1.8. A Model for Excess Mortality

Consider, as in Examples III.1.2 and III.1.3, independent non-negative random variables $X_1, \ldots, X_n$, but assume now that X_i has hazard function

$$\alpha_i(t, \theta) = \alpha_0(t, \theta) + \mu_i(t),$$

where $\mu_i(t)$ is known (cf. Andersen and Væth, 1989). In this model, the intensity process for the aggregated counting process $N_{\bullet}$ is

$$\lambda_{\bullet}(t) = \alpha_0(t, \theta)Y_{\bullet}(t) + Y_{\bullet}^{\mu}(t),$$

and the likelihood for the multivariate counting process $\mathbf{N}$ is

$$L(\theta) = \prod_{t \in \mathscr{T}} \left\{ \prod_{i=1}^{n} ((\alpha_0(t, \theta) + \mu_i(t))Y_i(t))^{\Delta N_i(t)} \right.$$
$$\left. \times (1 - (\alpha_0(t, \theta)Y_{\bullet}(t) + Y_{\bullet}^{\mu}(t))\,\mathrm{d}t)^{1 - \Delta N_{\bullet}(t)} \right\}.$$

It is seen that no sufficiency reduction is possible. □

Example III.1.9. An Additive Regression Model

Consider $(X_i, \mathbf{Z}_i)$, $i = 1, \ldots, n$, as in Example III.1.7 but assume that $\alpha_i(t, \theta)$ is given by

$$\alpha_i(t, \theta) = \alpha_0(t, \theta) + \boldsymbol{\beta}(t, \theta)^{\mathsf{T}} \mathbf{z}_i \tag{3.1.4}$$

rather than by the multiplicative form (3.1.3). The model (3.1.4) is a special case of the matrix version of the multiplicative intensity model suggested by Aalen (1980) to which we return in Section VII.4. Defining

$$\mathbf{Y}_i(t) = I(X_i \geq t)(1, z_{i1}, \ldots, z_{ip}), \quad i = 1, \ldots, n,$$

and

$$\boldsymbol{\alpha}(t) = (\alpha_0(t), \beta_1(t), \ldots, \beta_p(t))^{\mathsf{T}},$$

the multivariate counting process $\mathbf{N}$ has intensity process

$$\boldsymbol{\lambda}(t) = \mathbf{Y}(t)\boldsymbol{\alpha}(t),$$

where $\mathbf{Y}(t)$ is the $n \times (p + 1)$ matrix with row i equal to $\mathbf{Y}_i(t)$.

Setting up the likelihood as in Example III.1.7, it is easily seen that no sufficiency reduction is possible. Finally, it may be noted that a regression model for excess mortality can be obtained by combining Example III.1.8 with the present one. □

III.1.5. Other Counting Process Models

In the previous examples we have reformulated well-known models for complete life history data in terms of multivariate counting processes. We shall conclude this section by some models which are conveniently formulated directly as counting processes using Theorem II.7.5.

Example III.1.10. A Model for Matings of Drosophila Flies

The mating experiment was described in Example I.3.16. Let $N(t)$ be the number of matings initiated in the interval $[0, t]$ and $F(t)$ and $M(t)$ the number of female and male flies, respectively, not yet having initiated a mating just before time t. Thus, $F(t) = f_0 - N(t-)$ and $M(t) = m_0 - N(t-)$, where $f_0 = F(0)$ and $m_0 = M(0)$ are the number of female and male flies, respectively, in the pornoscope. Let $\mathscr{N}_t = \sigma(N(u), 0 \leq u \leq t)$ be generated by $N(\cdot)$ on an interval $\mathscr{T}$. Then a model for a univariate counting process $N(\cdot)$ can be set up by assuming that for a given locally integrable function $\alpha^\theta(t)$ parametrized by some θ the $(\mathrm{P}_\theta, (\mathscr{N}_t))$-intensity process for $N(t)$ is

$$\lambda^\theta(t) = \alpha^\theta(t) F(t) M(t).$$

By Theorem II.7.5, a counting process with this intensity process exists and is unique on $(\mathscr{N}_t)$. Thus, we have another example of Aalen's multiplicative intensity model. The interpretation of $\alpha^\theta(t)$ is that of an individual mating

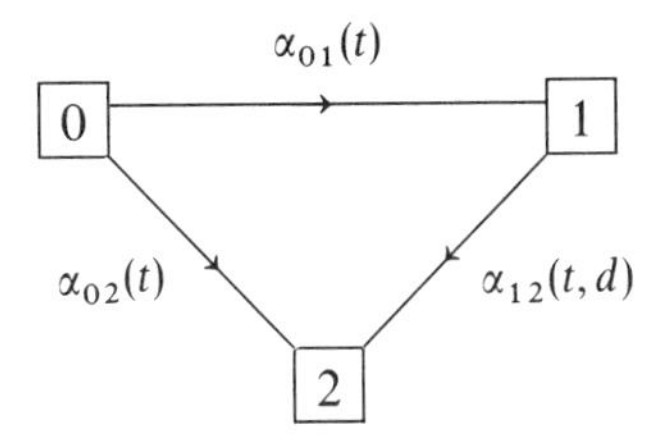

Figure III.1.1. An iliness–death process with duration dependence.

intensity because it is the intensity of $N(t)$ when $M(t) = F(t) = 1$. A more realistic and elaborate model for the mating pattern will be discussed in Example III.5.5. □

EXAMPLE III.1.11. An Illness–Death Process with Duration Dependence

Let states 0, 1, and 2 denote healthy, diseased, and dead, respectively, and define the counting process of transitions between these states by $\mathbf{N}(t) = (N_{01}(t), N_{02}(t), N_{12}(t))$, where $N_{0h}(t)$, $h = 1, 2$, has intensity process $\alpha_{0h}(t) Y_0(t)$ with $Y_0(t) = 1 - N_{01}(t-) - N_{02}(t-)$, whereas $N_{12}(t)$ has intensity process $\alpha_{12}(t, t - T) Y_1(t)$ with $Y_1(t) = N_{01}(t-) - N_{12}(t-)$ and $T = \inf\{t\colon N_{01}(t) = 1\}$; see Figure III.1.1. Thus, $Y_h(t)$ indicates that the individual is in state h at time $t-$, whereas T is the time of transition from 0 to 1 (if this transition ever occurs). It is seen that the intensity $\alpha_{12}(t, d)$ of dying while diseased depends on both *time* t and *duration* d. This is *not* a multiplicative intensity model because $\alpha_{12}(t, t - T) Y_1(t)$ cannot be written as a product of a deterministic function and a stochastic process independent of the parameter, except in the particular case when $\alpha_{12}(t, d)$ only depends on t, and the process corresponds to a *Markov* illness–death process; cf. Example III.1.4. When $\alpha_{12}(t, d)$ only depends on d, one has a special case of a *semi-Markov* or *Markov renewal* process to which we return in Examples X.1.7 and X.1.8. □

EXAMPLE III.1.12. Models for Software Reliability

A study on software reliability was described in Example I.3.19. A number of models have been proposed to analyze software reliability studies of this kind [see, e.g., Musa et al. (1987)]. We will consider two of these which may be described as follows. Let $N(t)$ be the number of software failures in $[0, t]$ and let F be the unknown true number of faults in the program at time 0.

For the Jelinski–Moranda model (Jelinski and Moranda, 1972) [or Musa model, cf. Musa (1975)], it is assumed that the failure rate of the system is proportional to the number of remaining faults and that each fault makes the same contribution to the failure rate of the system. Moreover, when a failure occurs, the corresponding fault is completely removed. Thus, the Jelinski–Moranda model may be specified by N having an intensity process of the form

$$\lambda(t) = \rho(F - N(t-)),$$

where ρ is the true failure rate per fault.

For the Littlewood model (Littlewood, 1980) each fault no longer makes the same contribution to the failure rate of the system. The argument for this is that "larger" faults may tend to produce failures earlier than "smaller" ones. The model may be given by specifying the intensity process of N to be

$$\lambda(t) = \frac{\alpha(F - N(t-))}{\beta + t}.$$

The model may alternatively be derived as a compound exponential failure model assuming that the failure rates for each of the original F faults form a sample from a gamma distribution with (inverse) scale parameter β and shape parameter α. By the reparametrization $\mu = \beta/\alpha$ and $\gamma = 1/\alpha$, the Littlewood model may be written

$$\lambda(t) = \frac{F - N(t-)}{\mu + \gamma t}.$$

Actually, this is an extension of the Littlewood model also including numerically small values of $\gamma \leq 0$. Note that when $\gamma = 0$, we get back to the Jelinski–Moranda model (with $\rho = 1/\mu$). □

Let us finally mention the most famous counting process!

EXAMPLE III.1.13. The Poisson Process

Let $(N(t), t \in \mathscr{T})$ be a homogeneous Poisson process with intensity θ (see Examples II.3.1 and II.3.2) and define $\mathscr{N}_t = \sigma(N(u), 0 \leq u \leq t)$. Then for any $r = 1, 2, \ldots; 0 = t_0 < t_1 < \cdots < t_r \leq \tau; n_1, n_2, \ldots, n_r = 0, 1, 2, \ldots,$

$$\mathrm{P}_\theta(N(t_j) - N(t_{j-1}) = n_j;\, j = 1, \ldots, r) = \prod_{j=1}^{r} \mathrm{P}_\theta(N(t_j) - N(t_{j-1}) = n_j)$$

and the distribution of $N(t_j) - N(t_{j-1})$ is Poisson with mean $\theta(t_j - t_{j-1})$ for $j = 1, \ldots, r$. Then it follows from Theorem II.7.1 that N has $(\mathrm{P}_\theta, \mathscr{N}_t)$-intensity process

$$\lambda^\theta(t) = \theta.$$

This is because the interevent times in N under P_θ are independent and exponentially distributed with intensity θ. The likelihood is given by

$$L(\theta) = \prod_{t \in \mathscr{T}} \theta^{\Delta N(t)}(1 - \theta\,\mathrm{d}t)^{1 - \Delta N(t)} = \theta^{N(\tau)} e^{-\theta\tau}.$$

Also, an inhomogeneous Poisson process and a doubly stochastic Poisson process are counting processes satisfying Aalen's multiplicative intensity model, cf. Jacobsen (1982, pp. 19–20). As Poisson processes play a minor role for the typical kind of applications that we have in mind in this book, we shall not go into further details here. □

III.2. Right-Censoring

As illustrated in the examples of Chapter I, incomplete observation in the form of various kinds of *censoring* can rarely be avoided in the study of life history data. In the previous section, we saw how several commonly used models for life history data could be based on multivariate counting processes when complete observation on $\mathscr{T}$ was possible. We shall now consider multivariate counting processes *composed of n individual processes* (each of which may be multivariate) and we shall see how the most common form of incomplete observation, *right-censoring*, may be "superimposed" onto such a model. Section III.2.1 provides an introduction to right-censoring and includes examples of models of right-censoring mechanisms. In Section III.2.2, the concept of *independent* right-censoring is introduced (cf. the introductory remarks in Section II.1), whereas in Section III.2.3, we discuss the statistical concept of *noninformative* right-censoring.

III.2.1. Introduction

We consider a multivariate counting process

$$\mathbf{N} = (N_{hi}; i = 1, \dots, n; h = 1, \dots, k),$$

where, as before, i indexes *individuals* and h *types* of events that the individuals may experience. For example, $h = 1, \dots, k$ may indicate the different causes of death in a competing risks model (cf. Example III.1.5).

Right-censoring of $\mathbf{N}$ is the situation where observation of $N_{hi}(\cdot)$, $h = 1, \dots, k$, is ceased after some (possibly random) time U_i, i.e., N_{hi} is only observed on the random set $E_i = \{t \leq U_i\} \subseteq \mathscr{T}$ or equivalently when the process

$$C_i(t) = I(t \in E_i) = I(t \leq U_i) \tag{3.2.1}$$

is unity. Thus, right-censoring is imposed onto $\mathbf{N}$ by *individual right-censoring processes* $C_1(\cdot), \dots, C_n(\cdot)$; see Aalen (1975, 1978b) and Aalen and Johansen (1978).

In the concrete examples of censoring to be discussed in the following, the censoring processes will typically depend only on i and not on h. It is easily seen, however, that the calculations will go through virtually unchanged with $C_i(\cdot)$ replaced by $C_{hi}(\cdot)$. Thus, different censoring mechanisms for the different types h of transitions can be handled within the framework in which we are working.

The introduction of right-censoring into the model will often bring in some extra random variation, i.e., the censoring process

$$\mathbf{C}(t) = (C_i(t), i = 1, \ldots, n)$$

need not be adapted to the "natural" filtration $(\mathscr{F}_t)$ in the examples of complete life history data considered in Section III.1. There, we had either $\mathscr{F}_t = \mathscr{N}_t$ or $\mathscr{F}_t = \mathscr{F}_0 \vee \mathscr{N}_t$ with $\mathscr{F}_0 = \sigma(\mathbf{X}_0)$. Thus, we now have to work with an *enlarged* filtration

$$(\mathscr{G}_t) \supseteq (\mathscr{F}_t),$$

and we shall make the assumption that

$$\mathbf{C}(\cdot) \text{ is } (\mathscr{G}_t)\text{-predictable.}$$

Since each $C_i(\cdot)$ is left-continuous, this is the case if $\mathbf{C}$ is *adapted* to $(\mathscr{G}_t)$, i.e., if the right-censoring times U_i are *stopping times* with respect to $(\mathscr{G}_t)$; see Section II.3.1. This is, for instance, the case if

$$\mathscr{G}_t = \mathscr{F}_t \vee \sigma(\mathbf{C}(u), u \leq t). \tag{3.2.2}$$

The interpretation is that *censoring may depend only on the past and not on future events.* The fact that we now consider another filtration may have the consequence that the compensator for $\mathbf{N}$ may change. We return to a discussion of this problem in Section III.2.2. After (possibly) having enlarged the filtrations from $(\mathscr{F}_t)$ to $(\mathscr{G}_t)$ to include any additional random variation in the right-censoring times, we shall now *reduce* the filtrations by specifying which data are available to the researcher at any time t after censoring. Thus, we do *not*, in general, assume $\mathscr{G}_t$ to represent the data at time t. First, the observable part of $\mathbf{N}$ or the *right-censored counting process* $\mathbf{N}^c = (N_{hi}^c)$ is given by

$$N_{hi}^c(t) = \int_0^t C_i(s)\, \mathrm{d}N_{hi}(s). \tag{3.2.3}$$

Next, as in Section III.1, we assume that $\mathbf{X}_0$ is observed (when relevant). We do *not* assume the whole censoring process $\mathbf{C}$ to be observed, however, and it is then a question for each individual i of whether or not the value of the censoring time U_i is observed. The situation is most easily thought of by introducing the concept of an *absorption time* by which we shall mean a (possibly random) time $\tau_i \in \overline{\mathscr{T}}$ with the property that the intensity process for N_{hi}, $h = 1, \ldots, k$, is zero for $t \geq \tau_i$. In the case of uncensored survival data (Examples III.1.2 and III.1.3), we have $\tau_i = X_i$. In an uncensored Markov process (Example III.1.4), we have $\tau_i = \inf_{t \in \mathscr{T}} \{t \colon X_i(t) \in A\}$, where $A \subseteq S$ is the subset of absorbing states. The idea is that typically, when the time of absorption for individual i precedes U_i, then $(N_{hi}(t), h = 1, \ldots, k;\ t \in \mathscr{T})$ is observed because in this case $N_{hi}(t) \equiv N_{hi}^c(t)$, but U_i itself is usually *not observed*. If there is no absorption of individual i before time U_i, then $(N_{hi}^c(t), h = 1, \ldots, k)$ is observed *together with the value of* U_i. We let $(\mathscr{F}_t^c)$ be the filtration generated by the observed data. In the next subsection, we shall see how this filtration may be represented by a *marked point process.*

Let us next study some examples of models for right-censoring mechanisms. For simplicity, we first consider the special case where a multivariate counting process $\mathbf{N}$ is defined from independent survival times $X_1, \ldots, X_n$ (Examples III.1.2 and III.1.3). Here, the observations typically consist of

$$(\tilde{X}_i, D_i; i = 1, \ldots, n),$$

where

$$\tilde{X}_i = X_i \wedge U_i$$

and

$$D_i = I(\tilde{X}_i = X_i).$$

Thus, for each individual an observation time $\tilde{X}_i$ and information on whether $\tilde{X}_i$ is a failure time or a censoring time is available. If $D_i = 1$, then the censoring time U_i is usually not observed; see, however, Example III.2.3. Thus, the data available at time t include $\mathbf{X}_0$ (if relevant) and

$$(\mathbf{N}^c(u), \mathbf{Y}^c(u); 0 \leq u \leq t)$$

where $\mathbf{N}^c = (N_i^c; i = 1, \ldots, n)$ and $\mathbf{Y}^c = (Y_i^c; i = 1, \ldots, n)$. Here

$$N_i^c(t) = I(\tilde{X}_i \leq t, D_i = 1)$$

is the right-censored counting process for individual i and

$$Y_i^c(t) = I(\tilde{X}_i \geq t)$$

indicates whether individual i is observed to be at risk just before time t.

We shall now see how some commonly used models for right-censoring fit into this setup. In Examples III.1.2 and III.1.3, we had a multivariate counting process with components defined by $N_i(t) = I(X_i \leq t)$ and we studied the compensator with respect to the self-exciting filtration $(\mathcal{N}_t)$. We first consider two examples of censoring processes $C_i(t) = I(t \leq U_i)$ predictable with respect to this filtration, i.e., examples where the original filtration $(\mathcal{N}_t)$ need not be enlarged to include the censoring.

Example III.2.1. Survival Data and Simple Type I Censorship

Here the observation of each individual is ceased at a common, *deterministic* time u_0 so $C_i(t) = I(t \leq u_0)$ is nonrandom and trivially predictable with respect to any filtration. This censoring scheme is most common in industrial life testing where n identical items are put on test simultaneously and observed on a fixed interval $[0, u_0]$. □

Example III.2.2. Survival Data and Simple Type II Censorship

In this case, the experiment is terminated at the time of the rth failure, $r \leq n$, i.e., $U_i = X_{(r)}, i = 1, \ldots, n$. Then $C_i = I(t \leq X_{(r)})$ is predictable with respect to

$(\mathcal{N}_t)$, $X_{(r)}$ being a *stopping time* with respect to this filtration. Type II censorship is also most common in industrial life testing experiments. Note that in this example the censored observation times $\tilde{X}_1, \ldots, \tilde{X}_n$ are *dependent*. □

In the next two examples, $\mathcal{G}_t$ is strictly larger than $\mathcal{F}_t$.

EXAMPLE III.2.3. Survival Data and Progressive Type I Censorship

In clinical trials, patients often enter the study consecutively ("staggered entry") while the study is closed at a particular date. When the interest focuses on the life time *from entry* (which might be the case if patients are *randomized* to some treatment at entry), the maximal time under study for patient i will be the time U_i from entry to the closing date. If the survival times from entry are *independent* of the entry times $\mathbf{X}_0$, say, and if we let $\mathcal{G}_0$ be generated by the entry times, then the censoring process $\mathbf{C}$ with components $C_i(t) = I(t \leq U_i)$ is predictable with respect to $(\mathcal{G}_t)$, where $\mathcal{G}_t = \mathcal{G}_0 \vee \mathcal{N}_t$. Another way of stating this is to say that *given* $\mathcal{G}_0$, the censoring times $U_1 = u_1, \ldots, U_n = u_n$ are *deterministic* and, thus, we have the generalization of Example III.2.1 known as progressive type I censorship. In this example, *all* the censoring times will be observable because it will be known when a patient would have left the study if he or she had not died before the closing date. This is, of course, a consequence of the assumption that *all* censoring is caused by patients being alive at the closing date. □

EXAMPLE III.2.4. Survival Data and Random Censorship

A generalization of Examples III.2.1 and III.2.3 is the *general* random censorship model where $\mathbf{U} = (U_1, \ldots, U_n)$ is independent of $\mathbf{X} = (X_1, \ldots, X_n)$, but where $\mathbf{U}$ may have an arbitrary distribution.

The *classical* or *simple* random censorship model in which $U_1, \ldots, U_n$ are assumed to be independent and identically distributed (i.i.d.) is the mathematically most tractable model for the censoring mechanism and it underlies the majority of papers on the analysis of survival data. In this case, we simply define $(\mathcal{G}_t)$ by (3.2.2). Alternatively, we might let $\mathcal{G}_0 = \sigma(\mathbf{U})$ and $\mathcal{G}_t = \mathcal{G}_0 \vee \mathcal{N}_t$, but except for the progressive type I censorship model in the previous example, it is usually intuitively very unnatural to model the censoring times as being realized at time 0, and this would also be contrary to our convention for which data are available at time t. □

The censoring schemes discussed in these examples generalize immediately to more general counting process models.

III.2.2. Independent Right-Censoring

So far we have not discussed the structure of the compensator of the basic counting process $\mathbf{N}$ and the right-censored counting process $\mathbf{N}^c$ with respect to the filtrations $(\mathscr{F}_t)$, $(\mathscr{G}_t)$, and $(\mathscr{F}_t^c)$. Assume, as in Section III.1, that our basic model is that $\mathbf{N}$ has $(\mathscr{F}_t)$-compensator $\mathbf{\Lambda}^\theta$ and intensity process $\boldsymbol{\lambda}^\theta$ with respect to a probability measure $\mathrm{P}_{\theta\phi}$. Thus, ϕ is a nuisance parameter and typically the distribution of the right-censoring times $\mathbf{U}$ may depend on ϕ. (This means that $\mathrm{P}_{\theta\phi}$ need not be the same probability measure as in Section III.1.)

A fundamental problem that we have to face is that the right-censoring may *alter the intensities for the events of interest*. If, for instance, in a clinical trial, patients who are particularly ill (or particularly well) are removed from study, then those patients who remain at risk are no longer "representative" for the sample of patients that we would have had, had there been no censoring. Right-censoring mechanisms which avoid this and similar problems and so to speak keep the risk sets representative for what they would have been without censoring are usually called *independent*. How to formalize this concept mathematically for general counting process models is not entirely obvious. We will use the following definition.

Definition III.2.1. Let $\mathbf{N}$ be a multivariate counting process with compensator $\mathbf{\Lambda}^\theta$ with respect to a given filtration $(\mathscr{F}_t)$ and probability measure $\mathrm{P}_{\theta\phi}$. Let $\mathbf{C}$ be a right-censoring process which is predictable with respect to a filtration $(\mathscr{G}_t) \supseteq (\mathscr{F}_t)$. Then we call the right-censoring of $\mathbf{N}$ generated by $\mathbf{C}$ *independent* if the $\mathrm{P}_{\theta\phi}$-compensator of $\mathbf{N}$ with respect to $(\mathscr{G}_t)$ is also $\mathbf{\Lambda}^\theta$.

In other words: The additional knowledge of the right-censoring times up until just before any time t should not alter the intensity process for $\mathbf{N}$ at time t. If, as in the discussion earlier, patients in a clinical trial who fared particularly well were censored, then knowledge about a patient still being uncensored would tell us that he or she had a higher death intensity compared to a situation without censoring. Hence, such a right-censoring scheme would not be independent unless the characteristics of the individual patients which tell us why he or she fared particularly well are included in the model as *covariates* (Example III.2.9).

Simple type I and simple type II censoring (Examples III.2.1 and III.2.2) are obviously independent because in these examples $(\mathscr{F}_t) = (\mathscr{G}_t)$. Also, random censorship as discussed in general in Example III.2.4, and in a special case in Example III.2.3, is independent, as we shall demonstrate in Example III.2.5. First, however, we study various consequences of independent censoring.

Because each N_{hi} has a decomposition

$$N_{hi}(t) = \Lambda_{hi}(t, \theta) + M_{hi}(t),$$

where M_{hi} is a local square integrable martingale with respect to $(\mathscr{G}_t)$ [see (2.4.1)], we have by (3.2.3) that

$$\begin{aligned}N_{hi}^c(t) &= \int_0^t C_i(s)\,\mathrm{d}N_{hi}(s)\\ &= \int_0^t C_i(s)\,\mathrm{d}\Lambda_{hi}(s,\theta) + \int_0^t C_i(s)\,\mathrm{d}M_{hi}(s)\\ &= \Lambda_{hi}^c(t,\theta) + M_{hi}^c(t).\end{aligned}$$

Here, the latter term by the predictability and boundedness of $C_i(\cdot)$ is again a local square integrable martingale with respect to $(\mathscr{G}_t)$, being a stochastic integral (Theorem II.3.1). Thus, under independent censoring, N_{hi}^c has $(\mathrm{P}_{\theta\phi}, (\mathscr{G}_t))$-compensator

$$\Lambda_{hi}^c(t,\theta) = \int_0^t C_i(s)\,\mathrm{d}\Lambda_{hi}(s,\theta)$$

for all $\phi \in \Phi$ (Section II.3.1). For the special case where $\mathbf{N}$ satisfies Aalen's *multiplicative intensity model* $\lambda_{hi}(t,\theta) = \alpha_{hi}(t,\theta)\,Y_{hi}(t)$ with respect to $(\mathscr{G}_t)$ (see Section III.1), it is seen that $\mathbf{N}^c$ also satisfies the multiplicative intensity model with respect to $(\mathscr{G}_t)$ with intensity process given by

$$\lambda_{hi}^c(t,\theta) = \alpha_{hi}(t,\theta)\,Y_{hi}^c(t),$$

where

$$Y_{hi}^c(t) = C_i(t)\,Y_{hi}(t).$$

That is, the observable counting process N_{hi}^c has the same "individual intensity" α_{hi}^θ as the uncensored process, but the random part Y_{hi}^c of the intensity process must in most examples be interpreted as the (predictable) indicator process for individual i being *observed to be at risk* for experiencing a type h event just before time t. In this case, the observations at time t can be described as

$$(\mathbf{X}_0, (\mathbf{N}^c(u), \mathbf{Y}^c(u)); 0 \le u \le t),$$

where $\mathbf{N}^c = (N_{hi}^c; h=1,\dots,k, i=1,\dots,n)$ and $\mathbf{Y}^c = (Y_{hi}^c; h=1,\dots,k, i=1,\dots,n)$. Equivalently, the observed data at time t can be specified as the σ-algebra

$$\mathscr{F}_t^c = \sigma(\mathbf{X}_0, (\mathbf{N}^c(u), \mathbf{Y}^c(u)); 0 \le u \le t).$$

Then it follows that the observed counting process $\mathbf{N}^c$ also satisfies a multiplicative intensity model with respect to the filtration $(\mathscr{F}_t^c)$ generated by the observed family of σ-algebras because this intensity process is given by

$$\mathrm{E}_{\theta\phi}(\lambda_{hi}^c(t,\theta)\,|\,\mathscr{F}_{t-}^c) = \alpha_{hi}(t,\theta)\,Y_{hi}^c(t)$$

by the innovation theorem (Section II.4.2). Here we use the fact that $Y_{hi}^c(t)$ (by definition of $\mathscr{F}_t^c$) is adapted to $\mathscr{F}_{t-}^c$. So, in this respect, independent right-censoring preserves the multiplicative intensity model.

In general, the compensator $\Lambda_{hi}^c(\cdot,\theta)$ may depend on θ and on the past observations in a more complicated way than specified by Aalen's model, for instance, in a censored regression model. We shall assume that the observations up to time t enable the researcher to calculate $\Lambda_{hi}^c(t, \theta)$ *for any given value of the unknown parameter* θ. We may, therefore, specify the observations available at time t as

$$(\mathbf{X}_0, (\mathbf{N}^c(u), 0 \leq u \leq t); \mathscr{L}^c(t)),$$

where

$$\mathscr{L}^c(t) = (\mathbf{\Lambda}^c(u, \theta); \theta \in \Theta, 0 \leq u \leq t) \tag{3.2.4}$$

is the *family* of $\mathrm{P}_{\theta\phi}$-compensators for $\mathbf{N}^c$ with respect to $(\mathscr{G}_t)$. Alternatively, the data at time t may be given as the σ-algebra

$$\mathscr{F}_t^c = \sigma(\mathbf{X}_0, (\mathbf{N}^c(u), 0 \leq u \leq t); \mathscr{L}^c(t)). \tag{3.2.5}$$

Then, by the innovation theorem, (3.2.4) is also the family of $(\mathrm{P}_{\theta\phi}, (\mathscr{F}_t^c))$-compensators for the right-censored counting process $\mathbf{N}^c$.

In nondegenerate cases, the two definitions of $(\mathscr{F}_t^c)$ for the multiplicative intensity model coincide, e.g., when the $\alpha_{hi}(\cdot, \theta)$ are positive on $\mathscr{T}$.

III.2.2.1. *Likelihood Constructions*

In the following, we shall use results from Section II.7 to construct the likelihood for $\mathbf{N}^c$ under independent right-censoring. We first consider the basic (completely observed) counting process $\mathbf{N}$. The $(\mathscr{F}_t)$-likelihood based on observation of $\mathbf{N}$ and $\mathbf{X}_0$, whose distribution may also depend on ϕ and maybe on θ too, is, by (2.7.2′), given by

$$\begin{aligned} L(\theta, \phi) &= L_0(\theta, \phi) \prod_{t \in \mathscr{T}} \left\{ \prod_{i=1}^{n} \prod_{h=1}^{k} \mathrm{d}\Lambda_{hi}^{\theta}(t)^{\Delta N_{hi}(t)} (1 - \mathrm{d}\Lambda_{\cdot\cdot}^{\theta}(t))^{1 - \Delta N_{\cdot\cdot}(t)} \right\} \\ &= L_0(\theta, \phi) L_\tau(\theta). \end{aligned} \tag{3.2.6}$$

Here $N_{\cdot\cdot} = \sum_{i=1}^n \sum_{h=1}^k N_{hi}$ and $\Lambda_{\cdot\cdot}^{\theta}$ is the $\mathrm{P}_{\theta\phi}$-compensator of $N_{\cdot\cdot}$ with respect to $(\mathscr{F}_t)$. If $\mathbf{X}_0$ does not depend on θ, then $L_\tau(\theta)$ is the full likelihood for θ based on the observation of $\mathbf{N}$; otherwise, $L_\tau(\theta)$ is a partial likelihood and the full conditional likelihood given $\mathscr{F}_0$. Consider now the right-censored counting process

$$\mathbf{N}^c = (N_{hi}^c, i = 1, \ldots, n; h = 1, \ldots, k).$$

As explained in Section III.2.1, we assume that the observations available at time t include $\mathbf{X}_0$ and $(\mathbf{N}^c(u), 0 \leq u \leq t)$ together with right-censoring times $U_i \leq t$ for individuals for which there is no time of absorption before U_i. As mentioned earlier, the observations may be formalized as $(\mathbf{X}_0, \mathbf{N}^c, \mathscr{L}^c(t))$ with $\mathscr{L}^c(t)$ defined by (3.2.4) as the family of $\mathrm{P}_{\theta\phi}$-compensators for $\mathbf{N}^c$ with respect to $(\mathscr{G}_t)$. It will be convenient to represent the observations instead as a *marked point process* $\mathbf{N}^*$ and derive the likelihood based on this marked point process along the lines of Section II.7.3. Thus, $\mathbf{N}^*$ plays the role of the process $\mathbf{N}$

in Section II.7.3 and has marks of the form $x = (y, u)$, say. Here, the first part y either carries the information that an event of type h happened to individual i or we may have $y = \varnothing$ (the "empty" mark) corresponding to "no event." The second part u tells which subset of the set of all individuals $\{1, \dots, n\}$ that was censored. Here, we may again have $u = \varnothing$, but $\mathbf{N}^*$ has no "completely empty" marks, i.e., marks of the form $x = (\varnothing, \varnothing)$. The observed counting process $\mathbf{N}^c$ then corresponds to the reduced process $\mathbf{N}^g$ in Section II.7.3 obtained by aggregation of $\mathbf{N}^*$ in the following way:

$$N_y^c = \sum_{x:\, g(x)=y} N_x^*,$$

where g is the coordinate mapping $g(y, u) = y$ and where we have written $\mathbf{N}^* = (N_x^*)$ and $\mathbf{N}^c = (N_y^c)$. The two ways of describing the data which we have now discussed are equivalent in the sense that for any $t \in \mathscr{T}$, $\mathscr{F}_t^c$ [defined in (3.2.5)] equals $\sigma(\mathbf{X}_0, \mathbf{N}^*(u), 0 \le u \le t)$.

We can now write down, parallel to (3.2.6), the $\mathrm{P}_{\theta\phi}$-likelihood $L^*(\theta, \phi)$ for $\mathbf{N}^*$ with respect to $(\mathscr{F}_t^c)$ informally as

$$L_\tau^*(\theta, \phi) = L_0(\theta, \phi) \prod_t \mathrm{P}_{\theta\phi}(\mathrm{d}\mathbf{N}^*(t) | \mathscr{F}_{t-}^c).$$

By the arguments leading to (2.7.15) [see also (2.1.14)], this may be rewritten as

$$L_\tau^*(\theta, \phi) = \prod_t \mathrm{P}_{\theta\phi}(\mathrm{d}\mathbf{N}^c(t) | \mathscr{F}_{t-}^c) L_0(\theta, \phi) \prod_t \mathrm{P}_{\theta\phi}(\mathrm{d}\mathbf{N}^*(t) | \mathrm{d}\mathbf{N}^c(t), \mathscr{F}_{t-}^c)$$

or

$$L_\tau^*(\theta, \phi) = L_\tau^c(\theta) L_\tau''(\theta, \phi), \tag{3.2.7}$$

where the contribution $L_0(\theta, \phi)$ from $\mathbf{X}_0$ has been absorbed in the second factor, $L_\tau''(\theta, \phi)$. In (3.2.7), the first factor, corresponding to the first factor in (2.7.15), equals

$$L_\tau^c(\theta) = \prod_t \left\{ \prod_{i,h} \mathrm{d}\Lambda_{hi}^c(t, \theta)^{\Delta N_{hi}^c(t)} (1 - \mathrm{d}\Lambda_{\cdot\cdot}^c(t, \theta))^{1 - \Delta N_{\cdot\cdot}^c(t)} \right\}, \tag{3.2.8}$$

where $N_{\cdot\cdot}^c = \sum_{h,i} N_{hi}^c$ and $\Lambda_{\cdot\cdot}^c(\cdot, \theta)$ is the $(\mathrm{P}_{\theta\phi}, (\mathscr{F}_t^c))$-compensator for $N_{\cdot\cdot}^c$. By our assumptions, $\Lambda_{\cdot\cdot}^c$ does not depend on ϕ, and as noted in Section II.7.3 [just below (2.7.15)], the *partial* likelihood function (3.2.8) based on the *innovative* events has the same form as the (partial) likelihood $L_\tau(\theta)$ in (3.2.6) based on the uncensored process $\mathbf{N}$. Thus, *independent right-censoring preserves the form of the (partial) likelihood.*

The fact that the form of the partial likelihood is preserved after independent right-censoring has the consequence that its *martingale properties* stay the same. For instance, the "score process" $((\partial/\partial\theta) \log L_t^c(\theta))$ is a $(\mathrm{P}_{\theta\phi}, (\mathscr{F}_t^c))$-martingale just as $((\partial/\partial\theta) \log L_t(\theta))$ is a $(\mathrm{P}_{\theta\phi}, (\mathscr{F}_t))$-martingale in the model without censoring; cf. Section II.7.2. This means that large sample statistical inference for independently censored data based on the partial likelihood will be much the same as that for uncensored data based on the full likelihood because this martingale structure plays such a central role in asymptotic theory as we shall see in Section VI.1.2 and in Chapter VIII.

III.2.2.2. *Examples*

After this general discussion of independent right-censoring, we turn to a series of examples starting with a continuation of Example III.2.4.

EXAMPLE III.2.5. Survival Data and Random Censorship

We first consider a single individual and let X and U be independent non-negative random variables, X having hazard rate α^θ while the distribution of U need not be absolutely continuous. We can represent the complete observation of X and U by a marked point process on the time interval $\mathscr{T}$ with marks $(1, \varnothing)$ at X and $(\varnothing, 1)$ at U. [Because $X = U$ with probability zero, a mark of the form $(1, 1)$ never occurs.] We let $(\mathscr{G}_t)$ be the filtration generated by this marked point process. Then the process $N(t) = I(X \leq t)$ has the same compensator

$$\Lambda(t, \theta) = \int_0^t \alpha(u, \theta) I(X \geq u)\, du$$

with respect to $(\mathscr{G}_t)$ as it has in the model without censoring, i.e., with respect to $(\mathscr{F}_t) = (\mathscr{N}_t)$. This is (by the independence of X and U) an immediate consequence of Theorem II.7.1 and shows that random censorship in this simple situation is independent.

That the observation of "the failure time" X may be prevented by right-censoring at U now corresponds to the situation where observation is terminated at the $(\mathscr{G}_t)$-*stopping time* $X \wedge U$ and that the corresponding mark is observed at that time, i.e., whether $X \wedge U$ is a failure time or a censoring time. Thus, the censoring process $C(t) = I(U \geq t)$ is $(\mathscr{G}_t)$-predictable and it follows that the censored process N^c has compensator

$$\Lambda^c(t, \theta) = \int_0^t \alpha(u, \theta) I(X \wedge U \geq u)\, du$$

with respect to $(\mathscr{G}_t)$ and, hence, also with respect to the filtration $(\mathscr{F}_t^c)$ generated by the *censored marked point process* $\mathbf{N}^*$ corresponding to the observation of $X \wedge U$ and the mark at that time.

In the case $n > 1$ with $X_1, \ldots, X_n$ i.i.d. and $\mathbf{U} = (U_1, \ldots, U_n)$ independent of $\mathbf{X} = (X_1, \ldots, X_n)$, one can go through the same arguments. From the "large" marked point process given by all the X_i and all the U_i, we can first define a multivariate counting process $\mathbf{N} = (N_1, \ldots, N_n)$, with the component $N_i(t) = I(X_i \leq t)$ counting the marks of the form $(i, \varnothing)$ corresponding to the failures as in the univariate case considered above. We can then calculate the $\mathrm{P}_{\theta\phi}$-compensator $\mathbf{\Lambda}(\cdot, \theta)$ for $\mathbf{N}$ with respect to the entire history $(\mathscr{G}_t)$ of the large process and note that it, by the independence of $\mathbf{X}$ and $\mathbf{U}$, coincides with the original $(\mathscr{F}_t)$-compensator in the model without censoring. This shows that general random censorship is independent. If $U_1, \ldots, U_n$ are mutually independent, this result can alternatively be derived from the univariate case and the product construction in Section II.4.3. □

EXAMPLE III.2.6. Censoring by Competing Risks

If interest in a study of survival data focuses on deaths from one specific cause, then one may wish to consider deaths due to other causes as right-censorings. As seen in Example III.1.5, the competing risks model is a simple random censorship model as just discussed. However, as mentioned in that example, the existence of the independent latent failure times is debatable. Hence, we shall now demonstrate how the situation can be modelled using marked point processes.

For simplicity, we consider just a single competing cause of failure and we let $\alpha_{01}(t, \theta)$ denote the hazard function for the cause of interest and let $\alpha_{02}(t, \theta, \phi)$ be the other cause-specific hazard. Consider a single individual, i, define $\mathbf{N}_i = (N_{0ji}, j = 1, 2)$ as in Example III.1.5, and consider the self-exciting filtration $(\mathscr{N}_t)$. We can then identify $\mathbf{N}_i$ with a marked point process with mark space $E = \{(1, \varnothing), (\varnothing, 2)\}$, where $(1, \varnothing)$ corresponds to death of cause 1 and $(\varnothing, 2)$ to death of the competing cause. Then Example III.1.5 shows that the $(\mathrm{P}_{\theta\phi}, (\mathscr{N}_t))$-compensator for the component $N_{01i}(t)$ counting the number of "marks of interest" in $[0, t]$ (i.e., marks of the form $(1, \varnothing)$] is

$$\Lambda_{1i}(t, \theta) = \int_0^t \alpha_{01}(u, \theta) Y_{0i}(u)\, du.$$

Here $Y_{0i}(t) = 1 - N_{0\cdot i}(t-)$ indicates whether individual i is alive (i.e., in state 0) at time $t-$. We can now define the right-censoring process $C_i(t) = I(t \le U_i)$, where U_i is the $(\mathscr{N}_t)$-stopping time:

$$U_i = \inf_{t \in \mathscr{T}} \{t: N_{02i}(t) = 1\}.$$

Observation of the *censored marked point process* now corresponds to observing $\tilde{X}_i = X_i \wedge U_i$, where

$$X_i = \inf_{t \in \mathscr{T}} \{t: N_{01i}(t) = 1\},$$

and the mark $(1, \varnothing)$ if $\tilde{X}_i = X_i$ or the mark $(\varnothing, 2)$ if $\tilde{X}_i = U_i$. The component $N_{01i}^c(t) = N_{01i}(t)$ counting the number of marks of the form $(1, \varnothing)$ in $[0, t]$ in the censored marked point process then has a $(\mathrm{P}_{\theta\phi}, (\mathscr{N}_t))$-compensator $\Lambda_{1i}(\cdot, \theta)$ which is adapted also to the filtration $(\mathscr{F}_t^c)$ generated by the censored marked point process [because $Y_{0i}(t) = I(\tilde{X}_i \ge t)$]. Thus, $\Lambda_{1i}(\cdot, \theta)$ is also the $(\mathrm{P}_{\theta\phi}, (\mathscr{F}_t^c))$-compensator for N_{01i}^c and this means that we can make inference on θ (and, hence, on the cause-specific hazard of interest, α_{01}^θ) in the presence of the competing risk by considering deaths from the other cause as independent censoring.

In the illness–death model defined in Example III.1.11, one may separately study $N_{01}(t)$, the healthy–diseased transition. This furnishes an example of (random) censoring by a competing risk: The transition $0 \rightarrow 2$ (death, while healthy) removes the individual from being at risk for the transition $0 \rightarrow 1$ (becoming ill). □

In the biostatistical literature, a main example of competing risks has been *bone marrow transplantation*. In Example VII.2.18, we shall briefly return to this point in connection with an analysis of the data introduced in Example I.3.14.

In the previous examples, we have demonstrated how some right-censoring mechanisms only depending on the previous history of $\mathbf{N}$ or on outside random variation are independent in the case of survival data. These models for right-censoring may be *combined* to more general models for the predictable process $\mathbf{C}$. Examples include right-censoring in a clinical trial with staggered entry (Example III.2.3) where censoring may, in addition to being caused by patients surviving until the closing date, also be a consequence of patients dying from causes unrelated to the one being studied (as in Example III.2.6). This was actually the case in the following example.

EXAMPLE III.2.7. Survival with Malignant Melanoma

In Example I.3.1, patients were followed from time of operation (in 1962–73) until the end of 1977. Thus, patients alive on 31 December 1977 were censored at that date corresponding to progressive type I censorship. However, 14 patients had died earlier from causes other than malignant melanoma, and in the study of the death intensity from the disease only, these patients are censored at the date of death from other causes. □

Next, we consider right-censoring in Markov processes (Example III.1.4).

EXAMPLE III.2.8. Right-Censoring in a Markov Process

It is easily seen going through Examples III.2.1–III.2.5 that the same arguments will apply starting with the uncensored Markov process model of Example III.1.4. Thus, for example, "censoring at the rth transition from state h to state j" would be an independent censoring scheme in this setting, being generated by an $(\mathscr{F}_t)$-stopping time.

We now return to the censoring patterns in some of the practical examples introduced in Section I.3 where Markov processes may be used as the basic modelling framework. In Example I.3.8, observation was complete: All mice were followed from time of radiation to death from either thymic lymphoma, reticulum cell sarcoma, or other causes. In Example I.3.9, where we studied both death from malignant melanoma and death from other causes, censoring was caused by patients being alive at the closing date (31 December 1977) as in Example III.2.3; cf. the discussion in Example III.2.7. In Example I.3.10, concerning psychiatric admissions and discharges, we pretended to have simple type I censorship because all women (neglecting death or emigration) were followed until 1 year after the date of birth.

It should also be mentioned that in a Markov process model, censoring may depend on the initial states generating $\mathscr{F}_0$. For example, one could have random censorship with different distributions according to the state in which the individuals were at time 0. □

The last remark in Example III.2.8 leads to another general class of censoring mechanisms relevant for regression models as exemplified in Examples III.1.7 and III.1.9.

Example III.2.9. Censoring Depending on Covariates

In the relative risk regression model (Example III.1.7) and in the additive regression model (Example III.1.9), the filtration considered was of the form $\mathscr{F}_t = \mathscr{F}_0 \vee \mathscr{N}_t$ with $\mathscr{F}_0$ generated by time-independent covariates $\mathbf{Z}_1, \ldots, \mathbf{Z}_n$. This means that the previously mentioned models for censoring mechanisms (Examples III.2.1–III.2.6) can be combined with censoring depending on the covariates generating $\mathscr{F}_0$. Thus, in a simple two-sample case, there may be different censoring distributions in the two samples. Also, a possible independent censoring scheme in a survival study with time since entry as the basic time scale and age at entry and sex included as covariates would be every year to censor, e.g., the oldest woman still alive.

Recalling that the censoring process has to be adapted to an extended filtration $(\mathscr{G}_t)$, it is crucial that the extension generated by the covariates upon which censoring depends does not change the compensator of $\mathbf{N}$. Therefore, an example of *dependent* right-censoring would be a case where the right-censoring process $\mathbf{C}$ depends on covariates which are *not* included in the model. We return to a discussion of this problem in Chapter IX. □

An example is the following:

Example III.2.10. CSL 1: A Randomized Clinical Trial in Liver Cirrhosis

In the randomized clinical trial on the effect of prednisone treatment versus placebo on survival in patients with liver cirrhosis, CSL 1 (Example I.3.4, see also Example I.3.12), most of the censoring was a consequence of patients still being alive at the closing date of the study (1 September 1974). Some patients left the study earlier, however, because of side effects of the prednisone treatment. Whether such a censoring scheme satisfies our requirements for "independence" may be difficult to judge in practice. Formally, the censoring process is *adapted* if all the covariates on which the withdrawal intensity depends (in particular, the treatment indicator) are included in the model for the death intensity (i.e., in $\mathscr{F}_0$) and if the withdrawal times U_i are included in $(\mathscr{G}_t)$. The crucial problem is whether the inclusion of the times of withdrawal *alters* the death intensities which basically amounts to whether or not X_i and U_i are *independent* given $\mathscr{F}_0$. This is, for instance, not the case if both the death and the withdrawal intensities depend on a common unobserved covariate. In such cases, the censoring due to withdrawal from side effects may bias the analysis of the death intensity and it is, therefore, strongly recommended in the clinical trial literature to keep track of all included patients whether or not they are still being treated and to base the treatment comparison on the randomized groups (the "intention to treat principle"), e.g., Pocock (1983, Section 12.3). □

In Definition III.2.1 of independent right-censoring, it is required that the censoring process $\mathbf{C}$ is predictable with respect to an extended filtration $(\mathscr{G}_t)$ while still preserving the compensator of $\mathbf{N}$. This means that examples of *dependent* right-censoring patterns include cases where $\mathbf{C}$ is not even *adapted* to the relevant $(\mathscr{G}_t)$; cf. Examples III.2.9 and III.2.10. Another example is the following.

Example III.2.11. Testing with Replacement

Suppose that objects (e.g., light bulbs) are life tested one at a time and at each failure time replaced by a new one. If observation is terminated at a fixed calendar time, the last object put on test will typically still be working and hence contribute with a censored lifetime. Thus, censoring of the last object (at lifetime t, say) depends on the lifetimes of previous objects which may well exceed t. A similar situation arises in the clinical trial Example III.2.3 if observation had not been terminated at a fixed calendar time but instead at the rth observed failure $X_{(r)}$ (as in type II censorship). With this stopping rule, patients with entry times later than that of the patient with failure time equal to $X_{(r)}$ may still be alive and, thus, censored after a time under study less than $X_{(r)}$. For a further discussion, the reader is referred to Sellke and Siegmund (1983), Slud (1984), and Arjas (1985). □

For the kind of censoring mechanisms in Example III.2.11, the methods to be discussed in the next chapters do not work, and in Sections X.1 and X.2, we return to a closer discussion on related topics; see also Gill (1980b, 1981).

We next consider the "pornoscope" model of Example I.3.16.

Example III.2.12. Mating of Drosophila Flies

Censoring of type I or type II (Examples III.2.1 and III.2.2) are also relevant for the pornoscope model in Example I.3.16. Thus, here one might choose to terminate the experiment at a fixed time u_0 or at the time of initiation of the rth mating. In the actual experiment, the following censoring was used. Let

$X_{(1)}$ denote the time of initiation of the first mating. Then the experiment was terminated at the first time after $X_{(1)} + 45$ min where no matings were going on—no later, however, than at time $X_{(1)} + 60$ min. It is seen that this censoring is independent, provided that the times of termination of matings are included in the filtration $(\mathscr{G}_t)$ and do not thereby alter the mating intensities. □

III.2.2.3. *Identifiability of Independent Right-Censoring Mechanisms*

In the next example, we shall relate our definition of independent right-censoring to other suggestions in the literature in the special case of i.i.d. survival times.

EXAMPLE III.2.13. Independent Censoring of i.i.d. Survival Times

Williams and Lagakos (1977) considered right-censoring of i.i.d. survival times $X_1, \ldots, X_n$ with hazard function $\alpha^\theta(t)$. They showed that if the model for the censored data satisfies a certain "constant sum" condition, then the likelihood for θ is proportional to (3.2.8). Kalbfleisch and MacKay (1979) showed that the constant sum condition is equivalent to another condition which is a *consequence* of our definition of independent censoring, namely, that the failure intensity at time t for an individual i at risk at that time (i.e., $\tilde{X}_i \geq t$) is $\alpha^\theta(t)$. Formulated in our notation, this condition simply states that the $(\mathrm{P}_{\theta\phi}, (\mathscr{F}_t^c))$-compensator for N_i^c is $\int \alpha^\theta Y_i^c$. This condition was verbally formulated by Cox (1975) and further discussed by Kalbfleisch and Prentice (1980, p. 120). It was given a precise mathematical formulation, not restricted to the (absolutely) continuous case, by Gill (1980a, Theorem 3.1.1). Thus, our requirement, being a condition on the larger filtration $(\mathscr{G}_t) \supseteq (\mathscr{F}_t^c)$, is a *stronger* requirement for independent censoring than those considered by these various authors. One may say that whereas our condition concerns both "survivors" and "diers," the classical conditions are only concerned with the "diers," but, as seen above, our condition does cover all the interesting models for right-censoring in the case of survival data; so, in this situation, we get the stronger requirements for free. Furthermore, our concept of independent censoring can be generalized to other models based on counting processes. □

So far we have discussed conditions on the censoring pattern which make the resulting observable processes tractable. The opposite wish is to see how much of the underlying structure is uniquely given if the observable processes are tractable. We conclude this section by indicating the results by Jacobsen (1989b), who studied i.i.d. survival data with hazard function $\alpha(t)$ (the restriction to identical distributions being made for convenience only).

Jacobsen's concept of independent censoring is also more restrictive than the one just considered (Example III.2.13) and differs slightly from ours. He also studied the marked point process $\mathbf{N}^*$, recording observed deaths and censorings, and considered the joint distribution of $\mathbf{N}^*$ and *all* the failure times $\mathbf{X} = (X_1, \ldots, X_n)$. He then showed first that if for all t the conditional distribution given $(X_1, \ldots, X_n)$ of the data available at time t (i.e., the process $\mathbf{N}^*$ stopped at t) only depends on the X_i through what we actually observe about them at time $t-$, i.e., that

$$\begin{aligned} X_i \geq t \quad & \text{if } X_i \wedge U_i \geq t, \\ X_i = x_i \quad & \text{if } X_i = x_i < t \quad \text{and} \quad U_i \geq x_i, \\ X_i > u_i \quad & \text{if } U_i = u_i < t \quad \text{and} \quad X_i \geq u_i, \end{aligned} \tag{3.2.9}$$

then N_i^c has $(\mathscr{F}_t^c)$-compensator

$$\Lambda_i^c(t) = \int_0^t \alpha(u) Y_i^c(u)\, du \tag{3.2.10}$$

with respect to the probability measure corresponding to the joint distribution of $\mathbf{N}^*$ and $\mathbf{X}$. Second, he showed that given an $(\mathscr{F}_t^c)$-compensator $\mathbf{\Lambda}^*$ for $\mathbf{N}^*$ satisfying (3.2.10) there exists one and only one joint distribution of $\mathbf{N}^*$ and $(X_1, \ldots, X_n)$ satisfying (3.2.9) such that $X_1, \ldots, X_n$ are i.i.d. with hazard $\alpha(\cdot)$ and $\mathbf{N}^*$ has compensator $\mathbf{\Lambda}^*$.

This distribution can be *simulated* in the way described in Example III.2.14 concerning a randomized version of what is known as *progressive type II censorship*.

EXAMPLE III.2.14. Survival Data and Randomized Progressive Type II Censorship

Suppose that n identical items are put on test simultaneously and let $X_1, \ldots, X_n$ be the i.i.d. lifetimes that would have been observed had there been no censoring. Next, generate n potential (possibly mutually dependent) right-censoring times $U_1^{(1)}, \ldots, U_n^{(1)}$ independent of the X_i and find $X_{(1)}$, the smallest X_i with $X_i \leq U_i^{(1)}$. Items j for which $U_j^{(1)} < X_{(1)}$ and $U_j^{(1)} < X_j$ are removed at the time points $U_j^{(1)}$. At time $X_{(1)}$, new potential right-censoring times $U_j^{(2)} > X_{(1)}$ are generated for each item j still on test. The joint distribution of the $U_j^{(2)}$ may depend on $X_{(1)}$ and the censoring times for those items actually removed in $[0, X_{(1)})$ (and on the labels of these items) but not on the lifetimes X_j of the items still on test. Next, $X_{(2)}$, the smallest X_i with $X_i \leq U_i^{(2)}$, is found and items j still on test and for which $U_j^{(2)} < X_{(2)}$ and $U_j^{(2)} < X_j$ are removed at $U_j^{(2)}$. At $X_{(2)}$, the censoring times for items still on test are once more updated and allowed to depend also on $X_{(2)}$ and censorings in $[X_{(1)}, X_{(2)})$ and so on. At every point in time, the decision to censor an item "still working" may depend arbitrarily on the past observations but not on the future. □

Jacobsen illustrated the concepts in the following example.

EXAMPLE III.2.15. "Counterexample"

Let X_1 and X_2 be i.i.d. with hazard function $\alpha(t)$ and let the censoring variables U_1 and U_2 be given by assuming that U_1 is independent of X_1 and X_2 and standard exponentially distributed and by defining

$$U_2 = \begin{cases} \infty & \text{if } X_1 \leq U_1 \\ X_1 + a & \text{if } U_1 < X_1. \end{cases}$$

One may check that this strange censoring pattern satisfies Jacobsen's as well as our definition of independent censoring. Of course, it is not *defined* by a simulation experiment as described in Example III.2.14, because the censoring time U_2 depends on the value of X_1 exactly when this is *not* observed. Jacobsen's result then tells us that there exists one and only one simulation experiment resulting in exactly the same observable process as in this example. The reason one feels unhappy about this example is that, though the censoring is *independent*, it is *informative* in the sense to be discussed in the next subsection. Thus, the problem with it is *statistical*, not *probabilistic*. □

In our view, the simulation experiment represents the canonical form of what one should understand by well-behaved independent right-censoring. It would be desirable to obtain a more abstract formulation (perhaps as a suitable stopping time condition on the $U_1, \ldots, U_n$) following up the introductory lines of Aalen (1982a) so that the concept may be defined for general counting processes.

III.2.3. Noninformative Right-Censoring

In this subsection, we consider a multivariate counting process

$$\mathbf{N} = (N_{hi}; i = 1, \ldots, n, h = 1, \ldots, k)$$

subject to independent right-censoring by a process $\mathbf{C}$. The observations are considered as a marked point process $\mathbf{N}^*$ as described above formula (3.2.7) and the $\mathrm{P}_{\theta\phi}$-likelihood for $\mathbf{N}^*$ with respect to the filtration $(\mathscr{F}_t^c)$ it generates is then, by (3.2.7), given by

$$L_\tau^*(\theta, \phi) = L_\tau^c(\theta) L_\tau''(\theta, \phi).$$

Here, the partial likelihood $L_\tau^c(\theta)$ for θ based on the observed counting process $\mathbf{N}^c$ [see (3.2.3)] by (3.2.8) is

$$L_\tau^c(\theta) = \prod_t \left\{ \prod_{i,h} \mathrm{d}\Lambda_{hi}^c(t, \theta)^{\Delta N_{hi}^c(t)} (1 - \mathrm{d}\Lambda_{\bullet\bullet}^c(t, \theta))^{1 - \Delta N_{\bullet\bullet}^c(t)} \right\}.$$

An important question is now whether $L_\tau^c(\theta)$ is the full likelihood for θ for each fixed $\phi \in \Phi$ based on observation of $\mathbf{N}^*$.

Definition III.2.2. If, for each fixed $\phi \in \Phi$, the likelihood (3.2.7) may be factored

$$L_\tau^*(\theta, \phi) = L_\tau^c(\theta) L_\tau''(\phi)$$

(so that L_τ'' does not depend on θ), then the independent right-censoring mechanism $\mathbf{C}$ is called *noninformative* for θ.

This precise definition (but without consideration of a nuisance parameter ϕ) is due to Arjas and Haara (1984) who made the discussion by Kalbfleisch and Prentice (1980, p. 126) rigorous. In fact, Arjas and Haara termed it *noninnovative* censoring and considered a more general situation with other kinds of censoring and with time-dependent covariates. Thus, their discussion included the concept of noninnovative covariates. We shall return to this in Sections III.4 and III.5. From the derivation of the likelihood in Section II.7.3 [see just above formula (2.7.15)], we may interpret noninformative right-censoring as follows: If, for each fixed $\phi \in \Phi$, the conditional distribution of $d\mathbf{N}^*(t)$ given $\mathscr{F}_{t-}^c$ and $d\mathbf{N}^c(t)$ does not depend on θ, then the right-censoring process $\mathbf{C}$ is noninformative for θ; or stated in other words, if, for each fixed $\phi \in \Phi$ and each $t \in \mathscr{T}$, the conditional intensity of certain individuals being censored at t given the past up until just before t and given a possible failure at t does not depend on θ, then the censoring is noninformative for θ.

In the special case of Aalen's multiplicative intensity model

$$\lambda_{hi}^c(t, \theta) = \alpha_{hi}(t, \theta) Y_{hi}^c(t, \theta),$$

where α_{hi} *does not depend on* i, i.e., $\alpha_{hi} = \alpha_h$, it is easily seen from (3.2.6)–(3.2.8) that if the censoring is noninformative for θ, then $(N_{h\cdot}^c, Y_{h\cdot}^c; h = 1, \ldots, k)$ is *sufficient for* θ. Here,

$$Y_{h\cdot}^c(t) = \sum_{i=1}^n Y_{hi}^c(t)$$

can often be interpreted as the total number of individuals *observed to be at risk* for experiencing a type h event just before time t. Thus, again, a process satisfying the multiplicative intensity model is obtained by aggregation. However, in contrast to Examples III.1.2 and III.1.4 with uncensored data where the aggregated counting processes $(N_{h\cdot}, h = 1, \ldots, k)$ (and $\mathbf{X}_0$) were themselves sufficient, $Y_{h\cdot}$ being a function of $N_{h\cdot}$ (and $\mathbf{X}_0$), it is now the *pairs* $(N_{h\cdot}^c, Y_{h\cdot}^c;\ h = 1, \ldots, k)$ which are sufficient under noninformative right-censoring.

In Example III.2.3, we have noninformative censoring provided that the entry time process does not depend on θ and in Examples III.2.4 and III.2.5 if the censoring distribution does not depend on θ. In Example III.2.9, we have noninformative censoring when there is random censorship with censoring distribution depending on covariates and if this distribution does not depend on θ.

Informative right-censoring may occur if censoring is due to competing causes of deaths with cause-specific intensities depending on θ (cf. Example III.2.6). One such example is the Koziol–Green model (Koziol and Green, 1976) where failure and censoring intensities are *proportional*. Statistical procedures based on $L_\tau^c(\theta)$ only will lose information if we have informative censoring and in such examples, more efficient methods may be applied. We have seen, however, that most sensible models for *right*-censoring mechanisms were noninformative. This is in contrast to other kinds of censoring, including *left*-censoring, to which we return in Section III.4.

III.3. Left-Truncation

The most common kind of incomplete information on life history data, *right-censoring*, was discussed in the previous section. To exemplify a different kind of incomplete observation, we consider the study of survival among insulin-dependent diabetics in Fyn county introduced in Example I.3.2.

EXAMPLE III.3.1. Survival Among Insulin Dependent Diabetics in the County of Fyn

Recall from Example I.3.2 that out of the about 450,000 inhabitants in Fyn county, Denmark, it was ascertained from prescriptions in the National Health Service files that $n = 1499$ suffered from insulin-dependent diabetes mellitus on 1 July 1973. They were all followed until 1 January 1982 with the purpose of assessing the mortality among diabetics. Suppose we want to study this mortality rate as a function of time since onset of diabetes, i.e., as a function of *disease duration*. Because a diabetic was only included in the sample conditionally on being alive on 1 July 1973, the relevant distribution to consider for the survival times X_i, $i = 1, \ldots, n$, is the conditional distribution of X given $X > V$ where the entry time, V, is the time since disease onset at 1 July 1973. The survival data are then said to be *left-truncated*. In Example III.3.6, we further discuss these data and present a model that allows us to study the mortality rate alternatively as a function of *age*. □

Left truncation is very common in fields like demography and epidemiology and is also present in several of the other practical examples introduced in Section I.3 (Examples I.3.3, I.3.5, I.3.11, and I.3.15). In fact, as recognized already by Halley (1693) [see also Kaplan and Meier (1958)], the construction of the classical life table involves following persons from an *entrance* (or left-truncation) age to an exit age and registering whether exit is due to death or right-censoring.

In this section, we shall consider in more generality counting process based models for left-truncated life history data. The setup is analogous to that studied in Section III.2. We consider a single individual, i, at a time and we drop the subscript i. We, therefore, let

$$\mathbf{N} = (N_h, h = 1, \ldots, k)$$

be a multivariate counting process on a space $(\Omega, \mathscr{F})$ with $\mathrm{P}_{\theta\phi}$-compensator $\mathbf{\Lambda}^\theta$ and intensity process $\boldsymbol{\lambda}^\theta$ with respect to a filtration $(\mathscr{F}_t)$ of the form $\mathscr{F}_t = \mathscr{F}_0 \vee \mathscr{N}_t$ (cf. Sections III.2.1 and III.2.2 where we noted that $\mathrm{P}_{\theta\phi}$ need not be the same probability measure as that considered for the complete data models in Section III.1). We shall restrict ourselves to a discussion of *independent left-truncation* (compare Section III.2.2), so we assume the existence of a larger filtration $(\mathscr{G}_t) \supseteq (\mathscr{F}_t)$ such that *the* $(\mathrm{P}_{\theta\phi}, (\mathscr{G}_t))$*-compensator of* $\mathbf{N}$ *is also* $\mathbf{\Lambda}^\theta$. The larger filtration is intended to carry the possible extra random variation involved in the truncation time, so we let V be a $(\mathscr{G}_t)$-stopping time and consider an event $A \in \mathscr{G}_V$ with $\mathrm{P}_{\theta\phi}(A) > 0$. The process $\mathbf{N}$ started at V is defined as

$$_V\mathbf{N}(t) = \mathbf{N}(t) - \mathbf{N}(t \wedge V), \tag{3.3.1}$$

cf. Section II.4.4. We want to study the process $\mathbf{N}$, starting from the time V, *given* that the event A (prior to V) has actually occurred. We call the process $_V\mathbf{N}$, *under this conditional distribution*, a *left-truncated* process. Proposition II.4.2 shows that left-truncation of $\mathbf{N}$ by the event A (before V) preserves the intensity of $\mathbf{N}$ after time V. Thus, the left-truncated counting process $_V\mathbf{N}$ has intensity process

$$_V\boldsymbol{\lambda}^\theta(t) = \boldsymbol{\lambda}^\theta(t) I(t > V) \tag{3.3.2}$$

with respect to the filtration $(_V\mathscr{G}_t)$ given by

$$_V\mathscr{G}_t = \mathscr{G}_t \vee \mathscr{G}_V$$

and the conditional probability $\mathrm{P}^A_{\theta\phi}$ given by

$$\mathrm{P}^A_{\theta\phi}(F) = \mathrm{P}_{\theta\phi}(F \cap A)/\mathrm{P}_{\theta\phi}(A), \quad F \in \mathscr{F}.$$

If $\mathscr{G}_t$ has the special form

$$\mathscr{G}_t = \mathscr{G}_0 \vee \mathscr{N}_t,$$

then

$$_V\mathscr{G}_t = \mathscr{G}_V \vee \sigma\{_V\mathbf{N}(u); 0 \le u \le t\},$$

and the $(\mathrm{P}^A_{\theta\phi}, (_V\mathscr{G}_t))$ conditional likelihood for $_V\mathbf{N}$ given $\mathscr{G}_V$, also the *partial* likelihood for $_V\mathbf{N}$, is

$$_V L(\theta) = \prod_{t > V} \left\{ \prod_{h=1}^{k} {}_V\lambda_h^\theta(t)^{\Delta_V N_h(t)} (1 - {}_V\lambda_{\cdot}^\theta(t)\,\mathrm{d}t)^{1 - \Delta_V N_{\cdot}(t)} \right\}. \tag{3.3.3}$$

We shall now discuss how the left-truncated data become available as time proceeds in parallel with the situation for right-censored data; cf. (3.2.4) and (3.2.5). Conditionally on A, we always assume that $_V\mathbf{N}$ is observed. Furthermore, we assume that the available data allow us to write down the intensity process ${}_V\boldsymbol{\lambda}^\theta$ for any $\theta \in \Theta$. We may then define the observed σ-algebra at time t, $t > V$, as

$${}_V\mathscr{F}_t^c = \sigma({}_V\mathbf{N}(u), V \le u \le t, {}_V\mathscr{L}(t)),$$

where

$${}_V\mathscr{L}(t) = ({}_V\boldsymbol{\lambda}^\theta(u); \theta \in \Theta; V \le u \le t).$$

Then ${}_V\boldsymbol{\lambda}^\theta$ is also (by the innovation theorem, Section II.4.2) the $({}_V\mathscr{F}_t^c, \mathrm{P}_{\theta\phi}^A)$-intensity process for ${}_V\mathbf{N}$ and (3.3.3) is a partial likelihood with respect to $({}_V\mathscr{F}_t^c)$. If $\mathbf{N}$ satisfies Aalen's multiplicative intensity model

$$\lambda_h^\theta(t) = \alpha_h^\theta(t) Y_h(t), \quad t \in \mathscr{T}, h = 1, \ldots, k,$$

with respect to $\mathrm{P}_{\theta\phi}$ and $(\mathscr{F}_t)$, then the left-truncated process ${}_V\mathbf{N}$ satisfies the multiplicative intensity model

$${}_V\lambda_h^\theta(t) = \alpha_h^\theta(t)\, {}_VY_h(t), \quad t > V,$$

with respect to $\mathrm{P}_{\theta\phi}^A$ and $({}_V\mathscr{G}_t)$, where

$${}_VY_h(t) = Y_h(t) I(t > V).$$

In this case, the data needed at time t, $t > V$, are $({}_V\mathbf{N}(u), {}_V\mathbf{Y}(u); V \le u \le t)$, where ${}_V\mathbf{Y} = ({}_VY_h, h = 1, \ldots, k)$, and we may define

$${}_V\mathscr{F}_t^c = \sigma({}_V\mathbf{N}(u), {}_V\mathbf{Y}(u); V \le u \le t).$$

In this case, we also have a multiplicative intensity model with respect to $({}_V\mathscr{F}_t^c)$.

We next consider some models for the generation of the left-truncation time V. In the most trivial example, $V = v_0$ is deterministic and we have $\mathscr{G}_t = \mathscr{F}_t$. More interesting is the following example.

Example III.3.2. Random Left-Truncation of a Survival Time

Let the random variable $X > 0$ have hazard function α_X^θ. Define $N(t) = N_X(t) = I(X \le t)$ and let $(\mathscr{N}_t)$ be the filtration generated by N_X. Assume that $V > 0$ is *independent* of X with distribution depending on parameters ϕ. Define the bivariate counting process (N_X, N_V) with components $N_X(t)$ and $N_V(t) = I(V \le t)$ and let $(\mathscr{G}_t)$ be the filtration it generates. The $(\mathscr{G}_t)$-intensity process $\lambda_X^{\theta\phi}$ for N_X with respect to the joint distribution $\mathrm{P}_{\theta\phi}$ of X and V is (as previously; cf. Example III.2.5)

$$\lambda_X^{\theta\phi}(t) = \lambda_X^\theta(t) = \alpha_X^\theta(t) I(t \le X)$$

which is the same as the intensity process with respect to $(\mathscr{N}_t)$. If the event

$$A = \{X > V\}$$

has positive probability, then it follows from the general discussion earlier that the intensity process for the left-truncated process ${}_VN_X(t) = N_X(t) - N_X(t \wedge V)$ with respect to the *conditional* joint distribution $\mathrm{P}_{\theta\phi}^A$ of X and V given A is

$${}_V\lambda_X^{\theta\phi}(t) = {}_V\lambda_X^\theta(t) = \alpha_X^\theta(t) I(V < t \le X).$$

We see that random left-truncation is independent and, hence, preserves the multiplicative structure of the intensity process. [For the left-truncation to be independent in this simple case, it suffices that the conditional joint density $h(x, v)$ given $X > V$ of (X, V) (when the density exists) can be written as a product of the form $h(x, v) = \tilde{f}(x)\tilde{g}(v)$; e.g., Tsai, 1990; Wellek, 1990.] In particular, ${}_V\lambda_X$ still only depends on θ. The conditional (given V) or partial likelihood for ${}_V N_X$ with respect to the conditional distribution given $X > V$ is, by (3.3.3),

$$
\begin{aligned}
{}_V L(\theta) &= \prod_{t > V} \{ {}_V\lambda_X^\theta(t)^{\Delta_V N_X(t)} (1 - {}_V\lambda_X^\theta(t)\,\mathrm{d}t)^{1 - \Delta_V N_X(t)} \} \\
&= \alpha_X^\theta(X) \exp\left(- \int_V^X \alpha_X^\theta(t)\,\mathrm{d}t \right) = S_X^\theta(X)\alpha_X^\theta(X)/S_X^\theta(V).
\end{aligned}
$$

It is seen that this is the conditional density of X given $X > V$, evaluated at (X, V). The random truncation model for survival data is further discussed in Example IV.1.7. □

When the distribution of X depends on covariates $\mathbf{Z}$ (Examples III.1.7 and III.1.9), the basic filtration is given by $\mathscr{F}_t = \mathscr{F}_0 \vee \mathscr{N}_t$ with $\mathscr{F}_0 = \sigma(\mathbf{Z})$. To write down the intensity process ${}_V\boldsymbol{\lambda}^\theta$ for the left-truncated process ${}_V\mathbf{N}$ we must include not only V but also $\mathbf{Z}$ in the "observed filtration" $({}_V\mathscr{F}_t^c)$. In this case, we have independent left-truncation if the conditional joint density of (X, V) given $X > V$ and $\mathbf{Z}$ can be written in a product form as described in Example III.3.2.

In Examples III.1.5 and III.2.6, the relationship between the random censorship model for survival data and a certain Markov process, the competing risks model, was studied. The next example establishes the relationship between the random truncation model and a certain Markov process; see Keiding and Gill (1990).

Example III.3.3. The Random Truncation Model for a Survival Time Viewed as a Markov Process

In the model of Example III.3.2, assume for convenience the distribution of V absolutely continuous with hazard $\alpha_V^\phi(t)$. Define the Markov process $U(t)$ by $U(0) = 0$ and transition intensities as specified in Figure III.3.1. The random

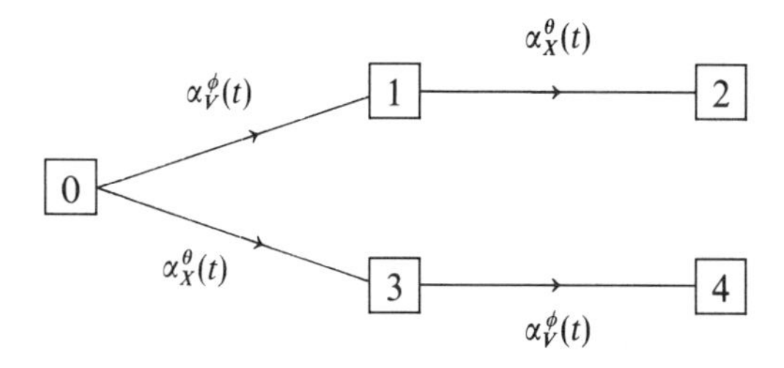

Figure III.3.1. Random left-truncation as a Markov process.

variables X and V correspond to the times of transition from 1 to 2 or 0 to 3, and from 0 to 1 or 3 to 4, respectively. Observation of (V, X) given $V < X$ is equivalent to observing $U(t)$ in the conditional distribution given $U(\tau) = 2$ (ultimate absorption in 2). The counting process

$$_V N_X(t) = N_X(t) - N_X(t \wedge V)$$

is identical to that counting transitions from state 1 to state 2,

$$_V N_X(t) = N_{12}(t),$$

and it is a standard result for Markov processes [see Hoem (1969b) for an explicit formulation] that the intensity $1 \to 2$ in the conditional Markov process is

$$\alpha_X^\theta(t)\frac{\mathrm{P}(U(\tau) = 2 | U(t) = 2)}{\mathrm{P}(U(\tau) = 2 | U(t) = 1)} = \alpha_X^\theta(t).$$

It now follows from Example III.1.4, used for the conditional Markov process given $\{X > V\}$, that $N_{12}(t)$ has intensity process

$$\alpha_X^\theta(t) Y_1(t) = \alpha_X^\theta(t) I(V < t \leq X)$$

with respect to the conditional distribution given $V < X$ and the filtration given by

$$_V\mathscr{F}_t = \sigma(I(V \leq u, V < X), I(V < X \leq u); 0 \leq u \leq t).$$

The likelihood with respect to θ consequently has the form (3.3.3) which may thus, in this case, be obtained without recourse to the proof given there. We return to this model in Example IV.3.10. □

Example III.3.4. Left-Truncation of a Markov Process

Let the Markov process $X = (X(t), t \in \mathscr{T})$, the counting process $\mathbf{N}$, and the filtration $(\mathscr{F}_t)$ be defined as in Example III.1.4. Let $V > 0$ be independent of X with distribution depending on parameters ϕ and define $N_V(t) = I(V \leq t)$. Then $\mathbf{N}$ has the same compensator both with respect to $(\mathscr{F}_t)$ and with respect to the filtration $(\mathscr{G}_t)$ defined by

$$\mathscr{G}_t = \mathscr{F}_t \vee \sigma(N_V(u), 0 \leq u \leq t),$$

and we can apply Proposition II.4.2 to any event $A \in \mathscr{G}_V = \sigma\{V, X_0; \mathbf{N}(t \wedge V), t \geq 0\}$ with $\mathrm{P}_{\theta\phi}(A) > 0$. Examples include events A indicating that a certain component N_h had at least or at most some specified number of jumps before V.

As an alternative to such partly *external* truncation (V independent of X), one might consider *internal* truncation, where V is an $(\mathscr{F}_t)$-stopping time. The obvious examples are the first arrival time to a particular state (cf. the illness–death process to be further discussed) or the first (or pth) time that the process (or one of its components) jumps. □

EXAMPLE III.3.5. The Illness–Death Process

Consider the illness–death process earlier discussed in Examples III.1.11 and III.2.6. A separate study of the transition $1 \to 2$ (death when diseased) may be performed by studying the counting process N_{12} which has intensity process $\alpha_{12}(t, t-T)Y_1(t)$, where $Y_1(t) = I$(the individual is in state 1 at time $t-$) and T is the entry time into state 1. Define, also, X to be the time of death and consider left-truncation by the event $A = \{T \leq V < X < \tau\}$ at some random time V.

Internal truncation is obtained by choosing, for instance, $V = T$, that is, follow the individual from the time of disease occurrence conditioning on it occurring: The truncated counting process would be

$$_VN_{12}(t) = N_{12}(t) - N_{12}(t \wedge V) = N_{12}(t)$$

with intensity process

$$\alpha_{12}(t, t-V)Y_1(t)I(t > V) = \alpha_{12}(t, t-T)Y_1(t).$$

In this case, except for the conditioning, left-truncation is equivalent to observation of the original process N_{12}.

External truncation is exemplified by choosing V independent of the illness–death process. On the conditioning event A, the truncated counting process $N_{12}(t) - N_{12}(t \wedge V)$ has intensity process

$$\alpha_{12}(t, t-T)Y_1(t)I(t > V) = \alpha_{12}(t, t-T)I(V < t \leq X).$$

It is seen that this is only observable (for given α_{12}) if we not only follow the diseased individual from time $V\,(> T)$ until death at time X, but also actually know the time T of disease occurrence. The latter condition (technically, that a nontrivial part of $\mathscr{G}_V$ is needed) is not always fulfilled and we shall return to that problem in Example VII.2.16.

Finally, in the particular case where $\alpha_{12}(t, d)$ does not depend on d (the original process is a Markov illness–death process), it is easily seen that both types of truncation (that is, following diseased individuals only, either from disease occurrence or from some random time, until death) are equivalent to studying a left-truncated random variable with hazard given by the death intensity of the diseased. □

So far we have only considered a (possibly multivariate) counting process corresponding to a single individual. We shall now briefly study independent left truncation of the processes corresponding to several independent individuals simultaneously. Let $\mathbf{N}_i = (N_{hi}, h = 1, \ldots, k)$, $i = 1, \ldots, n$ be independent and define $\mathbf{N} = (\mathbf{N}_i, i = 1, \ldots, n)$. Assume that $\mathbf{N}$ has the same $\mathrm{P}_{\theta\phi}$-compensator with respect to $(\mathscr{G}_t)$ and $(\mathscr{F}_t)$, where $(\mathscr{G}_t) \supseteq (\mathscr{F}_t)$ and $\mathscr{F}_t = \mathscr{F}_0 \vee \mathscr{N}_t$.

If V is a $(\mathscr{G}_t)$-stopping time, then we may apply the general result to the process

$$_V\mathbf{N}(t) = \mathbf{N}(t) - \mathbf{N}(t \wedge V)$$

corresponding to truncation of all individuals at the same time. Alternatively, we can let $(V_1, \mathbf{N}_1), \ldots, (V_n, \mathbf{N}_n)$ be mutually independent. In this case,

$$_{V_i}\mathbf{N}_i(t) = \mathbf{N}_i(t) - \mathbf{N}_i(t \wedge V_i), \quad i = 1, \ldots, n,$$

are independent and Proposition II.4.2 may be applied to each individual separately, after which the relevant intensity process for the multivariate counting process

$$({}_{V_i}\mathbf{N}_i, i = 1, \ldots, n)$$

can be found using the product construction in Section II.4.3.

More general cases with $V_1, \ldots, V_n$ being dependent seem to be more difficult to handle and the theory of left-truncation as a whole seems to be less rich than the theory of right-censoring.

Very frequently in practice, there will be both left-truncation and right-censoring. Here, we briefly indicate how the methods from this and the previous section can be combined. If, in the construction of our model, truncation precedes censoring, then conditionally on an event before a stopping time V_i the individual process $N_{hi}(\cdot)$ started at V_i is observed on a set of the form $(V_i, U_i]$ with $V_i \geq 0$ and $U_i \geq V_i$. If censoring precedes truncation, then the right-censored process N_{hi}^c started at V_i is only observed conditionally on an event before V_i.

In either case, one needs a specification of the conditional joint distribution of V_i and U_i given $U_i \geq V_i$ for $i = 1, \ldots, n$, possibly via a specification of their joint unconditional distribution. We shall return to this situation in the next section and conclude this one by a discussion of possible models for the left-truncation in two of the practical examples.

Example III.3.6. Diabetics in the County of Fyn

The problem and the data were presented in Example I.3.2 and we discussed in Example III.3.1 how the mortality rate might be studied as a function of disease duration. In this example, we want to model the mortality as a function of *age*. For simplicity, we consider just one sex. Furthermore, we first neglect migration both to and from the county of Fyn and consider the population of insulin-dependent diabetics diagnosed in Fyn county before 1 July 1973. We assume that the individuals behave independently and consider a single individual, *i*. Dropping the subscript *i*, we let

$$V = \text{age on 1 July 1973}$$

and

$$U = \text{age on 1 January 1982.}$$

Then $V < U$, and given the single covariate

$$Z = \text{age at diagnosis,}$$

$V = v_0$ and $U = u_0$ are deterministic. (There may, of course, be other covariates of interest which may easily be included in the model to be specified in the following.) We let

$$X = \text{age at death},$$

$\alpha_\theta(t; Z)$ be the hazard of X given Z (for t larger than age at diagnosis), and $N(t) = I(X \leq t)$. Then $\mathscr{F}_t = \mathscr{N}_t \vee \mathscr{F}_0 = \mathscr{G}_t$, where $\mathscr{F}_0 = \sigma(Z)$, and $N(t)$ has $(\mathrm{P}_{\theta\phi}, \mathscr{F}_t)$-intensity process

$$\lambda_\theta(t) = \alpha_\theta(t; Z) I(X \geq t),$$

where $\mathrm{P}_{\theta\phi}$ is the joint distribution of Z and X and ϕ parametrizes the marginal distribution of Z. We consider the event

$$A = \{X > v_0\} \in \mathscr{G}_{v_0}$$

and assume that $\mathrm{P}_{\theta\phi}(A) > 0$. Then the left-truncated counting process

$$_V N(t) = N(t) - N(t \wedge v_0)$$

has intensity process

$$_V \lambda_\theta(t) = \alpha_\theta(t; Z) I(v_0 < t \leq X)$$

with respect to the conditional distribution $\mathrm{P}_{\theta\phi}^A$ and the filtration $({}_V\mathscr{G}_t)$ where

$$_V \mathscr{G}_t = \sigma(Z; ({}_V N(u); v_0 < u \leq t)).$$

The *observed* right-censored and left-truncated counting process is

$$_V N^c(t) = \int_0^t C(u) \, \mathrm{d}\, {}_V N(u),$$

where $C(u) = I(u \leq u_0)$. It has intensity process

$$_V \lambda_\theta^c(t) = \alpha_\theta(t; Z) I(v_0 < t \leq X \wedge u_0)$$

with respect to $\mathrm{P}_{\theta\phi}^A$ and the *observed* filtration $({}_V\mathscr{F}_t^c)$, where

$$_V \mathscr{F}_t^c = \sigma(Z; ({}_V N^c(u); v_0 < u \leq t)).$$

In this simple case, truncation precedes censoring in the model construction.

If we no longer neglect migration *from* Fyn, the basic ("untruncated") model is a competing risks model with two cause-specific hazards: the death intensity $\alpha_\theta(t; Z)$ as before and an emigration intensity, say, $\gamma_{\theta\phi}(t; Z)$. We are then interested in the component $N_X(t)$ corresponding to death of the diabetic before age t; cf. Example III.2.6. This has intensity process

$$\lambda_X(t, \theta) = \alpha_\theta(t; Z) Y_0(t),$$

where $Y_0(t) = I(N_X(t-) = N_T(t-) = 0)$ indicates whether the diabetic is alive in Fyn at age $t-$ and $N_T(t)$ is the component of the basic counting process corresponding to emigration of the diabetic before age t. The event A is now defined as

$$A = \{Y_0(v_0) = 1\}$$

and we assume that $P_{\theta\phi}(A) > 0$. We let $\mathscr{F}_t = \mathscr{G}_t = \sigma(Z; (N_X(u), N_T(u); 0 \le u \le t))$ and ${}_V\mathscr{G}_t = \sigma(Z; (N_X(u), N_T(u); \ v_0 < u \le t))$. Then ${}_VN_X(t) = N_X(t) - N_X(t \wedge v_0)$ has $P^A_{\theta\phi}$-intensity process

$${}_V\lambda_X(t, \theta) = \alpha_\theta(t; Z) Y_0(t), \quad t > v_0,$$

with respect to $({}_V\mathscr{G}_t)$. The observed counting process

$${}_VN^c_X(t) = \int_0^t C(u) \, \mathrm{d}\, {}_VN_X(u)$$

then has $P^A_{\theta\phi}$-intensity process

$${}_V\lambda^c_X(t, \theta) = \alpha_\theta(t; Z) Y_0(t) I(t \le u_0), \quad t > v_0,$$

with respect to the observed filtration $({}_V\mathscr{F}^c_t)$, where

$${}_V\mathscr{F}^c_t = \sigma(Z; ({}_VN^c_X(u), {}_VN^c_T(u); v_0 < u \le t)).$$

In this case, some censoring (due to the competing risk: emigration) was present in the original model without truncation, and on top of the truncated model, some extra censoring was imposed due to the diabetic possibly being alive (in Fyn) on 1 January 1982. Note how we, to model the mortality as a function of age, explicitly included age at diagnosis as a covariate. In some preliminary analyses of the Fyn data, we shall, however, assume that the mortality only depends on age and not on age at diagnosis (see the examples using these data in Chapters IV–VI), whereas in Example VII.2.14, we return explicitly to the model studied above. □

EXAMPLE III.3.7. Mortality and Nephropathy Among Insulin-Dependent Diabetics at Steno Memorial Hospital

The problem and the data were presented in Example I.3.11. The basic (untruncated) model for a single individual is the three-state illness–death model discussed in Example III.1.11 with an intensity $\alpha^\theta_{01}(t)$ of developing diabetic nephropathy (DN), t referring to *time since onset of diabetes*, and death intensities $\alpha^\theta_{02}(t)$ before onset of DN, and $\alpha^\theta_{12}(t, d)$ after onset of DN, d referring to duration since onset of DN. These intensities may depend on covariates $\mathbf{Z}$ including, for instance, sex, age at diabetes onset, and calendar time at diabetes onset which have a distribution depending on nuisance parameters ϕ as in the previous example. We let $\mathscr{F}_t = \sigma(\mathbf{Z}; (N_{02}(u), N_{01}(u), N_{12}(u); 0 \le u \le t))$. With the usual definition of $Y_h(t)$, $h = 0, 1, 2$, $N_{0j}(t)$ then has intensity process

$$\lambda^\theta_{0j}(t) = \alpha^\theta_{0j}(t, \mathbf{Z}) Y_0(t), \quad j = 1, 2,$$

and $N_{12}(t)$ has intensity process

$$\lambda^\theta_{12}(t) = \alpha^\theta_{12}(t, t - T, \mathbf{Z}) Y_1(t)$$

with respect to $P_{\theta\phi}$ and $(\mathscr{F}_t)$. Here,

$$T = \inf_{t \in \mathscr{T}} \{Y_1(t) = 1\}$$

is the time (=disease duration) of onset of DN.

The (external) truncation time V (cf. Example III.3.6) is defined as time (=disease duration) of referral to the hospital ($V = \tau$ if the individual is never referred). We assume that V is independent of $\mathbf{N}$ with distribution also depending on ϕ. Another way of stating this is to say that from time 0 the diabetic is exposed to an intensity $\alpha_\phi(t)$ of being referred to Steno Memorial Hospital; cf. the discussion in Examples III.2.5 and III.3.3. This means that we assume the diabetics to be referred to the hospital independently of their later disease development; in particular, independently of possible later development of DN. If we now let

$$X = \inf_{t \in \mathscr{T}} \{Y_2(t) = 1\}$$

be the time of death, the truncation event is

$$A = \{V < X\}.$$

We assume that $P_{\theta\phi}(A) > 0$ and define $\mathscr{G}_t = \mathscr{F}_t \vee \sigma\{I(V \leq u), 0 \leq u \leq t\}$. Then $\mathbf{N}$ also has $(P_{\theta\phi}, \mathscr{G}_t)$-intensity process

$$\boldsymbol{\lambda}_\theta = (\lambda_{02}^\theta, \lambda_{01}^\theta, \lambda_{12}^\theta)$$

and the left-truncated counting process

$$_V\mathbf{N}(t) = \mathbf{N}(t) - \mathbf{N}(t \wedge V)$$

by Proposition II.4.2 has $(P_{\theta\phi}, \mathscr{G}_t)$-intensity process

$$_V\boldsymbol{\lambda}(t, \theta) = \boldsymbol{\lambda}(t, \theta) I(t > V).$$

Thus, both $\mathbf{Z}$, V, and $T \cdot I(T \leq V)$ must be included in the observed filtration to be able to calculate $_V\boldsymbol{\lambda}(t; \theta)$.

As in the previous example, a censoring mechanism may be superimposed onto $_V\mathbf{N}$ because the diabetic may still be alive on 31 December 1984. This censoring is deterministic for given value of $\mathbf{Z}$. □

III.4. General Censorship, Filtering, and Truncation

In Section III.2, we studied the case where the observation of the individual counting processes $\mathbf{N}_i = (N_{1i}, \ldots, N_{ki})$ was right-censored, i.e., the component i was observed not on $\mathscr{T}$ but only on a set of the form $E_i = [0, U_i]$. This is the most important example of incomplete observation, but there may be other observational plans of interest where observation of $\mathbf{N}_i$ is restricted to a subset $E_i \subseteq \mathscr{T}$.

Left-censoring corresponds to a set $E_i = (V_i, \tau]$, $V_i \geq 0$; as an example, we may consider the data regarding the time of descent from the trees of a baboon troop.

EXAMPLE III.4.1. Baboon Descent

The problem and the data were introduced in Example I.3.7. Recall that troops of baboons in the Amboseli Reserve, Kenya, sleep in the trees and descend for foraging at some time of the day. Observers often arrive later in the day (say, at time V_i) than this descent and, for such days, they can only ascertain that descent took place *before* V_i, so that the descent times are left-censored. □

When defined in terms of random variables, left-censoring is, of course, a concept symmetric to right-censoring, and, indeed, Ware and DeMets (1976) solved the baboon estimation problem by reversing time and using standard methods for right-censored data; see also Example IV.3.5. This trick, however, violates the basic role of the filtration in our framework, and, as we shall see presently, in more complicated models left-censoring presents special problems because of this.

Note the difference between left-censoring where $\mathbf{N}_i$ is only observed on $(V_i, \tau]$ and left-truncation (discussed in Section III.3) where $\mathbf{N}_i$, started at a stopping time V_i, is only observed on $(V_i, \tau]$ conditionally on an event prior to V_i. We shall return to a comparison of these two concepts later in this section.

Combination of left- and right-censoring corresponds to observing $\mathbf{N}_i$ on a set of the form $E_i = (V_i, U_i]$ and *censoring on intervals* more generally to a set

$$E_i = \bigcup_{j=1}^{r} (V_{ji}, U_{ji}], \tag{3.4.1}$$

where

$$0 \leq V_{1i} \leq U_{1i} \leq \cdots \leq V_{ri} \leq U_{ri} \leq \tau.$$

Thus, censoring corresponds to observing $\mathbf{N}$ on (possibly random) subintervals of $\mathcal{T}$ only.

EXAMPLE III.4.2. Mortality and Nephropathy Among Insulin-Dependent Diabetics

The problem and the data were introduced in Example I.3.11. Recall from that example that for some individuals developing diabetic nephropathy (DN), the exact time of onset of this complication was not observed and only a time U_1, last seen without DN, and a time V_2, first seen with DN, were known. For these patients observation is censored on the interval $(U_1, V_2]$. □

A concept closely related to censoring is *filtering* where the *jumps* of $\mathbf{N}$, i.e., $\mathrm{d}\mathbf{N}$, is observed on certain subintervals of $\mathscr{T}$ only. To exemplify the difference between these two concepts, we can consider a simple two-state Markov process model for a reversible disease (states *healthy* and *diseased*, transitions possible both ways, mortality disregarded; see, e.g., Example I.3.10). We can think of the case where occurrences of a recurrent disease are being studied in an individual on two intervals $[0, U_{1i}]$ and $(V_{2i}, U_{2i}]$, where $U_{1i} < V_{2i}$. If only new information is being collected during the two intervals, we are observing the disease process *via a filter* and the number of disease occurrences in $(U_{1i}, V_{2i}]$ will not be known. If, however, at time V_{2i} this number can be observed (via hospital records, interviews, or whatever), the observation of the disease process is *censored*. So, for a set E_i of the form (3.4.1), more information is available after a censored observation of the process than after observation via a filter, and only in the right-censoring case $E_i = [0, U_i]$ do the two concepts coincide. As we shall see presently, however, there may be cases where one deliberately throws away some information about the censored process and analyzes it as if it had been observed via a filter.

We shall now extend the method for handling right-censoring in Section III.2 to the more general plans of observation of $\mathbf{N}_i$ considered earlier. Corresponding to the set E_i, we define a *censoring* or *filtering* process $\mathbf{C}$ by

$$C_i(t) = I(t \in E_i)$$

and the *filtered* counting process by

$$N_{hi}^c(t) = \int_0^t C_i(u)\,\mathrm{d}N_{hi}(u).$$

Following the previous section, we restrict ourselves to *independent filtering* and assume the existence of a filtration $(\mathscr{G}_t) \supseteq (\mathscr{F}_t)$ such that $\mathbf{N} = (\mathbf{N}_i, i = 1, \ldots, n)$ *has the same* $\mathrm{P}_{\theta\phi}$*-compensator* $\mathbf{\Lambda}^\theta$ *with respect to both.* We also assume that the set E_i is such that C_i is $(\mathscr{G}_t)$*-predictable*, i.e., that the U_{ji} and V_{ji} are $(\mathscr{G}_t)$*-stopping times.* Then the $\mathrm{P}_{\theta\phi}$-compensator for N_{hi}^c with respect to $(\mathscr{G}_t)$ is

$$\Lambda_{hi}^c(t, \theta) = \int_0^t C_i(u)\,\mathrm{d}\Lambda_{hi}(u, \theta).$$

As in the cases of right-censoring, we assume that the available data include $\mathbf{X}_0$ and $\mathbf{N}^c$. For a set E_i of the form (3.4.1), time points V_{ji}, U_{ji} before the time τ_i of absorption are also observed. If τ_i is not observed, i.e., when τ_i belongs to some interval $(U_{ji}, V_{j+1,i}]$, then we may also observe the smallest V_{ji} such that $V_{ji} \geq \tau_i$. This is, for instance, the case when we have a left-censored survival time X_i and observe V_i when $V_i > X_i$; see Example III.4.4. In the case of censoring, we also observe the values of N_{hi} at the observed *entry times* V_{ji}. In the case of filtering, these values are not observed. We do, however, assume that the data at time t enable us to calculate $\mathbf{\Lambda}^c(t, \theta)$ for any given value of θ. Viewing these observed data as a marked point process $\mathbf{N}^*$, we can calculate the likelihood $L_\tau^*(\theta, \phi)$ corresponding to the filtration $(\mathscr{F}_t^c)$ generated by $\mathbf{N}^*$. The partial likelihood for $\mathbf{N}^c$ takes the same form

$$L_\tau^c(\theta) = \prod_{t \in \mathscr{T}} \prod_{h,i} \mathrm{d}\Lambda_{hi}^c(t, \theta)^{\Delta N_{hi}^c(t)}(1 - \mathrm{d}\Lambda_{\bullet\bullet}^c(t, \theta))^{1-\Delta N_{\bullet\bullet}^c(t)} \tag{3.4.2}$$

as in Section III.2.2; see (3.2.8).

When N_{hi} satisfies Aalen's multiplicative intensity model

$$\lambda_{hi}(t, \theta) = \alpha_{hi}^\theta(t)\, Y_{hi}(t)$$

with respect to $(\mathscr{F}_t)$, it follows that $N_{hi}^c(t)$ follows the multiplicative intensity model

$$\lambda_{hi}^c(t, \theta) = \alpha_{hi}^\theta(t)\, Y_{hi}^c(t)$$

with respect to $(\mathscr{F}_t^c)$. Here, $Y_{hi}^c(t) = Y_{hi}(t) C_i(t)$. Therefore, we also term the independent filtering process $\mathbf{C}$ an *Aalen filter*.

EXAMPLE III.4.3. Mau's Concept of Partitioned Counting Processes

Assume $k = 1$, let, for $i = 1, \ldots, n$ and some m, $0 \le S_{i0} \le S_{i1} \le \cdots \le S_{im} \le S_{i,m+1} = \tau$ be $(\mathscr{G}_t)$-stopping times and define stochastic processes

$$C_i^{(j)}(t) = I(S_{i,j-1} < t \le S_{ij}), \quad i = 1, \ldots, n;\ j = 1, \ldots, m+1.$$

Then to each component N_i there exists an $(m+1)$-variate *partitioned counting process* given by

$$N_{ij}(t) = \begin{cases} 0, & 0 \le t \le S_{ij} \\ N_i(t) - N_i(S_{ij}), & S_{ij} < t \le S_{i,j+1} \\ N_i(S_{i,j+1}) - N_i(S_{ij}), & S_{i,j+1} < t \le \tau, \end{cases}$$

$i = 1, \ldots, n$, $j = 1, \ldots, m+1$. Obviously, $N_{ij}(t)$ counts the events in the random interval $(S_{ij}, S_{i,j+1}]$. Mau (1985) noted that if each N_i satisfies the multiplicative intensity model, then also the partitioned counting process satisfies the multiplicative intensity model, now with intensity processes

$$\alpha_i^\theta(t)\, Y_{ij}(t) = \alpha_i^\theta(t) C_i^{(j)}(t)\, Y_i(t).$$

In our terminology, $\mathbf{C} = (C_i^{(j)})$ is an Aalen filter, so that the analysis of what happens to the counting process in particular random intervals may be performed using the powerful tools of the multiplicative intensity model. Mau (1987) showed how this allows monitoring of clinical trials, e.g., by separately analyzing the information from several calendar time intervals in a trial with staggered entry; cf. also Keiding et al. (1987) and Examples I.3.5 and V.2.5. □

When inference is based on the partial likelihood $L_\tau^c(\theta)$ alone using the Aalen filter, some factors in the full likelihood $L_\tau^*(\theta, \phi)$ for the marked point process $\mathbf{N}^*$ are disregarded. The marks for $\mathbf{N}^*$ which are discarded will, at time t, contain information on certain individuals either leaving the risk set or entering the risk set at that time and they may also be defined to carry information on occurrences of earlier events, the exact times of which are not observed. In that case, the marks will typically carry information on θ, the parameter of interest, and we then term the censoring mechanism (or the filter) $\mathbf{C}$ *informative* for θ; compare Section III.2.3. If $\mathbf{C}$ is *noninformative* for θ, $L_\tau^c(\theta)$ is the full likelihood (or at least the full conditional likelihood given $\mathbf{X}_0$) and no information is lost by basing the statistical inference on it. In the example mentioned earlier concerning a disease process observed on the set $E_i = [0, U_{1i}] \cup (V_{2i}, U_{2i}]$, the mark at V_{2i} may thus contain information of disease occurrences in $(U_{1i}, V_{2i}]$. In this case, some information is lost by only considering the process counting the number of disease occurrences filtered via the process $\mathbf{C}$ with components $C_i = I(t \in E_i)$. Thus, $\mathbf{C}$ is informative (for the parameters of the disease intensity) and it would be more efficient to base inference on the entire likelihood $L_\tau^*(\theta, \phi)$ than on the partial likelihood $L_\tau^c(\theta)$. On the other hand, the entire likelihood may depend on the nuisance parameter ϕ which is often inconvenient. In fact, one may not even be prepared to write down a full statistical model for $\mathbf{N}$, $\mathbf{C}$.

To return to a comparison of *left-truncation*, *left-censoring*, and *left-filtering*, note first the technical difference that the latter two keep the original sample space and probability measure, whereas left-truncation is a conditional procedure, restricted to a subset of the sample space and the corresponding conditional probability. Some further aspects are best considered in the simplest possible example.

EXAMPLE III.4.4. Random Left-Censoring of a Non-Negative Random Variable

We consider once more the setup from Examples III.2.5 and III.3.2: X and V are independent non-negative random variables, X has hazard function $\alpha_X(\cdot, \theta)$, and V has distribution function $F_V(\cdot, \phi)$ [and hazard function $\alpha_V(\cdot, \phi)$ if it exists]. Furthermore, complete observation of X and V is considered as observation of a marked point process as described in those examples. Right-censoring at V corresponds to only being able to observe $X \wedge V$ and the mark at that time, i.e., the counting process $N_X(t) = I(X \leq t)$ is only observed on the random interval $E = [0, V]$. Similarly, left-censoring corresponds to the case where $N_X(t)$ is only observed on a set $E = (V, \tau]$, and we shall assume that V is always observed. Thus, we observe the *filtered* counting process

$$N^c(t) = \int_0^t C(u)\,\mathrm{d}N(u) = (N(t) - N(V))I(t > V),$$

its $(\mathcal{G}_t)$-compensator (except for the value of the unknown θ)

$$\Lambda^c(t, \theta) = \int_0^t C(u)\,\mathrm{d}\Lambda(u, \theta) = \int_0^t \alpha(u, \theta)\,Y^c(u)\,\mathrm{d}u$$

[where $Y^c(t) = C(t)Y(t) = I(V < t \leq X)$], *and* the value $N(V) = I(X \leq V)$. That is, if $X > V$, then we observe *both* V and X, and if $X \leq V$, then we observe V *and* know that $X \leq V$. The partial likelihood $L_\tau^c(\theta)$ is in this case, according to (3.4.2),

$$L_\tau^c(\theta) = \alpha_X(X, \theta)^{I(X > V)} \prod_{V < t \leq X} (1 - \alpha_X(t, \theta)\,dt)$$

$$= \left(\frac{S_X(X, \theta)}{S_X(V, \theta)} \alpha_X(X, \theta) \right)^{I(X > V)}. \tag{3.4.3}$$

The second factor of the full likelihood $L_\tau^*(\theta, \phi)$ is

$$L_\tau''(\theta, \phi) = S_V(V, \phi)\alpha_V(V, \phi)S_X(V, \theta)^{I(V < X)}F_X(V, \theta)^{I(V \geq X)}$$

and it *does* depend on θ. So, obviously, observation of $N(V)$ gives us some information on θ, meaning that $C(\cdot)$ is informative for θ and that inference based on $L_\tau^c(\theta)$ only will not be fully efficient.

The fact that V is always observed is in contrast to the case of right-censoring (see, however, Example III.2.3). A more direct parallel definition of left-censoring would be to assume that $X \vee V$ were observed together with the mark at that time. That situation gives a different likelihood but it can be handled in a similar way. □

The example shows that left-censored data may be analyzed using the Aalen filter by treating the counting process as being observed with *delayed entry* or as being *left-filtered.* This way of analyzing left-censored data is, however, not fully efficient.

The partial likelihood (3.4.2) for the left-filtered process $\mathbf{N}^c$ is identical to the partial likelihood (3.3.3) based on the left-truncated process ${}_V\mathbf{N}$ in the sense that the data used in the two situations are the same and that the parameter θ enters into the two likelihoods in the same way, though, formally, the likelihood (3.3.3) is with respect to a conditional distribution $\mathrm{P}_{\theta\phi}^A$, whereas (3.4.2) is with respect to the original probability measure $\mathrm{P}_{\theta\phi}$. This shows that *left-truncated* counting processes can be correctly analyzed as counting processes observed with delayed entry: If individuals $i = 1, \ldots, n$ are observed, then individual i is included in the relevant risk set from the time V_i. Also, conditions for "independence" in the two situations are very much related. Thus, in Example III.3.2, we mentioned that, in the case of survival data, we have independent left-truncation if the conditional joint density of (X, V) given $X > V$ can be written in a certain product form and it is easily seen [e.g., Keiding (1992)] that the same condition ensures independent left-filtering.

For survival data, the basic difference between a left-censored and a left-truncated survival time X_i is that, in the latter case, individual i is only included in the sample conditional on its survival time exceeding the entry time V_i, whereas in the former case, individual i is always included in the sample, but observation of the exact failure time may be prevented for some reason. So, for n independent left-truncated observations, the partial likelihood will be a product of a *fixed number* (n) of factors of the form (3.4.3), whereas, for n independent left-censored observations, the partial likelihood is a product of a *random number* ($\leq n$) of factors of this form.

The situation with both left-truncation and right-censoring can also be handled using the Aalen filter. In this case, we can define the filtering process by

$$C_i(t) = I(V_i < t \leq U_i)$$

and base the inference on θ on the (partial) likelihood $L^c_\tau(\theta)$ with the form (3.4.2). In many realistic models for the distribution of $\mathbf{N}$, V, and U, the filtering process $\mathbf{C} = (C_1, \ldots, C_n)$ will be noninformative about θ and this analysis will be efficient. It should, however, be emphasized once more that for other incomplete plans of observation, a more efficient analysis can be carried out using the entire likelihood $L^*_\tau(\theta, \phi)$.

Right-truncation and more general types of truncation may be defined similarly to left-truncation, but none of these are conveniently dealt with in the present framework. As for left- versus right-censoring, the explanation is that the time direction given by the filtration destroys the symmetry between left and right, except for some simple cases where one may study right-truncation by reversal of time; see Example III.4.5. Keiding (1991) presented applications to the disease intensity ("incidence") in the illness–death model under special epidemiological sampling plans.

EXAMPLE III.4.5. Incubation Time of AIDS

The incubation time (duration from infection with the virus HIV until disease occurrence) of AIDS is difficult to measure because ordinarily the event of infection is unobserved. However, for some patients, the infection happened at a blood transfusion, say, at calendar time T_i. Only if the time when AIDS developed (say, U_i) is before the closing date (say, Z_i) is the patient observed. That is, the incubation time $X_i = U_i - T_i$ is observed conditional on $X_i < Z_i - T_i$, which is *right-truncation*.

As for left-censoring, reversal of time allows the explicit solutions of left-truncation to be imitated, the *retro-hazard* $\bar{\alpha}(t)$ given by

$$\bar{\alpha}(t)\,\mathrm{d}t = \mathrm{P}\{X > t - \mathrm{d}t \mid X \leq t\}$$

taking the place of the usual hazard. See Lagakos et al. (1988), Medley et al. (1988), Kalbfleisch and Lawless (1989), Keiding and Gill (1990), and Gross and Huber-Carol (1992). □

An extreme example of incomplete observation formally covered by the concept of interval censoring is observation of a *discrete skeleton* of the process, that is, N_h is observed at times $0 = \tau_0^{(h)} < \tau_1^{(h)} < \cdots < \tau_{r_h}^{(h)} \leq \tau$. Because one will then (with probability 1) never observe the exact time of a transition, the Aalen filter will reduce observation to nothing: the partial (filtered) likelihood $L_\tau^c(\theta) = 1$. This is, for instance, the case in connection with various kinds of grouped data from a Markov process as follows.

Let the Markov process X with state-space S, counting process $\mathbf{N}$, and filtration $(\mathscr{F}_t)$ be defined as in Example III.1.4. The statistical model is given by assuming some transitions impossible and the rest of the transition intensities [specified by the set $R \subseteq \{(h, j): h, j \in S, h \neq j\}$] arbitrarily varying. Intermittent observation of the counting process $\mathbf{N} = (N_{hj}, (h, j) \in R)$ is observation of $(N_{hj}(\tau_1^{(hj)}), \ldots, N_{hj}(\tau_{r_{hj}}^{(hj)}))$, where the τ's are assumed to be deterministic times unless otherwise specified. Note that $N_{hj}(\tau_{i+1}^{(hj)}) - N_{hj}(\tau_i^{(hj)})$ counts the number of transitions $h \to j$ in the time interval $(\tau_i^{(hj)}, \tau_{i+1}^{(hj)}]$, and, therefore, intermittent observation of the transition counts corresponds to *grouped observation* of the transition times. As indicated earlier, filtering removes all information in the sense that (with probability one) we have $\mathbf{N}^c = \mathbf{0}$ and $L_\tau^c(\theta) = 1$. In this case, the partial likelihood contains no information on θ. In connection with other kinds of grouped data, even less information may be available. Sometimes, only the state occupied at $\tau_0, \ldots, \tau_r$ is observed, i.e., $(X(\tau_j), j = 0, 1, \ldots, r)$ or equivalently $(\mathbf{Y}(\tau_j+), j = 0, 1, \ldots, r)$. Also, in this case, $L_\tau^c(\theta) = 1$ and one has to consider the full likelihood. Kalbfleisch and Lawless (1985) studied maximum likelihood estimation in the model with constant transition intensities based on *panel data*, i.e., observation of independent Markov processes $X_i(\cdot)$, $i = 1, \ldots, n$, at time points $\tau_0, \tau_1, \ldots, \tau_r$.

There may even be cases where the individual panel data are not available but only the number of individuals in each state. In this case, the *aggregated* data are $(Y_{h\cdot}(\tau_j); j = 0, 1, \ldots, r, h = 1, \ldots, k)$. We shall not discuss this kind of grouped data in this monograph except for some bibliographic remarks in this chapter and in Chapter VI.

III.5. Partial Model Specification. Time-Dependent Covariates

Under independent censoring or filtering, as we have seen in the previous sections, it is possible to write down a partial likelihood for θ, the parameter of interest, which has the same form as the likelihood for the full data and which does not depend on the nuisance parameter ϕ. Thus, the partial likelihood can be computed without actually specifying a model for the censoring mechanism, in fact as if censoring had been at fixed given times. Other examples of models which may be only partially specified are the Markov process model in Example III.1.4 and the Cox regression model in Example III.1.7. In the former, a specification of the distribution of the initial states is not needed for writing down the intensity process and thereby the partial likelihood $L_\tau(\theta)$, whereas, in the latter, inference can be performed conditionally on the covariates and without specifying a model for their distribution.

In the latter example, the covariates were *time-independent*, i.e., they were fixed given $\mathscr{F}_0$, but in several examples it is also of interest to study intensities conditionally on covariates which change in time. Some such time-dependent covariates may be deterministic or at least fixed, given $\mathscr{F}_0$.

EXAMPLE III.5.1. Survival Among Insulin-Dependent Diabetics in the County of Fyn

The problem and the data were introduced in Example I.3.2. In Example III.3.6, the basic time was taken to be the age of the patients, whereas the age at diagnosis was included as a time-independent covariate. The death intensity may also depend on the time-dependent covariate "disease duration" which can be computed for each age t knowing the age at diagnosis. Thus, the stochastic process $Z_i(t)$ = "disease duration for patient i at age t" is adapted to the filtration $({}_V\mathscr{F}_t^c)$ generated by the data. □

Similar remarks apply in the study of mortality and nephropathy among insulin-dependent diabetics (Examples I.3.11 and III.3.7).

EXAMPLE III.5.2. Psychiatric Admissions Among Women Having Just Given Birth

The problem and the data were introduced in Example I.3.10, and, as a basic model, we may use a Markov process as in Example III.1.4 with two states "in psychiatric ward" and "not in psychiatric ward." In this study, however, the admission intensity is likely to depend on the time since latest discharge from a psychiatric hospital (thereby, in fact, making the process *semi*-Markov). This time span may then be included as a time-dependent covariate and if information on the women's psychiatric history is included in $\mathscr{F}_0$, the covariate is adapted to $\mathscr{F}_0 \vee \mathscr{N}_t$. For a further discussion, see Example III.5.6. □

In such cases where the intensity depends on what Kalbfleisch and Prentice (1980, p. 123) termed a *defined time-dependent covariate*, the (partial) likelihood stays the same and inference based on the likelihood can be performed as if the covariate paths had been fixed in advance.

In other examples, there may be time-dependent covariates which are *truly random* in the sense that the processes $\mathbf{Z}_i(\cdot)$, $i = 1, \ldots, n$, are *not* automatically adapted to the filtration under consideration. Kalbfleisch and Prentice (1980, Section 5.3) distinguished between *ancillary covariates* and *internal covariates*, giving as an example of an *ancillary* covariate the level of air pollution in a study of the occurrence of asthma attacks. One important class of *internal* covariates is "disease complications" developing in a fashion unpredictable from the history of the process itself.

EXAMPLE III.5.3. CSL 1: A Randomized Clinical Trial in Liver Cirrhosis

The problem and the data were introduced in Example I.3.4. As mentioned there (see also Example I.3.12), several biochemical and clinical characteristics with potential prognostic information were recorded at the follow-up visits and it may be highly relevant to include these measurements as time-dependent covariates in the model for the death intensity. □

To include such covariates in the model, we must extend the filtration. One way of doing that is to consider the whole system of uncensored observations as developing according to a (very large) marked point process $\mathbf{N}_Z$ recording, with *innovative marks*, failures (and other transitions or events of interest) and, with *noninnovative marks*, changes of covariate values; see Arjas and Haara (1984) and Section II.7.3. To consider everything as a *point* process does pose some restrictions on the types of covariates considered in that (random) changes of covariate values have to be generated by an underlying process changing at discrete (possibly random) points in time and not continuously. So, if a continuously observed time-dependent covariate, which is not adapted, is to be included in the model, then its path has to be discretized in some way. This may, for instance, be done by defining its changes of values to happen at discrete points in time or at least to let its path vary deterministically except at a discrete set of points.

EXAMPLE III.5.4. The Feeding Pattern of a Rabbit

The problem and the data were presented in Example I.3.17. Each rabbit was observed continuously during a week; at times $T_1, T_2, \ldots$, it ate $x_1, x_2, \ldots$ grams of food.

Let $\mathbf{N}_Z$ be the marked point process with mark x_v at time T_v and let $(\mathscr{F}_t)$ be the filtration generated by this process. The $\mathscr{F}_t$-intensity process of the counting process $N(t) = \#\{v: T_v \leq t\}$ recording only the times of the meals (the "innovative" part of the mark), but not the marks $x_1, x_2, \ldots$ themselves (the "non-innovative" parts) may be postulated to have the form

$$\alpha(t, \gamma) \exp(\boldsymbol{\beta}^\top \mathbf{Z}(t)),$$

see Theorem II.7.5. Here the positive function $\alpha(t, \gamma)$ is *periodic* with period 24 h and $\mathbf{Z}(t)$ is assumed predictable with respect to $(\mathscr{F}_t)$, examples of components of $\mathbf{Z}(t)$ being the time since last meal, the weight of last meal, the number of meals during last 30 min, or the total intake during 60 min to 30 min prior to t. This provides another example of a partially specified model: If ϕ parametrizes the distribution of the x_v and if we let $\theta = (\gamma, \boldsymbol{\beta})$ be the parameter of interest, then the model is only specified to the extent that the dependence on θ of the $(P_{\theta\phi}, \mathscr{F}_t)$-intensity process for $N(t)$ is given. The model, proposed by Pons and Turckheim (1988a,b), is seen to be outside of Aalen's multiplicative intensity model, indeed being rather a generalization of the Cox regression model (cf. Example III.1.7) to point processes on the line.

The particular case $\boldsymbol{\beta} = \mathbf{0}$ reduces the model to the periodic Poisson process; cf. Jolivet et al. (1983). We return to an analysis of these data in Example VII.2.15. $\square$

In this example there was censoring of type I only: The rabbit was observed for 1 week. In other cases, one may wish to superimpose censoring or filtering onto the marked point process $\mathbf{N}_Z$ via an independent censoring or filtering process $\mathbf{C}$ which is predictable with respect to a filtration $(\mathscr{G}_t)$ larger than that generated by $\mathbf{N}_Z$. In this way, a *censored* or *filtered* marked point process, say, $\mathbf{N}_Z^*$, is obtained. As before, $\mathbf{N}_Z^*$ represents our observations and only a part of it, namely, $\mathbf{N}^c$, counts the observed transitions of interest. The process $\mathbf{N}^c$ carries the innovative parts of the marks, whereas the other marks of $\mathbf{N}_Z^*$ are noninnovative and include information on individuals entering or leaving the risk sets and on observed changes in covariate values. We assume that observation of $\mathbf{N}_Z^*$ enables us to calculate for each value of θ the $(\mathbf{P}_{\theta\phi}, (\mathscr{G}_t))$-compensator for $\mathbf{N}^c$.

EXAMPLE III.5.5. Mating of Drosophila Flies

The problem and the data were introduced in Example I.3.16. One possible model was set up in Example III.1.10, and possible censoring mechanisms discussed in Example III.2.12. In the latter example, an extended filtration $(\mathscr{G}_t)$ recording also (with noninnovative marks) the times of termination of matings was introduced. Because the biologists sometimes observe that the *male* flies mate more than once during one experiment, a more realistic model for the mating intensity process $\lambda(t, \theta)$ may be to write it as

$$\lambda(t, \theta) = \alpha(t, \theta) F(t) \tilde{M}(t).$$

where $\tilde{M}(t)$ is the number of male flies not engaged in a mating at time $t-$ and $F(t)$, as before, denotes the number of female flies not yet having initiated a mating before $t-$. It is seen that both $F(\cdot)$ and $\tilde{M}(\cdot)$ are $(\mathscr{G}_t)$-predictable.

Again, we have a partially specified model: Letting ϕ parametrize the distribution of the duration of the matings, we only have to specify how the $(\mathbf{P}_{\theta\phi}, \mathscr{G}_t)$-compensator of the process counting initiations of matings depends on θ. $\square$

We let $(\mathscr{F}_t^c)$ be the filtration generated by the data $\mathbf{N}_Z^*$. The full likelihood $L_\tau^*(\theta, \phi)$ for $\mathbf{N}_Z^*$ with respect to $(\mathscr{F}_t^c)$ can, as in Section III.4, be factored into the partial likelihood $L_\tau^c(\theta)$ based on $\mathbf{N}^c$ which does not depend on the nuisance parameter ϕ and a second factor $L_\tau''(\theta, \phi)$ which may or may not depend on θ. This means that inference on θ can be based on $L_\tau^c(\theta)$ only and it can be made without specifying the model for the censoring mechanism and the covariate processes. However, as before, a more efficient inference on θ may be obtained from the full likelihood $L_\tau^*(\theta, \phi)$ if the second factor does, in fact, depend on θ, i.e., if censoring or covariates are *informative*.

The partially specified model specifying only the $(\mathscr{F}_t^c)$-compensator $\mathbf{\Lambda}_\theta^c$ for $\mathbf{N}^c$ has some limitations due to the fact that only a small part of a big system is modelled. If one wants to make predictions on the basis of the model, then this is not directly possible if the model for $\mathbf{N}^c$ includes time-dependent covariates whose development in time is *not* modelled (Andersen, 1986; Andersen et al., 1991a). So, if prediction making is an important issue of a study, one has to either disregard time-dependent covariates or to model them. The latter possibility corresponds to labelling the marks for changes in these covariates innovative and to include the parameters for them in θ rather than in ϕ. We shall return to this in Examples VII.2.10 and VII.2.16 in connection with the CSL 1 survival data (Examples I.3.4 and I.3.12). It should be emphasized that the labelling of marks as innovative or noninnovative is up to the statistician and it depends on the purposes of the study. Another example of this problem was seen previously in that one may sometimes be interested in studying several cause-specific hazard functions in a competing risks model (Example III.1.5) and sometimes only deaths due to one cause are of interest, whereas deaths due to other causes are treated as censorings and the corresponding cause-specific hazard functions as nuisance parameters (Example III.2.6). This has also been discussed in connection with the melanoma survival data (Examples I.3.1, I.3.9, and III.2.7).

Another problem with censored or filtered observation of time-dependent covariates is that values of these covariates may not be observed, which may prevent one from computing even the partial likelihood $L_\tau^c(\theta)$. As discussed in Example III.3.5, this might be the case in the illness–death model with duration dependence introduced in Example III.1.11. Suppose that the $1 \rightarrow 2$ transition intensity is modelled as $\alpha_{12}(t, t - T) = \alpha_0(t)\exp(\beta(t - T))$ using the time-dependent covariate $Z(t) = t - T =$ "sojourn time in state 1 at time t," and suppose that at the entry time $V > T$, the value of T is unknown. Then the value of $Z(t)$ is unobservable. Another example is the following.

Example III.5.6. Psychiatric Admissions of Women Having Just Given Birth

The problem and the data were introduced in Example I.3.10. As mentioned there, only the psychiatric history from 1 October 1973 was known for the Danish women giving birth in 1975. This means that, for each woman, there may be insufficient information in $\mathscr{F}_0$ to assess the duration dependence discussed in Example III.5.2. We shall return to this in Example VII.2.17. □

III.6. Bibliographic Remarks

Sections III.1–III.3

The formulation of event history models based on counting processes was initiated by Aalen (1975, 1978b) who showed that his multiplicative intensity model covered, among other cases, the survival data (Examples III.1.1 and III.1.2), the Markov process (Examples III.1.4 and III.1.5), and Example III.1.10 on matings of Drosophila flies. Further examples and discussion were provided by Gill (1980a), Andersen et al. (1982), Jacobsen (1982), and Andersen and Borgan (1985); see also Section I.2. General likelihood constructions are found in the papers by Jacod (1975), Arjas and Haara (1984), Arjas (1989), and (using product-integrals) Gill and Johansen (1990). Also, the introduction into counting process models of right-censoring (Section III.2) and the more general kinds of incomplete observation (Section III.4) is due to Aalen (1975, 1978b) and was further discussed by Aalen and Johansen (1978) and Andersen et al. (1982). For survival data models, there has been quite an extensive discussion in the literature of independent right-censoring [see the references in Example III.2.13 and Lagakos and Williams (1978), Cox and Oakes (1984, Section 1.3), and Fleming and Harrington (1991, Section 1.3)] and noninformative right-censoring (Lagakos, 1979; Kalbfleisch and Prentice, 1980; Arjas and Haara, 1984; Arjas, 1989; Jacobsen, 1989b). For more general counting process models, the present discussion based on Arjas and Haara (1984) is new; cf. Andersen et al. (1988).

A general theory for inference with missing data was developed by Rubin (1976) and described in detail by Little and Rubin (1987). Recently, Heitjan and Rubin (1991) and Heitjan (1993) generalized this theory to include a concept called "coarsening at random" which includes as a special case a formalization of the property of independent right-censoring that "past observations do not affect the probabilities of future failures." Jacobsen and Keiding (1991) reconciled this development with the theory in the current chapter and that of Jacobsen (1989b).

Truncation and censoring in general was discussed by Hald (1949; 1952, p. 144), whereas early contributions on left-truncated survival data are the fundamental article by Kaplan and Meier (1958), the article on one-sample tests by Hyde (1977) [see also Hyde (1980)], and also Aalen (1978b) mentioned the concept of delayed entry. References to the demographic literature were given by Hoem (1976). In later years, the literature on left-truncated survival data has grown quite rapidly with main emphasis on the random truncation model; see Woodroofe (1985) and the references given by Keiding and Gill (1990).

Sections III.4 and III.5

The recent emphasis on modelling AIDS epidemics (Example III.4.5) has increased the interest in general censoring and truncation. But, nevertheless, the literature is still rather unsophisticated as regards concepts of independence and noninformativity of the censoring (and truncation) patterns, in effect assuming these to be deterministic. However, aspects of the iterative methods (versions of the EM algorithm) necessary to study the full likelihood and primarily developed by Turnbull (1974, 1976) and Dempster, Laird, and

Rubin (1977, Section 4.2) are instructive in the general modelling framework of this chapter. Whereas censoring is readily interpreted as being an example of incompletely observed data, it is at first sight more surprising that truncation may also be interpreted in this way. The idea is to consider among the unobserved data also the number of individuals who were never observed because their values are outside the relevant truncation set. Turnbull (1976) termed these the "ghosts" and Dempster, Laird, and Rubin (1977) gave a comprehensive discussion.

Later authors have primarily been concerned with the (difficult) task of providing asymptotic properties of estimators derived this way. An interesting modelling contribution was made by Samuelsen (1989), who suggested a stochastic process model for double censoring, generalizing the competing risks framework for random right-censoring (Example III.2.6) and the Markov process model for random left-truncation (Example III.3.3). We shall not discuss Samuelsen's work or these other kinds of incomplete observation in detail in this monograph but only briefly mention them in the Bibliographic Remarks of Chapter IV.

The literature on intermittent observation of a counting process is mainly concerned with the illness–death process earlier discussed in Examples III.1.11 and III.3.5. Whereas the time of death can usually be observed exactly, it is often difficult to assess exactly when disease occurred. One group of problems of this kind is *long-term animal carcinogenicity trials* where it is usually assumed that it can always be assessed after death whether disease had occurred or not. Often supplementary data are obtained by *serial sacrifice*, that is, animals are killed at prespecified times and it is assessed whether or not they already had the tumor. Most of the literature on designs of such trials [see, e.g., Borgan et al. (1984)] studies deterministic observational plans as opposed to plans determined adaptively by the development of the process. Exceptions are Mau (1986a) who formulated an explicitly random "associated design process" and Arjas and Haara (1992a, 1992b). The likelihood function usually becomes complicated and only certain functionals of the process are identifiable. Recent reviews well in tone with the approach taken here are by McKnight and Crowley (1984), McKnight (1985), and Dewanji and Kalbfleisch (1986); see also the monograph by Gart et al. (1986).

A somewhat different application of the simple illness–death process is to nonreversible complications of chronic diseases such as diabetes (Andersen, 1988) or cancer. Here, patients are examined at visits to the hospital and the determination of whether a transition $0 \to 1$ (onset of disease complication) has happened may only be performed at those times. It is, here, very important (though often overlooked in practice) to know whether the visits to the hospital are planned independently of the underlying disease process (as would be true for deterministic observational plans) or whether they may be triggered by the disease. Motivated by these problems, Grüger (1986) developed an interesting theory of *noninformative observational plans* for counting processes; see also Grüger et al. (1991).

Classification of time-dependent covariates into internal/external or innovative/noninnovative was discussed by Kalbfleisch and Prentice (1980), Arjas and Haara (1984), and Arjas (1989). The last two papers also discussed the important concept of partial model specification; see also Gill (1983b), Jacod (1987), Greenwood (1988), and Greenwood and Wefelmeyer (1990).

CHAPTER IV

Nonparametric Estimation

In the preceding chapter we discussed how the mathematical theory of Chapter II could be used to specify statistical models based on counting processes. In particular, Examples III.1.2–III.1.4 on identically distributed survival data, relative mortality, and finite-state Markov processes had a common structure. In all these examples, aggregation of the individual counting processes leads to a multiplicative model for the intensity process. Furthermore, in Sections III.2–III.4, we saw how this multiplicative structure was retained under independent right-censoring, filtering, and left-truncation. Another situation with multiplicative intensity process was provided in Examples III.1.10 and III.5.5 on matings for Drosophila flies.

In the present chapter, we will study nonparametric estimation for the multiplicative intensity model. Thus, our setup is the following. We fix a continuous-time interval $\mathscr{T}$ which may be of the form $[0, \tau)$ or $[0, \tau]$ for a given terminal time τ, $0 < \tau \leq \infty$ (cf. Section II.2). Note that the terminal point τ may or may not be included in $\mathscr{T}$. Let $(\Omega, \mathscr{F})$ be a measurable space equipped with a filtration $(\mathscr{F}_t, t \in \mathscr{T})$ satisfying the usual conditions (2.2.1) (except possibly completeness, cf. Section II.2) for each member of a family $\mathscr{P}$ of probability measures. Defined on $(\Omega, \mathscr{F})$ and adapted to the filtration, we have a multivariate counting process $\mathbf{N} = ((N_1(t), \ldots, N_k(t)); t \in \mathscr{T})$ satisfying the *multiplicative intensity model*, i.e., its $(\mathrm{P}, (\mathscr{F}_t))$-intensity process $\boldsymbol{\lambda} = (\lambda_1, \ldots, \lambda_k)$ is given by

$$\lambda_h(t) = \alpha_h(t) Y_h(t),$$

for $h = 1, \ldots, k$ and $\mathrm{P} \in \mathscr{P}$, cf. Sections II.4.1 and III.1.3. Here α_h is a nonnegative deterministic function (depending on P), whereas Y_h is a predictable process which is observable in the sense that it does not depend on P (cf. Section II.3.1). Often α_h is an individual force of transition, whereas Y_h counts the number at risk (cf. Examples III.1.2 and III.1.4), but other interpretations are also possible (cf. Example III.1.3). We will assume that α_h satisfies $\int_0^t \alpha_h(s)\,\mathrm{d}s < \infty$ for all h and all $t \in \mathscr{T}$. We consider nonparametric estimation in the present chapter, so no other assumptions are made on the α_h. Even if it is not made explicit in the notation, as opposed to what was the case in Chapter III, it must be stressed once more that the counting process $\mathbf{N} = (N_1, \ldots, N_k)$ as well as the Y_h will typically be derived by *aggregating* censored, filtered, or truncated observations as discussed in detail in Chapter III.

In Section IV.1, we define and study the Nelson–Aalen estimator for $A_h(t) = \int_0^t \alpha_h(s)\,\mathrm{d}s$, whereas in Section IV.2, we show how this estimator may be smoothed to obtain an estimator for α_h itself. In Sections IV.3 and IV.4, we consider the special cases of survival times (cf. Example III.1.2) and observations from finite-state Markov processes (cf. Example III.1.4). The Kaplan–Meier product-limit estimator is discussed in Section IV.3, whereas in Section IV.4, we define and study the product-limit estimator for the transition probabilities of a Markov process with finite state space proposed by Aalen and Johansen (1978). We return to a discussion of these estimators in Chapter VIII, where, in Sections VIII.2.4 and VIII.4.1, we prove that the Nelson–Aalen estimator is asymptotically efficient and show how this efficiency carries over to the Kaplan–Meier estimator and the product-limit estimator for Markov processes proposed by Aalen and Johansen.

Throughout the chapter, the counting processes are defined on a filtered probability space as described in Chapter III and outlined above, but we will only rarely make this explicit below.

IV.1. The Nelson–Aalen Estimator

IV.1.1. Definition and Basic Properties

We consider a multivariate counting process $\mathbf{N} = (N_1, \ldots, N_k)$ with intensity process $\boldsymbol{\lambda} = (\lambda_1, \ldots, \lambda_k)$ satisfying the multiplicative intensity model $\lambda_h(t) = \alpha_h(t) Y_h(t)$ as described in the introduction to this chapter.

To derive heuristically estimators for

$$A_h(t) = \int_0^t \alpha_h(s)\,\mathrm{d}s \tag{4.1.1}$$

for $h = 1, \ldots, k$, we use the fact that

$$M_h(t) = N_h(t) - \int_0^t \alpha_h(s) Y_h(s)\,\mathrm{d}s$$

is a local square integrable martingale [cf. Section II.4.1, in particular formulas (2.4.1) and (2.4.4)]. Thus we may write symbolically $dN_h(t) = \alpha_h(t)Y_h(t)\,dt + dM_h(t)$, where $dM_h(t)$ may be considered as a "random noise" component (cf. Section II.1, especially Figures II.1.4 and II.1.5). By this, a natural estimator of $A_h(t)$ is

$$\hat{A}_h(t) = \int_0^t Y_h(s)^{-1}\,dN_h(s), \tag{4.1.2}$$

as already noted in Section II.1 [cf. (2.1.9) and (2.1.12)]. Let $T_{h1} < T_{h2} < \cdots$ denote the successive jump times for N_h. Then N_h gives mass 1 to each of these jump times and mass 0 elsewhere, and it follows that we may write $\hat{A}_h(t)$ as the simple sum

$$\hat{A}_h(t) = \sum_{\{j:\, T_{hj} \le t\}} Y_h(T_{hj})^{-1}.$$

Thus, $\hat{A}_h$ is an increasing, right-continuous step-function with increment $1/Y_h(T_{hj})$ at the jump time T_{hj} of N_h.

The estimator $\hat{A}_h$ was introduced for counting process models by Aalen (1975, 1978b), and it generalizes the empirical cumulative intensity estimator, proposed independently by Nelson (1969, 1972) and Altshuler (1970) for the setup with censored failure time data. We will call $\hat{A}_h$ the *Nelson–Aalen estimator*. In Section IV.1.5, we will see how the Nelson–Aalen estimator may be derived as a nonparametric maximum likelihood estimator for certain situations.

To study the statistical properties of the Nelson–Aalen estimator, we introduce the indicator processes $J_h(t) = I(Y_h(t) > 0)$ and define

$$A_h^*(t) = \int_0^t \alpha_h(s)J_h(s)\,ds. \tag{4.1.3}$$

Note that $A_h^*(t)$ is almost the same as $A_h(t)$ when there is only a small probability that $Y_h(s) = 0$ for some $s \le t$. Because N_h may only jump when Y_h is positive, the Nelson–Aalen estimator may equivalently be written as

$$\hat{A}_h(t) = \int_0^t \frac{J_h(s)}{Y_h(s)}\,dN_h(s),$$

where $J_h(t)/Y_h(t)$ is interpreted as 0 whenever $Y_h(t) = 0$. Therefore,

$$\hat{A}_h(t) - A_h^*(t) = \int_0^t \frac{J_h(s)}{Y_h(s)}\,dM_h(s) \tag{4.1.4}$$

with M_h defined just below (4.1.1).

We note that J_h/Y_h is a predictable process for each h because this is the case for Y_h, and we assume that these processes are locally bounded (cf. Section II.2) as well. Then the $\hat{A}_h - A_h^*$ are stochastic integrals with respect to the local square integrable martingales M_h, and hence themselves local square integrable martingales (Theorem II.3.1). Thus, A_h^* is the compensator (cf. Section II.3.1) of $\hat{A}_h$. This fact is basic in the study of the statistical properties of the Nelson–Aalen estimator. Obviously, (4.1.4) takes the value zero when $t = 0$. Note that the assumption that J_h/Y_h is locally bounded is always fulfilled if there exists a constant $c > 0$ such that $Y_h(t) < c$ implies $Y_h(t) = 0$. Thus, in particular, the assumption holds when the Y_h take on integer values as is the case in many of the applications of the multiplicative intensity model.

By the orthogonality of the martingales M_h [cf. (2.4.3)] and (2.3.5), we, furthermore, get that the $\hat{A}_h - A_h^*$ have predictable covariation processes

$$\langle \hat{A}_h - A_h^*, \hat{A}_j - A_j^* \rangle(t) = \delta_{hj} \int_0^t \frac{J_h(s)}{Y_h(s)} \alpha_h(s)\,\mathrm{d}s \tag{4.1.5}$$

for $h, j = 1, \ldots, k$, where δ_{hj} is a Kronecker delta. Thus, the $\hat{A}_h - A_h^*$ are also orthogonal.

Now, by the result mentioned just after Example II.3.3, the local martingales (4.1.4) are square integrable on $[0, t]$ for a $t \in \mathscr{T}$ if and only if

$$\mathrm{E}\langle \hat{A}_h - A_h^* \rangle(t) = \int_0^t \mathrm{E}\left\{\frac{J_h(s)}{Y_h(s)}\right\} \alpha_h(s)\,\mathrm{d}s < \infty.$$

When we later talk about expected values, (co)variances, and correlations, it is tacitly understood that this condition holds for all h. This will make all these quantities well-defined. Because we have assumed $A_h(t) < \infty$ for all $t \in \mathscr{T}$, it is sufficient that $\mathrm{E}\{J_h(s)/Y_h(s)\}$ is bounded on $[0, t]$. As earlier, this is the case when the Y_h are integer-valued processes.

Then it follows that

$$\mathrm{E}\hat{A}_h(t) = \mathrm{E}A_h^*(t) = \int_0^t \alpha_h(s)\mathrm{P}(Y_h(s) > 0)\,\mathrm{d}s$$

for all $t \in \mathscr{T}$, so that the Nelson–Aalen estimator is, in general, biased downward, its bias being

$$\mathrm{E}\hat{A}_h(t) - A_h(t) = -\int_0^t \alpha_h(s)\mathrm{P}(Y_h(s) = 0)\,\mathrm{d}s.$$

However, when there is only a small probability that $Y_h(s) = 0$ for some $s \leq t$, the bias will be of little importance. An example of the evaluation of the bias is provided in Example IV.1.1. By the optional sampling theorem, the relation $\mathrm{E}\hat{A}_h(t) = \mathrm{E}A_h^*(t)$ continues to hold if t is replaced by a stopping time T satisfying the condition above with t replaced by T (cf. Sections II.3.1 and II.3.2).

By (4.1.5), it, furthermore, follows that the processes $\hat{A}_h - A_h^*$ have uncorrelated increments on $[0, t]$ and that $\hat{A}_h(s) - A_h^*(s)$ is uncorrelated with $\hat{A}_j(t) - A_j^*(t)$ for any s, t, and $h \neq j$. Therefore, plots of the Nelson–Aalen estimators for $h = 1, 2, \ldots, k$ may be judged independently of each other.

Following Aalen (1975, 1978b), we define the mean squared error function of $\hat{A}_h$ as

$$\tilde{\sigma}_h^2(t) = \mathrm{E}\{\hat{A}_h(t) - A_h^*(t)\}^2.$$

Using the fact that $\langle \hat{A}_h - A_h^* \rangle$ is the compensator of $(\hat{A}_h - A_h^*)^2$ [Section II.3.2 and (2.4.9)], we may write

$$\tilde{\sigma}_h^2(t) = \mathrm{E}\langle \hat{A}_h - A_h^* \rangle(t) = \int_0^t \mathrm{E}\left\{\frac{J_h(s)}{Y_h(s)}\right\} \alpha_h(s)\,\mathrm{d}s.$$

As an estimator for this mean squared error function, we use the optional variation process (cf. Section II.3.2)

$$\hat{\sigma}_h^2(t) = [\hat{A}_h - A_h^*](t) = \int_0^t J_h(s)(Y_h(s))^{-2}\,\mathrm{d}N_h(s), \tag{4.1.6}$$

where the integral may be written as a simple sum in a similar fashion as for $\hat{A}_h(t)$. It is seen that (4.1.6) is derived from (4.1.5) simply by replacing $\mathrm{d}A_h(s) = \alpha_h(s)\,\mathrm{d}s$ by $\mathrm{d}\hat{A}_h(s)$. Because, under the integrability condition stated below (4.1.5), $[\hat{A}_h - A_h^*] - \langle \hat{A}_h - A_h^* \rangle$ is a martingale over $[0, t]$, it follows that $\hat{\sigma}_h^2(t)$ is unbiased for $\tilde{\sigma}_h^2(t)$ (cf. Section II.3.2). When there is only a small probability that $Y_h(s) = 0$ for some $s \le t$, $A_h^*(t)$ is almost the same as the expected value of $\hat{A}_h(t)$. This suggests that (4.1.6) is a reasonable estimator for the variance of the Nelson–Aalen estimator, at least for large sample purposes. This will be confirmed by the asymptotic results in the next subsection.

An alternative to the variance estimator (4.1.6) may sometimes be useful in situations where the counting process N_h is obtained by aggregating n individual counting processes $N_{h1}, N_{h2}, \ldots, N_{hn}$. To be more specific, we assume that $N_h = \sum_{i=1}^n N_{hi}$, and that the N_{hi} have intensity processes λ_{hi} of the multiplicative form $\lambda_{hi}(t) = \alpha_h(t) Y_{hi}(t)$. Here α_h is the same for all i, and the Y_{hi} are assumed to be indicator processes, cf. Examples III.1.2 and III.1.4. The alternative variance estimator may then be derived by considering a natural "discrete" extension of this "continuous" model (cf. also Section IV.1.5). This is a model where the compensator of N_{hi} may be written as

$$\Lambda_{hi}(t) = \int_0^t Y_{hi}(s)\,\mathrm{d}A_h(s)$$

for A_h an arbitrary increasing function with $A_h(t) < \infty$ for $t \in \mathscr{T}$. Thus,

$$M_{hi}(t) = N_{hi}(t) - \int_0^t Y_{hi}(s)\,\mathrm{d}A_h(s)$$

is a local square integrable martingale for each i, and by (2.4.2), we have

$$\langle M_{hi} \rangle(t) = \int_0^t Y_{hi}(s)(1 - \Delta A_h(s))\,\mathrm{d}A_h(s).$$

Furthermore, we assume that the individual counting processes are conditionally independent given the past in the sense that

$$\langle M_{hi}, M_{hi'} \rangle = 0,$$

for $i \neq i'$, i.e., the martingales M_{hi} are orthogonal. For such an extended model, N_h may have multiple jumps, and it is, therefore, no longer a counting process. However, we still have that

$$M_h(t) = \sum_{i=1}^{n} M_{hi}(t) = N_h(t) - \int_0^t Y_h(s)\,\mathrm{d}A_h(s),$$

with $Y_h = \sum_{i=1}^n Y_{hi}$, is a local square integrable martingale. Its predictable variation process is by the assumed orthogonality of the M_{hi} and the bilinearity of the predictable variation process [cf. (2.3.4)] given by

$$\langle M_h \rangle(t) = \int_0^t Y_h(s)(1 - \Delta A_h(s))\,\mathrm{d}A_h(s).$$

Using (4.1.4) and Theorem II.3.1, this gives us

$$\langle \hat{A}_h - A_h^* \rangle(t) = \int_0^t \frac{J_h(s)}{Y_h(s)}(1 - \Delta A_h(s))\,\mathrm{d}A_h(s).$$

Now replacing A_h by the Nelson–Aalen estimator $\hat{A}_h$ in this expression, we arrive at the alternative variance estimator

$$\check{\sigma}_h^2(t) = \int_0^t J_h(s)(Y_h(s) - \Delta N_h(s))(Y_h(s))^{-3}\,\mathrm{d}N_h(s). \tag{4.1.7}$$

A comparison of this estimator with $\hat{\sigma}_h^2(t)$ given by (4.1.6) for censored survival data is provided in Example IV.1.1.

Example IV.1.1. Nelson–Aalen Estimator for Right-Censored Survival Data. Bias and Performance of Variance Estimators

Let $X_1, \ldots, X_n$ be i.i.d. non-negative random variables with absolutely continuous distribution function F, hazard rate function $\alpha = F'/(1 - F)$, and integrated hazard function $A(t) = \int_0^t \alpha(s)\,\mathrm{d}s$. We do not observe $X_1, \ldots, X_n$, only the right-censored sample $(\tilde{X}_i, D_i)$, $i = 1, \ldots, n$, where $\tilde{X}_i = X_i \wedge U_i$ and $D_i = I(\tilde{X}_i = X_i)$ for some censoring times $U_1, \ldots, U_n$. It was shown in Section III.2.2 that

$$N(t) = \sum_{i=1}^{n} I(\tilde{X}_i \leq t, D_i = 1)$$

is a univariate counting process with intensity process $\lambda(t) = \alpha(t)Y(t)$, with

$$Y(t) = \sum_{i=1}^{n} I(\tilde{X}_i \geq t),$$

when we have independent right-censoring. Now we have $J(t) = I(Y(t) > 0) = I(\tilde{X}_{(n)} \geq t)$, where $\tilde{X}_{(n)} = \max\{\tilde{X}_1, \ldots, \tilde{X}_n\}$, so that $J(t)/Y(t) \leq 1$ for all t.

(Remember that 0/0 is interpreted as 0.) Thus, the general conditions above are satisfied, and the Nelson–Aalen estimator $\hat{A}(t)$ takes the form

$$\hat{A}(t) = \int_0^t \frac{J(s)}{Y(s)} \mathrm{d}N(s) = \sum_{i:\tilde{X}_i \le t} \frac{D_i}{Y(\tilde{X}_i)}.$$

One use of the Nelson–Aalen estimator is to check graphically whether the lifetimes X_i appear to follow a certain parametric distribution. For example, the exponential distribution has constant hazard rate function α and cumulative hazard function $A(t) = \alpha t$. Therefore, the Nelson–Aalen estimator plotted against t should give an approximately straight line for the exponential model. The use of the Nelson–Aalen estimator for model checking was discussed in detail by Nelson (1982, Chap. 4), and we return to this in Section VI.3.1.

Klein (1988, 1991) studied the bias of the Nelson–Aalen estimator as well as the relative merits of the two estimators (4.1.6) and (4.1.7) for its variance. He considered the simple random censorship model (cf. Example III.2.4) where the $U_1, \ldots, U_n$ are i.i.d. random variables with distribution function G and where the X_i and the U_i are independent. The computations were done analytically, assuming a proportional hazards model for the survival and censoring distributions [cf. Chen, Hollander, and Langberg (1982)], and by Monte Carlo simulations. The simulations were performed with sample sizes 10 and 20 and with the survival distribution being the standard exponential distribution, whereas the censoring distribution was a Weibull distribution with shape parameter 0.5 or 2 and with a scale parameter adjusted so as to get a censoring percentage of 33%, 50%, or 67%. The same censoring percentages were considered in the analytic computations, but here the sample sizes of 50 and 100 were considered as well.

To a large extent Klein's results may be summarized by relating them to the value of $\mathrm{E}Y(t)$, the expected number still at risk just before time t. The bias of the Nelson–Aalen estimator was less than 0.25% of the integrated intensity for all situations reported by Klein as long as $\mathrm{E}Y(t) \ge 3$. When $\mathrm{E}Y(t) \ge 2$, this relative bias was less than 2%.

The variance estimator (4.1.6) was found to overestimate the true variance of the Nelson–Aalen estimator. However, as long as $\mathrm{E}Y(t) \ge 3$, the bias of this estimator did not exceed 5% of the true variance. For $\mathrm{E}Y(t) \ge 5$, the bias was of no practical importance. The variance estimator (4.1.7), on the other hand, tended to underestimate the true variance, and the bias was about 30% of the true variance when $\mathrm{E}Y(t)$ was around 3. Even for $\mathrm{E}Y(t)$ around 5, the bias was more than 15% of the true variance.

Of the two variance estimators, (4.1.7) always had the smallest mean squared error. For $\mathrm{E}Y(t) \ge 10$, the difference was rather small and not very important for practical purposes. However, this was not the case for small values of $\mathrm{E}Y(t)$. For $\mathrm{E}Y(t)$ around 5, the root of the mean squared error of (4.1.6) was approximately 50% larger than that of (4.1.7), whereas it for $\mathrm{E}Y(t)$ around 3 was approximately twice as large as that of (4.1.7).

We find it hard to give definitive recommendations on which of the variance estimators (4.1.6) and (4.1.7) to use based on Klein's (1988, 1991) results. In practice, variance estimators are often used in connection with confidence intervals and bands (cf. Section IV.1.3), and one would then like to know which of the two produce intervals and bands with the most correct coverage probabilities. In this connection, it may be a drawback with (4.1.7), which systematically underestimates the true variance, that it may give too low coverage probabilities. Further investigations are needed to get a definitive answer to this, however. On the other hand, a systematic handling of tied observations is only possible with (4.1.7).

Finally, it ought to be stressed that the difference between (4.1.6) and (4.1.7) is only of importance in connection with the handling of small risk sets. For many applications, therefore, it does not matter which of the two one applies. We have used (4.1.6) in connection with the practical examples in Section IV.1. □

We next turn to some of the practical examples introduced in Section I.3.

EXAMPLE IV.1.2. Survival with Malignant Melanoma. Nelson–Aalen Estimates for Male and Female Patients

The problem and the data were presented in Example I.3.1. For each individual i we study,

$$X_i = \text{time from operation until death from the disease}$$

with right-censoring variables

$U_i =$ time from operation until death from other causes, or the end of 1977, whichever comes first.

Of the 126 female patients, 28 were observed to die from the disease, and of the 79 male patients, 29 were observed to die from the disease.

The problem is seen to fit into the framework of Example IV.1.1 assuming the integrated hazard function of X_i to be A_s when individual i has sex s, $s =$ male, female. Thus, we study the *marginal distribution* of the X_i in the two groups assuming the survival times to be i.i.d. within groups. The Nelson–Aalen estimates $\hat{A}_{\text{male}}(t)$ and $\hat{A}_{\text{female}}(t)$ are plotted in Figure IV.1.1. Note that the slopes of the estimated integrated hazards are roughly constant (≈ 0.035 per year for women and ≈ 0.070 per year for men) indicating a constant intensity of death from melanoma about twice as large for men as for women. □

EXAMPLE IV.1.3. Mortality of Diabetics in the County of Fyn. Estimation of the Cumulative Death Intensity for Left-Truncated and Right-Censored Survival Data

The problem and the data were presented in Example I.3.2. For each diabetic i alive on 1 July 1973 we know

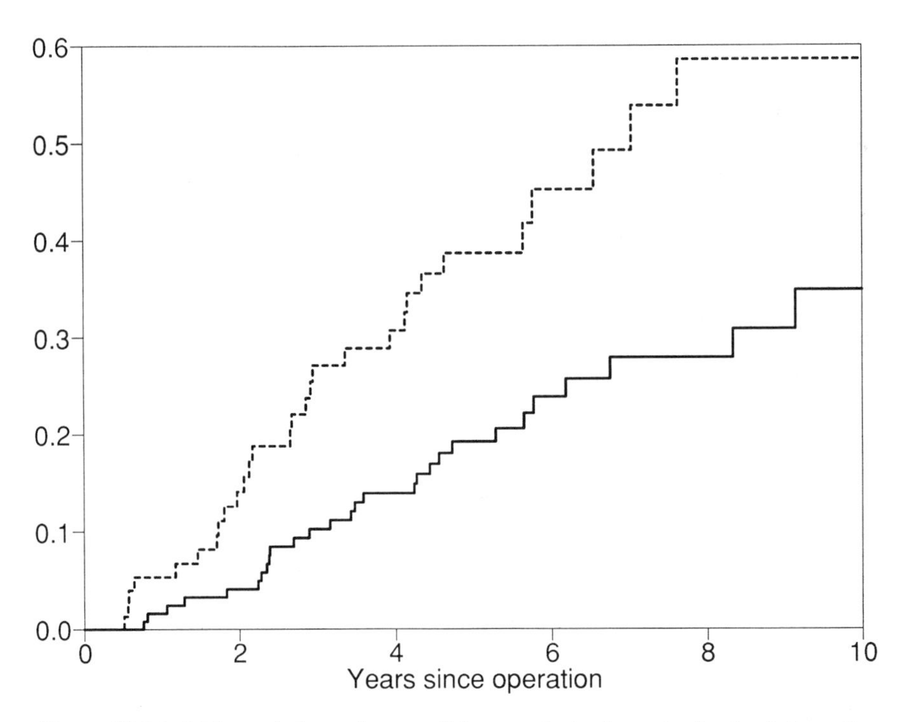

Figure IV.1.1. Nelson–Aalen estimates of the cumulative intensity for death from the disease for 126 female (—) and 79 male (- - -) patients with malignant melanoma.

$$X_i = \text{age at death},$$

if dead before 1 January 1982, or we just know that i was alive on 1 January 1982, or at an earlier emigration date. Define

$$U_i = \text{age at this date of no further follow-up}$$

and

$$V_i = \text{age on 1 July 1973}.$$

Of the 716 female diabetics in Fyn county on 1 July 1973, 237 were recorded to die during the follow-up period, whereas for the 783 males, 254 died. There were 1 male and 1 female emigrant.

We assume in this example that the hazard of X_i only depends on the sex of the patient: $\alpha_i(t) = \alpha_s(t)$, $s = f$ if i is a female, $s = m$ if i is a male. In particular, therefore, it is here assumed that the hazard does not depend on the duration of the disease. As discussed in detail in Example III.3.6, we have a situation with *left-truncated and right-censored survival data* and the counting processes (for $s = f, m$)

$$N_s(t) = \#\{i \text{ of sex } s\text{: } V_i < X_i \leq t \wedge U_i\},$$

Figure IV.1.2. The number of female diabetics at risk (Y_f) in the county of Fyn, 1973–81, as a function of their age.

counting the number of observed deaths at ages $\leq t$, have intensity processes $\alpha_s(t)\, Y_s(t)$ with

$$Y_s(t) = \#\{i \text{ of sex } s\colon V_i < t \leq X_i \wedge U_i\},$$

the number at risk at age $t-$. We note that in this example (in contrast to the previous one) the number of individuals at risk will both increase and decrease with t; cf. Figure IV.1.2 which gives the size of the risk set $Y_f(t)$ for female diabetics as a function of their age. The Nelson–Aalen estimates

$$\widehat{A}_s(t) = \int_0^t \frac{I(Y_s(u) > 0)}{Y_s(u)} \mathrm{d}N_s(u)$$

are shown in Figure IV.1.3.

We note that $\widehat{A}_f$ and $\widehat{A}_m$ are convex, corresponding to an increase with age of the death intensities α_f, α_m, and that $\widehat{A}_f(t) < \widehat{A}_m(t)$ with increasing difference, corresponding to $\alpha_f(t) < \alpha_m(t)$: lower death intensity for female than for male diabetics.

Note that, for left-truncated data, it may happen for some (typically small) ages t that $Y(t) = 0$, the risk set is *empty*; there is no evidence in the data on the death intensity $\alpha(t)$ over intervals with empty risk sets. In these data, no

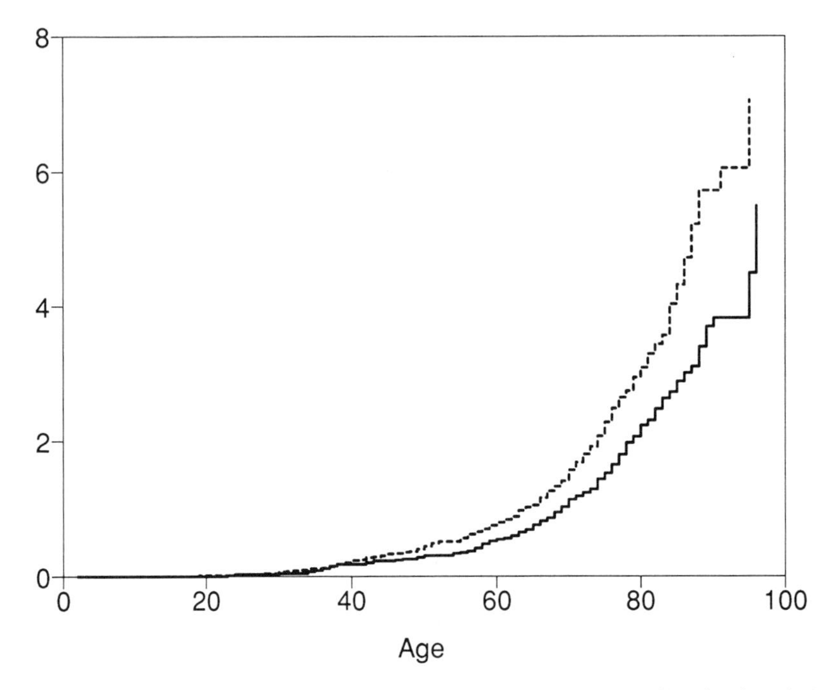

Figure IV.1.3. Nelson–Aalen estimates of integrated death intensities for female (—) and male (- - -) diabetics in the county of Fyn, 1973–81.

patient was under 2 years of age in 1973; hence, there is no information in the data on mortality for diabetics under the age of 2 years. □

Example IV.1.4. Psychiatric Admissions for Women Having Just Given Birth. Nelson–Aalen Estimates of the Integrated Admission and Discharge Intensities

The problem and the data were presented in Example I.3.10. As model for these data, we use a two-state Markov process $X_j(t)$ for each woman j, the states denoting "in psychiatric ward" (i) and "not in psychiatric ward" (o), respectively. Time t is time since birth. Following the discussion in Example III.1.4, we consider the bivariate counting process $(N_j^a(t), N_j^d(t))$ with

$$N_j^a(t) = \#\{\text{admissions between birth and time } t \text{ for woman } j\},$$

$$N_j^d(t) = \#\{\text{discharges between birth and time } t \text{ for woman } j\}.$$

The intensity process is $(\alpha_{aj}(t)\, Y_j^o(t), \alpha_{dj}(t)\, Y_j^i(t))$, where

$$Y_j^o(t) = I(\text{woman } j \text{ is alive and out of psychiatric ward at time } t-),$$

$$Y_j^i(t) = I(\text{woman } j \text{ is alive and in psychiatric ward at time } t-).$$

Neither deaths nor emigrations are recorded, but because less that 1% of these women are expected to die or emigrate during a 1-year period, the impact on the estimates will be negligible.

In the present example, we shall assume that the admission intensity $\alpha_{aj}(t) = \alpha_a(t)$ and the discharge intensity $\alpha_{dj}(t) = \alpha_d(t)$ do not differ between women. Aggregation then leads to the bivariate counting process $(N^a(t), N^d(t))$ counting all admissions and discharges within time t after giving birth for Danish women who gave birth in 1975. The intensity process is $(\alpha_a(t)Y^o(t), \alpha_d(t)Y^i(t))$ with

$$Y^o(t) = \#\{\text{women out of psychiatric ward at time } t-\},$$

$$Y^i(t) = \#\{\text{women in psychiatric ward at time } t-\}.$$

Nelson–Aalen estimation of the cumulative admission intensity $A_a(t)$ and cumulative discharge intensity $A_d(t)$ is now direct. Because $Y^o(t)$ is the number of Danish women giving birth in 1975 (71,378) minus those of these who were in psychiatric ward at time t after birth (a few hundred), one may approximate $Y^o(t)$ by 71,378.

The estimates $\hat{A}_a(t)$ and $\hat{A}_d(t)$ are shown in Figures IV.1.4 and IV.1.5. It is seen that $\alpha_a(t)$ decreases at least during the first couple of months and stabilizes, whereas $\alpha_d(t)$ seems to be approximately constant. □

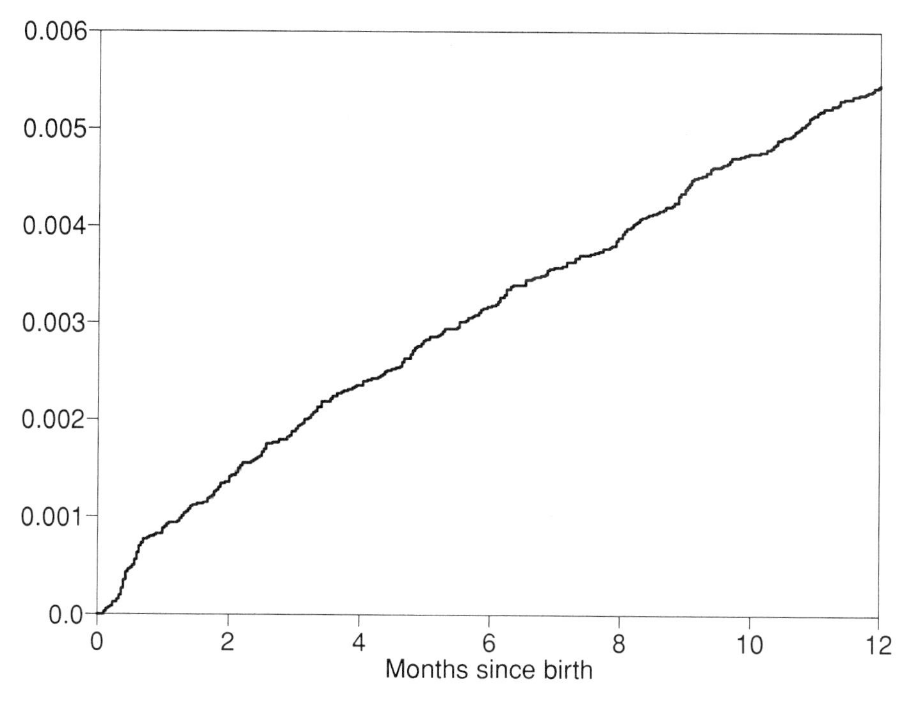

Figure IV.1.4. Estimated integrated admission intensity to psychiatric wards for Danish women giving birth in 1975, as function of time since birth.

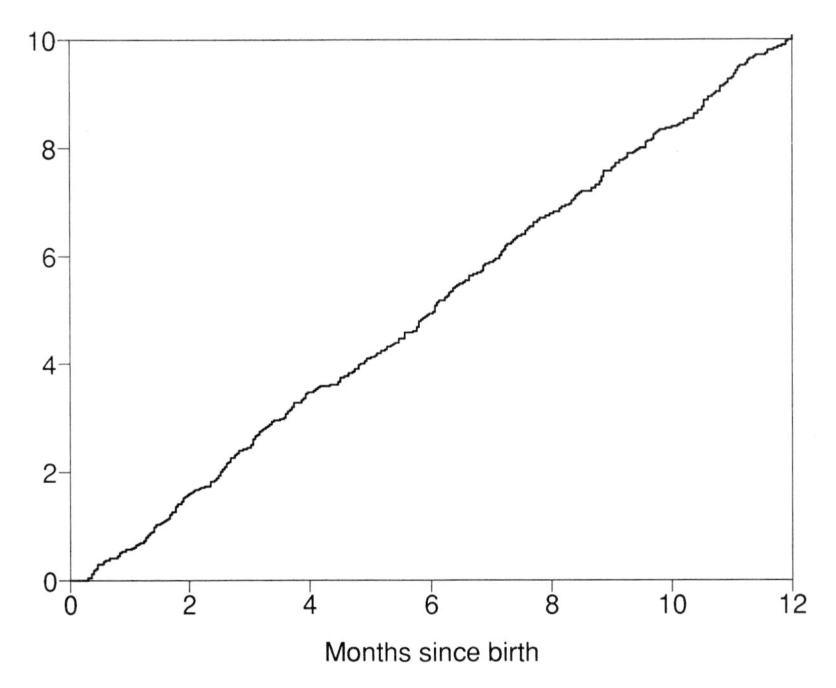

Figure IV.1.5. Estimated integrated discharge intensity from psychiatric wards for Danish women giving birth in 1975, as function of time since birth.

EXAMPLE IV.1.5. Mating of Drosophila Flies. Nelson–Aalen Estimates for Integrated Mating Intensities

The experiment was described and the data were quoted in Example I.3.16, and one possible model was discussed in Example III.1.10. With the notation introduced in the latter example, the Nelson–Aalen estimator of the integrated mating intensity $A(t) = \int_0^t \alpha(s)\,\mathrm{d}s$ is given by

$$\hat{A}(t) = \int_0^t \frac{I(F(s)M(s) > 0)}{F(s)M(s)}\,\mathrm{d}N(s),$$

with $N(t)$ denoting the number of matings initiated before time t, and $F(t)$ and $M(t)$ the number of females and males, respectively, not yet having initiated a mating just before time t.

The estimates are given in Figure IV.1.6 for the four combinations of ebony females/males and oregon females/males. The curves are plotted from time 0 until the time of termination of each experiment, cf. Examples I.3.16 and III.2.12. The most obvious patterns are that ebony females initiate matings sooner than oregon females and that ebony males mate sooner with ebony females than with oregon females, whereas oregon males do not discriminate so much. This suggests that a sexual selection is indeed taking place.

An alternative and more realistic model (model 2), assuming that males may mate more than once, was introduced in Example III.5.5. This amounts

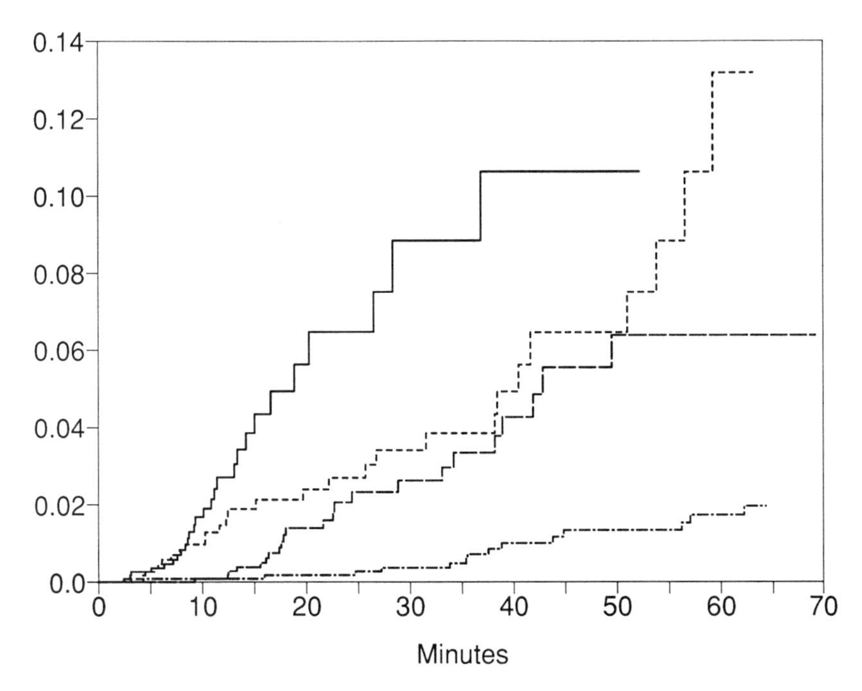

Figure IV.1.6. Integrated mating intensities (model 1) for ebony and oregon strains of Drosophila melanogaster. Female/male: ebony/ebony (—); ebony/oregon (- - -); oregon/oregon (— — —); oregon/ebony (- · · · -).

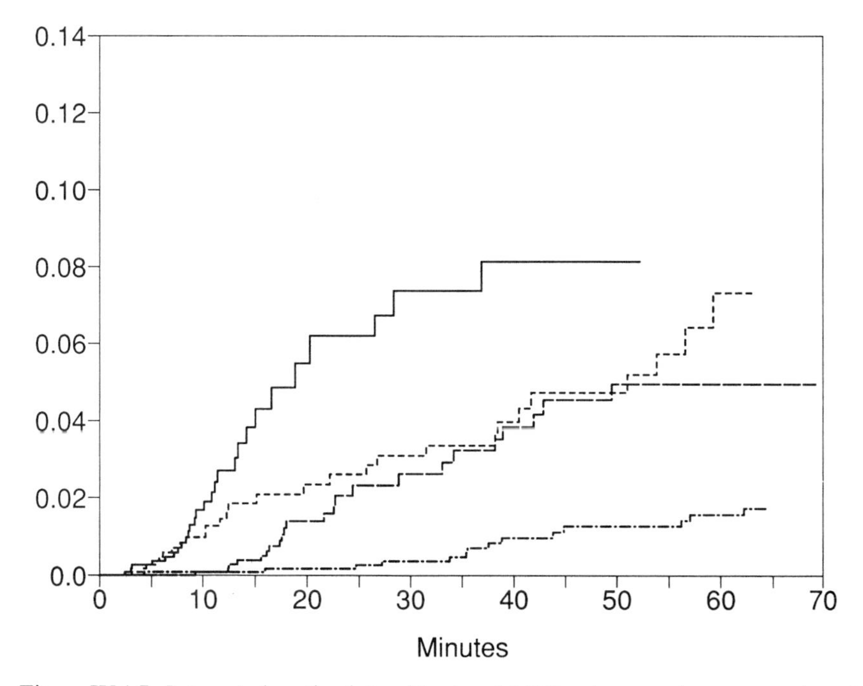

Figure IV.1.7. Integrated mating intensities (model 2) for ebony and oregon strains of Drosophila melanogaster. Female/male: ebony/ebony (—); ebony/oregon (- - -); oregon/oregon (— — —); oregon/ebony (- · · · -).

to changing the interpretation of $M(t)$ given earlier from "males not yet having initiated a mating before time t" into "males not being engaged in a mating just before time t." The estimates from model 2 are given in Figure IV.1.7 and the pattern is seen to be similar to that seen in Figure IV.1.6. □

IV.1.2. Large Sample Properties

To study the large sample properties of the Nelson–Aalen estimators, we consider a sequence of counting processes $\mathbf{N}^{(n)} = (N_1^{(n)}, \ldots, N_k^{(n)})$, $n = 1, 2, \ldots$, each satisfying the multiplicative intensity model $\lambda_h^{(n)}(t) = \alpha_h(t) Y_h^{(n)}(t)$, with the α_h being the same for all n. We introduce $J_h^{(n)}(t) = I(Y_h^{(n)}(t) > 0)$ and assume that $J_h^{(n)}/Y_h^{(n)}$ is locally bounded for each h and n. We will show that the Nelson–Aalen estimator is *uniformly consistent* on compact intervals.

Theorem IV.1.1. *Let $t \in \mathscr{T}$ and assume that, as $n \to \infty$,*

$$\int_0^t \frac{J_h^{(n)}(s)}{Y_h^{(n)}(s)} \alpha_h(s)\, \mathrm{d}s \xrightarrow{\mathrm{P}} 0 \tag{4.1.8}$$

and

$$\int_0^t (1 - J_h^{(n)}(s))\alpha_h(s)\, \mathrm{d}s \xrightarrow{\mathrm{P}} 0. \tag{4.1.9}$$

Then, as $n \to \infty$,

$$\sup_{s \in [0,t]} |\hat{A}_h^{(n)}(s) - A_h(s)| \xrightarrow{\mathrm{P}} 0.$$

PROOF. By the version (2.5.18) of Lenglart's inequality and (4.1.5), we get for any $\delta, \eta > 0$

$$\mathrm{P}\left\{ \sup_{s \in [0,t]} |\hat{A}_h^{(n)}(s) - A_h^{*(n)}(s)| > \eta \right\} \leq \frac{\delta}{\eta^2} + \mathrm{P}\left\{ \int_0^t \frac{J_h^{(n)}(s)}{Y_h^{(n)}(s)} \alpha_h(s)\, \mathrm{d}s > \delta \right\},$$

and it follows by (4.1.8) that

$$\sup_{s \in [0,t]} |\hat{A}_h^{(n)}(s) - A_h^{*(n)}(s)| \xrightarrow{\mathrm{P}} 0, \quad \text{as } n \to \infty.$$

Furthermore, $|A_h^{*(n)}(s) - A_h(s)| = \int_0^s (1 - J_h^{(n)}(u))\alpha_h(u)\, \mathrm{d}u$, and the theorem follows by (4.1.9). □

Because we have assumed that $A_h(t) < \infty$ for $t \in \mathscr{T}$, it is seen that a simple sufficient condition for (4.1.8) and (4.1.9) is that

$$\inf_{s \in [0,t]} Y_h^{(n)}(s) \xrightarrow{\mathrm{P}} \infty, \quad \text{as } n \to \infty. \tag{4.1.10}$$

Furthermore, if there exists a positive constant c not depending on n such that $Y_h^{(n)}(s) < c$ implies $Y_h^{(n)}(s) = 0$, a slightly weaker sufficient condition may be formulated; for, in this case, $\{J_h^{(n)}(s)/Y_h^{(n)}(s)\}\alpha_h(s)$ and $(1 - J_h^{(n)}(s))\alpha_h(s)$ are bounded by $\alpha_h(s)/c$ and $\alpha_h(s)$, respectively, and it follows by Proposition II.5.3 that a sufficient condition for (4.1.8) and (4.1.9) is that $Y_h^{(n)}(s) \xrightarrow{\text{P}} \infty$ as $n \to \infty$ for almost all $s \in [0, t]$.

The asymptotic distribution of the Nelson–Aalen estimators on compact intervals follows readily from the martingale central limit theorem. We introduce $\hat{\mathbf{A}}^{(n)} = (\hat{A}_1^{(n)}, \ldots, \hat{A}_k^{(n)})$ and define $\mathbf{A}$ in a similar manner. Then we have the following general result.

Theorem IV.1.2. *Let $t \in \mathscr{T}$ and assume that there exist a sequence of positive constants $\{a_n\}$, increasing to infinity as $n \to \infty$, and non-negative functions y_h such that α_h/y_h is integrable over $[0, t]$ for $h = 1, 2, \ldots, k$. Let*

$$\sigma_h^2(s) = \int_0^s \frac{\alpha_h(u)}{y_h(u)} \mathrm{d}u, \quad h = 1, 2, \ldots, k, \tag{4.1.11}$$

and assume that

(A) *For each $s \in [0, t]$ and $h = 1, 2, \ldots, k$,*

$$a_n^2 \int_0^s \frac{J_h^{(n)}(u)}{Y_h^{(n)}(u)} \alpha_h(u)\,\mathrm{d}u \xrightarrow{\text{P}} \sigma_h^2(s) \quad \text{as } n \to \infty. \tag{4.1.12}$$

(B) *For $h = 1, 2, \ldots, k$ and all $\varepsilon > 0$,*

$$a_n^2 \int_0^t \frac{J_h^{(n)}(u)}{Y_h^{(n)}(u)} \alpha_h(u) I\left\{\left| a_n \frac{J_h^{(n)}(u)}{Y_h^{(n)}(u)} \right| > \varepsilon\right\} \mathrm{d}u \xrightarrow{\text{P}} 0 \quad \text{as } n \to \infty. \tag{4.1.13}$$

(C) *For $h = 1, 2, \ldots, k$,*

$$a_n \int_0^t (1 - J_h^{(n)}(u))\alpha_h(u)\,\mathrm{d}u \xrightarrow{\text{P}} 0 \quad \text{as } n \to \infty. \tag{4.1.14}$$

Then

$$a_n(\hat{\mathbf{A}}^{(n)} - \mathbf{A}) \xrightarrow{\mathscr{D}} \mathbf{U} = (U_1, \ldots, U_k) \quad \text{as } n \to \infty$$

on $D[0, t]^k$, where $U_1, \ldots, U_k$ are independent Gaussian martingales with $U_h(0) = 0$ and $\operatorname{cov}(U_h(s_1), U_h(s_2)) = \sigma_h^2(s_1 \wedge s_2)$. Also, for $h = 1, 2, \ldots, k$,

$$\sup_{s \in [0, t]} |a_n^2 \hat{\sigma}_h^2(s) - \sigma_h^2(s)| \xrightarrow{\text{P}} 0 \quad \text{as } n \to \infty, \tag{4.1.15}$$

where $\hat{\sigma}_h^2(s)$ is defined by (4.1.6). Finally, (4.1.15) continues to hold when $\hat{\sigma}_h^2(s)$ is replaced by $\check{\sigma}_h^2(s)$ given by (4.1.7).

PROOF. By (4.1.4), we have

$$a_n(\hat{A}_h^{(n)}(s) - A_h^{*(n)}(s)) = a_n \int_0^t \frac{J_h^{(n)}(u)}{Y_h^{(n)}(u)} \mathrm{d}M_h^{(n)}(u),$$

$h = 1, \ldots, k$, so that Rebolledo's martingale central limit theorem (Theorem II.5.1) applies. With

$$H_{hj}^{(n)} = \frac{\delta_{hj} a_n J_h^{(n)}}{Y_h^{(n)}},$$

in (2.5.6) and (2.5.8) it follows immediately by Conditions A and B that

$$a_n(\hat{\mathbf{A}}^{(n)} - \mathbf{A}^{*(n)}) \xrightarrow{\mathscr{D}} \mathbf{U} \quad \text{as } n \to \infty.$$

Now Condition C ensures that

$$\sup_{s \in [0,t]} |a_n(A_h^{*(n)}(s) - A_h(s))| \xrightarrow{\mathrm{P}} 0$$

as $n \to \infty$, and the first assertion follows.

Furthermore, by (4.1.6) and Theorem II.5.1,

$$a_n^2 \hat{\sigma}_h^2(s) = [a_n(\hat{A}_h^{(n)} - A_h^{*(n)})](s) \to \operatorname{var} U_h(s) = \sigma_h^2(s)$$

uniformly on $[0, t]$ in probability as $n \to \infty$ and (4.1.15) follows.

Finally,

$$\hat{\sigma}_h^2(s) - \check{\sigma}_h^2(s) = \int_0^s J_h^{(n)}(u)(Y_h^{(n)}(u))^{-3} \Delta N_h^{(n)}(u) \, \mathrm{d}N_h^{(n)}(u),$$

so that

$$\begin{aligned} \sup_{s \in [0,t]} |a_n^2 \hat{\sigma}_h^2(s) - a_n^2 \check{\sigma}_h^2(s)| &\leq \sup_{s \in [0,t]} \left| \frac{\Delta N_h^{(n)}(s)}{Y_h^{(n)}(s)} \right| a_n^2 \int_0^t J_h^{(n)}(s)(Y_h^{(n)}(s))^{-2} \, \mathrm{d}N_h^{(n)}(s) \\ &= \sup_{s \in [0,t]} |\Delta \hat{A}_h^{(n)}(s)| \, a_n^2 \hat{\sigma}_h^2(t). \end{aligned}$$

Here the first term on the right-hand side converges to zero in probability by the uniform consistency of the Nelson–Aalen etimator (Theorem IV.1.1) and the continuity of A_h, whereas $a_n^2 \hat{\sigma}_h^2(t)$ converges to $\sigma_h^2(t)$ in probability according to what we just proved. Thus,

$$\sup_{s \in [0,t]} |a_n^2 \hat{\sigma}_h^2(s) - a_n^2 \check{\sigma}_h^2(s)| \xrightarrow{\mathrm{P}} 0$$

as $n \to \infty$, and it follows that (4.1.15) continues to hold when $\hat{\sigma}_h^2(s)$ is replaced by $\check{\sigma}_h^2(s)$. □

Because we have assumed that $A_h(t) < \infty$ for all h and $t \in \mathscr{T}$, Conditions A to C are satisfied provided that there exist functions y_h defined on $[0, t]$ with $\inf_{s \in [0,t]} y_h(s) > 0$ such that

$$\sup_{s \in [0,t]} |a_n^{-2} Y_h^{(n)}(s) - y_h(s)| \xrightarrow{\mathrm{P}} 0 \quad \text{as } n \to \infty. \tag{4.1.16}$$

Other sets of sufficient conditions may be found by Propositions II.5.2 and II.5.3. For an illustration, see Example IV.1.9.

In connection with Theorems IV.1.1 and IV.1.2, it is important to note that they continue to hold if $[0,t]$ is replaced everywhere by $[t_1,t_2]$, where $0 < t_1 < t_2$; $t_1, t_2 \in \mathscr{T}$ are fixed numbers, and at the same time $\widehat{A}_h^{(n)}$, A_h, $\hat{\sigma}_h^2$, $\breve{\sigma}_h^2$, and σ_h^2 are replaced by $\widehat{A}_h^{(n)} - \widehat{A}_h^{(n)}(t_1)$, $A_h - A_h(t_1)$, $\hat{\sigma}_h^2 - \hat{\sigma}_h^2(t_1)$, $\breve{\sigma}_h^2 - \breve{\sigma}_h^2(t_1)$, and $\sigma_h^2 - \sigma_h^2(t_1)$, respectively. Davidsen and Jacobsen (1991) showed how the results may be further extended to the situation where t_1 is a varying time parameter.

Also, as illustrated in the examples to follow, we use the normalizing factor $a_n = \sqrt{n}$, with n the number of "individual counting processes," in Theorem IV.1.2 in most applications of the multiplicative intensity model. But other choices may also occur, as will be seen for the model for mating of Drosophila flies in Example IV.1.10.

EXAMPLE IV.1.6. Right-Censored Survival Data

To illustrate the general results and conditions in this section, let us consider the random censorship model for censored survival data more closely; cf. Example III.2.4. Here, we have to suppose that for each n, $X_{1n}, X_{2n}, \ldots, X_{nn}$ and $U_{1n}, \ldots, U_{nn}$ are mutually independent with X_{in} having absolutely continuous distribution function F, hazard rate function α, and integrated hazard function A, while U_{in} has distribution function G_{in}. We define censored survival times $\widetilde{X}_{1n}, \ldots, \widetilde{X}_{nn}$ as in Example IV.1.1 and note that the $\widetilde{X}_{in}$ are independent with distribution function $H_{in} = 1 - (1 - F)(1 - G_{in})$. Finally, let $Y^{(n)}(t)$ and $\widehat{A}^{(n)}(t)$ be defined as in Example IV.1.1, where we have added a superscript (n) to indicate that these quantities are based on n censored survival times.

We assume that $F(t) < 1$, or equivalently $A(t) < \infty$. By (4.1.10), it is seen that

$$\sup_{s\in[0,t]} |\widehat{A}^{(n)}(s) - A(s)| \xrightarrow{\mathrm{P}} 0 \quad \text{as } n \to \infty$$

provided that

$$Y^{(n)}(t) \xrightarrow{\mathrm{P}} \infty \quad \text{as } n \to \infty. \tag{4.1.17}$$

Now

$$\mathrm{E}Y^{(n)}(t) = (1 - F(t)) \sum_{i=1}^{n} (1 - G_{in}(t-))$$

and

$$\begin{aligned}\operatorname{var} Y^{(n)}(t) &= (1 - F(t)) \sum_{i=1}^{n} (1 - G_{in}(t-))\{1 - (1 - F(t))(1 - G_{in}(t-))\}\\ &\le \mathrm{E}Y^{(n)}(t).\end{aligned}$$

Therefore, by Chebyshev's inequality,

$$\mathrm{P}(|Y^{(n)}(t) - \mathrm{E}Y^{(n)}(t)| < \varepsilon\sqrt{\mathrm{E}Y^{(n)}(t)}) \ge 1 - \frac{1}{\varepsilon^2},$$

for any $\varepsilon > 0$. So, in this case, (4.1.17) is equivalent to

$$\liminf_{n\to\infty} \sum_{i=1}^{n} (1 - G_{in}(t-)) = \infty,$$

a condition saying that the censoring must not be "too heavy" as n increases.

Let us then turn to the weak convergence of the Nelson–Aalen estimator. We will show that if there exists a (sub-)distribution function G with $G(t-) < 1$ such that

$$\sup_{s\in[0,t]} \left| n^{-1} \sum_{i=1}^{n} G_{in}(s) - G(s) \right| \to 0 \quad \text{as } n \to \infty, \tag{4.1.18}$$

then

$$\sqrt{n}(\hat{A}^{(n)} - A) \xrightarrow{\mathcal{D}} U \quad \text{as } n \to \infty.$$

Here U is a Gaussian martingale with $U(0) = 0$ and $\operatorname{cov}(U(s_1), U(s_2)) = \sigma^2(s_1 \wedge s_2)$, where $\sigma^2(s) = \int_0^s \{\alpha(u)/y(u)\}\, du$ with $y(s) = (1 - F(s))(1 - G(s-))$. To see that this is the case, we use the Glivenko–Cantelli theorem for independent, but not necessarily identically distributed random variables [e.g., Shorack and Wellner (1986, Theorem 3.2.1)] to get

$$\sup_{s\in[0,t]} \left| \frac{Y^{(n)}(s)}{n} - \frac{1}{n} \sum_{i=1}^{n} (1 - F(s))(1 - G_{in}(s-)) \right| \xrightarrow{\text{P}} 0$$

as $n \to \infty$. The result then follows by (4.1.16) and (4.1.18).

For the special case of simple random censorship, where $G_{in} = G$ for all i and n, we, in particular, have uniform consistency and weak convergence of the Nelson–Aalen estimator as stated earlier provided that $G(t-) < 1$.

We studied the bias and the mean squared error of $\hat{A}^{(n)}(s)$ for the simple random censorship model in Example IV.1.1. Hjort (1985b) showed that the skewness of the Nelson–Aalen estimator for this situation is

$$\frac{1}{\sqrt{n}} \frac{\int_0^s \{1/y^2(u)\}\alpha(u)\, du + \frac{3}{2}\sigma^4(s)}{\sigma^3(s)} + O(n^{-1}),$$

with $y(s) = (1 - F(s))(1 - G(s-))$, so that the distribution of $\hat{A}^{(n)}(s)$ tends to be skewed to the right. □

EXAMPLE IV.1.7. Left-Truncated and Right-Censored Survival Data

In this example, we will consider survival data with left-truncation and right-censoring. The reader is referred to Section III.3 for a general discussion of left-truncation and to Section III.4 for a discussion of left-truncated and right-censored data.

For simplicity, we will here restrict ourselves to the setup with random left-truncation and right-censoring. Thus, the situation is as follows. (See Examples III.3.2 and III.3.3 for a discussion of random left-truncated survival times, and Example III.3.6 for a particular illustration on how right-censoring may be introduced in a model for left-truncated data.) Let the lifetime X be absolutely continuously distributed with distribution function F, survival function $S = 1 - F$, density f, and hazard rate α. The truncation time V and the censoring time U have joint distribution function G, such that with probability 1, $V < U$, and they are independent of X. Moreover, we assume that there is a positive probability that $X > V$. Denote the probability measure corresponding to this setup by P^0. Now, we do not sample from P^0, but from the conditional distribution given the event $A = \{X > V\}$. Denote this conditional probability measure by P, i.e., $\mathrm{P}(\cdot) = \mathrm{P}^0(\cdot \mid X > V)$. Let $p = \mathrm{P}^0(X > V)$ and introduce for later use

$$\begin{aligned} y(s) &= \mathrm{P}(V < s \leq X \wedge U) \\ &= \mathrm{P}^0(V < s, X \geq s, U \geq s)/p \\ &= S(s)\{G(s-, \infty) - G(s-, s-)\}/p. \end{aligned}$$

Let now (V_i, X_i, U_i), $i = 1, \ldots, n$, be a sample of n i.i.d. triplets from this *conditional distribution.* Then our left-truncated and right-censored sample is given as $(V_1, \tilde{X}_1, D_1), \ldots, (V_n, \tilde{X}_n, D_n)$, where $\tilde{X}_i = X_i \wedge U_i > V_i$ and $D_i = I(\tilde{X}_i = X_i)$. We let, in the usual manner,

$$N^{(n)}(s) = \sum_{i=1}^{n} I(\tilde{X}_i \leq s, D_i = 1)$$

be the number of observed deaths up to time s and

$$Y^{(n)}(s) = \sum_{i=1}^{n} I(V_i < s \leq \tilde{X}_i)$$

be the number at risk just before time s. We also define $J^{(n)}(s) = I(Y^{(n)}(s) > 0)$. Then, as shown in Sections III.3 and III.4, $N^{(n)}$ is a univariate counting process with intensity process $\alpha Y^{(n)}$, and the integrated hazard function A may be estimated by the Nelson–Aalen estimator in the usual manner.

We assume that $F(t) < 1$, or, equivalently, $A(t) < \infty$, and that y defined earlier is positive on $(0, t]$, i.e., that $\mathrm{P}^0(X \geq s) > 0$ and $\mathrm{P}^0(V < s \leq U) > 0$ for all $0 < s \leq t$. Then, because $Y^{(n)}(s)$ is binomially distributed with parameters n and $y(s)$, it follows that $Y^{(n)}(s) \xrightarrow{\mathrm{P}} \infty$ as $n \to \infty$ for all $s \in (0, t]$. Therefore, by the remark following (4.1.10), the conditions of Theorem IV.1.1 are fulfilled and uniform consistency of $\hat{A}^{(n)}$ on $[0, t]$ follows.

We then turn to the weak convergence of the Nelson–Aalen estimator. In addition to the assumptions just mentioned, we now also assume that y is bounded away from zero on $[\varepsilon, t]$ for an $\varepsilon > 0$. Note that we may write

$$\frac{Y^{(n)}(s)}{n} = \frac{\sum_{i=1}^{n} I(V_i < s)}{n} - \frac{\sum_{i=1}^{n} I(\tilde{X}_i < s)}{n},$$

so that by the Glivenko–Cantelli theorem

$$\sup_{s \in [\varepsilon, t]} \left| \frac{Y^{(n)}(s)}{n} - \mathrm{P}(V < s) + \mathrm{P}(X \wedge U < s) \right| \xrightarrow{\mathrm{P}} 0 \quad \text{as } n \to \infty.$$

Now we have $y(s) = \mathrm{P}(V < s) - \mathrm{P}(X \wedge U < s)$, and it follows by (4.1.16) that

$$\sqrt{n}\{(\widehat{A}^{(n)} - \widehat{A}^{(n)}(\varepsilon)) - (A - A(\varepsilon))\} \xrightarrow{\mathscr{D}} U \quad \text{as } n \to \infty$$

on $D[\varepsilon, t]$, where U is a Gaussian martingale with $U(\varepsilon) = 0$ and $\operatorname{cov}(U(s), U(u)) = \sigma^2(s \wedge u)$ with $\sigma^2(s) = \int_\varepsilon^s \{\alpha(u)/y(u)\}\, du$. However, y is not bounded away from zero on the whole of $[0, t]$, so weak convergence of the Nelson–Aalen estimator on $[0, t]$ has to be derived directly by checking the Conditions A–C of Theorem IV.1.2 (using, e.g., Proposition II.5.3). We will not go into this here; see, however, Woodroofe (1985) and Keiding and Gill (1990) who considered the setup with random left-truncation only (cf. Example IV.3.10). By going through the arguments of Keiding and Gill (1990), or by the general result of Davidsen and Jacobsen (1991), one may, in fact, see that to get the weak convergence on the whole of $[0, t]$, we need the conditions given earlier in this example for each $\varepsilon > 0$ and, in addition, that $\int_0^t \{\alpha(u)/y(u)\}\, du < \infty$. □

EXAMPLE IV.1.8. Left-Filtered and Right-Censored Survival Data

Here we give a further discussion of the difference between left-truncation and left-filtering (delayed entry) (cf. Section III.4) in the setup of Example IV.1.7. In the notation of this example, we let (V_i, X_i, U_i), $i = 1, \ldots, n$, be i.i.d. triplets from the original *unconditional distribution* P^0. The observations are then of three types: For some individuals we observe the exact lifetimes $(V_i < X_i \leq U_i)$, for others we observe right-censored lifetimes $(V_i < U_i < X_i)$, whereas for the rest we only observe that their lifetimes are less than certain quantities $(X_i \leq V_i < U_i)$. This latter type of observation is said to be left-censored. If we deliberately throw away the information contained in these left-censored observations, the data are treated as left-filtered and right-censored. This amounts to considering the counting process

$$N^{(n)}(s) = \sum_{i=1}^{n} I(V_i < X_i \leq s, U_i \geq X_i)$$

counting the number of observed deaths until time s, and

$$Y^{(n)}(s) = \sum_{i=1}^{n} I(V_i < s \leq X_i \wedge U_i),$$

the number at risk at $s-$. Note that these quantities are almost the same as those of Example IV.1.7, and that by the results of Section III.4, $N^{(n)}$ is a counting process with intensity process $\alpha Y^{(n)}$ in the usual manner.

The difference between this example and the preceding one becomes clearer when we study the asymptotic properties of the Nelson–Aalen estimator for this situation. We concentrate on the weak convergence result and only remark that the uniform consistency result may be proved with the same argument as that of Example IV.1.7. We define

$$\begin{aligned} y^0(s) &= \mathrm{P}^0(V < s \leq X \wedge U) \\ &= \mathrm{P}^0(V < s) - \mathrm{P}^0(X \wedge U < s) \end{aligned}$$

and assume that this is bounded away from zero on $[\varepsilon, t]$ for an $\varepsilon > 0$. By the same argument as used in Example IV.1.7, we then get

$$\sup_{s \in [\varepsilon, t]} \left| \frac{Y^{(n)}(s)}{n} - y^0(s) \right| \xrightarrow{\mathrm{P}^0} 0 \quad \text{as } n \to \infty.$$

Thus, by (4.1.16),

$$\sqrt{n}\{(\hat{A}^{(n)} - \hat{A}^{(n)}(\varepsilon)) - (A - A(\varepsilon))\} \xrightarrow{\mathscr{D}} U^0 \quad \text{as } n \to \infty$$

on $D[\varepsilon, t]$, where U^0 is a Gaussian process with mean zero satisfying $U^0(\varepsilon) = 0$ and having covariance structure $\operatorname{cov}(U^0(s), U^0(u)) = \sigma_0^2(s \wedge u)$ with $\sigma_0^2(s) = \int_\varepsilon^s \{\alpha(u)/y^0(u)\}\, du$.

It is seen that the difference with Example IV.1.7 is that here the $y(u)$ of the previous example is replaced by $y^0(u) = py(u)$, where $p = \mathrm{P}^0(X > V)$. This corresponds to the fact that the expected number of observations that are not filtered, i.e., the expected "effective" sample size in the present setup equals np. (See also the discussion in Section III.4 just after Example III.4.4.) □

Example IV.1.9. Uncensored Observations from a Markov Process with Finite State Space

We consider the setup from Example III.1.4. Thus, $(X(t), t \in \mathscr{T})$ is a Markov process with finite state space S and transition intensities $\alpha_{hj}(t)$, $h \neq j$. We let $p_h = \mathrm{P}(X(0) = h)$ be the initial distribution ($\sum_{h \in S} p_h = 1$), and let the transition probabilities be denoted $P_{hj}(s, t)$. Then $P_j(s) = \sum_{h \in S} p_h P_{hj}(0, s)$ is the probability that the Markov process is in state j at time s.

Now we let $X_1(\cdot), X_2(\cdot), \ldots, X_n(\cdot)$ be n i.i.d. copies of $X(\cdot)$ and define an aggregated counting process $(N_{hj}^{(n)}, h \neq j)$, with $N_{hj}^{(n)}(t)$ counting the total number of direct transitions for $X_1(\cdot), \ldots, X_n(\cdot)$ from h to j in $[0, t]$. Furthermore, we let $Y_h^{(n)}(t)$ denote the total number of the $X_i(\cdot)$ which are in state h at time $t-$. Note that $Y_h(t)$ is binomially distributed with parameters n and $P_h(t)$. [In Example III.1.4, these processes were denoted $N_{hj.}(t)$ and $Y_{h.}(t)$, respectively. We do not write the subscript dot here for notational ease.] Then, by Theorem II.6.8 and Example III.1.4, $(N_{hj}^{(n)}, h \neq j)$ is a multivariate counting process with intensity process $(\lambda_{hj}^{(n)}, h \neq j)$ satisfying the multiplicative intensity model $\lambda_{hj}^{(n)}(t) = \alpha_{hj}(t) Y_h^{(n)}(t)$. It follows that the integrated transition intensities $A_{hj}(t) = \int_0^t \alpha_{hj}(u)\, du$, $h \neq j$, can be estimated by the Nelson–Aalen estimators

$$\hat{A}_{hj}^{(n)}(t) = \int_0^t J_h^{(n)}(u)(Y_h^{(n)}(u))^{-1}\, dN_{hj}^{(n)}(u),$$

where $J_h^{(n)}(t) = I(Y_h^{(n)}(t) > 0)$ as usual.

For this setup, we now want to check the conditions of Theorems IV.1.1 and IV.1.2. In order not to make things too complicated, we will restrict our attention to a subinterval $[t_1, t_2]$ of $\mathscr{T}$ on which $P_h(s) > 0$ for all $h \in S$ and almost all $s \in [t_1, t_2]$. Then obviously $Y_h^{(n)}(s) \xrightarrow{\mathrm{P}} \infty$ as $n \to \infty$, for all $h \in S$ and almost all $s \in [t_1, t_2]$. Then, by the remark just after (4.1.10), the conditions of Theorem IV.1.1 are satisfied, and it follows that the Nelson–Aalen estimators are uniformly consistent.

To prove weak convergence of the Nelson–Aalen estimators, here it is not straightforward to use (4.1.16) and the Glivenko–Cantelli theorem as in Examples IV.1.6 to IV.1.8. Therefore, we will check the conditions of Theorem IV.1.2 directly by using Proposition II.5.2. We start by showing in detail how (4.1.12) can be proved. Note first that

$$n\frac{J_h^{(n)}(s)}{Y_h^{(n)}(s)}\alpha_{hj}(s) \xrightarrow{\mathrm{P}} \frac{\alpha_{hj}(s)}{P_h(s)} \quad \text{as } n \to \infty,$$

for almost all $s \in [t_1, t_2]$. To check the conditions of Proposition II.5.2, we need the following technical result: If Z is binomially distributed with parameters n and p with $0 < p \leq 1$, then for each positive integer k

$$\mathrm{E}\left\{\frac{nI(Z > 0)}{Z}\right\}^k \leq \left(\frac{k+1}{p}\right)^k,$$

where 0/0 is interpreted as 0 [cf. McKeague and Utikal (1990a, Lemma 1)]. Using this result and the fact that $Y_h^{(n)}(s)$ is binomially distributed with parameters n and $P_h(s)$, we see that, for a fixed $s \in [t_1, t_2]$,

$$\mathrm{E}\{(nJ_h^{(n)}(s)(Y_h^{(n)}(s))^{-1}\alpha_{hj}(s))^2\}$$

is bounded uniformly in n. It follows that $\{nJ_h^{(n)}(s)(Y_h^{(n)}(s))^{-1}\alpha_{hj}(s); n = 1, 2, \ldots\}$ is uniformly integrable for each $s \in [t_1, t_2]$. Also, we see that

$$\mathrm{E}\left(n\frac{J_h^{(n)}(s)}{Y_h^{(n)}(s)}\alpha_{hj}(s)\right) \leq 2\frac{\alpha_{hj}(s)}{P_h(s)},$$

where the function on the right-hand side is integrable by the assumptions made earlier. Thus, the conditions of Proposition II.5.2 are fulfilled and it follows that

$$n\int_{t_1}^{s}\frac{J_h^{(n)}(u)}{Y_h^{(n)}(u)}\alpha_{hj}(u)\,\mathrm{d}u \xrightarrow{\mathrm{P}} \int_{t_1}^{s}\frac{\alpha_{hj}(u)}{P_h(u)}\,\mathrm{d}u,$$

for all $s \in [t_1, t_2]$, and (4.1.12) is proved. Now, (4.1.13) and (4.1.14) can be proved by a similar argument. Thus, we have that

$$(\sqrt{n}(\hat{A}_{hj}^{(n)} - A_{hj}); h \neq j) \xrightarrow{\mathscr{D}} (U_{hj}; h \neq j)$$

on $[t_1, t_2]$, where the U_{hj} are independent Gaussian martingales with $U_{hj}(0) = 0$ and $\mathrm{cov}(U_{hj}(s), U_{hj}(u)) = \int_{t_1}^{s \wedge u}\{\alpha_{hj}(v)/P_h(v)\}\,\mathrm{d}v$. □

Example IV.1.10. Matings of Drosophila Flies

In this example, we will discuss how one may derive the asymptotic properties of the Nelson–Aalen estimator for the simple statistical model for matings of Drosophila flies introduced in Example III.1.10. The more realistic model of Example III.5.5 will not be considered here.

The relevant large sample situation to consider for the model of Example III.1.10 is not one in which more and more flies are put in the same "pornoscope," but rather one in which more and more flies are put in larger and larger pornoscopes. We, therefore, consider a sequence of pornoscope experiments, indexed by n, and let s_n denote the size of the pornoscope in cm^2 in the nth experiment. Furthermore, we assume that the numbers of female and male flies per cm^2 are kept constant in all the experiments and equal to φ and μ. Thus, the total number of female and male flies in the pornoscope in the nth experiment are assumed to be $f_0^{(n)} = s_n\varphi$ and $m_0^{(n)} = s_n\mu$, respectively. In a similar manner, we assume that the chance of a mating of one male and one female fly, who have not yet initiated a mating, during a small time interval is inversely proportional to the size of the pornoscope. Formally, this is obtained by letting the individual mating intensity $\alpha^{(n)}(s)$ in the nth experiment satisfy $\alpha^{(n)}(s) = \theta(s)/s_n$. Here we will assume that θ, the individual mating intensity per cm^2, is a bounded function which is also bounded away from zero.

According to the model of Example III.1.10, we, therefore, consider a sequence of counting processes $N^{(n)}$, with $N^{(n)}(t)$ counting the number of matings initiated in the nth experiment in $[0, t]$ and having intensity processes of the form

$$\lambda^{(n)}(s) = \alpha^{(n)}(s)(f_0^{(n)} - N^{(n)}(s-))(m_0^{(n)} - N^{(n)}(s-)).$$

By introducing

$$Y^{(n)}(s) = s_n(\varphi - N^{(n)}(s-)/s_n)(\mu - N^{(n)}(s-)/s_n),$$

it is seen that $N^{(n)}$ has an intensity process of the multiplicative form $\lambda^{(n)}(s) = \theta(s)Y^{(n)}(s)$.

We now slightly reformulate our problem in that we will consider the asymptotic properties, over a fixed time interval $[0, t]$, of the Nelson–Aalen estimator

$$\hat{\Theta}^{(n)}(s) = \int_0^s (Y^{(n)}(u))^{-1}\, dN^{(n)}(u)$$

for $\Theta(s) = \int_0^s \theta(u)\, du$.

We will do this using Kurtz' law of large numbers (Theorem II.5.4). To this end, we let $\Lambda_n(s) = \int_0^s \lambda^{(n)}(u)\, du$ be the compensator of $N^{(n)}$, and note that we may write

$$\lambda^{(n)} = s_n\beta(\cdot, s_n^{-1}N^{(n)}),$$

where $\beta(u, x)$ is the nonanticipating non-negative function of $u \in \mathscr{T}$ and $x \in D(\mathscr{T})$ given (for $0 \le x(u-) \le \varphi \wedge \mu$) by

$$\beta(u, x) = \theta(u)(\varphi - x(u-))(\mu - x(u-)).$$

This function satisfies, for $u \le t$,

$$\beta(u, x) \le \theta(u)\varphi\mu \le \varphi\mu \sup_{u \le t} \theta(u)$$

and

$$\begin{aligned}
&|\beta(u, x_1) - \beta(u, x_2)| \\
&\quad \le \theta(u)\{(\varphi + \mu)|x_1(u-) - x_2(u-)| + |x_1(u-)^2 - x_2(u-)^2|\} \\
&\quad \le (\varphi + \mu + 2(\varphi \wedge \mu)) \sup_{u \le t} \theta(u) \sup_{u \le t} |x_1(u-) - x_2(u-)|.
\end{aligned}$$

Therefore, by Theorem II.5.4,

$$\sup_{s \le t} |s_n^{-1} N^{(n)}(s) - X(s)| \xrightarrow{\mathrm{P}} 0$$

as $n \to \infty$, where X is the solution of

$$\begin{aligned}
X(s) &= \int_0^s \beta(u, X)\, du \\
&= \int_0^s \theta(u)(\varphi - X(u-))(\mu - X(u-))\, du.
\end{aligned}$$

Clearly, $X(0) = 0$ and X is increasing and continuous.

We now introduce $y = (\varphi - X)(\mu - X)$ and consider an interval $[0, t]$ for which $y(t) = \inf_{s \in [0,t]} y(s) > 0$. Then by the above result

$$\sup_{s \in [0,t]} \left| \frac{Y^{(n)}(s)}{s_n} - y(s) \right| \xrightarrow{\mathrm{P}} 0$$

as $n \to \infty$. It follows using (4.1.10) and (4.1.16) that $\hat{\Theta}^{(n)}$ is uniformly consistent for Θ on $[0, t]$ and that

$$\sqrt{s_n}(\hat{\Theta}^{(n)} - \Theta) \xrightarrow{\mathscr{D}} U$$

as $n \to \infty$, on $D[0, t]$, where U is a Gaussian martingale with $U(0) = 0$ and covariance function given by

$$\operatorname{cov}(U(s), U(u)) = \sigma^2(s \wedge u) = \int_0^{s \wedge u} \frac{\theta(v)}{y(v)}\, dv.$$

Here, $\sigma^2(s)$ can be estimated uniformly consistently by $s_n \hat{\sigma}^2(s)$, where

$$\hat{\sigma}^2(s) = \int_0^s (Y^{(n)}(u))^{-2}\, dN^{(n)}(u);$$

cf. (4.1.6) and (4.1.15).

It is of some interest to reformulate these results in terms of the Nelson–Aalen estimator $\hat{A}^{(n)}(s)$ for the integrated individual mating intensity $A^{(n)}(s) = \int_0^s \alpha^{(n)}(u)\, du$. For this purpose we introduce, as in Example III.1.10, $F^{(n)}(s) = f_0^{(n)} - N^{(n)}(s-)$ and $M^{(n)}(s) = m_0^{(n)} - N^{(n)}(s-)$ for the number of female and male flies, respectively, that have not yet initiated a mating at time $t-$ (in the nth experiment). Then

$$\hat{A}^{(n)}(s) = \int_0^s \{F^{(n)}(u) M^{(n)}(u)\}^{-1}\, dN^{(n)}(u)$$

and its variance may be estimated by

$$\hat{\tau}^2(s) = \int_0^s \{F^{(n)}(u)M^{(n)}(u)\}^{-2}\,\mathrm{d}N^{(n)}(u)$$

in the usual manner.

Now $Y^{(n)} = s_n^{-1}F^{(n)}M^{(n)}$, and it follows that $\hat{\Theta}^{(n)} = s_n\hat{A}^{(n)}$, $\Theta = s_n A^{(n)}$, and $\hat{\sigma}^2 = s_n^2\hat{\tau}^2$. Therefore, the result derived above can also be written

$$s_n^{3/2}(\hat{A}^{(n)} - A^{(n)}) \xrightarrow{\mathscr{D}} U$$

as $n \to \infty$, where $\sigma^2(s) = \operatorname{var} U(s)$ can be estimated uniformly consistently by $s_n^3\hat{\tau}(s)$. □

In the next example, we study the model for relative mortality and the related excess mortality model introduced in Examples III.1.3 and III.1.8, respectively.

EXAMPLE IV.1.11. Models for Relative and Excess Mortality. Survival with Malignant Melanoma

In Example III.1.3, a model for relative mortality was discussed and we noted that if the individual counting process N_i had intensity process

$$\lambda_i(t) = \alpha(t)\mu_i(t)Y_i(t)$$

with $\mu_i(\cdot)$ known, then the aggregated process $N = N_1 + \cdots + N_n$ satisfied Aalen's multiplicative intensity model

$$\lambda(t) = \alpha(t)Y^{\mu}(t)$$

with

$$Y^{\mu}(t) = \sum_{i=1}^{n} \mu_i(t)Y_i(t).$$

This means that we can use the results of Section IV.1.1 and estimate the cumulative *relative* mortality

$$A(t) = \int_0^t \alpha(u)\,\mathrm{d}u$$

by the Nelson–Aalen estimator

$$\hat{A}(t) = \int_0^t \frac{J(u)}{Y^{\mu}(u)}\,\mathrm{d}N(u),$$

where $J(t) = I(Y^{\mu}(t) > 0) = I(Y(t) > 0)$. Here, the latter equality holds because all $\mu_i(t)$ are assumed positive. Furthermore, the asymptotic results of the present subsection hold provided that, e.g., there exists a positive function $\varphi(\cdot)$, bounded away from zero, such that, for all $t \in \mathscr{T}$,

$$\sup_{s\in[0,t]}\left|\frac{Y^{\mu}(s)}{n} - \varphi(s)\right| \xrightarrow{\mathrm{P}} 0. \tag{4.1.19}$$

The related model for *excess* mortality

$$\lambda_i(t) = (\gamma(t) + \mu_i(t))\,Y_i(t)$$

presented in Example III.1.8 does not give rise to a multiplicative intensity model for the aggregated counting process N, the latter having intensity process

$$\lambda(t) = \gamma(t)\,Y(t) + Y^{\mu}(t).$$

As we shall see presently, however, it is possible to apply the same ideas as in Section IV.1.1 to derive and study an estimator for the cumulative excess mortality

$$\Gamma(t) = \int_0^t \gamma(u)\,\mathrm{d}u$$

(Andersen and Væth, 1989). Thus, because

$$N(t) - \int_0^t Y^{\mu}(u)\,\mathrm{d}u - \int_0^t \gamma(u)\,Y(u)\,\mathrm{d}u = M(t),$$

where $M(t) = M_1(t) + \cdots + M_n(t)$ is a local square integrable martingale, a natural estimator for $\Gamma(t)$ is

$$\hat{\Gamma}(t) = \int_0^t J(u)\,Y(u)^{-1}\,\mathrm{d}N(u) - \int_0^t Y^{\mu}(u)\,Y(u)^{-1}\,\mathrm{d}u. \tag{4.1.20}$$

It is seen that $\hat{\Gamma}(t)$ is the difference between the ordinary Nelson–Aalen estimator for the integrated hazard and an integrated *average population hazard*

$$\int_0^t Y^{\mu}(u)\,Y(u)^{-1}\,\mathrm{d}u = \int_0^t \sum_{i=1}^n \mu_i(u)\frac{Y_i(u)}{Y(u)}\,\mathrm{d}u.$$

It follows that the difference

$$\hat{\Gamma}(t) - \int_0^t J(u)\gamma(u)\,\mathrm{d}u = \int_0^t J(u)\,Y(u)^{-1}\,\mathrm{d}M(u)$$

is a local square integrable martingale, and properties of the estimator can be derived following the lines of the present subsection provided that (4.1.16) (with $a_n = \sqrt{n}$) and (4.1.19) hold. Under these conditions, $\hat{\Gamma}(t)$ is uniformly consistent as $n \to \infty$ on compact subintervals of $\mathscr{T}$, and $n^{1/2}(\hat{\Gamma} - \Gamma)$ is asymptotically distributed as a mean zero Gaussian martingale with a variance

$$\int_0^t \left\{\frac{\gamma(u)}{y(u)} + \frac{\varphi(u)}{y^2(u)}\right\}\mathrm{d}u,$$

which we may estimate consistently by

$$n\int_0^t Y(u)^{-2}\,\mathrm{d}N(u).$$

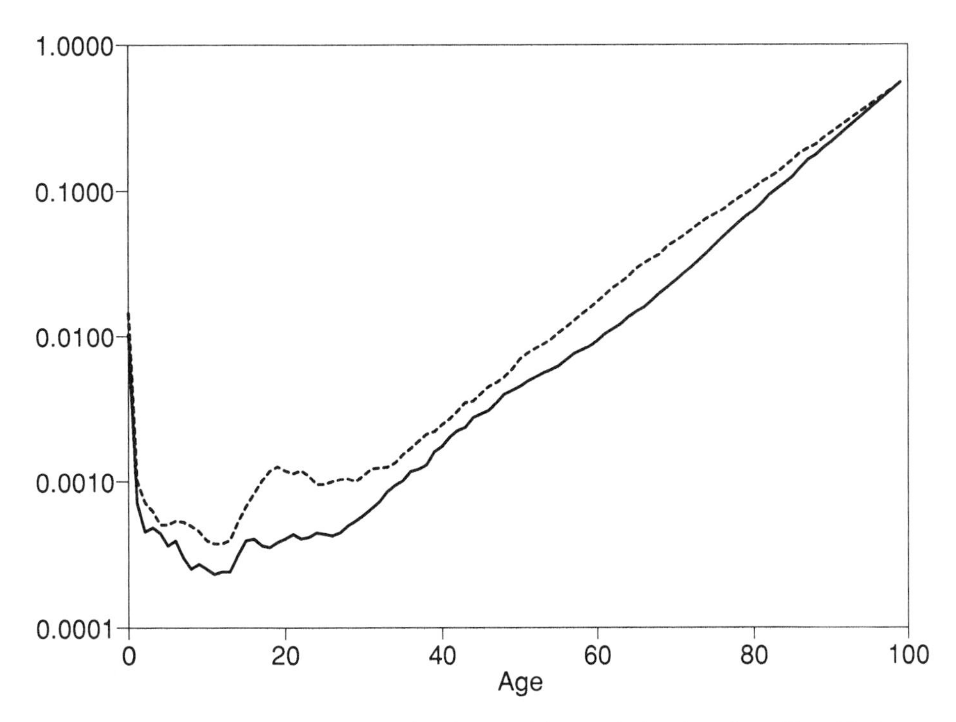

Figure IV.1.8. Mortality per year for the Danish population based on life table for the years 1971–75. Men: (- - -); women (—).

We shall now illustrate these models using the melanoma data introduced in Example I.3.1. As mentioned there, 71 patients died; 57 were recorded as deaths from malignant melanoma and 14 as deaths from other causes. The population mortality figures $\mu_i(\cdot)$ obtained from life tables published by the Danish Central Bureau of Statistics are given in Tables A.2 and A.3 in the Appendix and shown in Figure IV.1.8. The life table is from the period 1971–75 and the values are given for ages $x = 0, 1, \ldots, 99$ years separately for men and women. The curves are seen to have a characteristic "bathtub" shape increasing nearly log-linearly from age 30 and upward. Men have a higher mortality at all ages.

Considering first the multiplicative model, Figure IV.1.9 shows the estimated cumulative relative mortality for the melanoma patients based on these population mortality figures. The fact that the plot is approximately linear indicates that the relative mortality is roughly constant: $\alpha(t) \approx \alpha_0 \approx 3.4$.

Turning next to the additive model, Figure IV.1.10 shows the Nelson–Aalen estimator, the integrated average population hazard, and their difference: the integrated excess mortality $\hat{\Gamma}(t)$. All three curves look roughly linear indicating that also a model with a constant excess mortality $\gamma(t) \approx \gamma_0 \approx 0.04$ per year seems reasonable. The finding that the Nelson–Aalen estimator

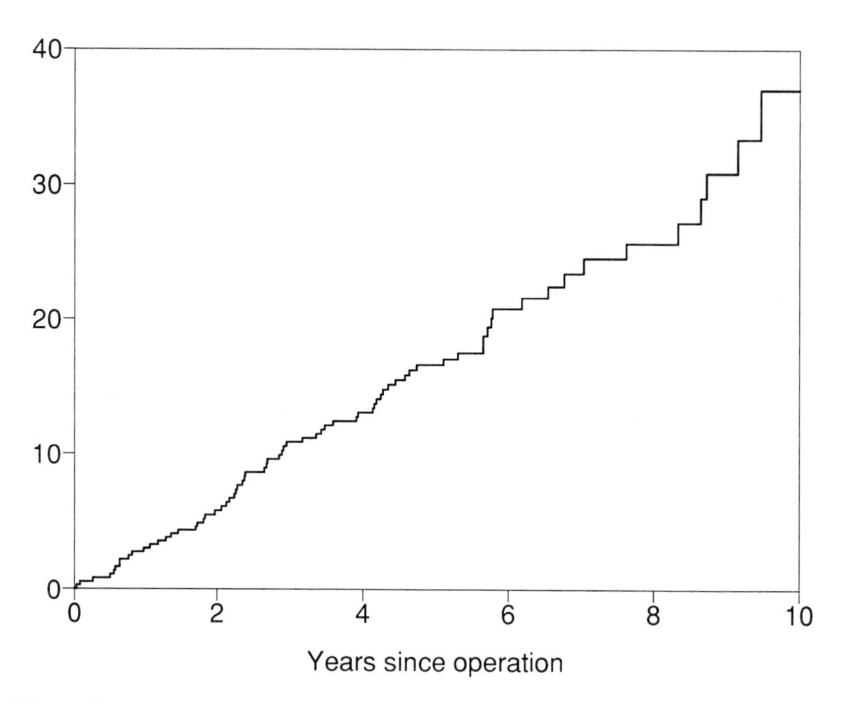

Figure IV.1.9. Integrated relative mortality for patients with malignant melanoma.

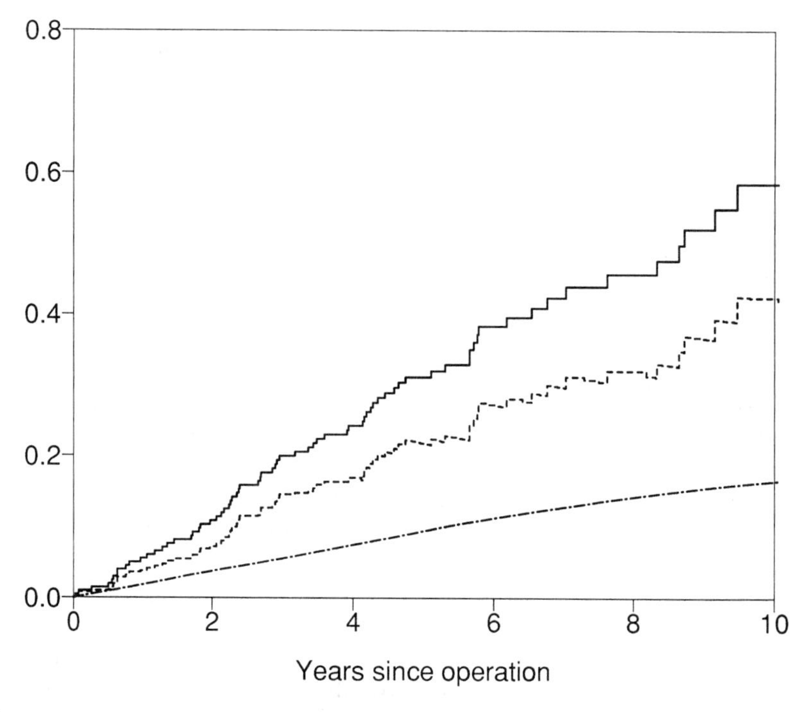

Figure IV.1.10. Survival with malignant melanoma. Nelson–Aalen estimate based on all 71 deaths (—); integrated average population hazard (-·-·-), and estimated integrated excess mortality (- - -).

is approximately linear is consistent with the findings in Example IV.1.2 (though we have in the present example considered all 71 deaths as events of interest and not only the 57 deaths from the disease). The fact that the average population hazard is approximately constant does *not* imply, however, that all $\mu_i(t)$ are identical—the population hazard certainly depends on both age and sex; see Figure IV.1.8.

To more closely study how the age of the patients may be accounted for, we now consider age as the basic time scale instead of time since operation. The survival time for patient i is then *left-truncated* at the age V_i at operation; following Example IV.1.7, we, therefore, redefine the counting processes and the risk indicator processes as follows:

$$N_i^*(a) = I(\text{patient } i \text{ is observed to die before } age\ a)$$

and

$$Y_i^*(a) = I(\text{patient } i \text{ is observed to be at risk at } age\ a-).$$

Then, with a slight abuse of notation for the population mortality, we can still look at the multiplicative model

$$N_i^*(a) = \int_0^a \alpha(u)\mu_i(u) Y_i^*(u)\,\mathrm{d}u + M_i^*(a)$$

or the additive model

$$N_i^*(a) = \int_0^a (\mu_i(u) + \gamma(u)) Y_i^*(u)\,\mathrm{d}u + M_i^*(a),$$

and we can estimate the age-specific relative mortality $\alpha(\cdot)$ or the age-specific excess mortality $\gamma(\cdot)$ as described above. Figure IV.1.11 shows the estimated integrated relative mortality in this time scale. The concave shape of the curve shows that the relative mortality decreases with age. Figure IV.1.12 shows the plots for the additive model. Both the Nelson–Aalen plot and the integrated average population mortality are convex, i.e., the corresponding hazards increase with age. However, their difference, the estimated integrated excess mortality, is still fairly linear (with slope ≈ 0.04 per year) suggesting an age-independent excess mortality.

Thus, a reasonable model for the mortality from all causes after an operation for malignant melanoma seems to be to write the hazard for patient i at time t after an operation as

$$\alpha_i(t) = \gamma_0 + \mu_i(t).$$

Note that this model has the same hazard structure as a competing risks model (Examples I.3.8, I.3.9, and III.1.5) with two causes of death: deaths from malignant melanoma with a time-independent cause-specific hazard function γ_0 and deaths from other causes with a cause-specific hazard function $\mu_i(t)$ which may be considered known from standard mortality tables. This interpretation is consistent with the finding that the estimated integrated excess mortality shown in Figure IV.1.10 is close to the Nelson–Aalen

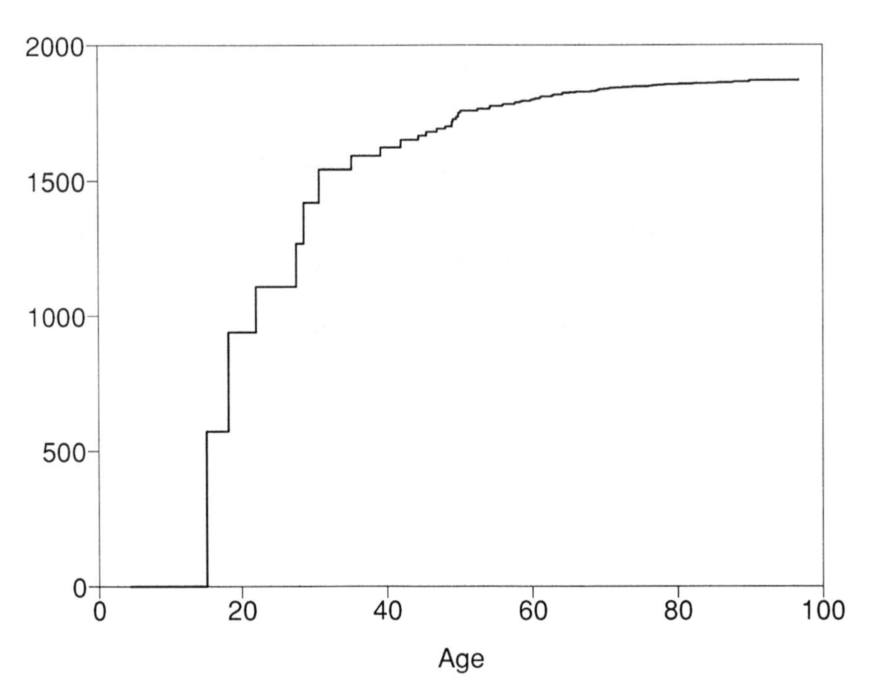

Figure IV.1.11. Integrated (age-specific) relative mortality for patients with malignant melanoma.

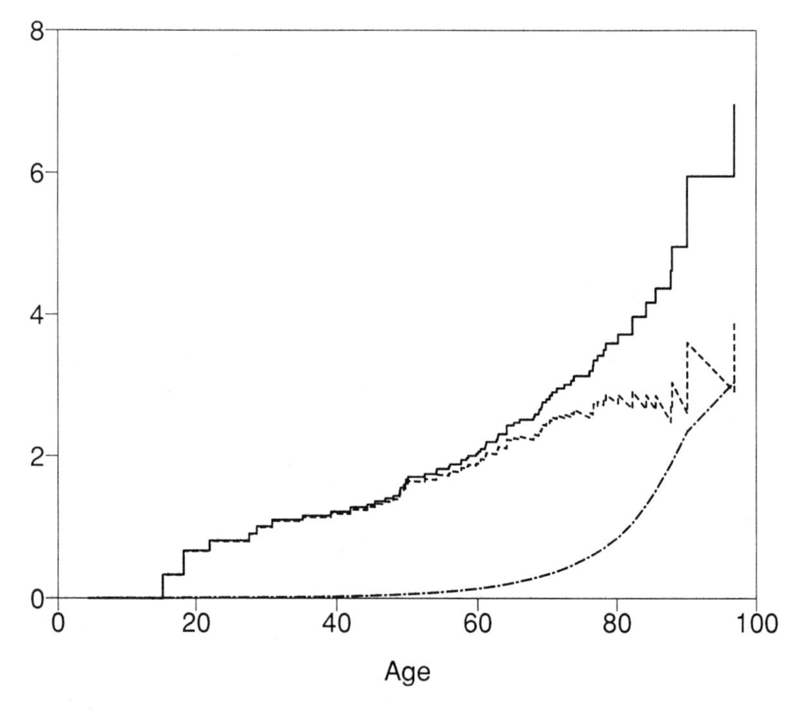

Figure IV.1.12. Survival with malignant melanoma. Estimated integrated (age-specific) mortality (—), integrated average population hazard (-·-·-), and estimated integrated excess mortality (- - -).

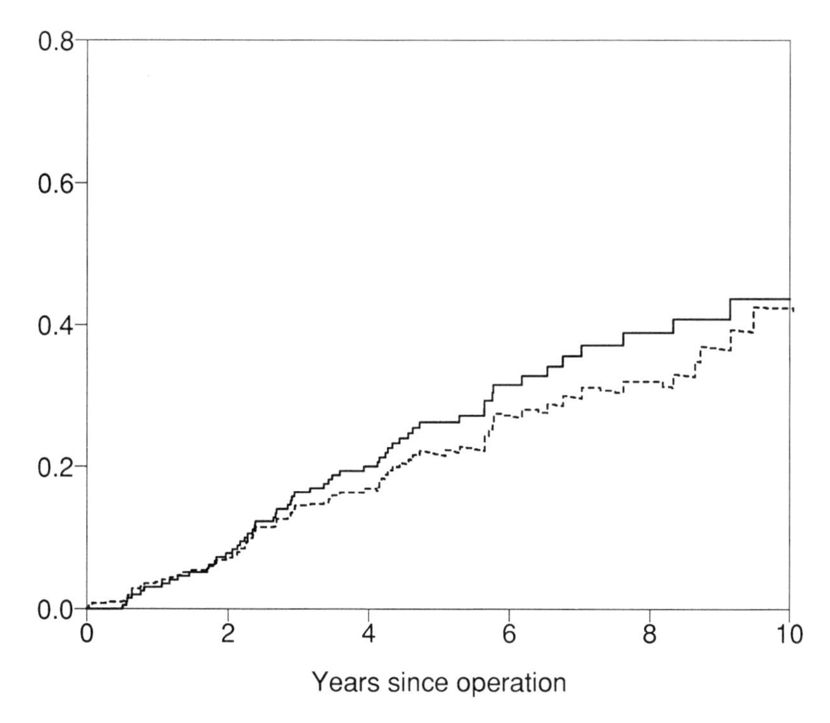

Figure IV.1.13. Survival with malignant melanoma. Estimated integrated excess mortality (- - -) and Nelson–Aalen estimate for the integrated mortality based on the 57 deaths from the disease (—).

estimator based on deaths from malignant melanoma only; cf. Figure IV.1.13. We shall return to a further discussion in Examples IV.4.1 and V.1.5 and to a further comparison of the multiplicative and additive models in Example VII.4.3. □

IV.1.3. Confidence Intervals and Confidence Bands

The asymptotic results of the preceding subsection may be used to provide pointwise confidence intervals and simultaneous confidence bands for $A_h(s) = \int_0^s \alpha_h(u)\,du$, $h = 1, \ldots, k$. In what follows, all formulas for intervals and bands are given with $\hat{\sigma}_h^2$ defined by (4.1.6). However, they are also true with $\check{\sigma}_h^2$ given by (4.1.7). We consider a fixed value of h in this subsection and omit the subscript h for notational ease.

IV.1.3.1. *Pointwise Confidence Intervals*

Let us first have a look at pointwise confidence intervals for $A(s)$ for a fixed $s \in \mathscr{T}$. The standard asymptotic $100(1-\alpha)\%$ interval ("linear" interval) is

$$\hat{A}^{(n)}(s) \pm c_{\alpha/2}\hat{\sigma}(s), \tag{4.1.21}$$

where $c_{\alpha/2}$ is the upper $\alpha/2$ fractile of the standard normal distribution. According to the Monte Carlo simulations for censored survival data of Bie, Borgan, and Liestøl (1987), this interval is not satisfactory for small sample sizes, however, and it may pay to consider transformations to improve the approximation to the asymptotic distribution.

This well-known idea works as follows. If g is a function which is differentiable in a neighborhood of $A(s)$, and $g'(x)$ is continuous and different from zero at $x = A(s)$, then by the ordinary delta-method (cf. the introduction to Section II.8)

$$\frac{g(\hat{A}^{(n)}(s)) - g(A(s))}{|g'(\hat{A}^{(n)}(s))|\,\hat{\sigma}(s)} \xrightarrow{\mathscr{D}} \mathscr{N}(0,1) \quad \text{as } n \to \infty.$$

It follows that an asymptotic $100(1-\alpha)\%$ confidence interval for $g(A(s))$ is

$$g(\hat{A}^{(n)}(s)) \pm c_{\alpha/2}|g'(\hat{A}^{(n)}(s))|\,\hat{\sigma}(s).$$

For a suitable choice of the function g, this interval may be inverted into an asymptotic $100(1-\alpha)\%$ confidence interval for $A(s)$ with better small sample properties than (4.1.21).

A systematic study of the best choice of g for various situations has not yet been performed. Bie et al. (1987) considered the transformations $g(x) = \log x$ and $g(x) = \arcsin(e^{-x/2})$. The latter of these is a variance-stabilizing transformation for the survival data situation with no censoring, and we will denote it the arcsin-transformation for short in the following. Assuming that $A(s) > 0$, the log-transformation gives the asymptotic $100(1-\alpha)\%$ confidence interval

$$\hat{A}^{(n)}(s)\exp\left\{\frac{\pm c_{\alpha/2}\hat{\sigma}(s)}{\hat{A}^{(n)}(s)}\right\}, \tag{4.1.22}$$

whereas the arcsin-transformation gives the interval

$$\begin{aligned} -2\log&\left(\sin\left\{\min\left(\frac{\pi}{2}, \arcsin(e^{-\hat{A}^{(n)}(s)/2}) + \tfrac{1}{2}c_{\alpha/2}\hat{\sigma}(s)(e^{\hat{A}^{(n)}(s)} - 1)^{-1/2}\right)\right\}\right) \\ &\le A(s) \\ &\le -2\log\left(\sin\left\{\max\left(0, \arcsin(e^{-\hat{A}^{(n)}(s)/2}) - \tfrac{1}{2}c_{\alpha/2}\hat{\sigma}(s)(e^{\hat{A}^{(n)}(s)} - 1)^{-1/2}\right)\right\}\right). \end{aligned} \tag{4.1.23}$$

The results of Bie et al. (1987) indicate that both of these confidence intervals are quite satisfactory also for sample sizes as low as 25, even with 50% censoring, for the survival data situation.

IV.1.3.2. *Simultaneous Confidence Bands*

We will next study simultaneous confidence bands for A on (subintervals of) $[0, t]$, where $t \in \mathscr{T}$. A class of confidence bands may be derived as follows [e.g., Doksum and Yandell (1984)]. Let q be a continuous and non-negative function on $[t_1, t_2]$ where $0 \le t_1 < t_2 \le t$. Then, by Theorem IV.1.2,

$$\left(\frac{a_n(\hat{A}^{(n)} - A)}{1 + a_n^2\hat{\sigma}^2}\right) q \circ \left(\frac{a_n^2\hat{\sigma}^2}{1 + a_n^2\hat{\sigma}^2}\right) \xrightarrow{\mathscr{D}} \left(\frac{U}{1 + \sigma^2}\right) q \circ \left(\frac{\sigma^2}{1 + \sigma^2}\right)$$

on $D[t_1, t_2]$ as $n \to \infty$, where $\circ$ denotes composition and U is a Gaussian martingale with $U(0) = 0$ and $\operatorname{cov}(U(s), U(u)) = \sigma^2(s \wedge u)$. Let W^0 denote the standard Brownian bridge. Then the processes

$$\left(\frac{U}{1 + \sigma^2}\right) q \circ \left(\frac{\sigma^2}{1 + \sigma^2}\right)$$

and

$$(qW^0) \circ \left(\frac{\sigma^2}{1 + \sigma^2}\right)$$

have the same distribution, both being zero mean Gaussian processes with the same covariance function. It follows that

$$\sup_{s \in [t_1, t_2]} \left| \frac{a_n(\hat{A}^{(n)}(s) - A(s))}{1 + a_n^2\hat{\sigma}^2(s)} q\left(\frac{a_n^2\hat{\sigma}^2(s)}{1 + a_n^2\hat{\sigma}^2(s)}\right) \right| \xrightarrow{\mathscr{D}} \sup_{x \in [c_1, c_2]} |q(x) W^0(x)| \quad (4.1.24)$$

as $n \to \infty$, where

$$c_i = \frac{\sigma^2(t_i)}{1 + \sigma^2(t_i)}. \quad (4.1.25)$$

This result may be inverted to yield simultaneous confidence bands. More precisely, the asymptotic $100(1 - \alpha)\%$ confidence band for A on $[t_1, t_2]$ is

$$\hat{A}^{(n)}(s) \pm a_n^{-1} K_{q,\alpha}(c_1, c_2)(1 + a_n^2\hat{\sigma}^2(s))/q\left(\frac{a_n^2\hat{\sigma}^2(s)}{1 + a_n^2\hat{\sigma}^2(s)}\right) \quad (4.1.26)$$

with $K_{q,\alpha}(c_1, c_2)$ the upper α fractile of the distribution of

$$\sup_{x \in [c_1, c_2]} |q(x) W^0(x)|.$$

For situations where one or both of c_1 and c_2 are not known, the unknown c_i may be replaced by

$$\hat{c}_i = \frac{a_n^2\hat{\sigma}^2(t_i)}{1 + a_n^2\hat{\sigma}^2(t_i)}$$

in (4.1.26).

Two choices of the weight function q in (4.1.26) seem to be of particular interest. The choice $q_1(x) = \{x(1 - x)\}^{-1/2}$ yields confidence bands proportional to the pointwise ones (4.1.21). With this choice, we may consider confidence bands on $[t_1, t_2]$, where $0 < c_1 < c_2 < 1$; cf. (4.1.25). The resulting asymptotic $100(1 - \alpha)\%$ confidence band for A on $[t_1, t_2]$ is

$$\hat{A}^{(n)}(s) \pm d_\alpha(\hat{c}_1, \hat{c}_2)\hat{\sigma}(s), \quad (4.1.27)$$

where $d_\alpha(c_1, c_2) = K_{q_1,\alpha}(c_1, c_2)$ is the upper α fractile of the distribution of

$$\sup_{c_1 \le x \le c_2} |W^0(x)\{x(1-x)\}^{-1/2}|.$$

This fractile may be found by the asymptotic approximation [Miller and Siegmund (1982, formula 8)]

$$\mathrm{P}\left\{\sup_{c_1 \le x \le c_2} |W^0(x)\{x(1-x)\}^{-1/2}| \ge d\right\}$$

$$= \frac{4\phi(d)}{d} + \phi(d)\left(d - \frac{1}{d}\right)\log\left\{\frac{c_2(1-c_1)}{c_1(1-c_2)}\right\} + o\{\phi(d)/d\}$$

as $d \to \infty$, where $\phi(d)$ is the standard normal density $(2\pi)^{-1/2}\exp(-\frac{1}{2}d^2)$. We note that (4.1.27) is equivalent to the confidence band proposed by Hjort (1985a) based on a transformation to the Ornstein–Uhlenbeck process. Following Nair (1984), who discussed a similar band for the survival distribution (cf. Section IV.3.3), we will call (4.1.27) a *equal precision band*, or *EP-band* for short.

The second choice of weight function of particular interest is $q_2(x) = 1$. The resulting asymptotic $100(1-\alpha)\%$ confidence band for A on $[t_1, t_2]$ is

$$\hat{A}^{(n)}(s) \pm a_n^{-1} e_\alpha(\hat{c}_1, \hat{c}_2)(1 + a_n^2 \hat{\sigma}^2(s)), \tag{4.1.28}$$

where $e_\alpha(c_1, c_2) = K_{q_2,\alpha}(c_1, c_2)$ is the upper α fractile of the distribution of

$$\sup_{c_1 \le x \le c_2} |W^0(x)|.$$

For this band, we will typically let $[t_1, t_2]$ be the whole of $[0, t]$, in which case tables of $e_\alpha(c_1, c_2) = e_\alpha(0, c_2)$ are given by Koziol and Byar (1975), Hall and Wellner (1980), and Schumacher (1984) for selected values of α and c_2. Chung (1986) gave tables of $e_\alpha(c_1, c_2)$ for general c_1 and c_2; see also the computer program WIENER PACK by Chung (1987). The band (4.1.28) is similar to a band for the survival distribution introduced by Hall and Wellner (1980) (cf. Section IV.3.3), and we will denote it the *Hall–Wellner band*, or *HW-band* for short.

A confidence band may be viewed as a one-sample test statistic in the sense that a hypothesis $A = A_0$ is rejected at significance level α if A_0 is not completely contained in the band. Such test statistics are termed one-sample maximal deviation (or Kolmogorov–Smirnov type) tests and will be further discussed in Section V.4.1. In particular, we will meet the tests corresponding to the EP- and HW-bands in Proposition V.4.1.

The confidence bands (4.1.27) and (4.1.28) are based directly on the weak convergence result of Theorem IV.1.2. These bands perform badly (for the survival data situation) even with sample sizes of 100–200 (Bie et al., 1987). However, as was the case for the pointwise confidence intervals, it may also, for confidence bands, be advantageous to consider transformations to improve the approximation to the asymptotic distribution. The idea is the same as the one used for the pointwise confidence intervals, and it works as follows. Let $0 \le t_1 < t_2 \le t$ and q be as earlier and let g be a function such that $g'(x)$ is different from zero and continuous on $(x_1 - \varepsilon, x_2 + \varepsilon)$ for an $\varepsilon > 0$, where $x_i = A(t_i)$. Then $a_n(g \circ \hat{A}^{(n)} - g \circ A)/g' \circ \hat{A}^{(n)}$ has the same asymptotic distribution as $a_n(\hat{A}^{(n)} - A)$ on $[t_1, t_2]$. By the argument leading to (4.1.26), it follows that an asymptotic $100(1 - \alpha)\%$ confidence band for $g \circ A$ on $[t_1, t_2]$ is

$$g(\hat{A}^{(n)}(s)) \pm a_n^{-1} K_{q,\alpha}(c_1, c_2) |g'(\hat{A}^{(n)}(s))| (1 + a_n^2 \hat{\sigma}^2(s)) / q\left(\frac{a_n^2 \hat{\sigma}^2(s)}{1 + a_n^2 \hat{\sigma}^2(s)}\right). \quad (4.1.29)$$

With a suitable function g, this band may be inverted into an asymptotic $100(1 - \alpha)\%$ confidence band for A. For the transformations $g(x) = \log x$ and $g(x) = \arcsin(e^{-x/2})$, we get bands similar to (4.1.22) and (4.1.23), respectively. For the weight function $q_1(x) = \{x(1 - x)\}^{-1/2}$, we get the transformed EP-bands. These are found by replacing $c_{\alpha/2}$ by $d_\alpha(\hat{c}_1, \hat{c}_2)$ in (4.1.22) and (4.1.23), and they are valid on $[t_1, t_2]$, with $0 < t_1 < t_2 \le t$ as given just above (4.1.27). For the weight function $q_2(x) = 1$, on the other hand, we get the transformed HW-bands. These are found by replacing $c_{\alpha/2}\hat{\sigma}(s)$ by $a_n^{-1} e_\alpha(\hat{c}_1, \hat{c}_2) \times (1 + a_n^2 \hat{\sigma}^2(s))$ in (4.1.22) and (4.1.23), and they are valid on $[t_1, t_2]$, where $0 < t_1 < t_2 \le t$ with $A(t_1) > 0$. In practice, one may use the (slightly) conservative approximation $e_\alpha(c_1, c_2) \approx e_\alpha(0, c_2)$ when c_1 is close to zero. Both of these transformed bands perform quite satisfactorily for sample sizes as low as 25, even with 50% censoring, for the survival data situation (Bie et al., 1987).

IV.1.3.3. *More on the Hall–Wellner Bands*

It is seen that the normalizing factors $\{a_n\}$ enter explicitly in the expressions for the general confidence band (4.1.26) and in the expression for the HW-band (4.1.28) as well as its transformed counterparts. In the results of Bie et al. (1987) on the HW-band referred to briefly earlier, the obvious choice $a_n = \sqrt{n}$ was used as a normalizing factor, where n is the number of individuals in the study; cf. Example IV.1.6. We will now consider more closely the impact of the choice of normalizing factors, and concentrate on the HW-band (4.1.28) on the interval $[0, t]$ for concreteness. Then, let $\{\hat{b}_n\}$ be a possibly random sequence and assume that $\hat{b}_n \xrightarrow{P} b \in (0, \infty)$ as $n \to \infty$. Then $a_n \hat{b}_n (\hat{A}^{(n)} - A) \xrightarrow{\mathscr{D}} bU$ as $n \to \infty$, where U is the Gaussian martingale defined some lines above (4.1.24), and $a_n^2 \hat{b}_n^2 \hat{\sigma}^2$ converges uniformly in probability to $b^2\sigma^2$ as n increases. By the same argument as was used to prove (4.1.28), we then get that an asymptotic $100(1 - \alpha)\%$ confidence band for A on $[0, t_2]$ is

$$\hat{A}^{(n)}(s) \pm (a_n \hat{b}_n)^{-1} e_\alpha(0, \tilde{c}_2)(1 + a_n^2 \hat{b}_n^2 \hat{\sigma}^2(s)), \quad (4.1.30)$$

where $e_\alpha(\cdot, \cdot)$ is as defined below (4.1.28) and

$$\tilde{c}_2 = \frac{a_n^2 \hat{b}_n^2 \hat{\sigma}^2(t_2)}{1 + a_n^2 \hat{b}_n^2 \hat{\sigma}^2(t_2)}.$$

If we now let $\hat{\gamma}_n = a_n^2 \hat{b}_n^2 \hat{\sigma}^2(t_2)$, then (4.1.30) may be rewritten as

$$\hat{A}^{(n)}(s) \pm \hat{\gamma}_n^{-1/2} e_\alpha\left(0, \frac{\hat{\gamma}_n}{1+\hat{\gamma}_n}\right)\hat{\sigma}(t_2)\left(1 + \hat{\gamma}_n \frac{\hat{\sigma}^2(s)}{\hat{\sigma}^2(t_2)}\right). \tag{4.1.31}$$

Because the sequence $\{\hat{b}_n\}$ may be chosen arbitrarily as long as $\hat{b}_n \xrightarrow{P} b \in (0, \infty)$ as $n \to \infty$, we have proved that (4.1.31) is a valid asymptotic $100(1-\alpha)\%$ confidence band for A on $[0, t_2]$ for any (possibly random) sequence $\{\hat{\gamma}_n\}$ satisfying $\hat{\gamma}_n \xrightarrow{P} \gamma$ as $n \to \infty$, for some $\gamma \in (0, \infty)$. In particular, one could choose $\hat{\gamma}_n \equiv \gamma$ for a prechosen γ, e.g., for $\gamma = 1$, this will give the band proposed by Andersen and Borgan (1985, p. 114).

Thus, the HW-band (4.1.28) is just one band, namely, the one with $\hat{\gamma}_n = a_n^2 \hat{\sigma}^2(t_2)$, in the collection of bands given by (4.1.31).

We will briefly consider the shape of the bands (4.1.31) for fixed n and varying $\hat{\gamma}_n$. To this end, we recall first that $\{W^0(t)\}$ is distributed as

$$\{W(t) - tW(1)\},$$

where W is a standard Brownian motion and that

$$\left\{\gamma^{-1/2} W\left(\frac{\gamma t}{1+\gamma}\right)\right\}$$

is distributed as $\{W(t/(1+\gamma))\}$ [e.g., Shorack and Wellner (1986, Section 2.2)]. From this, it follows that

$$\left\{\gamma^{-1/2} W^0\left(\frac{\gamma t}{1+\gamma}\right)\right\}$$

converges in distribution to a standard Brownian motion $\{W(t)\}$ as $\gamma \to 0$. Therefore,

$$\gamma^{-1/2} e_\alpha\left(0, \frac{\gamma}{1+\gamma}\right) \to u_\alpha$$

as $\gamma \to 0$, where u_α is the upper α fractile of the distribution of

$$\sup_{x \in [0,1]} |W(x)|.$$

Thus, for fixed n but $\hat{\gamma}_n \to 0$, the band (4.1.31) approaches the asymptotically valid band

$$\hat{A}^{(n)}(s) \pm u_\alpha \hat{\sigma}(t_2),$$

having constant width around $\hat{A}^{(n)}$. This band already was considered by Aalen (1976) for the special case of a competing risks model (cf. Example III.1.5). To get a rough understanding about how the width of the bands (4.1.31) changes with increasing $\hat{\gamma}_n$ (for fixed n), we begin by noting that $\hat{\gamma}_n^{-1/2}(1 + \hat{\gamma}_n \hat{\sigma}^2(s)/\hat{\sigma}^2(t_2))$ first decreases and obtains its minimum value when $\hat{\gamma}_n = \hat{\sigma}^2(t_2)/\hat{\sigma}^2(s)$ and then eventually increases at a rate of $\sqrt{\hat{\gamma}_n}$. Furthermore, $e_\alpha(0, \hat{\gamma}_n/(1+\hat{\gamma}_n))$ is increasing with $\hat{\gamma}_n$, but does not change very much with $\hat{\gamma}_n$

when $\hat{\gamma}_n \geq 3$ and $\alpha \leq 0.25$ (Hall and Wellner, 1980, Table 1). This indicates that as $\hat{\gamma}_n$ increases for fixed n, the width of the bands at small s decreases, whereas at large s it increases.

A further study of the performance of the bands (4.1.31) (and similar transformed versions) is certainly needed, but presently we do not know whether they will turn out to be of any *practical* importance.

IV.1.3.4. *Examples*

We next exemplify the use of confidence intervals and confidence bands using the melanoma data and the Fyn diabetics data.

EXAMPLE IV.1.12. Survival after Malignant Melanoma. Confidence Intervals and Confidence Bands

The data were introduced in Example I.3.1. In the group of 79 male patients, within which we assume the survival times to be i.i.d. (cf. Example IV.1.2), 29 died from the disease. Figure IV.1.14 shows the Nelson–Aalen estimate (see

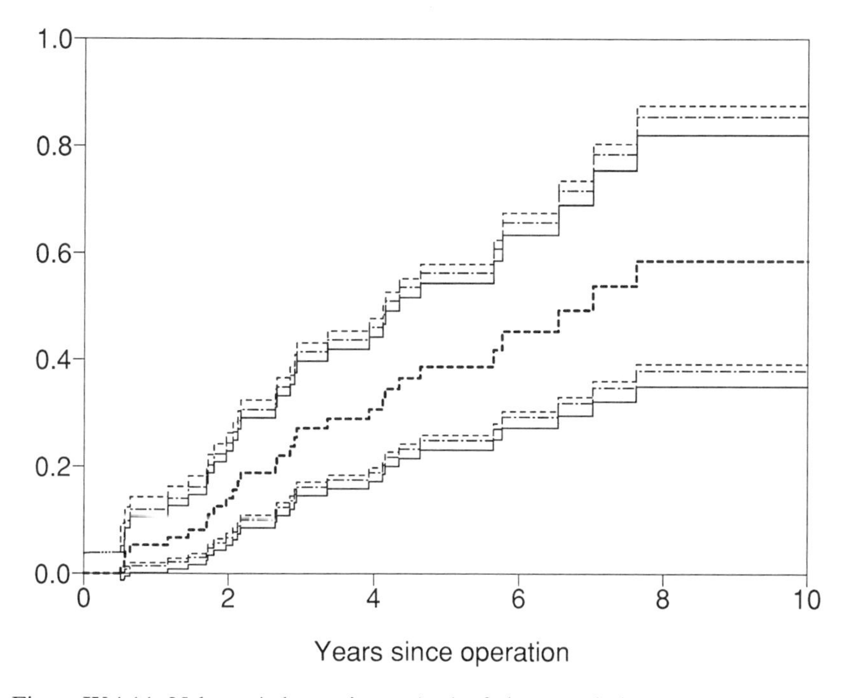

Figure IV.1.14. Nelson–Aalen estimate (- - -) of the cumulative intensity for death from the disease for 79 male patients with malignant melanoma, with approximate pointwise 95% confidence limits calculated according to three different formulas: Linear (—); log-transformation (- - -), and arcsin-transformation (- · · · · -). For $t < 185$ days, the "exact" formula (4.1.32) was used.

also Figure IV.1.1) as well as approximate pointwise 95% confidence limits calculated

(a) directly from (4.1.21) ("linear" limits),
(b) after log-transformation (4.1.22),
(c) after arscin-transformation (4.1.23).

It is seen that the lower linear limit (a) is negative just after the first failure time at 185 days. Further, the three limits are not too different, with (a) being symmetrical about the estimate and (a) $\leq$ (c) $\leq$ (b) for lower as well as higher limits.

Because both of the variance estimates (4.1.6) and (4.1.7) are zero before the first failure time, they are of little use when calculating confidence limits around an estimated cumulative hazard equal to zero. For $0 < t < 185$ days, we, therefore, calculated one-sided "exact" confidence limits for $A(t) = -\log S(t)$, where $S(t)$ is the survival probability, based on an observation of 0 in a binomial distribution $(Y(t), 1 - S(t))$. Thus, the upper limit $\bar{A}(t)$ in a $1 - \alpha$ confidence interval is given by

$$\{\exp(-\bar{A}(t))\}^{Y(t)} = \alpha, \tag{4.1.32}$$

i.e., $\bar{A}(t) = -\log \alpha^{1/Y(t)}$, which for $\alpha = 0.05$ gives

$$\bar{A}(t) = \frac{\log 20}{Y(t)} = \frac{2.996}{Y(t)}.$$

Figure IV.1.14 shows that this seems to be a reasonable procedure because the upper confidence limit is connected at the first failure time.

An alternative "exact" procedure for calculating confidence limits was discussed by Cox and Oakes (1984, p. 52).

In Figure IV.1.15, only the log-transformation is used. Pointwise 95% confidence limits according to (4.1.22) and (4.1.32) are given together with an EP (equal precision) 95% confidence band and an HW (Hall–Wellner) 95% confidence band derived as described below (4.1.29). The bands cover the interval $[t_1, t_2] = [1, 7]$ years. This leads to $\hat{c}_1 = 0.0539$ and $\hat{c}_2 = 0.4442$ [cf. (4.1.25)], so that the EP-band is given by replacing $c_{.025}$ in (4.1.22) by $d_{.05}(\hat{c}_1, \hat{c}_2)$, where

$$d_\alpha(c_1, c_2) = K_{q_1,\alpha}(c_1, c_2)$$

is the upper α fractile in the distribution of

$$\sup_{c_1 \leq x \leq c_2} |W^0(x)\{x(1-x)\}^{-1/2}|.$$

The EP-band has distance to the estimate proportional to the corresponding pointwise limits; in this example with proportionality factor

$$\frac{d_{.05}(\hat{c}_1, \hat{c}_2)}{c_{.025}} = \frac{2.887}{1.96} = 1.47,$$

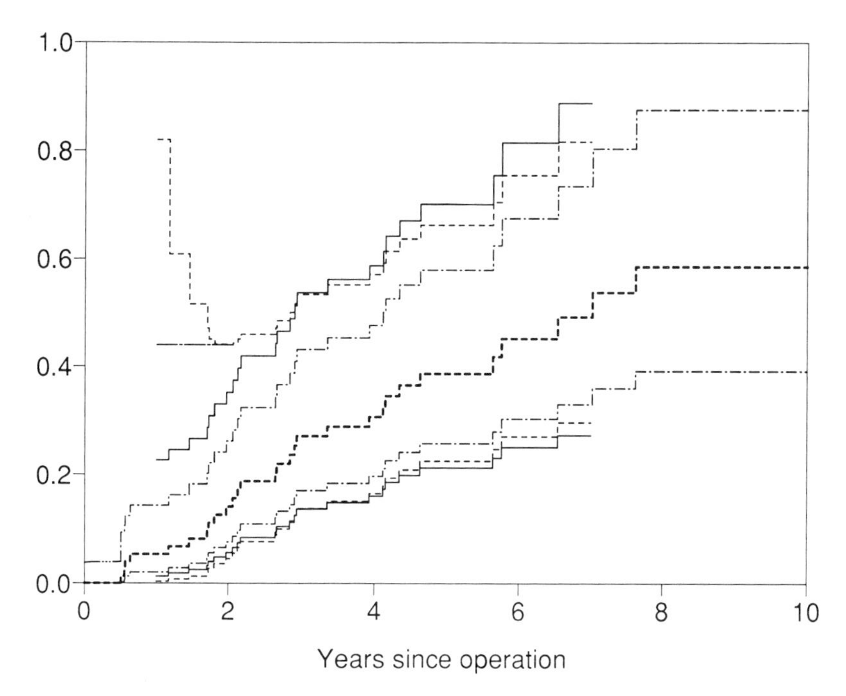

Figure IV.1.15. Nelson–Aalen estimate of the cumulative intensity for death from the disease for 79 male patients with malignant melanoma, with approximate pointwise 95% confidence limits (-····-) as well as EP- (—) and HW- (---) approximate 95% confidence bands on the interval [1, 7] years, all based on the log-transformation. The HW-band is "monotonized" up to $t \approx 2$ years.

where the fractile $d_{.05}(\hat{c}_1, \hat{c}_2) = 2.887$ is found by using the approximation of Miller and Siegmund (1982) given just below (4.1.27). The HW-band is obtained by replacing $c_{.025}\hat{\sigma}(s)$ in (4.1.22) by

$$\frac{e_{.05}(\hat{c}_1, \hat{c}_2)(1 + n\hat{\sigma}^2(s))}{\sqrt{n}}$$

with $e_\alpha(c_1, c_2)$ the upper α fractile in the distribution of

$$\sup_{c_1 \le x \le c_2} |W^0(x)|.$$

We apply the conservative approximation

$$e_{.05}(\hat{c}_1, \hat{c}_2) \approx e_{.05}(0, \hat{c}_2) = 1.231,$$

where the fractile is found by linear interpolation in the table of Koziol and Byar (1975). The same result $e_{.05}(\hat{c}_1, \hat{c}_2) = 1.231$ is obtained from the tables of Chung (1986).

It is noted that the HW-band is always wider than the pointwise confidence limits, but narrower than the EP-band for t larger than about 4 years. For small t, the HW-band is considerably wider, at $t = 1$ years attaining the value at the upper limit. Clearly, the nonincreasing upper limit is meaningless for a confidence band around an intrinsically increasing estimate. One might, therefore, in practice, replace the upper limit by the largest increasing function less than it, as indicated in Figure IV.1.15. □

EXAMPLE IV.1.13. Mortality of Diabetics in the County of Fyn. Confidence Intervals and Confidence Bands for the Integrated Death Intensity from Left-Truncated and Right-Censored Survival Data

The problem and the data were presented in Example I.3.2. Of the 716 female diabetics in Fyn county on 1 July 1973, 237 died before 1 January 1982.

Figure IV.1.16 shows the Nelson–Aalen estimate of the death intensity (cf. Example IV.1.3 and Figure IV.1.3) as well as approximate 95% confidence limits calculated

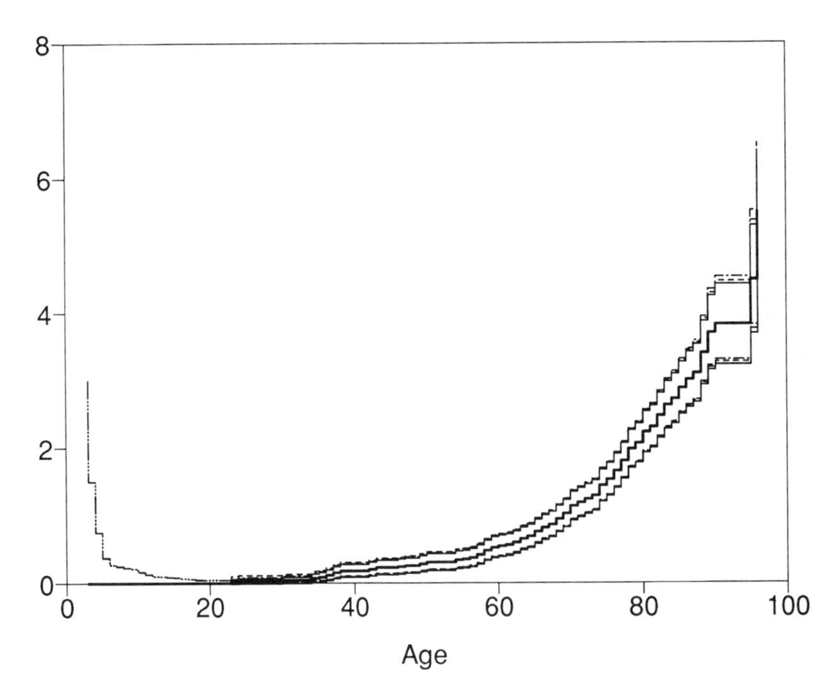

Figure IV.1.16. Nelson–Aalen estimate of the integrated death intensity for female diabetics in the county of Fyn, 1973–81, with approximate pointwise 95% confidence limits calculated according to three different formulas: Linear (—); log-transformation (- - -), and arcsin-transformation (- · · · -). The confidence limits for ages lower than 24 years (the lowest age of a death observed) are one-sided "exact" limits.

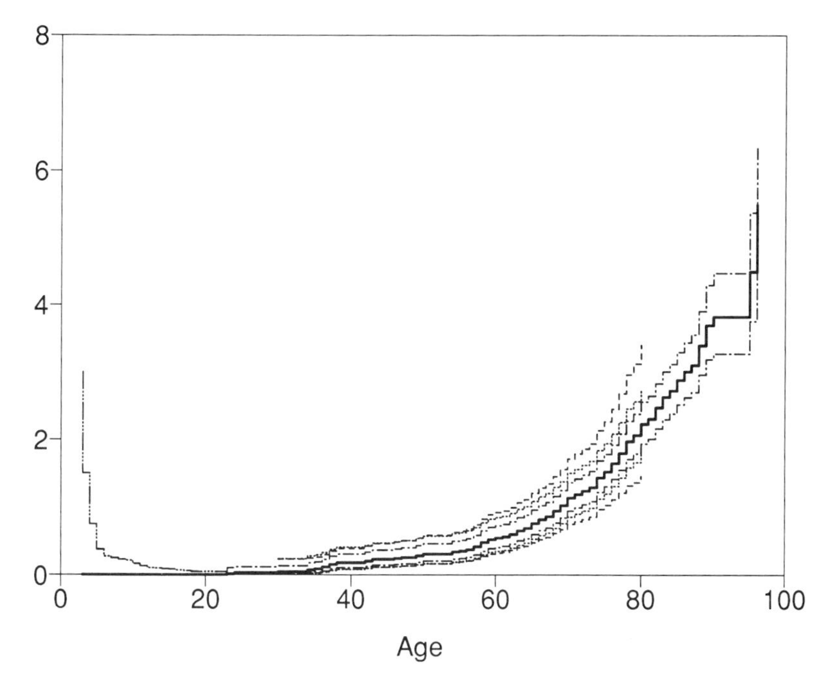

Figure IV.1.17. Nelson–Aalen estimates of the integrated death intensity for female diabetics in the county of Fyn, 1973–81, with approximate pointwise 95% confidence limits (-·--·-) as well as EP- (···), and HW- (---) approximate 95% confidence bands on the age interval [30, 80] years, all based on the log-transformation.

(a) directly from (4.1.21) ("linear" limits),
(b) after log-transformation (4.1.22),
(c) after arcsin-transformation (4.1.23).

The earliest death happened to a woman aged 24 years, so the confidence limits for ages from 2 (the lowest age with information) to 24 were constructed by the "exact" argument given in Example IV.1.12, modified to the varying risk set, so that at age t the confidence interval is $[0, 2.996/Y(t)]$, where $Y(t)$ is the number of patients at risk at age t (cf. Figure IV.1.2). Note that the very small risk set toward the lower age limit of 2 years of observed patients is reflected in widening confidence limits.

Figures IV.1.17 and IV.1.18 illustrate the dependence of simultaneous confidence bands on the age interval covered ($[t_1, t_2] = [30, 80]$ years and $[24, 90]$ years, respectively). The log-transformation is used throughout. The pointwise 95% confidence limits are compared to 95% EP-bands and 95% HW-bands, constructed as described below (4.1.29) and exemplified in Example IV.1.12. The values of $\hat{c}_1$, $\hat{c}_2$, $d_{.05}(\hat{c}_1, \hat{c}_2)$, and $e_{.05}(\hat{c}_1, \hat{c}_2)$ used in the two figures are the following:

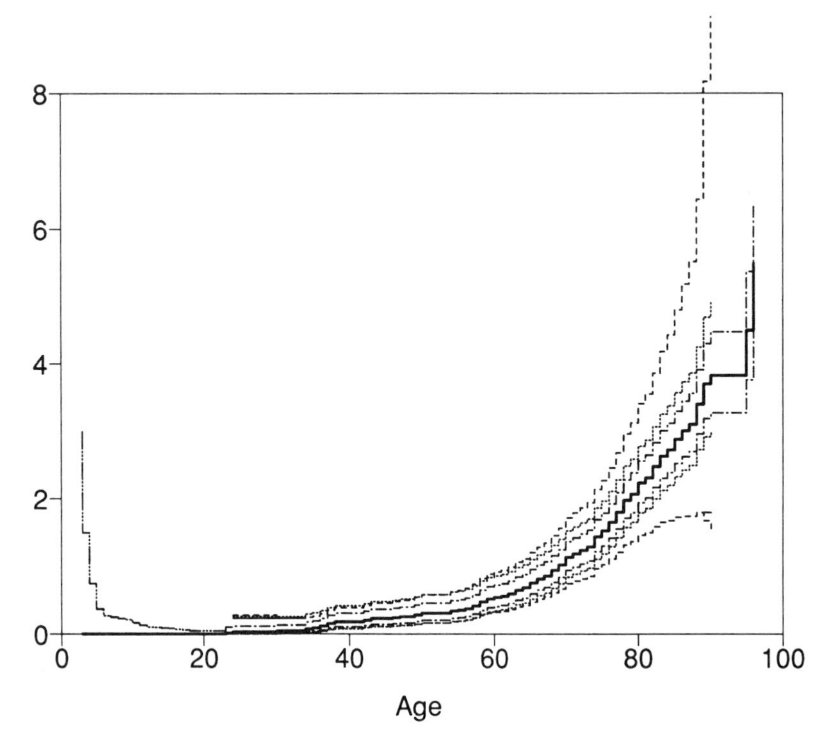

Figure IV.1.18. Nelson–Aalen estimate of the integrated death intensity for female diabetics in the county of Fyn, 1973–81, with approximate pointwise 95% confidence limits (-·-·-) as well as EP- (···), and HW- (---) approximate 95% confidence bands on the age interval [24, 90] years, all based on the log-transformation.

		Figure IV.1.17	Figure IV.1.18
Age interval:	$[t_1, t_2]$	[30, 80]	[24, 90]
	$\hat{c}_1$	.302	.231
	$\hat{c}_2$	.949	.985
	$d_{.05}(\hat{c}_1, \hat{c}_2)$	3.002	3.122
	$e_{.05}(\hat{c}_1, \hat{c}_2)$	1.358	1.358

In particular, $e_{.05}(\hat{c}_1, \hat{c}_2) = 1.358$ (both cases) is obtained by approximating $e_{.05}(\hat{c}_1, \hat{c}_2)$ by $e_{.05}(0, \hat{c}_2)$ and using Table 1 of Koziol and Byar (1975). The more elaborate tables of Chung (1986) yield $e_{.05}(.302, .949) = 1.347$ and $e_{.05}(.231, .985) = 1.355$ both very close to the approximate value of 1.358.

It is seen that the EP-bands are narrower than the HW-bands for high ages, and when the EP-band is based on a smaller interval (Fig. IV.1.17) indeed for most ages. The HW-band "explodes" for extreme ages, and this also happens for the EP-band at extreme high age. Monotonization (as indicated in Figure IV.1.18) might then be useful, although a sounder practical attitude in most cases might be to restrict attention to central parts of the age range (such as [30, 80] years) where there is sufficient information to produce reliable bands. □

IV.1.4. Bootstrapping the Nelson–Aalen Estimator

An alternative approach to confidence-band construction to that of the previous section is through the use of the bootstrap. This comes down to replacing the *asymptotic* distribution (given, e.g., by Theorem IV.1.2) of a process like $a_n(\hat{\mathbf{A}}^{(n)} - \mathbf{A})$, or a weighted or transformed version of this process, by a simulated *exact* distribution. When using the asymptotic approach, one may have to plug in estimates of parameters of the asymptotic distribution, for instance, the c_i of (4.1.25). In the bootstrap approach, one does not use asymptotic approximations but, on the other hand, instead of estimating one or two parameters like the c_i, one has to estimate all the parameters of the exact sampling situation (in particular, $\mathbf{A}$ itself).

The usual or nonparametric bootstrap (Efron, 1979) is based on the fact that in an i.i.d. setup, the empirical distribution of the data is an estimate of its probability law. So the bootstrap distribution is formed by taking random samples (with replacement) of size n from the empirical distribution (which puts mass $1/n$ on each of the observations). Parameters to be estimated, and their estimators, are seen as functionals of these distributions.

This approach is available in counting process models for n i.i.d. observations, for instance, for the Nelson–Aalen estimator of the cumulative hazard function A based on n randomly censored survival times in the simple random censorship model (Efron, 1981). The asymptotic correctness of the bootstrap [for estimating the distribution of the $n^{1/2}(\hat{A} - A)$] follows in this situation and similar ones from empirical process theory together with the functional delta-method, as shown by Gill (1989). It has also been checked by direct calculations and for various related problems by a number of other authors [the Kaplan–Meier estimator by Akritas (1986), Lo and Singh (1986), and Horváth and Yandell (1987); the Cox regression model by Hjort (1985b); the quantile product-limit estimator by Lo and Singh (1986), Horváth and Yandell (1987), and Doss and Gill (1992); cf. the bibliographic remarks in Section IV.5].

From the point of view that the bootstrap works because, for large samples, the empirical distribution is close to the true, and $a_n(\hat{\mathbf{A}}^{(n)} - \mathbf{A})$ has roughly the same asymptotic distribution under both, one may say that the idea of the bootstrap is just to estimate the asymptotic distribution of the quantity of interest by a simulation experiment. For our counting process models, this leads to a large number of alternative approaches, which, unlike the basic nonparametric bootstrap, are available for arbitrary counting process models and not just those based on i.i.d. samples.

For simplicity, we continue the discussion in the univariate case. The main feature of the asymptotic distribution of $a_n(\hat{A} - A)$ is that it has independent increments, with variances which correspond to the heuristic

$$\mathrm{d}N(t) \sim \text{binomial}(Y(t), \mathrm{d}A(t))$$

given $\mathscr{F}_{t-}$ because the conditional variance of $\mathrm{d}\hat{A}(t) = \mathrm{d}N(t)/Y(t)$ is then $(1/Y(t))\,\mathrm{d}A(t)(1 - \mathrm{d}A(t))$. The estimator (4.1.7) corresponds to plugging $\mathrm{d}\hat{A}^{(n)}(t)$ into this expression and adding over $0 \le t \le s$ to get an estimate of the variance of $\hat{A}^{(n)}(s)$. So consider now the following "weird bootstrap" (for which it is supposed that the process Y is integer-valued):

Definition IV.1.1 (The Weird Bootstrap). Given N, Y, and $\hat{A}$, let N^* be a process with independent binomial $(Y(t), \Delta\hat{A}(t))$ distributed increments at the jump times of N, constant between jump times. Let $\hat{A}^* = \int \mathrm{d}N^*/Y$. Estimate the distribution of $\hat{A} - A$ by the conditional distribution, given N and Y, of $\hat{A}^* - \hat{A}$.

Clearly, the distribution of $\hat{A}^* - \hat{A}$ can be easily simulated by independently sampling "failures at time t" from the risk set of $Y(t)$ individuals available at that time, giving each individual probability $1/Y(t)$ to fail. The strangeness of this bootstrap is that the same individuals at different times are not linked in any way, and the number at risk $Y(t)$ is not altered in the simulation either.

A multivariate version of the weird bootstrap is immediate: just follow the above recipe, componentwise. When there are tied jump times among different components of $\mathbf{N}$, then one has to decide whether the different components of $\Delta\mathbf{N}(t)$ should be taken to be independent binomial, or jointly multinomial, or something more complicated (for the Markov models of Section IV.4, we would take certain *groups* of components as independent multinomial).

One can show that under the conditions of Theorem IV.1.2, the conditional probability distribution (on $D[0,t]^k$) of $a_n(\hat{\mathbf{A}}^{(n)*} - \hat{\mathbf{A}}^{(n)})$, given the data $\mathbf{N}^{(n)}$, $\mathbf{Y}^{(n)}$, converges in probability to that of $\mathbf{U}$, and hence that confidence bands for the A_h based on the simulated distribution of $a_n(\hat{\mathbf{A}}^{(n)*} - \hat{\mathbf{A}}^{(n)})$ have asymptotically correct coverage probability. The idea of the proof is, first, to make an almost-sure convergence construction (Pollard, 1984, Section IV.3), and then to verify the conditions of the martingale central limit theorem for a sequence of versions of $a_n(\hat{\mathbf{A}}^{(n)*} - \hat{\mathbf{A}}^{(n)})$, for which $a_n(\hat{\mathbf{A}}^{(n)} - \mathbf{A})$ is converging almost surely, not just in distribution, to the process $\mathbf{U}$. See Gill (1989) and Doss and Gill (1992) for similar proofs, and Section VIII.1 for the same idea in a parametric context.

Many variants of this sampling scheme will give the same asymptotic distribution. For instance, because $\mathbf{A}$ is assumed continuous and $\mathbf{N}$ will only have single components jumping at a time, with a jump of size 1, one could reasonably replace the binomial $(Y_h(t), 1/Y_h(t))$ distribution with a Poisson (1) distribution, thus having the same mean and, asymptotically, the same variance.

In the simple case of this section, the weird bootstrap will be a lot harder to apply than the asymptotic methods described above. However, the method is also available for more complex models and more complex estimators such as those discussed in Sections IV.3 and IV.4 and even Chapter VII (with the obvious modifications) and may well become an attractive alternative to the calculation of a complicated asymptotic distribution.

IV.1.5. Maximum Likelihood Estimation

We motivated the Nelson–Aalen estimator at the beginning of this chapter by deriving it as the solution to a quite natural "martingale estimating equation"; in fact, one could say that the estimator is a method of moments estimator based directly on the counting process. The estimator turns out also to have a maximum likelihood interpretation, which adds further to its appeal and is, moreover, connected to its asymptotic optimality properties (discussed in Chapter VIII). A likelihood based derivation of the estimator also leads to likelihood-based estimated standard errors, tests, etc. However, this likelihood interpretation turns out not to be at all straightforward. We give here a general discussion on the principles involved in specifying a nonparametric maximum likelihood estimator (NPMLE), interwoven with the special case of deriving the Nelson–Aalen estimator $\hat{A}$ as NPMLE of the integrated hazard $A = \int \alpha$ for a univariate counting process N with intensity process $\lambda = \alpha Y$.

Assume we have noninformative censoring (see Section III.2.3) and let the nuisance parameters (involved in describing censoring, etc.) be fixed so that the likelihood depends only on the hazard α and can be written [cf. (2.7.2″)] as

$$\mathrm{dP}_A = \prod_t \{(Y(t)\alpha(t)\,\mathrm{d}t)^{\mathrm{d}N(t)}(1 - Y(t)\alpha(t)\,\mathrm{d}t)^{1-\mathrm{d}N(t)}\} \tag{4.1.33}$$

$$= \prod_t \{(Y(t)\,\mathrm{d}A(t))^{\mathrm{d}N(t)}(1 - Y(t)\,\mathrm{d}A(t))^{1-\mathrm{d}N(t)}\}, \tag{4.1.34}$$

where we have already in (4.1.34) suggestively rewritten the right-hand side fully in terms of the cumulative hazard $A = \int \alpha$ instead of α. Likelihood ratios are formed by taking ratios $\mathrm{dP}_A/\mathrm{dP}_{A'}$ of such expressions, forming Radon–Nikodym derivatives $(\mathrm{d}A/\mathrm{d}A')(t) = \mathrm{d}A(t)/\mathrm{d}A'(t)$ from the contributions at jump times, and taking product-integrals of the rest (cf. the discussion after Theorem II.7.2). These ratios are only well defined when $A \ll A'$; for instance, if $A = \int \alpha$ and $A' = \int \alpha'$, we require α to be zero where α' is and obtain $(\mathrm{d}A/\mathrm{d}A')(t) = \alpha(t)/\alpha'(t)$.

Now the true cumulative or integrated hazard function $A = \int \alpha$ is absolutely continuous, whereas the Nelson–Aalen estimator $\hat{A} = \int \mathrm{d}N/Y$ is a step-function, corresponding to a "hazard measure" which does not have a density with respect to Lebesgue measure but rather consists of discrete atoms of size $1/Y(t)$ at the jump times of N. If $\hat{A}$ is to be considered as maximum likelihood estimator, it seems we need to extend the model in some sense to allow such a discrete A as well; we cannot parametrize in terms of α at all.

The necessity of extending the model makes the status of $\hat{A}$ as NPMLE less unequivocable than would first appear. Conceivably for *any* estimator, even a very unreasonable one, one could construct an extension of the model, making it the NPMLE. So the extension has to be done in as natural a way as possible. However, there can be different criteria for this and they may not necessarily agree. From some points of view, the Nelson–Aalen estimator may not be the NPMLE at all. Already, in Section IV.1.1, we considered one particular discrete extension to motivate the variance estimator (4.1.7) for $\hat{A}$, and it was necessary to suppose that the multiplicative intensity model arose through aggregation of individual counting processes N_i satisfying the multiplicative intensity model with zero-one valued Y_i and with orthogonal counting process martingales M_i, leading to binomial-type expressions. This is reasonable when the multiplicative intensity model has arisen from the classical random censorship model of Example III.2.4, but not necessarily reasonable in other contexts.

If a well-developed theory of asymptotic behavior of the maximum likelihood estimator (MLE) for an infinite-dimensional parameter space existed, one could at least be confident in choosing that extension of the model which did guarantee nice asymptotic properties. However, such a theory does not exist at all and only some rudimentary results on consistency (Kiefer and Wolfowitz, 1956; Bahadur, 1967) and on efficiency (Gill, 1991b) are available. It is in any case very rare that such general theorems can actually be applied to specific models.

A further complication is that if the model has been extended to include discrete A as possible parameter values, a likelihood function in the usual sense typically does not exist: there is no single measure dominating all the probability measures in the model, and, hence, (4.1.33) or (4.1.34) cannot be interpreted as a probability density which, evaluated at the observed data, one can maximize over the parameter space. Fortunately, there is a nice solution to this problem, due to Kiefer and Wolfowitz (1956): Look at the maximum likelihood estimator for each submodel consisting of just two parameter values; this is well-defined since any two probability measures are dominated by a single measure, e.g., their sum. If there is one parameter value which comes out best when compared with any other, then call it the MLE. Clearly, this notion of MLE generalizes the usual definition. Moreover, it coincides with the following ad hoc solution, characterizing what is commonly meant by NPMLE (when available): if for each data point (N, Y) there is a parameter value giving that point positive probability, then choose the parameter value maximizing the (discrete) probability of the data.

Let us look now more closely at the problem in hand and see how these ideas work out. Formulas (4.1.33) and (4.1.34) gave an expression for the likelihood. Note that this result may possibly only have been arrived at after aggregation of component processes N_i satisfying the multiplicative intensity model with the same baseline hazard α, i.e., after a reduction by sufficiency (cf. Section III.2.3). Note also that the likelihood (4.1.33) for absolutely continuous A, equivalently the likelihood for α, is proportional to $\Pi_t \alpha(t)^{\Delta N(t)} \exp(-\int_0^\tau \alpha(t) Y(t)\,\mathrm{d}t)$, which can be made arbitrarily large by letting α be zero except very close to the jump times of N, where we let it peak higher and higher.

A possible way to arrive at a definition of NPMLE *without* extending the model in any way at all is to make use of the method of sieves. Suppose we maximize the likelihood over α with logarithmic derivative bounded by a constant, M say. This set of α is called a *sieve*. The new maximization problem does have a solution $\hat{\alpha}_M$, say, generating a cumulative hazard function $\hat{A}_M$. Now let M tend to infinity and we would find that $\hat{A}_M$ converges (in the sense of weak convergence) to the Nelson–Aalen estimator $\hat{A}$. As far as we know, no one has explored whether this idea can be made the basis of a general theory; i.e., whether all "reasonable" sequences of sieves lead to the same estimator and whether the idea works in general and not just for this and a few other special examples. References to the literature on the method of sieves and the related method of penalized maximum likelihood are given in the bibliographic remarks to Section IV.2.

To return to our main theme of constructing an NPMLE by extension of the model, note that when A is continuous, we can just as well, at least informally, rewrite (4.1.34) as

$$\mathrm{dP}_A = \prod_t ((Y(t)\,\mathrm{d}A(t))^{\mathrm{d}N(t)}) \exp\left(-\int_0^\tau Y(t)\,\mathrm{d}A(t)\right) \tag{4.1.35}$$

or as

$$\mathrm{dP}_A = \mathop{\pi}_t ((Y(t)\,\mathrm{d}A(t))^{\mathrm{d}N(t)}(1 - \mathrm{d}A(t))^{Y(t)-\mathrm{d}N(t)}). \tag{4.1.36}$$

This is true because, for continuous A, the product-integrals of (4.1.34) and (4.1.36) both evaluate to the exponential in (4.1.35). Now this immediately shows the problem we must face: All of the very different expressions (4.1.34)–(4.1.36) are meaningful when we let A be any cumulative hazard function, not just an absolutely continuous one, but they are definitely very different from one another.

Each corresponds intuitively to a different probabilistic extension to (4.1.33); in fact to a conditional Bernoulli, Poisson, or binomial model. One may interpret them as saying, conditional on $\mathscr{F}_{t-}$:

$$\mathrm{d}N(t) \sim \text{Bernoulli}(Y(t)\,\mathrm{d}A(t)), \tag{4.1.37}$$

$$\mathrm{d}N(t) \sim \text{Poisson}(Y(t)\,\mathrm{d}A(t)), \tag{4.1.38}$$

$$\mathrm{d}N(t) \sim \text{binomial}(Y(t), \mathrm{d}A(t)). \tag{4.1.39}$$

Other interpretations may also be possible (the likelihood does not determine the probability). Which extension seems most appealing will depend on more structure to the problem than that given by (4.1.33) alone.

All this makes it clear that we cannot just extend the model by writing down an expression for $\mathrm{dP}_A/\mathrm{dP}_{A'}$ in terms of A and A' and supposing this to be true for discontinuous as well as continuous A, A'; there are many ways to do this which possibly lead to different maximum likelihood estimators (though in this simple case, as we shall see now, all three lead to the Nelson–Aalen estimator) and to different likelihood-based variance estimators. In the multivariate case, even more possibilities present themselves.

We now show that (4.1.34)–(4.1.36) all lead to the same NPMLE $\hat{A}$ (in the sense of Kiefer and Wolfowitz, 1956), though to different likelihood-based (inverse information) variance estimators. Note first that a discrete A will beat a continuous A in any pairwise comparison. We consider, therefore, just A which have jumps where N does (though strictly speaking, $\mathrm{d}A$ may be chosen arbitrarily where Y is zero: the NPMLE is not unique when and where this happens). In the three likelihoods, we may then replace the differentials $\mathrm{d}N(t)$ and $\mathrm{d}A(t)$ everywhere by the differences or jumps $\Delta N(t)$ and $\Delta A(t)$; the integral $\int Y \mathrm{d}A$ in the Poisson case becomes the sum $\sum Y\Delta A$. The three expressions, up to factors involving factorials or binomial coefficients depending on the data, are now identical to the products of ordinary Bernoulli, Poisson, and binomial probabilities for the three interpretations (4.1.37)–(4.1.39) (where we also replace the differentials by differences). The likelihoods, being finite products with each term only depending on one unknown parameter $\Delta A(t)$, can now be easily maximized over these parameters, leading in each of the three cases (as we will check in a moment) to

$$\Delta\hat{A}(t) = \frac{\Delta N(t)}{Y(t)}.$$

Using the relation $A = \sum \Delta A$ for discrete cumulative hazard functions produces the Nelson–Aalen estimator as NPMLE for A.

One can continue by carrying out a formal calculation of observed Fisher information for each of these models, leading to variance estimators (inverse observed information). We pretend the jump times are fixed and that we have an ordinary likelihood function in the finite set of parameters $\Delta A(t)$. The matrix of negative second derivatives of the log-likelihood with respect to the $\Delta A(t)$ is inverted and evaluated at $\Delta\hat{A}(t) = \Delta N(t)/Y(t)$. This gives an estimated covariance matrix for the $\Delta\hat{A}(t)$ and, hence, on summation, for the covariance matrix of $\hat{A}(t)$. In each case, the likelihood is a product and the log-likelihood a sum of separate terms for each $\Delta A(t)$. The information matrix is, therefore, diagonal and the observed information easily inverted.

Consider just one term and write $D = \Delta N(t)$, $Y = Y(t)$, $a = \Delta A(t)$. The Bernoulli, Poisson, and binomial log-likelihoods (based on the distribution of D given Y) are

$$D\log(aY) + (1 - D)\log(1 - aY),$$
$$D\log(aY) - aY,$$
$$D\log a + (Y - D)\log(1 - a).$$

First derivatives are

$$\frac{D}{a} - \frac{(1 - D)Y}{1 - aY},$$
$$\frac{D}{a} - Y,$$
$$\frac{D}{a} - \frac{Y - D}{1 - a},$$

and negative second derivatives

$$\frac{D}{a^2} + \frac{(1 - D)Y^2}{(1 - aY)^2},$$
$$\frac{D}{a^2},$$
$$\frac{D}{a^2} + \frac{Y - D}{(1 - a)^2}.$$

Setting the first derivative to zero gives, in each case, $\hat{a} = D/Y$. Substituting in the negative second derivatives gives the observed information

$$\frac{Y^2}{D(1 - D)},$$
$$\frac{Y^2}{D},$$
$$\frac{Y^3}{D(Y - D)}.$$

Now because D is a zero-one variable the first, Bernoulli, case is degenerate. The observed information is infinite and its inverse is zero; the likelihood approach gives a foolish variance estimator for $\hat{A}$ identically equal to zero. However, the Poisson and binomial cases give sensible results: The inverse information is diagonal, with, on the diagonal, in the Poisson case, an estimated variance for $\Delta\hat{A}(t)$ equal to $\Delta N(t)/Y(t)^2$, and, in the binomial case, an estimated variance equal to $\Delta N(t)(Y(t) - \Delta N(t))/Y(t)^3$. Adding these diagonal elements gives the estimators (4.1.6) and (4.1.7) of variances, and also the same quantities as estimated covariances (the likelihood approach correctly estimates uncorrelated increments).

These calculations show that the "naive NPMLE" approach based on seeing (4.1.33) as a binomial or Poisson likelihood leads to both sensible estimators and to sensible variance estimators. This is important as the idea is easy to generalize to more complicated models and, with luck, it might still produce good results. Indeed, in Chapters VII and IX, we will use the Poisson likelihood to generate the estimators of the Cox regression model and to propose new estimators in so-called frailty models. We already saw in Section IV.1.4 that the binomial and Poisson interpretations were useful in inspiring bootstrap methodology.

We now return to the general discussion and to an appraisal of the significance of these discrete extensions of the multiplicative intensity model. We said expressions (4.1.34)–(4.1.36) were meaningful extensions allowing discrete A as parameter without saying in what sense meaningful. So far, we have only intended using them to form an expression for $\mathrm{dP}_A/\mathrm{dP}_{A'}$ for statistical purposes, i.e., to define an estimator $\hat{A}$ by the requirement $\mathrm{dP}_{\hat{A}}/\mathrm{dP}_A \geq 1$ for all A. Thus, we are only making use of the statistical interpretation of $\mathrm{dP}_A/\mathrm{dP}_{A'}$ as a likelihood ratio, and not of its probabilistic interpretation as a Radon–Nikodym derivative. Thus, to formally define an estimator $\hat{A}$, we do not actually need to construct P_A in the sense of exhibiting probability measures having the corresponding Radon–Nikodym derivatives on our sample space. This can be very difficult or even impossible. All of (4.1.34)–(4.1.36) have (nonunique) probabilistic interpretations: Bernoulli, Poisson, binomial. The last two interpretations seem to require the counting process N to be allowed to make multiple jumps at the fixed time points where $\Delta A(t) > 0$ (conditionally Poisson or binomial, respectively) and we may even need to extend the *sample space*, not just the parameter space, to make this possible.

Such proper and careful probabilistic extensions of the multiplicative intensity model have been made by Jacobsen (1984); see also Jacobsen (1982, Section 4.6). The general approach which he applies to a number of concrete examples is as follows. First, we must extend the sample space (e.g., the space of realizations of a multivariate counting process) to also include processes with multiple jumps. Second, we endow the enlarged sample space with a suitable topology. Now we can extend the model in a quite canonical way by adding to the collection of probability measures specified by the model all or some weak limits arising from sequences of probability measures in the model. In the extended model, if we are lucky, an MLE will exist as the probability measure which maximizes the probability of the observed data. This approach formalizes the idea we obtain from the method of sieves discussed above, of getting a sequence of parameter values giving higher and higher values to the likelihood function, such that the corresponding probability distributions approach a discrete distribution not in the model but on its boundary. To "complete" the model, we must first complete the sample space.

This is a very elegant approach but it has a number of drawbacks. The burden of choosing an analytically nice extension to a likelihood function has now shifted to choosing a nice extension to the sample space and fixing a nice topology on it. This is going to depend on more specific details of the model than just that it is "a multiplicative intensity model," as we have indicated above. Second, we get an estimated *model*, i.e., a distribution P, not an estimated *parameter* A in this case. The parametrization of the original model by α or by A was given in advance, but there is no reason why the extended model can be parametrized in a natural way (and certainly not in a unique way) by the "same" parameter (in an extended parameter space). (There is no canonical extension of the parameters.) Finally, if we believe in the original model, the statistical properties of an estimator of A cannot depend on hypothetical sampling distributions in an *extended* model. So the probabilistic nature of an extension to the model should be irrelevant, except perhaps if we worry about what it means to present the discrete measure $\hat{A}$ as an estimator of the continuous one A.

In view of this, Gill (1991b) proposed an almost opposite approach to the extension problem. In smooth parametric models, maximum likelihood estimators are interesting from a statistical point of view more because they are solutions of likelihood equations (score function, that is, derivative of log-likelihood, equals zero) than because they maximize likelihood functions. In particular, asymptotic optimality properties follow because (and if) the maximum likelihood estimator is a consistent solution of the likelihood equations. So, if one is trying to derive estimators with good large sample properties, it is more useful to try to extend score functions in an analytically smooth way than to extend likelihood functions or even Radon–Nikodym derivatives. This can be simpler: The differentiation involved in deriving a score function also often produces a simplification; irrelevant detail is thrown away, and we are not concerned with a probabilistic interpretation.

Now an immediate problem is that we seem to have no score functions in an infinite-dimensional case, let alone likelihood functions. However, consider a finite-dimensional parametric model: Then the score function is a vector, each of its components corresponding to a *one-dimensional submodel*, namely, that in which only one of the components of the parameter varies, the others being fixed. So the likelihood equations are a collection of score functions of one-dimensional submodels intersecting or passing through the given parameter value at which they are evaluated. Now this viewpoint is also available with an infinite-dimensional parameter space: Choose a convenient family of one-dimensional submodels, derive the score function for each, and try to solve the equations "score functions equal zero." If the family of submodels is rich enough, this may uniquely determine an estimator of, say, A.

To do this, we do have to write all these likelihood equations in terms of the same parameter A, and, moreover, if the original model allows only, e.g., absolutely continuous A, we still have an extension problem: there will usually not be a solution inside the original parameter space, so we still have to extend the score functions to discrete A too. To get an estimator of the parameter of the original model with nice statistical behavior, this should be done in as *analytically smooth* a way as possible, not as *probabilistically realistic* a way as possible. This approach does allow a generalization of asymptotic efficiency theory [see Section VIII.4.5 and Gill (1991b) or van der Vaart (1987)]: *if* the estimator obtained is consistent (it is at least Fisher consistent because the *expected* scores equal zero), and under further smoothness assumptions, the estimator is asymptotically efficient. Also we note that this approach is not in conflict with the probabilistic one, when it is available: If an estimator is the maximum likelihood estimator with respect to the Kiefer and Wolfowitz definition (whether or not the likelihood ratios concerned also have a probabilistic interpretation), then it is also a maximum likelihood estimator within any submodel including or passing through the estimated model. This is because the Kiefer and Wolfowitz approach coincides with the usual one when the latter is available. Thus, in particular, for any one-dimensional dominated submodel passing through the MLE, the corresponding likelihood equation is zero at that point.

Let us see how this works in our special case. Go back to the likelihood (4.1.33) for absolutely continuous A, and consider the following parametric submodels:

$$\alpha_h^\theta(t) = (1 + \theta h(t))\alpha(t),$$

where h is a fixed bounded function defining a direction in the parameter space, whereas θ is a real parameter in some interval about zero such that $1 + \theta h(t)$ stays non-negative for all t. At $\theta = 0$, we have $\alpha_h^\theta = \alpha$ so all these submodels pass, at $\theta = 0$, through the same point α.

Because, for given h, the intensity process is $\lambda^\theta(t) = (1 + \theta h(t))\alpha(t) Y(t)$, we have that

$$\frac{\partial}{\partial\theta} \log \lambda^\theta(t) = \frac{h(t)}{1 + \theta h(t)},$$

which reduces simply to $h(t)$ at $\theta = 0$. Therefore, the score for θ, at $\theta = 0$, is by (2.7.11) simply equal to $\int_0^\tau h\,\mathrm{d}M$ where $M = N - \int \alpha Y$. The score for the model in the direction of h can thus be written as

$$\int_0^\tau h\,\mathrm{d}N - \int_0^\tau h\alpha Y \tag{4.1.40}$$

or alternatively (and suggestively) as

$$\int_0^\tau h\,\mathrm{d}N - \int_0^\tau hY\,\mathrm{d}A. \tag{4.1.41}$$

We would, therefore, define an NPMLE of α by requiring that all score functions of models through $\hat{\alpha}$ are zero; in particular, (4.1.40) should be zero for all h. These equations however do not have a solution: Taking h to be the indicator functions $h = I_{[0,t]}$ gives the equations $N(t) = \int_0^t \alpha(s) Y(s)\,\mathrm{d}s$ for all t which cannot be satisfied. However, written in terms of A as (4.1.41), the equations have an obvious extension for A which are not absolutely continuous: the expression (4.1.41) itself. Choosing h to be the indicator function of intervals $[0, t]$ gives the equations $N(t) = \int_0^t Y \, \mathrm{d}A$ which has as solution $\hat{A} = \int \mathrm{d}N/Y$ (though $\mathrm{d}\hat{A}$ is arbitrary where Y is zero). This is also the solution when we let h vary completely arbitrarily.

Again we arrive at the Nelson–Aalen estimator. The extension problem has become the very easy problem of extending (4.1.40) to allow discrete A and this required much less imagination than obtaining one of (4.1.34)–(4.1.36) from (4.1.33). There was arbitrariness involved in choosing the parametric submodels, but any convenient but large enough set which uniquely identifies $\hat{A}$ will lead to the same answer.

Gill (1989, 1991b) showed how asymptotic properties of the NPMLE follow from properties of the scores (both in this special model and in general), under suitable regularity conditions of course, and in particular working in a situation with n independent and identically distributed observations, $n \to \infty$. In particular, the asymptotic covariance structure of the NPMLE can be deduced from the covariance structure of the scores (4.1.41); moreover, one may restrict attention to the special choices $h = I_{[0,t]}$ which were sufficient to determine the NPMLE.

Consider for a moment the two-parameter submodel $\alpha_{h,k}^{\theta,\phi} = (1 + \theta h)(1 + \phi k)\alpha$ for two fixed functions h and k. The score for θ at $\theta = 0$, $\phi = 0$ is given by (4.1.41), and the score for ϕ similarly. The second mixed derivative of the log-likelihood (at $\theta = 0$, $\phi = 0$) is, by (2.7.12), equal to $-\int_0^\tau hkY \, \mathrm{d}A$. So choosing the indicator functions $I_{[0,t]}$ and $I_{[0,s]}$ for the functions h and k, we find that the usual relations "expected score equals zero" and "expected second derivatives of log-likelihood equal minus covariance matrix of scores" reduce to the facts

$$\mathrm{E}N(t) - \int_0^t \mathrm{E}Y \, \mathrm{d}A = 0, \tag{4.1.42}$$

$$\operatorname{cov}\left(N(t) - \int_0^t Y \, \mathrm{d}A, N(s) - \int_0^s Y \, \mathrm{d}A\right) = \int_0^{s \wedge t} \mathrm{E}Y \, \mathrm{d}A \tag{4.1.43}$$

which are true if the counting process N is integrable. The point is that the asymptotic covariance structure of $\hat{A}$ *only* depends on relations (4.1.42) and (4.1.43) which themselves *only* depend on the form of the likelihood (4.1.33). Other models than the multiplicative intensity model which still lead to the same likelihood will have the same asymptotic distribution of the NPMLE $\hat{A}$, under sufficient regularity conditions.

IV.2. Smoothing the Nelson–Aalen Estimator

We consider a multivariate counting process $\mathbf{N} = (N_1, \ldots, N_k)$ with intensity process $\boldsymbol{\lambda} = (\lambda_1, \ldots, \lambda_k)$ of the multiplicative form $\lambda_h(t) = \alpha_h(t) Y_h(t)$. In the previous section, we derived and studied estimators for $A_h(t) = \int_0^t \alpha_h(s) \, \mathrm{d}s$, $h = 1$,

..., k. However, it is the α_h themselves which are the entities of real interest. Therefore, when studying a plot of the Nelson–Aalen estimator, one mainly focuses on the *slope* of the curve, cf. Examples IV.1.2–IV.1.5. Hence, it is useful to be able to directly estimate the α_h.

For the particular situation of censored survival data, a number of methods have been developed for nonparametric estimation of the death intensity α, and some of these have been extended to allow for estimation of the α_h in general counting process models. At the time of writing of this monograph, there is considerable research activity in this field, and it is, at present, impossible to give a definitive summary of this vast area.

Instead of trying to present a comprehensive (but nevertheless incomplete) review of nonparametric estimation of intensities, we have chosen to concentrate on *one* useful method in this section, namely, estimation of the α_h by kernel function smoothing of the Nelson–Aalen estimators. Other methods are, however, mentioned in the bibliographic remarks at the end of the chapter. Also, we will in the general discussion in this section only discuss some of the basic properties of the kernel function estimator, leaving a briefer discussion of other important problems to the examples. We consider a fixed value of h throughout the section and drop the subscript h for notational ease.

IV.2.1. The Kernel Function Estimator

The kernel function estimator for $\alpha(t)$ is derived by smoothing the increments of the Nelson–Aalen estimator $\hat{A}$. Formally, we define the estimator by

$$\hat{\alpha}(t) = b^{-1} \int_{\mathcal{T}} K\left(\frac{t-s}{b}\right) \mathrm{d}\hat{A}(s). \tag{4.2.1}$$

Here, the *kernel function* K is a bounded function which vanishes outside $[-1, 1]$ and has integral 1. The *bandwidth* or *window size* b is a positive parameter. The kernel function and the bandwidth both have to be chosen in concrete applications. One frequently used kernel function is the Epanechnikov kernel function $K(x) = 0.75(1 - x^2)$, $|x| \leq 1$. Other kernel functions are mentioned in Example IV.2.1.

If we let $T_1 < T_2 < \cdots$ denote the successive jump times of the counting process N, then $\hat{\alpha}(t)$ may be written equivalently as

$$\hat{\alpha}(t) = b^{-1} \sum_j K\left(\frac{t-T_j}{b}\right)(Y(T_j))^{-1}.$$

It should be realized that only indices j for which $t - b \leq T_j \leq t + b$ contribute to this sum. This implies that the integral in (4.2.1), as well as in similar expressions in what follows, may be restricted to the interval $[t - b, t + b]$. We also note that $\hat{\alpha}(t)$ is a weighted mean of the increments $Y(T_j)^{-1}$ of the Nelson–Aalen estimator over $[t - b, t + b]$.

Given a bandwidth b, we will in the general presentation in this and the following subsection only discuss estimation of $\alpha(t)$ for values of $t \in \mathscr{T}$ for which also $t \pm b \in \mathscr{T}$. The "tail problem," i.e., estimating $\alpha(t)$ for all $t \in \mathscr{T}$, is, however, considered briefly in Example IV.2.5.

The kernel function estimator (4.2.1) was proposed and studied in a counting process context by Ramlau-Hansen (1983a, b). His work was inspired by works on kernel function estimation of density functions; cf. the bibliographic comments in Section IV.5. The estimator (4.2.1) generalizes the kernel estimator for the intensity proposed by Watson and Leadbetter (1964a, b) for the situation with i.i.d. lifetimes.

To study the statistical properties of $\hat{\alpha}(t)$, it is convenient to introduce the quantity

$$\alpha^*(t) = b^{-1} \int_{\mathscr{T}} K\left(\frac{t-s}{b}\right) \mathrm{d}A^*(s), \tag{4.2.2}$$

cf. (4.1.3). Apart from the factor $J(s) = I(Y(s) > 0)$ included in A^*, this is just a smoothed version of α, namely,

$$\tilde{\alpha}(t) = b^{-1} \int_{\mathscr{T}} K\left(\frac{t-s}{b}\right) \alpha(s)\,\mathrm{d}s. \tag{4.2.3}$$

(Note that $\tilde{\alpha}$ depends on both the choice of kernel function and bandwidth.)

By (4.1.4), we get

$$\begin{aligned} \hat{\alpha}(t) - \alpha^*(t) &= b^{-1} \int_{\mathscr{T}} K\left(\frac{t-s}{b}\right) \mathrm{d}(\hat{A} - A^*)(s) \\ &= b^{-1} \int_{\mathscr{T}} K\left(\frac{t-s}{b}\right) J(s)(Y(s))^{-1}\,\mathrm{d}M(s), \end{aligned} \tag{4.2.4}$$

so that $\hat{\alpha}(t) - \alpha^*(t)$ is a stochastic integral with respect to the local square integrable martingale $M(t) = N(t) - \int_0^t \alpha(s) Y(s)\,\mathrm{d}s$. This fact provides the basis for studying the statistical properties of $\hat{\alpha}(t)$.

The first- and second-order moments of $\hat{\alpha}(t)$ exist provided that

$$\tilde{\tau}^2(t) = \mathrm{E}\{\hat{\alpha}(t) - \alpha^*(t)\}^2 < \infty.$$

By the definition of $\langle M \rangle$ as the compensator of M^2 (cf. Section II.3.2) and (4.2.4), we may write

$$\begin{aligned} \tilde{\tau}^2(t) &= b^{-2} \int_{\mathscr{T}} K^2\left(\frac{t-s}{b}\right) \mathrm{E}\left\{\frac{J(s)}{Y(s)}\right\} \alpha(s)\,\mathrm{d}s \\ &= b^{-1} \int_{-1}^{1} K^2(u) \mathrm{E}\left\{\frac{J(t-bu)}{Y(t-bu)}\right\} \alpha(t-bu)\,du. \end{aligned} \tag{4.2.5}$$

When we talk about expected values and variances, it is, therefore, tacitly understood that the right-hand side of (4.2.5) is finite. As noted in Section IV.1, this holds if there exists a constant $c > 0$ such that $Y(s) < c$ implies $Y(s) = 0$. In particular, this is the case when Y is an integer-valued process.

We then get

$$\begin{aligned}\mathrm{E}\hat{\alpha}(t) &= \mathrm{E}\alpha^*(t)\\ &= b^{-1}\int_{\mathscr{T}} K\left(\frac{t-s}{b}\right)\mathrm{P}(Y(s)>0)\alpha(s)\,\mathrm{d}s\\ &= \int_{-1}^{1} K(u)\mathrm{P}(Y(t-bu)>0)\alpha(t-bu)\,\mathrm{d}u, \end{aligned}\tag{4.2.6}$$

so that the kernel function estimator $\hat{\alpha}(t)$ is, in general, not even approximately unbiased, but rather estimates $\tilde{\alpha}$ defined in (4.2.3), i.e., a smoothed version of α. An asymptotic expression for the bias of the kernel function estimator is provided in formula (4.2.27).

By (4.2.6), $\alpha^*(t)$ is almost the same as $\mathrm{E}\hat{\alpha}(t)$ when there is only a small probability that $Y(s) = 0$ for some $s \in [t-b, t+b]$. Thus, the variance of $\hat{\alpha}(t)$ is approximately equal to $\tilde{\tau}^2(t)$ given by (4.2.5), at least in large samples.

In the next subsection, we will study the bias and the variance of the kernel estimator more closely, and also discuss the relative importance these two terms have on the performance of $\hat{\alpha}(t)$.

As an estimator for the variance of $\hat{\alpha}(t)$, we may use

$$\hat{\tau}^2(t) = b^{-2}\int_{\mathscr{T}} J(s)\left\{\frac{K((t-s)/b)}{Y(s)}\right\}^2 \mathrm{d}N(s), \tag{4.2.7}$$

which is an unbiased estimator for $\tilde{\tau}^2(t)$ [cf. the similar result for the Nelson–Aalen estimator given just below (4.1.6)].

An alternative variance estimator for $\hat{\alpha}(t)$ may be obtained by adopting the discrete extension model of Section IV.1.1. According to this model, $M = N - \int Y\,\mathrm{d}A$ has predictable variation process

$$\langle M\rangle(t) = \int_0^t Y(s)(1-\Delta A(s))\,\mathrm{d}A(s).$$

This suggests, using (4.2.4), that the variance of $\hat{\alpha}(t)$ may also be estimated by

$$\begin{aligned}\check{\tau}^2(t) &= b^{-2}\int_{\mathscr{T}} J(s)K^2\left(\frac{t-s}{b}\right)(Y(s))^{-1}(1-\Delta\hat{A}(s))\,\mathrm{d}\hat{A}(s)\\ &= b^{-2}\int_{\mathscr{T}} J(s)K^2\left(\frac{t-s}{b}\right)(Y(s)-\Delta N(s))(Y(s))^{-3}\,\mathrm{d}N(s),\end{aligned}$$

cf. (4.1.7).

We conclude this subsection by an example showing how the smoothed hazard estimates look for different choices of kernel function and bandwidth.

EXAMPLE IV.2.1. Mortality of Female Diabetics in the County of Fyn. Kernel Smoothing of the Death Intensity

This problem and the data were presented in Example I.3.2.

In general, the kernel function estimator $\hat{\alpha}(t)$ of an intensity function $\alpha(t)$ is a weighted mean of information from the interval $[t - b, t + b]$. If the counting process is defined on a bounded interval $[a, c]$, it follows that there are difficulties in estimating $\alpha(t)$ outside of $[a + b, c - b]$; we return to these *tail problems* in Example IV.2.5. Furthermore, one will usually want to avoid extending the weighted means to intervals of the time axis with no exposure, that is, where $Y(u) = 0$ [implying $\mathrm{d}A^*(u) = 0$ even though quite possibly $\mathrm{d}A(u) = \alpha(u)\,\mathrm{d}u > 0$].

For the female diabetics, we restrict attention to the age interval [10 years, 90 years] and estimate the intensity function over intervals [10 years + b, 90 years − b] depending on the bandwidth b (cf. Figure IV.1.2 which gives the size of the risk set).

Figure IV.2.1 contains estimates of the death intensity $\alpha(t)$ for female diabetics, using bandwidth $b = 15$ years and the following three kernels, all of which take the value 0 outside of $[-1, 1]$: the *uniform* kernel

$$K_U(x) = \tfrac{1}{2}, \quad -1 \leq x \leq 1,$$

the *Epanechnikov* kernel

$$K_E(x) = \tfrac{3}{4}(1 - x^2), \quad -1 \leq x \leq 1,$$

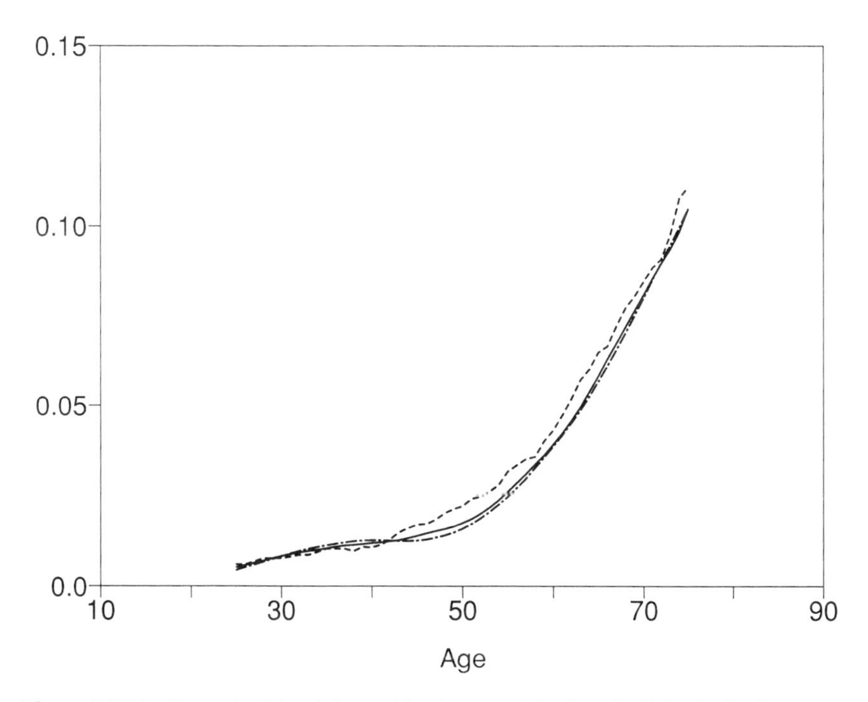

Figure IV.2.1. Smoothed death intensities (per year) for female diabetics in the county of Fyn, 1973–81, using the uniform kernel (- - -), the Epanechnikov kernel (—), and the biweight kernel (-·-·-), all with bandwidth 15 years.

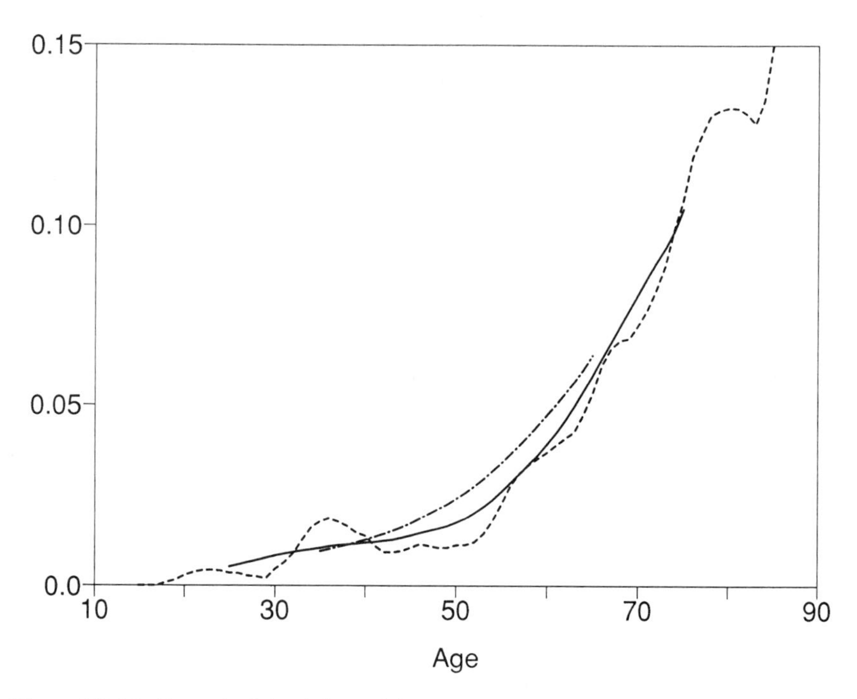

Figure IV.2.2. Smoothed death intensities (per year) for female diabetics in the county of Fyn, 1973–81, using the Epanechnikov kernel with bandwidth 5 years (- - -), 15 years (—), and 25 years (-·-·-).

and the *biweight* kernel

$$K_B(x) = \tfrac{15}{16}(1 - x^2)^2, \quad -1 \le x \le 1.$$

It is seen that the uniform kernel gives less smoothing than the other two kernel functions. But, broadly speaking, the three estimates are roughly identical, all displaying a clearly increasing death intensity with age, particularly after the age of 50 years. [This comparison of various kernels for fixed bandwidth b (here $= 15$ years) requires an additional comment. As is intuitively clear and also demonstrated in Figure IV.2.2, the more "spread out" the kernel, the smoother the estimate. However, the spread of the three kernels, measured, e.g., by the variance of the kernel (interpreted as a distribution), decreases from uniform ($\frac{1}{3}$) over Epanechnikov ($\frac{1}{5}$) to biweight ($\frac{1}{7}$). Thus, the uniform kernel is based on the "most spread out" information; nevertheless, the estimate is least smooth; see Figure IV.2.1.]

Figure IV.2.2 shows, for the case of the Epanechnikov kernel, how the estimate depends on the *choice of bandwidth* $b = 5$ years, 15 years, 25 years: the larger b, the smoother the estimate. We recall that this increased smoothness is only obtained at the expense of more biased estimates; cf. (4.2.6). □

IV.2.2. Large Sample Properties

In this subsection we will study the performance of the kernel function estimator in large samples. To this end we consider a sequence $\{N^{(n)}\}$ of counting processes, each with an intensity process of the multiplicative form $\lambda^{(n)}(t) = \alpha(t)Y^{(n)}(t)$, cf. Section IV.1.2, and we index all relevant quantities by n. As is usual in the kernel function estimation literature, we will let the bandwidth depend on n, i.e. $b = b_n$, and tend to zero as n increases, while the kernel function K remains the same for all n.

The plan of the subsection is as follows. First we give conditions for pointwise as well as uniform consistency of the kernel function estimator. Then we introduce the mean integrated squared error as a measure of the global performance of the kernel function estimator, and show how the bandwidth may be chosen in an optimal way in order to minimize the mean integrated squared error. Thereafter asymptotic normality of $\hat{\alpha}^{(n)}(t)$, for a fixed t, is considered. The subsection is concluded with four examples which illustrate the general theory, and also extends it by considering some topics not considered in the general treatment.

IV.2.2.1. *Consistency*

We first show that the kernel function estimator is pointwise consistent.

Theorem IV.2.1. *Let t be an interior point of $\mathscr{T}$ and assume that α is continuous at t. Let the bandwidth $b_n \to 0$ as $n \to \infty$. If there exists an $\varepsilon > 0$ such that*

$$\inf_{s \in [t-\varepsilon, t+\varepsilon]} b_n Y^{(n)}(s) \xrightarrow{\mathrm{P}} \infty \tag{4.2.8}$$

as $n \to \infty$, then $\hat{\alpha}^{(n)}(t) \xrightarrow{\mathrm{P}} \alpha(t)$ as $n \to \infty$.

Proof. It is enough to show that

$$|\hat{\alpha}^{(n)}(t) - \alpha^{*(n)}(t)| \xrightarrow{\mathrm{P}} 0, \tag{4.2.9}$$

$$|\alpha^{*(n)}(t) - \tilde{\alpha}^{(n)}(t)| \xrightarrow{\mathrm{P}} 0 \tag{4.2.10}$$

and

$$|\tilde{\alpha}^{(n)}(t) - \alpha(t)| \to 0 \tag{4.2.11}$$

as $n \to \infty$; cf. (4.2.1), (4.2.2), and (4.2.3). By version (2.5.18) of Lenglart's inequality and (4.2.4), we have for any η, $\delta > 0$

$$\mathrm{P}(|\hat{\alpha}^{(n)}(t) - \alpha^{*(n)}(t)| > \eta)$$

$$\leq \mathrm{P}\left(\sup_{t-b_n \leq s \leq t+b_n} \left| b_n^{-1} \int_{t-b_n}^{s} K\left(\frac{t-u}{b_n}\right) J^{(n)}(u)(Y^{(n)}(u))^{-1}\, \mathrm{d}M^{(n)}(u)\right| > \eta\right)$$

$$\leq \frac{\delta}{\eta^2} + \mathrm{P}\left(b_n^{-1} \int_{-1}^{1} K^2(u) J^{(n)}(t-b_n u)(Y^{(n)}(t-b_n u))^{-1} \alpha(t-b_n u)\, \mathrm{d}u > \delta\right).$$

Because α is continuous at t, and hence bounded in a neighborhood of t, and K is bounded, the last term on the right-hand side can be made arbitrarily small for n large enough using (4.2.8). Thus, (4.2.9) is proved. To prove (4.2.10), note that

$$|\alpha^{*(n)}(t) - \tilde{\alpha}^{(n)}(t)| \leq \int_{-1}^{1} |K(u)|\{1 - J^{(n)}(t - b_n u)\}\alpha(t - b_n u)\,du,$$

which, in fact, equals zero with probability converging to 1 by (4.2.8). Finally, because K has integral 1 over $[-1, 1]$,

$$|\tilde{\alpha}^{(n)}(t) - \alpha(t)| \leq \int_{-1}^{1} |K(u)||\alpha(t - b_n u) - \alpha(t)|\,du \to 0$$

as $n \to \infty$ by dominated convergence, and the proof is complete. □

By imposing stronger conditions, we are also able to prove uniform consistency of the kernel function estimator.

Theorem IV.2.2. *Let $t \in \mathscr{T}$ and let $0 < t_1 < t_2 < t$ be fixed numbers. Assume that the kernel K is of bounded variation and that α is continuous on $[0, t]$. Let the bandwidth $b_n \to 0$ as $n \to \infty$, and assume that, as $n \to \infty$,*

$$b_n^{-2} \int_0^t J^{(n)}(s)(Y^{(n)}(s))^{-1}\alpha(s)\,ds \xrightarrow{\mathrm{P}} 0 \tag{4.2.12}$$

and

$$\int_0^t (1 - J^{(n)}(s))\alpha(s)\,ds \xrightarrow{\mathrm{P}} 0. \tag{4.2.13}$$

Then, as $n \to \infty$,

$$\sup_{s \in [t_1, t_2]} |\hat{\alpha}^{(n)}(s) - \alpha(s)| \xrightarrow{\mathrm{P}} 0.$$

Proof. We have to prove that the convergence is uniform in (4.2.9), (4.2.10), and (4.2.11). Because K is assumed to be of bounded variation, we get for $s \in [t_1, t_2]$ and n large enough, using (4.2.4), that

$$|\hat{\alpha}^{(n)}(s) - \alpha^{*(n)}(s)| \leq 2b_n^{-1}V(K) \sup_{u \in [0, t]} |\hat{A}^{(n)}(u) - A^{*(n)}(u)|,$$

where $V(K)$ denotes the total variation of K. Just as in the proof of Theorem IV.1.1, the right-hand side converges to zero in probability as $n \to \infty$ by (4.2.12). This shows that

$$\sup_{s \in [t_1, t_2]} |\hat{\alpha}^{(n)}(s) - \alpha^{*(n)}(s)| \xrightarrow{\mathrm{P}} 0$$

as $n \to \infty$, so the convergence is uniform in (4.2.9). That this is also the case for (4.2.10) follows as in the proof of Theorem IV.2.1 using (4.2.13) and the boundedness of K. Finally, uniform convergence in (4.2.11) follows using the boundedness of K and the fact that α is continuous, and hence uniformly continuous, on $[0, t]$. □

We have assumed that $A(t) < \infty$ for $t \in \mathscr{T}$. Therefore, a sufficient condition for (4.2.12) and (4.2.13) is that

$$\inf_{s \in [0,t]} b_n^2 Y^{(n)}(s) \xrightarrow{\text{P}} \infty \quad \text{as } n \to \infty. \tag{4.2.14}$$

Comparing the conditions (4.2.8) and (4.2.14), we see that the bandwidth must tend toward zero more slowly to obtain uniform consistency than to obtain pointwise consistency. Moreover, by comparing these two conditions with (4.1.10), it is seen that the rate of increase of $Y^{(n)}$ must be larger to obtain consistency of the kernel function estimator than was the case for the Nelson–Aalen estimator itself.

IV.2.2.2. *Mean Integrated Squared Error and Optimal Bandwidth*

Theorems IV.2.1 and IV.2.2 show that the kernel function estimator is consistent provided that the bandwidth tends to zero sufficiently slowly as n increases. To be able to study more closely how the bandwidth should tend to zero, and even be able to pick an optimal bandwidth, we first have to select a measure for the global performance of the kernel function estimator. To this end, we restrict our attention to a fixed interval $[t_1, t_2]$ and assume that $c > 0$ is such that $t_i \pm c \in \mathscr{T}$ for $i = 1, 2$ and $[t_1 - b_n, t_2 + b_n] \subset [t_1 - c, t_2 + c]$. We then consider the mean integrated squared error (MISE)

$$\text{MISE}(\hat{\alpha}^{(n)}) = \text{E} \int_{t_1}^{t_2} \{\hat{\alpha}^{(n)}(t) - \alpha(t)\}^2 \, \text{d}t. \tag{4.2.15}$$

The mean integrated squared error may be written as a sum of a "squared bias term" and a "variance term." However, a little care is needed in doing so because of the possibility that $Y^{(n)}(s)$ may be zero for some s. We first note that by (4.2.6) and (4.2.3) we may write

$$\text{E}\hat{\alpha}^{(n)}(t) = \tilde{\alpha}^{(n)}(t) + R_1^{(n)}(t), \tag{4.2.16}$$

where by the boundedness of K and because $A(t_2 + c) < \infty$, we have

$$\begin{aligned} |R_1^{(n)}(t)| &= \left| \int_{-1}^{1} K(u) \text{P}(Y^{(n)}(t - b_n u) = 0) \alpha(t - b_n u) \, \text{d}u \right| \\ &\leq C_1 \sup_{t \in [t_1 - c, t_2 + c]} \text{P}(Y^{(n)}(t) = 0) \end{aligned} \tag{4.2.17}$$

for a constant C_1 not depending on n and t. Because the integrand in (4.2.15) is non-negative, we can reverse the order of integration and expectation to get

$$\mathrm{MISE}(\hat{\alpha}^{(n)}) = \int_{t_1}^{t_2} (\tilde{\alpha}^{(n)}(t) - \alpha(t))^2 \,\mathrm{d}t + \int_{t_1}^{t_2} \mathrm{E}(\hat{\alpha}^{(n)}(t) - \tilde{\alpha}^{(n)}(t))^2 \,\mathrm{d}t + R_2^{(n)}, \tag{4.2.18}$$

where, by (4.2.16) and (4.2.17),

$$|R_2^{(n)}| = \left| 2 \int_{t_1}^{t_2} R_1^{(n)}(t)(\tilde{\alpha}^{(n)}(t) - \alpha(t)) \,\mathrm{d}t \right|$$

$$\leq C_2 \sup_{t \in [t_1 - c, t_2 + c]} \mathrm{P}(Y^{(n)}(t) = 0), \tag{4.2.19}$$

and C_2 is another constant not depending on n and t. Formula (4.2.18) gives the desired decomposition of the mean squared error into a "squared bias term" and a "variance term" (plus a remainder term).

To further study the asymptotic behavior of the mean integrated squared error, we will restrict our attention to kernels satisfying

$$\int_{-1}^{1} K(t) \,\mathrm{d}t = 1, \quad \int_{-1}^{1} tK(t) \,\mathrm{d}t = 0, \quad \text{and} \quad \int_{-1}^{1} t^2 K(t) \,\mathrm{d}t = k_2 > 0, \tag{4.2.20}$$

which is the case for all symmetric non-negative functions with unit integral. We may then prove the following important result:

Theorem IV.2.3. *Assume that the kernel funtion satisfies* (4.2.20) *and that* α *is twice continuously differentiable on* $[t_1 - c, t_2 + c] \subset \mathcal{T}$ *for a* $c > 0$. *Assume, furthermore, that there exist a sequence* $\{a_n\}$, *increasing to infinity as* $n \to \infty$, *and a continuous function* y *such that*

$$\mathrm{E}\left\{\frac{a_n^2 J^{(n)}}{Y^{(n)}}\right\} \to \frac{1}{y}$$

uniformly on $[t_1 - c, t_2 + c]$ *as* $n \to \infty$. *Assume finally that*

$$\sup_{t \in [t_1 - c, t_2 + c]} \mathrm{P}(Y^{(n)}(t) = 0) = o(a_n^{-2}).$$

Then, as $n \to \infty$ *and* $b_n \to 0$, *the "squared bias term" of* (4.2.18) *may be written*

$$\int_{t_1}^{t_2} (\tilde{\alpha}^{(n)}(t) - \alpha(t))^2 \,\mathrm{d}t = \frac{1}{4} b_n^4 k_2^2 \int_{t_1}^{t_2} (\alpha''(t))^2 \,\mathrm{d}t + o(b_n^4), \tag{4.2.21}$$

where k_2 *is defined in* (4.2.20), *and the "variance term" in* (4.2.18) *may be written*

$$\int_{t_1}^{t_2} \mathrm{E}(\hat{\alpha}^{(n)}(t) - \tilde{\alpha}^{(n)}(t))^2 \,\mathrm{d}t = (a_n^2 b_n)^{-1} \int_{-1}^{1} K^2(t) \,\mathrm{d}t \int_{t_1}^{t_2} \frac{\alpha(t)}{y(t)} \,\mathrm{d}t + o((a_n^2 b_n)^{-1}). \tag{4.2.22}$$

PROOF. To prove (4.2.21), we use (4.2.3), (4.2.20), and a Taylor expansion to get

$$\begin{aligned}\tilde{\alpha}^{(n)}(t) - \alpha(t) &= \int_{-1}^{1} K(u)\{\alpha(t - b_n u) - \alpha(t)\}\,du \\ &= -b_n\alpha'(t)\int_{-1}^{1} uK(u)\,du + \frac{1}{2}b_n^2\alpha''(t)\int_{-1}^{1} u^2K(u)\,du + o(b_n^2) \\ &= \tfrac{1}{2}b_n^2\alpha''(t)k_2 + o(b_n^2).\end{aligned} \tag{4.2.23}$$

From this, (4.2.21) readily follows.

To prove (4.2.22), we first write

$$\begin{aligned}E(\hat{\alpha}^{(n)}(t) - \tilde{\alpha}^{(n)}(t))^2 &= E(\hat{\alpha}^{(n)}(t) - \alpha^{*(n)}(t))^2 + E(\alpha^{*(n)}(t) - \tilde{\alpha}^{(n)}(t))^2 \\ &\quad + 2E\{(\hat{\alpha}^{(n)}(t) - \alpha^{*(n)}(t))(\alpha^{*(n)}(t) - \tilde{\alpha}^{(n)}(t))\}\end{aligned} \tag{4.2.24}$$

Now, by (4.2.5),

$$\sup_{t\in[t_1,t_2]}\left| a_n^2 b_n E(\hat{\alpha}^{(n)}(t) - \alpha^{*(n)}(t))^2 - \{\alpha(t)/y(t)\}\int_{-1}^{1} K^2(u)\,du\right| \to 0$$

as $n \to \infty$, so that the leading term on the right-hand side of (4.2.24) may be written

$$(a_n^2 b_n)^{-1}\frac{\alpha(t)}{y(t)}\int_{-1}^{1} K^2(u)\,du + o((a_n^2 b_n)^{-1}).$$

Furthermore, by (4.2.2) and (4.2.3),

$$|\alpha^{*(n)}(t) - \tilde{\alpha}^{(n)}(t)| = \int_{-1}^{1} K(u)I(Y^{(n)}(t - b_n u) = 0)\alpha(t - b_n u)\,du,$$

and, therefore, by an argument similar to the one giving (4.2.17), it follows that the second term on the right-hand side of (4.2.24) is of the form $o(a_n^{-2})$. Finally, by the Cauchy–Schwarz inequality, the last term on the right-hand side of (4.2.24) is of the form $o((a_n^2 b_n^{1/2})^{-1})$, and (4.2.22) follows because b_n tends to zero as n increases. □

In connection with the assumptions of the theorem, it is worth pointing out that the sequence $\{a_n\}$ and the function y typically will be the same as in Theorem IV.1.2 and in Theorems IV.2.4 and IV.2.5 to be presented later. Thus, $a_n = \sqrt{n}$ in most applications, where n is the number of "individuals." Illustrations for particular situations are provided in Examples IV.1.6–IV.1.10. A further illustration, where the conditions of the present theorem are checked in detail, is provided in Example IV.2.2. We note here, however, that the condition made on the asymptotic behavior of $\sup_{t\in[t_1-c,t_2+c]} P(Y^{(n)}(t) = 0)$ is a very weak one because, typically, this quantity will tend exponentially fast to zero as the number of "individuals" increases.

A comparison of the approximations (4.2.21) and (4.2.22) for the two (main) components of the mean squared error (4.2.18) demonstrates one of the fundamental problems of kernel function smoothing, a problem that was illustrated empirically in Example IV.2.1: If, in an attempt to make the "squared bias term" (4.2.21) small, we choose a small bandwidth b_n, then the "variance term" (4.2.22) will become large. On the other hand, a large bandwidth will only reduce (4.2.22) at the expense of an increased "squared bias term."

The ideal bandwidth is, therefore, the one that balances the effect of these two terms of the mean integrated squared error in an optimal way. According to (4.2.18) and Theorem IV.2.3, we may express the mean integrated squared error by the asymptotic approximation

$$\mathrm{MISE}(\hat{\alpha}^{(n)}) = \frac{1}{4} b_n^4 k_2^2 \int_{t_1}^{t_2} (\alpha''(t))^2 \, \mathrm{d}t + a_n^{-2} b_n^{-1} \int_{-1}^{1} K^2(t) \, \mathrm{d}t \int_{t_1}^{t_2} \frac{\alpha(t)}{y(t)} \mathrm{d}t + o(b_n^4) + o((a_n^2 b_n)^{-1}). \tag{4.2.25}$$

If we minimize the sum of the two leading terms on the right-hand side of (4.2.25) with respect to b_n, we arrive at the following optimal bandwidth:

$$b_{n,\mathrm{opt}} = a_n^{-2/5} k_2^{-2/5} \left\{ \int_{-1}^{1} K^2(t) \, \mathrm{d}t \int_{t_1}^{t_2} \frac{\alpha(t)}{y(t)} \mathrm{d}t \right\}^{1/5} \left\{ \int_{t_1}^{t_2} (\alpha''(t))^2 \, \mathrm{d}t \right\}^{-1/5} \tag{4.2.26}$$

We note that when b_n is of the order $a_n^{-2/5}$ as in (4.2.26), then both of the remainder terms in (4.2.25) are of the form $o(a_n^{-8/5})$ so that a minimization of only the two leading terms is a reasonable procedure. Moreover, it is seen by (4.1.16) and the conditions of Theorem IV.2.3 that typically $Y^{(n)}(t)$ is of the order a_n^2 (in probability). Therefore, a bandwidth of the order $a_n^{-2/5}$ as in (4.2.26) will satisfy the conditions of the Theorems IV.2.1 and IV.2.2; cf. (4.2.8) and (4.2.14).

In Example IV.2.4, we will show how (4.2.26) may be used to estimate the optimal bandwidth in practice. As illustrated there, we then also have to estimate the integral of $(\alpha'')^2$. Another method for estimating the optimal bandwidth in practice is the "cross-validation" technique described in Example IV.2.3. Examples IV.2.4 and IV.2.5 offer some comparison on concrete data sets of the two methods for estimating the optimal bandwidth, whereas some references to theoretical results on these methods (for the case of censored survival data) are given in the bibliographic remarks.

We finally note that by combining (4.2.16) and (4.2.23) we have under the conditions of Theorem IV.2.3 the following asymptotic expression for the bias of the kernel estimator

$$\mathrm{E}(\hat{\alpha}^{(n)}(t)) - \alpha(t) = \tfrac{1}{2} b_n^2 \alpha''(t) k_2 + o(b_n^2) + o(a_n^{-2}). \tag{4.2.27}$$

An empirical study of the bias using this formula is provided in Example IV.2.4.

IV.2.2.3. *Asymptotic Normality*

We now turn to a study of the asymptotic distribution of the kernel function estimator. We will first give a general result where we do not need to be very specific about the rate of decrease of the bandwidth. In this result, we only have to assume that b_n tends to zero more slowly than a_n^{-2} as $n \to \infty$.

Theorem IV.2.4. *Let t be an interior point of $\mathscr{T}$ and assume that α is continuous at t. Assume that there exist a sequence of positive constants $\{a_n\}$, increasing to infinity as $n \to \infty$, and a function y, positive and continuous at t, such that*

$$\sup_{s \in [t-\varepsilon, t+\varepsilon]} |a_n^{-2} Y^{(n)}(s) - y(s)| \xrightarrow{\mathrm{P}} 0 \tag{4.2.28}$$

as $n \to \infty$, for an $\varepsilon > 0$. Then, as $n \to \infty$, $b_n \to 0$, and $a_n^2 b_n \to \infty$,

$$a_n b_n^{1/2}(\hat{\alpha}^{(n)}(t) - \tilde{\alpha}^{(n)}(t)) \xrightarrow{\mathscr{D}} \mathscr{N}(0, \tau^2(t)), \tag{4.2.29}$$

where $\tilde{\alpha}^{(n)}(t)$ is given by (4.2.3), *and*

$$\tau^2(t) = \frac{\alpha(t)}{y(t)} \int_{-1}^{1} K^2(u)\,du. \tag{4.2.30}$$

Also,

$$a_n^2 b_n \hat{\tau}^2(t) \xrightarrow{\mathrm{P}} \tau^2(t), \tag{4.2.31}$$

where $\hat{\tau}^2(t)$ is given by (4.2.7). *Finally, if the conditions above are fulfilled for $t = t_1$ and $t = t_2$ with $t_1 \neq t_2$, then $\hat{\alpha}^{(n)}(t_1)$ and $\hat{\alpha}^{(n)}(t_2)$ are asymptotically independent.*

We postpone the proof of this result.

Note first that if we, in addition to the conditions of the theorem, also assume that α has a bounded derivative in a neighborhood of t and that $a_n^2 b_n^3 \to 0$ as $n \to \infty$, then, by the mean value theorem,

$$a_n b_n^{1/2}(\tilde{\alpha}^{(n)}(t) - \alpha(t)) = a_n b_n^{1/2}\left(\int_{-1}^{1} K(u)\alpha(t - b_n u)\,du - \alpha(t)\right) \to 0$$

as $n \to \infty$. It then follows that $a_n b_n^{1/2}(\hat{\alpha}^{(n)}(t) - \alpha(t))$ also will have the asymptotic distribution described in (4.2.29). However, the above assumption that $a_n^2 b_n^3 \to 0$, as $n \to \infty$, requires that the bandwidth tends to zero faster than $a_n^{-2/3}$, and it is not fulfilled when b_n is of the order $a_n^{-2/5}$, as suggested by (4.2.25) and (4.2.26). A more useful asymptotic result, valid for symmetric kernel functions satisfying (4.2.20) and for bandwidths of the order $a_n^{-2/5}$, is therefore the following.

Theorem IV.2.5. *Let the conditions of Theorem* IV.2.4 *be fulfilled and assume, in addition, that α is twice continuously differentiable in a neighborhood of t and that the kernel function satisfies* (4.2.20). *Assume also that the bandwidth satisfies*

$$\limsup_{n\to\infty} a_n^{2/5} b_n < \infty.$$

Then

$$a_n b_n^{1/2}(\hat{\alpha}^{(n)}(t) - \alpha(t) - \tfrac{1}{2}b_n^2\alpha''(t)k_2) \xrightarrow{\mathscr{D}} \mathcal{N}(0, \tau^2(t))$$

as $n \to \infty$, *where* k_2 *and* $\tau^2(t)$ *are given by* (4.2.20) *and* (4.2.30).

This holds because, by (4.2.20) and a Taylor expansion (with ξ_n between $t - b_n u$ and t),

$$\begin{aligned}
a_n b_n^{1/2}&\{\tilde{\alpha}^{(n)}(t) - \alpha(t) - \tfrac{1}{2}b_n^2\alpha''(t)k_2\} \\
&= a_n b_n^{1/2}\left\{\int_{-1}^{1} K(u)\alpha(t - b_n u)\,\mathrm{d}u - \alpha(t) - \tfrac{1}{2}b_n^2\alpha''(t)k_2\right\} \\
&= \frac{1}{2} a_n b_n^{5/2}\left\{\int_{-1}^{1} u^2 K(u)\alpha''(\xi_n)\,\mathrm{d}u - \alpha''(t)k_2\right\} \\
&\to 0,
\end{aligned}$$

which, combined with (4.2.29), gives the stated result.

As mentioned in connection with Theorem IV.2.3, the normalizing factors $\{a_n\}$ in Theorems IV.2.4 and IV.2.5 are typically the same as considered in Theorem IV.1.2 on the Nelson–Aalen estimator. Thus, in particular, we have $a_n = \sqrt{n}$, with n the number of "individuals," for most applications of the multiplicative intensity model.

We finally prove Theorem IV.2.4.

PROOF OF THEOREM IV.2.4. By (4.2.4),

$$a_n b_n^{1/2}(\hat{\alpha}^{(n)}(t) - \alpha^{*(n)}(t)) = \int_{\mathscr{T}} H^{(n)}(s)\,\mathrm{d}M^{(n)}(s)$$

with $H^{(n)}(s) = a_n b_n^{-1/2} K((t - s)/b_n) J^{(n)}(s)(Y^{(n)}(s))^{-1}$, so that Rebolledo's martingale central limit theorem (Theorem II.5.1) may be applied. With $\lambda^{(n)}(s) = \alpha(s) Y^{(n)}(s)$ and the above choice of $H^{(n)}(s)$, we have

$$\begin{aligned}
\int_{\mathscr{T}} (H^{(n)}(s))^2 \lambda^{(n)}(s)\,\mathrm{d}s &= \int_{\mathscr{T}} a_n^2 b_n^{-1} K^2\left(\frac{t-s}{b_n}\right) J^{(n)}(s)(Y^{(n)}(s))^{-1}\alpha(s)\,\mathrm{d}s \\
&= \int_{-1}^{1} a_n^2 K^2(u) J^{(n)}(t - b_n u)(Y^{(n)}(t - b_n u))^{-1}\alpha(t - b_n u)\,\mathrm{d}u \\
&\xrightarrow{\mathrm{P}} \frac{\alpha(t)}{y(t)} \int_{-1}^{1} K^2(u)\,\mathrm{d}u
\end{aligned}$$

as $n \to \infty$, by (4.2.28) and the assumptions on α and y. Furthermore, for any $\varepsilon > 0$,

$$I(|H^{(n)}(s)| > \varepsilon) = I\left(\left|K\left(\frac{t-s}{b_n}\right) a_n^2 J^{(n)}(s)(Y^{(n)}(s))^{-1}\right| > \varepsilon a_n b_n^{1/2}\right)$$

converges to zero uniformly in probability because $a_n^2 b_n \to \infty$ as $n \to \infty$. It follows that

$$\int_{\mathcal{T}} (H^{(n)}(s))^2 \lambda^{(n)}(s) I(|H^{(n)}(s)| > \varepsilon)\, \mathrm{d}s \xrightarrow{\mathrm{P}} 0$$

as $n \to \infty$. Thus, by Theorem II.5.1, we have shown that

$$a_n b_n^{1/2}(\hat{\alpha}^{(n)}(t) - \alpha^{*(n)}(t)) \xrightarrow{\mathscr{D}} \mathscr{N}(0, \tau^2(t))$$

as $n \to \infty$ and that (4.2.31) holds. Furthermore, as in the proof of (4.2.10), we have

$$a_n b_n^{1/2}(\alpha^{*(n)}(t) - \tilde{\alpha}^{(n)}(t)) \xrightarrow{\mathrm{P}} 0$$

as $n \to \infty$, and (4.2.29) follows.

To prove the asymptotic independence of $\hat{\alpha}^{(n)}(t_1)$ and $\hat{\alpha}^{(n)}(t_2)$ for $t_1 \neq t_2$, we note that

$$a_n b_n^{1/2}(\hat{\alpha}^{(n)}(t_i) - \alpha^{*(n)}(t_i)) = \int_{\mathcal{T}} H_i^{(n)}(s)\, \mathrm{d}M^{(n)}(s),$$

for $i = 1, 2$, with $H_i^{(n)}(s) = a_n b_n^{-1/2} K((t_i - s)/b_n) J^{(n)}(s)(Y^{(n)}(s))^{-1}$. Therefore, as $n \to \infty$ and $b_n \to 0$,

$$\begin{aligned}
&\int_{\mathcal{T}} H_1^{(n)}(s) H_2^{(n)}(s) \lambda^{(n)}(s)\, \mathrm{d}s \\
&\quad = \int_{\mathcal{T}} a_n^2 b_n^{-1} K\left(\frac{t_1 - s}{b_n}\right) K\left(\frac{t_2 - s}{b_n}\right) J^{(n)}(s)(Y^{(n)}(s))^{-1} \alpha(s)\, \mathrm{d}s \\
&\quad \xrightarrow{\mathrm{P}} 0,
\end{aligned}$$

and from this the asymptotic independence follows using once more the martingale central limit theorem. □

IV.2.2.4. *Examples*

We will now give four examples. The first of these illustrates the conditions of Theorems IV.2.1–IV.2.5 for the competing risks model.

Example IV.2.2. The Competing Risks Model

The competing risks model is a continuous-time Markov process with one transient state labeled 0 and k absorbing states numbered from 1 to k (cf. Example III.1.5). In applications, we typically have that the state 0 corresponds to "alive", whereas transitions to the states 1 through k correspond to death according to k different causes (cf. Example I.3.8). Denote the transition probabilities of the model by $P_{hj}(s, t)$, and let the transition intensity from state 0 to state h be denoted $\alpha_{0h}(t)$, $h = 1, \ldots, k$. Then $P_{00}(0, t) =$

$\exp\{-\int_0^t \alpha(u)\,du\}$ with $\alpha = \sum_{h=1}^k \alpha_{0h}$. Consider n independent Markov processes of this kind and assume that each process starts in state 0 at time 0. Let $N_h^{(n)}$ count the number of transitions to state h during $[0, t]$. It was shown in Examples III.1.4 and III.1.5 that $N_h^{(n)}$ is a counting process with intensity process $\alpha_{0h} Y^{(n)}$, where $Y^{(n)}(t) = n - N^{(n)}(t-)$ with $N^{(n)} = \sum_{h=1}^k N_h^{(n)}$. Thus, the results of the present section may be used to estimate α_{0h}. We write b_{hn} for the bandwidth used when estimating α_{0h}.

We will here study the conditions of Theorems IV.2.1–IV.2.5 for this particular situation with $\mathscr{T}$ being the set of time points u for which $P_{00}(0, u) > 0$. We start out by considering the consistency of the kernel function estimator. Because $Y^{(n)}$ is decreasing and $P_{00}(0, t + \varepsilon) > 0$ for t an interior point of $\mathscr{T}$ and ε small enough,

$$\inf_{s \in [t-\varepsilon, t+\varepsilon]} b_{hn} Y^{(n)}(s) = n b_{hn} \frac{Y^{(n)}(t+\varepsilon)}{n} \xrightarrow{\mathrm{P}} \infty$$

if $nb_{hn} \to \infty$ as $n \to \infty$, i.e., if b_{hn} tends to zero more slowly than n^{-1}. Thus, (4.2.8) is fulfilled and $\hat{\alpha}_{0h}^{(n)}(t)$ is consistent provided that α_{0h} is continuous at t. Furthermore, if the kernel function is of bounded variation and the intensity is continuous, it is seen, using (4.2.14), that $\hat{\alpha}_{0h}^{(n)}$ is uniformly consistent in the manner described in Theorem IV.2.2 provided that $nb_{hn}^2 \to \infty$ as $n \to \infty$, i.e., provided that b_{hn} tends to zero more slowly than $n^{-1/2}$.

We will then prove that the conditions of Theorem IV.2.3 are fulfilled provided that $0 < P_{00}(0, t_2 + c) < P_{00}(0, t_1 - c) < 1$ for a $c > 0$, and all the α_{0h} are twice continuously differentiable on $[t_1 - c, t_2 + c]$. Using the result of Aalen (1976, Lemma 4.2), we first note that $\mathrm{E}\{nJ^{(n)}(t)/Y^{(n)}(t)\}$ converges uniformly to $1/P_{00}(0, t)$ on $[t_1 - c, t_2 + c]$ as $n \to \infty$. Furthermore,

$$\sup_{t \in [t_1 - c, t_2 + c]} \mathrm{P}(Y^{(n)}(t) = 0) = \{1 - P_{00}(0, t_2 + c)\}^n$$

tends exponentially fast to zero as $n \to \infty$, and it follows that the conditions of Theorem IV.2.3 are fulfilled with $a_n^2 = n$.

As for the asymptotic normality, the Glivenko–Cantelli theorem gives

$$\sup_{s \in \mathscr{T}} |n^{-1} Y^{(n)}(s) - P_{00}(0, s)| \xrightarrow{\mathrm{P}} 0$$

as $n \to \infty$, with $P_{00}(0, s) = \exp\{-\int_0^s \alpha(u)\,du\}$ a continuous function. Thus, (4.2.28) is fulfilled, and it follows (for example) by Theorem IV.2.5, for t an interior point of $\mathscr{T}$, that

$$(nb_{hn})^{1/2}(\hat{\alpha}_{0h}^{(n)}(t) - \alpha_{0h}(t) - \tfrac{1}{2} b_{hn}^2 \alpha_{0h}''(t) k_2) \xrightarrow{\mathscr{D}} \mathscr{N}(0, \tau_h^2(t))$$

with

$$\tau_h^2(t) = \alpha_{0h}(t) P_{00}(0, t)^{-1} \int_{-1}^{1} K^2(u)\,du,$$

when $n \to \infty$ provided that the kernel function satisfies (4.2.20), the bandwidth is of the order $n^{-1/5}$, and that α_{0h} has a continuous second-order derivative in a neighborhood of t. □

In the next example, a "cross-validation" technique is introduced for estimating the optimal bandwidth (4.2.26).

EXAMPLE IV.2.3. Mortality and Incidence of Diabetic Nephropathy (DN) in Insulin-Dependent Diabetics. Estimation of Optimal Bandwidth by Cross-Validation

The problem and the data were presented in Example I.3.11. As mentioned there, some of the transitions from the state 0: "alive without DN" to the state 1: "alive with DN" were *interval censored*; see also the discussion in Section III.5. Having predicted these transition times as described in Example I.3.11, the data for each of the 1503 male and 1224 female diabetics i allow us to "observe"

$$Y_{0i}(t) = I(i \text{ is observed to be in state 0 at time } t-)$$

and

$$N_{01i}(t) = I(i \text{ is observed to make a } 0 \rightarrow 1 \text{ transition in the interval } [0, t]).$$

We can then define

$$Y_0(t) = \sum_{i=1}^{n} Y_{0i}(t)$$

and

$$N_{01}(t) = \sum_{i=1}^{n} N_{01i}(t),$$

and letting $\alpha_{01}(t)$ be the intensity for developing DN, we may estimate the integrated transition intensity $A_{01}(t)$ by the Nelson–Aalen estimator.

To study the impact of having predicted some $0 \rightarrow 1$ transition times, we also calculated estimates of $A_{01}(t)$ where for the 97 males with a predicted time of occurrence of DN either the smallest possible time (i.e., the year after the year last seen without DN) or the largest possible time (i.e., the year before the year first seen with DN) of occurrence of DN was used instead of the corresponding predicted times.

Figure IV.2.3 shows the estimated $0 \rightarrow 1$ transition intensity $\hat{\alpha}_{01}(t)$ for males obtained by smoothing these three estimates of $A_{01}(t)$ using (4.2.1) with the uniform kernel $K_U(x) = 0.5I(|x| \leq 1)$ and the bandwidth $b = 8$ years.

The peak for the estimate based on the minimum transition times is clearly lower than for the other two estimates. This is partly due to the fact that the minimum times in a number of cases corresponded to unrealistically short diabetes durations when the patient was last seen without DN few years after the onset of the disease. The difference between the other two is of smaller

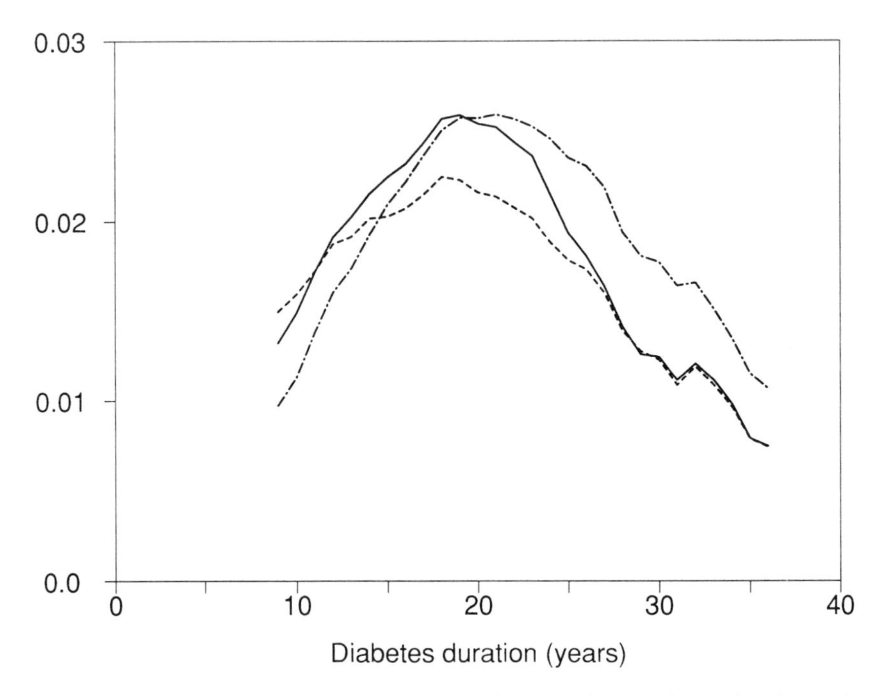

Figure IV.2.3. Estimated intensity per year of developing nephropathy for male insulin-dependent diabetics based on minimum transition times (- - -), maximum transition times (-·-·-), and predicted transition times (—) using the uniform kernel and bandwidth 8 years.

order of magnitude at least for diabetes durations up to about 25 years. We conclude that the procedure applied for predicting the unobserved transition times seems to give satisfactory results.

The bandwidth $b = 8$ years was chosen to minimize (an estimate of) the mean integrated squared error (4.2.15) as a function of b. Note that

$$\mathrm{MISE}(b) = \mathrm{E}\int_{t_1}^{t_2} \hat{\alpha}_{01}(t)^2\,\mathrm{d}t - 2\mathrm{E}\int_{t_1}^{t_2} \hat{\alpha}_{01}(t)\alpha_{01}(t)\,\mathrm{d}t + \int_{t_1}^{t_2} \alpha_{01}(t)^2\,\mathrm{d}t.$$

Here the first term is easy to estimate from $\hat{\alpha}_{01}(t)$, the last term does not depend on b, and Ramlau-Hansen (1981) showed that an approximately unbiased estimator of the second term is the "cross-validation" estimate

$$-2\sum_{i \neq j} \frac{1}{b} K\left(\frac{T_i - T_j}{b}\right) \frac{\Delta N_{01}(T_i)}{Y_0(T_i)} \frac{\Delta N_{01}(T_j)}{Y_0(T_j)};$$

where the sum is over i and j such that $i \neq j$ and $t_1 \leq T_i \leq t_2$; cf. also Rudemo (1982) and Bowman (1984).

Figure IV.2.4 shows the estimate for $\mathrm{MISE}(b) - \int_{t_1}^{t_2} \alpha_{01}(t)^2\,\mathrm{d}t$ for $b = 1$,

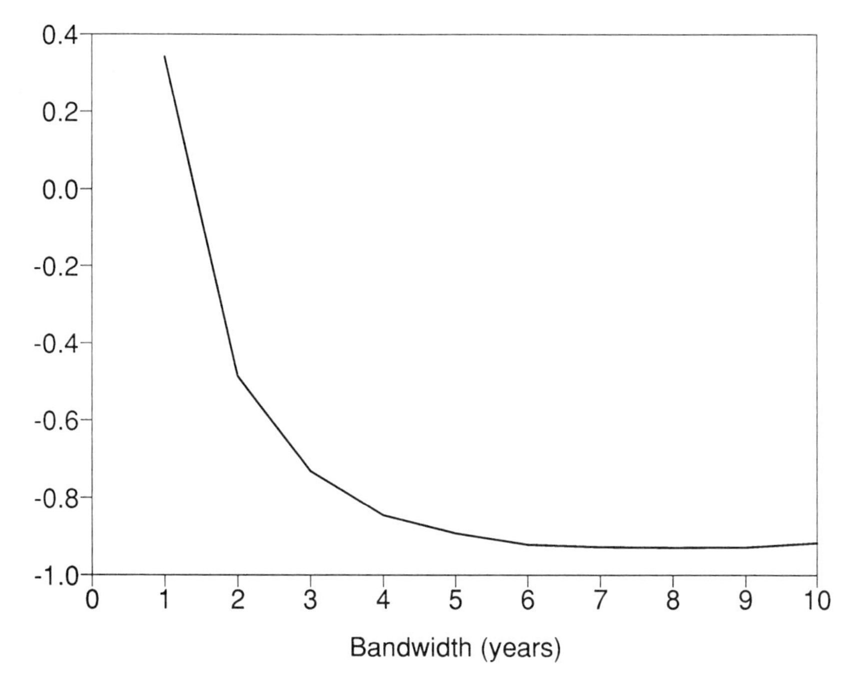

Figure IV.2.4. Estimate of the "risk function" $100(\text{MISE}(b) - \int_{11}^{34} \alpha_{01}(t)^2\,dt)$ for the uniform kernel for $b = 1, 2, \ldots, 10$ years.

2, ..., 10 years and for the uniform kernel choosing the interval $[t_1, t_2] =$ [11 years, 34 years]. It is seen that the minimum is attained for $b = 8$ years, but there seems to be little difference for values of b between 6 and 10 years. □

In the next example we extend the discussion of choice of bandwidth to also cover an estimate based on the theoretical expression (4.2.26). The possibilities of correcting the inherent bias [using (4.2.27)] are also discussed.

Example IV.2.4. Mortality of Female Diabetics in the County of Fyn. Optimal Bandwidth, Estimation of Local Bias, and Confidence Limits

The problem and the data were presented in Example I.3.2.

Let us return to the problem of smoothing the death intensity of female diabetics of Fyn introduced in Example IV.2.1. Using the age interval [25 years, 75 years] and the Epanechnikov kernel, the estimated mean integrated squared error was minimized for $b = 6$ years using the cross-validation procedure described in Example IV.2.3. The smoothed death intensity is shown in Figure IV.2.5. The local maximum around ages 35–40 years is interpreted as mortality for juvenile diabetics having developed diabetic nephropathy.

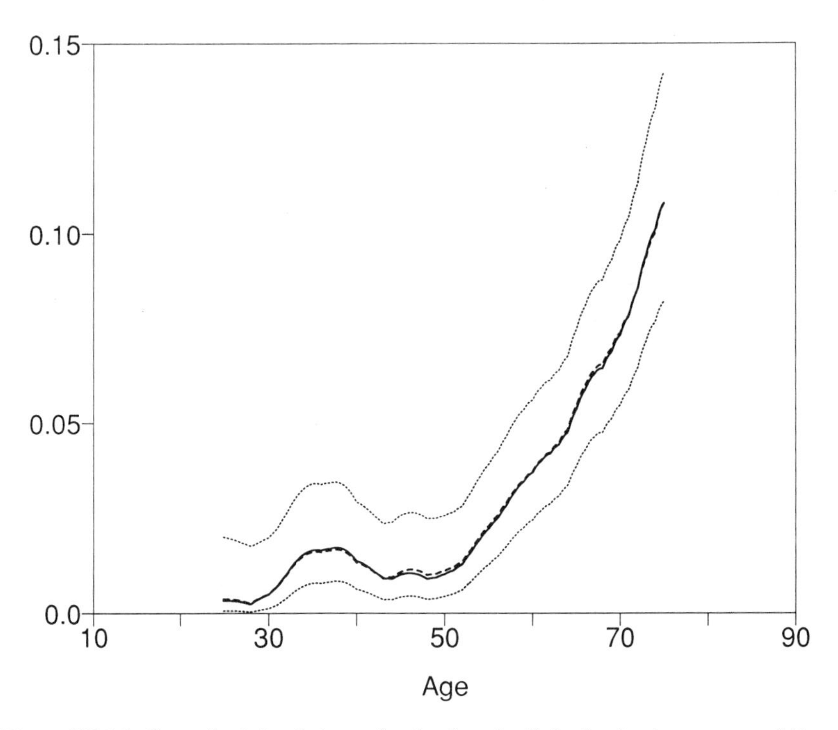

Figure IV.2.5. Smoothed death intensity for female diabetics in the county of Fyn, 1973–81, using the Epanechnikov kernel with the minimum MISE bandwidth estimated by cross-validation as $b = 6$ years (—). Correction for bias using a sixth-degree polynomial kernel K_1 and bandwidth $b_1 = 15$ years results in the estimate shown by a broken line (- - -). The dotted lines (···) are approximate pointwise 95% confidence limits about $\tilde{\alpha}(t)$, using the log-transformation.

We shall in this example explore somewhat further the inherent bias of $\hat{\alpha}$ as estimator of α. By (4.2.27),

$$E(\hat{\alpha}^{(n)}(t)) - \alpha(t) \approx \tfrac{1}{2} b_n^2 \alpha''(t) k_2(K)$$

with

$$k_2(K) = \int_{-1}^{1} t^2 K(t)\,\mathrm{d}t$$

which we assume to be positive. An estimate of the local bias using a particular kernel K and bandwidth b may, thus, be obtained as $\frac{1}{2} b^2 \hat{\alpha}''(t) k_2(K)$, where $\hat{\alpha}''$ is some estimate of the second derivative of α. To estimate α'', consider some kernel K_1 which is twice differentiable *on the whole line*. Then from

$$\hat{\alpha}(t) = \frac{1}{b_1} \int K_1\left(\frac{t-u}{b_1}\right) \mathrm{d}\hat{A}(u),$$

we get

$$\hat{\alpha}''(t) = \frac{1}{b_1^3} \int K_1''\left(\frac{t-u}{b_1}\right) \mathrm{d}\hat{A}(u).$$

The simplest symmetric twice differentiable kernel which is a polynomial times the indicator $I_{(-1,1)}$ is

$$K_1(x) = \tfrac{35}{32}(1 - x^2)^3 I_{(-1,1)}(x),$$

for which

$$K_1'(x) = -\tfrac{105}{16} x(1 - x^2)^2 I_{(-1,1)}(x)$$

and

$$K_1''(x) = -\tfrac{105}{16}(1 - 6x^2 + 5x^4) I_{(-1,1)}(x).$$

The broken line on Figure IV.2.5 shows the bias-corrected estimate

$$\hat{\alpha}(t) + \tfrac{1}{2} b^2 \hat{\alpha}''(t) k_2(K_E)$$

using $b_1 = 15$ years in the estimation of $\hat{\alpha}''$; in the present case, the influence of the bias correction is minor. [It would have been larger if (suboptimal) bandwidths b larger than 6 had been used.]

In (4.2.26), we derived the following asymptotically valid expression for the bandwidth minimizing MISE over the interval $t_1 < t < t_2$

$$b_{n,\mathrm{opt}} = k_2(K)^{-2/5} \left\{ \int_{-1}^{1} K^2(t)\,\mathrm{d}t \int_{t_1}^{t_2} \frac{\alpha(t)}{y(t)}\,\mathrm{d}t \right\}^{1/5} \left\{ \int_{t_1}^{t_2} (\alpha''(t))^2\,\mathrm{d}t \right\}^{-1/5} a_n^{-2/5}.$$

This may be estimated by noting that an estimator for

$$a_n^{-2} \int_{t_1}^{t_2} \frac{\alpha(t)}{y(t)}\,\mathrm{d}t$$

is

$$\hat{\sigma}^2(t_2) - \hat{\sigma}^2(t_1),$$

where $\hat{\sigma}^2(t)$ is the usual estimate for the variance of the Nelson–Aalen estimator; cf. (4.1.11) and (4.1.15). In our situation, using $t_1 = 25$ years, $t_2 = 75$ years, and kernels and bandwidths as before (in particular, $b_1 = 15$ years), we estimate b_{opt} as 10.5 years, somewhat larger than the cross-validation result of $b_{\mathrm{opt}} = 6$ years above.

Confidence limits for the *smoothed* intensity function $\tilde{\alpha}(t)$ [cf. (4.2.3)] are immediately obtained from Theorem IV.2.4 [by transforming symmetric confidence limits for $\log \tilde{\alpha}(t)$] as

$$\hat{\alpha}(t) \exp\left\{ \pm \frac{1.96 \hat{\tau}(t)}{\hat{\alpha}(t)} \right\};$$

cf. Figure IV.2.5. Care has to be exercised in the interpretation of these limits in view of the inherent bias of $\hat{\alpha}(t)$.

It may finally be noted that Hall and Marron (1987) pointed out (for density estimation) that $\int \hat{\alpha}''(t)^2\,dt$ is a biased estimate of $\int \alpha''(t)^2\,dt$. In a private communication, J. Perch Nielsen (March, 1991) suggested that an approximately unbiased estimator would be given by

$$\int_{t_1}^{t_2} \hat{\alpha}''(t)^2\,dt - \frac{c(K_1)}{b_1^5}(\hat{\sigma}^2(t_2) - \hat{\sigma}^2(t_1))$$

with

$$c(K_1) = \frac{1}{b}\int K_1''\left(\frac{t-u}{b}\right)^2 dt,$$

here assuming the value $c(K_1) = 35$. Use of this correction in the above situation would change the estimate of the optimal bandwidth from 10.5 to 11.6.

In our view, further clarification is required before correction for bias and estimation of optimal bandwidth may be based routinely on formulas like (4.2.27) and (4.2.26). □

The final example further illustrates choice of kernel and optimal bandwidth and goes on to suggest a way of handling the *tail problem*: extending the estimate to the boundary of the observation interval.

EXAMPLE IV.2.5. Psychiatric Admissions for Women Having Just Given Birth. Kernel Estimation of Admission and Discharge Intensities. Tail Problem and Optimal Bandwidth

The problem and the data were presented in Example I.3.10. The aim here is to obtain kernel smoothed estimates $\hat{\alpha}_a(t)$ and $\hat{\alpha}_d(t)$ of the admission intensity $\alpha_a(t)$ and the discharge intensity $\alpha_d(t)$, $0 \le t \le 365$ days. Thus, the desire is to obtain an estimate over the full observation interval, and we, therefore, have to address the *tail problem*. Furthermore, the choice of best bandwidth b for each of the kernels K_U (uniform) and K_E (Epanechnikov) defined in Example IV.2.1 will also be discussed here. Because of the extra complications connected to the tail problem, we first study bandwidth choice over a fixed subinterval $[c, 365 - c]$, using all observations on $[0, 365]$.

For the *admission intensity*, we used the interval $[80, 365 - 80] = [80, 285]$ days. Because the observations were available as an integer number of days, the uniform kernel was studied using bandwidths equal to an integer plus $\frac{1}{2}$ to avoid difficulties in interpreting jump points coinciding with the endpoints of the kernel. Using the cross-validation approach of Example IV.2.3, the optimal bandwidth was estimated as 58.5 days. For the Epanechnikov kernel, the estimated optimal bandwidth was 70 days.

It is interesting to relate these results to formula (4.2.26) on the optimal bandwidth. According to (4.2.26), the optimal bandwidth is asymptotically proportional to a product of a functional χ of the kernel and a functional of the intensity to be estimated, where

$$\chi^5 = \int_{-1}^{1} K(u)^2\,du \bigg/ \left(\int_{-1}^{1} u^2 K(u)\,du\right)^2.$$

We have $\chi_U^5 = 4.5$, $\chi_E^5 = 15$, yielding $\chi_E/\chi_U = 1.27$, which corresponds well with $b_{E,\text{opt}}/b_{U,\text{opt}} = 70/58.5 = 1.20$.

For the *discharge* intensity, we used the interval $[120, 365 - 120] = [120, 245]$ days and obtained an optimal $b = 111.5$ days for the uniform kernel, whereas the Epanechnikov kernel attained its minimum MISE at the maximum permissible value $b = 120$ days. We did not find it justified to use less than one-third ($= [120, 245]$) of the observations for estimation of MISE and so have not been able to identify an optimal b here.

As indicated earlier, the *tail problem* is to define the smoothed estimator $\hat{\alpha}_h(t)$, for $h = a, d$, for $0 \le t < b$ (and similarly for $365 - b < t \le 365$). We here proceed by defining a smooth family of nonsymmetric kernels $K_q(x)$ with support $[-1, q]$ and then using

$$\hat{\alpha}_h(t) = \frac{1}{b}\int_0^{365} K_q\left(\frac{t-u}{b}\right) d\hat{A}_h(u)$$

for $q = t/b$; it is clear that in this way the effective area of integration extends from 0 to $t + b$.

There are several ways of defining an extension of this kind; here we follow the suggestion by Gasser and Müller (1979) of multiplying $K(x)$ by a linear function and requiring that the product still have integral 1 and mean 0 over $[-1, q]$:

$$K_q(x) = K(x)(\alpha_q + \beta_q x), \qquad \int_{-1}^{q} K_q(x)\,dx = 1, \qquad \int_{-1}^{q} x K_q(x)\,dx = 0.$$

Obviously, these moment conditions specify two linear equations which define α_q and β_q.

For the uniform kernel, we get

$$K_{U,q}(x) = \frac{4(1+q^3)}{(1+q)^4} + \frac{6(1-q)}{(1+q)^3}x, \quad -1 \le x \le q,$$

and, for the Epanechnikov kernel, we get $K_{E,q}(x) = K_E(x)(\alpha_q + \beta_q x)$ with

$$\alpha_q = \left(\frac{2}{15} + \frac{q^3}{3} - \frac{q^5}{5}\right)\gamma_q, \qquad \beta_q = \frac{(1-q^2)^2\gamma_q}{4}$$

with

$$\gamma_q^{-1} = \frac{3}{4}\left\{\left(\frac{2}{15} + \frac{q^3}{3} - \frac{q^5}{5}\right)\left(\frac{2}{3} + q - \frac{q^3}{3}\right) - \frac{1}{16}(1-q^2)^4\right\}.$$

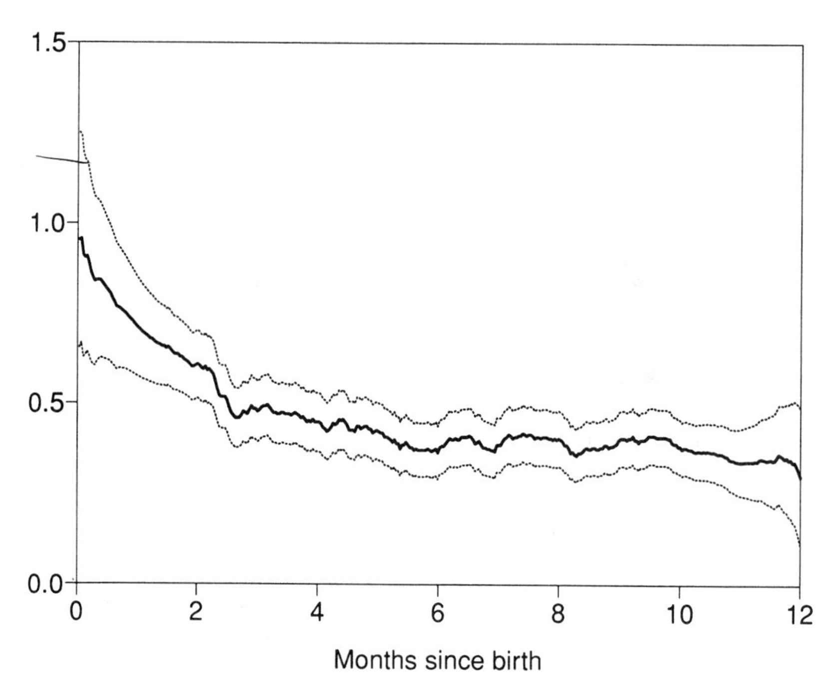

Figure IV.2.6. Estimated admission intensity per 10^3 months using the uniform kernel with the optimal bandwidth $b = 58.5$ days, with pointwise 95% confidence limits.

We note that $K_{E,q}(-x)$ is identical to the "Optimal 1" kernel quoted by Gasser and Müller (1979, Table 3) for $q = 0.0(0.2)1$. Also, note that for values of q close to 0, $\alpha_q + \beta_q x$ and, hence, $K_{E,q}(x)$ becomes negative for x close to -1.

Figures IV.2.6 and IV.2.7 show the estimated admission and discharge intensities, using the uniform kernel with optimal bandwidth (58.5 days, respectively 111.5 days) and extended in the tails as described. Approximate (linear) pointwise 95% confidence intervals are included.

For the Epanechnikov kernel, the optimal bandwidth for the admission intensity was 70 days. The corresponding (tail-amended) plots are shown in Figures IV.2.8–IV.2.10 based on $b = 15$, 70, and 120 days. For the discharge intensity, no optimal bandwidth could be identified (it is at least 120 days). Figure IV.2.11 shows the tail-amended plot based on $b = 120$ days.

The plots support the impression gained from the Nelson–Aalen plots of Example IV.1.4: For the admission intensity, there is an initial decrease, followed by a rather stable period from about 3 months to the end of the year. The discharge intensity seems to be constant. The confidence limits widen at the ends of the observation interval, reflecting the narrowing interval from which the tail kernel estimate derives its information. □

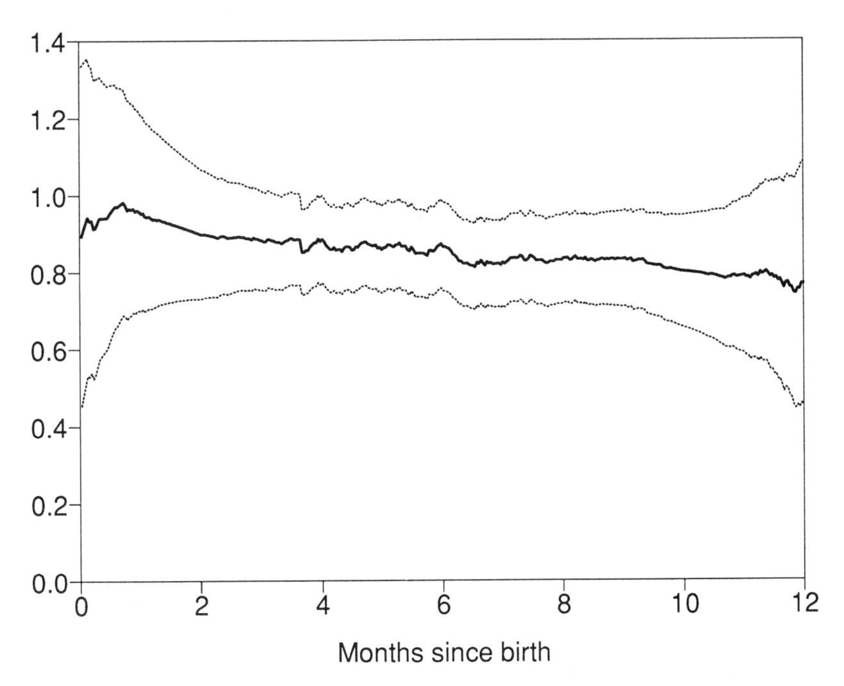

Figure IV.2.7. Estimated discharge intensity per month using the uniform kernel with the optimal bandwidth $b = 111.5$ days, with pointwise 95% confidence limits.

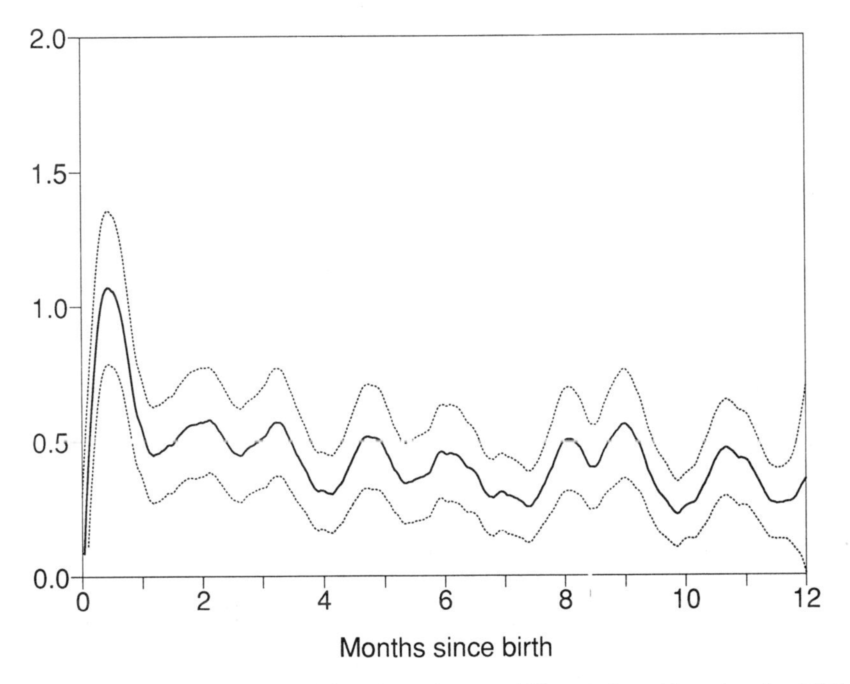

Figure IV.2.8. Estimated admission intensity per 10^3 months with pointwise 95% confidence limits using the Epanechnikov kernel and bandwidth $b = 15$ days.

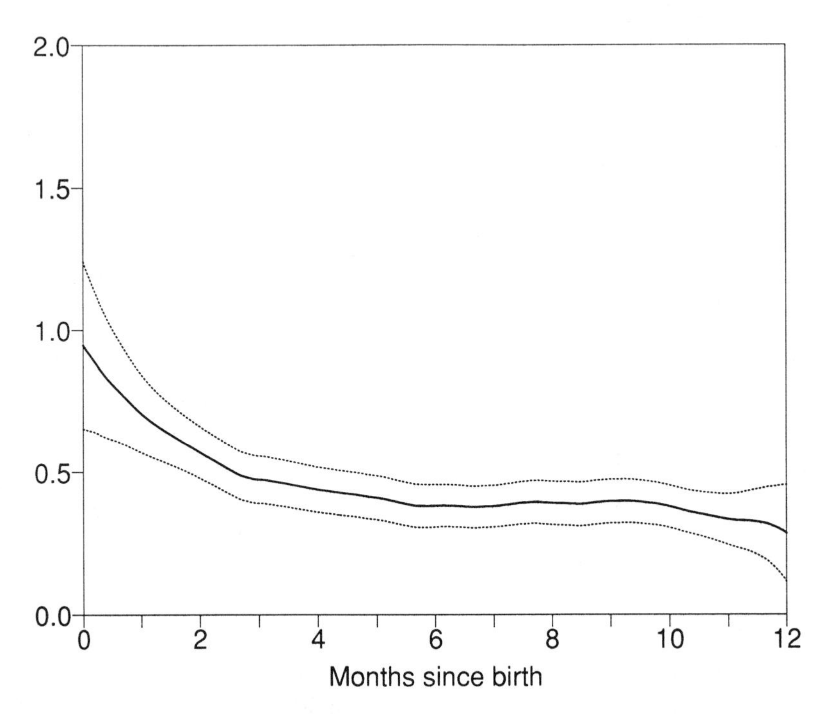

Figure IV.2.9. Estimated admission intensity per 10^3 months with pointwise 95% confidence limits using the Epanechnikov kernel and bandwidth $b = 70$ days (optimal).

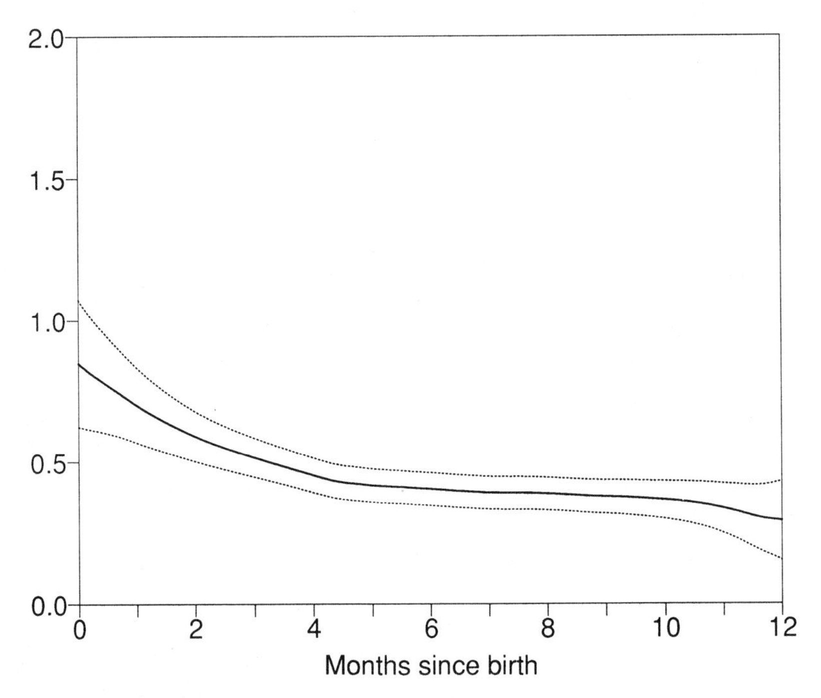

Figure IV.2.10. Estimated admission intensity per 10^3 months with pointwise 95% confidence limits using the Epanechnikov kernel and bandwidth $b = 120$ days.

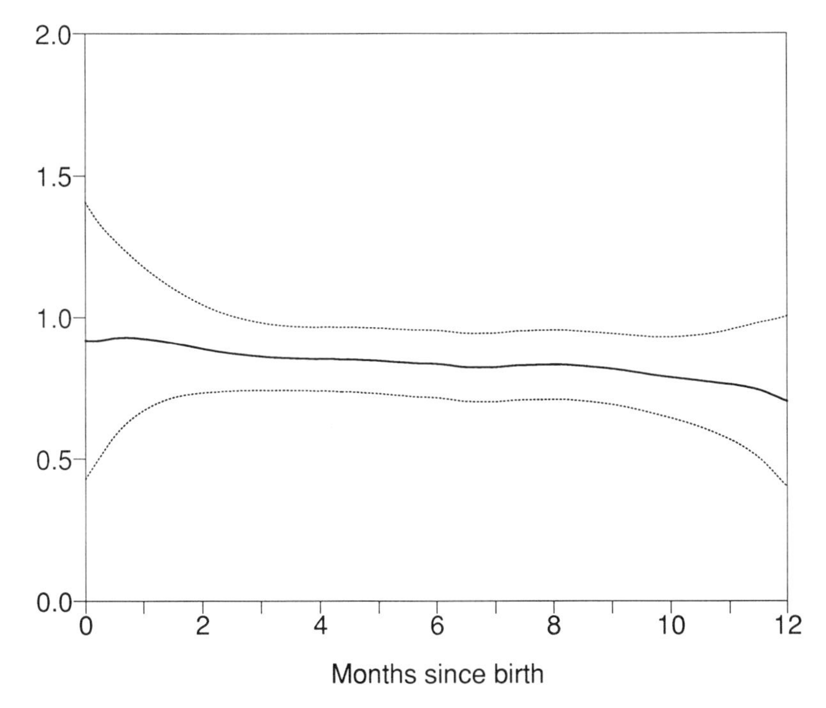

Figure IV.2.11. Estimated discharge intensity per month with pointwise 95% confidence limits using the Epanechnikov kernel and bandwidth $b = 120$ days.

IV.3. The Kaplan–Meier Estimator

In Sections IV.1 and IV.2, we considered nonparametric estimation in the multiplicative intensity model in its full generality. As illustrated by the examples there, and also by the examples of Chapter III, the results of Sections IV.1 and IV.2 may be applied to a number of situations, among which failure-time data is only one (prominent) example.

In this section, however, we will restrict attention to failure-time data. It was seen in Chapter III that for i.i.d. failure time data subject to independent right-censoring, as well as left-truncation and filtering, the univariate counting process $N(t)$, counting the number of observed failures in $[0, t]$, has intensity process of the multiplicative form $\alpha(t) Y(t)$. Here, α is the hazard rate function corresponding to the survival distribution S, assumed to be absolutely continuous, and $Y(t)$ is the number at risk just prior to time t. As before, we let A denote the integrated hazard function. Because $N(t) \le n$ for all t, where n is the number of individuals under study, the martingale $M(t) = N(t) - \int_0^t \alpha(s) Y(s)\,\mathrm{d}s$ is square integrable.

In Sections IV.3.1 and IV.3.2, we will derive and study the Kaplan–Meier estimator for the survival distribution

$$S(t) = \prod_{s \le t} (1 - \mathrm{d}A(s)) = \exp\left(-\int_0^t \alpha(u)\,\mathrm{d}u\right) \tag{4.3.1}$$

for $t \in \mathscr{T} = [0, \tau)$ with $\tau = \sup\{t: S(t) > 0\}$; cf. (2.6.12). Furthermore, the construction of pointwise confidence intervals and simultaneous confidence bands is considered in Section IV.3.3, whereas in Section IV.3.4, we consider estimation of quantiles and other functionals of S.

It should be emphasized that in the general derivations in this section, we *only* make use of the fact that N is a (bounded) counting process with intensity process αY. Therefore, our results are valid for any univariate counting process satisfying the multiplicative intensity model. But the estimated functional, $S(t) = \exp(-\int_0^t \alpha(u)\,\mathrm{d}u)$, is mainly of interest for the survival analysis setup we consider. Other applications of the results of this section are, however, to estimation of waiting time distributions and so-called partial transition probabilities of Markov processes with finite-state space. We return to this in Section IV.4.

IV.3.1. Definition and Basic Properties

The Kaplan–Meier estimator may be derived heuristically by partitioning the time axis into a number of small intervals, estimating the conditional survival probability in each interval by the proportion surviving, and multiplying these estimates together to get an estimate for the unconditional survival probability; cf. Kaplan and Meier's (1958) basic paper. (They also derived the estimator as a nonparametric maximum likelihood estimator.) We will derive the estimator directly from the Nelson–Aalen estimator using results on product-integrals, which can be considered to be a formalization of this idea. Our derivation will illustrate the close relation between these estimators, and it will also give a mathematical formulation which is very useful for deriving the statistical properties of the Kaplan–Meier estimator.

Let $\hat{A}$ be the Nelson–Aalen estimator (4.1.2) for the integrated hazard rate function. Then, (4.3.1) suggests that we may estimate the survival function by

$$\hat{S}(t) = \prod_{s \le t} (1 - \mathrm{d}\hat{A}(s)). \tag{4.3.2}$$

Because $\hat{A}$ is a step-function, we may, by (2.6.2), write

$$\hat{S}(t) = \prod_{s \le t} (1 - \Delta\hat{A}(s)) = \prod_{s \le t} \left(1 - \frac{\Delta N(s)}{Y(s)}\right)$$

which is the Kaplan–Meier estimator. It is well-known, and easily seen, that for uncensored failure-time data (Example III.1.2), $\hat{S}$ reduces to one minus the empirical distribution function. Note that our definition of $\hat{S}$ allows it to be strictly positive and constant to the right of the last observed failure, i.e., the last jump time of N. Alternative definitions of the Kaplan–Meier estimator for right-censored data are briefly reviewed after Example IV.3.1, where we also discuss how the Kaplan–Meier estimator may be interpreted as a nonparametric maximum likelihood estimator.

To study the statistical properties of the Kaplan–Meier estimator, we let $J(t) = I(Y(t) > 0)$ and $A^*(t) = \int_0^t J(u)\alpha(u)\,\mathrm{d}u$, as in Section IV.1, and define

$$S^*(t) = \exp(-A^*(t)) = \prod_{s \le t} (1 - \mathrm{d}A^*(s)). \tag{4.3.3}$$

Then, by version (2.6.5) of the Duhamel equation and (4.1.4), we get the basic relation

$$\begin{aligned} \frac{\hat{S}(t)}{S^*(t)} - 1 &= -\int_0^t \frac{\hat{S}(s-)}{S^*(s)} \mathrm{d}(\hat{A} - A^*)(s) \\ &= -\int_0^t \frac{\hat{S}(s-)J(s)}{S^*(s)Y(s)} \mathrm{d}M(s) \end{aligned} \tag{4.3.4}$$

for $t \in [0, \tau)$. Because $S^*(s) \ge S(s)$, the integrand in (4.3.4) is bounded by $1/S(s)$. Thus, $\hat{S}/S^* - 1$ is a local square integrable martingale on $[0, \tau)$ (Theorem II.3.1), which (by the result reviewed just below Example II.3.3) is square integrable on any proper subinterval of $[0, \tau)$.

It follows that

$$\mathrm{E}\{\hat{S}(t)/S^*(t)\} = 1 \tag{4.3.5}$$

and, therefore,

$$\mathrm{E}\hat{S}(t) \ge S(t)$$

for all $t \in [0, \tau)$, i.e., our version of the Kaplan–Meier estimator is, in general, biased upward. However, when there is only a small probability that $Y(s) = 0$ for some $s \le t$, the bias will be of little importance. A further study of the bias of the Kaplan–Meier estimator for right-censored failure-time data is provided in Example IV.3.1.

To arrive at an estimator for the variance of $\hat{S}$, we use (4.3.4) and the definition of $\langle M \rangle$ as the compensator of M^2 (cf. Section II.3.2) to get, for $t \in [0, \tau)$,

$$\mathrm{E}\{\hat{S}(t)/S^*(t) - 1\}^2 = \mathrm{E}\langle \hat{S}/S^* - 1 \rangle(t),$$

where, by (4.3.4) and (2.4.9),

$$\langle \hat{S}/S^* - 1 \rangle(t) = \int_0^t \left\{ \frac{\hat{S}(s-)}{S^*(s)} \right\}^2 \frac{J(s)}{Y(s)} \alpha(s)\,\mathrm{d}s.$$

By replacing $\hat{S}(s-)$ and $S^*(s)$ by $\hat{S}(s)$ (using the continuity of S) and $\mathrm{d}A(s) = \alpha(s)\,\mathrm{d}s$ by $\mathrm{d}\hat{A}(s)$ in the last expression, we arrive at the following estimator for the variance of $\hat{S}(t)/S(t)$:

$$\hat{\sigma}^2(t) = \int_0^t J(s)(Y(s))^{-2}\,\mathrm{d}N(s); \tag{4.3.6}$$

cf. (4.1.6). Thus, a reasonable estimator for the variance of $\hat{S}(t)$, at least for large sample purposes, is

$$\widehat{\text{var}}\,\hat{S}(t) = (\hat{S}(t))^2\hat{\sigma}^2(t). \tag{4.3.7}$$

Alternatively, we may, as in Section IV.1.1, consider an extension of the model to one involving an integrated hazard function A which is not necessarily continuous. For this model, $M = N - \int Y\,\mathrm{d}A$ is a martingale with

$$\langle M \rangle(t) = \int_0^t Y(s)(1 - \Delta A(s))\,\mathrm{d}A(s);$$

cf. the discussion above (4.1.7). In this extended model, we, therefore, have by (4.3.4) and (2.3.5)

$$\left\langle \frac{\hat{S}}{S^*} - 1 \right\rangle(t) = \int_0^t \left\{\frac{\hat{S}(s-)}{S^*(s)}\right\}^2 \frac{J(s)}{Y(s)}(1 - \Delta A(s))\,\mathrm{d}A(s).$$

Replacing S^* by $\hat{S}$ and A by $\hat{A}$ in this expression and noting that

$$\hat{S}(s) = \left(1 - \frac{\Delta N(s)}{Y(s)}\right)\hat{S}(s-),$$

we then arrive at the following alternative estimator for the variance of $\hat{S}(t)/S(t)$:

$$\hat{\hat{\sigma}}^2(t) = \int_0^t \{Y(s)(Y(s) - \Delta N(s))\}^{-1}\,\mathrm{d}N(s). \tag{4.3.8}$$

This yields the following estimator for the variance of $\hat{S}(t)$:

$$\widehat{\widehat{\text{var}}}\,\hat{S}(t) = (\hat{S}(t))^2\hat{\hat{\sigma}}^2(t), \tag{4.3.9}$$

which is the famous Greenwood's formula (Greenwood, 1926). If, for some time point s, $Y(s) = \Delta N(s)$, then $\hat{\hat{\sigma}}^2(t) = \infty$ for $t \geq s$. But in this case $\hat{S}(t) = 0$ for $t \geq s$ and (4.3.9) is interpreted as zero. For uncensored failure-time data, (4.3.9) reduces to $\hat{S}(t)(1 - \hat{S}(t))/n$ [e.g., Gill (1980a, p. 39)]. A discussion of the relative merits of these two estimators for right-censored survival data is included in Example IV.3.1. This discussion indicates that Greenwood's estimator (4.3.9) is to be preferred to (4.3.7).

Before we turn to this example, however, we will briefly consider a recursion formula for (4.3.9). This recursion formula is not of great importance here because Greenwood's formula itself is straightforward to use for computational purposes. The recursion formula is included here mainly because its extension to Markov processes considered in the next section is indeed very useful [cf. (4.4.20)]. The formula, which may be verified by a direct calculation, takes the form

$$\widehat{\widehat{\text{var}}}\,\hat{S}(t) = \widehat{\widehat{\text{var}}}\,\hat{S}(t-)(1 - \Delta\hat{A}(t))^2 + \hat{S}(t-)^2\,\widehat{\widehat{\text{var}}}(\Delta\hat{A}(t)),$$

where

$$\widehat{\widehat{\text{var}}}(\Delta\hat{A}(t)) = \frac{(Y(t) - \Delta N(t))\Delta N(t)}{Y(t)^3}$$

[cf. (4.1.7)]. We note that this formula corresponds to what one will get from a delta-method calculation based on the representation $\hat{S}(t) = \hat{S}(t-)(1 - \Delta\hat{A}(t))$.

EXAMPLE IV.3.1. Kaplan–Meier Estimator for Right-Censored Survival Data. Bias and Performance of Variance Estimators

We want to study the bias of our version of the Kaplan–Meier estimator for right-censored failure-time data. The situation is as described in Example IV.1.1, and we adopt the notation used there. Now $\hat{S}$, S^*, and M are constant to the right of $\tilde{X}_{(n)} = \max\{\tilde{X}_1, \dots, \tilde{X}_n\}$. Therefore, for all $t \in [0, \tau)$,

$$\frac{\hat{S}(t)}{S^*(t)} = \frac{\hat{S}(t)}{S(t)} + I(t \geq \tilde{X}_{(n)}) \left\{ \frac{\hat{S}(\tilde{X}_{(n)})}{S(\tilde{X}_{(n)})} - \frac{\hat{S}(\tilde{X}_{(n)})}{S(t)} \right\}.$$

Using (4.3.5), this gives

$$\mathrm{E}\hat{S}(t) = S(t) + \mathrm{E}\left\{ \frac{I(t \geq \tilde{X}_{(n)})\hat{S}(\tilde{X}_{(n)})(S(\tilde{X}_{(n)}) - S(t))}{S(\tilde{X}_{(n)})} \right\},$$

which shows that the absolute value of the bias increases with t. Because

$$\frac{S(\tilde{X}_{(n)}) - S(t)}{S(\tilde{X}_{(n)})} \leq 1 - S(t),$$

we also get the bounds

$$0 \leq \mathrm{E}\hat{S}(t) - S(t) \leq (1 - S(t))\mathrm{P}(\tilde{X}_{(n)} \leq t)$$

for all $t \in [0, \tau)$.

For the special case of simple random censorship, where $\tilde{X}_i = X_i \wedge U_i$ with $U_1, \dots, U_n$ i.i.d. with distribution functions G and independent of the X_i, the bounds take the form

$$0 \leq \mathrm{E}\hat{S}(t) - S(t) \leq (1 - S(t))\{1 - S(t)(1 - G(t))\}^n.$$

Because for the simple random censorship model,

$$\mathrm{E}Y(t) = n\mathrm{P}(\tilde{X}_i \geq t) = nS(t)(1 - G(t));$$

assuming the censoring distribution to be continuous, we further find

$$\begin{aligned} 0 \leq \mathrm{E}\hat{S}(t) - S(t) &\leq (1 - S(t))\left\{1 - \frac{\mathrm{E}Y(t)}{n}\right\}^n \\ &\leq (1 - S(t))\exp\{-\mathrm{E}Y(t)\}. \end{aligned}$$

Thus, e.g., for $\mathrm{E}Y(t) \geq 5$, the bias of the Kaplan–Meier estimator is at most $e^{-5} = 0.0067$. Studies of the bias of our version of the Kaplan–Meier estimator for small risk sets have been performed by Geurts (1987) and Klein (1988, 1991).

Klein (1988, 1991; cf. our Example IV.1.1) have also studied the performance of the two estimators (4.3.7) and (4.3.9) for the variance of the Kaplan–

Meier estimator. Both of these estimators were found to be negatively biased, but the Greenwood formula (4.3.9) was typically less biased than the estimator (4.3.7). For example, for $\mathrm{E}Y(t) \geq 5$, the bias of Greenwood's estimator was at most 15% of the true variance, whereas the relative bias of (4.3.7) was up to 25%. There were only rather small differences between the mean squared errors of (4.3.7) and (4.3.9). Klein also studied a suggestion of Wellner (1985) on how to estimate the variance of the Kaplan–Meier estimator when there is heavy censoring, as well as some modifications of Wellner's suggestion. All of these variance estimators were clearly more biased than the Greenwood formula (4.3.9), whereas their mean squared errors were close to that of the Greenwood estimator. Thus, it seems that for practical purposes one may always use the Greenwood formula (4.3.9) for estimating the variance of the Kaplan–Meier estimator. □

Before we turn to an investigation of the large sample properties of the Kaplan–Meier estimator, we conclude this subsection with some remarks on alternative definitions of the Kaplan–Meier estimator and discuss how the Kaplan–Meier estimator may be interpreted as a nonparametric maximum likelihood estimator.

Using the notation of Examples IV.1.1 and IV.3.1, all versions of the Kaplan–Meier estimator for right-censored data equal $\prod_{s \leq t}(1 - \Delta N(s)/Y(s))$ for $t \leq \tilde{X}_{(n)}$, and they are all equal to zero after $\tilde{X}_{(n)}$ if the event at this time is a failure. Thus, the various versions only differ when the last event is a censoring. In Kaplan and Meier's original paper, the estimator was left undefined for $t > \tilde{X}_{(n)}$ in such circumstances. Later, Efron (1967) slightly modified the estimator by setting it equal to zero after $\tilde{X}_{(n)}$ also when the event at this time is a censoring. Our version of the Kaplan–Meier estimator was proposed by Gill (1980a), and, as noted above, it satisfies $\hat{S}(t) = \hat{S}(\tilde{X}_{(n)})$ for $t > \tilde{X}_{(n)}$ also when the last event is a censoring.

Chen, Hollander, and Langberg (1982) and Geurts (1985) have studied the bias and the mean squared error of Efron's version of the Kaplan–Meier estimator. Comparisons of the bias and mean squared errors of Efron's and Gill's modifications have been performed by Geurts (1987) and Klein (1988, 1991). Not surprisingly, Gill's version (which we apply) was found to be best for large and moderate values of $S(t)$, whereas Efron's version performed best for small values of $S(t)$. But for practical purposes, the difference between the two is of no great importance.

In Section IV.1.5, we discussed a number of ways in which the Nelson–Aalen estimator could be interpreted as a nonparametric maximum likelihood estimator (NPMLE) of $A = \int \alpha$. The main problem there was how a discrete extension of the multiplicative intensity model should be performed, there being nothing canonical about one specific extension.

In the present section, however, we are considering survival data. So the survival function S and the cumulative hazard A for both continuous and discrete survival distributions satisfy the relations

$$S = \prod (1 - \mathrm{d}A),$$
$$A = -\int \frac{\mathrm{d}S}{S_-};$$

cf. (2.6.10) and (2.6.11). Thus, our model may be parametrized either by A or by S, and $\hat{A}$ is the NPMLE of A and $\hat{S}$ of S. Here the latter result can be obtained by the functional invariance of maximum likelihood estimators under 1–1 reparametrizations of the model, the proper likelihood here being of the binomial form (4.1.36), or directly by writing this likelihood as a function of S.

In the latter case, the approach of Gill (1991b) of defining the NPMLE as a solution of likelihood equations of smooth submodels works and gives the same answer. For a survival function S with density f, we would consider submodels f_h^θ defined by $f_h^\theta(t) = (1 + \theta h(t))f(t)$.

The binomial likelihood (4.1.36), which here is the proper one to use, leads through a standard information calculation to the Greenwood estimator (4.3.9). On the other hand, the Poisson likelihood (4.1.35), which here only is reasonable as a simple approximation, leads to the alternative estimator (4.3.7).

IV.3.2. Large Sample Properties

We will now study the large sample properties of the Kaplan–Meier estimator. We consider a sequence of univariate counting processes $N^{(n)}$, $n = 1, 2, \ldots$, each of the form described in the introduction to the present section, with α (and S) being the same for all n. Relevant quantities are indexed by n.

Let us first show that the Kaplan–Meier estimator is *uniformly consistent* on intervals $[0, t]$ for which $S(t) > 0$, i.e., on compact subintervals of $[0, \tau)$.

Theorem IV.3.1. *Let $t \in [0, \tau)$, and assume that, as $n \to \infty$,*

$$\int_0^t \frac{J^{(n)}(s)}{Y^{(n)}(s)} \alpha(s)\,\mathrm{d}s \xrightarrow{\mathrm{P}} 0 \tag{4.3.10}$$

and

$$\int_0^t (1 - J^{(n)}(s))\alpha(s)\,\mathrm{d}s \xrightarrow{\mathrm{P}} 0 \tag{4.3.11}$$

Then, as $n \to \infty$,

$$\sup_{s \in [0,t]} |\hat{S}^{(n)}(s) - S(s)| \xrightarrow{\mathrm{P}} 0.$$

Proof. This result is an immediate consequence of Theorem IV.1.1 and the continuity of product-integrals (cf. the remark below Proposition II.8.7). We also, however, present a simple direct proof: By version (2.5.18) of Lenglart's inequality and (4.3.4), we get, for any $\delta, \eta > 0$,

$$\mathrm{P}\left(\sup_{s \in [0,t]} \left| \frac{\hat{S}^{(n)}(s)}{S^{*(n)}(s)} - 1 \right| > \eta \right) \le \frac{\delta}{\eta^2} + \mathrm{P}\left(\int_0^t \left(\frac{\hat{S}^{(n)}(s-)}{S^{*(n)}(s)} \right)^2 \frac{J^{(n)}(s)}{Y^{(n)}(s)} \alpha(s)\,\mathrm{d}s > \delta \right).$$

Now, $\{\hat{S}^{(n)}(s)/S^{*(n)}(s)\}^2 \le (S(t))^{-2}$ for $s \in [0,t]$, and it follows by (4.3.10) that

$$\sup_{s \in [0,t]} \left| \frac{\hat{S}^{(n)}(s)}{S^{*(n)}(s)} - 1 \right| \xrightarrow{\mathrm{P}} 0 \quad \text{as } n \to \infty. \tag{4.3.12}$$

Furthermore, using (4.3.1), (4.3.3), and (2.6.5),

$$\begin{aligned} \left| \frac{S(s)}{S^{*(n)}(s)} - 1 \right| &= \int_0^s \frac{S(u)}{S^{*(n)}(u)} \mathrm{d}(A - A^{*(n)})(u) \\ &\le (S(t))^{-1} \int_0^t (1 - J^{(n)}(u))\alpha(u)\,\mathrm{d}u, \end{aligned} \tag{4.3.13}$$

so that, by (4.3.11),

$$\sup_{s \in [0,t]} \left| \frac{S(s)}{S^{*(n)}(s)} - 1 \right| \xrightarrow{\mathrm{P}} 0 \quad \text{as } n \to \infty.$$

Combining this with (4.3.12), we get

$$\sup_{s \in [0,t]} \frac{|\hat{S}^{(n)}(s) - S(s)|}{S^{*(n)}(s)} \xrightarrow{\mathrm{P}} 0$$

as $n \to \infty$, from which the result easily follows. □

We then turn to the problem of proving weak convergence of the Kaplan–Meier estimator on intervals $[0,t]$ for which $S(t) > 0$. The following general result holds.

Theorem IV.3.2. *Let $t \in [0, \tau)$, and assume that there exists a non-negative function y such that α/y is integrable over $[0,t]$. Let*

$$\sigma^2(s) = \int_0^s \frac{\alpha(u)}{y(u)}\,\mathrm{d}u \tag{4.3.14}$$

and assume that:

(A) *For each $s \in [0,t]$,*

$$n \int_0^s \frac{J^{(n)}(u)}{Y^{(n)}(u)} \alpha(u)\,\mathrm{d}u \xrightarrow{\mathrm{P}} \sigma^2(s) \quad \text{as } n \to \infty. \tag{4.3.15}$$

(B) *For all $\varepsilon > 0$,*

$$n \int_0^t \frac{J^{(n)}(s)}{Y^{(n)}(s)} \alpha(s) I\left(\left| \sqrt{n} \frac{J^{(n)}(s)}{Y^{(n)}(s)} \right| > \varepsilon \right) \mathrm{d}s \xrightarrow{\mathrm{P}} 0 \quad \text{as } n \to \infty. \tag{4.3.16}$$

(C)

$$\sqrt{n}\int_0^t (1 - J^{(n)}(u))\alpha(u)\,\mathrm{d}u \xrightarrow{\mathrm{P}} 0 \quad \text{as } n \to \infty. \tag{4.3.17}$$

Then

$$\sqrt{n}(\hat{S}^{(n)} - S) \xrightarrow{\mathscr{D}} -S \cdot U \quad \text{as } n \to \infty$$

on $D[0, t]$, *where* U *is a Gaussian martingale with* $U(0) = 0$ *and* $\operatorname{cov}(U(s_1), U(s_2)) = \sigma^2(s_1 \wedge s_2)$. *Moreover,*

$$\sup_{s \in [0,t]} |n\hat{\sigma}^2(s) - \sigma^2(s)| \xrightarrow{\mathrm{P}} 0 \quad \text{as } n \to \infty, \tag{4.3.18}$$

where $\hat{\sigma}^2(s)$ *is defined by* (4.3.6). *Finally,* (4.3.18) *continues to hold when* $\hat{\sigma}^2(s)$ *is replaced by* $\hat{\hat{\sigma}}^2(s)$ *given by* (4.3.8).

Proof. Because the survival function $S = S(A) = \prod (1 - \mathrm{d}A)$ is a functional of the integrated hazard A, Theorem IV.1.2 and the functional delta-method (Section II.8) may be used to derive the asymptotic distribution of the Kaplan–Meier estimator. According to Proposition II.8.7, S is compactly differentiable with derivative ($h \in D[0, t]$)

$$(\mathrm{d}S(A) \cdot h)(s) = -\int_0^s S(u-)\left(\frac{S(s)}{S(u)}\right)\mathrm{d}h(u) = -S(s)h(s)$$

(using the continuity of S). Therefore, by Theorem II.8.1 (and Lemma II.8.3), $\sqrt{n}(\hat{S}^{(n)} - S)$ is asymptotically equivalent to $-S\sqrt{n}(\hat{A}^{(n)} - A)$, and the weak convergence result follows using Theorem IV.1.2.

A direct martingale-based proof of this result is also possible, and we have chosen to present this as well. To this end, we note that, by (4.3.4),

$$\sqrt{n}\left\{\frac{\hat{S}^{(n)}(s)}{S^{*(n)}(s)} - 1\right\} = -\sqrt{n}\int_0^s \hat{S}^{(n)}(u-)J^{(n)}(u)\{S^{*(n)}(u)Y^{(n)}(u)\}^{-1}\,\mathrm{d}M^{(n)}(u).$$

By Conditions A and B, (4.3.12), and the fact that $\hat{S}^{(n)}(s)/S^{*(n)}(s) \leq (S(t))^{-1}$ for $s \in [0, t]$, we, therefore, get, by Rebolledo's martingale central limit theorem (Theorem II.5.1),

$$\sqrt{n}(\hat{S}^{(n)}/S^{*(n)} - 1) \xrightarrow{\mathscr{D}} -U,$$

as $n \to \infty$. Now Condition C and (4.3.13) ensure that

$$\sup_{s \in [0,t]} \left|\sqrt{n}\left(\frac{S(s)}{S^{*(n)}(s)} - 1\right)\right| \xrightarrow{\mathrm{P}} 0$$

as $n \to \infty$, so that

$$\frac{\sqrt{n}(\hat{S}^{(n)} - S)}{S^{*(n)}} \xrightarrow{\mathscr{D}} -U$$

as $n \to \infty$. From this, we again obtain the asymptotic distribution of the Kaplan–Meier estimator.

Note, furthermore, that (4.3.18) follows as in the proof of Theorem IV.1.2. It remains to prove that (4.3.18) continues to hold when $\hat{\sigma}^2$ is replaced by $\hat{\hat{\sigma}}^2$. To this end, note that

$$\hat{\hat{\sigma}}^2(s) - \hat{\sigma}^2(s) = \int_0^s \frac{\Delta N^{(n)}(u)}{(Y^{(n)}(u))^2(Y^{(n)}(u) - \Delta N^{(n)}(u))} \mathrm{d}N^{(n)}(u),$$

so that

$$\sup_{s \in [0,t]} |n\hat{\hat{\sigma}}^2(s) - n\hat{\sigma}^2(s)| \leq \sup_{s \in [0,t]} \left| \frac{\Delta N^{(n)}(s)}{Y^{(n)}(s) - \Delta N^{(n)}(s)} \right| n \int_0^t (Y^{(n)}(s))^{-2} \mathrm{d}N^{(n)}(s)$$

$$= \sup_{s \in [0,t]} \left| \frac{\Delta \hat{A}^{(n)}(s)}{1 - \Delta \hat{A}^{(n)}(s)} \right| n\hat{\sigma}^2(t).$$

Here, the first term on the right-hand side converges to zero in probability by the uniform consistency of the Nelson–Aalen estimator $\hat{A}^{(n)}$ (Theorem IV.1.1) and the continuity of A, whereas $n\hat{\sigma}^2(t)$ converges to $\sigma^2(t)$ in probability according to what we just proved. Thus,

$$\sup_{s \in [0,t]} |n\hat{\hat{\sigma}}^2(s) - n\hat{\sigma}^2(s)| \xrightarrow{\mathrm{P}} 0$$

as $n \to \infty$, and it follows that (4.3.18) continues to hold with $\hat{\sigma}^2(s)$ replaced by $\hat{\hat{\sigma}}^2(s)$. □

In connection with Theorems IV.3.1 and IV.3.2, it should be realized that the conditions of these theorems are exactly the same as the conditions of Theorems IV.1.1 and IV.1.2 (with $a_n = \sqrt{n}$ in Theorem IV.1.2). Therefore, the remarks on these conditions of Section IV.1.2 are still valid. It is also important to note that Theorems IV.3.1 and IV.3.2 continue to hold if $[0, t]$ is replaced everywhere by $[t_1, t_2]$ with $0 < t_1 < t_2 < \tau$, $\hat{S}^{(n)}(s)$ is replaced everywhere by $\hat{S}^{(n)}(s)/\hat{S}^{(n)}(t_1) = \prod_{(t_1,s]}(1 - \Delta N^{(n)}(u)/Y^{(n)}(u))$, and $S(s)$ is replaced everywhere by $S(s)/S(t_1) = \prod_{(t_1,s]}(1 - \mathrm{d}A(u)) = \exp(-\int_{t_1}^s \alpha(u)\,\mathrm{d}u)$, the conditional probability of surviving s given survival to t_1. This remark is especially important in connection with left-truncated and right-censored survival data, where it is often impossible to get sensible estimates of the survival function $S(\cdot)$ itself, and one has to be content with the conditional survival function $S(\cdot)/S(t_1)$ for a suitably chosen t_1; cf. Example IV.3.4.

In Examples IV.1.6–IV.1.8, we considered necessary conditions for uniform consistency and weak convergence of the Nelson–Aalen estimator for right-censored, as well as left-truncated and left-filtered, survival data. In light of the discussion above, the results of these examples carry over to the Kaplan–Meier estimator with only minor modifications.

Theorems IV.3.1 and IV.3.2 give weak uniform consistency and weak convergence of the Kaplan–Meier estimator on compact intervals $[0, t]$ for which $S(t) > 0$. The question whether these results may be extended to the whole positive real line was first addressed by Gill (1980a, 1983a), and his results were later improved by Wang (1987) and Ying (1989). We adopt the notation of Examples IV.1.1 and IV.3.1 and assume a simple random censorship model. Denote the censoring distribution by G and let $H = 1 - S(1 - G)$ be the distribution of the $\tilde{X}_i$. Then, Wang (1987) proved weak uniform consistency of the Kaplan–Meier estimator over the random interval $[0, \tilde{X}_{(n)}]$, whereas Ying (1989) proved the weak convergence result of Theorem IV.3.2 on $D[0, \tau_H]$, where $\tau_H = \sup\{t: H(t) > 1\}$.

IV.3.3. Confidence Intervals and Confidence Bands

The asymptotic results presented earlier may be used to provide pointwise confidence intervals and simultaneous confidence bands for the survival distribution. In what follows, all formulas for intervals and bands are given with $\hat{\sigma}^2$ defined by (4.3.6) for the nicety of the notation. However, they hold also with $\hat{\sigma}^2$ replaced by $\hat{\hat{\sigma}}^2$ given by (4.3.8).

Let us first consider pointwise confidence intervals for $S(s)$ for a fixed $s \in [0, \tau)$. The standard ("linear") asymptotic $100(1 - \alpha)\%$ interval is

$$\hat{S}^{(n)}(s) \pm c_{\alpha/2} \hat{S}^{(n)}(s) \hat{\sigma}(s), \tag{4.3.19}$$

where $c_{\alpha/2}$ is the upper $\alpha/2$ fractile of the standard normal distribution. This interval is not completely satisfactory for small sample sizes (Thomas and Grunkemeier, 1975). Using transformations as in Section IV.1.3, confidence intervals with better small sample properties may be derived. Kalbfleisch and Prentice (1980) and Thomas and Grunkemeier (1975) suggested the transformations $g(x) = \log(-\log x)$ and $g(x) = \arcsin \sqrt{x}$, respectively. These will be denoted the log-log-transformation and the arcsin-transformation in the following. (Note that arcsin-transformation had a slightly different meaning in Section IV.1.3.) Using these transformations, we arrive at the following $100(1 - \alpha)\%$ asymptotic confidence intervals:

$$\hat{S}^{(n)}(s)^{\exp\{\pm c_{\alpha/2} \hat{\sigma}(s) / \log \hat{S}^{(n)}(s)\}} \tag{4.3.20}$$

and

$$\begin{aligned}
&\sin^2\left\{\max\left(0, \arcsin(\hat{S}^{(n)}(s)^{1/2}) - \frac{1}{2} c_{\alpha/2} \hat{\sigma}(s) \left\{\frac{\hat{S}^{(n)}(s)}{1 - \hat{S}^{(n)}(s)}\right\}^{1/2}\right)\right\} \\
&\quad \le S(s) \\
&\quad \le \sin^2\left\{\min\left(\frac{\pi}{2}, \arcsin(\hat{S}^{(n)}(s)^{1/2}) + \frac{1}{2} c_{\alpha/2} \hat{\sigma}(s) \left\{\frac{\hat{S}^{(n)}(s)}{1 - \hat{S}^{(n)}(s)}\right\}^{1/2}\right)\right\};
\end{aligned} \tag{4.3.21}$$

cf. (4.1.22) and (4.1.23). The results of Borgan and Liestøl (1990) indicated that both of these intervals are quite satisfactory for right-censored survival data for sample sizes as low as 25, even with 50% censoring.

We then turn to the study of simultaneous confidence bands for S on (subintervals of) $[0,t]$, $t \in [0,\tau)$. By Theorems IV.1.2 and IV.3.2, $-\sqrt{n}(\hat{S}^{(n)} - S)/S$ has the same asymptotic distribution as the Nelson–Aalen estimator. Therefore, the arguments of Section IV.1.3 may be used directly to derive simultaneous confidence bands for S. Let $0 \leq t_1 \leq t_2 \leq t$ and define c_1 and c_2 by (4.1.25). We also introduce

$$\hat{c}_i = \frac{n\hat{\sigma}^2(t_i)}{1 + n\hat{\sigma}^2(t_i)},$$

$i = 1, 2$, as in Section IV.1.3.

If $[t_1, t_2]$ are such that $0 < c_1 < c_2 < 1$, then a $100(1-\alpha)\%$ asymptotic confidence band for S on $[t_1, t_2]$ is

$$\hat{S}^{(n)}(s) \pm d_\alpha(\hat{c}_1, \hat{c}_2)\hat{S}^{(n)}(s)\hat{\sigma}(s), \tag{4.3.22}$$

where $d_\alpha(c_1, c_2)$ is the upper α fractile in the distribution of

$$\sup_{c_1 \leq x \leq c_2} |W^0(x)\{x(1-x)\}^{-1/2}|;$$

cf. (4.1.27). This band was discussed by Nair (1981, 1984) and it is proportional to the pointwise interval (4.3.19). It is called the *equal precision band* or *EP-band* for short.

Another $100(1-\alpha)\%$ asymptotic confidence band for S on $[t_1, t_2]$ is

$$\hat{S}^{(n)}(s) \pm n^{-1/2}e_\alpha(\hat{c}_1, \hat{c}_2)(1 + n\hat{\sigma}^2(s))\hat{S}^{(n)}(s), \tag{4.3.23}$$

where $e_\alpha(c_1, c_2)$ is the upper α fractile in the distribution of

$$\sup_{c_1 \leq x \leq c_2} |W^0(x)|;$$

cf. (4.1.28). This band was first suggested by Hall and Wellner (1980) and it reduces to the well-known Kolmogorov band for completely observed survival data. We will denote it the *Hall–Wellner band* or *HW-band* for short.

As in Section IV.1.3, we may use transformations to try to improve the performance of the confidence bands for small and moderate sample sizes. For the transformations $g(x) = \log(-\log x)$ and $g(x) = \arcsin\sqrt{x}$ we get bands similar to (4.3.20) and (4.3.21), respectively. The transformed EP-bands are proportional to the pointwise intervals and are obtained by replacing $c_{\alpha/2}$ by $d_\alpha(\hat{c}_1, \hat{c}_2)$ in (4.3.20) and (4.3.21). These bands are valid on $[t_1, t_2]$, where $0 < t_1 \leq t_2 \leq t$ are such that $0 < c_1 < c_2 < 1$; cf. (4.1.25). If we, instead, replace $c_{\alpha/2}\hat{\sigma}(s)$ by $n^{-1/2}e_\alpha(\hat{c}_1, \hat{c}_2)(1 + n\hat{\sigma}^2(s))$ in (4.3.20) and (4.3.21), the transformed HW-bands result. For these, one must choose t_1 such that $c_1 > 0$. For the EP-bands, the transformations clearly improve the small sample performance, and the transformed bands are quite satisfactory for right-censored data for sample sizes as low as 25, even with 50% censoring. For HW-bands, a slight improvement is also obtained by using the transformations, but this is of less importance because the band (4.3.23) itself has reasonably good small sample behavior (Borgan and Liestøl, 1990).

By considering other sequences of normalizing factors than $\{\sqrt{n}\}$ as we did in Section IV.1.3, we also here arrive at a collection of $100(1-\alpha)\%$ asymptotic confidence bands for S on $[0, t_2]$ of the form

$$\widehat{S}^{(n)}(s) \pm \hat{\gamma}_n^{-1/2} e_\alpha\left(0, \frac{\hat{\gamma}_n}{1+\hat{\gamma}_n}\right)\hat{\sigma}(t_2)\left(1 + \hat{\gamma}_n \frac{\hat{\sigma}^2(s)}{\hat{\sigma}^2(t_2)}\right)\widehat{S}^{(n)}(s)$$

valid for any (possibly random) sequence $\{\hat{\gamma}_n\}$ satisfying $\hat{\gamma}_n \xrightarrow{P} \gamma$ as $n \to \infty$, for some $\gamma \in (0, \infty)$. Here, the HW-band (4.3.23) corresponds to the choice $\hat{\gamma}_n = n\hat{\sigma}^2(t_2)$.

Next, we illustrate the methods using the melanoma data.

EXAMPLE IV.3.2. Survival with Malignant Melanoma. Kaplan–Meier Estimates with Confidence Limits and Bands

The data were introduced in Example I.3.1, and the Kaplan–Meier plots for both male and female patients were given in Figure I.3.2. We shall here illustrate the calculation of confidence intervals and confidence bands using the data from the male patients. Figure IV.3.1 shows the standard error

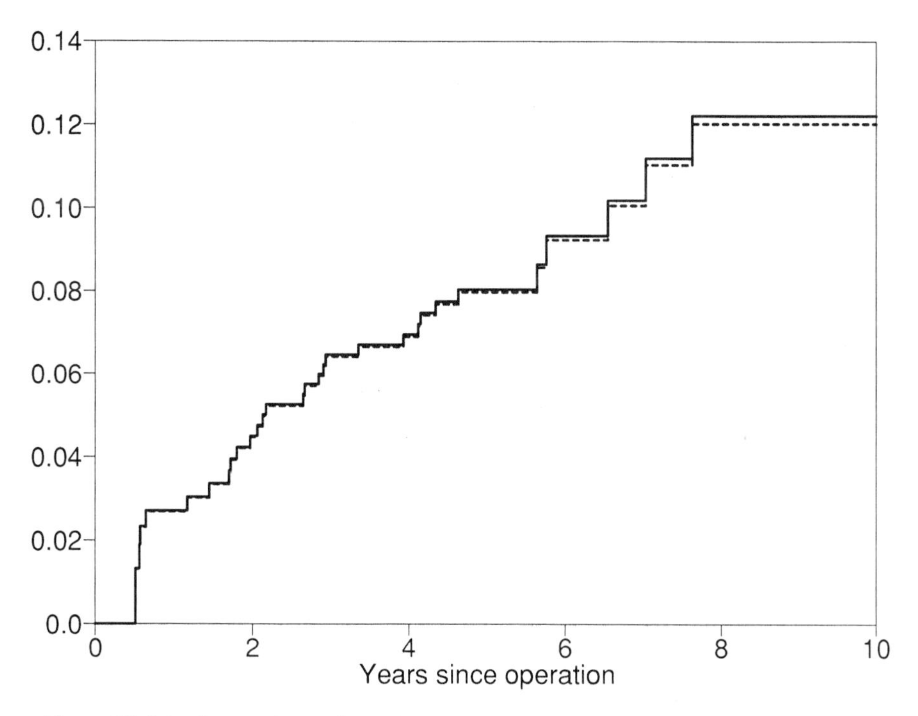

Figure IV.3.1. Comparison of standard error estimates for the survival probability for male patients with malignant melanoma: $\hat{\sigma}(t)$ given by (4.3.6) (- - -) and $\hat{\hat{\sigma}}(t)$ given by (4.3.8) (—).

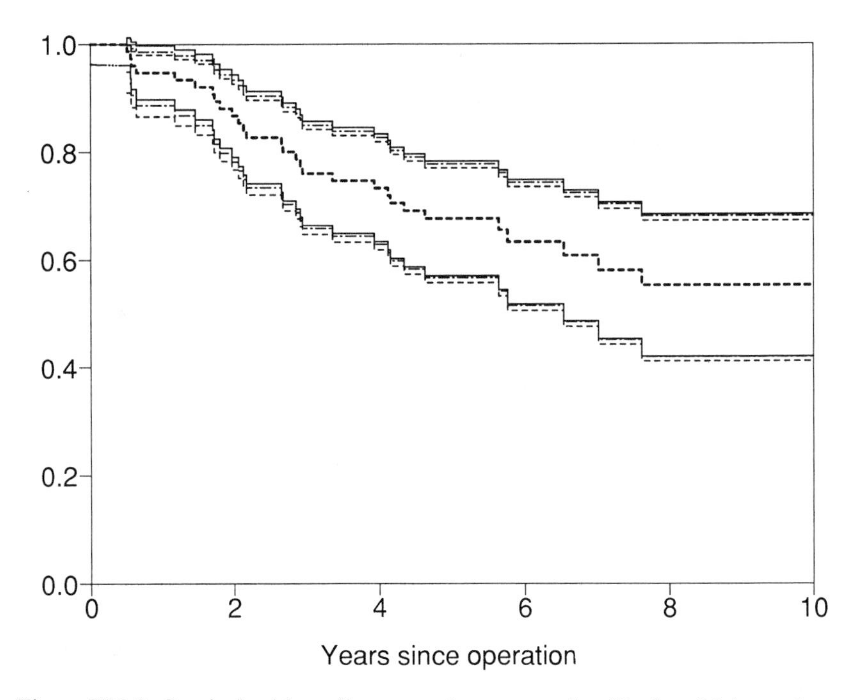

Figure IV.3.2. Survival with malignant melanoma, males: Kaplan–Meier estimate (- - -) with three different pointwise confidence limits: linear (—); log-log-transformed (- - -), and arcsin-transformed (- · · · -).

estimates $\hat{\sigma}(t)$, cf. (4.3.6), and $\hat{\hat{\sigma}}(t)$, cf. (4.3.8), for the male patients. The latter is obviously larger, but the difference is small, and, in the following plots, only the Greenwood estimator $\hat{\hat{\sigma}}(t)$ is used. Figure IV.3.2 shows the Kaplan–Meier estimate for males with approximate 95% pointwise confidence intervals using (4.3.19), (4.3.20), and (4.3.21). It is seen that the three intervals are roughly identical, but the linear confidence limits (4.3.19) include values greater than one just after the first failure time at 185 days (compare Figure IV.1.14). For $t < 185$ days, the same "exact" calculations as in Example IV.1.12 were used, i.e., the lower limits $\underline{S}(t)$ in a $1 - \alpha$ one-sided confidence interval was set to

$$\underline{S}(t) = \alpha^{1/Y(t)}.$$

Figure IV.3.3 shows the simultaneous EP-confidence bands valid for the interval $[1, 7]$ years ($\hat{c}_1 = 0.0539$, $\hat{c}_2 = 0.4442$; cf. Example IV.1.12). The bands based on the log-log-transformation are slightly broader than both those based on the arcsin-transformation and the linear bands. The latter include values greater than one and they are, therefore, not satisfactory to apply. Figure IV.3.4 shows the corresponding HW-bands valid for the same interval. As was the case for the bands based on the Nelson–Aalen estimator, some monotonization is needed for the transformed bands. As for the EP-

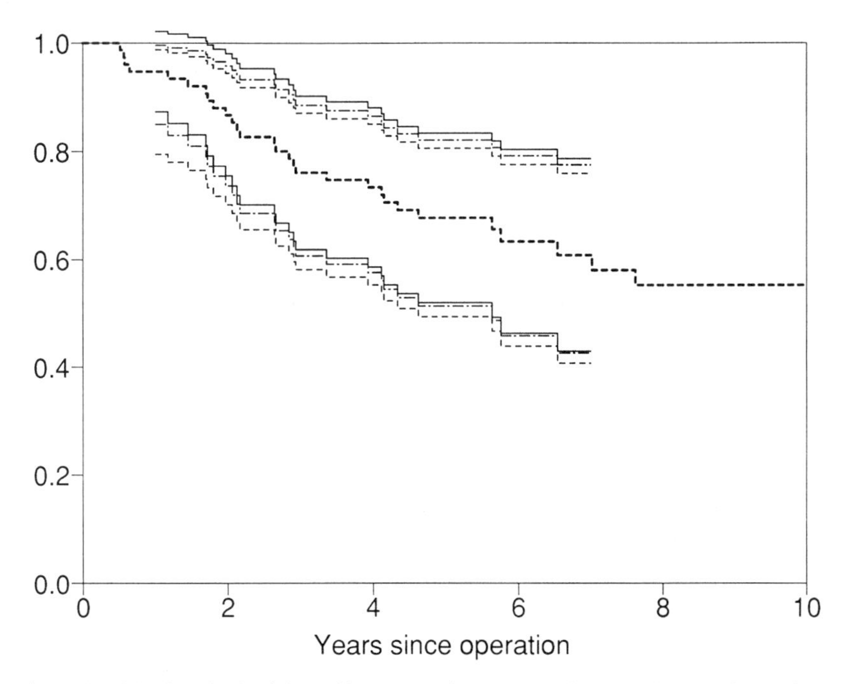

Figure IV.3.3. Survival with malignant melanoma, males: Kaplan–Meier estimate (- - -) with three different EP-confidence bands for the interval [1, 7] years: linear (—); log-log-transformed (- - -), and arcsin-transformed (-····-).

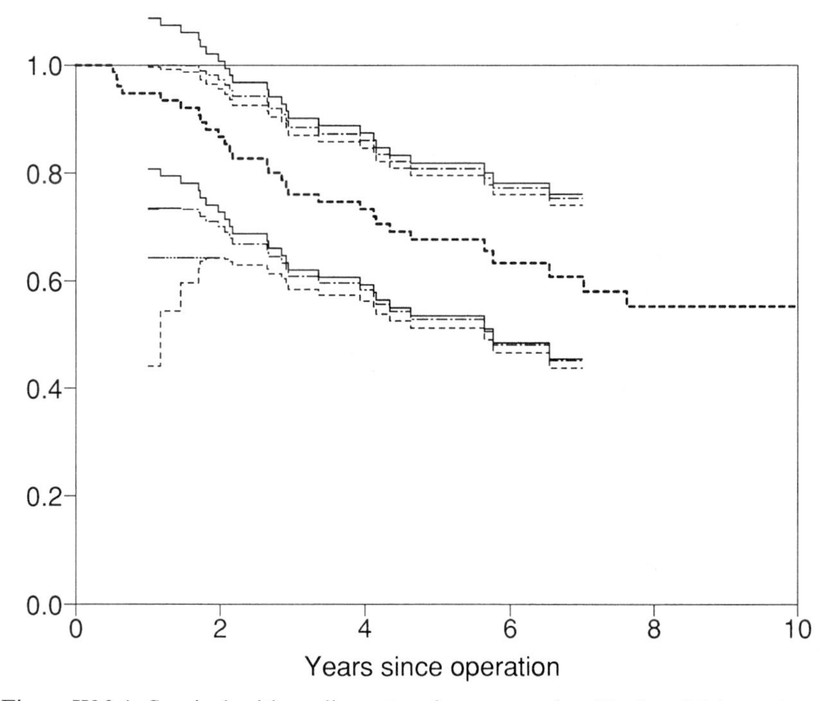

Figure IV.3.4. Survival with malignant melanoma, males: Kaplan–Meier estimate (- - -) with three different HW-confidence bands for the interval [1, 7] years: linear (—); log-log-transformed (- - -), and arcsin-transformed (-····-).

bands, the linear bands include values larger than one and the log-log-transformation gives broader bands than the arcsin-transformation. □

EXAMPLE IV.3.3. The Expected and Corrected Survival Functions. Survival with Malignant Melanoma

We now return to the additive hazard model considered in Example IV.1.11. Recall that the estimated integrated excess mortality $\hat{\Gamma}(t)$ was defined as

$$\hat{\Gamma}(t) = \int_0^t J(u)Y(u)^{-1}\,\mathrm{d}N(u) - \int_0^t \sum_{i=1}^n \mu_i(u)\frac{Y_i(u)}{Y(u)}\,\mathrm{d}u;$$

cf. (4.1.20). Taking product-integrals (cf. Section II.6) on both sides of this equation, we get

$$\prod_0^t (1 - \mathrm{d}\hat{\Gamma}(u)) = \hat{S}(t) \Big/ \exp\left\{-\int_0^t \sum_{i=1}^n \mu_i(u)\frac{Y_i(u)}{Y(u)}\,\mathrm{d}u\right\},$$

where $\hat{S}(t)$ is the Kaplan–Meier estimator. The denominator on the right-hand side is a continuous-time version of an *expected survival function* (Andersen and Væth, 1989) which is usually considered in connection with grouped survival data (Ederer et al., 1961). The left-hand side is, similarly, a continuous-time version of what is sometimes called a *corrected "survival function"* in spite of the fact that it needs neither be less than unity nor nonincreasing. The classical interpretation of the corrected survival function as the survival probability if deaths from other causes than the one under study had been eliminated is, however, appropriate as long as $\gamma(t) \geq 0$; cf. the discussion in connection with the competing risks model in Example IV.4.1. Large sample properties of the corrected survival function may be derived as in Section IV.3.2 under the conditions discussed in Example IV.1.11.

Returning now to the melanoma data introduced in Example I.3.1, Figure IV.3.5 shows the Kaplan–Meier estimate for these data based on all 71 deaths together with the expected and the corrected survival functions. The latter is actually increasing and, thus, greater than 1 before the first death time at $t = 10$ days.

For comparison, the Kaplan–Meier estimator based only on the 57 deaths from malignant melanoma is shown in Figure IV.3.6 together with the corrected survival curve. One should here note the close agreement between the two (compare Figure IV.1.13). □

The next example illustrates the necessary modifications for left-truncated (and right-censored) data.

EXAMPLE IV.3.4. Diabetics in the County of Fyn. Estimation of Conditional Survival Functions from Left-Truncated and Right-Censored Survival Data

The problem and the data were presented in Example I.3.2.

The data on mortality of diabetics in Fyn county are *left-truncated*, and,

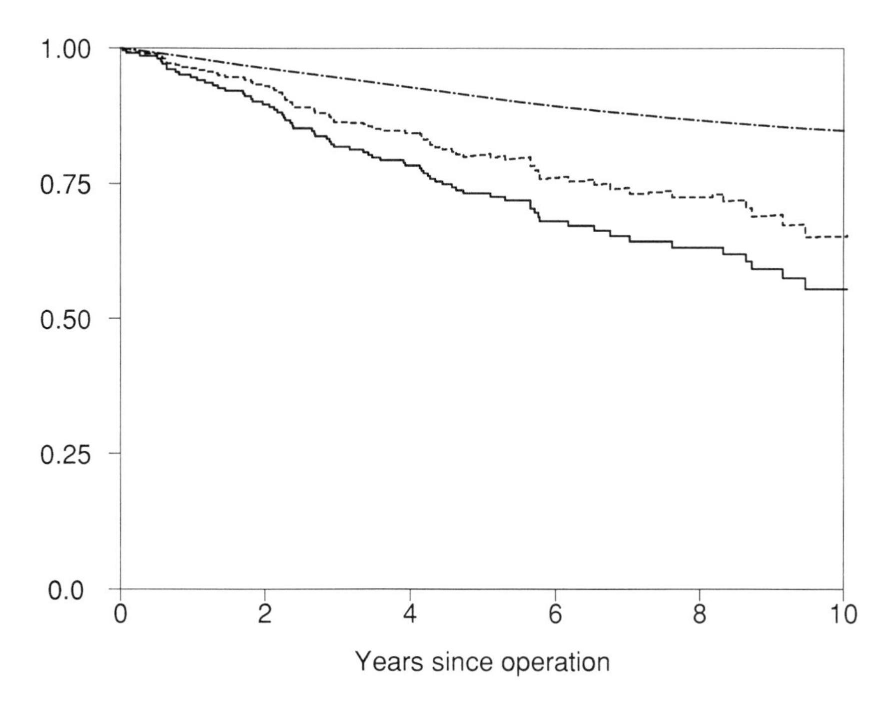

Figure IV.3.5. Survival with malignant melanoma: Kaplan–Meier estimate based on all 71 deaths (—), expected survival curve (-·-·-), and corrected survival curve (- - -).

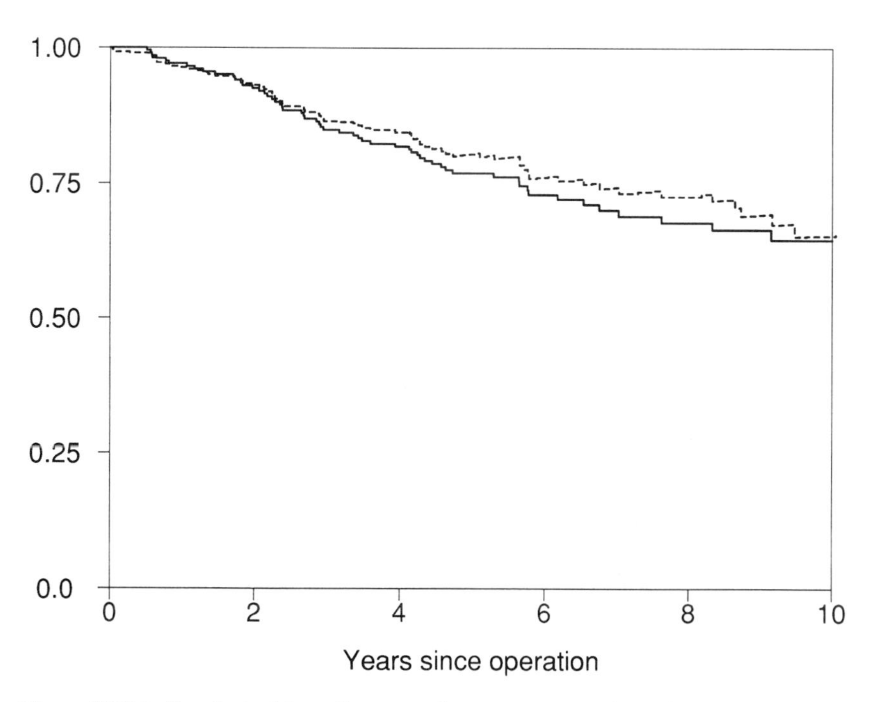

Figure IV.3.6. Survival with malignant melanoma: corrected survival curve (- - -) and Kaplan–Meier estimate (—) based on the 57 deaths from the disease.

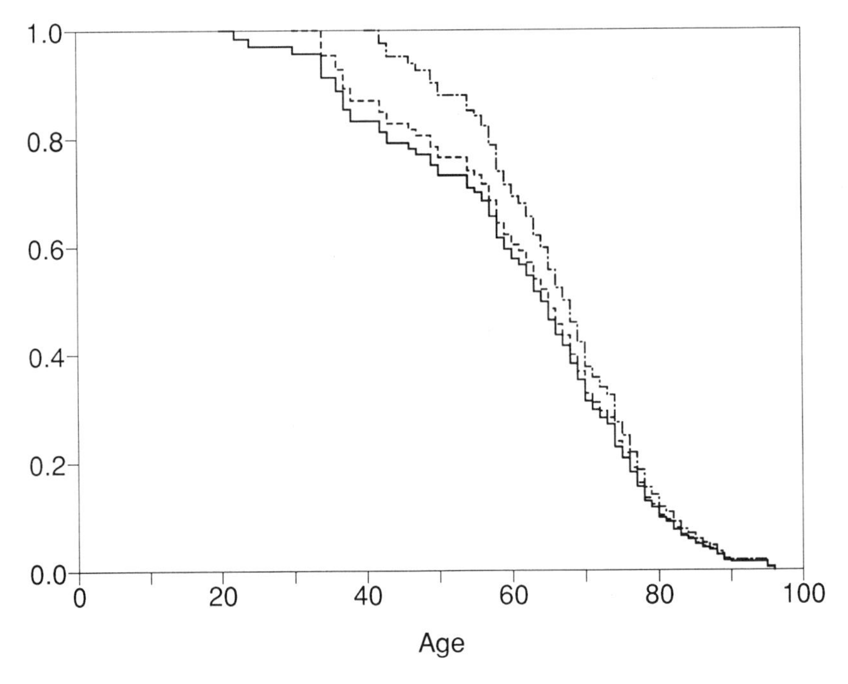

Figure IV.3.7. Estimated conditional survival probability for female diabetics in the county of Fyn, 1973–81, given survival to age 20 years (—), 30 years (- - -), and 40 years (- · · · · -).

as mentioned before (cf. Example IV.1.3), there is no information in the data on the death intensity for age intervals with empty risk sets (here: for ages ≤ 2 years). Moreover, the information is subject to large random variations when the risk sets are small, as will often be the case (such as it is here) toward the left end of the age scale.

The (product) integration of the local description of the mortality experience is, therefore, a more delicate issue under left-truncation. What can be usefully illustrated is the *conditional* survival distribution, given survival until some fixed age. Several curves will usually be necessary, there being nothing canonical about any particular age attained. Figure IV.3.7 shows estimated conditional survival functions given survival to ages 20 years, 30 years, and 40 years, respectively.

Confidence limits and bands may be obtained as before, interpreting $Y(s)$ in (4.3.8) as the number at risk at time (age) s. Figure IV.3.8 shows the estimated conditional survival function, given survival to age 30 years, with pointwise 95% confidence limits and a 95% EP confidence band, both based on the log-log-transformation. □

In the next example we illustrate how left-censoring may be handled by reversing time.

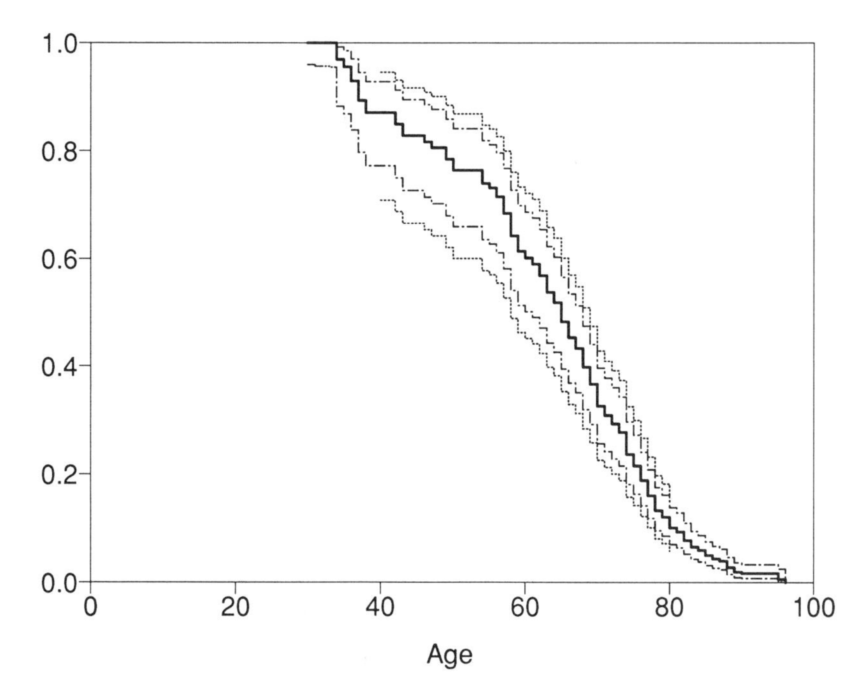

Figure IV.3.8. Estimated conditional survival probability for female diabetics in the county of Fyn, 1973–81, given survival to age 30 years (—), with pointwise 95% confidence limits (- · · · -) and 95% EP confidence bands over the age interval [40 years, 80 years] (· · ·) both based on the log-log-transformation.

EXAMPLE IV.3.5. Baboon Descent. Reversing Time for Left-Censored Data

The data and the problem were presented in Example I.3.7. In that example, times of arrival were given of the observer at the site of a baboon troop for 152 days as well as time of descent of the median baboon in the troop ("descent time" for short), if this event had not already happened when the observer arrived. For day i, denote by U_i the time of arrival of the observer and by X_i the descent time. Then U_i and $U_i \vee X_i$ are observed.

As mentioned in Example III.4.1, the descent times are usefully modelled as *left-censored*. However, by inverting time and studying 24 hours $- U_i$, 24 hours $- X_i$ ordinary right-censored data are obtained for which an ordinary Kaplan–Meier estimator may be calculated. The resulting estimate is given in Figure IV.3.9 in the original time scale.

Note that in this case (as will often be the case for left-censoring), the value of the censoring variable U_i is *always* observed, also on the days where X_i is observed.

Csörgő and Horváth (1985) went on to fit a parametric model to these data: a normal distribution with mean 8 h and 7 min and standard deviation 53 min. They also developed Kolmogorov–Smirnov- and Cramér–von

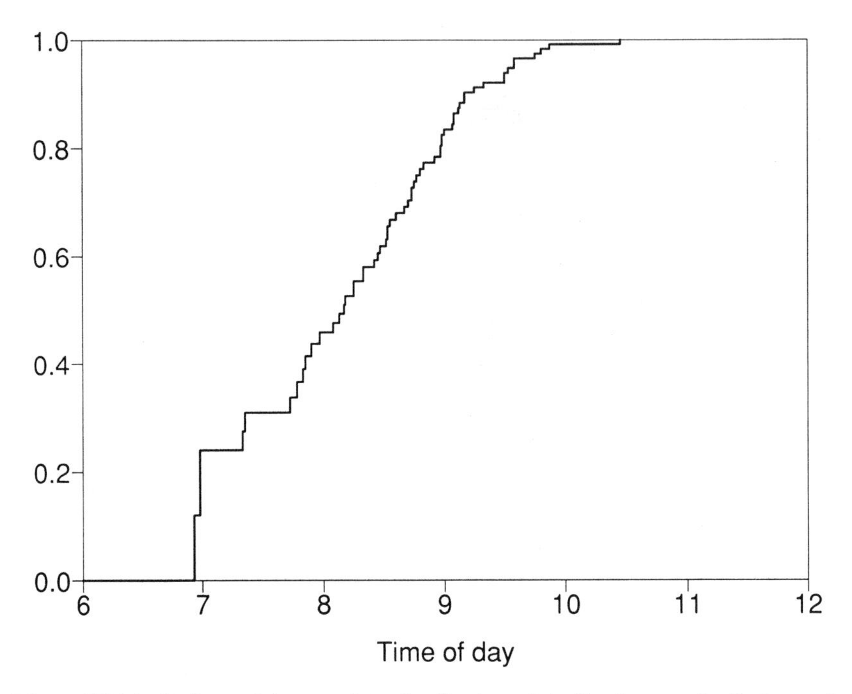

Figure IV.3.9. Estimated descent time distribution of the "main troop" of baboons in the Amboseli reserve, Kenya.

Mises-type goodness-of-fit statistics which did not indicate deviation from the hypothesis of a normal distribution. □

IV.3.4. Estimation of Quantiles and Other Functionals

Let us finally, in this section, mention how certain functionals of the survival function S, or equivalently of the distribution function $F = 1 - S$, may be estimated. We will mainly be interested in estimation of the quantiles because these are of prime importance in survival studies. But estimation of the mean, as well as some other functionals (to be described later), are also considered.

By a functional ϕ, we here understand a function from the space of distribution functions to the real line or to another function space. We are interested in estimating its value $\phi(F)$ when evaluated at the true distribution function. The pth quantile of F is, in particular, given by

$$F^{-1}(p) = \inf\{x\colon F(x) \geq p\},$$

i.e., it corresponds to ϕ being the right-continuous inverse of the distribution function evaluated at p. On the other hand, the mean is given as the functional

$$\int_0^\infty u\,\mathrm{d}F(u) = \int_0^\infty (1 - F(u))\,\mathrm{d}u,$$

whereas the expected time lived in the time interval $[0, t]$ is $\int_0^t (1 - F(u))\,\mathrm{d}u$. We will estimate $\phi(F)$ in the obvious way by $\phi(\hat{F}^{(n)})$, where $\hat{F}^{(n)} = 1 - \hat{S}^{(n)}$ with $\hat{S}^{(n)}$ being the Kaplan–Meier estimator. Because the Kaplan–Meier estimator is a nonparametric maximum likelihood estimator (as discussed toward the end of Section IV.3.1), the same will hold for the estimators considered in this subsection.

For some particular functionals, like the expected time lived in a certain time interval, large sample properties of $\phi(\hat{F}^{(n)})$ are almost immediate (cf. Example IV.3.8). For others, they may be derived using the theory of martingales and stochastic integrals. This was illustrated in connection with Theorem IV.3.2 where we were able to derive the properties of the Kaplan–Meier estimator considered as a functional of the Nelson–Aalen estimator $\hat{A}$ using martingale techniques.

In general, however, such techniques are not available. It is then often possible to study the properties of $\phi(\hat{F}^{(n)})$ using the functional delta-method reviewed in Section II.8: Provided that ϕ is (tangentially) compactly differentiable, we may approximate $\sqrt{n}(\phi(\hat{F}^{(n)}) - \phi(F))$ by $\mathrm{d}\phi(F)\cdot\sqrt{n}(\hat{F}^{(n)} - F)$, where $\mathrm{d}\phi(F)$ is the derivative of ϕ at F and it acts on $\sqrt{n}(\hat{F}^{(n)} - F)$ in a linear way; in fact, the two are asymptotically equivalent (Theorems II.8.1 and II.8.2). Therefore, by Theorem IV.3.2,

$$\sqrt{n}(\phi(\hat{F}^{(n)}) - \phi(F)) \xrightarrow{\mathscr{D}} \mathrm{d}\phi(F)\cdot Z \tag{4.3.24}$$

as $n \to \infty$, where $Z = (1 - F)U$, with U being the Gaussian martingale defined in Theorem IV.3.2.

It should be realized that the functional delta-method provides a general way of deriving large sample properties of functionals and that it will also often give the desired results in a more direct manner than the martingale approach when the latter is applicable. This was illustrated in the proof of Theorem IV.3.2 where we also used the functional delta-method to derive the large sample properties of the Kaplan–Meier estimator. In this connection, we, furthermore, note that the reason why the functional delta-method is not needed to derive the asymptotic distribution of the estimated expected time lived in the time interval $[0, t]$ (Example IV.3.8) is that the functional $\int_0^t (1 - F(u))\,\mathrm{d}u$ is linear, and, therefore, the need to approximate it by a linear functional (which is the whole purpose of the functional delta-method) does not arise.

We then turn to a closer study of the estimation of quantiles.

EXAMPLE IV.3.6. Estimation of Median and Other Quantiles. Lower Quartile in the Life Distribution of Male Patients with Malignant Melanoma

Let $0 < p < 1$, and define

$$\xi_p = F^{-1}(p) = \inf\{x\colon F(x) \geq p\} = \inf\{x\colon S(x) \leq 1 - p\}$$

as the pth quantile of the distribution function F. Because we have assumed that F is absolutely continuous with density $f = \alpha S$, we have $F(\xi_p) = p$, but this relation does not define ξ_p uniquely for time intervals where f vanishes. The quantiles are estimated by taking the right-continuous inverse of the empirical distribution function $\hat{F}^{(n)}$. Thus, we have

$$\hat{\xi}_p^{(n)} = \inf\{x\colon \hat{F}^{(n)}(x) \geq p\} = \inf\{x\colon \hat{S}^{(n)}(x) \leq 1 - p\}. \tag{4.3.25}$$

It is a simple exercise to show that $\hat{\xi}_p^{(n)}$ is consistent using Theorem IV.3.1.

To study the asymptotic distribution of the empirical quantile, we write $\xi_p = \phi(F)$ and $\hat{\xi}_p^{(n)} = \phi(\hat{F}^{(n)})$, where ϕ is the function from the space of distribution functions to the real line defined by

$$\phi(G) = G^{-1}(p) = \inf\{x\colon G(x) \geq p\}.$$

By Proposition II.8.4, we then have that ϕ is (tangentially) compactly differentiable at F with the derivative given by

$$\mathrm{d}\phi(F)\cdot h = -\frac{h(F^{-1}(p))}{f(F^{-1}(p))}$$

provided that $f(\xi_p) > 0$. It follows by (4.3.24) that

$$\sqrt{n}(\hat{\xi}_p^{(n)} - \xi_p) \xrightarrow{\mathscr{D}} -\frac{(1-p)U(F^{-1}(p))}{f(F^{-1}(p))}$$

as $n \to \infty$, so that

$$\sqrt{n}(\hat{\xi}_p^{(n)} - \xi_p) \xrightarrow{\mathscr{D}} \mathscr{N}\left(0, \frac{(1-p)^2\sigma^2(\xi_p)}{f^2(\xi_p)}\right), \tag{4.3.26}$$

where σ^2 is defined in (4.3.14).

To estimate the asymptotic variance of the empirical quantile, we insert one of the estimates (4.3.6) or (4.3.8) for σ^2 and (4.3.25) for ξ_p in the expression for the variance in (4.3.26). In addition, however, we have to estimate the density $f = -S'$. This may be done using the kernel function estimator

$$\hat{f}^{(n)}(t) = -b_n^{-1}\int_{\mathscr{T}} K\left(\frac{t-u}{b_n}\right)\mathrm{d}\hat{S}^{(n)}(u).$$

When K is the uniform kernel function (cf. Example IV.2.1), this amounts to estimating $f(t)$ by

$$\hat{f}^{(n)}(t) = \frac{1}{2b_n}(\hat{S}^{(n)}(t - b_n) - \hat{S}^{(n)}(t + b_n)).$$

For a particular illustration, we now consider the data on survival with malignant melanoma introduced in Example I.3.1. We restrict our attention to the male patients. For these, the Kaplan–Meier estimate for the survival distribution was shown in Figure IV.3.2 together with pointwise confidence intervals. We now find using (4.3.25) that $\xi_{0.25}$, i.e., the lower quartile in the life distribution of male melanoma patients, is estimated as $\hat{\xi}_{0.25} = 3.36$ years. To estimate its standard deviation, we first find $\hat{f}(3.36) = 0.0679$ using (rather arbitrarily) a kernel estimator with uniform kernel and bandwidth $b = 1.5$ years. By (4.3.26), we, therefore, find an estimated standard deviation of

$$\frac{0.75\hat{\sigma}(3.36)}{\hat{f}(3.36)} = \frac{0.75 \cdot 0.0671}{0.0679} = 0.741.$$

An approximate 95% confidence interval for $\xi_{0.25}$ is, therefore, $3.36 \pm 1.96 \cdot 0.741$, i.e., from 1.91 years to 4.81 years.

For the purpose of determining a confidence interval for a quantile ξ_p, it is, however, better to apply the idea of Brookmeyer and Crowley (1982a). Adapted to a transformed version of the Kaplan–Meier estimator, we then take as an approximate $100(1-\alpha)\%$ confidence interval for ξ_p all values ξ_p^0 which satisfy

$$\frac{|g(\hat{S}(\xi_p^0)) - g(1-p)|}{|g'(\hat{S}(\xi_p^0))|\,\hat{S}(\xi_p^0)\hat{\sigma}(\xi_p^0)} \le c_{\alpha/2},$$

i.e., all hypothesized values ξ_p^0 of ξ_p which are not rejected when testing the null hypothesis $\xi_p = \xi_p^0$ against the alternative hypothesis $\xi_p \neq \xi_p^0$ at the α level based on the asymptotic normality of $g(\hat{S}(t))$. Here, g is some given function like $g(x) = \log(-\log x)$ or $g(x) = \arcsin\sqrt{x}$; cf. Section IV.3.3. [Of course, the choice $g(x) = x$ corresponds to the case without transformation considered by Brookmeyer and Crowley (1982a).]

The advantage of such a procedure is that no estimation of the density $f(t)$ is needed. Note that the interval can be read directly from the lower and upper pointwise confidence limits for the survival distribution in exactly the same manner as $\hat{\xi}_p$ can be read from the Kaplan–Meier curve itself. This procedure is illustrated in Figure IV.3.10 based on the log-log-transformation, and the resulting approximate 95% confidence interval for $\xi_{0.25}$ for male melanoma patients is found to be from 2.13 years to 5.76 years. □

It is possible to extend the main result of the previous example to a result concerning the weak convergence of the whole quantile process (away from its endpoints). This will be considered in the next example.

EXAMPLE IV.3.7. The Quantile Process and Estimation of the Interquartile Range

We assume that the density f is continuous and positive on $[t_1 - \varepsilon, t_2 + \varepsilon]$ for given time points $0 < t_1 < t_2 < \tau$ and a given $\varepsilon > 0$, and consider the mapping $\phi(G) = G^{-1}$ as a mapping from the space $D[t_1, t_2]$ to the space $D[p_1, p_2]$, where $p_i = F(t_i)$ for $i = 1, 2$. By Proposition II.8.5, this mapping is (tangentially) compactly differentiable with derivative at F given by

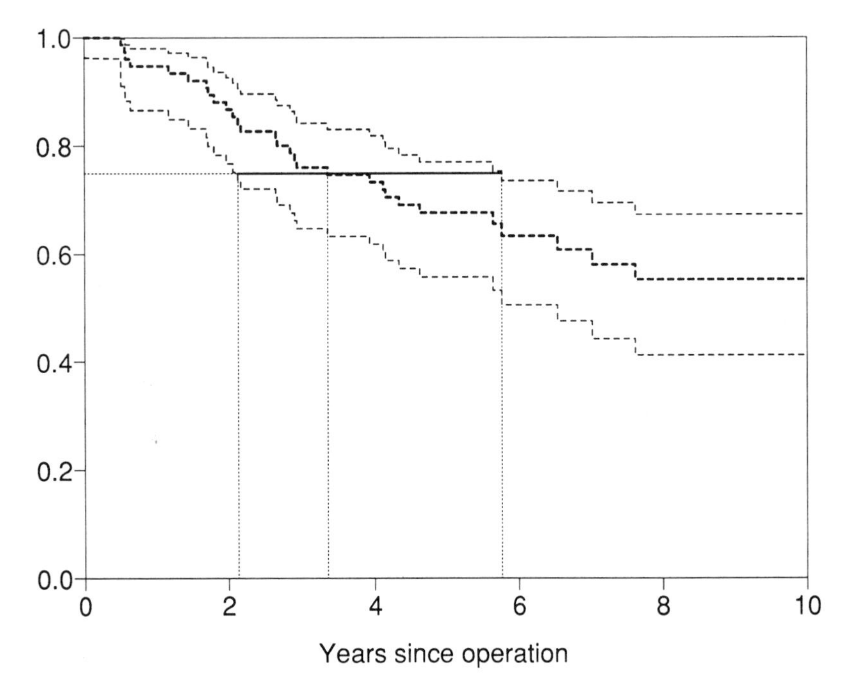

Figure IV.3.10. Survival with malignant melanoma, males: Estimation of lower quartile of the life distribution with approximate 95% confidence limits.

$$\mathrm{d}\phi(F)\cdot h = -\frac{h \circ F^{-1}}{f \circ F^{-1}}.$$

Therefore, by Theorem II.8.2, we have that

$$\sqrt{n}(\hat{F}^{(n)-1} - F^{-1}) \xrightarrow{\mathscr{D}} -\frac{(1-\imath)(U \circ F^{-1})}{f \circ F^{-1}}, \tag{4.3.27}$$

on $D[p_1, p_2]$ as $n \to \infty$, where $\imath$ denotes the identity mapping and U, as above, is the Gaussian martingale defined in Theorem IV.3.2.

In particular, we may use this result to derive the asymptotic distribution of the estimator

$$\hat{R}^{(n)} = \hat{\xi}^{(n)}_{0.75} - \hat{\xi}^{(n)}_{0.25}$$

for the interquartile range $\rho = \xi_{0.75} - \xi_{0.25}$. Note that by (4.3.27)

$$\sqrt{n}\left\{\begin{pmatrix}\hat{\xi}_{0.75}\\ \hat{\xi}_{0.25}\end{pmatrix} - \begin{pmatrix}\xi_{0.75}\\ \xi_{0.25}\end{pmatrix}\right\}$$

$$\xrightarrow{\mathscr{D}} \mathscr{N}\left[\mathbf{0}, \begin{pmatrix} \dfrac{\sigma^2(\xi_{0.75})}{16f^2(\xi_{0.75})} & \dfrac{3\sigma^2(\xi_{0.25})}{16f(\xi_{0.75})f(\xi_{0.25})} \\ \dfrac{3\sigma^2(\xi_{0.25})}{16f(\xi_{0.75})f(\xi_{0.25})} & \dfrac{9\sigma^2(\xi_{0.25})}{16f^2(\xi_{0.25})} \end{pmatrix}\right].$$

Thus, it is seen that

$$\sqrt{n}(\hat{R}^{(n)} - \rho) \xrightarrow{\mathscr{D}} \mathcal{N}\left(0, \frac{1}{16}\left\{\frac{\sigma^2(\xi_{0.75})}{f^2(\xi_{0.75})} + \frac{9\sigma^2(\xi_{0.25})}{f^2(\xi_{0.25})} - \frac{6\sigma^2(\xi_{0.25})}{f(\xi_{0.75})f(\xi_{0.25})}\right\}\right)$$

as $n \to \infty$. The asymptotic variance of $\hat{R}^{(n)}$ may be estimated as described just below (4.3.26). □

In survival studies, one will more often estimate the median lifetime than the mean lifetime, partly because it is hard to get reliable estimates of the latter. Nevertheless, let us also briefly consider how the mean, as well as the expected time lived in a restricted time interval, may be estimated. As noted above, no use of the functional delta-method will be needed to estimate the expected time lived in a restricted time interval because this is itself a linear functional.

EXAMPLE IV.3.8. Estimation of Mean Survival Time

Before we turn to the estimation of the mean itself, we first consider the simpler problem of estimating the expected time lived in the restricted time interval $[0, t]$, i.e.,

$$\mu_t = \int_0^t (1 - F(u))\,\mathrm{d}u = \int_0^t S(u)\,\mathrm{d}u,$$

for a fixed $t < \tau$. This is estimated by the area under the Kaplan–Meier curve

$$\hat{\mu}_t^{(n)} = \int_0^t \hat{S}^{(n)}(u)\,\mathrm{d}u,$$

and it follows immediately by Theorem IV.3.1 that $\hat{\mu}_t^{(n)}$ is consistent. Moreover, by Theorem IV.3.2,

$$\begin{aligned}
\sqrt{n}(\hat{\mu}_t^{(n)} - \mu_t) &= \int_0^t \sqrt{n}(\hat{S}^{(n)}(u) - S(u))\,\mathrm{d}u \\
&\xrightarrow{\mathscr{D}} -\int_0^t S(v)U(v)\,\mathrm{d}v \\
&= \int_0^t U(v)\,\mathrm{d}\bar{\mu}_t(v) \\
&= -\int_0^t \bar{\mu}_t(v)\,\mathrm{d}U(v)
\end{aligned}$$

as $n \to \infty$, where

$$\bar{\mu}_t(v) = \mu_t - \mu_v = \int_v^t S(u)\,\mathrm{d}u.$$

Thus, we have

$$\sqrt{n}(\hat{\mu}_t^{(n)} - \mu_t) \xrightarrow{\mathscr{D}} \mathcal{N}\left(0, \int_0^t \bar{\mu}_t^2(v)\,\mathrm{d}\sigma^2(v)\right).$$

The asymptotic variance of $\hat{\mu}_t^{(n)}$ may be estimated by inserting the proper estimators in the expression for the variance.

Gill (1983a) has discussed conditions under which this result may be extended to the unrestricted mean. We consider the simple random censorship model and adopt the notation of Examples IV.1.1 and IV.3.1. Denote by G the censoring distribution and let $H = 1 - S(1 - G)$ be the distribution of the $\tilde{X}_i$. We assume that the mean

$$\mu = \mu_\infty = \int_0^\infty S(u)\,du$$

and

$$\int_0^{\tau_H} (\mu - \mu_v)^2\,d\sigma^2(v)$$

are both finite, where τ_H is the (possibly infinite) time $\tau_H = \sup\{t: H(t) < 1\}$. As an estimator for μ Gill (1983a) suggested

$$\hat{\mu}^{(n)} = \hat{\mu}^{(n)}_{\tilde{X}_{(n)}} = \int_0^{\tilde{X}_{(n)}} \hat{S}^{(n)}(u)\,du,$$

and under the supplementary assumption

$$\sqrt{n}(\mu - \mu_{\tilde{X}_{(n)}}) \xrightarrow{\mathrm{P}} 0,$$

he proved that

$$\sqrt{n}(\hat{\mu}^{(n)} - \mu) \xrightarrow{\mathscr{D}} \mathscr{N}\left(0, \int_0^{\tau_H} (\mu - \mu_v)^2\,d\sigma^2(v)\right)$$

as $n \to \infty$. □

In reliability theory, a number of life distribution classes have been defined, e.g., distributions may be classified as having increasing failure rate (IFR), to be "new better than used" (NBU) or as having "decreasing mean residual life" (DMRL). According to the verbal descriptions, a life distribution is said to be NBU if

$$S(x + y) \leq S(x)S(y)$$

for all $x, y \geq 0$, and it is said to be DMRL if

$$\varepsilon(x) = \int_0^\infty \frac{S(x+t)}{S(x)}\,dt$$

is decreasing in x. As measures for departure from these two classes of life distributions, the following functionals have been suggested: For the NBU class

$$\begin{aligned}\gamma(F) &= \int_0^\infty \int_0^\infty \{S(x)S(y) - S(x+y)\}\,dF(x)\,dF(y)\\ &= \frac{1}{4} - \int_0^\infty \int_0^\infty S(x+y)\,dF(x)\,dF(y),\end{aligned}$$

and for the DMRL class

$$\nu(F) = \iint_{x<y} S(x)S(y)\{\varepsilon(x) - \varepsilon(y)\}\,\mathrm{d}F(x)\,\mathrm{d}F(y).$$

To test whether a life distribution belongs to one of these classes, one may, therefore, consider the statistics $\gamma(\hat{F})$ and $\nu(\hat{F})$, and their asymptotic distribution (properly normalized) may be derived using the functional delta-method. We will not go further into this, however. For a review on the above (and other) nonparametric concepts and methods in reliability, see Hollander and Proschan (1984).

Finally, in this subsection we consider two slightly different examples of functionals. The first of these gives an example of the estimation of so-called actuarial values.

Example IV.3.9. Estimation of Actuarial Values

In this example, we will consider a simple life insurance contract to illustrate how the basic concepts of life insurance fit into our framework. Further illustrations are provided in Examples IV.4.3 and VII.2.11.

An insurance company issues a life insurance policy to an x-year-old individual. The conditions of the contract are as follows: If the individual dies before age $x + m$, the spouse (or other relatives) of the insured receives a prespecified amount of money, B say, from the company. If the individual survives to age $x + m$, the company pays nothing. For this benefit, the individual has to pay a premium $\pi\,\mathrm{d}t$ to the company in each time interval $(t, t + \mathrm{d}t)$ after the issue of the policy as long as he or she is alive and below age $x + m$ years of age.

Let T denote the number of further years lived by the individual after the issue of the policy at age x. Denote, furthermore, by $S_x(t)$ the probability that T exceeds t years, i.e.,

$$S_x(t) = \frac{S(x+t)}{S(x)} = \exp\left\{-\int_x^{x+t} \alpha(u)\,\mathrm{d}u\right\}$$

is the conditional probability of surviving age $x + t$ given survival to age x. Here, $\alpha(s)$ is the death intensity for a s-year-old individual of the relevant sex. (For simplicity, we assume that the death intensity only depends on the age and sex of the individual and not on the time elapsed since the signing of the contract.) Finally, we let i be the annual interest rate and define the discount factor by $v = 1/(1 + i)$.

Then the cash value or discounted value of the benefit received by the spouse (or other relatives) of the insured is $v^T BI(T \leq m)$. The expected value of this random variable or its *actuarial value* is $B\bar{A}^{1}_{x:\overline{m}|}$, where

$$\bar{A}^{1}_{x:\overline{m}|} = \int_0^m v^t S_x(t)\alpha(x+t)\,\mathrm{d}t \tag{4.3.28}$$

using standard actuarial notation. On the other hand, the cash value of the premiums paid by the insured to the company is $\int_0^m v^t \pi I(T > t)\,dt$, and its actuarial value is $\pi \bar{a}_{x:\overline{m}|}$ with

$$\bar{a}_{x:\overline{m}|} = \int_0^m v^t S_x(t)\,dt. \tag{4.3.29}$$

Disregarding the company's expenses for simplicity in this example, the equivalence principle of insurance states that the actuarial values of the benefits and the premiums paid should be equal. Thus, the premium rate π is determined by the relation

$$B\bar{A}^1_{x:\overline{m}|} = \pi \bar{a}_{x:\overline{m}|},$$

which yields

$$\pi = B\frac{\bar{A}^1_{x:\overline{m}|}}{\bar{a}_{x:\overline{m}|}}.$$

To determine the premium rate π, the company, therefore, needs to estimate the actuarial values (4.3.28) and (4.3.29). In practical life insurance, this is usually accomplished assuming that the death intensity is of the Gompertz–Makeham form (cf. Examples VI.1.4 and VI.2.2). Following Præstgaard (1991), we will, however, illustrate how nonparametric estimation of the actuarial values $\bar{A}^1_{x:\overline{m}|}$ and $\bar{a}_{x:\overline{m}|}$ may be carried out. We concentrate on weak convergence of the marginal distributions of the estimators of the two actuarial values. Simultaneous weak convergence of the two may, however, also be derived in a similar manner.

As remarked toward the end of Section IV.3.2, $S_x(t)$ may be estimated by

$$\hat{S}_x^{(n)}(t) = \prod_{(x,x+t]} \left(1 - \frac{\Delta N^{(n)}(s)}{Y^{(n)}(s)}\right),$$

and its asymptotic distribution is given by an obvious modification of Theorem IV.3.2. Now we estimate $\bar{a}_{x:\overline{m}|}$ by

$$\hat{\bar{a}}_{x:\overline{m}|} = \int_0^m v^t \hat{S}_x^{(n)}(t)\,dt.$$

Like the expected time lived in a restricted time interval (Example IV.3.8), the actuarial value (4.3.29) is a linear functional. Therefore, no need for the functional delta-method arises, and we immediately have that

$$\sqrt{n}(\hat{\bar{a}}_{x:\overline{m}|} - \bar{a}_{x:\overline{m}|}) = \int_0^m v^t \sqrt{n}(\hat{S}_x^{(n)}(t) - S_x(t))\,dt$$

$$\xrightarrow{\mathscr{D}} -\int_0^m v^t S_x(t) U_x(t)\,dt, \tag{4.3.30}$$

where, by Theorem IV.3.2, U_x is a zero mean Gaussian martingale with covariance function $\operatorname{cov}(U_x(s), U_x(t)) = \sigma_x^2(s \wedge t)$ given by

$$\sigma_x^2(t) = \int_x^{x+t} \{\alpha(u)/y(u)\}\,du.$$

We introduce

$$\bar{a}_m(t) = \bar{a}_{x:\overline{m}|} - \bar{a}_{x:\overline{t}|} = \int_t^m v^u S_x(u)\,du,$$

and note that the right-hand side of (4.3.30) may be written

$$\int_0^m U_x(t)\,d\bar{a}_m(t) = -\int_0^m \bar{a}_m(t)\,dU_x(t).$$

Thus, we have proved that

$$\sqrt{n}(\hat{\bar{a}}_{x:\overline{m}|} - \bar{a}_{x:\overline{m}|}) \xrightarrow{\mathscr{D}} \mathcal{N}\left(0, \int_0^m (\bar{a}_{x:\overline{m}|} - \bar{a}_{x:\overline{t}|})^2\,d\sigma_x^2(t)\right)$$

as $n \to \infty$.

We then turn to the estimation of $\bar{A}^1_{x:\overline{m}|}$. By (4.3.28), this is estimated by

$$\hat{\bar{A}}^1_{x:\overline{m}|} = \int_0^m v^t \hat{S}_x^{(n)}(t)\,d\hat{A}_x^{(n)}(t),$$

where

$$\hat{A}_x^{(n)}(t) = \hat{A}^{(n)}(x+t) - \hat{A}^{(n)}(x)$$

with $\hat{A}^{(n)}$ being the Nelson–Aalen estimator. To derive the asymptotic distribution of this estimator, we first note that the actuarial value (4.3.28) may be written as the functional

$$\phi(V_x, A_x) = \int_0^m V_x(t)\,dA_x(t),$$

where

$$V_x(t) = v^t S_x(t)$$

and

$$A_x(t) = \int_x^{x+t} \alpha(u)\,du.$$

Moreover, its estimator takes the form $\phi(\hat{V}_x^{(n)}, \hat{A}_x^{(n)})$, where

$$\hat{V}_x^{(n)} = v^t \hat{S}_x^{(n)}(t).$$

Since the integrated intensity is assumed to be bounded on any compact subinterval of $\mathscr{T}$, it follows by Proposition II.8.6 that ϕ is compactly differentiable at (V_x, A_x) with derivative

$$d\phi(V_x, A_x)\cdot(h, k) = \int_0^m h\,dA_x + \int_0^m V_x\,dk.$$

To use this result to determine the asymptotic distribution of the estimated actuarial value, we need the joint asymptotic distribution of $\sqrt{n}(\hat{V}_x^{(n)} - V_x)$ and $\sqrt{n}(\hat{A}_x^{(n)} - A_x)$. By Theorems IV.1.2 and IV.3.2, we immediately have

$$\begin{pmatrix} \sqrt{n}(\hat{V}_x^{(n)} - V_x) \\ \sqrt{n}(\hat{A}_x^{(n)} - A_x) \end{pmatrix} \xrightarrow{\mathscr{D}} \begin{pmatrix} -V_x U_x \\ U_x \end{pmatrix}$$

as $n \to \infty$, where U_x is the Gaussian martingale defined earlier. Therefore, by the functional delta-method (Theorem II.8.1),

$$\begin{aligned}\sqrt{n}(\phi(\hat{V}_x^{(n)}, \hat{A}_x^{(n)}) - \phi(V_x, A_x)) &\xrightarrow{\mathscr{D}} -\int_0^m V_x(t) U_x(t) \, \mathrm{d}A_x(t) + \int_0^m V_x(t) \, \mathrm{d}U_x(t) \\ &= -\int_0^m U_x(t) \, \mathrm{d}\bar{A}_m^1(t) + \int_0^m V_x(t) \, \mathrm{d}U_x(t) \\ &= \int_0^m (V_x(t) - \bar{A}_m^1(t)) \, \mathrm{d}U_x(t),\end{aligned}$$

where

$$\bar{A}_m^1(t) = \bar{A}^1_{x:\overline{m}|} - \bar{A}^1_{x:\overline{t}|} = \int_t^m V_x(u) \, \mathrm{d}A_x(u).$$

Thus, we have proved that

$$\sqrt{n}(\hat{\bar{A}}^1_{x:\overline{m}|} - \bar{A}^1_{x:\overline{m}|}) \xrightarrow{\mathscr{D}} \mathscr{N}\left(0, \int_0^m (v^t S_x(t) - \bar{A}^1_{x:\overline{m}|} + \bar{A}^1_{x:\overline{t}|})^2 \, \mathrm{d}\sigma_x^2(t)\right)$$

as $n \to \infty$. For both the estimated actuarial values, the estimation of asymptotic variances is straightforward.

We have in this example looked at a simple life insurance contract. For a discussion of general actuarial values in a counting process framework, the reader is referred to Hoem and Aalen (1978). Præstgaard (1991) considered nonparametric estimators of general actuarial values combining the functional delta-method and the results of Section IV.4 on nonparametric estimation of the transition probabilities in a time inhomogeneous Markov process model (see also Example IV.4.3). □

The final example considers the problem of estimating the *unconditional* probability that the truncation time is smaller than the survival time based on a truncated sample.

EXAMPLE IV.3.10. Estimating the Unconditional Probability that $V \leq X$ from a Sample from the Conditional Distribution of (V, X) given $V \leq X$

In Example III.3.3, we outlined Keiding and Gill's (1990) construction of a five-state Markov process into which the random truncation model for a survival time (cf. Example III.3.2) may be imbedded, and in Example IV.1.7 we quoted Keiding and Gill's application of this embedding to obtain asymptotic results for the Nelson–Aalen estimator. We remarked toward the end of Section IV.3.2 that similar asymptotic results are valid for the Kaplan–Meier estimator. In this example, we briefly survey these results and apply them to the estimation of the *unconditional* probability p for the event $A = \{V \le X\}$ from the *conditional* distribution of (V, X) given A.

As in Examples III.3.2, III.3.3, and IV.1.7, we assume that V and X are independent random variables, and we denote their distribution functions G and F. Moreover, these are assumed to be absolutely continuous with support $[0, \infty)$ and hazard functions α_V and α_X, respectively. We denote the probability measure corresponding to this setup by P^0, and note that $p = \mathrm{P}^0(V \le X) > 0$. Now we do not sample from P^0, but from the conditional distribution given the event $A = \{V \le X\}$. We denote this conditional probability measure by P, i.e., $\mathrm{P}(\cdot) = \mathrm{P}^0(\cdot \mid X \ge V)$ (cf. Example IV.1.7).

Let now $(V_1, X_1), (V_2, X_2), \ldots, (V_n, X_n)$ be i.i.d. replicates from this *conditional distribution* and define

$$N_{01}(t) = \#\{V_i \le t\},$$
$$N_{12}(t) = \#\{V_i < X_i \le t\};$$

these are the transition counts over $(0, t]$ of the transitions $0 \to 1$ and $1 \to 2$, respectively, in the Markov process construction of Example III.3.3. Introduce also

$$Y_0(t) = \#\{V_i \ge t\},$$
$$Y_1(t) = \#\{V_i < t \le X_i\}.$$

Then (N_{01}, N_{12}) is a bivariate counting process with $N_{h,h+1}$ having intensity process of the multiplicative form $\phi_{h,h+1} Y_h$ for $h = 0, 1$. Here, ϕ_{01} and ϕ_{12} are the intensities for $0 \to 1$ and $1 \to 2$ transitions, respectively, in the conditional Markov process (with state space $\{0, 1, 2\}$) described in Example III.3.3. Thus,

$$\phi_{01}(t) = \alpha_V(t) \frac{(1 - F(t))(1 - G(t))}{\int_t^\infty (1 - F(v))\, \mathrm{d}G(v)}$$

and $\phi_{12}(t) = \alpha_X(t)$. Note that ϕ_{01} is the hazard rate of the conditional distribution of V given $V \le X$, whereas $\phi_{12} = \alpha_X$ is the hazard of the marginal distribution of X.

Our general results lead to uncorrelated Nelson–Aalen estimators

$$\hat{\Phi}_{01}(t) = \int_0^t \frac{\mathrm{d}N_{01}(u)}{Y_0(u)},$$

$$\hat{A}_X(t) = \hat{\Phi}_{12}(t) = \int_0^t \frac{\mathrm{d}N_{12}(u)}{Y_1(u)},$$

for which asymptotic distribution results are available along our standard routes (Theorems IV.1.1 and IV.1.2; cf. in particular, Example IV.1.7) although additional care [along the lines of Gill (1983a)] is required to extend the weak convergence from some subset $[\varepsilon, M]$, $0 < \varepsilon < M < \infty$, to the whole support $[0, \infty)$. Using the product-integral, estimators are obtained for $1 - F_1(t) = \mathrm{P}(V > t) = \mathrm{P}^0(V > t \mid V \le X)$ and $1 - F$; for F_1, we just get the empirical distribution function of the V_i, for $1 - F$ the Kaplan–Meier estimator

$$1 - \hat{F}(t) = \prod_{(0,t]} \left(1 - \frac{\mathrm{d}N_{12}(u)}{Y_1(u)}\right),$$

except that a new regularity condition is required: If there are $s < t < u$ such that $Y_1(t) = 0$ but $Y_1(s) > 0$, $Y_1(u) > 0$ (*empty inner risk sets*), then $1 - \hat{F}$ does not exist.

The basic asymptotic result may be phrased as follows. Assume

$$\int_0^\infty (1 - F)^{-1}\,\mathrm{d}G < \infty, \qquad \int_0^\infty G^{-1}\,\mathrm{d}F < \infty.$$

Then, as $n \to \infty$, $(\hat{\Phi}_{01}, \hat{\Phi}_{12})$ converges in $(D[0, \infty))^2$ to a pair (U_1, U_2) of independent Gaussian martingales with zero mean and variance functions given by

$$\operatorname{var} U_1(t) = \int_0^t \frac{\phi_{01}(u)\,\mathrm{d}u}{\mathrm{P}(V \ge u)} = \int_0^t \frac{p(1 - F(u))(1 - G(u))\alpha_V(u)\,\mathrm{d}u}{(\int_u^\infty (1 - F(v))\,\mathrm{d}G(v))^2},$$

$$\operatorname{var} U_2(t) = \int_0^t \frac{\phi_{12}(u)\,\mathrm{d}u}{\mathrm{P}(V < u \le X)} = \int_0^t \frac{p\alpha_X(u)\,\mathrm{d}u}{(1 - F(u))G(u)},$$

and $(\hat{F}_1, \hat{F})$ converges in $(D[0, \infty])^2$ to $((1 - F_1)U_1, (1 - F)U_2)$, where $1 - F_1(t) = p^{-1} \int_t^\infty (1 - F)\,\mathrm{d}G$.

By a delicate argument, including among other things the functional delta-method, it is then possible to derive the asymptotic distribution result for the nonparametric maximum likelihood estimator

$$\hat{p} = \int_0^\infty \hat{G}(t)\,\mathrm{d}\hat{F}(t)$$

of $p = \mathrm{P}^0(V \le X)$: As $n \to \infty$, $\sqrt{n}(\hat{p} - p)$ is asymptotically $\mathcal{N}(0, \sigma^2)$-distributed with

$$\sigma^2 = p^3 \int_0^\infty \frac{\mathrm{d}G}{1 - F} - p^2 + p^3 \int_0^\infty \left(\frac{1 - G}{1 - F}\right)^2 \frac{\mathrm{d}F}{G}.$$

Keiding and Gill (1988, 1990) gave Monte Carlo results on the small sample properties of this asymptotic distribution.

The nonparametric maximum likelihood estimator $\hat{G}$ of the distribution function of V may, in principle, be derived from $\hat{F}_1$ and $\hat{F}$, although an easier argument is to reverse time and reason by symmetry to obtain a Kaplan–Meier estimator in reversed time, using the same number-at-risk process Y_1 as for $1 - \hat{F}$. Keiding and Gill (1990) discussed the role of the "retro-hazard" $\bar{\alpha}_V(t)$ given by

$$\bar{\alpha}_V(t)\,\mathrm{d}t = \mathrm{P}^0(V \in (t - \mathrm{d}t, t] \mid V \leq t)$$

in the estimation under right-truncation and in the Markov process of Example III.3.3. Keiding (1991) gave examples of the use of $\bar{\alpha}_V(t)$ in epidemiology. □

IV.4. The Product-Limit Estimator for the Transition Matrix of a Nonhomogeneous Markov Process

In the previous section, we studied the Kaplan–Meier estimator for the survival distribution. Now this estimator will be generalized to an estimator of the transition matrix of a nonhomogeneous Markov process. Such a generalization was considered by Aalen (1978a) for the competing risks model (Example III.1.5) and independently by Aalen and Johansen (1978) and Fleming (1978a, 1978b) for arbitrary finite-state Markov processes. Our presentation is mainly based on Aalen and Johansen (1978), and we will denote the estimator the *Aalen–Johansen estimator.*

Consider a nonhomogeneous, time-continuous Markov process with finite state space $\{1, 2, \ldots, k\}$ having transition probabilities $P_{hj}(s, t)$ and transition intensities $\alpha_{hj}(t)$. It was seen in Chapter III (see, in particular, Examples III.1.4 and III.2.8) that for n conditionally independent replicates of such a Markov process (conditioning on the initial states), subject to independent right-censoring, left-truncation, and filtering, the multivariate counting process $\mathbf{N} = (N_{hj}; h \neq j)$, with N_{hj} counting the number of observed direct transitions from h to j in $[0, t]$, has intensity process $\boldsymbol{\lambda} = (\lambda_{hj}; h \neq j)$ of the multiplicative form $\lambda_{hj}(t) = \alpha_{hj}(t) Y_h(t)$. Here, $Y_h(t) \leq n$ is the number of sample paths observed to be in state h just prior to time t. We will derive and study a product-limit estimator for the $k \times k$ matrix of transition probabilities $\mathbf{P}(s, t) = \{P_{hj}(s, t)\}$ for $0 \leq s \leq t < \tau$, where $\tau = \sup\{u: \int_0^u \alpha_{hj}(v)\,\mathrm{d}v < \infty, h \neq j\}$.

Our general derivations are based *only* on the above-mentioned properties of the counting process $\mathbf{N} = (N_{hj})$ and on the product-integral representation of the matrix of transition probabilities given in Theorem II.6.7. Therefore, our results apply also to the estimation of so-called partial probabilities as commented upon at the end of Section IV.4.1 and as illustrated in some of the examples.

IV.4.1. Definition and Basic Properties

In this subsection, we derive and study the finite sample properties of the Aalen–Johansen estimator for the transition probabilities of a Markov process with finite state space in a manner very similar to that used in Section IV.3.1 for the Kaplan–Meier estimator. A study of its large sample properties is deferred to the next subsection.

The plan of the present subsection is as follows. First, we define the Aalen–Johansen estimator for the transition probability matrix and give a heuristic interpretation of the estimator. Then we derive a martingale representation of the estimator similar to (4.3.4) for the Kaplan–Meier estimator and use this to show that the Aalen–Johansen estimator is almost unbiased. The martingale representation is also used to derive estimators for the covariance matrix of the Aalen–Johansen estimator. Finally, we discuss how the Aalen–Johansen estimator may be interpreted as a nonparametric maximum likelihood estimator, discuss how it reduces when we have complete observation, consider estimation of so-called partial transition probabilities, and give a discussion (with practical applications) of four simple Markov process models.

IV.4.1.1. *The Estimator*

We consider the situation described in the introduction to this section. We define $\alpha_{hh} = -\sum_{j \neq h} \alpha_{hj}$ and let $A_{hj}(t) = \int_0^t \alpha_{hj}(s)\,\mathrm{d}s$ for all h, j. Then, by Theorem II.6.7, the transition probability matrix $\mathbf{P}(s,t) = (P_{hj}(s,t))$ is given as the product-integral

$$\mathbf{P}(s,t) = \prod_{(s,t]} (\mathbf{I} + \mathrm{d}\mathbf{A}(u)), \tag{4.4.1}$$

where $\mathbf{A} = \{A_{hj}\}$. We write, as usual, $J_h(t) = I(Y_h(t) > 0)$, and let

$$\hat{A}_{hj}(t) = \int_0^t J_h(s)(Y_h(s))^{-1}\,\mathrm{d}N_{hj}(s)$$

be the Nelson–Aalen estimator for A_{hj} $(h \neq j)$. Furthermore, we introduce $\hat{A}_{hh}(t) = -\sum_{j \neq h} \hat{A}_{hj}$. Relation (4.4.1) suggests that we estimate the matrix of transition probabilities by the $k \times k$ matrix

$$\hat{\mathbf{P}}(s,t) = \prod_{(s,t]} (\mathbf{I} + \mathrm{d}\hat{\mathbf{A}}(u)), \tag{4.4.2}$$

where $\hat{\mathbf{A}} = \{\hat{A}_{hj}\}$. Looking back at (4.3.1) and (4.3.2), we see that $\hat{\mathbf{P}}(s,t)$ is derived from the product-integral representation for $\mathbf{P}(s,t)$ exactly in the same manner as the Kaplan–Meier estimator was derived from the survival distribution. In fact, the results of Section IV.3, being valid for the Markov process with the two states "alive" and "dead," with "dead" absorbing (cf. Example III.1.5), are special cases of the results derived in the present section.

The Nelson–Aalen estimators are step-functions with a finite number of jumps on $(s,t]$. Therefore, the Aalen–Johansen estimator (4.4.2) is a finite product of matrices (cf. comment below Definition II.6.1). If one or more transitions are observed at time u (allowing for ties), then the contribution to (4.4.2) from this time point is a matrix $\mathbf{I} + \Delta\hat{\mathbf{A}}(u)$, where $\Delta\hat{\mathbf{A}}(u)$ is the $k \times k$ matrix with entry (h,j) equal to $\Delta N_{hj}(u)/Y_h(u)$ for $h \neq j$ and entry (h,h) equal to $-\Delta N_{h\cdot}(u)/Y_h(u)$ with $N_{h\cdot} = \sum_{j \neq h} N_{hj}$.

This interpretation of (4.4.2) and the Chapman–Kolmogorov equation

$$\mathbf{P}(s,t) = \prod_{m=1}^{M} \mathbf{P}(t_{m-1}, t_m),$$

where $s = t_0 < t_1 < \cdots < t_M = t$ and each interval contains at most one time point where transition(s) occur, also give a heuristic motivation for the Aalen–Johansen estimator: For the intervals where no transition occurs, the transition matrix is estimated by the identity matrix, whereas it is estimated by $\mathbf{I} + \Delta\hat{\mathbf{A}}(u)$ if one or more transitions are observed at time u in the interval.

We have formulated above the estimator in such a way that tied observations are allowed. However, under the model, there are with probability one no ties, and below we generally rule out the possibility of ties.

IV.4.1.2. *Martingale Representation and Approximate Unbiasedness*

To derive a martingale representation for the Aalen–Johansen estimator, which will be a key relation in our study of its statistical properties, we define

$$A_{hj}^*(t) = \int_0^t J_h(u)\alpha_{hj}(u)\,\mathrm{d}u,$$

as in Section IV.1, and introduce

$$\mathbf{P}^*(s,t) = \prod_{(s,t]} (\mathbf{I} + \mathrm{d}\mathbf{A}^*(u)), \tag{4.4.3}$$

where $\mathbf{A}^* = \{A_{hj}^*\}$ [cf. (4.3.3)]. Then, by Duhamel's equation (2.6.4),

$$\hat{\mathbf{P}}(s,t) - \mathbf{P}^*(s,t) = \int_s^t \hat{\mathbf{P}}(s,u-)\,\mathrm{d}(\hat{\mathbf{A}} - \mathbf{A}^*)(u)\mathbf{P}^*(u,t). \tag{4.4.4}$$

Now, by Theorem II.6.5, we have for $0 \le s \le t < \tau$

$$\begin{aligned} \det \mathbf{P}^*(s,t) &= \exp\left\{\int_s^t \sum_{h=1}^{k} J_h(u)\alpha_{hh}(u)\,\mathrm{d}u\right\} \\ &\ge \exp\left\{\int_s^t \sum_{h=1}^{k} \alpha_{hh}(u)\,\mathrm{d}u\right\} = \det \mathbf{P}(s,t) > 0, \end{aligned} \tag{4.4.5}$$

so that $\mathbf{P}^*(s,t)$ is nonsingular. Therefore, by multiplying both sides of (4.4.4) by the inverse of $\mathbf{P}^*(s,t) = \mathbf{P}^*(s,u)\mathbf{P}^*(u,t)$, we get

$$\hat{\mathbf{P}}(s,t)\mathbf{P}^*(s,t)^{-1} - \mathbf{I} = \int_s^t \hat{\mathbf{P}}(s,u-)\,\mathrm{d}(\hat{\mathbf{A}} - \mathbf{A}^*)(u)\mathbf{P}^*(s,u)^{-1}, \tag{4.4.6}$$

which is the desired martingale representation for the Aalen–Johansen estimator; cf. the similar representation (4.3.4) for the Kaplan–Meier estimator. To see that (4.4.6) actually defines a $k \times k$ matrix of square integrable martingales, we let $\mathbf{P}^*(s,u)^{-1} = \{P^{*mj}(s,u)\}$, and note that then the entry (h,j) of the matrix on the right-hand side of (4.4.6) takes the form

$$\sum_{l=1}^{k} \sum_{m=1}^{k} \int_s^t \hat{P}_{hl}(s, u-) P^{*mj}(s, u) \, \mathrm{d}(\hat{A}_{lm} - A^*_{lm})(u). \tag{4.4.7}$$

By (4.1.4), $\hat{A}_{lm} - A^*_{lm}$ is a square integrable martingale for each l, m. Moreover, the integrand is a predictable process. We will show that it is also bounded. We have $|\hat{P}_{hl}(s, u-)| \le 1$ and

$$|P^{*mj}(s, u)| = |(\det \mathbf{P}^*(s, u))^{-1} \operatorname{cof} P^*_{jm}(s, u)|,$$

where $\operatorname{cof} P^*_{jm}(s, t)$ is the cofactor of $P^*_{jm}(s, t)$. Since $|P^*_{hl}(s, u)| \le 1$ for all h, l, $|\operatorname{cof} P^*_{jm}(s, u)| \le (k-1)!$, and, therefore, by (4.4.5)

$$|P^{*mj}(s, u)| \le (k-1)! (\det \mathbf{P}^*(s, t))^{-1}. \tag{4.4.8}$$

It follows that the integrand in (4.4.7) is bounded and, therefore, that (4.4.6) in fact defines a $k \times k$ matrix of square integrable martingales.

Now (4.4.6) immediately gives

$$\mathrm{E}\{\hat{\mathbf{P}}(s, t) \mathbf{P}^*(s, t)^{-1}\} = \mathbf{I}, \tag{4.4.9}$$

which is a result for the Aalen–Johansen estimator similar to (4.3.5) for the Kaplan–Meier estimator. In particular, when there is only a small probability that $Y_h(u) = 0$ for some h and $s < u \le t$, $\hat{\mathbf{P}}(s, t)$ will be almost unbiased.

IV.4.1.3. *Estimators of the Covariance Matrix*

The relation (4.4.6) may also be used to derive estimators of the covariance matrix of the Aalen–Johansen estimator. The arguments are similar to those used in Section IV.3.1 for deriving the estimators (4.3.7) and (4.3.9) for the variance of the Kaplan–Meier estimator, but the algebra needed to carry out these arguments now becomes much more involved.

We will first derive an estimator for the covariance matrix of the Aalen–Johansen estimator corresponding to (4.3.7) for the Kaplan–Meier estimator. The resulting estimator is given in matrix form by (4.4.14) and (4.4.15) and elementwise by (4.4.16). We then derive an estimator of the "Greenwood type," i.e., corresponding to (4.3.9) for the Kaplan–Meier estimator. This estimator is given in matrix form by (4.4.17) and elementwise by (4.4.18). Finally, we derive a recursion formula for the "Greenwood-type" estimator corresponding to the one for the Kaplan–Meier estimator given just above Example IV.3.1. This recursion formula is given by (4.4.19) and (4.4.20). We recommend the use of the recursion formula for most numerical computations because it reduces the amount of computations considerably and also because it is able to handle tied observations [as is (4.4.17)].

Before we embark upon this program, we need to briefly review how one defines the covariance matrix of a $q \times r$ matrix-valued random variable $\mathbf{X}$. This is done by first arranging the columns of $\mathbf{X}$ on the top of each other into one vector (of dimension qr) denoted $\operatorname{vec} \mathbf{X}$. Then one defines the covariance matrix of $\mathbf{X}$ as the ordinary covariance matrix of $\operatorname{vec}\mathbf{X}$, i.e., as the matrix

$$\operatorname{cov}(\mathbf{X}) = \mathrm{E}\{(\operatorname{vec}\mathbf{X} - \operatorname{vec}\mathrm{E}\mathbf{X})(\operatorname{vec}\mathbf{X} - \operatorname{vec}\mathrm{E}\mathbf{X})^{\top}\}. \tag{4.4.10}$$

We note that $\operatorname{cov}(\mathbf{X})$ is a $qr \times qr$ matrix and that the (i,j)th block of it (when written as a partitioned matrix in the obvious way) is the $q \times q$ matrix of covariances between the elements of the ith and the jth column of $\mathbf{X}$. Furthermore, if $\mathbf{A}$ and $\mathbf{B}$ are $p \times q$ and $r \times s$ matrices, respectively, it follows using elementary properties of the vec-operator and Kronecker products of matrices [e.g., Mardia, Kent, and Bibby, (1979, Appendix A.2.5)] that

$$\operatorname{cov}(\mathbf{AXB}) = (\mathbf{B}^{\top} \otimes \mathbf{A})\operatorname{cov}(\mathbf{X})(\mathbf{B} \otimes \mathbf{A}^{\top}). \tag{4.4.11}$$

We also need to define the predictable covariation process of a $q \times r$ matrix $\mathbf{M} = \{M_{hj}\}$ of local square integrable martingales. In accordance with (4.4.10), this is defined as $\langle \mathbf{M} \rangle = \langle \operatorname{vec}\mathbf{M} \rangle$. (Remember the definition of the predictable covariation process of a vector-valued martingale in Section II.3.2.) Moreover, if $\mathbf{H} = \{H_{hj}\}$ and $\mathbf{K} = \{K_{hj}\}$ are $p \times q$ and $r \times s$ matrices, respectively, of locally bounded predictable processes, then the matrix-valued stochastic integral $\int \mathbf{H}\,\mathrm{d}\mathbf{MK}$ is the $p \times s$ matrix with (i,j)th entry $\sum_{l=1}^{q}\sum_{m=1}^{r}\int H_{il}K_{mj}\,\mathrm{d}M_{lm}$ (cf. Section II.3.3). In a similar manner as (4.4.11), the predictable covariation process of the local integrable martingale may be expressed as

$$\left\langle \int \mathbf{H}\,\mathrm{d}\mathbf{MK} \right\rangle = \int (\mathbf{K}^{\top} \otimes \mathbf{H})\,\mathrm{d}\langle \mathbf{M} \rangle (\mathbf{K} \otimes \mathbf{H}^{\top}). \tag{4.4.12}$$

We are now prepared to derive estimators for the covariance matrix of the Aalen–Johansen estimator. By (4.4.6), (4.4.9), and (4.4.10), the covariance matrix of $\hat{\mathbf{P}}(s,t)\mathbf{P}^*(s,t)^{-1}$ may be given as

$$\operatorname{cov}\{\hat{\mathbf{P}}(s,t)\mathbf{P}^*(s,t)^{-1}\} = \mathrm{E}\langle \hat{\mathbf{P}}(s,\cdot)\mathbf{P}^*(s,\cdot)^{-1} - \mathbf{I}\rangle(t),$$

where it follows using (4.4.6) and (4.4.12) that

$$\langle \hat{\mathbf{P}}(s,\cdot)\mathbf{P}^*(s,\cdot)^{-1} - \mathbf{I}\rangle(t)$$
$$= \int_s^t (\mathbf{P}^*(s,u)^{\top})^{-1} \otimes \hat{\mathbf{P}}(s,u-)\,\mathrm{d}\langle \hat{\mathbf{A}} - \mathbf{A}^*\rangle(u)\mathbf{P}^*(s,u)^{-1} \otimes \hat{\mathbf{P}}(s,u-)^{\top}. \tag{4.4.13}$$

To derive an estimator for the covariance matrix of $\hat{\mathbf{P}}(s,t)\mathbf{P}(s,t)^{-1}$ similar to (4.3.6) for the Kaplan–Meier estimator, we now replace $\hat{\mathbf{P}}(s,u-)$ and $\mathbf{P}^*(s,u)$ by $\hat{\mathbf{P}}(s,u)$ (using the continuity of $\mathbf{P}$) and $\mathrm{d}\langle \hat{\mathbf{A}} - \mathbf{A}^*\rangle(s)$ by $\mathrm{d}[\hat{\mathbf{A}} - \mathbf{A}^*](s)$ in (4.4.13), where $[\hat{\mathbf{A}} - \mathbf{A}^*] = [\operatorname{vec}(\hat{\mathbf{A}} - \mathbf{A}^*)]$ is the optional covariation process of the matrix-valued martingale $\hat{\mathbf{A}} - \mathbf{A}^*$ (cf. Section II.3.2). This suggests estimating the covariance matrix of $\hat{\mathbf{P}}(s,t)\mathbf{P}(s,t)^{-1}$ by

$$\int_s^t (\hat{\mathbf{P}}(s,u)^{\top})^{-1} \otimes \hat{\mathbf{P}}(s,u)\,\mathrm{d}[\hat{\mathbf{A}} - \mathbf{A}^*](u)\hat{\mathbf{P}}(s,u)^{-1} \otimes \hat{\mathbf{P}}(s,u)^{\top}.$$

Now, by (4.4.11),

$$\operatorname{cov}(\hat{\mathbf{P}}(s,t)) = (\mathbf{P}(s,t)^{\top} \otimes \mathbf{I}) \operatorname{cov}\{\hat{\mathbf{P}}(s,t)\mathbf{P}(s,t)^{-1}\}(\mathbf{P}(s,t) \otimes \mathbf{I}).$$

Using elementary properties of the vec-operator and Kronecker products [e.g., Mardia, Kent, and Bibby (1979, Appendix A.2.5)] and the Chapman–Kolmogorov equation $\mathbf{P}(s,t) = \mathbf{P}(s,u)\mathbf{P}(u,t)$ we, therefore, find that a reasonable estimator for the covariance matrix of $\hat{\mathbf{P}}(s,t)$, at least for large sample purposes, is

$$\widehat{\operatorname{cov}}(\hat{\mathbf{P}}(s,t)) = \int_s^t \hat{\mathbf{P}}(u,t)^{\top} \otimes \hat{\mathbf{P}}(s,u)\,\mathrm{d}[\hat{\mathbf{A}} - \mathbf{A}^*](u)\hat{\mathbf{P}}(u,t) \otimes \hat{\mathbf{P}}(s,u)^{\top} \tag{4.4.14}$$

which is a formula corresponding to (4.3.7) for the Kaplan–Meier estimator.

The $k^2 \times k^2$ matrix in (4.4.14) is written in a very compact form. It is, however, not very difficult to find an explicit expression for its elements. To this end, we first note that by (2.4.9)

$$[\hat{A}_{hh} - A^*_{hh}, \hat{A}_{hh} - A^*_{hh}](u) = \sum_{j \neq h} \int_0^u J_h(v)(Y_h(v))^{-2}\,\mathrm{d}N_{hj}(v);$$

$$[\hat{A}_{hh} - A^*_{hh}, \hat{A}_{hj} - A^*_{hj}](u) = -\int_0^u J_h(v)(Y_h(v))^{-2}\,\mathrm{d}N_{hj}(v), \quad \text{for } h \neq j;$$

$$[\hat{A}_{hj} - A^*_{hj}, \hat{A}_{hj} - A^*_{hj}](u) = \int_0^u J_h(v)(Y_h(v))^{-2}\,\mathrm{d}N_{hj}(v), \quad \text{for } h \neq j;$$

and

$$[\hat{A}_{hj} - A^*_{hj}, \hat{A}_{lm} - A^*_{lm}](u) = 0, \quad \text{otherwise.}$$

So if we let C_{gl} denote the $k \times k$ matrix with element (g,l) equal to 1, element (g,g) equal to -1, and the rest equal to zero, we may write

$$[\hat{\mathbf{A}} - \mathbf{A}^*](u) = \sum_{l=1}^{k} \sum_{g \neq l} \int_0^u \operatorname{vec} C_{gl} \operatorname{vec}^{\top} C_{gl} J_g(v)(Y_g(v))^{-2}\,\mathrm{d}N_{gl}(v).$$

Using this result, (4.4.14), and elementary properties of the vec-operator and Kronecker products, we find

$$\widehat{\operatorname{cov}}(\hat{\mathbf{P}}(s,t)) = \sum_{l=1}^{k} \sum_{g \neq l} \int_s^t \operatorname{vec}\{\hat{\mathbf{P}}(s,u)C_{gl}\hat{\mathbf{P}}(u,t)\} \operatorname{vec}^{\top}\{\hat{\mathbf{P}}(s,u)C_{gl}\hat{\mathbf{P}}(u,t)\} \times J_g(u)(Y_g(u))^{-2}\,\mathrm{d}N_{gl}(u). \tag{4.4.15}$$

Now the (h,j)th entry of $\hat{\mathbf{P}}(s,u)C_{gl}\hat{\mathbf{P}}(u,t)$ is $\hat{P}_{hg}(s,u)(\hat{P}_{lj}(u,t) - \hat{P}_{gj}(u,t))$, and it follows that the estimator for the covariance between $\hat{P}_{hj}(s,t)$ and $\hat{P}_{mr}(s,t)$ is given by

$$\widehat{\text{cov}}(\hat{P}_{hj}(s,t), \hat{P}_{mr}(s,t))$$
$$= \sum_{l=1}^{k} \sum_{g \neq l} \int_s^t \hat{P}_{hg}(s,u)\hat{P}_{mg}(s,u)\{\hat{P}_{lj}(u,t) - \hat{P}_{gj}(u,t)\}\{\hat{P}_{lr}(u,t) - \hat{P}_{gr}(u,t)\}$$
$$\times J_g(u)(Y_g(u))^{-2}\,\mathrm{d}N_{gl}(u). \tag{4.4.16}$$

In conclusion, the estimator for the covariance matrix of $\hat{\mathbf{P}}(s,t)$ similar to (4.3.7) for the Kaplan–Meier estimator may be given alternatively by (4.4.14) or (4.4.15), or elementwise by (4.4.16). Furthermore, by the martingale property of (4.4.6), we note that estimates for the covariances between $\hat{\mathbf{P}}(s,t)$ and $\hat{\mathbf{P}}(u,v)$ and their elements may be found by integrating over $(s,t] \cap (u,v]$ instead of $(s,t]$ in (4.4.14), (4.4.15), and (4.4.16).

As an alternative to the approach yielding (4.4.14), (4.4.15), and (4.4.16), we may, as in Sections IV.1.1 and IV.3.1, consider an extension of the model to one involving integrated transition intensity measures A_{hj} which are not necessarily continuous (cf. the discussion below of how the Aalen–Johansen estimator may be interpreted as a nonparametric maximum likelihood estimator). Such an extended model will lead to an estimator for the covariance matrix of $\hat{\mathbf{P}}(s,t)$ of the Greenwood type [cf. (4.3.9) for the Kaplan–Meier estimator]. For the extended model we have [cf. Theorem II.6.8 and (2.4.2)]

$$\langle M_{hj}, M_{hl}\rangle(t) = \int_0^t Y_h(s)(\delta_{jl} - \Delta A_{hj}(s))\,\mathrm{d}A_{hl}(s)$$

for $h \neq j$ and $h \neq l$, where δ_{jl} is a Kronecker delta, whereas

$$\langle M_{hj}, M_{gl}\rangle(t) = 0$$

for $g \neq h$. From this, we get, using (2.3.5) and the bilinearity of the predictable covariation process [cf. (2.3.4)],

$$\langle \hat{A}_{hj} - A^*_{hj}, \hat{A}_{hl} - A^*_{hl}\rangle(t) = \int_0^t \frac{J_h(s)}{Y_h(s)}(\delta_{jl} - \Delta A_{hj}(s))\,\mathrm{d}A_{hl}(s)$$

for $h \neq j$ and $h \neq l$,

$$\langle \hat{A}_{hh} - A^*_{hh}, \hat{A}_{hj} - A^*_{hj}\rangle(t) = -\int_0^t \frac{J_h(s)}{Y_h(s)}(1 - \Delta A_{h.}(s))\,\mathrm{d}A_{hj}(s)$$

for $h \neq j$,

$$\langle \hat{A}_{hh} - A^*_{hh}, \hat{A}_{hh} - A^*_{hh}\rangle(t) = \int_0^t \frac{J_h(s)}{Y_h(s)}(1 - \Delta A_{h.}(s))\,\mathrm{d}A_{h.}(s)$$

and

$$\langle \hat{A}_{hj} - A^*_{hj}, \hat{A}_{gl} - A^*_{gl}\rangle(t) = 0$$

for $g \neq h$. From (4.4.13) we may then derive an alternative estimator for the covariance matrix of $\hat{\mathbf{P}}(s,t)\mathbf{P}(s,t)^{-1}$ as follows. In (4.4.13), we replace $\mathbf{P}^*(s,u)$ by $\hat{\mathbf{P}}(s,u)$, whereas for the elements of $\langle \hat{\mathbf{A}} - \mathbf{A}^* \rangle$, we substitute $\hat{A}_{hj}$ for A_{hj} in the above expressions to get $\widehat{\widehat{\operatorname{cov}}}(\hat{\mathbf{A}}(t))$. The resulting estimator for the covariance matrix of $\hat{\mathbf{P}}(s,t)\mathbf{P}(s,t)^{-1}$ can then be converted into an estimator for the covariance matrix of $\hat{\mathbf{P}}(s,t)$ in a similar manner as above; cf. (4.4.14). The result is the Greenwood-type estimator

$$\widehat{\widehat{\operatorname{cov}}}(\hat{\mathbf{P}}(s,t)) = \int_s^t \hat{\mathbf{P}}(u,t)^{\mathsf{T}} \otimes \hat{\mathbf{P}}(s,u-)\,\widehat{\widehat{\operatorname{cov}}}(\mathrm{d}\hat{\mathbf{A}}(u))\hat{\mathbf{P}}(u,t) \otimes \hat{\mathbf{P}}(s,u-)^{\mathsf{T}} \tag{4.4.17}$$

which allows for ties in the data. However, when ties are present, we do not find any useful explicit expressions for the elements of (4.4.17). But for the special case where the data are truly continuous, so that $\Delta N_{\cdot\cdot}(t)$ equals 0 or 1 for all t, we may derive formulas for the elements of (4.4.17) in exactly the same manner as above. The resulting formula for the estimator for the covariance between $\hat{P}_{hj}(s,t)$ and $\hat{P}_{mr}(s,t)$ then becomes

$$\begin{aligned}
&\widehat{\widehat{\operatorname{cov}}}(\hat{P}_{hj}(s,t), \hat{P}_{mr}(s,t)) \\
&\quad = \sum_{l=1}^{k} \sum_{g \neq l} \int_s^t \hat{P}_{hg}(s,u-)\hat{P}_{mg}(s,u-)\{\hat{P}_{lj}(u,t) - \hat{P}_{gj}(u,t)\}\{\hat{P}_{lr}(u,t) - \hat{P}_{gr}(u,t)\} \\
&\qquad \times J_g(u)(Y_g(u) - 1)(Y_g(u))^{-3}\,\mathrm{d}N_{gl}(u).
\end{aligned} \tag{4.4.18}$$

Note that the difference between (4.4.16) and (4.4.18) is that, in the latter, we have replaced $\hat{P}_{hg}(s,u)$ and $\hat{P}_{mg}(s,u)$ by $\hat{P}_{hg}(s,u-)$ and $\hat{P}_{mg}(s,u-)$, whereas $(Y_g(u))^{-2}$ is replaced by $(Y_g(u) - 1)(Y_g(u))^{-3}$.

The explicit formulas (4.4.16) and (4.4.18) may be used directly for computational purposes. The computations, however, simplify considerably by using a *recursion formula* for the Greenwood-type estimator similar to the one for the Kaplan–Meier estimator given just above Example IV.3.1. This recursion formula may be derived heuristically as follows. On the right-hand side of (4.4.4), we replace $\hat{\mathbf{P}}$ and $\mathbf{P}^*$ by $\mathbf{P}$ to get

$$\hat{\mathbf{P}}(s,t) - \mathbf{P}(s,t) \approx \int_s^t \mathbf{P}(s,u-)\,\mathrm{d}(\hat{\mathbf{A}} - \mathbf{A}^*)(u)\mathbf{P}(u,t).$$

Since by (4.4.1)

$$\mathbf{P}(u,t) = \mathbf{P}(u,t-)(\mathbf{I} + \Delta\mathbf{A}(t)),$$

we have

$$\begin{aligned}
\hat{\mathbf{P}}(s,t) - \mathbf{P}(s,t) &\approx \int_s^{t-} \mathbf{P}(s,u-)\,\mathrm{d}(\hat{\mathbf{A}} - \mathbf{A}^*)(u)\mathbf{P}(u,t-)(\mathbf{I} + \Delta\mathbf{A}(t)) \\
&\quad + \mathbf{P}(s,t-)\Delta(\hat{\mathbf{A}} - \mathbf{A}^*)(t)\mathbf{P}(t,t) \\
&\approx (\hat{\mathbf{P}}(s,t-) - \mathbf{P}(s,t-))(\mathbf{I} + \Delta\mathbf{A}(t)) + \mathbf{P}(s,t-)\Delta(\hat{\mathbf{A}} - \mathbf{A}^*)(t).
\end{aligned}$$

Now, since $\hat{\mathbf{A}} - \mathbf{A}^*$ has uncorrelated increments, this suggests the recursion formula

$$\widehat{\widehat{\text{cov}}}(\hat{\mathbf{P}}(s,t)) = \{(\mathbf{I} + \Delta\hat{\mathbf{A}}(t))^{\mathsf{T}} \otimes \mathbf{I}\}\, \widehat{\widehat{\text{cov}}}(\hat{\mathbf{P}}(s,t-))\{(\mathbf{I} + \Delta\hat{\mathbf{A}}(t)) \otimes \mathbf{I}\}$$
$$+ \{\mathbf{I} \otimes \hat{\mathbf{P}}(s,t-)\}\, \widehat{\widehat{\text{cov}}}(\Delta\hat{\mathbf{A}}(t))\{\mathbf{I} \otimes \hat{\mathbf{P}}(s,t-)^{\mathsf{T}}\}, \tag{4.4.19}$$

which may, in fact, be verified by a direct calculation using (4.4.17). Written componentwise, the recursion formula becomes

$$\widehat{\widehat{\text{cov}}}(\hat{P}_{hj}(s,t), \hat{P}_{mr}(s,t)) = \sum_{g,l} \widehat{\widehat{\text{cov}}}(\hat{P}_{hg}(s,t-), \hat{P}_{ml}(s,t-))(\mathbf{I} + \Delta\hat{\mathbf{A}}(t))_{gj}(\mathbf{I} + \Delta\hat{\mathbf{A}}(t))_{lr}$$
$$+ \sum_{g,l} \hat{P}_{hg}(s,t-)\hat{P}_{ml}(s,t-)\, \widehat{\widehat{\text{cov}}}(\Delta\hat{A}_{gj}(t), \Delta\hat{A}_{lr}(t)), \tag{4.4.20}$$

where

$$\widehat{\widehat{\text{cov}}}(\Delta\hat{A}_{gj}(t), \Delta\hat{A}_{gr}(t)) = \frac{(\delta_{jr} Y_g(t) - \Delta N_{gj}(t))\Delta N_{gr}(t)}{Y_g(t)^3}$$

for $g \neq j$ and $g \neq r$,

$$\widehat{\widehat{\text{cov}}}(\Delta\hat{A}_{gg}(t), \Delta\hat{A}_{gr}(t)) = -\frac{(Y_g(t) - \Delta N_{g.}(t))\Delta N_{gr}(t)}{Y_g(t)^3}$$

for $g \neq r$,

$$\widehat{\widehat{\text{cov}}}(\Delta\hat{A}_{gg}(t), \Delta\hat{A}_{gg}(t)) = \frac{(Y_g(t) - \Delta N_{g.}(t))\Delta N_{g.}(t)}{Y_g(t)^3}$$

and

$$\widehat{\widehat{\text{cov}}}(\Delta\hat{A}_{gj}(t), \Delta\hat{A}_{lr}(t)) = 0$$

for $g \neq l$.

In Section IV.4.2, we show that the Aalen–Johansen etimator (suitably normalized) converges in distribution to a Gaussian process as the number of replicates of the Markov process tends to infinity. Therefore, the variance and covariance estimates just given may be used to set approximate pointwise confidence limits as illustrated in some of the examples to be presented.

IV.4.1.4. *Complete Observation*

In Section IV.3.1, we mentioned that the Kaplan–Meier estimator reduces to one minus the empirical distribution function when there is no censoring, truncation, or filtering, i.e., when we have complete observation, and moreover that for this situation Greenwood's formula (4.3.9) becomes just an ordinary binomial variance estimate. Let us now briefly see that similar results are valid for the Aalen–Johansen estimator as well. We first note that when we have complete observation

$$Y_h(t+) = Y_h(0+) + \sum_{j \neq h} N_{jh}(t) - \sum_{j \neq h} N_{hj}(t)$$

for all $h = 1, 2, \ldots, k$ and all $t \in \mathscr{T}$. Using this relation, it is fairly straightforward to check that (4.4.2) satisfies

$$\mathbf{Y}(t+) = \mathbf{Y}(s+)\hat{\mathbf{P}}(s, t),$$

where $\mathbf{Y}(t)$ is the row vector $(Y_1(t), \ldots, Y_k(t))$. Let $(p_1(0), p_2(0), \ldots, p_k(0))$ denote the initial distribution of the Markov process. Then, in particular, the Aalen–Johansen estimator for

$$p_h(t) = \sum_{j=1}^{k} p_j(0) P_{jh}(0, t),$$

the probability of being in state h at time t, is

$$\hat{p}_h(t) = \sum_{j=1}^{k} \frac{Y_j(0+)}{n} \hat{P}_{jh}(0, t) = \frac{Y_h(t+)}{n},$$

i.e., the fraction of sample paths found to be in state h at time t. Also the estimated covariance of $\hat{p}_h(t)$ and $\hat{p}_j(t)$ based on a multinomial distribution of $(Y_1(0+), \ldots, Y_k(0+))$ and the Greenwood-type covariance of $\hat{\mathbf{P}}(0, t)$ reduces to a usual estimated multinomial covariance matrix.

Illustrations of these relations are provided in Examples IV.4.1 and IV.4.2.

IV.4.1.5. *Nonparametric Maximum Likelihood Estimator*

As was the case in Section IV.3.1, where we discussed how the Kaplan–Meier estimator could be interpreted an a nonparametric maximum likelihood estimator (NPMLE), there is also for Markov processes a canonical discrete extension of the model. For both the continuous and the discrete cases, the connection between the matrix of integrated transition intensity measures $\mathbf{A}$ and the transition probability matrix $\mathbf{P}$ of a Markov process is given by (2.6.13) and (2.6.14). Thus, our model may be parametrized either by $\mathbf{P}$ or $\mathbf{A}$ (together with the initial distribution).

Moreover, thinking of $\mathbf{P}$ as the matrix of transition probabilities also for discrete-time Markov chains, the sensible extension of the likelihood is [cf. (4.1.36)]

$$\prod_t \prod_h \left\{ \prod_{j \neq h} (Y_h(t)\, \mathrm{d}A_{hj}(t))^{\mathrm{d}N_{hj}(t)} (1 - \mathrm{d}A_{h.}(t))^{Y_h(t) - \mathrm{d}N_{h.}(t)} \right\}.$$

This has the interpretation of conditionally independent multinomially distributed number of jumps: Given $\mathscr{F}_{t-}$, $(\mathrm{d}N_{hj}, j \neq h)$ is independent over $h = 1, \ldots, k$ and multinomially $(Y_h(t), \mathrm{d}A_{hj}(t){:}\, j \neq h)$ distributed. The likelihood is then fully correct for discrete-time Markov chains, and (as described in Section IV.1.5) it reduces to the correct likelihood for the absolutely continuous case. From the extended likelihood, it follows as in Section IV.1.5 that the Nelson–Aalen estimators $\hat{A}_{hj}$ are NPMLE, and, therefore, by the functional invariance of maximum likelihood estimators under 1–1 reparametrizations of the model, the Aalen–Johansen estimator $\hat{\mathbf{P}}$ is NPMLE as well.

An information calculation for $\hat{\mathbf{A}}$ based on the above likelihood, together with the delta-method and the relation $\mathbf{P} = \prod(\mathbf{I} + \mathrm{d}\mathbf{A})$ leads to the Greenwood-type estimator (4.4.17). Alternatively, first reparametrization by $\mathbf{P}(t, t + \mathrm{d}t) = \mathbf{I} + \mathrm{d}\mathbf{A}(t)$ and then calculation of the observed information gives the same answer. On the other hand, a 'Poisson likelihood' [cf. (4.1.35)]

$$\left\{\prod_t \prod_{j \neq h} (Y_h(t)\,\mathrm{d}A_{hj}(t))^{\mathrm{d}N_{hj}(t)}\right\} \exp\left\{-\sum_{j \neq h} \int_0^\tau Y_h(t)\,\mathrm{d}A_{hj}(t)\right\},$$

which here only is reasonable as a simple approximation, leads to the alternative estimator (4.4.16).

IV.4.1.6. *Partial Transition Probabilities and Waiting time Distributions*

From a Markov process with transition intensities $\alpha_{hj}(t)$ and transition probabilities $P_{hj}(s, t)$, one may construct a new Markov process by substituting zero for some of the α_{hj}. Following Hoem (1969b), we will denote this new Markov process a *partial* process and its transition probabilities $\bar{P}_{hj}(s, t)$ for *partial transition probabilities* in contrast to the *influenced transition probabilities* of the original Markov process.

One important example of such a partial Markov process is a competing risks model where one or more of the causes of death are "eliminated" by substituting zero for the cause-specific intensities for these particular causes of death. For a concrete example of estimation and interpretation of partial transition probabilities in such a competing risks model, see Example IV.4.1.

We let (in the general setup) $\bar{\mathbf{A}}$ be the $k \times k$ matrix of the remaining A_{hj} (with zeros substituted at the proper places). Then the $k \times k$ matrix of partial transition probabilities $\bar{\mathbf{P}}(s, t) = \{\bar{P}_{hj}(s, t)\}$ (which may have a number of structural zeros and ones) is given by the product-integral $\bar{\mathbf{P}}(s, t) = \prod_{(s,t]}(\mathbf{I} + \mathrm{d}\bar{\mathbf{A}}(u))$ (see Example IV.4.1 for an illustration for the competing risks model). This may, of course, be estimated by the Aalen–Johansen estimator $\hat{\bar{\mathbf{P}}}(s, t) = \prod_{(s,t]}(\mathbf{I} + \mathrm{d}\hat{\bar{\mathbf{A}}}(u))$, where $\hat{\bar{\mathbf{A}}}$ is the $k \times k$ matrix of Nelson–Aalen estimators for those (h, j) for which α_{hj} is not set equal to zero. Since all our derivations are based only on the properties of the counting processes (N_{hj}) and on the product-integral representation (4.4.1), all the results in the present, as well as the following subsection, continue to hold for the estimation of partial transition probabilities.

We finally note that the waiting time distribution in state j,

$$\begin{aligned} S_j(s, t) &= \exp\left\{-\sum_{l \neq j} \int_s^t \alpha_{jl}(u)\,\mathrm{d}u\right\} \\ &= \prod_{(s,t]} (1 + \mathrm{d}A_{jj}(u)), \end{aligned}$$

may be estimated by the Kaplan–Meier estimator

$$\hat{S}_j(s,t) = \prod_{(s,t]} (1 + \mathrm{d}\hat{A}_{jj}(u)) = \prod_{(s,t]} \left(1 - \frac{\Delta N_j(u)}{Y_j(u)}\right),$$

where $N_j = \sum_{l \neq j} N_{jl}$ counts the number of exits from state j. The statistical properties of this estimator follow directly from the results of Section IV.3 and need no further discussion here.

IV.4.1.7. *Examples*

We end this subsection with a detailed discussion of four simple Markov process models. The purpose of these examples is both to illustrate and elaborate on the general theory and to provide concrete practical illustrations of its use. We start with a discussion of the competing risks model.

EXAMPLE IV.4.1. Competing Risks. Radiation-Exposed Mice and Survival with Malignant Melanoma

We consider the competing risks model, cf. Example III.1.5 and Figure I.3.5, and we will see how the general theory applies to this simple Markov process model.

Let $s < T_1 < T_2 < \cdots < T_m \leq t$ be the times of the observed deaths, due to any cause, between s and t. Then, by (4.4.2), the transition probability matrix $\mathbf{P}(s,t) = (P_{hj}(s,t))$ is estimated by

$$\hat{\mathbf{P}}(s,t) = \prod_{i=1}^{m} (\mathbf{I} + \Delta\hat{\mathbf{A}}(T_i)),$$

with

$$\mathbf{I} + \Delta\hat{\mathbf{A}}(T_i) = \begin{pmatrix} 1 - \dfrac{\Delta N_0(T_i)}{Y_0(T_i)} & \dfrac{\Delta N_{01}(T_i)}{Y_0(T_i)} & \dfrac{\Delta N_{02}(T_i)}{Y_0(T_i)} & \cdots & \dfrac{\Delta N_{0k}(T_i)}{Y_0(T_i)} \\ 0 & 1 & 0 & \cdots & 0 \\ 0 & 0 & 1 & \cdots & 0 \\ \vdots & \vdots & \vdots & \ddots & \vdots \\ 0 & 0 & 0 & \cdots & 1 \end{pmatrix},$$

where $N_0 = \sum_{j=1}^{k} N_{0j}$ counts the total number of observed deaths.

For computational purposes, a direct application of (4.4.2), as explained above, is very simple to use. It may, nevertheless, be of some interest also to derive explicit expressions for the elements of $\hat{\mathbf{P}}(s,t)$. Straightforward matrix multiplication shows that

$$\hat{P}_{00}(s,t) = \prod_{i=1}^{m} \left(1 - \frac{\Delta N_0(T_i)}{Y_0(T_i)}\right),$$

i.e., the Kaplan–Meier estimator based on the total number of deaths, whereas for $j = 1, 2, \ldots, k$,

$$\hat{P}_{0j}(s,t) = \sum_{i=1}^{m} \left\{ \prod_{h<i} \left(1 - \frac{\Delta N_0(T_h)}{Y_0(T_h)}\right)\right\} \frac{\Delta N_{0j}(T_i)}{Y_0(T_i)}$$

$$= \int_{(s,t]} \hat{P}_{00}(s,u-)\,\mathrm{d}\hat{A}_{0j}(u).$$

For the competing risks model, it is well-known that

$$P_{0j}(s,t) = \int_s^t P_{00}(s,u)\alpha_{0j}(u)\,\mathrm{d}u$$

for $j = 1, 2, \ldots, k$. It is seen that the estimator is of exactly the same form. This reflects the fact that the empirical transition probabilities are derived from the Nelson–Aalen estimators in the same manner as the transition probabilities are constructed from the (integrated) transition intensities; cf. (4.4.1) and (4.4.2). In Examples IV.4.2 and IV.4.3 we will elaborate a little more on the relation between the transition probabilities and their empirical counterparts. The argument used in these examples to derive explicit expressions for the elements of $\hat{\mathbf{P}}(s,t)$ could also have been used here as an alternative to the direct matrix multiplications.

The variances and covariances of the estimators may, according to (4.4.16), be estimated by

$$\widehat{\operatorname{cov}}(\hat{P}_{0j}(s,t), \hat{P}_{0r}(s,t))$$

$$= \sum_{l=1}^{k} \int_s^t \hat{P}_{00}(s,u)^2 \{\delta_{lj} - \hat{P}_{0j}(u,t)\}\{\delta_{lr} - \hat{P}_{0r}(u,t)\} J_0(u)(Y_0(u))^{-2}\,\mathrm{d}N_{0l}(u)$$

for all $j, r = 0, 1, 2, \ldots, k$. The alternative Greenwood-type estimator (4.4.18) results by substituting $\hat{P}_{00}(s,u-)$ for $\hat{P}_{00}(s,u)$ and $(Y_0(u)-1)(Y_0(u))^{-3}$ for $(Y_0(u))^{-2}$ in this expression. In particular, (4.4.16) yields

$$\widehat{\operatorname{var}}\,\hat{P}_{00}(s,t) = (\hat{P}_{00}(s,t))^2 \int_s^t J_0(u)(Y_0(u))^{-2}\,\mathrm{d}N_0(u)$$

[cf. (4.3.6) and (4.3.7)], whereas (4.4.18) gives an estimator of the Greenwood form [cf. (4.3.8) and (4.3.9)]. Furthermore, for $j = 1, 2, \ldots, k$, we get, by (4.4.16),

$$\widehat{\operatorname{var}}\,\hat{P}_{0j}(s,t) = \int_s^t \{\hat{P}_{00}(s,u)\hat{P}_{0j}(u,t)\}^2 J_0(u)(Y_0(u))^{-2}\,\mathrm{d}N_0(u)$$

$$+ \int_s^t \{\hat{P}_{00}(s,u)\}^2\{1 - 2\hat{P}_{0j}(u,t)\} J_0(u)(Y_0(u))^{-2}\,\mathrm{d}N_{0j}(u),$$

whereas the alternative estimator (4.4.18) may be derived from this expression as explained earlier.

From the competing risks model, one may derive a partial model by substituting zero for α_{0h} for $h \neq j$. This partial model will have transition probabilities $\bar{P}_{00}(s,t) = \exp\{-\int_s^t \alpha_{0j}(u)\,du\}$ and $\bar{P}_{0j}(s,t) = 1 - \bar{P}_{00}(s,t)$. Now $\bar{P}_{0j}(s,t)$ is estimated by

$$\hat{\bar{P}}_{0j}(s,t) = 1 - \hat{\bar{P}}_{00}(s,t) = 1 - \prod_{(s,t]}\left(1 - \frac{\Delta N_{0j}(u)}{Y_0(u)}\right),$$

and we have, by (4.4.16),

$$\widehat{\operatorname{var}}\,\hat{\bar{P}}_{0j}(s,t) = (\hat{\bar{P}}_{00}(s,t))^2 \int_{(s,t]} J_0(u)(Y_0(u))^{-2}\,dN_{0j}(u),$$

whereas (4.4.18) gives an estimator of the Greenwood form. The difference between $\hat{\bar{P}}_{0j}(s,t)$ and $\hat{P}_{0j}(s,t)$ will be illustrated later.

We apply this to causes of death in radiation-exposed male mice as introduced in Example I.3.8. Some comparisons of the temporal patterns of the causes of death (thymic lymphoma, reticulum cell sarcoma, other causes) may be performed using Nelson–Aalen plots of the cause-specific integrated intensities as demonstrated in detail by Aalen (1982b). Thus, Figure IV.4.1 compares the three causes of death for germ-free mice, showing that mortality from thymic lymphoma happens early, that from reticulum cell sarcoma and

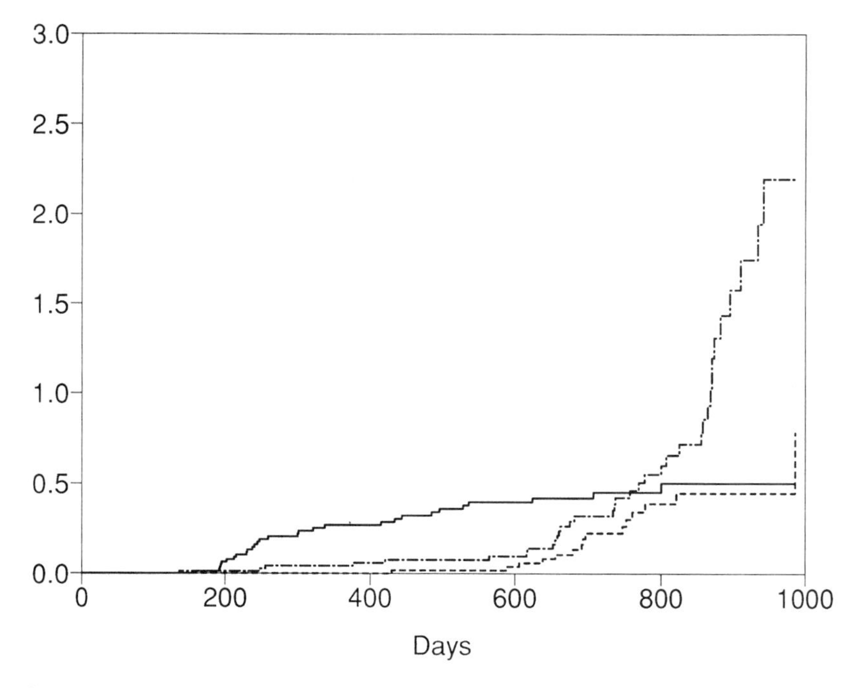

Figure IV.4.1. Estimated integrated cause-specific death intensities for mice in a germ-free environment: Thymic lymphoma (—), reticulum cell sarcoma (- - -), and other causes (-·-·-).

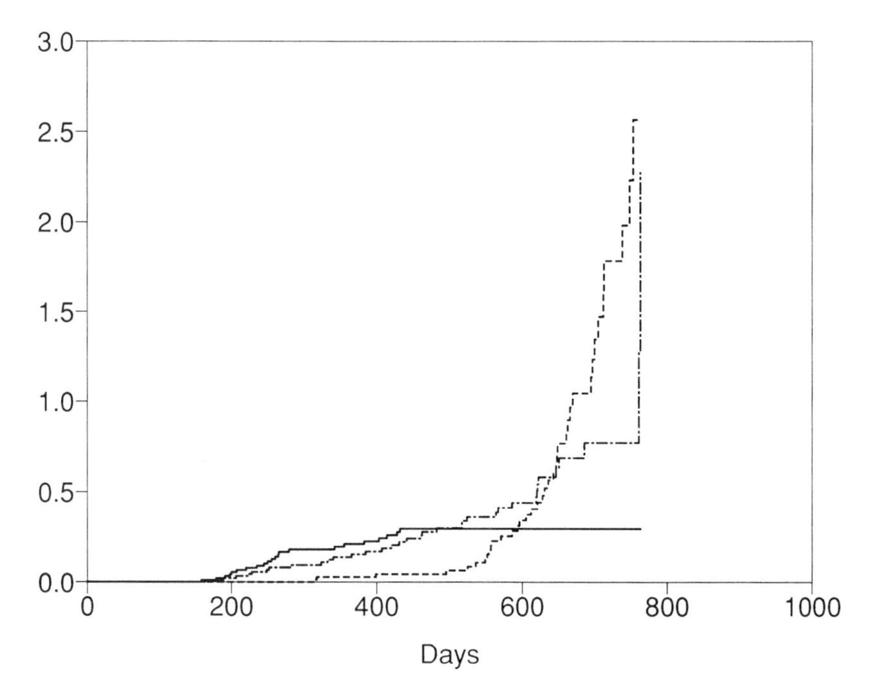

Figure IV.4.2. Estimated integrated cause-specific death intensities for mice in a conventional environment: Thymic lymphoma (—), reticulum cell sarcoma (- - -), and other causes (-····-).

that from other causes later. Figure IV.4.2 shows that, in a conventional environment, deaths from other causes have an initially similar development as deaths from thymic lymphoma. Figure IV.4.3 shows that the mortality from thymic lymphoma is similar in the two environments.

In addition, however, it is sometimes of interest to estimate quantities like $P_{0j}(0, t)$, the probability of dying of cause j before time t, sometimes called the "cumulative incidence function." This is plotted in Figure IV.4.4 for germ-free mice, cause of death j = thymic lymphoma, with pointwise approximate 95% confidence limits

$$\exp\left(\log \hat{P}_{0j}(0, t) \pm 1.96\{\widehat{\widehat{\operatorname{var}}}\, \hat{P}_{0j}(0, t)\}^{1/2}/\hat{P}_{0j}(0, t)\right).$$

Since in this example there is no censoring, we are here considering a closed system. Therefore, the Aalen–Johansen estimate for the probability of dying from a given cause is just the fraction of the individuals observed to have died of that cause, and its variance estimate is just the usual binomial variance estimate (cf. the general discussion earlier).

The probability $P_{0j}(0, t)$ describes the risk of dying from cause j before time t *in an environment where other causes of death are working* (in the example, reticulum cell sarcoma and other causes). In the competing risks

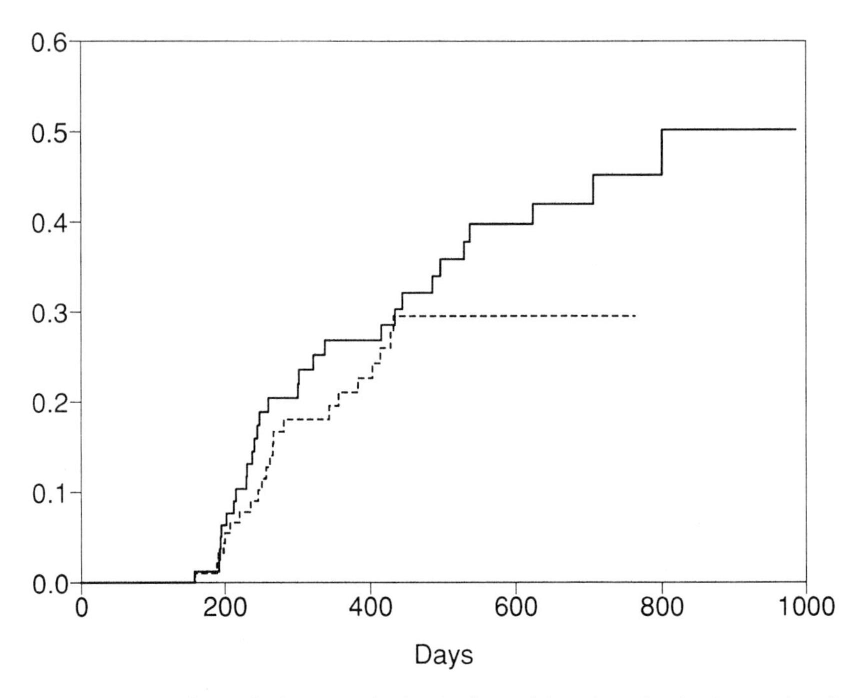

Figure IV.4.3. Estimated integrated death intensities for death from thymic lymphoma in germ-free (—) and conventional (- - -) environments.

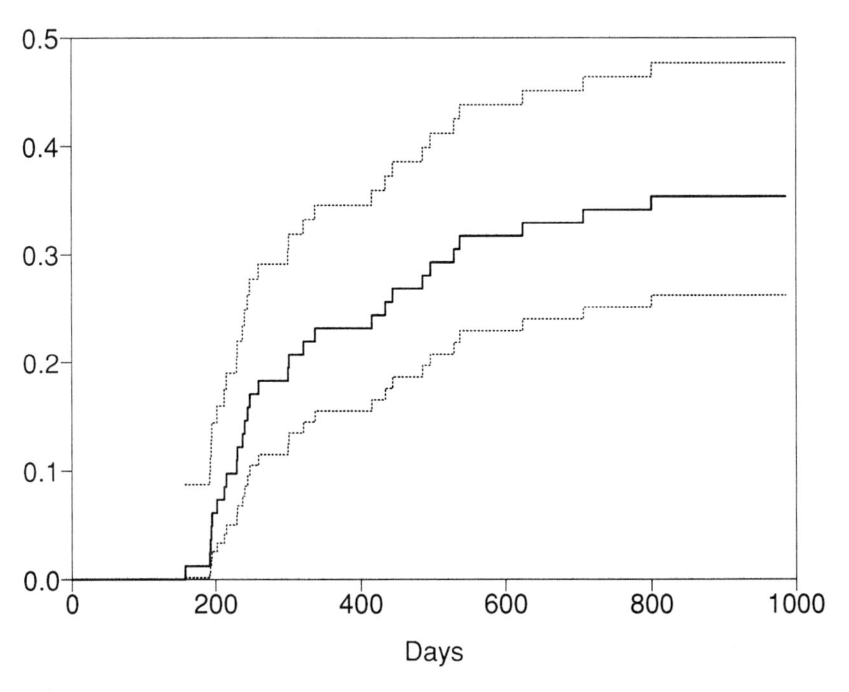

Figure IV.4.4. Estimated probability of dying from thymic lymphoma with pointwise 95% confidence limits for mice in germ-free environment based on the log-transformation.

literature, there has been an extensive discussion of how to model the risk of dying from a given cause j if all or some other causes of death were removed (cf. Example III.1.5). Under the empirically unverifiable condition of *independent competing risks* [cf. Gail (1975), David and Moeschberger (1978), Kalbfleisch and Prentice (1980, Chapter 7)], the probability of dying from cause j if all other causes were removed is given by

$$\begin{aligned}\bar{P}_{0j}(0,t) &= 1 - \bar{P}_{00}(0,t)\\ &= 1 - \prod_{0}^{t}(1 - \mathrm{d}A_{0j}(u))\\ &= 1 - \exp\left\{-\int_0^t \alpha_{0j}(u)\,\mathrm{d}u\right\}\end{aligned}$$

as discussed earlier. We call this the partial transition probability in contrast to the influenced transition probability $P_{0j}(0,t)$ which depends also on $\alpha_{0h}(u)$ for $h \neq j$.

Figure IV.4.5 shows $\hat{P}_{0j}(0,t)$ and $\hat{\bar{P}}_{0j}(0,t)$ for $j =$ thymic lymphoma for both environments. It is seen that $\hat{P}_{0j}(0,t) \leq \hat{\bar{P}}_{0j}(0,t)$, in accordance with the inequality

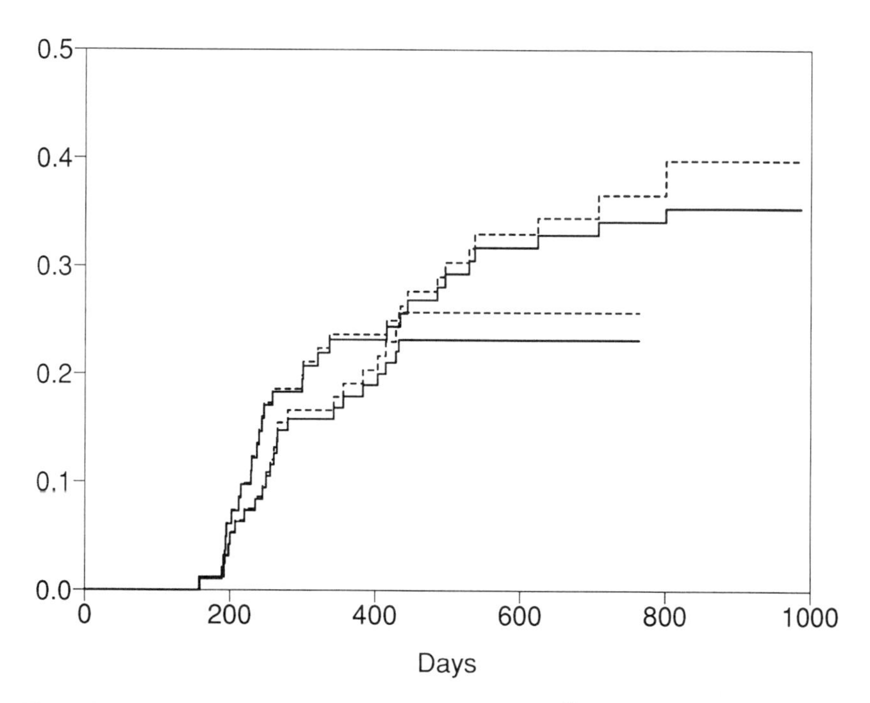

Figure IV.4.5. The influenced transition probability $\hat{P}_{0j}(0,t)$ (—) and the partial transition probability $\hat{\bar{P}}_{0j}(0,t)$ (- - -) for the cause $j =$ thymic lymphoma. Upper curves: mice from a germ-free environment. Lower curves: mice from a conventional environment.

$$P_{0j}(0,t) = \int_0^t P_{00}(0,u-)\,\mathrm{d}A_{0j}(u)$$
$$\leq \int_0^t \bar{P}_{00}(0,u-)\,\mathrm{d}A_{0j}(u) = \bar{P}_{0j}(0,t).$$

Formally, $\bar{P}_{0j}(0,t)$ corresponds to a hypothetical situation where only cause j is active, and with the same intensity as before. Although this situation is usually hard to interpret in biological contexts, it is instructive to realize that the inequality correponds to mice that "should have died of thymic lymphoma, but were killed off by other causes before then."

In cancer survival studies, one sometimes considers deaths from cancer (say, cause j) and deaths from other causes separately (see, for instance, Examples IV.1.11 and IV.3.3 on survival with malignant melanoma). An analysis of the cause-specific hazard α_{0j} for cancer then enables one to calculate $\hat{\bar{P}}_{00}(0,t) = 1 - \hat{\bar{P}}_{0j}(0,t)$, which is then interpreted as an estimate of the probability of surviving up to time t were cancer the only possible cause of death. The relevance of such an approach is often debatable, but in examples where, e.g., the hazard due to other causes is close to the hazard for the general population (Examples IV.1.11 and IV.3.3), it seems possible to give $\hat{\bar{P}}_{0j}$ a firm interpretation.

This is actually what we have done so far in the examples concerning survival with malignant melanoma by regarding deaths from other causes than the disease as censorings. The survival curves in Figures I.3.2 and IV.3.2–4 should, therefore, be interpreted accordingly. In Example V.1.5, a comparison of the mortality from other causes with that of the Danish population will be performed. □

EXAMPLE IV.4.2. The Two-State Illness–Death Model. Psychiatric Admissions for Women Having Just Given Birth

Let us next see how the general theory applies to the two-state Markov illness–death model with states 0 and 1, corresponding to "healthy" and "diseased," respectively, and with transition intensities α_{01} and α_{10} between these two states (cf. Figure I.3.6 with state 2 deleted).

By (4.4.2), $\mathbf{P}(s,t) = (P_{hj}(s,t))$ may be estimated as follows. Let $s < T_1 < T_2 < \cdots < T_m \leq t$ be the observed times of transitions from state 0 to state 1 or vice versa. Then

$$\hat{\mathbf{P}}(s,t) = \prod_{i=1}^{m} (\mathbf{I} + \Delta\hat{\mathbf{A}}(T_i)),$$

where now

$$\mathbf{I} + \Delta\hat{\mathbf{A}}(T_i) = \begin{pmatrix} 1 - \dfrac{\Delta N_{01}(T_i)}{Y_0(T_i)} & \dfrac{\Delta N_{01}(T_i)}{Y_0(T_i)} \\ \dfrac{\Delta N_{10}(T_i)}{Y_1(T_i)} & 1 - \dfrac{\Delta N_{10}(T_i)}{Y_1(T_i)} \end{pmatrix}.$$

This gives a simple algorithm for the computation of the empirical transition probabilities.

As in the previous example, it is illustrative to derive explicit expressions for the elements of $\hat{\mathbf{P}}(s,t)$ and to compare them with the formulas for the transition probabilities. Using the Kolmogorov forward differential equation (2.6.14) and the fact that $P_{00}(s,t) + P_{01}(s,t) = 1$, we get

$$\begin{aligned} \mathrm{d}P_{01}(s,t) &= P_{00}(s,t-)\,\mathrm{d}A_{01}(t) - P_{01}(s,t-)\,\mathrm{d}A_{10}(t) \\ &= \mathrm{d}A_{01}(t) - P_{01}(s,t-)\,\mathrm{d}(A_{10} + A_{01})(t), \end{aligned}$$

which is valid for any two-state Markov model with integrated intensity measures A_{01} and A_{10}. Using (2.6.7), this differential equation is seen to have the solution

$$P_{01}(s,t) = \int_{(s,t]} \prod_{(u,t]} \{1 - \mathrm{d}(A_{10} + A_{01})(v)\}\,\mathrm{d}A_{01}(u).$$

For the absolutely continuous case, where $A_{01}(t) = \int_0^t \alpha_{01}(u)\,\mathrm{d}u$ and $A_{10}(t) = \int_0^t \alpha_{10}(u)\,\mathrm{d}u$, the well-known formula

$$P_{01}(s,t) = \int_s^t \exp\left\{-\int_u^t (\alpha_{10}(v) + \alpha_{01}(v))\,\mathrm{d}v\right\}\alpha_{01}(u)\,\mathrm{d}u$$

for the transition probability results. On the other hand, for the purely discontinuous case where A_{01} and A_{10} are the Nelson–Aalen estimators $\hat{A}_{01}$ and $\hat{A}_{10}$, we get, for the Aalen–Johansen estimator,

$$\begin{aligned} \hat{P}_{01}(s,t) &= \int_s^t \prod_{(u,t]} \{1 - \mathrm{d}(\hat{A}_{01} + \hat{A}_{10})(v)\}\,\mathrm{d}\hat{A}_{01}(u) \\ &= \sum_{i=1}^m \left\{\prod_{h=i+1}^m \left(1 - \frac{\Delta N_{01}(T_h)}{Y_0(T_h)} - \frac{\Delta N_{10}(T_h)}{Y_1(T_h)}\right)\right\}\frac{\Delta N_{01}(T_i)}{Y_0(T_i)}. \end{aligned}$$

The symmetric expression for $\hat{P}_{10}(s,t)$ is derived by interchanging 0 and 1 in this formula.

As for the estimators for the variances and covariances, we get, by (4.4.16),

$$\begin{aligned} \widehat{\operatorname{var}}\,\hat{P}_{01}(s,t) &= \int_s^t (\hat{P}_{00}(s,u))^2\{\hat{P}_{11}(u,t) - \hat{P}_{01}(u,t)\}^2 J_0(u)(Y_0(u))^{-2}\,\mathrm{d}N_{01}(u) \\ &\quad + \int_s^t (\hat{P}_{01}(s,u))^2\{\hat{P}_{01}(u,t) - \hat{P}_{11}(u,t)\}^2 J_1(u)(Y_1(u))^{-2}\,\mathrm{d}N_{10}(u), \end{aligned}$$

whereas the estimator for $\operatorname{var}\hat{P}_{10}(s,t)$ is given by interchanging 0 and 1. Moreover,

$$\widehat{\text{cov}}(\hat{P}_{01}(s,t), \hat{P}_{10}(s,t))$$
$$= \int_s^t \hat{P}_{00}(s,u)\hat{P}_{10}(s,u)\{\hat{P}_{11}(u,t) - \hat{P}_{01}(u,t)\}\{\hat{P}_{10}(u,t) - \hat{P}_{00}(u,t)\}$$
$$\times J_0(u)(Y_0(u))^{-2}\,\mathrm{d}N_{01}(u)$$
$$+ \int_s^t \hat{P}_{01}(s,u)\hat{P}_{11}(s,u)\{\hat{P}_{01}(u,t) - \hat{P}_{11}(u,t)\}\{\hat{P}_{00}(u,t) - \hat{P}_{10}(u,t)\}$$
$$\times J_1(u)(Y_1(u))^{-2}\,\mathrm{d}N_{10}(u).$$

The alternative variance and covariance estimators of the Greenwood type may be found from these expressions as explained below (4.4.18).

The quantity $\hat{P}_{00}(s,t) = 1 - \hat{P}_{01}(s,t)$ estimates the probability that an individual who is in state 0 at time s will also be in this state at time t. One may also be interested in

$$\hat{S}_0(s,t) = \prod_{(s,t]} \left(1 - \frac{\Delta N_{01}(u)}{Y_0(u)}\right),$$

which estimates the probability of being continuously in state 0 between time s and t. We immediately have, by (4.4.16),

$$\widehat{\text{var}}\, \hat{S}_0(s,t) = (\hat{S}_0(s,t))^2 \int_s^t J_0(u)(Y_0(u))^{-2}\,\mathrm{d}N_{01}(u),$$

cf. (4.3.6) and (4.3.7), whereas (4.4.18) gives Greenwood's formula.

We now go on to consider the example concerning psychiatric admissions for women having just given birth introduced in Example I.3.10. In the above two-state Markov process, we let state 0 be out of psychiatric care (denoted by o) and state 1 be in psychiatric care (denoted by i), and let (p_o, p_i) denote the initial distribution of the process (cf. Example III.1.4).

Table IV.4.1 contains estimates of the probabilities of being in psychiatric care: given in psychiatric care at day of birth: $\hat{P}_{ii}(0,t)$; given not in psychiatric care at day of birth: $\hat{P}_{oi}(0,t)$; the correlation between these as well as the marginal probability of being in psychiatric care at time t: $\widehat{\Pr}(t) = \hat{p}_o\hat{P}_{oi}(0,t) +$

Table IV.4.1. Estimates of Transition Probabilities, Their Correlation, and The Prevalence.

day t	$\hat{P}_{ii}(0,t)$	$\hat{P}_{oi}(0,t)$	$\widehat{\widehat{\text{corr}}}(\hat{P}_{oi}(0,t), \hat{P}_{ii}(0,t))$	$\widehat{\Pr}(t)$
90	0.082425	0.000562	0.557	0.000574
180	0.007366	0.000517	0.480	0.000518
270	0.000942	0.000490	0.638	0.000490
360	0.000460	0.000420	0.974	0.000420

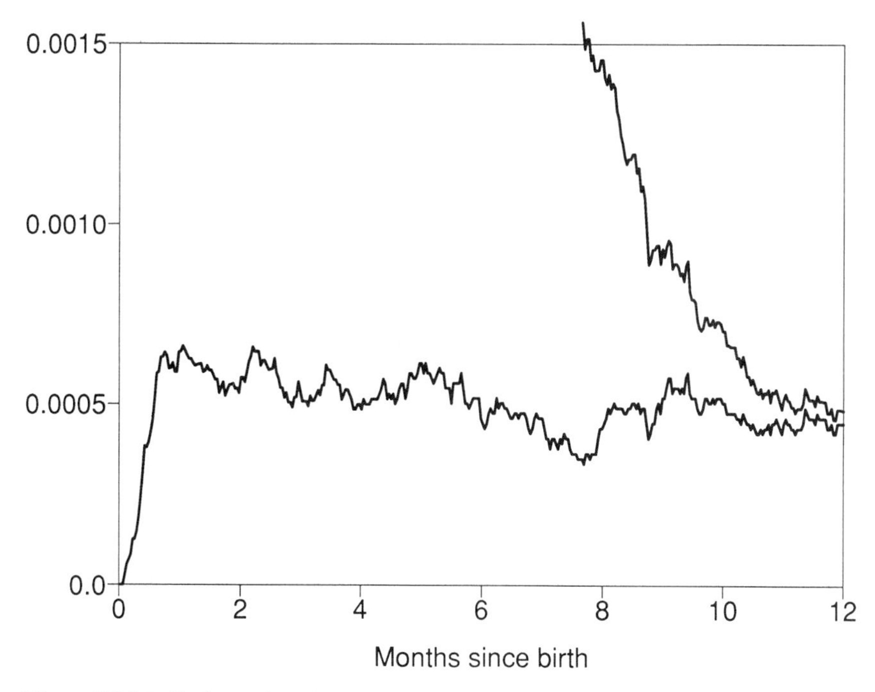

Figure IV.4.6. Estimated probabilities of being in psychiatric care at day t given in psychiatric care at day of birth $[\hat{P}_{ii}(0, t)]$ (upper curve) and not in psychiatric care at day of birth $[\hat{P}_{oi}(0, t)]$ (lower curve).

$\hat{p}_i \hat{P}_{ii}(0, t)$ taking $\hat{p}_i = 1 - \hat{p}_o = 11/71{,}378$ corresponding to the fraction of women in psychiatric care at day of birth. Since our calculations essentially assume the system to be closed (disregarding deaths and migrations), it follows from the remarks above that $\widehat{\Pr}(t)$ equals the fraction of women in psychiatric care at time t, the so-called *prevalence*.

Figures IV.4.6 and IV.4.7 show $\hat{P}_{oi}(0, t)$ and $\hat{P}_{ii}(0, t)$ for all t and, in Figure IV.4.8, the use of the standard error estimates for setting approximate confidence limits is illustrated for the case of $P_{oi}(0, t)$.

A striking feature of the transition probability estimates is that $\hat{P}_{oi}(0, t)$ and $\hat{P}_{ii}(0, t)$ are almost identical at the end of the one-year follow-up period. At the same time, their correlation approaches 1. This *ergodic* property states that after about a year, the process has *lost memory* in the sense that, according to *the predictions of the model*, it is equally likely for a woman to be in psychiatric care whether or not she was in psychiatric care at the day of birth. Certainly, this is a consequence of the size of the transition intensities, but it is not directly available from the figures in Examples IV.1.4 and IV.2.5 of the estimated transition intensities. This is one way in which the transition probability estimates usefully supplement the exploratory analysis based on (cumulative or smoothed) transition *intensity* estimates. □

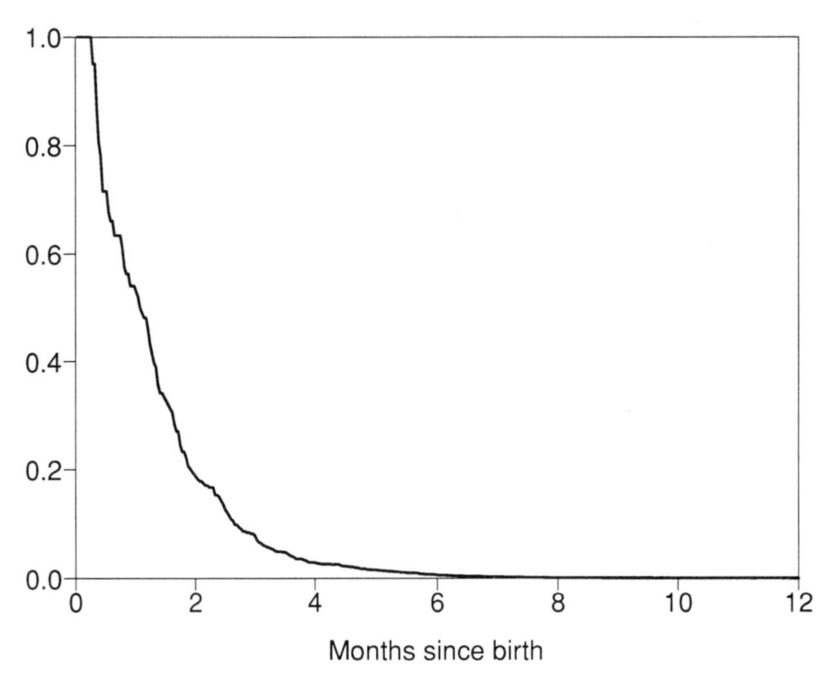

Figure IV.4.7. Estimated probability $\hat{P}_{ii}(0, t)$ of being in psychiatric care at day t given in psychiatric care at day 0 (day of birth).

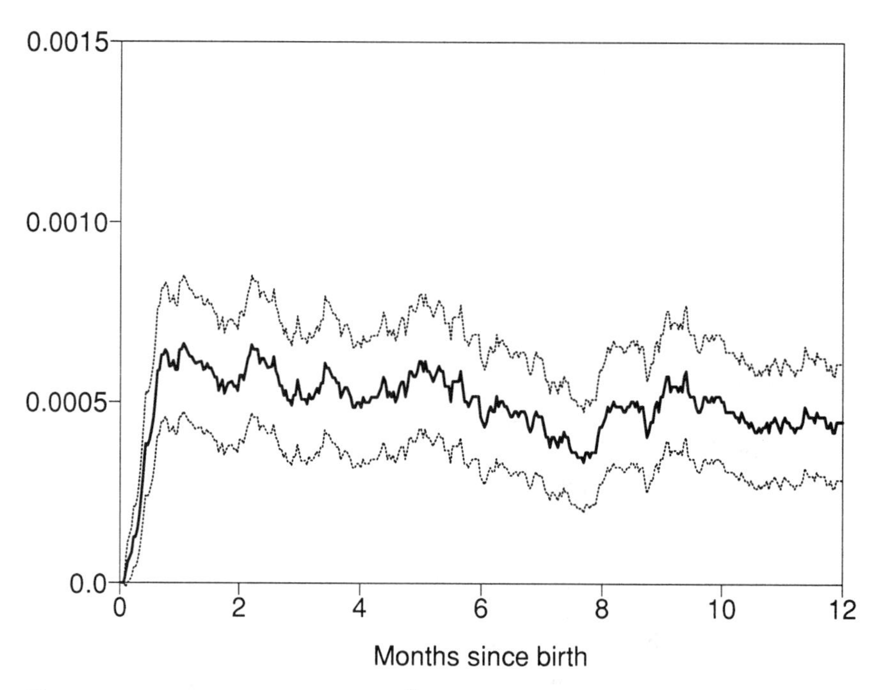

Figure IV.4.8. Estimated probability $\hat{P}_{oi}(0, t)$ of being in psychiatric care at day t given not in psychiatric care at day of birth with (linear) pointwise 95% confidence limits.

EXAMPLE IV.4.3. The Three-State Illness–Death Model Without Recovery. Nephropathy for Diabetics

As a third example, let us consider the three-state Markov illness–death model with states 0, 1, and 2 corresponding to "healthy," "diseased" and "dead," respectively, and where no recovery is possible (i.e., $\alpha_{10} = 0$), cf. Example III.1.11 and Figure I.3.6 (with no arrow from "diseased" to "healthy"). As in the preceding examples, we let $s < T_1 < T_2 < \cdots < T_m \leq t$ be the times of observed transitions between any two states, and note that $\mathbf{P}(s,t) = (P_{hj}(s,t)))$ is estimated by

$$\hat{\mathbf{P}}(s,t) = \prod_{i=1}^{m} (\mathbf{I} + \Delta\hat{\mathbf{A}}(T_i)),$$

with

$$\mathbf{I} + \Delta\hat{\mathbf{A}}(T_i) = \begin{pmatrix} 1 - \dfrac{\Delta N_0(T_i)}{Y_0(T_i)} & \dfrac{\Delta N_{01}(T_i)}{Y_0(T_i)} & \dfrac{\Delta N_{02}(T_i)}{Y_0(T_i)} \\ 0 & 1 - \dfrac{\Delta N_{12}(T_i)}{Y_1(T_i)} & \dfrac{\Delta N_{12}(T_i)}{Y_1(T_i)} \\ 0 & 0 & 1 \end{pmatrix},$$

where $N_0 = N_{01} + N_{02}$. Again, this provides the best algorithm for computing the Aalen–Johansen estimator.

As in the previous example, we will also derive explicit expressions for the elements of $\hat{\mathbf{P}}(s,t)$ and compare them to the similar formulas for the transition probabilities. By the Kolmogorov forward differential equation (2.6.14), we get

$$dP_{00}(s,t) = -P_{00}(s,t-)\,d(A_{01} + A_{02})(t),$$
$$dP_{11}(s,t) = -P_{11}(s,t-)\,dA_{12}(t),$$

and

$$dP_{01}(s,t) = P_{00}(s,t-)\,dA_{01}(t) - P_{01}(s,t-)\,dA_{12}(t).$$

These equations are valid for any three-state Markov illness–death model without recovery having integrated intensity measures A_{01}, A_{02}, and A_{12}. Using (2.6.7), these differential equations are seen to have the solutions

$$P_{00}(s,t) = \prod_{(s,t]} \{1 - d(A_{01} + A_{02})(u)\},$$
$$P_{11}(s,t) = \prod_{(s,t]} \{1 - dA_{12}(u)\}$$

and

$$P_{01}(s,t) = \int_{(s,t]} P_{00}(s,u-)\,dA_{01}(u) \prod_{(u,t]} (1 - dA_{12}(v))$$
$$= \int_{(s,t]} P_{00}(s,u-)\,dA_{01}(u) P_{11}(u,t).$$

For the absolutely continuous case, the well-known formulas

$$P_{00}(s,t) = \exp\left\{-\int_s^t (\alpha_{01}(u) + \alpha_{02}(u))\,du\right\},$$

$$P_{11}(s,t) = \exp\left\{-\int_s^t \alpha_{12}(u)\,du\right\},$$

and

$$P_{01}(s,t) = \int_s^t P_{00}(s,u)\alpha_{01}(u)P_{11}(u,t)\,du$$

result. For the purely discontinuous case, where A_{01}, A_{02}, and A_{12} are the Nelson–Aalen estimators $\hat{A}_{01}$, $\hat{A}_{02}$, and $\hat{A}_{12}$, we get the empirical transition probabilities

$$\hat{P}_{00}(s,t) = \prod_{(s,t]} \{1 - d(\hat{A}_{01} + \hat{A}_{02})(u)\} = \prod_{i=1}^{m}\left(1 - \frac{\Delta N_0(T_i)}{Y_0(T_i)}\right),$$

$$\hat{P}_{11}(s,t) = \prod_{(s,t]} (1 - d\hat{A}_{12}(u)) = \prod_{i=1}^{m}\left(1 - \frac{\Delta N_{12}(T_i)}{Y_1(T_i)}\right),$$

and

$$\hat{P}_{01}(s,t) = \int_s^t \hat{P}_{00}(s,u-)\,d\hat{A}_{01}(u)\hat{P}_{11}(u,t)$$

$$= \sum_{i=1}^{m}\left\{\prod_{h=1}^{i-1}\left(1 - \frac{\Delta N_0(T_h)}{Y_0(T_h)}\right)\frac{\Delta N_{01}(T_i)}{Y_0(T_i)}\prod_{h=i+1}^{m}\left(1 - \frac{\Delta N_{12}(T_h)}{Y_1(T_h)}\right)\right\}.$$

Since $\hat{P}_{00}(s,t)$ and $\hat{P}_{11}(s,t)$ are Kaplan–Meier-type estimators, their variances may be estimated as described in Section IV.3.1, whereas by (4.4.16),

$$\widehat{\operatorname{var}}\,\hat{P}_{01}(s,t) = \int_s^t \{\hat{P}_{00}(s,u)(\hat{P}_{11}(u,t) - \hat{P}_{01}(u,t))\}^2 J_0(u)(Y_0(u))^{-2}\,dN_{01}(u)$$

$$+ \int_s^t \{\hat{P}_{00}(s,u)\hat{P}_{01}(u,t)\}^2 J_0(u)(Y_0(u))^{-2}\,dN_{02}(u)$$

$$+ \int_s^t \{\hat{P}_{01}(s,u)\hat{P}_{11}(u,t)\}^2 J_1(u)(Y_1(u))^{-2}\,dN_{12}(u).$$

The alternative Greenwood-type estimator for the variance may again be found from this expression as explained below (4.4.18).

Various interpretations of the transition probabilities are available in concrete applications of the model. In a study involving treatment of cancer, state 0 could correspond to "no response to treatment," state 1 to "response to treatment," and state 2 to "relapse." The probability $P_{01}(0,t)$ is then the so-called "probability of being in response function" suggested by Temkin (1978) and sometimes used as an outcome measure when studying the efficacy of cancer chemotherapy.

In studies of disability and life insurance, state 0 is the "active" state, state 1 corresponds to "disabled," and state 2 to "dead." In this context, the transition probabilities enter into the calculation of life insurance premiums as follows: If an insured person pays the premium $\pi\,\mathrm{d}t$ when alive and active in the time interval $(t, t + \mathrm{d}t)$ after the issue of the policy at age x, then the actuarial value (cf. Example IV.3.9) of the premiums paid by the insured is

$$\pi \bar{a}^{aa}_{x:\overline{m}|} = \pi \int_0^m v^t P_{00}(x, x+t)\,\mathrm{d}t.$$

Here the insurance runs for m years and $v = 1/(1+i)$, where i is the interest rate. If the individual dies before age $x + m$, the spouse (or other relatives) of the insured gets a prespecified amount B. The actuarial value of this benefit is

$$B\bar{A}^{\,1}_{x:\overline{m}|} = B \int_0^m v^t \{P_{00}(x, x+t)\alpha_{02}(x+t) + P_{01}(x, x+t)\alpha_{12}(x+t)\}\,\mathrm{d}t.$$

The net premium π must then satisfy $\pi \bar{a}^{aa}_{x:\overline{m}|} = B\bar{A}^{\,1}_{x:\overline{m}|}$.

From disability and mortality data, one can now estimate $\bar{a}^{aa}_{x:\overline{m}|}$ and $\bar{A}^{\,1}_{x:\overline{m}|}$ using the Aalen–Johansen estimators for $P_{00}(x, x+t)$ and $P_{01}(x, x+t)$ combined with the Nelson–Aalen estimators for (the integrals of) α_{02} and α_{12}. As shown by Præstgaard (1991), large sample properties of such estimators can be studied along the lines of Example IV.3.9 using the functional delta-method (Section II.8).

In Example VII.2.11, we return to life insurance aspects of the Steno Memorial hospital data introduced in Example I.3.11. Here, we only apply the Aalen–Johansen estimator for studying the survival probability $P_{00}(s,t) + P_{01}(s,t) = 1 - P_{02}(s,t)$, after t years of diabetes, given no sign of diabetic nephropathy (DN) at diabetes duration $s < t$.

Figure IV.4.9 shows these estimates for $s = 5$ years for patients with age at onset of diabetes between 16 and 30 years and separately for males and females. It is seen that females have a better prognosis than males, mainly owing to the fact that females have a lower intensity of developing DN.

It is interesting to compare the estimate for females with that based on the Fyn diabetics survival data (Example IV.3.4). The seemingly better survival seen in the present example than that in Figure IV.3.7 (compare with the estimates corresponding to conditioning on being alive at age 20 or 30 years) is a consequence of the conditioning of being without DN at diabetes duration $s = 5$ years, whereas some unknown fraction of the Fyn diabetics will have developed DN at age 20 or 30 years. □

Example IV.4.4. The Three-State Markov Illness–Death Model with Recovery. Abnormal Prothrombin Levels in Liver Cirrhosis

Let us now extend the model of the previous example by allowing for recovery. Thus, our model is as illustrated in Figure I.3.6. Again, the transition probability matrix may be estimated by

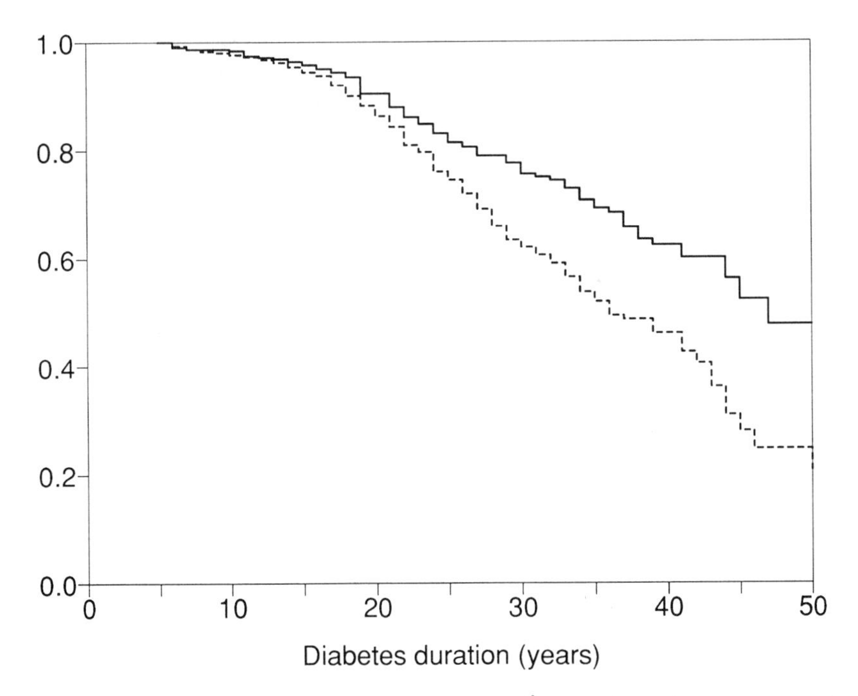

Figure IV.4.9. Estimated survival probabilities $1 - \hat{P}_{02}(5, t)$ for male (- - -) and female (—) insulin-dependent diabetics with age at disease onset 16–30 years given no sign of diabetic nephropathy after 5 years of diabetes duration.

$$\hat{\mathbf{P}}(s,t) = \prod_{i=1}^{m} (\mathbf{I} + \Delta\hat{\mathbf{A}}(T_i)),$$

where now ($N_0 = N_{01} + N_{02}$ and $N_1 = N_{10} + N_{12}$)

$$\mathbf{I} + \Delta\hat{\mathbf{A}}(T_i) = \begin{pmatrix} 1 - \dfrac{\Delta N_0(T_i)}{Y_0(T_i)} & \dfrac{\Delta N_{01}(T_i)}{Y_0(T_i)} & \dfrac{\Delta N_{02}(T_i)}{Y_0(T_i)} \\ \dfrac{\Delta N_{10}(T_i)}{Y_1(T_i)} & 1 - \dfrac{\Delta N_1(T_i)}{Y_1(T_i)} & \dfrac{\Delta N_{12}(T_i)}{Y_1(T_i)} \\ 0 & 0 & 1 \end{pmatrix}$$

and $s < T_1 < T_2 < \cdots < T_m \leq t$ as usual denote the times of the observed transitions between any two states.

For this model, no explicit expressions for the transition probabilities $P_{hj}(s, t)$ or the corresponding Aalen–Johansen estimates are available.

We shall illustrate this using the data from the randomized clinical trial CSL 1 on the effect of prednisone treatment versus placebo on survival in patients with liver cirrhosis (Examples I.3.4 and I.3.12). As explained in the latter example, we shall here assume that observation of the prothrombin

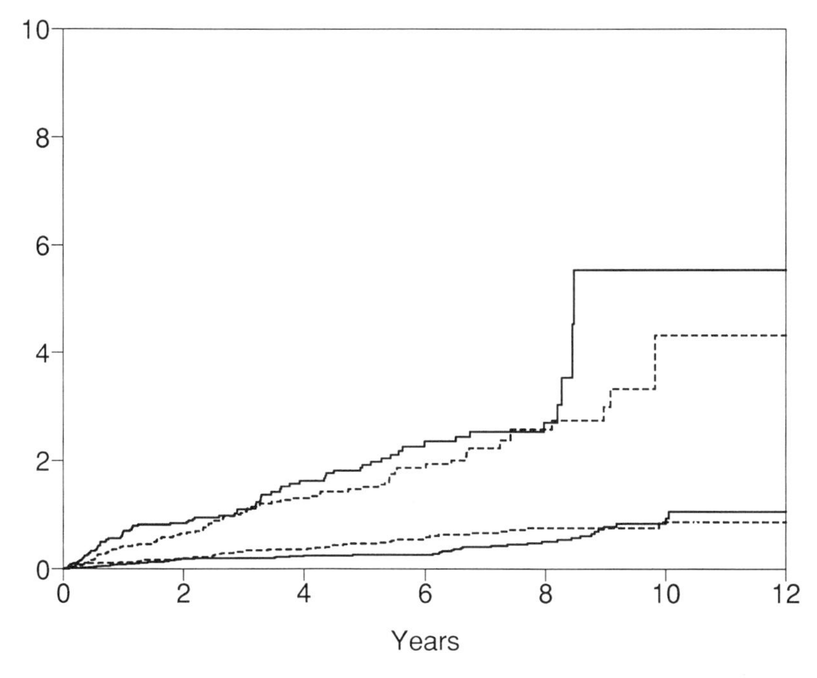

Figure IV.4.10. Nelson–Aalen estimates for the integrated death intensities $A_{02}(t)$ (lower pair of curves) and $A_{12}(t)$ (upper pair of curves). Prednisone (—); placebo (- - -).

index is *continuous*, and we define changes of values to take place at the time points where patients are examined at follow-up visits to the hospital. We let state 0 correspond to "normal prothrombin level," state 1 to "low (or abnormal) prothrombin level," and state 2 to "dead."

Figure IV.4.10 shows the Nelson–Aalen estimates for the integrated death intensities $\hat{A}_{02}(t)$ and $\hat{A}_{12}(t)$ for prednisone- and placebo-treated patients, respectively. It is seen that the death intensity for both treatments is higher for patients with low prothrombin, the *ratio* between the intensities for low and normal values being larger for prednisone-treated patients.

Figure IV.4.11 shows $\hat{A}_{01}(t)$ and $\hat{A}_{10}(t)$, for both prednisone- and placebo-treated patients. It seems as if placebo-treated patients are more likely to go from the "normal" to the "abnormal" (low) state, whereas prednisone treatment seems to slightly increase the intensity $\alpha_{10}(t)$, the difference, however, being of a smaller order of magnitude. These tendencies confirm what we saw in Table I.3.7.

It is not obvious how to summarize the treatment effect on the prognosis based on these intensity plots. For this purpose a more satisfactory outcome measure to use is the *survival probability* $1 - P_{h2}(0, t)$, *given the initial state* $h = 0$ or 1. Figures IV.4.12 and IV.4.13 show the Aalen–Johansen estimates

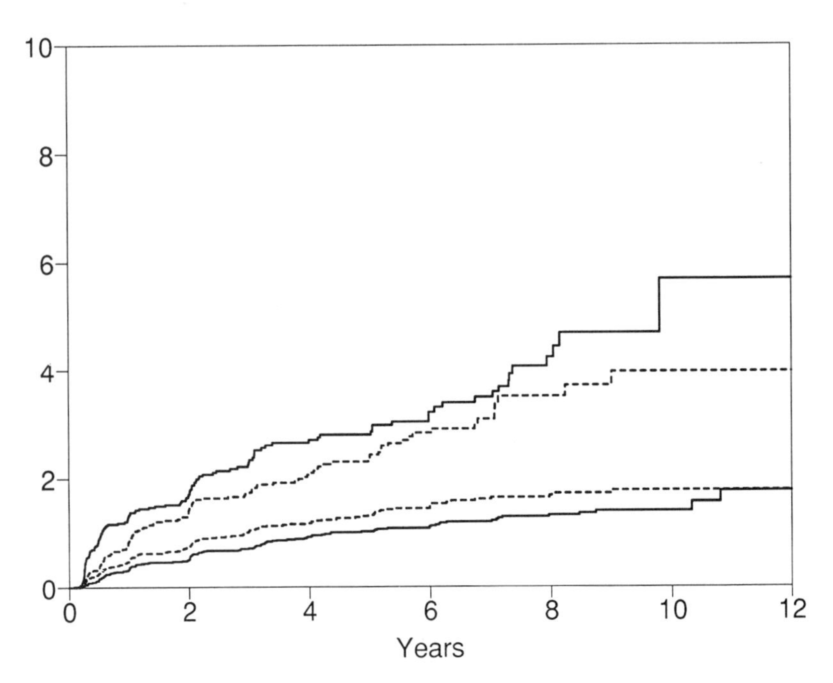

Figure IV.4.11. Estimated integrated transition intensities for prednisone- (—) and placebo- (- - -) treated patients. Upper pair of curves: low to normal prothrombin [$\hat{A}_{10}(t)$]. Lower pair of curves: normal to low prothrombin [$\hat{A}_{01}(t)$].

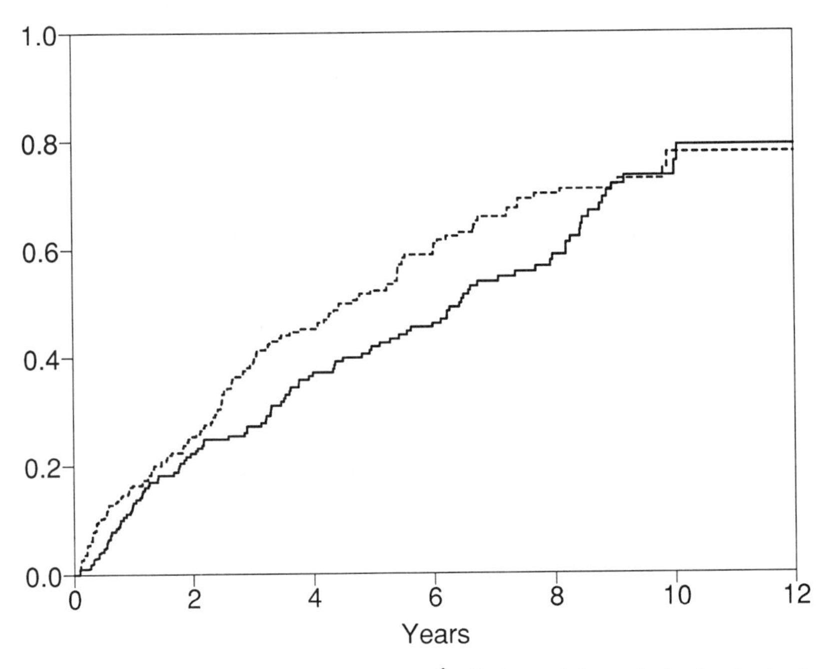

Figure IV.4.12. Aalen–Johansen estimates $\hat{P}_{02}(0, t)$: prednisone (—); placebo (- - -).

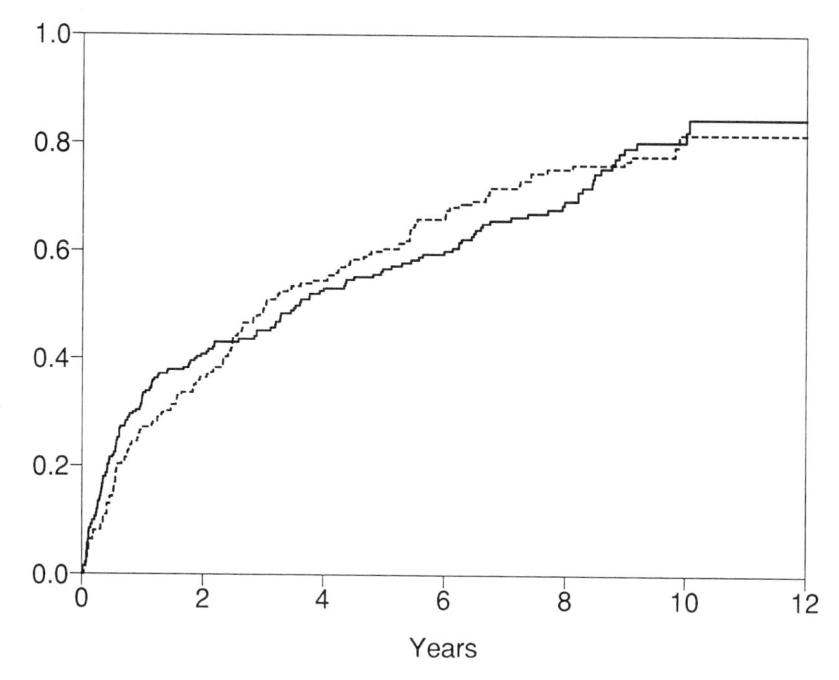

Figure IV.4.13. Aalen–Johansen estimates $\hat{P}_{12}(0, t)$: prednisone (—); placebo (- - -).

$\hat{P}_{02}(0, t)$ and $\hat{P}_{12}(0, t)$ for both treatment groups. There is a tendency that the estimated death probability $\hat{P}_{02}(0, t)$ is higher for placebo-treated patients, whereas the two estimates for $\hat{P}_{12}(0, t)$ are roughly similar.

Figure IV.4.14 shows $\hat{P}_{02}(0, t)$ for the prednisone group with approximate 95% pointwise confidence intervals computed in this way.

When the main outcome measure is the survival probability, it is of interest to compare the Aalen–Johansen estimates with the simple Kaplan–Meier estimates based on those patients in the relevant treatment group who are in state 0 or state 1 at time 0. As an example of such a comparison Figure IV.4.15 shows estimates of the survival probability $1 - P_{02}(0, t)$ for patients with normal prothrombin index at time 0. It is seen that the Kaplan–Meier estimate and Aalen–Johansen estimate are very similar. However, the *variance* of the Kaplan–Meier estimate is larger than that of the Aalen–Johansen estimate; cf. Figure IV.4.16. This is probably a consequence of the fact that the former estimate is based only on patients with normal prothrombin index at time 0, whereas all patients contribute information to the latter. We may, therefore, conclude that there may be a considerable gain in precision by using the procedures discussed in this section, but it should be kept in mind that the (Markov) assumption behind the Aalen–Johansen estimates is stronger than the assumption for the Kaplan–Meier estimate. □

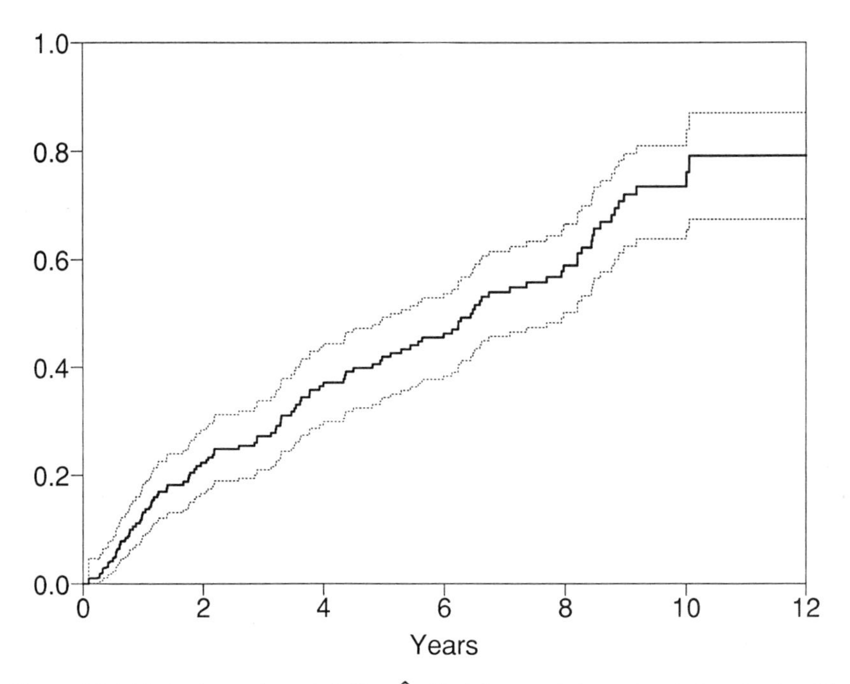

Figure IV.4.14. Estimated probability $\hat{P}_{02}(0, t)$ for prednisone-treated patients with approximate 95% confidence limits calculated using the log-transform.

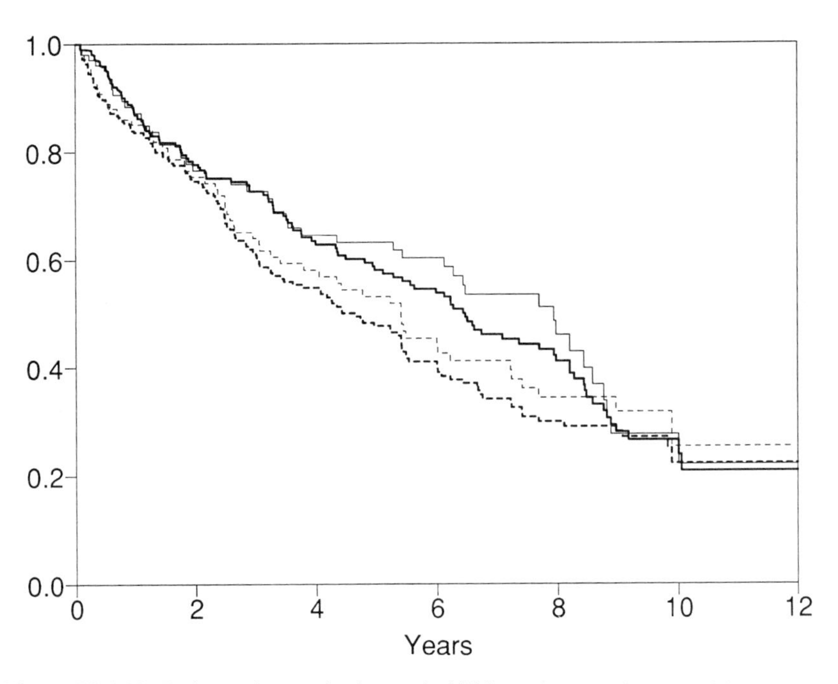

Figure IV.4.15. Estimated survival probabilities for patients with normal prothrombin index at time 0. Comparison of Aalen–Johansen estimates (thick curves) and Kaplan–Meier estimates (thin curves). Prednisone (—); placebo (- - -).

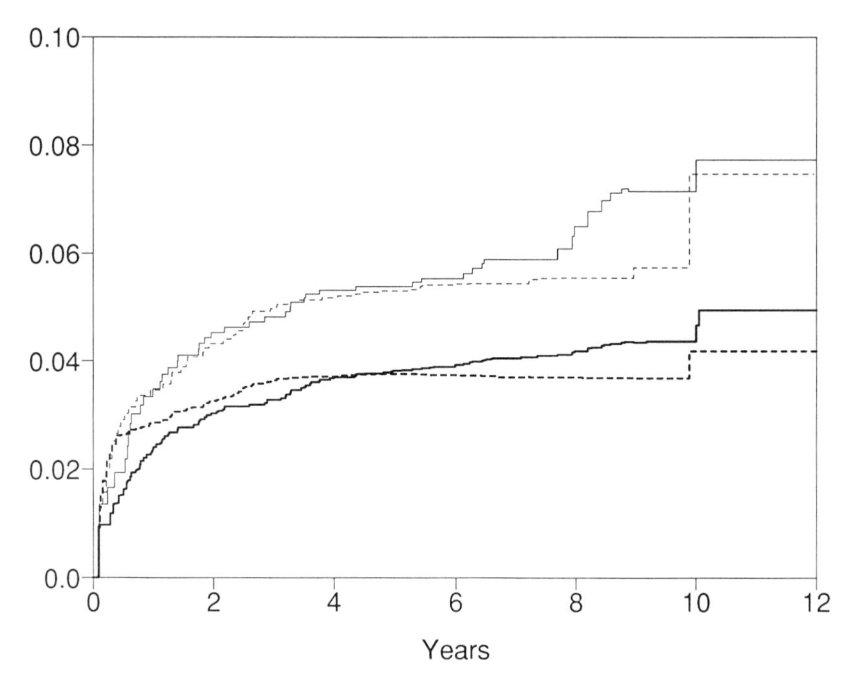

Figure IV.4.16. Estimated standard deviations of Aalen–Johansen estimates (thick curves) and Kaplan–Meier estimates (thin curves) for the survival probability $1 - P_{02}(0, t)$ given normal prothrombin index at time 0. Prednisone (—); placebo (- - -).

IV.4.2. Large Sample Properties

We will now study the asymptotic properties of $\hat{\mathbf{P}}(s, t)$ when the number n of observed processes increases to infinity. Both $\hat{\mathbf{P}}(s, t)$ and $\mathbf{P}^*(s, t)$ will depend on n, but we do not make this explicit in the notation.

Let us first prove that the estimator is uniformly consistent under a set of conditions similar to those needed for the Nelson–Aalen estimator. These conditions were discussed further in Example IV.1.9 for the case of no censoring. We define the norm of a matrix $\mathbf{B} = (b_{hj})$ by $\|\mathbf{B}\| = \sup_h \sum_j |b_{hj}|$ and recall that $\tau = \sup\{u: \int_0^u \alpha_{hj}(v)\,dv < \infty, h \neq j\}$. We then have the following result.

Theorem IV.4.1. *Let $s, v \in [0, \tau)$ with $s < v$ and assume that, as $n \to \infty$, for all $h \neq j$*

$$\int_s^v \frac{J_h^{(n)}(u)}{Y_h^{(n)}(u)} \alpha_{hj}(u)\,du \xrightarrow{\mathrm{P}} 0 \tag{4.4.21}$$

and

$$\int_s^v (1 - J_h^{(n)}(u))\alpha_{hj}(u)\,du \xrightarrow{P} 0. \tag{4.4.22}$$

Then, as $n \to \infty$,

$$\sup_{t \in [s,v]} \|\hat{\mathbf{P}}(s,t) - \mathbf{P}(s,t)\| \xrightarrow{P} 0.$$

PROOF. This result is an immediate consequence of Theorem IV.1.1 and the continuity of product-integrals (cf. the remark below Proposition II.8.7). We, however, also give a direct proof: Since $\mathbf{P}^*(s,t)$ is a stochastic matrix, we have $\|\mathbf{P}^*(s,t)\| = 1$, and it follows that

$$\begin{aligned}\|\hat{\mathbf{P}}(s,t) - \mathbf{P}(s,t)\| &= \|(\hat{\mathbf{P}}(s,t)\mathbf{P}^*(s,t)^{-1} - \mathbf{I})\mathbf{P}^*(s,t) + \mathbf{P}^*(s,t) - \mathbf{P}(s,t)\| \\ &\le \|\hat{\mathbf{P}}(s,t)\mathbf{P}^*(s,t)^{-1} - \mathbf{I}\| + \|\mathbf{P}^*(s,t) - \mathbf{P}(s,t)\|.\end{aligned}$$

We will treat these two terms separately. By (4.1.4) and (4.4.7), the (h,j)th entry of $\hat{\mathbf{P}}(s,t)\mathbf{P}^*(s,t)^{-1} - \mathbf{I}$ is a sum of stochastic integrals of the form

$$\int_s^t \hat{P}_{hl}(s,u-)P^{*mj}(s,u)J_l^{(n)}(u)(Y_l^{(n)}(u))^{-1}\,dM_{lm}^{(n)}(u)$$

with $M_{ll}^{(n)} = -\sum_{j \ne l} M_{lj}^{(n)}$. Now,

$$\langle M_{ll}^{(n)} \rangle(t) = -\int_0^t \alpha_{ll}(u)\,Y_l^{(n)}(u)\,du,$$

and version (2.5.18) of Lenglart's inequality gives

$$\begin{aligned}&\mathrm{P}\left\{\sup_{t\in[s,v]}\left|\int_s^t \hat{P}_{hl}(s,u-)P^{*mj}(s,u)J_l^{(n)}(u)(Y_l^{(n)}(u))^{-1}\,dM_{lm}^{(n)}(u)\right| > \eta\right\} \\ &\quad \le \frac{\delta}{\eta^2} + \mathrm{P}\left\{\varepsilon_{lm}\int_s^v \{\hat{P}_{hl}(s,u-)P^{*mj}(s,u)\}^2 J_l^{(n)}(u)(Y_l^{(n)}(u))^{-1}\alpha_{lm}(u)\,du > \delta\right\}\end{aligned}$$

for all $\eta, \delta > 0$, where $\varepsilon_{lm} = 1$ for $l \ne m$ and $\varepsilon_{ll} = -1$. Using (4.4.8), the right-hand side is bounded by

$$\frac{\delta}{\eta^2} + \mathrm{P}\left\{c\varepsilon_{lm}\int_s^v \frac{J_l^{(n)}(u)}{Y_l^{(n)}(u)}\alpha_{lm}(u)\,du > \delta\right\}$$

for some constant c. By (4.4.21), it, therefore, follows that

$$\sup_{t\in[s,v]} \|\hat{\mathbf{P}}(s,t)\mathbf{P}^*(s,t)^{-1} - \mathbf{I}\| \xrightarrow{P} 0$$

as $n \to \infty$.

We then use Duhamel's equation (2.6.4) to get

$$\mathbf{P}^*(s,t) - \mathbf{P}(s,t) = \int_s^t \mathbf{P}^*(s,u)\,d(\mathbf{A}^{*(n)} - \mathbf{A})(u)\mathbf{P}(u,t).$$

Since $\|\mathbf{P}^*(s,u)\| = \|\mathbf{P}(u,t)\| = 1$, it follows that

$$\|\mathbf{P}^*(s,t) - \mathbf{P}(s,t)\| \le \int_s^t \|\mathrm{d}(\mathbf{A}^{*(n)} - \mathbf{A})(u)\|$$

$$\le 2\sum_h \int_s^v (1 - J_h^{(n)}(u))|\alpha_{hh}(u)|\,\mathrm{d}u. \tag{4.4.23}$$

By (4.4.22), we, therefore, have

$$\sup_{t\in[s,v]} \|\mathbf{P}^*(s,t) - \mathbf{P}(s,t)\| \xrightarrow{\mathrm{P}} 0$$

as $n \to \infty$, and the proof is complete. □

We will now prove a weak convergence result for $\hat{\mathbf{P}}(s,\cdot)$ under a set of conditions similar to those needed for the Nelson–Aalen estimator. The conditions were discussed for the situation with no censoring in Example IV.1.9.

Theorem IV.4.2. *Let $s, v \in [0,\tau)$ with $s < v$, and assume that there exist non-negative functions y_h defined on $[s,v]$ such that α_{hj}/y_h is integrable over $[s,v]$ for all $h \ne j$. Let*

$$\sigma_{hj}(t) = \int_s^t \{\alpha_{hj}(u)/y_h(u)\}\,\mathrm{d}u \tag{4.4.24}$$

for $h \ne j$ and assume that:

(A) *For each $t \in [s,v]$ and all $h \ne j$,*

$$n\int_s^t \frac{J_h^{(n)}(u)}{Y_h^{(n)}(u)}\alpha_{hj}(u)\,\mathrm{d}u \xrightarrow{\mathrm{P}} \sigma_{hj}(t) \quad \text{as } n \to \infty. \tag{4.4.25}$$

(B) *For all $\varepsilon > 0$ and all $h \ne j$,*

$$n\int_s^v \frac{J_h^{(n)}(u)}{Y_h^{(n)}(u)}\alpha_{hj}(u) I\left(\left|\sqrt{n}\frac{J_h^{(n)}(u)}{Y_h^{(n)}(u)}\right| > \varepsilon\right)\mathrm{d}u \xrightarrow{\mathrm{P}} 0 \quad \text{as } n \to \infty. \tag{4.4.26}$$

(C) *For all $h \ne j$,*

$$\sqrt{n}\int_s^v (1 - J_h^{(n)}(u))\alpha_{hj}(u)\,\mathrm{d}u \xrightarrow{\mathrm{P}} 0 \quad \text{as } n \to \infty. \tag{4.4.27}$$

Then

$$\sqrt{n}(\hat{\mathbf{P}}(s,\cdot) - \mathbf{P}(s,\cdot)) \xrightarrow{\mathscr{D}} \int_s^{\cdot} \mathbf{P}(s,u)\,\mathrm{d}\mathbf{U}(u)\mathbf{P}(u,\cdot),$$

where $\mathbf{U} = (U_{hj})$ is a $k \times k$ matrix-valued process, where, for $h \ne j$, the U_{hj} are independent Gaussian martingales with $U_{hj}(0) = 0$ and $\operatorname{cov}(U_{hj}(t_1), U_{hj}(t_2)) = \sigma_{hj}(t_1 \wedge t_2)$ and $U_{hh} = -\sum_{j\ne h} U_{hj}$.

PROOF. The transition probability matrix $\mathbf{P}(s,t) = \prod_{(s,t]}(\mathbf{I} + \mathrm{d}\mathbf{A})$ is by Proposition II.8.7 a compactly differentiable functional of the integrated intensity measure $\mathbf{A}$. The weak convergence result is, therefore, an immediate consequence of Theorem IV.1.2 and the functional delta-method (Theorem II.8.1).

It is, however, also possible to give a martingale-based proof, and we have chosen to present this as well. To this end, we note that we may write

$$\sqrt{n}(\hat{\mathbf{P}}(s,t) - \mathbf{P}(s,t)) = \sqrt{n}(\hat{\mathbf{P}}(s,t)\mathbf{P}^*(s,t)^{-1} - \mathbf{I})\mathbf{P}^*(s,t) + \sqrt{n}(\mathbf{P}^*(s,t) - \mathbf{P}(s,t)).$$

By (4.4.23) and Assumption C, it now follows that

$$\sup_{t \in [s,v]} \|\sqrt{n}(\mathbf{P}^*(s,t) - \mathbf{P}(s,t))\| \xrightarrow{\mathrm{P}} 0 \tag{4.4.28}$$

as $n \to \infty$. Therefore, it suffices to prove that

$$\sqrt{n}(\hat{\mathbf{P}}(s,\cdot)\mathbf{P}^*(s,\cdot)^{-1} - \mathbf{I}) \xrightarrow{\mathscr{D}} \int_s^{\cdot} \mathbf{P}(s,u)\,\mathrm{d}\mathbf{U}(u)\mathbf{P}(s,u)^{-1}. \tag{4.4.29}$$

By (4.1.4) and (4.4.7), the (h,j)th entry of the left-hand side of (4.4.29) is the stochastic integral

$$\sum_{l=1}^{k}\sum_{m=1}^{k} \sqrt{n}\int_s^{\cdot} \hat{P}_{hl}(s,u-)P^{*mj}(s,u)J_l^{(n)}(u)(Y_l^{(n)}(u))^{-1}\,\mathrm{d}M_{lm}^{(n)}(u).$$

Now the conditions of the present theorem imply those of Theorem IV.4.1, and (4.4.29) follows by the martingale central limit theorem (Theorem II.5.1) using (4.4.8), (4.4.28), and Conditions A and B. □

We will now have a closer look at the covariance structure of the limiting Gaussian process

$$\mathbf{Z}(s,t) = \int_s^t \mathbf{P}(s,u)\,\mathrm{d}\mathbf{U}(u)\mathbf{P}(u,t).$$

By (4.4.12), the covariance matrix of $\mathbf{Z}(s,t)$ may be given as

$$\begin{aligned}\operatorname{cov}(\mathbf{Z}(s,t)) &= \langle \mathbf{Z}(s,\cdot)\rangle(t) \\ &= \int_s^t \mathbf{P}(u,t)^{\top} \otimes \mathbf{P}(s,u)\,\mathrm{d}\langle \mathbf{U}\rangle(u)\mathbf{P}(u,t) \otimes \mathbf{P}(s,u)^{\top},\end{aligned}$$

where

$$\langle \mathbf{U}\rangle(u) = \sum_{l=1}^{k}\sum_{g\neq l}\int_0^u \operatorname{vec} C_{gl}\operatorname{vec}^{\top} C_{gl}\{\alpha_{gl}(v)/y_g(v)\}\,\mathrm{d}v,$$

and, therefore [cf. (4.4.15)],

$$\operatorname{cov}(\mathbf{Z}(s,t)) = \sum_{l=1}^{k} \sum_{g \neq l} \int_s^t \operatorname{vec}\{\mathbf{P}(s,u) C_{gl} \mathbf{P}(u,t)\} \operatorname{vec}^{\mathsf{T}}\{\mathbf{P}(s,u) C_{gl} \mathbf{P}(u,t)\}$$
$$\times \{\alpha_{gl}(u)/y_g(u)\}\, \mathrm{d}u.$$

In the same manner as we derived (4.4.16) from (4.4.15), we get that the covariances of the limiting Gaussian process $\mathbf{Z} = (Z_{hj})$ may be given by

$$\operatorname{cov}(Z_{hj}(s,t), Z_{mr}(s,t)) = \sum_{l=1}^{k} \sum_{g \neq l} \int_s^t P_{hg}(s,u) P_{mg}(s,u) \{P_{lj}(u,t) - P_{gj}(u,t)\}$$
$$\times \{P_{lr}(u,t) - P_{gr}(u,t)\} \frac{\alpha_{gl}(u)}{y_g(u)}\, \mathrm{d}u. \tag{4.4.30}$$

By (4.1.15),

$$\sup_{t \in [s,v]} \left| n \int_s^t J_h^{(n)}(u) (Y_h^{(n)}(u))^{-2}\, \mathrm{d}N_{hj}^{(n)}(u) - \sigma_{hj}(t) \right| \xrightarrow{\mathrm{P}} 0$$

as $n \to \infty$ for all $h \neq j$, under the assumptions of Theorem IV.4.2. Moreover, by Theorem IV.4.1, $\hat{\mathbf{P}}(s,t)$ is uniformly consistent under the same assumptions, and it follows that n times (4.4.16) is a uniformly consistent estimator for (4.4.30). Thus,

$$\sup_{t \in [s,v]} \| n \widehat{\operatorname{cov}}(\hat{\mathbf{P}}(s,t)) - \operatorname{cov}(\mathbf{Z}(s,t)) \| \xrightarrow{\mathrm{P}} 0$$

as $n \to \infty$, where $\widehat{\operatorname{cov}}(\hat{\mathbf{P}}(s,t))$ is given by (4.4.14) or, equivalently, (4.4.15). By the final result of Theorem IV.1.2, this also holds true when $\widehat{\operatorname{cov}}(\hat{\mathbf{P}}(s,t))$ is replaced by the estimator $\widehat{\widehat{\operatorname{cov}}}(\hat{\mathbf{P}}(s,t))$ with elements given by (4.4.18).

IV.5. Bibliographic Remarks

Section IV.1

Nelson (1969, 1972) introduced what we term the Nelson–Aalen estimator as a graphical technique for hazard plotting of censored failure-time data to obtain engineering information on the form of the failure-time distribution. Independently of Nelson, Altshuler (1970) derived the Nelson–Aalen estimator in the context of competing risks animal experiments. He proved that the estimator is unbiased over time intervals where it is guaranteed (by means of an experimental design with equal-age replacement of failed animals) that the risk set never becomes empty.

The large sample properties of the Nelson–Aalen estimator were first derived by Breslow and Crowley (1974) for censored survival data in the simple random censorship model as a main step in their study of the large sample properties of the Kaplan–Meier estimator. They derived our Theorem IV.1.2 for this particular situation using a Skorohod construction and results on weak convergence of empirical distribution functions. Gill (1989) showed that this proof is essentially a functional delta-method proof; its steps are contained in the differentiability propositions of Section II.8. Independently of Breslow and Crowley (1974), Aalen (1976) (based on a 1973 Technical Report from the University of Oslo) studied the Nelson–Aalen estimators for the cause-specific death intensities of a competing risks model. He derived approximate expressions for their expected values and variances, and he proved almost sure uniform consistency for the estimators as well as for the estimators for their variances. Asymptotic normality for finite-dimensional distributions was established by means of characteristic functions, and weak convergence for the Nelson–Aalen estimators considered as processes by applying the theory of Billingsley (1968).

As mentioned in Sections I.1 and I.2, Aalen (1975, 1978b) in his path-breaking work on the multiplicative intensity model for counting processes suggested the Nelson–Aalen estimator for this class of models and studied its small and large sample properties using martingale methods. Our presentation in Sections IV.1.1 and IV.1.2 is based directly on Aalen's work, but is updated according to later theoretical developments (cf. Section I.2).

The literature on confidence intervals and confidence bands for the integrated intensity functions is fairly small, but parallels the more abundant literature on confidence intervals and bands for the survival curve (cf. the Bibliographic Remarks to Section IV.3). Section IV.1.3, which is mainly based on Bie, Borgan, and Liestøl (1987), gives references to work in the area. Nair (1981) also explicitly considered confidence bands for the cumulative hazard function based on censored data.

For the competing risks model, Burke, Csörgő, and Horváth (1981) and Csörgő and Horváth (1982a) showed how the Nelson–Aalen estimator and certain transforms of it can be approximated strongly by Gaussian processes. The latter used the approximation results to derive conservative (as opposed to approximate) confidence bands for the integrated intensity function, but these seem to be of little practical importance.

Bootstrapping procedures for the simple random censorship model for censored survival data was first discussed by Efron (1981). His suggestion was to obtain independent bootstrap samples of the (uncensored) survival times and the censoring times from the Kaplan–Meier estimates for the survival and censoring distributions, respectively, and then combine these in the usual manner to obtain a censored bootstrap sample. Moreover, he showed that this bootstrap algorithm is identical to the one in which we resample with replacement from the observed pairs $(\tilde{X}_i, D_i)$, $i = 1, 2, \ldots, n$. Independently, Akritas (1986), Lo and Singh (1986), and Horváth and Yandell (1987) proved that Efron's suggestion is (asymptotically) correct for estimating the (normalized) distribution of the Kaplan–Meier estimator. The lines of proofs taken in these papers were quite different. Akritas (1986) used martingale theory along the lines of Gill (1980a), Lo and Singh (1986) established a representation of the Kaplan–Meier estimator as an i.i.d. mean process and gave a corresponding bootstrap version of the representation, whereas Horváth and Yandell (1987) used strong approximations of the bootstrapped Kaplan–Meier process to Wiener processes. The latter two papers also provided rates of convergence of their approximations and established parallel results for the bootstrapped Kaplan–Meier quantile process.

Doss and Gill (1992) discussed a modification [proposed by Hjort (1985b) in the context of Cox's regression model] in which the bootstrap sample of the censoring times is obtained conditionally on what is actually known about the true censoring times. They also discussed a similar conditional bootstrap algorithm for the situation where the censoring is obtained by a combination of progressive type I and simple random censoring. For both of these algorithms, the (asymptotic) correctness of the bootstrap for estimating the distribution of the Kaplan–Meier estimator was shown by Kim (1990), and based on this, Doss and Gill (1992) showed the correctness of the bootstrap for estimating the distribution of the Kaplan–Meier quantile process.

The "weird bootstrap" of Definition IV.1.1 is new here. It is not tailored to fit a particular censoring pattern as was the case with the above-mentioned bootstrap procedures. Moreover, it may, as noted in Section IV.1.4, find useful applications for the more complex estimators and models of Sections IV.3. and IV.4. In density estimation problems, however, the use of the bootstrap is problematic and our approach is not immediately available for the problems of Section IV.2. The classical i.i.d. bootstrap already fails for ordinary density estimation as Efron (1979) pointed out in his first paper on the subject. The reason for this is that the large sample distribution of, e.g., a kernel density estimator, depends on the true *density* and even on its second-order *derivative*, whereas the empirical distribution has no density, let alone such a smooth one. Implicitly [as in the proposal by Hall (1990)] or explicitly, one has to bootstrap from a preliminarily estimated smooth distribution, in fact, oversmoothed so as to estimate second-order derivatives correctly as well as the density itself. It is not clear how our weird bootstrap could be adapted to cope with such problems.

Section IV.1.5 gives a discussion of various proposals which have been put forward on how a nonparametric maximum likelihood estimator (NPMLE) may be defined, and Bibliographic comments on the literature are given along the way. We also mention that, for the special cases of censored survival data and observations from finite-state Markov process, Johansen (1978), following the lines of Kaplan and Meier (1958), showed that the Nelson–Aalen estimator and the corresponding product-limit estimators (i.e., the Kaplan–Meier and Aalen–Johansen estimators) are NPMLE in the sense of Kiefer and Wolfowitz (1956). Moreover, Johansen (1983) proposed the Poisson extension (4.1.38) of the multiplicative intensity model for counting processes and gave (4.1.35) a probabilistic interpretation. Again, he derived the

Nelson–Aalen estimator as the NPMLE in the Kiefer–Wolfowitz sense, cf. our derivation below (4.1.38). Scholz (1980) pointed out that *all* definitions of maximum likelihood estimators depend on which choice of probability density is made (only almost everywhere determined). He showed that the wrong choice can result in a complete breakdown of the method and proposed taking a limit of ratios of probabilities to make the right choice. Applied to censored survival data, his careful analysis again yields the Kaplan–Meier estimator.

Hjort (1990b) derived a nonparametric Bayes estimator for the integrated intensities in survival and Markov process models subject to right-censoring using a particular class of prior processes, called by him beta processes. His Bayes estimator is easy to interpret and to compute, and it gives the Nelson–Aalen estimator as the limiting case of a vague prior.

Doubly censored survival data, i.e., data which are censored both to the left and to the right, can be analyzed as left-filtered and right-censored as described in Section III.4 and Example IV.1.8. However, by such an analysis, one deliberately throws away the information contained in the left-censored observations. Turnbull (1974) derived an iterative estimation procedure (for estimating the survival curve) which also uses the information contained in the left-censored data (cf. the Bibliographic Remarks to Chapter III). Samuelsen (1989) described this procedure with a special emphasis on the estimation of the integrated intensity. More importantly, by embedding the model for doubly censored data in a six-state Markov process model, he was able to derive the asymptotic properties of the "one-step estimator" for the integrated intensity obtained by iterating only once by the Turnbull algorithm (starting from the Nelson–Aalen estimator obtained by treating the left-censored data as left-filtered).

Section IV.2

The problem of smoothing the Nelson–Aalen estimator of the integral of α in the multiplicative intensity model for counting processes is in many respects closely related to nonparametric density estimation and nonparametric curve estimation (or regression smoothing). A nice survey of density estimation (for uncensored data), with a view toward applications, was given by Silverman (1986). His main emphasis was on the kernel smoothing method, but other methods like the nearest neighbor method, kernel smoothing with variable bandwidth, maximum penalized likelihood, and orthogonal series expansions were also discussed. Izenman (1991) reviewed recent developments in this field and gave an extensive bibliography. Other recent surveys by Müller (1988) and Härdle (1990) were phrased in the context of regression smoothing but are equally relevant for our context.

In these Bibliographic Remarks, we concentrate on works on nonparametric estimation of the hazard function for censored survival data, or more generally on estimation of α in the multiplicative intensity model for counting processes. It is important to realize, however, that the developments within nonparametric density estimation and regression smoothing will often have implications for the estimation of α in the multiplicative intensity model.

The kernel function estimator for the hazard function was first proposed and studied by Watson and Leadbetter (1964a, 1964b) for the uncensored data case, while Rice and Rosenblatt (1976) further investigated its properties.

Ramlau-Hansen (1983a) considered the kernel function estimator within the framework of the multiplicative intensity model for counting processes, thereby extending its use to censored survival data, Markov process models, etc. (cf. Section III.1). Within this framework, he studied small sample properties (expectation, variance) as well as large sample properties (consistency, asymptotic normality) of the estimator. In a companion paper, Ramlau-Hansen (1983b), among other things, showed how the analysis of the mean integrated squared error from density estimation readily extends to the multiplicative intensity model. Our Section IV.2 is to a large extent based on Ramlau-Hansen's work.

Independently of Ramlau-Hansen, Tanner and Wong (1983) and Yandell (1983) studied the kernel function estimator for the hazard function for randomly censored survival data by more classical methods. Tanner and Wong's (1983) results are to a large extent contained as special cases in those of Ramlau-Hansen (1983a). However, they did not assume a compact support of the kernel function. Yandell (1983), on the other hand, used strong approximations to derive the asymptotic distribution of the maximal deviation of the kernel estimate from the true hazard. Using this result, he developed simultaneous confidence bands for the hazard function, and he studied the performance of these bands by stochastic simulation.

In kernel function estimation, both the bandwidth and the kernel must be chosen by the user. Of these two choices, the first one is the most important, and except for an empirical illustration of the performance of various kernels in Example IV.2.1, we have in Section IV.2 concentrated on the choice of the bandwidth.

The cross-validation method for determining the bandwidth based on the (mean) integrated squared error was proposed by Rudemo (1982) and Bowman (1984) for density estimation with uncensored data. It has since then been investigated intensively by a number of authors and is known to possess certain optimality properties [cf. the review by Izenman (1991)]. For randomly censored survival data, Marron and Padgett (1987) showed that the cross-validation method gives the optimal bandwidth (for estimating the density) in the sense that the ratio between the integrated squared error obtained by the "cross-validation bandwidth" and the infimum of the integrated squared error obtained by any bandwidth converges to one almost surely. Inspired by Rudemo (1982), Ramlau-Hansen (1981) (in an unpublished report in Danish) suggested the cross-validation method reviewed in Example IV.2.3 for the multiplicative intensity model for counting processes. We expect the cross-validation method to possess similar optimality properties in the counting process setup as for censored survival data, but to our knowledge no study exists which supports this conjecture.

The literature on the choice of the kernel function is far less extensive than that on the choice of the bandwidth. An argument for the optimality of the Epanechnikov kernel was reviewed by Silverman (1986). Ramlau-Hansen (1983b), Müller (1984), and Gasser, Müller, and Mammitzsch (1985) discussed various criteria for optimality of the kernel function and derived classes of kernel functions having specific optimality properties.

Like most of the literature on density estimation, nonparametric regression, and hazard estimation, we have adopted "the L_2-view" and concentrated on the squared error between the estimator $\hat{\alpha}$ and the true value of α. Devroye and Györfi (1985) and Devroye (1987) gave arguments for focusing on the absolute error ("the L_1-view") and presented a theory for density estimation (for uncensored data) based on this approach.

We have, in Section IV.2, concentrated on the kernel function estimator (4.2.1) for estimating α in the multiplicative intensity model. A number of the other methods from density estimation (for uncensored data) have, however, also been modified to allow for estimation of the hazard function for censored survival data or α in the multiplicative intensity model for counting processes.

For randomly censored survival data, various proposals for kernel estimators for the hazard function with variable bandwidths (depending on the data) have been put forward and studied by Tanner (1983), Tanner and Wong (1984), Schäfer (1985), and Müller and Wang (1990). Bartoszyński et al. (1981) proposed a variable bandwidth method for estimating the intensity of a nonstationary Poisson process based on censored replicates of the process, and Hjort (1985a, 1992b) indicated how an estimator for α based on orthogonal series expansions could be implemented for the multiplicative intensity model for counting processes. Further suggestions were given by Hjort (1992b).

As discussed in Section IV.1.5, a nonparametric maximum likelihood estimator for α does not exist since the likelihood (4.1.33) is unbounded if α is allowed to be any non-negative function. There are essentially two ways in which the likelihood method can be salvaged to produce a nonparametric estimate. One method is to subtract a "penalty term" from the log-likelihood function which penalizes rough behavior of the estimate. The resulting penalized log-likelihood function can then be maximized to produce an estimate for α. An alternative is to adopt Grenander's method of sieves where the log-likelihood function is maximized over a constrained subspace of the parameter space and the constraint is relaxed as sample size grows. A penalized maximum likelihood estimate for the hazard function based on censored survival data was first suggested by Anderson and Senthilselvan (1980), whereas Bartoszyński et al. (1981) proposed such an estimate for the intensity of a nonstationary Poisson process. Senthilselvan (1987) and O'Sullivan (1988) proposed estimates for randomly censored survival data based on a reparametrization of the penalized log-likelihood function by the square root and the logarithm of the hazard function, respectively. These papers were mainly concerned with problems concerning the existence and the numerical evaluation of the estimates. Antoniadis (1989) and Antoniadis and Grégoire (1990) suggested estimators in the multiplicative intensity model (based on a reparametrization by $\log \alpha$ and $\sqrt{\alpha}$, respectively) obtained by penalizing a modification of the log-likelihood function, and they studied their asymptotic properties using results on counting processes, martingales, etc. Karr (1987) studied maximum likelihood estimation of α in the multiplicative intensity model using the method of sieves, and he derived consistency of the estimator and weak convergence of its integral toward the same limit as the Nelson–Aalen estimator.

Section IV.3

The actuarial estimator for the survival function has been used for hundreds of years. The product-limit estimator, obtained as a limiting case of the actuarial estimator, seems to have been first proposed by Böhmer (1912). It was, however, lost sight of by later researchers and not further investigated until Kaplan and Meier's (1958) important paper appeared. The collaboration of this paper resulted when Kaplan and Meier almost simultaneously prepared separate manuscripts that were in many respects similar. Kaplan and Meier (1958) derived the estimator for right-censored survival data by a heuristic argument based on the chain rule for conditional probabilities, but they also showed how it could be interpreted as a nonparametric maximum likelihood estimator. Furthermore, they showed that the estimator is almost unbiased and derived an approximate expression for its variance. Renewal testing as well as the necessary modification of the estimator for left-truncated and right-censored data were also discussed.

Efron (1967) showed that (his version of) the Kaplan–Meier estimator is a so-called "self-consistent" estimator. His self-consistency equations are, in fact, the generalized score equations discussed toward the end of Section IV.1.5 (Gill, 1989). Peterson (1977) expressed the Kaplan–Meier estimator as a function of two empirical subsurvival functions. This idea was taken up by Gill (1981) in connection with testing with replacement.

The large sample properties of the Kaplan–Meier estimator for the random censorship model were first conjectured by Efron (1967) and later proved by Breslow and Crowley (1974) (cf. the Bibliographic Remarks to Section IV.1). Meier (1975) derived parallel results for fixed censoring times.

As a by-product of their study of the product-limit estimator for the transition probabilities of a Markov process (Section IV.4), Aalen and Johansen (1978) derived small as well as large sample properties of the Kaplan–Meier estimator for quite general censoring patterns using product-integration and results on counting processes, martingales, and stochastic integrals. Along the same lines, Gill (1980a) performed a detailed study of the Kaplan–Meier estimator for arbitrary survival distributions (not necessarily continuous) under independent right-censoring (his concept of independent censoring differs slightly from ours, cf. Example III.2.13). [Although product-integrals are not explicitly mentioned by Gill (1980a), product-integral-based relations like our (4.3.4) are central for his treatment.] Some of the results of Gill (1980a) were summarized and extended by Gill (1983a). Our treatment in Sections IV.3.1 and IV.3.2 is mainly based on Gill's work, but we have restricted our attention to absolutely continuous survival distributions. Moreover, unlike Gill (1980a, 1983a), we only consider the asymptotic properties of the Kaplan–Meier estimator on compact time intervals $[0, t]$ for which $S(t) > 0$. Remarks as to how the results extend to the whole support of S are, however, provided at the end of Section IV.3.2.

Additional large sample results for the Kaplan–Meier estimator have been provided by Csörgő and Horváth (1983) who showed it to be strongly uniformly consistent under random censoring and studied its rate of convergence to the true distribution function. They also gave a review of earlier related works; in fact, Winter, Földes, and Rejtő (1978) seem to be the first to have proved strong uniform consistency of the estimator. Furthermore, Wellner (1985) showed that for the simple random censorship model (N, Y) converges jointly to a Poisson process (with dependent components) under "heavy censoring," and he used this to derive Poisson-type limit theorems for the Kaplan–Meier estimator (and the Nelson–Aalen estimator). For the Kaplan–Meier estimator, he used this to propose an alternative approximate variance formula (cf. Example IV.3.1).

As mentioned earlier, already Kaplan and Meier (1958) described how their estimator could be adapted to handle left-truncated data, and the practical use of the product-limit estimator under left-truncation has flourished since then. Nevertheless, recent years have seen a renewed interest in the study of the theoretical properties of the Kaplan–Meier estimator under left-truncation. Most of these works have used a classical approach, like Woodroofe (1985), who was motivated from problems in astronomy, and Wang, Jewell, and Tsai (1986). Keiding and Gill (1990), however, showed that by a simple reparametrization of the left-truncation model as a five-state Markov process, the results of Aalen and Johansen (1978) can be used directly to study the properties of the Kaplan–Meier estimator under left-truncation (cf. Examples III.3.3 and IV.3.10).

As illustrated in Example IV.3.4, it may be difficult to get sensible estimates for the whole survival curve based on left-truncated and right-censored data. In this connection, it is worth noting that Lai and Ying (1991a) proposed a modified Kaplan–Meier estimator which discards the contributions to the product (4.3.2) for the (random) time intervals where there are small risk sets, and they proved uniform consistency and weak convergence of their modified estimator over the entire observable range of the distribution function under a general model for random left-truncation and right-censoring.

Moreover, a small simulation study indicated that their modified estimator becomes more stable than the ordinary Kaplan–Meier estimator especially for left-truncated data.

Confidence bands for the survival curve have been proposed by Gillespie and Fisher (1979), Hall and Wellner (1980), and Nair (1981, 1984). Csörgő and Horváth (1986) and Hollander and Peña (1989) introduced classes of bands which show the relationship between most of these proposals and also include some new bands. In Section IV.3.3, we have concentrated on the bands discussed by Hall and Wellner (1980) and Nair (1981, 1984), as well as their transformed counterparts studied by Borgan and Liestøl (1990). We also mention that Thomas and Grunkemeier (1975), based on a restricted product-limit estimator, gave two alternatives to the "linear" confidence interval (4.3.19) having small sample properties comparable to our transformed intervals (4.3.20) and (4.3.21).

Burke, Csörgő, and Horváth (1981) and Csörgő and Horváth (1982b) showed for the competing risks model how the Kaplan–Meier estimator and certain transforms of it can be approximated strongly by Gaussian processes [see also Major and Rejtő (1988)].

Nonparametric Bayesian estimation of the survival curve for right-censored failure times has been considered by a number of authors. Susarla and Van Ryzin (1976) derived a nonparametric Bayesian estimator (giving the Kaplan–Meier estimator as the limiting case of a vague prior) using a Dirichlet process prior [cf. Ferguson (1973)]. Ferguson and Phadia (1979) extended their results to a more general class of prior distributions, namely, the processes "neutral to the left" [cf. Doksum (1974)]. Another nonparametric Bayesian estimator for the survival function was obtained by Hjort (1990b) by product-integrating his nonparametric Bayes estimator for the cumulative hazard function (cf. Bibliographic Remarks to Section IV.1).

As mentioned in connection with the Bibliographic Remarks to Chapter III and Section IV.1, Turnbull (1974), based on the idea of self-consistency due to Efron (1967), derived an iterative procedure for estimating the survival curve nonparametrically from doubly censored data. Also for such missing data problems the score functions of suitably chosen parametric submodels (cf. Section IV.1.5) coincide exactly with the self-consistency equations and also have an interpretion in terms of the EM algorithm (Gill, 1989).

By product-integrating his "one-step estimator" for the integrated intensity (cf. Bibliographic Remarks to Section IV.1), Samuelsen (1989) derived a (suboptimal) "one-step estimator" for the survival curve and described its asymptotic properties, including an explicit expression for its asymptotic covariance function which may be estimated in a simple way. Chang and Yang (1987), on the other hand, proved that the Turnbull estimator itself (i.e., iterating to the limit) is strongly consistent, whereas Chang (1990) showed that it (properly normalized) converges weakly to a Gaussian process. However, his suggested estimator of the covariance function is only given as the solution to a set of equations which may be complicated to evaluate even numerically.

Extensions of the Turnbull algorithm to the illness–death model of Example IV.4.3 were studied by Frydman (1991, 1992).

In the special case where each individual is seen only *once* and where the observation of every lifetime thus is either right- or left-censored, an estimation algorithm equivalent to that of Turnbull's was derived by Ayer et al. (1955), whereas large sample results were later given by Groeneboom and Wellner (1992).

A completely different kind of two-sided censoring occurs when we only observe the parts of the life histories of individuals which overlap a given "observational window"; in fact, only the length of the overlap is observed. Though one can distinguish "censored" and "uncensored" observations, the Kaplan–Meier estimator is quite inappropriate (though often used) in this situation. Wijers (1991) showed consistency of the nonparametric maximum likelihood estimator (computed by the EM algorithm) for this problem. The problem also turns up in spatial statistics when a line segment process is observed through a finite window. Baddeley and Gill (1992) showed on the other hand how the Kaplan–Meier estimator itself could successfully be used to estimate interpoint distance distributions for a spatial point process observed through a finite window.

Reid (1981) performed a general study of functionals of the Kaplan–Meier estimator using influence functions and derived the large sample properties of the mean, the trimmed mean, the median as well as other functionals as special cases. The argument given by Reid (1981) is wrong, however, since Frechet differentiability fails here. But luckily, compact differentiability works, so her conclusions are correct. Mauro (1985) proved consistency of a class of functionals using combinatorial arguments. Stute and Wang (1993) used more combinatorics, and an elegant backward supermartingale argument, to greatly improve these results. Our treatment in Section IV.3.4 is based on the functional delta-method, and bibliographic comments on this are given in Chapter II.

The performance of the "test-based" confidence interval for the median proposed by Brookmeyer and Crowley (1982a) (cf. Example IV.3.6) was further investigated by Slud, Byar, and Green (1984) and compared to the class of "reflected" confidence intervals. Jennison and Turnbull (1985) suggested a confidence interval for the median similar to the one of Brookmeyer and Crowley (1982a), but with the Greenwood estimate for the variance of the Kaplan–Meier estimator replaced by an estimate based on the restricted product-limit estimator of Thomas and Grunkemeier (1975).

Weak convergence of the quantile process for the Kaplan–Meier estimator seems to have been first studied by Sander (1975). A detailed study of the asymptotic properties of this process, including its rate of convergence to the true quantiles, was provided by Aly, Csörgő, and Horváth (1985) using strong approximations in the random censorship model.

Many authors have discussed estimation of the mean lifetime on the basis of the Kaplan–Meier estimator; in fact, such a discussion was included already in Kaplan and Meier's original paper. Results on the asymptotic properties of the estimated mean have been proved by Yang (1977), Susarla and Van Ryzin (1980), and Gill (1983a), among others. Of these, Gill's result reviewed in Example IV.3.8 is the most general, and Kumazawa (1987) based his study of the "mean residual life process" on Gill's work.

Section IV.4

Aalen (1978a) was the first to extend the Kaplan–Meier estimator beyond the censored survival data setup. As a continuation of his 1976 paper on the Nelson–Aalen estimators for the integrated transition intensities in the competing risks model (cf. Bibliographic Remarks to Section IV.1), he derived what we term the Aalen–Johansen estimators for the transition probabilities of this model and studied their properties using the techniques of Breslow and Crowley (1974) and his 1976 paper. He gave the estimators as well as their variance estimates in the explicit form of our Example IV.4.1.

The paper by Aalen (1978a) was soon to be followed by the important work of Aalen and Johansen (1978). They introduced the Aalen–Johansen estimators for general finite-state Markov processes, allowing quite general censoring patterns, and used product-integration combined with results on counting processes, martingales, and stochastic integrals to study their small and large sample properties. Unfortunately, they gave neither theoretical nor practical examples to illustrate their results, and this may help explain why it has taken so long before the importance of this work has been widely recognized. The presentation in Section IV.4 is based on Aalen and Johansen's work, but it extends their treatment on a number of occasions. In particular, Aalen and Johansen (1978) only gave an estimator for the covariance matrix corresponding to (4.4.14).

Independently of Aalen and Johansen (1978), Fleming (1978a, 1978b) defined the Aalen–Johansen estimators as the implicit solution to the Kolmogorov (or Volterra) equation (with the integrated intensities replaced by the Nelson–Aalen estimators), cf. (2.6.14) and Theorem II.6.1. Using essentially the functional delta-method as well as results on martingales (but not product-integrals), he was then able (in an i.i.d. setup) to derive similar large sample results as Aalen and Johansen (1978), including proofs of uniform consistency and weak convergence of the estimator. He also proposed the estimator (4.4.16) for the covariances and showed it to be uniformly consistent.

Aalen (1982b), in an unpublished report, illustrated the use of the Aalen–Johansen estimator for the competing risks model using the same data set as in Example IV.4.1, whereas Keiding and Andersen (1989) presented a case study of the two-state Markov process (cf. Example IV.4.2).

CHAPTER V

Nonparametric Hypothesis Testing

The present chapter takes the same starting point as Chapter IV, although rather than *estimation* in the multiplicative intensity model, we now consider *hypothesis testing*. The general remarks made in the Introduction to Chapter IV concerning the underlying model and the specification of the counting process under aggregation of censored, truncated, or filtered observations, still apply.

Thus, we consider a multivariate counting process $\mathbf{N} = ((N_1(t), \ldots, N_k(t)); t \in \mathscr{T})$ with intensity process $\boldsymbol{\lambda} = (\lambda_1, \ldots, \lambda_k)$ given by

$$\lambda_h(t) = \alpha_h(t) Y_h(t)$$

for $h = 1, \ldots, k$; it is assumed throughout that the cumulative intensity

$$A_h(t) = \int_0^t \alpha_h(s) \, \mathrm{d}s$$

is finite for all $h = 1, \ldots, k$ and all $t \in \mathscr{T}$.

Our main interest is in a general linear k-sample statistic for the hypothesis $\alpha_1 = \cdots = \alpha_k$ obtained as follows. The Nelson–Aalen estimator of the, under the hypothesis common, cumulative intensity is compared to the Nelson–Aalen estimator of the hth component, using general predictable locally bounded non-negative weight processes K_h:

$$Z_h(t) = \int_0^t K_h(s) J_h(s) \left(\frac{\mathrm{d}N_h(s)}{Y_h(s)} - \frac{\mathrm{d}N_{\bullet}(s)}{Y_{\bullet}(s)} \right).$$

Suitable quadratic forms in the $Z_h(t)$, $h = 1, \ldots, k$, form our general test statistic, which covers almost all known "linear" nonparametric test statistics for censored data. The specification and exemplification of this statistic form the bulk of Section V.2, for general k as well as for $k = 2$.

One-sample tests may be derived along similar lines; cf. Section V.1. Some other topics are mentioned more briefly in Section V.3: tests based on intensity ratio estimates, tests for trend, and tests based on stratification in the multiplicative intensity model. In Section V.4, we briefly discuss test statistics based on nonlinear functionals of the estimated cumulative intensity (Kolmogorov–Smirnov, Cramér–von Mises, etc.) as well as indicate how the present framework is immediately suited for the development of *stopping rules* allowing termination of the experiment (trial) as soon as conclusive evidence has been reached.

V.1. One-Sample Tests

Consider a univariate counting process $(N(t); t \in \mathcal{T})$ with intensity process given by $\alpha(t)Y(t)$, $t \in \mathcal{T}$. The null hypothesis $\alpha = \alpha_0$ is to be tested, where α_0 is a known intensity function. Here, $A(t) < \infty$, $t \in \mathcal{T}$, and similarly for the integral A_0 of the hypothesized α_0.

V.1.1. A General Linear One-Sample Test Statistic

The idea behind the test statistic is to compare the increments $\mathrm{d}\hat{A}(t)$ of the Nelson–Aalen estimator (4.1.2),

$$\hat{A}(t) = \int_0^t \{J(s)/Y(s)\}\,\mathrm{d}N(s)$$

with $\alpha_0(t)\,\mathrm{d}t$, allowing for a stochastic weight function $K(t)$. Formally, let $J(t) = I(Y(t) > 0)$ and define, as in (4.1.3),

$$A_0^*(t) = \int_0^t \alpha_0(s)\,J(s)\,\mathrm{d}s.$$

Then, under the null hypothesis, $\hat{A} - A_0^*$ is a local square integrable martingale with predictable variation process

$$\langle \hat{A} - A_0^* \rangle(t) = \int_0^t J(s)\alpha_0(s)/Y(s)\,\mathrm{d}s.$$

Our general test statistic is to be based on the stochastic process

$$Z(t) = \int_0^t K(s)\,\mathrm{d}\{\hat{A}(s) - A_0^*(s)\}, \tag{5.1.1}$$

where K is any locally bounded predictable non-negative stochastic process. It is assumed throughout that $K(s) = 0$ whenever $Y(s) = 0$. It follows immediately from the definition and Section II.3.3 that, under the null hypothesis, Z is a local square integrable martingale with predictable variation process

$$\langle Z \rangle(t) = \int_0^t K^2(s)\,\mathrm{d}\langle \hat{A} - A_0^* \rangle(s) = \int_0^t K^2(s)\alpha_0(s)/Y(s)\,\mathrm{d}s$$

and optional variation process

$$[Z](t) = \int_0^t K^2(s)Y^{-2}(s)\,\mathrm{d}N(s).$$

The hypothesis $\alpha = \alpha_0$ may now be tested using the standardised test statistic $Z_0(t)\{\langle Z_0 \rangle(t)\}^{-1/2}$ which may be expected to be approximately standard normally distributed; this is, indeed, proved below. Usually t is chosen as τ or some similar "large" time. One may as an alternative use $[Z](t)$ rather than $\langle Z \rangle(t)$ in the denominator.

EXAMPLE V.1.1. The Standardized Mortality Ratio (SMR) and the One-Sample Log-Rank Test

In (5.1.1), let $K = Y$, then (using $YJ = Y$)

$$Z(t) = \int_0^t Y(s)\left\{\frac{J(s)}{Y(s)}\mathrm{d}N(s) - J(s)\alpha_0(s)\,\mathrm{d}s\right\} = N(t) - E(t),$$

where

$$E(t) = \int_0^t \alpha_0(s)Y(s)\,\mathrm{d}s = \langle Z \rangle(t)$$

is often denoted "the expected number of events in $[0, t]$ under $\alpha = \alpha_0$" because $\mathrm{E}\{E(t)\} = \mathrm{E}\{N(t)\}$ under the null hypothesis. The ratio $N(t)/E(t)$ is called the *standardized mortality ratio* (SMR) in the survival data example. The test statistic $Z(t)\{\langle Z \rangle(t)\}^{-1/2}$ here becomes $\{N(t) - E(t)\}E(t)^{-1/2}$ which is called the *one-sample log-rank statistic* for reasons to be explained in Section V.2. Furthermore, we have

$$[Z](t) = \int_0^t Y^2(s)J(s)Y^{-2}(s)\,\mathrm{d}N(s) = N(t)$$

so that the test statistic $Z(t)\{[Z](t)\}^{-1/2}$ here specializes to $\{N(t) - E(t)\}N(t)^{-1/2}$.

Consider now an alternative α_1 with $\alpha_1(s) > \alpha_0(s)$, $0 \leq s \leq t$. We get

$$\begin{aligned}\mathrm{E}_{\alpha_1}\{[Z](t) - \langle Z \rangle(t)\} &= \mathrm{E}_{\alpha_1}\left\{N(t) - \int_0^t \alpha_0(s)Y(s)\,\mathrm{d}s\right\} \\ &= \mathrm{E}_{\alpha_1}\left\{\int_0^t [\alpha_1(s) - \alpha_0(s)]Y(s)\,\mathrm{d}s\right\} > 0.\end{aligned}$$

We will, therefore, expect to find $[Z](t) > \langle Z \rangle(t)$ at such alternatives, leading to test statistics $Z/[Z]^{1/2}$ that are numerically smaller, i.e., less powerful than $Z/\langle Z \rangle^{1/2}$.

In most cases, one will use $t = \tau$ in the calculation of the test statistic. We return to the SMR in Example VI.1.2. □

EXAMPLE V.1.2. Censored Survival Data

Continuing the discussion in Section III.2 and Examples IV.1.1 and IV.1.6, let $X_1, \ldots, X_n$ be i.i.d. non-negative random variables with absolutely continuous distribution function F, hazard rate function $\alpha = F'/(1 - F)$ and integrated hazard function $A(t) = \int_0^t \alpha(s)\,\mathrm{d}s$. For individual i, we observe $\tilde{X}_i = X_i \wedge U_i$ and $D_i = I(\tilde{X}_i = X_i)$ for some censoring times $U_1, \ldots, U_n$. Then under independent censoring (cf. Section III.2.2)

$$N(t) = \sum_{i=1}^{n} I(\tilde{X}_i \leq t, D_i = 1)$$

specifies a univariate counting process with intensity process $\alpha(t)Y(t)$ with

$$Y(t) = \sum_{i=1}^{n} I(\tilde{X}_i \geqq t).$$

We shall again be concerned with the hypothesis H: $\alpha = \alpha_0$ (or, equivalently, $F = F_0$) and we shall now study several choices of the weight process K. Let first $K = Y$; then (using $YJ = Y$)

$$\begin{aligned} Z(t) &= \int_0^t Y(s)\left(\frac{J(s)}{Y(s)}\mathrm{d}N(s) - J(s)\alpha_0(s)\,\mathrm{d}s\right) \\ &= N(t) - \int_0^t Y(s)\alpha_0(s)\,\mathrm{d}s. \end{aligned}$$

Thus, specializing Example V.1.1, $Z(t) = N(t) - E(t)$, with $N(t)$ the observed number of deaths in $[0, t]$, and

$$E(t) = \sum_{i=1}^{n} A_0(\tilde{X}_i \wedge t),$$

the classical expected number of deaths in $[0, t]$. The test statistic

$$Z(t)\{\langle Z\rangle(t)\}^{-1/2} = \{N(t) - E(t)\}E(t)^{-1/2}$$

was discussed in this context by Breslow (1975) and generalized to left-truncation (still allowing right-censoring) by Hyde (1977); Woolson (1981) motivated this "one-sample log-rank statistic" as a large-sample approximation of the two-sample so-called log-rank statistic to be discussed in Example V.2.1.

A family of test statistics generalizing this was suggested by Harrington and Fleming (1982) by choosing $K(t) = Y(t)[S_0(t)]^\rho$, where $S_0(t) = \exp\{-A_0(t)\}$ is the survival function under the hypothesis. Although the general test statistic above was given (with asymptotic theory) by Andersen et al. (1982), further particular cases were suggested by Gatsonis et al. (1985) with complicated specific derivations of expressions for the variance and no asymptotic theory. Indeed, besides $K = Y$ as already studied here, these authors suggested $K = S_0 Y$ and $K = (1 + \{\log(1 - S_0)\}/S_0)Y$.

Finally, a slightly different one-sample linear test statistic was studied by Hollander and Proschan (1979) motivated by a two-sample statistic of Efron (1967); cf. Example V.2.1. Hollander and Proschan studied the Kaplan–Meier estimator modified to be zero after the last (possibly censored) observation, cf. the discussion after Example IV.3.1, in our notation

$$\hat{S}^*(t) = \hat{S}(t)I(Y(t+) > 0).$$

They then studied the statistic

$$C = -\int_0^\infty S_0(s)\,\mathrm{d}\hat{S}^*(s)$$

which may be interpreted as $P(X > Y)$ where X and Y are independent and X is distributed according to the hypothetical distribution S_0 and Y according to the "empirical distribution" $\hat{S}^*$.

By partial integration

$$C = [-S_0(s)\hat{S}^*(s)]_0^\infty + \int_0^\infty \hat{S}^*(s-)\,\mathrm{d}S_0(s)$$
$$= 1 + \int_0^\infty \hat{S}^*(s-)\,\mathrm{d}S_0(s),$$

so by adding the two representations one obtains, with

$$\hat{A}^*(t) = -\int_0^t \mathrm{d}\hat{S}^*(s)/\hat{S}^*(s-),$$

that

$$2C = 1 + \int_0^\infty \hat{S}^*(s-)\,\mathrm{d}S_0(s) - \int_0^\infty S_0(s)\,\mathrm{d}\hat{S}^*(s)$$
$$= 1 - \int_0^\infty \hat{S}^*(s-)S_0(s)\,\mathrm{d}A_0(s) + \int_0^\infty S_0(s)\hat{S}^*(s-)\,\mathrm{d}\hat{A}^*(s)$$
$$= 1 + \int_0^\infty S_0(s)\hat{S}^*(s-)\{\mathrm{d}\hat{A}^*(s) - \mathrm{d}A_0(s)\}.$$

Note that $\mathrm{d}\hat{A}^* = \mathrm{d}\hat{A}$, except at the largest time $T = \max_i \tilde{X}_i$ where we have $\Delta\hat{A}^*(T) = 1$, $\Delta\hat{A}(T) = I(T \text{ uncensored})$. Furthermore, A_0^* and A_0 are both continuous, with $\mathrm{d}A_0^*(s) = \mathrm{d}A_0(s)$, $0 \le s < T$. Therefore,

$$2C - 1 - I(T \text{ is censored})\hat{S}(T-)S_0(T) = \int_0^\infty S_0(s)\hat{S}^*(s-)\{\mathrm{d}\hat{A}(s) - \mathrm{d}A_0^*(s)\},$$

where the right-hand side is of the form (5.1.1) with $K = S_0\hat{S}_-^*$ and very close to $2C - 1$. The difference is a tail effect, and, indeed, Hollander and Proschan explicitly assumed $n^{1/2}S_0(T)$ and $n^{1/2}\hat{S}(T-)$ asymptotically negligible—an assumption that is unnecessary for our slight modification. □

EXAMPLE V.1.3. Mortality of Female Diabetics in the County of Fyn. Test of the Hypothesis That the Mortality of Diabetics Equals That of the General Population

The problem and the data were presented in Example I.3.2.

We illustrate the test statistics of the previous example by applying them—in the obvious modification for left-truncated data—to testing the hypothesis that the mortality of female diabetics in Fyn county, 1973–81, equals that of official Danish vital statistics, 1976–80, as specified in 1-year age groups. Taking $\tau = 100$ years and using the notation $O = N(\tau)$, $E = E(\tau)$, there were $O = 237$ observed deaths among the 716 female diabetics against $E = 80.1$ expected, yielding an estimate $O/E = 2.96$ (often termed SMR = 296) of the relative mortality. The test statistics become $(O - E)E^{-1/2} = 17.5$ and $(O - E)O^{-1/2} = 10.2$, respectively, both to be judged highly significant when compared to the standard normal distribution.

As mentioned earlier, Harrington and Fleming (1982) considered the family of weight processes $K(t) = Y(t)[S_0(t)]^\rho$. As an illustration, we calculate the test statistics for $\rho = \frac{1}{2}$ and $\rho = 1$ ($\rho = 0$ corresponding to the one-sample log-rank test already discussed).

For $\rho = \frac{1}{2}$, we get

$$Z(100) = \int_0^{100} Y(t)[S_0(t)]^{1/2}\,\mathrm{d}\{\hat{A}(t) - A_0^*(t)\} = 192.53 - 59.54 = 132.99,$$

$$\langle Z\rangle(100) = \int_0^{100} Y^2(t)S_0(t)\alpha_0(t)/Y(t)\,\mathrm{d}t = 46.98,$$

$$[Z](100) = \int_0^{100} Y^2(t)S_0(t)Y^{-2}(t)\,\mathrm{d}N(t) = 161.68,$$

yielding test statistics $132.99/\sqrt{46.98} = 19.40$ and $132.99/\sqrt{161.68} = 10.46$, respectively.

For $\rho = 1$, we get $Z(100) = 161.68 - 46.98 = 114.70$, $\langle Z\rangle(100) = 32.43$, $[Z](100) = 121.16$ yielding test statistics 20.14 and 10.42, respectively.

We note a very clear difference between the two choices of variance estimates—in the present case the use of $\langle Z\rangle$ yielded a smaller variance, hence a more significant test statistic, as noted in Example V.1.1 for the special case $\rho = 0$. □

EXAMPLE V.1.4. Aggregation: Models for Relative Mortality

In the model for relative mortality discussed in Examples III.1.3 and IV.1.11, the individual counting process $N_i(t)$, $i = 1, \ldots, n$ had intensity process $\lambda_i(t) = \alpha(t)\mu_i(t)Y_i(t)$, where the "population mortality" $\mu_i(t)$ is assumed known. It was remarked that the aggregated process $N = N_1 + \cdots + N_n$ still satisfies Aalen's multiplicative intensity model: $\lambda(t) = \alpha(t) Y^{\mu}(t)$ with

$$Y^{\mu}(t) = \sum_{i=1}^{n} \mu_i(t) Y_i(t).$$

Simple hypotheses on the relative mortality $\alpha(t)$ may, therefore, be tested by a direct application of the results above. In particular, one may test H_0: $\alpha(t) = 1$, that is, that the mortality of the study population equals that of the general population, even if the mortality of each single individual differs. However, the occurrence of the "survival function" $S_0(t) = \exp\{-A_0(t)\}$ in the Harrington–Fleming and Hollander–Proschan families of test statistics is no longer so clearly motivated [for $\alpha_0(t) = 1$, an exponential weight function appears]. □

EXAMPLE V.1.5. Malignant Melanoma

The problem and the data were introduced in Example I.3.9.

In Example IV.4.1, we discussed the possibility of interpreting the partial transition probability

$$\bar{P}^i_{02}(0, t) = \exp\left\{-\int_0^t \alpha^i_{02}(s)\,ds\right\}$$

of individual i dying from other causes than melanoma as an estimate of the mortality "had melanoma been eliminated." In this connection, it is interesting to test the hypothesis that $\alpha^i_{02}(t)$ is equal to the population mortality, which, in this case, we shall take as the then current population mortality $\mu_s(t)$ for the same sex s as the patient. As explained in Example V.1.4, this may be reformulated as a test of unit relative mortality. Using the classical "one-sample log-rank" test, there are $O = 7 + 7 = 14$ observed male + female deaths versus $E = 12.315 + 8.929 = 21.244$ expected, yielding a relative mortality estimate $O/E = 14/21.24 = 0.66$ (SMR = 66) and test statistics of the above hypothesis of $(O - E)E^{-1/2} = -1.57$ and $(O - E)O^{-1/2} = -1.94$, so that the mortality from other causes of the melanoma patients is only two-thirds of that of the general population; owing to the small number of deaths, this decrease is, however, at most marginally statistically significant. □

V.1.2. Asymptotic Null Distribution

To study the asymptotic properties of the test statistic, we consider, in a similar way as in Section IV.1.2, a sequence of counting processes $N^{(1)}$, $N^{(2)}$, ... with intensity processes

$$\lambda^{(n)} = \alpha Y^{(n)}, \quad n = 1, 2, \ldots,$$

where α is the same for all n. Other quantities from Section V.1.1 with superscript n are similarly defined.

Theorem V.1.1. *Assume that there exist a sequence of positive constants $\{c_n\}$ and non-negative functions k and y on $\mathscr{T}$ such that $k^2\alpha_0/y$ is well-defined and integrable on $\mathscr{T}$. Further, assume under $\mathrm{H}_0\colon \alpha = \alpha_0$ that*

(A) *For all $t \in \mathscr{T}$,*

$$c_n^2 \int_0^t \frac{\{K^{(n)}(s)\}^2}{Y^{(n)}(s)} \alpha_0(s)\,\mathrm{d}s \xrightarrow{\mathrm{P}} \int_0^t \frac{k^2(s)\alpha_0(s)}{y(s)}\,\mathrm{d}s \tag{5.1.2}$$

as $n \to \infty$;

(B) *For all $t \in \mathscr{T}$ and all $\varepsilon > 0$,*

$$c_n^2 \int_0^t \frac{\{K^{(n)}(s)\}^2}{Y^{(n)}(s)} \alpha_0(s) I\left(\left|\frac{c_n K^{(n)}(s)}{Y^{(n)}(s)}\right| > \varepsilon\right) \mathrm{d}s \xrightarrow{\mathrm{P}} 0 \tag{5.1.3}$$

as $n \to \infty$.

Then under $H_0\colon \alpha = \alpha_0$, as $n \to \infty$, the process $c_n Z^{(n)}$ will converge to a mean-zero Gaussian martingale U with covariance function

$$r(s,t) = \operatorname{cov}(U(s), U(t)) = \int_0^{s \wedge t} k^2(s)\alpha_0(s)/y(s)\,\mathrm{d}s.$$

PROOF. The theorem follows directly from Rebolledo's theorem (Theorem II.5.1). □

Condition A is seen to be equivalent to

$$\langle c_n Z^{(n)} \rangle(t) \xrightarrow{\mathrm{P}} r(t,t)$$

and it, therefore, also follows from Theorem II.5.1 that

$$[c_n Z^{(n)}](t) \xrightarrow{\mathrm{P}} r(t,t).$$

Hence, each of the two suggested test statistics $Z^{(n)}(t)\{\langle Z^{(n)}\rangle(t)\}^{-1/2}$ and $Z^{(n)}(t)\{[Z^{(n)}](t)\}^{-1/2}$ is asymptotically standard normal, for each $t \in \mathscr{T}$ with $r(t,t) > 0$. Note that, in general, to prove the asymptotic distribution of the test statistic for a particular t, one needs only verify conditions A and B for that particular t (cf. Theorem II.5.1). Usually one will want to take t as the right endpoint τ of $\mathscr{T}$; in this case, τ will have to belong to $\mathscr{T}$.

When checking conditions A and B in practice it is often useful to invoke "Gill's conditions" (Proposition II.5.3), as follows.

Proposition V.1.2. *Assume, under the hypothesis:*

(i) *There exist a sequence of constants $\{a_n\}$ increasing to infinity as $n \to \infty$ and a sequence of constants $\{d_n\}$ such that for almost all $s \in [0, \tau]$ we have*

$$a_n^{-2} Y^{(n)}(s) \xrightarrow{\text{P}} y(s) < \infty \tag{5.1.4}$$

and

$$d_n^{-1} K^{(n)}(s) \xrightarrow{\text{P}} k(s) < \infty \tag{5.1.5}$$

as $n \to \infty$.

(ii) *For all* $\delta > 0$, *there exists a function* k_δ *such that*

$$\liminf_{n\to\infty} \text{P}\{(a_n/d_n)^2[K^{(n)}(s)]^2/Y^{(n)}(s) \leq k_\delta(s), \text{ all } s \in \mathscr{T}\} \geq 1 - \delta. \tag{5.1.6}$$

Then, provided that $\alpha_0 k_\delta$ *is integrable on* $\mathscr{T}$ *for all* $\delta > 0$, *Conditions* A *and* B *of Theorem* V.1.1 *are satisfied with* $c_n = a_n/d_n$.

Proof. Proposition II.5.3 is directly applicable because the integrands in (5.1.2) and (5.1.3) are both bounded by $c_n^2[K^{(n)}(s)]^2\alpha_0(s)/Y^{(n)}(s)$. □

In the following example, we document explicitly the verification of conditions A and B for the one-sample log-rank and the Harrington–Fleming test statistics under random right-censoring in the survival analysis situation.

Example V.1.6. Asymptotic Distribution of One-Sample Statistics for Censored Survival Data

Continuing the discussion of Example V.1.2, consider the general random censorship model (cf. Examples III.2.4 and IV.1.6), where, for each n, $(X_1^{(n)}, U_1^{(n)}, \ldots, X_n^{(n)}, U_n^{(n)})$ are all independent, with $X_i^{(n)}$ having distribution function F, density $f = F'$, and hazard rate $\alpha = f/(1 - F)$, where $U_i^{(n)}$ has distribution function $G_i^{(n)}$. The censored survival times $\tilde{X}_i^{(n)} = X_i^{(n)} \wedge U_i^{(n)}$ are then independent, $i = 1, \ldots, n$, with distribution function $H_i^{(n)} = 1 - (1 - F)(1 - G_i^{(n)})$.

We shall assume that

$$n^{-1} \sum_{i=1}^n G_i^{(n)}(s) \to G(s), \tag{5.1.7}$$

where G is a (sub-) distribution function, for $n \to \infty$ and all $s \in [0, \infty]$, and we will prove that conditions A and B of Theorem V.1.1 hold using Proposition V.1.2.

First note that

$$\text{E}(Y^{(n)}(s)) = \text{E}\left(\sum_{i=1}^n I(\tilde{X}_i^{(n)} \geq t)\right) = [1 - F(s)] \sum_{i=1}^n [1 - G_i^{(n)}(s-)]$$

and

$$\text{var}(Y^{(n)}(s)) \leq \text{E}(Y^{(n)}(s));$$

cf. Example IV.1.6. Therefore, by (5.1.7) and Chebyshev's inequality,

$$n^{-1} Y^{(n)}(s) \xrightarrow{\text{P}} y(s) = [1 - F(s)][1 - G(s-)] \tag{5.1.8}$$

as $n \to \infty$. For the log-rank test, we choose $K = Y$, so that it immediately follows that (5.1.4) and (5.1.5) hold with $a_n^2 = n$, $d_n = n$, $k = y$, and, hence, $c_n = n^{-1/2}$. To prove (5.1.6), we need the following "in probability linear bound" by Daniels (1945). Let H_n be the empirical distribution function based on a random sample of size n from the continuous distribution function H. Then, for any $\delta \in (0, 1)$,

$$P(1 - H_n(s-) \leq \delta^{-1}(1 - H(s)) \quad \text{for all } s \in [0, \infty]) = 1 - \delta. \quad (5.1.9)$$

We first use this result on the empirical distribution function of the uncensored $X_i^{(n)}$ to obtain

$$P\left(n^{-1} \sum_{i=1}^{n} I(X_i^{(n)} \geq s) \leq \delta^{-1}(1 - F(s)) \text{ for all } s \in [0, \infty]\right) = 1 - \delta;$$

since $Y^{(n)}(s) = \sum I(\tilde{X}_i^{(n)} \geq s) \leq \sum I(X_i^{(n)} \geq s)$, we get

$$P(n^{-1} Y^{(n)}(s) \leq \delta^{-1}(1 - F(s)) \text{ for all } s \in [0, \infty]) \geq 1 - \delta$$

and (5.1.6) is satisfied with $k_\delta = \delta^{-1}(1 - F)$. The result then follows noting that $\delta^{-1}(1 - F(s))\alpha(s) = \delta^{-1} f(s)$ is integrable. Note that this proof immediately generalizes to the Harrington–Fleming family of statistics obtained from $K = YS_0^\rho$, $0 < \rho \leq 1$. Indeed, (5.1.4) is the same as before and (5.1.5) is direct, with $k = y\, S_0^\rho$. That (5.1.6) holds follows from the result for the log-rank test, choosing the same $k_\delta(s)$, because $K = Y\, S_0^\rho \leq Y$ so that $K^2/Y \leq Y$.

Hollander and Proschan's choice $K = S_0 \hat{S}^*_-$ can also be dealt with under the natural integrability condition needed to ensure finiteness of the asymptotic variance by using Gill's [(1980a, Theorem 3.2.1) or (1983a, Lemma 2.6)] extension of Daniels' in probability linear bound of the Kaplan–Meier estimator. □

V.1.3. Local Asymptotic Power

To compare the virtues of the different members of the class of tests so far defined, it is useful to study their *local power*, that is, power against interesting sequences of alternatives converging toward the null hypothesis.

At the end of Section VIII.2.3, based on the theory of contiguity and local asymptotic normality, we will mention optimality properties of these tests within the class of all possible tests. Because of the extra structure assumed there, we will be able to dispense with some of the more technical assumptions needed below.

Consider a sequence $N^{(1)}, N^{(2)}, \ldots$ of counting processes with intensity processes $\lambda^{(n)} = \alpha^{(n)} Y^{(n)}$, $n = 1, 2, \ldots$, where $\alpha^{(n)}$ is no longer independent of n but now converges to α_0 as $n \to \infty$. It is useful to choose

$$\alpha^{(n)}(t) = \alpha_0(t) + a_n^{-1}\theta(t) + \rho^{(n)}(t), \quad (5.1.10)$$

where $\{a_n\}$ is the sequence of (5.1.4), $\theta(t)$ is a given function on $\mathscr{T}$, and $\rho^{(n)}(t)$

is a sequence of functions satisfying $\sup_{[0,\tau]} |a_n \rho^{(n)}(t)| \to 0$ as $n \to \infty$. When $N^{(n)}$ is distributed according to $\alpha^{(n)}$, $M^{(n)}(t) = N^{(n)}(t) - \int_0^t \alpha^{(n)}(u) Y^{(n)}(u)\,du$ is still a martingale [although $\hat{A}^{(n)}(t) - A_0^{*(n)}(t)$ is not]. Hence, we have the decomposition

$$\begin{aligned} Z^{(n)}(t) &= \int_0^t K^{(n)}(s)\{d\hat{A}^{(n)}(s) - dA_0^{*(n)}(s)\} \\ &= \int_0^t K^{(n)}(s)/Y^{(n)}(s)\,dM^{(n)}(s) + \int_0^t K^{(n)}(s)\{a_n^{-1}\theta(s) + \rho^{(n)}(s)\}\,ds \\ &= Z_1^{(n)}(t) + Z_2^{(n)}(t). \end{aligned}$$

It is seen that $Z_1^{(n)}(t)$ is a local square integrable martingale with predictable variation process

$$\langle Z_1^{(n)} \rangle (t) = \int_0^t \frac{\{K^{(n)}(s)\}^2}{Y^{(n)}(s)} \{\alpha_0(s) + a_n^{-1}\theta(s) + \rho^{(n)}(s)\}\,ds.$$

Now assume that (i) and (ii) of Proposition V.1.2 hold also under the sequence of local alternatives $\alpha^{(n)}$ and, moreover, that

(iii) For all $\delta > 0$, there exists a function g_δ such that

$$\liminf_{n\to\infty} \mathrm{P}(d_n^{-1} K^{(n)}(s) \le g_\delta(s) \text{ for all } s \in \mathcal{T}) \ge 1 - \delta. \tag{5.1.11}$$

Furthermore assume that k_δ, $\alpha_0 k_\delta$, θk_δ, g_δ, and θg_δ are integrable on $\mathcal{T}$ for all $\delta > 0$. It then follows as in Theorem V.1.1 (cf. Proposition V.1.2) that, with $c_n = a_n/d_n$, $c_n Z_1^{(n)}$ will converge in distribution to a mean-zero Gaussian martingale with variance function

$$\int_0^t k^2(s)(\alpha_0(s)/y(s))\,ds \tag{5.1.12}$$

as $n \to \infty$ (and hence $\alpha^{(n)} \to \alpha_0$ along the sequence of local alternatives). Also,

$$\sup_{t\in\mathcal{T}} \left| c_n Z_2^{(n)}(t) - \int_0^t k(s)\theta(s)\,ds \right| \xrightarrow{\mathrm{P}} 0,$$

so that $c_n Z^{(n)}$ converges in distribution to a Gaussian process with the same variance function (5.1.12) but with mean function $\int k\theta$.

The normalized test statistic $Z^{(n)}(t)\{\langle Z^{(n)} \rangle (t)\}^{-1/2}$, therefore, converges in distribution (along the sequence of local alternatives) to a normal distribution with unit variance and mean

$$\Pi(k, \theta, t) = \int_0^t k(s)\theta(s)\,ds \Bigg/ \left(\int_0^t k^2(s)[\alpha_0(s)/y(s)]\,ds \right)^{1/2}.$$

This expression may be used to find the approximate power of a specific test and calculate Pitman efficiencies as demonstrated for k-sample tests in Section V.2.3. Here, we only discuss how optimal tests may be derived.

The greatest local power at a given θ is obtained by choosing k to maximize $\Pi(k, \theta, t)$. Note here that by the Cauchy–Schwarz inequality for functions $f \geq 0$, $g \geq 0$,

$$\int_0^t f(s)g(s)\,ds \leq \left[\int_0^t f^2(s)\,ds \int_0^t g^2(s)\,ds\right]^{1/2}$$

with equality only when f and g are proportional. Hence, by writing the numerator of $\Pi(k, \theta, t)$ above as

$$\int_0^t k(s)[\alpha_0(s)/y(s)]^{1/2}\theta(s)[y(s)/\alpha_0(s)]^{1/2}\,ds,$$

we obtain

$$\Pi(k, \theta, t)^2 \leq \int_0^t \theta^2(s)[y(s)/\alpha_0(s)]\,ds$$

with equality only when $k[\alpha_0/y]^{1/2}$ and $\theta[y/\alpha_0]^{1/2}$ are proportional, that is, when

$$k(t) = c\theta(t)y(t)/\alpha_0(t). \tag{5.1.13}$$

For *proportional hazards alternatives*

$$\alpha^{(n)}(t) = (1 + a_n^{-1}\gamma)\alpha_0(t),$$

we have $\theta(t) = \gamma\alpha_0(t)$, and the optimal k is

$$k(t) = c\gamma\alpha_0(t)y(t)/\alpha_0(t) = (\text{const.})y(t),$$

leading to the choice $K^{(n)}(t) = Y^{(n)}(t)$ of the one-sample log-rank test. The regularity conditions may be checked for the censored survival data example along a similar path as in Example V.1.6, using the "in probability linear bounds" of van Zuijlen; see Shorack and Wellner (1986, Inequality 25.3.1).

For testing against *constant alternatives*, we have $\alpha^{(n)}(t) = \alpha_0(t) + a_n^{-1}\gamma$ with $\theta(t) = \gamma$, leading to the choice $K^{(n)}(t) = Y^{(n)}(t)/\alpha_0(t)$.

EXAMPLE V.1.7. Characterization of Some Tests for Censored Survival Data from Their Local Power Properties

In the random censorship model of Example V.1.6, consider a sequence of distribution functions $F^{(1)}, F^{(2)}, \ldots$ of $X^{(1)}, X^{(2)}, \ldots$ given by the generalized local location parameter family

$$F^{(n)}(t) = \Psi(g(t) + n^{-1/2}\varphi),$$

where Ψ is a fixed absolutely continuous distribution function with positive, continuously differentiable density ψ on $(-\infty, \infty)$ and g is a fixed nondecreasing differentiable function from $(0, \infty)$ onto $(-\infty, \infty)$. Then $F^{(n)}$ has hazard function

$$\alpha^{(n)}(t) = h\{g(t) + n^{-1/2}\varphi\}g'(t)$$

with $h = \psi/(1 - \Psi)$ the hazard function corresponding to Ψ. Let $a_n = \sqrt{n}$, $\alpha_0(t) = h\{g(t)\}g'(t)$ and $\theta(t) = \varphi h'\{g(t)\}g'(t)$ and assume that ψ and g are sufficiently regular so that (5.1.10) is satisfied with these functions.

We now choose $g(t) = \Psi^{-1}\{F_0(t)\}$ corresponding to the null hypothesis, and in view of (5.1.13), which gave the optimal limiting weight function

$$k(t) = c\theta(t)y(t)/\alpha_0(t),$$

we put

$$K(t) = Y(t)\theta(t)/\alpha_0(t) = Y(t)\frac{h'\{\Psi^{-1}[F_0(t)]\}}{h\{\Psi^{-1}[F_0(t)]\}}.$$

For $h(x) = e^x$, we have the extreme value distribution $\Psi(x) = 1 - \exp(-e^x)$; here $h' = h$ and $K = Y$, leading us to the log-rank test once again. We have, thus, seen that the one-sample *log-rank test is locally most powerful* (within this class of statistics) against location (shift) alternatives on the scale given by the extreme value distribution. (Note that these alternatives correspond to proportional hazards alternatives on the scale given by the exponential distribution.)

Next, choosing the logistic distribution

$$\Psi(x) = e^x/(1 + e^x), \qquad \psi(x) = e^x/(1 + e^x)^2, \qquad h(x) = \Psi(x)$$

so that $h' = \Psi(1 - \Psi)$, we obtain, from $\Psi^{-1}(y) = \log\{y/(1 - y)\}$,

$$K(t) = Y(t)\frac{F_0(t)\{1 - F_0(t)\}}{F_0(t)} = Y(t)S_0(t),$$

the $\rho = 1$ specialization of the Harrington–Fleming test statistics. This statistic is seen to be *locally most powerful* (within this class of statistics) *against logistic location alternatives* and can, therefore, be regarded as a generalized one-sample Wilcoxon test statistic.

The full Harrington–Fleming family may be generated from the distribution functions

$$\Psi_\rho(t) = 1 - (1 + \rho e^x)^{-1/\rho}, \quad 0 < \rho \leq 1$$

(which has the above-mentioned extreme value distribution as its limit for $\rho \to 0$). In fact, we get

$$\psi_\rho(t) = e^x(1 + \rho e^x)^{-(1/\rho)-1}, \qquad h_\rho(t) = e^x(1 + \rho e^x)^{-1},$$

and $\Psi^{-1}(y) = \log\{\rho^{-1}[1 - (1 - y)^{-\rho}]\}$, so that the optimal choice is

$$K_\rho(t) = Y(t)S_0^\rho(t).$$

The optimality of these tests in larger classes is briefly discussed at the end of Section VIII.2.3. One could also consider more general models for the local alternatives; see (8.4.10). □

V.2. k-Sample Tests

In the previous section we considered the one-sample testing problem. We will now extend the ideas to the two-sample and the k-sample problems.

V.2.1. The General Test Statistic

We consider a k-variate counting process $\mathbf{N} = (N_1, \ldots, N_k)$ with $k \geq 2$ having an intensity process $\boldsymbol{\lambda} = (\lambda_1, \ldots, \lambda_k)$ of the multiplicative form $\lambda_h(t) = \alpha_h(t) Y_h(t)$. We will derive a class of test statistics for the hypothesis

$$\mathrm{H}_0\colon \alpha_1 = \alpha_2 = \cdots = \alpha_k. \tag{5.2.1}$$

The common value of $\alpha_1, \alpha_2, \ldots, \alpha_k$ will be denoted α. Our derivations are based on Aalen (1978b) who considered the two-sample situation and Andersen et al. (1982) who extended Aalen's results to the general k-sample problem.

As in Section V.1, the construction of a test statistic is based on the simple idea of comparing the Nelson–Aalen estimators [cf. (4.1.2)]

$$\hat{A}_h(t) = \int_0^t J_h(s)(Y_h(s))^{-1}\,\mathrm{d}N_h(s),$$

where J_h/Y_h is assumed to be locally bounded for each h, with an estimator of the hypothesized common value

$$A(t) = \int_0^t \alpha(s)\,\mathrm{d}s.$$

This latter quantity may be estimated by

$$\hat{A}(t) = \int_0^t J(s)(Y_{\bullet}(s))^{-1}\,\mathrm{d}N_{\bullet}(s),$$

where $N_{\bullet} = \Sigma_h N_h$; $Y_{\bullet} = \Sigma_h Y_h$ and $J(t) = I(Y_{\bullet}(t) > 0)$. This follows as in Section IV.1.1, because under the hypothesis, $N_{\bullet}(t)$ is a univariate counting process with intensity process $\alpha(t) Y_{\bullet}(t)$.

Now it is reasonable to compare $\hat{A}_h(t)$ and $\hat{A}(t)$ only for values of t for which $Y_h(t) > 0$. Therefore, we introduce

$$\tilde{A}_h(t) = \int_0^t J_h(s)\,\mathrm{d}\hat{A}(s) = \int_0^t J_h(s)(Y_{\bullet}(s))^{-1}\,\mathrm{d}N_{\bullet}(s).$$

We note that, when the hypothesis (5.2.1) holds true,

$$\hat{A}_h(t) - \tilde{A}_h(t) = \int_0^t \frac{J_h(s)}{Y_h(s)}\mathrm{d}M_h(s) - \int_0^t \frac{J_h(s)}{Y_{\bullet}(s)}\mathrm{d}M_{\bullet}(s),$$

where $M_{\bullet} = \Sigma_h M_h$. Thus, except for random variations, $\hat{A}_h$ and $\tilde{A}_h$ are equal under the hypothesis, their difference being a local square integrable martingale. In a similar manner as in Section V.1, we now introduce non-negative locally bounded predictable weight processes K_h and define stochastic processes

$$Z_h(t) = \int_0^t K_h(s)\,\mathrm{d}(\hat{A}_h - \tilde{A}_h)(s),$$

$h = 1, 2, \ldots, k$. These processes accumulate the (weighted) differences in the increments of $\hat{A}_h$ and $\tilde{A}_h$ and, therefore, provide a reasonable basis for constructing a test statistic for the hypothesis.

It turns out that a special choice of weight process covers the examples of interest, namely, the case where the K_h processes have the form

$$K_h(t) = Y_h(t)K(t),$$

where K is a non-negative, locally bounded predictable process that only depends on the process $(N_{\bullet}, Y_{\bullet})$. It is assumed that K is zero where $Y_{\bullet}$ is zero, and we interpret $K/Y_{\bullet}$ as zero where $Y_{\bullet}$ is zero. For this particular choice of the K_h processes, we may write

$$Z_h(t) = \int_0^t K(s)\,\mathrm{d}N_h(s) - \int_0^t K(s)\frac{Y_h(s)}{Y_{\bullet}(s)}\,\mathrm{d}N_{\bullet}(s), \tag{5.2.2}$$

$h = 1, 2, \ldots, k$. (Since $J_h Y_h = Y_h$, the factor J_h may be omitted.) We note that (5.2.2) yields

$$\sum_{h=1}^{k} Z_h(t) = 0.$$

In the sequel, we will restrict our attention to weight processes of the form $K_h(t) = Y_h(t)K(t)$. It should be realized, however, that most of the arguments to be presented may be modified to cover completely general weight processes.

Under hypothesis (5.2.1), we may now write

$$\begin{aligned} Z_h(t) &= \int_0^t K(s)\,\mathrm{d}M_h(s) - \int_0^t K(s)\frac{Y_h(s)}{Y_{\bullet}(s)}\,\mathrm{d}M_{\bullet}(s) \\ &= \sum_{l=1}^{k} \int_0^t K(s)\left(\delta_{hl} - \frac{Y_h(s)}{Y_{\bullet}(s)}\right)\mathrm{d}M_l(s), \end{aligned} \tag{5.2.3}$$

where δ_{hl} is a Kronecker delta. It follows that the Z_h are local square integrable martingales. Using (2.4.9), their predictable covariation processes take the form

$$\begin{aligned} \langle Z_h, Z_j\rangle(t) &= \sum_{l=1}^{k} \int_0^t K^2(s)\left(\delta_{hl} - \frac{Y_h(s)}{Y_{\bullet}(s)}\right)\left(\delta_{jl} - \frac{Y_j(s)}{Y_{\bullet}(s)}\right)\alpha(s)\,Y_l(s)\,\mathrm{d}s \\ &= \int_0^t K^2(s)\frac{Y_h(s)}{Y_{\bullet}(s)}\left(\delta_{hj} - \frac{Y_j(s)}{Y_{\bullet}(s)}\right)\alpha(s)\,Y_{\bullet}(s)\,\mathrm{d}s. \end{aligned} \tag{5.2.4}$$

By the result mentioned just after Example II.3.3, the local martingales (5.2.3) are square integrable over $[0, t]$ provided that $\mathrm{E}\langle Z_h \rangle(t) < \infty$. Therefore, a sufficient condition for the first- and second-order moments of the $Z_h(t)$ to exist is by (5.2.4) that $\int_0^t \mathrm{E}\{K^2(s) Y_{\bullet}(s)\} \alpha(s)\,\mathrm{d}s < \infty$. When this is the case, we have under the hypothesis that $\mathrm{E}Z_h(t) = 0$ and that

$$\operatorname{cov}(Z_h(t), Z_j(t)) = \mathrm{E}\langle Z_h, Z_j \rangle(t)$$

may be estimated unbiasedly by

$$\hat{\sigma}_{hj}(t) = \int_0^t K^2(s) \frac{Y_h(s)}{Y_{\bullet}(s)} \left(\delta_{hj} - \frac{Y_j(s)}{Y_{\bullet}(s)} \right) \mathrm{d}N_{\bullet}(s). \tag{5.2.5}$$

The last fact follows because the difference between (5.2.5) and (5.2.4) is a local square integrable martingale under the hypothesis. The $k \times k$ matrix with elements given by (5.2.5) is denoted $\hat{\mathbf{\Sigma}}(t)$.

Let $\mathbf{Z}(t) = (Z_1(t), \ldots, Z_k(t))^\mathsf{T}$. Then a reasonable test statistic for the hypothesis (5.2.1) is the quadratic form

$$X^2(t) = \mathbf{Z}(t)^\mathsf{T} \hat{\mathbf{\Sigma}}(t)^- \mathbf{Z}(t), \tag{5.2.6}$$

where t is chosen as τ, the upper endpoint of $\mathscr{T}$, or some similar "large" time, and $\hat{\mathbf{\Sigma}}(t)^-$ is a generalized inverse. Now, $\hat{\mathbf{\Sigma}}(t)$ has rank $k - 1$ (its maximum rank) provided that for each h, j in a chain of pairs connecting $\{1, 2, \ldots, k\}$, there exists a time point where $N_{\bullet}$ jumps and such that K, Y_h, and Y_j are positive (e.g., Gill, 1986a, p. 114 and Appendix 1). Therefore, if we delete, for instance, the last row and column of $\hat{\mathbf{\Sigma}}(t)$, to give $\hat{\mathbf{\Sigma}}_0(t)$, say, and let $\mathbf{Z}_0(t) = (Z_1(t), \ldots, Z_{k-1}(t))^\mathsf{T}$, then (5.2.6) may alternatively be given as

$$X^2(t) = \mathbf{Z}_0(t)^\mathsf{T} \hat{\mathbf{\Sigma}}_0(t)^{-1} \mathbf{Z}_0(t),$$

where $\hat{\mathbf{\Sigma}}_0(t)^{-1}$ is an ordinary inverse. It will be shown in the following subsection that $X^2(t)$ is asymptotically χ^2 distributed, under hypothesis (5.2.1), provided that some mild regularity conditions are fulfilled.

For the two-sample case, i.e., $k = 2$, (5.2.6) reduces to

$$X^2(t) = (Z_1(t))^2 / \hat{\sigma}_{11}(t).$$

In this case, one may equivalently base the test on the fact that

$$U(t) = Z_1(t) \hat{\sigma}_{11}(t)^{-1/2}$$

is asymptotically standard normally distributed under the hypothesis that $\alpha_1 = \alpha_2$. Note also that a straightforward reformulation of (5.2.2) and (5.2.5) gives

$$Z_1(t) = \int_0^t L(s)\,\mathrm{d}\hat{A}_1(s) - \int_0^t L(s)\,\mathrm{d}\hat{A}_2(s) \tag{5.2.7}$$

and

$$\hat{\sigma}_{11}(t) = \int_0^t L^2(s) \{Y_1(s) Y_2(s)\}^{-1}\,\mathrm{d}(N_1 + N_2)(s), \tag{5.2.8}$$

where we have introduced the weight process

$$L(t) = K(t)Y_1(t)Y_2(t)\{Y_1(t) + Y_2(t)\}^{-1}. \tag{5.2.9}$$

It is seen that L is zero when at least one of Y_1 and Y_2 is zero. This formulation of the two-sample statistic was the one originally proposed by Aalen (1978b), and it corresponds closely to the way we defined the one-sample statistic in (5.1.1).

Before we derive the results concerning the asymptotic distribution of the test statistics, we will consider a number of examples. First, we will see how the test statistic (5.2.6), as well as its two-sample analogue, generalize well-known tests for censored survival data. Then we discuss alternative (co)variance estimators for censored survival data, and finally six practical examples are presented.

EXAMPLE V.2.1. Censored Survival Data

Let X_{hi} for $i = 1, \ldots, n_h$, $h = 1, \ldots, k$ be independent non-negative random variables with absolutely continuous distribution function F_h and hazard rate function α_h in group number h. We do not observe the X_{hi}, however, but only the right-censored samples $(\tilde{X}_{hi}, D_{hi})$, $h = 1, \ldots, k$, where $\tilde{X}_{hi} = X_{hi} \wedge U_{hi}$ and $D_{hi} = I(\tilde{X}_{hi} = X_{hi})$ for some censoring times U_{hi}. In this setup, we want to test the hypothesis $F_1 = F_2 = \cdots = F_k$, or equivalently the hypothesis that the α_h are identical; cf. (5.2.1).

It was shown in Section III.2 that $\mathbf{N} = (N_1, \ldots, N_k)$ with

$$N_h(t) = \sum_{i=1}^{n_h} I(\tilde{X}_{hi} \le t, D_{hi} = 1),$$

is a k-variate counting process with intensity process $\boldsymbol{\lambda} = (\lambda_1, \ldots, \lambda_k)$ of the multiplicative form $\lambda_h(t) = \alpha_h(t) Y_h(t)$, with

$$Y_h(t) = \sum_{i=1}^{n_h} I(\tilde{X}_{hi} \ge t),$$

when we have independent right-censoring. Thus, the situation with censored failure-time data fits into the setup of the present section. We will illustrate how most well-known tests for censored failure-time data are special cases of (5.2.6) obtained by particular choices of the weight process K.

With the choice $K(t) = I(Y_{\cdot}(t) > 0)$, we may write

$$Z_h(\infty) = O_h - E_h,$$

where

$$O_h = N_h(\infty)$$

is the observed number of failures in group number h and

$$E_h = \int_0^\infty \{Y_h(s)/Y_{\cdot}(s)\}\,\mathrm{d}N_{\cdot}(s) = \int_0^\infty Y_h(s)\,\mathrm{d}\hat{A}(s).$$

The E_h are often denoted "the expected numbers of failures" under the hypothesis since when H_0 holds true, we have $\mathrm{E}Z_h(\infty) = 0$ and, therefore, $\mathrm{E}(E_h) = \mathrm{E}(O_h)$. Thus, for this choice of weight process, the test statistic (5.2.6) (with $t = \infty$) is nothing but the well-known log-rank test for censored failure-time data (Mantel, 1966; Cox, 1972; Peto and Peto, 1972). The name log-rank test was proposed by Peto and Peto (1972) and is due to the fact that, in the absence of censoring, the log-rank test reduces to the Savage test which have scores which are approximately linearly related to the logarithm of the rank of the observations when they are ranked from the largest to the smallest (cf. Example V.2.9). The log-rank test is optimal for proportional hazards alternatives (cf. Example V.2.14).

Furthermore, with the choice $K(t) = Y_{\bullet}(t)$, we get

$$Z_h(\infty) = \int_0^\infty Y_{\bullet}(s)\,\mathrm{d}N_h(s) - \int_0^\infty Y_h(s)\,\mathrm{d}N_{\bullet}(s).$$

The test statistic (5.2.6) then is the generalization of the Wilcoxon and Kruskal–Wallis tests to right-censored data of Gehan (1965) and Breslow (1970).

Tarone and Ware (1977) suggested a family of test statistics which contains the log-rank test and the Gehan–Breslow generalization of the Wilcoxon and Kruskal–Wallis tests as special cases. Their family is obtained by choosing the weight process $K(t) = g(Y_{\bullet}(t))$ for a fixed function g. In particular, they proposed the choice $g(y) = \sqrt{y}$.

Prentice (1978) [see also Kalbfleisch and Prentice (1980, Chapter 6)], suggested another generalization of the Wilcoxon and Kruskal–Wallis tests to censored survival data. To describe this test, we introduce

$$\tilde{S}(t) = \prod_{s \le t}\left(1 - \frac{\Delta N_{\bullet}(s)}{Y_{\bullet}(s) + 1}\right),$$

which is close to the Kaplan–Meier estimator (see Section IV.3) based on the combined sample. Then Prentice's suggestion is obtained by letting $K(t) = \tilde{S}(t-)Y_{\bullet}(t)/(Y_{\bullet}(t) + 1)$. It is seen that (5.2.2) with this choice of K may be rewritten as

$$Z_h(t) - \int_0^t \tilde{S}(s)\,\mathrm{d}N_h(s) - \int_0^t \tilde{S}(s)\frac{Y_h(s)}{Y_{\bullet}(s)}\,\mathrm{d}N_{\bullet}(s).$$

[The reason for defining $K(t)$ as above and not just as $\tilde{S}(t)$ is to make it predictable, cf. Andersen et al. (1982, Section 3.4) for a general discussion of this point.] A generalization of the Wilcoxon test to censored data very close to the one of Prentice (1978) was earlier proposed by Peto and Peto (1972).

Intuitively speaking, the difference between the Gehan–Breslow and the Peto–Prentice generalizations of the Wilcoxon and Kruskal–Wallis tests is that the former uses a weight process ($K = Y_{\bullet}$) that depends on failures as well as censorings, whereas the weight process ($K \approx \tilde{S}$) of the latter essentially depends on the survival experience in the combined sample. The dependence on the censoring for the Gehan–Breslow test can lead to anomalous results when the censoring patterns greatly differ in the k groups as illustrated by the case study of Prentice and Marek (1979). Moreover, we will, in Example V.2.14, find that the Peto–Prentice test is optimal for a time-transformed logistic location family, whereas the Gehan-Breslow test has no such simple optimality properties [see, however, the discussion by Gill (1980a, Chapter 5)].

A general theory for linear rank tests for censored survival data was developed by Prentice (1978); see also Kalbfleisch and Prentice (1980, Chapter 6). His generalization of the Wilcoxon and Kruskal–Wallis tests is one particular member of this class. The basic test statistic in this class is expressed as a linear combination of scores associated with the censored and uncensored observations in such a way that the same score applies to all censored observations between two adjacent failure times. It is explained in detail by Andersen et al. (1982, Section 3.4) how this general test statistic may also be written in the form (5.2.2) with an appropriate choice of the weight process $K(t)$.

Since this embedding of Prentice's approach into the counting process framework in some sense contains censored data versions of most classical linear rank tests, let us briefly indicate how Prentice's formulation of the test statistics look. In the classical theory of rank tests for uncensored data [cf. Hájek and Sĭdák, (1967); see also Example V.2.9], a test for the hypothesis $F_1 = F_2 = \cdots = F_k$ is based on the statistics

$$\sum_{i=1}^{n_h} a_n(R_{hi}), \quad h = 1, 2, \ldots, k.$$

Here, the R_{hi} are the ranks of the X_{hi} in the combined sample, and the scores $a_n(j)$ are obtained from a *score function* ϕ on $(0, 1]$ by $a_n(j) = \mathrm{E}\phi(T_{(j)})$, where $T_{(j)}$ is the jth-order statistic in a uniform $(0, 1)$ sample of size n, or by the approximation $a_n(j) = \phi(j/(n + 1))$, which gives rise to an asymptotically equivalent test. Prentice proposed generalizations of both of these sets of scores to censored data. His main proposal, which we shall not consider here, corresponds to the first set of scores mentioned above and was based on generalizing the concept of the rank vector. His secondary proposal was to suggest a set of approximate scores, which he conjectured would yield an asymptotically equivalent test. This conjecture was later shown to be correct by Cuzick (1985). For the set of approximate scores, the classical scores $\phi(R_{hi}/(n + 1))$ were replaced by their analogues for censored data as follows. The uncensored observations X_{hi} received the scores

$$\phi(1 - \tilde{S}(X_{hi})),$$

with $\tilde{S}$ as defined above, which are easily seen to reduce to the classical ones when there is no censoring. A censored observation $\tilde{X}_{hi}$ indicates that the survival time is larger than $\tilde{X}_{hi}$. However, only the information that it is larger than the largest observed failure time $X_{h'i'}$, say, less than or equal to

$\widetilde{X}_{hi}$, is used in Prentice's construction of the test statistic. Therefore, the relevant score associated with the censored observation $\widetilde{X}_{hi}$ is

$$\Phi(u) = \frac{1}{1-u} \int_u^1 \phi(v)\, dv$$

evaluated at $u = \phi(1 - \widetilde{S}(X_{h'i'}))$.

Harrington and Fleming (1982) introduced a class of test statistics for censored survival data which contains the log-rank test and a test very close to (and asymptotically equivalent to) Peto and Peto's and Prentice's generalization of the Wilcoxon and Kruskal–Wallis tests as special cases. Their test statistic is obtained by letting $K(t) = (\hat{S}(t-))^\rho I(Y_\bullet(t) > 0)$ in (5.2.2). Here,

$$\hat{S}(t) = \prod_{s \le t} \left(1 - \frac{\Delta N_\bullet(s)}{Y_\bullet(s)}\right)$$

is the Kaplan–Meier estimator (cf. Section IV.3) based on the combined sample and ρ is a fixed number between zero and one. It is seen that $\rho = 0$ gives the log-rank test, whereas $\rho = 1$ gives a test similar to Peto and Peto's and Prentice's generalization of the Wilcoxon and Kruskal–Wallis tests.

For the two-sample case, Efron (1967) studied another generalization of the Wilcoxon test to censored data than the ones mentioned above. Let us conclude this example by showing how Efron's test statistic may be written in the form (5.2.7). The calculations are similar to those of Example V.1.2 for the one-sample modification of Efron's statistic proposed by Hollander and Proschan (1979). We introduce

$$\hat{S}_h^*(t) = \hat{S}_h(t) J_h(t)$$

for $h = 1, 2$, where $\hat{S}_h(t)$ is the Kaplan–Meier estimator for the hth sample, and let

$$J_h(t) = I\left(\max_i \widetilde{X}_{hi} \ge t\right)$$

as usual. Thus, $\hat{S}_h^*(t)$ is the Kaplan–Meier estimator modified to be zero after $T_h = \max_i \widetilde{X}_{hi}$; cf. the discussion after Example IV.3.1. Then, Efron's statistic may be written as

$$\hat{W} = -\int_0^\infty \hat{S}_1^*(s-)\, d\hat{S}_2^*(s).$$

If we let $T = T_1 \wedge T_2$, then

$$\begin{aligned}\hat{W} &= -\int_0^\infty \hat{S}_1(s-) J_1(s)\, d\hat{S}_2(s) + I(T_2 \le T_1) \hat{S}_1(T-) \hat{S}_2(T-) \\ &= \int_0^\infty \hat{S}_1(s-) \hat{S}_2(s-) J_1(s) J_2(s)\, d\hat{A}_2(s) + I(T_2 \le T_1) \hat{S}_1(T-) \hat{S}_2(T-),\end{aligned}$$

where $\hat{A}_2$ is the Nelson–Aalen estimator for the second sample. By integrating the first expression for $\hat{W}$ by parts, and supposing that there are no ties among the $\tilde{X}_{hi}$, we also find that

$$\hat{W} = 1 + \int_0^\infty \hat{S}_2^*(s-)\, \mathrm{d}\hat{S}_1^*(s).$$

If we, therefore, repeat the previous calculations and add the results, we find

$$2\hat{W} - 1 = -Z_1(\infty) + (I(T_2 \le T_1) - I(T_1 \le T_2))\hat{S}_1(T-)\hat{S}_2(T-),$$

where $Z_1(t)$ is given by (5.2.7) with $L(t) = \hat{S}_1(t-)\hat{S}_2(t-)J_1(t)J_2(t)$. Here, the last term is negligible compared to the first one when there are no ties among the $\tilde{X}_{hi}$. However, if we use the above expression to extend the definition of $\hat{W}$ to tied $\tilde{X}_{hi}$ (even if F_1 and F_2 are continuous), this last term can cause disastrous behavior of $\hat{W}$. So, following Gill (1980a), it seems better to redefine Efron's (1967) statistic as $\frac{1}{2} - \frac{1}{2}Z_1(\infty)$. Thus, this slightly modified version of Efron's statistic is of the form (5.2.7) with L as given above.

In Example VII.2.4, we discuss how the log-rank test as well as the other tests considered in this example may be interpreted as score tests in semiparametric regression models. □

Example V.2.2. Alternative (Co)variance Estimators for Censored Survival Data

For all the test statistics mentioned in Example V.2.1, except Efron's test statistic, variances and covariances may be estimated unbiasedly by inserting the proper weight process K in (5.2.5). For Efron's test, we may insert the L process given in Example V.2.1 in (5.2.8). These estimates are based on the model with absolutely continuous distribution functions F_h.

To be able to treat ties in a systematic manner, it is of interest to consider an extension of the model, where it is no longer assumed that the distribution functions are continuous. Thus, hazard rates α_h do not exist, but cumulative hazard rates $A_h(t) = \int_0^t (1 - F_h(s-))^{-1}\, \mathrm{d}F_h(s)$ may still be defined.

We, furthermore, assume that the censoring mechanism is such that the counting processes $N_{hi}(t) = I(\tilde{X}_{hi} \le t, D_{hi} = 1)$ have integrated intensity processes $\Lambda_{hi}(t) = \int_0^t I(\tilde{X}_{hi} \ge s)\, \mathrm{d}A_h(s)$ and that the local martingales $M_{hi} = N_{hi} - \Lambda_{hi}$ for $i = 1, 2, \ldots, n_h$, $h = 1, 2, \ldots, k$ are orthogonal; cf. a similar extension in Section IV.1.1. This is the case for all the common types of right-censoring like general random censorship (Example III.2.4) and progressive censoring of types I and II (Gill, 1980a, Corollary 3.1.1). Essentially the assumption means that the remaining lifetimes for individuals still at risk are conditionally independent given the "past" and that their distribution is the same as it would have been had there been no censoring.

Then, in this extended model,

$$M_h(t) = N_h(t) - \int_0^t Y_h(s)\, \mathrm{d}A_h(s)$$

for $h = 1, \ldots, k$ are orthogonal martingales with

$$\langle M_h \rangle(t) = \int_0^t Y_h(s)(1 - \Delta A_h(s))\,\mathrm{d}A_h(s).$$

Using (5.2.3), we then get

$$\langle Z_h, Z_j \rangle(t) = \int_0^t K^2(s) \frac{Y_h(s)}{Y_{\bullet}(s)} \left(\delta_{hj} - \frac{Y_j(s)}{Y_{\bullet}(s)} \right) Y_{\bullet}(s)(1 - \Delta A(s))\,\mathrm{d}A(s),$$

under the hypothesis, where A is the common value of the A_h. This is an expression parallel to (5.2.4).

Now, by Gill (1980a, Proposition 3.2.2),

$$\int_0^t (\Delta N_{\bullet}(s) - 1)\,\mathrm{d}N_{\bullet}(s) - \int_0^t Y_{\bullet}(s)(Y_{\bullet}(s) - 1)\Delta A(s)\,\mathrm{d}A(s)$$

is a martingale under the hypothesis. Using this fact, it is straightforward to check that with

$$\hat{\hat{\sigma}}_{hj}(t) = \int_0^t K^2(s) \frac{Y_h(s)}{Y_{\bullet}(s)} \left(\delta_{hj} - \frac{Y_j(s)}{Y_{\bullet}(s)} \right) \frac{Y_{\bullet}(s) - \Delta N_{\bullet}(s)}{Y_{\bullet}(s) - 1}\,\mathrm{d}N_{\bullet}(s), \qquad (5.2.10)$$

we have

$$\mathrm{E}\hat{\hat{\sigma}}_{hj}(t) = \mathrm{E}\langle Z_h, Z_j \rangle(t) = \operatorname{cov}(Z_h(t), Z_j(t))$$

under the hypothesis. Thus, $\hat{\hat{\sigma}}_{hj}(t)$ is an unbiased estimator in the extended model; in fact, it is the commonly used estimator for censored survival data. Of course, when there are no tied observations, $\hat{\sigma}_{hj}$ and $\hat{\hat{\sigma}}_{hj}$ coincide.

For censored survival data, one may consider the model with simple random censorship (Example III.2.4) with equal censoring in each group. More precisely, in this model, the censoring times U_{hi}, $i = 1, \ldots, n_h$, $h = 1, \ldots, k$, are i.i.d. with distribution function G and independent of the X_{hi}. For such a situation, under the hypothesis, the conditional distribution of the observations, obtained by holding the $n = \Sigma n_h$ observed values of $(\tilde{X}_{hi}, D_{hi})$ fixed, but allocating them at random among the k groups of size $n_1, n_2, \ldots, n_k$, is the permutation distribution. Since we have assumed that the weight process K only depends on $N_{\bullet}$ and $Y_{\bullet}$, it remains fixed under the permutations. Now the Z_h are still mean-zero (discrete-time) martingales under the permutation distribution (with respect to an appropriate choice of $\mathscr{F}_t$ specifying the allocation of observations to samples for those observations which are smaller than or equal to t). Their predictable covariation processes can be shown to be of the form (5.2.10) when we allow for ties in the data. Then, taking (permutation) expectations (using results on the moments of the multihypergeometric distribution), one finds that the permutation covariance between $Z_h(\infty)$ and $Z_j(\infty)$ is

$$\tilde{\sigma}_{hj}(\infty) = \frac{n_h}{n-1} \left(\delta_{hj} - \frac{n_j}{n} \right) \int_0^\infty K^2(s)(Y_{\bullet}(s) - \Delta N_{\bullet}(s))(Y_{\bullet}(s))^{-1}\,\mathrm{d}N_{\bullet}(s). \quad (5.2.11)$$

These permutation-based covariances offer an alternative to (5.2.10) (both allow ties in the data). One should be careful, however, only to use the permutation covariances when the censoring patterns are similar in all groups; cf. Example V.2.4. When the permutation covariances are used, the test statistic (5.2.6) may be rewritten in a simpler form not involving matrix inversion; cf. Example V.2.10. □

In Example V.2.1, we discussed a number of test statistics for the setup with censored survival data. Of these, only the log-rank test, corresponding to $K(t) = I(Y_{\cdot}(t) > 0)$, and the Gehan–Breslow generalization of the Wilcoxon and Kruskal–Wallis tests, corresponding to $K(t) = Y_{\cdot}(t)$, are immediately usable for other applications of the multiplicative intensity model. Also the tie-corrected (co)variance estimator (5.2.10) is especially designed for censored survival data and is not applicable in general.

An extended model of the form discussed in Example V.2.2 is, however, also appropriate if one wants to compare the transition intensities for the same type of transition in k different Markov process models of the same structure (see Example V.2.6 for a concrete situation). This is also the case when one compares two or more transition intensities of the same Markov process provided that the intensities to be compared are relative to transitions going *from* different states. Thus, for these two situations, the tie-corrected estimator (5.2.10) is still appropriate.

We then turn to the six practical examples.

Example V.2.3. Malignant Melanoma. Comparison of Mortality Rates for Men and Women

The data were presented in Example I.3.1. In Example IV.1.2, it was noted that the mortality from melanoma (as a function of duration since operation) appeared to be about twice as large for males as for females. We want to test the hypothesis of equal mortality for males and females, using the two-sample versions of some of the test statistics derived here. We choose $\tau = 14$ years corresponding to the maximum follow-up time.

For the weight process $K(t) = I(Y_{\cdot}(t) > 0)$, we obtain the log-rank test, where (5.2.7) and (5.2.8) yield

$$Z_1(\tau) = -9.143, \qquad \hat{\sigma}_{11}(\tau) = 12.923$$

and, thus, a test statistic

$$U(\tau) = Z_1(\tau)\hat{\sigma}_{11}(\tau)^{-1/2} = u = -2.543,$$

corresponding to a (two-sided) $P = .011$. The Gehan–Wilcoxon test is obtained by choosing $K(t) = Y_{\cdot}(t)$; (5.2.7) and (5.2.8) then yield

$$Z_1(\tau) = -1563, \qquad \hat{\sigma}_{11}(\tau) = 329754,$$

and test statistic -2.722, $P = .0065$.

Finally, the generalization of the Wilcoxon test due to Prentice is obtained by letting $K(t) = \tilde{S}(t-)Y_{\bullet}(t)/\{Y_{\bullet}(t)+1\}$, where $\tilde{S}(t-)$ is the left-continuous modification of a modified Kaplan–Meier estimator of the survivor function based on the combined data from males and females. We get a test statistic of

$$U(\tau) = Z_1(\tau)\hat{\sigma}_{11}(\tau)^{-1/2} = u = -8.188/\sqrt{9.4445} = -2.664,$$

$P = .0077$.

As will be seen in Example V.2.14, the log-rank test is locally most powerful against proportional hazards alternatives, whereas the Peto–Prentice generalization of the Wilcoxon test is optimal for a time-transformed logistic location family and therefore, puts more weight on early deviations. Although a first glance at Figure IV.1.1 indicates near proportionality between the intensities for males and females, closer inspection reveals jumps for males at times 200–300 days before anything has happened to the females.

There are no ties in the present data, so the tie-corrected variance estimate (5.2.10) is identical to the uncorrected (5.2.8). □

EXAMPLE V.2.4. Occurrence of Chronic Graft-versus-Host Disease after Bone Marrow Transplantation

The problem and the data were presented in Example I.3.14. Here, we shall consider as response the time until development of chronic graft-versus-host disease (GvHD), taking death, relapse, and loss to follow-up (e.g., due to end of study) as censorings. We are particularly interested in testing homogeneity across the five participating Nordic centers, which had the following total numbers of patients and cases of chronic GvHD:

Center	1	2	3	4	5	Total
All patients	68	93	9	15	5	190
Chronic GvHD	14	23	3	0	4	44

We use as τ the maximum follow-up time (=7.5 years). We first study the log-rank test, obtained from the weight process $K(t) = I(Y_{\bullet}(t) > 0)$; this yields the vector $(Z_1(\tau), \ldots, Z_5(\tau))$ [cf. (5.2.2)]

$$-3.57 \quad 2.47 \quad 1.98 \quad -4.40 \quad 3.52$$

with estimated covariance matrix $(\hat{\sigma}_{hj}(\tau))$ [cf. (5.2.5)]

10.54	−8.19	−0.40	−1.76	−0.19
−8.19	10.94	−0.48	−2.05	−0.23
−0.40	−0.48	1.00	−0.10	−0.014
−1.76	−2.05	−0.10	3.95	−0.042
−0.19	−0.23	−0.014	−0.042	0.47

The inverse of the upper left-hand 4×4 submatrix is

$$\begin{matrix} 2.175 & 2.117 & 2.111 & 2.120 \\ 2.117 & 2.165 & 2.111 & 2.119 \\ 2.111 & 2.111 & 3.086 & 2.113 \\ 2.120 & 2.119 & 2.113 & 2.350 \end{matrix}$$

and the test statistic becomes 35.75, which should be compared to the χ^2 distribution, d.f. = 4, $P < .0001$, indicating a clear difference between centers in the occurrence of chronic GvHD.

The time until chronic GvHD was recorded in days, and at each of the days 99, 100, 101, and 132, two patients got GvHD, whereas at each of the days 123 and 176, three patients got GvHD. If the correction for these ties indicated by the variance estimate (5.2.10) is introduced, the test statistic is change to 35.89.

For the Breslow generalization of the Kruskal–Wallis test, we obtain a test statistic of 34.22 using the variance estimate (5.2.5) and 34.35 using the tie-corrected version (5.2.10).

Finally, for Prentice's Kruskal–Wallis generalization, straightforward use of (5.2.5) yields 34.77 and tie-correction gives 34.90.

We have used this example to test various common standard statistical packages.

First, using PROC LIFETEST of SAS (1985), two different results may be obtained. If numerical covariates are defined as indicators of the centers, and these are tested for association with the response using the TEST statement, the statistic obtained from the "log-rank scores" was 35.75, exactly equal to our un-tie-corrected log-rank result. The test using "Wilcoxon scores" yielded a statistic of 24.94, which has no resemblance to our results. The other procedure in this SAS program is to use the STRATA statement, which yielded log-rank and "Wilcoxon" test statistics of 35.89 and 34.35, respectively, exactly our tie-corrected results.

The latter (i.e., tie-corrected) results were also obtained from program P1L of BMDP (1988).

Finally, SPSS (1981) uses the "Lee–Desu" statistic (Lee and Desu, 1972) which here resulted in a value of 23.45. The "Lee–Desu" statistic corresponds to the Breslow generalization of the Kruskal–Wallis test, but it uses the permutation covariances (5.2.11), only valid for equal censoring, instead of one of the covariance estimators (5.2.5) or (5.2.10), which are valid in general. This illustrates that one should be careful only to use the permutation approach (and, therefore, also the "Lee–Desu" test of SPSS) when the censoring patterns are similar in all groups. □

Example V.2.5. Prognostic Significance of Residual Cancer Tissue after Diagnostic Biopsy in Breast Cancer

The problem and the data were presented in Example I.3.5. Briefly, breast cancer patients were divided into two groups according to whether residual cancer tissue (RCT) was left in the biopsy cavity (RCT+) or not (RCT−) after the diagnostic biopsy.

We here only study the pre- (and peri-)menopausal patients. For the patients entering in 1979–81, an interim analysis was performed concerning recurrence-free survival through 1981. There were no ties and the log-rank statistic (with $\tau = 3$ years, the maximum follow-up time) was $u = 3.38$, $P = .0007$, indicating a (highly surprising) decreased recurrence-free survial for RCT + patients.

The relevant protocol was open for accrual until the end of November 1982. A confirmatory analysis based on new patients only (experience from areas B and C in Figure I.3.4) gave (letting τ be the maximum follow-up time until the closing date of 28 February 1986, that is, 4 years and 2 months) the value $u = 1.43$, $P = .15$ of the log-rank statistic.

Using the modification of the log-rank statistic to left-truncated data allows inclusion of the experience from areas D and E of Figure I.3.4. The value of the log-rank test statistic based on B, C, D, and E (with τ equal to the maximum follow-up time of 7 years and 2 months) is $u = 3.95$, $P = .0001$, confirming the preliminary finding.

It may be remarked that this conclusion would also have been reached had the confirmatory analysis been based on follow-up only through 1982 (areas B and D): $u = 3.17$, $P = .0015$.

Note that the "reuse" of the patients through the left-truncation argument rests heavily on the model, which here assumes the patients to form a homogeneous group, with mortality depending only on time since entry. Hidden heterogeneities ("frailties"; cf. Chapter IX) would invalidate the test. However, inclusion of several known risk factors through a Cox regression model (cf. Chapter VII) is permitted in this framework, and this did not change the qualitative conclusions (unpublished results). □

EXAMPLE V.2.6. Abnormal Prothrombin Levels in Liver Cirrhosis. Comparison of the Transition Intensities Between Treatment Groups

The data were introduced in Example I.3.12. In Example IV.4.4, we presented estimates for integrated intensities and transition probabilities for these data based on a Markov process model with the three states 0 = "normal prothrombin level," 1 = "abnormal prothrombin level," and 2 = "dead." The test statistics discussed in this section may now be used to test whether each of the separate intensities $\alpha_{01}(t)$, $\alpha_{10}(t)$, $\alpha_{02}(t)$, $\alpha_{12}(t)$ for transitions between the three states differ between prednisone- and placebo-treated patients.

We use as τ the maximal follow-up time 12.75 years. The two test statistics that allow immediate generalization beyond the classical censored survival data problem are the log-rank and Gehan–Wilcoxon statistics, which we shall use here. As remarked just below Example V.2.2, the tie corrections are applicable for the test of the mentioned homogeneity hypotheses, and we shall only quote the results of these.

For the transition $0 \rightarrow 1$ (normal $\rightarrow$ low prothrombin; cf. Figure IV.4.11), we get

log-rank	$u = 2.12$	$P = .034$
Gehan–Wilcoxon	$u = 2.33$	$P = .020$,

indicating an increased risk of the normal → low transition for placebo-treated patients over prednisone-treated patients.

The reverse transition $1 \to 0$ (cf. Figure IV.4.11) yields

log-rank	$u = -2.73$	$P = .006$
Gehan–Wilcoxon	$u = -3.78$	$P = .0002$,

that is, an increased change of recovery for prednisone-treated patients over placebo-treated patients.

Concerning mortality of patients with normal level of prothrombin (transition $0 \to 2$; cf. Figure IV.4.10), we get

log-rank	$u = 1.27$	$P = .20$
Gehan–Wilcoxon	$u = 1.76$	$P = .08$,

indicating that the slightly increased risk of dying for placebo-treated over prednisone-treated patients with normal prothrombin is not statistically significant.

Finally, for the mortality of patients with low level of prothrombin (transition $1 \to 2$; cf. Figure IV.4.10), we get

log-rank	$u = -1.69$	$P = .09$
Gehan–Wilcoxon	$u = -1.97$	$P = .05$,

so that the increased mortality of prednisone-treated patients over placebo-treated patients with low prothrombin is close to statistical significance at conventional levels. □

Example V.2.7. Matings of Drosophila Flies. Comparison of Mating Intensities

The mating experiment and the data were described in Example I.3.16, whereas possible statistical models were discussed in Examples III.1.10 and III.5.5.

The test statistics described in this section allow comparison of the mating intensities of the four combinations of ebony females/males and oregon females/males. We only quote the results for the more realistic model 2, described in Example III.5.5; cf. the estimated integrated intensities of Figure IV.1.7. As τ was used the time of termination of the experiment, which as explained in Example I.3.16, was between 45 and 60 min after the start, according to when after the 45 min there were no matings going on.

The only test statistics relevant here are the log-rank and Breslow–

Kruskal–Wallis statistics, and because the assumptions underlying the derivations of the tie corrections are clearly not fulfilled, the uncorrected versions are quoted.

We get

log-rank	$X^2 = 53.1$	d.f. $= 3$	$P < .0001$
Gehan–Wilcoxon	$X^2 = 54.3$	d.f. $= 3$	$P < .0001$,

showing a highly significant diference in mating intensity between the four combinations, as one would probably expect judged from Figure IV.1.7.

Note that these test statistics do not utilize the special "factorial" structure of the intensities given by the female/male types. We return to this in Example V.3.3. □

EXAMPLE V.2.8. Pustulosis Palmo-Plantaris and Menopause

The problem and the data were introduced in Example I.3.13. As indicated there, Aalen et al. (1980) posited a Markov process model (Example III.1.4) such as illustrated in Figure I.3.7 for the development of natural menopause, induced menopause, and pustulosis palmo-plantaris. A basic complication is that sampling takes place *only among the diseased*, that is, from patients in states ID, D, and MD. Various models for how sampling happens were studied by Aalen et al. (1980) and Borgan (1980); in all cases extending to a Markov process with more states, some of the additional states indicating that not only had the particular event happened, but the patient had also been sampled. The observed process is then the conditional stochastic process, given that the patient is sampled. This is again a Markov process, with intensities γ_{0D}, $\gamma_{M,MD}$, and $\gamma_{I,ID}$ of acquiring the disease before menopause and after natural and induced menopause, respectively. On the basis of the data, we may test the hypothesis H_γ: $\gamma_{0D} = \gamma_{M,MD} = \gamma_{I,ID}$, and since (for the sampling models considered) it can be shown that the hypothesis H_α: $\alpha_{0D} = \alpha_{M,MD} = \alpha_{I,ID}$ implies H_γ, in this way we obtain a valid (though possibly conservative) test of H_α.

For the data of Table I.3.8, a three-sample log-rank test yields $X^2 = 22.4$, d.f. $= 2$, $P < 0.0001$, indicating a highly significant change of pustulosis incidence at menopause. Although the evidence is somewhat convoluted, it is clear that this incidence is considerably *increased* after both natural and induced menopause. □

V.2.2. Asymptotic Null Distribution and Conservative Tests

In this subsection we first derive the asymptotic null distribution for the general test statistic (5.2.6). We also mention explicitly the two-sample case. We then consider the particular situation where the risk sets are asymptotically proportional. For this situation alternative versions of the test statistic exist, one of which is conservative.

V.2.2.1. *Asymptotic Null Distribution*

We now turn to the study of the asymptotic distribution of $(Z_1, \ldots, Z_k)$ and the test statistic (5.2.6) under the hypothesis (5.2.1). Assume for this purpose that we have a sequence of k-variate counting processes $\mathbf{N}^{(n)} = (N_1^{(n)}, \ldots, N_k^{(n)})$, $n = 1, 2, \ldots$, with intensity processes $\boldsymbol{\lambda}^{(n)} = (\lambda_1^{(n)}, \ldots, \lambda_k^{(n)})$ of the multiplicative form $\lambda_h^{(n)}(t) = \alpha_h(t) Y_h^{(n)}(t)$, where the α_h are the same for all n. All relevant quantities are indexed by n. We may prove the following theorem.

Theorem V.2.1. *Assume that there exist a sequence of positive constants* $\{c_n\}$ *and non-negative functions* $y_1, y_2, \ldots, y_k, \kappa$ *such that* $\kappa^2 \alpha y_{\bullet}$, *with* $y_{\bullet} = \Sigma_h y_h$, *is integrable on* $\mathscr{T}$. *Let*

$$\sigma_{hj}(t) = \int_0^t \kappa^2(s) \frac{y_h(s)}{y_{\bullet}(s)} \left(\delta_{hj} - \frac{y_j(s)}{y_{\bullet}(s)} \right) \alpha(s) y_{\bullet}(s) \, \mathrm{d}s \tag{5.2.12}$$

for all $t \in \mathscr{T}$ *and all* h, j, *and assume that under the hypothesis* (5.2.1), *we have*

(A) *For all* h, j *and all* $t \in \mathscr{T}$,

$$c_n^2 \int_0^t (K^{(n)}(s))^2 Y_h^{(n)}(s) Y_j^{(n)}(s) (Y_{\bullet}^{(n)}(s))^{-1} \alpha(s) \, \mathrm{d}s$$

$$\xrightarrow{\mathrm{P}} \int_0^t \kappa^2(s) y_h(s) y_j(s) (y_{\bullet}(s))^{-1} \alpha(s) \, \mathrm{d}s \quad \textit{as } n \to \infty.$$

(B) *For all* $t \in \mathscr{T}$ *all* $\varepsilon > 0$,

$$c_n^2 \int_0^t (K^{(n)}(s))^2 I(|c_n K^{(n)}(s)| > \varepsilon) \alpha(s) Y_{\bullet}^{(n)}(s) \, \mathrm{d}s \xrightarrow{\mathrm{P}} 0 \quad \textit{as } n \to \infty.$$

Then, under hypothesis (5.2.1),

$$c_n(Z_1^{(n)}, \ldots, Z_k^{(n)}) \xrightarrow{\mathscr{D}} (U_1, \ldots, U_k) \quad \textit{as } n \to \infty,$$

in $D(\mathscr{T})^k$, *where the* U_h *are mean-zero Gaussian martingales with* $U_h(0) = 0$ *and* $\operatorname{cov}(U_h(s), U_j(t)) = \sigma_{hj}(s \wedge t)$. *Furthermore, under the hypothesis, for all* h, j,

$$\sup_{t \in \mathscr{T}} |c_n^2 \hat{\sigma}_{hj}(t) - \sigma_{hj}(t)| \xrightarrow{\mathrm{P}} 0 \quad \textit{as } n \to \infty,$$

where $\hat{\sigma}_{hj}(t)$ *is defined by* (5.2.5).

Proof. We first note that $\sigma_{hj}(t)$ of (5.2.12) is well-defined by the assumed integrability of $\kappa^2 \alpha y_{\bullet}$ and the Cauchy-Schwarz inequality. By (5.2.3) we may write $c_n Z_h^{(n)}(t) = \sum_{l=1}^k \int_0^t H_{hl}^{(n)}(s) \, \mathrm{d}M_l^{(n)}(s)$ with

$$H_{hl}^{(n)}(s) = c_n K^{(n)}(s) \{\delta_{hl} - Y_h^{(n)}(s)/Y_{\bullet}^{(n)}(s)\},$$

and the weak convergence result follows from the martingale central limit theorem (Theorem II.5.1) using (5.2.4) and conditions A and B.

Applying the same theorem to the martingales

$$c_n W_h^{(n)}(t) = c_n \sum_{l=1}^{k} \int_0^t K^{(n)}(s) Y_h^{(n)}(s) (Y_{\bullet}^{(n)}(s))^{-1} \, \mathrm{d}M_l^{(n)}(s),$$

$h = 1, 2, \ldots, k$; we, furthermore, have

$$c_n^2 [W_h^{(n)}, W_j^{(n)}](t) = c_n^2 \int_0^t (K^{(n)}(s))^2 Y_h^{(n)}(s) Y_j^{(n)}(s) (Y_{\bullet}(s))^{-2} \, \mathrm{d}N_{\bullet}^{(n)}(s)$$

$$\to \int_0^t \kappa^2(s) y_h(s) y_j(s) (y_{\bullet}(s))^{-1} \alpha(s) \, \mathrm{d}s \tag{5.2.13}$$

uniformly on $\mathcal{T}$ in probability as $n \to \infty$ for all h, j. From this, the uniform consistency of $c_n^2 \hat{\sigma}_{hj}$ follows using (5.2.5) and (5.2.12). □

When checking conditions A and B in practice, it is often useful to use Proposition II.5.3 as follows.

Proposition V.2.2. *Assume that*

(i) *There exist a sequence of constants* $\{a_n\}$, *increasing to infinity as* $n \to \infty$, *and a sequence of constants* $\{d_n\}$ *such that for all* h *and almost all* $s \in [0, \tau]$, *we have, under the hypothesis,*

$$Y_h^{(n)}(s)/a_n^2 \xrightarrow{P} y_h(s) < \infty \tag{5.2.14}$$

and

$$K^{(n)}(s)/d_n \xrightarrow{P} \kappa(s) < \infty \tag{5.2.15}$$

as $n \to \infty$.

(ii) *For all* $\delta > 0$, *there exists a function* k_δ *such that*

$$\liminf_{n \to \infty} \mathrm{P}((a_n d_n)^{-2} (K^{(n)}(s))^2 Y_{\bullet}^{(n)}(s) \leq k_\delta(s) \text{ all } s \in \mathcal{T}) \geq 1 - \delta \tag{5.2.16}$$

under the hypothesis.

Then, provided that αk_δ *is integrable on* $\mathcal{T}$ *for all* $\delta > 0$, *conditions* A *and* B *of Theorem* V.2.1 *are satisfied with* $c_n = (a_n d_n)^{-1}$.

Proof. The integrands in conditions A and B are both bounded by $c_n^2 (K^{(n)}(s))^2 \alpha(s) Y_{\bullet}(s)$. Therefore, Proposition II.5.3 is directly applicable and the result follows. □

Examples V.2.9 and V.2.10 will illustrate how conditions (5.2.14)–(5.2.16) may be checked in practice.

Writing $\mathbf{Z}^{(n)}(t) = (Z_1^{(n)}(t), \ldots, Z_k^{(n)}(t))^\mathsf{T}$, we, therefore, especially have for any fixed $t \in \mathscr{T}$ that

$$c_n \mathbf{Z}^{(n)}(t) \xrightarrow{\mathscr{D}} \mathscr{N}(\mathbf{0}, \mathbf{\Sigma}(t)), \tag{5.2.17}$$

as $n \to \infty$, under the assumptions of Theorem V.2.1, where $\mathbf{\Sigma}(t)$ is the (singular) $k \times k$ matrix with elements $\sigma_{hj}(t)$ defined by (5.2.12). [In fact, we only need to check conditions A and B of Theorem V.2.1 for this particular $t \in \mathscr{T}$ to get (5.2.17); cf. Theorem II.5.1.] Now, $\mathbf{\Sigma}(t)$ has rank $k - 1$ (its maximum rank) under mild conditions, such as the existence for each h, j in a chain of pairs connecting $\{1, 2, \ldots, k\}$ of a set $B_{hj} \subset [0, t]$ with positive Lebesgue measure such that $\kappa^2(s) y_h(s) y_j(s) \alpha(s) / y_{\cdot}(s) > 0$ for $s \in B_{hj}$ (e.g., Gill, 1986a, p. 114 and Appendix 1). Moreover, by (5.2.13), $c_n^2 \hat{\mathbf{\Sigma}}(t)$ is a consistent estimator for $\mathbf{\Sigma}(t)$, and, therefore, in particular the probability that $c_n^2 \hat{\mathbf{\Sigma}}(t)$ has rank $k - 1$ increases to unity as n tends to infinity. Hence, under hypothesis (5.2.1), the test statistic $X^2(t)$ given by (5.2.6) is asymptotically χ^2 distributed with $k - 1$ degrees of freedom as already stated in Section V.2.1. Often $\tau \in \mathscr{T}$ and the test statistic is calculated for $t = \tau$ as illustrated in Section V.2.1.

For the two-sample case, we, in particular, get by Theorem V.2.1 that $c_n Z_1^{(n)} \xrightarrow{\mathscr{D}} U_1$ as $n \to \infty$, on $D(\mathscr{T})$, where $Z_1^{(n)}$ is given by (5.2.2) or (5.2.7), and U_1 is a Gaussian martingale with $U_1(0) = 0$ and $\operatorname{cov}(U_1(s), U_1(t)) = \sigma_{11}(s \wedge t)$ given by (5.1.12). If we introduce

$$l(t) = \kappa(t) y_1(t) y_2(t) \{y_1(t) + y_2(t)\}^{-1}$$

[cf. (5.2.9)], we may alternatively write

$$\sigma_{11}(t) = \int_0^t l^2(s)(y_1(s) + y_2(s)) \{y_1(s) y_2(s)\}^{-1} \alpha(s) \, \mathrm{d}s.$$

In Section V.4, we discuss how this result on weak convergence of the whole $Z_1^{(n)}$ process may be used to derive Kolmogorov–Smirnov- and Cramér–von Mises-type tests, as well as certain sequential tests, for the hypothesis that $\alpha_1 = \alpha_2$. Here, we only note that we especially have

$$Z_1^{(n)}(t) \{\hat{\sigma}_{11}(t)\}^{-1/2} \xrightarrow{\mathscr{D}} \mathscr{N}(0, 1)$$

as $n \to \infty$, a result already stated in Section V.2.1.

V.2.2.2. *Asymptotically Proportional Risk Sets and Conservative Approximations*

An important special structure of the assumptions of Theorem V.2.1 occurs when

$$y_h(s) / y_{\cdot}(s) = p_h \tag{5.2.18}$$

for all h and all $s \in \mathscr{T}$ for which $y_{\cdot}(s) > 0$. Obviously, the p_h have to sum to one. When (5.2.18) holds true, we say that we have *asymptotically proportional risk sets*. Then some modifications of the test statistic (5.2.6) are possible. For, in this situation, $\mathbf{\Sigma}(t)$ of (5.2.17) has elements of the form

$$\sigma_{hj}(t) = p_h(\delta_{hj} - p_j) \int_0^t \kappa^2(s)\alpha(s)y_{\bullet}(s)\,\mathrm{d}s, \tag{5.2.19}$$

i.e., it is proportional to the covariance matrix of a multinomial random variable.

Moreover, by (5.2.13), the integral on the right-hand side of (5.2.19) may be estimated consistently by

$$c_n^2 \int_0^t (K^{(n)}(s))^2\,\mathrm{d}N_{\bullet}^{(n)}(s).$$

If we, therefore, assume the existence of constants or random variables $\hat{p}_h^{(n)}$ such that $\hat{p}_h^{(n)} \xrightarrow{\mathrm{P}} p_h$ as $n \to \infty$, then we may estimate $\sigma_{hj}(t)$ consistently by $c_n^2\sigma_{hj}^*(t)$, where

$$\sigma_{hj}^*(t) = \hat{p}_h^{(n)}(\delta_{hj} - \hat{p}_j^{(n)}) \int_0^t (K^{(n)}(s))^2\,\mathrm{d}N_{\bullet}^{(n)}(s). \tag{5.2.20}$$

[For survival data, (5.2.20) is close to the permutation estimator (5.2.11) using $\hat{p}_h^{(n)} = n_h/n$; cf. Examples V.2.9 and V.2.10.] It follows that an alternative version of the test statistic (5.2.6) is

$$X_p^2(t) = \left\{\sum_{h=1}^{k} \frac{(Z_h^{(n)}(t))^2}{\hat{p}_h^{(n)}}\right\}\left\{\int_0^t (K^{(n)}(s))^2\,\mathrm{d}N_{\bullet}^{(n)}(s)\right\}^{-1}. \tag{5.2.21}$$

It is seen, using (5.2.13) and (5.2.18), that one possible choice of the $\hat{p}_h^{(n)}$ is

$$\int_0^t (K^{(n)}(s))^2 \frac{Y_h^{(n)}(s)}{Y_{\bullet}^{(n)}(s)}\,\mathrm{d}N_{\bullet}^{(n)}(s) \Big/ \int_0^t (K^{(n)}(s))^2\,\mathrm{d}N_{\bullet}^{(n)}(s). \tag{5.2.22}$$

Then (5.2.21) takes the form

$$X_c^2(t) = \sum_{h=1}^{k} \left\{(Z_h^{(n)}(t))^2 \Big/ \int_0^t (K^{(n)}(s))^2 \frac{Y_h^{(n)}(s)}{Y_{\bullet}^{(n)}(s)}\,\mathrm{d}N_{\bullet}^{(n)}(s)\right\}. \tag{5.2.23}$$

By direct computations similar to those of Crowley and Breslow (1975), it may be verified that the difference between the $k \times k$ matrix $\mathbf{\Sigma}^*(t)$, with elements given by (5.2.20) with the $\hat{p}_h^{(n)}$ as in (5.2.22), and $\hat{\mathbf{\Sigma}}(t)$ given by (5.2.5) is always positive semidefinite. Thus, (5.2.23) always takes on smaller values than (5.2.6), whereas (5.2.6) always has the correct asymptotic distribution. A common terminology is to state that (5.2.23) is *conservative* except in the case where we have asymptotically proportional risk sets.

V.2.2.3. *Examples*

Let us then illustrate the conditions of Theorem V.2.1, as well as the alternative versions (5.2.21) and (5.2.23) of the test statistic, by some examples.

EXAMPLE V.2.9. Classical Nonparametric Tests

Consider the classical k sample situation where the object is to test the hypothesis $F_1 = F_2 = \cdots = F_k$ in the setup with X_{hi} for $i = 1, 2, \ldots, n_h$, $h = 1, 2, \ldots, k$ (with $n = \Sigma n_h$) independent random variables with absolutely continuous distribution function F_h with support $[0, \infty)$, density f_h, and hazard rate α_h in group number h. By Example III.1.2, the k-dimensional counting process $\mathbf{N}^{(n)} = (N_1^{(n)}, \ldots, N_k^{(n)})$, with

$$N_h^{(n)}(t) = \sum_{i=1}^{n_h} I(X_{hi} \le t)$$

counting the number of observations in the hth group in $[0, t]$, has an intensity process $\boldsymbol{\lambda}^{(n)} = (\lambda_1^{(n)}, \ldots, \lambda_k^{(n)})$ of the multiplicative form

$$\lambda_h^{(n)}(t) = \alpha_h(t)\, Y_h^{(n)}(t)$$

with

$$Y_h^{(n)}(t) = n_h - N_h^{(n)}(t-).$$

We want to test the hypothesis that the α_h are identical; cf. (5.2.1).

We restrict our attention to the situation where the weight process $K^{(n)}$ in (5.2.2) may be given as $K^{(n)}(t) = K_0(Y_\bullet^{(n)}(t))$ for some fixed function K_0 (that is the same for all n) with $K_0(0) = 0$. Since we have

$$Y_h^{(n)}(t) = \int_{[t,\infty)} \mathrm{d}N_h^{(n)}(u),$$

we may, after a change of the order of integration, rewrite $Z_h^{(n)}(\infty)$ as

$$Z_h^{(n)}(\infty) = \int_0^\infty K_0(Y_\bullet^{(n)}(s))\, \mathrm{d}N_h^{(n)}(s)$$
$$- \int_0^\infty \int_{(0,u]} K_0(Y_\bullet^{(n)}(s))(Y_\bullet^{(n)}(s))^{-1}\, \mathrm{d}N_\bullet^{(n)}(s)\, \mathrm{d}N_h^{(n)}(u);$$

cf. (5.2.2). If, furthermore, $R_{hi}^{(n)}$ denotes the rank of X_{hi} in the combined sample of size n, then $Y_\bullet^{(n)}(X_{hi}) = n + 1 - R_{hi}^{(n)}$, and it is seen that

$$Z_h^{(n)}(\infty) = \sum_{i=1}^{n_h} a_n(R_{hi}^{(n)})$$

is a linear rank statistic with scores given by

$$a_n(j) = K_0(n + 1 - j) - \sum_{v=n+1-j}^{n} K_0(v)/v.$$

Especially for $K_0(y) = y$ and $K_0(y) = I_{(0,\infty)}(y)$, one gets

$$Z_h^{(n)}(\infty) = n_h(n + 1) - 2 \sum_{i=1}^{n_h} R_{hi}^{(n)}$$

and

$$Z_h^{(n)}(\infty) = n_h - \sum_{i=1}^{n_h} \sum_{v=n+1-R_{hi}^{(n)}}^{n} 1/v,$$

respectively, corresponding to rank statistics of the Wilcoxon and Savage type [e.g., Hájek and Sĭdák (1967, pp. 87 and 97)].

As mentioned in Example V.2.1, the generalization of the Savage test to censored data is commonly denoted as the log-rank test. As indicated in that example, the reason for this name is that for the Savage test the score associated with X_{hi} is

$$1 - \sum_{v=n+1-R_{hi}^{(n)}}^{n} \frac{1}{v} \approx 1 + \sum_{v=n+1-R_{hi}^{(n)}}^{n} \log\left(1 - \frac{1}{v}\right) = 1 + \log\left\{\frac{n - R_{hi}}{n}\right\},$$

i.e., the score is approximately linearly related to $\log(n - R_{hi})$.

Let us then consider the situation with $K_0(y) = y$ more specifically. We will first see how the conditions of the Theorem V.2.1 may be verified by using Proposition V.2.2. With $K^{(n)}(s) = Y_{\bullet}^{(n)}(s)$, $a_n = n^{1/2}$, $d_n = n$, and, hence, $c_n = n^{-3/2}$, we have for all $s \in [0, \infty]$, under the hypothesis that all F_h equal F, that

$$\frac{Y_h^{(n)}(s)}{a_n^2} = \frac{n_h}{n} \frac{Y_h^{(n)}(s)}{n_h} \xrightarrow{\mathrm{P}} p_h(1 - F(s)) < \infty$$

and

$$\frac{K^{(n)}(s)}{d_n} = \frac{Y_{\bullet}^{(n)}(s)}{n} \xrightarrow{\mathrm{P}} (1 - F(s)) < \infty$$

as $n \to \infty$, provided that $\hat{p}_h^{(n)} = n_h/n \to p_h$ as $n \to \infty$. Thus, (5.2.14) and (5.2.15) hold with $y_h(s) = p_h(1 - F(s))$ and $\kappa(s) = y_{\bullet}(s) = 1 - F(s)$.

To prove (5.2.16), we use the result by Daniels (1945) reviewed in connection with Example V.1.6. Since, under the hypothesis that all F_h equal F, $1 - Y_{\bullet}^{(n)}/n$ is the (left continuous version of) the empirical distribution for F based on a sample of size $n = \Sigma n_h$, (5.1.9) gives

$$\mathrm{P}(Y_{\bullet}^{(n)}(s)/n \leq \delta^{-1}(1 - F(s)) \text{ for all } s \in [0, \infty]) = 1 - \delta.$$

Thus, we have, for any $\delta \in (0, 1)$ with $k_\delta(s) = \delta^{-3}(1 - F(s))^3$, that

$$\begin{aligned} &\mathrm{P}((a_n d_n)^{-2}(K^{(n)}(s))^2 Y_{\bullet}^{(n)}(s) \leq k_\delta(s) \text{ for all } s \in [0, \infty]) \\ &= \mathrm{P}\left(\left(\frac{Y_{\bullet}^{(n)}(s)}{n}\right)^3 \leq \frac{1}{\delta^3}(1 - F(s))^3 \text{ for all } s \in [0, \infty]\right) = 1 - \delta, \end{aligned}$$

so that (5.2.16) is satisfied. The result then follows noting that $\alpha(s)k_\delta(s) = \delta^{-3}(1 - F(s))^3\alpha(s) \leq \delta^{-3}f(s)$ is integrable.

Since we found $y_h(s) = p_h(1 - F(s))$, (5.2.18) holds true, and versions (5.2.21) and (5.2.23) of the test statistic apply. Let us first consider a slightly modified version of (5.2.21). Since we are in a situation with no censoring, we may take the permutation distribution approach and estimate the covariances by (5.2.11), i.e., by

$$\tilde{\sigma}_{hj}(\infty) = \hat{p}_h^{(n)}(\delta_{hj} - \hat{p}_j^{(n)})\frac{n}{n-1}\int_0^\infty Y_{\bullet}^{(n)}(s)(Y_{\bullet}^{(n)}(s) - 1)\,\mathrm{d}N_{\bullet}(s),$$

where $\hat{p}_h^{(n)} = n_h/n$. It is easy to see that this is a consistent estimator for (5.2.12). Then, since we have

$$\int_0^\infty Y^{(n)}(s)(Y^{(n)}(s) - 1)\,\mathrm{d}N_{\bullet}(s) = \sum_{i=1}^n i(i-1) = \tfrac{1}{3}(n-1)n(n+1),$$

a modified version of the test statistic (5.2.21) is

$$\left\{\sum_{h=1}^k \frac{Z_h^{(n)}(\infty)^2}{\hat{p}_h^{(n)}}\right\}\left\{\frac{n}{n-1}\int_0^\infty Y_{\bullet}^{(n)}(s)(Y_{\bullet}^{(n)}(s) - 1)\,\mathrm{d}N_{\bullet}(s)\right\}^{-1}$$
$$= \frac{3}{n(n+1)}\sum_{h=1}^k \frac{1}{n_h}\left\{n_h(n+1) - 2\sum_{i=1}^{n_h} R_{hi}\right\}^2,$$

which is exactly the Kruskal–Wallis test [e.g., Hájek and Sĭdák (1967, p. 104)].

Using instead the estimators (5.2.20) (with $\hat{p}_h^{(n)} = n_h/n$) for the covariances (which are close to the permutation based estimators), the test statistic (5.2.21) turns out to be smaller than the Kruskal–Wallis statistic by a factor of $n/(n + \frac{1}{2})$. The test statistic defined by (5.2.23) is not related in such a simple way to the Kruskal–Wallis test. □

EXAMPLE V.2.10. Tests for Censored Survival Data

In this example, we will study further the tests for censored survival data from Example V.2.1. We consider the random censorship model (cf. Examples III.2.4 and IV.1.6), where, for each n, $X_{hi}^{(n)}$, $U_{hi}^{(n)}$ for $i = 1, 2, \ldots, n_h$, $h = 1, 2, \ldots, k$, $n = \Sigma_h n_h$, are mutually independent with the $X_{hi}^{(n)}$ having absolutely continuous distribution function F_h with support $[0, \infty)$, density f_h, and hazard rate α_h, whereas $U_{hi}^{(n)}$ has distribution function $G_{hi}^{(n)}$. We define censored survival times $\tilde{X}_{hi}^{(n)}$ as in Example V.2.1 and note that the $\tilde{X}_{hi}^{(n)}$ are independent with distribution function $H_{hi}^{(n)} = 1 - (1 - F_h)(1 - G_{hi}^{(n)})$. Finally, we let $N_h^{(n)}(t)$, $Y_h^{(n)}(t)$, etc., be defined as in Example V.2.1, where we have added a superscript n to indicate the dependence on the sample size.

We will first show that the condition (5.2.14) is satisfied provided that $n_h/n \to p_h$ as $n \to \infty$ and that there exist (sub-)distribution functions $G_1, \ldots, G_k$ such that

$$n_h^{-1}\sum_{i=1}^{n_h} G_{hi}^{(n)}(s) \to G_h(s) \tag{5.2.24}$$

as $n \to \infty$ for all h and all $s \in [0, \infty]$. We note that, under the hypothesis that all F_h equal F, we have

$$\mathrm{E}Y_h^{(n)}(s) = (1 - F(s)) \sum_{i=1}^{n_h} (1 - G_{hi}^{(n)}(s-))$$

and

$$\operatorname{var} Y_h^{(n)}(s) \leq \mathrm{E}Y_h^{(n)}(s);$$

cf. Example IV.1.6. Therefore, by the assumption above and Chebyshev's inequality,

$$\frac{Y_h^{(n)}(s)}{n} = \frac{n_h}{n}\frac{Y_h^{(n)}(s)}{n_h} \xrightarrow{\mathrm{P}} y_h(s) = p_h(1 - F(s))(1 - G_h(s-)) \tag{5.2.25}$$

as $n \to \infty$ under the hypothesis that the F_h are identical. Thus, (5.2.14) holds true with $a_n = n^{1/2}$.

The other two conditions, (5.2.15) and (5.2.16) of Proposition V.2.2, depend on the choice of the weight process and, therefore, have to be checked separately for each particular choice of $K^{(n)}$.

For the log-rank test $K^{(n)}(t) = I(Y_{\bullet}(t) > 0)$, and (5.2.15) is trivially fulfilled with $d_n = 1$ and $\kappa(t) = I(y_{\bullet}(t) > 0)$ with

$$y_{\bullet}(s) = (1 - F(s))(1 - \Sigma_h p_h G_h(s-)).$$

To prove (5.2.16), we use (5.1.9) to get, for any $\delta \in (0, 1)$, under the hypothesis,

$$\mathrm{P}\left(\frac{1}{n}\sum_{h=1}^{k}\sum_{i=1}^{n_h} I\{X_{hi}^{(n)} \geq s\} \leq \frac{1}{\delta}(1 - F(s)) \text{ for all } s \in [0, \infty]\right) = 1 - \delta.$$

From this, it follows that

$$\mathrm{P}\left(\frac{Y_{\bullet}^{(n)}(s)}{n} \leq \frac{1}{\delta}(1 - F(s)) \text{ for all } s \in [0, \infty]\right) \geq 1 - \delta,$$

so that (5.2.16) is satisfied with $k_\delta = \delta^{-1}(1 - F)$. Furthermore,

$$\delta^{-1}(1 - F(s))\alpha(s) = \delta^{-1}f(s)$$

is integrable, and it follows by Proposition V.2.2 that the conditions of Theorem V.2.1 are fulfilled for the log-rank test for the general random censorship model provided that (5.2.24) holds true.

For the generalization of the Wilcoxon and Kruskal–Wallis tests of Gehan (1965) and Breslow (1970), we have $K^{(n)}(t) = Y_{\bullet}^{(n)}(t)$. It then follows by (5.2.25) that (5.2.15) holds true with $\kappa(t) = y_{\bullet}(t)$ and $d_n = n$. Also, (5.2.16) can be shown to hold true by combining the argument used for the log-rank test above with that of Example V.2.9 for the case without censoring.

Let us also have a look at the test statistic of Harrington and Fleming (1982). (Prentice's generalization of the Wilcoxon and Kruskal–Wallis tests may be treated in a similar manner.) For this test statistic, we have $K^{(n)}(t) =$

$(\hat{S}^{(n)}(t-))^\rho I(Y_\bullet^{(n)}(t) > 0)$, where $\hat{S}^{(n)}$ is the Kaplan–Meier estimator based on the combined sample and ρ is a fixed number between 0 and 1. Then, by Theorem IV.3.1, (5.2.15) holds true with $\kappa(t) = S(t)^\rho I(y_\bullet(t) > 0)$ and $d_n = 1$. Moreover, since $(\hat{S}^{(n)}(t-))^\rho \le 1$, the argument used to prove (5.2.16) for the log-rank test works here as well, and the conditions of Theorem V.2.1 are fulfilled for the Harrington and Fleming test statistic under random censorship.

By (5.2.25), it is seen that, in general, $y_h(t)/y_\bullet(t)$ depends on t. However, when there is asymptotically equal censorship in the sense that $G_1 = G_2 = \cdots = G_k$, then (5.2.18) is satisfied and versions (5.2.21) and (5.2.23) of the test statistic also apply. In particular, for the log-rank test, (5.2.23) takes the form

$$X_c^2 = \sum_{h=1}^{k} \frac{(O_h - E_h)^2}{E_h},$$

where O_h and E_h were introduced in connection with Example V.2.1. This is a version of the log-rank test, known to be conservative in the case of unequal censoring patterns in the k groups (Peto and Pike, 1973; Crowley and Breslow, 1975).

Finally, when the censoring patterns are equal in all groups (not only asymptotically equal), i.e., all $G_{hi}^{(n)}$ are equal, the permutation covariances (5.2.11) may be used. These give the statistic

$$(n-1)\left\{\sum_{h=1}^{k} \frac{(Z_h^{(n)}(\infty))^2}{n_h}\right\}\left\{\int_0^\infty K^2(s)(Y_\bullet(s) - \Delta N_\bullet(s))(Y_\bullet(s))^{-1}\,\mathrm{d}N_\bullet(s)\right\}^{-1}$$

which is close to (5.2.21) (with $\hat{p}_h^{(n)} = n_h/n$ and $t = \infty$). For the Breslow–Gehan choice of weight process $K = Y_\bullet$, this test is in SPSS (1981), known as the "Lee–Desu" test. □

EXAMPLE V.2.11. Survival with Malignant Melanoma. Conservative Approximations to Test Statistics

The data were introduced in Example I.3.1. As explained there, the patients with malignant melanoma were censored at the time (= duration since operation) at end of study 31 December 1977 or at time of death of other causes. This censoring pattern is very compatible with the concept of equal censoring underlying the conservative approximation (5.2.23) of the test statistics. In this example, we compare the results of Example V.2.3 with the ones we get by using this conservative approximation.

For the problem of comparing death from melanoma for male and female melanoma patients, we quote the following results:

test statistic	(5.2.7)–(5.2.8)	(5.2.23)
log-rank	−2.543	−2.542
Gehan–Wilcoxon	−2.722	−2.721
Prentice–Wilcoxon	−2.664	−2.663.

In this case, the effect is negligible. Note that in this two-sample situation, the gain in computational ease by using the conservative approximations is rather marginal. In fact, if O_h and E_h denote observed and "expected" deaths, respectively, in group h, the log-rank test statistic given by (5.2.7) and (5.2.8) may be represented as

$$U = V^{-1/2}(O_1 - E_1),$$

where $V = \hat{\sigma}_{11}(\tau)$[cf. (5.2.8)] may be easily calculated in parallel with E_h (though it is a separate calculation). The conservative approximation, in χ^2 form, is the slightly easier and perhaps more familiar

$$(O_1 - E_1)^2/E_1 + (O_2 - E_2)^2/E_2.$$

The parameters p_h represent the (assumed constant) fraction of the individuals at risk that belong to group h. The estimates $\hat{p}_h$ of (5.2.22), therefore, represent an average fraction of group h individuals at risk, weighted over time according to the particular test statistic, specified by the weight process K.

In the present example, all three test statistics yielded the value $\hat{p}_{\text{female}} = 0.65$, compared to the fact that at time 0, there were 126/79 = 0.61/0.39 females/males (since the females "die more slowly," there are on the average more of them at risk). □

EXAMPLE V.2.12. Bone Marrow Transplantation

The problem and the data were presented in Example I.3.14. In this example, we compare the log-rank test result of Example V.2.4 with the one we get using the conservative version of the log-rank test presented in Example V.2.10. We may calculate the conservative approximation without or with the tie-modified variance estimate. The results are as follows:

	no tie correction	tie correction
original test	35.75	35.89
conservative approximation	35.28	35.42.

The estimates $\hat{p}_h$ of (5.2.22) for the five centers are (whether or not tie corrected)

$$0.399 \quad 0.466 \quad 0.023 \quad 0.100 \quad 0.011$$

which compare to the following initial frequencies

$$0.358 \quad 0.489 \quad 0.047 \quad 0.079 \quad 0.026.$$

Even with this rather clear deviation from the hypothesis of asymptotically proportional risk sets, the error in using the conservative approximation is practically insignificant. □

EXAMPLE V.2.13. A Markov Illness–Death Model. Nephropathy and Mortality for Insulin-Dependent Diabetics

Let us consider the three-state Markov illness–death model without recovery having states 0 = "healthy," 1 = "diseased," and 2 = "dead" and transition intensities α_{01}, α_{02}, and α_{12} for the three possible types of transition.

For this model, the transition probabilities take the form

$$P_{00}(s,t) = \exp\left\{-\int_s^t (\alpha_{01}(u) + \alpha_{02}(u))\,du\right\},$$

$$P_{11}(s,t) = \exp\left\{-\int_s^t \alpha_{12}(u)\,du\right\},$$

and

$$P_{01}(s,t) = \int_s^t P_{00}(s,u)\alpha_{01}(u)P_{11}(u,t)\,du;$$

cf. Example IV.4.3.

Assume now that we want to test the hypothesis of nondifferential mortality, i.e., the hypothesis that $\alpha_{02}(t) = \alpha_{12}(t)$ for all t, based on observation of n independent copies, all in state 0 at time 0, of the Markov process over $[0,\tau]$ with $\int_0^\tau \alpha_{hj}(u)\,du < \infty$ for $(h,j) = (0,1), (0,2), (1,2)$. It is known from Example III.1.4 that for this case $(N_{02}^{(n)}, N_{12}^{(n)})$, with $N_{hj}^{(n)}(t)$ counting the total number of transitions from state h to state j in $[0,t]$, forms a bivariate counting process with intensity process $(\lambda_{02}^{(n)}, \lambda_{12}^{(n)})$ of the form

$$\lambda_{hj}^{(n)}(t) = \alpha_{hj}(t)\,Y_h^{(n)}(t),$$

where $Y_h^{(n)}(t)$ is the number of copies of the Markov process found to be in state h just prior to time t.

Thus, the situation fits into the framework of this section, and the test statistic (5.2.6) may be used, e.g., with $K^{(n)}(s) = Y_\cdot^{(n)}(s) = Y_0^{(n)}(s) + Y_1^{(n)}(s)$ or $K^{(n)}(s) = I(Y_\cdot^{(n)}(s) > 0)$. By applying the technique of Examples V.2.9 and V.2.10 (and noting that $Y_1^{(n)}/n$ is the difference between two empirical distribution functions), we may prove that the conditions of Theorem V.2.1 are satisfied with

$$y_0(t) = P_{00}(0,t)$$

and, since under the hypothesis $\alpha_{02} = \alpha_{12}$,

$$y_1(t) = P_{01}(0,t) = \exp\left\{-\int_0^t \alpha_{02}(u)\,du\right\}\left\{1 - \exp\left(-\int_0^t \alpha_{01}(u)\,du\right)\right\}.$$

Thus, writing $y_{\cdot}(t) = y_0(t) + y_1(t)$, it is seen that

$$\frac{y_0(t)}{y_{\cdot}(t)} = 1 - \frac{y_1(t)}{y_{\cdot}(t)} = \exp\left(-\int_0^t \alpha_{01}(u)\,du\right)$$

does depend on t, and version (5.2.23) of the test statistic is (strictly) conservative.

We remind the reader that the multiplicative structure of the intensity process is preserved under independent right-censoring and left-truncation (Sections III.2.2 and III.3) and go on to illustrate the above results using the data on nephropathy (DN) and mortality among insulin-dependent diabetics introduced in Example I.3.11. Here, it was noted that out of a total number of person-years at risk of 44,561 in state 0: "alive without DN", 451 diabetics died, whereas as many as 267 died from state 1: "alive with DN", where the number of person-years at risk was only 5024. This very high mortality with DN is also seen from the Nelson–Aalen plots (Figures V.2.1 and V.2.2) of the integrated hazard estimates $\hat{A}_{02}(t)$ and $\hat{A}_{12}(t)$ for men and women, respectively, and confirmed by any statistic (that one hardly needs to calculate). Thus, the values of the log-rank test statistics are 29.4 for men and 27.3 for women and those of the Wilcoxon-type statistic, choosing $K = Y_0 + Y_1$ (or $L = Y_0 Y_1$), are 28.5 for men and 25.6 for women all highly significant when

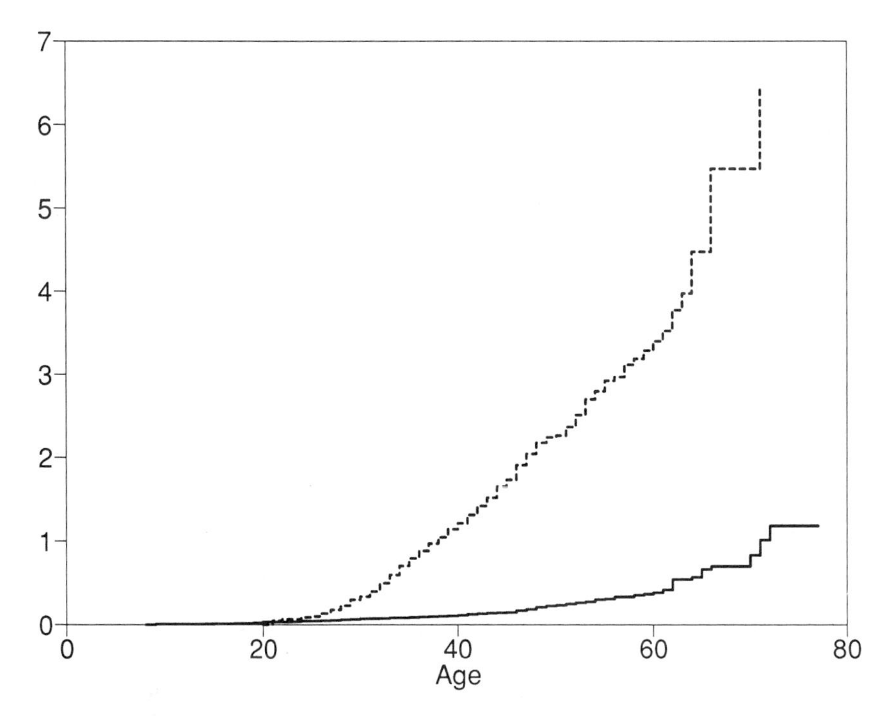

Figure V.2.1. Nelson–Aalen estimates of integrated death intensities for male diabetics without DN (——) and with DN (- - -).

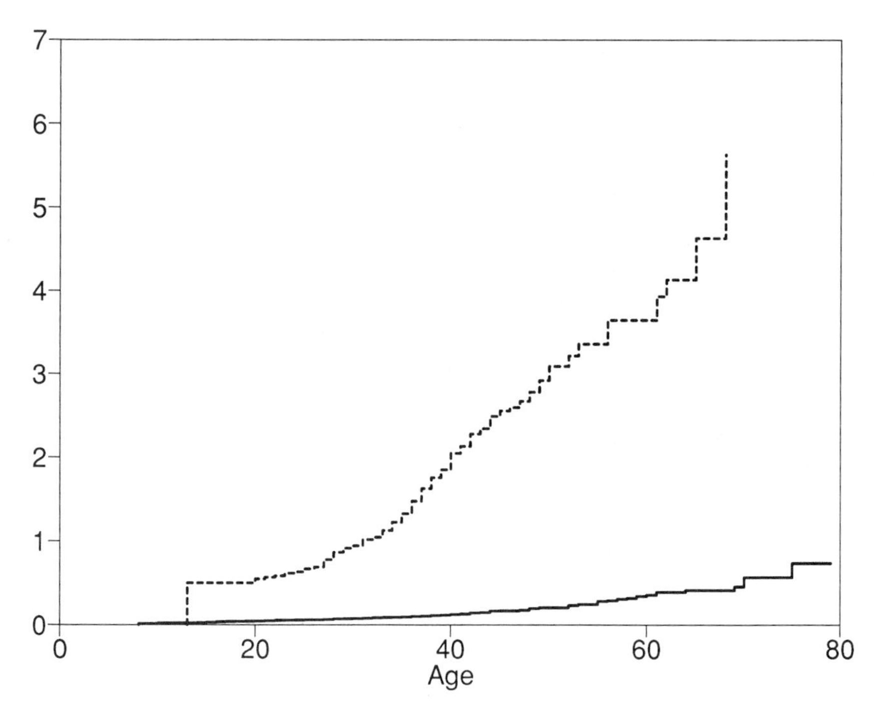

Figure V.2.2. Nelson–Aalen estimates of integrated death intensities for female diabetics without DN (——) and with DN (- - -).

referred to a standard normal distribution. This is also the case for the corresponding conservative approximations (5.2.23) taking the values $829.49 = (28.8)^2$ and $722.32 = (26.9)^2$ for men and women, respectively, for the log-rank test and $792.66 = (28.2)^2$ and $639.90 = (25.3)^2$ for the Wilcoxon test. So even if there are no theoretical justifications for using the conservative approximation (5.2.23) in this example, the difference between (5.2.23) and the standard test statistic (5.2.6) is of little importance. □

V.2.3. Local Asymptotic Power

In the two preceding subsections, we have derived and studied a class of test statistics for hypothesis (5.2.1). So far, all our derivations have been done assuming the hypothesis (5.2.1) to hold true. In this subsection, we will study the behavior of the test statistics outside the hypothesis and derive optimal tests. As in Section V.1.3, we will do this by considering the properties of the test statistics under a sequence of local alternatives. A further study of the optimality properties of these tests, using the theory of contiguity and local asymptotic normality, is provided in Sections VIII.2.3 and VIII.4.2.

Consider, therefore, a sequence of counting processes $\mathbf{N}^{(n)} =$

$(N_1^{(n)}, \ldots, N_k^{(n)})$, $n = 1, 2, \ldots$, as in Section V.2.2, but assume now that $\lambda_h^{(n)}(t) = \alpha_h^{(n)}(t) Y_h^{(n)}(t)$ with

$$\alpha_h^{(n)}(t) = \alpha(t) + \theta_h(t)/a_n + \rho_h^{(n)}(t) \tag{5.2.26}$$

for $\theta_1, \ldots, \theta_k$ fixed functions on $\mathscr{T}$, $\{a_n\}$ a sequence of constants increasing to infinity as $n \to \infty$, and $\{\rho_h^{(n)}\}$, $h = 1, \ldots, k$ sequences of functions satisfying $\sup_{[0,\tau]} |a_n \rho_h^{(n)}(t)| \to 0$ for each h as $n \to \infty$. As will be seen below, the normalizing sequence $\{a_n\}$ will be the same as the one defined in connection with (5.2.14). Moreover, the functions θ_h will usually be of the form $\theta_h(t) = \varphi_h \gamma(t) \alpha(t)$ for constants φ_h and a fixed function γ [cf. (5.2.32)].

With the $\alpha_h^{(n)}(t)$ as in (5.2.26), the $Z_h^{(n)}(t)$ of (5.2.2) may be written as

$$Z_h^{(n)}(t) = X_h^{(n)}(t) + V_h^{(n)}(t) \tag{5.2.27}$$

with

$$X_h^{(n)}(t) = \sum_{l=1}^{k} \int_0^t K^{(n)}(s) \left(\delta_{hl} - \frac{Y_h^{(n)}(s)}{Y_{\bullet}^{(n)}(s)} \right) \mathrm{d}M_l^{(n)}(s)$$

and

$$V_h^{(n)}(t) = a_n^{-1} \int_0^t K^{(n)}(s) Y_h^{(n)}(s) \left\{ \theta_h(s) - \sum_{l=1}^{k} \frac{Y_l^{(n)}(s)}{Y_{\bullet}^{(n)}(s)} \theta_l(s) \right\} \mathrm{d}s + R_{1h}^{(n)}(t),$$

where for all h

$$|R_{1h}^{(n)}(t)| = o(a_n^{-1}) \int_0^t K^{(n)}(s) Y_{\bullet}^{(n)}(s) \, \mathrm{d}s;$$

cf. (5.2.3). It is seen that the $X_h^{(n)}$ are local square integrable martingales under the sequence of local alternatives (5.2.26) having predictable covariation processes

$$\begin{aligned}
&\langle X_h^{(n)}, X_j^{(n)} \rangle (t) \\
&\quad = \int_0^t (K^{(n)}(s))^2 \frac{Y_h^{(n)}(s)}{Y_{\bullet}^{(n)}(s)} \left(\delta_{hj} - \frac{Y_j^{(n)}(s)}{Y_{\bullet}^{(n)}(s)} \right) \alpha(s) Y_{\bullet}^{(n)}(s) \, \mathrm{d}s \\
&\qquad + a_n^{-1} \sum_{l=1}^{k} \int_0^t (K^{(n)}(s))^2 \left(\delta_{hl} - \frac{Y_h^{(n)}(s)}{Y_{\bullet}^{(n)}(s)} \right) \left(\delta_{jl} - \frac{Y_j^{(n)}(s)}{Y_{\bullet}^{(n)}(s)} \right) \theta_l(s) Y_l^{(n)}(s) \, \mathrm{d}s \\
&\qquad + R_{2hj}^{(n)}(t),
\end{aligned}$$

where for all h, j

$$|R_{2hj}^{(n)}(t)| = o(a_n^{-1}) \int_0^t (K^{(n)}(s))^2 Y_{\bullet}^{(n)}(s) \, \mathrm{d}s;$$

cf. (5.2.4).

Assume now that (i) and (ii) of Proposition V.2.2 continue to hold under the sequence of local alternatives (5.2.26) and, moreover, that

(iii) For all $\delta > 0$, there exists a function g_δ such that

$$\liminf_{n\to\infty} \mathrm{P}(a_n^{-2} d_n^{-1} K^{(n)}(s) Y_{\bullet}^{(n)}(s) \leq g_\delta(s) \text{ for all } s \in \mathscr{T}) \geq 1 - \delta. \tag{5.2.28}$$

Furthermore, we assume that k_δ, αk_δ, $k_\delta \sum_{l=1}^{k} |\theta_l|$, g_δ, and $g_\delta \sum_{l=1}^{k} |\theta_l|$ are integrable on $\mathscr{T}$ for all $\delta > 0$. Then it follows as in Theorem V.2.1 and Proposition V.2.2 that with $c_n = (a_n d_n)^{-1}$ we have

$$c_n(X_1^n, \ldots, X_k^{(n)}) \xrightarrow{\mathscr{D}} (U_1, \ldots, U_k)$$

as $n \to \infty$, along the sequence of local alternatives (5.2.26), with the U_h as defined in Theorem V.2.1. Furthermore,

$$\sup_{t \in \mathscr{T}} \left| c_n V_h^{(n)}(t) - \int_0^t \kappa(s) y_h(s) \{\theta_h(s) - \bar{\theta}(s)\} \, \mathrm{d}s \right| \xrightarrow{\mathrm{P}} 0$$

as $n \to \infty$, where

$$\bar{\theta}(s) = \sum_{l=1}^{k} \{y_l(s)/y_{\bullet}(s)\} \theta_l(s). \tag{5.2.29}$$

Also, the uniform consistency of $c_n^2 \hat{\sigma}_{hj}$ continues to hold along the sequence of local alternatives.

In particular, we, therefore, have for any fixed $t \in \mathscr{T}$

$$c_n \mathbf{Z}^{(n)}(t) \xrightarrow{\mathscr{D}} \mathscr{N}(\boldsymbol{\xi}(t), \boldsymbol{\Sigma}(t))$$

as $n \to \infty$, where we have introduced $\boldsymbol{\xi}(t) = (\xi_1(t), \ldots, \xi_k(t))^{\mathsf{T}}$ given by

$$\xi_h(t) = \int_0^t \kappa(s) y_h(s) \{\theta_h(s) - \bar{\theta}(s)\} \, \mathrm{d}s. \tag{5.2.30}$$

It follows that, under the sequence of local alternatives (5.2.26), the test statistic $X^2(t)$ given by (5.2.6) asymptotically has a noncentral χ^2 distribution with $k - 1$ degrees of freedom and noncentrality parameter

$$\zeta(t) = \boldsymbol{\xi}(t)^{\mathsf{T}} \boldsymbol{\Sigma}(t)^{-} \boldsymbol{\xi}(t). \tag{5.2.31}$$

Therefore, the local asymptotic power [versus the sequence of alternatives (5.2.26)] of an α-level test based on $X^2(t)$ becomes $1 - \Gamma_{k-1}(\gamma_\alpha; \zeta(t))$, where $\Gamma_{k-1}(\cdot; \zeta(t))$ is the cumulative noncentral χ^2 distribution with $k - 1$ degrees of freedom and noncentrality parameter $\zeta(t)$ and γ_α is the upper α point of the central χ^2 distribution with $k - 1$ degrees of freedom.

This result may be used to pick an optimal choice of weight process $K^{(n)}$ for a particular situation. It is seen that one should choose $K^{(n)}$ such that the corresponding κ [cf. (5.2.15)] maximizes the noncentrality parameter (5.2.31).

We are not able to find an explicit expression for this optimal κ in general [see (8.2.21) for an implicit solution]. However, for the two-sample case, i.e., when $k = 2$, we have by (5.2.12), (5.2.30), and (5.2.31) that

$$\zeta(t) = \left\{ \int_0^t \kappa(s) y_1(s) y_2(s) (y_{\bullet}(s))^{-1} \delta(s)\, ds \right\}^2 \times \left\{ \int_0^t \kappa^2(s) y_1(s) y_2(s) (y_{\bullet}(s))^{-1} \alpha(s)\, ds \right\}^{-1}$$

with $\delta = \theta_1 - \theta_2$. Now by the Cauchy–Schwarz inequality,

$$\zeta(t) \leq \int_0^t y_1(s) y_2(s) \delta^2(s) \{ y_{\bullet}(s) \alpha(s) \}^{-1}\, ds$$

with equality if and only if κ is proportional to δ/α. If we, therefore, assume that the θ_h in (5.2.26) are of the form

$$\theta_h(s) = \varphi_h \gamma(s) \alpha(s), \tag{5.2.32}$$

where the φ_h are constants and γ a fixed function, then it is optimal to have κ proportional to γ. Thus, in particular, the choice of weight process $K^{(n)}(s) = I(Y_{\bullet}^{(n)}(s) > 0)$, corresponding to the log-rank test, is optimal when $\gamma \equiv 1$, i.e., for the alternative of (local) proportional α_h. Other applications of this result to the particular case of censored survival data are provided in Example V.2.14.

When we have asymptotically proportional risk sets and when, moreover, (5.2.32) holds true, we are able to derive the optimal κ analytically also when $k > 2$. For, in this case, $\bar{\theta}(s) = \bar{\varphi}\gamma(s)\alpha(s)$ in (5.2.29), with $\bar{\varphi} = \sum_l \varphi_l p_l$, and $\boldsymbol{\Sigma}(\tau)$ has elements given by (5.2.19). It then follows that the noncentrality parameter (5.2.31) takes the form

$$\begin{aligned} \zeta(t) &= \sum_{h=1}^k p_h^{-1} \xi_h^2(t) \Big/ \int_0^t \kappa^2(s) \alpha(s) y_{\bullet}(s)\, ds \\ &= \left(\int_0^t \kappa(s) \gamma(s) \alpha(s) y_{\bullet}(s)\, ds \right)^2 \left(\int_0^t \kappa^2(s) \alpha(s) y_{\bullet}(s)\, ds \right)^{-1} \sum_{h=1}^k p_h (\varphi_h - \bar{\varphi})^2. \end{aligned} \tag{5.2.33}$$

Using the Cauchy–Schwarz inequality as above, we have

$$\zeta(t) \leq \int_0^t \gamma^2(s) \alpha(s) y_{\bullet}(s)\, ds \sum_{h=1}^k p_h (\varphi_h - \bar{\varphi})^2 \tag{5.2.34}$$

with equality if and only if κ is proportional to γ. Thus, again, a test with a weight process $K^{(n)}$ which makes κ proportional to γ [cf. (5.2.15)] is optimal.

In relation to (5.2.34), it should be stressed that typically we will only have asymptotically proportional risk sets in the survival data setup with (asymptotically) equal censoring in the k groups. This was illustrated in the Examples V.2.10 and V.2.13. An explanation of why we are only able to determine explicitly the optimal choice of weight process when $k = 2$ or when we have asymptotically proportional risk sets is given in connection to (8.2.21) in

Section VIII.2.3. In Sections VIII.2.3 and VIII.4.2, we, furthermore, show that a choice of weight process which makes κ proportional to γ has certain optimality properties (to be defined in Section VIII.2.3) also when the risk sets are not asymptotically proportional.

In addition to providing a basis for deriving optimal tests for a particular situation, (5.2.31) may also be used to approximate the power of a specific test versus a fixed alternative. Given a particular value of n and of the alternative $\alpha_h^{(n)}$ against which the power is to be computed, one identifies $\alpha_h^{(n)}(s)$ with $\alpha(s) + \theta_h(s)/a_n$ and, hence, obtains $\theta_h(s) = a_n(\alpha_h^{(n)}(s) - \alpha(s))$. Inserting these into (5.2.30), one finds that the power of the test is approximately $1 - \Gamma_{k-1}(\gamma_\alpha; a_n^2 \zeta^{(n)}(t))$, where $\zeta^{(n)}(t)$ is given as in (5.2.31), but with $\xi(t)$ now given by (5.2.29) and (5.2.30) with $\theta_h(s)$ replaced by $\alpha_h^{(n)}(s)$.

Finally, (5.2.31) may be used to compute Pitman efficiencies. To this end, consider two sequences of test statistics for hypothesis (5.2.1) having weight processes K and $\tilde{K}$, respectively. Remember that a_n^2, in most applications (cf. the remark just above Example IV.1.6), corresponds to the number of "individual counting processes" on which a test is based. Therefore, the Pitman efficiency of the test with weight process K relative to the one with weight process $\tilde{K}$ is defined as

$$e(K, \tilde{K}, t) = \lim \frac{\tilde{a}_n^2}{a_n^2},$$

where $\{a_n\}$ and $\{\tilde{a}_n\}$ are the normalizing sequences required for the two tests to achieve the same limiting power. Then if $\zeta(t)$ and $\tilde{\zeta}(t)$ denote the values of (5.2.31) for the two tests in question, it can be seen that the Pitman efficiency may be given as

$$e(K, \tilde{K}, t) = \frac{\zeta(t)}{\tilde{\zeta}(t)},$$

i.e., as the ratio between the noncentrality parameters, or, in this context, the efficacies for the two tests. A detailed study of the efficacies for some common two-sample tests are provided by Gill (1980a, Chapter 5.2–3).

EXAMPLE V.2.14. Censored Survival Data

To further illustrate the general results on local asymptotic power of the test statistic (5.2.6), we consider the situation with censored survival data more closely. The derivations will be quite parallel to those of Example V.1.7 for the one-sample case. As in Example V.2.10, we adopt the random censorship model where for each n, $X_{hi}^{(n)}$, $U_{hi}^{(n)}$ for $i = 1, 2, \ldots, n_h$, $h = 1, 2, \ldots, k$, $n = \Sigma n_h$, are mutually independent with the $X_{hi}^{(n)}$ having absolutely continuous distribution function $F_h^{(n)}$ with support $[0, \infty)$, density $f_h^{(n)}$, and hazard rate $\alpha_h^{(n)}$, whereas $U_{hi}^{(n)}$ has distribution function $G_{hi}^{(n)}$. The censored survival times $\tilde{X}_{hi}^{(n)}$, the counting processes $N_h^{(n)}$, etc., are defined as in Examples V.2.1 and V.2.10.

We will assume that the distribution functions $F_1^{(n)}, \ldots, F_k^{(n)}$ form a generalized local location family in the sense that

$$F_h^{(n)}(t) = \Psi(g(t) + n^{-1/2}\varphi_h), \tag{5.2.35}$$

where Ψ is a fixed absolutely continuous distribution function with positive continuously differentiable density ψ on $(-\infty, \infty)$, and g is a fixed nondecreasing differentiable function from $(0, \infty)$ onto $(-\infty, \infty)$. Note that, then $F_h^{(n)}$ has hazard function

$$\alpha_h^{(n)}(t) = h(g(t) + n^{-1/2}\varphi_h)g'(t), \tag{5.2.36}$$

where $h = \psi/(1 - \Psi)$ is the hazard function corresponding to Ψ. Therefore, as in Example V.1.7, a Taylor expansion gives that (5.2.26) is fulfilled with $a_n = \sqrt{n}$, $\alpha(t) = h(g(t))g'(t)$, and $\theta_h(t) = \varphi_h h'(g(t))g'(t)$. Thus, (5.2.32) also holds true with

$$\begin{aligned}\gamma(t) &= h'(g(t))/h(g(t)) \\ &= h'(\Psi^{-1}(F(t)))/h(\Psi^{-1}(F(t))),\end{aligned} \tag{5.2.37}$$

where $F(t) = \Psi(g(t))$ is the common distribution function under the null hypothesis.

Now the general theory suggests that we get an optimal test for the two-sample situation quite generally, and for the k-sample situation ($k > 2$) under asymptotically equal censorship (cf. Example V.2.10), if we use the weight process

$$K^{(n)}(t) = I(Y_\bullet^{(n)}(t) > 0)\frac{h'(\Psi^{-1}(\hat{F}^{(n)}(t-)))}{h(\Psi^{-1}(\hat{F}^{(n)}(t-)))}, \tag{5.2.38}$$

where $\hat{F}^{(n)} = 1 - \hat{S}^{(n)}$ with

$$\hat{S}^{(n)}(t) = \prod_{s \le t}\left(1 - \frac{\Delta N_\bullet^{(n)}(s)}{Y_\bullet^{(n)}(s)}\right),$$

the Kaplan–Meier estimator based on the combined sample. Note that $K^{(n)}(t)$ does not depend on $g(t)$.

Let us see what (5.2.38) looks like for the choices of Ψ considered in Example V.1.7. [Further examples are provided by Gill (1980a, Chapter 5.3).] When $\Psi(x) = 1 - e^{-e^x}$, i.e., an extreme value distribution, then $h(x) = e^x$ and the log-rank test is optimal. This reflects the optimality of the log-rank test for proportional hazards alternatives or Lehmann alternatives, where

$$(1 - F_h^{(n)}) = (1 - F)^{\exp(\beta_h^{(n)})}.$$

For, in this situation,

$$\begin{aligned}F_h^{(n)} &= 1 - \exp\{e^{\beta_h^{(n)}}\log(1 - F)\} \\ &= \Psi(\log(-\log(1 - F)) + \beta_h^{(n)});\end{aligned}$$

so that by taking $g = \log(-\log(1 - F))$ and letting $\beta_h^{(n)} = n^{-1/2}\varphi_h$, we arrive at (5.2.35).

For the logistic distribution $\Psi(x) = h(x) = e^x/(1+e^x)$, $h'(x) = \Psi(x)(1-\Psi(x))$, and $\Psi^{-1}(y) = \log(y/(1-y))$. The optimal choice of weight process is, therefore,

$$K^{(n)}(t) = I(Y_{\cdot}^{(n)}(t) > 0)\hat{S}^{(n)}(t-),$$

corresponding to the Peto–Prentice generalization of the Wilcoxon and Kruskal–Wallis test to censored data; cf. Example V.2.1. When there is no censoring, $\hat{S}^{(n)}(t-) = Y_{\cdot}^{(n)}(t)/n$ and the Wilcoxon and Kruskal–Wallis tests result.

Finally, for the family of distributions

$$\Psi_\rho(x) = \begin{cases} 1 - \exp(e^{-x}), & \rho = 0 \\ 1 - (1+\rho e^x)^{-1/\rho}, & 0 < \rho \le 1, \end{cases}$$

we have $h_\rho(x) = e^x/(1+\rho e^x)$ and $\Psi_\rho^{-1}(y) = \log\{\rho^{-1}(1-(1-y)^{-\rho})\}$, for $\rho > 0$, and the optimal choice of weight process becomes

$$K^{(n)}(t) = I(Y_{\cdot}^{(n)}(t) > 0)\hat{S}^{(n)}(t-)^\rho.$$

This is the weight process suggested by Harrington and Fleming (1982); cf. Example V.2.1. □

It should be noted that it is straightforward to extend the argument of this example to general counting process models, where $\mathbf{N} = (N_1^{(n)}, \ldots, N_k^{(n)})$, $n = 1, 2, \ldots$, have intensity processes of the multiplicative form with $\alpha_h^{(n)}(t)$ given by (5.2.36). For then the common value of the $\alpha_h^{(n)}(t)$ under the hypothesis takes the form $\alpha(t) = h(g(t))g'(t)$ and the corresponding integrated intensity is, therefore, $A(t) = H(g(t))$, where H is the integrated intensity of h. Thus, $g(t) = H^{-1}(A(t))$, and by (5.2.37) we get an optimal test (when $k = 2$ or when we have asymptotically proportional risk sets) by choosing the weight process

$$K^{(n)}(t) = I(Y_{\cdot}^{(n)} > 0)\frac{h'(H^{-1}(\hat{A}^{(n)}(t))}{h(H^{-1}(\hat{A}^{(n)}(t))},$$

where

$$\hat{A}^{(n)}(t) = \int_0^t \frac{\mathrm{d}N_{\cdot}(s)}{Y_{\cdot}^{(n)}(s)}$$

is the Nelson–Aalen estimator based on $N_{\cdot} = \Sigma_h N_h$. A further discussion of optimal tests for general counting process models satisfying (a slight generalization of) (5.2.36) is provided in Section VIII.4.2.

Our discussion on the optimal choice of weight process in the present subsection only assumes that we have a sequence of k-variate counting process models satisfying the multiplicative intensity model with the deterministic part of the intensity process given by (5.2.26) and (5.2.32). No other assumptions are made on how the counting process models relate to the underlying sequence of probability models. In particular, it suffices that the censoring is independent; no assumption of noninformative censoring is needed.

Assuming that the censoring is noninformative, however, more can be said on the optimality of these test satistics. For then, as shown in Sections VIII.2.3 and VIII.4.2 (using the theory of contiguity and local asymptotic normality), the tests are asymptotically efficient, i.e., they are optimal within the class of (almost) all tests, not only within the restricted class of tests given by (5.2.6) and not just in the two-sample or asymptotically proportional risk sets cases.

V.3. Other Linear Nonparametric Tests

In this section, we consider nonparametric tests for some more special situations than those studied in Section V.2. In Section V.3.1, tests based on *intensity ratio estimates* are derived; in Section V.3.2, we study *stratified tests*, whereas Section V.3.3 contains a discussion of *trend tests*.

V.3.1. Tests Based on Intensity Ratio Estimates

The setup is now the same as in Section V.2.1: $\mathbf{N} = (N_1, \dots, N_k)$ is a k-variate counting process ($k \geq 2$) with a multiplicative intensity process $\boldsymbol{\lambda} = (\lambda_1, \dots, \lambda_k)$ with $\lambda_h(t) = \alpha_h(t) Y_h(t)$, $h = 1, \dots, k$, and we consider the null hypothesis

$$\mathrm{H}_0\colon \alpha_1 = \alpha_2 = \cdots = \alpha_k;$$

cf. (5.2.1). In the present subsection, we shall discuss tests for H_0 against a *proportional hazards alternative*:

$$\mathrm{H}_1\colon \alpha_h(t) = \theta_h \alpha_1(t), \quad h = 1, \dots, k \tag{5.3.1}$$

(where $\theta_1 = 1$) in which case the null hypothesis is

$$\mathrm{H}_0\colon \theta_1 = \theta_2 = \cdots = \theta_k = 1. \tag{5.3.2}$$

The common shape of the hazards, $\alpha_1(t)$, is still left unspecified and, in that sense, our setup is nonparametric or rather *semiparametric*. We shall return to this semiparametric model in more generality in Sections VII.2 and VII.3, whereas the present subsection, following Andersen (1983a), mostly deals with the two-sample case ($k = 2$), adding only a couple of remarks on the general k-sample situation.

In the two-sample case we denote by θ the parameter θ_2 in (5.3.1) and instead of testing equality of $\alpha_1(t)$ as in (5.3.2), we consider the slightly more general null hypothesis

$$\mathrm{H}_0\colon \theta = \theta_0 \tag{5.3.3}$$

which we test against the alternative

$$\mathrm{H}_1\colon \alpha_2(t) = \theta\alpha_1(t), \quad \theta \neq \theta_0. \tag{5.3.4}$$

As in Section V.2.1, we introduce a whole class of test statistics for H_0 each given by a predictable weight process $L(\cdot)$ as in (5.2.7) and (5.2.9) and (as we shall see presently) each corresponding to an estimate $\hat{\theta}_L$ of θ which is consistent and asymptotically normally distributed under the conditions of Theorem V.2.1. The estimator is defined by

$$\hat{\theta}_L = \hat{\theta}_L(\tau) = \frac{\int_0^\tau L(s)\,\mathrm{d}\hat{A}_2(s)}{\int_0^\tau L(s)\,\mathrm{d}\hat{A}_1(s)}, \tag{5.3.5}$$

where $\hat{A}_h(\cdot)$ is the Nelson–Aalen estimator for the integrated α_h, $h = 1, 2$, and it follows that under H_0

$$\hat{\theta}_L(t) - \theta_0 = \int_0^t L(s)\left(\frac{\mathrm{d}M_2(s)}{Y_2(s)} - \theta_0\frac{\mathrm{d}M_1(s)}{Y_1(s)}\right)\left(\int_0^t L(s)\frac{\mathrm{d}N_1(s)}{Y_1(s)}\right)^{-1} \tag{5.3.6}$$

for all $t \in (0, \tau]$. We now consider weight processes given by (5.2.9), i.e.,

$$L(t) = K(t)\,Y_1(t)\,Y_2(t)\,Y_{\bullet}(t)^{-1}$$

and assume that conditions (5.2.14), (5.2.15), and (5.2.16) hold under hypothesis (5.3.3). In this case, y_1, y_2, κ, and k_δ all depend on θ_0. We write $y_1(t, \theta_0)$, $y_2(t, \theta_0)$, etc., and we let

$$l(s, \theta_0) = \frac{\kappa(s, \theta_0)}{y_1(s, \theta_0)^{-1} + y_2(s, \theta_0)^{-1}}$$

be the limit in P_{θ_0} probability of $c_n L(s)$. Under these conditions, $\hat{\theta}_L$ is consistent and $a_n(\hat{\theta}_L(\cdot) - \theta_0)$ converges weakly in $D(\mathscr{T})$ to a mean-zero Gaussian martingale U_l with $U_l(0) = 0$ and

$$\operatorname{cov}(U_l(u), U_l(t))$$
$$= \theta_0^2 \int_0^{t \wedge u} l^2(s, \theta_0)\left(\frac{1}{y_1(s, \theta_0)} + \frac{1}{\theta_0 y_2(s, \theta_0)}\right)\alpha_1(s)\,\mathrm{d}s \left\{\int_0^{t \wedge u} l(s, \theta_0)\alpha_1(s)\,\mathrm{d}s\right\}^{-2}. \tag{5.3.7}$$

The weak convergence of the numerator of (5.3.6) (normalized by $a_n c_n$) follows from the martingale central limit theorem exactly as in Theorem V.2.1. Convergence in P_{θ_0} probability of c_n times the denominator to

$$\int_0^t l(s, \theta_0)\alpha_1(s)\,\mathrm{d}s$$

is a simple exercise using Lenglart's inequality (2.5.18) and, finally, the consistency follows from the weak convergence result and the fact that $a_n^{-1} \to 0$.

A consistent estimate of the asymptotic variance $\sigma_l^2(t, \theta_0) = \mathrm{var}_{\theta_0} U_l(t)$ in (5.3.7) is given by a_n^2 times

$$\hat{\sigma}_L^2(t, \theta_0) = \theta_0^2 \left(\int_0^t \frac{L^2(u)}{Y_1(u)\theta_0 Y_2(u)} \mathrm{d}N_{\cdot}(u) \right) \left(\int_0^t L(u) \frac{\mathrm{d}N_1(u)}{Y_1(u)} \right)^{-2} \tag{5.3.8}$$

and the test statistic

$$(\hat{\theta}_L - 1)^2 / \hat{\sigma}_L^2(\tau, 1)$$

for the hypothesis $\theta = 1$ of *identical* hazards for N_1 and N_2 is, therefore, exactly the same as the X^2 we considered in Section V.2.1.

We next illustrate the methods using the melanoma data. Further examples are provided in Example VII.3.5.

EXAMPLE V.3.1. Survival with Malignant Melanoma. Estimation of the Ratio Between the Death Intensities for Men and Women

The melanoma data were introduced in Example I.3.1.

The Nelson–Aalen plots in Figure IV.1.1 suggested that the intensity of dying from the disease was fairly constant for both men and women and, therefore, a model with a *constant ratio* between the death intensities seems reasonable. The estimator corresponding to the log-rank test, i.e., the weight process $L(t) = Y_1(t) Y_2(t) / Y_{\cdot}(t)$, takes the value $\hat{\theta}_L = 1.95$ with an estimated standard error under the null hypothesis of $\hat{\sigma}_L(\infty, 1) = 0.373$. This agrees with the log-rank test statistic quoted in Example V.2.3 since

$$(1.95 - 1)^2/(0.373)^2 = (-2.543)^2 = 6.47.$$

Since the null hypothesis is rejected it is of particular interest to report a confidence interval for θ. This may be obtained as the interval

$$\left\{ \theta_0 : \frac{(\hat{\theta}_L - \theta_0)^2}{\hat{\sigma}_L^2(\infty, \theta_0)} \le \chi_{1-\alpha,1}^2 \right\} = (1.16, 3.27)$$

for $\alpha = 0.05$. It is worth noting that since $\hat{\theta}_L$ is equivalent to the log-rank test statistic, it may be useful to report the estimate and its standard error in addition to (or instead of) the value of the log-rank test statistic in situations where the proportional hazards assumption seems tenable. □

A natural question is which L to choose to get the best test statistic for H_0 or, equivalently, to get the smallest asymptotic variance of $\hat{\theta}_L$. By the Cauchy–Schwarz inequality

$$\sigma_l^2(t, \theta) \ge \left\{ \theta^{-2} \int_0^t \left(\frac{1}{y_1(s, \theta)} + \frac{1}{\theta y_2(s, \theta)} \right)^{-1} \alpha_1(s) \, \mathrm{d}s \right\}^{-1}, \tag{5.3.9}$$

where the right-hand side is the asymptotic variance corresponding to the choice of weight process

$$L(t) = \theta^{-1} \frac{Y_1(t)Y_2(t)}{Y_1(t) + \theta Y_2(t)}. \tag{5.3.10}$$

Therefore, the best weight process, in general, depends on the unknown θ. When θ is close to 1, however, the *log-rank* test which is obtained with $L = Y_1 Y_2 / Y_{\cdot}$ [cf. Example V.2.1] is nearly optimal within the class considered. This result is in agreement with the local asymptotic power discussion in Section V.2.3 and was also noted by Crowley et al. (1982) and Andersen (1983a). The result will be further explored in Section VIII.4. If $\tilde{\theta}$ is a (preliminary) consistent estimator for θ, then (5.3.10) with $\theta = \tilde{\theta}$ yields an optimal estimator attaining the lower bound for the asymptotic variance. For instance, $\tilde{\theta}$ may be chosen as $\tilde{\theta} = \hat{A}_2(\tau)/\hat{A}_1(\tau)$ which is (5.3.5) with the choice $L(t) \equiv 1$. [Inserting $\tilde{\theta}$ into (5.3.10), the process $L(t)$ is, of course, no longer predictable. By a Taylor expansion argument as in Section VI.3.3, one can, however, show that the test statistic is still asymptotically normally distributed.] This is the *two-step estimator* introduced by Begun and Reid (1983) for the case of right-censored survival data. In Example VII.2.6, we shall see that the *Cox partial likelihood estimator* is also optimal in this sense.

In the general k-sample case (5.3.1), we can define estimators

$$\hat{\theta}_{Lh}(\tau) = \hat{\theta}_{Lh} = \frac{\int_0^\tau L_h(s)\,\mathrm{d}\hat{A}_h(s)}{\int_0^\tau L_h(s)\,\mathrm{d}\hat{A}_1(s)}, \quad h = 2, \ldots, k,$$

as in (5.3.5). For each h, we now have a representation of $\hat{\theta}_{Lh}(t) - \theta_{h0}$ as in (5.3.6) when the true hazard ratio $\alpha_h(t)/\alpha_1(t)$ is θ_{h0} and the asymptotic joint distribution of the $k - 1$ estimators can be derived using the martingale central limit theorem (and Lenglart's inequality) under conditions similar to those in the two-sample case.

V.3.2. Stratified Tests

We assume that we have m multivariate counting processes of the type discussed in Section V.2, i.e., $\mathbf{N}_s = (N_{1s}, \ldots, N_{ks})$ for $s = 1, \ldots, m$ are k-variate counting processes with $\mathbf{N}_s$ having an intensity process $\boldsymbol{\lambda}_s = (\lambda_{1s}, \ldots, \lambda_{ks})$ of the multiplicative form $\lambda_{hs}(t) = \alpha_{hs}(t) Y_{hs}(t)$. We want to test the hypothesis

$$\mathrm{H}_0\colon \alpha_{1s} = \alpha_{2s} = \cdots = \alpha_{ks} \quad \text{for all } s = 1, \ldots, m. \tag{5.3.11}$$

This situation occurs when the $\mathbf{N}_s$ count the events of interest in each of m strata, and we want to test the hypothesis of equal α_{hs} within each stratum, allowing for heterogeneity between strata. Practical illustrations are provided in Examples V.3.2 and V.3.3.

It is simple to develop a test for the hypothesis (5.3.11) using the results of Section V.2. To this end, we define for each stratum the same quantities as in Section V.2, putting on an extra index s to indicate the dependence of these quantities on the actual stratum. In particular, we let $N_{\bullet s} = \sum_h N_{hs}$ and $Y_{\bullet s} = \sum_h Y_{hs}$ and define

$$Z_{hs}(t) = \int_0^t K_s(u)\,\mathrm{d}N_{hs}(u) - \int_0^t K_s(u)\frac{Y_{hs}(u)}{Y_{\bullet s}(u)}\mathrm{d}N_{\bullet s}(u) \tag{5.3.12}$$

and

$$\hat{\sigma}_{hjs}(t) = \int_0^t K_s^2(u)\frac{Y_{hs}(u)}{Y_{\bullet s}(u)}\left(\delta_{hj} - \frac{Y_{js}(u)}{Y_{\bullet s}(u)}\right)\mathrm{d}N_{\bullet s}(u), \tag{5.3.13}$$

for $s = 1, 2, \dots, m$. Furthermore, we let $\mathbf{Z}_s(t)$ and $\hat{\mathbf{\Sigma}}_s(t)$, for $s = 1, \dots, m$, be the vectors and matrices with elements given by (5.3.12) and (5.3.13), respectively. Then all the results of Section V.2 remain valid within each stratum. We assume that the conditions of Theorem V.2.1 are fulfilled for each stratum with normalizing constants c_{sn} in stratum s and write $c_{\bullet n} = \sum_s c_{sn}$. Then, provided that $c_{\bullet n}/c_{sn} \to \pi_s$ as $n \to \infty$, it follows by (5.2.17) that

$$c_{\bullet n} \sum_{s=1}^{m} \mathbf{Z}_s(t) \xrightarrow{\mathscr{D}} \mathscr{N}\left(\mathbf{0}, \sum_{s=1}^{m} \pi_s^2 \mathbf{\Sigma}_s(t)\right),$$

for any fixed $t \in \mathscr{T}$ as $n \to \infty$. Here the $k \times k$ matrix $\sum_s \pi_s^2 \mathbf{\Sigma}_s(t)$ of rank $k - 1$ may be estimated consistently by $c_{\bullet n}^2 \sum_s \hat{\mathbf{\Sigma}}_s(t)$.

As a test statistic for the hypothesis (5.3.11), we may, therefore, use the quadratic form

$$\left(\sum_{s=1}^{m} \mathbf{Z}_s(t)\right)^{\mathsf{T}} \left(\sum_{s=1}^{m} \hat{\mathbf{\Sigma}}_s(t)\right)^{-} \left(\sum_{s=1}^{m} \mathbf{Z}_s(t)\right), \tag{5.3.14}$$

which is asymptotically χ^2 distributed with $k - 1$ degrees of freedom under the hypothesis. The test statistic may be evaluated as described below (5.2.6) by deleting the last element of the vector $\sum_s \mathbf{Z}_s(t)$ and the last row and column of the matrix $\sum_s \hat{\mathbf{\Sigma}}_s(t)$ and performing an ordinary matrix inversion. When appropriate, the tie-corrected estimator defined by (5.2.10) (within each stratum) may be used in (5.3.14) instead of (5.3.13).

For the two-sample case, one may equivalently base the test on the fact that

$$\left(\sum_{s=1}^{m} Z_{1s}(t)\right)\left(\sum_{s=1}^{m} \hat{\sigma}_{11s}(t)\right)^{-1/2}$$

is asymptotically standard normally distributed when $\alpha_{1s} = \alpha_{2s}$ for $s = 1, \dots, m$.

To actually compute the test statistic (5.3.14), a choice of weight process K_s in (5.3.12) has to be made. For this purpose, any of the weight processes discussed in Section V.2 (see, in particular, Example V.2.1) may be used. In particular, $K_s = I(Y_{\bullet s} > 0)$ will give the stratified log-rank test.

By the way the test statistic (5.3.14) is constructed, it is seen that one can only expect the stratified tests to have good power against alternatives where the deviations from the hypothesis (5.3.11) go in the same direction in all strata. Normally this is also the alternatives one has in mind when performing a stratified test. A situation where this is not the case, however, is illustrated in Example V.3.3.

Example V.3.2. Survival with Malignant Melanoma. Effect of Sex and Tumor Thickness

The melanoma data were introduced in Example I.3.1. In malignant melanoma, one of the strongest prognostic factors is the tumor thickness. Figure V.3.1 shows the Nelson–Aalen plots of the integrated death intensities in three groups obtained by stratification by thickness using the cut points 2 mm and 5 mm. The hazard seems to increase with tumour thickness, and the log-rank test statistic for the hypothesis of no effect of thickness takes the value 31.58 (d.f. = 2, $P < .0001$). In Example V.2.3, we saw that the death intensity was significantly related to the sex of the patients. Since in this data set men tend to have thicker tumors than women (the mean thickness for men is 3.6 mm and for women it is 2.5 mm), it may be the case that the sex

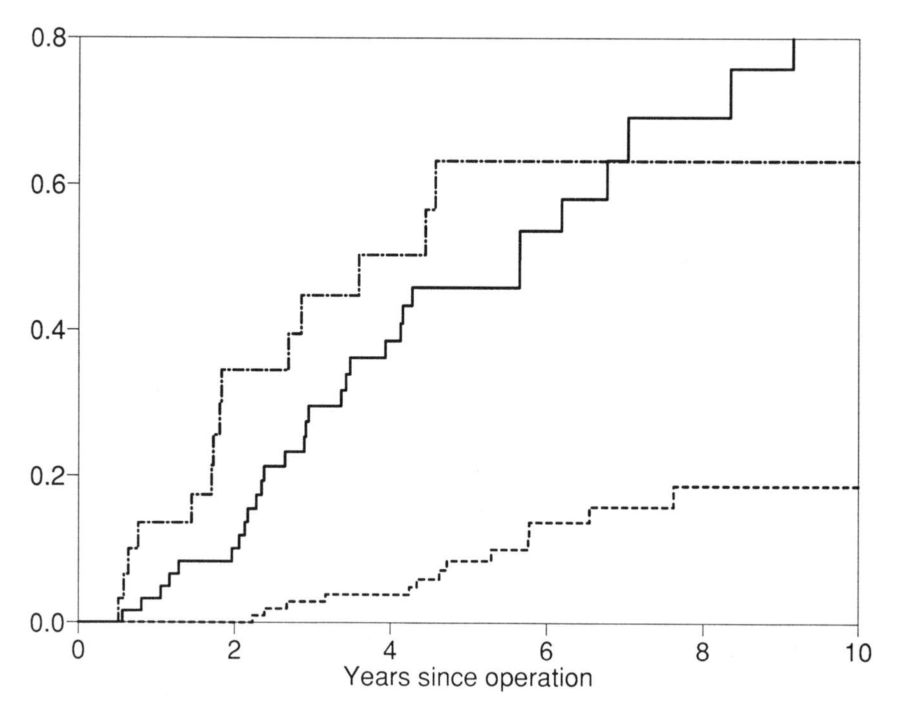

Figure V.3.1. Malignant melanoma. Nelson–Aalen estimates for patients with thickness below 2 mm (- - -); between 2 mm and 5 mm (——), and above 5 mm (-·-·-).

effect is due to differences in thickness. We, therefore, want to test the hypothesis H_0: $\alpha_{1s} = \alpha_{2s}$, $s = 1, 2, 3$, where the stratum index s refers to the three thickness groups, $h = 1$ corresponds to women and $h = 2$ to men; cf. (5.3.11). The stratified log-rank test statistic takes the value 3.22 (d.f. = 1, $P = .07$) which is considerably smaller than the value $6.47 = (-2.543)^2$ of the log-rank test statistic for sex found in Example V.2.3. This result suggests that the difference between the mortality for men and women can partly be ascribed to differences in tumor thickness. We shall return to this point in Examples VII.2.5 and VII.3.2. □

EXAMPLE V.3.3. Matings of Drosophila Flies

The problem and the data were introduced in Example I.3.16. In Example V.2.7, the mating intensities for the four combinations of ebony(e)/oregon (o) and female/male Drosophila flies were compared and found to be highly significantly different. It may be of interest to examine whether these differences can be explained by different mating activity for the two types of males whereas the female type may have no effect. Thus, we want to test the hypothesis $\alpha_{ee} = \alpha_{oe}$ and $\alpha_{eo} = \alpha_{oo}$, where the first index specifies the female type. The "symmetric" hypothesis $\alpha_{ee} = \alpha_{eo}$ and $\alpha_{oe} = \alpha_{oo}$ of no effect of male type may also be of interest. Both situations are seen to fit into the structure of the present subsection. We will use the stratified log-rank test which for the two sample case considered here takes the form

$$\left(\sum_s O_{1s} - \sum_s E_{1s}\right)^2 \Big/ \sum_s V_s = \left(\sum_s O_{2s} - \sum_s E_{2s}\right)^2 \Big/ \sum_s V_s$$

with $O_{hs} = N_{hs}(\infty)$, $E_{hs} = \int_0^\infty (Y_{hs}/Y_{\bullet s})\,dN_{\bullet s}$, and $V_s = \hat{\sigma}_{11s}(\infty)$. Table V.3.1 gives the summary statistics for the calculation of the relevant stratified log-rank-type test statistics. For the former hypothesis, the statistic is $(55 - 31.658)^2/18.846 = 28.91 = (37 - 60.342)^2/18.846$, whereas for the latter, we get $(41 - 45.388)^2/19.952 = 0.97 = (51 - 46.612)^2/19.952$. Thus, the female type has a significant effect on the mating intensities, but the hypothesis that only the female type affects the intensities is not rejected. From Figure IV.1.7, it seems as if the ebony females have a higher mating intensity than the oregon females. Within each stratum defined by the female type, however, it is seen that the "homogametic" mating intensity (i.e., same type of female and male) is higher than the "heterogametic." This is exactly the type of alternative hypothesis against which the stratified statistics have poor power and explains why the second hypothesis was not rejected. □

Above, we have proved that (5.3.14) is asymptotically χ^2 distributed with $k - 1$ degrees of freedom under the hypothesis assuming that the number of strata remains fixed as $n \to \infty$. This made it possible to apply directly the results of Section V.2. Under suitable regularity conditions, it is, however, not difficult to prove the same result when the number of strata is allowed to

Table V.3.1. Calculation of Stratified Log-rank Test Statistics for Drosophila Flies

(a) Hypothesis: $\alpha_{ee} = \alpha_{oe}$ and $\alpha_{eo} = \alpha_{oo}$

	$(f, m) = (e, e)$	$(f, m) = (e, o)$	Sum
Obs. (O_{hs})	27	28	55
"Exp." (E_{hs})	9.888	21.770	31.658
V_s	6.627	12.219	18.846
	$(f, m) = (o, e)$	$(f, m) = (o, o)$	Sum
Obs.	14	23	37
"Exp."	31.112	29.230	60.342
V_s	6.627	12.219	18.846

(b) Hypothesis: $\alpha_{ee} = \alpha_{eo}$ and $\alpha_{oe} = \alpha_{oo}$

	$(f, m) = (e, e)$	$(f, m) = (o, e)$	Sum
Obs.	27	14	41
"Exp."	20.670	24.718	45.388
V_s	12.008	7.944	19.952
	$(f, m) = (e, o)$	$(f, m) = (o, o)$	Sum
Obs.	28	23	51
"Exp."	34.330	12.282	46.612
V_s	12.008	7.944	19.952

increase with n (cf. the proofs in Section VII.2.2). A practical consequence of this is that one may expect (5.3.14) to be approximately χ^2 distributed under the hypothesis, not only when the amount of information in each stratum is large but also when the number of strata is large even if we have only a small or moderate amount of information in each stratum.

In particular, the above results remain valid for the case of pair-matched censored survival data (with "stratum" corresponding to "pair") as illustrated in the final example of this subsection.

EXAMPLE V.3.4. Pair-Matched Censored Survival Data

Let X_{hs} for $h = 1, 2$ and $s = 1, 2, \ldots, m$ be independent non-negative random variables with absolutely continuous distribution functions F_{hs} and hazard rate functions α_{hs}. As usual, we do not observe X_{hs}, but only the right-censored samples $(\tilde{X}_{hs}, D_{hs})$, where $\tilde{X}_{hs} = X_{hs} \wedge U_{hs}$ and $D_{hs} = I(\tilde{X}_{hs} = X_{hs})$ for some censoring times U_{hs}. We want to test the hypothesis $F_{1s} = F_{2s}$, $s = 1, 2, \ldots, m$, or equivalently $\alpha_{1s} = \alpha_{2s}$, $s = 1, 2, \ldots, m$ [cf. (5.3.11)].

We introduce $N_{hs}(t) = I(\tilde{X}_{hs} \leq t, D_{hs} = 1)$ and $Y_{hs}(t) = I(\tilde{X}_{hs} \geq t)$. With $K_s = L_s(Y_{1s} + Y_{2s})/(Y_{1s} Y_{2s})$, the basic term (5.3.12) of the test statistic may be written

$$Z_{1s}(t) = \int_0^t L_s(u)\frac{dN_{1s}(u)}{Y_{1s}(u)} - \int_0^t L_s(u)\frac{dN_{2s}(u)}{Y_{2s}(u)}$$

[cf. (5.2.7) and (5.2.9)]. Here, the weight process L_s has the property of being equal to zero when $Y_{1s}Y_{2s} = 0$, that is, it is only nonzero before the smaller of $\tilde{X}_{1s}$ and $\tilde{X}_{2s}$. Thus, $Z_{1s}(t)$ equals

$$\begin{aligned} &L_s(X_{1s})I(X_{1s} \leq t) \quad \text{if } X_{1s} = \tilde{X}_{1s} < \tilde{X}_{2s}, \\ -&L_s(X_{2s})I(X_{2s} \leq t) \quad \text{if } X_{2s} = \tilde{X}_{2s} < \tilde{X}_{1s}, \end{aligned}$$

and it is zero for all t if $\tilde{X}_{1s} \wedge \tilde{X}_{2s}$ corresponds to a censoring. The basic term (5.3.13) of the variance estimate may be written

$$\hat{\sigma}_{11s}(t) = \int_0^t L_s^2(u)\{Y_{1s}(u)Y_{2s}(u)\}^{-1}\,d(N_{1s} + N_{2s})(u)$$

[cf. (5.2.8)], and, similarly, this equals

$$\begin{aligned} &L_s^2(X_{1s})I(X_{1s} \leq t) \quad \text{if } X_{1s} = \tilde{X}_{1s} < \tilde{X}_{2s}, \\ &L_s^2(X_{2s})I(X_{2s} \leq t) \quad \text{if } X_{2s} = \tilde{X}_{2s} < \tilde{X}_{1s}, \end{aligned}$$

and it is zero for all t if $\tilde{X}_{1s} \wedge \tilde{X}_{2s}$ corresponds to a censoring.

The particular choice $K_s = I(Y_s > 0)$, or equivalently

$$L_s = (Y_{1s}Y_{2s})/(Y_{1s} + Y_{2s}),$$

leads to a log-rank-type test with

$$\sum_{s=1}^{m} Z_{1s}(t) = \tfrac{1}{2}(D_1(t) - D_2(t)),$$

where $D_h(t)$ is the number of matched pairs where the individual from group h died while the other individual was still at risk, and

$$\sum_{s=1}^{m} \hat{\sigma}_{11s}(t) = \tfrac{1}{4}(D_1(t) + D_2(t)).$$

Thus, for testing the hypothesis we may refer the test statistic

$$\frac{\sum_{s=1}^{m} Z_{1s}(t)}{(\sum_{s=1}^{m} \hat{\sigma}_{11s}(t))^{1/2}} = \frac{D_1(t) - D_2(t)}{(D_1(t) + D_2(t))^{1/2}}$$

to the standard normal distribution. In the literature on censored matched-pair survival analysis, this is usually referred to as a sign test. This test will be rederived as a score test in a regression model in Example VII.2.13, where an application to the data introduced in Example I.3.6 will also be given.

Note that pairs where the first event is a censoring do not contribute information to the test. Perhaps for this reason, almost all of the literature on censored matched-pair survival analysis (cf. bibliographic remarks in Section

V.5) attempts to utilize "interpair information" by phrasing the discussion in terms of a random effects model where X_{1s} and X_{2s} are dependent for each s, usually without an explicit distributional form of this dependence. In Examples IX.1.1 and IX.4.2, we comment briefly on models with parametric dependence. □

V.3.3. Trend Tests

We consider the setup of Section V.2.1 and adopt the notation used there. Thus, $\mathbf{N} = (N_1, \ldots, N_k)$ is a k-variate counting process having an intensity process $\boldsymbol{\lambda} = (\lambda_1, \ldots, \lambda_k)$ of the multiplicative form $\lambda_h(t) = \alpha_h(t) Y_h(t)$. The test statistics (5.2.6) are an omnibus type of tests which test the hypothesis (5.2.1) of equal α_h against the general alternative that the α_h are not all equal.

However, it may sometimes be the case that the h indexing the components of the counting process refers to a natural ordering and that the relevant alternatives to consider satisfy

$$\alpha_1 \leq \alpha_2 \leq \cdots \leq \alpha_k. \tag{5.3.15}$$

An example of such a situation is provided in Example V.3.5. Then the omnibus tests (5.2.6) will have poor power, and more powerful tests, so-called tests for trend, should be applied.

The basic idea in deriving a test for trend is to associate with the hth component of the counting process a *score* x_h, $h = 1, \ldots, k$. The scores should be chosen such that $x_1 < x_2 < \cdots < x_k$, i.e., such that large values of the scores are associated with large α_h under the alternative. A common choice in practice is simply to let $x_h = h$ to indicate the order of the components, but sometimes the x_h may be chosen to be some other quantities which characterize the components (cf. Example V.3.5).

Following Tarone (1975) and Tarone and Ware (1977) we may then, for any fixed $t \in \mathscr{T}$, define a general trend test statistic as

$$\sum_{h=1}^{k} x_h Z_h(t) = \mathbf{x}^\mathsf{T} \mathbf{Z}(t),$$

where $\mathbf{x} = (x_1, \ldots, x_k)^\mathsf{T}$. Under the assumptions of Theorem V.2.1, it immediately follows by (5.2.17) that

$$c_n \mathbf{x}^\mathsf{T} \mathbf{Z}(t) \xrightarrow{\mathscr{D}} \mathscr{N}(0, \mathbf{x}^\mathsf{T} \boldsymbol{\Sigma}(t) \mathbf{x})$$

as $n \to \infty$. Thus, the trend statistic

$$\frac{\sum_{h=1}^{k} x_h Z_h(t)}{\sqrt{\mathbf{x}^\mathsf{T} \hat{\boldsymbol{\Sigma}}(t) \mathbf{x}}}, \tag{5.3.16}$$

with $\hat{\boldsymbol{\Sigma}}(t)$ defined by (5.2.5), is asymptotically standard normally distributed under the hypothesis (5.2.1). When appropriate the tie-corrected estimator (5.2.10) may be used in (5.3.16) instead of (5.2.5).

It should be noted that (5.3.16) does not change its value if we use the scores $x_h' = ax_h + b$, with $a > 0$, instead of the scores x_h. Thus, the trend test is unaffected by an affine transformation of the scores. Moreover, under suitable regularity conditions, it is simple to prove that (5.3.16) is asymptotically standard normally distributed also when k increases with n (cf. Section VII.2.2).

If we replace $\hat{\boldsymbol{\Sigma}}(t)$ in (5.3.16) by $\boldsymbol{\Sigma}^*(t)$ which have elements given by (5.2.20) with the $\hat{p}_h^{(n)}$ as given in (5.2.22), a conservative version of the trend test statistic results. This takes the form

$$\frac{\sum_{h=1}^k x_h Z_h(t)}{\sqrt{V^*(t)}}, \tag{5.3.17}$$

where

$$V^*(t) = \sum_{h=1}^k x_h^2 \int_0^t K(s)^2 \frac{Y_h(s)}{Y_\cdot(s)} \mathrm{d}N_\cdot(s) - \frac{\left(\sum_{h=1}^k x_h^2 \int_0^t K(s)^2 Y_h(s)(Y_\cdot(s))^{-1} \mathrm{d}N_\cdot(s)\right)^2}{\int_0^t K(s)^2 \mathrm{d}N_\cdot(s)}.$$

In particular, for the choice $K(s) = I(Y_\cdot(s) > 0)$ of weight process, the conservative version of the log-rank test for trend becomes

$$\frac{\sum_h x_h(N_h(t) - E_h(t))}{\sqrt{\sum_h x_h^2 E_h(t) - (\sum_h x_h E_h(t))^2/(\sum_h E_h(t))}},$$

where $E_h(t) = \int_0^t Y_h(Y_\cdot)^{-1} \mathrm{d}N_\cdot$ (remember that $\sum_h N_h(t) = \sum_h E_h(t)$).

Tarone (1975) showed that under hypothesis (5.2.1) the difference between the omnibus test (5.2.6) and the square of the trend test (5.3.16) is asymptotically χ^2 distributed with $k - 2$ degrees of freedom, and he suggested using this difference as a test for departure from trend. This may be a useful procedure to get a feeling for how much of the deviation from the hypothesis may be explained by a trend. But the result does not make the difference usable as a formal test statistic for trend since what is needed then is not its distribution under hypothesis (5.2.1), but rather its distribution under some model for trend in the α_h. A rigorous approach for testing departure from trend, therefore, necessitates use of the regression models of Chapter VII, and we shall return to this point in Example VII.2.4.

EXAMPLE V.3.5. Survival with Malignant Melanoma. Test for Trend for Tumor Thickness

The melanoma data were introduced in Example I.3.1. In Example V.3.2, Figure V.3.1 suggested that the intensity of dying from malignant melanoma increases with increasing tumor thickness. Since this is also what one would expect a priori, it may be appropriate to apply a test for trend in this situation. Using the score $x_1 = 1$, $x_2 = 2$, $x_3 = 3$ for the three thickness strata, the log-rank test for trend is $5.28 = \sqrt{27.93}$, whereas the conservative version (5.3.17) is $5.27 = \sqrt{27.80}$. The corresponding values of the three-sample log-rank test statistics (5.2.6) and (5.2.23) are 31.58 and 31.48, respectively, suggesting that "the trend explains most of the variation" between the three strata. Using instead the score given by the mean tumor thickness (in mm) in each group ($x_1 = 1.05$, $x_2 = 3.34$, $x_3 = 8.48$), the log-rank trend statistics become somewhat smaller: (5.3.16) gives 4.65, whereas (5.3.17) takes the value 4.63. In this example, the latter and perhaps more natural choice of scores gives a less significant test statistic than the simple one studied in the first place. We return to a closer examination of the effect of tumor thickness in Example VII.3.2. This will explain the different performance of the two sets of scores. □

V.4. Using the Complete Test Statistic Process

For the k-sample test problems ($k = 1, 2, \ldots$) studied in Sections V.1 and V.2, the basic tool was the k-dimensional martingale $Z(t)$ of (5.1.1) or $\mathbf{Z}(t) = (Z_1(t), \ldots, Z_k(t))$ of (5.2.2). However, we always evaluated $Z(t)$ at one (large) value of t, typically at the end of the study. The purpose of this final section of the hypothesis testing chapter is to demonstrate two areas where the use of the complete sample path of Z (or $\mathbf{Z}$) opens new possibilities. Section V.4.1 discusses tests based on nonlinear functionals of the estimated cumulative intensity, where our approach unifies the results on Kolmogorov–Smirnov, Cramér–von Mises, and Anderson–Darling tests for censored data. Section V.4.2 briefly indicates how the asymptotic results on Z and $\mathbf{Z}$ connect the literature on sequential tests for censored data with well-known results for sequential analysis of Brownian motion.

The discussion will be phrased in the context of one- and two-sample tests.

V.4.1. Tests Based on Nonlinear Functionals of the Estimated Cumulative Intensity

So far, we have considered generalizations of classical linear rank tests to the counting process framework, including the possibility of censoring, filtering, and truncation. We have seen that for censored survival data, the log-rank-type tests have some local optimality properties against proportional hazards alternatives, whereas the Peto–Prentice generalizations of the Wilcoxon- or Kruskal–Wallis-type tests have similar properties against logistic location alternatives.

It is well-known in the classical situation (i.e., uncensored survival data) that these tests can have very low power against other important alternative hypotheses such as "crossing hazards." A further battery of tests based on the empirical distribution function (therefore, sometimes called EDF tests) has been developed with this motivation. The latter tests generally have some power against all alternatives.

The purpose of this section is to discuss how versions of such test statistics for the one- and two-sample problems in the general counting process model may be based directly on the basic stochastic processes $Z(t)$ in (5.1.1) and $Z_1(t)$ in (5.2.7). The asymptotic theory necessary to obtain approximate distribution results for the test statistics is also contained in direct corollaries of the basic Theorems V.1.1 and V.2.1. We shall restrict ourselves to a brief survey; indeed, a number of details concerning specific implementation still require further investigation.

Most of our discussion will be in the one-sample framework of Section V.1. We there considered a (one-dimensional) counting process $(N(t), t \in \mathscr{T})$ with intensity process $\alpha(t)Y(t)$; we wanted to test the hypothesis $\alpha = \alpha_0$, α_0 known. The basic tool was the stochastic process (5.1.1)

$$Z(t) = \int_0^t K(s)\,\mathrm{d}\{\hat{A}(s) - A_0^*(s)\}.$$

Based on this, a test statistic was derived by selecting one value t_0 of t (often $t_0 = \tau$) and using $Z(t_0)$ normalized by an estimate of its standard deviation.

The following two classes of nonlinear functionals of $\{Z(t)\}$ will be considered in this section: First, *maximal deviation* (or Kolmogorov–Smirnov-type) statistics such as $\sup_{t \in \mathscr{T}_0} |Z(t)|$ for some $\mathscr{T}_0 \subseteq \mathscr{T}$; and second, generalizations to the general counting process model of *integrated squared deviation* (or Cramer–von Mises-type) statistics of the type

$$\int_0^{t_0} Z^2(s)\,\mathrm{d}H(s)$$

for a suitable weight process H.

The exposition has the aim of generating asymptotic expressions for test statistics which enable us to draw on the elaborate existing literature for nonlinear EDF tests, in particular so that no new derivations of approximate percentage points of test statistics are required.

For the *maximal deviation statistics*, the possibilities may be summarized in the following proposition, which is a direct corollary of Theorem V.1.1 and elementary results on transformations of Brownian motion as already exploited in the discussions of confidence limits and bands in Sections IV.1.3 and IV.3.3 [see, in particular, the arguments leading to (4.1.24)].

Proposition V.4.1. *In the notation of Theorem* V.1.1, *let* $\hat{\sigma}^2(t)$ *denote a uniformly consistent estimator of*

$$\sigma^2(t) = \int_0^t \frac{k^2(s)\alpha_0(s)}{y(s)}\,\mathrm{d}s, \tag{5.4.1}$$

such as $\hat{\sigma}^2(t) = \langle c_n Z^{(n)} \rangle(t)$ *or* $[c_n Z^{(n)}](t)$. *Then, under the hypothesis* $\alpha = \alpha_0$:

(a) *The asymptotic distribution of*

$$\sup_{t_1 \le t \le t_2} |c_n Z^{(n)}(t)|$$

is that of $\sup_{e_1 \le x \le e_2} |W(x)|$, *where* W *is standard Brownian motion and* $e_i = \sigma^2(t_i)$, $i = 1, 2$.

(b) *The asymptotic distribution of*

$$\sup_{t_1 \le t \le t_2} |c_n Z^{(n)}(t)/\{1 + \hat{\sigma}^2(t)\}|$$

is that of $\sup_{d_1 \le x \le d_2} |W^0(x)|$, *where* W^0 *is a standard Brownian bridge and* $d_i = \sigma^2(t_i)/\{1 + \sigma^2(t_i)\}$, $i = 1, 2$.

Finally,

(c) *the asymptotic distribution of*

$$\sup_{t_1 \le t \le t_2} |c_n Z^{(n)}(t)/\hat{\sigma}(t)|$$

is that of $\sup_{d_1 \le x \le d_2} |W^0(x)\{x(1 - x)\}^{-1/2}|$.

From these results, one may connect to ordinary distribution results for maximal deviation statistics; for useful concise tables, see, e.g., Schumacher (1984) or Shorack and Wellner (1986). For the special weight function $K = J$, the procedure was discussed at length in Section IV.1.3, where we noted that simultaneous confidence bands for A are equivalent to maximal deviation statistics based on $\sup_{t \in \mathcal{T}_0} |\hat{A}(t) - A(t)|$. In particular, the HW- and EP-bands correspond to classes (b) and (c) of tests of Proposition V.4.1.

The generality allows for many different choices of weight function K, and since the existing literature contains no results for the multiplicative intensity model for general counting processes, we can get no general guidance from there. Indeed, all previous work seems to be tied to the example of censored survival data. For this example, however, the tradition from the classical one-sample Kolmogorov–Smirnov statistic is strong: It is taken for granted that the test statistic should be based on the maximal absolute deviation between empirical and hypothesized distribution function. Aalen (1976, p. 24) briefly indicated how Kolmogorov–Smirnov tests may be obtained (in essentially the way outlined here) for the multiple decrement model; Gill (1980a, p. 80) also indicated how one-sample Kolmogorov–Smirnov tests (and the equivalent confidence band for the survival function) may be obtained as a corollary of the weak convergence results. The question of weight function was taken up by Fleming et al. (1980) and Fleming and Harrington (1981) who based their detailed discussion on the wish first to obtain something that would reduce (at least approximately) to the traditional Kolmogorov–Smirnov test for uncensored survival data, and next to generate a family of weight functions enabling the statistician to tune the sensitivity of the test toward alternatives of particular interest.

Two-sample maximal deviation test statistics may similarly be defined by replacing Z by Z_1, where [from (5.2.7)]

$$Z_1(t) = \int_0^t L(s)\,\mathrm{d}\{\hat{A}_1(s) - \hat{A}_2(s)\},$$

and by replacing $\hat{\sigma}^2(t)$ by $c_n^2 \hat{\sigma}_{11}(t)$ where $\hat{\sigma}_{11}(t)$ is given in (5.2.8). Brownian motion or Brownian bridge asymptotics are then available from Theorem V.2.1. The same authors as above also considered the two-sample problem for censored survival data. A further most useful survey, with tables and a comparative simulation study, is due to Schumacher (1984), who also gave arguments for basing the test statistic on the difference $|\log \hat{A}_1(t) - \log \hat{A}_2(t)|$ rather than $|\hat{A}_1(t) - \hat{A}_2(t)|$ as above. (We again refer to the parallel discussion in Section IV.1.3 on the choice of transformation for deriving confidence bands.) Fleming et al. (1987) surveyed various maximal deviation statistics for the two-sample censored survival data problem and gave many results from Monte Carlo studies of the small sample properties.

Turning next to the *integrated squared deviation statistics*, we restrict ourselves to presenting versions of the Cramér–von Mises and Anderson–Darling statistics for the one-sample counting process problem. The following heuristics underlie the derivations. Classically, integrated squared deviation statistics were derived for uncensored survival data as the Cramér–von Mises statistic

$$-\int_0^\infty \{\hat{S}(t) - S_0(t)\}^2 \, \mathrm{d}S_0(t)$$

with asymptotic distribution (under the hypothesis $\alpha = \alpha_0$, that is, $S = S_0$); after suitable normalization,

$$\int_0^1 W^0(t)^2 \, \mathrm{d}t,$$

where W^0 is standard Brownian bridge on $[0, 1]$. Similarly, the Anderson–Darling statistic was defined as

$$-\int_0^\infty \frac{\{\hat{S}(t) - S_0(t)\}^2}{S_0(t)\{1 - S_0(t)\}} \, \mathrm{d}S_0(t)$$

with asymptotic distribution given by

$$\int_0^1 \frac{W^0(t)^2}{t(1-t)} \, \mathrm{d}t.$$

In our situation, we have $c_n Z(t)$ approximately distributed as $W(\sigma^2(t))$, where $\sigma^2(t)$ was defined in Proposition V.4.1 and W (as usual) is standard Brownian motion. To obtain similar asymptotic results as in the classical case, we first recall that for W^0 a standard Brownian bridge,

$$W(t) = (1 + t) W^0\left(\frac{t}{1+t}\right)$$

is standard Brownian motion. Thus, we have to show that the asymptotic distribution of the Cramér–von Mises statistic is that of

$$\int_{u=\sigma^2(t_1)}^{\sigma^2(t_2)} W^0\left(\frac{u}{1+u}\right)^2 \mathrm{d}\left(\frac{u}{1+u}\right) = \int_{u=\sigma^2(t_1)}^{\sigma^2(t_2)} \frac{W(u)^2}{(1+u)^2} \mathrm{d}\left(\frac{u}{1+u}\right)$$

which brings us back to our asymptotic theory, setting $u = \hat{\sigma}^2(t)$ in the approximation.

Proposition V.4.2. *Under the hypothesis $\alpha = \alpha_0$, the asymptotic distribution of the Cramér–von Mises statistic*

$$\int_{t_1}^{t_2} c_n^2 \left(\frac{Z^{(n)}(s)}{1 + \hat{\sigma}^2(s)} \right)^2 \mathrm{d}\left(\frac{\hat{\sigma}^2(s)}{1 + \hat{\sigma}^2(s)} \right)$$

is that of

$$\int_{d_1}^{d_2} W^0(t)^2 \,\mathrm{d}t, \tag{5.4.2}$$

where W^0 is standard Brownian bridge and $d_i = \sigma^2(t_i)/\{1 + \sigma^2(t_i)\}$, $i = 1, 2$.

Under the hypothesis, if $0 < d_1 < d_2 < 1$, the asymptotic distribution of the Anderson–Darling statistic

$$\int_{t_1}^{t_2} c_n^2 \frac{\{Z^{(n)}(s)\}^2}{\hat{\sigma}^2(s)} \mathrm{d}\left(\frac{\hat{\sigma}^2(s)}{1 + \hat{\sigma}^2(s)} \right)$$

is that of

$$\int_{d_1}^{d_2} \frac{W^0(t)^2}{t(1-t)} \mathrm{d}t. \tag{5.4.3}$$

PROOF. The first result essentially holds by the continuous mapping theorem applied to the pair

$$c_n^2 \left(\frac{Z^{(n)}}{1 + \hat{\sigma}^2} \right)^2, \qquad \frac{\hat{\sigma}^2}{1 + \hat{\sigma}^2}$$

and to the mapping $(x, y) \rightarrow \int x \,\mathrm{d}y$ discussed in Proposition II.8.6 and the remark following it. To see that this is applicable, note that weak convergence of the first element of this pair to a continuous limit, in the Skorohod sense, implies weak convergence in the sense of Dudley with respect to the supremum norm (see Section II.8). We also have uniform convergence, in probability, of the second element of the pair to a deterministic limit. Together we, therefore, have joint weak convergence in $(D[t_1, t_2])^2$, supremum norm, to a random element of $(D[t_1, t_2])^2$ whose second component is, in fact, fixed. Moreover, along the whole sequence the variation of the second element of the pair, $\hat{\sigma}^2(t_2)/\{1 + \hat{\sigma}^2(t_2)\}$, is always less than 1. Now apply the continuous mapping theorem and the remark after Proposition II.8.6.

The second result holds by the same argument applied to

$$c_n^2 \frac{(Z^{(n)})^2}{\hat{\sigma}^2}, \qquad \frac{\hat{\sigma}^2}{1 + \hat{\sigma}^2}$$

in $(D[t_1, t_2])^2$. □

Again, similar results may be obtained for the two-sample case. The already mentioned very useful survey by Schumacher (1984) also covers integrated squared derivation statistics.

To sum up this brief exposition: The counting process framework allows direct generalization of the classical Kolmogorov–Smirnov, Cramér–von Mises, and Anderson–Darling statistics. The required asymptotic results are corollaries of previous theorems of the present chapter, and the test statistics, though new in this generality, extend previous suggestions for censored survival data. There is a close connection between one-sample Kolmogorov–Smirnov statistics and confidence bands (Sections IV.1.3 and IV.3.3).

V.4.2. Sequential Tests

The theory developed so far is based on the concept of continuous observation of the counting processes involved for $0 \leq t \leq \tau$. Nevertheless, we have only discussed statistical inference procedures to be performed after the observation has been concluded, and, in particular, we have not built the possibility of data-dependent stopping rules into the statistical estimation and testing procedures. The closest we have come to this area is the generality of the censoring process, where the theory of Chapter III is rich enough to let censoring patterns depend adaptively on the realized counting process, while still preserving the general martingale structure.

This generality invites more sophisticated statistical inference procedures. Thus, it may be desirable to monitor the process continuously, perform a test at every time t, and stop the process as soon as a prescribed significance level has been attained. It is immediately clear that this so-called *nominal* significance level is not the actual significance level of the test specified. For testing that the drift parameter of a Brownian motion is zero, there exists an elaborate literature calculating such significance levels under various postulated stopping rules. The purpose of the present section is to indicate that many of the results in this literature are available for direct implementation into our framework via the basic stochastic processes $Z(t)$ in (5.1.1) and $(Z_h(t))$ in (5.2.2) and their asymptotic approximations by Gaussian martingales; cf. Theorem V.1.1 and V.2.1.

We shall concentrate on the two-sample test as an example, but one- and k-sample tests would work similarly. For the k-sample test, the stochastic process $X^2(t)$ of (5.2.6) is approximated by a similar quadratic form (a so-called *Bessel process*) $\mathbf{U}(t)^\mathsf{T}\boldsymbol{\Sigma}(t)^-\mathbf{U}(t)$ of the limiting Gaussian martingale of Theorem V.2.1, where $\boldsymbol{\Sigma}(t) = (\sigma_{hj}(t))$ is given by (5.2.12).

As usual, the results may be specialized to censored survival data, but we refrain from doing so here. The reason is that the present framework only covers the relatively uncommon situation where all patients are put on trial simultaneously. (In reliability applications, simultaneous entry in life testing is common.) To handle *staggered entry*, one has to involve calendar time and time on study simultaneously, and this will be discussed in Section X.2.

Thus, we consider a bivariate counting process $(N_1(t), N_2(t))$ with intensity process $(\alpha_1(t)Y_1(t), \alpha_2(t)Y_2(t))$. The hypothesis we want to test is $H_0: \alpha_1 = \alpha_2$. For the predictable weight process $(L(t))$ we define [cf. (5.2.7)] the basic stochastic process

$$Z(t) = \int_0^t L(s)\,d\{\hat{A}_1(s) - \hat{A}_2(s)\}$$

and, from (5.2.8),

$$\hat{\sigma}^2(t) = \int_0^t L^2(s)\{Y_1(s)Y_2(s)\}^{-1}\,d\{N_1 + N_2\}(s).$$

Let [cf. (5.2.12) and the remarks following Proposition V.2.2]

$$\sigma^2(t) = \int_0^t l^2(s)\{y_1(s) + y_2(s)\}\{y_1(s)y_2(s)\}^{-1}\alpha(s)\,ds,$$

and assume that there exist constants $c_1, c_2, \ldots$ such that for a sequence of such problems (not indicated in the notation), $c_n^2\hat{\sigma}^2(t) \xrightarrow{P} \sigma^2(t)$ as $n \to \infty$; also assume a suitable Lindeberg condition and integrability conditions regarding l and y. Then by Theorem V.2.1, under the null hypothesis, $c_n Z \xrightarrow{\mathscr{D}} U$, where U is a mean-zero Gaussian martingale with covariance function $\sigma^2(t)$, that is, $U \overset{\mathscr{D}}{=} W \circ \sigma^2$, where W is a standard Brownian motion. Furthermore, Section V.2.3 considers contiguous alternatives of the form (5.2.26), $\alpha_h(t) = \alpha(t) + \theta_h(t)/a_n + o(a_n^{-1})$, where the sequence a_n normalizes $Y_h(t)$: $Y_h(t)/a_n^2 \xrightarrow{P} y_h(t)$ as $n \to \infty$. It is shown there under natural regularity conditions that then $c_n^2\hat{\sigma}^2 \xrightarrow{P} \sigma^2$ still holds along the sequence of local alternatives and that $c_n Z \xrightarrow{\mathscr{D}} U$, where U now is a Gaussian martingale with covariance function $\sigma^2(t)$ and expectation

$$\xi(t) = \int_0^t l(s)\{\theta_1(s) - \theta_2(s)\}\,ds.$$

If W is a Brownian motion with drift μ, we have $E\{W(t)\} = \mu t$, $E\{W(\sigma^2(t))\} = \mu\sigma^2(t)$, so it is of particular interest to check under which circumstances one has $\xi(t) = \mu\sigma^2(t)$ or

$$l(s)\{\theta_1(s) - \theta_2(s)\} = \mu l^2(s)\frac{y_1(s) + y_2(s)}{y_1(s)y_2(s)}\alpha(s).$$

It is seen that this will hold if $\theta_1(s) - \theta_2(s)$ is proportional to $l(s)\alpha(s)\{y_1(s) + y_2(s)\}/\{y_1(s)y_2(s)\}$. However, this is exactly the optimal choice of weight process for the fixed-sample test, developed toward the end of Section V.2.3; see, in particular, formula (5.2.32).

With these preparations, we are now ready to take advantage of the rich literature for sequential analysis of Brownian motion. To recapitulate: Let

$\kappa(s) = l(s)\{y_1(s) + y_2(s)\}/\{y_1(s)y_2(s)\}$; see the discussion after Proposition V.2.2. Consider the local family of statistical models with

$$\alpha_h(t) = \alpha(t) + \varphi_h \kappa(t)\alpha(t)/a_n + \rho_h^{(n)}(t),$$

where $\sup_{[0,\tau]} |a_n \rho_h^{(n)}(t)| \to 0$ as $n \to \infty$. The basic stochastic process

$$Z(t) = \int_0^t L(s)\,\mathrm{d}\{\hat{A}_1(s) - \hat{A}_2(s)\}$$

is asymptotically a time-transformed Brownian motion: $c_n Z \xrightarrow{\mathscr{D}} U$ with $U \stackrel{\mathscr{D}}{=} W \circ \sigma^2$, where the Brownian motion W has variance one per unit time and drift $\varphi_1 - \varphi_2$. The test of the hypothesis H_0: $\alpha_1(t) = \alpha_2(t)$ transforms to a test of H_0: $\varphi_1 = \varphi_2$, that is, of zero drift of the Brownian motion.

Consider first the so-called repeated significance test [for an excellent introduction; see Siegmund (1985, Section IV.2)]. For any fixed t, we would base a test of H_0 on the statistic $|Z(t)|/\hat{\sigma}(t)$, rejecting if this is greater than some fractile b. Let T be the random time

$$T = \inf\{t > t_0 \colon |Z(t)|/\hat{\sigma}(t) > b\}.$$

Clearly, T is a stopping time with respect to the basic filtration $(\mathscr{F}_t)$. Define the repeated significance test of H_0 as the procedure that stops sampling at $\min(T, t_1)$, where $t_1 > t_0$, rejects H_0 if $T \leq t_1$, and accepts H_0 if $T > t_1$. Thus, the value of the power function in the alternative μ is $\mathrm{P}_\mu\{T \leq t_1\}$; and we define the so-called average sample size (ASN) $\mathrm{E}_\mu(T \wedge t_1)$. Siegmund also provided a detailed discussion of the validity of various approximations to the power and ASN functions, assuming the Brownian motion model is exact; by our general approximation results, these may be implemented as approximations to the counting process model.

A further example of a sequential test is the adaptation of the classical sequential probability ratio test, where the stopping boundaries are two horizontal lines in the $(t, W(t))$ [that is, $(u, Z(u))$] plane. Although it may be proved that the limiting sample size $T < \infty$ a.s., in practice an upper bound on T may be useful here as well. See Siegmund (1985, Sections II.2 and II.3).

V.5. Bibliographic Remarks

Sections V.1 and V.2

For counting processes, two-sample tests came first historically; the work of Aalen (1975) on the log-rank test and Aalen (1978b) on the general two-sample test included version (5.2.7)–(5.2.9) of the test statistic and showed how the asymptotic distribution of the test statistic may be derived with the same tools as those for the Nelson–Aalen estimator. Gill (1980a) gave a detailed study of the two-sample test statistics for censored survival data using counting process methodology. See Example V.2.1 regarding the relation to the survival analysis literature on two-sample tests. Asymptotic optimality of these tests will be studied in Chapter VIII.

The counting process formulations of one- and k-sample tests were surveyed by Andersen et al. (1982); references to the survival analysis literature are given in Examples V.1.2 and V.2.1. Our exposition on the local power calculations is based on the results by Gill (1980a) for the two-sample case and Hjort (1984, 1985a) for the k-sample case.

Prentice (1978) gave an influential framework for rank tests for censored survival data, generalizing the classical work of Hájek and Sĭdák (1967) for the uncensored case. These tests are based on the marginal probability of a suitable generalized rank vector as if the data had been obtained under progressive type II censoring. Further work along these lines, including Pitman efficiency calculations, was done by Leurgans (1984), Struthers (1984), and Cuzick (1985), in part based on the counting process and martingale framework. Leurgans (1983) defined three classes of rank tests for censored survival data: the asymptotically efficient, the approximately distribution-free, and the approximately unbiased. Various specific tests were studied, and among the results were that the Gehan generalization of the Wilcoxon test belongs to none of the classes. A useful comparison of the various approaches was given by Gu et al. (1991).

Small sample results exist primarily for two-sample tests for censored survival data, comparing size and power of various tests under specified parametric models by Monte Carlo simulation (Lee et al., 1975; Lininger et al., 1979; Latta, 1981; Kellerer and Chmelevsky, 1983). Öhman (1990) compared various forms of generalized two-sample Wilcoxon tests obtained by different choices of the variance estimator, including one motivated by a jackknife argument.

Section V.3

Among the many further nonparametric tests for censored data, we presented in this section three examples: tests based on intensity ratio estimates, cf. Andersen (1983a), generalizing Crowley et al. (1982); stratified tests (Peto et al., 1977); and trend tests, where additional early work was done by Pons (1981). As mentioned in Example V.3.4, the special theme of tests for censored matched-pair survival data has been studied at length, primarily in the context of random effects models, where the two survival times in a pair are correlated. O'Brien and Fleming (1987) gave a pragmatic survey of this literature, whereas Dabrowska (1986, 1989a, 1990) obtained asymptotic results. See Dabrowska (1990) for a useful discussion and survey. These results are not directly accessible from the simple counting process framework, which requires adaptation to handle dependent lifetimes; cf. Chapter IX.

The literature contains several other types of nonparametric tests for censored data; we mention here Brookmeyer and Crowley's (1982b) k-sample median test, a class of two-sample tests based on the weighted difference between the Kaplan–Meier estimators: $\int w(t)[\hat{S}_1(t) - \hat{S}_2(t)]\,dt$; see Pepe and Fleming (1989, 1991) and Gray's (1988) test for comparing the "cumulative incidences" $\hat{P}_{0j}(s,t)$ (cf. Examples IV.4.1) of some fixed cause j between two or more groups. Jones and Crowley (1989, 1990) proposed a general class of linear tests, incorporating many of the above, motivated by the Mantel–Haenszel tradition discussed earlier; see further Example VII.2.4.

A somewhat different approach to rank tests for censored survival data was taken by Albers and Akritas (1987) who assumed equal censoring and then ordered the censored and uncensored observations separately, in effect producing two classical rank tests, one for the uncensored and one for the censored observations, which were then combined. Albers (1988a, 1988b) extended the idea from two-sample tests to k-sample tests (and regression) and to matched pairs. Neuhaus (1988) and Janssen (1989) also assumed equal censoring and derived a family of rank tests, optimal against specific alternatives, in the framework of modern asymptotic decision theory. The approach to censoring taken by these authors is principally different from ours: They essentially study the distribution of the censored survival times directly, whereby the assumption of equal censoring distribution enters directly into the statistical inference.

Tests based on the complete test statistic process are covered in Section V.4.

Section V.4.1

For a general introduction to nonlinear tests based on the empirical distribution function, Durbin's (1973) concise text is still unsurpassed. The standard comprehensive theoretical reference is now the treatise by Shorack and Wellner (1986), who also presented many details, such as tables of percentage points of the various statistics.

For the maximal deviation statistics, the equivalence with confidence bands has been mentioned in the text; this means that many of the references quoted in connection with Sections IV.1.3 and IV.3.3 are also relevant here. As important recent surveys of the censored survival data problems, we mention Nair (1984), Doksum and Yandell (1984), Schumacher (1984), Csörgő and Horváth (1986), and Fleming et al. (1987), who also connected to the classification of Leurgans (1983) mentioned earlier. Further interesting work on Kolmogorov–Smirnov-type two-sample tests has been done by Janssen and Milbrodt (1991), who showed that to maximize the power of such omnibus tests against (semi)parametric alternatives, the optimal weight function for the parametric alternative has to be modified slightly.

A very interesting combinatorial approach to *exact* Kolmogorov type one-sample tests for left-truncated and right-censored survival data is due to Guilbaud (1988). His results showed essentially that the critical region from the uncensored case is adaptively reduced or enlarged as the risk set is reduced or enlarged. Guilbaud's results thus provided important additional insight into the effects of censoring on these statistics.

The literature on integrated squared deviation statistics for censored survival data is limited. Early work by Pettitt and Stephens (e.g., 1976) essentially assumed censoring in a fixed point. The attempt by Koziol and Green (1976) of a direct generalization of the classical Cramér–von Mises statistic by replacing the empirical survival function with the Kaplan–Meier estimator has attracted much interest in the literature, although we consider the restriction on the censoring distribution ($1 - G = S_0^\beta$ for some β) to limit the applicability severely.

The Cramér–von Mises statistic of Proposition V.4.2 was suggested by Koziol (1978), and the corresponding two-sample generalization by Koziol and Yuh (1982).

The problem of goodness-of-fit tests when parameters are estimated is discussed in Section VI.3.3, and further references are given in the bibliographic remarks to Chapter VI.

Section V.4.2

The basic structure of the current theory of sequential tests and confidence intervals was admirably presented by Siegmund (1985), who, furthermore, gave many thoughtful comments on the practical range of these theoretical results.

Except for a preliminary report by Mau (1986c), the simplicity of the results for general counting processes as presented here has not been recorded in the literature before.

Sequential rank tests for comparison of survival distributions were developed by Sen and co-workers; see the surveys by Sen (1981, 1985). Their main results concerned simultaneous entry and so-called progressive censoring, that is, observation is discontinued at some specified time t_1 for all individuals, where t_1 is either chosen in advance (type I progressive censoring) or as the time $T_{(r)}$ of the rth failure, where r is chosen in advance (type II progressive censoring). S. Monty, in a Copenhagen cand. act. (M.Sc.) thesis, carefully documented how these results are covered by our master theorems, such as Theorem V.2.1.

In medical applications, it will rarely be relevant to have simultaneous entry, and we return to sequential methods for clinical trials with staggered entry in Section X.2.

CHAPTER VI

Parametric Models

In the preceding two chapters, we have discussed nonparametric estimation and testing in the multiplicative intensity model for counting processes. In the present chapter, we will study parametric inference for this model and also for other counting process models with parametrically specified intensity processes.

Our setup is as follows. Let $\mathbf{N} = ((N_1(t), \ldots, N_k(t));\ t \in \mathscr{T})$ be a k-variate counting process with intensity process $\boldsymbol{\lambda} = (\lambda_1, \ldots, \lambda_k)$. We assume that the intensity process $\lambda_h(t) = \lambda_h(t; \boldsymbol{\theta})$ is a predictable process [with respect to some filtration $(\mathscr{F}_t)$] which may be specified by a q-dimensional parameter $\boldsymbol{\theta} = (\theta_1, \ldots, \theta_q)$. In this chapter, we will often consider the multiplicative intensity model. For this model, it is assumed that $\lambda_h(t; \boldsymbol{\theta}) = \alpha_h(t; \boldsymbol{\theta}) Y_h(t)$, where $\alpha_h(t; \boldsymbol{\theta})$ is a function specified by $\boldsymbol{\theta}$, whereas Y_h, as in Chapters III–V, is a predictable process. A number of examples of models having an intensity process of this multiplicative form have been discussed in the preceding three chapters; the difference from Chapters IV and V being that we will now consider parametric models for the $\alpha_h(t; \boldsymbol{\theta})$.

In Section VI.1, we consider maximum likelihood estimation for k-variate counting process models with parametrically specified intensity processes. Test statistics closely related to this method of estimation, like Wald's test, the score test, and the likelihood ratio test are also discussed. In Section VI.2, we consider a wider class of estimators, namely, the class of "M-estimators," and show that the maximum likelihood estimator is optimal within this class. The optimality of the maximum likelihood estimator in general is studied in Section VIII.2.2. Various techniques for checking the appropriateness of a specified form of the $\lambda_h(t; \boldsymbol{\theta})$ are discussed in Section VI.3. These include graphical techniques such as the use of the Nelson–Aalen estimator and the

Total Time on Test plot (TTT-plot), as well as more formal goodness-of-fit tests.

In Section VII.6, we will return to a discussion of parametric regression models, whereas parametric "frailty" models are studied in Chapter IX. Finally, in Example X.1.9, we indicate how the methodology of the present chapter continues to be valid also for some situations (e.g., semi-Markov models, "staggered entry") where the nonparametric methods of Chapters IV and V break down by adopting Arjas's (1985, 1986) idea of letting calendar time ("real time") be the basic time variable.

VI.1. Maximum Likelihood Estimation

VI.1.1. The Maximum Partial Likelihood Estimator and Related Tests

We consider a multivariate counting process $\mathbf{N} = (N_1, \ldots, N_k)$ with intensity process $\boldsymbol{\lambda} = (\lambda_1, \ldots, \lambda_k)$ specified by a q-dimensional parameter $\boldsymbol{\theta} = (\theta_1, \ldots, \theta_q) \in \Theta$, an open subset of the q-dimensional Euclidean space. Thus, we may write $\lambda_h(t) = \lambda_h(t; \boldsymbol{\theta})$. Our aim in this subsection is to derive the maximum (partial) likelihood estimator for $\boldsymbol{\theta}$, discuss test statistics related to this method of estimation, and illustrate the methods by a number of examples. A formal proof of the large sample properties is deferred to the next subsection.

By (2.7.4′) and the results of Section II.7.3 [cf. also (3.2.8) and (3.4.2)] the partial likelihood may be written

$$L_\tau(\boldsymbol{\theta}) = \left\{ \prod_{t \in \mathscr{T}} \prod_{h=1}^{k} \lambda_h(t; \boldsymbol{\theta})^{\Delta N_h(t)} \right\} \exp\left\{ -\int_0^\tau \sum_{h=1}^{k} \lambda_h(t; \boldsymbol{\theta}) \, \mathrm{d}t \right\}. \tag{6.1.1}$$

Thus, the log-partial-likelihood function takes the form

$$C_\tau(\boldsymbol{\theta}) = \int_0^\tau \sum_{h=1}^{k} \log \lambda_h(t; \boldsymbol{\theta}) \, \mathrm{d}N_h(t) - \int_0^\tau \sum_{h=1}^{k} \lambda_h(t; \boldsymbol{\theta}) \, \mathrm{d}t, \tag{6.1.2}$$

and, assuming that we may interchange the order of differentiation and integration, the vector $\mathbf{U}_\tau(\boldsymbol{\theta})$ of *score statistics* $U_\tau^j(\boldsymbol{\theta})$, $j = 1, \ldots, q$, is given by

$$U_\tau^j(\boldsymbol{\theta}) = \int_0^\tau \sum_{h=1}^{k} \frac{\partial}{\partial \theta_j} \log \lambda_h(t; \boldsymbol{\theta}) \, \mathrm{d}N_h(t) - \int_0^\tau \sum_{h=1}^{k} \frac{\partial}{\partial \theta_j} \lambda_h(t; \boldsymbol{\theta}) \, \mathrm{d}t. \tag{6.1.3}$$

As for the classical case of maximum likelihood estimation for i.i.d. random variables, the log-partial-likelihood function (6.1.2) may have a number of local maxima. Thus, the equation $\mathbf{U}_\tau(\boldsymbol{\theta}) = \mathbf{0}$ may have multiple solutions. Here, we will not discuss conditions for a unique solution to this "partial likelihood equation," but rather consider the maximum partial likelihood estimator $\hat{\boldsymbol{\theta}}$ given as a solution to the equation $\mathbf{U}_\tau(\boldsymbol{\theta}) = \mathbf{0}$. If more than one solution is found in a concrete situation, one could then check which of these gives the largest value of (6.1.2).

In Section VI.1.2, we will give formal proofs showing that $\hat{\boldsymbol{\theta}}$ enjoys "the usual" properties for maximum likelihood estimators, well-known from the case of i.i.d. random variables, under similar regularity conditions. More precisely, in the next subsection, we will prove that (with a probability tending to one) there exists a consistent solution $\hat{\boldsymbol{\theta}}$ of the equation $\mathbf{U}_\tau(\boldsymbol{\theta}) = \mathbf{0}$ and that this solution is asymptotically multinormally distributed around the true parameter value and with a certain covariance matrix. This asymptotic covariance matrix may, in the usual manner, be estimated by $\mathscr{I}_\tau(\hat{\boldsymbol{\theta}})^{-1}$, where $-\mathscr{I}_\tau(\boldsymbol{\theta})$ is the matrix of second-order partial derivatives of the log-partial-likelihood function (6.1.2). Thus, writing $\mathscr{I}_\tau^{jl}(\boldsymbol{\theta})$ for the (j, l)th element of the matrix $\mathscr{I}_\tau(\boldsymbol{\theta})$, and assuming that we may interchange the order of integration and differentiation in (6.1.3), we have that

$$\mathscr{I}_\tau^{jl}(\boldsymbol{\theta}) = \int_0^\tau \sum_{h=1}^k \frac{\partial^2}{\partial\theta_j\,\partial\theta_l}\lambda_h(t;\boldsymbol{\theta})\,\mathrm{d}t - \int_0^\tau \sum_{h=1}^k \frac{\partial^2}{\partial\theta_j\,\partial\theta_l}\log\lambda_h(t;\boldsymbol{\theta})\,\mathrm{d}N_h(t). \quad (6.1.4)$$

This result on the asymptotic distribution of $\hat{\boldsymbol{\theta}}$ immediately gives that the *Wald test statistic*

$$(\hat{\boldsymbol{\theta}} - \boldsymbol{\theta}_0)^{\mathsf{T}}\mathscr{I}_\tau(\hat{\boldsymbol{\theta}})(\hat{\boldsymbol{\theta}} - \boldsymbol{\theta}_0) \quad (6.1.5)$$

is asymptotically χ^2 distributed with q degrees of freedom under the simple hypothesis H_0: $\boldsymbol{\theta} = \boldsymbol{\theta}_0$. Also the *score test statistic*

$$\mathbf{U}_\tau(\boldsymbol{\theta}_0)^{\mathsf{T}}\mathscr{I}_\tau(\boldsymbol{\theta}_0)^{-1}\mathbf{U}_\tau(\boldsymbol{\theta}_0) \quad (6.1.6)$$

and the partial *likelihood ratio test statistic*

$$2(C_\tau(\hat{\boldsymbol{\theta}}) - C_\tau(\boldsymbol{\theta}_0)) \quad (6.1.7)$$

have the same asymptotic distribution as the Wald test statistic. For the special case of $q = 1$, one may, of course, consider the equivalent versions $(\hat{\theta} - \theta_0)\mathscr{I}_\tau(\hat{\theta})^{1/2}$ and $U_\tau(\theta_0)\mathscr{I}_\tau(\theta_0)^{-1/2}$ for the Wald test and the score test statistics, both being asymptotically standard normally distributed when $\theta = \theta_0$.

These results are proved in Section VI.1.2, where we also comment upon the fact that the partial likelihood ratio test statistic for a composite null hypothesis will retain the same asymptotic distribution as in the classical i.i.d. case.

We, furthermore, note that for situations where (6.1.1) is proportional to the full likelihood, (i.e., when we have noninformative censoring; cf. Section III.2.3) $\hat{\boldsymbol{\theta}}$ will be an ordinary maximum likelihood estimator. Otherwise, it is only a maximum partial likelihood estimator. In any case, the asymptotic properties of $\hat{\boldsymbol{\theta}}$ and the related test statistics are as outlined earlier. In the following, therefore, we will not make a point out of the possibility that (6.1.1) may be just a *partial* likelihood. Consequently, we will for the rest of this chapter drop the word "partial" when we talk about maximum likelihood estimators, likelihood ratio tests, etc. The reader should, however, keep in mind that whether we really do have ordinary maximum likelihood estimators, likelihood ratio tests, etc., or just partial ones depends on whether or not we have noninformative censoring.

We will, in this subsection, illustrate the general formulations by a number of examples. In most of these examples, we will consider the multiplicative intensity model where the intensity process may be written in the form

$$\lambda_h(t;\boldsymbol{\theta}) = \alpha_h(t;\boldsymbol{\theta})\,Y_h(t) \tag{6.1.8}$$

as described in the introduction to the chapter.

As remarked in Section III.1.3 and illustrated a number of times in the preceding three chapters, a multiplicative intensity model for (N_h) is often obtained by aggregating individual counting processes N_{hi} $(i = 1, 2, \ldots, n,$ $h = 1, 2, \ldots, k)$ each having intensity processes of the form $\lambda_{hi}(t;\boldsymbol{\theta}) = \alpha_h(t;\boldsymbol{\theta})\,Y_{hi}(t)$. We note that for such a situation the likelihood (6.1.1) based on (N_{hi}) is proportional to the one based on (N_h); in fact, we noted in Section III.2.3 that $(N_h, Y_h; h = 1, \ldots, k)$ is sufficient (when censoring is noninformative). Thus, in most examples to be presented we can (and will) base our inference on such aggregated processes. An exception is Example VI.1.5 where no such sufficiency reduction is available.

EXAMPLE VI.1.1. Occurrence/Exposure Rates. Suicides Among Danish Nonmanual Workers

The situation with a q-variate counting process $(N_1, \ldots, N_q)$ satisfying the multiplicative intensity model (6.1.8) with all the $\alpha_h(t;\boldsymbol{\theta})$ constant is very simple from a mathematical point of view, but nevertheless deserves special attention because of its practical importance, in particular, in demography and epidemiology. This situation arises, for example, when $N_1, \ldots, N_q$ count the number of deaths in q age groups and the force of mortality is assumed to be constant for each of these age groups. More generally, any Markov process with finite state-space and piecewise constant intensities can be reformulated to fit into this framework (cf. Example VI.1.7).

Let the constant value of $\alpha_h(\cdot;\boldsymbol{\theta})$ be denoted θ_h, $h = 1, \ldots, q$. Then the score functions (6.1.3) may be written

$$U_\tau^j(\boldsymbol{\theta}) = \frac{N_j(\tau)}{\theta_j} - R_j(\tau),$$

where

$$R_j(\tau) = \int_0^\tau Y_j(t)\,\mathrm{d}t$$

is the "exposure" or "total time on test," corresponding to θ_j. It follows that the maximum likelihood estimators are the *occurrence/exposure* rates

$$\hat{\theta}_j = N_j(\tau)/R_j(\tau).$$

These are asymptotically normally distributed around the true parameter values. Furthermore, by (6.1.4),

$$\mathscr{I}_\tau^{jl}(\boldsymbol{\theta}) = \delta_{jl} N_j(\tau)/\theta_j^2,$$

where δ_{jl} is a Kronecker delta, so that the occurrence/exposure rates are asymptotically independent with an asymptotic variance for $\hat{\theta}_j$ that can be estimated by $\hat{\theta}_j^2/N_j(\tau) = \hat{\theta}_j/R_j(\tau)$.

As an illustration, we consider the data introduced in Example I.3.3 on suicides of Danish nonmanual workers, 1970–80. Assume that the mortality is constant in each 5-year age group; then the estimated age-specific suicide intensities for male and female academics (group I) are given in Table VI.1.1. In particular, for 20–24-year-old males, we get $\hat{\theta} = 2/13439 = 0.00015$ per year with estimated variance $\hat{\theta}/13439 = 0.00011^2$.

Confidence intervals are conveniently based on the log-transformation

$$\operatorname{var}(\log \hat{\theta}_j) \approx \hat{\theta}_j^{-2} \operatorname{var}(\hat{\theta}_j) \approx 1/N_j(\tau),$$

so that for 20–24-year-old males, we get the approximate 95% confidence interval

$$0.00015\, e^{\pm 1.96/\sqrt{2}} = [0.00004, 0.00060].$$

For a careful study of confidence intervals for "rare events," see Schou and Væth (1980).

For 20–24-year-old females, there are no observed suicides, yielding an

Table VI.1.1. Estimated Age-Specific Suicide Intensities (with Approximate Standard Deviations) for Male and Female Academics. Values per 10^4 Person Years

Age 9 Nov. 1970	Males	Females
20–24	1.5 (1.1)	0 (0[a])
25–29	1.9 (0.6)	3.6 (1.6)
30–34	4.9 (1.0)	4.9 (1.9)
35–39	4.0 (0.9)	7.6 (2.9)
40–44	5.2 (1.0)	6.4 (2.6)
45–49	4.1 (0.9)	4.9 (2.2)
50–54	5.1 (1.1)	4.6 (2.3)
55–59	5.8 (1.3)	9.0 (3.7)
60–64	2.6 (1.1)	4.1 (2.9)

[a] see text for construction of approximate 95% confidence interval.

estimate $\hat{\theta} = 0/2366 = 0$ with a standard error estimate of 0. The latter is clearly unsatisfactory and may, for the purpose of setting approximate confidence intervals, be replaced by the following consideration. The likelihood for θ is proportional to that obtained from a Poisson distribution with parameter 2366 θ. A (one-sided) exact test of $\theta = \theta_0$ in this distribution when $D = 0$ suicides are observed will accept the hypothesis at level 5% when

$$0.05 \leq P_{\theta_0}(D = 0) = e^{-2366\,\theta_0}$$

or $\theta_0 \leq 0.0013$, so that a 95% confidence interval is [0, 0.0013].

Table VI.1.1 gives the impression of a somewhat higher suicide intensity for females than for males. Further analysis of these data follows in the next two examples. □

EXAMPLE VI.1.2. The Standardized Mortality Ratio. Suicides Among Danish Nonmanual Workers

We consider a univariate counting process $N(t)$ with intensity process of the multiplicative form $\lambda(t) = \theta\alpha_0(t)Y(t)$, where $\alpha_0(t)$ is a known function. Then, by (6.1.3), the score function $U_\tau(\theta)$ takes the form

$$U_\tau(\theta) = \frac{N(\tau)}{\theta} - E(\tau),$$

where

$$E(\tau) = \int_0^\tau \alpha_0(t)Y(t)\,dt$$

may be interpreted as the "expected number" of events when $\theta = 1$ (cf. Example V.1.1). Therefore, the maximum likelihood estimator of θ is

$$\hat{\theta} = N(\tau)/E(\tau),$$

which in the survival analysis setup is known as the *standardized mortality ratio* (SMR). By the results reviewed above, $\hat{\theta}$ is asymptotically normally distributed around the true parameter value and with an asymptotic variance that may be estimated by

$$\mathscr{I}_\tau(\hat{\theta})^{-1} = \hat{\theta}^2/N(\tau);$$

cf. (6.1.4).

For testing the hypothesis that $\theta = 1$, the Wald test statistic is given by

$$(\hat{\theta} - 1)\mathscr{I}_\tau(\hat{\theta})^{1/2} = \{N(\tau) - E(\tau)\}N(\tau)^{-1/2}.$$

This could also have been obtained as the score test statistic $U_\tau(1)\mathscr{I}_\tau(1)^{-1/2}$. This test statistic, which is standard normally distributed when $\theta = 1$, was discussed already in Example V.1.1.

On the other hand, if the parametrization $\beta = \log\theta$ is used, we have, by (6.1.3),

$$U_\tau(\beta) = N(\tau) - e^\beta E(\tau),$$

and, therefore,

$$\mathscr{I}_\tau(\beta) = e^\beta E(\tau),$$

yielding the score test statistic

$$U_\tau(0)\mathscr{I}_\tau(0)^{-1/2} = \{N(\tau) - E(\tau)\}E(\tau)^{-1/2},$$

which is the "one-sample log-rank test" of Example V.1.1.

Finally, by (6.1.2), the likelihood ratio test statistic, which is independent of the parametrization, is

$$\begin{aligned} 2\{C_\tau(\hat{\theta}) - C_\tau(1)\} &= 2\{N(\tau)\log\hat{\theta} - \hat{\theta}E(\tau) + E(\tau)\} \\ &= 2\{N(\tau)\log(N(\tau)/E(\tau)) - N(\tau) + E(\tau)\}. \end{aligned}$$

This test statistic is asymptotically χ^2 distributed with one degree of freedom under the hypothesis that $\theta = 1$.

We now continue the analysis from the previous example of suicides among Danish nonmanual workers, 1970–80, in particular academics. A common method for analyzing such data is to regard the age-specific mortality (here, suicide rate) of the combined data as known [$=\alpha_0(t)$] and calculate the standardized mortality ratio of the study group. In this example, we use all Danish (male and female) nonmanual workers as the standard population. (Although this is standard practice, it is obviously debatable to consider the standard population "known" when the study group constitutes such a substantial fraction of it.) Let D_{px} and R_{px} be the number of deaths (here, suicides) and person-years lived in age group $x = 1, \dots, 9$ in the study population ($p = a$) and standard population ($p = s$). We then put

$$\alpha_0(t) = D_{sx}/R_{sx}, \quad 20 + 5(x-1) \le t < 20 + 5x$$

(see Example VI.1.1) and obtain the expected number of deaths (suicides) as

$$\int_{20}^{65} \alpha_0(t)Y(t)\,\mathrm{d}t = \sum_{x=1}^{9} \frac{D_{sx}}{R_{sx}} \int_{20+5(x-1)}^{20+5x} Y(t)\,\mathrm{d}t = \sum_{x=1}^{9} \frac{D_{sx}}{R_{sx}} R_{ax}.$$

Consider, in particular, female academics, of whom $O = 41$ commited suicide in 1970–80. The expected number of suicides is

$$E = \frac{200}{1351884}(2366) + \cdots + \frac{83}{314434}(4866) = 26.07,$$

so that the SMR is $O/E = 1.57$. The standard error may be estimated by $\mathrm{SMR}/\sqrt{O} = 1.57/\sqrt{41} = 0.25$. The Wald or score test statistic of no excess suicides in this group takes the value $(O - E)/\sqrt{O} = (41 - 26.07)/\sqrt{41} = 2.33$, $P = .02$, whereas the one-sample log-rank test yields $(O - E)/\sqrt{E} = (41 - 26.07)/\sqrt{26.07} = 2.92$, $P = .004$. Finally, the likelihood ratio statistic gives $2\{O\log(O/E) - O + E\} = 7.13 = 2.67^2$, $P = .007$. (All P values are two-sided.) As usual (cf. Examples V.1.1 and V.1.3), we get a stronger test for the alternative of increased mortality in the study group by estimating the variance by E rather than O; the likelihood ratio statistic is in between.

For male academics, we similarly get SMR = 1.20 with an estimated standard error of 0.10. Hence, a test for equal excess suicide rate for male and female academics may be carried out as

$$u = \frac{1.57 - 1.20}{\sqrt{0.25^2 + 0.10^2}} = 1.37, \qquad P = .17.$$

There exists a huge literature about the standardized mortality ratio and other ways of standardizing vital rates; for general reviews, see, e.g., Fleiss (1981, Chapter 14) and Breslow and Day (1987, Chapters 3 and 4); Keiding (1987) surveyed the historical development. Hoem (1987) reviewed the theory of standardization under a multiplicative (or log-linear) intensity model of the form considered in the following example. □

EXAMPLE VI.1.3. Log-Linear Intensity Models. Suicides Among Danish Nonmanual Workers

In many practical situations, e.g., in demography or epidemiology, there is a meaningful specification of a multivariate counting process $\mathbf{N}$ into, say, $\{N_{i_1 i_2 \cdots i_l}; i_j = 1, 2, \ldots, k_j; j = 1, 2, \ldots, l\}$. This may correspond to any multidimensional grouping of events, such as cross-classification of deaths by age group, sex, occupation, marital status, region of residence, and so on. We assume that the intensity process is of the multiplicative form (6.1.8) and, moreover, that all $\alpha_{i_1 i_2 \cdots i_l}(\cdot; \boldsymbol{\theta})$ are constant, i.e., $\alpha_{i_1 i_2 \cdots i_l}(t; \boldsymbol{\theta}) = \theta_{i_1 i_2 \cdots i_l}$ for all $t \in \mathscr{T}$.

For such situations, one may consider log-linear (or multiplicative) models for $\theta_{i_1 i_2 \cdots i_l}$ in a similar manner as for contingency table models (e.g., Bishop, Fienberg, and Holland, 1975). Thus, for example, a model with only main effects may be specified by

$$\theta_{i_1 i_2 \cdots i_l} = \alpha_{i_1} \beta_{i_2} \gamma_{i_3} \delta_{i_4} \cdots \varepsilon_{i_l},$$

whereas a model with first-order interaction between the second and the third "factor" takes the form

$$\theta_{i_1 i_2 \cdots i_l} = \alpha_{i_1} \phi_{i_2 i_3} \delta_{i_4} \cdots \varepsilon_{i_l}.$$

One may of course also consider models with more interaction terms of the first and higher orders, all up to the "saturated" model which leaves the $\theta_{i_1 i_2 \cdots i_l}$ unspecified. The likelihood ratio test can be used to test the appropriateness of the various model specifications just as for log-linear models for contingency tables.

By (6.1.1) and (6.1.8), the likelihood may now be written

$$\prod_{i_1 i_2 \cdots i_l} \left\{ \theta_{i_1 i_2 \cdots i_l}^{N_{i_1 i_2 \cdots i_l}(\tau)} e^{-\theta_{i_1 i_2 \cdots i_l} R_{i_1 i_2 \cdots i_l}(\tau)} \right\}$$

with

$$R_{i_1 i_2 \cdots i_l}(\tau) = \int_0^\tau Y_{i_1 i_2 \cdots i_l}(t)\,dt,$$

the "exposure" corresponding to the cross-classification (or "cell") $i_1 i_2 \cdots i_l$. It is seen that the likelihood is proportional to the likelihood of a product Poisson model. Therefore, log-linear intensity models for counting processes may be fitted by GLIM (e.g., Aitkin et al., 1989) and similar statistical packages by treating the $N_{i_1 i_2 \cdots i_l}(\tau)$ as "independent Poisson variables" with "parameters" $\theta_{i_1 i_2 \cdots i_l} R_{i_1 i_2 \cdots i_l}(\tau)$. Thus, for large sample inference, it is not necessary to assume that the $N_{i_1 i_2 \cdots i_l}(\tau)$ actually are Poisson distributed as is often done in the literature on such models. The important assumption is that of a constant intensity within each "cell"; see Breslow and Day (1987, Section 4.2) for a further discussion of the Poisson assumption.

In the example concerning suicides among Danish nonmanual workers, several variables were registered at entrance into study on the census date 9 November 1970. In addition to the three variables, age group, sex, and job status group, already used in Examples VI.1.1 and VI.1.2, we here also consider marital status and geographical region.

Log-linear intensity models were studied for the suicide intensity θ_{asgmr} with

- a = age group = 20–24, ..., 60–64;
- s = sex = m, f;
- g = group among nonmanual workers = academics, with advanced but nonacademic training, with extensive practical training, other nonmanual workers;
- m = marital status = unmarried (i.e., never married), married, widowhood, divorced (including separated);
- r = geographical region = capital, other urban areas, rural areas.

As always with such models, rational choices have to be made to reduce the number of parameters of the model. Using likelihood ratio tests, the full model θ_{asgmr} was first easily reduced to $\lambda_{as}\mu_{sgmr}$ and then to $\lambda_{as}\phi_{sg}\delta_m\varepsilon_r$, ($\chi^2 = 96.8$, d.f. = 83, $P = .14$), so that the effects of marital status and of geographical region are the same for all combinations of all other variables, whereas, for each sex, the effects of age and job status are multiplicative. Furthermore, the hypothesis $\lambda_{as} = \alpha_a\kappa_s$ of the same relative age effects for males and females is accepted ($\chi^2 = 3.6$, d.f. = 8, $P = .89$), whereas the hypothesis $\phi_{sg} = \beta_s\gamma_g$ of no interaction between sex and job status group is rejected ($\chi^2 = 25.5$, d.f. = 3, $P = .00001$). This means that the relative suicide intensities in the four job status groups display different patterns for males and females. No further reduction of this model

$$\theta_{asgmr} = \alpha_a\phi_{sg}\delta_m\varepsilon_r$$

is possible, and the parameter estimates are given in Table VI.1.2. (Note that

Table VI.1.2. Estimates in Final Model for Suicide Intensities among Danish Nonmanual Workers 1970–80. Variables as Registered 9 November 1970. Standard Deviations of Estimates in Brackets

a	Age	$\hat{\alpha}_a$ (per Year)	
1	20–24	0.00029 (.00003)	
2	25–29	0.00041 (.00004)	
3	30–34	0.00068 (.00007)	
4	35–39	0.00080 (.00008)	
5	40–44	0.00088 (.00009)	
6	45–49	0.00098 (.00010)	
7	50–54	0.00086 (.00009)	
8	55–59	0.00086 (.00009)	
9	60–64	0.00053 (.00007)	
g	Job Status	$\hat{\phi}_{mg}$ (Male)	$\hat{\phi}_{fg}$ (Female)
1	Academics	1 (—)	1.00 (.18)
2	Advanced nonacad.	0.89 (.09)	0.55 (.06)
3	Extensive practical	0.82 (.07)	0.74 (.06)
4	Other	1.12 (.11)	0.47 (.06)
m	Marital Status	$\hat{\delta}_m$	
1	Unmarried	1 (—)	
2	Married	0.58 (.03)	
3	Widowhood	0.95 (.13)	
4	Divorced	1.46 (.10)	
r	Geographical Region	$\hat{\varepsilon}_r$	
1	Capital	1 (—)	
2	Other urban	0.76 (.04)	
3	Rural	0.70 (.05)	

we have normalized by setting $\phi_{m1} = \delta_1 = \varepsilon_1 = 1$.) The final model, thus, specifies multiplicativity between age and all other factors, and within these, marital status and geographical region factor out, whereas sex and occupational group interact. It is remarkable that the difference between males and females is found to lie not in the relative age pattern (as is common for many other causes of death) but in the job status. Also note that contrary to the case for most causes of death, the intensity of suicides is maximal around ages 45–49. As perhaps expected, the suicide intensity increases with increasing urbanization and is high for divorced and low for married with unmarried and widowed in between. Finally, the sex and job status interaction shows that the suicide rates for male and female academics are equal, in contrast

Table VI.1.3. Relative Distributions of Person Years for Academics

	Urbanization			
	Capital	Other Urban	Rural	Total
Males	0.43	0.32	0.25	1
Females	0.58	0.27	0.15	1

	Marital Status				
	Unmarried	Married	Widowhood	Divorced	Total
Males	0.12	0.84	0.01	0.03	1
Females	0.32	0.55	0.04	0.09	1

to the indication from Example VI.1.1 and the conclusion from the SMR analysis in Example VI.1.2. The reason is that female academics are much more concentrated in the high-risk groups on urbanization and marital status than male academics; cf. Table VI.1.3. Therefore, the further "standardization" with respect to these confounders eliminates the earlier observed sex difference. □

EXAMPLE VI.1.4. Left-Truncated and Right-Censored Survival Data. Diabetics in the County of Fyn

We consider the situation with left-truncated and right-censored survival data. Thus, we have lifetimes $X_1, \ldots, X_n$, with common hazard rate function $\alpha(\cdot;\boldsymbol{\theta})$, sampled conditionally on $X_i > V_i$ for each i, where $V_1, \ldots, V_n$ are truncation times. This left-truncated sample is then subject to right-censoring at the censoring times $U_1, \ldots, U_n$. Our sample of left-truncated and right-censored lifetimes, therefore, consists of the triplets $(V_i, \tilde{X}_i, D_i)$, where $\tilde{X}_i = X_i \wedge U_i > V_i$ and $D_i = I(\tilde{X}_i = X_i)$; cf. Examples III.3.6 and IV.1.7.

Now, as shown in Sections III.3 and III.4, the process

$$N(t) = \sum_{i=1}^{n} I(\tilde{X}_i \leq t, D_i = 1),$$

recording the number of observed deaths up to time t, is a counting process with intensity process of the multiplicative form $\alpha(t;\boldsymbol{\theta})Y(t)$, where

$$Y(t) = \sum_{i=1}^{n} I(V_i < t \leq \tilde{X}_i)$$

is the number at risk just before time t.

Thus, by (6.1.1), the (partial) likelihood may now be written (with $\tau = \infty$)

$$\left\{\prod_t (\alpha(t;\boldsymbol{\theta})Y(t))^{\Delta N(t)}\right\} \exp\left\{-\int_0^\infty \alpha(t;\boldsymbol{\theta})Y(t)\,dt\right\}$$

$$= \left\{\prod_{i=1}^{n} (\alpha(\tilde{X}_i;\boldsymbol{\theta})Y(\tilde{X}_i))^{D_i}\right\} \exp\left\{-\sum_{i=1}^{n}\int_{V_i}^{\tilde{X}_i} \alpha(t;\boldsymbol{\theta})\,dt\right\}.$$

This is proportional to

$$\prod_{i=1}^{n} \left\{ \alpha(\tilde{X}_i; \boldsymbol{\theta})^{D_i} \exp\left(- \int_{V_i}^{\tilde{X}_i} \alpha(t; \boldsymbol{\theta})\, dt \right) \right\},$$

an expression well-known from classical survival analysis. It follows that the essential part of the log-likelihood function (6.1.2) takes the form

$$\sum_{i=1}^{n} \left\{ D_i \log \alpha(\tilde{X}_i; \boldsymbol{\theta}) - \int_{V_i}^{\tilde{X}_i} \alpha(t; \boldsymbol{\theta})\, dt \right\},$$

and that the maximum likelihood estimator $\hat{\boldsymbol{\theta}}$ is given as a solution to the equations

$$\sum_{i=1}^{n} \left\{ D_i \frac{(\partial/\partial\theta_j)\alpha(\tilde{X}_i; \boldsymbol{\theta})}{\alpha(\tilde{X}_i; \boldsymbol{\theta})} - \int_{V_i}^{\tilde{X}_i} \frac{\partial}{\partial\theta_j} \alpha(t; \boldsymbol{\theta})\, dt \right\} = 0,$$

$j = 1, \ldots, q$; cf. (6.1.3).

As a particular illustration, let us consider the data on mortality of diabetics in the county of Fyn introduced in Example I.3.2 concentrating on the female diabetics. As described in Example III.3.6, we have that V_i is the age of the ith female diabetic on 1 July 1973. Furthermore, $\tilde{X}_i$ is her age at death if she is observed to die before 1 January 1982; otherwise, it is her age at the date of no further follow-up (1 January 1982 or date of emigration). We will consider a model where the death intensity for the female diabetics is of the Gompertz–Makeham form, i.e.,

$$\alpha(t; \boldsymbol{\theta}) = a + bc^t,$$

with $\boldsymbol{\theta} = (a, b, c)$, a model which is often used for modelling human mortality at adult ages. (Only two of the female diabetics are observed to die at an age below 30 years.) Then the likelihood equations take the form

$$\sum_{i=1}^{n} \frac{D_i}{a + bc^{\tilde{X}_i}} - \sum_{i=1}^{n} (\tilde{X}_i - V_i) = 0,$$

$$\sum_{i=1}^{n} \frac{D_i c^{\tilde{X}_i}}{a + bc^{\tilde{X}_i}} - \frac{1}{\log c} \sum_{i=1}^{n} (c^{\tilde{X}_i} - c^{V_i}) = 0,$$

$$\sum_{i=1}^{n} \frac{D_i \tilde{X}_i b c^{\tilde{X}_i - 1}}{a + bc^{\tilde{X}_i}} + \frac{b}{c(\log c)^2} \sum_{i=1}^{n} (c^{\tilde{X}_i} - c^{V_i}) - \frac{b}{\log c} \sum_{i=1}^{n} (\tilde{X}_i c^{\tilde{X}_i - 1} - V_i c^{V_i - 1}) = 0.$$

These equations have to be solved by an iterative procedure.

For the data on female diabetics, we have subtracted 50 years from the $\tilde{X}_i$ and the V_i before fitting the model, corresponding to the reparametrization

$$\alpha(t; \boldsymbol{\theta}) = a + bc^{t-50},$$

to avoid an extremely high correlation between the estimates for b and c. We then get the maximum likelihood estimates

$$\hat{a} = -0.00245, \qquad \hat{b} = 0.0234, \qquad \hat{c} = 1.060$$

with estimated standard deviations

$$\widehat{\text{s.d.}}(\hat{a}) = 0.00326, \qquad \widehat{\text{s.d.}}(\hat{b}) = 0.0054, \qquad \widehat{\text{s.d.}}(\hat{c}) = 0.009,$$

and with estimated correlations

$$\widehat{\text{corr}}(\hat{a}, \hat{b}) = -0.920, \qquad \widehat{\text{corr}}(\hat{a}, \hat{c}) = 0.880, \qquad \widehat{\text{corr}}(\hat{b}, \hat{c}) = -0.948.$$

The constant a in the Gompertz–Makeham formula is far from being significantly different from zero (likelihood ratio test statistic 0.59, d.f. = 1, $P = .44$), so that the Gompertz model

$$\alpha(t; \boldsymbol{\theta}) = bc^{t-50}$$

gives an equally good fit to the data. For this model, the estimates become, with estimated standard deviations in brackets,

$$\hat{b} = 0.0199\ (0.0023) \qquad \hat{c} = 1.066\ (0.005)$$

and with estimated correlation

$$\widehat{\text{corr}}(\hat{b}, \hat{c}) = -0.820.$$

The estimated value of c is for both models around 1.06 which is a lower value than that found in the general Danish female population where c is around 1.10 based on official vital statistics (cf. Figure IV.1.8). This reflects the fact that the relative mortality of diabetics is decreasing with age.

We return to an assessment of the goodness of fit of the Gompertz model in Examples VI.3.3, VI.3.7, and VI.3.9. M-estimators for the Gompertz–Makeham model are discussed in Example VI.2.3. □

Example VI.1.5. Excess and Relative Mortality. Survival with Malignant Melanoma

In Examples III.1.3 and III.1.8, we introduced models for relative and excess mortality. These models were further studied in Example IV.1.11, where we considered nonparametric inference. Here, we will consider parametric inference for models where the relative mortality and the excess mortality are assumed to be constant.

Let us first consider the model for constant *relative* mortality. Then we have individual counting processes $N_i(t)$, $i = 1, \ldots, n$, with intensity processes

$$\lambda_i(t; \theta) = \theta \mu_i(t) Y_i(t),$$

with $\mu_i(t)$ known and $Y_i(t)$ an indicator for the ith individual to be at risk at $t-$. As noted in Example III.1.3, the aggregated process $N = \sum_{i=1}^{n} N_i$ then satisfies the multiplicative intensity model (6.1.8) with

$$\lambda(t; \theta) = \theta Y^{\mu}(t),$$

where

$$Y^{\mu}(t) = \sum_{i=1}^{n} \mu_i(t) Y_i(t)$$

as in Example IV.1.11. Thus, the present situation is only a slight generalization of Example VI.1.2, and it follows that the maximum likelihood estimator for θ is

$$\hat{\theta} = N(\tau) \Big/ \int_0^\tau Y^\mu(t)\,\mathrm{d}t.$$

This estimator for the relative mortality is asymptotically normally distributed around the true parameter value and with an asymptotic variance that may be estimated by $\hat{\theta}^2/N(\tau)$.

For the melanoma data introduced in Example I.3.1, we saw in Figure IV.1.9 that it seemed reasonable to consider a model with a constant relative mortality when follow-up time was considered the basic time scale (but not if age were considered the basic time scale, cf. Figure IV.1.11). The maximum likelihood estimate for θ is

$$\hat{\theta} = \frac{71}{21.24} = 3.34$$

with an estimated standard deviation of

$$\widehat{\text{s.d.}}(\hat{\theta}) = \frac{\hat{\theta}}{\sqrt{N(\tau)}} = \frac{3.34}{\sqrt{71}} = 0.40.$$

(Note that $\hat{\theta}$ is the same in the two different models specified by taking the basic time scale to be follow-up time or to be age.) The value of $\hat{\theta}$ is close to the approximate slope that could be read off Figure IV.1.9. Both the Wald test statistic and the likelihood ratio test statistic for the hypothesis $\theta = 1$ of identical mortalities for the melanoma patients and the general Danish population are highly significant. In particular, the Wald test statistic takes the value $(\hat{\theta} - 1)/\widehat{\text{s.d.}}(\hat{\theta}) = 5.90$.

For the related model with constant *excess* mortality, we have intensity process

$$\lambda_i(t;\gamma) = \{\gamma + \mu_i(t)\}\, Y_i(t),$$

for the ith individual counting process N_i. It follows that the intensity process for the aggregated process N takes the form

$$\lambda(t;\gamma) = \gamma Y(t) + Y^\mu(t),$$

where as usual $Y = \sum_{i=1}^n Y_i$. Thus, the aggregated process has an intensity process which is not of the multiplicative form (6.1.8). As noted just below (6.1.8), we do not have the simple sufficiency reduction of the former examples, and the inference should be based on the individual counting processes N_i. By (6.1.3), the maximum likelihood estimator $\hat{\gamma}$ is given as the solution to the equation

$$\int_0^\tau \sum_{i=1}^n \frac{\mathrm{d}N_i(t)}{\hat{\gamma} + \mu_i(t)} - \int_0^\tau Y(t)\,\mathrm{d}t = 0.$$

This equation has to be solved by an iterative procedure. The estimator $\hat{\gamma}$ for the excess mortality is asymptotically normally distributed with the proper expectation and with an asymptotic variance that according to (6.1.4) may be estimated by

$$\hat{\sigma}^2 = \left\{ \int_0^\tau \sum_{i=1}^n \frac{\mathrm{d}N_i(t)}{(\hat{\gamma} + \mu_i(t))^2} \right\}^{-1}.$$

For the melanoma data (introduced in Example I.3.1), a model with a constant excess mortality seemed acceptable both compared to a model where the excess mortality depends on follow-up time and compared to a model where it depends on age; cf. Figures IV.1.10 and IV.1.12. The maximum likelihood estimate for γ is in both cases

$$\hat{\gamma} = 0.0400 \text{ (per year)}$$

with an estimated standard deviation

$$\hat{\sigma} = 0.0066 \text{ (per year).}$$

Again both the Wald test statistic and the likelihood ratio test statistic for the hypothesis of identical mortalities for the melanoma patients and the general Danish population ($\gamma = 0$) are highly significant. Thus, $\hat{\gamma}/\hat{\sigma} = 6.10$ for the Wald test statistic.

A model generalizing both of the above models is the one for which the N_i have intensity processes of the form

$$\lambda_i(t; \gamma, \theta) = \{\gamma + \theta\mu_i(t)\} Y_i(t).$$

This model is, in fact, a special case of the linear parametric regression models of Example VII.6.2. Again the inference has to be based on the individual processes, and it is seen by (6.1.3) that the maximum likelihood estimators are given as solutions to the equations

$$\int_0^\tau \sum_{i=1}^n \frac{\mathrm{d}N_i(t)}{\hat{\gamma} + \hat{\theta}\mu_i(t)} - \int_0^\tau Y(t)\,\mathrm{d}t = 0,$$

$$\int_0^\tau \sum_{i=1}^n \frac{\mu_i(t)\,\mathrm{d}N_i(t)}{\hat{\gamma} + \hat{\theta}\mu_i(t)} - \int_0^\tau Y^\mu(t)\,\mathrm{d}t = 0.$$

It follows from these equations that the maximum likelihood estimators satisfy the simple equation

$$N(\tau) = \hat{\gamma}R(\tau) + \hat{\theta}E(\tau),$$

where $R(t) = \int_0^t Y(u)\,\mathrm{d}u$ and $E(t) = \int_0^t Y^\mu(u)\,\mathrm{d}u$. The estimates must be found by an iterative procedure. An estimate of the asymptotic covariance matrix of $\hat{\gamma}$ and $\hat{\theta}$ may be found in the usual manner using (6.1.4).

For this linear model, the maximum likelihood estimates for the melanoma data are $\hat{\theta} = 1.19$ (0.38) and $\hat{\gamma} = 0.0378$ (0.0077) (per year) with an estimated correlation of -0.54. These estimates do not depend on the choice of time scale either. Here, the hypothesis $\theta = 1$ can be accepted (likelihood ratio test statistic 0.26, d.f. = 1, $P = .61$), but the hypothesis $\gamma = 0$ cannot (likelihood ratio test statistic 52.3). Thus, within the linear model the simple additive model is acceptable, but not the simple multiplicative one.

M-estimators for the excess model and the linear model are studied in Example VI.2.4. In Example VII.4.3, we shall return to a discussion of a nonparametric linear intensity model obtained by replacing γ and θ in the parametric linear model by unknown time-varying functions. □

EXAMPLE VI.1.6. The Dynamics of a Natural Baboon Troop. Maximum Likelihood Estimation

To describe the data presented in Example I.3.18, we suggested in Example III.1.6 the Markov birth–immigration–death–emigration process $\{X(t): 0 \le t \le \tau\}$, with $X(t)$ specifying the size of the troop at time t, and with constant intensities λ (birth), μ (death), ρ (emigration), and ν (immigration). We noted that this corresponds to the multivariate counting process $\mathbf{N}(t) = (N_1(t), N_2(t), N_3(t), N_4(t))$ counting the number of births, deaths, emigrations, and immigrations in $[0, t]$, respectively. This is a multiplicative intensity model with the constant intensity functions λ, μ, ρ, and ν. The likelihood based on complete observation over $[0, \tau]$ is

$$\lambda^{N_1(\tau)}\mu^{N_2(\tau)}\rho^{N_3(\tau)}\nu^{N_4(\tau)}e^{-(\lambda+\mu+\rho)R(\tau)-\nu\tau},$$

where $R(t) = \int_0^t X(u)\,du$, i.e., total time at risk. The maximum likelihood estimates are the occurrence/exposure rates

$$\begin{aligned}
\hat{\lambda} &= N_1(\tau)/R(\tau), & \widehat{\text{s.d.}}(\hat{\lambda}) &= \hat{\lambda}/N_1(\tau)^{1/2},\\
\hat{\mu} &= N_2(\tau)/R(\tau), & \widehat{\text{s.d.}}(\hat{\mu}) &= \hat{\mu}/N_2(\tau)^{1/2},\\
\hat{\rho} &= N_3(\tau)/R(\tau), & \widehat{\text{s.d.}}(\hat{\rho}) &= \hat{\rho}/N_3(\tau)^{1/2},\\
\hat{\nu} &= N_4(\tau)/\tau, & \widehat{\text{s.d.}}(\hat{\nu}) &= \hat{\nu}/N_4(\tau)^{1/2},
\end{aligned}$$

and $\hat{\lambda}$, $\hat{\mu}$, $\hat{\rho}$, and $\hat{\nu}$ are asymptotically independent; cf. Example VI.1.1. For these data, we have, for $\tau = 373$ days,

$$N_1(\tau) = 10, \qquad N_2(\tau) = 12, \qquad N_3(\tau) = 3, \qquad N_4(\tau) = 3, \qquad R(\tau) = 15407,$$

that is (per 10^4 days), with standard deviations in brackets,

$$\hat{\lambda} = (10/15407) \times 10^4 = 6.49\ (2.05)$$

$$\hat{\mu} = 7.79\ (2.25), \qquad \hat{\rho} = 1.95\ (1.13), \qquad \hat{\nu} = 80.0\ (46.2).$$

We return to assessments of the goodness of fit in Examples VI.3.2 and VI.3.4. □

EXAMPLE VI.1.7. Abnormal Prothrombin Levels in Liver Cirrhosis. Models with Piecewise Constant Transition Intensities

As an example of a Markov process model with piecewise constant intensities, we shall now return to the prothrombin data introduced in Example

Table VI.1.4. Prothrombin Data: Estimates (per Year) in Model with Piecewise Constant Transition Intensities (Estimated Standard Deviations in Brackets)

	Prednisone		Placebo	
Transition	0–1 Year	1 Year and Up	0–1 Year	1 Year and Up
From low to normal ($1 \to 0$)	1.32 (0.14)	0.45 (0.056)	0.82 (0.095)	0.44 (0.049)
From normal to low ($0 \to 1$)	0.35 (0.052)	0.15 (0.016)	0.48 (0.067)	0.19 (0.020)
From low to dead ($1 \to 2$)	0.66 (0.097)	0.31 (0.046)	0.42 (0.067)	0.32 (0.041)
From normal to dead ($0 \to 2$)	0.087 (0.026)	0.069 (0.011)	0.12 (0.033)	0.085 (0.013)

I.3.12. The Nelson–Aalen plots in Figures IV.4.10 and IV.4.11 indicate that the transition intensities might well be considered constant or perhaps rather piecewise constant with a breakpoint after about 1 year of treatment. As explained in Example VI.1.1 the maximum likelihood estimates in such a model are simple occurrence/exposure rates and these are shown in Table VI.1.4. The estimates in the simpler model assuming all intensities to be *constant* are shown in Table I.3.7, but for both treatment groups, the likelihood ratio test statistic for the reduction to that model is highly significant (72.77 for prednisone and 52.25 for placebo, both with d.f. = 4). Judging from Table VI.1.4 the intensities seem to *decrease* with time. Comparing the intensities in the two treatment groups, the tendencies are the same as seen in Example IV.4.4: The transition intensities between the two transient states and that from "normal" to "dead" suggest that prednisone treatment has a beneficial effect, whereas prednisone increases the death intensity from the state "low." To summarize these effects, it is, therefore, of interest to calculate the *transition probabilities* $P_{hj}(s,t)$, $s < t$, h, $j = 0, 1, 2$, based on the parametric model. When $t \leq 1$ year or $s \geq 1$ year, the $P_{hj}(s,t)$ are given by simple closed form functions of the transition intensities (e.g., Chiang, 1968, Section 4.2), and when $s < 1$ year $< t$, the Chapman–Kolmogorov equations

$$P_{hj}(s,t) = \sum_{l=0,1,2} P_{hl}(s,1)P_{lj}(1,t)$$

and the simple formulas for $P_{hl}(s,1)$ and $P_{lj}(1,t)$ may be applied. As an example, Figure VI.1.1 shows $\hat{P}_{02}(0,t)$ for both treatment groups together with the corresponding Aalen–Johansen estimates (see also Figure IV.4.12). It is seen that the parametric and the nonparametric estimates are very close and that the probability of having died by time t is lower for the prednisone-treated patients. □

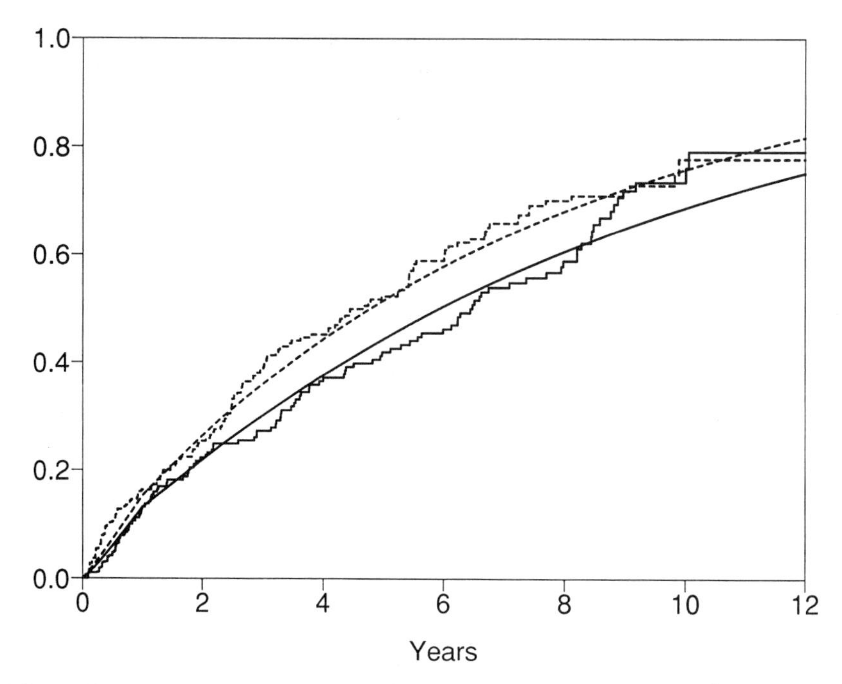

Figure VI.1.1. Prothrombin data. Estimated transition probabilities $\hat{P}_{02}(0, t)$ based on a nonparametric analysis and on a model with piecewise constant transition intensities. Prednisone (—); placebo (- - -).

EXAMPLE VI.1.8. Software Reliability. Maximum Likelihood Estimation

In Example III.1.12, two models for the study of software reliability were introduced. In this example, we discuss maximum likelihood estimation for these models and present an analysis of the data introduced in Example I.3.19.

For the Jelinski–Moranda model, the counting process N, counting the number of failures of the system, has an intensity process of the form

$$\lambda(t) = \rho(F - N(t-)),$$

where the failure rate per fault, ρ, and the initial number of faults in the system, F, are the unknown parameters. Although F can only take integer values, we choose to treat it as a continuously varying parameter in the following. By (6.1.3), the maximum likelihood estimators $\hat{\rho}$ and $\hat{F}$ are then given as the solutions to the equations

$$N(\tau)/\rho - \int_0^\tau (F - N(t-))\,dt = 0$$

and

$$\int_0^\tau (F - N(t-))^{-1}\,\mathrm{d}N(t) - \rho\tau = 0.$$

Thus, $\hat{F}$ is given by

$$\int_0^\tau \frac{\mathrm{d}N(t)}{\hat{F} - N(t-)} - \frac{\tau N(\tau)}{\int_0^\tau (\hat{F} - N(t-))\,\mathrm{d}t} = 0,$$

whereas

$$\hat{\rho} = \frac{N(\tau)}{\int_0^\tau (\hat{F} - N(t-))\,\mathrm{d}t}$$

is of the occurrence/exposure type, but with estimated "exposure." By the results reviewed earlier (see also Example VI.1.12), $(\hat{\rho}, \hat{F})$ is approximately bivariate nomally distributed around the true parameter value with a covariance matrix which may be estimated by $\hat{\mathscr{I}}_\tau^{-1}$, where, according to (6.1.4),

$$\hat{\mathscr{I}}_\tau = \begin{pmatrix} N(\tau)/\hat{\rho}^2 & \tau \\ \tau & \int_0^\tau (F - N(t-))^{-2}\,\mathrm{d}N(t) \end{pmatrix}.$$

For the data presented in Example I.3.19, the maximum likelihood estimates become $\hat{\rho} = 5.40$ (per Ms CPU-time) and $\hat{F} = 44.5$ when $\tau = 0.57657$ Ms is chosen as the time of the last (i.e., 43rd) observed failure. The estimated standard deviations are $\widehat{\text{s.d.}}(\hat{\rho}) = 1.14$ (per ms) and $\widehat{\text{s.d.}}(\hat{F}) = 1.98$, and the estimated correlation is 0.69.

We then consider the Littlewood model. For this model, the intensity process has the form

$$\lambda(t) = \frac{\alpha(F - N(t-))}{\beta + t} = \frac{F - N(t-)}{\mu + \gamma t},$$

where $\mu = \beta/\alpha$ and $\gamma = 1/\alpha$. For the data of Example I.3.19, we found (using a downhill simplex method) the maximum likelihood estimates $\hat{\mu} = 0.185$, $\hat{\gamma} = 0.000$, and $\hat{F} = 44.5$ (corresponding to the Jelinski–Moranda model) when the likelihood was maximized subject to the constraints $\mu > 0$, $\gamma \geq 0$, and $F \geq 43$ and CPU-time was measured in Ms. (As above, $\tau = 0.57657$ Ms was taken as the time of the 43rd failure.) In Example III.1.12, we noted, however, that the reparametrized Littlewood model remains valid for negative values of γ satisfying $\gamma \geq -\mu/\tau$. Maximizing the likelihood on the larger parameter space given by $\mu > 0$, $\gamma \geq -\mu/\tau$, and $F \geq 43$, we obtained the maximum likelihood estimates $\hat{\mu} = 0.209$, $\hat{\gamma} = -0.255$, and $\hat{F} = 43.0$.

For both of these situations, the maximum likelihood estimates for the Littlewood model are found on the boundary of the parameter space. Therefore, the usual large sample estimates for standard deviations and correlations do not apply. Also, the approximate χ^2 distribution of the likelihood ratio test statistic for testing the hypothesis that $\gamma = 0$, corresponding to the Jelinski–Moranda model, is not justified. Nevertheless, the value 156.86 for the maximized log-likelihood for the Littlewood model, compared to the value 156.39 for the Jelinski–Moranda model, indicates a satisfactory fit of the latter.

In Example VI.2.5, we present M-estimators for the Littlewood model.

□

VI.1.2. Large Sample Properties

We shall derive the large sample properties of the maximum likelihood estimator $\hat{\boldsymbol{\theta}}$ as well as the test statistics mentioned in the preceding subsection. To avoid confusion we will now denote by $\boldsymbol{\theta}_0 = (\theta_{10}, \dots, \theta_{q0})$ the true value of the parameter and reserve $\boldsymbol{\theta}$ for the free parameter in the log-likelihood function (6.1.2), the score function (6.1.3), etc. Consider therefore a sequence of counting processes $\mathbf{N}^{(n)} = (N_1^{(n)}, \dots, N_{k_n}^{(n)})$; $n = 1, 2, \dots$; with intensity processes $\boldsymbol{\lambda}^{(n)} = (\lambda_1^{(n)}, \dots, \lambda_{k_n}^{(n)})$ of the parametric form $\lambda_h^{(n)}(t) = \lambda_h^{(n)}(t; \boldsymbol{\theta}_0)$; $h = 1, 2, \dots, k_n$; where $\boldsymbol{\theta}_0$ is assumed to be the same for all n. Note that we do allow the dimension k_n of the counting process to depend on n. This is done in order to make the results general enough to cover the models of Example VI.1.5 and more generally the parametric regression models of Section VII.6. But for most applications of the general theory in the present chapter we have $k_n = k$, i.e. independent of n. Of course the counting process and its intensity process as well as the log-likelihood function (6.1.2) and its derivatives with respect to $\boldsymbol{\theta}$ now depend on n, but in order to simplify the notation we will not make this dependence explicit in the following.

We do not aim at proving the results under a minimal set of conditions. Rather we state sufficient conditions in the spirit of Cramér (1945) which are general enough to cover the applications we have in mind in the present chapter as well as in Sections VII.6 and VIII.2.2. Our conditions and proofs follow the lines of the paper by Borgan (1984), who studied the maximum likelihood estimator for the multiplicative intensity model (6.1.8). We adopt the notation $(\partial/\partial\theta_j)g(\boldsymbol{\theta}_0)$ for $(\partial/\partial\theta_j)g(\boldsymbol{\theta})|_{\boldsymbol{\theta}=\boldsymbol{\theta}_0}$, and state the following condition.

Condition VI.1.1.

(A) There exists a neighborhood Θ_0 of $\boldsymbol{\theta}_0$ such that for all n, h and $\boldsymbol{\theta} \in \Theta_0$, and almost all $t \in \mathscr{T}$, the partial derivatives of $\lambda_h(t; \boldsymbol{\theta})$ and $\log \lambda_h(t; \boldsymbol{\theta})$ of the first, second, and third order with respect to $\boldsymbol{\theta}$ exist and are continuous in $\boldsymbol{\theta}$ for $\boldsymbol{\theta} \in \Theta_0$. Moreover, the log-likelihood function (6.1.2) may be differentiated three times with respect to $\boldsymbol{\theta} \in \Theta_0$ by interchanging the order of integration and differentiation.

(B) There exist a sequence $\{a_n\}$ of non-negative constants increasing to infinity as $n \to \infty$ and finite functions $\sigma_{jl}(\boldsymbol{\theta})$ defined on Θ_0 such that for all j, l

$$a_n^{-2} \int_0^\tau \sum_{h=1}^{k_n} \left\{ \frac{\partial}{\partial \theta_j} \log \lambda_h(t; \boldsymbol{\theta}_0) \right\} \left\{ \frac{\partial}{\partial \theta_l} \log \lambda_h(t; \boldsymbol{\theta}_0) \right\} \lambda_h(t; \boldsymbol{\theta}_0)\, \mathrm{d}t \xrightarrow{\mathrm{P}} \sigma_{jl}(\boldsymbol{\theta}_0)$$

as $n \to \infty$.

(C) For all h, j and all $\varepsilon > 0$, we have that

$$a_n^{-2} \int_0^\tau \sum_{h=1}^{k_n} \left\{ \frac{\partial}{\partial \theta_j} \log \lambda_h(s; \boldsymbol{\theta}_0) \right\}^2 I\left(\left| a_n^{-1} \frac{\partial}{\partial \theta_j} \log \lambda_h(s; \boldsymbol{\theta}_0) \right| > \varepsilon \right) \lambda_h(s; \boldsymbol{\theta}_0)\, \mathrm{d}s \xrightarrow{\mathrm{P}} 0$$

as $n \to \infty$.

(D) The matrix $\boldsymbol{\Sigma} = \{\sigma_{jl}(\boldsymbol{\theta}_0)\}$ with $\sigma_{jl}(\boldsymbol{\theta}_0)$ defined in Condition B is positive definite.

(E) For any n and each $h = 1, 2, \ldots, k_n$ there exist predictable processes G_{hn} and H_{hn} not depending on $\boldsymbol{\theta}$ such that for all $t \in \mathscr{T}$

$$\sup_{\boldsymbol{\theta} \in \Theta_0} \left| \frac{\partial^3}{\partial \theta_j\, \partial \theta_l\, \partial \theta_m} \lambda_h(t, \boldsymbol{\theta}) \right| \leq G_{hn}(t),$$

and

$$\sup_{\boldsymbol{\theta} \in \Theta_0} \left| \frac{\partial^3}{\partial \theta_j\, \partial \theta_l\, \partial \theta_m} \log \lambda_h(t, \boldsymbol{\theta}) \right| \leq H_{hn}(t),$$

for all j, l, m. Moreover

$$a_n^{-2} \int_0^\tau \sum_{h=1}^{k_n} G_{hn}(t)\, \mathrm{d}t, \qquad a_n^{-2} \int_0^\tau \sum_{h=1}^{k_n} H_{hn}(t) \lambda_h(t; \boldsymbol{\theta}_0)\, \mathrm{d}t$$

as well as (for all j, l)

$$a_n^{-2} \int_0^\tau \sum_{h=1}^{k_n} \left\{ \frac{\partial^2}{\partial \theta_j\, \partial \theta_l} \log \lambda_h(t; \boldsymbol{\theta}_0) \right\}^2 \lambda_h(t; \theta_0)\, \mathrm{d}t$$

all converge in probability to finite quantities as $n \to \infty$, and, for all $\varepsilon > 0$,

$$a_n^{-2} \int_0^\tau \sum_{h=1}^{k_n} H_{hn}(t) I(a_n^{-1}(H_{hn}(t))^{1/2} > \varepsilon) \lambda_h(t; \boldsymbol{\theta}_0)\, \mathrm{d}t \xrightarrow{\mathrm{P}} 0.$$

Before we embark on deriving the statistical properties of the maximum likelihood estimator $\hat{\boldsymbol{\theta}}$, let us briefly comment upon these conditions. As will become evident later, the essential point in the derivation of the large sample properties of $\hat{\boldsymbol{\theta}}$ is that the score statistics (6.1.3) evaluated at $\boldsymbol{\theta}_0$ are the stochastic integrals $\int_0^\tau \sum_h (\partial/\partial\theta_j) \log \lambda_h(s; \boldsymbol{\theta}_0)\, \mathrm{d}M_h(s)$, where $M_h(t) = N_h(t) - \int_0^t \lambda_h(s; \boldsymbol{\theta}_0)\, \mathrm{d}s$; cf. (2.7.11) and (6.1.13). Conditions B and C are just condition (2.5.1) on convergence in probability of the predictable covariation processes and the Lindeberg-type condition (2.5.3) for proving weak convergence of (normalized versions of) these stochastic integrals using the martingale central limit theorem (Theorem II.5.1). Moreover, to convert the asymptotic normality of the stochastic integrals into a result on the asymptotic normality of $\hat{\boldsymbol{\theta}}$, Taylor expansions are used in exactly the same manner as for the classical i.i.d. case. By Condition A, such Taylor expansions are valid, whereas Condition E ensures the remainder terms in these Taylor expansions to behave properly.

Furthermore, note that the assumption in Condition B also may be written

$$a_n^{-2} \int_0^\tau \sum_{h=1}^{k_n} \frac{\partial^2}{\partial\theta_j\,\partial\theta_l} \lambda_h(t;\boldsymbol{\theta}_0)\,dt - a_n^{-2} \int_0^\tau \sum_{h=1}^{k_n} \left\{ \frac{\partial^2}{\partial\theta_j\,\partial\theta_l} \log \lambda_h(t;\boldsymbol{\theta}_0) \right\} \lambda_h(t;\boldsymbol{\theta}_0)\,dt \xrightarrow{\mathrm{P}} \sigma_{jl}(\boldsymbol{\theta}_0)$$

as $n \to \infty$. Therefore, the intuitive content of the conditions is that $a_n^{-2} \sum_{h=1}^{k_n} \lambda_h(t;\boldsymbol{\theta})$ and $a_n^{-2} \sum_{h=1}^{k_n} \log \lambda_h(t;\boldsymbol{\theta})\lambda_h(t;\boldsymbol{\theta}_0)$ both converge (sufficiently uniformly) in probability to nice functions and also that their partial derivatives with respect to $\boldsymbol{\theta}$ of the first, second, and third order converge in probability to the corresponding partial derivatives of the limiting functions. For the multiplicative intensity model (6.1.8) (with $k_n = k$ fixed), this essentially amounts to requiring that the $\alpha_h(t;\boldsymbol{\theta})$ are nice, differentiable functions and that $a_n^{-2} Y_h(t)$ converges in probability to functions $y_h(t)$. Comparing this with (4.1.16), we see that the normalizing factor a_n typically will be the same as the one used in Chapters IV and V. Thus, we will usually have $a_n = \sqrt{n}$, with n the number of *individual* counting processes from which the N_h are composed; see, however, Example VI.1.12 for an instance where this is not the case.

We will return to a discussion of Condition VI.1.1 in connection with some examples later. First, we will prove that with a probability tending to one, there exists a solution of the likelihood equation and this solution is consistent. Note, however, that this does not rule out the possibility of the likelihood equation having other (inconsistent) solutions.

Theorem VI.1.1. *Assume that Condition* VI.1.1 *holds. Then, with a probability tending to one, the equation* $\mathbf{U}_\tau(\boldsymbol{\theta}) = \mathbf{0}$ *has a solution* $\hat{\boldsymbol{\theta}}$ *and* $\hat{\boldsymbol{\theta}} \xrightarrow{\mathrm{P}} \boldsymbol{\theta}_0$ *as* $n \to \infty$.

Proof. By a Taylor series expansion, we get, for any $\boldsymbol{\theta} \in \Theta_0$,

$$U_\tau^j(\boldsymbol{\theta}) = U_\tau^j(\boldsymbol{\theta}_0) - \sum_{l=1}^{q} (\theta_l - \theta_{l0})\mathscr{I}_\tau^{jl}(\boldsymbol{\theta}_0) + \tfrac{1}{2} \sum_{l=1}^{q} \sum_{m=1}^{q} (\theta_l - \theta_{l0})(\theta_m - \theta_{m0}) R_\tau^{jlm}(\boldsymbol{\theta}^*),$$

where $\boldsymbol{\theta}^*$ is on the line segment joining $\boldsymbol{\theta}$ and $\boldsymbol{\theta}_0$, and

$$R_\tau^{jlm}(\boldsymbol{\theta}) = \frac{\partial^3 C_\tau(\boldsymbol{\theta})}{\partial\theta_j\,\partial\theta_l\,\partial\theta_m} = \int_0^\tau \sum_{h=1}^{k_n} \frac{\partial^3}{\partial\theta_j\,\partial\theta_l\,\partial\theta_m} \log \lambda_h(t;\boldsymbol{\theta})\,dN_h(t) - \int_0^\tau \sum_{h=1}^{k_n} \frac{\partial^3}{\partial\theta_j\,\partial\theta_l\,\partial\theta_m} \lambda_h(t;\boldsymbol{\theta})\,dt. \tag{6.1.9}$$

We need to show that

$$a_n^{-2} U_\tau^j(\boldsymbol{\theta}_0) \xrightarrow{\mathrm{P}} 0 \tag{6.1.10}$$

and

$$a_n^{-2} \mathscr{I}_\tau^{jl}(\boldsymbol{\theta}_0) \xrightarrow{\mathrm{P}} \sigma_{jl}(\boldsymbol{\theta}_0) \tag{6.1.11}$$

as $n \to \infty$, and that there exists a finite constant M, not depending on $\boldsymbol{\theta}$, such that

$$\lim_{n\to\infty} \mathrm{P}(|a_n^{-2} R_\tau^{jlm}(\boldsymbol{\theta})| < M \text{ for all } j, l, m, \text{ and all } \boldsymbol{\theta} \in \Theta_0) = 1. \tag{6.1.12}$$

From (6.1.10)–(6.1.12), the theorem will follow by the same argument as is used for proving consistency for the maximum likelihood estimator in the case of i.i.d. random variables and in other situations as well. This argument was given in detail by Billingsley (1961, pp. 12 and 13).

We now introduce the notation $C_t(\boldsymbol{\theta})$, $U_t^j(\boldsymbol{\theta})$, $\mathscr{I}_t^{jl}(\boldsymbol{\theta})$, and $R_t^{jlm}(\boldsymbol{\theta})$ for the stochastic processes we get when τ is replaced by t in (6.1.2), (6.1.3), (6.1.4), and (6.1.9), respectively.

Let us first prove (6.1.10). By (2.4.1) and (2.4.4), (6.1.3) evaluated at the true parameter $\boldsymbol{\theta}_0$ equals

$$U_t^j(\boldsymbol{\theta}_0) = \int_0^t \sum_{h=1}^{k_n} \frac{\partial}{\partial \theta_j} \log \lambda_h(s; \boldsymbol{\theta}_0)\, \mathrm{d}M_h(s), \tag{6.1.13}$$

where

$$M_h(t) = N_h(t) - \int_0^t \lambda_h(s; \boldsymbol{\theta}_0)\, \mathrm{d}s, \tag{6.1.14}$$

$h = 1, \ldots, k_n$, by (2.4.3) are orthogonal local square integrable martingales with predictable variation processes

$$\langle M_h \rangle(t) = \int_0^t \lambda_h(s; \boldsymbol{\theta}_0)\, \mathrm{d}s. \tag{6.1.15}$$

By Condition B, the integral

$$a_n^{-2} \int_0^\tau \sum_{h=1}^{k_n} \left\{ \frac{\partial}{\partial \theta_j} \log \lambda_h(s; \boldsymbol{\theta}_0) \right\}^2 \mathrm{d}\langle M_h \rangle(s)$$

is finite with a probability tending to one as n increases. Thus, it suffices to prove (6.1.10) assuming this integral to be finite. But then, by Theorem II.3.1, $a_n^{-1} U_t^j(\boldsymbol{\theta}_0)$ is a stochastic integral and, therefore, a local square integrable martingale. By (6.1.15) and (2.4.9), it has predictable variation process

$$a_n^{-2} \langle U^j(\boldsymbol{\theta}_0) \rangle(t) = a_n^{-2} \int_0^t \sum_{h=1}^{k_n} \left\{ \frac{\partial}{\partial \theta_j} \log \lambda_h(s; \boldsymbol{\theta}_0) \right\}^2 \lambda_h(s; \boldsymbol{\theta}_0)\, \mathrm{d}s.$$

By Condition B, $a_n^{-2} \langle U^j(\boldsymbol{\theta}_0) \rangle(\tau) \xrightarrow{\mathrm{P}} \sigma_{jj}(\boldsymbol{\theta}_0)$ as $n \to \infty$. Therefore, an application of version (2.5.18) of Lenglart's inequality gives that, for all $\delta, \eta > 0$, we have

$$\mathrm{P}\left(\sup_{t\in\mathscr{T}}|a_n^{-2}U_t^j(\boldsymbol{\theta}_0)| > \eta\right) \le \frac{\delta}{\eta^2} + \mathrm{P}(a_n^{-4}\langle U^j(\boldsymbol{\theta}_0)\rangle(\tau) > \delta), \quad (6.1.16)$$

which proves (6.1.10).

To prove (6.1.11), note that, by (6.1.4) and (6.1.14), we may write

$$a_n^{-2}\mathscr{I}_\tau^{jl}(\boldsymbol{\theta}_0) = a_n^{-2}\int_0^\tau \sum_{h=1}^{k_n}\left\{\frac{\partial}{\partial\theta_j}\log\lambda_h(s;\boldsymbol{\theta}_0)\right\}\left\{\frac{\partial}{\partial\theta_l}\log\lambda_h(s;\boldsymbol{\theta}_0)\right\}\lambda_h(s;\boldsymbol{\theta}_0)\,\mathrm{d}s$$

$$-a_n^{-2}\int_0^\tau \sum_{h=1}^{k_n}\frac{\partial^2}{\partial\theta_j\,\partial\theta_l}\log\lambda_h(s;\boldsymbol{\theta}_0)\,\mathrm{d}M_h(s).$$

By Condition E and an argument similar to the one used to prove (6.1.10), it follows that the last of these terms converges to zero in probability. By Condition B, (6.1.11) is then an immediate consequence.

Finally, to prove (6.1.12), we note that (6.1.9) and Condition E give, for any n and all j, l, m and all $\boldsymbol{\theta}\in\Theta_0$,

$$|a_n^{-2}R_\tau^{jlm}(\boldsymbol{\theta})| \le a_n^{-2}\int_0^\tau\sum_{h=1}^{k_n}H_{hn}(s)\,\mathrm{d}N_h(s) + a_n^{-2}\int_0^\tau\sum_{h=1}^{k_n}G_{hn}(s)\,\mathrm{d}s.$$

Here, the last term converges in probability to a finite quantity by Condition E, whereas the first one by (2.4.9) is the optional variation process (evaluated at τ) of the (with a probability tending to one) local square integrable martingale

$$a_n^{-1}\int_0^{\cdot}\sum_{h=1}^{k_n}(H_{hn}(s))^{1/2}\,\mathrm{d}M_h(s).$$

This martingale has predictable variation process

$$a_n^{-2}\int_0^{\cdot}\sum_{h=1}^{k_n}H_{hn}(s)\lambda_h(s;\boldsymbol{\theta}_0)\,\mathrm{d}s.$$

By Theorem II.5.1, the optional and predictable variation processes have the same limits in probability, so it follows by Condition E that

$$a_n^{-2}\int_0^\tau\sum_{h=1}^{k_n}H_{hn}(s)\,\mathrm{d}N_h(s)$$

also converges in probability to some finite quantity as $n\to\infty$, and (6.1.12) follows. □

We will also prove the following result about the asymptotic distribution of the maximum likelihood estimator.

Theorem VI.1.2. *Assume that Condition* VI.1.1 *holds, and let* $\hat{\boldsymbol{\theta}}$ *be a consistent solution of the equation* $\mathbf{U}_\tau(\boldsymbol{\theta}) = \mathbf{0}$. *Then*

$$a_n(\hat{\boldsymbol{\theta}} - \boldsymbol{\theta}_0) \xrightarrow{\mathcal{D}} \mathcal{N}(\mathbf{0}, \boldsymbol{\Sigma}^{-1}),$$

where $\boldsymbol{\Sigma} = \{\sigma_{jl}(\boldsymbol{\theta}_0)\}$ *defined in Condition D may be estimated consistently by* $a_n^{-2}\mathcal{I}_\tau(\hat{\boldsymbol{\theta}})$.

PROOF. Taylor-expanding $U_\tau^j(\hat{\boldsymbol{\theta}})$ around $\boldsymbol{\theta}_0$ gives, when $\hat{\boldsymbol{\theta}} \in \Theta_0$,

$$0 = a_n^{-1} U_\tau^j(\hat{\boldsymbol{\theta}}) = a_n^{-1} U_\tau^j(\boldsymbol{\theta}_0) - \sum_{l=1}^{q} a_n(\hat{\theta}_l - \theta_{l0}) a_n^{-2} \mathcal{I}_\tau^{jl}(\boldsymbol{\theta}^*),$$

where $\boldsymbol{\theta}^*$ is on the line segment between $\hat{\boldsymbol{\theta}}$ and $\boldsymbol{\theta}_0$. Therefore, the result about the weak convergence of $\hat{\boldsymbol{\theta}}$ will follow by the standard argument used to prove the asymptotic distribution of the maximum likelihood estimator if we can prove that

$$a_n^{-1} \mathbf{U}_\tau(\boldsymbol{\theta}_0) \xrightarrow{\mathcal{D}} \mathcal{N}(\mathbf{0}, \boldsymbol{\Sigma}), \tag{6.1.17}$$

and

$$a_n^{-2} \mathcal{I}_\tau^{jl}(\boldsymbol{\theta}^*) \xrightarrow{\mathrm{P}} \sigma_{jl}(\boldsymbol{\theta}_0) \tag{6.1.18}$$

for all j, l and any random $\boldsymbol{\theta}^*$ such that $\boldsymbol{\theta}^* \xrightarrow{\mathrm{P}} \boldsymbol{\theta}_0$ as $n \to \infty$ [cf. Billingsley (1961, Theorems 2.2 and 10.1)].

Let us first prove (6.1.17). As in the proof of Theorem VI.1.1, we may assume that the $a_n^{-1} U_t^j(\boldsymbol{\theta}_0)$ are local square integrable martingales (since this is the case with a probability tending to one). Then, by (6.1.13), (6.1.15), and Condition B,

$$\begin{aligned} &\langle a_n^{-1} U^j(\boldsymbol{\theta}_0), a_n^{-1} U^l(\boldsymbol{\theta}_0) \rangle(\tau) \\ &\quad = a_n^{-2} \int_0^\tau \sum_{h=1}^{k_n} \left\{ \frac{\partial}{\partial \theta_j} \log \lambda_h(s; \boldsymbol{\theta}_0) \right\} \left\{ \frac{\partial}{\partial \theta_l} \log \lambda_h(s; \boldsymbol{\theta}_0) \right\} \lambda_h(s; \boldsymbol{\theta}_0) \, \mathrm{d}s \\ &\quad \xrightarrow{\mathrm{P}} \sigma_{jl}(\boldsymbol{\theta}_0) \end{aligned}$$

as $n \to \infty$, for all j, l. From this and Condition C, (6.1.17) follows by the martingale central limit theorem (Theorem II.5.1).

Let us then turn to (6.1.18). By a Taylor series expansion, we have, when $\boldsymbol{\theta}^* \in \Theta_0$,

$$a_n^{-2} \mathcal{I}_\tau^{jl}(\boldsymbol{\theta}^*) = a_n^{-2} \mathcal{I}_\tau^{jl}(\boldsymbol{\theta}_0) - a_n^{-2} \sum_{m=1}^{q} (\theta_m^* - \theta_{m0}) R_\tau^{jlm}(\check{\boldsymbol{\theta}}),$$

where $R_\tau^{jlm}(\boldsymbol{\theta})$ is defined by (6.1.9) and $\check{\boldsymbol{\theta}}$ is on the line segment joining $\boldsymbol{\theta}^*$ and $\boldsymbol{\theta}_0$. By (6.1.11), the first term converges in probability to $\sigma_{jl}(\boldsymbol{\theta}_0)$ as $n \to \infty$, whereas, by (6.1.12), the second term is bounded in probability by $qM\|\boldsymbol{\theta}^* - \boldsymbol{\theta}_0\|$ for some finite constant M not depending on $\boldsymbol{\theta}^*$. Here $\|\cdot\|$ is the ordinary q-dimensional Euclidean norm. This proves (6.1.18) and also establishes that $a_n^{-2}\mathcal{I}_\tau(\hat{\boldsymbol{\theta}})$ is a constant estimator for $\boldsymbol{\Sigma}$. □

We note that in the proofs of these two theorems the theory of counting processes is only used to prove the results (6.1.10), (6.1.11), (6.1.12), and

(6.1.17). From these four properties of the derivatives of the log-likelihood function, the theorems follow by exactly the same arguments as are used to prove the similar results for the classical case of i.i.d. random variables. Compared to the i.i.d. case, therefore, the main difference is that in the present context Lenglart's inequality is used to establish the convergence in probability results (6.1.10) and (6.1.11), instead of the law of large numbers used in the classical case, whereas we have to use the martingale central limit theorem to establish the weak convergence result (6.1.17) which in the classical case is proved by the central limit theorem for i.i.d. random variables.

Now the classical proofs of the asymptotic properties of the test statistics (6.1.5)–(6.1.7) are based solely on the above-mentioned properties of the derivatives of the log-likelihood function. We may, therefore, copy the standard argument as follows. By Theorem VI.1.2, it is immediately clear that the Wald test statistic (6.1.5) is asymptotically χ^2 distributed with q degrees of freedom when the true parameter value is $\boldsymbol{\theta}_0$. Also, by (6.1.11) and (6.1.17), the same holds for the score test statistic (6.1.6). The similar result for the likelihood ratio test statistic (6.1.7) follows by Taylor-expanding $C_\tau(\boldsymbol{\theta}_0)$ around $C_\tau(\hat{\boldsymbol{\theta}})$ (assuming $\hat{\boldsymbol{\theta}} \in \Theta_0$) and then using the fact that $U_\tau^j(\hat{\boldsymbol{\theta}}) = 0$ and (6.1.12) to get

$$2\{C_\tau(\hat{\boldsymbol{\theta}}) - C_\tau(\boldsymbol{\theta}_0)\} = \sum_{j=1}^{q} \sum_{l=1}^{q} \mathscr{I}_\tau^{jl}(\hat{\boldsymbol{\theta}})(\theta_{j0} - \hat{\theta}_j)(\theta_{l0} - \hat{\theta}_l) + o_p(1).$$

From this, it follows that the likelihood ratio test statistic is asymptotically equivalent to the Wald test statistic and, therefore, also asymptotically χ^2 with q degrees of freedom when the true parameter value is $\boldsymbol{\theta}_0$.

Also the usual asymptotic properties for the likelihood ratio test statistic for a *composite* null hypothesis may be proved as for the case of i.i.d. random variables using (6.1.10), (6.1.11), (6.1.12), and (6.1.17). For details, see, e.g., Serfling (1980, Section 4.4.4).

In connection with Theorems VI.1.1 and VI.1.2, it is important to note that if Condition VI.1.1 as it stands is not fulfilled, we still have analogous results for a time interval $[t_1, t_2] \subset \mathscr{T}$ for which the condition does hold.

We will now illustrate how Condition VI.1.1 may be checked in a few simple examples.

EXAMPLE VI.1.9. Occurrence/Exposure Rates

We will consider the asymptotic distribution of the occurrence/exposure rates of Example VI.1.1. Therefore, let $\mathbf{N}^{(n)} = (N_1^{(n)}, \ldots, N_q^{(n)})$, $n = 1, 2, \ldots$, be a sequence of q-variate counting processes (with q not depending on n) with intensity process

$$\lambda_h^{(n)}(t) = \lambda_h^{(n)}(t; \boldsymbol{\theta}_0) = \theta_{h0} Y_h^{(n)}(t)$$

for the hth component and assume that $\theta_{h0} > 0$ for all h. We suppress the dependence on n in the notation as we did above.

As in Example VI.1.1, we let

$$R_j(\tau) = \int_0^\tau Y_j(t)\,\mathrm{d}t.$$

Assume now that there exist a sequence of constants $\{a_n\}$ tending to infinity as $n \to \infty$ and constants $r_j(\tau) > 0$ such that

$$a_n^{-2} R_j(\tau) \xrightarrow{\mathrm{P}} r_j(\tau) \tag{6.1.19}$$

as $n \to \infty$ for all j, a condition stating that the "average" exposure should stabilize as n increases. (Remember that $a_n^{-2} = 1/n$ in most applications.)

Under these rather weak assumptions, Condition VI.1.1 is seen to hold as follows. Condition A follows immediately since $(\partial/\partial\theta_j)\lambda_h(t;\boldsymbol{\theta}) = \delta_{hj} Y_h(t)$, and $(\partial/\partial\theta_j)\log\lambda_h(t;\boldsymbol{\theta}) = \delta_{hj}/\theta_j$, where δ_{hj} is a Kronecker delta. That Conditions B and C hold are immediate consequences of (6.1.19), whereas Condition D is fulfilled because, here, $\sigma_{jl}(\boldsymbol{\theta}_0) = \delta_{jl} r_j(\tau)/\theta_{j0}$. For Condition E, we have, for example,

$$\sup_{\boldsymbol{\theta}\in\Theta_0}\left|\frac{\partial^3}{\partial\theta_j\,\partial\theta_l\,\partial\theta_m}\log\lambda_h(t;\boldsymbol{\theta})\right| \le \sup_{\boldsymbol{\theta}\in\Theta_0}\left|\frac{2}{\theta_h^3}\right|,$$

which is bounded by a positive constant independent of h, j, l, m (since we have assumed that all θ_{h0} are positive). The other parts of Condition E follow in a similar manner.

Thus, Condition VI.1.1 holds, and it follows by Theorem VI.1.1 that the occurrence/exposure rates

$$\hat{\theta}_j = N_j(\tau)/R_j(\tau)$$

are consistent. Moreover, since $\sigma_{jl}(\boldsymbol{\theta}_0) = \delta_{jl} r_j(\tau)/\theta_{j0}$, the matrix $\boldsymbol{\Sigma}$ defined in Condition D is diagonal, and it follows by Theorem VI.1.2 that the random vector with components $a_n(\hat{\theta}_j - \theta_{j0})$ converges in distribution to a vector with components which are independent and normally distributed with means zero and variances $\theta_{j0}/r_j(\tau)$ for $j = 1, \ldots, q$. □

EXAMPLE VI.1.10. Standardized Mortality Ratio

We will consider the one-dimensional counting process model of Example VI.1.2, for which the standardized mortality ratio (SMR) is the maximum likelihood estimator. Since we are now studying the asymptotic properties, we assume that $N^{(n)}$, $n = 1, 2, \ldots$, is a sequence of counting processes with intensity process

$$\lambda^{(n)}(t) = \lambda^{(n)}(t;\theta_0) = \theta_0\alpha_0(t)\,Y^{(n)}(t)$$

in the nth model. As earlier, we suppress the dependence on n in the following. Moreover, we assume that $\theta_0 > 0$.

As in Example VI.1.2, we define

$$E(\tau) = \int_0^\tau \alpha_0(t) Y(t)\,dt,$$

and we assume that there exist a sequence of constants $\{a_n\}$ tending to infinity as $n \to \infty$ and a constant $e(\tau) > 0$ such that

$$a_n^{-2} E(\tau) \xrightarrow{P} e(\tau) \tag{6.1.20}$$

as $n \to \infty$. This condition essentially says that the "average expected" number of events should stabilize as n increases.

Then it may be proved in a similar manner to Example VI.1.9 that Condition VI.1.1 holds. Thus, it follows by Theorems VI.1.1 and VI.1.2 that the SMR

$$\hat{\theta} = N(\tau)/E(\tau)$$

is consistent and that

$$a_n(\hat{\theta} - \theta_0) \xrightarrow{\mathscr{D}} \mathscr{N}(0, \theta_0/e(\tau))$$

as $n \to \infty$. □

Example VI.1.11. Right-Censored Survival Data

Let us consider maximum likelihood estimation for right-censored survival data. We adopt the general random censorship model (cf. Examples III.2.4, IV.1.6, and V.1.6), so we assume that for each n, $X_1^{(n)}, X_2^{(n)}, \ldots, X_n^{(n)}$ and $U_1^{(n)}, U_2^{(n)}, \ldots, U_n^{(n)}$ are mutually independent with $X_i^{(n)}$ having absolutely continuous distribution function F with support $[0, \infty)$ and hazard function $\alpha = F'/(1 - F)$, whereas $U_i^{(n)}$ has distribution function $G_i^{(n)}$. The censored survival times $\tilde{X}_i^{(n)} = X_i^{(n)} \wedge U_i^{(n)}$ are then independent with distribution function $H_i^{(n)} = 1 - (1 - F)(1 - G_i^{(n)})$. As usual, we let $D_i^{(n)} = I(\tilde{X}_i^{(n)} = X_i^{(n)})$.

The distribution of the uncensored $X_i^{(n)}$ is assumed to depend on a q-dimensional parameter $\boldsymbol{\theta} = (\theta_1, \ldots, \theta_q)$, and to indicate this, we write $F(\cdot;\boldsymbol{\theta})$ and $\alpha(\cdot;\boldsymbol{\theta})$ for the distribution and hazard functions. As usual, the true parameter value is denoted by $\boldsymbol{\theta}_0$. The counting process

$$N^{(n)}(t) = \sum_{i=1}^{n} I(\tilde{X}_i^{(n)} \le t, D_i^{(n)} = 1)$$

then has intensity process of the multiplicative form

$$\lambda^{(n)}(t;\boldsymbol{\theta}_0) = \alpha(t;\boldsymbol{\theta}_0) Y^{(n)}(t)$$

with

$$Y^{(n)}(t) = \sum_{i=1}^{n} I(\tilde{X}_i^{(n)} \ge t)$$

(cf. Example IV.1.1). Omitting the superscript (n) in the following, the log-likelihood (6.1.2) (apart from a constant) takes the form

$$C(\boldsymbol{\theta}) = \int_0^\infty \log \alpha(t; \boldsymbol{\theta}) \, \mathrm{d}N(t) - \int_0^\infty \alpha(t; \boldsymbol{\theta}) Y(t) \, \mathrm{d}t$$
$$= \sum_{i=1}^n D_i \log \alpha(\tilde{X}_i; \boldsymbol{\theta}) - \sum_{i=1}^n \int_0^{\tilde{X}_i} \alpha(t; \boldsymbol{\theta}) \, \mathrm{d}t.$$

We will assume that there exists a neighborhood Θ_0 of $\boldsymbol{\theta}_0$ such that the derivatives of $\alpha(t; \boldsymbol{\theta})$ of the first, second, and third order with respect to $\boldsymbol{\theta}$ exist and are continuous in $\boldsymbol{\theta}$ for $\boldsymbol{\theta} \in \Theta_0$. Moreover, we assume that $\int_0^x \alpha(t; \boldsymbol{\theta}) \, \mathrm{d}t$ for any finite x may be differentiated three times with respect to $\boldsymbol{\theta} \in \Theta_0$ by interchanging the order of integration and differentiation. Then part A of Condition VI.1.1 is fulfilled and the maximum likelihood estimator $\hat{\boldsymbol{\theta}}$ is the solution to the equations

$$\sum_{i=1}^n D_i \frac{\partial}{\partial \theta_j} \log \alpha(\tilde{X}_i; \boldsymbol{\theta}) - \sum_{i=1}^n \int_0^{\tilde{X}_i} \frac{\partial}{\partial \theta_j} \alpha(t; \boldsymbol{\theta}) \, \mathrm{d}t = 0$$

for $j = 1, 2, \ldots, q$. We will discuss under which supplementary conditions parts B–E of Condition VI.1.1 are satisfied.

To this end, we must impose certain boundedness and integrability conditions on the derivatives of $\alpha(t; \boldsymbol{\theta})$ with respect to $\boldsymbol{\theta}$. We assume that

$$\int_0^\infty \left\{ \frac{\partial}{\partial \theta_j} \log \alpha(t; \boldsymbol{\theta}_0) \right\}^2 \alpha(t; \boldsymbol{\theta}_0)(1 - F(t; \boldsymbol{\theta}_0)) \, \mathrm{d}t < \infty,$$
$$\int_0^\infty \left\{ \frac{\partial^2}{\partial \theta_j \, \partial \theta_l} \log \alpha(t; \boldsymbol{\theta}_0) \right\}^2 \alpha(t; \boldsymbol{\theta}_0)(1 - F(t; \boldsymbol{\theta}_0)) \, \mathrm{d}t < \infty$$

for all j, l, and that there exist functions γ and ρ such that

$$\sup_{\boldsymbol{\theta} \in \Theta_0} \left| \frac{\partial^3}{\partial \theta_j \, \partial \theta_l \, \partial \theta_m} \alpha(t; \boldsymbol{\theta}) \right| \leq \gamma(t),$$
$$\sup_{\boldsymbol{\theta} \in \Theta_0} \left| \frac{\partial^3}{\partial \theta_j \, \partial \theta_l \, \partial \theta_m} \log \alpha(t; \boldsymbol{\theta}) \right| \leq \rho(t) \tag{6.1.21}$$

for all $t \in [0, \infty]$ and

$$\int_0^\infty \gamma(t)(1 - F(t; \boldsymbol{\theta}_0)) \, \mathrm{d}t < \infty,$$
$$\int_0^\infty \rho(t) \alpha(t; \boldsymbol{\theta}_0)(1 - F(t; \boldsymbol{\theta}_0)) \, \mathrm{d}t < \infty. \tag{6.1.22}$$

We must also assume that the censoring stabilizes as n grows, in the sense that

$$n^{-1} \sum_{i=1}^n G_i^{(n)}(t) \to G(t),$$

as $n \to \infty$ for all $s \in [0, \infty]$ for some (sub-)distribution function G; cf. (5.1.7).

Then, by Example V.1.6,

$$n^{-1}Y(t) \xrightarrow{P} y(t) = (1 - F(t;\boldsymbol{\theta}_0))(1 - G(t-)) \tag{6.1.23}$$

and for any $\delta \in (0, 1)$,

$$\mathrm{P}(n^{-1}Y(t) \leq \delta^{-1}(1 - F(t;\boldsymbol{\theta}_0)) \text{ for all } t \in [0, \infty]) \geq 1 - \delta; \tag{6.1.24}$$

cf. (5.1.8) and the result just below (5.1.9).

Then parts B, C, and E of Condition VI.1.1 can also be seen to hold. Consider, for illustration, the condition on the third-order partial derivative of $\log \lambda(t, \boldsymbol{\theta})$ in part E. By (6.1.21), this is bounded uniformly for $\boldsymbol{\theta} \in \Theta_0$ by ρ. Moreover, by (6.1.23),

$$n^{-1}\rho(t)\alpha(t;\boldsymbol{\theta}_0)Y(t) \xrightarrow{P} \rho(t)\alpha(t;\boldsymbol{\theta}_0)y(t),$$

and by (6.1.24),

$$\mathrm{P}(n^{-1}\rho(t)\alpha(t;\boldsymbol{\theta}_0)Y(t) \leq \delta^{-1}\rho(t)\alpha(t;\boldsymbol{\theta}_0)(1 - F(t;\boldsymbol{\theta}_0)) \text{ for all } t \in [0, \infty]) \geq 1 - \delta,$$

where $\rho(t)\alpha(t;\boldsymbol{\theta}_0)(1 - F(t;\boldsymbol{\theta}_0))$ is integrable by (6.1.22). Proposition II.5.3, therefore, gives that

$$n^{-1}\int_0^\infty \rho(t)\alpha(t;\boldsymbol{\theta}_0)Y(t)\,dt \xrightarrow{P} \int_0^\infty \rho(t)\alpha(t;\boldsymbol{\theta}_0)y(t)\,dt < \infty$$

as $n \to \infty$. The other conditions in part E as well as part B and C are seen to hold by similar arguments (and by the use of the Cauchy–Schwarz inequality for the integrability in part B). In particular, we get from part B

$$\sigma_{jl}(\boldsymbol{\theta}_0) = \int_0^\infty \left\{\frac{\partial}{\partial\theta_j}\log\alpha(t;\boldsymbol{\theta}_0)\right\}\left\{\frac{\partial}{\partial\theta_l}\log\alpha(t;\boldsymbol{\theta}_0)\right\}\alpha(t;\boldsymbol{\theta}_0)y(t)\,dt,$$

and provided that $\boldsymbol{\Sigma} = \{\sigma_{jl}(\boldsymbol{\theta}_0)\}$ is positive definite (part D), Condition VI.1.1 holds and Theorems VI.1.1 and VI.1.2 apply. □

Example VI.1.12. Software Reliability Models

In this example, we will discuss how one may derive formally the asymptotic properties of the maximum likelihood estimators for the software reliability models considered in Example VI.1.8, concentrating, in particular, on the Jelinski–Moranda model. For a general treatment of large sample results for software reliability models using similar techniques, see van Pul (1992a).

In software reliability models, one of the parameters is F, the number of faults in the system. It is obviously not possible to derive any sensible large sample results by considering a sequence of models with F fixed. A more relevant large sample situation to consider is one in which there are more and more faults in larger and larger programs observed over a finite time period. This is the same type of reasoning as was used in the study of the large sample properties of the Nelson–Aalen estimator for matings of Drosophila flies in Example IV.1.10.

We, therefore, consider a sequence of software reliability models, indexed by n, and let s_n denote the size of the program, e.g., measured in thousands of lines of programming code, in the nth model. We assume that $s_n \to \infty$ as $n \to \infty$ and that the number of faults per 1000 lines of code is equal to ϕ_0 in all models. Thus, the total number of faults in the nth model is $F^{(n)} = s_n \phi_0$. For simplicity, we will let the observation period be $[0, \tau]$ for all models, where τ is a finite number. It is, however, not very difficult to modify the proofs to be presented to allow for an observation period $[0, T^{(n)}]$ in the nth model, with the $T^{(n)}$ being stopping times tending to τ in probability as n increases. In particular, one may decide to stop the observation after a fixed number of failures as we did in Example VI.1.8.

For the Jelinski–Moranda model, the counting process $N^{(n)}$, counting the number of failures in the nth model, will have the intensity process

$$\lambda^{(n)}(t; \boldsymbol{\theta}_0) = \rho_0(F^{(n)} - N^{(n)}(t-)) = s_n \rho_0(\phi_0 - N^{(n)}(t-)/s_n),$$

where we have written $\boldsymbol{\theta}_0 = (\rho_0, \phi_0)$. We assume that $\rho_0 > 0$ and $\phi_0 > 0$ and will derive the large sample properties of $\hat{\boldsymbol{\theta}} = (\hat{\rho}, \hat{\phi})$, the maximum likelihood estimators for the parameters of this reformulated Jelinski–Moranda model, by checking Condition VI.1.1.

The key step in doing this is to use the result of Kurtz reviewed in Section II.5.2. We write $\Lambda_n(t) = \int_0^t \lambda^{(n)}(u; \boldsymbol{\theta}_0)\,\mathrm{d}u$ for the compensator of $N^{(n)}$, and note that we may write

$$\lambda^{(n)} = s_n \beta(\cdot, s_n^{-1} N^{(n)}),$$

where $\beta(u, x)$ is the nonanticipating non-negative function of $u \in \mathscr{T}$ and $x \in D(\mathscr{T})$ given (for $0 \le x(u-) \le \phi_0$) by

$$\beta(u, x) = \rho_0(\phi_0 - x(u-)).$$

For $u \le \tau$, this function satisfies $\beta(u, x) \le \rho_0 \phi_0$ and

$$|\beta(u, x_1) - \beta(u, x_2)| \le \rho_0 \sup_{u \le \tau} |x_1(u-) - x_2(u-)|.$$

Therefore, by Kurtz's law of large numbers (Theorem II.5.4),

$$\sup_{t \le \tau} |s_n^{-1} N^{(n)}(t) - X(t)| \xrightarrow{\mathrm{P}} 0 \tag{6.1.25}$$

as $n \to \infty$, where X is the solution of

$$\begin{aligned} X(t) &= \int_0^t \beta(u, X)\,\mathrm{d}u \\ &= \int_0^t \rho_0(\phi_0 - X(u-))\,\mathrm{d}u. \end{aligned}$$

Clearly, $X(0) = 0$, and it is easily seen that $X(t) = \phi_0(1 - e^{-\rho_0 t})$.

In particular, we have that

$$N^{(n)}(\tau-)/s_n \xrightarrow{\mathrm{P}} \phi_0(1 - e^{-\rho_0 \tau})$$

as $n \to \infty$. It follows that there exists a $\phi' < \phi_0$ such that the probability of the event $\{N^{(n)}(\tau-)/s_n < \phi'\}$ tends to one as n tends to infinity. Thus, it suffices to check parts A–E of Condition VI.1.1 assuming $N^{(n)}(\tau-)/s_n < \phi'$. Then A is easily seen to hold. Furthermore, writing $\boldsymbol{\theta} = (\theta_1, \theta_2) = (\rho, \phi)$ and using (6.1.25), part B is fulfilled with $a_n = \sqrt{s_n}$,

$$\sigma_{11}(\boldsymbol{\theta}_0) = \frac{1}{\rho_0}\int_0^\tau (\phi_0 - X(t))\,dt = \frac{\phi_0}{\rho_0^2}(1 - e^{-\rho_0\tau}),$$

$$\sigma_{12}(\boldsymbol{\theta}_0) = \sigma_{21}(\boldsymbol{\theta}_0) = \tau,$$

and

$$\sigma_{22}(\boldsymbol{\theta}_0) = \rho_0 \int_0^\tau \frac{1}{\phi_0 - X(t)}\,dt = \frac{1}{\phi_0}(e^{\rho_0\tau} - 1).$$

On the event $\{N^{(n)}(\tau-)/s_n < \phi'\}$, the derivatives $(\partial/\partial\theta_j)\log\lambda^{(n)}(t;\boldsymbol{\theta})$ are bounded for $j = 1, 2$. Thus part C follows since $s_n \to \infty$ as $n \to \infty$. Part D is an immediate consequence of the expressions for $\sigma_{jl}(\boldsymbol{\theta}_0)$ just derived. Finally, for the boundedness condition, part E, we only show the result for the third-order partial derivatives of $\log\lambda^{(n)}(t;\boldsymbol{\theta})$. The other conditions in part E may be checked in a similar manner. Now we find

$$\frac{\partial^3}{\partial\rho^3}\log\lambda(t;\boldsymbol{\theta}) = \frac{2}{\rho^3}$$

and

$$\frac{\partial^3}{\partial\phi^3}\log\lambda(t;\boldsymbol{\theta}) = \frac{2}{(\phi - N^{(n)}(t-)/s_n)^3},$$

whereas the mixed third-order partial derivatives vanish. Thus, choosing an $\varepsilon > 0$ satisfying $\rho_0 - \varepsilon > 0$ and $\phi_0 - \phi' - \varepsilon > 0$, the absolute value of these two derivatives are bounded in a neighborhood of $\boldsymbol{\theta}_0 = (\rho_0, \phi_0)$ by $2(\rho_0 - \varepsilon)^{-3}$ and $2(\phi_0 - \phi' - \varepsilon)^{-3}$, respectively. The boundedness condition, part E, for the third-order partial derivatives of $\log\lambda^{(n)}(\boldsymbol{\theta})$ then follows using (6.1.25).

Thus, we have proved that

$$\sqrt{s_n}(\hat{\boldsymbol{\theta}} - \boldsymbol{\theta}_0) \xrightarrow{\mathscr{D}} \mathscr{N}(\mathbf{0}, \boldsymbol{\Sigma}^{-1})$$

as $n \to \infty$, where $\boldsymbol{\Sigma}$ is the 2×2 matrix with elements $\sigma_{jl}(\boldsymbol{\theta}_0)$ as given above. The matrix $\boldsymbol{\Sigma}$ may be estimated consistently by $s_n^{-1}\mathscr{I}_\tau(\hat{\boldsymbol{\theta}})$ in the usual manner or, alternatively, by inserting $\hat{\rho}$ and $\hat{\phi}$ in the above expressions.

The (extended) Littlewood model may be handled along similar lines. With $\boldsymbol{\theta} = (\alpha, \beta, \phi)$, the intensity process for $N^{(n)}$ now becomes

$$\lambda^{(n)}(t;\boldsymbol{\theta}_0) = \frac{\alpha_0(F^{(n)} - N^{(n)}(t-))}{\beta_0 + t} = \frac{s_n\alpha_0(\phi_0 - N^{(n)}(t-)/s_n)}{\beta_0 + t}.$$

We may then use the same argument as for the Jelinski–Moranda model to find that (6.1.25) holds, with X being the solution of

$$X(t) = \int_0^t \frac{\alpha_0(\phi_0 - X(u-))}{\beta_0 + u} \mathrm{d}u.$$

This gives

$$X(t) = \phi_0\left(1 - \left(\frac{\beta_0}{\beta_0 + t}\right)^{\alpha_0}\right).$$

Using this result, one may then check Condition VI.1.1 in a similar manner to the Jelinski–Moranda model. We omit further details and refer the interested readers to van Pul (1992a) for a thorough treatment of the Littlewood model.

Having derived the asymptotic distribution for the parameters of the reformulated Jelinski–Moranda and Littlewood models, we may now go back to the original parametrizations. It follows that the maximum likelihood estimator $\hat{F}^{(n)} = s_n\hat{\phi}$ for the number of faults in the nth model and the maximum likelihood estimators for the other model parameters are jointly approximately multinormally distributed around the true parameter values with a covariance matrix that may be estimated as described in Example VI.1.8, i.e., by $\hat{\mathscr{I}}_\tau^{-1}$, where now the elements of the matrix $\hat{\mathscr{I}}_\tau$ are found by differentiating minus the log-likelihood two times with respect to the parameters of the original model; cf. (6.1.4). □

VI.2. M-Estimators

VI.2.1. Introduction and Examples

In the preceding section, we studied maximum likelihood estimation for counting process models with parametrically specified intensity processes. In the present section, we will study a wider class of estimators, namely, what we may call M-estimators. To this end, we first note that (6.1.3) may be written

$$U_\tau^j(\boldsymbol{\theta}) = \int_0^\tau \sum_{h=1}^k \left\{\frac{\partial}{\partial \theta_j} \log \lambda_h(t;\boldsymbol{\theta})\right\}\{\mathrm{d}N_h(t) - \lambda_h(t;\boldsymbol{\theta})\,\mathrm{d}t\}.$$

As realized by Hjort (1985a), in the context of the multiplicative intensity model, an extension is now possible. We can define an estimator $\tilde{\boldsymbol{\theta}}$ as a solution to the equations

$$U_\tau^j(\boldsymbol{\theta}) = \int_0^\tau \sum_{h=1}^k K_{hj}(t;\boldsymbol{\theta})\{\mathrm{d}N_h(t) - \lambda_h(t;\boldsymbol{\theta})\,\mathrm{d}t\} = 0, \tag{6.2.1}$$

$j = 1, \ldots, q$. Here the $K_{hj}(t; \boldsymbol{\theta})$ may be deterministic functions or, more generally, predictable, locally bounded processes. Different choices of these will produce different estimators in the whole class of M-estimators. It is seen that the maximum likelihood estimator corresponds to the particular choice of

$$K_{hj}(t; \boldsymbol{\theta}) = \frac{\partial}{\partial \theta_j} \log \lambda_h(t; \boldsymbol{\theta}) \tag{6.2.2}$$

in (6.2.1).

An examination of the proofs of the Theorems VI.1.1 and VI.1.2 will reveal that the arguments used there carry over almost step-by-step to the present situation provided that the $K_{hj}(t; \boldsymbol{\theta})$ fulfill regularity conditions similar to the ones for (6.2.2) stated in Condition VI.1.1. We postpone the details to the next subsection. The result is that $\tilde{\boldsymbol{\theta}}$ is asymptotically multinormally distributed around the true parameter value with a certain covariance matrix. This asymptotic covariance matrix may be estimated by $\tilde{\boldsymbol{\Xi}} = \tilde{\boldsymbol{\Psi}}^{-1} \tilde{\boldsymbol{\Delta}} (\tilde{\boldsymbol{\Psi}}^{\mathsf{T}})^{-1}$ where $\tilde{\boldsymbol{\Psi}}$ and $\tilde{\boldsymbol{\Delta}}$ have elements

$$\tilde{\psi}_{jl} = \int_0^\tau \sum_{h=1}^k K_{hj}(t; \tilde{\boldsymbol{\theta}}) \left\{ \frac{\partial}{\partial \theta_l} \log \lambda_h(t; \tilde{\boldsymbol{\theta}}) \right\} \mathrm{d}N_h(t) \tag{6.2.3}$$

and

$$\tilde{\delta}_{jl} = \int_0^\tau \sum_{h=1}^k K_{hj}(t; \tilde{\boldsymbol{\theta}}) K_{hl}(t; \tilde{\boldsymbol{\theta}}) \, \mathrm{d}N_h(t), \tag{6.2.4}$$

respectively. Alternatively, $\mathrm{d}N_h(t)$ may be replaced by $\lambda_h(t; \tilde{\boldsymbol{\theta}}) \, \mathrm{d}t$ in these expressions. Note that the proposed estimator $\tilde{\boldsymbol{\Xi}}$ does not reduce to the inverse of the observed information, i.e., $\mathscr{I}(\hat{\boldsymbol{\theta}})^{-1}$ [see (6.1.4)], when the weight process is given by (6.2.2). So the estimator also offers an alternative for estimating the covariance matrix of the maximum likelihood estimator.

M-estimators are useful in situations where the maximum likelihood estimator is difficult (or even impossible) to compute. For such situations, the M-estimators may give satisfactory final estimates or they may provide good starting values for the evaluation of the maximum likelihood estimator by an iterative procedure. It is also possible to use a "one-step estimator" by Newton–Raphson-iterating once toward the maximum likelihood estimator starting from an M-estimator. This will give an estimator with the same asymptotic distribution as the maximum likelihood estimator. Finally, it should be possible to develop M-estimators for particular situations with better robustness properties than the maximum likelihood estimator, but such studies still have to be done for counting process models.

We will consider five examples. The first of these is concerned with uncensored failure-time data, and it shows that the M-estimators defined by (6.2.1) deserve their name since they generalize the classical M-estimators for uncensored data.

EXAMPLE VI.2.1. M-Estimators for Uncensored Failure-Time Data

For i.i.d. non-negative random variables $X_1, X_2, \ldots, X_n$ with distribution function $F(\cdot;\theta)$ and hazard function $\alpha(\cdot;\theta)$, for the sake of simplicity assumed to depend on a one-dimensional parameter θ, the counting process

$$N(t) = \sum_{i=1}^{n} I(X_i \leq t)$$

has intensity process

$$\lambda(t;\theta) = \alpha(t;\theta) Y(t)$$

with

$$Y(t) = \sum_{i=1}^{n} I(X_i \geq t)$$

(Example III.1.2). Thus, for this situation the M-estimator $\tilde{\theta}$ is, by (6.2.1), the solution to the equation

$$\int_0^\infty K(t;\theta)\{\mathrm{d}N(t) - \alpha(t;\theta)Y(t)\,\mathrm{d}t\} = \sum_{i=1}^{n} \left\{K(X_i;\theta) - \int_0^{X_i} K(t;\theta)\alpha(t;\theta)\,\mathrm{d}t\right\} = 0.$$

If we, therefore, introduce

$$\rho(t;\theta) = K(t;\theta) - \int_0^t K(u;\theta)\alpha(u;\theta)\,\mathrm{d}u, \tag{6.2.5}$$

it is seen that $\tilde{\theta}$ may be solved from

$$\sum_{i=1}^{n} \rho(X_i;\theta) = 0,$$

i.e., it is an M-estimator in the classical sense [cf. Huber (1981)].

Moreover, given a function $\rho(t;\theta)$ there exists by Theorem II.6.3 a unique function $K(t;\theta)$ satisfying (6.2.5), and this function is given by

$$K(t;\theta) = \rho(t;\theta) + \int_0^t \rho(s;\theta)\exp\left\{\int_s^t \alpha(u;\theta)\,\mathrm{d}u\right\}\alpha(s;\theta)\,\mathrm{d}s.$$

In the classical situation of uncensored data it is therefore no restriction to consider M-estimators as the solution to (6.2.1). □

The relation between the classical M-estimators for uncensored data and the M-estimators for counting process models discussed in the above example connects the latter to the vast literature on robust M-estimators for uncensored data and, thus, paves the way for a study of robust M-estimators for censored survival data and other counting process models. We will not pursue this here, however, but rather consider four examples where in each case the important point is to derive M-estimators which are simple to calculate.

The first of these essentially concerns occurrence/exposure rates, but in the noval setting of stochastic epidemic models. The other three relate to situations already encountered in Section VI.1.

EXAMPLE VI.2.2. Stochastic Epidemic Models

The simple epidemic model is the Markov process on $[0, \infty)$ specified as follows (Becker, 1989, Chapter 7). At time $t = 0$, a (closed) community of size $s + i$ consists of s susceptibles and i infectives. Individuals who are infected become infectious immediately and remain infectious for the duration of the epidemic. Let $N(t)$ denote the number of individuals infected during $(0, t]$ so that $S(t) = s - N(t)$ and $I(t) = i + N(t)$ are the number of susceptibles and infectives, respectively, at time t. We assume that

$$P(N(t+h) = N(t) + 1 \mid S(t), I(t)) = \beta I(t)S(t)h + o(h),$$

$$P(N(t+h) = N(t) \mid S(t), I(t)) = 1 - \beta I(t)S(t)h + o(h).$$

It is seen that, with respect to the filtration generated by $\{S(t), I(t)\}$, $N(t)$ is then a counting process with intensity process $\beta I(t-)S(t-)$. According to Section VI.1.1, the maximum likelihood estimator based on observation over $[0, \tau]$ is then given by the occurrence/exposure rate

$$\hat{\beta} = N(\tau) \Big/ \int_0^\tau I(t)S(t)\,dt$$

with standard deviation estimate

$$\widehat{\text{s.d.}}(\hat{\beta}) = \hat{\beta}/\sqrt{N(\tau)}.$$

In practice, it may be unrealistic to assume that $S(t)$ and $I(t)$ are known at each time t. Define, therefore, the predictable process

$$K(t) = J(t)/[I(t-)S(t-)],$$

where $J(t) = I(S(t-) > 0)$ and $K(t)$ is defined to be zero when $J(t) = 0$.

The estimating equation (6.2.1) with this weight process then takes the form

$$\begin{aligned} U_\tau(\beta) &= \int_0^\tau \frac{J(t)}{I(t-)S(t-)}\{dN(t) - \beta I(t-)S(t-)\,dt\} \\ &= \int_0^\tau \frac{J(t)\,dN(t)}{I(t-)S(t-)} - \beta \int_0^\tau J(t)\,dt = 0 \end{aligned}$$

so that the M-estimator is

$$\tilde{\beta} = \int_0^\tau \frac{J(t)\,dN(t)}{I(t-)S(t-)} \Big/ \int_0^\tau J(t)\,dt.$$

The numerator takes the form

$$\int_0^\tau \frac{J(t)}{I(t-)S(t-)} \mathrm{d}N(t) = \frac{1}{si} + \frac{1}{(s-1)(i+1)} + \cdots + \frac{1}{\{S(\tau)+1\}\{I(\tau)-1\}}$$

and the denominator is

$$\int_0^\tau J(t)\,\mathrm{d}t = \inf\{t: S(t) = 0\} \wedge \tau,$$

which, in particular, equals τ when $S(\tau) > 0$. The resulting M-estimator of β is seen to depend only on the number of infectives and susceptibles at time τ and the time $\inf\{t: S(t) = 0\}$ of the last infection, if that happened before τ.

To estimate the asymptotic variance of $\tilde{\beta}$, we note that by (6.2.3)

$$\tilde{\psi} = \int_0^\tau J(t)\,\mathrm{d}t$$

and by (6.2.4)

$$\tilde{\delta} = \int_0^\tau \frac{J(t)\,\mathrm{d}N(t)}{\{I(t-)S(t-)\}^2}.$$

Thus, we may use

$$\tilde{\xi} = \frac{\tilde{\delta}}{\tilde{\psi}^2} = \int_0^\tau \frac{J(t)\,\mathrm{d}N(t)}{\{I(t-)S(t-)\}^2} \left(\int_0^\tau J(t)\,\mathrm{d}t \right)^{-2}$$

as an estimator for the variance of $\tilde{\beta}$ [as also suggested by Becker (1989, Section 7.3)].

Becker (1989, Chapter 7) suggested several applications of this approach to more sophisticated epidemic models. □

EXAMPLE VI.2.3. Mortality of Diabetics in the County of Fyn. Preliminary Estimates for the Gompertz–Makeham Survival Distribution

In Example VI.1.4, we showed how the maximum likelihood estimators may be derived as a solution of three rather complicated nonlinear equations for left-truncated and right-censored survival data having a death intensity of the Gompertz–Makeham form

$$\alpha(t;\boldsymbol{\theta}) = a + bc^t.$$

To solve these equations iteratively, it may be important to have good starting values for the iterative procedure. Such starting values may be provided by the M-estimators studied in this example.

We concentrate on a fixed interval $[t_1, t_2] \subset \mathcal{T}$ and choose a d satisfying $(t_2 - t_1)/3 \le d < t_2 - t_1$. On $[0, d]$, we then specify a function h. As we will soon see, various choices of h will correspond to various choices of M-estimators (within a subclass of all M-estimators). We then let $l = (t_2 - t_1 - d)/2$ and define the functions H_1, H_2, H_3 by

$$H_j(t) = \begin{cases} h(t - t_1 - (j-1)l) & \text{for } t \in I_j = [t_1 + (j-1)l, t_2 - (3-j)l] \\ 0 & \text{otherwise.} \end{cases}$$

Thus, H_1, H_2, and H_3 are just the h shifted to the intervals $[t_1, t_1 + d]$, $[t_1 + l, t_1 + l + d]$, and $[t_1 + 2l, t_1 + 2l + d]$, respectively. The M-estimators to be considered in this example are then defined by (6.2.1) (with $k = 1$) using weight processes $K_j = H_j/Y$ for $j = 1, 2, 3$, where Y is defined as in Example VI.1.4.

Thus, $(\tilde{a}, \tilde{b}, \tilde{c})$ is given as the solution to the equations

$$\int_{I_j} \frac{H_j(t)}{Y(t)} \mathrm{d}N(t) - \int_{I_j} H_j(t)(a + bc^t)\,\mathrm{d}t = 0,$$

$j = 1, 2, 3$. Now

$$\int_{I_j} H_j(t)(a + bc^t)\,\mathrm{d}t = a \int_0^d h(t)\,\mathrm{d}t + bc^{t_1 + (j-1)l} \int_0^d c^t h(t)\,\mathrm{d}t,$$

and it follows that

$$\tilde{c} = \left(\frac{G_3 - G_2}{G_2 - G_1} \right)^{1/l},$$

where

$$G_j = \int_{I_j} \frac{H_j(t)}{Y(t)} \mathrm{d}N(t).$$

Furthermore,

$$\tilde{a} = \frac{\tilde{c}^l G_1 - G_2}{(\tilde{c}^l - 1) \int_0^d h(t)\,\mathrm{d}t}$$

and

$$\tilde{b} = \frac{G_2 - G_1}{\tilde{c}^{t_1}(\tilde{c}^l - 1) \int_0^d \tilde{c}^t h(t)\,\mathrm{d}t}.$$

The asymptotic covariance matrix of $(\tilde{a}, \tilde{b}, \tilde{c})$ may be estimated using (6.2.3) and (6.2.4) (with $k = 1$).

These M-estimators are essentially "continuous" versions of the moment methods used by actuaries for graduating occurrence/exposure rates for the death intensity by analytic graduation. Corresponding to the suggestion by King and Hardy (1880), we may let $h(t) = kt$ for $0 \le t \le d/2$ and $h(t) = k(d - t)$ for $d/2 \le t \le d$ with $d = (t_2 - t_1)/2$ and $k = d/2$. Other choices of constants k and $d \ge (t_2 - t_1)/3$ are also considered in the actuarial tradition. [Within this tradition the name "King–Hardy's method" is often (for reasons unknown to us) used for the case with $d = l = (t_2 - t_1)/3$ and $h(t) = 1$ for $0 \le t \le d$.] Forsén (1979) gave references to the actuarial literature and discussed other choices of h (within the framework of graduating occurrence/exposure rates).

We will not go into a further discussion of the best choice of h. That is a question open for further research. Here, we will be content with illustrating the use of the M-estimators by applying the King–Hardy method to the data on diabetics in the county of Fyn (cf. Examples I.3.2 and VI.1.4). We consider only the female diabetics and, as in Example VI.1.4, we subtract 50 years from the entry and (censored) survival times before fitting the model. This corresponds to the reparametrization.

$$\alpha(t;\boldsymbol{\theta}) = a + bc^{t-50}.$$

Choosing $t_1 = 25$ and $t_2 = 85$, we then get the estimates

$$\tilde{a} = 0.00798, \qquad \tilde{b} = 0.00926, \qquad \tilde{c} = 1.099$$

with estimated standard deviations

$$\widehat{\text{s.d.}}(\tilde{a}) = 0.00448, \qquad \widehat{\text{s.d.}}(\tilde{b}) = 0.00508, \qquad \widehat{\text{s.d.}}(\tilde{c}) = 0.024$$

and with correlations

$$\widehat{\text{corr}}(\tilde{a},\tilde{b}) = -0.855, \qquad \widehat{\text{corr}}(\tilde{a},\tilde{c}) = 0.810, \qquad \widehat{\text{corr}}(\tilde{b},\tilde{c}) = -0.983.$$

It is seen that the M-estimate for c is higher than the maximum likelihood estimate given in Example VI.1.4 and that its estimated standard deviation is more than twice as large as the estimated standard deviation for the maximum likelihood estimate. Due to the high positive correlation between the estimators for a and c and the high negative correlation between the estimators for b and c, the M-estimate for a is somewhat higher and the M-estimate for b somewhat lower than the corresponding maximum likelihood estimates. □

EXAMPLE VI.2.4. Excess Mortality and Linear Intensity Models. Survival with Malignant Melanoma

We consider the model for excess mortality of Example VI.1.5 where the individual counting processes N_i have intensity processes of the form

$$\lambda_i(t;\gamma) = \{\gamma + \mu_i(t)\}\, Y_i(t).$$

We saw in the above-mentioned example that the maximum likelihood estimator for this model is given implicitly as the solution to a nonlinear equation. By using the theory of the present section, it is possible to find a (not fully efficient) estimator by solving

$$\int_0^\tau \sum_{i=1}^n K_i(t)\{\mathrm{d}N_i(t) - \lambda_i(t;\gamma)\,\mathrm{d}t\} = 0$$

for γ, the $K_i(t)$ being suitably chosen locally bounded predictable processes not depending on γ. Thus, we find the estimator

$$\tilde{\gamma} = \frac{\int_0^\tau \sum_{i=1}^n K_i(t)\,\mathrm{d}N_i(t) - \int_0^\tau \sum_{i=1}^n K_i(t)\,Y_i(t)\mu_i(t)\,\mathrm{d}t}{\int_0^\tau \sum_{i=1}^n K_i(t)\,Y_i(t)\,\mathrm{d}t}.$$

This estimator is asymptotically normally distributed around γ. Its asymptotic variance may be estimated by a direct application of (6.2.3) and (6.2.4) or by replacing $dN_i(t)$ by $\{\tilde{\gamma} + \mu_i(t)\} Y_i(t)\,dt$ in these expressions. The latter approach yields

$$\tilde{\psi} = \int_0^\tau \sum_{i=1}^n K_i(t) Y_i(t)\,dt$$

and

$$\tilde{\delta} = \int_0^\tau \sum_{i=1}^n K_i^2(t)\{\tilde{\gamma} + \mu_i(t)\} Y_i(t)\,dt$$

with a corresponding estimator

$$\tilde{\xi} = \frac{\tilde{\delta}}{\tilde{\psi}^2} = \frac{\int_0^\tau \sum_{i=1}^n K_i^2(t)\{\tilde{\gamma} + \mu_i(t)\} Y_i(t)\,dt}{\{\int_0^\tau \sum_{i=1}^n K_i(t) Y_i(t)\,dt\}^2}$$

for the asymptotic variance of $\tilde{\gamma}$.

A simple choice of weight processes is $K_i(t) = Y_i(t)$ which gives the M-estimator

$$\tilde{\gamma} = \frac{N(\tau) - E(\tau)}{R(\tau)}$$

with $R(\tau) = \int_0^\tau Y(u)\,du$ and $E(\tau) = \int_0^\tau Y^\mu(u)\,du$ as in Example VI.1.5 (Buckley, 1984; Pocock et al., 1982; Andersen and Væth, 1989.) Since for this estimator $\tilde{\psi} = R(\tau)$ and $\tilde{\delta} = \int_0^\tau \sum_{i=1}^n \{\tilde{\gamma} + \mu_i(t)\} Y_i(t)\,dt = N(\tau)$ its variance may be estimated simply by $\tilde{\xi} = N(\tau)/R(\tau)^2$. Another possible choice of weight processes is to let $K_i(t)$ be the age of the ith individual at follow-up time t. A third possibility worth mentioning is to fix a time point $t_0 \leq \tau$ and (independent of i) let $K_i(t) = J(t)/Y(t)$ for $t \leq t_0$ and $K_i(t) = 0$ for $t > t_0$. Here, $J(t) = I(Y(t) > 0)$ as usual. With this choice, it is seen that $\tilde{\gamma}$ is the average slope of the estimator $\hat{\Gamma}$ of Example IV.1.11 for the integrated excess mortality over the interval $[0, t_0]$.

Again, the methods will be illustrated using the melanoma data introduced in Example I.3.1. The estimator based on $K_i = Y_i$ gives

$$\tilde{\gamma} = \frac{71 - 21.24}{1209.11} = 0.0412 \text{ (per year)}$$

with an estimated standard deviation of 0.0070 per year. The choice $K_i(t) =$ "age of ith individual at $t-$" gives the estimate $\tilde{\gamma} = 0.0427$ with estimated standard deviation [derived by (6.2.3) and (6.2.4)] 0.0083. Finally, the estimator based on the average slope of $\hat{\Gamma}$ of Example IV.1.11 over 0–10 years, i.e., the choice $K_i(t) = 1/Y(t)$ for $t \leq 10$ and $K_i(t) = 0$ otherwise, gives $\tilde{\gamma} = \hat{\Gamma}(10)/10 = 0.0418$ per year with an estimated standard deviation of 0.0084 per year [using the estimated standard deviation of $\hat{\Gamma}(10)$ from Example IV.1.11].

Comparing with the maximum likelihood estimate $\hat{\gamma} = 0.0400$ per year and its estimated standard deviation $\hat{\sigma} = 0.0066$ per year, it is seen that estimates of the same order of magnitude are obtained here, whereas the M-estimators are less precise.

Let us also briefly consider the linear intensity model of Example VI.1.5 where the N_i have intensity processes

$$\lambda_i(t; \gamma, \theta) = \{\gamma + \theta\mu_i(t)\} Y_i(t).$$

By (6.2.1), we may find estimators $\tilde{\gamma}$ and $\tilde{\theta}$ as solutions to the system of linear equations

$$\int_0^\tau \sum_{i=1}^n K_{ij}(t)\, \mathrm{d}N_i(t) - \tilde{\gamma} \int_0^\tau \sum_{i=1}^n K_{ij}(t) Y_i(t)\, \mathrm{d}t - \tilde{\theta} \int_0^\tau \sum_{i=1}^n K_{ij}(t) Y_i(t)\mu_i(t)\, \mathrm{d}t = 0$$

for $j = 1, 2$, where the K_{ij} are locally bounded predictable processes. A possible choice is $K_{i1}(t) = Y_i(t)$ and $K_{i2}(t) =$ "age of ith individual at $t-$," the choice of K_{i1} leading to the equation

$$N(\tau) - \tilde{\gamma} R(\tau) - \tilde{\theta} E(\tau) = 0$$

which is also satisfied for the maximum likelihood estimators $(\hat{\gamma}, \hat{\theta})$; cf. Example VI.1.5.

For the melanoma data, this gives the M-estimates

$$\tilde{\theta} = 1.25, \qquad \tilde{\gamma} = 0.0368 \text{ per year}$$

with estimated standard deviations

$$\widehat{\text{s.d.}}(\tilde{\theta}) = 0.42, \qquad \widehat{\text{s.d.}}(\tilde{\gamma}) = 0.0081 \text{ per year}$$

and a correlation of -0.60. These estimates are of the same order of magnitude as the maximum likelihood estimates of Example VI.1.5, and as for the excess model they are more variable. □

Example VI.2.5. Software Reliability Models

We consider the Littlewood model for software reliability; cf. Examples III.1.12, VI.1.8, and VI.1.12. For this model, the counting process N has intensity process of the form

$$\lambda(t) = \frac{\alpha(F - N(t-))}{\beta + t} = \frac{F - N(t-)}{\mu + \gamma t}$$

with $\mu = \beta/\alpha$ and $\gamma = 1/\alpha$. Following a suggestion of K.O. Dzhaparidze reported by Geurts, Hasselaar, and Verhagen (1988) simple M-estimators for α, β, and F may be found by solving the equations

$$\int_0^\tau K_j(t) \left\{ \mathrm{d}N(t) - \frac{\alpha(F - N(t-))}{\beta + t} \mathrm{d}t \right\} = 0,$$

$j = 1, 2, 3$, with $K_j(t) = (\beta + t)t^{j-1}$. Since, by integration by parts,

$$\int_0^\tau N(t-)t^{j-1}\,\mathrm{d}t = \frac{\tau^j}{j}N(\tau) - \frac{1}{j}\int_0^\tau t^j\,\mathrm{d}N(t),$$

we get the following linear system of equations for determining $\tilde{\beta}$, $\tilde{\alpha}\tilde{F}$, and $\tilde{\alpha}$:

$$\tilde{\beta}\int_0^\tau t^{j-1}\,\mathrm{d}N(t) - \tilde{\alpha}\tilde{F}\frac{\tau^j}{j} + \int_0^\tau t^j\,\mathrm{d}N(t) + \tilde{\alpha}\left(\frac{\tau^j}{j}N(\tau) - \frac{1}{j}\int_0^\tau t^j\,\mathrm{d}N(t)\right) = 0,$$

$j = 1, 2, 3$. Solving these we get the M-estimates (with estimated standard deviations in brackets) $\tilde{\alpha} = 11.2$ (12.8), $\tilde{\beta} = 2.20$ (2.55), and $\tilde{F} = 46.4$ (1.06) based on all the $n = 43$ failures reported in Example I.3.19. The estimated correlations are $\widehat{\text{corr}}(\tilde{\alpha}, \tilde{\beta}) = 0.9995$, $\widehat{\text{corr}}(\tilde{\alpha}, \tilde{F}) = -0.932$, and $\widehat{\text{corr}}(\tilde{\beta}, \tilde{F}) = -0.924$. We note that the M-estimate for the number of faults in the system is somewhat higher than the maximum likelihood estimate found in Example VI.1.8. However, the extremely high correlation between $\tilde{\alpha}$ and $\tilde{\beta}$ indicates that this parametrization of the Littlewood model may be numerically unstable.

For the alternative parameterization of the model, we get the M-estimates (with estimated standard deviations in brackets) $\tilde{\mu} = 0.196$ (0.007) and $\tilde{\gamma} = 0.089$ (0.102) with estimated correlations $\widehat{\text{corr}}(\tilde{\mu}, \tilde{\gamma}) = -0.359$, $\widehat{\text{corr}}(\tilde{\mu}, \tilde{F}) = -0.094$, and $\widehat{\text{corr}}(\tilde{\gamma}, \tilde{F}) = 0.932$. Thus, the correlations are much lower for this parametrization of the model. Also it is seen that the estimate for μ is close to the maximum likelihood estimate found in Example VI.1.8, whereas γ is not significantly different from zero. □

VI.2.2. Large Sample Properties

As explained below, consistency of the M-estimators does not follow by an argument along the lines of Theorem VI.1.1. The existence of a consistent solution to Eq. (6.2.1) may probably be proved under conditions which mimic those for the classical case of i.i.d. random variables (Huber, 1967; 1981, Section 6.2). We will not pursue this here, however, but rather assume that the M-estimators we are considering are consistent. That this actually is the case must then be checked for each particular situation (and this is straightforward in Examples VI.2.2–5). Once consistency is assumed, the large sample distribution of the M-estimators may be derived in a similar manner as for the maximum likelihood estimator. To start with, we list the required conditions, assuming the general setup of Section VI.1.2:

Condition VI.2.1. Assume that parts A, C, and E of Condition VI.1.1 hold with $(\partial/\partial\theta_j)\log\lambda_h(t;\boldsymbol{\theta})$ replaced by $K_{hj}(t;\boldsymbol{\theta})$ [and, therefore, $(\partial/\partial\theta_j)\lambda_h(t;\boldsymbol{\theta})$ replaced by $K_{hj}(t;\boldsymbol{\theta})\lambda_h(t;\boldsymbol{\theta})$]. Moreover, assume that

(B′) There exist a sequence $\{a_n\}$ of non-negative constants increasing to infinity as $n \to \infty$ and finite functions $\psi_{jl}(\boldsymbol{\theta})$ and $\delta_{jl}(\boldsymbol{\theta})$ defined on Θ_0 such that for all j, l

$$a_n^{-2}\int_0^\tau \sum_{h=1}^{k_n} K_{hj}(t;\boldsymbol{\theta}_0)K_{hl}(t;\boldsymbol{\theta}_0)\lambda_h(t;\boldsymbol{\theta}_0)\,dt \xrightarrow{P} \delta_{jl}(\boldsymbol{\theta}_0)$$

and

$$a_n^{-2}\int_0^\tau \sum_{h=1}^{k_n} K_{hj}(t;\boldsymbol{\theta}_0)\left\{\frac{\partial}{\partial\theta_l}\log\lambda_h(t;\boldsymbol{\theta}_0)\right\}\lambda_h(t;\boldsymbol{\theta}_0)\,dt \xrightarrow{P} \psi_{jl}(\boldsymbol{\theta}_0)$$

as $n \to \infty$.

(D′) The matrices $\boldsymbol{\Psi} = \{\psi_{jl}(\boldsymbol{\theta}_0)\}$ and $\boldsymbol{\Delta} = \{\delta_{jl}(\boldsymbol{\theta}_0)\}$ with elements defined in Condition B′ are, respectively, nonsingular and positive definite.

Let now $\mathscr{I}_\tau^{jl}(\boldsymbol{\theta})$ denote the negative of the first-order partial derivative with respect to θ_l of $U_\tau^j(\boldsymbol{\theta})$ defined in (6.2.1) and let $R_\tau^{jlm}(\boldsymbol{\theta})$ denote its second-order partial derivative with respect to θ_l and θ_m. Then, by going through the arguments of Theorem VI.1.1, it is seen that (6.1.10) and (6.1.12) remain true under the new and more general definitions of the quantities involved. Also [cf. (6.1.11)],

$$a_n^{-2}\mathscr{I}_\tau^{jl}(\boldsymbol{\theta}_0) \xrightarrow{P} \psi_{jl}(\boldsymbol{\theta}_0)$$

as $n \to \infty$, where $\psi_{jl}(\boldsymbol{\theta}_0)$ is defined in Condition B′. For the maximum likelihood estimator, (6.1.10)–(6.1.12) are sufficient to show that there exists a consistent solution to the likelihood equations. However, the argument used in this proof (e.g., Billingsley, 1961, pp. 12 and 13) exploits the fact that $\boldsymbol{\Sigma}$ defined in part D of Condition VI.1.1 is positive definite and, therefore, cannot be copied here.

Instead of (6.1.17), we now get

$$a_n^{-1}\mathbf{U}_\tau(\boldsymbol{\theta}_0) \xrightarrow{\mathscr{D}} \mathscr{N}(\mathbf{0}, \boldsymbol{\Delta}),$$

where $\boldsymbol{\Delta} = \{\delta_{jl}(\boldsymbol{\theta}_0)\}$ with $\delta_{jl}(\boldsymbol{\theta}_0)$ defined in Condition B′.

From these results, the asymptotic normality of the M-estimators follows exactly as for the maximum likelihood estimator provided we assume consistency. Thus, we have proved:

Theorem VI.2.1. *Assume that Condition* VI.2.1 *holds and that* $\tilde{\boldsymbol{\theta}}$ *is a consistent solution to Eq.* (6.2.1). *Then*

$$a_n(\tilde{\boldsymbol{\theta}} - \boldsymbol{\theta}_0) \xrightarrow{\mathscr{D}} \mathscr{N}(\mathbf{0}, \boldsymbol{\Psi}^{-1}\boldsymbol{\Delta}(\boldsymbol{\Psi}^{\mathsf{T}})^{-1})$$

as $n \to \infty$. *The matrices* $\boldsymbol{\Psi}$ *and* $\boldsymbol{\Delta}$ *defined in Condition D′ may be estimated consistently by* $a_n^{-2}\tilde{\boldsymbol{\Psi}}$ *and* $a_n^{-2}\tilde{\boldsymbol{\Delta}}$ *with elements as given by* (6.2.3) *and* (6.2.4) (*with k replaced by* k_n).

In a similar manner to the classical case of i.i.d. random variables, we may prove that the maximum likelihood estimator $\hat{\boldsymbol{\theta}}$ is optimal within the class of

M-estimators $\tilde{\boldsymbol{\theta}}$ in the sense that

$$\boldsymbol{\Psi}^{-1}\boldsymbol{\Delta}(\boldsymbol{\Psi}^{\mathsf{T}})^{-1} - \boldsymbol{\Sigma}^{-1}, \tag{6.2.6}$$

with $\boldsymbol{\Sigma}$ defined in part D of Condition VI.1.1, is always non-negative definite. This may be proved by first noting that

$$\begin{bmatrix} a_n^{-1} \int_0^\tau \sum_{h=1}^{k_n} K_{h1}(t;\boldsymbol{\theta}_0)\,\mathrm{d}M_h(t) \\ \vdots \\ a_n^{-1} \int_0^\tau \sum_{h=1}^{k_n} K_{hq}(t;\boldsymbol{\theta}_0)\,\mathrm{d}M_h(t) \\ a_n^{-1} \int_0^\tau \sum_{h=1}^{k_n} \frac{\partial}{\partial\theta_1} \log\lambda_h(t;\boldsymbol{\theta}_0)\,\mathrm{d}M_h(t) \\ \vdots \\ a_n^{-1} \int_0^\tau \sum_{h=1}^{k_n} \frac{\partial}{\partial\theta_q} \log\lambda_h(t;\boldsymbol{\theta}_0)\,\mathrm{d}M_h(t) \end{bmatrix} \xrightarrow{\mathscr{D}} \mathscr{N}\left(\mathbf{0}, \begin{pmatrix} \boldsymbol{\Delta} & \boldsymbol{\Psi} \\ \boldsymbol{\Psi}^{\mathsf{T}} & \boldsymbol{\Sigma} \end{pmatrix}\right)$$

so that the matrix

$$\begin{pmatrix} \boldsymbol{\Delta} & \boldsymbol{\Psi} \\ \boldsymbol{\Psi}^{\mathsf{T}} & \boldsymbol{\Sigma} \end{pmatrix}$$

is non-negative definite. From this, it follows as in Rao (1973, p. 327) that

$$\boldsymbol{\Delta} - \boldsymbol{\Psi}\boldsymbol{\Sigma}^{-1}\boldsymbol{\Psi}^{\mathsf{T}}$$

is non-negative definite, and from this it is an immediate consequence that the matrix in (6.2.6) is non-negative definite.

VI.3. Model Checking

A common and useful technique for checking the validity of a parametric model is to embed it in a larger parametric model and use, e.g., the likelihood ratio test to check whether the reduction to the actual model is valid; for applications in survival analysis, see, e.g., Kalbfleisch and Prentice (1980, Chapter 3) and Cox and Oakes (1984, Chapter 3). We will illustrate this technique in Example VI.3.1. Also, the appropriateness of the Gompertz model (relative to the Gompertz–Makeham model) for the mortality of female diabetics was checked in this manner in Example VI.1.4.

Here, we will not discuss this technique any further, but rather concentrate on a number of special techniques which are valuable for model checking. First, we illustrate use of the Nelson–Aalen estimator for model checking, then the Total Time on Test plot from reliability as well as some related tests are discussed, and, finally, we consider some possibilities for constructing formal goodness-of-fit tests versus general alternatives.

VI.3.1. The Nelson–Aalen Estimator

As mentioned in Example IV.1.1, one use of the Nelson–Aalen estimator for survival data is to check graphically whether the lifetimes appear to follow a certain parametric distribution; in fact, this was the rationale behind the estimator in Nelson's (1969) original paper. For example, the exponential distribution has constant hazard rate function α and cumulative hazard function $A(t) = \alpha t$. Therefore, the Nelson–Aalen estimator plotted versus t should give an approximately straight line for the exponential model. As another example, consider the Weibull distribution with hazard rate function $\alpha(t) = \alpha\theta(\alpha t)^{\theta-1}$ and cumulative hazard function $A(t) = (\alpha t)^{\theta}$. Here, $\log A(t) = \theta \log \alpha + \theta \log t$, so that $\log \hat{A}(t)$ plotted versus $\log t$ should yield an approximately straight line for the Weibull model.

This approach may, of course, be used for graphical checks, not only for survival data but also for other parametric counting process models. Furthermore, by adding confidence bands to the plot (cf. Section IV.1.3), more formal testing procedures are obtained. This is illustrated in Example VI.3.1 concerning the melanoma data and Example VI.3.2 on the dynamics of a baboon troop.

EXAMPLE VI.3.1. Survival with Malignant Melanoma. Check of Exponential Model

In Example IV.1.2, Figure IV.1.1 showed the Nelson–Aalen estimates for the integrated hazard for male and female melanoma patients. Both curves (as also mentioned in that example) look roughly linear, suggesting that a model with a constant hazard may be appropriate. In Figure VI.3.1, the corresponding integrated hazard estimates (straight lines) based on the exponential model are added and they are seen to approximate the Nelson–Aalen estimates quite well. The slopes are

$$\hat{\alpha}_0 = 28/786.9 = 0.0356 \text{ (per year)}$$

and

$$\hat{\alpha}_1 = 29/420.8 = 0.0689 \text{ (per year)}$$

for women and men, respectively, with estimated standard deviations $\widehat{\text{s.d.}}(\hat{\alpha}_0) = 0.0067$ (per year) and $\widehat{\text{s.d.}}(\hat{\alpha}_1) = 0.0128$ (per year). The likelihood ratio test statistic for the hypothesis $\alpha_0 = \alpha_1$ takes the value 6.15 ($P = .01$ when referred to the χ^2 distribution with one degree of freedom).

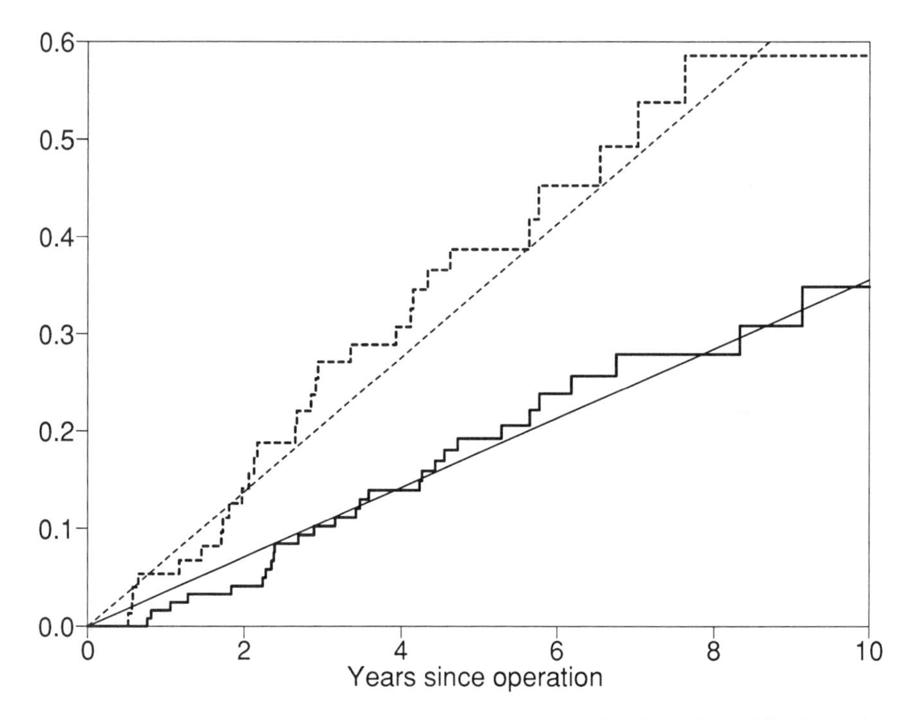

Figure VI.3.1. Malignant melanoma. Estimated cumulative death intensities for male (- - -) and female (—) patients based on a nonparametric analysis (step-functions) and on an exponential model (straight lines).

Figure IV.1.15 showed the Nelson–Aalen estimate for males with 95% confidence bands for the interval [1 year, 7 years]. In Figure VI.3.2, the line with slope $\hat{\alpha}_1$ is added and it is seen to lie well within the bands, again speaking in favor of the exponential model.

As indicated above, it is also possible to check the appropriateness of the exponential model by fitting a Weibull model $\alpha\theta(\alpha t)^{\theta-1}$ for the death intensity. The maximum likelihood estimates for θ become (with estimated standard deviations in brackets) 1.197 (0.202) for women and 1.017 (0.166) for men. Thus, for both sexes the Wald test for the hypothesis $\theta = 1$ gives quite insignificant results. The likelihood ratio test gives similar results. This further indicates that an exponential model is a reasonable one. □

Example VI.3.2. The Dynamics of a Natural Baboon Troop. Nelson–Aalen Plots

The problem and the data were introduced in Example I.3.18. Nelson–Aalen plots based on $N_1(t)$ and $N_2(t)$ (counting the births and deaths, respec-

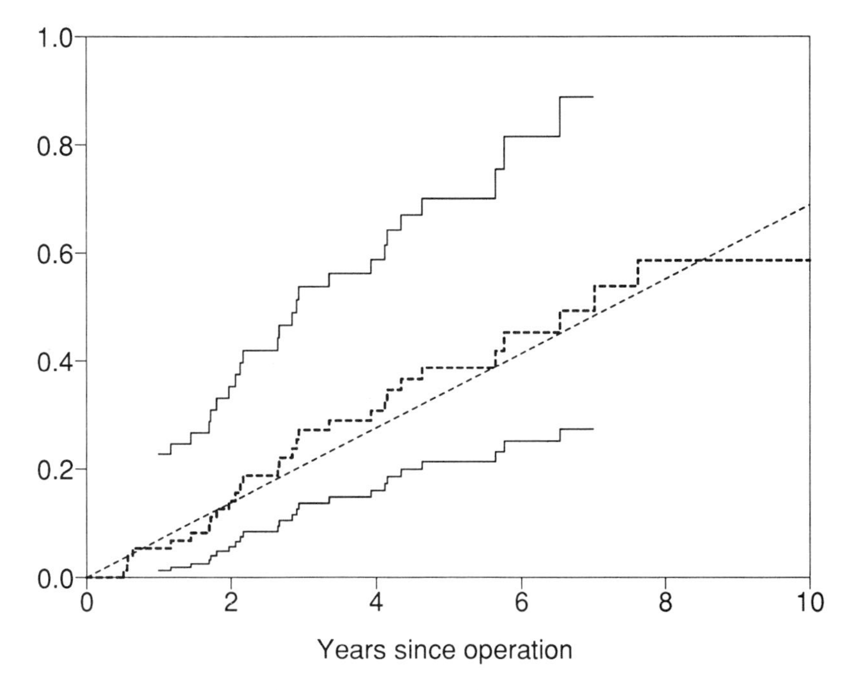

Figure VI.3.2. Nelson–Aalen estimate for male patients with malignant melanoma with approximate log-transformed 95% EP-confidence band for the interval [1, 7] years (step-functions). Also shown is the estimate based on an exponential model (straight line).

tively) of Example VI.1.6 are given in Figures VI.3.3 and VI.3.4 with 95% EP-confidence bands for the interval [100, 360] days based on the log-transformation. Furthermore, the lines with slope $\hat{\lambda}$ and $\hat{\mu}$, respectively, are plotted to obtain a visual comparison with the integrated intensity under the hypothesis of constant intensities. It seems as if the uncertainty is too large to rule out this hypothesis. In Examples VI.3.4 and VI.3.8, we return to the problem of testing the hypothesis of constant intensities. □

Example VI.3.3. Mortality of Female Diabetics in the County of Fyn. Goodness-of-Fit of the Gompertz Model Assessed by the Nelson–Aalen Estimates of the Integrated Death Intensity

The problem and the data were introduced in Example I.3.2.

In Figure VI.3.5 we give the Nelson-Aalen plot for the integrated death intensity for female diabetics together with a curve representing the integrated death intensity based on the Gompertz model fitted by maximum

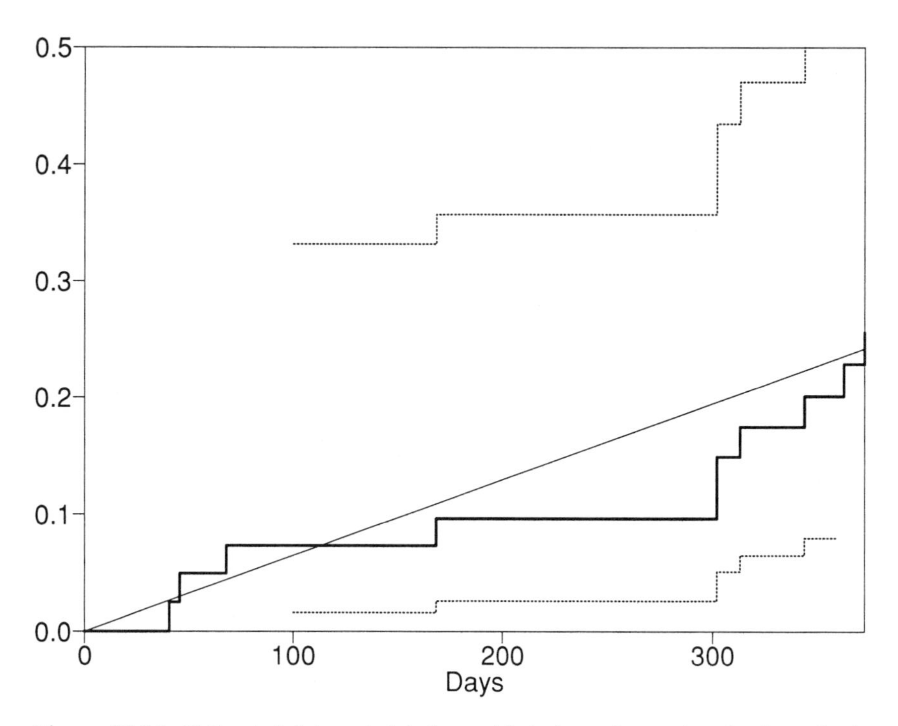

Figure VI.3.3. Estimated integrated baboon birth intensity under the hypothesis of constant birth intensity (straight line) and Nelson–Aalen plot of births with 95% EP-confidence band over the interval [100, 360] days based on the log-transformation (step-functions).

likelihood; cf. Example VI.1.4. The two estimates for the integrated death intensity are seen to follow each other closely, indicating that the Gompertz model gives a satisfactory description of the mortality of insulin-dependent female diabetics. □

One may, in fact, go one step further and actually estimate the parameters and make formal hypothesis tests directly from the Nelson–Aalen plots using regression techniques. For some fixed time points $t_1, t_2, \ldots, t_m$, the m-vector

$$\hat{\mathbf{A}}_m = (\hat{A}(t_1), \ldots, \hat{A}(t_m))^\top$$

has, according to Theorem IV.1.2, an asymptotic multinormal distribution with mean $(A(t_1), \ldots, A(t_m))^\top$ and a covariance matrix, say $\mathbf{\Gamma}$, whose element $(j, l), j, l = 1, \ldots, m$, we may estimate by $\hat{\sigma}^2(t_j \wedge t_l)$; cf. (4.1.6) and (4.1.7). Thus, parameters $\boldsymbol{\theta}$ for a given specification $A(t; \boldsymbol{\theta})$ of $A(t)$ can be estimated using weighted least squares by minimizing

$$(\hat{\mathbf{A}}_m - \mathbf{A}_m(\boldsymbol{\theta}))^\top \hat{\mathbf{\Gamma}}^{-1} (\hat{\mathbf{A}}_m - \mathbf{A}_m(\boldsymbol{\theta})),$$

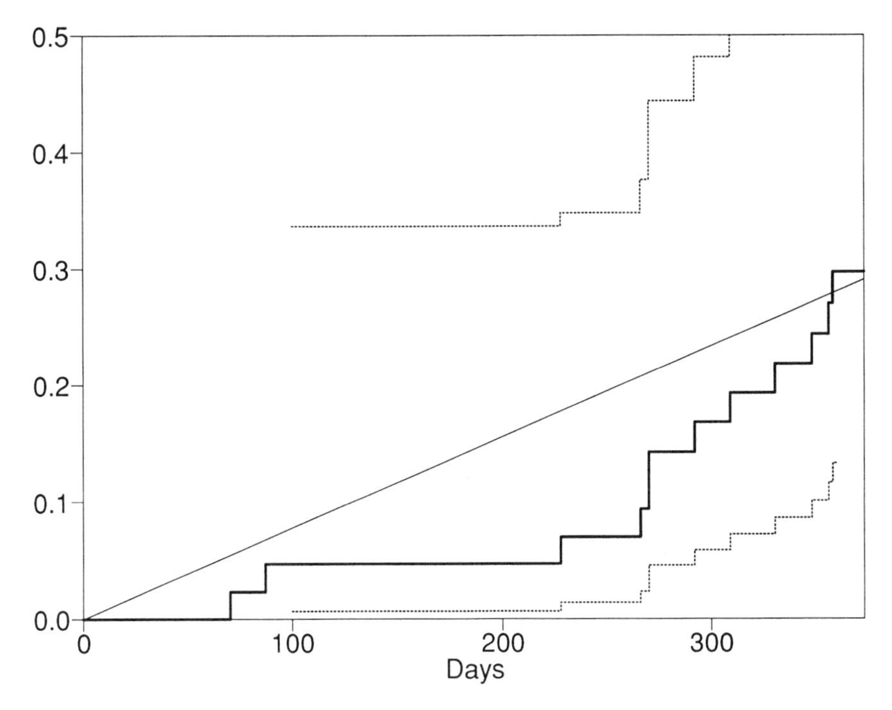

Figure VI.3.4. Estimated integrated baboon death intensity under the hypothesis of constant death intensity (straight line) and Nelson–Aalen plot of deaths with 95% EP-confidence band over the interval [100, 360] days based on the log-transformation (step-functions).

where $\mathbf{A}_m(\boldsymbol{\theta}) = (A(t_1;\boldsymbol{\theta}),\ldots,A(t_m;\boldsymbol{\theta}))^\mathsf{T}$, with respect to $\boldsymbol{\theta}$. Moreover, if the model holds, this quantity, with the estimated parameter inserted, is asymptotically χ^2 distributed with degrees of freedom equal to m minus the number of estimated parameters. This procedure is essentially a "continuous" version of the curve-fitting technique (analytic graduation) used by actuaries and demographers to smooth a sequence of occurrence/exposure rates; see the Bibliographic Remarks in Section VI.4. If the time points are numerous and fill the whole interval $[0, \tau]$, the resulting estimator is close to being efficient ("the Nelson–Aalen estimator is asymptotically sufficient").

One final possible use of the Nelson–Aalen estimator for goodness-of-fit checking of parametric survival data models is based on the fact that for i.i.d. lifetimes $X_1, \ldots, X_n$ with hazard $\alpha(t;\boldsymbol{\theta})$, the cumulative hazards $A(X_1;\boldsymbol{\theta}), \ldots,$ $A(X_n;\boldsymbol{\theta})$ evaluated at the observed lifetimes (assuming no censoring) are i.i.d. with a unit exponential distribution. Use of the corresponding "residuals" $A(X_1;\hat{\boldsymbol{\theta}}), \ldots, A(X_n;\hat{\boldsymbol{\theta}})$ for checking the specification of a model is discussed in Section VII.3.4 in connection with regression models.

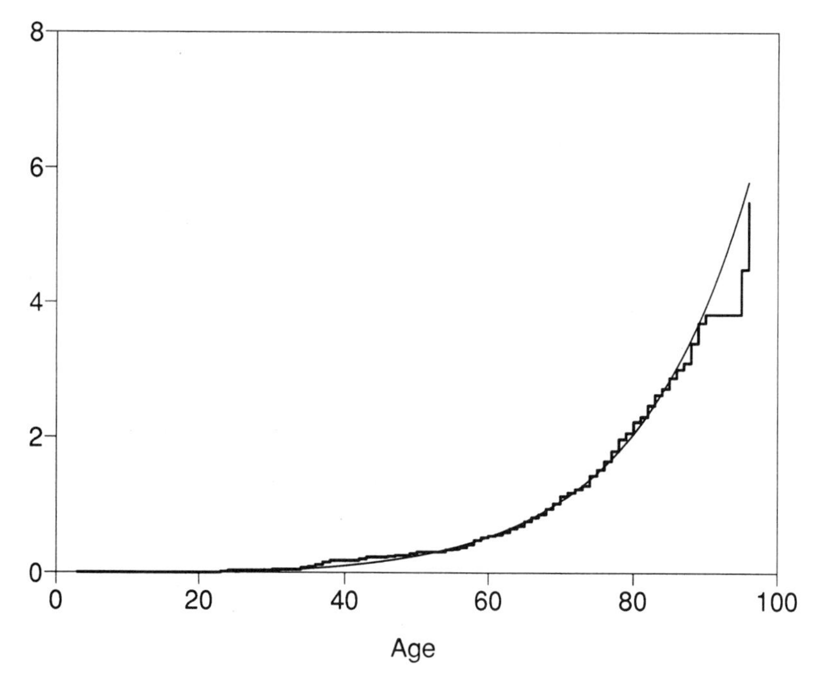

Figure VI.3.5. Estimated integrated death intensity for female diabetics in Fyn under the Gompertz model (smooth line) and the corresponding Nelson–Aalen plot (step-function).

VI.3.2. The Total Time on Test Plot

We consider a univariate counting process $N(t)$ satisfying the multiplicative intensity model, i.e., it has intensity process of the form

$$\lambda(t) = \alpha(t)Y(t). \tag{6.3.1}$$

In the present section, we will consider a graphical technique, the so-called Total Time on Test plot, or TTT-plot for short, for checking the assumption that $\alpha(t)$ is constant. We will also study related test statistics for formally testing the hypothesis that there exists a θ such that

$$\alpha(t) = \theta \tag{6.3.2}$$

for all $t \in \mathscr{T}$.

It is simple to motivate the TTT-plot using the usual decomposition (2.4.1) of N into its compensator and a martingale. To this end, we assume that (6.3.2) holds and introduce

$$R(t) = \int_0^t Y(s)\,\mathrm{d}s; \tag{6.3.3}$$

cf. Example VI.1.1. Then

$$N(t) = \theta R(t) + M(t),$$

with M a local square integrable martingale. Therefore, we have

$$\mathrm{E}R(t) = \mathrm{E}N(t)/\theta$$

(provided that the expectations exist), and it follows that a plot of $R(t)$ against $N(t)$ should give approximately a straight line with slope θ^{-1}. This plot may be transformed to the unit square by choosing a (possibly random) time $T \in \mathscr{T}$ and plotting $R(t)/R(T)$ against $N(t)/N(T)$. This is the TTT-plot, the name being due to the fact that $R(t)$ given by (6.3.3) measures the "exposure" or "total time on test" when $Y(t)$ is the size of the risk set. The plot should approximate a straight line with unit slope when (6.3.2) holds.

Let $T_1 < T_2 < T_3 < \cdots$ be the jump times of N and assume that $T_m < T \leq T_{m+1}$. It is then quite common to plot $R(T_i)/R(T)$ against $N(T_i)/N(T) = i/N(T)$ for $i = 1, 2, \ldots, m$ and connect the points with straight lines. We will, however, not adopt this convention, but rather plot $R(t)/R(T)$ against $N(t)/N(T)$ for all $t \in [0, T]$ and connect the vertical lines one gets in this manner by horizontal lines.

The TTT-plot is especially well-suited for situations where the alternative to the hypothesis (6.3.2) of special interest is that $\alpha(t)$ is monotone. Since we have

$$\mathrm{E}N(t) = \mathrm{E}\int_0^t \alpha(s)Y(s)\,\mathrm{d}s = \int_0^t \alpha(s)\,\mathrm{d}\mathrm{E}R(s)$$

under (6.3.1), it is seen that $\mathrm{dE}R(t) = \mathrm{dE}N(t)/\alpha(t)$. Therefore, the TTT-plot will tend to be concave for an increasing $\alpha(t)$ and convex when $\alpha(t)$ is decreasing.

Closely related to the TTT-plot is the cumulative total time on test statistic defined as $N(T)$ times the area under the TTT-plot. According to the just-mentioned properties of the TTT-plot, this statistic for testing hypothesis (6.3.2) tends to take on large values when $\alpha(t)$ is increasing and small values when it is decreasing. One may also use a Kolmogorov–Smirnov-type test, i.e., reject the hypothesis when the maximum distance between the TTT-plot and the diagonal line $y = x$ is large.

To study formally the properties of the TTT-plot, we follow Aalen and Hoem (1978) and note that in the new (random) time scale measured by "the total time on test" (6.3.3), N is transformed into a counting process N^* given by

$$N^*(u) = N(R^{-1}(u)) \tag{6.3.4}$$

on the (random) interval $[0, R(\tau)]$. Here, $R^{-1}(u) = \inf\{t\colon R(t) \geq u\}$. According to the results of Section II.5.2, this counting process has intensity process

$$\lambda^*(u) = \alpha(R^{-1}(u)). \tag{6.3.5}$$

Especially under hypothesis (6.3.2), it is a counting process with a constant intensity process θ on $[0, R(\tau)]$, i.e., a randomly stopped Poisson process with constant intensity.

Now, if there exists an m such that $N(\tau) \geq m + 1$ with probability one (which is the case for completely observed failure-time data and failure-time data subject to simple type II censoring), the statistical properties of the TTT-plot are easily derived for the case when we let T equal the $(m + 1)$th jump time T_{m+1} of N. For then under hypothesis (6.3.2), N^* is a Poisson process with constant intensity θ stopped at the $(m + 1)$th event, and $R(T_i)/R(T)$, $i = 1, 2, \ldots, m$, will have the same distribution as the order statistics based on m i.i.d. uniform [0, 1] random variables. From this, some exact distributional results for the TTT-plot and the test statistics may be derived [e.g., Barlow et al. (1972), Barlow and Campo (1975)]. Moreover, it follows that the TTT-plot (properly normalized) converges in distribution to a standard Brownian bridge when (6.3.2) holds. From this, we also get asymptotic results for the test statistics as reviewed below.

In most applications of the multiplicative intensity model (6.3.1), there does not exist an m such that $N(\tau) \geq m + 1$ with probability one, and then the above results do not apply directly. Using the random time change (6.3.4) and standard results on weak convergence of processes, however, Gill (1986b) proved that under hypothesis (6.3.2) the TTT-plot has the same asymptotic distribution as the empirical distribution function from a sample of i.i.d. uniform [0, 1] random variables, taking $N(T)$ as the number of observations. [For the special case of $T = \tau$, this result will follow also from (6.3.21).] It follows immediately that the asymptotic distribution of the signed area between the TTT-plot and the diagonal line $y = x$, times $\sqrt{N(T)}$, is the same as that of $\int_0^1 W^0(x)\,dx$, W^0 being the standard Brownian bridge. Therefore, the normalized cumulative Total Time on Test statistic

$$\sqrt{N(T)}\left(\frac{1}{N(T)} \sum_{i=1}^{N(T)} \frac{R(T_i)}{R(T)} - \frac{1}{2}\right) \tag{6.3.6}$$

is asymptotically normally distributed with mean zero and variance 1/12 under the hypothesis. Furthermore, the Kolmogorov–Smirnov-type test for hypothesis (6.3.2) is to reject at the level α when

$$\sqrt{N(T)} \sup_{0 \leq t \leq T} \left| \frac{R(t)}{R(T)} - \frac{N(t)}{N(T)} \right| > e_\alpha, \tag{6.3.7}$$

where e_α is the upper α fractile in the distribution of $\sup_{0 \leq x \leq 1} |W^0(x)|$.

Example VI.3.4. The Dynamics of a Natural Baboon Troop. TTT-Plot and Statistics

The problem and the data were introduced in Example I.3.18.

Figures VI.3.6 and 7 contain TTT-plots for births and deaths, respectively, with the identity line indicating the null hypothesis of a constant intensity and boundaries $\pm e_\alpha/\sqrt{N_h(T)}$ around the identity line, where we use

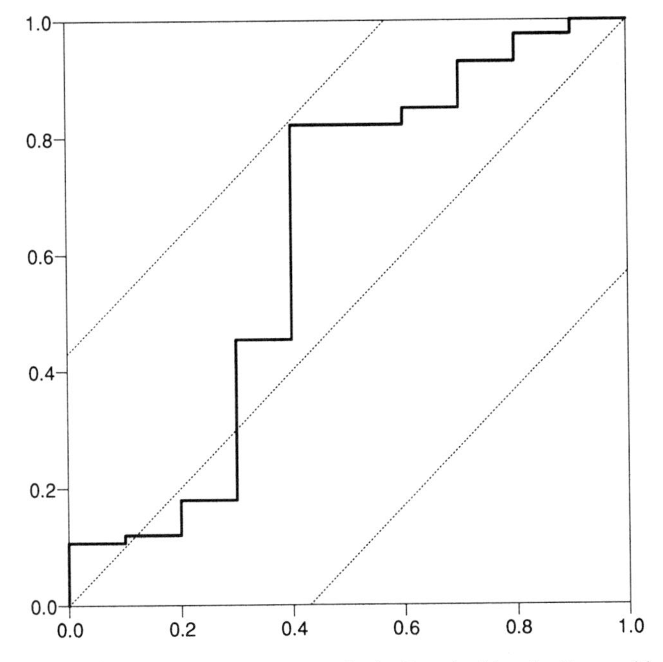

Figure VI.3.6. TTT-plot for baboon births, including the identity line and boundaries corresponding to the Kolmogorov–Smirnov test (6.3.7).

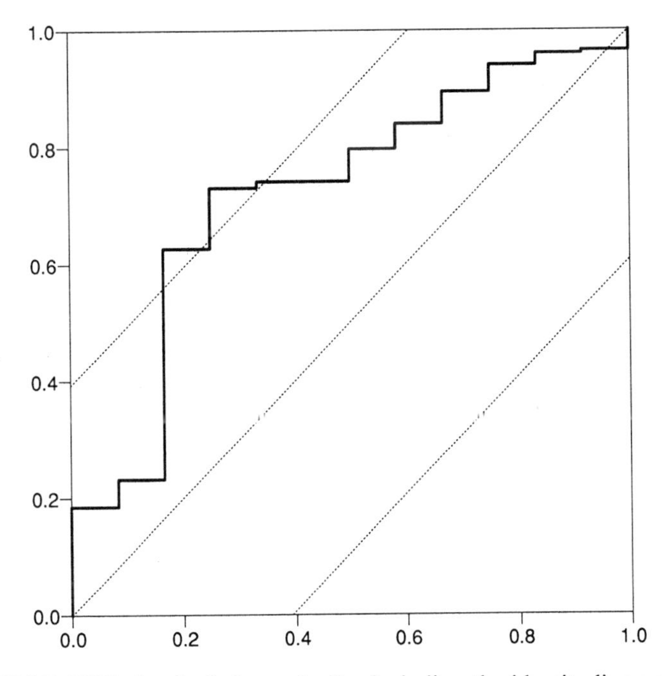

Figure VI.3.7. TTT-plot for baboon deaths, including the identity line and boundaries corresponding to the Kolmogorov–Smirnov test (6.3.7).

$e_{0.05} = 1.36$ [e.g., Shorack and Wellner (1986, Table 1)] corresponding to the Kolmogorov–Smirnov test (6.3.7). Furthermore, $N_1(T) = 10$ is the total number of births and $N_2(T) = 12$ the total number of deaths over the $T = \tau = 373$ days of observation.

It is seen that the TTT-plot for births (Figure VI.3.6) stays just within its boundaries, whereas that for deaths (Figure VI.3.7) exceeds one of them. For deaths, the value of the maximal vertical distance between the TTT-plot and the diagonal is 0.48, corresponding to a value of the Kolmogorov–Smirnov statistic of $\sqrt{12} \cdot 0.48 = 1.67$, i.e., $P = .008$. For the births, we similarly get a Kolmogorov–Smirnov statistic of $\sqrt{10} \cdot 0.42 = 1.33$; $P = .06$.

The cumulative TTT statistic (6.3.6) based on the signed area between the TTT-plot and the diagonal takes the values 1.053 for deaths and 0.398 for births, corresponding to $P = .003$ and .17, respectively.

Note that both tests based on the TTT-plot give significant deviation from the null hypothesis of constant death intensity and that a marginally significant result is obtained for the birth intensity when the Kolmogorov–Smirnov test is used. These results are in contrast to the confidence bands of Figures VI.3.3 and VI.3.4, which were far from crossing the estimated integrated intensities under the hypothesis. As mentioned earlier, the tests based on the TTT-plot are particularly useful for testing "exponentiality" (i.e., constant intensity) against "increasing failure rate" alternatives. According to Figure VI.3.4, we have exactly such an alternative for the death intensity, whereas the omnibus EP-confidence band is not sufficiently sensitive against this particular departure from the hypothesis of constant intensity. □

EXAMPLE VI.3.5. Survival with Malignant Melanoma: A Model with Constant Relative Mortality

The melanoma data were introduced in Example I.3.1. We now turn to the problem of testing whether a model with a constant *relative* mortality fits these data; cf. Examples IV.1.11 and VI.1.5. Figure VI.3.8 shows the TTT-plot for testing against a relative mortality that varies with follow-up time: $E(t)/E(T)$ is plotted against $N(t)/N(T)$ where we have chosen $T = 10$ years, and

$$E(t) = \int_0^t Y^{\mu}(u)\,du$$

is the "expected" number of failures before time t. The plot also shows the boundaries $\pm e_\alpha/\sqrt{71}$ around the diagonal, where $e_\alpha = 1.36$ corresponding to the level $\alpha = 0.05$ of the Kolmogorov–Smirnov test (6.3.7). Since the curve does not cross these boundaries, the Kolmogorov–Smirnov test is not significant at the 5% level; in fact, the test statistic takes the value 0.49 corresponding to $P = .97$.

To test against a model where the relative mortality varies with age, the counting processes are redefined as in Example IV.1.11. Figure VI.3.9 shows the corresponding TTT-plot with significance boundaries for $\alpha = 0.05$. The

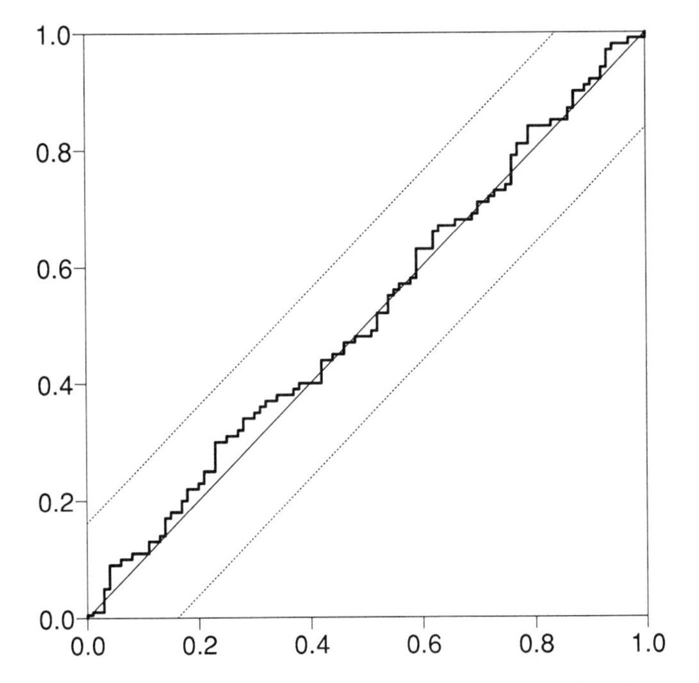

Figure VI.3.8. TTT-plot for a constant relative mortality model for the melanoma data using follow-up time as the basic time scale.

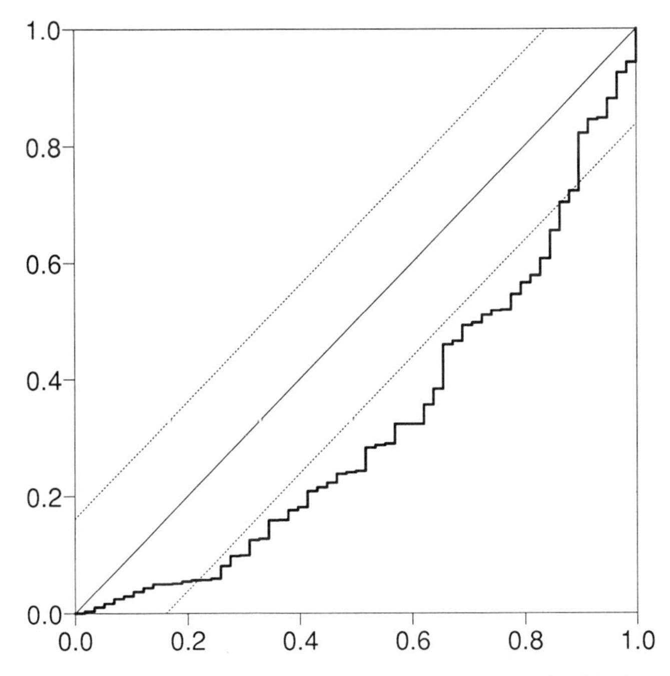

Figure VI.3.9. TTT-plot for a constant relative mortality model for the melanoma data using age as the basic time scale.

plot indicates a very poor fit of the model with a constant relative mortality, a fact which is confirmed by the value 2.26 of the Kolmogorov–Smirnov test statistic ($P < .001$). The *convex* shape of the curve suggests that the relative mortality *decreases* with age; cf. also Figure IV.1.11. (Since few patients contribute information outside the age interval from 30 to 80 years, the plot is based only on this interval.) □

VI.3.3. Goodness-of-Fit Tests

In this subsection, we consider a univariate counting process $N(t)$ with intensity process $\lambda(t)$. Our object is to derive formal goodness-of-fit tests for the hypothesis

$$\lambda(t) = \lambda(t; \boldsymbol{\theta}), \tag{6.3.8}$$

i.e., the hypothesis that $\lambda(t)$, in addition to observable processes, may be specified by a q-dimensional parameter $\boldsymbol{\theta} = (\theta_1, \ldots, \theta_q)$. For the particular case of the multiplicative intensity model (6.1.8), i.e., when $\lambda(t) = \alpha(t) Y(t)$, this amounts to testing the hypothesis that $\alpha(t) = \alpha(t; \boldsymbol{\theta})$.

In Section VI.3.1, we saw how the Nelson–Aalen estimator could be used for graphically checking whether the $\alpha(t)$ of the multiplicative intensity model had a certain form by comparing the Nelson–Aalen estimator

$$\hat{A}(t) = \int_0^t \{J(s)/Y(s)\}\, \mathrm{d}N(s)$$

with

$$A^*(t; \boldsymbol{\theta}) = \int_0^t J(s)\alpha(s; \boldsymbol{\theta})\, \mathrm{d}s,$$

where, as usual, $J(t) = I(Y(t) > 0)$. If the specified parametric model is correct, these two functions should agree reasonably well for a suitably chosen $\boldsymbol{\theta}$ value, e.g., the maximum likelihood estimator $\hat{\boldsymbol{\theta}}$. To construct formal goodness-of-fit tests, therefore, it is reasonable to consider the process

$$\hat{A}(t) - A^*(t; \hat{\boldsymbol{\theta}}) = \int_0^t \frac{J(s)}{Y(s)} \{\mathrm{d}N(s) - \alpha(s; \hat{\boldsymbol{\theta}}) Y(s)\, \mathrm{d}s\}$$

or, more generally, a weighted version of this (cf. Section V.1).

To test hypothesis (6.3.8) for the general model, it is in a similar manner reasonable to compare $N(t)$ and $\int_0^t \lambda(s; \hat{\boldsymbol{\theta}})\, \mathrm{d}s$. Goodness-of-fit tests may, therefore, be based on the process

$$Z(t) = \int_0^t K(s; \hat{\boldsymbol{\theta}}) \{\mathrm{d}N(s) - \lambda(s; \hat{\boldsymbol{\theta}})\, \mathrm{d}s\}, \tag{6.3.9}$$

where $K(t; \hat{\boldsymbol{\theta}})$ is a suitably chosen process which we allow to depend on the maximum likelihood estimator $\hat{\boldsymbol{\theta}}$ for greater flexibility.

In this subsection, we will consider various tests for (6.3.8) based on the goodness-of-fit process (6.3.9). Following Hjort (1990a), we first derive the limiting distribution of (6.3.9) (properly normalized) under the hypothesis. This limiting distribution is, in general, quite intractable. However, for the one-parameter case with weight process $K(t;\hat{\theta}) = (\partial/\partial\theta)\log\lambda(t;\hat{\theta})$ Brownian bridge asymptotics are available, and we show how this may be used to construct Kolmogorov–Smirnov-type tests and generalized TTT-plots. Alternatively, the limiting distribution of (6.3.9) may be used to construct χ^2-type tests.

Another approach is to apply the idea of Khmaladze (1981, 1988) by basing goodness-of-fit tests on a transformation of (6.3.9) to a process with a nice limiting distribution, and we indicate how Khmaladze's ideas may be implemented in our framework.

In the main body of this section, we consider, for the ease of presentation, a univariate counting process as outlined earlier. For some situations (e.g., regression models), we need to consider multivariate counting processes, however, and we conclude our discussion of goodness-of-fit tests by indicating how our results may be extended to cover the multivariate case.

VI.3.3.1. *Weak Convergence of the Goodness-of-Fit Process*

We study the asymptotic properties of process (6.3.9) under the hypothesis. Thus, we consider a sequence of models of the above form, as described at the beginning of Section VI.1.2, all having intensity process of the form (6.3.8). As before, we denote by $\boldsymbol{\theta}_0$ the true value of the parameter. We assume that the intensity process $\lambda(t;\boldsymbol{\theta})$ and the weight process $K(t;\boldsymbol{\theta})$ satisfy Conditions VI.1.1 and VI.2.1, with $k_n = 1$ and with $K_{hj}(t;\boldsymbol{\theta})$ replaced by $K(t;\boldsymbol{\theta})$ in Condition VI.2.1, and not only when the integrals are taken over $[0,\tau]$ but when they are taken over any interval $[0,t]$ with $t \leq \tau$. [It should be realized, however, that the $K(t;\boldsymbol{\theta})$ considered here otherwise has nothing to do with the $K_{hj}(t;\boldsymbol{\theta})$ processes considered in connection with the M-estimators in Section VI.2.] This will ensure that all the Taylor expansions to be presented are valid, and that the remainder terms behave properly. Formal proofs may be written out using arguments similar to the proofs of Theorems VI.1.1 and VI.1.2 (see also Hjort, 1990a).

Moreover, to be able to express our results in the nicest possible way, we assume that there exist functions $f(t;\boldsymbol{\theta})$ and $\kappa(t;\boldsymbol{\theta})$ such that

$$a_n^{-2}\lambda(t;\boldsymbol{\theta}_0) \xrightarrow{\mathrm{P}} f(t;\boldsymbol{\theta}_0) \tag{6.3.10}$$

and

$$K(t;\boldsymbol{\theta}_0) \xrightarrow{\mathrm{P}} \kappa(t;\boldsymbol{\theta}_0) \tag{6.3.11}$$

as $n \to \infty$. [More precisely, we assume that the convergence is sufficiently uniform so that the limits on the right-hand sides in Conditions VI.1.1.B and VI.2.1.B′ may be expressed by substituting $f(t;\boldsymbol{\theta}_0)$ for $a_n^{-2}\lambda(t;\boldsymbol{\theta}_0)$, $(\partial/\partial\theta_l)\log f(s;\boldsymbol{\theta}_0)$ for $(\partial/\partial\theta_l)\log\lambda(s;\boldsymbol{\theta}_0)$ and $\kappa(t;\boldsymbol{\theta}_0)$ for $K(t;\boldsymbol{\theta}_0)$ in the corresponding left-hand expressions.]

Then we get from (6.3.9), by a Taylor series expansion around $\boldsymbol{\theta}_0$,

$$\begin{aligned} Z(t) &= \int_0^t K(s;\hat{\boldsymbol{\theta}})\,\mathrm{d}N(s) - \int_0^t K(s;\hat{\boldsymbol{\theta}})\lambda(s;\hat{\boldsymbol{\theta}})\,\mathrm{d}s \\ &= \int_0^t K(s;\boldsymbol{\theta}_0)\,\mathrm{d}M(s) - \sum_{j=1}^q \left\{(\hat{\theta}_j - \theta_{0j})\int_0^t K(s;\boldsymbol{\theta}_0)\frac{\partial}{\partial\theta_j}\lambda(s;\boldsymbol{\theta}_0)\,\mathrm{d}s\right\} \\ &\quad + \sum_{j=1}^q \left\{(\hat{\theta}_j - \theta_{0j})\int_0^t \frac{\partial}{\partial\theta_j}K(s;\boldsymbol{\theta}_0)\,\mathrm{d}M(s)\right\} + o_{\mathrm{P}}(a_n), \end{aligned} \tag{6.3.12}$$

where $M(t)$ is the local square integrable martingale

$$M(t) = N(t) - \int_0^t \lambda(s;\boldsymbol{\theta}_0)\,\mathrm{d}s.$$

Furthermore, it follows from Lenglart's inequality (see Section II.5.2) as in (6.1.16) that

$$\sup_{t\in\mathscr{T}}\left|a_n^{-2}\int_0^t \frac{\partial}{\partial\theta_j}K(s;\boldsymbol{\theta}_0)\,\mathrm{d}M(s)\right| \xrightarrow{\mathrm{P}} 0$$

as $n\to\infty$. By Theorem VI.1.2, therefore, the third term on the right-hand side of (6.3.12) is of the order $o_{\mathrm{P}}(a_n)$. Finally, from (6.1.13) and the proof of Theorem VI.1.2, we have that

$$a_n(\hat{\boldsymbol{\theta}} - \boldsymbol{\theta}_0) = \boldsymbol{\Sigma}^{-1}a_n^{-1}\int_0^\tau \left(\frac{\partial}{\partial\theta_1}\log\lambda(s;\boldsymbol{\theta}_0),\ldots,\frac{\partial}{\partial\theta_q}\log\lambda(s;\boldsymbol{\theta}_0)\right)^{\mathsf{T}}\mathrm{d}M(s) + o_{\mathrm{P}}(1),$$

where $\boldsymbol{\Sigma}$ is given by part D of Condition VI.1.1.

We denote by $\xi(t;\boldsymbol{\theta}_0)$ the vector of partial derivatives of $\log f(t;\boldsymbol{\theta})$ [cf. (6.3.10)] with respect to $\boldsymbol{\theta}$ evaluated at the true parameter value $\boldsymbol{\theta}_0$, i.e.,

$$\xi(t;\boldsymbol{\theta}_0) = \left(\frac{\partial}{\partial\theta_1}\log f(s;\boldsymbol{\theta}_0),\ldots,\frac{\partial}{\partial\theta_q}\log f(s;\boldsymbol{\theta}_0)\right)^{\mathsf{T}}. \tag{6.3.13}$$

Then

$$\boldsymbol{\Sigma} = \int_0^\tau \xi(s;\boldsymbol{\theta}_0)\xi^{\mathsf{T}}(s;\boldsymbol{\theta}_0)f(s;\boldsymbol{\theta}_0)\,\mathrm{d}s. \tag{6.3.14}$$

Furthermore, we introduce the vector

$$\boldsymbol{\Psi}(t;\boldsymbol{\theta}_0) = \int_0^t \kappa(s;\boldsymbol{\theta}_0)\xi(s;\boldsymbol{\theta}_0)f(s;\boldsymbol{\theta}_0)\,\mathrm{d}s. \tag{6.3.15}$$

Combining this, we obtain from (6.3.12)

$$a_n^{-1}Z(t) = a_n^{-1}\int_0^t K(s;\boldsymbol{\theta}_0)\,\mathrm{d}M(s)$$

$$-\boldsymbol{\Psi}(t;\boldsymbol{\theta}_0)^{\mathsf{T}}\boldsymbol{\Sigma}^{-1}a_n^{-1}\int_0^\tau \left(\frac{\partial}{\partial\theta_1}\log\lambda(s;\boldsymbol{\theta}_0),\ldots,\frac{\partial}{\partial\theta_q}\log\lambda(s;\boldsymbol{\theta}_0)\right)^{\mathsf{T}}\mathrm{d}M(s)$$

$$+\, o_{\mathrm{P}}(1)$$

as $n \to \infty$. By the martingale central limit theorem (Theorem II.5.1), it follows that $a_n^{-1}Z$ converges weakly in $D(\mathscr{T})$ to the process U given by

$$U(t) = \int_0^t \kappa(s;\boldsymbol{\theta}_0)\,\mathrm{d}V(s) - \boldsymbol{\Psi}(t;\boldsymbol{\theta}_0)^{\mathsf{T}}\boldsymbol{\Sigma}^{-1}\int_0^\tau \boldsymbol{\xi}(s;\boldsymbol{\theta}_0)\,\mathrm{d}V(s), \qquad (6.3.16)$$

where V is a Gaussian martingale with $V(0) = 0$ and

$$\operatorname{cov}(V(s), V(t)) = \int_0^{s\wedge t} f(s;\boldsymbol{\theta}_0)\,\mathrm{d}s.$$

By (6.3.14), it follows that $\int_0^\tau \boldsymbol{\xi}(s;\boldsymbol{\theta}_0)\,\mathrm{d}V(s)$ has covariance matrix $\boldsymbol{\Sigma}$, and, therefore, we have

$$\operatorname{cov}(U(s), U(t)) = \int_0^{s\wedge t} \kappa^2(s;\boldsymbol{\theta}_0)f(s;\boldsymbol{\theta}_0)\,\mathrm{d}s - \boldsymbol{\Psi}(s;\boldsymbol{\theta}_0)^{\mathsf{T}}\boldsymbol{\Sigma}^{-1}\boldsymbol{\Psi}(t;\boldsymbol{\theta}_0) \qquad (6.3.17)$$

for the covariance structure of the limiting process.

VI.3.3.2. *The One-Parameter Case*

In general, the limiting process (6.3.16) is quite intractable. However, for the special case of $q = 1$ and when the weight process is chosen to be

$$K(t;\hat{\theta}) = \frac{\partial}{\partial\theta}\log\lambda(t;\hat{\theta}), \qquad (6.3.18)$$

we see by comparing (6.1.3) and (6.3.9) that $Z(\tau) = 0$. Moreover, by (6.3.13)–(6.3.15)

$$\operatorname{cov}(U(s), U(t)) = \sigma^2(\tau;\theta_0)\left\{\frac{\sigma^2(s\wedge t;\theta_0)}{\sigma^2(\tau;\theta_0)} - \frac{\sigma^2(s;\theta_0)}{\sigma^2(\tau;\theta_0)}\frac{\sigma^2(t;\theta_0)}{\sigma^2(\tau;\theta_0)}\right\},$$

where

$$\sigma^2(t;\theta_0) = \int_0^t \left\{\frac{\partial}{\partial\theta}\log f(s;\theta_0)\right\}^2 f(s;\theta_0)\,\mathrm{d}s.$$

If we, therefore, let $p(t) = \sigma^2(t;\theta_0)/\sigma^2(\tau;\theta_0)$ and remember that $\sigma^2(\tau;\theta_0)$ is estimated consistently by $a_n^{-2}\mathscr{I}_\tau(\hat{\theta})$ given by (6.1.4) (with no superscripts j and l), it is seen that

$$\mathscr{I}_\tau(\hat{\theta})^{-1/2} Z \xrightarrow{\mathscr{D}} W^0 \circ p, \tag{6.3.19}$$

i.e., it behaves asymptotically as a time-transformed Brownian bridge.

Thus, in particular, a Kolmogorov–Smirnov-type test for hypothesis (6.3.8) is to reject at the α level when

$$\mathscr{I}_\tau(\hat{\theta})^{-1/2} \sup_{t \in [0,\tau]} \left| \int_0^t \frac{\partial}{\partial\theta} \log \lambda(s;\hat{\theta})\, \mathrm{d}N(s) - \int_0^t \frac{\partial}{\partial\theta} \lambda(s;\hat{\theta})\, \mathrm{d}s \right| > e_\alpha, \tag{6.3.20}$$

where e_α is the upper α fractile in the distribution of $\sup_{0 \le x \le 1} |W^0(x)|$.

It is of some interest to focus on the special case $\lambda(t;\theta) = \theta Y(t)$ already considered in Section VI.3.2. For, in this case, we have $\hat{\theta} = N(\tau)/R(\tau)$, with $R(t)$ given by (6.3.3), and $\mathscr{I}_\tau(\hat{\theta}) = N(\tau)/\hat{\theta}^2$, so that (6.3.19) gives

$$\sqrt{N(\tau)} \left\{ \frac{N(t)}{N(\tau)} - \frac{R(t)}{R(\tau)} \right\} \xrightarrow{\mathscr{D}} W^0 \circ p. \tag{6.3.21}$$

Thus, (6.3.19) provides an alternative proof of the asymptotic properties of the TTT-plot of Section VI.3.2.

Moreover, this also suggests how one may generalize the TTT-plot to the present situation. To this end, we first note that since $Z(\tau) = 0$, we have

$$\int_0^\tau \frac{\partial}{\partial\theta} \log \lambda(s;\hat{\theta})\, \mathrm{d}N(s) = \int_0^\tau \frac{\partial}{\partial\theta} \lambda(s;\hat{\theta})\, \mathrm{d}s.$$

Therefore, we get a generalized TTT-plot for one-parameter models by plotting

$$\int_0^t \frac{\partial}{\partial\theta} \lambda(s;\hat{\theta})\, \mathrm{d}s \bigg/ \int_0^\tau \frac{\partial}{\partial\theta} \lambda(s;\hat{\theta})\, \mathrm{d}s$$

versus

$$\int_0^t \frac{\partial}{\partial\theta} \log \lambda(s;\hat{\theta})\, \mathrm{d}N(s) \bigg/ \int_0^\tau \frac{\partial}{\partial\theta} \log \lambda(s;\hat{\theta})\, \mathrm{d}N(s).$$

Alternatively, one may perform the plot on a transformed time scale to get exactly the same asymptotic distribution as the ordinary TTT-plot under the hypothesis. We do not, however, consider this option any further here.

EXAMPLE VI.3.6. Survival with Malignant Melanoma: A Model with Constant Excess Mortality

The melanoma data were introduced in Example I.3.1. In Example VI.3.5, we studied models for these data with a constant *relative* mortality. In the present example, we shall similarly study the model with a constant *excess* mortality. We will base our analysis on the aggregated counting process $N = \sum_{i=1}^n N_i$ which by Example VI.1.5 has intensity process

$$\lambda(t;\gamma) = \gamma Y(t) + Y^\mu(t).$$

As noted in Example VI.1.5, the aggregated processes N, Y, and Y^μ are not sufficient, however, so some information is lost by this aggregation. At the end of this section, we show how an analysis similar to the one of the present example may be based on the individual counting processes N_i.

Now we have

$$\frac{\partial}{\partial\gamma}\log\lambda(t;\gamma) = \frac{Y(t)}{\gamma Y(t) + Y^\mu(t)},$$

and the suggested generalized TTT-plot consists of plotting

$$\int_0^t Y(u)\,\mathrm{d}u \Big/ \int_0^\tau Y(u)\,\mathrm{d}u$$

versus

$$\int_0^t \frac{Y(u)}{\hat\gamma Y(u) + Y^\mu(u)}\,\mathrm{d}N(u) \Big/ \int_0^\tau \frac{Y(u)}{\hat\gamma Y(u) + Y^\mu(u)}\,\mathrm{d}N(u).$$

Here, $\hat\gamma$ is the (not fully efficient) maximum likelihood estimator *based on the aggregated process*, i.e., the (M-) estimator solving

$$\int_0^\tau \frac{Y(t)}{\hat\gamma Y(t) + Y^\mu(t)}\,\mathrm{d}N(t) - \int_0^\tau Y(t)\,\mathrm{d}t = 0.$$

For the melanoma data, we get $\hat\gamma = 0.0410$ per year with an estimated standard deviation of 0.0070 per year.

The generalized TTT-plot is shown in Figure VI.3.10 using follow-up time as the time scale. Also shown are the 95% boundaries $\pm 1.36/\{\mathscr{I}_\tau(\hat\gamma)^{-1/2}R(\tau)\}$ around the diagonal. Here, $\mathscr{I}_\tau(\hat\gamma)^{-1/2} = \widehat{\text{s.d.}}(\hat\gamma)$, and we have used that $\hat\gamma$ solves the equation

$$R(\tau) = \int_0^\tau Y(u)\,\mathrm{d}u = \int_0^\tau \frac{Y(u)}{\hat\gamma Y(u) + Y^\mu(u)}\,\mathrm{d}N(u).$$

It is seen that the curve nowhere crosses the boundaries and the Kolmogorov–Smirnov-type test statistic (6.3.20) based on the generalized TTT-plot takes the nonsignificant value

$$\mathscr{I}_\tau(\hat\gamma)^{-1/2} \sup_t \left| \int_0^t \frac{Y(u)}{\hat\gamma Y(u) + Y^\mu(u)}\,\mathrm{d}N(u) - \int_0^t Y(u)\,\mathrm{d}u \right| = 0.53.$$

Thus, we cannot reject the hypothesis that the excess mortality is independent of follow-up time. The same conclusion is reached when age is used as the basic time scale.

Alternative TTT-plots and statistics based on the M-estimator

$$\tilde\gamma = (N(\tau) - E(\tau)) \Big/ \int_0^\tau Y(t)\,\mathrm{d}t$$

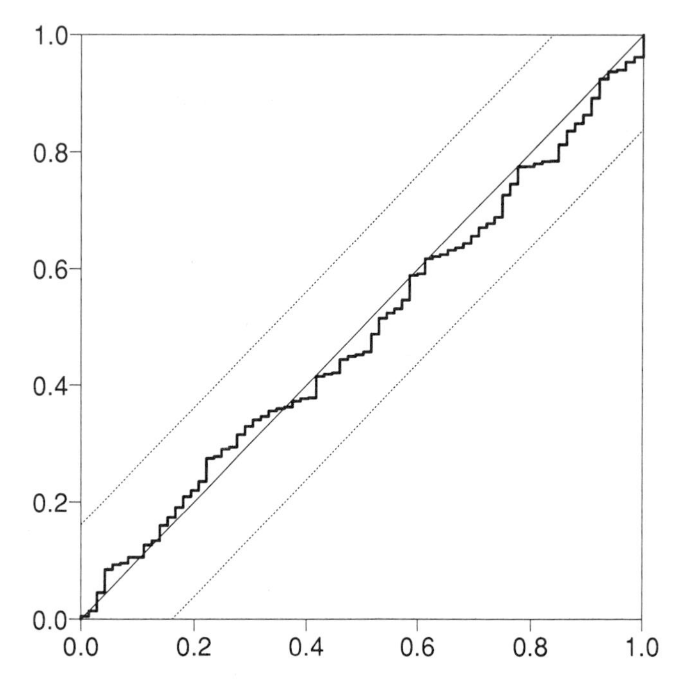

Figure VI.3.10. Generalized TTT-plot for a constant excess mortality model for the melanoma data using follow-up time as the basic time scale.

were suggested by Andersen and Væth (1989) and the same conclusions as above were obtained. □

VI.3.3.3. χ^2 *Tests*

The limiting distribution of $a_n^{-1} Z$ may, in any case, be used to derive χ^2-type tests for hypothesis (6.3.8). To this end, we let $0 = t_0 < t_1 < \cdots < t_m = \tau$ be a partition of $\mathscr{T}$ into m subintervals $I_i = [t_{i-1}, t_i)$, and define

$$Z_i = \int_{I_i} \mathrm{d}Z(s) = \int_{I_i} K(s; \hat{\boldsymbol{\theta}}) \{\mathrm{d}N(s) - \lambda(s; \hat{\boldsymbol{\theta}}) \, \mathrm{d}s\}$$

and $\mathbf{Z} = (Z_1, \ldots, Z_m)^{\mathsf{T}}$. Then, by (6.3.16),

$$a_n^{-1} \mathbf{Z} \xrightarrow{\mathscr{D}} \mathscr{N}(\mathbf{0}, \boldsymbol{\Phi})$$

as $n \to \infty$ for a certain covariance matrix $\boldsymbol{\Phi}$ which may be derived from (6.3.17). If $a_n^{-2} \hat{\boldsymbol{\Phi}}$ is a consistent estimator of this matrix (e.g., defined by (6.3.24)) and $\hat{\boldsymbol{\Phi}}^-$ denotes a (possibly) generalized inverse, we may consider the test statistic

$$X^2 = \mathbf{Z}^{\mathsf{T}} \hat{\boldsymbol{\Phi}}^- \mathbf{Z}. \tag{6.3.22}$$

If $\mathbf{\Phi}$ is nonsingular (as will be the case in many applications), then (6.3.22) will be approximately χ^2 distributed with m degrees of freedom under the hypothesis. An important special case where this is not the case is when the weight process is of the form

$$K(t;\hat{\boldsymbol{\theta}}) = \sum_{j=1}^{q} c_j \left\{ \frac{\partial}{\partial \theta_j} \log \lambda(t;\hat{\boldsymbol{\theta}}) \right\} \tag{6.3.23}$$

for some constants $c_1, c_2, \ldots, c_q$. For then $\sum_{i=1}^{m} Z_i = 0$ by the definition of the maximum likelihood estimator, and (6.3.22) typically will be approximately χ^2 distributed with $m - 1$ degrees of freedom. [However, as pointed out by Li and Doss (1991), care may have to be exercised when the rank of $\mathbf{\Phi}$ is less than m.]

As an estimator for $\mathbf{\Phi}$, we may take $\hat{\mathbf{\Phi}} = \{\hat{\phi}_{ij}\}$ with entry (i, j) given as

$$\hat{\phi}_{ij} = \delta_{ij}\hat{d}_i - \hat{\boldsymbol{\psi}}_i^{\mathsf{T}} \mathscr{I}_\tau(\hat{\boldsymbol{\theta}})^{-1} \hat{\boldsymbol{\psi}}_j. \tag{6.3.24}$$

Here

$$\hat{d}_i = \int_{I_i} K^2(s;\hat{\boldsymbol{\theta}})\, \mathrm{d}N(s),$$

$$\hat{\boldsymbol{\psi}}_i = \int_{I_i} K(s;\hat{\boldsymbol{\theta}}) \boldsymbol{\xi}(s;\hat{\boldsymbol{\theta}})\, \mathrm{d}N(s),$$

with

$$\boldsymbol{\xi}(s;\hat{\boldsymbol{\theta}}) = \left(\frac{\partial}{\partial \theta_1} \log \lambda(s;\hat{\boldsymbol{\theta}}), \ldots, \frac{\partial}{\partial \theta_q} \log \lambda(s;\hat{\boldsymbol{\theta}}) \right)^{\mathsf{T}}, \tag{6.3.25}$$

and $\mathscr{I}_\tau(\boldsymbol{\theta})$ is given by (6.1.4).

If, in particular, we take $K(s;\hat{\boldsymbol{\theta}}) = 1$, X^2 given by (6.3.22) simply compares the observed and "expected" number of events [i.e., $\int_{I_i} \mathrm{d}N(t)$ and $\int_{I_i} \lambda(t;\hat{\boldsymbol{\theta}})\,\mathrm{d}t$] over the m intervals. This gives a dynamic generalization of the classical χ^2 test of Karl Pearson. Details on the general test statistic (6.3.22), further special cases, and some preliminary power calculations were provided by Hjort (1990a).

EXAMPLE VI.3.7. Generalized Pearson χ^2 Test for the Gompertz Model. Diabetics on Fyn.

We consider the Gompertz model of Example VI.1.4 using the parametrization $\lambda(t;\boldsymbol{\theta}) = \alpha(t;\boldsymbol{\theta}) Y(t) = bc^t Y(t)$ with $Y(t)$ being "the number at risk process." We will consider the generalized Pearson test for this situation, i.e., the test (6.3.22) with $K(t;\hat{\boldsymbol{\theta}}) = 1$. [Note that this choice of weight process is of the form (6.3.23) for the Gompertz model.] Then

$$Z_i = O_i - E_i$$

with

$$O_i = \int_{I_i} \mathrm{d}N(s)$$

and

$$E_i = \int_{I_i} \lambda(s; \hat{\boldsymbol{\theta}})\,\mathrm{d}s = \hat{b} \int_{I_i} (\hat{c})^s Y(s)\,\mathrm{d}s$$

being the observed and "expected" number of deaths in the ith interval, respectively. The covariance matrix of the Z_i may be estimated by (6.3.24) with $\hat{d}_i = O_i$, and

$$\hat{\boldsymbol{\psi}}_i = \int_{I_i} \hat{\boldsymbol{\xi}}(s; \hat{\boldsymbol{\theta}})\,\mathrm{d}N(s) = \begin{pmatrix} O_i/\hat{b} \\ \int_{I_i} s\,\mathrm{d}N(s)/\hat{c} \end{pmatrix}.$$

For the data on female diabetics in the county of Fyn introduced in Example I.3.2, maximum likelihood estimates for the present parametrization of the Gompertz model are easily obtained from those given in Example VI.1.4 (using a different parametrization). Using the age intervals 0–44, 45–59, 60–69, 70–79, and 80+ the observed numbers of deaths are 17, 26, 55, 86, and 53, respectively, whereas the corresponding "expected" numbers become 14.45, 31.70, 57.32, 77.46 and 56.09, respectively. The generalized Pearson statistic takes the value 2.76 (d.f. = 4, $P = .60$) indicating a nice fit of the Gompertz model. □

VI.3.3.4. *Khmaladze's Goodness-of-Fit Idea*

We observed above that the limiting distribution (6.3.16) of the goodness-of-fit process (6.3.9) was complicated when $q \geq 2$ and, therefore, not well-suited to derive goodness-of-fit tests for hypothesis (6.3.8). In the context of empirical processes, Khmaladze (1981) suggested that rather than to base goodness-of-fit tests directly on the complicated limiting distribution of (6.3.9), such tests should be based on a transformation of (6.3.9) to a process with a nice limiting distribution. This is obtained by (asymptotically) replacing (6.3.9) by this process minus its compensator.

To see how this can be done, we consider the asymptotic situation where the limiting process U corresponding to the goodness-of-fit process Z is given by (6.3.16). We want to determine the compensator of U relative to the filtration

$$\mathscr{G}_t = \sigma\left\{V(s), s \leq t; \int_0^\tau \boldsymbol{\xi}(s; \boldsymbol{\theta}_0)\,\mathrm{d}V(s)\right\},$$

i.e., the filtration generated by V [cf. (6.3.16)] with $\int_0^\tau \boldsymbol{\xi}(s; \boldsymbol{\theta}_0)\,\mathrm{d}V(s)$ "added at time zero." Note that U is adapted to $\mathscr{G}_t$, but not to the filtration generated by V. Now, in (6.3.16),

$$\mathbf{\Psi}(t;\boldsymbol{\theta}_0)^{\mathsf{T}}\mathbf{\Sigma}^{-1}\int_0^\tau \xi(s;\boldsymbol{\theta}_0)\,\mathrm{d}V(s)$$

is $\mathscr{G}_0$-measurable, so to find the compensator of U, we just need to find the $\mathscr{G}_t$-compensator of $\int_0^t \kappa(s;\boldsymbol{\theta}_0)\,\mathrm{d}V(s)$. Arguing informally, we have

$$\begin{aligned}
&\mathrm{E}\left(\mathrm{d}\int_0^t \kappa(s;\boldsymbol{\theta}_0)\,\mathrm{d}V(s)\,\middle|\,\mathscr{G}_t\right)\\
&\quad= \kappa(t;\boldsymbol{\theta}_0)\mathrm{E}\left(\mathrm{d}V(t)\,\middle|\,V(s), s \le t; \int_0^\tau \xi(s;\boldsymbol{\theta}_0)\,\mathrm{d}V(s)\right)\\
&\quad= \kappa(t;\boldsymbol{\theta}_0)\mathrm{E}\left(\mathrm{d}V(t)\,\middle|\,\int_t^\tau \xi(s;\boldsymbol{\theta}_0)\,\mathrm{d}V(s)\right),
\end{aligned}$$

where the last equality follows since V is a Gaussian martingale and, therefore, has independent increments. Since $\mathrm{d}V(t)$ and $\int_t^\tau \xi(s;\boldsymbol{\theta}_0)\,\mathrm{d}V(s)$ are jointly multinormally distributed with mean zero, we get

$$\begin{aligned}
&\mathrm{E}\left(\mathrm{d}V(t)\,\middle|\,\int_t^\tau \xi(s;\boldsymbol{\theta}_0)\,\mathrm{d}V(s)\right)\\
&\quad= \mathrm{cov}\left(\mathrm{d}V(t), \int_t^\tau \xi(s;\boldsymbol{\theta}_0)\,\mathrm{d}V(s)\right)\left\{\mathrm{cov}\left(\int_t^\tau \xi(s;\boldsymbol{\theta}_0)\,\mathrm{d}V(s)\right)\right\}^{-1}\int_t^\tau \xi(s;\boldsymbol{\theta}_0)\,\mathrm{d}V(s)\\
&\quad= \xi^{\mathsf{T}}(t;\boldsymbol{\theta}_0)f(t;\boldsymbol{\theta}_0)\,\mathrm{d}t\left\{\int_t^\tau \xi(s;\boldsymbol{\theta}_0)\xi^{\mathsf{T}}(s;\boldsymbol{\theta}_0)f(s;\boldsymbol{\theta}_0)\,\mathrm{d}s\right\}^{-1}\int_t^\tau \xi(s;\boldsymbol{\theta}_0)\,\mathrm{d}V(s).
\end{aligned}$$

These informal calculations suggest that

$$\begin{aligned}
\tilde{U}(t) = &\int_0^t \left\{\kappa(s;\boldsymbol{\theta}_0)\xi^{\mathsf{T}}(s;\boldsymbol{\theta}_0)\left(\int_s^\tau \xi(u;\boldsymbol{\theta}_0)\xi^{\mathsf{T}}(u;\boldsymbol{\theta}_0)f(u;\boldsymbol{\theta}_0)\,\mathrm{d}u\right)^{-1}\right.\\
&\times\left.\int_s^\tau \xi(u;\boldsymbol{\theta}_0)\,\mathrm{d}V(u)\right\}f(s;\boldsymbol{\theta}_0)\,\mathrm{d}s\\
&- \mathbf{\Psi}(t;\boldsymbol{\theta}_0)^{\mathsf{T}}\mathbf{\Sigma}^{-1}\int_0^\tau \xi(u;\boldsymbol{\theta}_0)\,\mathrm{d}V(u)
\end{aligned}\tag{6.3.26}$$

is the compensator of U and, therefore, that

$$\begin{aligned}
&U(t) - \tilde{U}(t)\\
&\quad= \int_0^t \kappa(s;\boldsymbol{\theta}_0)\,\mathrm{d}V(s) - \int_0^t \left\{\kappa(s;\boldsymbol{\theta}_0)\xi^{\mathsf{T}}(s;\boldsymbol{\theta}_0)\left(\int_s^\tau \xi(u;\boldsymbol{\theta}_0)\xi^{\mathsf{T}}(u;\boldsymbol{\theta}_0)f(u;\boldsymbol{\theta}_0)\,\mathrm{d}u\right)^{-1}\right.\\
&\quad\times\left.\int_s^\tau \xi(u;\boldsymbol{\theta}_0)\,\mathrm{d}V(u)\right\}f(s;\boldsymbol{\theta}_0)\,\mathrm{d}s
\end{aligned}\tag{6.3.27}$$

is a zero-mean Gaussian martingale.

In fact, by a direct calculation, one may show that

$$\operatorname{cov}(U(s) - \tilde{U}(s), U(t) - \tilde{U}(t)) = \int_0^{s \wedge t} \kappa^2(s;\boldsymbol{\theta}_0) f(s;\boldsymbol{\theta}_0)\,\mathrm{d}s,$$

which proves that $U - \tilde{U}$ is, indeed, a Gaussian martingale having the same distribution as the leading term on the right-hand side of (6.3.16). Thus, (6.3.27) is an "untied" version of the "tied down" process (6.3.16) and, therefore, it has a much more tractable distribution.

The idea now is to construct a transform of the goodness-of-fit process (6.3.9) which has the same asymptotic distribution as $U - \tilde{U}$ by substituting the empirical counterparts for the quantities entering in $U - \tilde{U}$. A little care is needed here, however, since the empirical version of V is $a_n^{-1}M$, and this quantity is not observable. What is observable is not M, but the "estimated residual process"

$$\hat{M}(t) = N(t) - \int_0^t \lambda(s;\hat{\boldsymbol{\theta}})\,\mathrm{d}s. \tag{6.3.28}$$

Now by (6.3.14), (6.3.15), and (6.3.16) (with $\kappa = 1$), the asymptotic counterpart of $a_n^{-1}\hat{M}$ is

$$\hat{V}(t) = V(t) - \left(\int_0^t \boldsymbol{\xi}^{\mathsf{T}}(s;\boldsymbol{\theta}_0) f(s;\boldsymbol{\theta}_0)\,\mathrm{d}s\right) \times \left(\int_0^\tau \boldsymbol{\xi}(s;\boldsymbol{\theta}_0)\boldsymbol{\xi}^{\mathsf{T}}(s;\boldsymbol{\theta}_0) f(s;\boldsymbol{\theta}_0)\,\mathrm{d}s\right)^{-1} \int_0^\tau \boldsymbol{\xi}(s;\boldsymbol{\theta}_0)\,\mathrm{d}V(s).$$

Moreover, it is seen that (6.3.26) remains true when V is replaced by $\hat{V}$ on the right-hand side. Thus, we may in this modified version of (6.3.26) replace $\hat{V}$ by $a_n^{-1}\hat{M}$, $\kappa(s;\boldsymbol{\theta}_0)$ by $K(s;\hat{\boldsymbol{\theta}})$, $\boldsymbol{\xi}(s;\theta_0)$ by $\hat{\boldsymbol{\xi}}(s;\hat{\boldsymbol{\theta}})$ given by (6.3.25), and $f(s;\boldsymbol{\theta}_0)\,\mathrm{d}s$ by $a_n^{-2}\,\mathrm{d}N(s)$ to obtain the empirical counterpart of $\tilde{U}$. (Note that, by this substitution, the last term vanishes by the definition of the maximum likelihood estimator.) Thus, one will expect $a_n^{-1}Z^*$, with

$$Z^*(t) = Z(t) - \int_0^t \Bigg\{K(s;\hat{\boldsymbol{\theta}})\hat{\boldsymbol{\xi}}^{\mathsf{T}}(s;\hat{\boldsymbol{\theta}}) \times \left(\int_s^\tau \hat{\boldsymbol{\xi}}(u;\hat{\boldsymbol{\theta}})\hat{\boldsymbol{\xi}}^{\mathsf{T}}(u;\hat{\boldsymbol{\theta}})\,\mathrm{d}N(u)\right)^{-1} \int_s^\tau \hat{\boldsymbol{\xi}}(u;\hat{\boldsymbol{\theta}})\,\mathrm{d}\hat{M}(u)\Bigg\}\,\mathrm{d}N(s), \tag{6.3.29}$$

to be asymptotically distributed as (6.3.27). (The integrals from s to τ should be interpreted as being over $[s,\tau]$.) Weak convergence of $a_n^{-1}Z^*$ on $[0,\tau-\varepsilon]$ for a fixed $\varepsilon > 0$ may easily be proved, but weak convergence on the whole of $[0,\tau]$ is a very delicate matter by the inverse term in (6.3.29) which will be small for s close to τ.

It follows that a plot of

$$Z^*(t)\Big/\sqrt{\int_0^\tau K^2(s;\hat{\boldsymbol{\theta}})\,\mathrm{d}N(s)}$$

versus

$$\int_0^t K^2(s;\hat{\boldsymbol{\theta}})\,\mathrm{d}N(s) \Big/ \int_0^\tau K^2(s;\hat{\boldsymbol{\theta}})\,\mathrm{d}N(s)$$

should resemble a realization of a standard Brownian motion W on the interval [0, 1] if the model specification (6.3.8) is correct. In a similar manner, a Kolmogorov–Smirnov-type test rejects the hypothesis (6.3.8) at level α if

$$\sup_{0\le t\le\tau}\left| Z^*(t) \Big/ \sqrt{\int_0^\tau K^2(s;\hat{\boldsymbol{\theta}})\,\mathrm{d}N(s)}\right| \ge w_\alpha,$$

where w_α is the upper fractile in the distribution of $\sup_{0\le x\le 1}|W(x)|$.

We will illustrate how these general results specialize in two particular situations. The first example is a one-parameter model. Thus, for this example, Khmaladze's idea is not really needed, but it is illuminating to see what the actual processes look like for a particularly simple example.

EXAMPLE VI.3.8. Khmaladze's Goodness-of-Fit Test for the Dynamics of a Natural Baboon Troop

For the data on the natural baboon troop introduced in Example I.3.18, we study deaths, for which the parameter $\boldsymbol{\theta}$ is the scalar μ. The process $N(t)$ counting the number of deaths in $[0, t]$ has, under the assumed model, intensity process $\lambda(t;\mu) = \mu X(t-)$, with $X(t)$ specifying the size of the troop at time t (cf. Example III.1.6). Therefore, $(\partial/\partial\mu)\log\lambda(t;\mu) = 1/\mu$, and using the weight process $K(t;\hat{\mu}) = 1$, $Z^*(t)$ given by (6.3.29) takes the form

$$Z^*(t) = \hat{M}(t) - \int_0^t \frac{\hat{M}(\tau) - \hat{M}(s-)}{N(\tau) - N(s-)}\,\mathrm{d}N(s)$$

with $\hat{M}$ given in (6.3.28).

Figure VI.3.11 shows $(N(t), \hat{M}(t))$ as well as $(N(t), Z^*(t))$. The maximal deviation of $Z^*(t)$ from 0 is 6.82 [attained at $N(t) = 10$], so a Kolmogorov–Smirnov test value is obtained as $6.82/\sqrt{N(\tau)} = 6.82/\sqrt{12} = 1.97$, which, compared to the 5% critical value of 2.25, does not indicate a strong deviation from the hypothesis, in contrast to the more specific tests of Example VI.3.4 based on the TTT-plot. □

EXAMPLE VI.3.9. Khmaladze Goodness-of-Fit Test for the Gompertz Model. Diabetics in Fyn

The situation is as described in connection with Example VI.3.7, and we choose $K(t;\boldsymbol{\theta}) = 1$ here as well to allow comparison with the results of that example.

Now

$$\hat{\boldsymbol{\xi}}(t;\hat{\boldsymbol{\theta}}) = \begin{pmatrix} 1/\hat{b} \\ t/\hat{c} \end{pmatrix} = \begin{pmatrix} 1/\hat{b} & 0 \\ 0 & 1/\hat{c} \end{pmatrix}\begin{pmatrix} 1 \\ t \end{pmatrix}.$$

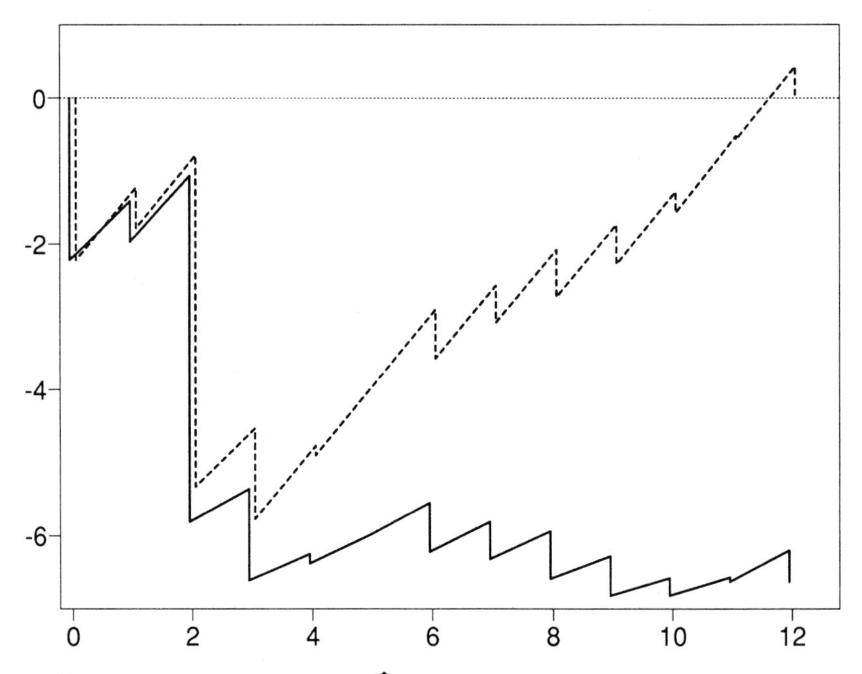

Figure VI.3.11. Baboon deaths: $\widehat{M}(t)$ (- - -) and $Z^*(t)$ (—) plotted against $N(t)$.

It is seen that when this expression is substituted into (6.3.29), the fixed matrix on the right-hand side will cancel, and, therefore, we may for $\hat{\boldsymbol{\xi}}(t;\hat{\boldsymbol{\theta}})$ in (6.3.29) simply take $(1,t)^{\mathsf{T}}$. Thus, we get

$$Z^*(t) = \widehat{M}(t)$$
$$-\int_0^t \left\{(1,s)\begin{pmatrix} N(\tau)-N(s-) & \int_s^\tau u\,\mathrm{d}N(u) \\ \int_s^\tau u\,\mathrm{d}N(u) & \int_s^\tau u^2\,\mathrm{d}N(u)\end{pmatrix}^{-1}\begin{pmatrix}\widehat{M}(\tau)-\widehat{M}(s-) \\ \int_s^\tau u\,\mathrm{d}\widehat{M}(u)\end{pmatrix}\right\}\mathrm{d}N(s),$$

where

$$\widehat{M}(\tau) - \widehat{M}(s-) = N(\tau) - N(s-) - \hat{b}\int_s^\tau (\hat{c})^u Y(u)\,\mathrm{d}u$$

and

$$\int_s^\tau u\,\mathrm{d}\widehat{M}(u) = \int_s^\tau u\,\mathrm{d}N(u) - \hat{b}\int_s^\tau u(\hat{c})^u Y(u)\,\mathrm{d}u.$$

Figure VI.3.12 shows $(N(t),\widehat{M}(t))$ and $(N(t),Z^*(t))$. The maximal absolute value of $Z^*(t)$ is 44.25 [obtained at $t = 95$ years, $N(t) = 236$], so the Kolmogorov–Smirnov-type test statistic takes the significant value $44.25/\sqrt{237} = 2.87$ ($P < .0001$). Thus, in contrast to the result of the χ^2 test of Example VI.3.7, the Kolmogorov–Smirnov-type test seems to be able to

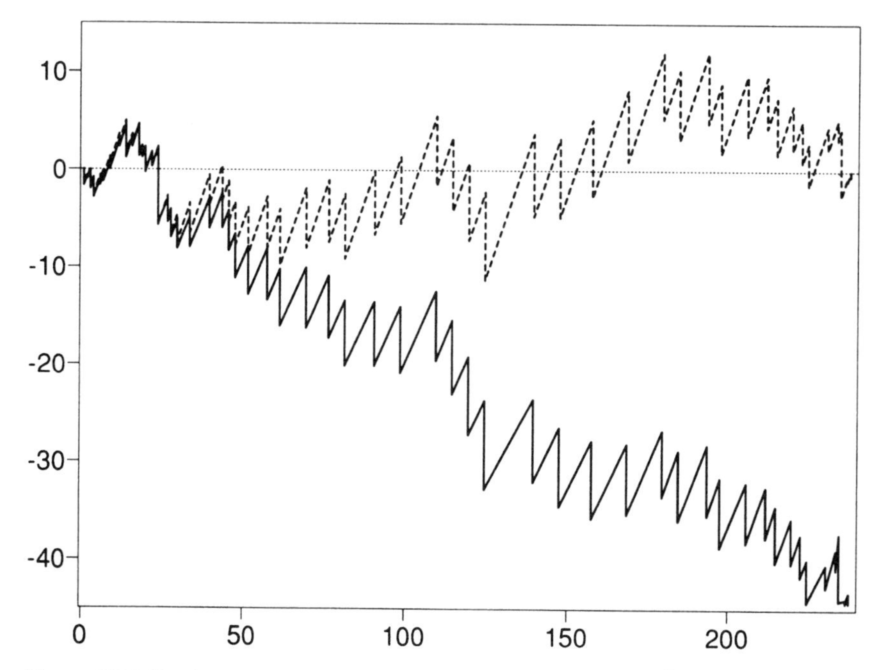

Figure VI.3.12. Gompertz model for female diabetics in Fyn: $\hat{M}(t)$ (- - -) and $Z^*(t)$ (—) plotted against $N(t)$.

detect that the mortality of the female diabetics increases more steeply than predicted by the Gompertz model before approximately 40 years of age and less steep thereafter (cf. Figure VI.3.5). (But we have to admit that, at present, little is known about how various alternatives to the Gompertz model will turn up in the plot.) □

VI.3.3.5. *Extension to Multivariate Counting Processes*

Earlier, we have considered goodness-of-fit tests based on a univariate counting process. As illustrated in Example VI.3.6, this is sometimes too restrictive, however, and there is a need to generalize the results to multivariate counting processes. Such a generalization is also needed to cover the parametric regression models of Section VII.6.

Therefore, let $(N_1(t), N_2(t), \ldots, N_k(t))$ be a multivariate counting process with intensity process $(\lambda_1(t), \lambda_2(t), \ldots, \lambda_k(t))$. We want to test the hypothesis

$$\lambda_h(t) = \lambda_h(t; \boldsymbol{\theta}), \quad h = 1, 2, \ldots, k, \tag{6.3.30}$$

i.e., the hypothesis that all the $\lambda_h(t)$, in addition to observable processes, may be specified by a q-dimensional parameter $\boldsymbol{\theta} = (\theta_1, \ldots, \theta_q)$. As in Sections VI.1 and VI.2, we will allow the number of components of the multivariate counting process to increase with n, but we do not make this possibility explicit in the notation.

We will base tests for hypothesis (6.3.30) on the goodness-of-fit process

$$Z(t) = \int_0^t \sum_{h=1}^k K_h(s;\hat{\boldsymbol{\theta}})\{\mathrm{d}N_h(s) - \lambda_h(s;\hat{\boldsymbol{\theta}})\,\mathrm{d}s\}, \tag{6.3.31}$$

where the $K_h(t;\hat{\boldsymbol{\theta}})$ are suitably chosen processes which may depend on the maximum likelihood estimator $\hat{\boldsymbol{\theta}}$. The $K_h(t;\boldsymbol{\theta})$ are assumed to satisfy Condition VI.2.1 in a similar manner to that described for the univariate case earlier.

Then, by exactly the same argument as was used in the univariate case, we find that $a_n^{-1}Z$ converges weakly in $D(\mathscr{T})$ to a Gaussian process U with covariance function

$$\operatorname{cov}(U(s), U(t)) = \delta(s \wedge t;\boldsymbol{\theta}_0) - \boldsymbol{\Psi}(s;\boldsymbol{\theta}_0)^{\mathsf{T}}\boldsymbol{\Sigma}^{-1}\boldsymbol{\Psi}(t;\boldsymbol{\theta}_0). \tag{6.3.32}$$

Here $\boldsymbol{\Sigma}$ is defined in part D of Condition VI.1.1, whereas $\delta(t;\boldsymbol{\theta}_0)$ and

$$\boldsymbol{\Psi}(t;\boldsymbol{\theta}_0) = (\psi_1(t;\boldsymbol{\theta}_0), \ldots, \psi_q(t;\boldsymbol{\theta}_0))^{\mathsf{T}}$$

are given by (cf. Condition VI.2.1)

$$a_n^{-2}\int_0^t \sum_{h=1}^k K_h(u;\boldsymbol{\theta}_0)^2 \lambda_h(u;\boldsymbol{\theta}_0)\,\mathrm{d}u \xrightarrow{\mathrm{P}} \delta(t;\boldsymbol{\theta}_0)$$

and

$$a_n^{-2}\int_0^t \sum_{h=1}^k K_h(u;\boldsymbol{\theta}_0)\left\{\frac{\partial}{\partial\theta_j}\log\lambda_h(u;\boldsymbol{\theta}_0)\right\}\lambda_h(u;\boldsymbol{\theta}_0)\,\mathrm{d}u \xrightarrow{\mathrm{P}} \psi_j(t;\boldsymbol{\theta}_0)$$

as $n \to \infty$.

This limiting distribution may be used to derive χ^2 tests exactly as in the univariate case.

Moreover, for the one-parameter case, and when the weight processes are chosen to be

$$K_h(t;\hat{\theta}) = \frac{\partial}{\partial\theta}\log\lambda_h(t;\hat{\theta}),$$

Brownian bridge asymptotics are available as in (6.3.19). In particular, a Kolmogorov–Smirnov-type test for hypothesis (6.3.30) is to reject at the α level when

$$\mathscr{I}_\tau(\hat{\theta})^{-1/2}\sup_{t\in[0,\tau]}\left|\int_0^t \sum_{h=1}^k \frac{\partial}{\partial\theta}\log\lambda_h(s;\hat{\theta})\,\mathrm{d}N_h(s) - \int_0^t \sum_{h=1}^k \frac{\partial}{\partial\theta}\lambda_h(s;\hat{\theta})\,\mathrm{d}s\right| > e_\alpha,$$

where e_α is the upper α fractile in the distribution of $\sup_{0\le x\le 1}|W^0(x)|$.

The generalized TTT-plot from the univariate case is also easily obtained for the multivariate case as illustrated in the final example in this section.

Example VI.3.10. Survival with Malignant Melanoma: A Model with Constant Excess Mortality—Continued

If in the situation of Example VI.3.6 we base our analysis on the individual counting processes N_i with intensity processes

$$\lambda_i(t;\gamma) = \{\gamma + \mu_i(t)\}\, Y_i(t),$$

the generalized TTT-plot consists of plotting

$$\int_0^t Y(u)\,du \bigg/ \int_0^\tau Y(u)\,du$$

versus

$$\int_0^t \sum_{i=1}^n \frac{dN_i(u)}{\hat{\gamma} + \mu_i(u)} \bigg/ \int_0^\tau \sum_{i=1}^n \frac{dN_i(u)}{\hat{\gamma} + \mu_i(u)},$$

where, now, $\hat{\gamma}$ is the maximum likelihood estimator of Example VI.1.5. It is seen that the ordinate values obtained for this generalized TTT-plot and for the one in Example VI.3.6 are the same, whereas the obtained abscissa values differ. The maximum difference of the abscissa values is only 0.018, however, so the two generalized TTT-plots are not easily distinguished, and we do not show the plot of the present example. The Kolmogorov–Smirnov-type test statistic now takes the value

$$\mathscr{I}_\tau(\hat{\gamma})^{-1/2} \sup_t \left| \int_0^t \sum_{i=1}^n \frac{dN_i(u)}{\hat{\gamma} + \mu_i(u)} - \int_0^t Y(u)\,du \right| = 0.056,$$

giving a P value of approximately 90%. Thus, the results of the present example differ only marginally from those of Example VI.3.6. □

VI.4. Bibliographic Remarks

Sections VI.1 and VI.2

For censored failure-time data, there exists a huge literature on maximum likelihood estimation and other methods of estimation in parametric models; see, e.g., the bibliographic comments in the books of Gross and Clark (1975, Chapters 3 and 4), Kalbfleisch and Prentice (1980, Chapter 3), and Lawless (1982, Chapters 3–5).

For simple Markov process models like the competing risks model and the three-state illness–death model (or disability model), there is a long tradition in demography and actuarial science for estimating the transition intensities, assumed (sometimes implicitly) to be piecewise constant, by occurrence/exposure rates. These estimates are often considered to be preliminary (or "raw") estimates and a subsequent fitting of a nice parametric function to them is then performed. Hoem (1972, 1976) reviewed and gave a statistical interpretation of this "two-step" approach to parameter estimation (which is often denoted "analytic graduation"). Simple Markov processes with (piecewise) constant transition intensities have also for a long time been used to model and analyze medical follow-up studies, two important early contributions being Fix and Neyman (1951) and Sverdrup (1965).

The first study of maximum likelihood estimators for parametric counting process models seems to have been performed by Aalen (1975) [for a journal publication, see Aalen and Hoem (1978)]. Aalen considered counting processes satisfying the multiplicative intensity model with constant α_h and following Keiding (1975), he used random time changes of the counting processes to independent Poisson processes to derive the asymptotic properties of the occurrence/exposure rates (Examples VI.1.1 and VI.1.9)

Borgan (1984) studied maximum likelihood estimation for general parametric versions of the multiplicative intensity model for counting processes, whereas Hjort (1985a) noted that Borgan's results could be extended to the class of M-estimators. The theory and proofs presented in our Sections VI.1 and VI.2 are based on these publications, but we have extended the results to cover general parametric counting process models not necessarily satisfying the multiplicative intensity model.

The assumptions in our Condition VI.1.1 are similar to those adopted by Cramér (1945) for the setup with i.i.d. random variables. LeCam and Yang (1990, Sections 6.2 and 6.7) discussed these Cramér-type conditions, as well as other conditions used in the literature, and reviewed the "differentiability in quadratic mean" condition. It would be of interest to try to imitate the latter weaker condition for asymptotic normality in the counting process case.

We have, in the present chapter, concentrated on maximum likelihood estimators and M-estimators. Other estimation methods known from the classical setup of i.i.d. random variables may also be considered within the counting process framework. In particular, Aven (1986) derived Bayes estimators for parametric counting process models, whereas Hjort (1986) showed that (within the traditional nonBayesian framework), the Bayes estimators and the maximum likelihood estimators have the same asymptotic distribution; in fact, he showed them to be asymptotically equivalent.

The asymptotic results presented in this chapter are suitable for situations where we either have an increasing number of components (typically corresponding to n individuals) of the multivariate counting process or where (for a fixed number of components) the intensity process is increasing with n (typically since we have aggregated over n "individual" processes although this need not be the case; cf. Example VI.1.12). Svensson (1990), on the other hand, considered a multivariate counting process with a fixed number of components and a parametric intensity process observed over an interval

$[0, u]$, and he derived strong consistency and asymptotic normality of the maximum likelihood estimator when $u \to \infty$. As particular illustrations, he considered renewal counting processes and birth-and-death processes.

In this monograph, we restrict our attention to counting process models and other stochastic process models (like Markov process models) which may be expressed by counting processes. There exists, however, a theory for maximum likelihood estimators and M-estimators (or estimating equations) in stochastic process models more generally which resembles the one for counting processes presented in Sections VI.1 and VI.2. See Barndorff-Nielsen and Sørensen (1992) for a readable review with a lot of examples and further references.

All the results in Sections VI.1 and VI.2 are derived under the assumption that the specified parametric model is correct. Hjort (1992a) adopted the model agnostic point of view and studied the asymptotic properties of the maximum likelihood estimators and the M-estimators for censored survival data outside model conditions. Along the same lines, he also discussed model-based and model-robust bootstrapping for parametric censored survival data models.

Toward the end of Section III.4, we indicated why the methodology presented in the present chapter (and in the rest of this monograph) does not apply to various kinds of grouped and aggregated data from Markov processes. For some such situations, maximum likelihood estimation may be performed, however, when the transition intensities are assumed (piecewise) constant (but perhaps depend on a parameter $\boldsymbol{\theta}$). For then the likelihood function, which typically will depend on the transition probabilities, may be expressed (without integration) by means of the transition intensities. In particular, Kalbfleisch and Lawless (1985) studied estimation based on panel data along these lines, whereas Kay (1986) discussed the same methodology for analyzing cancer markers in survival studies.

Section VI.3

We mentioned in the introductory remarks to Section VI.3 that a common and useful technique for checking the validity of a parametric model is to embed it in a larger parametric model and then test whether the reduction to the actual model is valid. This technique for goodness-of-fit testing for censored survival data was reviewed, e.g., by Kalbfleisch and Prentice (1980, Chapter 3), Lawless (1982, Chapter 9), and Cox and Oakes (1984, Chapter 3). Along these lines, Gray and Pierce (1985) showed how Neyman's "smooth tests" (derived by embedding the model in a "rich" parametric family and using the score test) may be adapted to censored survival data.

Nelson (1969, 1972) introduced the Nelson–Aalen estimator for censored data mainly as a graphical tool for assessing the fit of specific parametric failure-time models much in the same manner as we have described in Section VI.3.1. This use of the Nelson–Aalen estimator for model checking was reviewed in detail by Nelson (1982, Chapter 4) who also described how special hazard papers may be used for checking models like the normal and the lognormal.

A huge literature exists on tests for exponentiality for uncensored and censored failure-time data; see, e.g., Lawless (1982, Chapter 9) and Doksum and Yandell (1984). In Section VI.3.2, we have concentrated on the extension to counting process models of the TTT-plot with related tests.

The TTT-plot was proposed for failure-time data by Barlow and Campo (1975), and it has proven to be a very useful tool in a number of reliability applications; see, e.g., the review paper by Bergman (1985). Aalen and Hoem (1978) generalized the TTT-plot to counting process models with a multiplicative intensity process and illustrated its use outside the censored failure-time data setup. (In fact, they applied the plot to show that the mating intensity of Drosophila flies is not constant using the first 25 mating times for the experiment with ebony males and ebony females reported in Example I.3.16.) Our treatment of the TTT-plot follows Aalen and Hoem (1978) and Gill (1986b).

The tests based on the TTT-plot have a much longer history than the plot itself; in fact, Epstein and Sobel (1953) already considered the cumulative total time on test statistic for type II censored failure-time data. See Doksum and Yandell (1984) for a review on the literature of these tests for uncensored and censored failure-time data.

Durbin (1973) reviewed results on the distribution of the empirical distribution function for uncensored data when parameters are estimated. Fairly few papers exist extending these results to the censored data situation. Hjort (1990a) considered this problem for counting process models having a multiplicative intensity process and gave references to earlier related (but independent) work for censored survival data. [The main results of Hjort (1990a) were given already by Hjort (1985a).] Our derivation of the asymptotic distribution of the "goodness-of-fit process" (6.3.9) is based on Hjort (1990a), but we have extended his results to cover general parametric counting process models not necessarily satisfying the multiplicative intensity model.

Hjort (1990a) also noted that the "goodness-of-fit process" has a nice limiting distribution in the one-parameter case and that this can be used to construct Kolmogorov–Smirnov-type goodness-of-fit tests and the like (cf. Section V.4.1). The generalized TTT-plot is new here. We do not know how useful this will turn out to be, however, and further studies are needed to get an understanding for how specific deviations from given models will be reflected in the plot.

Our presentation of χ^2 tests are also based on Hjort's (1990a) work. χ^2 tests for censored survival data have also been studied by Habib and Thomas (1986), Akritas (1988), and Li and Doss (1991). In particular, the test of Akritas (1988, Section 4) based on the maximum likelihood estimator corresponds to the test (6.3.22) with weight process $K(t; \hat{\boldsymbol{\theta}}) = 1$. Akritas (1988,

Section 3) also studied a χ^2 test of the simple form $\sum (O_i - E_i)^2/E_i$ based on a modified minimum χ^2 estimator.

To our knowledge, nothing has been published on real applications of Khmaladze's (1981, 1988) goodness-of-fit tests and on how his idea (from empirical processes) may be adapted to censored survival data and counting process models. However, in a technical report, Geurts et al. (1988) applied the methodology to assess the fit of the Jelinski–Moranda model for the data presented in Example I.3.19, and they concluded that this model gives an adequate fit to the data. The plots based on (6.3.29) given in Figures VI.3.11 and VI.3.12 are new here. As for the generalized TTT-plot for the one-parameter situation, further studies are needed to understand how specific deviations from given models will be reflected in the plot.

CHAPTER VII

Regression Models

In the preceding three chapters, we have mainly studied various kinds of statistical inference applicable for the analysis of life history data from individuals stratified into *homogeneous groups* (e.g., Examples III.1.2, III.1.4, and III.1.5) though other applications were also possible (e.g., Examples III.1.3, III.1.8, and III.1.10). One of the main examples was Aalen's multiplicative intensity model. In this chapter, we shall return to a study of *regression models* as exemplified for uncensored data in Examples III.1.7 and III.1.9, for their right-censored counterparts in Section III.2 (e.g., Example III.2.9), and for the more general models in Sections III.3, III.4, and III.5.

VII.1. Introduction. Regression Model Formulation

We consider a multivariate counting process

$$\mathbf{N} = (N_{hi}, h = 1, \ldots, k; i = 1, \ldots, n)$$

with compensator

$$\boldsymbol{\Lambda}^{\theta} = (\Lambda^{\theta}_{hi}, h = 1, \ldots, k; i = 1, \ldots, n)$$

with respect to some filtration $(\mathscr{F}_t)$. As discussed in Chapter III, there may be nuisance parameters φ, but the compensator is assumed to depend only on θ, the parameter of interest. (For ease of notation, we have dropped the superscript c introduced in Section III.5 to distinguish between the underlying uncensored process and the observed, censored or filtered process.) We may then write the partial likelihood as

$$L_\tau(\theta) = \prod_t \left\{ \prod_{h,i} \mathrm{d}\Lambda_{hi}(t,\theta)^{\Delta N_{hi}(t)} (1 - \mathrm{d}\Lambda_{\cdot\cdot}(t,\theta))^{1-\Delta N_{\cdot\cdot}(t)} \right\}; \tag{7.1.1}$$

cf. (3.2.8) and (3.4.2).

In this chapter, we shall study models relating $\Lambda_{hi}(\cdot,\theta)$ to a vector of *covariates* $\mathbf{Z}_i(t)$, $i = 1, \ldots, n$, observed for individual i. A fundamental assumption is that the processes $\mathbf{Z}_i(\cdot)$ are $(\mathscr{F}_t)$*-predictable*, meaning that the value of the covariate at time t should be known just before time t; cf. Sections II.1 and II.3.1. An important special case is models with fixed (i.e., time-independent) covariates; cf. Examples III.1.7, III.1.9, and III.2.9. Furthermore, the $\mathbf{Z}_i(\cdot)$ are assumed to be *locally bounded* (Section II.2).

We shall consider models where Λ_{hi}^θ is absolutely continuous, i.e., where an *intensity process* λ_{hi}^θ exists, in which case the compensator is given by

$$\Lambda_{hi}^\theta(t) = \int_0^t \lambda_{hi}^\theta(u)\,\mathrm{d}u, \quad h = 1, \ldots, k, i = 1, \ldots, n.$$

Also, the models that will be discussed are characterized by a multiplicative structure of the individual intensity processes, i.e.,

$$\lambda_{hi}^\theta(t) = Y_{hi}(t)\alpha_{hi}^\theta(t; \mathbf{Z}_i(t)), \quad h = 1, \ldots, k, i = 1, \ldots, n, \tag{7.1.2}$$

where the $Y_{hi}(\cdot)$ are *predictable* and do not depend on θ. Usually, $Y_{hi}(t)$ contains information on whether individual i is observed to be at risk for experiencing a type h event just before time t. The function $\alpha_{hi}^\theta(\cdot)$ specifies the dependence on the covariates $\mathbf{Z}_i$ and on the parameter θ and it can usually be interpreted as an individual hazard rate function, relative hazard, or type h transition intensity; cf. the examples in Chapter III.

We shall consider two main classes of models for α_{hi}^θ, *multiplicative hazard models* and *additive hazard models*. In Sections VII.2 and VII.3, we discuss semiparametric multiplicative hazard models of the form

$$\alpha_{hi}^\theta(t; \mathbf{Z}_i(t)) = \alpha_{h0}(t,\gamma) r(\boldsymbol{\beta}_h^\mathsf{T} \mathbf{Z}_i(t)), \quad h = 1, \ldots, k, i = 1, \ldots, n, \tag{7.1.3}$$

where $\theta = (\gamma, \boldsymbol{\beta}_1, \ldots, \boldsymbol{\beta}_k)$. This class of models is also known as *relative risk regression models* since they specify how the *ratio* between the hazard functions for events of type h for two individuals (1 and 2),

$$\frac{\alpha_{h1}^\theta(t; \mathbf{Z}_1(t))}{\alpha_{h2}^\theta(t; \mathbf{Z}_2(t))} = \frac{r(\boldsymbol{\beta}_h^\mathsf{T} \mathbf{Z}_1(t))}{r(\boldsymbol{\beta}_h^\mathsf{T} \mathbf{Z}_2(t))},$$

depends on the covariates for these individuals and on the vector $\boldsymbol{\beta}_h$ of regression coefficients. This dependence is specified via the *relative risk function* $r(x)$, which must be positive and which is often chosen to be the exponential function $r(x) = \exp(x)$ (Cox, 1972). The *underlying* or *baseline hazard functions* $\alpha_{h0}(t, \gamma)$ are non-negative and they are left completely unspecified.

In Section VII.4, we discuss *nonparametric additive* hazard models of the form

$$\alpha_{hi}^{\theta}(t; \mathbf{Z}_i(t)) = \beta_0^{(h)}(t, \theta) + \boldsymbol{\beta}^{(h)}(t, \theta)^{\mathsf{T}} \mathbf{Z}_i(t), \tag{7.1.4}$$

(cf. Example III.1.9), where the shapes of the *regression function* vectors $\boldsymbol{\beta}^{(h)}(t, \theta)$ are left completely unspecified. Other non- and semiparametric regression models are discussed in Section VII.5, whereas Section VII.6 deals with *parametric* regression models.

We have formulated the multiplicative model (7.1.3) with separate regression coefficient vectors $\boldsymbol{\beta}_h$ for each type of transition $h = 1, \ldots, k$ [and similarly for the additive model (7.1.4)]. There may be examples, however, where some parameters are common to several types of transition and, therefore, it is convenient to formulate the multiplicative model as

$$\alpha_{hi}^{\theta}(t; \mathbf{Z}_i(t)) = \alpha_{h0}(t, \gamma) r(\boldsymbol{\beta}^{\mathsf{T}} \mathbf{Z}_{hi}(t)), \quad h = 1, \ldots, k, \tag{7.1.5}$$

with one vector of regression coefficients $\boldsymbol{\beta} = (\beta_1, \ldots, \beta_p)$ containing all the *different* parameters in the $\boldsymbol{\beta}_h$ vectors and, therefore, *not depending on the type* h and with *type-specific covariates*

$$\mathbf{Z}_{hi}(t) = (Z_{hi1}(t), \ldots, Z_{hip}(t)), \quad h = 1, \ldots, k, i = 1, \ldots, n.$$

[The additive model (7.1.4) can be rewritten in a similar manner.] Here, $\mathbf{Z}_{hi}(t)$ should be computable from the vector $\mathbf{Z}_i(t)$ of basic covariates for individual i and from the history $\mathscr{F}_{t-}$ up until just before time t and, technically, the $\mathbf{Z}_{hi}(t)$, $h = 1, \ldots, n$, are everywhere assumed to be *predictable* and *locally bounded*. The basic trick is that if in the entire model (i.e., for all types together) there are p regression coefficients to be estimated, then each of the type-specific covariate vectors $\mathbf{Z}_{hi}$ for individual i should be made p-variate possibly by including extra components equal to zero for all i. It may seem intuitively more natural to consider models with *separate regression coefficients* for each type of transition as in (7.1.3) and (7.1.4), but we have chosen for several reasons to use the formulation (7.1.5) with type-specific covariates. First, it does allow models where some parameters are common to several transitions; second, as we shall see in Example VII.1.1 and Example VII.2.5, the interesting models can be formulated in this way, and finally, the formulations and proofs of the large sample properties in Section VII.2.2 simplify when there is only one regression coefficient vector.

The following example illustrates some of the flexibility available when formulating models as (7.1.2) and (7.1.5) using type-specific covariates.

EXAMPLE VII.1.1. Mortality and Nephropathy for Insulin-Dependent Diabetics. Description of Possible Models with Type-Specific Covariates.

The data were introduced in Example I.3.11. There are three basic types of transitions (Example III.3.7):

$h = 02 \sim$ death without diabetic nephropathy (DN),
$h = 01 \sim$ occurrence of DN,
$h = 12 \sim$ death after occurrence of DN.

One way of studying the effect of sex on the corresponding transition intensities would be to have proportional death intensities for males and females without DN, i.e.,

$$\alpha_{02,i}(t)\,Y_{02,i}(t) = \begin{cases} \alpha_{02,0}(t)\,Y_{0i}(t) & \text{if } i \text{ is a female} \\ \alpha_{02,0}(t)e^{\beta_{02}}\,Y_{0i}(t) & \text{if } i \text{ is a male.} \end{cases} \tag{7.1.6}$$

Here, time t could refer to diabetes duration as in Example IV.4.3, $\alpha_{02,0}(t)$ is the death intensity for females without DN, and $Y_{02,i}(t) = Y_{0i}(t)$ indicates whether patient i is observed to be in state 0: "alive without DN" at time $t-$. For the DN intensity $\alpha_{01,i}(t)$, we may look at the "stratified" model

$$\alpha_{01,i}(t)\,Y_{01,i}(t) = \begin{cases} \alpha_{01F,0}(t)\,Y_{0i}(t) & \text{if } i \text{ is a female} \\ \alpha_{01M,0}(t)\,Y_{0i}(t) & \text{if } i \text{ is a male,} \end{cases}$$

where $\alpha_{01F,0}(t)$ is the DN intensity for females, $\alpha_{01M,0}(t)$ the DN intensity for males, and $Y_{01,i}(t) = Y_{0i}(t)$. Finally, for the mortality after occurrence of DN, we may again consider a model with proportional intensities

$$\alpha_{12,i}(t)\,Y_{12,i}(t) = \begin{cases} \alpha_{12,0}(t)\,Y_{1i}(t) & \text{if } i \text{ is a female} \\ \alpha_{12,0}(t)e^{\beta_{12}}\,Y_{1i}(t) & \text{if } i \text{ is a male.} \end{cases}$$

Here, $Y_{12,i}(t) = Y_{1i}(t)$ indicates whether patient i is observed to be in state 1: "alive with DN" at time $t-$ and $\alpha_{12,0}(t)$ is the death intensity for females with DN.

This model can be formulated as (7.1.2), (7.1.5) as follows. We split the type $h = 01$ into two types $h = 01F$ and $h = 01M$ and define the indicators

$$Y_{01F,i}(t) = Y_{0i}(t)I(i \text{ is a female})$$

and

$$Y_{01M,i}(t) = Y_{0i}(t)I(i \text{ is a male})$$

and the type-specific covariates

$$Z_{02,i,1} = I(i \text{ is a male}), \qquad Z_{02,i,2} \equiv 0,$$

$$Z_{01M,i,1} = Z_{01M,i,2} = Z_{01F,i,1} = Z_{01F,i,2} \equiv 0,$$

$$Z_{12,i,1} \equiv 0, \qquad Z_{12,i,2} = I(i \text{ is a male}).$$

Then we may write for each of the four types $h = 02, 01F, 01M, 12$

$$\alpha_{hi}(t) = \alpha_{h0}(t)\exp(\boldsymbol{\beta}^{\mathsf{T}}\mathbf{Z}_{hi});$$

cf. (7.1.5) with $\boldsymbol{\beta} = (\beta_{02}, \beta_{12})$ and the above definitions of the type-specific covariates. In this model where each of the transition intensities has been modelled separately, it may be examined whether sex affects the two death intensities in the same way by replacing the two regression coefficients β_{02} and β_{12} by a single β, resulting in a model where the two death intensities are not modelled separately. This amounts to including the type-specific covariates $Z_{02,i} = Z_{12,i} = I$ (i is a male) and $Z_{01M,i} = Z_{01F,i} \equiv 0$ instead of those mentioned above.

Alternatively, one might consider modelling the death intensity in state 0 by means of a regression model for the *relative mortality*; cf. Example III.1.3. This can be done by replacing $Y_{02,i}(t) = Y_{0i}(t)$ in (7.1.6) by $Y_{02,i}(t) = \mu_i(t) Y_{0i}(t)$, $\mu_i(t)$ being the population mortality at time t for an individual born the same year as patient i and having the same sex as patient i. In this model, $\exp(\beta_{02})$ is the ratio between the relative mortalities for male and female diabetics without DN and $\alpha_{02,0}(t)$ is the *relative* mortality for female diabetics without DN.

Instead of letting t denote diabetes duration, one can consider age the basic time scale and include age at onset of the disease as a time-fixed covariate or diabetes duration as a time-dependent covariate. Also, for the $1 \rightarrow 2$ transition intensity, the sojourn time spent in state 1 at time t could be included as a time-dependent covariate; cf. Examples III.5.1 and III.5.2.

When the transition intensities are modelled separately, it is even possible to consider different basic time scales for the different types of transition; thus, the death intensities $\alpha_{02,i}(\cdot)$ and $\alpha_{12,i}(\cdot)$ might be considered primarily a function of age and $\alpha_{01F,i}(\cdot)$ and $\alpha_{01M,i}(\cdot)$ primarily a function of diabetes duration.

We return to analyses of these data in Example VII.2.11. □

The next example shows how a new time-dependent covariate may be computed from basic time-fixed and time-dependent covariates.

Example VII.1.2. CSL 1—A Randomized Clinical Trial in Patients with Liver Cirrhosis. A Possible Model for the Effect of Alcohol Consumption

The data were introduced in Example I.3.4. Some of the interesting variables have to do with the alcohol consumption reported by the individual patients. So, the basic covariates $Z_i(t)$ may include the time-independent covariates

$$Z_{i1} = \text{average alcohol consumption per day before entry}$$

and

$$Z_{i2} = \text{duration of alcohol abuse before entry}$$

and the time-dependent covariate

$$Z_{i3}(t) = \text{alcohol consumption per day at time } t-.$$

From these three basic covariates, a new time-dependent covariate

$$Z_{i4}(t) = Z_{i1} Z_{i2} + \int_0^t Z_{i3}(u)\,du,$$

measuring the cumulative amount of alcohol consumed before t can be computed, and a model for the hazard function of the form

$$\alpha_i(t, \theta) = \alpha_0(t) \exp(\beta_4 Z_{i4}(t))$$

might be considered.

These data will be further analyzed in Example VII.2.16. □

VII.2. Semiparametric Multiplicative Hazard Models

This section and the next deal with semiparametric multiplicative hazard regression models of the form (7.1.5). In Section VII.2.1, we consider estimation of the vector of regression coefficients $\boldsymbol{\beta}$ and the underlying intensities $\alpha_{h0}(\cdot)$, $h = 1, \ldots, k$, as well as testing hypotheses about $\boldsymbol{\beta}$. In Section VII.2.2, large sample properties of the estimators are derived for the case where the relative risk function $r(x)$ is the exponential function $\exp(x)$. Section VII.2.3 discusses estimation of various functionals, whereas Section VII.2.4 contains some special cases not covered by the general theory. Some special problems in models with time-dependent covariates are discussed in Section VII.2.5 in connection with some practical examples. The following section (VII.3) is concerned with goodness-of-fit problems for this model.

VII.2.1. Estimation and Large Sample Properties

We consider the model given by the multiplicative hazard function

$$\alpha_{hi}^{\theta}(t; \mathbf{Z}_i(t)) = \alpha_{h0}(t, \gamma) r(\boldsymbol{\beta}^{\mathsf{T}} \mathbf{Z}_{hi}(t)),$$

where the predictable and locally bounded type-specific covariates $\mathbf{Z}_{hi}(\cdot)$ are defined from a vector $\mathbf{Z}_i(\cdot)$ of basic covariates; cf. (7.1.5). We assume that $\boldsymbol{\beta}$ has dimension p, whereas γ is infinite dimensional; in fact, as mentioned in Section VII.1, the underlying hazard functions are left completely unspecified and we write $\alpha_{h0}(t)$ for $\alpha_{h0}(t, \gamma)$. Following Chapters IV and V, however, we do assume that the $\alpha_{h0}(\cdot)$ are non-negative with

$$\int_0^t \alpha_{h0}(u)\, \mathrm{d}u < \infty, \quad h = 1, \ldots, k, \tag{7.2.1}$$

for all $t \in \mathscr{T}$.

In this case, the partial likelihood (7.1.1) is proportional to the product integral

$$\prod_{t \in \mathscr{T}} \left\{ \prod_{h,i} (\mathrm{d}A_{h0}(t) r(\boldsymbol{\beta}^{\mathsf{T}} \mathbf{Z}_{hi}(t)))^{\Delta N_{hi}(t)} \left(1 - \sum_{h=1}^{k} \mathrm{d}A_{h0}(t) S_h^{(0)}(\boldsymbol{\beta}, t)\right)^{1 - \Delta N_{\cdot\cdot}(t)} \right\}, \tag{7.2.2}$$

where we have defined

$$S_h^{(0)}(\boldsymbol{\beta}, t) = \sum_{i=1}^{n} r(\boldsymbol{\beta}^\mathsf{T}\mathbf{Z}_{hi}(t))Y_{hi}(t), \quad h = 1, \ldots, k, \tag{7.2.3}$$

and

$$A_{h0}(t) = \int_0^t \alpha_{h0}(u)\,du, \quad h = 1, \ldots, k. \tag{7.2.4}$$

If $Y_{hi}(t)$ is the indicator for individual i being at risk for a type h transition at time $t-$, then (7.2.3) is the sum over the type h *risk set* at time t of the quantities $r(\boldsymbol{\beta}^\mathsf{T}\mathbf{Z}_{hi}(t))$.

Estimation of the integrated underlying hazard functions $A_{h0}(t)$ for given $\boldsymbol{\beta}$ can now be performed from (7.2.2) following the lines of Section IV.1.5. For continuous A_{h0}, the "multinomial" likelihood (7.2.2) can be rewritten in the "Poisson" form

$$\prod_t \left\{ \prod_{h,i} (\mathrm{d}A_{h0}(t) r(\boldsymbol{\beta}^\mathsf{T}\mathbf{Z}_{hi}(t)))^{\Delta N_{hi}(t)} \right\} \exp\left[-\sum_{h=1}^{k} \int_0^\tau S_h^{(0)}(\boldsymbol{\beta}, u)\,\mathrm{d}A_{h0}(u) \right]; \tag{7.2.2'}$$

cf. (4.1.35). Now, for fixed value of $\boldsymbol{\beta}$, maximization of (7.2.2) and of (7.2.2′) with respect to $\Delta A_{h0}(t)$ leads to

$$\Delta \hat{A}_{h0}(t, \boldsymbol{\beta}) = \frac{\Delta N_h(t)}{S_h^{(0)}(\boldsymbol{\beta}, t)}$$

and, therefore, still for a fixed value of $\boldsymbol{\beta}$, we would estimate $A_{h0}(t)$ by the Nelson–Aalen estimator

$$\hat{A}_{h0}(t, \boldsymbol{\beta}) = \int_0^t \frac{J_h(u)}{S_h^{(0)}(\boldsymbol{\beta}, u)}\,\mathrm{d}N_h(u), \tag{7.2.5}$$

where $J_h(u) = I(Y_h(u) > 0)$, $Y_h = Y_{h1} + \cdots + Y_{hn}$, and $N_h = N_{h1} + \cdots + N_{hn}$; cf. (4.1.2). The estimate $\hat{A}_{h0}(t, \boldsymbol{\beta})$ can be written as a sum over observed times of type h transitions in $[0, t]$.

Inserting (7.2.5) into (7.2.2) or (7.2.2′) we obtain, respectively, the following partially maximized (partial) likelihoods (likelihood profile) only depending on $\boldsymbol{\beta}$ (assuming no ties):

$$L(\boldsymbol{\beta}) \mathop{\pi}_{t \in \mathcal{T}} \prod_h \Delta N_h(t)^{\Delta N_h(t)} (1 - \mathrm{d}N_{\cdot\cdot}(t))^{1-\Delta N_{\cdot\cdot}(t)}$$

or

$$L(\boldsymbol{\beta}) \prod_{t \in \mathcal{T}} \prod_h \Delta N_h(t)^{\Delta N_h(t)} \exp(-N_{\cdot\cdot}(\tau)).$$

Here

$$L(\boldsymbol{\beta}) = \prod_{t \in \mathcal{T}} \prod_{h,i} \left(\frac{r(\boldsymbol{\beta}^\mathsf{T}\mathbf{Z}_{hi}(t))}{S_h^{(0)}(\boldsymbol{\beta}, t)} \right)^{\Delta N_{hi}(t)} \tag{7.2.6}$$

is the *Cox partial likelihood* derived by (Cox 1972, 1975) for the case of censored survival data via entirely different routes; cf. Example VII.2.1. We shall base estimation of $\boldsymbol{\beta}$ on (7.2.6), or equivalently on the log Cox partial likelihood

$$C_\tau(\boldsymbol{\beta}) = \log L(\boldsymbol{\beta})$$
$$= \sum_{h=1}^{k} \left[\sum_{i=1}^{n} \int_0^\tau \log r(\boldsymbol{\beta}^\mathsf{T}\mathbf{Z}_{hi}(t))\, \mathrm{d}N_{hi}(t) - \int_0^\tau \log S_h^{(0)}(\boldsymbol{\beta}, t)\, \mathrm{d}N_h(t) \right], \qquad (7.2.7)$$

and the value of $\boldsymbol{\beta}$ which maximizes $C_\tau(\boldsymbol{\beta})$ (if such a value exists) will be denoted $\hat{\boldsymbol{\beta}}$. We then estimate $A_{h0}(t)$ by $\hat{A}_{h0}(t, \hat{\boldsymbol{\beta}})$ [cf. (7.2.5)] which is often called the *Breslow estimator*. From $\hat{A}_{h0}(t, \hat{\boldsymbol{\beta}})$ estimates of the underlying hazard functions, $\alpha_{h0}(t)$ can be obtained, for instance, by kernel function smoothing as in Section IV.2, i.e., we let

$$\hat{\alpha}_{h0}(t) = b_h^{-1} \int_{\mathscr{T}} K_h\left(\frac{t-u}{b_h}\right) \mathrm{d}\hat{A}_{h0}(u, \hat{\boldsymbol{\beta}}) \qquad (7.2.8)$$

for some bandwidths $b_h > 0$, $h = 1, \ldots, k$.

Efficiency of the (nonparametric maximum likelihood) estimators $\hat{\boldsymbol{\beta}}$ and $\hat{A}_{h0}(t, \hat{\boldsymbol{\beta}})$ is discussed in Section VIII.4.3.

Example VII.2.1. Cox's Partial Likelihood

We consider (independently) right-censored survival data

$$(\tilde{X}_1, D_1), \ldots, (\tilde{X}_n, D_n),$$

where $\tilde{X}_i = X_i \wedge U_i$ and $D_i = I(\tilde{X}_i = X_i)$; see, for instance, Example IV.1.1. Assume that the hazard function for X_i is $\alpha_i(t) = \alpha_0(t) r(\boldsymbol{\beta}^\mathsf{T}\mathbf{Z}_i(t))$, where $\mathbf{Z}_i(\cdot)$ is deterministic; for instance, time-independent though this need not be the case. Cox (1975) studied likelihood constructions in this case introducing the concept of a *partial likelihood*. Letting

$$X_{(1)} < \cdots < X_{(m)}$$

denote the *observed and ordered* failure times (assumed distinct), where $m \le n$, letting C_j represent information on censoring in the interval $[X_{(j-1)}, X_{(j)})$ plus the fact that one individual fails at $X_{(j)}$, and letting (j) specify the individual failing at $X_{(j)}$, the sequences

$$(\tilde{X}_1, D_1), \ldots, (\tilde{X}_n, D_n) \quad \text{and} \quad (C_1, (1), \ldots, C_m, (m), C_{m+1})$$

contain the same information. The partial likelihood based on $((1), \ldots, (m))$ in the sequence $(C_1, (1), \ldots, C_m, (m), C_{m+1})$:

$$\prod_{j=1}^{m} \Pr((j) | C_j, (C_l, (l)), l = 1, \ldots, j-1),$$

(see Section II.7.3) is then

$$\prod_{j=1}^{m} \frac{r(\boldsymbol{\beta}^{\mathsf{T}}\mathbf{Z}_{(j)})}{\sum_{l \in R_{(j)}} r(\boldsymbol{\beta}^{\mathsf{T}}\mathbf{Z}_{(j)}^{l})}. \tag{7.2.9}$$

Here $R_{(j)} = \{i: \tilde{X}_i \geq X_{(j)}\}$ is the risk set at time $X_{(j)}-$, $\mathbf{Z}_{(j)}$ is the covariate vector at $X_{(j)}$ for the individual failing at that time, and $\mathbf{Z}_{(j)}^{i} = \mathbf{Z}_i(X_{(j)})$ when $i \in R_{(j)}$. Equation (7.2.9) is a consequence of the fact that

$$\Pr((j) \text{ fails at } X_{(j)} | \text{one failure at } X_{(j)}, R_{(j)}) = \frac{r(\boldsymbol{\beta}^{\mathsf{T}}\mathbf{Z}_{(j)})}{\sum_{l \in R_{(j)}} r(\boldsymbol{\beta}^{\mathsf{T}}\mathbf{Z}_{(j)}^{l})}.$$

Obviously, (7.2.6) reduces to (7.2.9) in this special case.

For later use, we define

$$p_i(t, \boldsymbol{\beta}) = \frac{Y_i(t) r(\boldsymbol{\beta}^{\mathsf{T}}\mathbf{Z}_i(t))}{S^{(0)}(\boldsymbol{\beta}, t)}, \tag{7.2.10}$$

where $Y_i(t) = I(\tilde{X}_i \geq t)$ and we may then also write Cox's partial likelihood as

$$L(\boldsymbol{\beta}) = \prod_{i=1}^{n} p_i(\tilde{X}_i, \boldsymbol{\beta})^{D_i}.$$

As explained in more detail in Section II.7.3, this way of deriving Cox's partial likelihood is also available for general counting processes [see also the arguments of Self and Prentice (1982)]. Kalbfleisch and Prentice (1973) derived (7.2.9) as the marginal likelihood of the ranks in the case of (untied) survival data with time-fixed covariates and (staggered) type II censoring.

□

In Examples VII.2.2 and VII.2.5, we shall see how (7.2.6) reduces in some concrete models.

Example VII.2.2. Mortality and Nephropathy for Insulin-Dependent Diabetics. Partial Likelihoods

The data were introduced in Example I.3.11. In the first model considered in Example VII.1.1 with different effects of sex on the two death intensities and with different underlying DN intensities for males and females, the partial likelihood (7.2.6) becomes

$$L(\beta_{02}, \beta_{12}) = \prod_{t \in \mathcal{T}} \prod_{i=1}^{n} \prod_{h \in H} \left(\frac{r(\beta_{02} Z_{hi1} + \beta_{12} Z_{hi2})}{\sum_{j=1}^{n} Y_{hj}(t) r(\beta_{02} Z_{hj1} + \beta_{12} Z_{hj2})} \right)^{\Delta N_{hi}(t)},$$

where $H = \{02, 01M, 01F, 12\}$ is the type set. This further reduces to a product of two factors

$$L(\beta_{02}, \beta_{12}) = \prod_{t \in \mathcal{T}} \prod_{i=1}^{n} \left(\frac{r(\beta_{02} I(i \text{ is a male}))}{\sum_{j=1}^{n} Y_{0j}(t) r(\beta_{02} I(j \text{ is a male}))} \right)^{\Delta N_{02i}(t)}$$
$$\times \prod_{t \in \mathcal{T}} \prod_{i=1}^{n} \left(\frac{r(\beta_{12} I(i \text{ is a male}))}{\sum_{j=1}^{n} Y_{1j}(t) r(\beta_{12} I(j \text{ is a male}))} \right)^{\Delta N_{12i}(t)},$$

each of which depends on one β only. Note that because no effects are estimated for the two types $h = 01M$, $h = 01F$, these two types do not enter the partial likelihood. Assuming the same effect (β) of sex on the $0 \to 2$ and $1 \to 2$ transition intensities, the partial likelihood is

$$L(\beta) = \prod_{t \in \mathscr{T}} \prod_{i=1}^{n} \left(\frac{r(\beta I(i \text{ is a male}))}{\sum_{j=1}^{n} Y_{0j}(t) r(\beta I(j \text{ is a male}))} \right)^{\Delta N_{02i}(t)} \times \left(\frac{r(\beta I(i \text{ is a male}))}{\sum_{j=1}^{n} Y_{1j}(t) r(\beta I(j \text{ is a male}))} \right)^{\Delta N_{12i}(t)}.$$

Considering regression models for the relative mortality as in Example VII.1.1 does not entail any further difficulties. In the above likelihood expressions, $Y_{0j}(t)$ and $Y_{1j}(t)$ should just be multiplied by the relevant value $\mu_j(t)$ of the known population mortality. □

In the following, attention is restricted to the case where the relative risk function $r(\cdot)$ is the exponential function $r(x) = \exp(x)$ (Andersen and Gill, 1982; Andersen and Borgan, 1985). First, the existence of $\hat{\boldsymbol{\beta}}$ is discussed for this case.

Other choices of $r(\cdot)$ (Prentice and Self, 1983) are briefly discussed in Example VII.2.12.

Example VII.2.3. Existence of $\hat{\boldsymbol{\beta}}$

Using the same notation as in Example VII.2.1, the Cox partial likelihood (7.2.9) may be rewritten as

$$L(\boldsymbol{\beta}) = \prod_{j=1}^{m} \left\{ 1 + \sum_{l \in R_{(j)} \backslash (j)} \exp(\boldsymbol{\beta}^{\mathsf{T}} (\mathbf{Z}_{(j)}^{l} - \mathbf{Z}_{(j)})) \right\}^{-1}.$$

Jacobsen (1989b) showed that $C_\tau(\cdot) = \log L(\cdot)$ is *strictly* concave if and only if the *contrast covariate vectors*

$$\mathbf{Z}_{(j)}^{l} - \mathbf{Z}_{(j)} \quad \text{for } j = 1, \ldots, m, \text{ and } l \in R_{(j)} \backslash (j)$$

span the entire p-dimensional space. Furthermore, he showed that $\hat{\boldsymbol{\beta}}$ exists and is unique if and only if $\mathbf{0}$ belongs to the *interior of the convex hull* of the contrast covariate vectors. Thus, $\hat{\boldsymbol{\beta}}$ does *not* exist if, for all risk sets $R_{(j)}$, the covariate vector $\mathbf{Z}_{(j)}$ for the individual (j) failing at time $X_{(j)}$ is extreme "in the same direction" compared to the covariate vectors $\mathbf{Z}_{(j)}^{l}$ for $l \in R_{(j)} \backslash (j)$, e.g., if age is included as covariate and if it is always the oldest individual at risk who fails. □

When $r(\cdot) = \exp(\cdot)$, the expression (7.2.3) becomes

$$S_h^{(0)}(\boldsymbol{\beta}, t) = \sum_{i=1}^{n} \exp(\boldsymbol{\beta}^{\mathsf{T}} \mathbf{Z}_{hi}(t)) Y_{hi}(t), \quad h = 1, \ldots, k, \tag{7.2.11}$$

and we define the p-vectors

$$\mathbf{S}_h^{(1)}(\boldsymbol{\beta}, t) = \sum_{i=1}^{n} \mathbf{Z}_{hi}(t) \exp(\boldsymbol{\beta}^{\mathsf{T}} \mathbf{Z}_{hi}(t)) Y_{hi}(t), \quad h = 1, \ldots, k, \tag{7.2.12}$$

and the $p \times p$ matrices

$$\mathbf{S}_h^{(2)}(\boldsymbol{\beta}, t) = \sum_{i=1}^{n} \mathbf{Z}_{hi}^{\otimes 2}(t) \exp(\boldsymbol{\beta}^{\mathsf{T}} \mathbf{Z}_{hi}(t)) Y_{hi}(t), \quad h = 1, \ldots, k \tag{7.2.13}$$

(where for a p-vector $\mathbf{a}$, $\mathbf{a}^{\otimes 2}$ is the $p \times p$ matrix $\mathbf{a}\mathbf{a}^{\mathsf{T}}$). Expressions (7.2.12) and (7.2.13) are the first- and second-order partial derivatives of $S_h^{(0)}(\boldsymbol{\beta}, t)$ with respect to $\boldsymbol{\beta}$. Furthermore, we define the p-vector

$$\mathbf{E}_h(\boldsymbol{\beta}, t) = (E_{hj}(\boldsymbol{\beta}, t), j = 1, \ldots, p) = \frac{\mathbf{S}_h^{(1)}(\boldsymbol{\beta}, t)}{S_h^{(0)}(\boldsymbol{\beta}, t)} \tag{7.2.14}$$

and the $p \times p$ matrix

$$\mathbf{V}_h(\boldsymbol{\beta}, t) = (V_{hjl}(\boldsymbol{\beta}, t), j, l = 1, \ldots, p) = \frac{\mathbf{S}_h^{(2)}(\boldsymbol{\beta}, t)}{S_h^{(0)}(\boldsymbol{\beta}, t)} - \mathbf{E}_h(\boldsymbol{\beta}, t)^{\otimes 2}. \tag{7.2.15}$$

The quantities $\mathbf{E}_h(\boldsymbol{\beta}, t)$ and $\mathbf{V}_h(\boldsymbol{\beta}, t)$ are the expectation and the covariance matrix, respectively, of the type-specific covariate vector $\mathbf{Z}_{hi}(t)$ if an individual i is selected with probability

$$p_{hi}(t, \boldsymbol{\beta}) = \frac{Y_{hi}(t) \exp(\boldsymbol{\beta}^{\mathsf{T}} \mathbf{Z}_{hi}(t))}{S_h^{(0)}(\boldsymbol{\beta}, t)}$$

[see (7.2.10)], i.e., the individual is selected at time t from the type h risk set with a probability proportional to his or her intensity.

With these definitions, the vector of *score statistics*

$$\frac{\partial}{\partial \beta_j} C_\tau(\boldsymbol{\beta}) = U_\tau^j(\boldsymbol{\beta}), \quad j = 1, \ldots, p$$

[cf. (7.2.7)] can be written as

$$\mathbf{U}_\tau(\boldsymbol{\beta}) = \sum_{h=1}^{k} \left[\sum_{i=1}^{n} \int_0^\tau \mathbf{Z}_{hi}(t) \, \mathrm{d}N_{hi}(t) - \int_0^\tau \mathbf{E}_h(\boldsymbol{\beta}, t) \, \mathrm{d}N_h(t) \right] \tag{7.2.16}$$

and the matrix of second-order partial derivatives of $C_\tau(\boldsymbol{\beta})$ is $-\mathscr{I}_\tau(\boldsymbol{\beta})$, where

$$\mathscr{I}_\tau(\boldsymbol{\beta}) = \sum_{h=1}^{k} \int_0^\tau \mathbf{V}_h(\boldsymbol{\beta}, t) \, \mathrm{d}N_h(t). \tag{7.2.17}$$

Standard likelihood theory would suggest that the distribution in large samples of $\sqrt{n}(\hat{\boldsymbol{\beta}} - \boldsymbol{\beta}_0)$, where $\boldsymbol{\beta}_0$ is the true parameter vector, is approximately multinormal with mean zero and a covariance matrix which may be estimated consistently by $(n^{-1}\mathscr{I}_\tau(\hat{\boldsymbol{\beta}}))^{-1}$. In Section VII.2.2, we shall show that this is, indeed, the case under some regularity conditions. From this result, it immediately follows that the asymptotic distribution of the *Wald test statistic*

$$(\hat{\boldsymbol{\beta}} - \boldsymbol{\beta}_0)^{\mathsf{T}} \mathscr{I}_\tau(\hat{\boldsymbol{\beta}})(\hat{\boldsymbol{\beta}} - \boldsymbol{\beta}_0)$$

for the simple hypothesis $\boldsymbol{\beta} = \boldsymbol{\beta}_0$ is χ^2 with p degrees of freedom. Standard likelihood theory would also suggest that both the (Cox partial) *likelihood ratio test statistic*

$$-2 \log Q = 2(C_\tau(\hat{\boldsymbol{\beta}}) - C_\tau(\boldsymbol{\beta}_0))$$

and *the score test statistic*

$$\mathbf{U}_\tau(\boldsymbol{\beta}_0)^\mathsf{T} \mathscr{I}_\tau(\boldsymbol{\beta}_0)^{-1} \mathbf{U}_\tau(\boldsymbol{\beta}_0) \tag{7.2.18}$$

have the same asymptotic distribution as the Wald test statistic. Again, this turns out to be true under the same regularity conditions.

Composite hypotheses concerning $\boldsymbol{\beta}$ can be dealt with similarly [Section VI.1.2; Serfling (1980, Subsection 4.4.4)].

The next example shows how these test statistics simplify in the case of survival data.

EXAMPLE VII.2.4. Test Statistics for Survival Data: The $(p + 1)$-Sample Case and Trend Tests

Suppose the n individuals consist of $p + 1$ groups labelled $0, 1, \ldots, p$ among which we want to compare the survival time distributions as in the setup of Example V.2.1. For each individual, we let

$$Z_{ij} = I(i \text{ belongs to group } j), \quad j = 1, \ldots, p,$$

i.e., the hazard for individual i is

$$\alpha_i(t) = \alpha_0(t) \exp\left\{ \sum_{j=1}^{p} \beta_j Z_{ij} \right\}.$$

Under the null hypothesis that the corresponding regression coefficients vanish, $\beta_1 = \cdots = \beta_p = 0$, the score statistic (7.2.16) has components

$$U_\tau^j(0) = N_{\bullet}^{(j)}(\tau) - \int_0^\tau \frac{Y_{\bullet}^{(j)}(t)}{Y_{\bullet}(t)} \, \mathrm{d}N_{\bullet}(t), \quad j = 1, \ldots, p.$$

Here, $N_{\bullet}^{(j)} = \sum_i Z_{ij} N_i$, $Y_{\bullet}^{(j)} = \sum_i Z_{ij} Y_i$, $N_{\bullet} = \sum_j N_{\bullet}^{(j)}$, and $Y_{\bullet} = \sum_j Y_{\bullet}^{(j)}$. Furthermore, the second-order partial derivative matrix has elements

$$\mathscr{I}_\tau^{jl}(0) = \int_0^\tau \frac{Y_{\bullet}^{(j)}(t)}{Y_{\bullet}(t)} \left(\delta_{jl} - \frac{Y_{\bullet}^{(l)}(t)}{Y_{\bullet}(t)} \right) \mathrm{d}N_{\bullet}(t), \quad j = 1, \ldots, p;$$

cf. (7.2.17). This shows that the score test statistic (7.2.18) is simply the *log-rank* test statistic introduced in Example V.2.1; see also Cox (1972).

Also the other test statistics for this situation discussed in Example V.2.1 are obtained as special cases of (7.2.18) by appropriate choices of time-dependent covariates (Lustbader, 1980; Oakes, 1981). Thus, the choice

$$Z_{ij}(t) = I(i \text{ belongs to group } j) K(t), \quad j = 1, \ldots, p,$$

yields the test statistic corresponding to the predictable weight process $K(t)$;

cf. Section V.2.1. Optimality properties of these test statistics will be discussed in Sections VIII.2.3 and VIII.4.2. For $\boldsymbol{\beta} \neq \mathbf{0}$, this is a rather curious (unrealistic) model, but a test of $\boldsymbol{\beta} = \mathbf{0}$ is still a test of the null hypothesis of interest, and when K is almost deterministic (large sample sizes), the alternatives do become meaningful.

If a *score* x_j, $j = 0, 1, \ldots, p$, where $x_0 < x_1 < \cdots < x_p$, is associated with each group, then we may consider the one-dimensional covariate

$$Z_i = x_j \quad \text{if } i \text{ belongs to group } j.$$

In this case, the score test for the hypothesis $\beta = 0$ in the model $\alpha_i(t) = \alpha_0(t)\exp(\beta Z_i)$ is simply the *log-rank test for trend* defined in Section V.3.3, and, more generally, the covariate $Z_i(t) = x_j K(t)$ (if i belongs to group j) yields as score test the general trend test considered there. Having specified the trend hypothesis, it is also possible to test for *departures from trend*. Thus, the reduction from the general proportional hazards model

$$\alpha_i(t) = \alpha_0(t)\exp\left(\sum_{j=1}^{p} \beta_j Z_{ij}\right)$$

to the trend model

$$\alpha_i(t) = \alpha_0(t)\exp(\beta Z_i)$$

can be tested using, for instance, the partial likelihood ratio test statistic with $p - 1$ degrees of freedom. Furthermore, the sum of this test statistic and the partial likelihood ratio test statistic for the hypothesis $\beta = 0$ in the trend model is equal to the partial likelihood ratio $(p + 1)$-sample test statistic for the hypothesis $\beta_1 = \cdots = \beta_p = 0$ in the general proportional hazards model.

If the individuals are not grouped but rather to each individual there is associated a one-dimensional *continuous* (*possibly time-dependent*) *covariate* $Z_i(t)$, then one may again consider the model $\alpha_i(t) = \alpha_0(t)\exp(\beta Z_i(t))$. In this case, the score test statistic for the hypothesis $\beta = 0$ is

$$\frac{U_\tau^2(0)}{\mathscr{I}_\tau(0)} = \frac{\left\{\int_0^\tau \sum_{i=1}^n (Z_i(t) - \overline{Z^1}(t))\,\mathrm{d}N_i(t)\right\}^2}{\int_0^\tau (\overline{Z^2}(t) - (\overline{Z^1}(t))^2)\,\mathrm{d}N_.(t)},$$

where $\overline{Z^l}(t) = \sum_{i=1}^n Z_i^l(t)Y_i(t)/Y_.(t)$, $l = 1, 2$. Test statistics of this type were studied by Jones and Crowley (1989, 1990) who also suggested using truncated versions of the statistic obtained by downweighting extreme values of $Z_i(t)$. □

We shall conclude this section by an example using the melanoma data. The main purpose of the example is to illustrate how the Cox partial likelihood (7.2.6) reduces in some concrete models and the use of the test statistics and of type-specific covariates. We return to a more definitive analysis of these data in Section VII.3.

EXAMPLE VII.2.5. Survival with Malignant Melanoma.
Simple Multiplicative Regression Models

The data were introduced in Example I.3.1. There is basically one type of event of interest "death from malignant melanoma" and for each of the $n = 205$ patients we now consider the covariates,

$$Z_{i1} = \begin{cases} 1 & \text{if patient } i \text{ is a male} \\ 0 & \text{if patient } i \text{ is a female} \end{cases}$$

and

Z_{i2} = thickness in mm of the tumor for patient i minus the mean tumor thickness 2.92 mm.

If we consider the model with only the covariate Z_{i1} included, then we are in the simple two-sample case which is a special case of Example VII.2.4 and the hazard function for individual i is

$$\alpha_i(t) = \alpha_0(t)\exp(\beta_1 Z_{i1}),$$

time t referring to time since operation. Here we find $\hat{\beta}_1 = 0.662$ with an estimated standard deviation of 0.265 corresponding to an estimated hazard ratio between men and women of $\hat{\theta}_1 = \exp(\hat{\beta}_1) = 1.94$ and an approximate 95% confidence interval for θ_1 equal to (1.15, 3.26), compare Example V.3.1. This yields a Wald test statistic of $(0.662/0.265)^2 = 6.24$ which we may compare with the log-rank test statistic (=the score test statistic; cf. Example VII.2.4) 6.47 and the partial likelihood ratio test statistic 6.15, all giving P-values around 0.01. Figure VII.2.1 shows the estimated integrated hazards in the two groups, i.e., the estimated integrated underlying hazard $\hat{A}_0(t, \hat{\beta}_1)$ for females and $\hat{\theta}_1 \hat{A}_0(t, \hat{\beta}_1)$ for males. The estimates are close to the Nelson–Aalen plots in Figure IV.1.1.

Considering next a model with only the covariate Z_{i2} included, we find $\hat{\beta}_2 = 0.160$ (0.031). Thus, assuming that the log hazard function increases linearly with tumor thickness, the estimated increase in hazard when the tumor thickness increases by 1 mm is $\exp(\hat{\beta}_2) = 1.17$ with an approximate 95% confidence interval (1.10, 1.25). In this case, the Wald test statistic is $(0.160/0.031)^2 = 26.28$ and the partial likelihood ratio test statistic 19.19. The score test statistic, closely connected with the log-rank test for trend (Examples V.3.5 and VII.2.4), takes the value 28.70.

The simplest possible model for the death intensity $\alpha_i(t)$ for patient i including both covariates is

$$\alpha_i(t) = \alpha_0(t)\exp(\beta_1 Z_{i1} + \beta_2 Z_{i2}), \quad i = 1, \ldots, n. \tag{7.2.19}$$

In this model, $\exp(\beta_1)$ is the ratio of the death intensities for male and female patients for *given tumor thickness*; $\exp(\beta_2)$ is the increase (or decrease) in the death intensity comparing two patients *with the same sex* and differing 1 mm in tumor thickness.

Thus, also in this model it is assumed that the log hazard function in-

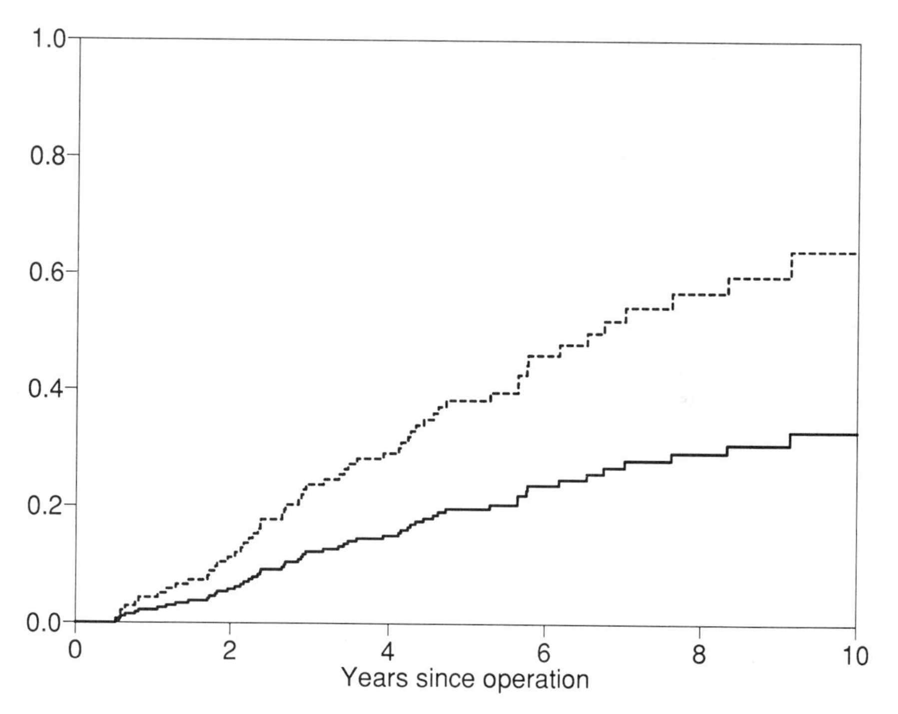

Figure VII.2.1. Estimated integrated hazards for men (---) and women (—) with malignant melanoma based on a proportional hazards model.

creases linearly with the tumor thickness. The underlying hazard function $\alpha_0(t)$ is the death intensity for patients with $Z_{i1} = Z_{i2} = 0$, i.e., females with tumor thickness equal to the mean tumor thickness.

In this model, the partial likelihood (7.2.6) is

$$L(\beta_1, \beta_2) = \prod_{i=1}^{n} \left(\frac{\exp(\beta_1 Z_{i1} + \beta_2 Z_{i2})}{\sum_{j \in R_i} \exp(\beta_1 Z_{j1} + \beta_2 Z_{j2})} \right)^{D_i}$$

with D_i and R_i as defined in Example VII.2.1. The estimates are

$$\hat{\beta}_1 = 0.574\ (0.265)$$

and

$$\hat{\beta}_2 = 0.159\ (0.033)$$

with an estimated correlation of -0.018. Hence, a slight positive correlation between Z_{i1} and Z_{i2} has reduced the marginal significance of both covariates in that the Wald test statistic for the hypothesis $\beta_1 = 0$ is 4.68 and the partial likelihood ratio test statistic is 4.64, whereas the corresponding test statistics for the hypothesis $\beta_2 = 0$ are 23.71 and 17.67, respectively (all statistics to be referred to the χ_1^2 distribution).

One possible model with *interaction* between sex and tumor thickness is obtained by replacing the covariate tumor thickness (Z_{i2}) by the two covariates

$$Z_{i3} = \begin{cases} Z_{i2} & \text{if patient } i \text{ is a male} \\ 0 & \text{otherwise} \end{cases}$$

and

$$Z_{i4} = \begin{cases} Z_{i2} & \text{if patient } i \text{ is a female} \\ 0 & \text{otherwise.} \end{cases}$$

Here β_1 is still a sex effect, $\exp(\beta_1)$ now being the ratio between the hazards for male and female patients *with tumor thickness equal to the mean tumor thickness*, whereas the coefficients β_3 and β_4 corresponding to Z_{i3} and Z_{i4} are effects of tumor thickness for male and female patients, respectively. In this model, the estimates are

$$\hat{\beta}_1 = 0.532\ (0.283),$$

$$\hat{\beta}_3 = 0.178\ (0.053),$$

and

$$\hat{\beta}_4 = 0.148\ (0.043)$$

and the Wald test statistic for interaction is

$$\frac{(0.178 - 0.148)^2}{0.053^2 + 0.043^2 - 2(0.053)(0.043)(0.009)} = 0.19$$

(0.009 being the estimated correlation between $\hat{\beta}_3$ and $\hat{\beta}_4$). The corresponding partial likelihood ratio test statistic is 0.20 both giving a P-value of 0.65.

In the models considered so far, it has been assumed that male and female patients have *proportional hazards*—an assumption that one would probably like to check in the analysis. We shall consider model checking in Section VII.3 and just mention here that such checks are frequently based on *the stratified Cox regression model*; cf. Kalbfleisch and Prentice (1980, Section 4.4). Here, the model (7.2.19) is replaced by the model

$$\alpha_i(t) = \begin{cases} \alpha_{10}(t)\exp(\beta_2 Z_{i2}) & \text{if patient } i \text{ is a female} \\ \alpha_{20}(t)\exp(\beta_2 Z_{i2}) & \text{if patient } i \text{ is a male,} \end{cases} \tag{7.2.20}$$

where there is still the same effect, β_2, of tumor thickness for males and females but where there are now separate underlying hazards: $\alpha_{10}(t)$ for females and $\alpha_{20}(t)$ for males. In this model, it can be investigated whether the estimates are compatible with the assumption of proportional hazards for males and females, i.e., that the ratio $\alpha_{20}(t)/\alpha_{10}(t)$ is constant over time.

The stratified model (7.2.20) may be written as (7.1.2) and (7.1.5) by defining

$$Y_{1i}(t) = I(i \text{ is a female at risk at time } t-)$$

and

$$Y_{2i}(t) = I(i \text{ is a male at risk at time } t-).$$

Thus, in (7.2.20) sex enters by defining a *type* $h = 1, 2$ where females are only at risk for type 1 events and males only for type 2 events, but no extra type-specific covariates are needed to specify the model.

We now let D_{hi} be the indicator $D_i I(i \text{ belongs to stratum } h)$, $h = 1, 2$, and

$$R_{hi} = \{j: Y_{hj}(\tilde{X}_i) = 1\}$$

the type h risk set just before time $\tilde{X}_i$. Then the partial likelihood (7.2.6) is

$$L(\beta_2) = \prod_{i=1}^{n} \prod_{h=1}^{2} \left(\frac{\exp(\beta_2 Z_{i2})}{\sum_{j \in R_{hi}} \exp(\beta_2 Z_{j2})} \right)^{D_{hi}} \tag{7.2.21}$$

which cannot be further reduced. In this model, we find $\hat{\beta}_2 = 0.158$ (0.033) and the estimated underlying hazards $\hat{A}_{10}(t, \hat{\beta}_2)$ and $\hat{A}_{20}(t, \hat{\beta}_2)$ are shown in Figure VII.2.2. The curve for females is close to that in Figure VII.2.1, whereas that for males is somewhat lower. This is due to the fact that we have now taken the effect of tumor thickness into account and "standardized" the estimates to the value 2.92 mm which is lower than the corresponding average

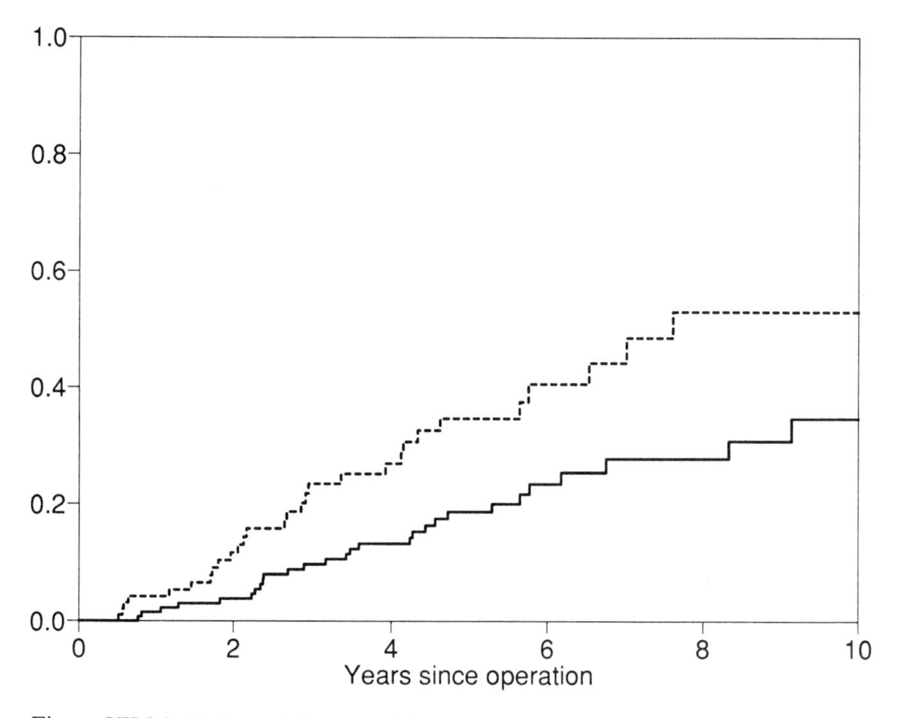

Figure VII.2.2. Estimated integrated hazards for men (---) and women (—) with malignant melanoma and tumor thickness equal to 2.92 mm based on a stratified model.

tumor thickness for men (3.61 mm). The fact that both integrated hazard estimates are roughly linear suggests that the proportional hazards assumption is reasonable.

Alternatively, completely separate models for males and females may be considered. This can be done by replacing the expression $\beta_2 Z_{i2}$ in (7.2.20) including the common covariate Z_{i2} for tumor thickness by $\beta_3 Z_{i3} + \beta_4 Z_{i4}$ with the two separate covariates for thickness defined earlier: Z_{i3} for men and Z_{i4} for women. Now, Z_{i3} and Z_{i4} play the role of type-specific covariates and letting $\boldsymbol{\beta} = (\beta_3, \beta_4)$, $Z_{hi1} = Z_{i4}$, $Z_{hi2} \equiv 0$ for $h = 1$, and $Z_{hi1} \equiv 0$ and $Z_{hi2} = Z_{i3}$ for $h = 2$, the model has the form (7.1.2) and (7.1.5). Here, the partial likelihood (7.2.6) reduces to a product

$$L(\beta_3)L(\beta_4) = \prod_{i=1}^{n} \left(\frac{\exp(\beta_3 Z_{i3})}{\sum_{j \in R_{1i}} \exp(\beta_3 Z_{j3})} \right)^{D_{1i}} \prod_{i=1}^{n} \left(\frac{\exp(\beta_4 Z_{i4})}{\sum_{j \in R_{2i}} \exp(\beta_4 Z_{j4})} \right)^{D_{2i}} \tag{7.2.22}$$

of two factors each depending on only one of the β. The estimators for β_3 and β_4 are asymptotically *independent* and the Wald test statistic for interaction based on the estimates

$$\hat{\beta}_3 = 0.172\ (0.052)$$

and

$$\hat{\beta}_4 = 0.149\ (0.043)$$

is simply

$$\frac{(0.172 - 0.149)^2}{(0.052)^2 + (0.043)^2} = 0.12.$$

This is very close to the partial likelihood ratio test statistic 0.12 based on (7.2.21) and (7.2.22) both giving a P-value of 0.73.

It may also be of interest to analyze the intensity of dying from other causes in a competing risks model; cf. Examples I.3.9, III.1.5, and IV.4.1. Here, two types $h = 1, 2$ may be defined as

$$h = 1 \sim \text{death from malignant melanoma,}$$

$$h = 2 \sim \text{death from other causes.}$$

A model with different effects of sex on the two cause-specific intensities (and disregarding for a moment the tumor thickness) can be written in the form (7.1.2) and (7.1.5) by letting $Y_{1i}(t) = Y_{2i}(t) = I$ (patient i is at risk at time $t-$), by defining the type-specific covariates, $Z_{hi1} = Z_{i1}$ and $Z_{hi2} \equiv 0$ for $h = 1$ and $Z_{hi1} \equiv 0$ and $Z_{hi2} = Z_{i1}$ for $h = 2$, and letting $\boldsymbol{\beta} = (\beta_1, \beta_2)$, where β_1 is now the effect of sex on transitions of type $h = 1$ (death from malignant melanoma) and β_2 is the effect of sex of transitions of type $h = 2$ (death from other causes). Again, this corresponds to completely separate models for the two cause-specific hazard functions and the partial likelihood (7.2.6) reduces to a product

$$L(\beta_1)L(\beta_2) = \prod_{i=1}^{n} \left(\frac{\exp(\beta_1 Z_{i1})}{\sum_{j \in R_i} \exp(\beta_1 Z_{j1})} \right)^{D_{1i}} \prod_{i=1}^{n} \left(\frac{\exp(\beta_2 Z_{i1})}{\sum_{j \in R_i} \exp(\beta_2 Z_{j1})} \right)^{D_{2i}} \quad (7.2.23)$$

of two factors each depending on one β only (D_{hi} indicates whether individual i was observed to die from cause h, $h = 1, 2$). The estimates become $\hat{\beta}_1 = 0.662$ (0.265) as in the first model considered in this example and $\hat{\beta}_2 = 0.630$ (0.536), and it seems likely that we can accept a model (which, however, may be somewhat difficult to interpret) with the *same effect* of sex on the two cause-specific hazards. This model can also be written in the form (7.1.5)

$$\alpha_{hi}(t) = \alpha_{h0}(t)\exp(\beta Z_{hi1}), \quad h = 1, 2, i = 1, \ldots, n, \quad (7.2.24)$$

where $Z_{hi1} = Z_{i1}$ for both $h = 1$ and $h = 2$, and we get the partial likelihood

$$L(\beta) = \prod_{i=1}^{n} \prod_{h=1}^{2} \left(\frac{\exp(\beta Z_{i1})}{\sum_{j \in R_i} \exp(\beta Z_{j1})} \right)^{D_{hi}} = \prod_{i=1}^{n} \left(\frac{\exp(\beta Z_{i1})}{\sum_{j \in R_i} \exp(\beta Z_{j1})} \right)^{D_i}$$

since $D_{1i} + D_{2i} = D_i$. Thus, assuming the same effect of sex on $\alpha_{1i}(t)$ and $\alpha_{2i}(t)$, the partial likelihood is the same as if there was only one cause of death under consideration. There are still, however, different underlying cause-specific hazards $\alpha_{h0}(t)$, $h = 1, 2$, yielding different estimates

$$\hat{A}_{10}(t, \hat{\beta}) = \sum_{\tilde{X}_i \le t} \frac{D_{1i}}{\sum_{j \in R_i} \exp(\hat{\beta} Z_{j1})}$$

and

$$\hat{A}_{20}(t, \hat{\beta}) = \sum_{\tilde{X}_i \le t} \frac{D_{2i}}{\sum_{j \in R_i} \exp(\hat{\beta} Z_{j1})},$$

the sum of which equals the estimated integrated underlying hazard function

$$\hat{A}_0(t, \hat{\beta}) = \sum_{\tilde{X}_i \le t} \frac{D_i}{\sum_{j \in R_i} \exp(\hat{\beta} Z_{j1})}$$

in the model, disregarding causes of death. In this model, we find

$$\hat{\beta} = 0.656 \ (0.238)$$

and the Wald test statistic

$$\frac{(0.662 - 0.630)^2}{(0.265)^2 + (0.536)^2} = 0.001,$$

and the partial likelihood ratio test statistic 0.003 for the hypothesis $\beta_1 = \beta_2$ are both extremely small. The estimated integrated hazard functions $\hat{A}_{10}(t, \hat{\beta})$, $\hat{A}_{20}(t, \hat{\beta})$, and $\hat{A}_0(t, \hat{\beta})$ are shown in Figure VII.2.3. It is seen that the hazard of dying from other causes is higher than the hazard of dying from the disease shortly after operation, but otherwise it is very low.

If, in the model (7.2.24), one wants to adjust for the effect of tumor thick-

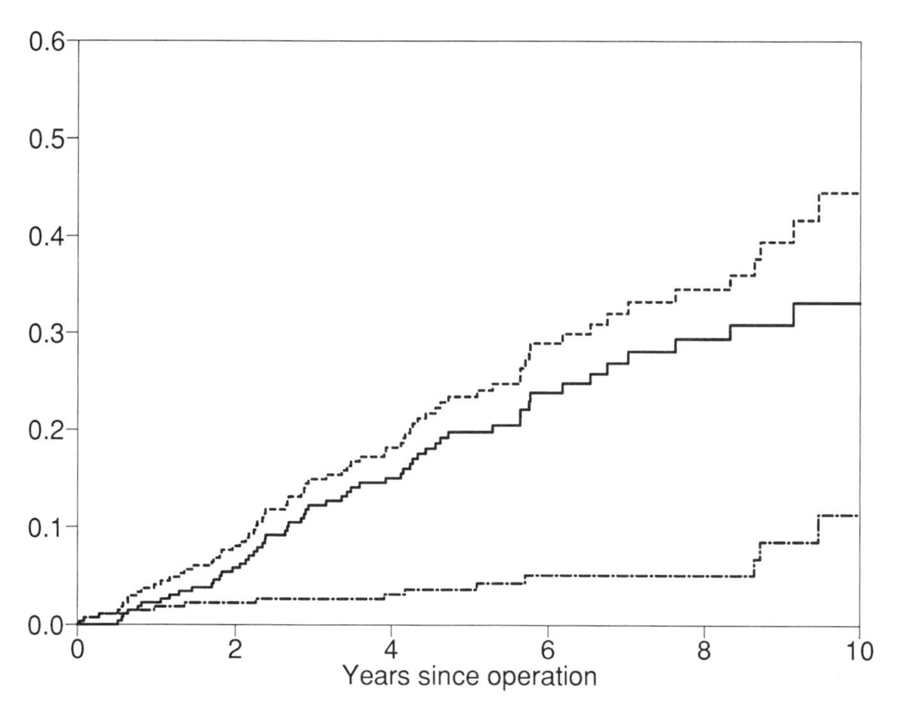

Figure VII.2.3. Estimated integrated hazards for women with malignant melanoma: death from the disease (—), death from other causes (-·-·-), death from all causes (- - -).

ness, then this covariate is likely to have no effect on the intensity of dying from other causes. Such a model can be formulated using a type-specific covariate

$$Z_{hi2} = \begin{cases} Z_{i2} & \text{if } h = 1 \\ 0 & \text{if } h = 2 \end{cases} \tag{7.2.25}$$

and it has neither the form of a simple stratified Cox regression model nor does it correspond to separate models for males and females. But it may still be written in the general form (7.1.5) and the partial likelihood is

$$\begin{aligned} L(\beta, \beta_3) &= \prod_{i=1}^{n} \prod_{h=1}^{2} \left(\frac{\exp(\beta Z_{hi1} + \beta_3 Z_{hi2})}{\sum_{j \in R_i} \exp(\beta Z_{hj1} + \beta_3 Z_{hj2})} \right)^{D_{hi}} \\ &= \prod_{i=1}^{n} \left(\frac{\exp(\beta Z_{i1} + \beta_3 Z_{i2})}{\sum_{j \in R_i} \exp(\beta Z_{j1} + \beta_3 Z_{j2})} \right)^{D_{1i}} \left(\frac{\exp(\beta Z_{i1})}{\sum_{j \in R_i} \exp(\beta Z_{j1})} \right)^{D_{2i}} \end{aligned} \tag{7.2.26}$$

with β_3 representing the effect of tumor thickness on $\alpha_{1i}(t)$. It cannot be further reduced but it may be maximized using standard software by creating a data file with $2n = 410$ records and defining a stratum indicator h which is 1 for the first $n = 205$ records and 2 for the last n records. Furthermore, the failure indicator should be D_{1i} in stratum 1 and D_{2i} in stratum 2 and the covariate for tumor thickness should be defined by (7.2.25). Apart from that, the records in the two strata are pairwise identical.

The estimates based on (7.2.26) become

$$\hat{\beta} = 0.585\ (0.238)$$

and

$$\hat{\beta}_3 = 0.159\ (0.033).$$

Thus, the estimated effect of sex decreases when tumor thickness is taken into account, whereas the effect of thickness is very similar to that obtained in earlier models. □

VII.2.2. Large Sample Properties

In this section, we study the large sample properties (i.e., $n \to \infty$) of the estimator $\hat{\boldsymbol{\beta}}$ and of the processes $\hat{A}_{h0}(\cdot, \hat{\boldsymbol{\beta}}) - A_{h0}(\cdot)$. We, therefore, consider a sequence of models indexed by n with processes $N_{hi}^{(n)}$, $Y_{hi}^{(n)}$, and $\mathbf{Z}_{hi}^{(n)}$ defined on the nth sample space $(\Omega^{(n)}, \mathscr{F}^{(n)}, \mathscr{P}^{(n)})$ and adapted to a filtration $(\mathscr{F}_t^{(n)})$. We will, however, drop the superscript (n) from the notation and here just remind the reader that these quantities do depend on n, whereas the true parameter values $\boldsymbol{\beta}_0$ and $\alpha_{h0}(\cdot)$ are the same for all n. Convergence in probability and convergence in distribution are always as n tends to infinity. Often, n is the sample size, though this need not always be the case, but for ease of notation we have chosen to use (n) as the sequence of rate constants rather than the more general (a_n^2) used elsewhere in this volume.

The reader should recall definitions (7.2.11)–(7.2.17).

The key fact in the proof of the large sample properties of $\hat{\boldsymbol{\beta}}$ is that the score statistics (7.2.16) evaluated at $\boldsymbol{\beta}_0$ and considered as processes in t

$$U_t^j(\boldsymbol{\beta}_0) = \sum_{h=1}^{k} \sum_{i=1}^{n} \int_0^t \{Z_{hij}(u) - E_{hj}(\boldsymbol{\beta}_0, u)\}\, \mathrm{d}M_{hi}(u), \quad j = 1, \ldots, p,$$

by Theorem II.3.1 are local square integrable martingales, being linear combinations of stochastic integrals of predictable and locally bounded processes with respect to the local square integrable martingales

$$M_{hi}(t) = N_{hi}(t) - \int_0^t \alpha_{h0}(u) \exp(\boldsymbol{\beta}_0^\mathsf{T} \mathbf{Z}_{hi}(u))\, Y_{hi}(u)\, \mathrm{d}u.$$

The M_{hi} are orthogonal with

$$\langle M_{hi} \rangle (t) = \int_0^t \alpha_{h0}(u) \exp(\boldsymbol{\beta}_0^\mathsf{T} \mathbf{Z}_{hi}(u))\, Y_{hi}(u)\, \mathrm{d}u; \tag{7.2.27}$$

cf. (2.4.3).

In the proofs, we need the following regularity condition where the norm of a vector $\mathbf{a} = (a_i)$ or a matrix $\mathbf{A} = (a_{ij})$ is $\|\mathbf{a}\| = \sup_i |a_i|$ and $\|\mathbf{A}\| = \sup_{i,j} |a_{ij}|$, respectively. We use $|\mathbf{a}|$ for the Euclidean norm of $\mathbf{a}$.

Condition VII.2.1. There exist a neighborhood $\mathscr{B}$ of $\boldsymbol{\beta}_0$ and scalar, p-vector and $p \times p$ matrix functions $s_h^{(0)}$, $\mathbf{s}_h^{(1)}$, and $\mathbf{s}_h^{(2)}$, respectively, $h = 1, \ldots, k$, defined on $\mathscr{B} \times \mathscr{T}$ such that for $m = 0, 1, 2$, $h = 1, \ldots, k$:

(a)
$$\sup_{\boldsymbol{\beta} \in \mathscr{B}, t \in \mathscr{T}} \left\| \frac{1}{n} \mathbf{S}_h^{(m)}(\boldsymbol{\beta}, t) - \mathbf{s}_h^{(m)}(\boldsymbol{\beta}, t) \right\| \xrightarrow{\mathrm{P}} 0 \quad \text{as } n \to \infty;$$

(b) $\mathbf{s}_h^{(m)}(\cdot)$ is a continuous function of $\boldsymbol{\beta} \in \mathscr{B}$ uniformly in $t \in \mathscr{T}$ and bounded on $\mathscr{B} \times \mathscr{T}$;

(c) $s_h^{(0)}(\boldsymbol{\beta}_0, \cdot)$ is bounded away from zero on $\mathscr{T}$;

(d)
$$\mathbf{s}_h^{(1)}(\boldsymbol{\beta}, t) = \frac{\partial}{\partial \boldsymbol{\beta}} s_h^{(0)}(\boldsymbol{\beta}, t), \qquad \mathbf{s}_h^{(2)}(\boldsymbol{\beta}, t) = \frac{\partial^2}{\partial \boldsymbol{\beta}^2} s_h^{(0)}(\boldsymbol{\beta}, t) \quad \text{for } \boldsymbol{\beta} \in \mathscr{B}, t \in \mathscr{T};$$

(e) $\boldsymbol{\Sigma}_\tau = \sum_{h=1}^k \int_0^\tau \mathbf{v}_h(\boldsymbol{\beta}_0, t) s_h^{(0)}(\boldsymbol{\beta}_0, t) \alpha_{h0}(t)\,\mathrm{d}t$ is positive definite, where $\mathbf{v}_h = (v_{hjl}) = \mathbf{s}_h^{(2)}/s_h^{(0)} - \mathbf{e}_h^{\otimes 2}$ and $\mathbf{e}_h = (e_{hj}) = \mathbf{s}_h^{(1)}/s_h^{(0)}$.

We are now able to prove the following consistency theorem.

Theorem VII.2.1. *Assume Condition* VII.2.1 *and that* (7.2.1) *holds for* $t = \tau$. *Then the probability that the equation* $\mathbf{U}_\tau(\boldsymbol{\beta}) = \mathbf{0}$ *has a unique solution* $\hat{\boldsymbol{\beta}}$ *tends to* 1 *and* $\hat{\boldsymbol{\beta}} \xrightarrow{\mathrm{P}} \boldsymbol{\beta}_0$ *as* $n \to \infty$.

PROOF. The process

$$X(\boldsymbol{\beta}, t) = \frac{1}{n}(C_t(\boldsymbol{\beta}) - C_t(\boldsymbol{\beta}_0))$$
$$= \frac{1}{n} \sum_{h=1}^k \left[\sum_{i=1}^n \int_0^t (\boldsymbol{\beta} - \boldsymbol{\beta}_0)^\mathsf{T} \mathbf{Z}_{hi}(s)\,\mathrm{d}N_{hi}(s) - \int_0^t \log\left(\frac{S_h^{(0)}(\boldsymbol{\beta}, s)}{S_h^{(0)}(\boldsymbol{\beta}_0, s)} \right) \mathrm{d}N_h(s) \right]$$

has compensator

$$\tilde{X}(\boldsymbol{\beta}, t) = \frac{1}{n} \sum_{h=1}^k \int_0^t \left[(\boldsymbol{\beta} - \boldsymbol{\beta}_0)^\mathsf{T} \mathbf{S}_h^{(1)}(\boldsymbol{\beta}_0, s) - \log\left(\frac{S_h^{(0)}(\boldsymbol{\beta}, s)}{S_h^{(0)}(\boldsymbol{\beta}_0, s)} \right) S_h^{(0)}(\boldsymbol{\beta}_0, s) \right] \alpha_{h0}(s)\,\mathrm{d}s,$$

their difference being a local square integrable martingale. The predictable variation process of this martingale is by (7.2.27) and (2.4.9)

$$\langle X(\boldsymbol{\beta}, \cdot) - \tilde{X}(\boldsymbol{\beta}, \cdot) \rangle(t) = \frac{1}{n^2} \sum_{h=1}^k \sum_{i=1}^n \int_0^t \left[(\boldsymbol{\beta} - \boldsymbol{\beta}_0)^\mathsf{T} \mathbf{Z}_{hi}(s) - \log\left(\frac{S_h^{(0)}(\boldsymbol{\beta}, s)}{S_h^{(0)}(\boldsymbol{\beta}_0, s)} \right) \right]^2$$
$$\times \alpha_{h0}(s) \exp(\boldsymbol{\beta}_0^\mathsf{T} \mathbf{Z}_{hi}(s)) Y_{hi}(s)\,\mathrm{d}s.$$

It now follows from (7.2.1) and Condition VII.2.1(a), (b), (c) that $\tilde{X}(\boldsymbol{\beta}, \tau)$ for $\boldsymbol{\beta} \in \mathscr{B}$ converges in probability to some function

$$f(\boldsymbol{\beta}) = \sum_{h=1}^k \int_0^\tau \left[(\boldsymbol{\beta} - \boldsymbol{\beta}_0)^\mathsf{T} \mathbf{s}_h^{(1)}(\boldsymbol{\beta}_0, s) - \log\left(\frac{s_h^{(0)}(\boldsymbol{\beta}, s)}{s_h^{(0)}(\boldsymbol{\beta}_0, s)} \right) s_h^{(0)}(\boldsymbol{\beta}_0, s) \right] \alpha_{h0}(s)\,\mathrm{d}s,$$

whereas $n\langle X(\boldsymbol{\beta}, \cdot) - \tilde{X}(\boldsymbol{\beta}, \cdot) \rangle(\tau)$ for $\boldsymbol{\beta} \in \mathscr{B}$ converges in probability to a (finite) function of $\boldsymbol{\beta}$. Also, by the inequality of Lenglart (Section II.5.2),

$$X(\boldsymbol{\beta}, \tau) \xrightarrow{\mathrm{P}} f(\boldsymbol{\beta}).$$

Now, by (7.2.1) and Condition VII.2.1(b), (c), (d), we have for $\boldsymbol{\beta} \in \mathscr{B}$

$$\frac{\partial}{\partial \boldsymbol{\beta}} f(\boldsymbol{\beta}) = \int_0^\tau \sum_{h=1}^k (\mathbf{e}_h(\boldsymbol{\beta}_0, s) - \mathbf{e}_h(\boldsymbol{\beta}, s)) s_h^{(0)}(\boldsymbol{\beta}_0, s) \alpha_{h0}(s)\, \mathrm{d}s$$

which is zero for $\boldsymbol{\beta} = \boldsymbol{\beta}_0$. Furthermore,

$$-\frac{\partial^2}{\partial \boldsymbol{\beta}^2} f(\boldsymbol{\beta}) = \int_0^\tau \sum_{h=1}^k \mathbf{v}_h(\boldsymbol{\beta}, s) s_h^{(0)}(\boldsymbol{\beta}_0, s) \alpha_{h0}(s)\, \mathrm{d}s,$$

which is positive semidefinite and, by Condition VII.2.1(e), positive definite for $\boldsymbol{\beta} = \boldsymbol{\beta}_0$. Thus, $X(\boldsymbol{\beta}, \tau)$ converges pointwise in probability to a concave function $f(\boldsymbol{\beta})$ on $\mathscr{B}$ with a unique maximum at $\boldsymbol{\beta} = \boldsymbol{\beta}_0$. The random function $X(\boldsymbol{\beta}, \tau)$ is also concave and has a maximum at $\boldsymbol{\beta} = \hat{\boldsymbol{\beta}}$ when the maximum exists. We now apply Theorem II.1 of Andersen and Gill (1982, Appendix II) saying that if a sequence of random *concave* functions converges *pointwise* in probability to a real function on an open convex set E, then the convergence is *uniform* on compact subsets of E, in probability. This shows that the convergence is uniform over $\mathscr{B}$ and, therefore, the maximizing value(s) $\hat{\boldsymbol{\beta}}$ of $X(\boldsymbol{\beta}, \tau)$ converge(s) in probability to the maximizing value $\boldsymbol{\beta}_0$ of $f(\boldsymbol{\beta})$. □

Note that the crucial assumption for the (asymptotic) existence of $\hat{\boldsymbol{\beta}}$ is (e) in Condition VII.2.1. Sufficient conditions for this condition to hold have been discussed in an i.i.d. situation by Næs (1982); see also Example VII.2.7. Exact conditions for the existence of $\hat{\boldsymbol{\beta}}$ were outlined in Example VII.2.3 in the previous subsection.

To prove asymptotic normality of $\hat{\boldsymbol{\beta}}$, some boundedness condition on the covariates is needed. Ordinary boundedness, as assumed by Næs (1982), is sufficient. Boundedness by a random variable with finite rth moment ($r > 2$), by a Chebychev-type inequality, also implies the following Lindeberg-type condition which is exactly what is needed in the proof:

Condition VII.2.2. There exists a $\delta > 0$ such that

$$n^{-1/2} \sup_{h,i,t} |\mathbf{Z}_{hi}(t)|\, Y_{hi}(t) I\big(\boldsymbol{\beta}_0^\mathsf{T} \mathbf{Z}_{hi}(t) > -\delta |\mathbf{Z}_{hi}(t)|\big) \xrightarrow{\mathrm{P}} 0 \quad \text{as } n \to \infty.$$

The matrices $\mathscr{I}_\tau(\boldsymbol{\beta})$ and $\boldsymbol{\Sigma}_\tau$ were defined in (7.2.17) and Condition VII.2.1(e), respectively. We define $\mathscr{I}_t(\boldsymbol{\beta})$ and $\boldsymbol{\Sigma}_t$ similarly by integrating up to t.

Theorem VII.2.2. *Assume Conditions* VII.2.1 *and* VII.2.2 *and that* (7.2.1) *holds for* $t = \tau$. *Then, as* $n \to \infty$,

$$n^{1/2}(\hat{\boldsymbol{\beta}} - \boldsymbol{\beta}_0) \xrightarrow{\mathscr{D}} \mathscr{N}(\mathbf{0}, \boldsymbol{\Sigma}_\tau^{-1})$$

and

$$\sup_{t \in \mathscr{T}} \|n^{-1} \mathscr{I}_t(\hat{\boldsymbol{\beta}}) - \boldsymbol{\Sigma}_t\| \xrightarrow{\mathrm{P}} 0. \tag{7.2.28}$$

In particular,

$$n^{-1}\mathscr{I}_\tau(\hat{\boldsymbol\beta}) \xrightarrow{\mathrm{P}} \boldsymbol\Sigma_\tau \quad \textit{as } n \to \infty.$$

Proof. The idea in the proof is first to use the martingale central limit theorem (Theorem II.5.1) to prove asymptotic normality of the score statistic (7.2.16) evaluated at $\boldsymbol\beta_0$:

$$n^{-1/2}\mathbf{U}_\tau(\boldsymbol\beta_0) \xrightarrow{\mathscr{D}} \mathscr{N}(\mathbf{0}, \boldsymbol\Sigma_\tau); \tag{7.2.29}$$

next to Taylor-expand $U_\tau^j(\boldsymbol\beta)$ around $\boldsymbol\beta_0$:

$$U_\tau^j(\boldsymbol\beta) - U_\tau^j(\boldsymbol\beta_0) = -\mathscr{I}_\tau^j(\boldsymbol\beta_j^*)(\boldsymbol\beta - \boldsymbol\beta_0), \quad j = 1, \ldots, p,$$

where each $\boldsymbol\beta_j^*$ is on the line segment between $\boldsymbol\beta$ and $\boldsymbol\beta_0$ and $\mathscr{I}_\tau^j(\boldsymbol\beta)$ is the row vector with elements

$$-\frac{\partial}{\partial\beta_l}U_\tau^j(\boldsymbol\beta), \quad l = 1, \ldots, p;$$

and, finally, to insert $\boldsymbol\beta = \hat{\boldsymbol\beta}$ to get

$$n^{-1/2}U_\tau^j(\boldsymbol\beta_0) = \{n^{-1}\mathscr{I}_\tau^j(\boldsymbol\beta_j^*)\}n^{1/2}(\hat{\boldsymbol\beta} - \boldsymbol\beta_0) \tag{7.2.30}$$

since $U_\tau^j(\hat{\boldsymbol\beta}) = 0$. Then the asymptotic normality of $\hat{\boldsymbol\beta}$ follows from (7.2.28), (7.2.29), and Theorem VII.2.1.

To prove (7.2.29), we note that

$$n^{-1/2}\mathbf{U}_t(\boldsymbol\beta_0) = \sum_{h=1}^{k}\sum_{i=1}^{n}\int_0^t \mathbf{H}_{hi}^{(n)}(u)\,\mathrm{d}M_{hi}(u),$$

where $\mathbf{H}_{hi}^{(n)}(t) = (H_{hij}^{(n)}(t))$ and

$$H_{hij}^{(n)}(t) = n^{-1/2}(Z_{hij}(t) - E_{hj}(\boldsymbol\beta_0, t)), \quad j = 1, \ldots, p.$$

Now (using (2.4.9))

$$\begin{aligned}
&\langle n^{-1/2}U^j(\boldsymbol\beta_0), n^{-1/2}U^l(\boldsymbol\beta_0)\rangle(t) \\
&= \sum_{h=1}^{k}\sum_{i=1}^{n}\int_0^t H_{hij}^{(n)}(u)H_{hil}^{(n)}(u)\alpha_{h0}(u)Y_{hi}(u)\exp(\boldsymbol\beta_0^\mathsf{T}\mathbf{Z}_{hi}(u))\,\mathrm{d}u \\
&\xrightarrow{\mathrm{P}} \sum_{h=1}^{k}\int_0^t v_{hjl}(\boldsymbol\beta_0, u)s_h^{(0)}(\boldsymbol\beta_0, u)\alpha_{h0}(u)\,\mathrm{d}u, \quad j, l = 1, \ldots, p,
\end{aligned}$$

by (7.2.1) and Condition VII.2.1(a), (b), (c), and it remains to be proved that for all $\varepsilon > 0$

$$\sum_{h=1}^{k}\sum_{i=1}^{n}\int_0^\tau \{H_{hij}^{(n)}(t)\}^2\alpha_{h0}(t)Y_{hi}(t)\exp(\boldsymbol\beta_0^\mathsf{T}\mathbf{Z}_{hi}(t))I(|H_{hij}^{(n)}(t)| > \varepsilon)\,\mathrm{d}t \xrightarrow{\mathrm{P}} 0,$$

$j = 1, \ldots, p$. Here we use the elementary inequality

$$|\mathbf{a} - \mathbf{b}|^2 I(|\mathbf{a} - \mathbf{b}| > \varepsilon) \le 4|\mathbf{a}|^2 I(|\mathbf{a}| > \varepsilon/2) + 4|\mathbf{b}|^2 I(|\mathbf{b}| > \varepsilon/2).$$

First, we note that

$$\sum_{h=1}^{k} \int_0^\tau |\mathbf{E}_h(\boldsymbol{\beta}_0, t)|^2 I(n^{-1/2}|\mathbf{E}_h(\boldsymbol{\beta}_0, t)| > \varepsilon)\alpha_{h0}(t)\frac{1}{n}S_h^{(0)}(\boldsymbol{\beta}_0, t)\,\mathrm{d}t \xrightarrow{\mathrm{P}} 0$$

by (7.2.1) and Condition VII.2.1(a), (b), (c). To prove that

$$\sum_{h=1}^{k} \int_0^\tau \frac{1}{n}\sum_{i=1}^{n} |\mathbf{Z}_{hi}(t)|^2 I(n^{-1/2}|\mathbf{Z}_{hi}(t)| > \varepsilon)\, Y_{hi}(t)\exp(\boldsymbol{\beta}_0^\mathsf{T}\mathbf{Z}_{hi}(t))\alpha_{h0}(t)\,\mathrm{d}t \xrightarrow{\mathrm{P}} 0,$$

we note that

$$\begin{aligned}\sum_{h=1}^{k} \int_0^\tau \frac{1}{n}\sum_{i=1}^{n} &|\mathbf{Z}_{hi}(t)|^2 I(n^{-1/2}|\mathbf{Z}_{hi}(t)| > \varepsilon, \boldsymbol{\beta}_0^\mathsf{T}\mathbf{Z}_{hi}(t) \le -\delta|\mathbf{Z}_{hi}(t)|)\\ &\times Y_{hi}(t)\exp(\boldsymbol{\beta}_0^\mathsf{T}\mathbf{Z}_{hi}(t))\alpha_{h0}(t)\,\mathrm{d}t\\ \le \sum_{h=1}^{k} \int_0^\tau \frac{1}{n}\sum_{i=1}^{n} &|\mathbf{Z}_{hi}(t)|^2 \exp(-\delta|\mathbf{Z}_{hi}(t)|)I(|\mathbf{Z}_{hi}(t)| > \varepsilon n^{1/2})\, Y_{hi}(t)\alpha_{h0}(t)\,\mathrm{d}t \xrightarrow{\mathrm{P}} 0\end{aligned}$$

by (7.2.1) and Condition VII.2.1(c) since $x^2\exp(-\delta x) \to 0$ as $x \to \infty$. Finally, by Condition VII.2.2,

$$\begin{aligned}\sum_{h=1}^{k} \int_0^\tau \frac{1}{n}\sum_{i=1}^{n} &|\mathbf{Z}_{hi}(t)|^2 I(n^{-1/2}|\mathbf{Z}_{hi}(t)| > \varepsilon,\ \boldsymbol{\beta}_0^\mathsf{T}\mathbf{Z}_{hi}(t) > -\delta|\mathbf{Z}_{hi}(t)|)\\ &\times Y_{hi}(t)\exp(\boldsymbol{\beta}_0^\mathsf{T}\mathbf{Z}_{hi}(t))\alpha_{h0}(t)\,\mathrm{d}t \xrightarrow{\mathrm{P}} 0\end{aligned}$$

and (7.2.29) follows by Theorem II.5.1. To prove (7.2.28) for any $\boldsymbol{\beta}^*$ such that $\boldsymbol{\beta}^* \xrightarrow{\mathrm{P}} \boldsymbol{\beta}_0$, we note that

$$\begin{aligned}\|n^{-1}\mathscr{I}_t(\boldsymbol{\beta}^*) - \boldsymbol{\Sigma}_t\| \le \sum_{h=1}^{k}\Bigg\{&\left\|\int_0^t (\mathbf{V}_h(\boldsymbol{\beta}^*, u) - \mathbf{v}_h(\boldsymbol{\beta}^*, u))\frac{\mathrm{d}N_h(u)}{n}\right\|\\ &+ \left\|\int_0^t (\mathbf{v}_h(\boldsymbol{\beta}^*, u) - \mathbf{v}_h(\boldsymbol{\beta}_0, u))\frac{\mathrm{d}N_h(u)}{n}\right\|\\ &+ \left\|\int_0^t \mathbf{v}_h(\boldsymbol{\beta}_0, u)\left\{\frac{\mathrm{d}N_h(u)}{n} - \frac{1}{n}S_h^{(0)}(\boldsymbol{\beta}_0, u)\alpha_{h0}(u)\,\mathrm{d}u\right\}\right\|\\ &+ \left\|\int_0^t \mathbf{v}_h(\boldsymbol{\beta}_0, u)\left(\frac{1}{n}S_h^{(0)}(\boldsymbol{\beta}_0, u) - s_h^{(0)}(\boldsymbol{\beta}_0, u)\right)\alpha_{h0}(u)\,\mathrm{d}u\right\|\Bigg\}.\end{aligned} \tag{7.2.31}$$

We now deal with the four terms on the right-hand side of (7.2.31) separately and consider a fixed value of h. From Condition VII.2.1(a), (b), (c), it follows that

$$\sup_{t \in \mathscr{T},\, \boldsymbol{\beta} \in \mathscr{B}} \|\mathbf{V}_h(\boldsymbol{\beta}, t) - \mathbf{v}_h(\boldsymbol{\beta}, t)\| \xrightarrow{\mathrm{P}} 0,$$

and since $\boldsymbol{\beta}^* \xrightarrow{\mathrm{P}} \boldsymbol{\beta}_0$, $\sup_t \|\mathbf{V}_h(\boldsymbol{\beta}^*, t) - \mathbf{v}_h(\boldsymbol{\beta}^*, t)\| \xrightarrow{\mathrm{P}} 0$ also. By the inequality of Lenglart (2.5.17), we have

$$\mathrm{P}\left\{\frac{1}{n}N_h(\tau) > \eta\right\} \le \frac{\delta}{\eta} + \mathrm{P}\left\{\int_0^\tau \frac{1}{n}S_h^{(0)}(\boldsymbol{\beta}_0, t)\alpha_{h0}(t)\,\mathrm{d}t > \delta\right\},$$

and, by choosing $\delta > \int_0^\tau s_h^{(0)}(\boldsymbol{\beta}_0, t)\alpha_{h0}(t)\,dt$, the latter probability tends to 0 as $n \to \infty$. Thus, $(1/n)N_h(\tau)$ is bounded in probability and the first of the terms in (7.2.31) converges to 0 in probability uniformly in t. The second term can be dealt with similarly because $\boldsymbol{\beta}^* \xrightarrow{P} \boldsymbol{\beta}_0$. The integral in the third term is with respect to the local square integrable martingale $M_h = M_{h1} + \cdots + M_{hn}$, and by using the version (2.5.18) of the inequality of Lenglart,

$$P\left\{\sup_{t \in \mathcal{T}} \left| \int_0^t v_{hjl}(\boldsymbol{\beta}_0, u)\frac{1}{n}\,dM_h(u)\right| > \eta\right\}$$

$$\leq \frac{\delta}{\eta^2} + P\left\{\frac{1}{n^2}\int_0^\tau v_{hjl}^2(\boldsymbol{\beta}_0, t)S_h^{(0)}(\boldsymbol{\beta}_0, t)\alpha_{h0}(t)\,dt > \delta\right\}.$$

This, together with (7.2.1) and Condition VII.2.1(a), (b), (c), shows that also the third term vanishes uniformly in probability as $n \to \infty$. The same conditions deal with the fourth term and the proof is complete. □

The proofs for Theorem VII.2.1 and VII.2.2 are taken from Andersen and Gill (1982) and extended to the k-variate case (Andersen and Borgan, 1985). For bounded covariates, Næs (1982) showed asymptotic normality of $\hat{\boldsymbol{\beta}}$ in an i.i.d. situation using results on discrete-time martingales. Similar techniques were used by Arjas and Haara (1988a) to show the same result under weaker boundedness conditions on the covariates; see Example VII.2.7.

The reason why one should not impose too strong boundedness conditions on the covariates is that if the asymptotic results only hold under such strong conditions, then the large sample *approximations* using the limiting normal distribution for $\hat{\boldsymbol{\beta}}$ are likely to be poor in concrete examples, including some large covariate values. Along these lines, Bednarski (1989) showed that $\hat{\boldsymbol{\beta}}$ may be inconsistent if the "nice" distribution of the covariates is contaminated with some very large values for which the model fails, even if the proportion of such covariate values vanishes as n tends to infinity. One way of overcoming such difficulties may be to filter away individuals with extreme covariate values from the risk set (cf. Example III.2.9)—in fact, Bednarski (1991) has suggested a robust estimator of $\boldsymbol{\beta}$ based on this idea—but more practical experience with such a procedure is needed.

An example where all possible boundedness conditions on $\mathbf{Z}$ are certainly satisfied is the following.

Example VII.2.6. The Two-Sample Case

Consider the univariate case $k = 1$, $p = 1$ with $Z_i \in \{0, 1\}$, $i = 1, \ldots, n$, and assume that $N_i(t)$ has intensity process

$$\lambda_i(t) = Y_i(t)\exp(\beta Z_i)\alpha_0(t).$$

Then, $\hat{\beta}$ is the solution to the equation $U_\tau(\beta) = 0$, i.e.,

$$N_\cdot^{(1)}(\tau) = \int_0^\tau \frac{Y_\cdot^{(1)}(t)\exp(\beta)}{Y_\cdot^{(0)}(t) + Y_\cdot^{(1)}(t)\exp(\beta)}\,dN_\cdot(t),$$

where $N_{\bullet}^{(j)}(t) = \sum_i I(Z_i = j) N_i(t)$, $j = 0, 1$, $N_{\bullet}(t) = N_{\bullet}^{(0)}(t) + N_{\bullet}^{(1)}(t)$, and $Y_{\bullet}^{(j)}(t) = \sum_i I(Z_i = j) Y_i(t)$, $j = 0, 1$. This equation has the interpretation that the observed number of events $N_{\bullet}^{(1)}(\tau)$ in the group with $Z_i = 1$ should equal an "expected" number of events in the same group under the assumption of proportional intensities; cf. the discussion in Example V.2.1.

Suppose now that there exist bounded, continuous, and positive functions $y_j(t)$, $j = 0, 1$, bounded away from zero such that

$$\sup_{t \in \mathscr{T}} \left| \frac{1}{n} Y_{\bullet}^{(j)}(t) - y_j(t) \right| \xrightarrow{\text{P}} 0.$$

Then Condition VII.2.2(a)–(d) is fulfilled and since

$$\Sigma_\tau = \int_0^\tau \frac{\exp(\beta) y_0(t) y_1(t)}{y_0(t) + \exp(\beta) y_1(t)} \alpha_0(t)\, dt$$

is positive, $\hat{\beta}$ is consistent and asymptotically normal if only $A_0(\tau) < \infty$. (As we shall see in the next example, the latter condition can actually be relaxed in an i.i.d. situation when the covariates are bounded.) From this expression for Σ_τ it is seen that the limiting variance of $\exp(\hat{\beta})$ is the same as the right-hand side of (5.3.9) corresponding to the optimal two-step estimator considered in Section V.3.1. □

Next, we discuss the regularity Conditions VII.2.1 and VII.2.2 in the i.i.d. case.

EXAMPLE VII.2.7. The i.i.d. Case

Andersen and Gill (1982, Theorem 4.1) showed (when $k = 1$) that if

$$(N_i(t), Y_i(t), \mathbf{Z}_i(t), t \in \mathscr{T}), \quad i = 1, \ldots, n,$$

are i.i.d., Conditions VII.2.1 and VII.2.2 hold provided that (i) Σ_τ is positive definite; (ii) $P(Y_i(t) = 1 \text{ for all } t \in \mathscr{T}) > 0$; and (iii) there exists a neighborhood $\mathscr{B}$ of $\boldsymbol{\beta}_0$ such that

$$\mathrm{E}\left(\sup_{t \in \mathscr{T}, \boldsymbol{\beta} \in \mathscr{B}} Y_i(t) |\mathbf{Z}_i(t)|^2 \exp(\boldsymbol{\beta}^\mathsf{T} \mathbf{Z}_i(t)) \right) < \infty. \tag{7.2.32}$$

Their proof, using functional forms of the law of large numbers, immediately extends to the case $k > 1$.

As a special case, we now consider the situation in Examples III.1.7 and III.2.9. We let $\mathbf{Z}_1, \ldots, \mathbf{Z}_n$ be i.i.d. p-dimensional covariates and $X_1, \ldots, X_n$ independent *survival times* where the conditional distribution of X_i given $(\mathbf{Z}_1, \ldots, \mathbf{Z}_n) = (\mathbf{z}_1, \ldots, \mathbf{z}_n)$ has hazard function

$$\alpha_i(t) = \alpha_0(t) \exp(\boldsymbol{\beta}^\mathsf{T} \mathbf{z}_i).$$

Furthermore, we let $U_1, \ldots, U_n$ be mutually independent right-censoring times, we let $\tilde{X}_i = X_i \wedge U_i$, and we assume that X_i and U_i are conditionally independent given $\mathbf{Z}_i$. If, for instance, the conditional distribution of U_i given $\mathbf{Z} = \mathbf{z}$ has hazard function of the form $\alpha_{0,u}(t)\exp(\boldsymbol{\beta}_u^\mathsf{T}\mathbf{z}_i)$ (where $\boldsymbol{\beta}_u$ may be $\mathbf{0}$ in which case $U_1, \ldots, U_n$ are i.i.d.) or if (almost surely) $U_i = t_0 \le \tau$, then we are in an i.i.d. case and Conditions VII.2.1 and VII.2.2 hold if (i) $\boldsymbol{\Sigma}_\tau$ is positive definite; (ii) $\mathrm{P}(\tilde{X}_i > t) > 0$ for all $t \in \mathscr{T}$; and (iii) there exists a neighborhood $\mathscr{B}$ of $\boldsymbol{\beta}_0$ such that

$$\mathrm{E}\left(\sup_{\boldsymbol{\beta} \in \mathscr{B}} \max |\mathbf{Z}_i|^2 \exp(\boldsymbol{\beta}^\mathsf{T}\mathbf{Z}_i) \right) < \infty.$$

Note that the last condition [as well as (7.2.32)] is trivially fulfilled if the $\mathbf{Z}_i$ are *bounded*. By Condition VII.2.1(e),

$$\boldsymbol{\Sigma}_\tau = \int_0^\tau \mathbf{v}(\boldsymbol{\beta}_0, t) s^{(0)}(\boldsymbol{\beta}_0, t)\alpha_0(t)\,\mathrm{d}t,$$

where, in the i.i.d. case, $\mathbf{v}(\boldsymbol{\beta}_0, t)$ is the covariance matrix of $\mathbf{Z}$ with respect to the probability measure P_t given by

$$\mathrm{dP}_t \propto Y(t)\exp(\boldsymbol{\beta}^\mathsf{T}\mathbf{Z})\,\mathrm{dP}$$

(Andersen and Gill, 1982). Therefore, $\boldsymbol{\Sigma}_\tau$ is positive definite, provided that there exists a set $\mathscr{T}' \subseteq \mathscr{T}$ with $\int_{\mathscr{T}'} \alpha_0(t)\,\mathrm{d}t > 0$ such that the components of $\mathbf{Z}_i$ are linearly independent with P_t probability 1 for $t \in \mathscr{T}'$.

So far, we have worked under the somewhat unsatisfactory assumption (7.2.1) that $A_{h0}(t)$ is finite also for $t = \tau$ which will often only be the case if τ is finite. Under stronger boundedness conditions on the covariates than (7.2.32), the results of, e.g., Theorems VII.2.1 and VII.2.2 can be shown without assuming that (7.2.1) holds for $t = \tau$. Andersen and Gill (1982) indicated how the proofs could be modified in the i.i.d. case assuming $\mathbf{Z}$ to be *bounded*, $\mathrm{N}(\tau)$ to have finite mean, and that for all $t < \tau$, the probability that $\mathrm{Y}(u) = 1$ for all $u \le t$ is strictly positive. In all cases, what has to be shown is that contributions to integrals over $[t, \tau)$ can be made arbitrarily small, uniformly in n, by choosing t close enough to τ. (For an example of such a technique, see Example V.1.6.)

Also, as mentioned above, Arjas and Haara (1988a) showed consistency and asymptotic normality of $\hat{\boldsymbol{\beta}}$ based on all data from the interval $[0, \tau)$. Their boundedness condition on the covariates was

$$\sup_{t \in [0,\tau)} \{n^{-1/2}|\mathbf{Z}_i(t)|\, Y_i(t), i = 1, \ldots, n\} \xrightarrow{\mathrm{P}} 0. \qquad \square$$

We now turn to a study of the large sample properties of the estimated integrated underlying hazards $\hat{A}_{h0}(t, \hat{\boldsymbol{\beta}})$, $h = 1, \ldots, k$. The following theorem provides a succinct description of the asymptotic joint distribution of these estimates and of $\hat{\boldsymbol{\beta}}$:

Theorem VII.2.3. *Assume Conditions* VII.2.1 *and* VII.2.2 *and that* (7.2.1) *holds with* $t = \tau$. *Then* $n^{1/2}(\hat{\boldsymbol{\beta}} - \boldsymbol{\beta}_0)$ *and the processes*

$$W_h(\cdot) = n^{1/2}(\hat{A}_{h0}(\cdot, \hat{\boldsymbol\beta}) - A_{h0}(\cdot)) + n^{1/2}(\hat{\boldsymbol\beta} - \boldsymbol\beta_0)^{\mathsf T} \int_0^{\cdot} \mathbf{e}_h(\boldsymbol\beta_0, u)\alpha_{h0}(u)\,\mathrm{d}u,$$

$h = 1, \ldots, k$, are asymptotically independent. The limiting distribution of W_h is that of a zero-mean Gaussian martingale with variance function

$$\omega_h^2(t) = \int_0^t \frac{\alpha_{h0}(u)}{s_h^{(0)}(\boldsymbol\beta_0, u)}\,\mathrm{d}u.$$

Proof. We write $n^{1/2}(\hat{A}_{h0}(t, \hat{\boldsymbol\beta}) - A_{h0}(t))$ as

$$n^{1/2} \int_0^t J_h(u)\{S_h^{(0)}(\hat{\boldsymbol\beta}, u)^{-1} - S_h^{(0)}(\boldsymbol\beta_0, u)^{-1}\}\,\mathrm{d}N_h(u)$$

$$+ n^{1/2} \int_0^t J_h(u)\{S_h^{(0)}(\boldsymbol\beta_0, u)^{-1}\,\mathrm{d}N_h(u) - \alpha_{h0}(u)\,\mathrm{d}u\}$$

$$+ n^{1/2} \int_0^t (J_h(u) - 1)\alpha_{h0}(u)\,\mathrm{d}u.$$

Here the third term converges in probability to zero by Conditions VII.2.1(a), (b), (c). The second term, say $\hat{W}_h(t)$, is a stochastic integral with respect to a local square integrable martingale, namely,

$$\hat{W}_h(t) = n^{1/2} \int_0^t J_h(u) S_h^{(0)}(\boldsymbol\beta_0, u)^{-1}\,\mathrm{d}M_h(u).$$

By Taylor expansion around $\boldsymbol\beta_0$, the first term equals

$$-n^{1/2}(\hat{\boldsymbol\beta} - \boldsymbol\beta_0)^{\mathsf T} \int_0^t J_h(u)\mathbf{E}_h(\boldsymbol\beta_h^*, u) S_h^{(0)}(\boldsymbol\beta_h^*, u)^{-1}\,\mathrm{d}N_h(u)$$

with $\boldsymbol\beta_h^*$ on the line segment between $\hat{\boldsymbol\beta}$ and $\boldsymbol\beta_0$ and, following the proof of Theorem VII.2.2, it can be shown that

$$\sup_{t \in \mathscr{T}} \left\| -\int_0^t J_h(u)\mathbf{E}_h(\boldsymbol\beta_h^*, u) S_h^{(0)}(\boldsymbol\beta_h^*, u)^{-1}\,\mathrm{d}N_h(u) + \int_0^t \mathbf{e}_h(\boldsymbol\beta_0, u)\alpha_{h0}(u)\,\mathrm{d}u \right\| \xrightarrow{\mathrm{P}} 0$$

for any $\boldsymbol\beta_h^* \xrightarrow{\mathrm{P}} \boldsymbol\beta_0$.

Since, by (7.2.30), $\hat{\boldsymbol\beta} - \boldsymbol\beta_0 = \mathscr{I}_\tau(\boldsymbol\beta_0)^{-1}\mathbf{U}_\tau(\boldsymbol\beta_0) + \boldsymbol\varepsilon$, where $\boldsymbol\varepsilon \xrightarrow{\mathrm{P}} \mathbf{0}$ and because the local martingales $\mathbf{U}_t(\boldsymbol\beta_0)$ and $\hat{W}_h(t)$, $h = 1, \ldots, k$, are orthogonal,

$$\begin{aligned}
&\langle \mathbf{U}(\boldsymbol\beta_0), \hat{W}_h \rangle(t) \\
&= n^{1/2} \int_0^t \sum_{i=1}^n (\mathbf{Z}_{hi}(u) - \mathbf{E}_h(u)) J_h(u) S_h^{(0)}(\boldsymbol\beta_0, u)^{-1} Y_{hi}(u) \exp(\boldsymbol\beta_0^{\mathsf T}\mathbf{Z}_{hi}(u))\alpha_{h0}(u)\,\mathrm{d}u \\
&= 0,
\end{aligned}$$

it follows that $n^{1/2}(\hat{\boldsymbol\beta} - \boldsymbol\beta_0)$ and $\hat{W}_h(\cdot)$ are asymptotically independent. Also,

$\hat{W}_h$ and $\hat{W}_{h'}$ are orthogonal for $h \neq h'$ and, hence, asymptotically independent. It now only remains to find the limiting distribution of $\hat{W}_h(\cdot)$. This, however, follows from the martingale central limit theorem and from Condition VII.2.1(a), (b), (c), exactly as in the proof for Theorem IV.1.2, and the proof is complete. □

It is interesting to note that a formal information calculation based on the "Poisson" likelihood (7.2.2′) gives rise to the same variance estimates as stated in the following three corollaries to Theorem VII.2.3.

Corollary VII.2.4. *Under the same conditions as in Theorem* VII.2.3, *the process* $(n^{1/2}(\hat{A}_{h0}(\cdot, \hat{\boldsymbol\beta}) - A_{h0}(\cdot)), h = 1, \ldots, k)$ *converges weakly to a k-variate Gaussian process with mean zero and covariance function*

$$\delta_{hh'}\omega_h^2(s \wedge t) + \int_0^s \mathbf{e}_h^\mathsf{T}(\boldsymbol\beta_0, u)\alpha_{h0}(u)\,du\, \boldsymbol\Sigma_\tau^{-1} \int_0^t \mathbf{e}_{h'}(\boldsymbol\beta_0, u)\alpha_{h'0}(u)\,du,$$
$$h, h' = 1, \ldots, k,$$

which can be estimated uniformly consistently by

$$n\left\{\delta_{hh'} \int_0^{s \wedge t} S_h^{(0)}(\hat{\boldsymbol\beta}, u)^{-2}\, dN_h(u) \right.$$
$$\left. + \int_0^s \mathbf{E}_h^\mathsf{T}(\hat{\boldsymbol\beta}, u) S_h^{(0)}(\hat{\boldsymbol\beta}, u)^{-1}\, dN_h(u) \mathscr{I}_\tau(\hat{\boldsymbol\beta})^{-1} \int_0^t \mathbf{E}_{h'}(\hat{\boldsymbol\beta}, u) S_{h'}^{(0)}(\hat{\boldsymbol\beta}, u)^{-1}\, dN_{h'}(u)\right\}.$$

Proof. For the last part of the theorem, see how (7.2.31) was dealt with in the proof of Theorem VII.2.2. □

Corollary VII.2.5. *Under the same conditions as in Theorem* VII.2.3, *the asymptotic covariance of* $n^{1/2}(\hat{\boldsymbol\beta} - \boldsymbol\beta_0)$ *and* $n^{1/2}(\hat{A}_{h0}(t, \hat{\boldsymbol\beta}) - A_{h0}(t))$ *is*

$$-\boldsymbol\Sigma_\tau^{-1} \int_0^t \mathbf{e}_h(\boldsymbol\beta_0, u)\alpha_{h0}(u)\,du$$

which can be estimated uniformly consistently by

$$-n\mathscr{I}_\tau(\hat{\boldsymbol\beta})^{-1} \int_0^t \mathbf{E}_h(\hat{\boldsymbol\beta}, u) S_h^{(0)}(\hat{\boldsymbol\beta}, u)^{-1}\, dN_h(u).$$

The integrated hazard for an individual with (fixed) basic covariates $\mathbf{Z}_0$ and type-specific covariates $\mathbf{Z}_{h0}$ can be estimated by $\exp(\hat{\boldsymbol\beta}^\mathsf{T}\mathbf{Z}_{h0})\hat{A}_{h0}(t, \hat{\boldsymbol\beta})$, and, by a standard use of the delta-method, we can find its asymptotic variance.

Corollary VII.2.6. *Under the same conditions as in Theorem* VII.2.3, *the asymptotic variance of* $n^{1/2}(\exp(\hat{\boldsymbol\beta}^\mathsf{T}\mathbf{Z}_{h0})\hat{A}_{h0}(t, \hat{\boldsymbol\beta}) - \exp(\boldsymbol\beta_0^\mathsf{T}\mathbf{Z}_{h0})A_{h0}(t))$ *can be estimated uniformly consistently by*

$$n(\exp(\hat{\boldsymbol{\beta}}^{\mathsf{T}}\mathbf{Z}_{h0}))^2\left\{\int_0^t S_h^{(0)}(\hat{\boldsymbol{\beta}},u)^{-2}\,\mathrm{d}N_h(u)\right.$$
$$\left.+\int_0^t (\mathbf{E}_h(\hat{\boldsymbol{\beta}},u)-\mathbf{Z}_{h0})^{\mathsf{T}}\,\mathrm{d}\hat{A}_{h0}(u,\hat{\boldsymbol{\beta}})\mathscr{I}_\tau(\hat{\boldsymbol{\beta}})^{-1}\int_0^t (\mathbf{E}_h(\hat{\boldsymbol{\beta}},u)-\mathbf{Z}_{h0})\,\mathrm{d}\hat{A}_{h0}(u,\hat{\boldsymbol{\beta}})\right\}. \tag{7.2.33}$$

Another use of the delta-method shows that the asymptotic variance of the logarithm of this quantity, i.e., of $n^{1/2}\{(\hat{\boldsymbol{\beta}}-\boldsymbol{\beta}_0)^{\mathsf{T}}\mathbf{Z}_{h0}+\log\hat{A}_{h0}(t,\hat{\boldsymbol{\beta}})-\log A_{h0}(t)\}$ can be estimated uniformly consistently by

$$n\{\hat{A}_{h0}(t,\hat{\boldsymbol{\beta}})^{-2}\int_0^t S_h^{(0)}(\hat{\boldsymbol{\beta}},u)^{-2}\,\mathrm{d}N_h(u)+(\overline{\mathbf{Z}}_h(t)-\mathbf{Z}_{h0})^{\mathsf{T}}\mathscr{I}_\tau(\hat{\boldsymbol{\beta}})^{-1}(\overline{\mathbf{Z}}_h(t)-\mathbf{Z}_{h0})\},$$

where the "average covariate vector" $\overline{\mathbf{Z}}_h(t)$ over $[0,t]$ is given by

$$\overline{\mathbf{Z}}_h(t)=\hat{A}_{h0}(t,\hat{\boldsymbol{\beta}})^{-1}\int_0^t \mathbf{E}_h(\hat{\boldsymbol{\beta}},u)\,\mathrm{d}\hat{A}_{h0}(u,\hat{\boldsymbol{\beta}}).$$

The interpretation is that the uncertainty of the estimated integrated intensity for a given individual depends on the "distance" of the covariate vector $\mathbf{Z}_{h0}$ for this individual to the average covariate vector $\overline{\mathbf{Z}}_h(\cdot)$ [see Altman and Andersen (1986)], where the distance is measured by means of the estimated covariance matrix for $\hat{\boldsymbol{\beta}}$.

Next we illustrate the use of Corollaries VII.2.4–VII.2.6 on the melanoma data.

EXAMPLE VII.2.8. Survival with Malignant Melanoma: Confidence Limits for the Integrated Hazard

The data were introduced in Example I.3.1 and in Example VII.2.5 several simple multiplicative regression models for the melanoma data were considered. In Figure VII.2.1, we saw that the estimated integrated hazards for men and women in a proportional hazards model including no other covariates were close to the Nelson–Aalen estimates. Figure VII.2.4 shows the estimate for men with approximate 95% pointwise confidence limits calculated from Corollary VII.2.6 (using the log-transformation as described above). The confidence limits are seen to be slightly more narrow than those based on the Nelson–Aalen estimate shown in Figure IV.1.14.

The estimated hazards for men and women in a stratified model adjusting for the effect of tumor thickness were shown in Figure VII.2.2. Figure VII.2.5 shows the estimate for males with approximate 95% confidence limits calculated in the same way as in Figure VII.2.4. The confidence limits seem to be slightly broader than in Figure VII.2.4 probably partly owing to the fact that only the 29 deaths among the male patients contribute to the estimation here, whereas all 57 deaths contribute to the estimate based on the proportional hazards model shown in Figure VII.2.4. □

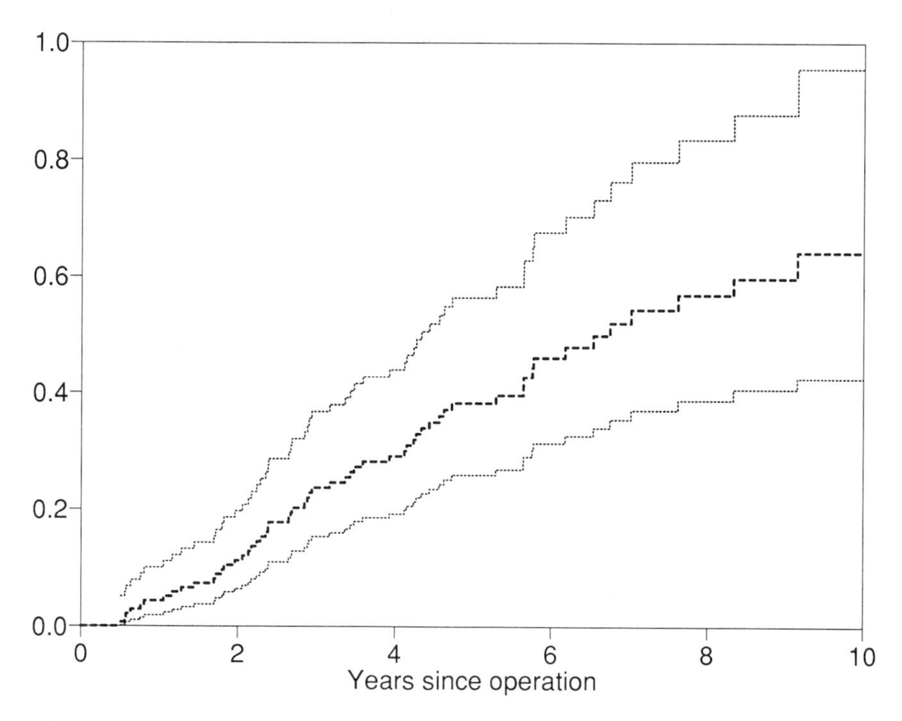

Figure VII.2.4. Estimated integrated hazard for men with malignant melanoma based on a proportional hazards model. Approximate 95% pointwise confidence limits are obtained using the log-transformation.

Finally, we consider large sample properties of the smoothed estimate (7.2.8), i.e.,

$$\hat{\alpha}_{h0}(t) = b_h^{-1} \int_{\mathscr{T}} K_h\left(\frac{t-u}{b_h}\right) \mathrm{d}\hat{A}_{h0}(u, \hat{\boldsymbol{\beta}})$$

using results similar to those in Section IV.2. As in Theorem IV.2.5, we assume that the kernel function $K_h(t)$ satisfies (4.2.20), i.e.,

$$\int_{-1}^{1} K_h(t)\,\mathrm{d}t = 1, \qquad \int_{-1}^{1} tK_h(t)\,\mathrm{d}t = 0, \qquad \int_{-1}^{1} t^2 K_h(t)\,\mathrm{d}t = k_{2h} > 0.$$

Then the following theorem holds:

Theorem VII.2.7. *Let t be an interior point in $\mathscr{T}$ and assume that, for each $h = 1, \ldots, k$, α_{h0} is twice continuously differentiable in a neighborhood of t. Assume that Conditions* VII.2.1 *and* VII.2.2 *hold and that $b_h = b_h^{(n)} \to 0$ as $n \to \infty$ in such a way that $nb_h^{(n)} \to \infty$* and $\limsup n^{1/5} b_h^{(n)} < \infty$. *Then $(nb_h)^{1/2}(\hat{\alpha}_{h0}(t) - \alpha_{h0}(t) - \frac{1}{2} b_h^2 \alpha''_{h0}(t) k_{2h})$, $h = 1, \ldots, k$, are asymptotically independent and*

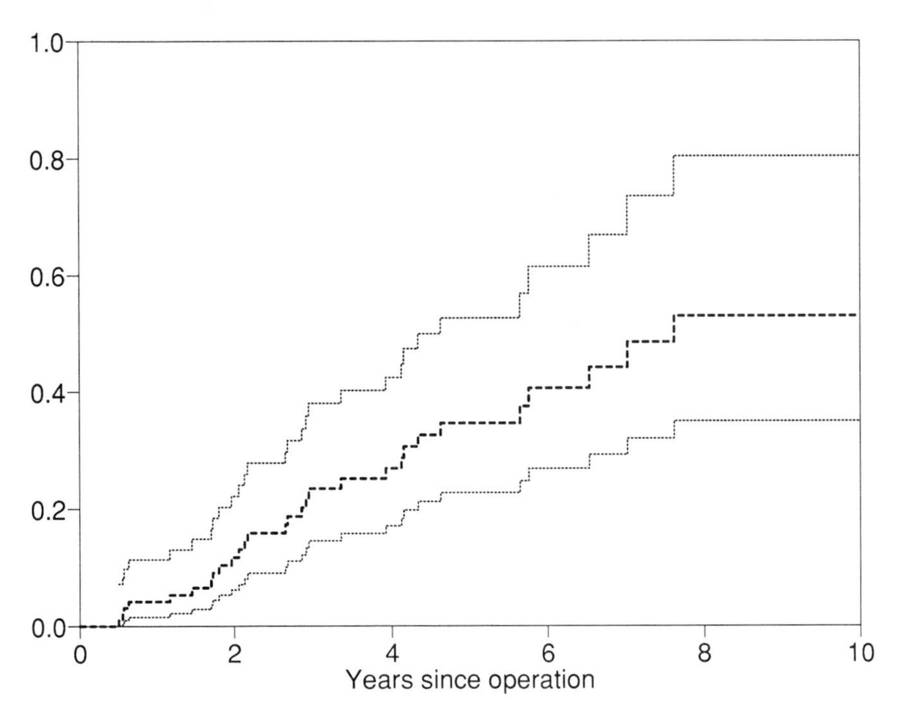

Figure VII.2.5. Estimated integrated hazard for men with malignant melanoma based on a stratified model. Approximate 95% pointwise confidence limits are obtained using the log-transformation.

$$(nb_h)^{1/2}(\hat{\alpha}_{h0}(t) - \alpha_{h0}(t) - \tfrac{1}{2}b_h^2\alpha''_{h0}(t)k_{2h}) \xrightarrow{\mathscr{D}} \mathscr{N}(0, \tau_h^2(t)) \quad \textit{as } n \to \infty,$$

where the asymptotic variance

$$\tau_h^2(t) = \frac{\alpha_{h0}(t)}{s_h^{(0)}(\boldsymbol{\beta}_0, t)} \int_{-1}^{1} K_h^2(u)\,\mathrm{d}u$$

can be estimated consistently by

$$nb_h^{-1} \int_0^\tau K_h^2\left(\frac{t-u}{b_h}\right) S_h^{(0)}(\hat{\boldsymbol{\beta}}, u)^{-2}\,\mathrm{d}N_h(u).$$

Proof. The proof consists in combining the proofs of Theorems IV.2.4 and IV.2.5 with the proof for Theorem VII.2.3. We write

$$(nb_h)^{1/2}(\hat{\alpha}_{h0}(t) - \alpha_{h0}(t) - \tfrac{1}{2}b_h^2\alpha''_{h0}(t)k_{2h})$$

$$= \frac{(nb_h)^{1/2}}{b_h}\int_0^\tau K_h\left(\frac{t-u}{b_h}\right) J_h(u)\left\{\frac{\mathrm{d}N_h(u)}{S_h^{(0)}(\hat{\boldsymbol{\beta}}, u)} - \alpha_{h0}(u)\,\mathrm{d}u\right\}$$

$$+ (nb_h)^{1/2}\left\{\frac{1}{b_h}\int_0^\tau K_h\left(\frac{t-u}{b_h}\right) J_h(u)\alpha_{h0}(u)\,\mathrm{d}u - \alpha_{h0}(t) - \frac{1}{2}b_h^2\alpha''_{h0}(t)k_{2h}\right\}$$

and note that the last term converges in probability to 0 as $n \to \infty$ by precisely the same arguments as in the proofs of Theorems IV.2.4 and IV.2.5. Following the proof of Theorem VII.2.3, the first term can be rewritten as

$$-b_h^{1/2} n^{1/2} (\hat{\boldsymbol{\beta}} - \boldsymbol{\beta}_0)^{\mathsf{T}} \frac{1}{b_h} \int_0^{\tau} K_h\left(\frac{t-u}{b_h}\right) \mathbf{E}_h(\boldsymbol{\beta}_h^*, u) J_h(u) S_h^{(0)}(\boldsymbol{\beta}_h^*, u)^{-1} \, \mathrm{d}N_h(u)$$

$$+ (nb_h)^{1/2} \frac{1}{b_h} \int_0^{\tau} K_h\left(\frac{t-u}{b_h}\right) J_h(u) S_h^{(0)}(\boldsymbol{\beta}_0, u)^{-1} \, \mathrm{d}M_h(u)$$

with $\boldsymbol{\beta}_h^*$ on the line segment between $\hat{\boldsymbol{\beta}}$ and $\boldsymbol{\beta}_0$. Now, $n^{1/2}(\hat{\boldsymbol{\beta}} - \boldsymbol{\beta}_0)$ converges in distribution, the integral in the first term converges in probability [to $\mathbf{e}_h(\boldsymbol{\beta}_0, t)\alpha_{h0}(t)$] and, since $b_h^{1/2} \to 0$, the first term vanishes asymptotically. The weak convergence of the last term follows exactly as in the proof of Theorem IV.2.4 and the asymptotic independence for different values of h follows from the orthogonality of the local martingales $M_1, \ldots, M_k$. □

It should be noted that the asymptotic independence (and the fact that the variance of $\hat{\boldsymbol{\beta}}$ does not influence the asymptotic variance) is a consequence of $b_h \to 0$. As discussed just after Theorem IV.2.4, the bias term vanishes only under additional stronger conditions like $nb_h^3 \to 0$.

In Examples VII.2.16 and VII.2.17 we shall calculate smoothed baseline hazards.

VII.2.3. Estimation of Functionals

In the two preceding subsections, estimation of the *intensities* and large sample properties of the estimators have been considered. From the point of view of *presentation* of the results from survival analyses or analyses of Markov process models, survival- or other transition probabilities (or other functionals of the intensities) are often easier to interpret; cf. Sections IV.3 and IV.4. In this subsection, we shall discuss how such estimates can be obtained from multiplicative hazard regression models. We consider the special cases of survival data and inhomogeneous Markov processes, and, finally, a more general disability model is studied with the purpose of estimating life insurance premiums for insulin-dependent diabetics.

From a Cox regression model for survival data with fixed (i.e., time-independent) covariates, one is often interested in estimating survival probabilities $S(t, \mathbf{Z}_0)$ for individuals with given fixed covariates $\mathbf{Z}_0$. This can be done from the estimated regression coefficients $\hat{\boldsymbol{\beta}}$ and the integrated baseline hazard estimate $\hat{A}_0(t, \hat{\boldsymbol{\beta}})$ by the product integral

$$\hat{S}(t, \mathbf{Z}_0) = \prod_{u \le t} \{1 - \exp(\hat{\boldsymbol{\beta}}^{\mathsf{T}} \mathbf{Z}_0) \, \mathrm{d}\hat{A}_0(u, \hat{\boldsymbol{\beta}})\}; \tag{7.2.34}$$

cf. (4.3.2). Defining

$$A^*(t,\mathbf{Z}_0) = \exp(\boldsymbol{\beta}_0^\mathsf{T}\mathbf{Z}_0)\int_0^t J(s)\alpha_0(s)\,\mathrm{d}s$$

and

$$S^*(t,\mathbf{Z}_0) = \prod_{s\le t}(1 - \mathrm{d}A^*(s,\mathbf{Z}_0)) = \exp\{-A^*(t,\mathbf{Z}_0)\},$$

it follows from the version (2.6.5) of Duhamel's equation that

$$\frac{\hat{S}(t,\mathbf{Z}_0)}{S^*(t,\mathbf{Z}_0)} - 1 = -\int_0^t \frac{\hat{S}(s-,\mathbf{Z}_0)}{S^*(s,\mathbf{Z}_0)}\{\exp(\hat{\boldsymbol{\beta}}^\mathsf{T}\mathbf{Z}_0)\,\mathrm{d}\hat{A}_0(s,\hat{\boldsymbol{\beta}}) - \mathrm{d}A^*(s,\mathbf{Z}_0)\}.$$

As in Theorems IV.3.1 and IV.3.2, this equation or the delta-method can be used to show that $n^{1/2}(\hat{S}(t,\mathbf{Z}_0) - S(t,\mathbf{Z}_0))$ behaves asymptotically like

$$-n^{1/2}S(t,\mathbf{Z}_0)\{\hat{A}_0(t,\hat{\boldsymbol{\beta}})\exp(\hat{\boldsymbol{\beta}}^\mathsf{T}\mathbf{Z}_0) - A_0(t)\exp(\boldsymbol{\beta}_0^\mathsf{T}\mathbf{Z}_0)\}$$

and the variance of $n^{1/2}(\hat{S}(t,\mathbf{Z}_0) - S(t,\mathbf{Z}_0))$ can, therefore, by Corollary VII.2.6, be estimated by $\hat{S}(t,\mathbf{Z}_0)^2$ times expression (7.2.33).

This result may be used for calculating pointwise confidence limits for $S(t,\mathbf{Z}_0)$. Note that the estimates $\hat{S}(t,\mathbf{Z}_0)$ may be negative if there exists a t_0 such that

$$\Delta N_\cdot(t_0)\exp(\hat{\boldsymbol{\beta}}^\mathsf{T}\mathbf{Z}_0) > \sum_{i=1}^n Y_i(t_0)\exp(\hat{\boldsymbol{\beta}}^\mathsf{T}\mathbf{Z}_i).$$

This is obviously most likely to happen for values of $\mathbf{Z}_0$ which are *extreme* compared to the covariates for the individuals on which the estimation was based. It is also most likely to happen at time points t_0 where the size of the risk set $Y_\cdot(t_0)$ is relatively small. In both cases, the estimated variance of $\hat{S}(t,\mathbf{Z}_0)$ will be large as seen above. The fact that meaningless estimates may result from heavy extrapolation from a regression model is, of course, well-known from, e.g., ordinary linear regression.

Alternatively, one may consider (asymptotically equivalent) estimates of $S(t,\mathbf{Z}_0)$ of the form

$$\tilde{S}(t,\mathbf{Z}_0) = \tilde{S}_0(t)^{\exp(\hat{\boldsymbol{\beta}}^\mathsf{T}\mathbf{Z}_0)}$$

for some estimate $\tilde{S}_0(t)$ of the "underlying survival function" $S_0(t) = \exp\{-A_0(t)\}$. Possibilities are

$$\tilde{S}_0(t) = \prod_{u\le t}(1 - \mathrm{d}\hat{A}_0(u,\hat{\boldsymbol{\beta}})) \tag{7.2.35}$$

and

$$\tilde{S}_0(t) = \exp\{-\hat{A}_0(t,\hat{\boldsymbol{\beta}})\}, \tag{7.2.36}$$

and when $\Delta N_\cdot(t) \le 1$, i.e., when there are no ties among the observed survival times, one may also use

$$\tilde{S}_0(t) = \prod_{\substack{\tilde{X}_i\le t\\ D_i=1}}\left(1 - \frac{\exp(\hat{\boldsymbol{\beta}}^\mathsf{T}\mathbf{Z}_i)}{S^{(0)}(\hat{\boldsymbol{\beta}},\tilde{X}_i)}\right)^{\exp(-\hat{\boldsymbol{\beta}}^\mathsf{T}\mathbf{Z}_i)}, \tag{7.2.37}$$

where $S^{(0)}(\boldsymbol{\beta}, \cdot)$ is given by (7.2.11). The relative merits of these estimates have not been studied systematically. One should note that only the estimate based on (7.2.35) is *not* invariant under translation of the covariates. The estimate based on (7.2.35) may be negative, whereas (7.2.36) is always positive and (7.2.37) is always non-negative. The latter estimate (with an extension to tied data) was suggested by Kalbfleisch and Prentice (1973) [cf. Kalbfleisch and Prentice (1980, Section 4.3)] and was later studied by Bailey (1983) and by Jacobsen (1984). We prefer the estimate (7.2.34) since, because of the simple way in which it is obtained from the integrated hazard estimate using the product integral, it immediately generalizes to other finite-state Markov processes as we shall see later. For survival data, we first consider the melanoma data as illustration.

Example VII.2.9. Survival with Malignant Melanoma. Estimation of Survival Probabilities

The data were introduced in Example I.3.1 and we now continue the discussion of regression models from Example VII.2.5, considering only deaths from the disease. In addition to the two covariates, sex ($Z_{i1} = 1$ for males and 0 for females) and tumor thickness in mm (Z_{i2}), we now consider the covariate

$$Z_{i3} = \begin{cases} 1 & \text{if ulceration is present in the tumor of patient } i \\ 0 & \text{otherwise.} \end{cases}$$

The estimates in this model are as follows:

j	Covariate Z_j	$\hat{\beta}_j$	$\widehat{\text{s.d.}}(\hat{\beta}_j)$	$(\hat{\beta}_j/\widehat{\text{s.d.}}(\hat{\beta}_j))^2$
1	Sex	0.459	0.267	2.97
2	Thickness	0.113	0.038	8.94
3	Ulceration	1.17	0.311	14.03

Here $\widehat{\text{s.d.}}(\hat{\beta}_j)$ is the estimated standard error of $\hat{\beta}_j$ and $\hat{\beta}_j/\widehat{\text{s.d.}}(\hat{\beta}_j)$ [or $(\hat{\beta}_j/\widehat{\text{s.d.}}(\hat{\beta}_j))^2$] is the Wald test statistic for the effect of Z_j taking the other covariates into account. It is seen that when both tumor thickness and ulceration (both of which are positively correlated to sex) are taken into account, the significance of sex decreases ($P = .08$). Also the estimated effect of thickness decreases when ulceration is entered into the model.

From this model, we can estimate the survival probability for patients with given values of the three covariates as described above. Figure VII.2.6 shows the estimate (7.2.34) for males ($Z_1 = 1$) and females ($Z_1 = 0$) with tumor thickness 2.92 mm ($Z_2 = 0$) and no ulceration ($Z_3 = 0$). The difference between the two curves is much smaller than between the Kaplan–Meier estimates in Figure I.3.1 as expected. The other estimates (7.2.35)–(7.2.37) were identical to (7.2.34) to three decimal places. Figure VII.2.7 shows the estimate for males with pointwise 95% confidence limits calculated as described above

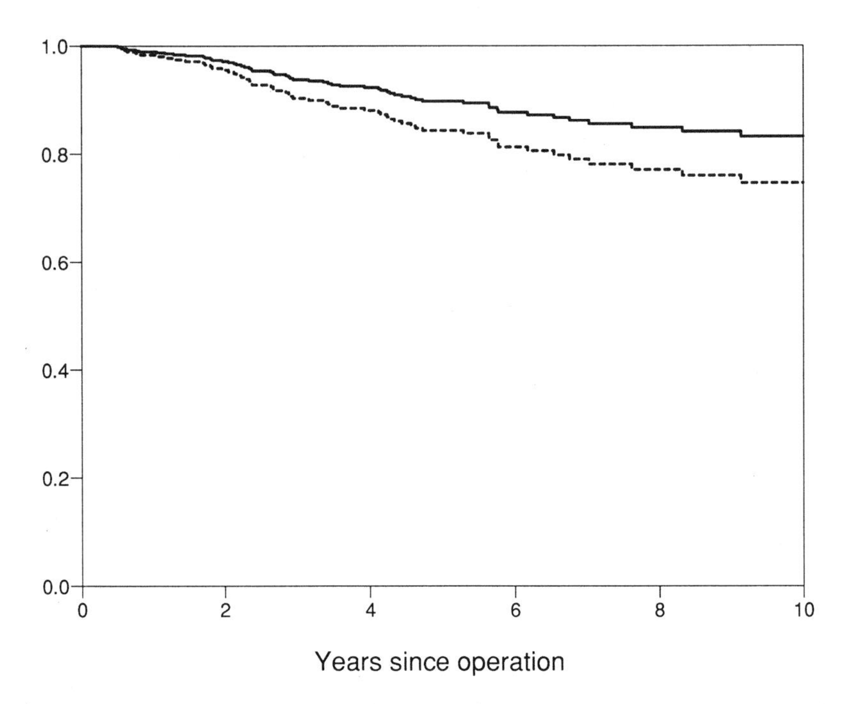

Figure VII.2.6. Estimated survival functions for men (---) and women (—) with malignant melanoma with tumor thickness 2.92 mm and no uleration.

using the log(−log)-transformation. Once an estimate of the survival curve and the corresponding confidence interval for a given covariate pattern has been obtained, the median or other quantiles and their respective confidence intervals can be estimated as described in Example IV.3.6. For instance, from Figure VII.2.7, a lower 0.10 quantile of 3.17 years can be read off with an approximate 95% confidence interval ranging from 2.13 to 5.65 years, whereas the estimated lower 0.20 quantile is 6.54 years [approximate 95% confidence interval (3.59 years, $+\infty$)] □

We now return to the situation studied in Section IV.4 and assume that the transition intensity $\alpha_{hji}(t)$ from state h to state j in a Markov process for individual i is given by

$$\alpha_{hji}(t) = \alpha_{hj0}(t)\exp(\boldsymbol{\beta}_0^{\mathsf{T}}\mathbf{Z}_{hji}), \quad h, j = 1, \ldots, k, h \neq j, i = 1, \ldots, n,$$

where the type-specific time-fixed covariate $\mathbf{Z}_{hji}$ is computed from a vector $\mathbf{Z}_i$ of basic covariates for individual i. Estimates $\hat{\boldsymbol{\beta}}$ and $\hat{A}_{hj0}(t, \hat{\boldsymbol{\beta}})$ can then be obtained as described in Section VII.2.1 and from these estimates one may wish to estimate probabilities $P_{hj}(s, t; \mathbf{Z}_0)$ for individuals with given (fixed) basic covariates $\mathbf{Z}_0$ and corresponding type-specific covariates $\mathbf{Z}_{hj0}$.

This can be done, as in Section IV.4, by letting

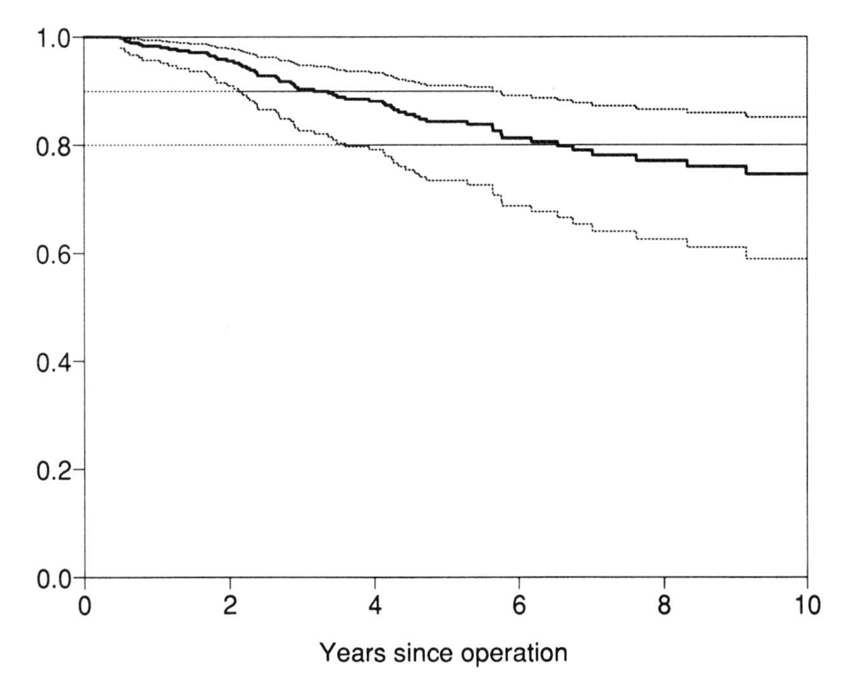

Figure VII.2.7. Estimated survival function for men with malignant melanoma with tumor thickness 2.92 mm and no ulceration. Approximate 95% pointwise confidence intervals are calculated using the log(−log)-transformation and the horizontal lines are used for calculating approximate 95% confidence intervals for the lower 0.10 and 0.20 quantiles.

$$\hat{A}_{hj}(t; \mathbf{Z}_0) = \hat{A}_{hj0}(t, \hat{\boldsymbol{\beta}}) \exp(\hat{\boldsymbol{\beta}}^\mathsf{T} \mathbf{Z}_{hj0}), \quad h \neq j,$$

and

$$\hat{A}_{hh}(t; \mathbf{Z}_0) = -\sum_{j \neq h} \hat{A}_{hj}(t, \mathbf{Z}_0), \quad h = 1, \ldots, k.$$

We then estimate $\mathbf{P}(s, t; \mathbf{Z}_0) = (P_{hj}(s, t; \mathbf{Z}_0); h, j = 1, \ldots, k)$ by the product integral

$$\hat{\mathbf{P}}(s, t; \mathbf{Z}_0) = \prod_{(s,t]} (\mathbf{I} + d\hat{\mathbf{A}}(u; \mathbf{Z}_0)), \tag{7.2.38}$$

where $\hat{\mathbf{A}} = (\hat{A}_{hj})$; cf. (4.4.2). The estimate (7.2.38) is meaningful as long as

$$\Delta \hat{A}_{hh}(t; \mathbf{Z}_0) \geq -1;$$

see Theorem II.6.7.

We now show how large sample properties of $\hat{\mathbf{P}}(s, t; \mathbf{Z}_0)$ can be derived from properties of $(\hat{\boldsymbol{\beta}}, \hat{\mathbf{A}}_0(t, \hat{\boldsymbol{\beta}}))$ (Theorem VII.2.3) using the functional delta-method (Section II.8). [see Andersen et al. (1991a) for a similar derivation.]

Using Taylor expansion as in the proof of Theorem VII.2.3, it is seen that the (h, j) element, $h \neq j$, of the $k \times k$ matrix $n^{1/2}(\hat{\mathbf{A}}(t; \mathbf{Z}_0) - \mathbf{A}(t, \mathbf{Z}_0))$ is asymptotically equivalent to

$$\exp(\boldsymbol{\beta}_0^{\mathsf{T}} \mathbf{Z}_{hj0}) \left\{ n^{1/2}(\hat{\boldsymbol{\beta}} - \boldsymbol{\beta}_0)^{\mathsf{T}} \int_0^t (\mathbf{Z}_{hj0} - \mathbf{e}_{hj}(\boldsymbol{\beta}_0, u)) \alpha_{hj0}(u) \, du + \int_0^t J_h(u) n S_{hj}^{(0)}(\boldsymbol{\beta}_0, u)^{-1} n^{-1/2} \, dM_{hj}(u) \right\}$$
$$= X_{1hj}^{(n)}(t) + X_{2hj}^{(n)}(t), \tag{7.2.39}$$

say, where $\mathbf{e}_{hj}(\boldsymbol{\beta}, u)$ is defined in Condition VII.2.1. (Asymptotically equivalent means convergence in probability to zero of the supremum norm of the difference.) By the compact differentiability of the product-integral (Proposition II.8.7),

$$\sqrt{n}\left(\prod (\mathbf{I} + d\hat{\mathbf{A}}(\cdot\,; \mathbf{Z}_0)) - \prod (\mathbf{I} + d\mathbf{A}(\cdot\,; \mathbf{Z}_0)) \right)$$

is asymptotically equivalent to the process equal in the point t to

$$\int_s^t \prod_{(s,u)} (\mathbf{I} + d\mathbf{A}(\cdot\,; \mathbf{Z}_0))(d\mathbf{X}_1^{(n)}(u) + d\mathbf{X}_2^{(n)}(u)) \prod_{(u,t)} (\mathbf{I} + d\mathbf{A}(\cdot\,; \mathbf{Z}_0)). \tag{7.2.40}$$

Moreover, this process considered as a functional of $\mathbf{X}_1^{(n)} + \mathbf{X}_2^{(n)}$ is continuous with respect to the supremum norm.

By Theorem VII.2.3, $\mathbf{X}_1^{(n)}$ and $\mathbf{X}_2^{(n)}$ are asymptotically independent. Therefore, the covariance matrix for $n^{1/2}(\hat{\mathbf{P}}(s, t; \mathbf{Z}_0) - \mathbf{P}(s, t; \mathbf{Z}_0))$ is a sum of two $k^2 \times k^2$ matrices. Since the elements $X_{2hj}^{(n)}$, $h \neq j$, of $\mathbf{X}_2^{(n)}$ are orthogonal local square integrable martingales, it follows directly, as in (4.4.14), that the second term may be estimated by the matrix

$$\int_s^t \hat{\mathbf{P}}(u, t; \mathbf{Z}_0)^{\mathsf{T}} \otimes \hat{\mathbf{P}}(s, u; \mathbf{Z}_0) \, d[\mathbf{X}_2^{(n)}](u) \hat{\mathbf{P}}(u, t; \mathbf{Z}_0) \otimes \hat{\mathbf{P}}(s, u, \mathbf{Z}_0)^{\mathsf{T}}$$

with elements n times

$$\widehat{\operatorname{cov}}_2(\hat{P}_{hj}(s, t; \mathbf{Z}_0), \hat{P}_{mr}(s, t; \mathbf{Z}_0))$$
$$= \sum_{g \neq l} \int_s^t \hat{P}_{hg}(s, u; \mathbf{Z}_0) \hat{P}_{mg}(s, u; \mathbf{Z}_0) \{\hat{P}_{lj}(u, t; \mathbf{Z}_0) - \hat{P}_{gj}(u, t; \mathbf{Z}_0)\}$$
$$\times \{\hat{P}_{lr}(u, t; \mathbf{Z}_0) - \hat{P}_{gr}(u, t; \mathbf{Z}_0)\} J_g(u) (\exp(\hat{\boldsymbol{\beta}}^{\mathsf{T}} \mathbf{Z}_{gl0}))^2 S_{gl}^{(0)}(\hat{\boldsymbol{\beta}}, u)^{-2} \, dN_{gl}(u).$$

cf. (4.4.16). Alternatively, a recursion formula like (4.4.19) can be used.

Since each element (h, j) of the first term of (7.2.40) has the form

$$n^{1/2}(\hat{\boldsymbol{\beta}} - \boldsymbol{\beta}_0)^{\mathsf{T}} \sum_{g,l} \int_s^t P_{hg}(s, u; \mathbf{Z}_0) \, d\mathbf{W}_{gl}(u) P_{lj}(u, t; \mathbf{Z}_0),$$

the first term for the estimated covariance of $\hat{P}_{hj}(s, t; \mathbf{Z}_0)$ and $\hat{P}_{mr}(s, t; \mathbf{Z}_0)$ is

$$\widehat{\text{cov}}_1(\hat{P}_{hj}(s,t;\mathbf{Z}_0), \hat{P}_{mr}(s,t;\mathbf{Z}_0))$$
$$= \int_s^t \sum_{g,l} \hat{P}_{hg}(s,u;\mathbf{Z}_0)\,\mathrm{d}\hat{\mathbf{W}}_{gl}^{\mathsf{T}}(u)\hat{P}_{lj}(u,t;\mathbf{Z}_0)\mathscr{I}_\tau(\hat{\boldsymbol{\beta}})^{-1}$$
$$\times \int_s^t \sum_{g,l} \hat{P}_{mg}(s,u;\mathbf{Z}_0)\,\mathrm{d}\hat{\mathbf{W}}_{gl}(u)\hat{P}_{lr}(u,t;\mathbf{Z}_0),$$

where for $g \neq l$

$$\hat{\mathbf{W}}_{gl}(t) = \exp(\hat{\boldsymbol{\beta}}^{\mathsf{T}}\mathbf{Z}_{gl0}) \int_0^t (\mathbf{Z}_{gl0} - \mathbf{E}_{gl}(\hat{\boldsymbol{\beta}}, u))J_g(u)S_{gl}^{(0)}(\hat{\boldsymbol{\beta}}, u)^{-1}\,\mathrm{d}N_{gl}(u),$$

whereas $\hat{\mathbf{W}}_{gg} = -\sum_{l \neq g} \hat{\mathbf{W}}_{gl}$.

Next we shall see how these results may be used in connection with the prothrombin data.

Example VII.2.10. Abnormal Prothrombin Levels in Liver Cirrhosis. Estimation of Transition Probabilities in Finite-State Markov Processes

The data were introduced in Example I.3.12. We now continue the discussion from Example IV.4.4 by considering models where the transition intensities of the four types are *proportional* for prednisone- and placebo-treated patients. Thus, for each of the 488 patients, we consider a single basic covariate

$$Z_i = \begin{cases} 1 & \text{if patient } i \text{ had prednisone treatment} \\ 0 & \text{if patient } i \text{ had placebo treatment.} \end{cases}$$

From Z_i, four type-specific covariates are defined,

$$\mathbf{Z}_{01i} = \begin{bmatrix} Z_i \\ 0 \\ 0 \\ 0 \end{bmatrix}, \quad \mathbf{Z}_{10i} = \begin{bmatrix} 0 \\ Z_i \\ 0 \\ 0 \end{bmatrix}, \quad \mathbf{Z}_{02i} = \begin{bmatrix} 0 \\ 0 \\ Z_i \\ 0 \end{bmatrix}, \quad \mathbf{Z}_{12i} = \begin{bmatrix} 0 \\ 0 \\ 0 \\ Z_i \end{bmatrix},$$

and the estimates for the elements of the corresponding regression coefficient vector $\boldsymbol{\beta} = (\beta_{01}, \beta_{10}, \beta_{02}, \beta_{12})$ are (with estimated standard deviations in brackets)

$$\hat{\beta}_{01} = -0.256\ (0.121),$$
$$\hat{\beta}_{10} = 0.308\ (0.114),$$
$$\hat{\beta}_{02} = -0.249\ (0.197),$$
$$\hat{\beta}_{12} = 0.249\ (0.148).$$

(Note that in this simple case, the estimation of the four parameters can be made separately.) The estimates confirm what was seen in Figures IV.4.10 and

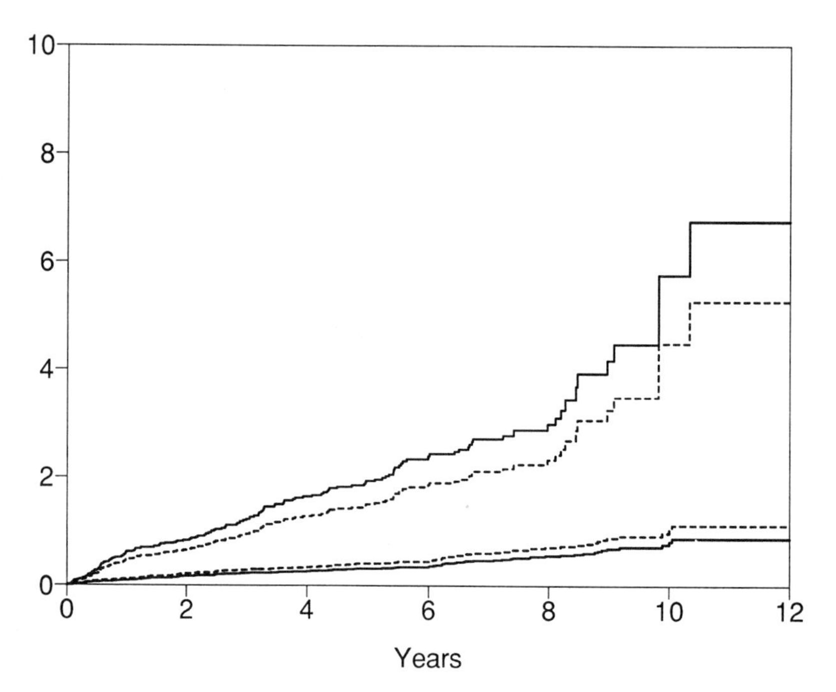

Figure VII.2.8. Estimated integrated death intensities for prednisone- (—) and placebo-treated (- - -) patients. Upper pair of curves: $1 \to 2$; lower pair of curves: $0 \to 2$.

IV.4.11: prednisone (significantly) increases $\alpha_{10}(t)$ and decreases $\alpha_{01}(t)$ and for the death intensities, there are (insignificant) tendencies that prednisone increases $\alpha_{12}(t)$ and decreases $\alpha_{02}(t)$. The estimated integrated intensities

$$\hat{A}_{hj0}(t, \hat{\boldsymbol{\beta}}) \quad \text{for placebo}$$

and

$$\exp(\hat{\beta}_{hj})\hat{A}_{hj0}(t, \hat{\boldsymbol{\beta}}) \quad \text{for prednisone,}$$

$h = 0, 1, j = 0, 1, 2, h \neq j$, are shown in Figures VII.2.8 and VII.2.9 and they are seen to resemble the corresponding Nelson–Aalen estimates from Example IV.4.4.

Under the hypothesis H_0: $\beta_{02} = \beta_{12}$ of identical treatment effects on the death intensities, there are four type-specific covariates in the model which is now three-dimensional:

$$\mathbf{Z}_{01i} = \begin{bmatrix} Z_i \\ 0 \\ 0 \end{bmatrix}, \quad \mathbf{Z}_{10i} = \begin{bmatrix} 0 \\ Z_i \\ 0 \end{bmatrix}, \quad \mathbf{Z}_{02i} = \mathbf{Z}_{12i} = \begin{bmatrix} 0 \\ 0 \\ Z_i \end{bmatrix}.$$

This hypothesis can be tested using the statistic

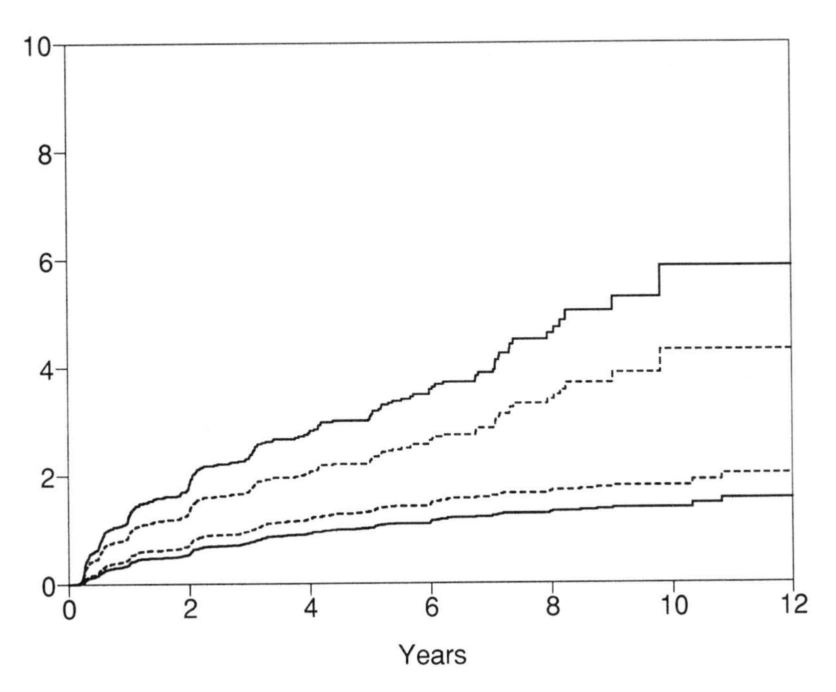

Figure VII.2.9. Estimated integrated transition intensities for prednisone- (—) and placebo-treated (- - -) patients. Upper pair of curves: $1 \to 0$; lower pair of curves: $0 \to 1$.

$$\frac{\hat{\beta}_{02} - \hat{\beta}_{12}}{\{\widehat{\text{s.d.}}(\hat{\beta}_{02})^2 + \widehat{\text{s.d.}}(\hat{\beta}_{12})^2\}^{1/2}} = -2.02,$$

giving a P-value of 0.04 when referred to the standard normal distribution.

For comparison with Figures IV.4.12 and IV.4.13, the estimated death probabilities $\hat{P}_{02}(0, t; Z_0)$ and $\hat{P}_{12}(0, t; Z_0)$ are shown in Figures VII.2.10 and VII.2.11, respectively, both for prednisone- ($Z_0 = 1$) and placebo-treated ($Z_0 = 0$) patients. Again, the estimates are close to the Aalen–Johansen estimates but for the probabilities $P_{02}(0, t)$ given normal prothrombin at start of treatment, prednisone treatment now seems to be uniformly superior. The main differences from the previous analyses are seen in the estimated standard deviations of the estimates. Figure VII.2.12 shows the estimated standard deviation of the estimate for prednisone treated patients (Figure VII.2.10) calculated as described above and the corresponding quantity for the Markov process without assuming proportional hazards (Figure IV.4.16). There seems to be some further gain in precision using the Cox regression model, but the reliability of this finding, of course, depends on the fit of the proportional hazards model. This may be investigated as discussed in Section VII.3. □

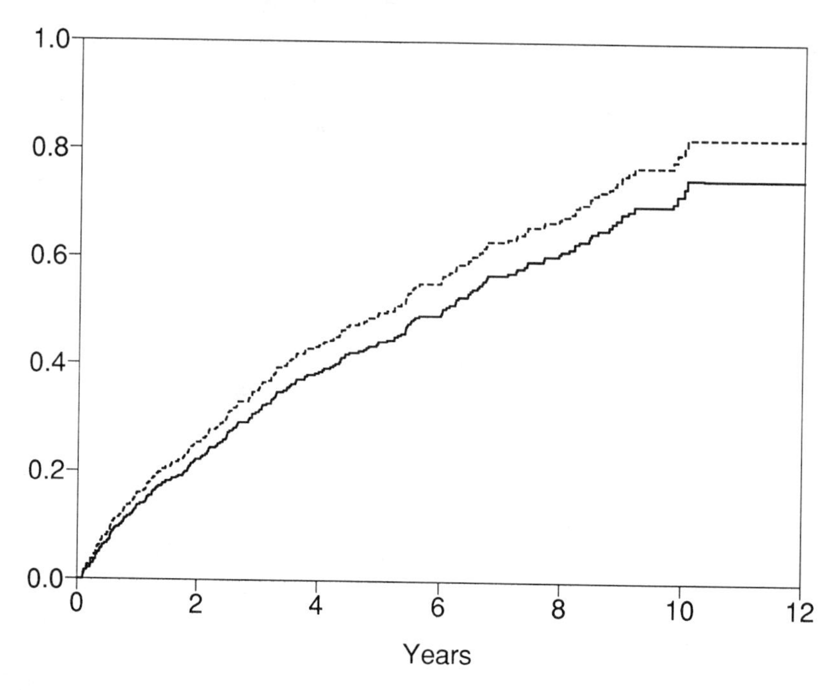

Figure VII.2.10. Estimated probabilities $\hat{P}_{02}(0, t; Z_0)$ of being dead by time t given normal prothrombin at time 0. Prednisone, $Z_0 = 1$: (—); placebo, $Z_0 = 0$: (- - -).

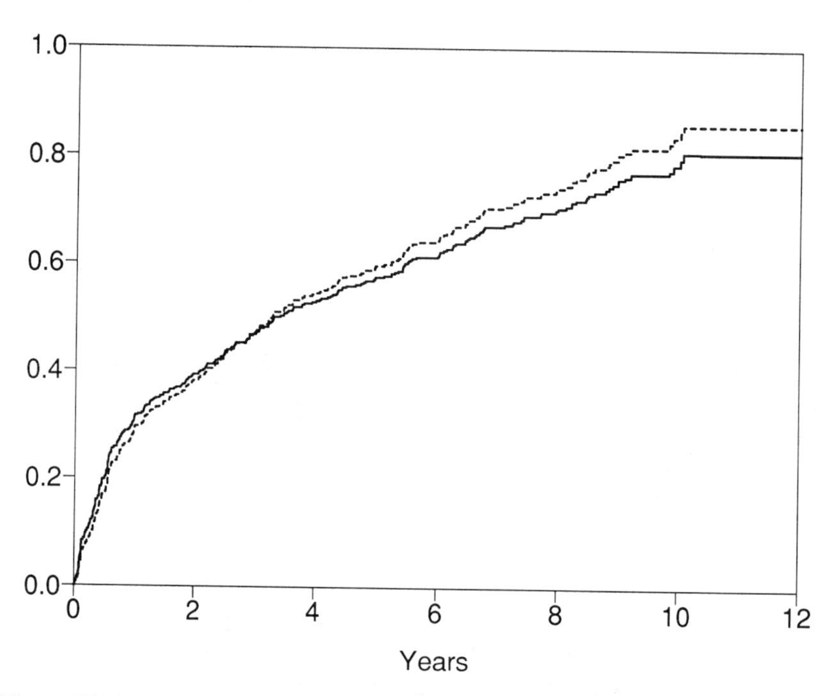

Figure VII.2.11. Estimated probabilities $\hat{P}_{12}(0, t; Z_0)$ of being dead by time t given low prothrombin at time 0. Prednisone, $Z_0 = 1$: (—); placebo, $Z_0 = 0$: (- - -).

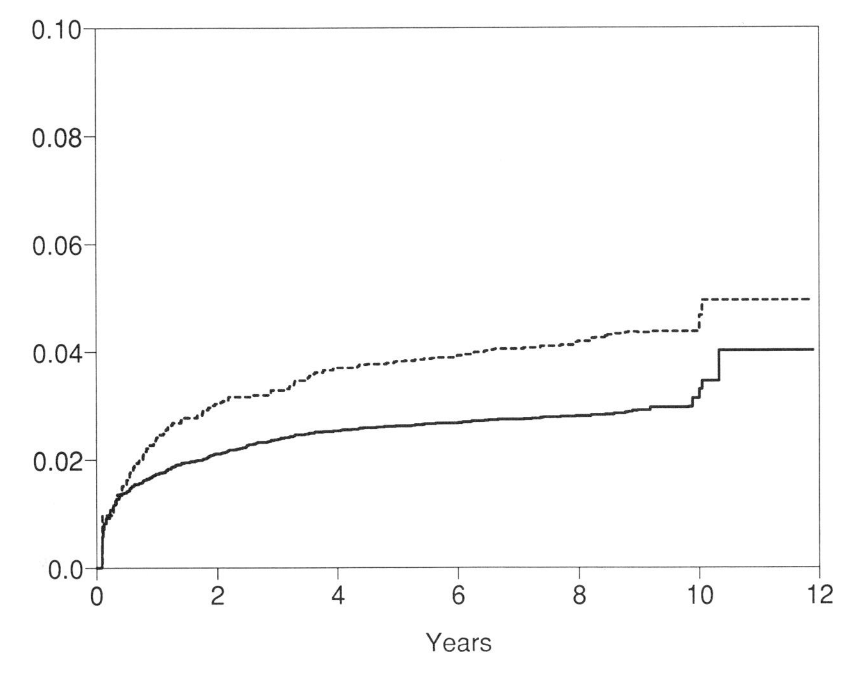

Figure VII.2.12. Estimated standard deviations of estimates of $P_{02}(0, t)$ for prednisone-treated patients: $\hat{P}_{02}(0, t; Z_0 = 1)$ based on a proportional hazards model (—), $\hat{P}_{02}(0, t)$ based on a Markov process model (- - -) (see Figure IV.4.16).

In the next example, we study a general disability model where functionals of interest are *life insurance premiums* (Examples IV.3.9 and IV.4.3).

EXAMPLE VII.2.11. Nephropathy and Mortality among Insulin-Dependent Diabetics. Calculation of Life Insurance Premiums

The data were introduced in Example I.3.11. In this example, regression models for the transition intensities will be developed with the purpose of combining the estimates into life insurance premiums for insulin-dependent diabetics (Ramlau-Hansen et al., 1987).

Andersen (1988) studied regression models for a part of the present data set (individuals diagnosed between 1933 and 1952) modelling the effects of some of the basically continuous covariates like calendar year of disease onset and duration of diabetes at the time of onset of diabetic nephropathy (DN) using piecewise constant functions. To avoid a too irregular behavior of the life insurance premiums calculated from the entire data set, Ramlau-Hansen et al. (1987) approximated these effects using linear functions. The models (to be discussed in the following) are given by

Table VII.2.1. Estimated Regression Coefficients $\hat{\beta}$ in the Models for $\alpha_{01}(\cdot)$, $\alpha_{02}(\cdot)$, and $\alpha_{12}(\cdot)$

Transition	Covariate	$\hat{\beta}$	$\widehat{\text{s.d.}}(\hat{\beta})$
From "no DN" to "DN": $0 \to 1$	Time of disease onset, t_0 (years)	-0.0259	0.0050
	Age at disease onset, a_0 (years)	-0.0121	0.0041
From "no DN" to "death": $0 \to 2$	Time of disease onset, t_0 (years)	-0.0148	0.0062
	I (males, $a \leq 17$ years)	-1.450	0.541
	I (males, $18 \leq a \leq 35$ years)	-0.264	0.220
	I (males, $a \geq 36$ years)	-0.125	0.160
From "DN" to "death": $1 \to 2$	Time of disease onset, t_0 (years)	-0.0308	0.0056
	Disease duration, $a_1 - a_0$ at onset of DN (years)	-0.0373	0.0108
	I ($a - a_1 \leq 4$ years)	-0.356	0.212
	I ($5 \leq a - a_1 \leq 14$ years)	0.548	0.186
	I ($a - a_1 > 14$ years)	0	—

$$\alpha_{02,i}(a) = \alpha_{02,0}(a)\mu_i(a)\exp\left(\beta_1^{02} t_0^i + \sum_{j=2}^{4} \beta_j^{02} Z_j^{02,i}(a)\right), \quad a \geq a_0^i,$$

$$\alpha_{12,i}(a) = \alpha_{12,0}(a)\exp\left(\beta_1^{12} t_0^i + \beta_2^{12}(a_1^i - a_0^i) + \sum_{j=3}^{4} \beta_j^{12} Z_j^{12,i}(a)\right), \quad a \geq a_1^i,$$

$$\alpha_{01,i}(a) = \alpha_{01s_i,0}(a - a_0)\exp(\beta_1^{01} t_0^i + \beta_2^{01} a_0^i), \quad a \geq a_0^i,$$

and estimated regression coefficients from these models are shown in Table VII.2.1. Here, a denotes age, a_0 is age at onset of diabetes, a_1 is age at onset of DN, and t_0 is the calendar year of onset of diabetes. It is seen that all three intensities decrease with t_0, indicating a general medical improvement over time. The intensity $\alpha_{01}(\cdot)$ of developing DN also decreases with increasing values of a_0 and the death intensity $\alpha_{12}(\cdot)$ after onset of DN decreases with increasing values of $a_1 - a_0$, the duration of diabetes at the time of onset of DN. Furthermore, $\alpha_{12}(\cdot)$ depends on $a - a_1$, the time since onset of DN through the time-dependent covariates $Z_3^{12,i}(a) = I(a - a_1^i \leq 4 \text{ years})$ and $Z_4^{12,i}(a) = I(5 \leq a - a_1^i \leq 14 \text{ years})$. Hence, the model is a *semi-Markov* process rather than a Markov process. The model for $\alpha_{01}(\cdot)$ is stratified by sex, s (Examples VII.1.1 and VII.2.2), since men and women are suspected to have different DN intensities and, therefore, the estimation of missing $0 \to 1$ transition times was carried out separately for males and females (Example I.3.11). The model for $\alpha_{12}(\cdot)$ is a "standard" multiplicative model for the absolute mortality, whereas that for $\alpha_{02}(\cdot)$ is a regression model for the *relative mortality*, denoting by $\mu_i(t)$ the known population mortality (Examples III.1.7, VII.1.1, and VII.2.2). The choice between such models is further discussed in Example VII.2.14 in the next subsection. The last three regression coefficients in the model for $\alpha_{02}(\cdot)$ corresponding to the time-dependent covariates $Z_2^{02,i}(a) = I(s_i = 1, a \leq 17 \text{ years})$, $Z_3^{02,i}(a) = I(s_i = 1, 18 \leq a \leq 35 \text{ years})$, and $Z_4^{02,i}(a) = I(s_i = 1,\ a \geq 36 \text{ years})$ suggest that men $(s = 1)$ and women $(s = 0)$ have nonproportional relative mortalities before onset of DN. In fact, the ratio between the relative mortality for males and that for females is less than one, but it increases with age.

For the death intensities $\alpha_{02}(\cdot)$ and $\alpha_{12}(\cdot)$, the basic time scale was chosen to be age, a, whereas, for the DN intensity, $\alpha_{01}(\cdot)$ the disease duration, $a - a_0$, was chosen as basic time scale (Andersen, 1988). The rationale behind these particular choices was to use as basic time scale the one that had the largest effect on the intensities and then model the effect of other time variables parametrically.

From the estimates of the regression coefficients and the integrated baseline transition intensities, the transition probabilities for given patient characteristics can be estimated. This is in spite of the fact that the model as mentioned above is not a Markov model.

For given values $\mathbf{Z}_0 = (a_0, s_0, t_0)$ of the fixed covariates, we denote by $\alpha_{01}(a; \mathbf{Z}_0)$ and $\alpha_{02}(a; \mathbf{Z}_0)$ the transition intensities from state 0 at age a $(a \geq a_0)$. Furthermore, for given $\mathbf{Z}_0$ and given age a_1 at time of DN, we denote by $\alpha_{12}(a; \mathbf{Z}_0, a_1)$ the death intensity at age a $(a \geq a_1)$ from state 1. Then (for $x > a$)

$$P_{00}(a, x; \mathbf{Z}_0) = \exp\left(- \int_a^x (\alpha_{01}(y; \mathbf{Z}_0) + \alpha_{02}(y; \mathbf{Z}_0))\, dy \right),$$

$$P_{01}(a, x; \mathbf{Z}_0) = \int_a^x P_{00}(a, y; \mathbf{Z}_0)\alpha_{01}(y; \mathbf{Z}_0) P_{11}(y, x; \mathbf{Z}_0, y)\, dy,$$

$$P_{02}(a, x; \mathbf{Z}_0) = 1 - P_{00}(a, x; \mathbf{Z}_0) - P_{01}(a, x; \mathbf{Z}_0),$$

$$P_{11}(a, x; \mathbf{Z}_0, a_1) = \exp\left(- \int_a^x \alpha_{12}(y; \mathbf{Z}_0, a_1)\, dy \right),$$

$$P_{12}(a, x; \mathbf{Z}_0, a_1) = \int_a^x P_{11}(a, y; \mathbf{Z}_0, a_1)\alpha_{12}(y; \mathbf{Z}_0, a_1)\, dy,$$

and once these formulas are established, the life insurance premiums can be calculated following the lines of Examples IV.3.9 and IV.4.3. This was done by Ramlau-Hansen et al. (1987), and Table VII.2.2 shows examples of how many times the premiums calculated in this way are larger than premiums calculated using standard life tables. It is everywhere assumed that the diabetic shows no sign of DN at the time of issue of the insurance. It is seen that the relative premiums decrease both with increasing age at issue and increas-

Table VII.2.2. Relative Life Insurance Premiums for Diabetics Compared with Insurances at Standard Rates (a_0 = Age at Disease Onset). Term of Insurance: 20 years

Age at Issue (years)	Males		Females	
	$a_0 = 10$ years	$a_0 = 20$ years	$a_0 = 10$ years	$a_0 = 20$ years
20	6.1	3.8	5.3	3.5
30	3.9	3.9	3.8	4.2
40	2.4	2.4	2.5	2.5
50	1.6	1.5	1.5	1.3

ing age at onset of diabetes. Furthermore, the relative premiums are in most cases smaller for females than for males. Finally, the order of magnitude (1–6) of the relative premiums shows that it is possible to offer life insurances for insulin-dependent diabetics, and the work of Ramlau-Hansen et al. (1987) has, in fact, resulted in a new way of rating life insurance policies for Danish diabetics without sign of DN. □

VII.2.4. Some Special Models

In this subsection, we discuss some special cases which are not directly covered by the general results in Section VII.2.2. In that section, attention was restricted to the exponential function as relative risk function and here we first briefly present the similar results for a general $r(\cdot)$.

EXAMPLE VII.2.12. Other Relative Risk Functions $r(x)$

In Section VII.2.2, we studied the case where $r(x) = \exp(x)$, corresponding to the classical Cox regression model. In this case, the partial derivatives of $S_h^{(0)}(\boldsymbol{\beta}, t)$ with respect to the components of $\boldsymbol{\beta}$ again yielded sums $\mathbf{S}_h^{(1)}(\boldsymbol{\beta}, t)$ and $\mathbf{S}_h^{(2)}(\boldsymbol{\beta}, t)$ involving the exponential function [cf. (7.2.11)–(7.2.13)], and the process $X(\boldsymbol{\beta}, t)$ (Theorem VII.2.1) was automatically concave as a function of $\boldsymbol{\beta}$. Prentice and Self (1983) studied general relative risk functions $r(\cdot)$, for instance, $r(x) = 1 + x$, and discussed conditions to be added in such cases for the required asymptotic properties to hold. We shall now briefly review their results.

We still define $S_h^{(0)}(\boldsymbol{\beta}, t)$ by (7.2.3). Letting $u(\cdot) = \log r(\cdot)$, we further define

$$\mathbf{S}_h^{(1)}(\boldsymbol{\beta}, t) = \frac{\partial}{\partial \boldsymbol{\beta}} S_h^{(0)}(\boldsymbol{\beta}, t), \qquad \mathbf{S}_h^{(3)}(\boldsymbol{\beta}, t) = \frac{\partial^2}{\partial \boldsymbol{\beta}^2} S_h^{(0)}(\boldsymbol{\beta}, t),$$

$$\mathbf{S}_h^{(2)}(\boldsymbol{\beta}, t) = \sum_{i=1}^{n} Y_{hi}(t) \mathbf{Z}_{hi}^{\otimes 2}(t) \frac{\partial}{\partial \boldsymbol{\beta}} u(\boldsymbol{\beta}^\mathsf{T} \mathbf{Z}_{hi}(t)) r(\boldsymbol{\beta}^\mathsf{T} \mathbf{Z}_{hi}(t)),$$

$$S_h^{(4)}(\boldsymbol{\beta}, t) = \sum_{i=1}^{n} Y_{hi}(t)[u(\boldsymbol{\beta}^\top \mathbf{Z}_{hi}(t)) - u(\boldsymbol{\beta}_0^\top \mathbf{Z}_{hi}(t))] r(\boldsymbol{\beta}_0^\top \mathbf{Z}_{hi}(t)),$$

$$\mathbf{S}_h^{(5)}(\boldsymbol{\beta}, t) = \frac{\partial}{\partial \boldsymbol{\beta}} S_h^{(4)}(\boldsymbol{\beta}, t), \qquad \mathbf{S}_h^{(6)}(\boldsymbol{\beta}, t) = \frac{\partial^2}{\partial \boldsymbol{\beta}^2} S_h^{(4)}(\boldsymbol{\beta}, t),$$

and we define $\mathbf{E}_h(\boldsymbol{\beta}, t)$ and $\mathbf{V}_h(\boldsymbol{\beta}, t)$ by (7.2.14) and (7.2.15) as in the case $r(\cdot) = \exp(\cdot)$. The processes $\mathbf{S}_h^{(m)}(\boldsymbol{\beta}, t)$, $m = 1, \ldots, 6$, simplify considerably in this case.

Prentice and Self (1983) then basically required Condition VII.2.1 to hold for $m = 0, 1, \ldots, 6$ and assumed that $r(\boldsymbol{\beta}^\top \mathbf{Z}_{hi})$ is locally bounded away from zero for $\boldsymbol{\beta}$ in a neighborhood $\mathscr{B}$ of $\boldsymbol{\beta}_0$. Then the proofs follow the lines of the proofs in Section VII.2.2 with some extra difficulties involved mainly in the proof of consistency since $X(\cdot, t)$ is only concave with a probability tending to one inside a neighborhood of $\boldsymbol{\beta}_0$. □

In Section VII.2.2, we assumed k, the number of types (or strata) to be fixed as n, the number of individuals, tends to infinity. In models for survival in matched pairs, however (Holt and Prentice, 1974), we have $k = n/2$ and the asymptotic results presented in Section VII.2.2 do not apply. Gross and Huber (1987) considered that problem and we shall now briefly discuss their results.

Example VII.2.13. Survival in Matched Pairs. Length of Remission in Leukemia Patients

For each member $j = 1, 2$ in the ith pair, $i = 1, \ldots, n$, a possibly right-censored survival time $\tilde{X}_{ij}$ and a vector of covariates $\mathbf{Z}_{ij}(t)$ are observed. We define the processes $N_{ij}(t) = I(\tilde{X}_{ij} \le t, D_{ij} = 1)$ and $Y_{ij}(t) = I(\tilde{X}_{ij} \ge t)$ as previously and assume that the counting process $N_{ij}(t)$ has intensity process

$$\lambda_{ij}(t) = Y_{ij}(t) \exp(\boldsymbol{\beta}^\top \mathbf{Z}_{ij}(t)) \alpha_{0i}(t), \quad i = 1, \ldots, n, j = 1, 2. \qquad (7.2.41)$$

We can then estimate $\boldsymbol{\beta}$ from the log Cox partial likelihood (7.2.7),

$$C_\tau(\boldsymbol{\beta}) = \int_0^\tau \sum_{i=1}^{n} \sum_{j=1}^{2} \log \left\{ \frac{\exp(\boldsymbol{\beta}^\top \mathbf{Z}_{ij}(t))}{\sum_{j=1}^{2} \exp(\boldsymbol{\beta}^\top \mathbf{Z}_{ij}(t)) Y_{ij}(t)} \right\} dN_{ij}(t).$$

It is seen that pairs where the smaller observation time is a censoring time contribute no information to $C_\tau(\boldsymbol{\beta})$ since the term from the larger observation time is always $\log(1)$. Consistency and asymptotic normality of $\hat{\boldsymbol{\beta}}$ do not follow from Theorems VII.2.1 and VII.2.2 and even though the proofs for these theorems may be modified to cover the case where both the number of strata and the number of individuals in each stratum tend to infinity, they do not immediately generalize to the present situation where the number of individuals in each stratum is fixed.

Gross and Huber considered a more restrictive model assuming one underlying hazard function $\alpha_0(t)$ and proportional hazards between pairs with proportionality factors given by nuisance parameters $\exp(\gamma_i)$, $i = 1, \ldots, n$.

Under the assumption that (7.2.1) holds for $t = \tau$ and that the $\mathbf{Z}_{ij}(t)$ are bounded, they were able to prove consistency and asymptotic normality provided, furthermore, that certain $\mathbf{S}^{(m)}$ processes involving the nuisance parameters γ converge uniformly in probability to some nondegenerate limiting functions $\mathbf{s}^{(m)}$. Going through the proofs of Gross and Huber (1987), it is seen that these results hold also for the model (7.2.41) if one in the definition of their $\mathbf{S}^{(m)}$ processes replaces $\exp(\gamma_i)$ by $\alpha_{0i}(t)$ and assume (7.2.1) to hold for $t = \tau$ and for all $i = 1, \ldots, n$.

A particularly simple example arises in, for instance, controlled and paired experiments. Here, $Z_{ij}(t)$ may include only a single time-independent indicator with $Z_{ij} = 0$ for the placebo-treated member of a pair and $Z_{ij} = 1$ for the other. Then $C_\tau(\beta)$ reduces to

$$C_\tau(\beta) = D_1 \log\theta - (D_0 + D_1)\log(1 + \theta)$$

with $\theta = \exp(\beta)$, D_0 = number of pairs where the smaller observation time is a failure time for the placebo-treated member, and D_1 is defined similarly. Thus,

$$\hat{\theta} = \frac{D_1}{D_0},$$

and $\hat{\beta} = \log\hat{\theta}$ is asymptotically normal with a variance which can be estimated by $D_0^{-1} + D_1^{-1}$, provided that the regularity conditions of Gross and Huber hold. These, however, simplify considerably in this simple case and what we need is $A_{0i}(\tau) < \infty$, $i = 1, \ldots, n$, and the existence of a neighborhood $\mathscr{B}$ of β_0 and bounded, continuous, and positive functions $y_1(t)$, $y_2(t)$, and $y_{12}(t)$ such that as $n \to \infty$

$$\sup_{t \in \mathscr{T}} \left| \frac{1}{n} \sum_{i=1}^{n} \alpha_{0i}(t) Y_{i1}(t) - y_1(t) \right| \xrightarrow{\text{P}} 0,$$

$$\sup_{t \in \mathscr{T}} \left| \frac{1}{n} \sum_{i=1}^{n} \alpha_{0i}(t) Y_{i2}(t) - y_2(t) \right| \xrightarrow{\text{P}} 0,$$

$$\sup_{t \in \mathscr{T}} \left| \frac{1}{n} \sum_{i=1}^{n} \alpha_{0i}(t) Y_{i1}(t) Y_{i2}(t) - y_{12}(t) \right| \xrightarrow{\text{P}} 0.$$

For the model considered by Gross and Huber (1987), $\alpha_{0i}(t)$ in these sums should be replaced by $\exp(\gamma_i)$. In both cases, the conditions may be verified using versions of the law of large numbers for nonidentically distributed random variables if we assume the pairs $(\tilde{X}_{i1}, \tilde{X}_{i2})$, $i = 1, \ldots, n$, mutually independent.

Also, the score test for the hypothesis $\beta = 0$ takes a very simple form in this case. From (7.2.7), (7.2.16), and (7.2.17), it follows that

$$\frac{U_\tau(0)^2}{\mathscr{I}_\tau(0)} = \frac{(D_1 - D_0)^2}{D_1 + D_0};$$

see Downton (1972) for the case with no censoring and Example V.3.4. Under the above assumptions, this test statistic is approximately χ^2 distributed with one degree of freedom when $\beta = 0$.

For the data of Freireich et al. (1963) concerning remission lengths in leukemia patients and introduced in Example I.3.6, we have $D_0 = 18$ and $D_1 = 3$ giving the estimate

$$\hat{\theta} = 3/18 = 0.167 = \exp(-1.792)$$

with estimated standard deviation of $\hat{\beta}\,(1/3 + 1/18)^{1/2} = 0.624$. Thus, the Wald test statistic for the hypothesis $\beta = 0$ takes the value $(1.792/0.624)^2 = 8.25$ and the score test statistic is $(3 - 18)^2/(3 + 18) = 10.71$. Evaluated in the χ_1^2 distribution, both test statistics give P-values below 0.5%.

Here, we have disregarded the sequential stopping rule mentioned in Example I.3.6. As also mentioned there, these data have been analyzed several times in the survival data literature without consideration of the basic paired design of the study. Using a standard Cox regression model for the two-sample case (Example VII.2.6), we find $\hat{\beta} = -1.51 = \log(0.22)$ with estimated standard deviation 0.40 both fairly close to the values obtained above from the matched analysis. □

We next consider the regression model for the relative mortality mentioned in Example III.1.7.

Example VII.2.14. A Regression Model for the Relative Mortality. Mortality of Diabetics in the County of Fyn.

Nonparametric models for relative mortality were studied in Examples III.1.3 and IV.1.11 and, as mentioned in Examples III.1.7, VII.1.1, and VII.2.2, it is easy to incorporate regression terms into such a model. Thus, for survival data, the intensity process for N_i is simply written as

$$\lambda_i(t) = Y_i(t)\alpha_0(t)\mu_i(t)\exp(\boldsymbol{\beta}^{\mathsf{T}}\mathbf{Z}_i(t)),$$

where $\mu_i(\cdot)$ is the (known) population mortality corresponding to individual i and $Y_i(t)$ is the usual indicator for individual i being at risk at time $t-$. Thus, the model fits into the framework of Section VII.2.2, but to verify the conditions for consistency and asymptotic normality of $\hat{\boldsymbol{\beta}}$, some regularity conditions on the $\mu_i(\cdot)$ are needed.

We shall not pursue that any further here but rather, as an example, consider the survival data from the 716 female and 783 male diabetics from the county of Fyn introduced in Example I.3.2, discussing both regression models for the relative mortality and the absolute mortality. One regression model for the absolute mortality including information on sex and age at disease onset would be to specify the age-specific hazard function as

$$\alpha_i(a) = \alpha_0(a)\exp(\beta_1 Z_{i1} + \beta_2 Z_{i2}), \quad i = 1, \ldots, 1499,$$

with $Z_{i1} = I$ (i is a male) and Z_{i2} is the age a_{i0} (in years) at disease onset. In the estimation, the individuals, as usual, enter the risk set at their age, V_i, on 1 July 1973 and leave it at their age $\tilde{X}_i$ at death, emigration, or 1 January 1982, i.e.,

$$Y_i(a) = I(V_i < a \leq \tilde{X}_i).$$

Technically, the estimation of β_1 and β_2 may be performed, e.g., using the notion of "time-dependent strata" in BMDP 2L (BMDP, 1988). The estimates are

$$\hat{\beta}_1 = 0.384\ (0.093),$$

$$\hat{\beta}_2 = -0.0095\ (0.0039)$$

with standard deviation given in brackets. Thus, female diabetics have a lower mortality than males (see also Fig. IV.1.3) and the mortality decreases with increasing age at disease onset.

Instead of including the time-fixed covariate "age at disease onset," one might consider including the time-dependent covariate "disease duration at age a," i.e., $Z_{i3}(a) = a - a_{i0}$. Since age is the basic time scale and since the age at disease onset entered linearly into the log hazard function, this amounts to a reparametrization of the model where the regression coefficient for $Z_{i3}(a)$ is $\beta_3 = -\beta_2$ and the sex effect β_1 is unchanged. The underlying hazard function, however, changes since its interpretation changes.

As an alternative to regression models for the *absolute* mortality, models for the *relative* mortality can be studied as described above. In the model including Z_1 and Z_2, the estimates are

$$\hat{\beta}_1 = -0.145\ (0.094),$$

$$\hat{\beta}_2 = -0.0098\ (0.0039).$$

Note that the effect of sex has changed because of the higher population mortality for males. Thus, men and women seem to have the same relative mortality, whereas the effect of age at onset is almost the same as in the previous model. One way of possibly choosing between the two classes of models would be to consider an intermediate model

$$\alpha_i(a) = \mu_i(a)^{\beta_0} \exp(\beta_1 Z_{i1} + \beta_2 Z_{i2})\alpha_0(a) = \exp(\beta_0 \log \mu_i(a) + \beta_1 Z_{1i} + \beta_2 Z_{i2})\alpha_0(a),$$

including the time-dependent covariate $\log \mu_i(a)$. Here, $\beta_0 = 0$ corresponds to a model for the absolute mortality and $\beta_0 = 1$ to a model for the relative mortality (Andersen et al., 1985). The estimates are

$$\hat{\beta}_0 = -0.393\ (0.629),$$

$$\hat{\beta}_1 = 0.592\ (0.345),$$

$$\hat{\beta}_2 = -0.0094\ (0.0039),$$

indicating that the model for the absolute mortality ($\beta_0 = 0$) could be preferable to the model for the relative mortality. The partial likelihood ratio statistic and the Wald test statistic for the former hypothesis are 0.38 and 0.39, respectively, and for the latter hypothesis, the corresponding test statistics are 4.78 and 4.90, respectively (all statistics to be referred to the χ_1^2 distribution).

The fact that age at onset significantly affects the mortality has the consequence that the analyses of these data presented in earlier chapters have to be interpreted with some caution since it is only for *given* age at onset that we have a mortality which only depends on age; see Example III.3.6. □

In the asymptotics considered so far, the time interval $\mathscr{T}$ has been fixed while the number of individual counting processes observed on $\mathscr{T}$ increased. We conclude this subsection by an example where a single process is observed on a time interval the length of which tends to infinity assuming the baseline intensity to be *periodic*.

EXAMPLE VII.2.15. Cox's Periodic Regression Model. The Feeding Pattern of a Rabbit

The motivation for studying a periodic version of the Cox regression model was given in Examples I.3.17 and III.5.4 concerning models for the feeding pattern of a rabbit. The data consists of the rabbit's eating times $0 < T_1 < T_2 < \cdots$ and the corresponding quantities $x_1, x_2, \ldots$ eaten and we define the filtration $(\mathscr{F}_t)$ by

$$\mathscr{F}_t = \sigma\{(T_v, x_v);\, v = 1, 2, \ldots, T_v \leq t\}.$$

The $(\mathscr{F}_t)$ intensity process of the counting process

$$N(t) = \#\{v\colon T_v \leq t\}$$

is then assumed to have the form

$$\lambda(t) = \alpha_0(t)\exp(\boldsymbol{\beta}^{\mathsf{T}}\mathbf{Z}(t))$$

for some p-vector of $(\mathscr{F}_t)$-predictable *covariate processes* $\mathbf{Z}(t)$. Leaving α_0 unspecified except for the basic assumption of periodicity with period $\Delta(=1$ day), $\alpha_0(t) = \alpha_0(t + \Delta)$, the model is semiparametric.

Pons and Turckheim (1988a) studied the estimation of $A_0 = \int \alpha_0$ and $\boldsymbol{\beta}$ based on observation of the marked point process $(\mathbf{T}, \mathbf{Z})$ on $[0, n\Delta]$ and they derived large sample properties of the estimators as $n \to \infty$. Following the same routes as in Section VII.2.1 and letting $S_v = T_v$, modulo Δ, the estimators are given by

$$\hat{A}_0(t) = \sum_{S_v \leq t} \frac{1}{\sum_{j=0}^{n-1} \exp(\hat{\boldsymbol{\beta}}^{\mathsf{T}}\mathbf{Z}(S_v + j\Delta))},$$

where $\hat{\boldsymbol{\beta}}$ maximizes

$$L(\boldsymbol{\beta}) = \prod_{v} \frac{\exp(\boldsymbol{\beta}^{\mathsf{T}}\mathbf{Z}(T_v))}{\sum_{j=0}^{n-1}\exp(\boldsymbol{\beta}^{\mathsf{T}}\mathbf{Z}(S_v + j\Delta))},$$

i.e., $\hat{\boldsymbol{\beta}}$ is solution to the score equation $\mathbf{U}_n(\boldsymbol{\beta}) = (\partial/\partial\boldsymbol{\beta})C_n(\boldsymbol{\beta}) = \mathbf{0}$, where

$$C_n(\boldsymbol{\beta}) = \sum_{j=0}^{n-1}\left\{\int_0^{\Delta} \boldsymbol{\beta}^{\mathsf{T}}\mathbf{Z}(u+j\Delta)\,\mathrm{d}N(u+j\Delta) - \int_0^{\Delta}\log\sum_{k=0}^{n-1} e^{\boldsymbol{\beta}^{\mathsf{T}}\mathbf{Z}(u+k\Delta)}\,\mathrm{d}N(u+j\Delta)\right\}.$$

Since $\mathbf{U}_n(\boldsymbol{\beta}_0)$ has no obvious $(\mathscr{F}_t)$-martingale properties, the methods of proofs used in Section VII.2.2 no longer apply. Pons and Turckheim (1988a) showed that the estimators $\hat{A}_0$ and $\hat{\boldsymbol{\beta}}$ enjoy the usual large sample properties assuming both conditions like our Conditions VII.2.1 and VII.2.2 to hold (their Condition C) and some additional ergodicity (their Condition A) and mixing conditions (their Condition B) on the covariates.

We refer the reader to Pons and Turckheim (1988a) for further details and conclude this example by quoting some results from their analysis of the data presented in Example I.3.17. Including the two time-dependent covariates,

$$Z_1(t) = \text{time (in hours) since last meal before } t$$

and

$$Z_2(t) = \text{weight (in grams) of last meal before } t,$$

the estimates become (with estimated standard deviations in brackets)

$$\hat{\beta}_1 = 1.62\ (0.20)$$

and

$$\hat{\beta}_2 = -2.3\ (0.3).$$

As expected, the intensity increases with the time since last meal and it decreases with the weight of the last meal. Both covariates have highly significant effects, but inclusion of information on earlier meals does not improve the fit of the model. □

VII.2.5. Time-Dependent Covariates

In the basic formulation of the regression models [cf. (7.1.2)], time-dependent covariates have been allowed. One of the strengths of using the counting process and martingale approach in the study of the Cox model and related regression models (compared to more classical i.i.d. techniques) is, in fact, the ease with which time-dependent covariates can be dealt with. In Examples VII.2.11 (life insurance for diabetics) and VII.2.14 (diabetics on Fyn), we have also seen examples of what in Section III.5 was termed *defined* time-dependent covariates, typically durations since some observed events. However, as also mentioned in Section III.5, in other examples, other types of time-dependent covariates may be relevant. One such case, the randomized clinical trial in liver cirrhosis, CSL 1, was advertised in Example III.5.3. In the following example that discussion will be expanded both with the purpose of comparing analyses with and without inclusion of time-dependent covariates and to illustrate some of the methodological difficulties one may encounter when time-dependent covariates are used.

EXAMPLE VII.2.16. CSL 1: A Randomized Clinical Trial in Liver Cirrhosis. Regression Analyses with Time-Fixed and Time-Dependent Covariates

In this example we shall summarize some Cox regression analyses of the data introduced in Example I.3.4. Schlichting et al. (1983) and Christensen et al. (1985) developed a multiplicative model for the death intensity including a treatment indicator (prednisone versus placebo) and 12 clinical, biochemical, and histological variables recorded at the time of entry into the trial. Since the model is a standard Cox regression model with only time-fixed covariates, we shall not discuss it any further here but refer the reader to Schlichting et al. (1983, Tables 4 and 5) and Christensen et al. (1985, Table 1) for details. The estimated baseline intensity $\hat{\alpha}_0(t; \hat{\boldsymbol{\beta}})$, shown in Figure VII.2.13, is seen to be decreasing initially. The estimate was obtained from $\hat{A}_0(t; \hat{\boldsymbol{\beta}})$ using Epanechnikov's kernel function and $b = 1.5$ years.

As mentioned in Example I.3.12, a number of variables were also mea-

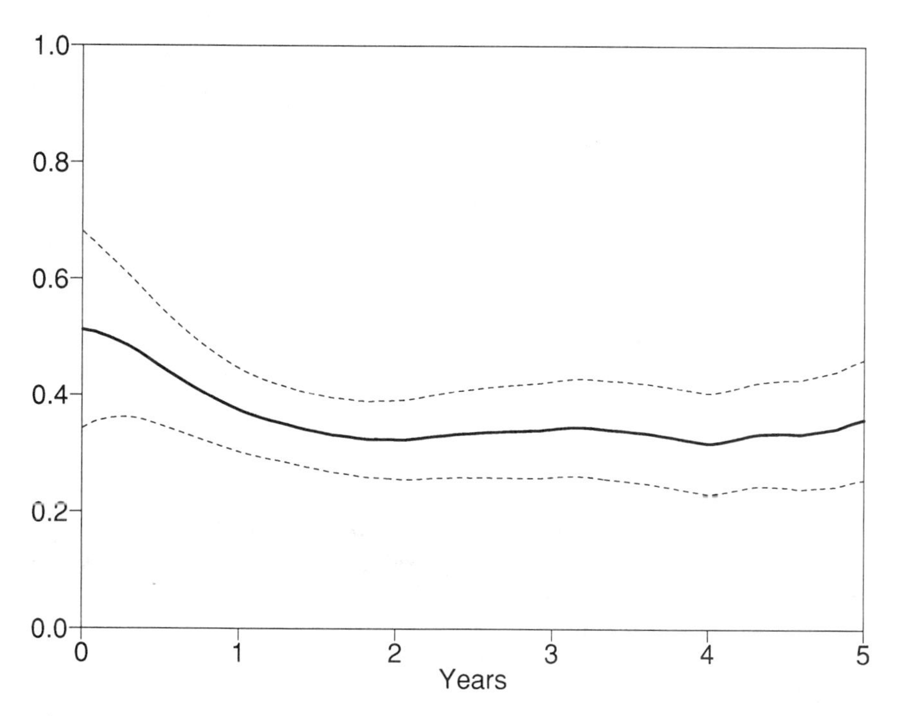

Figure VII.2.13. Estimated baseline hazard function and approximate 95% pointwise confidence interval in a final model for the CSL 1 data with only time-fixed covariates.

sured at follow-up visits during the trial. The visits were scheduled to take place 3, 6, and 12 months after entry and, thereafter, once a year. When assessing the death intensity for a given patient during the course of the trial, it is likely that more recent values of the covariates are more influential than those recorded at the time of entry, and, therefore, a model for the death intensity taking the follow-up measurements into account is of considerable interest (Example III.5.3). Such a model may, however, cause some computational difficulties and also some problems in interpretation. It is seen from (7.2.7) that to compute the contribution to the log Cox partial likelihood at any death time, the covariate values for *all* individuals at risk at that time are needed ($S_h^{(0)}(\boldsymbol{\beta}, t)$). We, therefore, have a missing data problem trying to analyze a model for which sufficiently detailed observations are not available. An operational way of attacking the problem would be to perform some interpolation between the observed covariate values, and Christensen et al. (1986) in their analyses used the convention that the values of the covariates for a patient still alive and under observation at some given time t are the values recorded at the *last* follow-up examination *before* time t. This seems the most natural choice, but other approaches like using the measurement from the follow-up closest to time t would be possible when approximating the likelihood.

The estimates using this convention in a model similar to that presented by Christensen et al. (1986) are shown in Table VII.2.3. It is seen that high levels of bilirubin, the presence of marked gastrointestinal bleeding, and a bad nutritional status are associated with a bad prognosis and the presence of inflammation in the liver connective tissue with a good one. Prognosis worsens with increasing age and increasing alkaline phosphatase and with decreasing albumin. Higher alcohol consumption is also associated with higher mortality, whereas inclusion of the duration of alcohol abuse did not improve the fit of the model (Example VII.1.2). There was a significant interaction between ascites and treatment and between prothrombin index and treatment suggesting that prednisone should not be given to patients with ascites but might be given to patients with a high prothrombin index; cf. also Examples IV.4.4 and VII.2.10. Age was included as age a_0 (in years) at entry, i.e., as a time-fixed covariate. Because of this *linear* score, however, the same estimated regression coefficient would have been obtained for age had it instead been included as the time-dependent covariate current age, $a_0 + t$ (in years) at time t.

Figure VII.2.14 shows the estimated baseline hazard function which appears roughly *constant*. The fact that the estimated hazard function decreases initially when no time-dependent covariates are included in the analyses (Figure VII.2.13) may reflect that patients are often accrued to a clinical trial in a somewhat critical phase of their disease and that it is not entirely possible to adjust for such within individual variation using the covariates recorded at entry. Taking into account, however, the course of the disease by conditioning at every time t on the present value of the variable makes it

Table VII.2.3. Estimated Regression Coefficients $\hat{\beta}$ in a Final Model for the CSL 1 Data with Time-Dependent Covariates

Covariate	Score	$\hat{\beta}$	$\widehat{\text{s.d.}}(\hat{\beta})$
Age at entry	years—60	0.0483	0.0086
Treatment	1: placebo 0: prednisone	−0.206	0.226
Inflammation in liver connective tissue	1: moderate or marked 0: none or slight	−0.479	0.135
Bilirubin	1: ≥ 70 μmol/L 0: < 70 μmol/L	1.124	0.175
Alkaline phosphatase	log (value in KA units)—4	0.315	0.108
Albumin	log (value in g/L)—4	−1.450	0.251
Nutritional status	1: meager/or extremely meager 0: otherwise	0.442	0.162
Alcohol consumption	0: < 10 g/day, 3: 10–50 g/day, 9: > 50 g/day	0.134	0.0245
Marked gastrointestinal bleeding	1: present 0: absent	1.509	0.180
Prothrombin index (pred.)	1: $> 70\%$ and prednisone 0: otherwise	−1.120	0.223
Prothrombin index (plac.)	1: $> 70\%$ and placebo 0: otherwise	−0.518	0.202
Ascites (slight, pred.)	1: present and prednisone 0: otherwise	1.112	0.244
Ascites (slight, plac.)	1: present and placebo 0: otherwise	0.325	0.270
Ascites (marked, pred.)	1: present and prednisone 0: otherwise	1.747	0.250
Ascites (marked, plac.)	1: present and placebo 0: otherwise	1.178	0.223

possible to perform this adjustment in a more efficient way and the conclusion from the analysis based on the time-dependent covariates that the death intensity is constant for given current covariate values is clinically more interpretable. This is because time since randomization/start of treatment should have little influence in a clinical trial where there is no marked treatment effect and where time 0 is at the first contact to the hospital and not, for instance, time of disease onset or time of operation.

There may, however, be other problems in *interpreting* the results for the model with time-dependent covariates. This has to do with the fact that the *survival probability* is not a functional only of the intensity when (non-deterministic) time-dependent covariates are included in the model; cf. the discussion in Section III.5. The survival probability also depends on the stochastic structure of the future development of the covariates and, hence, a model for this is needed if survival probabilities, which are often the goal

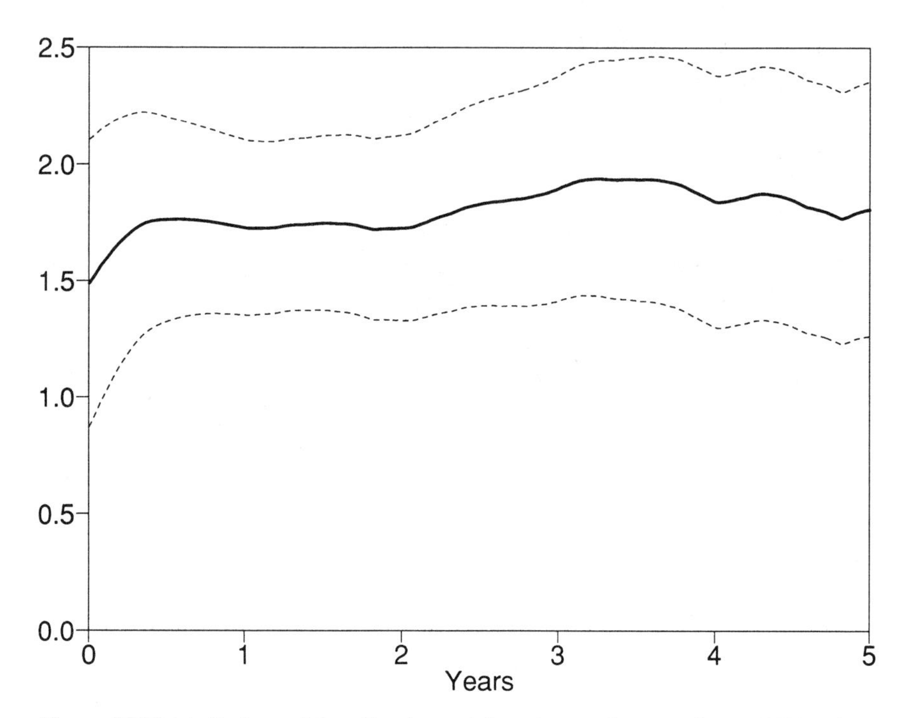

Figure VII.2.14. Estimated baseline hazard function and approximate 95% pointwise confidence interval in a final model for the CSL 1 data with time-dependent covariates.

of the analysis, are to be estimated. An example of how that may be achieved was given in Example VII.2.10. Alternatively, one may keep to models with time-fixed covariates or approximate short-term survival probabilities by functions only depending on the death intensity and the current covariate values. Thus, Christensen et al. (1986) and Andersen (1986) used the approximation

$$\exp\{-\hat{\alpha}_0 \Delta t \exp(\hat{\boldsymbol{\beta}}^{\mathsf{T}} \mathbf{Z}_i(t))\}$$

for the conditional probability of surviving time $t + \Delta t$ given survival until time t for $\Delta t = 0.5$ years. As an estimate of the constant baseline hazard, they used the M-estimator (see Section VI.2)

$$\hat{\alpha}_0 = N_{\bullet}(\infty)\left\{\int_0^\infty S^{(0)}(\hat{\boldsymbol{\beta}}, u)\,\mathrm{d}u\right\}^{-1},$$

where $S^{(0)}$ is given by (7.2.11). Alternatively, the computationally more complicated maximum likelihood estimate (to be discussed in Section VII.6) could have been used. □

The next example concerns psychiatric admissions among women who have just given birth and illustrates how a *semi-Markov* model can be analyzed using time-dependent covariates to model the duration dependence; see Examples III.5.2 and VII.2.11. Potential problems with incomplete information on the duration dependence were announced in Example III.5.6.

In Section X.1, we return to nonparametric analyses of semi-Markov models.

EXAMPLE VII.2.17. Psychiatric Admissions for Women Having Just Given Birth. The Influence of Covariates on the Admission Intensity

The problem and the data were introduced in Example I.3.10. In Chapter IV (Examples IV.1.4, IV.2.5, and IV.4.2), the intensities of admission to and discharge from psychiatric hospitals among Danish women giving birth in 1975 were analyzed under the somewhat unrealistic assumption that this group of women is homogeneous. In the present example, some regression analyses of the admission intensity are offered, following Andersen and Rasmussen (1986). The included covariates and the estimated regression coefficients are shown in Table VII.2.4. If only the four demographic variables,

Table VII.2.4. Estimated Regression Coefficients $\hat{\beta}$ in a Model for the Admission Intensity

Covariate	$\hat{\beta}$	$\widehat{\text{s.d.}}(\hat{\beta})$
Marital status		
Not married	0.279	0.129
Married	0	—
Age		
Age $\leq$ 18 years	0.226	0.319
18 years $<$ age $\leq$ 34 years	0	—
Age $>$ 34 years	0.311	0.217
Parity		
No prior children	−0.117	0.133
1 prior child	0	—
2 prior children	0.517	0.165
$\geq$3 prior children	0.428	0.224
Geographical region		
Capital	0	—
Urban areas	−0.204	0.124
Rural areas	−0.678	0.147
Prior admissions		
Latest discharge in $(t-30$ days, $t)$	6.480	0.160
Latest discharge in $(t-180$ days, $t-30$ days$)$	5.108	0.168
Latest discharge in $(-\infty, t-180$ days$)$	3.018	0.220
No prior admissions	0	—

age, parity, marital status, and geographical region, were included in the model (and if the discharge intensity, which we do not analyze further here, were still the same for all women and only depended on the time since birth), then we would still have a Markov process. As seen in the table, however, the effects of variables describing the duration dependence are highly significant and, therefore, the Markov assumption is totally unjustified. For given values of the demographic variables, women with a recent contact to a psychiatric ward have a much higher admission intensity than women without. It is seen that married women have a lower admission intensity and that the same is the case for women in the "normal" childbearing age interval from 19 to 34 years. Furthermore, the admission intensity seems roughly to increase with parity and with the degree of urbanization, the highest intensity being in the capital. Returning to the influence of prior admissions, it seems as if the effect *decreases* with time. Here, it should be recalled, however, that information on admissions before 1 October 1973 is not available (Example III.5.6) and, therefore, the estimated effect of admissions more than 180 days ago should be interpreted with caution since women with no prior admissions and some women with latest discharge in the interval $(-\infty, t - 180 \text{ days})$ [namely, those (unknown) women whose latest discharge was before 1 October 1973] are being pooled. If we leave out the latter covariate and keep the demographic factors and the other two time-dependent covariates for which

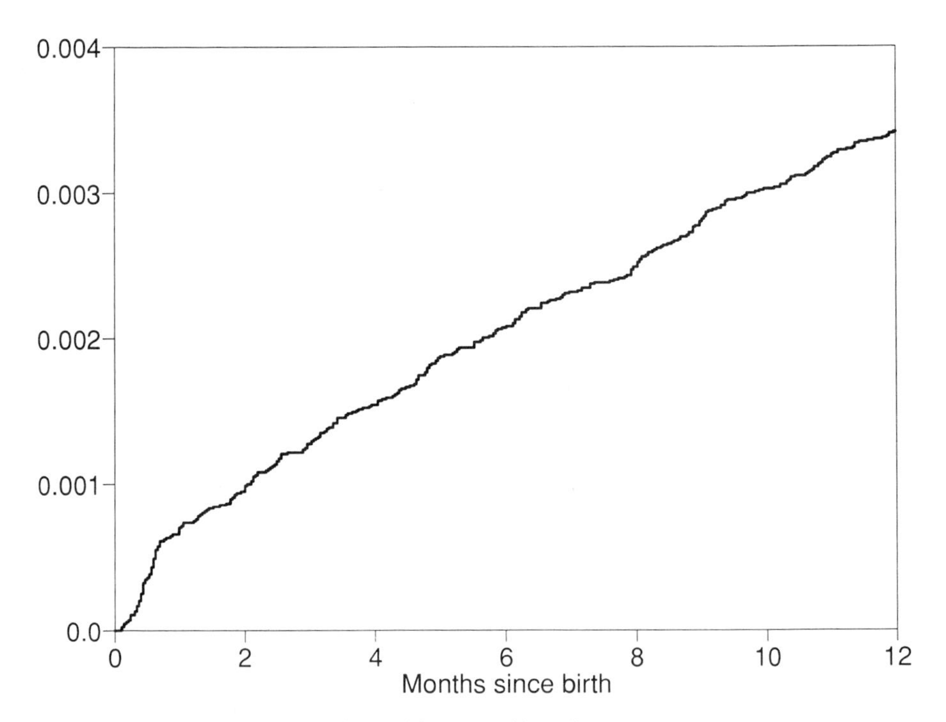

Figure VII.2.15. Estimated integrated baseline admission intensity.

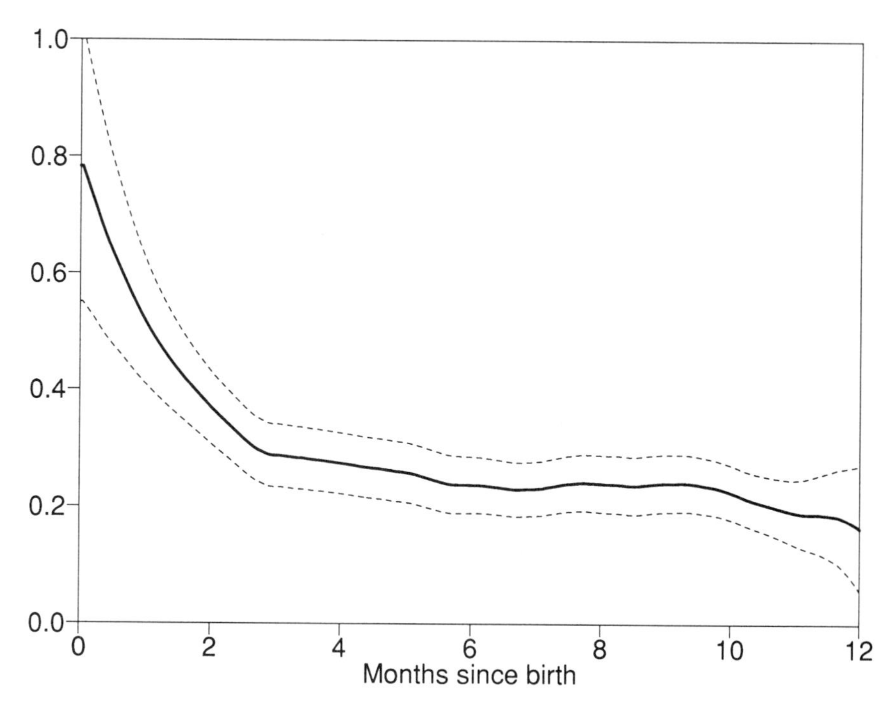

Figure VII.2.16. Estimated baseline admission intensity per 10^3 months with approximate 95% pointwise confidence intervals.

we do have complete information, neither the estimated regression coefficients nor the estimated underlying admission intensity change very much and, therefore, it is plausible that these effects are estimated "correctly."

Figure VII.2.15 shows the integrated baseline admission intensity, i.e., the integrated intensity for married women, aged 19–34 years with 1 prior birth, living in the Copenhagen area and with no prior admissions. The shape resembles what we saw in Example IV.1.4. Figure VII.2.16 shows a smoothed version of this using the same smoothing procedure as in Example IV.2.5 (i.e., Epanechnikov's kernel with bandwidth 70 days and extended to cover the two tails). Again, the shape is very similar to what we saw earlier, the estimated intensity decreasing with time. But the actual level of the intensity has decreased since we have now standardized to a group of women with a relatively low admission intensity. □

We finally return to the bone marrow transplantation data (see the discussion just after Example III.2.6).

Example VII.2.18. Bone Marrow Transplantation

The problem and the data were introduced in Example I.3.14; cf., in particular, Figure I.3.8. In this example, we shall use proportional intensity models with time-dependent covariates to analyze the transitions between (1) the initial state after transplantation, (2) development of acute (stage II–IV) graft-versus-host disease (GvHD), (3) chronic GvHD, (4) leukemia relapse, and (5) death. We should mention that a fifth state, of development of cytomegalovirus (CMV) infection after transplantation, was of particular interest to Jacobsen et al. (1990), whose data we reanalyze. However, for simplicity, we disregard posttransplant CMV infection. Of the original 190 patients, we include only the 163 suffering from acute myeloid leukemia, acute lymphoblastic leukemia, or lymphoblastic lymphoma.

The strategy for estimating the intensities of all transitions in Figure I.3.8 (except that from relapse to death) is to formulate four proportional intensity models, each handling all transitions into one of the four states, acute GvHD, chronic GvHD, relapse, and death (except from relapse), and with time-dependent covariates taking care of the transitions between previous states.

Tables VII.2.5–VII.2.8 show the estimated regression coefficients in the final models for development of acute GvHD and chronic GvHD and for relapse and death, respectively. In each case, all other possible transitions are scored as censorings when necessary. For example, when estimating the intensity for developing chronic GvHD, we interpret relapse and death as censorings.

Table VII.2.5 shows that the intensity of developing acute GvHD (this takes place only during the first 3 or 4 months after transplantation) is increased by patient CMV immunity at transplantation and HLA mismatch between donor and patient; in addition, the risk is increased if transplantation takes place during relapse.

Development of chronic GvHD (this takes place later than 2 to 3 months after transplantation) is rather more likely for patients who earlier developed acute GvHD, and, in addition, older donors carry an increased intensity (Table VII.2.6).

Table VII.2.5. Estimated Regression Coefficients $\hat{\beta}$ in a Model for Development of Acute GvHD

Covariate	$\hat{\beta}$	$\widehat{\text{s.d.}}(\hat{\beta})$
Patient CMV immune at transplantation		
Yes	0.907	0.308
No	0	—
HLA type		
Incomplete match	0.713	0.326
Complete match	0	—
Transplantation during		
Relapse	0.960	0.349
Remission	0	—

Table VII.2.6. Estimated Regression Coefficients $\hat{\beta}$ in a Model for Development of Chronic GvHD

Covariate	$\hat{\beta}$	$\widehat{\text{s.d.}}(\hat{\beta})$
Fixed		
Donor age		
>20 years	0.878	0.402
≤ 20 years	0	—
Donor → patient age/sex		
Female > 20 years to male	0.856	0.402
Other	0	—
Time dependent		
Transition from		
Acute GvHD	0.704	0.339
Initial state	0	—

Table VII.2.7. Estimated Regression Coefficients $\hat{\beta}$ in a Model for Relapse

Covariate	$\hat{\beta}$	$\widehat{\text{s.d.}}(\hat{\beta})$
Fixed		
Donor CMV status		
Immune	−1.611	0.451
Not immune	0	—
Transplantation		
During first remission	0	—
Later	2.232	0.494
Donor age		
>20 years	1.813	0.436
≤ 20 years	0	—
Time dependent		
Transition from		
Chronic GvHD	−2.516	0.795
Acute GvHD (only) or initial	0	—

Table VII.2.7 shows that the intensity of relapse (*censored at death*) is decreased if the donor was CMV immune and if the patient had had chronic GvHD. The latter phenomenon is called an "antileukemic effect" of chronic GvHD and has attracted considerable clinical interest.

The separate interpretation of cause-specific intensities in competing risks models was briefly discussed in Example IV.4.1, where we pointed out that in biological contexts it is usually rather speculative to discuss what would have happened if a particular risk was removed (here: what would the relapse intensity be if death from other causes were impossible?). Indeed, bone marrow transplantation has been a main example in the biostatistical literature on competing risks (Prentice et al., 1978; Kalbfleisch and Prentice, 1980, Chapter 7; Pepe et al., 1991).

In the present situation, it is obviously unnatural to regard death from other causes as an independent censoring in the sense of Section III.2.2, and a natural suspicion is that the decreased relapse intensity with chronic GvHD may be explained by increased mortality from the latter complication. Jacobsen et al. (1990) argued, however, as follows. Analysis of the intensity of death in remission (that is, before relapse) [cf. Table VII.2.8] showed that whereas acute GvHD does increase the intensity of death in remission, chronic GvHD does not add to this intensity. (The regression coefficient 0.241 for "chronic only" is insignificant and that for "acute and chronic" of 2.187 is insignificantly larger than that for "acute only" of 1.926; the standard deviation of the difference $2.187 - 1.926 = 0.261$ being estimated as 0.522.)

Table VII.2.8. Estimated Regression Coefficients $\hat{\beta}$ in a Model for Death in Remission

	Model Including Chronic GvHD		Final Model	
Covariate	$\hat{\beta}$	$\widehat{\text{s.d.}}(\hat{\beta})$	$\hat{\beta}$	$\widehat{\text{s.d.}}(\hat{\beta})$
Fixed				
MTX prophylaxis against acute GvHD				
Given	0.935	0.322	0.949	0.321
Not given	0	—	0	—
Transplantation				
During first remission	0	—	0	—
Later	0.747	0.356	0.760	0.355
During remission	0	—	0	—
During relapse	0.940	0.425	0.914	0.418
Donor age				
>20 years	0.787	0.336	0.827	0.327
≤ 20 years	0	—	0	—
Time dependent				
GvHD				
None	0	—	0	—
Acute only	1.926	0.398		
Chronic only	0.241	0.835		
Acute and chronic	2.187	0.550		
Acute $\pm$ chronic			1.958	0.354

This makes it less likely that the antileukemic effect of chronic GvHD may be explained by a selective excess mortality in remission for patients with chronic GvHD. We have added an extra column where only acute GvHD is included as covariate. □

VII.3. Goodness-of-Fit Methods for the Semi-parametric Multiplicative Hazard Model

In this section, we turn to a discussion of methods for examining the basic assumptions of the multiplicative hazard model. For simplicity, we shall study only a single type of event and in the basic model only time-fixed covariates will be included, though most of the methods to be discussed do work for more general counting processes and they are extendable to deal with time-dependent covariates. Furthermore, the only relative risk function under consideration will be the exponential function.

Thus, our basic model is that $N_1(t), \ldots, N_n(t)$, $t \in \mathscr{T}$, are *univariate* counting processes, N_i, having intensity process

$$\lambda_i(t) = \alpha_i(t) Y_i(t) = \alpha_0(t) \exp(\boldsymbol{\beta}^\mathsf{T} \mathbf{Z}_i) Y_i(t), \quad i = 1, \ldots, n. \tag{7.3.1}$$

In this model, the two basic assumptions are the *proportional hazards* assumption: the ratio

$$\frac{\alpha_i(t)}{\alpha_j(t)} = \exp\{\boldsymbol{\beta}^\mathsf{T}(\mathbf{Z}_i - \mathbf{Z}_j)\}, \qquad i, j = 1, \ldots, n, \tag{7.3.2}$$

between the hazards for any two individuals i and j does not depend on t; and the assumption of *log-linearity*: the log hazard

$$\log \alpha_i(t) - \log \alpha_0(t) = \boldsymbol{\beta}^\mathsf{T} \mathbf{Z}_i, \qquad i = 1, \ldots, n, \tag{7.3.3}$$

for any individual i, depends linearly on $\mathbf{Z}_i$. We assume that all relevant *interaction terms* between basic covariates are included in (7.3.1). This assumption of *additivity* will not be further discussed in this section as we have seen earlier (Example VII.2.5) how tests for interaction could be performed under the assumptions of proportional hazards and log-linearity using, e.g., the partial likelihood ratio test.

VII.3.1. Simple Graphical Checks of Proportional Hazards

The way in which we shall first approach methods to check (7.3.2) is by examining whether there are proportional hazards *between strata* defined from a single covariate (Z_p) assuming proportional hazards *within each stratum.* This strategy of treating one covariate (Z_p) at a time and assuming a lot of structure concerning the effects of the remaining covariates $(Z_1, \ldots, Z_{p-1})$

in the model is, of course, not optimal. This is because a misspecification of the model for the effect of $(Z_1, \ldots, Z_{p-1})$ may blur the examination of Z_p. It does, however, give a systematic procedure for examining all covariates of interest, and if the assumptions for some covariates turn out to be suspect, then some of the checks of the rest may be repeated; see, for instance, Example VII.3.4.

The notation will be as follows. The covariate vector for individual i will be written $\mathbf{Z}_i = (\mathbf{Z}_i^0, Z_{ip}) = (Z_{i1}, \ldots, Z_{i,p-1}, Z_{ip})$ and we assume for simplicity that $\mathbf{Z}_i^0 = (Z_{i1}, \ldots, Z_{i,p-1})$ does not include any interaction terms between Z_{ip} and other covariates. We consider a stratification of the range of Z_p into disjoint strata $I_1, \ldots, I_k$ and define the stratum $h(i)$ for individual i by $h(i) = h$ when $Z_{ip} \in I_h$, $i = 1, \ldots, n$, $h = 1, \ldots, k$.

We then want to compare the *stratified model* with separate underlying hazards $\alpha_{10}(t), \ldots, \alpha_{k0}(t)$ in each stratum where the hazard is given by

$$\alpha_i(t) = \alpha_{h(i)0}(t)\exp(\boldsymbol{\beta}^\mathsf{T}\mathbf{Z}_i^0) \tag{7.3.4}$$

(see, for instance, Example VII.2.5) with *the proportional hazards model* where there is one underlying hazard $\alpha_0(t)$ and proportionality factors $\exp(\beta_{p+1})$, $\ldots$, $\exp(\beta_{p+k})$, $(\beta_{p+k} \equiv 0)$ such that

$$\alpha_i(t) = \exp(\beta_{p+h(i)})\alpha_0(t)\exp(\boldsymbol{\beta}^\mathsf{T}\mathbf{Z}_i^0), \quad h = 1, \ldots, k. \tag{7.3.5}$$

Defining $Y_{hi}(t) = \delta_{hh(i)} Y_i(t)$, the model (7.3.4) has the form considered in the two previous sections and we may, therefore, estimate

$$A_{h0}(t) = \int_0^t \alpha_{h0}(u)\,du$$

by $\hat{A}_{h0}(t, \hat{\boldsymbol{\beta}})$ defined in (7.2.5) with $\hat{\boldsymbol{\beta}}$ found from (7.2.7). Introducing now, as in Example VII.2.4, $k - 1$ covariates $Z_{i,p+1}, \ldots, Z_{i,p+k-1}$ to describe the effect of Z_{ip} by the stratum indicators

$$Z_{i,p+h} = I(Z_{ip} \in I_h), \quad h = 1, \ldots, k - 1,$$

the model (7.3.5) can be written as

$$\alpha_i(t) = \alpha_0(t)\exp\left(\sum_{\substack{j=1 \\ j \neq p}}^{p+k-1} \beta_j Z_{ij}\right).$$

This model also has the form (7.1.5) and estimates of the regression coefficients and the integrated baseline hazard can be based on (7.2.5) and (7.2.7). Due to the different numbers of "nuisance functions," however, the models (7.3.4) and (7.3.5) are not directly comparable using, for instance, a partial likelihood ratio test.

We shall now discuss two simple graphical methods for comparison of these two models.

One may plot

$$\log \hat{A}_{10}(t, \hat{\boldsymbol{\beta}}), \ldots, \log \hat{A}_{k0}(t, \hat{\boldsymbol{\beta}})$$

versus t [or some other fixed function of t such as $\log(t)$]. Under the proportional hazards model (7.3.5), these curves should be approximately parallel, the constant vertical distance between $\log \hat{A}_{h0}(t, \hat{\boldsymbol{\beta}})$ and $\log \hat{A}_{k0}(t, \hat{\boldsymbol{\beta}})$ being approximately $\hat{\beta}_{p+h}$, $h = 1, \ldots, k-1$. Alternatively, one may directly plot the differences

$$\log \hat{A}_{h0}(t, \hat{\boldsymbol{\beta}}) - \log \hat{A}_{k0}(t, \hat{\boldsymbol{\beta}}), \quad h = 1, \ldots, k-1,$$

versus t [or $(\log(t)]$. Two drawbacks to this approach should be mentioned. First, by Corollary VII.2.6, the variances of these curves depend on t. Second, the kind of discrepancies from the curves being parallel that one may spot on such a plot may not reflect in any simple fashion the kind of discrepancies from proportionality that one may be interested in. Thus, a decreasing distance between $\log \hat{A}_{h0}(t, \hat{\boldsymbol{\beta}})$ and $\log \hat{A}_{k0}(t, \hat{\boldsymbol{\beta}})$ indicates that

$$\frac{\alpha_{h0}(t)}{\alpha_{k0}(t)} < \frac{A_{h0}(t)}{A_{k0}(t)} \tag{7.3.6}$$

and *not* necessarily that the hazard ratio $\alpha_{10}(t)/\alpha_{k0}(t)$ is decreasing. However, for the special case of *Weibull* distributed survival data $A_{h0}(t) = \alpha_h t^{\theta_h}$, $h = 1, \ldots, k$ [i.e., $\log \hat{A}_{h0}(t, \hat{\boldsymbol{\beta}})$ plotted against $\log t$ should be approximately *linear*; see Section VI.3.1], (7.3.6) is equivalent to $\theta_h < \theta_k$, which does imply that $\alpha_{h0}(t)/\alpha_{k0}(t)$ is decreasing.

This second drawback may be avoided by considering, instead, the smoothed estimates $\hat{\alpha}_{h0}(t)$, $h = 1, \ldots, k$ (cf. Theorem VII.2.7), but the price that is paid in increasing bias and variance is unclear (Gray, 1990).

Another graphical method based on the simple cumulative hazard estimates is to plot

$$\hat{A}_{h0}(t, \hat{\boldsymbol{\beta}}), \quad h = 1, \ldots, k-1, \quad \text{against } \hat{A}_{k0}(t, \hat{\boldsymbol{\beta}})$$

for all or some selected values of t. Under the proportional hazards model (7.3.5), these curves should approximate straight lines with intercepts zero and slopes close to $\exp(\hat{\beta}_{p+h})$, $h = 1, \ldots, k-1$. Again, the variances of the curves are nonconstant (Corollary VII.2.4), but some discrepancies from linearity may be easier to interpret (Gill and Schumacher, 1987). Thus, if $\hat{A}_{h0}(t, \hat{\boldsymbol{\beta}})$ plotted against $\hat{A}_{k0}(t, \hat{\boldsymbol{\beta}})$ gives a *convex* (concave) curve, then this suggests that the corresponding hazard ratio $\alpha_{h0}(t)/\alpha_{k0}(t)$ is *increasing* (decreasing). Furthermore, a piecewise linear curve indicates that $\alpha_{h0}(t)$ and $\alpha_{k0}(t)$ are piecewise proportional.

Comparing with Section V.3.1, it is seen that this is equivalent to plotting the numerator of the estimate $\hat{\theta}_{hL}(t)$ against its denominator choosing $L(t) \equiv 1$. Therefore, alternative plots are obtained using different weight processes $L(\cdot)$.

Dabrowska et al. (1989) studied this procedure in the special two-sample case with no covariates. Among other things, they derived confidence bands for the *relative trend function* $A_2 \circ A_1^{-1}$ and also for the *relative change function* $(A_2 - A_1)/A_1$ which is constant under the proportional hazards model.

Later, Dabrowska et al. (1992) extended the procedure to the general Cox regression model.

Next, we illustrate the use of some of these plots on the melanoma data.

EXAMPLE VII.3.1. Survival with Malignant Melanoma. Graphical Checks of Proportional Hazards

The data were introduced in Example I.3.1. We now continue Example VII.2.9 by stratifying, in turn, after sex, tumor thickness, and ulceration and considering some of the plots discussed earlier. Figures VII.3.1–VII.3.3 show plots of the estimates $\log \hat{A}_{h0}(t, \hat{\boldsymbol{\beta}})$ against t for

1. $h =$ females, males,
2. $h =$ thickness < 2 mm, $2 \leq$ thickness < 5 mm, thickness ≥ 5 mm,
3. $h =$ no ulceration, ulceration,

respectively, adjusting in all three cases for the effects of the other two covariates. While the curves for males and females in Figure VII.3.1 look roughly parallel, the vertical distance between the curves for thickness and ulceration seem to decrease with time. Thus, the proportional hazards assumption for thickness and/or ulceration seems questionable judged from these figures, whereas men and women do seem to have proportional death intensities.

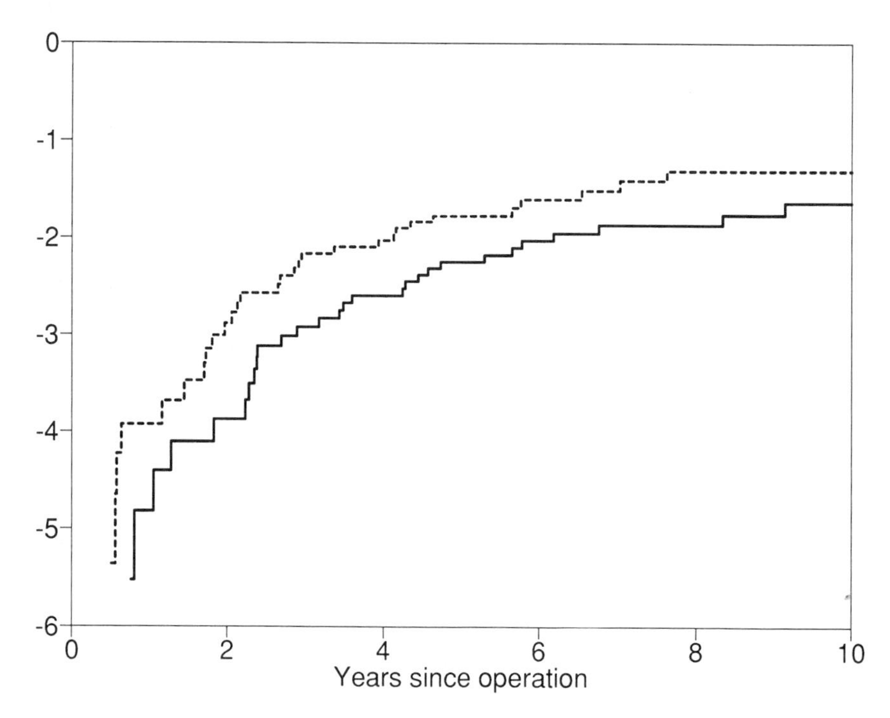

Figure VII.3.1. The log estimated integrated baseline hazards for men (- - -) and women (—).

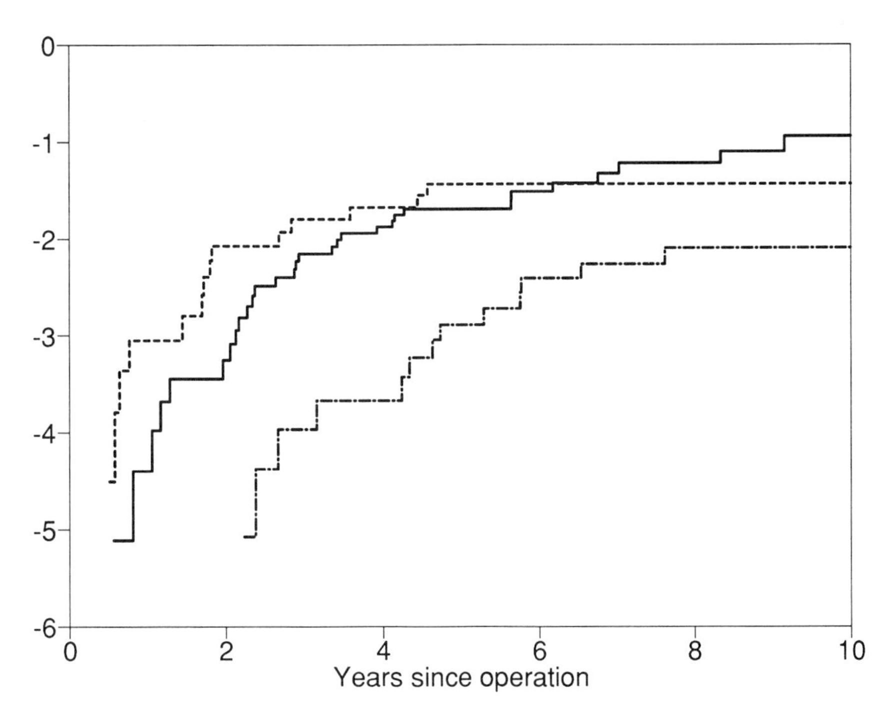

Figure VII.3.2. The log estimated integrated baseline hazards for patients with tumor thickness < 2 mm (-·-·-·-), 2 ≤ tumor thickness < 5 mm (—), and tumor thickness ≥ 5 mm (- - -).

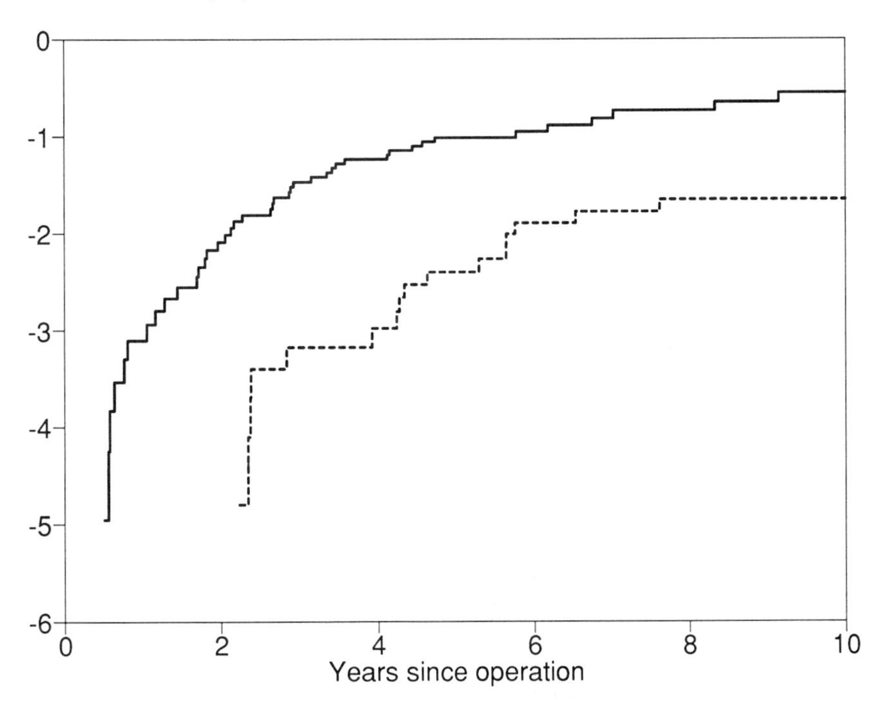

Figure VII.3.3. The log estimated integrated baseline hazards for patients without (- - -) and with (—) ulceration.

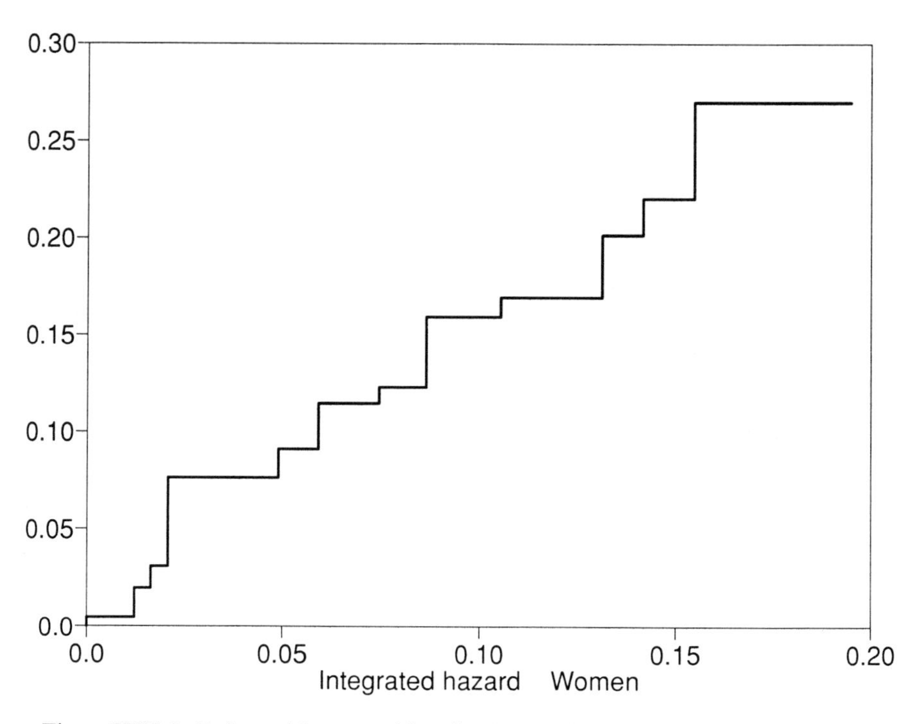

Figure VII.3.4. Estimated integrated baseline hazard for men plotted against that for women.

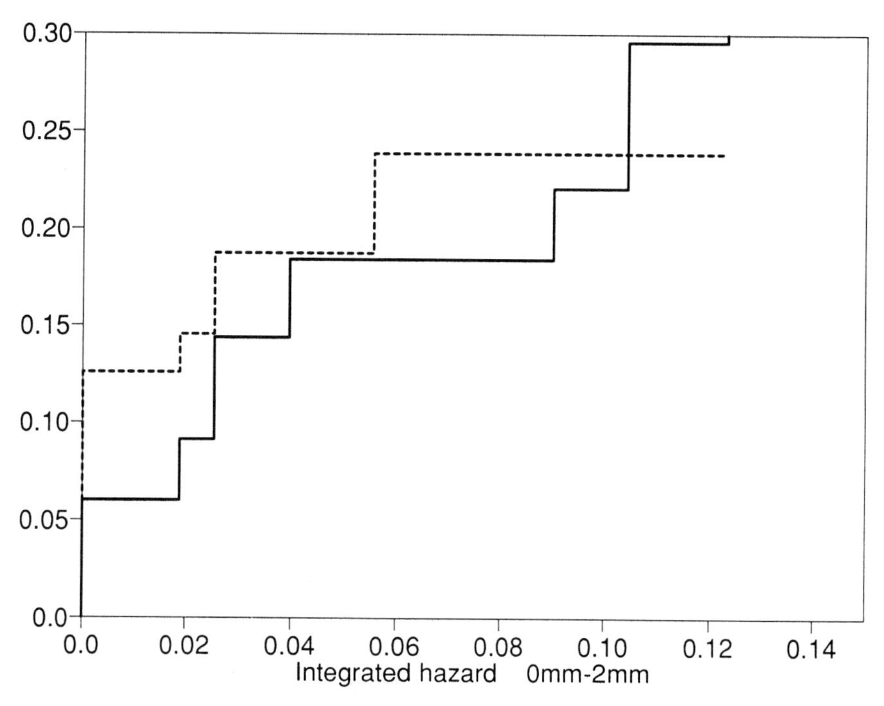

Figure VII.3.5. Estimated integrated baseline hazards for patients with $2 \leq$ tumor thickness < 5 mm (—) and tumor thickness ≥ 5 mm (- - -) plotted against that for patients with tumor thickness < 2 mm.

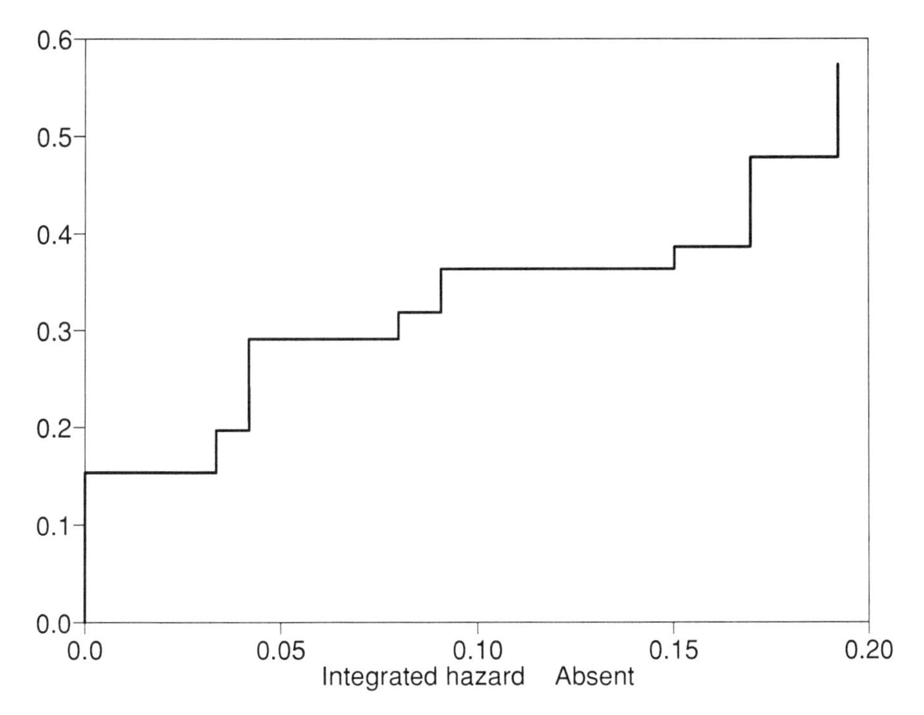

Figure VII.3.6. Estimated integrated baseline hazard for patients with ulceration plotted against that for patients without.

The same tendencies are seen in Figures VII.3.4–VII.3.6 where the estimated integrated hazards in the different strata are plotted against each other. Figure VII.3.4 for the stratification by sex shows approximately a straight line with slope ≈ 1.6 close to the estimate $\exp(\hat{\beta}_1)$ in Example VII.2.9. But both in Figure VII.3.5 and Figure VII.3.6, we get concave looking curves indicating that the hazard ratios decrease with time. We return to *tests* for proportional hazards based on these data in Example VII.3.4. □

VII.3.2. Tests for Log-Linearity

In this subsection, we turn to a brief discussion of the simpler problem of examining the assumption (7.3.3) of log-linearity once the proportional hazards assumption (7.3.2) has been investigated. We, therefore, assume that a *score* x_h of Z_p can be connected to each stratum, $h = 1, \ldots, k$, and it is then a question of whether the estimates $\hat{\beta}_{p+h}$ corresponding to the stratum indicators (where $\hat{\beta}_{p+k} \equiv 0$) are compatible with the hypothesis

$$\beta_{p+h} = \beta_p x_h, \quad h = 1, \ldots, k. \tag{7.3.7}$$

There are several ways of examining the hypothesis (7.3.7). The estimates $\hat{\beta}_{p+h}$

may be plotted against x_h, $h = 1, \ldots, k$, to see whether a straight line is obtained. Alternatively, it can be tested whether $\beta_{p+h} = 0, h = 1, \ldots, k - 1$, in a model also including the covariate Z_p. Tests for hypothesis (7.3.7) against other alternatives were considered by Thomsen (1988) and we shall now illustrate these techniques using the melanoma data.

EXAMPLE VII.3.2. Survival with Malignant Melanoma. Tests for Log-Linearity of the Effect of Tumor Thickness

The data were introduced in Example I.3.1. In all previous examples in the present chapter using these data, it has been assumed that tumor thickness has a linear effect on the log hazard function. To investigate this hypothesis, the patients are once more stratified into three groups using the cutpoints 2 mm and 5 mm for thickness. Considering a model also including sex ($Z_{i1} = 1$ for males and 0 for females) and ulceration ($Z_{i3} = 1$ if present and 0 if absent), we get the estimates (where Z_{i2} is the tumor thickness in millimeters):

Covariate	$\hat{\beta}$	$\widehat{\text{s.d.}}(\hat{\beta})$
Sex (Z_{i1})	0.356	0.272
Ulceration (Z_{i3})	0.958	0.324
$I(2 \leq Z_{i2} < 5$ mm)	1.113	0.361
$I(Z_{i2} \geq 5$ mm)	1.129	0.418

The max log partial likelihood is -261.31 and it is seen that the relation between the estimates

$$0, \qquad 1.113, \qquad 1.129$$

and the mean tumor thickness in the three groups:

$$1.05 \text{ mm}, \qquad 3.34 \text{ mm}, \qquad 8.48 \text{ mm}$$

does not correspond very well to (7.3.7). The inappropriateness of scoring the tumor thickness simply using Z_{i2} is also seen by adding Z_{i2} to this model:

Covariate	$\hat{\beta}$	$\widehat{\text{s.d.}}(\hat{\beta})$
Sex (Z_{i1})	0.399	0.273
Thickness (Z_{i2})	0.127	0.072
Ulceration (Z_{i3})	0.898	0.326
$I(2 \leq Z_{i2} < 5$ mm)	0.847	0.394
$I(Z_{i2} \geq 5$ mm)	0.226	0.699

Here the max log partial likelihood is -259.97 and the partial likelihood ratio test statistic for elimination of the last two covariates is 7.08 (2 d.f.) corresponding to $P = 0.03$.

Another way of illustrating the score of tumor thickness would be to add to the model discussed in Example VII.2.9 (including sex, thickness, and ulceration) variables of the form

$$(Z_{i2} - x_j)I(Z_{i2} \geq x_j)$$

for some cutpoints $x_1, x_2, \ldots$ (Thomsen, 1988). Choosing again the cutpoints $x_1 = 2$ mm and $x_2 = 5$ mm, we get the estimates:

Covariate	$\hat{\beta}$	$\widehat{\text{s.d.}}(\hat{\beta})$
Sex (Z_{i1})	0.457	0.289
Thickness (Z_{i2})	1.006	0.440
Ulceration (Z_{i3})	0.884	0.326
$(Z_{i2} - 2)I(Z_{i2} \geq 2$ mm$)$	−0.968	0.530
$(Z_{i2} - 5)I(Z_{i2} \geq 5$ mm$)$	0.042	0.205

The max log partial likelihood is −260.83 and the partial likelihood ratio test statistic for the reduction to the model in Example VII.2.9 takes the value 5.36, which compared to a χ_2^2 distribution gives $P = 0.07$. The estimates indicate a strong log-linear effect for tumors thinner than 2 mm and a much smaller effect for thicker tumors, i.e., the influence of thickness on the log hazard seems to be a *concave* function. Replacing, therefore, Z_{i2} by Z_{i4}, defined as log(tumor thickness for patient i), we get the estimates:

Covariate	$\hat{\beta}$	$\widehat{\text{s.d.}}(\hat{\beta})$	$\hat{\beta}$	$\widehat{\text{s.d.}}(\hat{\beta})$
Sex (Z_{i1})	0.401	0.285	0.381	0.271
Ulceration (Z_{i2})	0.884	0.328	0.939	0.324
log (thickness) (Z_{i4})	0.965	0.525	0.576	0.179
$(Z_{i4} - \log(2))I(Z_{i4} \geq \log(2))$	−0.536	0.894	—	—
$(Z_{i4} - \log(5))I(Z_{i4} \geq \log(5))$	−0.077	1.029	—	—

The max log partial likelihood is −261.44 and the partial likelihood ratio test statistic for the reduction to the model assuming a log-linear effect on the hazard of log(thickness) is 0.86, corresponding to $P = 0.65$.

The latter model (with estimates shown in the right hand side of the table) is, therefore, preferable to the model with a linear effect of thickness. It can finally be noted that the inclusion of the covariates $I(2 \leq Z_{i2} < 5$ mm$)$ and $I(Z_{i2} \geq 5$ mm$)$ does not significantly improve the fit of this last model (partial likelihood ratio test statistic 3.22, 2 d.f., $P = 0.20$). □

Another way of modelling the effect of a continuous covariate is by means of *threshold values*. Testing the significance of a single, *fixed* threshold at some value γ simply amounts to including the covariate $I(Z_{ip} \leq \gamma)$ in the model. In the univariate case $p = 1$, Jespersen (1986) considered the problem of testing against a single, *unknown* breakpoint, and, in the following example, we shall briefly review his results.

EXAMPLE VII.3.3. Dichotomizing a Single Continuous Covariate. Effect of Tumor Thickness on Survival with Malignant Melanoma

Consider the following model for the intensity process:

$$\lambda_i(t) = Y_i(t)\alpha_0(t)\exp\{\beta I(Z_i \le \gamma)\}$$

with a univariate continuous covariate Z_i, $i = 1, \ldots, n$. For a fixed breakpoint γ, the regression parameter β may be estimated by the value $\hat{\beta}_\gamma$ maximizing

$$C_\tau(\beta, \gamma) = \beta \sum_{i=1}^{n} N_i(\tau)I(Z_i \le \gamma) - \int_0^\tau \log\left(\sum_{i=1}^{n} Y_i(t)\exp\{\beta I(Z_i \le \gamma)\}\right)dN_\cdot(t)$$

[cf. (7.2.7)] and γ can then be estimated by the value $\hat{\gamma}$ maximizing the function $L(\hat{\beta}_\gamma, \gamma)$ [see (7.2.6)] which is piecewise constant between the observed covariate values. To test whether Z_i has no effect against a model with a breakpoint at $Z = \hat{\gamma}$ amounts to testing the hypothesis H_0: $\beta = 0$. Jespersen (1986) studied the score test statistic which for fixed γ is

$$S_\gamma = \frac{\partial}{\partial\beta} C_\tau(0, \gamma) = \sum_{i=1}^{n} N_i(\tau)I(Z_i \le \gamma) - \int_0^\tau \frac{\sum_{i=1}^{n} Y_i(t)I(Z_i \le \gamma)}{Y_\cdot(t)}\, dN_\cdot(t),$$

see (7.2.16), and suggested using $\sup_\gamma |S_\gamma|$ (properly normalized) for testing H_0. Because the parameter γ is only present in the model under the alternative hypothesis $\beta \neq 0$, the large sample distributional properties of the score statistic under H_0 do not follow from the results of Section VII.2.2.

For right-censored survival data, Jespersen (1986) showed (among other things) that, in an i.i.d. case (Example VII.2.7) with G being the distribution of Z_i, the process

$$(N_\cdot(\tau)^{-1/2} S_{G^{-1}(s)}; s \in [0, 1])$$

converges weakly in the space $D[0, 1]$ to the standard Brownian bridge $W^0(\cdot)$ under H_0 provided that $\mathrm{P}(\tilde{X}_i > \tau) > 0$ and that the distribution of censoring times $U_1, \ldots, U_n$ has at most a finite number of discontinuities.

Thus, the significance of the "best" breakpoint $\hat{\gamma}$ can be tested referring the value of $N_\cdot(\tau)^{-1/2} \sup_\gamma |S_\gamma|$ to the distribution of $\sup_{0 \le s \le 1} |W^0(s)|$ (Koziol and Byar, 1975; Hall and Wellner, 1980, cf. also Section IV.1.3).

As an example, we once more consider the melanoma data introduced in Example I.3.1 and the covariate tumor thickness. Figure VII.3.7 shows the "profile" log-likelihood $C_\tau(\hat{\beta}_\gamma, \gamma)$ as a function of the threshold value γ for tumor thickness and it is seen that the "best" breakpoint is at $\hat{\gamma} = 2.1$ mm. Figure VII.3.8 shows $N_\cdot(\tau)^{-1}(S_\gamma)^2$ as a function of γ. The curve is similar in shape to Figure VII.3.7 and again the maximum value is obtained for $\hat{\gamma} = 2.1$ mm. The maximum value is $(20.84)^2/57 = (2.76)^2$ and corresponds to a P

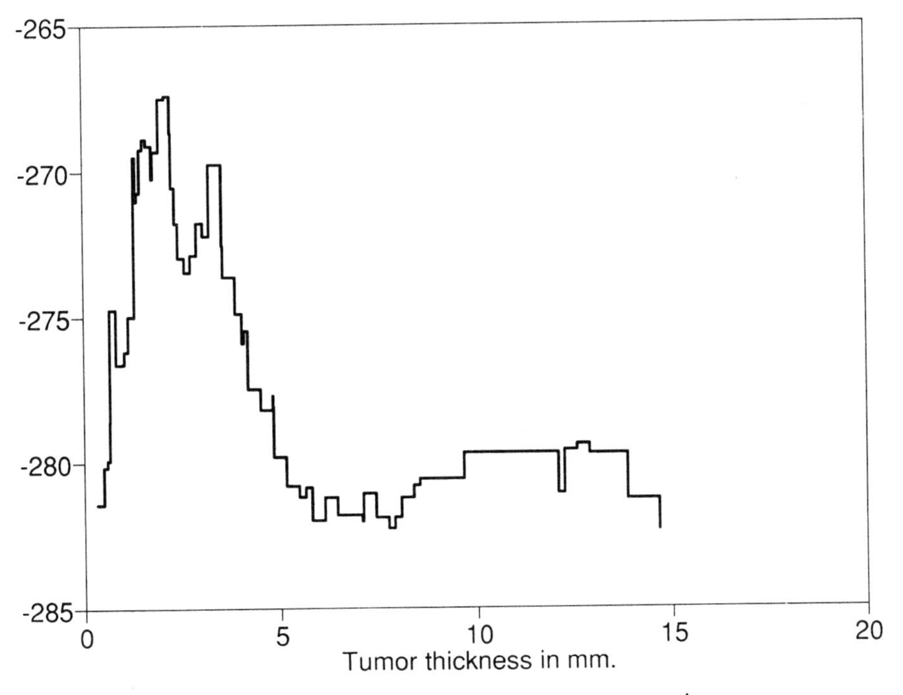

Figure VII.3.7. "Profile" log partial likelihood $C_\tau(\hat{\beta}_\gamma, \gamma)$.

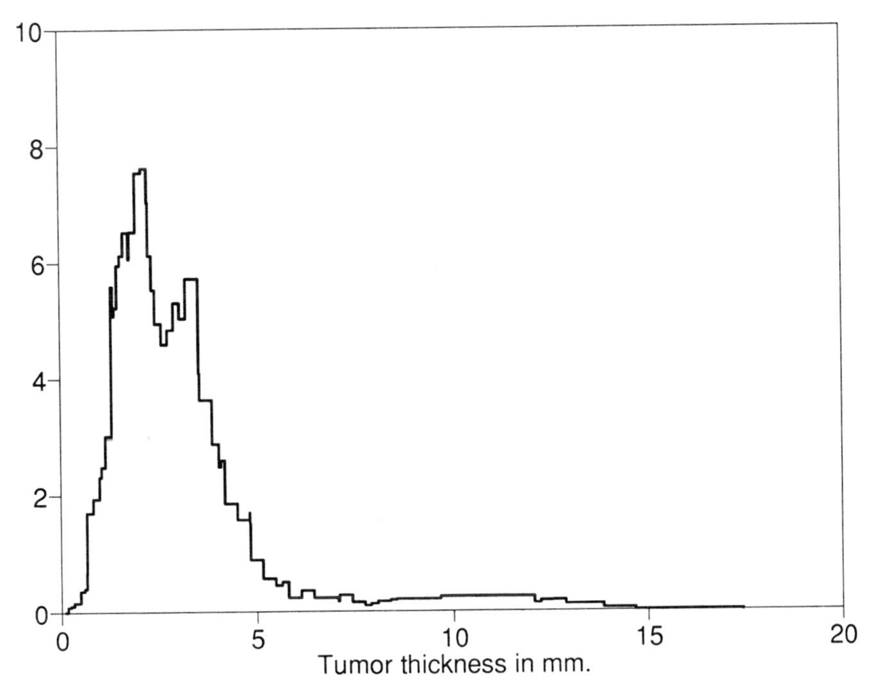

Figure VII.3.8. Normalized score test statistic $N_{\cdot}(\tau)^{-1}(S_\gamma)^2$.

value less than 0.001, reflecting the strong prognostic value of tumor thickness. (The upper 0.001 fractile in the distribution of $\sup_{0 \le s \le 1} |W^0(s)|$ is 1.95.) □

VII.3.3 Tests for Proportional Hazards

We shall now discuss *tests* for the proportional hazards assumption (7.3.2). For tests against specific alternatives, Cox (1972) suggested including some deterministic time-dependent covariate in the model (7.3.5) or in the corresponding model with the stratum indicators $Z_{p+1}, \ldots, Z_{p+k-1}$ replaced by another suitably scored version of Z_p. Thus, including, for instance, the covariate $Z_{ip}g(t)$ for some fixed function of time such as $g(t) = t$, $g(t) = \log t$ or $g(t) = I(t \le t_0)$ gives a test for particular deviations from the proportional hazards assumption. We shall illustrate it using the melanoma data.

EXAMPLE VII.3.4. Survival with Malignant Melanoma. Tests for Proportional Hazards Using Time-Dependent Covariates

The data were introduced in Example I.3.1. As starting point, we take the model from Example VII.3.2 including sex (Z_{i1}), ulceration (Z_{i3}), and log(tumor thickness) (Z_{i4}). To that model we add, in turn, the time-dependent covariates

1. $Z_{i5}(t) = Z_{i1}(\log(t) - 7)$,
2. $Z_{i6}(t) = Z_{i3}(\log(t) - 7)$,
3. $Z_{i7}(t) = I(2 \le Z_{i2} < 5 \text{ mm})(\log(t) - 7)$,
   $Z_{i8}(t) = I(Z_{i2} \ge 5 \text{ mm})(\log(t) - 7)$,

where t is in days and $7 \simeq \log(3 \cdot 365)$ is subtracted from $\log(t)$ to avoid too large numerical covariate values [cf. Cox (1972)]. In the three models, we get the following estimates:

Covariate	$\hat\beta$ ($\widehat{\text{s.d.}}(\hat\beta)$)	$\hat\beta$ ($\widehat{\text{s.d.}}(\hat\beta)$)	$\hat\beta$ ($\widehat{\text{s.d.}}(\hat\beta)$)
Sex (Z_{i1})	0.352 (0.276)	0.372 (0.270)	0.419 (0.271)
Ulceration (Z_{i3})	0.932 (0.324)	1.048 (0.360)	0.960 (0.326)
log(thickness)(Z_{i4})	0.582 (0.180)	0.576 (0.181)	0.547 (0.191)
$Z_{i5}(t)$	−0.408 (0.394)		
$Z_{i6}(t)$		−1.189 (0.589)	
$Z_{i7}(t)$			−0.677 (0.594)
$Z_{i8}(t)$			−1.513 (0.600)

The partial likelihood ratio test statistics for proportional hazards are

1. sex: 1.12, 1 d.f., $P = 0.29$,
2. ulceration: 5.20, 1 d.f., $P = 0.02$,
3. thickness: 8.28, 2 d.f., $P = 0.02$,

confirming what we saw on the plots in Example VII.3.1. Thus, the proportional hazards assumption for ulceration and/or thickness is unjustified. Considering, however, a model *stratified* by ulceration and including sex, log(thickness), $Z_{i7}(t)$, and $Z_{i8}(t)$; the effect of the latter two is no longer significant at the 5% level ($P = 0.06$). The estimates are

Covariate	$\hat{\beta}$ ($\widehat{\text{s.d.}}(\hat{\beta})$)	$\hat{\beta}$ ($\widehat{\text{s.d.}}(\hat{\beta})$)	$\hat{\beta}$ ($\widehat{\text{s.d.}}(\hat{\beta})$)
Sex (Z_{i1})	0.402 (0.271)	0.360 (0.270)	
log(thickness)(Z_{i4})	0.538 (0.191)	0.560 (0.178)	0.589 (0.175)
$Z_{i7}(t)$	−0.367 (0.663)		
$Z_{i8}(t)$	−1.201 (0.636)		

In the stratified model including log(thickness) and sex, the effect of the latter is insignificant ($P = 0.18$) and, eliminating sex from the model, the estimated effect of log(thickness) is 0.589 (0.175). In this model, none of the other covariates mentioned in Example I.3.1 are significant and, therefore, a "final model" for the intensity of dying from malignant melanoma has been obtained. The estimated integrated hazards in the two strata are shown in Figure VII.3.9. □

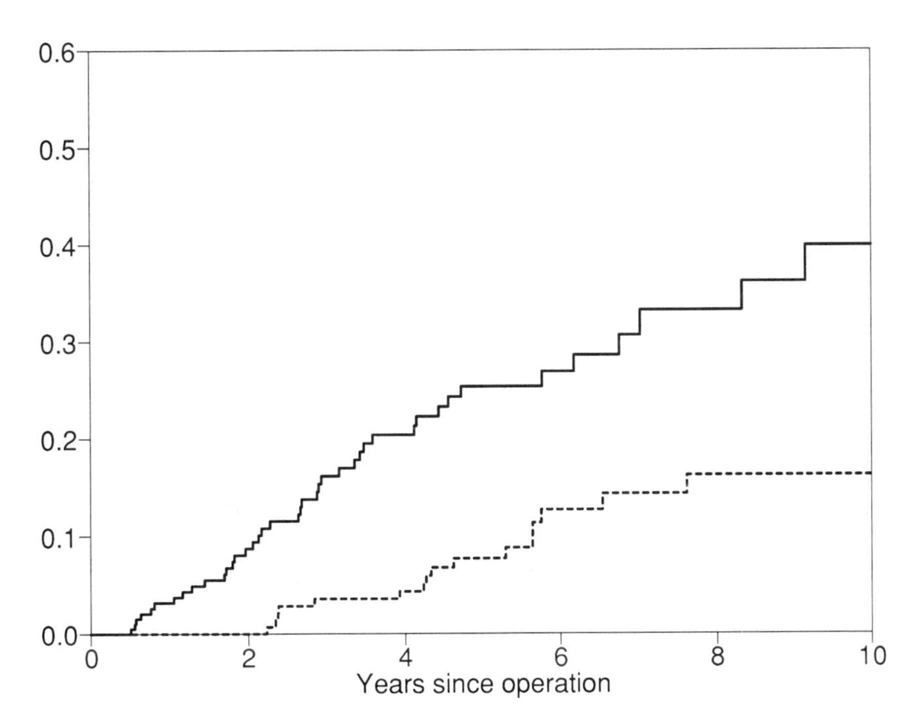

Figure VII.3.9. Estimated integrated baseline hazards for patients without (---) and with (—) ulceration in the "final" model for the melanoma data.

This class of tests can of course also be used in the simple two-sample model (Example VII.2.6) where $p = 1$ and Z_1 is an indicator variable. In this simple model, alternative test statistics against less specific alternatives are available as demonstrated in the following example.

EXAMPLE VII.3.5. Goodness-of-Fit Methods for the Two-Sample Proportional Hazards Model. Survival with Malignant Melanoma

Consider the two-sample proportional hazards model discussed in Example VII.2.6 and denote by $\theta = \exp(\beta) = \alpha_1(t)/\alpha_0(t)$, the true ratio between the hazards in group 1 and group 0. Following the idea of constructing nonparametric two-sample tests (Section V.2), one may base tests for hypothesis H_0 of proportional hazards on processes of the form

$$X(t) = \int_0^t L_1(s)(\mathrm{d}\tilde{A}_1(s) - \tilde{\theta}\,\mathrm{d}\tilde{A}_0(s)), \quad t \in \mathscr{T}.$$

Here, $L_1(\cdot)$ is a (predictable) weight process and $\tilde{A}_1(\cdot)$, $\tilde{A}_0(\cdot)$, and $\tilde{\theta}$ are estimates for the cumulative hazards $A_1(\cdot)$ and $A_0(\cdot)$ and their ratio θ (Andersen, 1983b). As one special case of this, we may let $\hat{A}_h(t)$, $h = 0, 1$, be the Nelson–Aalen estimators, and

$$\tilde{\theta} = \hat{\theta}_{L_2} = \frac{\int_0^\tau L_2(t)\,\mathrm{d}\hat{A}_1(t)}{\int_0^\tau L_2(t)\,\mathrm{d}\hat{A}_0(t)},$$

an estimator of the kind discussed in Section V.3.1. Then the resulting test statistic $X(\tau)$ (properly normalized by a standard deviation estimate under H_0) is identical to the test statistic discussed by Gill and Schumacher (1987). Thus, Gill and Schumacher defined

$$\hat{K}_{jh} = \int_0^\tau L_j(t)\,\mathrm{d}\hat{A}_h(t), \quad j = 1, 2, h = 0, 1$$

(i.e., $\hat{\theta}_{L_2} = \hat{K}_{21}/\hat{K}_{20}$) and

$$\hat{V}_{jj'} = \int_0^\tau L_j(t)L_{j'}(t)\{Y_1(t)Y_0(t)\}^{-1}\,\mathrm{d}N_{\cdot}(t), \quad j, j' = 1, 2,$$

and showed that the distribution of the statistic

$$\frac{\hat{K}_{10}\hat{K}_{21} - \hat{K}_{20}\hat{K}_{11}}{\sqrt{\hat{K}_{20}\hat{K}_{21}\hat{V}_{11} - \hat{K}_{20}\hat{K}_{11}\hat{V}_{12} - \hat{K}_{10}\hat{K}_{21}\hat{V}_{21} + \hat{K}_{10}\hat{K}_{11}\hat{V}_{22}}}$$

[which is just a normalized version of $X(\tau)$ under H_0] is asymptotically standard normal. Choosing $L_1(t)$ and $L_2(t)$ such that $L_1(t)/L_2(t)$ is *monotone*, they also showed that the test statistic is consistent again the alternative that the ratio $\alpha_1(t)/\alpha_0(t)$ is *monotone*. This is, for instance, satisfied in the case of right-censored survival data if $L_1 = Y_1 Y_0$ (corresponding to the Gehan–Wilcoxon test) and $L_2 = Y_1 Y_0 Y_{\cdot}^{-1}$ (corresponding to the log-rank test); cf. (5.2.9) and Example V.2.1.

Lin (1991) studied an extension of the Gill–Schumacher test to the general Cox regression model for survival data.

If one, instead, chooses $L_1(t) = L_2(t)$ to be

$$L(t) = \tilde{\theta}^{-1} \frac{Y_1(t) Y_0(t)}{Y_0(t) + \tilde{\theta} Y_1(t)},$$

where $\tilde{\theta}$ is some (preliminary) consistent estimate [i.e., $\hat{\theta}_L$ is the *two-step* estimator introduced by Begun and Reid (1983), see also Section V.3.1], it can be seen that the asymptotic distribution of

$$n^{-1/2} \left\{ n^{-1} \int_0^\tau L^2(u) \left(\frac{dN_1(u)}{Y_1^2(u)} + \hat{\theta}_L^2 \frac{dN_0(u)}{Y_0^2(u)} \right) \right\}^{-1/2} X(\cdot)$$

is that of a time-transformed Brownian bridge $W^0(h(\cdot)/h(\tau))$. Here, $W^0(\cdot)$ is the standard Brownian bridge on [0, 1] and

$$h(t) = \theta^{-1} \int_0^t \frac{y_1(u) y_0(u)}{y_0(u) + \theta y_1(u)} \alpha_0(u) \, du$$

is the limit in probability of

$$n^{-1} \int_0^t L^2(u) \left(\frac{dN_1(u)}{Y_1^2(u)} + \hat{\theta}_L^2 \frac{dN_0(u)}{Y_0^2(u)} \right).$$

Thus, the supremum over $t \in \mathcal{T}$ of the absolute value of this statistic can be referred to the distribution of $\sup_{t \in [0,1]} |W^0(t)|$.

The same limiting distribution was obtained by Wei (1984) for the statistic

$$\sup_{t \in \mathcal{T}} I_\tau(\hat{\beta})^{-1/2} |U_t(\hat{\beta})|$$
$$= \frac{\sup_{t \in \mathcal{T}} |N_1(t) - \int_0^t \exp(\hat{\beta}) Y_1(u)/[Y_0(u) + \exp(\hat{\beta}) Y_1(u)] \, dN_.(u)|}{\sqrt{\int_0^\tau \exp(\hat{\beta}) Y_1(u) Y_0(u)/(Y_0(u) + \exp(\hat{\beta}) Y_1(u))^2 \, dN_.(u)}}$$

[cf. (7.2.16) and (7.2.17)], the latter requiring the additional (minor) computational effort of calculating $\hat{\beta}$, whereas no iterative procedure is needed for the former. Wei showed that the test is consistent against the general alternative that $\alpha_1(t)$ and $\alpha_0(t)$ are not proportional. Since

$$U_t(\hat{\beta}) = \int_0^t Y_1(u) \left\{ \frac{dN_1(u)}{Y_1(u)} - \exp(\hat{\beta}) \, d\hat{A}_0(u, \hat{\beta}) \right\},$$

the test statistic of Wei can also be written in the general form considered above.

For the melanoma data introduced in Example I.3.1, we found in Example V.3.1 that the hazard ratio estimate for males versus females corresponding to the log-rank test was $\hat{\theta}_{L_2} = 1.95$. Similar calculations for the Gehan–Wilcoxon test yields $\hat{\theta}_{L_1} = 2.10$ and the Gill–Schumacher test statistic for proportional hazards takes the value -0.81, giving $P = 0.42$ when referred to a standard normal distribution. For ulceration, we find $\hat{\theta}_{L_2} = 4.33$ and $\hat{\theta}_{L_1} = 5.27$ and the test statistic becomes -2.33 ($P = 0.02$). These results are in accordance with those of Example VII.3.4 where we also accepted the proportional hazards hypothesis for sex but not for ulceration.

For Wei's suggestion, the test statistic for sex becomes 0.94 ($P = 0.34$) and that for ulceration 1.19 ($P = 0.11$). In the case of ulceration where a reasonable alternative is a monotone hazard ratio, the more omnibus type test of Wei does not detect departures from proportionality. □

Even though the test of Wei (1984) discussed in Example VII.3.5 works for any *one-dimensional covariate* Z_i [e.g., for the melanoma data, we get a test statistic of 0.82 ($P = 0.50$) corresponding to tumor thickness], similar methods for the general p-variate proportional hazards model are not directly available. Haara (1987) showed that under the conditions of Theorem VII.2.2, the process

$$n^{-1/2}\mathbf{U}_t(\hat{\boldsymbol{\beta}})$$

converges weakly on $[0, \tau]$ to a zero-mean p-variate Gaussian process $\mathbf{U}(\cdot)$ with covariance

$$\operatorname{cov}(\mathbf{U}(s), \mathbf{U}(t)) = \boldsymbol{\Sigma}_s - \boldsymbol{\Sigma}_s\boldsymbol{\Sigma}_\tau^{-1}\boldsymbol{\Sigma}_t, \quad s \leq t;$$

cf. Condition VII.2.1(e). This result which obviously contains the result of Wei (1984) as a special case is, however, not directly applicable for testing the goodness-of-fit of the Cox regression model as properties of the limiting process $\mathbf{U}(\cdot)$ are not well-known. Some rather informal attempts in this direction were made by Nagelkerke et al. (1984) and also the so-called "innovations approach" of Khmaladze (1981, 1988), [cf. Section VI.3.3], may be applied. The latter approach amounts to considering the empirical version of the limiting process $\mathbf{U}(\cdot)$ minus its compensator; say,

$$\tilde{\mathbf{U}}(t) = -\int_0^t \mathrm{d}\boldsymbol{\Sigma}_s(\boldsymbol{\Sigma}_\tau - \boldsymbol{\Sigma}_s)^{-1}\mathbf{U}(s)$$

since $\mathbf{W} = \mathbf{U} - \tilde{\mathbf{U}}$ is a Gaussian martingale with $\operatorname{cov}(\mathbf{W}(s), \mathbf{W}(t)) = \boldsymbol{\Sigma}_s$, $s \leq t$. Therefore, in large samples, the components of

$$\hat{\mathbf{W}}(t) = n^{-1/2}\left\{\mathbf{U}_t(\hat{\boldsymbol{\beta}}) + \int_0^t \mathrm{d}\mathscr{I}_s(\hat{\boldsymbol{\beta}})(\mathscr{I}_\tau(\hat{\boldsymbol{\beta}}) - \mathscr{I}_s(\hat{\boldsymbol{\beta}}))^{-1}\mathbf{U}_s(\hat{\boldsymbol{\beta}})\right\}$$

behave like Gaussian martingales with variances and covariances given by the relevant elements of $\boldsymbol{\Sigma}_t$ which we may estimate consistently by $n^{-1}\mathscr{I}_t(\hat{\boldsymbol{\beta}})$. Since, however, the components are dependent, it may be less obvious which functionals of $\hat{\mathbf{W}}(t)$ to use as test statistics and we shall not pursue this approach any further here.

VII.3.4. Methods Based on Residuals

In classical *linear normal models* residuals, defined as the difference between observations and their estimated expected values, are often useful when assessing goodness-of-fit of a given model. For the class of *generalized linear models*, wider classes of residuals were discussed by McCullagh and Nelder (1989, Section 2.4). For the models treated in the present section, residuals can be defined via the basic counting process decomposition where

$$M_i(t) = N_i(t) - \int_0^t Y_i(u)\exp(\boldsymbol{\beta}_0^{\mathsf{T}}\mathbf{Z}_i)\alpha_0(u)\,\mathrm{d}u,\ i = 1, \ldots, n,$$

are orthogonal local square integrable martingales. Inserting the estimated parameter values into the compensator, we get the *residuals*

$$\begin{aligned}\widehat{M}_i(t) &= N_i(t) - \int_0^t Y_i(u)\exp(\hat{\boldsymbol{\beta}}^{\mathsf{T}}\mathbf{Z}_i)\,\mathrm{d}\hat{A}_0(u,\hat{\boldsymbol{\beta}})\\ &= N_i(t) - \int_0^t p_i(u,\hat{\boldsymbol{\beta}})\,\mathrm{d}N_{\bullet}(u), \quad i = 1, \ldots, n,\end{aligned} \tag{7.3.8}$$

with $p_i(u,\hat{\boldsymbol{\beta}}) = Y_i(u)\exp(\hat{\boldsymbol{\beta}}^{\mathsf{T}}\mathbf{Z}_i)/S^{(0)}(\hat{\boldsymbol{\beta}}, u)$; cf. (7.2.10). We, furthermore, define $\hat{r}_i = N_i(\tau) - \widehat{M}_i(\tau)$.

For right-censored survival data, we get, for $t = \tau$,

$$\widehat{M}_i(\tau) = D_i - \exp(\hat{\boldsymbol{\beta}}^{\mathsf{T}}\mathbf{Z}_i)\hat{A}_0(\tilde{X}_i, \hat{\boldsymbol{\beta}}),$$

i.e., $\hat{r}_i = D_i - \widehat{M}_i(\tau)$ is the residual in the sense of Cox and Snell (1968) further discussed by Kay (1977). Kay then noted that in the uncensored case the true values $r_i = \exp(\boldsymbol{\beta}_0^{\mathsf{T}}\mathbf{Z}_i)A_0(\tilde{X}_i)$, $i = 1, \ldots, n$, constitute a sample of n independent unit exponential variables (Section VI.3.1). If some of the $\tilde{X}_i$ are right-censored, one may treat the corresponding residuals $\hat{r}_i$ as right-censored and use the methods discussed in Section VI.3 for checking the (approximate) exponentiality of $\hat{r}_1, \ldots, \hat{r}_n$. Thus, Kay (1977) [see also Crowley and Hu (1977)] suggested using the *Nelson–Aalen plot*—possibly within a number of strata. It is, however, not obvious how to interpret departures from linearity of this plot (Crowley and Storer, 1983) and also the closeness of the distribution of $\hat{r}_1, \ldots, \hat{r}_n$ to the unit exponential distribution in small samples is unclear (Lagakos, 1981).

Another graphical procedure based on the residuals (7.3.8) was suggested by Arjas (1988). It is based on the fact that under the model (7.3.5) with proportional hazards both within and between k strata, the differences

$$N_h(t) - \int_0^t \sum_{h(i)=h} p_i(u,\boldsymbol{\beta}_0)\,\mathrm{d}N_{\bullet}(u), \quad h = 1, \ldots, k, \tag{7.3.9}$$

where $N_h(t) = \sum_{h(i)=h} N_i(t)$ and $\boldsymbol{\beta}_0 = (\beta_1^0, \ldots, \beta_{p-1}^0, \beta_{p+1}^0, \ldots, \beta_{p+k-1}^0)$ is the true parameter vector, are (local) martingales. Therefore, plots of

$$\int_0^{X^h_{(m)}} \sum_{h(i)=h} p_i(u, \hat{\boldsymbol{\beta}}) \, \mathrm{d}N_{\bullet}(u), \quad m = 1, \ldots, N_h(\tau), h = 1, \ldots, k, \tag{7.3.10}$$

versus m, where $X^h_{(m)}$, $m = 1, \ldots, N_h(\tau)$, are the ordered jump times in stratum h, should be approximately straight lines with unit slope. This is because the martingale property of (7.3.9) implies that

$$\mathrm{E}N_h(t) = \mathrm{E} \int_0^t \sum_{h(i)=h} p_i(u, \boldsymbol{\beta}_0) \, \mathrm{d}N_{\bullet}(u),$$

and, inserting $t = X^h_{(m)}$ such that $N_h(t) = m$, it is seen that

$$\int_0^{X^h_{(m)}} \sum_{h(i)=h} p_i(u, \hat{\boldsymbol{\beta}}) \, \mathrm{d}N_{\bullet}(u) \approx m.$$

This plot is a Total Time on Test plot (cf. Section VI.3.2) for the residuals $\hat{r}_i$. Arjas (1988) noted that if the strata are ordered so that the ratios $\alpha_{h0}(t)/\alpha_{h'0}(t)$ are increasing in t when $h < h'$, then the curve for stratum 1 is likely to be concave with the "expected" failure counts (7.3.10) being first larger than the observed and then smaller. For stratum k, the situation is reversed: The curve is likely to be convex with "expected" failure counts (7.3.10) being first too small and then too large compared to the observed failure counts.

Tests for the proportional hazards model based on these residuals were briefly discussed by Arjas (1988) and Arjas and Haara (1988b).

EXAMPLE VII.3.6. Survival with Malignant Melanoma. Residual Plots

The data were introduced in Example I.3.1. In Examples VII.3.4 and VII.3.5, it was seen that the proportional hazards assumption was not reasonable for the covariate ulceration and an alternative model stratified by that variable seemed more appropriate. Figure VII.3.10 shows the Nelson–Aalen plot partitioned by ulceration for the $\hat{r}_i$ from the model assuming proportional hazards for ulceration. The plot certainly suggests a poor fit of that model, both curves deviating rather clearly from the line through the origin with unit slope. Figure VII.3.11, showing the equivalent plot for the model *stratified* by ulceration, is much nicer (disregarding the last parts of the curves). The example shows that this plotting procedure does have some power to detect departures from a given model. What will be difficult, however, is to imagine what kind of departures from the model would lead to curves like those in Figure VII.3.10.

This seems to be easier based on Figure VII.3.12 showing the plot suggested by Arjas (1988) of the "expected failure counts" (7.3.10) against the observed for the model assuming proportional hazards for ulceration. The figure certainly suggests a *monotone* hazard ratio between the two strata because we have a concave curve for the "no ulceration" stratum and a convex curve for

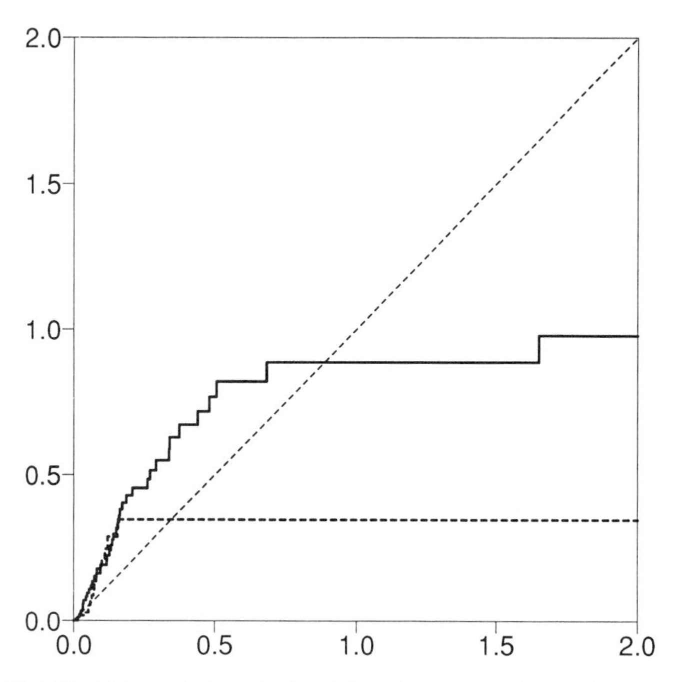

Figure VII.3.10. Nelson–Aalen plot for $\hat{r}_i$ based on a model assuming proportional hazards for ulceration. (- - -): no ulceration; (—): ulceration.

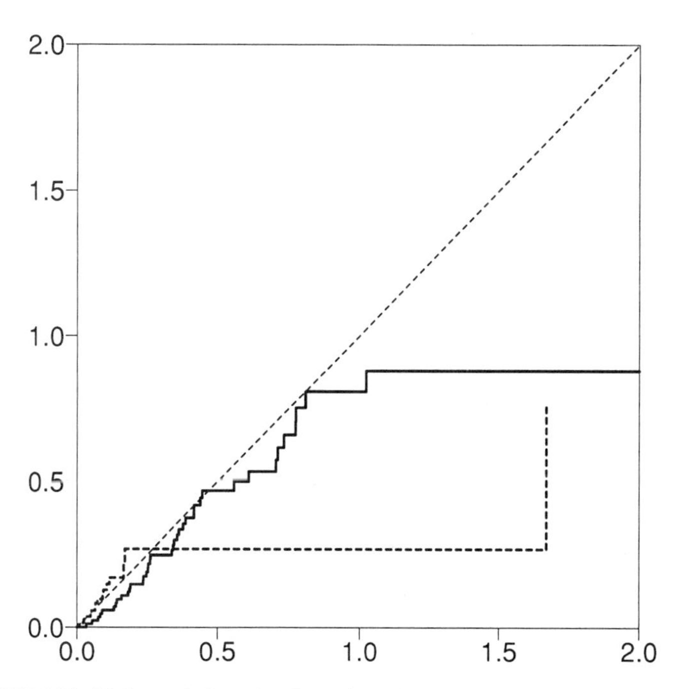

Figure VII.3.11. Nelson–Aalen plot for $\hat{r}_i$ based on a model stratified by ulceration. (- - -): no ulceration; (—): ulceration.

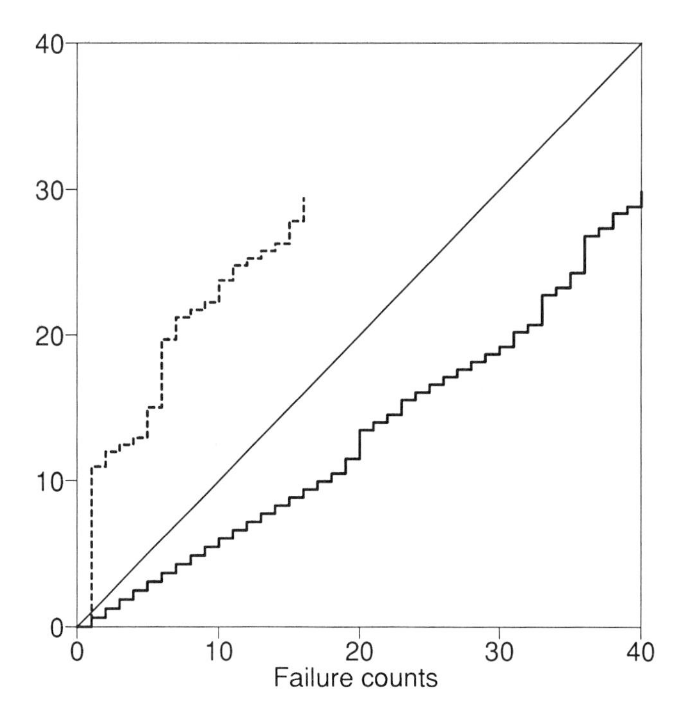

Figure VII.3.12. Expected failure counts (7.3.10) plotted against the observed for the model assuming proportional hazards for ulceration. (---): no ulceration; (—): ulceration.

the "ulceration" stratum. This conclusion is in accordance with the results of Examples VII.3.1 and VII.3.4. □

Since M_i, $i = 1, \ldots, n$, defined above (7.3.8) are local martingales, the same will hold for $\int_0^t \mathbf{K}_i(s)\,\mathrm{d}M_i(s)$ for predictable processes $\mathbf{K}_i(\cdot)$, $i = 1, \ldots, n$. Here, $\mathbf{K}_i$ may be a vector, in which case the integration is componentwise in the obvious way. Thus, a broader class of "residuals" can be defined as

$$\int_0^t \mathbf{K}_i(s)\,\mathrm{d}\hat{M}_i(s) = \int_0^t \mathbf{K}_i(s)\,\mathrm{d}N_i(s) - \int_0^t \mathbf{K}_i(s)p_i(s, \hat{\boldsymbol{\beta}})\,\mathrm{d}N_{\bullet}(s), \quad i = 1, \ldots, n; \tag{7.3.11}$$

cf. Barlow and Prentice (1988).

One possible choice is $\mathbf{K}_i(s) = \mathbf{Z}_i$. For the model (7.3.5), one may, in particular, let $\mathbf{K}_i(s)$ be the vector with components $Z_{i,p+h} = I(i$ belongs to stratum $h)$, $h = 1, \ldots, k-1$. In this case, component h of (7.3.11) is

$$\int_0^\tau Z_{i,p+h}\,\mathrm{d}N_i(t) - \int_0^\tau Z_{i,p+h}p_i(t, \hat{\boldsymbol{\beta}})\,\mathrm{d}N_{\bullet}(t),$$

and, summing over individuals in stratum h, we get

$$\int_0^\tau \mathrm{d}N_h(t) - \int_0^\tau \sum_{h(i)=h} p_i(t, \hat{\boldsymbol{\beta}})\,\mathrm{d}N_{\boldsymbol{\cdot}}(t).$$

For any partition $0 = \tau_0 < \tau_1 < \cdots < \tau_m = \tau$ of $\mathscr{T}$, this is the sum

$$\sum_{j=1}^{m} (O_{jh} - E_{jh})$$

of the observed number of deaths in $[\tau_{j-1}, \tau_j)$

$$O_{jh} = \int_{\tau_{j-1}}^{\tau_j} \mathrm{d}N_h(t)$$

minus the "expected" number of deaths in $[\tau_{j-1}, \tau_j)$

$$E_{jh} = \int_{\tau_{j-1}}^{\tau_j} \sum_{h(i)=h} p_i(t, \hat{\boldsymbol{\beta}})\,\mathrm{d}N_{\boldsymbol{\cdot}}(t)$$

in stratum h, $h = 1, \ldots, k-1$, as defined by Schoenfeld (1980a). He suggested using the statistic obtained by normalizing the vector $(O_{jh} - E_{jh}, j = 1, \ldots, m;$ $h = 1, \ldots, k-1)$ by its estimated covariance matrix as a $\chi^2_{m(k-1)}$ goodness-of-fit statistic for the model (7.3.5); cf. also Section VI.3.3. Other partitions of the range of $\mathbf{Z}$ may yield more omnibus type goodness-of-fit test statistics.

A second choice of weight to be discussed is $\mathbf{K}_i(t) = \mathbf{Z}_i - \mathbf{E}(\boldsymbol{\beta}_0, t)$ with $\mathbf{E}(\cdot, \cdot)$ defined by (7.2.14). In this case, the residual (7.3.11) inserting $\hat{\boldsymbol{\beta}}$ instead of $\boldsymbol{\beta}_0$ is, for survival data,

$$(\mathbf{Z}_i - \mathbf{E}(\hat{\boldsymbol{\beta}}, \tilde{X}_i))D_i - \int_0^\tau (\mathbf{Z}_i - \mathbf{E}(\hat{\boldsymbol{\beta}}, t))p_i(t, \hat{\boldsymbol{\beta}})\,\mathrm{d}N_{\boldsymbol{\cdot}}(t). \tag{7.3.12}$$

The first term in (7.3.12) is the vector of *partial residuals* introduced by Schoenfeld (1982), expressing for each failure time $\tilde{X}_i$ (i.e., $D_i = 1$) the difference between the covariates $\mathbf{Z}_i$ for the failing individual and the corresponding "expected value" $\mathbf{E}(\hat{\boldsymbol{\beta}}, \tilde{X}_i)$; cf. (7.2.14). The sum over all individuals of the partial residuals yields the score statistic (7.2.16) evaluated at $\hat{\boldsymbol{\beta}}$, i.e., the sum is zero. Partial sums over the ordered failure times give the score statistics evaluated at successive failure times on which the test statistic of Wei (1984) is based in the case $p = 1$; cf. Example VII.3.5.

It was shown by Cain and Lange (1984) and Reid and Crépeau (1985) that $\mathscr{I}_\tau(\hat{\boldsymbol{\beta}})^{-1}$ times (7.3.12) estimates the difference $\hat{\boldsymbol{\beta}} - \hat{\boldsymbol{\beta}}_{-i}$ between the regression parameter estimate based on the whole sample of n individuals and the estimate $\hat{\boldsymbol{\beta}}_{-i}$ obtained by deleting individual i from the sample. That this is the case can also be seen from (7.2.30) which implies that

$$\hat{\boldsymbol{\beta}} - \boldsymbol{\beta}_0 \simeq \mathscr{I}_\tau^{-1}(\boldsymbol{\beta}_0)\mathbf{U}_\tau(\boldsymbol{\beta}_0) = \mathscr{I}_\tau^{-1}(\boldsymbol{\beta}_0)\left\{\sum_{j=1}^{n} \int_0^\tau \mathbf{Z}_j\,\mathrm{d}N_j(u) - \int_0^\tau \mathbf{E}(\boldsymbol{\beta}_0, u)\,\mathrm{d}N_{\boldsymbol{\cdot}}(u)\right\}$$

and similarly (with obvious notation)

$$\hat{\boldsymbol{\beta}}_{-i} - \boldsymbol{\beta}_0 \simeq \mathscr{I}_\tau^{-1}(\boldsymbol{\beta}_0)\left\{\sum_{j\neq i}\int_0^\tau \mathbf{Z}_j\, \mathrm{d}N_j(u) - \int_0^\tau \mathbf{E}_{-i}(\boldsymbol{\beta}_0, u)\, \mathrm{d}N_{\cdot -i}(u)\right\}.$$

Therefore, by a simple calculation

$$\hat{\boldsymbol{\beta}} - \hat{\boldsymbol{\beta}}_{-i} \simeq \mathscr{I}_\tau^{-1}(\hat{\boldsymbol{\beta}})\left\{\int_0^\tau (\mathbf{Z}_i - \mathbf{E}_{-i}(\hat{\boldsymbol{\beta}},u))\, \mathrm{d}N_i(u) - \int_0^\tau (\mathbf{Z}_i - \mathbf{E}_{-i}(\hat{\boldsymbol{\beta}}, u))p_i(\hat{\boldsymbol{\beta}}, u)\, \mathrm{d}N_{\cdot}(u)\right\}$$

which is (almost) the desired equation. Other approximations to $\hat{\boldsymbol{\beta}} - \hat{\boldsymbol{\beta}}_{-i}$ were obtained by Storer and Crowley (1985) and Lustbader (1986) generalizing diagnostics from linear regression [e.g., Belsley, Kuh, and Welsch (1980); Cook and Weisberg (1982)]. Usually $(\hat{\boldsymbol{\beta}} - \hat{\boldsymbol{\beta}}_{-i})_j$; $j = 1, , \ldots, p$ (possibly suitably scaled by the estimated standard error of $\hat{\beta}_j$) is plotted against the case indicator i, against Z_{ij} or (for survival data) against $\tilde{X}_i$ to get a visual impression of how each individual affects $\hat{\boldsymbol{\beta}}$, as we shall see in the following example.

Example VII.3.7. Survival with Malignant Melanoma. Regression Diagnostics

This example illustrates the use of regression diagnostics for the melanoma data introduced in Example I.3.1. We shall consider the "final" model from

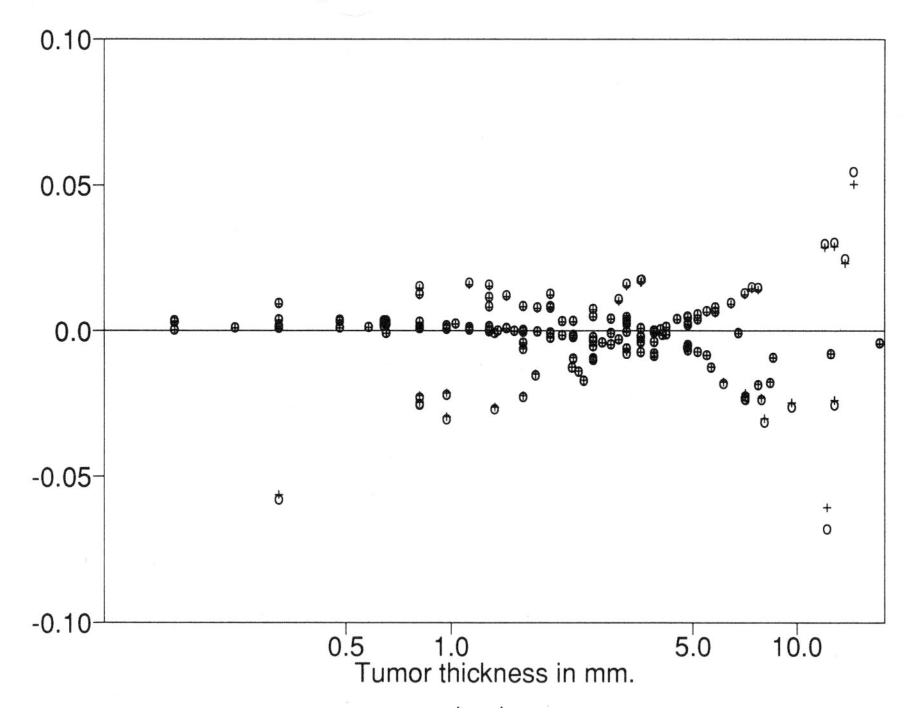

Figure VII.3.13. Regression diagnostic $\hat{\beta} - \hat{\beta}_{-i}$ (o) and approximation based on (7.3.12) (+) for tumor thickness Z_i plotted against Z_i.

Example VII.3.4 stratified by ulceration and including log(thickness) as the only covariate. We study how each of the patients influences the estimate $\hat{\beta}$ for this covariate, Z_i. The estimate based on all patients is 0.589 (0.175) (Example VII.3.4). For each patient ($i = 1, \ldots, 205$), $\hat{\beta}_{-i}$ was calculated directly by fitting the model of the above type for the remaining 204 patients. Furthermore, $\hat{\beta} - \hat{\beta}_{-i}$ was approximated using (7.3.12). Figure VII.3.13 shows both the "exact" $\hat{\beta} - \hat{\beta}_{-i}$ and the approximation plotted against the covariate Z_i. We, first, note the close agreement between the exact values and the approximation. However, the latter is systematically numerically smaller with a bias that seems to increase with the absolute values of $\hat{\beta} - \hat{\beta}_{-i}$. Figure VII.3.14 shows the approximation based on (7.3.12) plotted against the survival/censoring time $\tilde{X}_i$ using different symbols for censored and uncensored observations. In both figures, three outliers seem identifiable: One with a very large value of tumor thickness (14.66 mm) and a relatively short observation time (a failure at 2.85 years), one with a large value of tumor thickness (12.24 mm) and a long observation time (a censoring at 10.60 years), and one with a very small value of tumor thickness (0.32 mm) and a short observation time (a failure at 2.24 years). The first of these would lead to a smaller estimated effect of tumor thickness if excluded and the last two to a larger $\hat{\beta}$. Since even these largest values of the diagnostic are small (compared, e.g., to the esti-

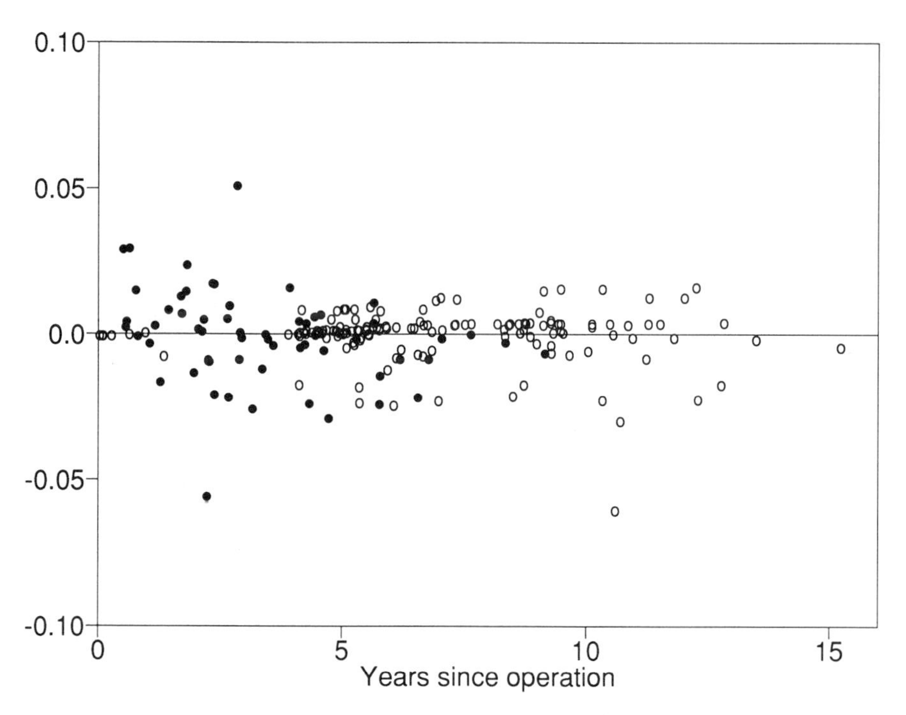

Figure VII.3.14. Approximation to regression diagnostic $\hat{\beta} - \hat{\beta}_{-i}$ for tumor thickness plotted against survival/censoring time $\tilde{X}_i$: censored observations (o); uncensored observations (●).

mated standard deviation 0.175 of $\hat{\beta}$), one would probably not be very suspicious about the model. □

Therneau et al. (1990) [see also Fleming and Harrington (1991, Chapter 4)] gave a detailed discussion on how to use both the simple residual (7.3.8), the "score residual" (7.3.12), and a third "deviance residual" to study other aspects of the multiplicative hazard model, including the functional form of the covariate effect (Section VII.3.2) and the accuracy of prediction for single individuals.

VII.3.5. Concluding Remarks

Methods for assessing goodness-of-fit of a given statistical model are an important part of any statistical analysis since investigators may be seriously mislead if conclusions are drawn on erroneous assumptions. The process of goodness-of-fit examination is a combination of graphical displays of the data and more formal hypothesis tests, the former usually being the more important part. It is, therefore, crucial that the structure that one looks for in the plot is simple and that various kinds of discrepancies from that structure suggest certain deviations from the model.

We, therefore, believe that plots like Figures VII.3.4–VII.3.6 (plotting one integrated baseline hazard estimate against another) and Figure VII.3.12 (Arjas's "residual" plot) will be more useful for checking proportionality than figures like Figures VII.3.1–VII.3.3 (plotting the log integrated hazard against time) and Figures VII.3.10–VII.3.11 (the Nelson–Aalen "residual" plot).

Tests for proportionality against certain alternatives should be easy and flexible to perform as an integrated part of the analysis of the model. Therefore, the tests based on time-dependent covariates are likely to be superior to the alternative suggestions.

Finally, for continous covariates, the method for checking log-linearity suggested by Thomsen (1988) and the regression diagnostics for detecting influential points discussed in Example VII.3.7 seem useful.

VII.4. Nonparametric Additive Hazard Models

VII.4.1. Introduction and Basic Properties

The previous two sections have dealt with the semiparametric multiplicative hazard model where the effect of a vector of possibly time-dependent covariates $(Z_{i1}(t), \ldots, Z_{ip}(t))$ for individual i, $i = 1, \ldots, n$, was described by unknown regression coefficients $(\beta_1, \ldots, \beta_p)$. In particular, the effect on the intensity was assumed to be constant over time. In this section, we discuss the nonparametric regression model introduced by Aalen (1980) where the type h intensity is given by (7.1.4); see also Example III.1.9. For simplicity, we formulate the basic model only in the univariate case ($k = 1$), so

$$\mathbf{N}(t) = (N_i(t); i = 1, \dots, n)$$

is a multivariate counting process, the individual (univariate) process N_i having $(\mathscr{F}_t)$-intensity process given by

$$\lambda_i(t) = \alpha_i(t; \mathbf{Z}_i(t)) Y_i(t) \tag{7.4.1}$$

with [dropping the superscript θ in (7.1.4)]

$$\alpha_i(t, \mathbf{Z}_i(t)) = \beta_0(t) + \beta_1(t) Z_{i1}(t) + \cdots + \beta_p(t) Z_{ip}(t). \tag{7.4.2}$$

The general k-variate case can easily be written in this way with appropriate choices of type-specific covariates. In (7.4.2), $\alpha_i(\cdot, \cdot)$ must be non-negative, whereas the regression functions $\beta_j(t)$, $j = 0, 1, \dots, p$, are left completely unspecified except for the assumption that

$$\int_0^t |\beta_j(u)| \, du < \infty, \quad j = 0, 1, \dots, p, \tag{7.4.3}$$

for all $t \in \mathscr{T}$. In this subsection, first, we shall consider estimation of the *integrated* regression functions

$$B_j(t) = \int_0^t \beta_j(u) \, du, \quad j = 0, 1, \dots, p,$$

and their variances and also of $\beta_0(t), \dots, \beta_p(t)$. Testing hypotheses about the regression functions is also briefly addressed. Section VII.4.2 discusses large sample properties of the estimators.

The model given by (7.4.1)–(7.4.2) can be written in matrix form

$$\mathbf{N}(t) = \int_0^t \mathbf{Y}(u) \boldsymbol{\beta}(u) \, du + \mathbf{M}(t) \tag{7.4.4}$$

and is sometimes denoted the *matrix multiplicative intensity model.* In (7.4.4), $\mathbf{N}(t)$ is the n-vector of individual counting processes, $\mathbf{M}(t)$ is an n-vector of local square integrable martingales, $\boldsymbol{\beta}(t) = (\beta_0(t), \beta_1(t), \dots, \beta_p(t))$, and $\mathbf{Y}(t)$ is the $n \times (p+1)$ matrix with ith row, $i = 1, \dots, n$, given by

$$Y_i(t)(1, Z_{i1}(t), \dots, Z_{ip}(t)) - (Y_{ij}(t), j - 0, 1, \dots, p),$$

say; see also Example III.1.9. Equation (7.4.4), written symbolically as

$$\mathrm{d}\mathbf{N}(t) = \mathbf{Y}(t) \, \mathrm{d}\mathbf{B}(t) + \mathrm{d}\mathbf{M}(t),$$

[compare (2.1.6) and Section IV.1.1] motivates "generalized Nelson–Aalen estimators" for $\mathbf{B}(t) = (B_0(t), B_1(t), \dots, B_p(t))$ of the form

$$\hat{\mathbf{B}}(t) = \int_0^t J(u) \mathbf{Y}^-(u) \, \mathrm{d}\mathbf{N}(u). \tag{7.4.5}$$

Here, $\mathbf{Y}^{-}(t)$ is a predictable generalized inverse of $\mathbf{Y}(t)$, i.e., a $(p+1) \times n$ matrix satisfying

$$\mathbf{Y}^{-}(t)\mathbf{Y}(t) = \mathbf{I},$$

the $(p+1) \times (p+1)$ identity matrix, and

$$J(t) = I(\text{rank } \mathbf{Y}(t) = p+1) \tag{7.4.6}$$

is the predictable indicator of $\mathbf{Y}(t)$ having full rank (assuming $p + 1 \leq n$). Assuming $J\mathbf{Y}^{-}$ to be locally bounded and defining

$$\mathbf{B}^{*}(t) = \int_{0}^{t} J(u)\boldsymbol{\beta}(u)\,\mathrm{d}u$$

[compare (4.1.3)], we have by (7.4.4) that

$$(\hat{\mathbf{B}} - \mathbf{B}^{*})(t) = \int_{0}^{t} J(u)\mathbf{Y}^{-}(u)\,\mathrm{d}\mathbf{M}(u) \tag{7.4.7}$$

is a $(p+1)$-variate local martingale with predictable variation process

$$\langle \hat{\mathbf{B}} - \mathbf{B}^{*} \rangle(t) = \int_{0}^{t} J(u)\mathbf{Y}^{-}(u)\,\mathrm{diag}\,\boldsymbol{\lambda}(u)(\mathbf{Y}^{-}(u))^{\mathsf{T}}\,\mathrm{d}u \tag{7.4.8}$$

and optional variation process

$$[\hat{\mathbf{B}} - \mathbf{B}^{*}](t) = \int_{0}^{t} J(u)\mathbf{Y}^{-}(u)\,\mathrm{diag}\,\mathrm{d}\mathbf{N}(u)(\mathbf{Y}^{-}(u))^{\mathsf{T}} \tag{7.4.9}$$

[see (2.4.8)]. Here, diag $\mathbf{v}$ for an n-vector $\mathbf{v}$ is the $n \times n$ diagonal matrix with the elements of $\mathbf{v}$ in the diagonal.

Equations (7.4.8) and (7.4.9) suggest that the mean squared error function

$$\boldsymbol{\Sigma}(t) = \mathrm{E}(\hat{\mathbf{B}} - \mathbf{B}^{*})(\hat{\mathbf{B}} - \mathbf{B}^{*})^{\mathsf{T}}(t)$$

be estimated by

$$\hat{\boldsymbol{\Sigma}}(t) = [\hat{\mathbf{B}} - \mathbf{B}^{*}](t).$$

We shall, indeed, use $\hat{\mathbf{B}}(t)$ and $\hat{\boldsymbol{\Sigma}}(t)$ as estimators of $\mathbf{B}(t)$ and $\boldsymbol{\Sigma}(t)$, but to calculate the estimates, a choice of generalized inverse must be made. An unweighted least squares principle applied to the linear equations (7.4.4) would suggest using

$$\mathbf{Y}^{-}(t) = (\mathbf{Y}(t)^{\mathsf{T}}\mathbf{Y}(t))^{-1}\mathbf{Y}(t)^{\mathsf{T}} \tag{7.4.10}$$

(Aalen, 1980; Huffer and McKeague, 1991; McKeague, 1988a). With this choice, $\hat{\boldsymbol{\Sigma}}(t)$ can be written as

$$\hat{\boldsymbol{\Sigma}}(t) = \int_{0}^{t} J(u)(\mathbf{Y}^{\mathsf{T}}(u)\mathbf{Y}(u))^{-1}\,\mathrm{d}\mathbf{H}(u)(\mathbf{Y}^{\mathsf{T}}(u)\mathbf{Y}(u))^{-1},$$

where $\mathrm{d}\mathbf{H}(u)$ is the $(p+1) \times (p+1)$ matrix with elements

$$\mathrm{d}H_{jl}(u) = \sum_{i=1}^{n} Y_{ij}(u) Y_{il}(u) \,\mathrm{d}N_i(u).$$

In the univariate ($p = 0$) case, (7.4.10) reduces to the $1 \times n$ matrix

$$\left(\sum_{i=1}^{n} Y_i(t)^2\right)^{-1} (Y_1(t), \ldots, Y_n(t))$$

and the estimate (7.4.5) becomes

$$\hat{B}_0(t) = \int_0^t I(Y_{.}(u) > 0) \left(\sum_{i=1}^{n} Y_i(u)^2\right)^{-1} \sum_{i=1}^{n} Y_i(u) \,\mathrm{d}N_i(u).$$

When all the $Y_i(t)$ are *indicator processes* (as in models for censored or filtered survival data or Markov processes), the estimator $\hat{B}_0(t)$ is the Nelson–Aalen estimator (4.1.2).

Since the unweighted least squares principle leading to (7.4.10) does not take into account the fact that the individual $M_i(t)$ may have unequal variances, one cannot expect this choice of generalized inverse to be optimal in any sense. We return to the question of optimality in Section VIII.4.4 and just note now that to achieve some kind of optimality, a weighted least squares principle should be preferred. This leads to choices of the form

$$\mathbf{Y}^{-}(t) = (\mathbf{Y}(t)^{\mathsf{T}} \hat{\mathbf{W}}(t) \mathbf{Y}(t))^{-1} \mathbf{Y}(t)^{\mathsf{T}} \hat{\mathbf{W}}(t) \tag{7.4.11}$$

(McKeague, 1988a; Huffer and McKeague, 1991), $\hat{\mathbf{W}}(t)$ being an $n \times n$ diagonal matrix with element (i, i), an estimate of λ_i^{-1} "proportional to the inverse variance $(\alpha_i(t, \mathbf{Z}_i(t)) Y_i(t) \,\mathrm{d}t)^{-1}$ of $\mathrm{d}M_i(t)$." In fact, we show in Section VIII.4.4 that this choice *is* optimal. In the univariate case, we have

$$\alpha_i(t, \mathbf{Z}_i(t)) = \beta_0(t) Y_i(t)$$

and the ith diagonal element can be chosen as $w_i(t) = I(Y_i(t) > 0) Y_i(t)^{-1}$, this choice leading to the Nelson–Aalen estimator as an estimator for $B_0(t)$.

To calculate the weighted least squares estimate $\hat{B}_j^{w}(t)$ for $B_j(t)$ in the general case, a two-stage procedure is used. First, a preliminary estimate $\hat{B}_j(t)$ is obtained using (7.4.5) and, for instance, the generalized inverse (7.4.10). Next, $\hat{B}_j(t)$ is smoothed to get an estimate of $\beta_j(t)$,

$$\hat{\beta}_j(t) = b_j^{-1} \int_{\mathscr{T}} K((t - s)/b_j) \,\mathrm{d}\hat{B}_j(s), \quad j = 1, \ldots, p \tag{7.4.12}$$

[where, as in (4.2.1), K is a kernel function and $b_j > 0$ the bandwidth] and (7.4.11) is then computed with $\hat{W}(t)$ being the diagonal matrix with element (i, i) equal to $\{Y_i(t) \hat{\alpha}_i(t, \mathbf{Z}_i(t))\}^{-1} I(\hat{\alpha}_i(t, \mathbf{Z}_i(t)) Y_i(t) > 0)$, where

$$\hat{\alpha}_i(t, \mathbf{Z}_i(t)) = \hat{\beta}_0(t) + \sum_{j=1}^{p} \hat{\beta}_j(t) Z_{ij}(t), \quad i = 1, \ldots, n. \tag{7.4.13}$$

To stabilize this two-stage procedure, it is probably advisable to apply a relatively large bandwidth in (7.4.12), that is, to *oversmooth* in the first stage.

Furthermore, to use the usual techniques in the derivation of the large sample properties of the estimate $\hat{B}_j^w(t)$ (Huffer and McKeague, 1991), the generalized inverse (7.4.11) and, hence, also the estimates (7.4.12) and (7.4.13) need to be *predictable*, leading to the application in (7.4.12) of a kernel with support on $(0, 1]$ rather than on $[-1, 1]$. It may, however, be possible to relax the assumption of predictability using alternative techniques of proof. At any rate, a *final* estimate of $\beta_j(t)$ can be obtained using (7.4.12) with a kernel having support on $[-1, 1]$ as in Section IV.2.

Before considering some examples we shall briefly see how some of the ideas of Chapter V may be used for the construction of tests for the hypothesis of certain, say q, covariates having no effect on the intensity, i.e.,

$$\mathrm{H}_0\colon \beta_j(t) = 0 \quad \text{for all } t \in \mathscr{T}$$

for $j \in J \subseteq \{1, \ldots, p\}$ with $|J| = q$. For each fixed j, one simply considers a process of the form

$$X_j(t) = \int_0^t L_j(s)\,\mathrm{d}\hat{B}_j(s),$$

where $L_j(s)$ is a predictable weight process. Under H_0, $B_j^*(t) \equiv 0$ for $j \in J$ and $\mathbf{X}_J = (X_j, j \in J)$ is a vector of mean-zero local square integrable martingales with predictable and optional variation processes given by the $q \times q$ matrices

$$\langle \mathbf{X}_J \rangle(t) = \int_0^t \operatorname{diag} \mathbf{L}(s)\,\mathrm{d}\langle \hat{\mathbf{B}}_J \rangle(s) \operatorname{diag} \mathbf{L}(s)$$

and

$$[\mathbf{X}_J](t) = \int_0^t \operatorname{diag} \mathbf{L}(s)\,\mathrm{d}[\hat{\mathbf{B}}_J](s) \operatorname{diag} \mathbf{L}(s),$$

respectively, where $\hat{\mathbf{B}}_J$ is the q-vector $(\hat{\mathbf{B}}_j, j \in J)$; see (2.4.8). Thus, the estimated covariance under H_0 between $X_j(t)$ and $X_l(t)$, $j, l \in J$, is

$$[X]_{jl}(t) = \int_0^t L_j(s) L_l(s)\,\mathrm{d}\hat{\Sigma}_{jl}(s),$$

where $\hat{\Sigma}_{jl}(s) = \widehat{\operatorname{cov}}(\hat{B}_j(s), \hat{B}_l(s))$. It follows from the asymptotic results in the next subsection that the distribution in large samples under H_0 of $X_j(t)[X]_{jj}(t)^{-1/2}$ is approximately normal and that of $\mathbf{X}_J(t)^\mathsf{T}[\mathbf{X}_J](t)^{-1}\mathbf{X}_J(t)$ is approximately χ^2 with q degrees of freedom. Choices of L_j to achieve reasonable power against specific classes of alternatives were considered by Aalen (1980, 1989) who argued that for alternatives of the form $\beta_j(t) > 0$ or $\beta_j(t) < 0$ for all $t \in \mathscr{T}$, $L_j(t)$ could advantageously be chosen as the reciprocal of the jth diagonal element of the matrix $(\mathbf{Y}(t)^\mathsf{T}\mathbf{Y}(t))^{-1}$.

Obviously, one could go further in the direction of choosing alternative weight processes or considering other classes of tests. Thus, Huffer and McKeague (1991) considered Kolmogorov–Smirnov-type tests (Section V.4.1).

We saw above how the estimator (7.4.5) reduces in the univariate case. Next, we consider the two-sample case.

EXAMPLE VII.4.1. The Two-Sample Case

Assume that we have two groups of individuals labelled 0 and 1 and a single time-fixed covariate $Z_{i1} = I(i \text{ belongs to group } 1)$. Let the hazard function be given by (7.4.2), i.e.,

$$\alpha_i(t, Z_i) = \beta_0(t) + \beta_1(t) Z_{i1}$$

and define $Y_\cdot^{(j)}$ and $N_\cdot^{(j)}$, $j = 0, 1$, as in Example VII.2.6. A simple calculation shows that the least squares estimators given by (7.4.5) and (7.4.10) in this case reduce to

$$\hat{B}_0(t) = \int_0^t J(u) \frac{\mathrm{d}N_\cdot^{(0)}(u)}{Y_\cdot^{(0)}(u)}$$

and

$$\hat{B}_1(t) = \int_0^t J(u) \left(\frac{\mathrm{d}N_\cdot^{(1)}(u)}{Y_\cdot^{(1)}(u)} - \frac{\mathrm{d}N_\cdot^{(0)}(u)}{Y_\cdot^{(0)}(u)} \right),$$

that is, the Nelson–Aalen estimator in group 0 and the difference between that of group 1 and that of group 0, respectively, both calculated up to the time

$$\tau_0 = \inf\{t\colon Y_\cdot^{(0)}(t) Y_\cdot^{(1)}(t) = 0\},$$

where the rank of $\mathbf{Y}(t)$ becomes 1.

Also, the entries in the estimated covariance matrix $\hat{\boldsymbol{\Sigma}}(t)$ reduce to the corresponding quantities for the Nelson–Aalen estimators

$$\hat{\Sigma}_{00}(t) = \int_0^t J(u) \frac{\mathrm{d}N_\cdot^{(0)}(u)}{Y_\cdot^{(0)}(u)^2},$$

$$\hat{\Sigma}_{11}(t) = \int_0^t J(u) \left(\frac{\mathrm{d}N_\cdot^{(0)}(u)}{Y_\cdot^{(0)}(u)^2} + \frac{\mathrm{d}N_\cdot^{(1)}(u)}{Y_\cdot^{(1)}(u)^2} \right),$$

$$\hat{\Sigma}_{01}(t) = \hat{\Sigma}_{10}(t) = -\int_0^t J(u) \frac{\mathrm{d}N_\cdot^{(0)}(u)}{Y_\cdot^{(0)}(u)^2}.$$

Finally, let us study the proposed test for hypothesis $\mathrm{H}_0\colon \beta_1(t) \equiv 0$ of no effect of the covariate.

This leads to choosing

$$L_1(t) = Y_\cdot^{(0)}(t) Y_\cdot^{(1)}(t) Y_\cdot^{(\cdot)}(t)^{-1}$$

as in the *log-rank test* (Example V.2.1). The proposed estimate of the variance of the test statistic $X_1(\tau)$, however, becomes

$$\int_0^\tau L_1(t)^2 \left(\frac{\mathrm{d}N_\cdot^{(0)}(t)}{Y_\cdot^{(0)}(t)^2} + \frac{\mathrm{d}N_\cdot^{(1)}(t)}{Y_\cdot^{(1)}(t)^2} \right)$$

instead of

$$\int_0^\tau L_1(t)^2 (Y_\bullet^{(0)}(t) Y_\bullet^{(1)}(t))^{-1} (\mathrm{d}N_\bullet^{(0)}(t) + \mathrm{d}N_\bullet^{(1)}(t))$$

as in (5.2.8). This difference is due to the fact that we do not now reestimate the variance under H_0. □

Practical examples of the use of the additive model were presented by Aalen (1980, 1989) and by Mau (1986b). We shall once more use the melanoma data as illustration.

EXAMPLE VII.4.2. Survival with Malignant Melanoma. Effect of Sex, Tumor Thickness, and Ulceration in an Additive Hazard Model

This example continues, first of all, Examples VII.2.5 and VII.2.9 where multiplicative hazard regression models for the data first introduced in Example I.3.1 were discussed. In the present example, additive models will be considered and, as an example of the two-sample case (cf. Examples VII.2.6 and VII.4.1), we consider the covariate

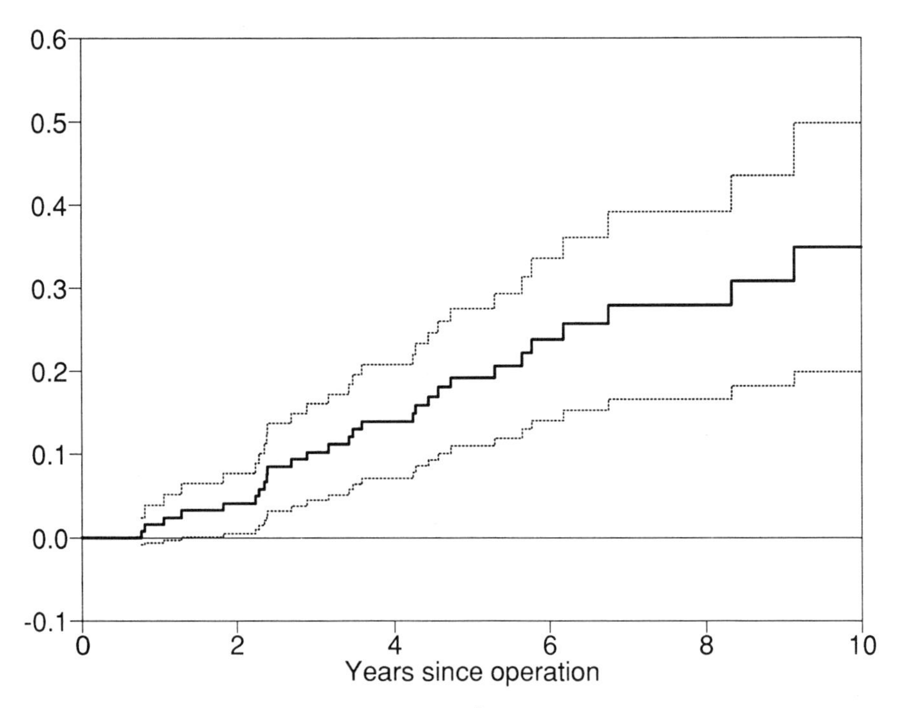

Figure VII.4.1. Estimated "baseline" hazard $\hat{B}_0(t)$ in an additive model including only sex as covariate. Approximate pointwise 95% confidence limits obtained as "estimate $\pm 1.96\widehat{\text{s.d.}}$"

$$Z_{i1} = \begin{cases} 1 & \text{if patient } i \text{ is a man} \\ 0 & \text{if patient } i \text{ is a woman.} \end{cases}$$

In this case, $\hat{B}_0(t)$ is the estimated integrated hazard for women and it is shown in Figure VII.4.1 with approximate 95% pointwise confidence limits calculated using (7.4.9) and the generalized inverse (7.4.10). The estimate $\hat{B}_1(t)$, the difference between the Nelson–Aalen estimates for men and women (cf. Figure IV.1.1), is shown in Figure VII.4.2. The pointwise confidence limits suggest that men have a higher hazard than women (Examples V.2.3 and VII.2.5) and this tendency is confirmed by the log-rank-type test statistic discussed in the previous example, taking the value 5.41 ($P = 0.02$) which is slightly smaller than the ordinary log-rank test statistic (6.47).

Turning next to a model like that considered in Example VII.2.9 including also the covariates

$$Z_{i2} = \text{“tumor thickness for patient” } i - 2.92 \text{ mm}$$

and

$$Z_{i3} = \begin{cases} 1 & \text{if ulceration is present in the tumor of patient } i \\ 0 & \text{otherwise,} \end{cases}$$

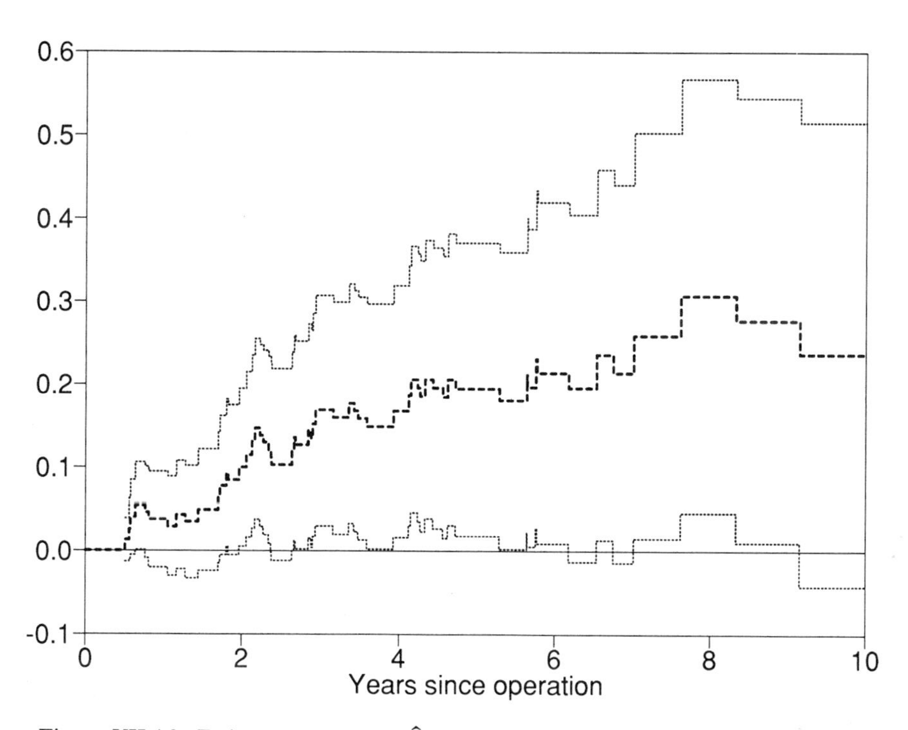

Figure VII.4.2. Estimated sex effect $\hat{B}_1(t)$ in an additive model including no other covariates.

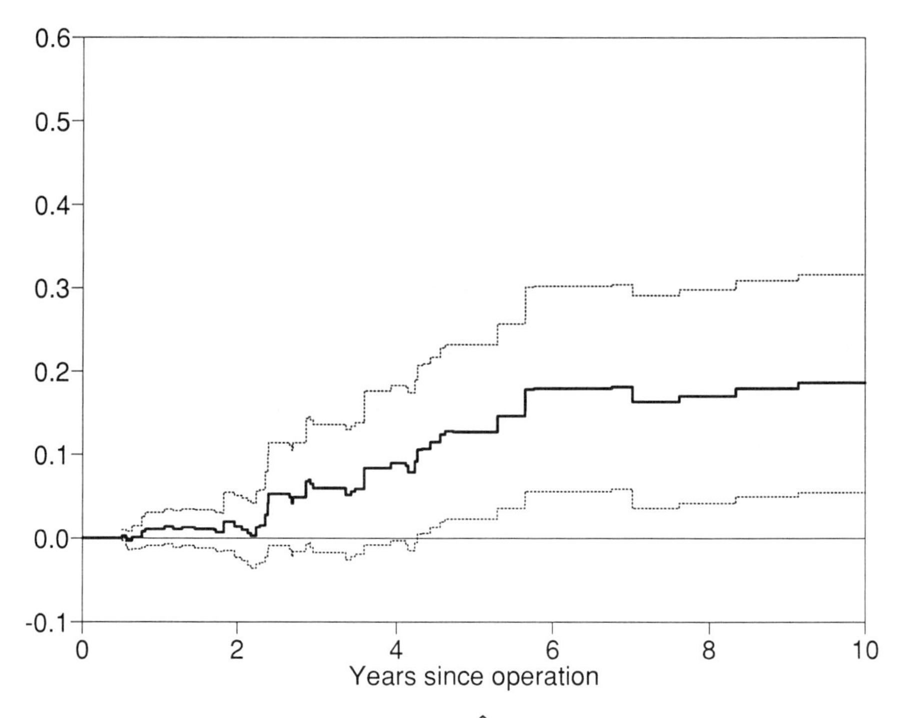

Figure VII.4.3. Estimated "baseline" hazard $\hat{B}_0(t)$ in an additive model including sex, thickness, and ulceration.

the estimates $\hat{B}_j(t)$, $j = 0, 1, 2, 3$, are shown in Figures VII.4.3–VII.4.6. As in Example VII.2.9, the effect of sex seems to vanish once the other variables are included in the model, whereas ulceration and thickness still have clear effects, the effect of the latter, however, tends to decrease with time. In fact, one of the strengths of the nonparametric model is that temporal patterns of covariate effects may be studied. The test statistic for the effect of sex adjusting for the effects of ulceration and thickness is 2.40 ($P = 0.12$).

EXAMPLE VII.4.3. A Combined Model for Excess and Relative Mortality. Survival with Malignant Melanoma

In Example IV.1.11, models for relative mortality $\alpha_i(t) = \alpha(t)\mu_i(t)$ and excess mortality $\alpha_i(t) = \gamma(t) + \mu_i(t)$ were considered, $\mu_i(t)$ being a known hazard function. A model containing both of these as special cases is the following (Andersen and Væth, 1989):

$$\alpha_i(t) = \gamma(t) + \alpha(t)\mu_i(t), \tag{7.4.14}$$

which is seen to be an example of Aalen's additive hazard model (7.4.2) with a single time-dependent covariate $Z_{i1}(t) = \mu_i(t)$. Fitting the model (7.4.14) therefore provides a means for possibly choosing between the two simpler models since both of the hypotheses $\gamma(t) \equiv 0$ and $\alpha(t) \equiv 1$ can be tested along

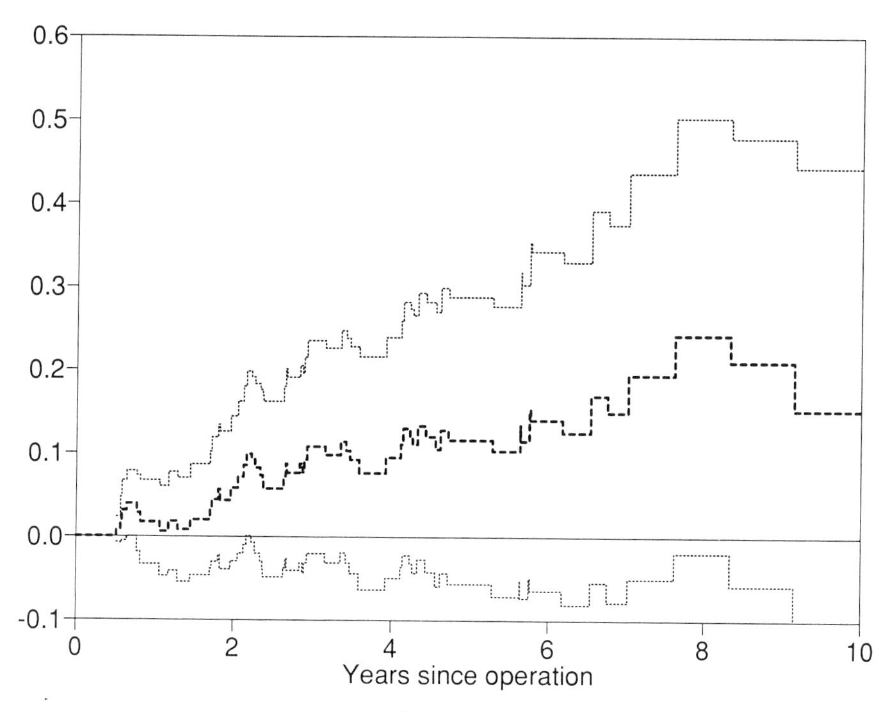

Figure VII.4.4. Estimated sex effect $\hat{B}_1(t)$ in an additive model including thickness and ulceration.

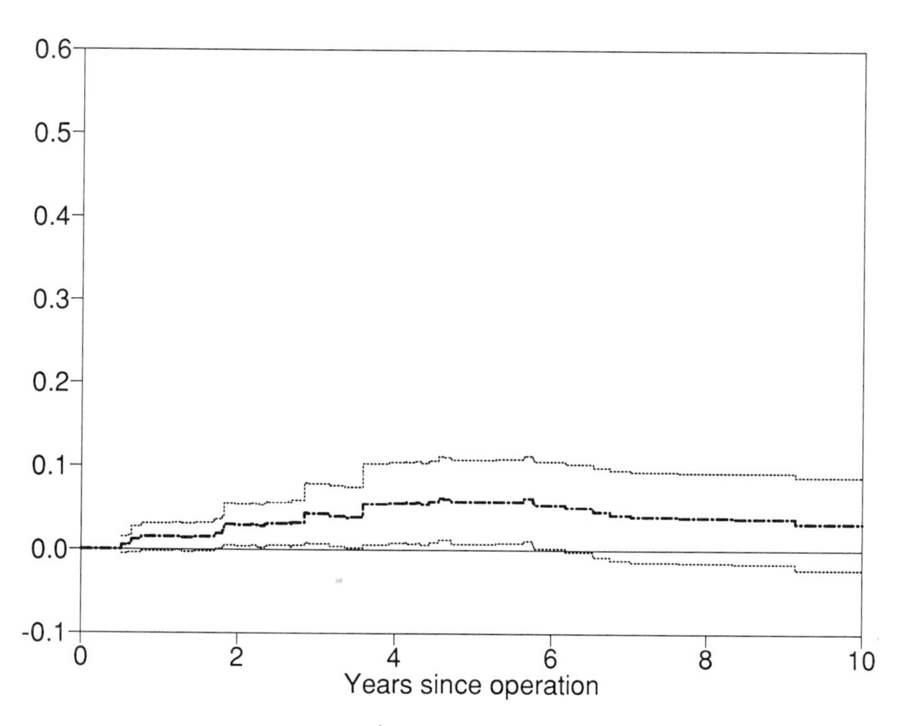

Figure VII.4.5. Estimated effect $\hat{B}_2(t)$ of tumor thickness in an additive model including sex and ulceration.

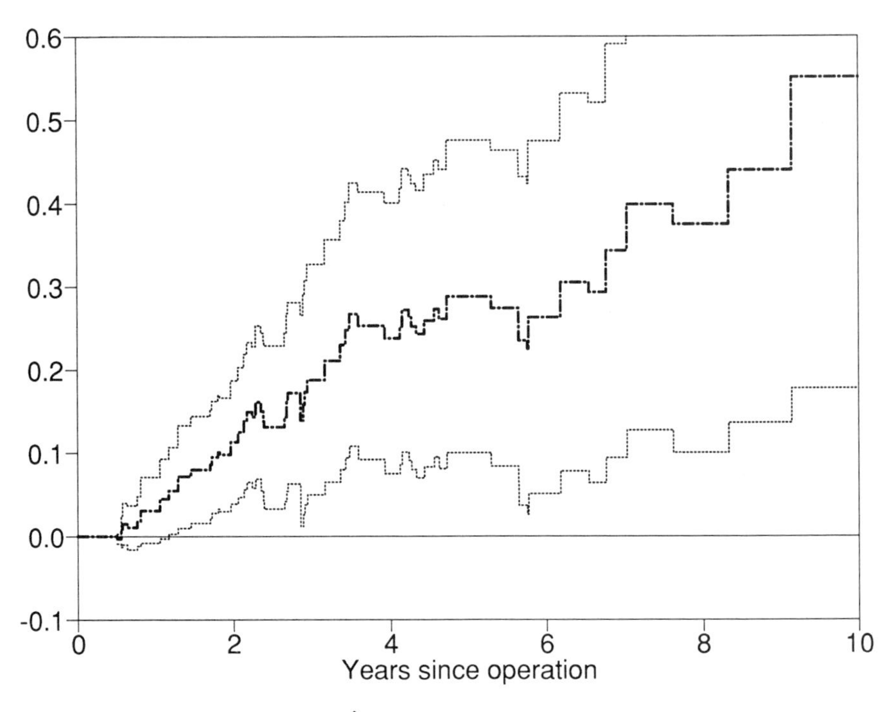

Figure VII.4.6. Estimated effect $\hat{B}_3(t)$ of ulceration in an additive model including sex and tumor thickness.

the lines of this subsection. [For a discussion of (7.4.14) with *constant* regression functions γ and α, see Example VI.1.5.]

For the melanoma data introduced in Example I.3.1, we estimated both the integral of $\alpha(\cdot)$ and that of $\gamma(\cdot)$ in Example IV.1.11 considering either time since operation or age as the basic time scale. Figures VII.4.7 and VII.4.8 show the corresponding estimates based on the combined model (7.4.14) as functions of time since operation. The estimate $\hat{\Gamma}(t)$ looks roughly linear (with a slope ≈ 0.03 per year) suggesting that $\gamma(t)$ is approximately constant. The hypothesis $\gamma(t) \equiv 0$ is clearly rejected when tested as outlined above (e.g., $P < 0.001$ when calculated at $\tau = 9$ years). The estimate $\hat{A}(t)$ is first almost constant and from 2 years and upward, it increases roughly linearly (with a slope close to unity). The test statistic for the hypothesis $\alpha(t) \equiv 1$ is insignificant at $\tau = 9$ years. The model $\alpha_i(t) = \gamma_0 + \mu_i(t)$ considered in Examples IV.1.11 and VI.1.5, therefore, seems reasonable, whereas the joint hypothesis $\gamma(t) \equiv 0$ and $\alpha(t) \equiv 1$ corresponding to the mortality for the melanoma patients being identical to that of the general Danish population is clearly rejected, as expected.

Turning, as in Example IV.1.11, to considering instead models with age as the basic time scale, Figures VII.4.9 and VII.4.10 show the estimated integrated regression functions $\hat{\Gamma}(a)$ and $\hat{A}(a)$, respectively, in the combined model

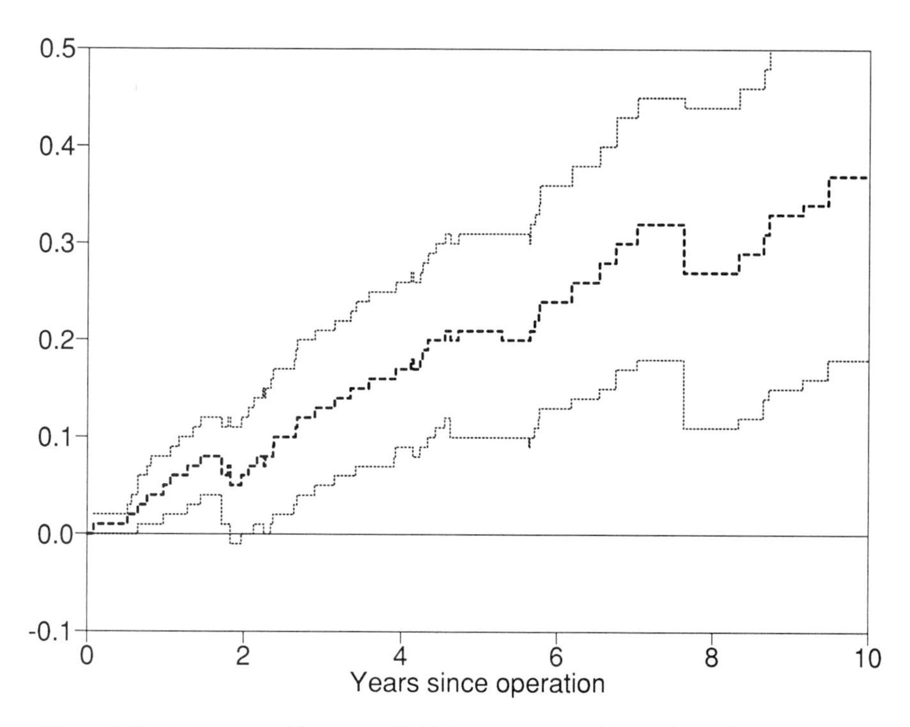

Figure VII.4.7. Estimated integral of $\gamma(t)$ in the model $\alpha_i(t) = \gamma(t) + \alpha(t)\mu_i(t)$ with approximate pointwise 95% confidence limits.

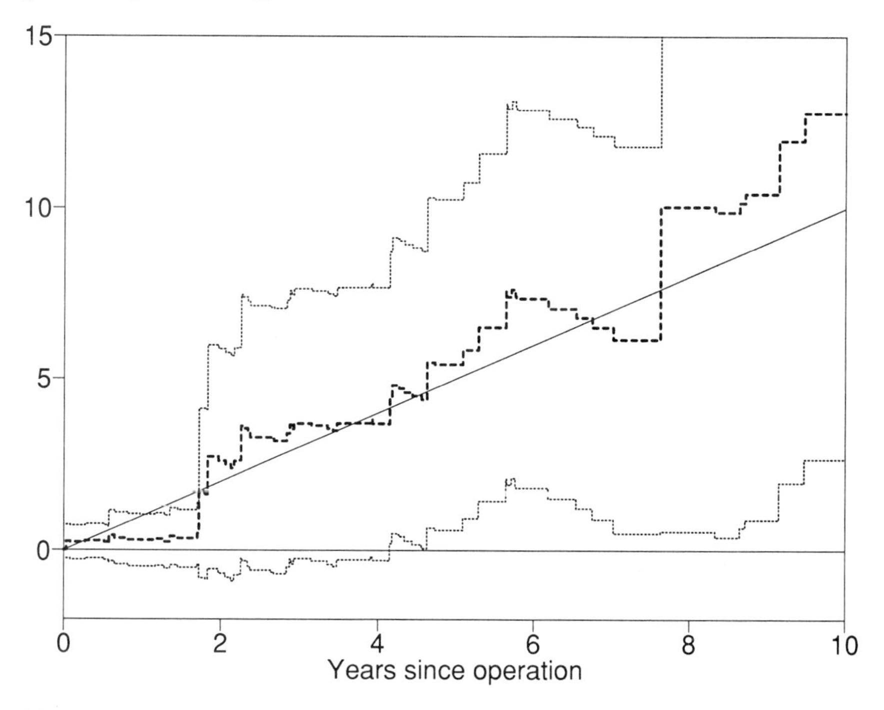

Figure VII.4.8. Estimated integral of $\alpha(t)$ in the model $\alpha_i(t) = \gamma(t) + \alpha(t)\mu_i(t)$ with approximate pointwise 95% confidence limits.

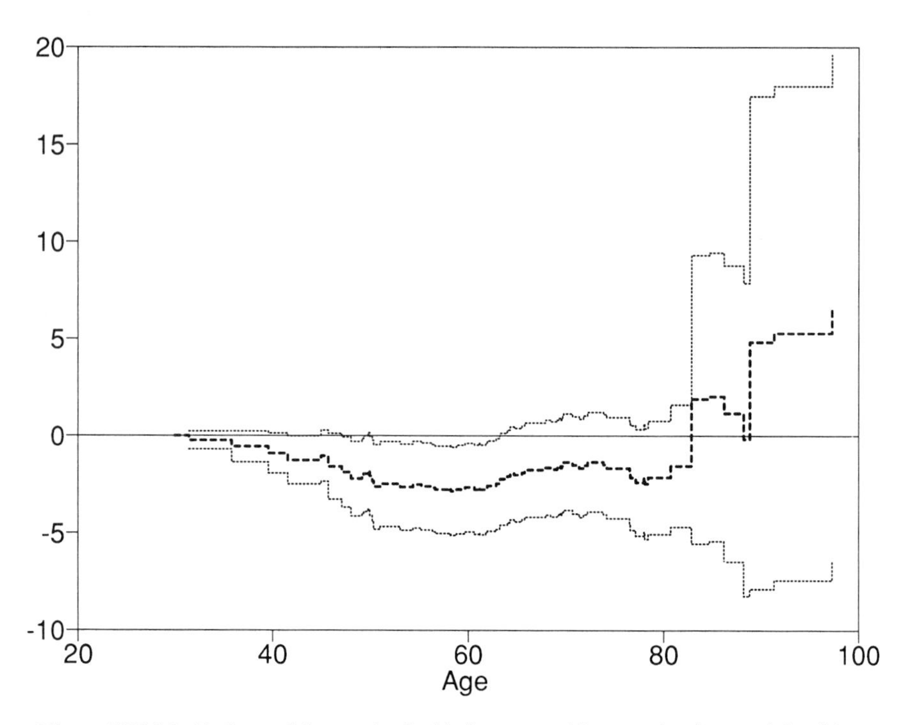

Figure VII.4.9. Estimated integral of $\gamma(a)$ from age 30 years in the model $\alpha_i(a) = \gamma(a) + \alpha(a)\mu_i(a)$ with approximate pointwise 95% confidence limits.

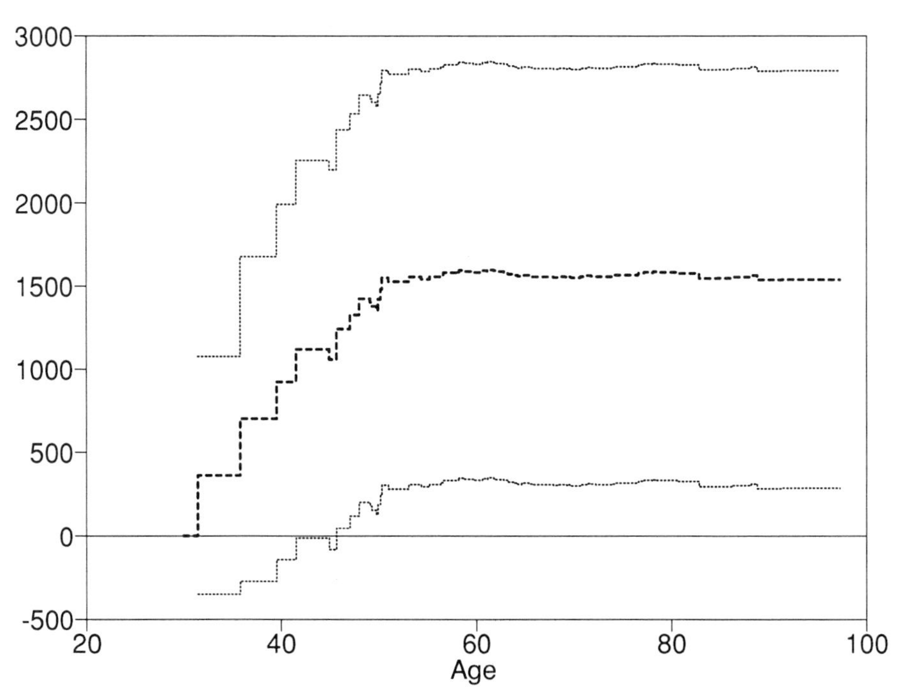

Figure VII.4.10. Estimated integral of $\alpha(a)$ from age 30 years in the model $\alpha_i(a) = \gamma(a) + \alpha(a)\mu_i(a)$ with approximate pointwise 95% confidence limits.

(7.4.14). Comparing Figure VII.4.10 with Figure IV.1.10, it is seen that the slope of $\hat{A}(a)$ for small ages is much higher in the combined model and the strong negative correlation between $\hat{A}(a)$ and $\hat{\Gamma}(a)$ then "forces" the latter to be negative; cf. Figure VII.4.9.

Again test statistics for the hypotheses $\gamma(a) \equiv 0$ or $\alpha(a) \equiv 1$ can be considered. Since the departures from these null hypotheses do not seem to be of the form $\gamma(a) > 0$ or $\gamma(a) < 0$ (Figure VII.4.9) and similarly for $\alpha(a)$ (Figure VII.4.10), the values of the test statistics vary considerably with time. The joint hypothesis $\gamma(a) \equiv 0$ and $\alpha(a) \equiv 1$ is, however, consistently rejected. □

VII.4.2. Large Sample Properties

For any choice of generalized inverse $\mathbf{Y}^{-}(t)$ one can, as indicated by Aalen (1980), formulate conditions on the elements of $J(t)\mathbf{Y}^{-}(t)$ ensuring that the martingale central limit theorem applies to the local square integrable martingales $\hat{\mathbf{B}} - \mathbf{B}^*$; cf. (7.4.7). Following Huffer and McKeague (1991) and McKeague (1988a), here we shall mainly consider the least squares choice (7.4.10), in which case the conditions can be given directly in terms of the matrix $\mathbf{Y}(t)$. As for the semiparametric multiplicative hazard model discussed in the earlier sections of this chapter [see Condition VII.2.1] some asymptotic stability and boundedness conditions are needed for the covariate processes. Recall from Section VII.4.1 that $Y_{i0}(t) = Y_i(t)$ and $Y_{ij}(t) = Y_i(t)Z_{ij}(t)$ for $j = 1, \ldots, p$ and define for $j, k, l = 0, 1, \ldots, p$, $t \in \mathcal{T}$,

$$\begin{aligned} R_j^{(1)}(t) &= \sum_{i=1}^{n} Y_{ij}(t), \\ R_{jk}^{(2)}(t) &= \sum_{i=1}^{n} Y_{ij}(t)\,Y_{ik}(t), \\ R_{jkl}^{(3)}(t) &= \sum_{i=1}^{n} Y_{ij}(t)\,Y_{ik}(t)\,Y_{il}(t). \end{aligned} \tag{7.4.15}$$

Then the following theorem holds.

Theorem VII.4.1. *Let $t \in \mathcal{T}$ and assume that for $j, k, l = 0, 1, \ldots, p$ there exist continuous functions $r_j^{(1)}$, $r_{jk}^{(2)}$, and $r_{jkl}^{(3)}$ such that as $n \to \infty$*

(A)
$$\sup_{s \in [0,t]} \left| \frac{1}{n} R_j^{(1)}(s) - r_j^{(1)}(s) \right| \xrightarrow{\mathrm{P}} 0,$$

$$\sup_{s \in [0,t]} \left| \frac{1}{n} R_{jk}^{(2)}(s) - r_{jk}^{(2)}(s) \right| \xrightarrow{\mathrm{P}} 0,$$

$$\sup_{s \in [0,t]} \left| \frac{1}{n} R_{jkl}^{(3)}(s) - r_{jkl}^{(3)}(s) \right| \xrightarrow{\mathrm{P}} 0.$$

Assume, furthermore, that for $j = 0, 1, \ldots, p$

(B) $$n^{-1/2} \sup_{\substack{i=1,\ldots,n \\ s \in [0,t]}} |Y_{ij}(s)| \xrightarrow{\mathrm{P}} 0 \quad \text{as } n \to \infty$$

and that for all $s \in [0, t]$

(C) *the matrix* $\mathbf{r}^{(2)}(s) = (r_{jk}^{(2)}(s))$ *is nonsingular.*

Then

$$\sqrt{n}(\hat{\mathbf{B}} - \mathbf{B}) \xrightarrow{\mathscr{D}} \mathbf{U} = (U_0, U_1, \ldots, U_p) \quad \text{as } n \to \infty$$

in $D[0, t]^{p+1}$, *where* $\hat{\mathbf{B}}$ *is given by* (7.4.5), $\mathbf{Y}^-$ *by* (7.4.10), *and* $\mathbf{U}$ *is a mean-zero Gaussian martingale with*

$$\operatorname{cov}(U_j(s), U_k(s)) = \sum_{g,l,m=0}^{p} \int_0^s r_{glm}^{(3)}(u)(\mathbf{r}^{(2)}(u))_{jl}^{-1}(\mathbf{r}^{(2)}(u))_{km}^{-1}\beta_g(u)\,\mathrm{d}u.$$

Furthermore,

$$\sup_{s \in [0,t]} \left| \sum_{g,l,m} n \int_0^s R_{glm}^{(3)}(u)(\mathbf{R}^{(2)}(u))_{jl}^{-1}(\mathbf{R}^{(2)}(u))_{km}^{-1}\,\mathrm{d}\hat{B}_g(u) - \operatorname{cov}(U_j(s), U_k(s)) \right| \xrightarrow{\mathrm{P}} 0.$$

Proof. We rewrite $\sqrt{n}(\hat{\mathbf{B}} - \mathbf{B})$ as

$$\begin{aligned}\sqrt{n}(\hat{\mathbf{B}} - \mathbf{B})(t) = {} & \frac{1}{\sqrt{n}} \int_0^t J(s)\left\{\left(\frac{1}{n}\mathbf{R}^{(2)}(s)\right)^{-1} - (\mathbf{r}^{(2)}(s))^{-1}\right\}\mathbf{Y}^{\mathsf{T}}(s)\,\mathrm{d}\mathbf{M}(s) \\ & + \frac{1}{\sqrt{n}} \int_0^t J(s)(\mathbf{r}^{(2)}(s))^{-1}\mathbf{Y}^{\mathsf{T}}(s)\,\mathrm{d}\mathbf{M}(s) \\ & + \sqrt{n} \int_0^t (J(s) - 1)\boldsymbol{\beta}(s)\,\mathrm{d}s \end{aligned} \tag{7.4.16}$$

and deal with the three terms on the right-hand side separately. We want to show that the first and the last term vanish asymptotically while the second converges weakly to $\mathbf{U}$. From conditions (A) and (C), it follows that the last term converges uniformly to zero in probability; see, e.g., McKeague (1988b, Lemma 4.2) for a similar argument. The jth component of the first term,

$$\tilde{X}_j^{(n)}(t) = \frac{1}{\sqrt{n}} \int_0^t J(s) \sum_{i=1}^{n} \sum_{l=0}^{p} \left\{\left(\frac{1}{n}\mathbf{R}^{(2)}(s)\right)^{-1} - (\mathbf{r}^{(2)}(s))^{-1}\right\}_{jl} Y_{il}(s)\,\mathrm{d}M_i(s),$$

$j = 0, 1, \ldots, p$, is a local square integrable martingale. By the inequality of Lenglart (2.5.18),

$$\mathrm{P}\left(\sup_{s \in [0,t]} |\tilde{X}_j^{(n)}(s)| > \varepsilon\right) \leq \frac{\eta}{\varepsilon^2} + \mathrm{P}(\langle \tilde{X}_j^{(n)} \rangle(t) > \eta)$$

for any $\eta, \varepsilon > 0$. Here

$$\langle \tilde{X}_j^{(n)} \rangle(t) = \frac{1}{n} \int_0^t J(s) \sum_{i=1}^n \left\{ \sum_{l=0}^p \left(\left(\frac{1}{n} \mathbf{R}^{(2)}(s) \right)^{-1} - (\mathbf{r}^{(2)}(s))^{-1} \right)_{jl} Y_{il}(s) \right\}^2 \lambda_i(s) \, \mathrm{d}s$$

and, by conditions (A) and (C), there exist random variables C_l, $l = 0, 1, \ldots, p$, with $C_l \xrightarrow{\mathrm{P}} 0$ such that

$$\langle \tilde{X}_j^{(n)} \rangle(t) \le \sum_{k=0}^p \sum_{l=0}^p C_k C_l \int_0^t \frac{1}{n} \sum_{i=1}^n Y_{ik}(s) Y_{il}(s) \lambda_i(s) \, \mathrm{d}s$$

and, therefore, by condition (A), the first term on the right-hand side of (7.4.16) converges uniformly to zero in probability. The second term is also a local square integrable martingale with component j given by

$$X_j^{(n)}(t) = \frac{1}{\sqrt{n}} \sum_{i=1}^n \int_0^t \sum_{l=0}^p (\mathbf{r}^{(2)}(s))_{jl}^{-1} Y_{il}(s) \, \mathrm{d}M_i(s),$$

$j = 0, 1, \ldots, p$, and condition (A) immediately gives that $\langle X_j^{(n)}, X_k^{(n)} \rangle(t)$ converges in probability to the expression for $\operatorname{cov}(U_j(t), U_k(t))$ stated in the theorem. Finally, the Lindeberg condition in the martingale central limit theorem follows in a straightforward manner from conditions (B) and (C) and the proof is complete. □

The proof is taken from Huffer and McKeague (1991); see also McKeague (1988a). The latter paper also discusses the asymptotic properties of the weighted least squares estimator given by (7.4.11). Obviously, stronger conditions are needed for this choice of generalized inverse; cf. also Section IV.2. We shall not study this case in detail but only briefly discuss the extra regularity conditions and refer to McKeague (1988a) for further information. As in Theorem IV.2.2, $\beta_0(t), \ldots, \beta_p(t)$ must be continuous, the kernel $\tilde{K}$ of bounded variation and the bandwidth $\tilde{b} = \tilde{b}^{(n)}$ must tend to zero in such a way that $n\tilde{b}^{(n)} \to \infty$. This ensures uniform consistency of (7.4.13) on some interval $[t_0, t]$. Furthermore, functions $v_{jl}(\cdot)$, $j, l = 0, 1, \ldots, p$, must exist such that

$$\sup_{s \in [0,t]} \left| \frac{1}{n} \sum_{i=1}^n Y_{ij}(s) Y_{il}(s) / \lambda_i(s) - v_{jl}(s) \right| \xrightarrow{\mathrm{P}} 0,$$

and the limiting matrix $\mathbf{v}(s) = (v_{jl}(s))$ should be nonsingular for all $s \in [0, t]$. Then under some further regularity conditions on the covariates [stronger than condition (B) in Theorem VII.4.1], $\sqrt{n}(\hat{\mathbf{B}}^w - \mathbf{B})$ converges weakly on $D[t_0, t]^{p+1}$ to a zero-mean $(p+1)$-variate Gaussian martingale with covariance between components j and l given by

$$\int_{t_0}^s (\mathbf{v}^{-1}(u))_{jl} \, \mathrm{d}u.$$

A uniformly consistent estimator of the latter is

$$n \sum_{i=1}^{n} \int_{t_0}^{s} (\mathbf{Y}^{-}(u))_{ji} (\mathbf{Y}^{-}(u))_{li} \, \mathrm{d}N_i(u),$$

where $\mathbf{Y}^{-}(u)$ is given by (7.4.11).

As a special case, we finally consider the following example.

EXAMPLE VII.4.4. The i.i.d. Case

As was the case for the Cox regression model (Example VII.2.7) it is possible to give more explicit conditions for Theorem VII.4.1 to hold if the processes

$$((N_i, Y_i, (Z_{i1}, \dots, Z_{ip}))(t), t \in \mathscr{T}, i = 1, \dots, n)$$

are assumed to be i.i.d. [and the $Z_{ij}(\cdot)$ to be left-continuous with right-hand limits]. Using functional forms of the strong law of large numbers [e.g., Andersen and Gill (1982, Appendix III)], it can be seen that a sufficient condition for conditions (A) and (B) to hold is that

$$\mathrm{E}\left(\sup_{s \in [0,t]} Y_i(s) |Z_{ij}^3(s)| < \infty \right)$$

for $j = 1, \dots, p$. Furthermore, the matrix $\mathbf{r}^{(2)}$ now has elements

$$r_{jl}^{(2)}(t) = \mathrm{E}\, Y_i(t) Z_{ij}(t) Z_{il}(t),$$

$j, l = 0, 1, \dots, p$, and condition (C), therefore, corresponds to a linear independence condition on the covariates. □

VII.5. Other Non- and Semiparametric Regression Models

Sections VII.2–VII.4 have dealt with some particular semiparametric and nonparametric specifications of how the hazard part $\alpha_{hi}(t; \mathbf{Z}_i(t))$ of the intensity process in (7.1.2) may depend on the individual covariate processes $\mathbf{Z}_i(\cdot)$, $i = 1, \dots, n$. In Section VII.5.1, we shall briefly present some alternative, more general nonparametric models putting most emphasis on results where counting process and martingale techniques have played a central role (McKeague and Utikal, 1990a, 1990b, 1991). Section VII.5.2 deals with two linear, semiparametric alternatives to the Cox regression model for survival data. In the *accelerated failure-time models*, the log survival time is linear in the covariates except for an error term with some unknown distribution [e.g., Ritov (1990)]. In *transformation models*, some unknown function of the survival time (like the log cumulative hazard as in the Cox regression model) is linear in the covariates with a known distribution [e.g., Dabrowska and Doksum (1988)].

VII.5.1. General Nonparametric Regression Models

Consider the model where the individual (and for convenience univariate) counting processes N_i, $i = 1, \dots, n$, have intensity processes given by

$$\lambda_i(t) = Y_i(t)\alpha(t; \mathbf{Z}_i(t)), \quad i = 1, \dots, n; \tag{7.5.1}$$

cf. (7.1.2). Following Beran (1981) and McKeague and Utikal (1990a), we make no further assumptions about the function α which we then want to estimate. We shall assume that the individual covariate processes $\mathbf{Z}_i(\cdot)$ are univariate though the resūlts in theory may be generalized to multivariate covariates.

Let z be fixed and consider an interval I_z containing z and having length w. Now define

$$N_i(t, z) = \int_0^t I(Z_i(s) \in I_z)\, \mathrm{d}N_i(s)$$

and (7.5.2)

$$Y_i(t, z) = I(Z_i(t) \in I_z)\, Y_i(t).$$

Then (for the special case of survival data) $N(t, z) = \sum_i N_i(t, z)$ counts the number of failures in $[0, t]$ among individuals with covariate values close to z and $Y(t, z) = \sum_i Y_i(t, z)$ is the number at risk at time $t-$ with covariate values close to z. The integrated hazard

$$A(t, z) = \int_0^t \alpha(s, z)\, \mathrm{d}s \tag{7.5.3}$$

is now estimated by the Nelson–Aalen estimator

$$\hat{A}(t, z) = \int_0^t \frac{N(\mathrm{d}s, z)}{Y(s, z)}. \tag{7.5.4}$$

From (7.5.4) estimates, $\tilde{\alpha}(t, z)$ of $\alpha(t, z)$ for any t, z can then be obtained by first smoothing $\hat{A}(t, z)$ in the t direction, getting

$$\hat{\alpha}(t, z) = b^{-1} \int_{\mathscr{T}} K\left(\frac{t - s}{b}\right) \hat{A}(\mathrm{d}s, z) \tag{7.5.5}$$

and then in the z direction,

$$\tilde{\alpha}(t, z) = \tilde{b}^{-1} \int \tilde{K}\left(\frac{z - x}{\tilde{b}}\right) \hat{\alpha}(t, x)\, \mathrm{d}x. \tag{7.5.6}$$

Here K and $\tilde{K}$ are kernel functions and b and $\tilde{b}$ bandwidths (Section IV.2). McKeague and Utikal (1990a) [see also Dabrowska (1987)] derived large sample properties as $n \to \infty$ of (7.5.4), (7.5.5), and (7.5.6) when $w \to 0$, $b \to 0$, $\tilde{b} \to 0$ are suitably balanced. In particular, they found conditions under which the process

$$(nw_n)^{1/2}(\hat{A}(\cdot, z) - A(\cdot, z))$$

converges weakly and

$$(nw_nb_n)^{1/2}(\hat{\alpha}(t, z) - \alpha(t, z))$$

converges in distribution to certain Gaussian limits with zero mean and covariances that can be estimated consistently. Also, uniform consistency of (7.5.6) was proved.

Keiding (1990) suggested using instead a bivariate smoothing technique to estimate $\alpha(\cdot, \cdot)$ and presented some practical illustrations of the methods in connection with the so-called *Lexis diagram*. We return to this in Section X.1.

Again taking (7.5.1) as their starting point, McKeague and Utikal (1990a) estimated

$$\mathscr{A}(t, z) = \int_0^z A(t, x)\,\mathrm{d}x = \int_0^t \int_{-\infty}^z \alpha(s, x)\,\mathrm{d}x\,\mathrm{d}s \tag{7.5.7}$$

[cf. (7.5.3)] by

$$\hat{\mathscr{A}}(t, z) = \int_{-\infty}^z \hat{A}(t, x)\,\mathrm{d}x. \tag{7.5.8}$$

From the limiting distribution of (7.5.8), which they derived under certain regularity conditions, McKeague and Utikal (1990a, 1990b) proposed tests for the hypothesis $\alpha(t, z) = \alpha(t)$ of no effect of the covariate or the hypothesis

$$\alpha(t, z) = \alpha_1(t)\alpha_2(z) \tag{7.5.9}$$

of proportionality. Also, tests for the reduction of (7.5.1) to Aalen's (1980) additive model

$$\alpha(t, z) = \beta_0(t) + \beta_1(t)z,$$

(Section VII.4) were studied as well as tests for the further reduction of (7.5.9) to the Cox proportional hazards model $\alpha_2(z) = \exp(\beta z)$.

Various special cases of the general multiplicative model (7.5.9) have been studied in the literature. Thus, Thomas (1983) estimated $\alpha_2(z)$ assuming it to be monotone, Hastie and Tibshirani (1986) used local likelihoods methods for estimation, whereas O'Sullivan (1989) used a penalized likelihood approach. Finally, Zucker and Karr (1990) studied a model of the form

$$\alpha(t, z) = \exp\left(\sum\nolimits_j z_j\alpha_j(t)\right)$$

with $\alpha_j(\cdot)$ unknown.

Common to all the suggestions discussed in this section is that more experience is needed with the application of the model, some of which are suspected to require very large samples to work properly.

VII.5.2. Other Semiparametric Linear Regression Models for Survival Data

In the Cox (1972) proportional hazards regression model for survival data with time-fixed covariates, the hazard is $\alpha(t, \mathbf{z}) = \alpha_0(t)\exp(\boldsymbol{\beta}^\mathsf{T}\mathbf{z})$ (Sections VII.2 and VII.3). Since the cumulative hazard evaluated at the failure time X is unit exponentially distributed (Section VI.3.1), the log cumulative hazard evaluated at X can be written as

$$\log A_0(X) = -\boldsymbol{\beta}^\mathsf{T}\mathbf{z} + \varepsilon, \tag{7.5.10}$$

where $\mathrm{P}(\varepsilon > t) = e^{-e^t}$, i.e., ε has an extreme value distribution. A general *transformation model* is a model where for some unknown monotone function h

$$h(X) = \boldsymbol{\beta}^\mathsf{T}\mathbf{z} + \varepsilon, \tag{7.5.11}$$

where ε has some known distribution. This is one generalization of the Cox model and Dabrowska and Doksum (1988) studied partial likelihood methods for estimating $\boldsymbol{\beta}$ based on a (possibly right-censored) sample of independent survival times $\tilde{X}_i$, $i = 1, \ldots, n$, and their corresponding covariates $\mathbf{Z}_i$, $i = 1, \ldots, n$.

Accelerated failure-time models can also be written as (7.5.11), but here h is known, $h(X) = \log X$. If the distribution, say F_0, of the error terms $\varepsilon_1, \ldots, \varepsilon_n$ is also known (up to a finite number of parameters), then the model is fully parametric (Example VII.6.3), whereas we have a semiparametric model if this distribution is unknown. This model has been studied by various authors including Miller (1976), Buckley and James (1979), Koul et al. (1981), Louis (1981), Wei and Gail (1983), James and Smith (1984), Ritov and Wellner (1988), Lai and Ying (1991b), Wei et al. (1990), Tsiatis (1990), and Ritov (1990), and, in the following, the results of the latter two papers will be briefly presented.

Let $V_i = \log \tilde{X}_i$ and assume that the covariates have been centered by subtracting their means. Then the score equation for $\boldsymbol{\beta}$ is

$$\sum_{i=1}^{n} \mathbf{Z}_i \left\{ D_i \frac{f_0'(V_i - \boldsymbol{\beta}^\mathsf{T}\mathbf{Z}_i)}{f_0(V_i - \boldsymbol{\beta}^\mathsf{T}\mathbf{Z}_i)} - (1 - D_i) \frac{f_0(V_i - \boldsymbol{\beta}^\mathsf{T}\mathbf{Z}_i)}{S_0(V_i - \boldsymbol{\beta}^\mathsf{T}\mathbf{Z}_i)} \right\} = \mathbf{0}, \tag{7.5.12}$$

where $f_0 = F_0'$, $S_0 = 1 - F_0$, and D_i, as usual, is the indicator $I(\tilde{X}_i = X_i)$. For known F_0, this equation may be solved to get an estimate of $\boldsymbol{\beta}$. When F_0 is unknown, one may replace the score $-f_0'/f_0$ by a suitable function s, for instance, $s(t) = t$ as for the standard normal distribution to get the equation $\boldsymbol{\Psi}(\boldsymbol{\beta}, s) = \mathbf{0}$, where

$$\boldsymbol{\Psi}(\boldsymbol{\beta}, s) = \sum_{i=1}^{n} \mathbf{Z}_i \left\{ D_i s(V_i - \boldsymbol{\beta}^\mathsf{T}\mathbf{Z}_i) - (1 - D_i) \frac{\int_{V_i - \boldsymbol{\beta}^\mathsf{T}\mathbf{Z}_i}^{\infty} s(t)\,\mathrm{d}S_0(t)}{S_0(V_i - \boldsymbol{\beta}^\mathsf{T}\mathbf{Z}_i)} \right\}. \tag{7.5.13}$$

Note that for $s(t) = t$, equation (7.5.13) reduces to the ordinary least squares equation when there are no censored observations. By partial integration, it follows that the equation can also be interpreted as an ordinary least squares equation where to each *censored* residual $V_i - \boldsymbol{\beta}^\mathsf{T}\mathbf{Z}_i$ is added the conditional expectation

$$\frac{\int_{V_i - \boldsymbol{\beta}^\mathsf{T}\mathbf{Z}_i}^{\infty} S_0(t)\,\mathrm{d}t}{S_0(V_i - \boldsymbol{\beta}^\mathsf{T}\mathbf{Z}_i)},$$

given that the residual exceeds the observed value. Equation (7.5.13) may be solved iteratively (if it has a solution) from a starting value $\hat{\boldsymbol{\beta}}^{(0)}$ by replacing the unknown S_0 by the Kaplan–Meier estimator $\hat{S}_0^{\hat{\boldsymbol{\beta}}^{(0)}}$ (Section IV.3) based on the (possibly censored) residuals $V_i - \hat{\boldsymbol{\beta}}^{(0)\mathsf{T}}\mathbf{Z}_i$. The large sample properties of the resulting estimator, originally proposed by Buckley and James (1979), were studied by Ritov (1990) using (among other tools) martingale methods.

A different approach to the estimation of $\boldsymbol{\beta}$ was taken by Tsiatis (1990) motivated by linear rank tests (Prentice, 1978). Thus, a linear rank statistic for testing $\boldsymbol{\beta} = \boldsymbol{\beta}_0$ is

$$\boldsymbol{\Gamma}(\boldsymbol{\beta}_0, w) = \sum_{i=1}^{n} \int_{-\infty}^{\infty} \{\mathbf{Z}_i - \overline{\mathbf{Z}}^{\boldsymbol{\beta}_0}(y)\} w(y)\,\mathrm{d}N_i(y + \boldsymbol{\beta}_0^\mathsf{T}\mathbf{Z}_i), \qquad (7.5.14)$$

where $\overline{\mathbf{Z}}^{\boldsymbol{\beta}}$ is the average

$$\overline{\mathbf{Z}}^{\boldsymbol{\beta}}(y) = \frac{\sum_{j=1}^{n} \mathbf{Z}_j I(V_j - \boldsymbol{\beta}^\mathsf{T}\mathbf{Z}_j \geq y)}{\sum_{j=1}^{n} I(V_j - \boldsymbol{\beta}^\mathsf{T}\mathbf{Z}_j \geq y)}$$

and $w(y)$ is a weight function of the "residual time scale" $y = v - \boldsymbol{\beta}_0^\mathsf{T}\mathbf{z}$. An estimate for $\boldsymbol{\beta}$ is then obtained as a solution to the equation $\boldsymbol{\Gamma}(\boldsymbol{\beta}, w) = \mathbf{0}$ and Tsiatis (1990) studied its large sample properties using to a certain extent methods based on counting processes. One of the main technical difficulties in both approaches has to do with the fact that it is necessary to deal with several time scales (several values of $\boldsymbol{\beta}$) simultaneously.

Ritov (1990) established an asymptotic equivalence between the two classes of estimators. More specifically, he showed that for any score $s(\cdot)$ in (7.5.13) there exists a weight $w(\cdot)$ in (7.5.14) [depending on $s(\cdot)$ and the Kaplan–Meier estimator $\hat{S}_0^{\boldsymbol{\beta}}$ for the residuals] such that $\boldsymbol{\Psi}(\boldsymbol{\beta}, s) = \boldsymbol{\Gamma}(\boldsymbol{\beta}, w)$. This $w(\cdot)$ is given by

$$w(t; s, \hat{S}_0^{\boldsymbol{\beta}}) = s(t) + \frac{\int_t^\infty s(u)\,\mathrm{d}\hat{S}_0^{\boldsymbol{\beta}}(u)}{\hat{S}_0^{\boldsymbol{\beta}}(t)}.$$

Conversely, for a given weight $w(\cdot)$ in (7.5.14), there is a score $s(\cdot)$ in (7.5.13) such that $\boldsymbol{\Psi}(\boldsymbol{\beta}, s) = \boldsymbol{\Gamma}(\boldsymbol{\beta}, w)$, namely,

$$s(t; w) = w(t) + \int_{-\infty}^{t} w(u)\hat{S}_0^{\boldsymbol{\beta}}(u-)^{-1}\,\mathrm{d}\hat{S}_0^{\boldsymbol{\beta}}(u).$$

The operators W_F and S_F corresponding to a distribution function F given by

$$W_F s(t) = s(t) - \frac{\int_t^\infty s(u)\,\mathrm{d}F(u)}{1 - F(t)}$$

and

$$S_F w(t) = w(t) - \int_{-\infty}^{t} w(u)\frac{\mathrm{d}F(u)}{1 - F(u-)}$$

used by Ritov to establish the equivalence are, in fact, well-known: if $F = F_\theta$ depends on a real parameter θ and if we let $f_\theta = F'_\theta$ be the density function and $\alpha_\theta = f_\theta/(1 - F_\theta)$ the hazard function, then

$$W_F\left\{\frac{\partial}{\partial\theta}\log f_\theta\right\} = \frac{\partial}{\partial\theta}\log\alpha_\theta$$

and, conversely,

$$S_F\left\{\frac{\partial}{\partial\theta}\log\alpha_\theta\right\} = \frac{\partial}{\partial\theta}\log f_\theta.$$

Further properties of these operators were studied by Ritov and Wellner (1988); see also Efron and Johnstone (1990).

VII.6. Parametric Regression Models

VII.6.1. Examples and Large Sample Properties

We now turn to models where the intensity process for N_{hi},

$$\lambda_{hi}^{\boldsymbol{\theta}}(t) = Y_{hi}(t)\alpha_{hi}^{\boldsymbol{\theta}}(t; \mathbf{Z}_i(t)) \tag{7.6.1}$$

[cf. (7.1.2)] is specified by a q-dimensional parameter $\boldsymbol{\theta} = (\theta_1, \ldots, \theta_q)$; see also the introductory remarks in Chapter VI. In the present section, we first consider some examples and then we outline the large sample properties of the model (7.6.1). Section VII.6.2 contains a brief discussion of goodness-of-fit methods.

EXAMPLE VII.6.1. Multiplicative Models

Our main example of a parametric regression model will be the *multiplicative model* given by

$$\alpha_{hi}^{\boldsymbol{\theta}}(t; \mathbf{Z}_i(t)) = \alpha_{h0}(t, \gamma)r(\boldsymbol{\beta}^{\mathsf{T}}\mathbf{Z}_{hi}(t)) \tag{7.6.2}$$

[cf. (7.1.5)], where $\boldsymbol{\theta} = (\gamma, \boldsymbol{\beta})$ is finite-dimensional. We shall restrict ourselves to the case where the relative risk function is the exponential function, in which case (for survival data) (7.6.2) covers among other models:

(1) exponential regression: $\alpha_{h0}(t, \gamma) = \gamma_h$,
(2) Weibull regression: $\alpha_{h0}(t, \gamma) = \gamma_h t^{\rho_h}$,
(3) Gompertz–Makeham regression: $\alpha_{h0}(t, \gamma) = a_h + b_h c_h^t$,
(4) piecewise exponential regression: $\alpha_{h0}(t, \gamma)$ piecewise constant,
(5) relative mortality regression: $\alpha_{hi}(t, \boldsymbol{\theta}) = \mu_i(t) \exp(\boldsymbol{\beta}^\mathsf{T} \mathbf{Z}_{hi}(t))$.

Note that the piecewise exponential regression model (4) is a generalization of the log-linear intensity models studied in Example VI.1.3 in that the covariates are now arbitrary (and do not necessarily only take a finite number of different values). This model [and, hence, (1)] is also immediately applicable for more general counting process models, i.e., not necessarily models for survival data, but also (2) or (3) may be relevant models for mortality rates, for instance, in a disability model. In model (5), the $\mu_i(\cdot)$ are (assumed known) population mortalities (Breslow et al., 1983); see, e.g., Examples III.1.3, III.1.7, and IV.1.11.

Below we shall study in some detail conditions for the usual asymptotic properties of (7.6.2) to hold as well as some numerical examples. Section VII.6.2 deals with goodness-of-fit methods for this model. □

EXAMPLE VII.6.2. Additive Models

Parametric versions of the *additive models* considered (for the nonparametric case) in Section VII.4 may be of interest. Here

$$\alpha_{hi}^{\boldsymbol{\theta}}(t; \mathbf{Z}_i(t)) = \sum_{j=0}^{p} \beta_j(t, \boldsymbol{\theta}) \mathbf{Z}_{hi}(t) \tag{7.6.3}$$

[cf. (7.1.4)], where $\boldsymbol{\theta}$ is q-dimensional. In particular, the β_j may be *constant*, $\beta_j(t, \boldsymbol{\theta}) = \theta_j$, as considered in a very special case in the linear intensity model of Example VI.1.5, or they may include population mortality as in the model studied by Pocock et al. (1982). □

EXAMPLE VII.6.3. Log-Linear Survival Time Models

In the spirit of classical linear normal models, one sometimes, for survival data, studies models where the log survival time, $\log(X_i)$, itself is linear in the covariates; cf. Section VII.5.2. Thus,

$$\log(X_i) = \sum_{j=0}^{p} \beta_j Z_{ij} + \varepsilon_i, \tag{7.6.4}$$

where the error terms $\varepsilon_1, \ldots, \varepsilon_n$ are i.i.d. with mean zero and a distribution belonging to some parametric class like the normal, the logistic, or one of the extreme value distributions. Here, the hazard function of X_i has the form

$$\alpha^{\theta}(t; \mathbf{Z}_i) = \alpha_0(t \exp(-\boldsymbol{\beta}^\mathsf{T} \mathbf{Z}_i)) \exp(-\boldsymbol{\beta}^\mathsf{T} \mathbf{Z}_i), \tag{7.6.5}$$

where the shape of α_0 depends on the distribution of the error terms. Models with such a hazard function are known as *accelerated failure-time models* and include the Weibull regression models (2) mentioned in Example VII.6.1. Semiparametric inference in this model was discussed in Section VII.5.2. □

Inference from parametric regression models will be based on the (partial) likelihood (7.1.1) which gives a log (partial) likelihood of the following form:

$$C_\tau(\boldsymbol{\theta}) = \sum_{h,i} \left(\int_0^\tau \log\{\lambda_{hi}^{\boldsymbol{\theta}}(t)\}\, \mathrm{d}N_{hi}(t) - \int_0^\tau \lambda_{hi}^{\boldsymbol{\theta}}(t)\, \mathrm{d}t \right). \tag{7.6.6}$$

Furthermore, assuming enough regularity to interchange the order of differentiation and integration, the vector of score statistics ($U_\tau^j(\boldsymbol{\theta}) = (\partial/\partial\theta_j)C_\tau(\boldsymbol{\theta})$, $j = 1,\ldots,q$) is given by

$$U_\tau^j(\boldsymbol{\theta}) = \sum_{h,i} \left(\int_0^\tau \frac{\partial}{\partial\theta_j} \log \lambda_{hi}^{\boldsymbol{\theta}}(t)\, \mathrm{d}N_{hi}(t) - \int_0^\tau \frac{\partial}{\partial\theta_j} \lambda_{hi}^{\boldsymbol{\theta}}(t)\, \mathrm{d}t \right); \tag{7.6.7}$$

cf. also (6.1.3). The maximum (partial) likelihood estimator is now defined as the solution $\hat{\boldsymbol{\theta}}$ to the equation $\mathbf{U}_\tau(\boldsymbol{\theta}) = \mathbf{0}$.

The general results in Section VI.1.2 are formulated in such a way that the condition to be checked for the consistency and asymptotic normality to hold is exactly Condition VI.1.1. We are not going to discuss this condition in more detail, but, following Borgan (1984), we shall rather see how they are translated for our main example: the multiplicative model (7.6.2) with $r(\cdot) = \exp(\cdot)$ discussed in Example VII.6.1. For this model, the expressions for the score vector $\mathbf{U}_\tau(\boldsymbol{\gamma}, \boldsymbol{\beta})$ reduce to

$$\begin{aligned}
&\frac{\partial}{\partial\boldsymbol{\gamma}} C_\tau(\boldsymbol{\gamma}, \boldsymbol{\beta}) \\
&\quad = \sum_{h,i} \left\{ \int_0^\tau \frac{\partial}{\partial\boldsymbol{\gamma}} \log \alpha_{h0}(t, \boldsymbol{\gamma})\, \mathrm{d}N_{hi}(t) - \int_0^\tau \frac{\partial}{\partial\boldsymbol{\gamma}} \alpha_{h0}(t, \boldsymbol{\gamma}) Y_{hi}(t) \exp(\boldsymbol{\beta}^\mathsf{T} \mathbf{Z}_{hi}(t))\, \mathrm{d}t \right\} \\
&\quad = \sum_{h=1}^k \int_0^\tau \frac{\partial}{\partial\boldsymbol{\gamma}} \log \alpha_{h0}(t, \boldsymbol{\gamma}) \{ \mathrm{d}N_{h\cdot}(t) - \alpha_{h0}(t, \boldsymbol{\gamma}) S_h^{(0)}(\boldsymbol{\beta}, t)\, \mathrm{d}t \},
\end{aligned} \tag{7.6.8}$$

$$\begin{aligned}
\frac{\partial}{\partial\boldsymbol{\beta}} C_\tau(\boldsymbol{\gamma}, \boldsymbol{\beta}) &= \sum_{h,i} \left\{ \int_0^\tau \mathbf{Z}_{hi}(t)\, \mathrm{d}N_{hi}(t) - \int_0^\tau \alpha_{h0}(t, \boldsymbol{\gamma}) \mathbf{Z}_{hi}(t) Y_{hi}(t) \exp(\boldsymbol{\beta}^\mathsf{T} \mathbf{Z}_{hi}(t))\, \mathrm{d}t \right\} \\
&= \int_0^\tau \left\{ \sum_{h,i} \mathbf{Z}_{hi}(t)\, \mathrm{d}N_{hi}(t) - \sum_{h=1}^k \alpha_{h0}(t, \boldsymbol{\gamma}) \mathbf{S}_h^{(1)}(\boldsymbol{\beta}, t)\, \mathrm{d}t \right\},
\end{aligned} \tag{7.6.9}$$

where the sums $S_h^{(0)}$ and $\mathbf{S}_h^{(1)}$ are defined in (7.2.11) and (7.2.12), respectively. Inserting the true parameter value $(\boldsymbol{\gamma}_0, \boldsymbol{\beta}_0)$ into (7.6.8) and (7.6.9) evaluated at t instead of τ, we get the local martingales

$$\frac{\partial}{\partial\boldsymbol{\gamma}} C_t(\boldsymbol{\gamma}_0, \boldsymbol{\beta}_0) = \sum_{h=1}^k \int_0^t \frac{\partial}{\partial\boldsymbol{\gamma}} \log \alpha_{h0}(u, \boldsymbol{\gamma}_0)\, \mathrm{d}M_{h\cdot}(u) = \mathbf{M}^{\boldsymbol{\gamma}}(t)$$

and

$$\frac{\partial}{\partial \boldsymbol{\beta}} C_t(\gamma_0, \boldsymbol{\beta}_0) = \sum_{h=1}^{k} \sum_{i=1}^{n} \int_0^t \mathbf{Z}_{hi}(u)\, \mathrm{d}M_{hi}(u) = \mathbf{M}^{\boldsymbol{\beta}}(t),$$

respectively. These have predictable covariation processes

$$\langle \mathbf{M}^{\boldsymbol{\gamma}} \rangle (t) = \sum_{h=1}^{k} \int_0^t \left\{ \frac{\partial}{\partial \boldsymbol{\gamma}} \log \alpha_{h0}(u, \gamma_0) \right\}^{\otimes 2} S_h^{(0)}(\boldsymbol{\beta}_0, u) \alpha_{h0}(u, \gamma_0)\, du,$$

$$\langle \mathbf{M}^{\boldsymbol{\beta}} \rangle (t) = \sum_{h=1}^{k} \int_0^t \mathbf{S}_h^{(2)}(\boldsymbol{\beta}_0, u) \alpha_{h0}(u, \gamma_0)\, du,$$

$$\langle \mathbf{M}^{\boldsymbol{\gamma}}, \mathbf{M}^{\boldsymbol{\beta}} \rangle (t) = \sum_{h=1}^{k} \int_0^t \left\{ \frac{\partial}{\partial \boldsymbol{\gamma}} \log \alpha_{h0}(u, \gamma_0) \right\} \otimes \mathbf{S}_h^{(1)}(\boldsymbol{\beta}_0, u) \alpha_{h0}(u, \gamma_0)\, du,$$

where $\mathbf{S}_h^{(2)}$ is defined in (7.2.13). Therefore, the main assumption for (B) in Condition VI.1.1 to hold is (a) in Condition VII.2.1. For the Lindeberg condition (C) in Condition VI.1.1, we need the Lindeberg condition VII.2.2, whereas for the first part of (E) in Condition VI.1.1, it is sufficient that a condition like Condition VII.2.1(a) also holds for the third-order partial derivatives of $S_h^{(0)}(\boldsymbol{\beta}, t)$ with respect to $\boldsymbol{\beta}$. Finally, (A) in Condition VI.1.1 should hold for $\alpha_{h0}(t, \gamma)$ which must also be bounded away from zero on $\Theta_0 \times \mathscr{T}$ [see also Condition VII.2.1(c)].

As a simple numerical example of a parametric regression model for survival data, we once more study the melanoma data.

EXAMPLE VII.6.4. Survival with Malignant Melanoma. Regression Models with Piecewise Constant Hazard

We consider the melanoma data introduced in Example I.3.1 and study a model with a piecewise constant baseline hazard function (Example VII.6.1). Following the results of Section VII.3 (e.g., Examples VII.3.1 and VII.3.4), possibly nonproportional hazards for patients with and without ulceration are allowed. With the covariates sex ($Z_{i1} = 1$ for males and 0 for females) and log (thickness in mm) (Z_{i4}) defined as in Example VII.3.2, the model is

$$\alpha_i(t) = \alpha_0^{(i)}(t) \exp(\beta_1 Z_{i1} + \beta_4 Z_{i4}),$$

where

$$\alpha_0^{(i)}(t) = \begin{cases} \alpha_{00} & \text{if } t < 3 \text{ years and patient } i \text{ does not have ulceration} \\ \alpha_{01} & \text{if } t < 3 \text{ years and patient } i \text{ has ulceration} \\ \alpha_{10} & \text{if } t \geq 3 \text{ years and patient } i \text{ does not have ulceration} \\ \alpha_{11} & \text{if } t \geq 3 \text{ years and patient } i \text{ has ulceration.} \end{cases}$$

In this model, $\hat{\beta}_1 = 0.357$ (0.270) and $\hat{\beta}_4 = 0.531$ (0.177), close to the estimates found for the Cox regression model in Example VII.3.4.

Furthermore,

$$\log \hat{\alpha}_{00} = -12.870\,(1.02),$$
$$\log(\hat{\alpha}_{10}/\hat{\alpha}_{00}) = 0.540\,(0.529),$$
$$\log(\hat{\alpha}_{01}/\hat{\alpha}_{00}) = 1.451\,(0.501),$$
$$\log\{\hat{\alpha}_{11}\hat{\alpha}_{00}/(\hat{\alpha}_{10}\hat{\alpha}_{01})\} = -0.886\,(0.618).$$

Thus, for $t < 3$ years, ulceration increases the hazard by a factor $\exp(1.451) = 4.27$, whereas the effect for $t \geq 3$ years is $\exp(1.451 - 0.886) = 1.76$. This difference is, however, insignificant, the Wald test statistic being $(-0.886/0.618)^2 = 2.04$, whereas the likelihood ratio test statistic is 2.10 (both giving $P \approx 0.15$ when referred to a χ_1^2 distribution) and the conclusion is that, within the piecewise exponential model, the proportional hazards assumption for ulceration cannot be rejected.

In the reduced model, ulceration is, therefore, included as a covariate ($Z_{i3} = 1$ if present and 0 if absent), whereas the baseline hazard is

$$\alpha_0(t) = \begin{cases} \alpha_{0.} & \text{if } t < 3 \text{ years} \\ \alpha_{1.} & \text{if } t \geq 3 \text{ years,} \end{cases}$$

and the estimates are $\hat{\beta}_1 = 0.360$ (0.270), $\hat{\beta}_4 = 0.533$ (0.177), $\hat{\beta}_3 = 0.953$ (0.324). Furthermore,

$$\log \hat{\alpha}_{0.} = -12.493\,(0.960)$$
$$\log(\hat{\alpha}_{1.}/\hat{\alpha}_{0.}) = -0.103\,(0.265)$$

and we cannot reject the hypothesis of a constant hazard function [Wald test statistic $(-0.103/0.265)^2 = 0.15$, likelihood ratio test statistic 0.10, both $P > 0.70$]. In the model with a constant hazard function, the effect of sex is insignificant ($P = 0.25$) and we end up with a model where $\log \hat{\alpha}_0 = -12.622$ (0.942), whereas $\hat{\beta}_4 = 0.570$ (0.173) and $\hat{\beta}_3 = 0.995$ (0.319), in close agreement with the results of the Cox regression models. It should, however, be emphasized that the test for a nonconstant effect of ulceration using time-dependent covariates (Example VII.3.4) seemed to be more sensitive than the test based on a piecewise exponential model used in the present example. □

VII.6.2. Goodness-of-Fit Methods for the Multiplicative Model

In this section, we briefly consider some graphical and numerical ways of checking the multiplicative model (7.6.2). Following Section VI.3.1, a given parametric specification $\alpha_{h0}(t, \boldsymbol{\gamma})$ of the baseline intensity function in (7.6.2) may be checked by graphical inspection of the estimate $\hat{A}_{h0}(t, \hat{\boldsymbol{\beta}})$ in the "corresponding" semiparametric model; cf. (7.2.5). For example, an approximately *linear* integrated baseline hazard estimate $\hat{A}_{h0}(t, \hat{\boldsymbol{\beta}})$ suggests that a model with a constant baseline hazard $\alpha_{h0}(t, \boldsymbol{\gamma}) = \gamma_h$ might be appropriate. This is, for instance, the case for the melanoma data; cf. Examples VII.3.4 and

VII.6.4. Also, "residuals" as introduced at the end of Section VI.3.1 and further discussed in Section VII.3.4 may be used in connection with parametric regression models but we shall not go into further details here.

Tests for a parametric specification of a regression model may be carried out by embedding it in a larger model class and testing the significance of the reduction using, for instance, the (partial) likelihood ratio test statistic. This was exemplified in Example VII.6.4 where a multiplicative regression model with a constant baseline hazard function was tested against a model with a piecewise constant baseline hazard.

Let us further mention that the approach of Hjort (1990a) discussed in Section VI.3.3 may also be applied in a regression context. Thus, for the case $k = 1$, Hjort (1990a) studied processes of the form

$$Z(t) = \int_0^t K(s, \hat{\gamma}, \hat{\boldsymbol{\beta}}) \{ \mathrm{d}\hat{A}_0(s, \hat{\boldsymbol{\beta}}) - J(s)\alpha_0(s, \hat{\gamma})\, \mathrm{d}s \}; \tag{7.6.10}$$

cf. (6.3.9). Here, $(\hat{\boldsymbol{\beta}}, \hat{\gamma})$ are the maximum likelihood estimates solving the equations $(\partial/\partial\boldsymbol{\beta})C_\tau(\gamma, \boldsymbol{\beta}) = \mathbf{0}$ and $(\partial/\partial\boldsymbol{\gamma})C_\tau(\gamma, \boldsymbol{\beta}) = \mathbf{0}$ [cf. (7.6.8) and (7.6.9)], $K(\cdot)$ is a suitably chosen weight process, and $\hat{A}_0(s, \hat{\boldsymbol{\beta}})$ is defined in (7.2.5). Weak convergence of (7.6.10) (properly normalized) was derived by Hjort (1990a) following the lines of Section VI.3.3. He also studied formal χ^2-type goodness-of-fit tests.

We finally mention that also the approach of Khmaladze (1981, 1988) (see Section VI.3.3) may be used for goodness-of-fit examination of a parametric regression model.

VII.7. Bibliographic Remarks

Section VII.1

The first regression model formulated for counting processes was the additive model of Aalen (1980) who also remarked that the Cox regression model could be discussed in this framework. This was later done by Næs (1982) and Andersen and Gill (1982) (for the univariate case, $k = 1$) and by Andersen and Borgan (1985). See also Prentice et al. (1981), Andersen et al. (1988), and the bibliographic remarks for Chapter III.

The fundamental paper by Cox (1972) gave rise to an extensive discussion in the literature on the interpretation of (7.2.6) as a "likelihood function" (Kalbfleisch and Prentice, 1973, 1980; Breslow, 1974; Cox, 1975; Jacobsen, 1982, 1984; Bailey, 1983; Johansen, 1983) and on the estimation of the cumulative baseline hazard function (Breslow, 1972, 1974; Kalbfleisch and Prentice, 1973, 1980; Jacobsen, 1982, 1984; Bailey, 1983; Link, 1984; Pons and Turckheim, 1987). In our treatment of this topic in Section VII.2.1, following Johansen (1983) we derived the Breslow estimator for the cumulative baseline hazard as a nonparametric maximum likelihood estimator and the Cox partial likelihood as a profile likelihood.

Section VII.2

The large sample properties of the estimators in the Cox regression model were derived using counting process techniques by Andersen and Gill (1982) [see also Prentice and Self (1983) for a discussion of relative risk functions other than the exponential function, Example VII.2.12]. Alternative proofs have been presented by Tsiatis (1981), Næs (1982), and Bailey (1983); see Section 4 in the paper by Andersen and Gill (1982).

Small sample properties of $\hat{\boldsymbol{\beta}}$ and the related test statistics for the special case of censored survival data have been investigated by Peace and Flora (1978), Johnson et al. (1982), and Lee et al. (1983) using simulations. The general tendencies found in these studies seem to be that one should not be too pessimistic about small samples; in many of the situations studied, the asymptotics work quite well even for sample sizes about 50. The performance certainly depends on features like amount of censoring and balancedness and correlation of covariates. Efficiency of $\hat{\boldsymbol{\beta}}$ compared to a parametric model for the baseline hazard was discussed by Efron (1977) and Oakes (1977); see also Kalbfleisch and Prentice (1980, Section 4.7), Cox and Oakes (1984, Section 8.5), and our Sections VIII.4.3 and VIII.5. Also, the bootstrap investigations of Chen and George (1985) and Altman and Andersen (1989) point to a satisfactory performance of the asymptotics (for the sample sizes they studied). Further bootstrap results were given by Hjort (1985b). Results on existence and robustness of $\hat{\boldsymbol{\beta}}$ were given by Jacobsen (1989a) (see Example VII.2.3) and by Bednarski (1989, 1991), respectively.

In our discussion of the asymptotic results for the Cox regression model, we have concentrated on what may be termed a "classical cohort design" where it is assumed that covariate informaton $Z_{hi}(t)$ is available for all individuals at risk at time $t-$, i.e., for all i with $Y_{hi}(t) = 1$. Less data demanding designs have been considered in the literature, the basic idea being that covariates are assembled for all individuals who experience one of the events under study (the *cases*) and for a random sample of the remaining individuals. Thus, Thomas (1977), see also Prentice and Breslow (1978), discussed the possibility of taking, at every event time t, a random sample of the individuals at risk at time t. This "retrospective" or "nested or synthetic case-control" design was further discussed by, among others, Oakes (1981), Breslow et al. (1983), and Prentice (1986b). Goldstein and Langholz (1992) derived large sample results for this procedure using counting process theory, and their approach was simplified and extended to more general sampling designs by Borgan, Goldstein, and Langholz (1992) using marked point processes. Another possibility suggested by Prentice (1986a) [with large sample theory by Self and Prentice (1988)] is the "case-cohort" design where, again, data from all individuals experiencing an event are analyzed together with data from a subcohort randomly selected at the beginning of the study.

Bayesian analysis was discussed by Kalbfleisch and Prentice (1980, Section 8.4), based on the paper by Kalbfleisch (1978b), and later by Hjort (1990b), and (in the context of so-called *frailty models*, see Chapter IX) by Clayton (1991).

Section VII.3

In Section VII.3, we have put most emphasis on goodness-of-fit methods for the Cox regression model where counting processes play a role and which are, therefore, applicable outside the classical censored survival data situation. Development of goodness-of-fit methods for that situation attracted a lot of attention in the 1980s. Early contributions on *graphical methods* were the papers by Kay (1977), Crowley and Hu (1977), Cox (1979), Lagakos (1981), Schoenfeld (1982), and Andersen (1982), whereas various *tests* for the model were discussed by Schoenfeld (1982), Andersen (1982), and Nagelkerke et al. (1984). Aranda-Ordaz (1983), O'Quigley and Moreau (1984, 1986), and Moreau et al. (1985, 1986) developed tests for the model by imbedding it in larger classes of models. Kay (1984) gave a summary of these various proposals for graphical and numerical methods which seem to highlight different aspects of the model. For more recent methods based on residuals and other regression diagnostics, the reader is referred to the references in Section VII.3.4.

A case study with a nice review of the Cox regression model was presented by Kardaun (1983).

Sections VII.4–VII.6

The nonparametric additive model discussed in Section VII.4 was introduced by Aalen (1980) but received little attention until Mau (1986b) demonstrated its use as an exploratory graphical tool for detection of a possible time-dependent effect of a covariate. A review of the model was given by Aalen (1989), whereas the large sample properties of the estimator were studied by McKeague (1988a) and by Huffer and McKeague (1991) following up on the early work of Aalen (1980). McKeague (1986, 1988b) discussed this model in connection with sieve estimation and grouped survival data, respectively.

References for the general nonparametric regression models and the accelerated failure-time model are given in Sections VII.5.1 and VII.5.2, respectively. Various versions of the transformation model (7.5.11) have been studied using rank or partial likelihood methods (Pettitt, 1982, 1983, 1984; Cuzick, 1985, 1988; Clayton and Cuzick, 1985a, 1985b; Doksum,

1987; Dabrowska and Doksum, 1988) building partly on earlier work by Kalbfleisch (1978a) and Prentice (1978).

Our discussions of parametric regression models in Sections VII.6.1 and VII.6.2 are based on the papers by Borgan (1984) and Hjort (1990a), respectively. There exists substantial literature on various special parametric regression models for survival data. A nice review of this literature is found in Chapter 6 in the book by Lawless (1982); see also Section 10.3 in the book by Lee (1980) and the bibliographic remarks to Chapter 3 in the book by Kalbfleisch and Prentice (1980).

Further Topics

As mentioned in Section I.1, there are other models for survival data (and more general life history data) than those based on counting processes which are relevant in practice. Among such techniques, we here mention regression analysis of *grouped survival data* discussed by Prentice and Gloeckler (1978) and Pierce et al. (1979) and refer the reader to Section 7.3 in the book by Lawless (1982) for further details.

In the presentation in this chapter, we have everywhere assumed that we have been analyzing the correct model. There exists a fairly large literature, however, on *misspecified models*. The misspecification could consist in analyzing a proportional hazards model when the true model is an accelerated failure-time model (Solomon, 1984; Struthers and Kalbfleisch, 1986). Another misspecification could be the omission of important covariates in a Cox regression model (Lagakos and Schoenfeld, 1984; Struthers and Kalbfleisch, 1986; Morgan, 1986; Hougaard, 1986a, 1986b, 1987; Bretagnolle and Huber–Carol, 1988) or in other types of regression models (Gail et al., 1984, 1988; Gail, 1986; Aalen, 1989), or the effect of a covariate could be specified using a wrong functional form (Lagakos, 1988). Models misspecified by omitting important covariates are related to the so-called *frailty models* to be discussed in Chapter IX.

CHAPTER VIII
Asymptotic Efficiency

In the counting process models we have studied in previous chapters, likelihood considerations have often lead to the construction of specific test statistics and estimators, in both parametric and nonparametric problems. At the same time, the martingale central limit theorem provided large sample distributional properties of these statistical procedures. This has also enabled us to make an asymptotic comparison between specific procedures and to choose the best within some special classes of procedures: for instance, the "martingale-based" k-sample tests of Section V.2 and the M-estimators of Section VI.2.

However, by combining likelihood and central limit theory from the start, we can, using LeCam's theory of local asymptotic normality, get much more: We can derive asymptotic bounds to the effectiveness of all statistical procedures within very broad classes and thereby show that the best test statistic or best estimator of the just mentioned rather special classes actually have much wider and stronger optimality properties. With this general theory, we can show, for instance, that tests of hypotheses, within a class only restricted by asymptotic requirements of size and perhaps also asymptotic unbiasedness or invariance, cannot asymptotically have more than the power of the best test of Section V.2 against certain alternatives; that estimators within a class of reasonable estimators defined by a notion of regularity (a kind of asymptotic equivariance) cannot have asymptotic distributions more concentrated about the true parameter value than a certain normal distribution; the same asymptotic distribution as that of the maximum likelihood estimator or best M-estimator of Section VI.2. The variance of this optimal limiting distribution corresponds to the inverse of the Fisher information and is commonly called the information bound.

The classical theory of local asymptotic normality—as treated in such books as those of Ibragimov and Khas'minskii (1981), Fabian and Hannan

(1985), Strasser (1985), LeCam (1986), and LeCam and Yang (1990)—is most fully developed in the context of finite-dimensional models, i.e., models with a Euclidean parameter of small, fixed dimension. It even includes constructive methods for achieving these bounds which work under the weakest possible assumptions, though they are not much used in practice. The general idea here (in estimation) is to use a one-step Newton–Raphson or scoring method starting from a square root-n consistent initial estimate, itself constructively defined by a discretization procedure. More often used maximum likelihood and Bayes methods only work under stronger conditions, but then share these optimality properties.

In recent years, much effort has been spent in building an infinite-dimensional version of the theory; important contributions being due to Levit (1973, 1975, 1978), Beran (1977), Pfanzagl (1982), Millar (1979), and Begun et al. (1983). For recent syntheses of this work, we refer especially to the monograph of van der Vaart (1988) and the book of Bickel et al. (1993). It turns out that this infinite-dimensional theory can be quite fully described in finite-dimensional terms—in fact, in terms of various finite-dimensional submodels. These considerations of parametric submodels are closely connected to the approach to nonparametric maximum likelihood estimation based on score functions of parametric submodels of Section IV.1.5, and to the local alternatives of Section V.2.3. Asymptotic efficiency in an infinite-dimensional testing or estimation problem is equivalent to asymptotic efficiency in the "hardest" parametric submodel. Determining the hardest parametric submodel typically becomes a problem in the calculus of variations, more abstractly but equivalently formulated as a problem of inverting an "information" operator in a Hilbert space, or of calculating a projection into a subspace. All the formulations result in practice in the need to solve some integral or differential equation. The feasibility of these theoretical calculations is generally closely connected with the possibility of actually constructing efficient procedures.

Of course, not only is the theory concerned with "nice" estimators, "nice" test statistics, etc. (restrictions of regularity, unbiasedness, invariance); it is also only about "nice" models, in the rough sense of models to which asymptotic normality is relevant at all. This corresponds, in the classical case of n independent and identically distributed observations (the i.i.d. case), to models where Fisher information is finite and positive and varies continuously with the parameters of the model. The higher the dimensionality of the parameter space, the easier it is for these requirements to be violated. As we depart from i.i.d. models, the amount of statistical information about different parameters may accumulate at different rates (as the amount of observations increases) and may not even become asymptotically nonrandom. But despite these possibilities, the theory has important practical consequences for many of the counting process models we have studied and we proceed to describe it in more detail.

The combination of likelihood plus central limit theorem finds expression in the concept of *local asymptotic normality* (LAN) of a sequence of statistical models or experiments. In the i.i.d. case, we write likelihoods as products of terms for independent random variables and use the classical central limit theorem (applied to the log-likelihood); in the counting process case, we write likelihoods as product-integrals of (conditional) terms for infinitesimal time intervals and use the martingale central limit theorem. But apart from this, results and methods of proof run completely in parallel. In fact, both are special cases of the more general theory of local asymptotic normality for filtered spaces, based on Hellinger processes; see Jacod and Shiryayev (1987).

In Section VIII.1, we give an overview of the theory of local asymptotic normality and asymptotic efficiency theory, concentrating on parametric models. In Section VIII.2, we show how local asymptotic normality may be verified in models for counting processes, and give applications to parametric and nonparametric models. Asymptotic efficiency is proved of the maximum likelihood estimators of Section VI.1. More strikingly, for the multiplicative intensity model, very strong nonparametric optimality properties are demonstrated of the "best" counting process based k-sample tests of Section V.2 (in particular, the log-rank test, the Peto–Prentice Wilcoxon generalization, the Harrington and Fleming tests) and of the nonparametric estimators (Nelson–Aalen, Kaplan–Meier, Aalen–Johansen) of Chapter IV. This is done by consideration of "hardest" parametric submodels. Section VIII.3 returns to the general theory and gives a more powerful and more abstract theory on optimal estimation and testing in semiparametric (infinite-dimensional) models. Section VIII.4 gives counting process applications: nonparametric estimation (VII.4.1) and testing (VIII.4.2) in the multiplicative intensity model are reconsidered from the abstract point of view, and then the Cox regression model and the Aalen linear regression model are analyzed in Sections VIII.4.3 and VIII.4.4. Nonparametric optimality of the Cox maximum partial likelihood estimator and of the best generalized least squares type estimator in the respective models is demonstrated. Section VIII.4.5 connects these optimality results to the nonparametric maximum likelihood interpretations of some of these statistics and also briefly discusses optimality considerations for partially specified models. Section VIII.5 contains further bibliographic and historical remarks.

Sections VIII.2 and VIII.4 start with more detailed overviews of the results of the sections, in particular summarizing the conclusions for the models of Chapters IV to VII in more detail.

VIII.1. Contiguity and Local Asymptotic Normality

In this section, we review the basic concepts of local asymptotic normality and of one of its ingredients, contiguity (both originating in the work of LeCam, 1953), briefly describing their application to the theory of asymptotically efficient tests and estimators.

VIII.1.1. Contiguity

Let $(\mathrm{P}^{(n)})$, $(\mathrm{Q}^{(n)})$ be two sequences of probability measures where, for each $n = 1, 2, \ldots$, $\mathrm{P}^{(n)}$ and $\mathrm{Q}^{(n)}$ are defined on the same measurable space (sample space) $(\Omega^{(n)}, \mathscr{F}^{(n)})$. Consider the basic statistical problem of choosing between $\mathrm{P}^{(n)}$ and $\mathrm{Q}^{(n)}$ (a simple null hypothesis versus a simple alternative). The theory of contiguity gives conditions under which this problem does not degenerate as $n \to \infty$. We say that the sequence $(\mathrm{Q}^{(n)})$ is contiguous to $(\mathrm{P}^{(n)})$, written $\mathrm{Q}^{(n)} \lhd \mathrm{P}^{(n)}$, if any of the following (equivalent) hypotheses hold:

$$\text{For any sequence of events } A^{(n)} \in \mathscr{F}^{(n)} \text{ for each } n,$$
$$\mathrm{P}^{(n)}(A^{(n)}) \to 0 \quad \Rightarrow \quad \mathrm{Q}^{(n)}(A^{(n)}) \to 0. \tag{8.1.1}$$

$$\text{For any sequence of random variables } T^{(n)},$$
$$T^{(n)} \overset{\mathrm{P}^{(n)}}{\to} 0 \quad \Rightarrow \quad T^{(n)} \overset{\mathrm{Q}^{(n)}}{\to} 0. \tag{8.1.2}$$

The sequence of distributions (under $\mathrm{P}^{(n)}$)
of the (possibly infinite) likelihood ratio $\mathrm{dQ}^{(n)}/\mathrm{dP}^{(n)}$
is bounded in probability as $n \to \infty$. (8.1.3)

If $\mathrm{Q}^{(n)} \lhd \mathrm{P}^{(n)}$ and $\mathrm{P}^{(n)} \lhd \mathrm{Q}^{(n)}$, written $P^{(n)} \lhd\rhd Q^{(n)}$, we say $\mathrm{P}^{(n)}$ and $\mathrm{Q}^{(n)}$ are (mutually) contiguous.

To explain these hypotheses, we recall that two probability measures P and Q are said to be *equivalent* if they are mutually absolutely continuous; their Radon–Nikodym derivative (the likelihood ratio) is then nowhere zero or infinite. If two probability measures are not equivalent, then one posseses a singular component with respect to the other. This leads to the possibility of a degenerating statistical testing problem—one has, for instance, tests of zero size with positive power. From (8.1.3), also holding with the roles of $\mathrm{P}^{(n)}$ and $\mathrm{Q}^{(n)}$ interchanged, we see that mutual contiguity is a kind of asymptotic equivalence of the probability measures $\mathrm{P}^{(n)}$ and $\mathrm{Q}^{(n)}$. Contiguous sequences have asymptotically nondegenerating likelihood ratios so that the statistical problem of choosing between $\mathrm{P}^{(n)}$ and $\mathrm{Q}^{(n)}$ is in the limit nontrivial. Condition (8.1.1) also describes an asymptotic equivalence between $\mathrm{P}^{(n)}$ and $\mathrm{Q}^{(n)}$; recall that P and Q are equivalent if and only if $\mathrm{P}(A) = 0 \Leftrightarrow \mathrm{Q}(A) = 0$. Finally, condition (8.1.2) (with "$\Rightarrow$" replaced by "$\Leftrightarrow$") describes a powerful characterization of contiguity: Convergence in probability of a sequence of random variables under $\mathrm{P}^{(n)}$ and under $\mathrm{Q}^{(n)}$ are equivalent and the limits are the same.

If the likelihood ratio $\mathrm{dQ}^{(n)}/\mathrm{dP}^{(n)}$ actually converges in distribution, much more can be said and a large theory originated by LeCam exists, called "the theory of convergence of statistical experiments," concerned with statistical consequences (in terms of risk functions of general statistical procedures) of this. Of special interest is the case when the likelihood ratio converges in distribution to the likelihood ratio for a normal shift experiment.

Suppose X is a random variable with the $\mathscr{N}(0, 1)$ distribution under P and the $\mathscr{N}(\sigma, 1)$ distribution, $\sigma \neq 0$, under Q. Equivalently, dividing X by the constant σ, the problem is to decide whether we have an observation from the $\mathcal{N}(0, 1/\sigma^2)$ or the $\mathcal{N}(1, 1/\sigma^2)$ distribution. The statistical information in the observation is seen to depend on σ^2. Consider the likelihood ratio dQ/dP based on one observation of X: We have

$$\frac{\mathrm{dQ}}{\mathrm{dP}} = \frac{(2\pi)^{-1/2}\exp(-\frac{1}{2}(X-\sigma)^2)}{(2\pi)^{-1/2}\exp(-\frac{1}{2}X^2)} = \exp(\sigma X - \tfrac{1}{2}\sigma^2). \tag{8.1.4}$$

We see that under P, $\log(\mathrm{dQ}/\mathrm{dP}) = \sigma X - \frac{1}{2}\sigma^2$ is $\mathcal{N}(-\frac{1}{2}\sigma^2, \sigma^2)$ distributed, whereas under Q, it has the $\mathcal{N}(+\frac{1}{2}\sigma^2, \sigma^2)$ distribution.

Suppose T is another random variable, under P jointly normally distributed with log(dQ/dP). It turns out we can easily describe the distribution of T under Q. By taking its regression on the log-likelihood ratio, T can be written as a regression coefficient (covariance divided by variance σ^2) times log(dQ/dP), plus a residual, a random variable independent of log(dQ/dP), and normally distributed. Under Q, by its independence with the likelihood ratio, the distribution of the residual does not change (and remains independent). So, by the regression representation, under Q the distribution of T stays jointly normal with the log-likelihood ratio, with the same variance and covariance as before; only its mean changes by an amount equal to the change σ^2 in the mean of the log-likelihood ratio times the regression coefficient: itself equal to covariance divided by σ^2. The net result is, thus, a shift in the mean of T equal to its covariance with the log-likelihood ratio.

These observations clarify the meaning of LeCam's famous "third lemma," which gives a condition for contiguity with a "normal limit" in the sense just described together with a means of transforming asymptotic normality of a test statistic or estimator under the null hypothesis sequence $(\mathrm{P}^{(n)})$ to the alternative hypothesis sequence $(\mathrm{Q}^{(n)})$:

Theorem VIII.1.1 (LeCam's Third Lemma). *Let* $(\mathrm{P}^{(n)})$, $(\mathrm{Q}^{(n)})$ *be sequences of probability measures on* $(\Omega^{(n)}, \mathscr{F}^{(n)})$ *and let* $(\mathbf{T}^{(n)})$ *be a sequence of random vectors. Suppose*

$$\begin{bmatrix} \log\dfrac{\mathrm{dQ}^{(n)}}{\mathrm{dP}^{(n)}} \\ \mathbf{T}^{(n)} \end{bmatrix} \xrightarrow{\mathscr{D}(\mathrm{P}^{(n)})} \mathcal{N}\left(\begin{bmatrix} -\dfrac{1}{2}\sigma^2 \\ \boldsymbol{\mu} \end{bmatrix}, \begin{pmatrix} \sigma^2 & \boldsymbol{\gamma}^{\mathrm{T}} \\ \boldsymbol{\gamma} & \boldsymbol{\Sigma} \end{pmatrix}\right). \tag{8.1.5}$$

Then $\mathrm{P}^{(n)} \lhd\rhd \mathrm{Q}^{(n)}$ *and*

$$\begin{bmatrix} \log\dfrac{\mathrm{dQ}^{(n)}}{\mathrm{dP}^{(n)}} \\ \mathbf{T}^{(n)} \end{bmatrix} \xrightarrow{\mathscr{D}(\mathrm{Q}^{(n)})} \mathcal{N}\left(\begin{bmatrix} +\dfrac{1}{2}\sigma^2 \\ \boldsymbol{\mu}+\boldsymbol{\gamma} \end{bmatrix}, \begin{pmatrix} \sigma^2 & \boldsymbol{\gamma}^{\mathrm{T}} \\ \boldsymbol{\gamma} & \boldsymbol{\Sigma} \end{pmatrix}\right). \tag{8.1.6}$$

To illustrate the statistical application of this theorem, suppose $T^{(n)}$ is a random variable used as a test statistic for one-sided testing of the null hypothesis $\mathrm{P}^{(n)}$ against the alternative $\mathrm{Q}^{(n)}$ at size α. Suppose $T^{(n)}$ is asymptotically normal under the null hypothesis $\mathrm{P}^{(n)}$ and is, in fact, standardized to be asymptotically $\mathcal{N}(0, 1)$. Moreover, suppose $T^{(n)}$ converges jointly in distribution with $\log(dQ^{(n)}/dP^{(n)})$ under $P^{(n)}$ to a bivariate normal distribution, so that we also have contiguity with a normal limit. Taking the "asymptotic information" to be σ^2 leaves only the asymptotic correlation coefficient ρ (supposed positive) between test statistic and log-likelihood ratio to be specified to determine the joint limiting distribution, which is then

$$\mathcal{N}\left(\begin{pmatrix} -\frac{1}{2}\sigma^2 \\ 0 \end{pmatrix}, \begin{pmatrix} \sigma^2 & \rho\sigma \\ \rho\sigma & 1 \end{pmatrix}\right), \quad \sigma > 0, \rho \geq 0.$$

Under the alternative $Q^{(n)}$, we, therefore, have joint convergence in distribution to the bivariate normal distribution:

$$\mathcal{N}\left(\begin{pmatrix} \frac{1}{2}\sigma^2 \\ \rho\sigma \end{pmatrix}, \begin{pmatrix} \sigma^2 & \rho\sigma \\ \rho\sigma & 1 \end{pmatrix}\right).$$

Now $\sigma^{-1}(\log(dQ^{(n)}/dP^{(n)}) + \frac{1}{2}\sigma^2)$ is asymptotically $\mathcal{N}(0, 1)$ under $(P^{(n)})$ and $\mathcal{N}(\sigma, 1)$ under $(Q^{(n)})$. The most powerful test is based on this statistic, by the Neyman–Pearson lemma. So, letting Φ be the standard normal distribution function, and defining the critical value $u_\alpha = \Phi^{-1}(1 - \alpha)$, we find that the asymptotic power of the most powerful size-α test is $1 - \Phi(u_\alpha - \sigma)$; whereas the asymptotic power of the one-sided size-α test based on $T^{(n)}$ is $1 - \Phi(u_\alpha - \rho\sigma)$. We say that $T^{(n)}$ has a Pitman asymptotic relative efficiency of ρ^2 relative to the optimal test. Moreover, $T^{(n)}$ is asymptotically efficient (asymptotically most powerful) if and only if $\rho = 1$, under which condition $T^{(n)}$ is actually asymptotically equivalent to the Neyman–Pearson likelihood ratio test: We then have

$$T^{(n)} - \frac{1}{\sigma}\left(\log\left(\frac{dQ^{(n)}}{dP^{(n)}}\right) + \frac{1}{2}\sigma^2\right) \xrightarrow{P^{(n)}} 0,$$

and this holds under $Q^{(n)}$ too.

Note that σ^2 tells us how easily $P^{(n)}$ and $Q^{(n)}$ can be asymptotically distinguished; we will see it presently appearing exactly in the role of the asymptotic Fisher information for this experiment.

VIII.1.2. Local Asymptotic Normality

We now extend the discussion to the case of a (finite-dimensional) parametric model. Consider a sequence of models

$$(\Omega^{(n)}, \mathscr{F}^{(n)}, \mathscr{P}^{(n)} = \{P_{\boldsymbol{\theta}}^{(n)}: \boldsymbol{\theta} \in \Theta\}), \quad n = 1, 2, \ldots,$$

each with the same parameter space $\Theta \subseteq \mathbb{R}^p$. Let $a_n \to \infty$ be a given sequence of numbers; typically $a_n = n^{1/2}$ if n can be interpreted as a sample size in the nth model. We look in the neighborhood of a fixed point $\boldsymbol{\theta}_0$ in the interior of Θ and suppose that, as $n \to \infty$, the information about $\boldsymbol{\theta}$ increases in such a way that for each vector $\mathbf{h} \in \mathbb{R}^p$, defining

$$\boldsymbol{\theta}_n(\mathbf{h}) = \boldsymbol{\theta}_0 + a_n^{-1}\mathbf{h} \tag{8.1.7}$$

or, more generally,

$$\boldsymbol{\theta}_n(\mathbf{h}) = \boldsymbol{\theta}_0 + a_h^{-1}\mathbf{h} + o(a_n^{-1}); \tag{8.1.7'}$$

see Section VIII.4,

$$(\mathrm{P}_{\boldsymbol{\theta}_0}^{(n)}) = (\mathrm{P}_{\boldsymbol{\theta}_n(\mathbf{0})}^{(n)}) \lhd\rhd (\mathrm{P}_{\boldsymbol{\theta}_n(\mathbf{h})}^{(n)}) \quad \text{for each } \mathbf{h}.$$

Such sequences $\boldsymbol{\theta}_n = \boldsymbol{\theta}_n(\mathbf{h})$ are called *local sequences*. [Note that if consistent estimators of $\boldsymbol{\theta}$ exist, we cannot have $(\mathrm{P}_{\boldsymbol{\theta}_0}^{(n)}) \lhd\rhd (\mathrm{P}_{\boldsymbol{\theta}_1}^{(n)})$ for $\boldsymbol{\theta}_0 \neq \boldsymbol{\theta}_1$ since, under contiguity, convergence in probability to *different* limits is impossible.] If we actually have a nontrivial ($\sigma > 0$) normal limit, as in LeCam's third lemma, the discussion of the testing problem we gave there suggests that it will not be possible to find estimators of $\boldsymbol{\theta}$ with a better convergence rate than a_n. We will see that this is indeed true under the hypothesis of *local asymptotic normality* (LAN): We say that the sequence of experiments has the LAN property at $\boldsymbol{\theta}_0$ with asymptotic information matrix $\mathscr{I}$ (a positive definite symmetric matrix) and relative to the rate sequence a_n if for every $r \geq 1$ and $\mathbf{h}_1, \ldots, \mathbf{h}_r \in \mathbb{R}^p$ (thought of as column vectors) we have

$$\left(\log \frac{d\mathrm{P}_{\boldsymbol{\theta}_n(\mathbf{h}_1)}^{(n)}}{d\mathrm{P}_{\boldsymbol{\theta}_0}^{(n)}}, \ldots, \log \frac{d\mathrm{P}_{\boldsymbol{\theta}_n(\mathbf{h}_r)}^{(n)}}{d\mathrm{P}_{\boldsymbol{\theta}_0}^{(n)}}\right)^{\mathsf{T}} \xrightarrow{\mathscr{D}(\mathrm{P}_{\boldsymbol{\theta}_0}^{(n)})} \mathscr{N}(\boldsymbol{\mu}, \boldsymbol{\Sigma}) \tag{8.1.8}$$

with $\boldsymbol{\Sigma} = \mathbf{H}^{\mathsf{T}}\mathscr{I}\mathbf{H}$, $\mathbf{H}$ being the $p \times r$ matrix $(\mathbf{h}_1 \cdots \mathbf{h}_r)$, $\boldsymbol{\mu} = -\frac{1}{2}\operatorname{diag}(\boldsymbol{\Sigma})$, where here $\operatorname{diag}(\boldsymbol{\Sigma})$ means the vector of diagonal elements of the matrix $\boldsymbol{\Sigma}$.

The log-likelihood ratios here are extended real valued. For small n, $\boldsymbol{\theta}_n(\mathbf{h}) = \boldsymbol{\theta}_0 + a_n^{-1}\mathbf{h}$ may not be in Θ, but for all sufficiently large n the left-hand side of (8.1.8) is meaningful.

Note that LeCam's third lemma now gives mutual contiguity of all sequences $(\mathrm{P}_{\boldsymbol{\theta}_n(\mathbf{h})}^{(n)})$ as well as the possibility of transferring (joint) asymptotic normality under one sequence to another.

LAN is often formulated in a different, but equivalent way, which is often useful: We have LAN if a sequence of random vectors $\mathbf{U}^{(n)}$ exists such that, for each $\mathbf{h} \in \mathbb{R}^p$,

$$\log \frac{d\mathrm{P}_{\boldsymbol{\theta}_n(\mathbf{h})}^{(n)}}{d\mathrm{P}_{\boldsymbol{\theta}_0}^{(n)}} - \mathbf{h}^{\mathsf{T}}\mathbf{U}^{(n)} + \frac{1}{2}\mathbf{h}^{\mathsf{T}}\mathscr{I}\mathbf{h} \xrightarrow{\mathrm{P}_{\boldsymbol{\theta}_0}^{(n)}} 0, \tag{8.1.9}$$

where

$$\mathbf{U}^{(n)} \xrightarrow{\mathscr{D}(\mathrm{P}_{\boldsymbol{\theta}_0}^{(n)})} \mathscr{N}(\mathbf{0}, \mathscr{I}), \tag{8.1.10}$$

$\mathscr{I}$ and a_n as before. (Other versions of LAN are discussed in Section VIII.1.5.) The components of $\mathbf{U}^{(n)}$ can be recovered (up to asymptotic equivalence) from (8.1.9) by choosing $\mathbf{h}$ as each of the unit vectors in $\mathbb{R}^p$ in turn. Calling the ith unit vector $\mathbf{e}_i$, one could take $U_i^{(n)} = \log(d\mathrm{P}_{\boldsymbol{\theta}_n(\mathbf{e}_i)}^{(n)}/d\mathrm{P}_{\boldsymbol{\theta}_0}^{(n)}) + \frac{1}{2}\mathscr{I}_{ii}$.

However, much more useful is to identify $\mathbf{U}^{(n)}$ (up to asymptotic equivalence) as the derivative of the log-likelihood ratio $\log(\mathrm{dP}^{(n)}_{\boldsymbol{\theta}_n(\mathbf{h})}/\mathrm{dP}^{(n)}_{\boldsymbol{\theta}_0})$ with respect to $\mathbf{h}$, evaluated at $\mathbf{h} = \mathbf{0}$; to see this, differentiate with respect to $\mathbf{h}$ and then put $\mathbf{h} = \mathbf{0}$ in (8.1.9). So, considering (8.1.7) or (8.1.7′) as a reparametrization, depending on n, from $\boldsymbol{\theta}$ to $\mathbf{h}$ by the substitution $\boldsymbol{\theta} = \boldsymbol{\theta}_n(\mathbf{h})$, the components of the vector $\mathbf{U}^{(n)}$ are approximately the set of *score functions* for the components of $\mathbf{h}$, equal to a_n^{-1} times the score functions for those for $\boldsymbol{\theta}$. (The information for $\boldsymbol{\theta}$ grows at rate a_n^2, whereas that for $\mathbf{h}$ remains bounded). Thus, typically, writing $\log \mathrm{lik}^{(n)}_{\boldsymbol{\theta}}(\boldsymbol{\theta}) = \log \mathrm{dP}^{(n)}_{\boldsymbol{\theta}}/\mathrm{dP}^{(n)}_{\boldsymbol{\theta}_0}$ and $\log \mathrm{lik}^{(n)}_{\mathbf{h}}(\mathbf{h}) = \log \mathrm{dP}^{(n)}_{\boldsymbol{\theta}_n(\mathbf{h})}/\mathrm{dP}^{(n)}_{\boldsymbol{\theta}_0}$, we may take in (8.1.9) and (8.1.10)

$$\mathbf{U}^{(n)} = \frac{\partial}{\partial \mathbf{h}} \log \mathrm{lik}_{\mathbf{h}}(\mathbf{0}) = a_n^{-1} \frac{\partial}{\partial \boldsymbol{\theta}} \log \mathrm{lik}_{\boldsymbol{\theta}}(\boldsymbol{\theta}_0). \tag{8.1.11}$$

In a parametric model for a counting process, Theorem VIII.2.2 will show that LAN holds in the form given by (8.1.9)–(8.1.11) under precisely the same conditions as we needed to prove asymptotic normality of maximum likelihood estimators in Section VI.1.2. In Sections VIII.3 and VIII.4, we will extend the discussion to non- and semiparametric models. However the main ingredient will simply be the LAN assumption for conveniently chosen parametric submodels of a similar local nature to (8.1.7′), and the assumption could again be verified by checking the conditions of Theorem VI.1.2. However, before making these connections to models based on counting processes, we will proceed with the general theory.

VIII.1.3. Asymptotic Optimality in Estimation and Testing

Suppose $\mathbf{U}$ is a random vector with the $\mathcal{N}(\mathcal{I}\mathbf{h}, \mathcal{I})$ distribution under $\mathrm{P}_{\mathbf{h}}$, and define $\mathbf{X} = \mathcal{I}^{-1}\mathbf{U}$. Then, just as in the discussion of (8.1.4), we see that $\mathrm{dP}_{\mathbf{h}}/\mathrm{dP}_{\mathbf{0}} = \exp(\mathbf{h}^{\mathsf{T}}\mathbf{U} - \frac{1}{2}\mathbf{h}^{\mathsf{T}}\mathcal{I}\mathbf{h})$, where $\mathbf{U} \sim \mathcal{N}(\mathbf{0}, \mathcal{I})$ under $\mathrm{P}_{\mathbf{0}}$, is the likelihood ratio based on one observation of $\mathbf{X}$. Note that the score functions for the components of $\mathbf{h}$ at $\mathbf{h} = \mathbf{0}$ are exactly the components of $\mathbf{U}$. Thus, LAN says that for n large, statistical inference about $\boldsymbol{\theta}$ in a $1/a_n$ neighborhood of $\boldsymbol{\theta}_0$ is approximately the same as statistical inference about $\mathbf{h} = a_n(\boldsymbol{\theta} - \boldsymbol{\theta}_0)$ in the multivariate normal shift experiment: Under $\mathrm{P}_{\mathbf{h}}$, observe $\mathbf{X} \sim \mathcal{N}(\mathbf{h}, \mathcal{I}^{-1})$. We should not expect to get a better asymptotic variance of an estimator of $\boldsymbol{\theta}$, when rescaled by a_n, than $\mathcal{I}^{-1}$, the inverse of the information matrix. Similarly, testing problems cannot be expected to be easier than in the corresponding multivariate normal problems.

As far as a simple testing problem is concerned, we have already seen some elaboration of this idea. For composite hypotheses, we must be aware that optimal tests in the asymptotic situation only exist within classes restricted by unbiasedness or invariance (or one considers best average power over certain ellipsoids, or minimax tests), so the asymptotic optimality theory we describe later has to have similar qualifications. We come back to this in a moment. For estimation, the situation is delicate for different reasons. The phenomenon of "superefficiency" means that estimators can be constructed which can behave on subspaces of Θ of measure zero far better than what we have just described. Such estimators are not really as pathological as they are usually made out to be: Think of the common procedure of first testing a null hypothesis and then estimating, using a better estimator (under the null hypothesis) if the null hypothesis is accepted. However, the resulting procedure is going to have rather erratic behavior at contiguous alternatives to the null hypothesis.

Based on LAN, two types of theorems are commonly given in the literature, which take account of the just mentioned difficulties in two different ways. A large literature, starting with Hájek (1972) and LeCam (1972), is devoted to *local asymptotic minimax* theorems, giving lower bounds to the maximum local risk of an estimator sequence in terms of the same quantity for the limiting experiment ($\mathbf{X} = \mathscr{I}^{-1}\mathbf{U}$ is minimax for estimating $\mathbf{h}$ under "bowl-shaped" loss functions—symmetric about zero with convex level sets). These results are, however, rather technical and to check that the lower bounds are actually achieved in practice is quite difficult, requiring *uniform* convergence in distribution over local neighborhoods. The other type of theorem, originating with Hájek (1970), is the so-called *convolution* theorem, which restricts attention to a nicely behaved subclass of estimators, and then states a perhaps more appealing property. Here, we concentrate on this type.

Suppose $\mathbf{T}^{(n)}$ is an estimator sequence for $\boldsymbol{\theta}$ in our model. We say that $\mathbf{T}^{(n)}$ is *regular* if, for all $\mathbf{h} \in \mathbb{R}^p$,

$$a_n(\mathbf{T}^{(n)} - \boldsymbol{\theta}_n(\mathbf{h})) \xrightarrow{\mathscr{D}(\mathbf{P}^{(n)}_{\boldsymbol{\theta}_n(\mathbf{h})})} \mathbf{Z} \tag{8.1.12}$$

for some random vector $\mathbf{Z}$, the same for all $\mathbf{h}$. This is a kind of local unbiasedness or equivariance: Under a small change in $\boldsymbol{\theta}$, the limiting distribution of $\mathbf{T}_n$ shifts by the same amount: Rewrite (8.1.12) as

$$a_n(\mathbf{T}^{(n)} - \boldsymbol{\theta}_0) \xrightarrow{\mathscr{D}(\mathbf{P}^{(n)}_{\boldsymbol{\theta}_n(\mathbf{h})})} \mathbf{Z} + \mathbf{h}. \tag{8.1.12$'$}$$

We can now state the famous Hájek (1970) convolution theorem:

Theorem VIII.1.2. *Suppose* $\mathbf{T}^{(n)}$ *is a regular estimator of* $\boldsymbol{\theta}$ *in a* LAN *model. Then the limiting random vector* $\mathbf{Z}$ *in* (8.1.12) *can be written as the sum* $\mathbf{Z} = \mathbf{X} + \mathbf{Y}$ *of two independent random vectors, one of them,* $\mathbf{X}$*, having the* $\mathscr{N}(\mathbf{0}, \mathscr{I}^{-1})$ *distribution.*

This result says that the limiting distribution of a regular estimator is *more spread out* or *less concentrated* about $\boldsymbol{\theta}_0$ than the $\mathscr{N}(\mathbf{0}, \mathscr{I}^{-1})$ distribution. The variance cannot be smaller and there may also be bias. The probability that $\mathbf{Z}$ lies in any convex region symmetric about $\mathbf{0}$ is smaller than that for the $\mathscr{N}(\mathbf{0}, \mathscr{I}^{-1})$ distribution [Anderson's (1955) lemma]. If a regular estimator sequence achieves the lower bound of Theorem VIII.1.2 (i.e., the component $\mathbf{Y}$ in the limiting distributing is not present), we say it is *asymptotically efficient* or *asymptotically optimal.*

It turns out that $\mathbf{T}^{(n)}$ is regular and asymptotically efficient if and only if it is asymptotically equivalent to a particular linear combination of the components an approximate score $\mathbf{U}^{(n)}$ [satisfying (8.1.9) and (8.1.10)] for $\mathbf{h}$. We say, in general, that an estimator $\mathbf{T}^{(n)}$ is *asymptotically linear* if the difference between $a_n(\mathbf{T}^{(n)} - \boldsymbol{\theta}_0)$ and $\mathbf{AU}^{(n)}$ converges in $\mathrm{P}_{\boldsymbol{\theta}_0}^{(n)}$ probability to zero as $n \to \infty$, where $\mathbf{A}$ is any fixed $p \times p$ matrix. The approximating linear combination of the approximate scores, $\mathbf{AU}^{(n)}$, is called the *influence function* of the estimator. Then $\mathbf{T}^{(n)}$ is regular and asymptotically efficient if and only if $a_n(\mathbf{T}^{(n)} - \boldsymbol{\theta}_0)$ is asymptotically equivalent to the *optimal influence function* $\mathscr{I}^{-1}\mathbf{U}^{(n)}$. The proof that asymptotic linearity with optimal influence function implies regularity and, hence, efficiency is just a question of using LeCam's third lemma to get the asymptotic distribution of $a_n(\mathbf{T}^{(n)} - \boldsymbol{\theta}_0)$ under $\mathrm{P}_{\boldsymbol{\theta}_n(\mathbf{h})}^{(n)}$ and comparing with the alternative version of regularity, (8.1.12′).

The lower bound $\mathscr{I}^{-1}$ to the variance of a regular estimator of $\boldsymbol{\theta}$ is called the *information bound.*

In many situations, we are only interested in estimating a function of $\boldsymbol{\theta}$, $\boldsymbol{\kappa}(\boldsymbol{\theta}) \in \mathbb{R}^k$ say; maybe one of the components of $\boldsymbol{\theta}$. Regularity of an estimator of $\boldsymbol{\kappa}(\boldsymbol{\theta})$ is defined in just the same way as for $\boldsymbol{\theta}$. We assume that $\boldsymbol{\kappa}(\boldsymbol{\theta})$ is differentiable at $\boldsymbol{\theta} = \boldsymbol{\theta}_0$ and denote by $\mathrm{d}\boldsymbol{\kappa}$ the $k \times p$ matrix of partial derivatives at that point.

Theorem VIII.1.3. *Suppose $\mathbf{T}^{(n)}$ is a regular estimator of the differentiable function $\boldsymbol{\kappa}$ of $\boldsymbol{\theta}$ in a LAN model. Then the limiting distribution $\mathbf{Z}$ of $a_n(\mathbf{T}^{(n)} - \boldsymbol{\kappa}(\boldsymbol{\theta}_0))$ under $\mathrm{P}_{\boldsymbol{\theta}_0}^{(n)}$ is the convolution of the $\mathscr{N}(\mathbf{0}, \mathrm{d}\boldsymbol{\kappa}\mathscr{I}^{-1}\mathrm{d}\boldsymbol{\kappa}^\top)$ distribution with another distribution.*

Again, we say that we have asymptotic efficiency when the limiting distribution is just $\mathscr{N}(\mathbf{0}, \mathrm{d}\boldsymbol{\kappa}\mathscr{I}^{-1}\mathrm{d}\boldsymbol{\kappa}^\top)$. The optimal asymptotic variance is again called the information bound. An estimator is regular and asymptotically efficient if and only if it is asymptotically linear with optimal influence function, i.e., $a_n(\mathbf{T}^{(n)} - \boldsymbol{\kappa}(\boldsymbol{\theta}_0))$ is asymptotically equivalent to $\mathrm{d}\boldsymbol{\kappa}\mathscr{I}^{-1}\mathbf{U}^{(n)}$. One easily sees that taking a differentiable function $\boldsymbol{\kappa}$ of an efficient estimator for $\boldsymbol{\theta}$ yields an efficient estimator for $\boldsymbol{\kappa}(\boldsymbol{\theta})$.

A further very important result is the following: An asymptotically linear, regular estimator is automatically asymptotically efficient; thus, its influence function, some linear transformation of $\mathbf{U}^{(n)}$, is forced by regularity to be the *optimal* influence function $\mathrm{d}\boldsymbol{\kappa}\mathscr{I}^{-1}\mathbf{U}^{(n)}$. The proof of this result is a nice exercise in contiguity theory. Suppose $\mathbf{T}^{(n)}$ is asymptotically linear, that is, $\mathbf{Y}^{(n)} = a_n(\mathbf{T}^{(n)} - \boldsymbol{\kappa}(\boldsymbol{\theta}_0))$ is asymptotically equivalent under $\mathrm{P}_{\boldsymbol{\theta}_0}^{(n)}$ to $\mathbf{AU}^{(n)}$ for some matrix $\mathbf{A}$. It follows that $\mathbf{Y}^{(n)}$ is asymptotically jointly normally distributed with $\log(\mathrm{dP}_{\boldsymbol{\theta}_n(\mathbf{h})}/\mathrm{dP}_{\boldsymbol{\theta}_0})$ under $\mathrm{P}_{\boldsymbol{\theta}_0}^{(n)}$; its asymptotic mean is zero and its asymptotic covariance with the log-likelihood ratio is $\mathbf{A}\mathscr{I}\mathbf{h}^\top$. Therefore, its asymptotic mean under $\mathrm{P}_{\boldsymbol{\theta}_n(\mathbf{h})}^{(n)}$ is $\mathbf{A}\mathscr{I}\mathbf{h}^\top$, whereas its asymptotic variance remains unaltered. But, by regularity of $\mathbf{T}^{(n)}$ as estimator of $\boldsymbol{\kappa}(\boldsymbol{\theta})$ and differentiability of $\boldsymbol{\kappa}$, this asymptotic mean must equal $\mathrm{d}\boldsymbol{\kappa} \cdot \mathbf{h}$. Since we have equality for all $\mathbf{h}$, it follows that $\mathbf{A}\mathscr{I} = \mathrm{d}\boldsymbol{\kappa}$ or $\mathbf{A} = \mathrm{d}\boldsymbol{\kappa}\mathscr{I}^{-1}$.

Let us now formulate some theorems on testing problems, not often discussed in the literature from the point of view of the theory of Local Asymptotic Normality. The material here is based on Choi (1989).

Consider the null hypothesis $\boldsymbol{\theta} \in \Theta_0$, where Θ_0 is a subspace of Θ of the form

$$\Theta_0 = \{\boldsymbol{\theta} \in \Theta : \boldsymbol{\kappa}(\boldsymbol{\theta}) = \boldsymbol{\kappa}_0\} \quad \text{or} \quad \Theta_0 = \{\boldsymbol{\theta} \in \Theta : \kappa(\boldsymbol{\theta}) \leq \kappa_0\},$$

where the differentiable function $\boldsymbol{\kappa}$ takes values in $\mathbb{R}^k$; $k \geq 1$ in the first case but $\boldsymbol{\kappa} = \kappa$ is real, $k = 1$ in the second case. We suppose we have LAN at $\boldsymbol{\theta}_0$, where $\boldsymbol{\kappa}(\boldsymbol{\theta}_0) = \boldsymbol{\kappa}_0$, so $\boldsymbol{\theta}_0 \in \Theta_0$. Consider an *asymptotically size-α test* for the given testing problem; that is, a sequence $\psi^{(n)}$ of critical functions such that for all local sequences of parameter values in the null hypothesis, $\boldsymbol{\theta}_n(\mathbf{h}) = \boldsymbol{\theta}_0 + a_n^{-1}\mathbf{h}$, $\boldsymbol{\theta}_n(\mathbf{h}) \in \Theta_0$ for all n, $\limsup \mathrm{E}_{\boldsymbol{\theta}_n(\mathbf{h})}\psi^{(n)} \leq \alpha$.

Theorem VIII.1.4. *Consider the testing problem for the null hypothesis $\kappa(\boldsymbol{\theta}) \leq \kappa_0 \in \mathbb{R}$ (κ real-valued and differentiable). Then the sequence of tests $\psi^{(n)}$ is asymptotically uniformly most powerful size-α with respct to local sequences $\boldsymbol{\theta}_n = \boldsymbol{\theta}_0 + a_n^{-1}\mathbf{h}$ if and only if it is asymptotically equivalent to the sequence of tests: Reject the null hypothesis if and only if*

$$\frac{\mathrm{d}\kappa \mathscr{I}^{-1}\mathbf{U}^{(n)}}{(\mathrm{d}\kappa \mathscr{I}^{-1}\,\mathrm{d}\kappa^{\top})^{1/2}} \geq u_\alpha.$$

Here, that a test is *asymptotically uniformly most powerful* means that the asymptotic probability of rejecting the null hypothesis, for any local sequence in the alternative, is not smaller than that for any other asymptotically size-α test. Suppose now $\mathbf{T}^{(n)}$ is an efficient regular estimator of $\boldsymbol{\theta} \in \Theta$, then an asymptotically size-α test can be obtained as: Reject if $a_n(\kappa(\mathbf{T}^{(n)}) - \kappa_0)/(\mathrm{d}\kappa \mathscr{I}^{-1}\,\mathrm{d}\kappa^{\top})^{1/2} \geq u_\alpha$. We see that this test is asymptotically uniformly most powerful or, in other words, that an efficient (or optimal) test of $\kappa(\boldsymbol{\theta}) \leq \kappa_0$ is obtained by using an efficient (or optimal) estimator of $\kappa(\boldsymbol{\theta})$ as test statistic.

This theorem describes optimality in a one-sided, one-dimensional testing problem very nicely, even in the presence of nuisance parameters. For the other kinds of null hypotheses just mentioned, we can only obtain optimality within a restricted class of test statistics. For the null hypothesis $\kappa(\boldsymbol{\theta}) = \kappa_0$, κ real-valued (thus, a two-sided alternative, nuisance parameters allowed), we can only prove: The equal-tails two-sided test based on an efficient regular estimator of $\kappa(\boldsymbol{\theta})$ is asymptotically efficient within the class of asymptotically unbiased tests. (Asymptotically unbiased tests have limiting power against local alternatives nowhere less than their size.) Dropping the restriction that $\boldsymbol{\kappa}$ is one-dimensional, even less can be said. Choi (1989) shows that the natural asymptotic χ^2-type test of $\boldsymbol{\kappa}(\boldsymbol{\theta}) = \boldsymbol{\kappa}_0$ based on an asymptotically efficient regular estimator of $\boldsymbol{\kappa}$ is asymptotically uniformly most powerful among all asymptotically invariant tests, where invariance is with respect to "information-preserving" transformations of $\mathbf{h}$. This test is typically asymptotically equivalent to such well-known and much used tests as the likelihood ratio test, the Wald test, and the score test. The invariance concept used here is, however, difficult to motivate. Just because the asymptotic testing problem is invariant under some set of transformations seems a poor reason to restrict attention to asymptotically invariant tests.

There is another way of showing that these very natural tests have an appealing optimality property, by considering in more detail the importance of detecting various alternatives to the null hypothesis [the ideas here go back to Wald (1943)]. When testing against a highly composite alternative hypothesis, the kind of test one should use obviously should depend on the kind of alternative which it is most important to detect. If one does not have any special reason to demand high power against particular alternatives, one could look at a collection of points in the alternative hypothesis which, each on their own, are equally difficult to statistically distinguish from the null hypothesis, and demand that an omnibus test does equally well against all these, statistically, equally important alternatives. This leads to a consideration of the envelope power function (the maximal power at a given point in the alternative of all tests of a given size). Asymptotically, the envelope power turns out to be constant on certain local ellipsoids determined by the information matrix [in fact, alternatives where $(\mathrm{d}\boldsymbol{\kappa}\cdot\mathbf{h})^{\mathsf{T}}(\mathrm{d}\boldsymbol{\kappa}\mathscr{I}^{-1}\,\mathrm{d}\boldsymbol{\kappa}^{\mathsf{T}})^{-1}(\mathrm{d}\boldsymbol{\kappa}\cdot\mathbf{h})$ is constant]. The natural tests described above are optimal in the sense of having the best average power over these ellipsoids; or if you prefer in the sense of having the best minimum power over these ellipsoids (an *asymptotically minimax test* with respect to the ellipsoids). They have constant power on these ellipsoids and are best among all such tests (this is the property of being the *best asymptotically invariant test* mentioned above). Finally, they are optimal in the sense of having the smallest maximum shortcoming from the envelope power. Such tests are called *asymptotically most stringent*; note that this property makes no special reference to the ellipsoids any more. (It is proved via construction of the ellipsoids because on a region where the envelope power is constant, the properties of minimax and most stringent test coincide.)

It is important that the dimension of the hypothesis being tested, k, is finite for this notion of testing optimality to be meaningful. It one tests a truly infinite-dimensional hypothesis (e.g., when carrying out an omnibus goodness-of-fit test), it turns out that the above optimality notions become useless: For instance, the minimum power over an infinite-dimensional ellipsoid of any test cannot exceed its size, so *all tests* are asymptotically minimax, asymptotically most stringent, etc., with regard to alternatives defined in terms of the envelope power.

A final technical aside is that we will claim the optimality of certain tests by just showing they have the best asymptotic power function. Strictly speaking, to establish such a property as asymptotically most stringent, one must calculate the maximum shortcoming first, and then let n tend to infinity, not vice versa. This would require an analysis of the uniformity of convergence of the power function of the test to its limit. Tests based on regular estimators will satisfy the required uniformity.

VIII.1.4. Toward an Infinite Dimensional Theory

We conclude Section VIII.1 with some remarks on a generalization and reformulation of Theorem VIII.1.3 which pave the way for a discussion of convolution theorems for infinite-dimensional parameters in Section VIII.3. In particular, we give a geometrical interpretation of the information bounds in terms of the geometry of the (asymptotic) score functions $\mathbf{h}^{\mathsf{T}}\mathbf{U}$. This material is not needed for Section VIII.2, which contains applications of the finite-dimensional theory presented earlier.

We note, first, that to establish the convolution theorem it suffices only to assume the LAN and regularity property for all $\mathbf{h}$ in a cone $C \subseteq \mathbb{R}^p$ (i.e., $\mathbf{h} \in C$ and $\alpha \in \mathbb{R}$, $\alpha \geq 0 \Rightarrow \alpha\mathbf{h} \in C$) provided the linear span of C, $\text{lin}(C)$, equals all of $\mathbb{R}^p$; in this context, this just means that C^0, the interior of C, must be nonempty. This means that we can also consider $\boldsymbol{\theta}_0$ not necessarily in the interior of Θ.

What happens if $\text{lin}(C)$ is a proper subset of $\mathbb{R}^p$? Consider the case when $\text{lin}(C) = \{\mathbf{h}\colon h_{q+1} = \cdots = h_p = 0\}$ for some $1 \leq q < p$. This means that the last $p - q$ coordinates of $\boldsymbol{\theta}$ remain fixed. Thus, working with a smaller cone C corresponds to restricting attention to a submodel. The cone gives the smaller collection of directions in which the parameter can now vary. Partition $\boldsymbol{\theta}$ into $\boldsymbol{\theta}_1$ and $\boldsymbol{\theta}_2$. For the submodel $\boldsymbol{\theta}_2$ fixed, we can obviously apply the convolution theorem to regular estimators of $\boldsymbol{\theta}_1$, getting the lower bound for the asymptotic variance $(\mathscr{I}_{11})^{-1}$, where

$$\mathscr{I} = \begin{pmatrix} \mathscr{I}_{11} & \mathscr{I}_{12} \\ \mathscr{I}_{21} & \mathscr{I}_{22} \end{pmatrix}$$

is also partitioned according to the first q and last $p - q$ elements. We have

$$(\mathscr{I}_{11})^{-1} = (\mathscr{I}^{-1})_{11} - (\mathscr{I}^{-1})_{12}(\mathscr{I}^{-1})_{22}^{-1}(\mathscr{I}^{-1})_{21},$$

showing that $(\mathscr{I}^{-1})_{11}$—the lower bound for estimation of $\boldsymbol{\theta}_1$ when $\boldsymbol{\theta}_2$ varies too—equals $(\mathscr{I}_{11})^{-1}$ plus a positive semidefinite matrix. Thus, the lower bound for estimation of $\boldsymbol{\theta}_1$, when $\boldsymbol{\theta}_2$ is known, is smaller than the lower bound when it is unknown. Similarly, for estimating $\boldsymbol{\kappa}(\boldsymbol{\theta})$, when working in the smaller model, we get the lower bound $(d\boldsymbol{\kappa})_1(\mathscr{I}_{11})^{-1}(d\boldsymbol{\kappa})_1^{\mathsf{T}}$, where $(d\boldsymbol{\kappa})_1$ denotes the first q columns of $d\boldsymbol{\kappa}$. This lower bound is again by some algebra smaller than the lower bound $d\boldsymbol{\kappa}\mathscr{I}^{-1}\, d\boldsymbol{\kappa}^{\mathsf{T}}$ for estimation of $\boldsymbol{\kappa}(\boldsymbol{\theta})$ in the larger model.

This simple result has a geometrical interpretation which becomes important for the infinite-dimensional extension of these results, to be presented in Section VIII.3. Let $\mathbb{H}$ denote the Hilbert space obtained by giving $\mathbb{R}^p$ the inner product $(\mathbf{h}, \mathbf{h}')_{\mathbb{H}} = \mathbf{h}^{\mathsf{T}}\mathscr{I}\mathbf{h}'$. Thus, we give $\mathbb{R}^p$ a new metric measuring (statistical) information distance for our limiting experiment. Under LAN, we have joint weak convergence of the log-likelihood ratios on the left-hand side of (8.1.8) to a multivariate normal with means $-\frac{1}{2}(\mathbf{h}_i, \mathbf{h}_i)_{\mathbb{H}} = -\frac{1}{2}\|\mathbf{h}_i\|_{\mathbb{H}}^2$ and with covariance $(\mathbf{h}_i, \mathbf{h}_j)_{\mathbb{H}}$. Note that $\|\mathbf{h}\|_{\mathbb{H}}^2 = \mathrm{E}(\mathbf{h}^\mathsf{T}\mathbf{U})^2$ where $\mathbf{U} \sim \mathcal{N}(\mathbf{0}, \mathcal{I})$. So this Hilbert space can be identified with the space of (limiting) score functions for $\mathbf{h}$.

The derivative $\mathrm{d}\boldsymbol{\kappa}$, acting on $\mathbf{h}$ by matrix multiplication, can also be represented, coordinatewise, as k linear maps on $\mathbb{H}$. To be specific, we can write for the ith coordinate $(\mathrm{d}\boldsymbol{\kappa}\cdot\mathbf{h})_i = \mathrm{d}\kappa_i\mathcal{I}^{-1}\mathcal{I}\mathbf{h} = (\mathcal{I}^{-1}\,\mathrm{d}\kappa_i^\mathsf{T}, \mathbf{h})_{\mathbb{H}}$, where $\mathrm{d}\kappa_1, \ldots, \mathrm{d}\kappa_k$ are the k rows of the matrix $\mathrm{d}\boldsymbol{\kappa}$ [do not confuse the $1 \times p$ row vector $\mathrm{d}\kappa_1$ with the $k \times q$ matrix $(\mathrm{d}\boldsymbol{\kappa})_1$]. Let $\kappa_1', \ldots, \kappa_k'$ be given by $\mathcal{I}^{-1}(\mathrm{d}\kappa_1)^\mathsf{T}, \ldots, \mathcal{I}^{-1}(\mathrm{d}\kappa_k)^\mathsf{T} \in \mathbb{H}$. We can, therefore, write

$$\mathrm{d}\boldsymbol{\kappa}\cdot\mathbf{h} = \begin{bmatrix} (\kappa_1', \mathbf{h})_{\mathbb{H}} \\ \vdots \\ (\kappa_k', \mathbf{h})_{\mathbb{H}} \end{bmatrix},$$

showing that the κ_i' are the derivatives (at $\boldsymbol{\theta}_0$) of $\kappa_i(\boldsymbol{\theta})$ with respect to $\boldsymbol{\theta} \in \mathbb{H}$.

The covariance matrix $\mathrm{d}\boldsymbol{\kappa}\mathcal{I}^{-1}(\mathrm{d}\boldsymbol{\kappa})^\mathsf{T}$ equals the matrix of the inner products of $\kappa_1', \ldots, \kappa_k'$ with one another; moreover, by some easy algebra, $(\mathrm{d}\boldsymbol{\kappa})_1(\mathcal{I}_{11})^{-1}(\mathrm{d}\boldsymbol{\kappa})_1^\mathsf{T}$ equals the matrix of inner products of the *projections* of $\kappa_1', \ldots, \kappa_k'$ (in $\mathbb{H}$) into $\mathrm{lin}(C)$. To see this, let $\mathbf{e}_1, \mathbf{e}_2, \ldots$ be the q-vectors $(1,0,0,\ldots)^\mathsf{T}$, $(0,1,0,\ldots)^\mathsf{T}$, $\ldots$, which span $\mathrm{lin}(C)$. Note that the $p \times k$ matrix $(\mathbf{e}_1, \ldots, \mathbf{e}_q)(\mathcal{I}_{11})^{-1}(\mathrm{d}\boldsymbol{\kappa})_1^\mathsf{T}$ represents k linear combinations of the column vectors $\mathbf{e}_1, \ldots, \mathbf{e}_q$ which have the same inner products (in $\mathbb{H}$) with $\mathbf{e}_1, \ldots, \mathbf{e}_q$ as $\kappa_1', \ldots, \kappa_k'$. Thus, these linear combinations are the required projections into $\mathrm{lin}(C)$, and their inner products can be written down directly.

This last result remains true for any cone C, not just the special choice just studied: If we have LAN for $\mathbf{h} \in C$, a cone, then the lower bound for the asymptotic covariance matrix of a regular estimator of $\boldsymbol{\kappa}(\boldsymbol{\theta})$ is given by the matrix of inner products of the projections (in $\mathbb{H}$) of the derivatives (in $\mathbb{H}$) of the coordinates of $\boldsymbol{\kappa}$ into $\mathrm{lin}(C)$. In Section VIII.3, we will see the same result when both $\boldsymbol{\kappa}$ and $\boldsymbol{\theta}$ are infinite-dimensional.

This geometrical point of view produces another useful result. Suppose our model gives rise to a cone C and that $\mathrm{lin}(C)$ is a subspace of $\mathbb{H}$ of dimension $q \le p$. The projections of $\kappa_1', \ldots, \kappa_k'$ into $\mathrm{lin}(C)$ span a smaller subspace still, typically of dimension k; think of the case when k is much smaller than q, itself much smaller than p. We think of this new subspace as corresponding to a k-dimensional submodel of our original q-dimensional model, itself embedded in a still larger p-dimensional model. (In Section VIII.3, we will consider estimating a k-dimensional parameter in a semi-parametric model, though infinite-dimensional still a submodel of a "full" or completely nonparametric model which makes no assumptions about the distribution of the data at all.) Clearly, the projections of the κ_i' into the k-dimensional subspace spanned by their very own projections into $\mathrm{lin}(C)$ are the same as these projections into $\mathrm{lin}(C)$ and, therefore, the information bound for estimation of $\boldsymbol{\kappa}(\boldsymbol{\theta})$ is the same for the model corresponding to C as for the much smaller model corresponding to this subspace. For all other

k-dimensional subspaces of lin(C) however, the projections are not the same and the lengths of the projections will be shorter.

This means that when estimating a k-dimensional parameter we can find a k-dimensional submodel of the original model for which the information bound is exactly the same. This must be the "hardest" k-dimensional submodel since it gives the same bound as for the complete model, whereas for all other submodels the information bound decreases (projections into subspaces are always shorter). In fact, the calculation of projections is essentially the same (at least, locally) as the construction of a hardest submodel. If we can exhibit a submodel, and a regular estimator in the whole model, such that the information bound in the submodel corresponds with the asymptotic variance of the estimator, then this must be the hardest submodel and the estimator is efficient for the whole model.

VIII.1.5. Technical Remarks on LAN

As given by LeCam (1970), the definition of the LAN property requires a strengthened form of (8.1.9) and a corresponding strengthening of (8.1.8), with $\mathbf{h}$ on the left-hand side of (8.1.9) replaced by elements $\mathbf{h}_n$ of an arbitrary bounded sequence of elements of $\mathbb{R}^p$; we call this property S-LAN. Regularity of an estimator and differentiability of a function of a parameter are usually also given in the corresponding stronger forms (S-regularity and S-differentiability). This strengthening is not needed to prove either convolution or local asymptotic minimax theorem (see van der Vaart, 1988), but it does play an essential role in showing the invariance of the results under different parametrizations and rate sequences, as well as in showing the attainability of the bounds. We emphasize (8.1.8) rather than the more elegant (8.1.9) because it is the former which gives the best extension to infinite-dimensional parameter spaces. The natural extension of (8.1.9) to the infinite-dimensional case would require the existence of a Hilbert space valued (infinite-dimensional) analogue to $\mathbf{U}^{(n)}$ (see Ibragimov and Khas'minskii, 1991), whereas the extension of (8.1.8) remains firmly finite-dimensional but is all that is needed to extend the convolution and asymptotic local minimax theorems.

Here we have only introduced scalar rate sequences a_n; this is often replaced by a sequence of positive definite matrices M_n. Such a generalization is important when statistical information about different parameters accrues at different rates as $n \to \infty$. For a careful treatment of these issues, see Fabian and Hannan (1982, 1985).

There is a nice connection between the theory presented in this section and the parametric bootstrap. Suppose one is interested in estimating the parameter $\boldsymbol{\theta}$ and, moreover, in computing confidence intervals for components of it, or functions of it. One way to proceed is to simply plug in the MLE of $\boldsymbol{\theta}$ into the expression for the asymptotic information and then use an approximate multivariate normal distribution of $a_n(\hat{\boldsymbol{\theta}}_n - \boldsymbol{\theta})$. Other methods based on the observed information and on the likelihood ratio test are also available. Another less standard possibility is to use the parametric bootstrap. This means one estimates $\boldsymbol{\theta}$ by $\hat{\boldsymbol{\theta}}_n$, then simulates new data from the estimated model $\mathrm{P}^{(n)}_{\hat{\boldsymbol{\theta}}_n}$, computing a new estimate $\hat{\boldsymbol{\theta}}^*_n$. The simulated (and observable) distribution of $\hat{\boldsymbol{\theta}}^*_n - \hat{\boldsymbol{\theta}}_n$ ($\hat{\boldsymbol{\theta}}_n$ fixed) is now used as an estimate of the distribution of $\hat{\boldsymbol{\theta}}_n - \boldsymbol{\theta}$ and hence confidence regions can be calculated. (It is not actually necessary to use the same estimate of $\boldsymbol{\theta}$ under which to do the simulations as the value of the estimator of interest, though it would be unusual to use a different one.)

Suppose now the model satisfies the S-LAN property and that $\hat{\boldsymbol{\theta}}_n$ is an S-regular estimator of $\boldsymbol{\theta}$. Since $a_n(\hat{\boldsymbol{\theta}}_n - \boldsymbol{\theta}_0) \xrightarrow{\mathcal{D}} \mathbf{Z}$ for some random variable $\mathbf{Z}$ (under $\mathrm{P}^{(n)}_{\boldsymbol{\theta}_0}$), we may invoke the Skorohod–Dudley almost sure convergence construction [see Pollard (1984)] to guarantee the existence of a sequence $\hat{\boldsymbol{\theta}}'_n$ having the same distribution as $\hat{\boldsymbol{\theta}}_n$, such that $a_n(\hat{\boldsymbol{\theta}}'_n - \boldsymbol{\theta}_0) \to \mathbf{Z}'$ almost surely, where $\mathbf{Z}'$ has the same distribution as $\mathbf{Z}$. Thus, $\hat{\boldsymbol{\theta}}'_n = \boldsymbol{\theta}_0 + a_n^{-1}\mathbf{Z}' + o(a_n)$ almost surely.

The sequence $\hat{\boldsymbol{\theta}}'_n$ thus satisfies (with probability one) the S-LAN property, and therefore, by S-regularity, the distribution of $a_n(\hat{\boldsymbol{\theta}}'^*_n - \hat{\boldsymbol{\theta}}'_n)$ converges (almost surely) to that of $\mathbf{Z}$ along the sequence $\mathrm{P}^{(n)}_{\hat{\boldsymbol{\theta}}'_n}$. Since convergence almost surely implies convergence in probability and sampling under $\mathrm{P}^{(n)}_{\hat{\boldsymbol{\theta}}'_n}$ and under $\mathrm{P}^{(n)}_{\hat{\boldsymbol{\theta}}_n}$ is, for each n, distributionally the same, the distribution of $a_n(\hat{\boldsymbol{\theta}}^*_n - \hat{\boldsymbol{\theta}}_n)$ converges in probability under the original sequence of models to that of $\mathbf{Z}$ too. The bootstrap, thus, works "in probability"; this is enough to guarantee convergence of the coverage probability of bootstrap-based confidence regions to their nominal level. For more details, see van Pul (1992b).

VIII.2. Local Asymptotic Normality in Counting Process Models

The general theory of LAN will now be specialized to counting process models. Section VIII.2.1 represents a general theorem due to Jacod and Shiryayev (1987) giving essentially the weakest possible conditions for LAN; it closely imitates classical theorems for LAN in parametric models for independent, identically distributed random variables and proves LAN in the form (8.1.8), stating joint asymptotic normality of the log-likelihood ratios for any finite collection of local alternatives. The classical theory gives local asymptotic normality under the condition of mean square differentiability of the root density of one observation with respect to the parameter. For counting process models, it turns out that a completely analogous result can be given in terms of mean square differentiability in a certain sense of the root intensity process.

Though this general theorem is very elegant, in our applications we will mainly prove the LAN property via the definition (8.1.9), which approximates the log-likelihood ratio at a single local alternative in terms of some approximate asymptotically normally distributed score $\mathbf{U}^{(n)}$; in fact, we will usually attempt to prove this for the exact local score $\mathbf{U}^{(n)}$ defined by (8.1.11). In Section VIII.2.2, we show, taking this approach, that LAN holds in parametric models under exactly the same assumptions as used in Section VI.1.2 to derive large sample properties of maximum likelihood estimators; these estimators are then asymptotically efficient and the usual likelihood-based testing procedures asymptotically optimal too.

In the rest of the chapter, we will usually leave the checking of the LAN property (or even the exact formulation of conditions for it to hold) to the reader, who now has the choice between direct verification of the definitions (8.1.8) or (8.1.9), verification of the conditions of the Jacod and Shiryayev Theorem VIII.2.1, or verification of the conditions of Theorem VI.1.2. The latter conditions are, of course, much stronger than the others, but in well-behaved models rather easy to check.

In the last two subsections, we obtain optimality results on the nonparametric k-sample tests of Chapter V and the nonparametric estimators of Chapter IV, through the device of constructing parametric submodels (for which the LAN property may be verified by checking the asymptotic normality of the score of local parameters). In both cases, we are working in the multiplicative intensity model for a k-variate counting process.

For the testing problem (5.2.1) of equality of the unknown functions α_h, it turns out we can find $2k$-dimensional parametric submodels for which the optimal tests of the corresponding parametric hypothesis are asymptotically equivalent to the best tests of the type (5.2.2) and (5.2.6). This result can be phrased as a dramatic extension of the optimality results of Section V.2.3 where only some partial results were obtained ($k = 2$ or asymptotically proportional risk sets) and where the comparison was *within* the class of tests of the form (5.2.2). Now our result says that the best tests of the form (5.2.2), also for $k \neq 2$ and for asymptotically nonproportional risk sets for alternatives of the type (5.2.26) and (5.2.32), are in fact the best of *all* tests if α is completely unknown.

In particular, for censored survival data (Example V.2.14), the k-sample log-rank test is the asymptotically optimal test against Lehmann alternatives (proportional hazards with baseline hazard unknown), and the Peto–Prentice generalization of the Wilcoxon and Kruskal–Wallis test is the best test against unknown time transformations of logistic location alternatives. These two models are invariant under monotone time transformations, essentially forcing one to restrict attention to rank tests, so we may say: The log-rank test is the optimal rank test for Lehmann alternatives; the Peto–Prentice test is the optimal rank rest for logistic location alternatives. Optimality is here asymptotic, and in any of the senses described in Section VIII.1.3 (most stringent; or, with regard to local alternatives equally distant from the null hypothesis: minimax, best average power, or best constant power).

Section VIII.2.4 considers nonparametric estimation for the multiplicative intensity model. A one-dimensional parametric submodel can be found making the Nelson–Aalen estimator of one of the integrated hazard functions at a given time point, $\hat{A}_h(t)$ say, asymptotically equally good as the best parametric estimator of the same quantity. This result can be phrased more dramatically as giving asymptotic optimality of $\hat{A}_h(t)$ as an estimator of $A_h(t)$ in the completely *nonparametric* multiplicative intensity model. We show how the result can be extended to the joint estimation of cumulative hazards for various t and h, and also to finite-dimensional functionals of the $\mathbf{A}$ such as quantiles or product-integrals (or quantiles of product-integrals) at a finite number of time points. In particular, therefore, we have shown the asymptotic optimality of the Kaplan–Meier estimator of the survival probability at a particular time point when the survival function is completely unknown; more generally, the same holds for the Aalen–Johansen estimator of transition probabilities in an inhomogenous Markov process with completely unknown intensities.

Though these results are very nice, one still would like to get stronger results of some kind of optimality, for instance, of the Nelson–Aalen estimator considered as an estimator of a whole unknown *function*. For this purpose, Section VIII.3 will present a more abstract optimality theory allowing infinite-dimensional parameters. This theory also removes the ad hoc nature of the "hardest parametric models" given in Sections VIII.2.3 and VIII.2.4: The theory will implicitly include a construction of these hardest models. The problems treated in Sections VIII.2.3 and VIII.2.4 will be discussed again from the point of view of the general abstract theory in Section VIII.4.

VIII.2.1. Jacod and Shiryayev's Theorem

In this section, we present a version of a general theorem of Jacod and Shiryayev, giving conditions for local asymptotic normality in the form (8.1.8) to hold when the statistical model under consideration is a counting process model. The theorem is given to show the elegance of the LAN theory and to show how one of the main theorems in the theory of LAN in i.i.d. models has an exact analogue for counting processes. In our applications, we will usually take a different route to LAN, via (8.1.9)–(8.1.11).

Suppose $\mathbf{N} = \mathbf{N}^{(n)}$ is a k_n-variate counting process—from now on we suppress the upper index (n)—having intensity process $\boldsymbol{\lambda}^{\boldsymbol{\theta}}$ under $\mathbf{P}_{\boldsymbol{\theta}}$ with respect to a filtration satisfying the usual conditions (2.2.1) (except possibly completeness). Here the parameter $\boldsymbol{\theta}$ varies in a parameter space $\Theta \subseteq \mathbb{R}^p$, and $\boldsymbol{\theta}_0$ is a fixed point in its interior. Suppose $\mathscr{F}_t = \mathscr{F}_0 \vee \sigma\{\mathbf{N}(s)\colon s \leq t\}$ and suppose all $\mathbf{P}_{\boldsymbol{\theta}}$, $\boldsymbol{\theta} \in \Theta$ coincide on $\mathscr{F}_0$; or, more generally, that we have *noninformative censoring* as defined in Section III.2.3, with all $\mathbf{P}_{\boldsymbol{\theta}}$ again coinciding on $\mathscr{F}_0$. Then the partial likelihood for $\boldsymbol{\theta}$ based on the counting process $\mathbf{N}$ coincides with the full likelihood generated by a larger marked point process generating the filtration. Suppose $\mathscr{F} = \mathscr{F}_\tau$. Let $\mathscr{I}$ be a $p \times p$ symmetric, positive definite matrix and $a_n \to \infty$ a sequence of constants; define $\boldsymbol{\theta}_n(\mathbf{h}) = \boldsymbol{\theta}_0 + a_n^{-1}\mathbf{h}$, $\mathbf{h} \in \mathbb{R}^p$.

Theorem VIII.2.1 (Jacod and Shiryayev). *Suppose, for each* $\mathbf{h}, \mathbf{h}' \in \mathbb{R}^p$,

$$\sum_{j=1}^{k_n} \int_0^\tau \left(\left(\frac{\lambda_j^{\boldsymbol{\theta}_n(\mathbf{h})}}{\lambda_j^{\boldsymbol{\theta}_0}}\right)^{1/2} - 1\right)\left(\left(\frac{\lambda_j^{\boldsymbol{\theta}_n(\mathbf{h}')}}{\lambda_j^{\boldsymbol{\theta}_0}}\right)^{1/2} - 1\right)\lambda_j^{\boldsymbol{\theta}_0} \xrightarrow{\mathrm{P}_{\boldsymbol{\theta}_0}} \frac{1}{4}\mathbf{h}^\mathsf{T}\mathscr{I}\mathbf{h}', \quad (8.2.1)$$

$$\sum_{j=1}^{k_n} \int_0^\tau \left(\left(\frac{\lambda_j^{\boldsymbol{\theta}_n(\mathbf{h})}}{\lambda_j^{\boldsymbol{\theta}_0}}\right)^{1/2} - 1\right)^2 I\left(\left|\left(\frac{\lambda_j^{\boldsymbol{\theta}_n(\mathbf{h})}}{\lambda_j^{\boldsymbol{\theta}_0}}\right)^{1/2} - 1\right| > \varepsilon\right)\lambda_j^{\boldsymbol{\theta}_0} \xrightarrow{\mathrm{P}_{\boldsymbol{\theta}_0}} 0. \quad (8.2.2)$$

Then we have local asymptotic normality and, for each $\mathbf{h} \in \mathbb{R}^p$,

$$\log\frac{\mathrm{dP}_{\boldsymbol{\theta}_n(\mathbf{h})}}{\mathrm{dP}_{\boldsymbol{\theta}_0}} + \frac{1}{2}\mathbf{h}^\mathsf{T}\mathscr{I}\mathbf{h} - 2\sum_{j=1}^{k_n} \int_0^\tau \left(\left(\frac{\lambda_j^{\boldsymbol{\theta}_n(\mathbf{h})}}{\lambda_j^{\boldsymbol{\theta}_0}}\right)^{1/2} - 1\right)\mathrm{d}M_j^{\boldsymbol{\theta}_0} \xrightarrow{\mathrm{P}_{\boldsymbol{\theta}_0}} 0, \quad (8.2.3)$$

where $\mathbf{M}^{\boldsymbol{\theta}} = \mathbf{N} - \int \boldsymbol{\lambda}^{\boldsymbol{\theta}}$.

Write $U^{(n)}(\mathbf{h})$ for the stochastic integral part of (8.2.3) (including the factor 2), which therefore states that the log-likelihood ratio for comparing $\boldsymbol{\theta}_n(\mathbf{h})$ to $\boldsymbol{\theta}_0$ is asymptotically equivalent, under $\mathrm{P}_{\boldsymbol{\theta}_0}$, to $U^{(n)}(\mathbf{h}) - \frac{1}{2}\mathbf{h}^\mathsf{T}\mathscr{I}\mathbf{h}$. Conditions (8.2.1) and (8.2.2) are the conditions needed in the martingale central limit theorem to guarantee asymptotic multivariate normality, with means zero and covariances $\mathbf{h}^\mathsf{T}\mathscr{I}\mathbf{h}'$, for such stochastic integrals $U^{(n)}(\mathbf{h})$, $U^{(n)}(\mathbf{h}')$. So, once we have established the asymptotic equivalence, we will indeed have local asymptotic normality in the sense of (8.1.8). The beautiful thing is that conditions (8.2.1) and (8.2.2) are also exactly what is needed to show the asymptotic negligibility of the remainder term in (8.2.3), i.e., to prove asymptotic equivalence.

To interpret the theorem in terms of the asymptotic normality of an approximate vector of scores for the local parameter $\mathbf{h}$ [see (8.1.9)–(8.1.11)], note that, by a first-order Taylor expansion of the root intensity process at $\boldsymbol{\theta}_n(\mathbf{h})$ about $\mathbf{h} = \mathbf{0}$,

$$2\left(\left(\frac{\lambda_j^{\boldsymbol{\theta}_n(\mathbf{h})}}{\lambda_j^{\boldsymbol{\theta}_0}}\right)^{1/2} - 1\right) \approx a_n^{-1}\mathbf{h}^\mathsf{T}\left(\frac{\partial}{\partial\boldsymbol{\theta}}\log\lambda_j^{\boldsymbol{\theta}_0}\right). \quad (8.2.4)$$

Recall from Sections VI.1.1 and VI.1.2, in particular formula (6.1.13), that the vector of score functions for the components of $\boldsymbol{\theta}$, at $\boldsymbol{\theta} = \boldsymbol{\theta}_0$, can be written as

$$\sum_{j=1}^{k_n} \int_0^\tau \frac{\partial}{\partial\boldsymbol{\theta}}\log\lambda_j^{\boldsymbol{\theta}_0}\,\mathrm{d}M_j^{\boldsymbol{\theta}_0}.$$

Therefore,

$$U^{(n)}(\mathbf{h}) \approx a_n^{-1}\mathbf{h}^\mathsf{T}\sum_{j=1}^{k_n} \int_0^\tau \frac{\partial}{\partial\boldsymbol{\theta}}\log\lambda_j^{\boldsymbol{\theta}_0}\,\mathrm{d}M_j^{\boldsymbol{\theta}_0} = a_n^{-1}\mathbf{h}^\mathsf{T}\frac{\partial}{\partial\boldsymbol{\theta}}\log\left(\frac{\mathrm{dP}_{\boldsymbol{\theta}}}{\mathrm{dP}_{\boldsymbol{\theta}_0}}\right)$$

$$= \mathbf{h}^\mathsf{T}\frac{\partial}{\partial\mathbf{h}}\log\left(\frac{\mathrm{dP}_{\boldsymbol{\theta}_n(\mathbf{h})}}{\mathrm{dP}_{\boldsymbol{\theta}_0}}\right).$$

We may identify the components of the vector $\mathbf{U}^{(n)}$ of (8.1.9), the alternative definition of LAN, with each $U^{(n)}(\mathbf{e}_i)$, $\mathbf{e}_i$ being the ith unit vector in $\mathbb{R}^p$. The above approximation shows, therefore, that $U^{(n)}(\mathbf{e}_i)$ is as we expect an approximate score function for the ith component of the local parameter $\mathbf{h}$ after the reparametrization $\boldsymbol{\theta}_n(\mathbf{h}) = \boldsymbol{\theta}_0 + a_n^{-1}\mathbf{h}$ from $\boldsymbol{\theta}$ to $\mathbf{h}$. Section VIII.2.2 shows that under suitable smoothness conditions, one may take $\mathbf{U}^{(n)}$ as *exactly* the score for $\mathbf{h}$.

To illustrate Theorem VIII.2.1, we verify its conditions in the most simple model, Example VI.1.2 (the standardized mortality ratio), already considered from a large sample point of view in Example VI.1.10.

EXAMPLE VIII.2.1. Standardized Mortality Ratio

Consider a univariate counting process $N = N^{(n)}$ with intensity process

$$\lambda^\theta(t) = \theta\alpha_0(t)Y^{(n)}(t), \quad \theta \in \Theta = (0, \infty),$$

where $\alpha_0(t)$ is a known function of t (the baseline mortality function) and $Y^{(n)}(t)$ is an observable, predictable process, the number at risk. Suppressing dependence on n, the maximum likelihood estimator of θ is $\hat{\theta} = N(\tau)/E(\tau)$, where $E(\tau) = \int_0^\tau \alpha_0(t)Y(t)\,dt$. Let $\theta_0 \in \Theta$ be a fixed parameter value. We suppose a sequence of constants $a_n \to \infty$ exists such that

$$a_n^{-2}E(\tau) \xrightarrow{P_{\theta_0}} e(\tau) > 0.$$

By Example VI.1.10,

$$a_n(\hat{\theta} - \theta_0) \xrightarrow{\mathscr{D}(\theta_0)} \mathcal{N}(0, \theta_0/e(\tau)).$$

We will discover later that $\hat{\theta}$ is actually a regular estimator of θ.

First, we verify local asymptotic normality. Since θ is one-dimensional, we consider $h \in \mathbb{R}$ and rewrite conditions (8.2.1) and (8.2.2) as

$$\int_0^\tau \left(\left(\frac{\lambda^{\theta_0 + a_n^{-1}h}(s)}{\lambda^{\theta_0}(s)}\right)^{1/2} - 1\right)^2 \lambda^{\theta_0}(s)\,ds \xrightarrow{P_{\theta_0}} \frac{1}{4}h^\mathsf{T}\mathscr{I}h, \tag{8.2.5}$$

$$\int_0^\tau \left(\left(\frac{\lambda^{\theta_0 + a_n^{-1}h}(s)}{\lambda^{\theta_0}(s)}\right)^{1/2} - 1\right)^2 I\left(\left|\left(\frac{\lambda^{\theta_0 + a_n^{-1}h}(s)}{\lambda^{\theta_0}(s)}\right)^{1/2} - 1\right| > \varepsilon\right)\lambda^{\theta_0}(s)\,ds \xrightarrow{P_{\theta_0}} 0, \tag{8.2.6}$$

where we note that

$$\left(\frac{\lambda^{\theta_0 + a_n^{-1}h}(s)}{\lambda^{\theta_0}(s)}\right)^{1/2} = \left(\frac{\theta_0 + a_n^{-1}h}{\theta_0}\right)^{1/2} = \left(1 + \frac{a_n^{-1}h}{\theta_0}\right)^{1/2}$$

and

$$\lambda^{\theta_0}(s) = \theta_0\alpha_0(s)Y(s).$$

Thus, (8.2.6) holds trivially. Moreover, we can rewrite the left-hand side of (8.2.5) as

$$\left(\frac{(1 - a_n^{-1}h/\theta_0)^{1/2} - 1}{a_n^{-1}}\right)^2 \frac{\theta_0 E(\tau)}{a_n^2}$$

which converges in probability to $(\frac{1}{2}h/\theta_0)^2\theta_0 e(\tau) = \frac{1}{4}h^2 e(\tau)/\theta_0$. Thus, if the preliminary conditions $\mathscr{F}_t = \mathscr{F}_0 \vee \sigma\{N(s): s \le t\}$ and P_θ is constant on $\mathscr{F}_0$ hold, we have local asymptotic normality with information matrix (scalar) $e(\tau)/\theta_0$.

By the convolution theorem (Theorem VIII.1.2), for every regular estimator $\hat{\theta}$ of θ, the asymptotic distribution of $a_n(\hat{\theta} - \theta_0)$ is less concentrated about zero than the $\mathcal{N}(0, \mathscr{I}^{-1})$ distribution, in this case, the $\mathcal{N}(0, \theta_0/e(\tau))$ distribution. This coincides with the asymptotic distribution of the maximum likelihood estimator. To show efficiency of the latter, it remains to show that this estimator is regular.

A direct proof of regularity is very easy in this case. Note that $a_n(\hat{\theta} - \theta_n(h)) = a_n^{-1}M^{\theta_n(h)}(\tau)/(a_n^{-2}E(\tau))$. Since $a_n^{-2}E(\tau)$ converges in probability under θ_0, it converges in probability under the contiguous sequence $\theta_n(h)$ to the same limit. Under this sequence, $\langle a_n^{-1}M^{\theta_n(h)}\rangle(\tau) = \theta_n(h)a_n^{-2}E(\tau)$ which, therefore, converges in probability also, to the same limit as under θ_0. The Lindeberg condition for $a_n^{-1}M^{\theta_n(h)}$ is trivially satisfied since the jumps of this martingale are all equal to a_n^{-1}. Combining, we get asymptotically normality of $a_n(\hat{\theta} - \theta_n(h))$ under $\theta_n(h)$ with the same limit as under θ_0.

A more subtle proof is to note that, from (8.2.3), the log-likelihood ratio [for comparing $\theta_n(h)$ and θ_0] is asymptotically equivalent to $2((1 + a_n^{-1}h/\theta_0)^{1/2} - 1)M^{\theta_0}(\tau) - \frac{1}{2}h^{\mathsf{T}}\mathscr{I}h$. Since $a_n^{-1}M^{\theta_0}(\tau)/\theta_0$ is asymptotically $\mathcal{N}(0, \mathscr{I})$ distributed, the log-likelihood ratio is also asymptotically equivalent to $ha_n^{-1}M^{\theta_0}(\tau)/\theta_0 - \frac{1}{2}h^{\mathsf{T}}\mathscr{I}h$, whereas $a_n(\hat{\theta} - \theta_0) = M^{\theta_0}(\tau)/E(\tau) = a_n^{-1}M^{\theta_0}(\tau)/(a_n^{-2}E(\tau))$. So we can take the approximate score for h, $U^{(n)}$ of (8.1.9), to be $a_n^{-1}M^{\theta_0}(\tau)/\theta_0$; and $a_n(\hat{\theta} - \theta_0)$ is asymptotically equivalent to $U^{(n)}/\mathscr{I}$. Thus, $\hat{\theta}$ is asymptotically linear with the optimal influence function and, therefore, regular and efficient. In this model, $(\partial/\partial\theta)\log\lambda^\theta$ is the constant $1/\theta$, so the score for θ is actually exactly equal to $M^\theta(\tau)/\theta$ and that for h is exactly $a_n^{-1}M^\theta(\tau)/\theta$.

The fact that this maximum likelihood estimator is asymptotically linear with optimal influence function is no coincidence; in the next section, we will derive this result, in general, for maximum likelihood estimators in parametric models. □

VIII.2.2. LAN in Parametric Models

This section reconsiders the general parametric model of Chapter VI. Thus, we suppose $\mathbf{N}$ is a k_n-variate counting process with intensity process $\boldsymbol{\lambda}(\cdot; \boldsymbol{\theta}) = \boldsymbol{\lambda}_{\boldsymbol{\theta}}$ (a subscript is more convenient here). It turns out that the conditions of Section VI.1.2 for asymptotic normality of the maximum likelihood estimator $\hat{\boldsymbol{\theta}}$ give (by an equally simple proof) local asymptotic normality of the model and asymptotic efficiency of the estimator, as well asymptotic optimality of the usual likelihood-based test procedures. Efficiency theory in semiparametric models is either explicitly (as in the rest of Section VIII.2) or implicity (as in Sections VIII.3 and VIII.4) built on the LAN property of suitable parametric submodels, so the following theorem can be used again and again in the rest of the chapter.

Theorem VIII.2.2. *Under Condition* VI.1.1 *together with the assumption of noninformative censoring* (*so the partial likelihood based on the observed counting process is actually the full likelihood for the parameter of interest*), *we have local asymptotic normality at* $\boldsymbol{\theta}_0$ *with rate sequence* a_n *the same as in part* (B) *of Condition* VI.1.1, *and with asymptotic information matrix* $\mathscr{I} = \boldsymbol{\Sigma}$ *given in part* (D) *of the conditions. Moreover, in* (8.1.9), *one may take*

$$\mathbf{U}^{(n)} = a_n^{-1}\mathbf{U}(\boldsymbol{\theta}_0),$$

the score for $\mathbf{h}$ *at* $\mathbf{h} = \mathbf{0}$, *in the reparametrization* (8.1.7), *or the scaled score for* $\boldsymbol{\theta}$ *at* $\boldsymbol{\theta}_0$. *The maximum likelihood estimator* $\hat{\boldsymbol{\theta}}$ *is regular and asymptotically efficient.*

For testing, one may conclude that the (asymptotically equivalent) score test, Wald test, and likelihood ratio test of the null hypothesis $\boldsymbol{\theta} = \boldsymbol{\theta}_0$ are asymptotically most stringent, have asymptotically best average power, are asymptotically minimax, and asymptotically most powerful invariant, with respect to the local ellipsoids defined by a constant value of $\mathbf{h}^{\mathsf{T}}\mathscr{I}\mathbf{h}$ where, asymptotically, the envelope power is constant too. These are alternatives equally far from the null in a statistical sense. Similarly the standard likelihood-based tests of a hypothesis $\boldsymbol{\kappa}(\boldsymbol{\theta}) = \boldsymbol{\kappa}_0$, for any differentiable function $\boldsymbol{\kappa}$ with matrix of partial derivatives $\mathbf{d}\boldsymbol{\kappa}$ at $\boldsymbol{\theta}_0$, have the same optimality properties with respect to local ellipsoids where $(\mathbf{d}\boldsymbol{\kappa}\cdot\mathbf{h})^{\mathsf{T}}(\mathbf{d}\boldsymbol{\kappa}\mathscr{I}^{-1}\,\mathbf{d}\boldsymbol{\kappa}^{\mathsf{T}})^{-1}(\mathbf{d}\boldsymbol{\kappa}\cdot\mathbf{h})$ takes a fixed value and, hence, there is asymptotically constant envelope power. When κ is scalar, the tests are asymptotically uniformly most powerful unbiased. Also in that case, the one-sided tests of $\kappa(\boldsymbol{\theta}) \le \kappa_0$ are asymptotically most powerful, without any qualification.

PROOF. For simplicity, we will take $k_n = 1$ for all n, so the counting process N is univariate, and we will take θ to be scalar too. The general case is proved in a similar manner, along the lines of Theorems VI.1.1 and VI.1.2. Dependence on θ will now be expressed by a subscript. Write $M_\theta = N - \int \lambda_\theta$. By the prime we denote a derivative with respect to θ. Recall that $\theta_n(h) = \theta_0 + a_n^{-1}h$. By noninformative censoring, we have that the log-likelihood ratio for $\theta_n(h)$ against θ_0 is

$$\begin{aligned} C(h) &= \log\frac{\mathrm{d}P_{\theta_n(h)}}{\mathrm{d}P_{\theta_0}} \\ &= \int_0^\tau \log(\lambda_{\theta_n(h)})\,\mathrm{d}N - \int_0^\tau \lambda_{\theta_n(h)}\,\mathrm{d}t - \int_0^\tau \log(\lambda_{\theta_0})\,\mathrm{d}N + \int_0^\tau \lambda_{\theta_0}\,\mathrm{d}t. \end{aligned}$$

Of course, $C(0) = 0$, and the first, second, and third derivatives of C with respect to h are

$$a_n^{-1}\left(\int_0^\tau \frac{\lambda'_{\theta_n(h)}}{\lambda_{\theta_n(h)}}\,\mathrm{d}N - \int_0^\tau \lambda'_{\theta_n(h)}\right),$$

$$a_n^{-2}\left(\int_0^\tau \left(\frac{\lambda''_{\theta_n(h)}}{\lambda_{\theta_n(h)}} - \left(\frac{\lambda'_{\theta_n(h)}}{\lambda_{\theta_n(h)}}\right)^2\right)\mathrm{d}N - \int_0^\tau \lambda''_{\theta_n(h)}\right),$$

$$a_n^{-3}\left(\int_0^\tau (\log \lambda_{\theta_n(h)})'''\,\mathrm{d}N - \int_0^\tau \lambda'''_{\theta_n(h)}\right).$$

So we get the Taylor expansion

$$\begin{aligned} C(h) &= h a_n^{-1}\int_0^\tau (\log \lambda_{\theta_0})'\,\mathrm{d}M_{\theta_0} \\ &\quad - \tfrac{1}{2}h^2 a_n^{-2}\int_0^\tau ((\log \lambda_{\theta_0})')^2\lambda_{\theta_0} + \tfrac{1}{2}h^2 a_n^{-2}\int_0^\tau (\log \lambda_{\theta_0})''\,\mathrm{d}M_{\theta_0} \\ &\quad + \tfrac{1}{6}h^3 a_n^{-3}\int_0^\tau (\log \lambda_{\theta_n(h^*)})'''\,\mathrm{d}N - \tfrac{1}{6}h^3 a_n^{-3}\int_0^\tau \lambda'''_{\theta_n(h^*)} \\ &= hU_n - \tfrac{1}{2}h^2 I_n + \tfrac{1}{2}h^2 R_{1n} + \tfrac{1}{6}h^3 R_{2n} - \tfrac{1}{6}h^3 R_{3n}, \end{aligned}$$

say, where h^* is on the line segment between 0 and h. Now parts (B) and (C) of Condition VI.1.1 are, respectively, the condition of convergence in probability of the predictable variation process and the Lindeberg condition for proving the weak convergence of the stochastic integral U_n to the $\mathcal{N}(0, \mathcal{I})$ distribution by the martingale central limit theorem. Term I_n converges in probability to the asymptotic variance $\mathcal{I}$ by part (B). Recalling that the score function for θ at θ_0, denoted by $U(\theta_0)$, was equal to $\int_0^\tau (\log \lambda_{\theta_0})'\,\mathrm{d}M_{\theta_0}$, we see that we will have the local asymptotic normality property (8.1.9) with $U^{(n)} = a_n^{-1}U(\theta_0)$ if only we can show that terms R_{1n}, R_{2n}, and R_{3n} converge in probability to zero. Moreover, since Theorem VI.1.2 shows that $a_n(\hat{\theta} - \theta_0)$ is asymptotically equivalent to $\mathcal{I}^{-1}a_n^{-1}U(\theta_0)$, regularity and asymptotic efficiency of $\hat{\theta}$ hold by Theorem VIII.1.2 and the remarks following it.

Now the predictable variation process of $a_n^{-1}\int(\log \lambda_{\theta_0})''\,\mathrm{d}M_{\theta_0}$ is $a_n^{-2}\int((\log \lambda_{\theta_0})'')^2\lambda_{\theta_0}$, so the convergence assumption of part (E) and Lenglart's inequality deal with term R_{1n} (we even have another factor a_n^{-1} to use). Term R_{3n} is bounded, also by part (E), by $a_n^{-3}\int_0^\tau G(t)\,\mathrm{d}t$, so it is asymptotically negligible too since a_n^{-2} times the integral of G converges in probability to a finite quantity. Finally, also using part (E), term R_{2n} is bounded by $a_n^{-3}\int_0^\tau H(t)\,\mathrm{d}N(t)$. Since H is predictable and non-negative, this integral is the optional variation process (evaluated at τ) of the local square integrable martingale $a_n^{-3/2}\int H^{1/2}\,\mathrm{d}M_{\theta_0}$. This martingale has predictable variation process $a_n^{-3}\int H\lambda_{\theta_0}$. By the corollary of the martingale central limit theorem, the optional and predictable variation processes have the same limit in probability. Thus, R_{2n} converges in probability to zero by part (E), thanks to the factor $a_n^{-1} \to 0$. □

VIII.2.3. Efficiency of Nonparametric k-Sample Tests

In the previous subsection, we considered a genuinely parametric model. Now we turn to some nonparametric models, establishing efficiency of the nonparametric k-sample tests of Chapter V and (in the next subsection) of the estimators of Chapter IV by embedding into the corresponding nonparametric models some carefully chosen parametric submodels.

Consider the multiplicative intensity model with completely unknown hazard functions $\alpha_1, \ldots, \alpha_k$, whose equality is to be tested. Since the notation h will now stand for one of the k samples, we will use a symbol ϕ for local parameters. In Section V.2.3, we discussed the power of counting process based tests by considering the "simple alternatives"

$$\alpha_h^{(n)}(t) = \alpha(t) + a_n^{-1}\phi_h\gamma(t)\alpha(t) + o(a_n^{-1}). \tag{8.2.7}$$

[see (5.2.26) and (5.2.32)] to the "simple null hypothesis"

$$\alpha_h^{(n)}(t) = \alpha(t), \quad h = 1, \ldots, k.$$

The "little o" here is uniform in $t \in \mathscr{T}$.

This actually corresponds exactly to a local reparametrization of a global, but parametric, submodel

$$\alpha_h^{(n)}(t) = \alpha(t; \theta_h^{(n)}) \tag{8.2.8}$$

with the reparametrization

$$\theta_h^{(n)} = \theta_0 + a_n^{-1}\phi_h. \tag{8.2.9}$$

This can be seen by substituting (8.2.9) into (8.2.8) and then carrying out a first-order Taylor expansion which gives precisely, for each t, the expansion (8.2.7) with

$$\gamma(t) = \frac{\partial}{\partial\theta}\log\alpha(t; \theta_0). \tag{8.2.10}$$

The submodel therefore restricts the α_h to be members of a one-dimensional parametric family of hazard functions. In the nonparametric or infinite-dimensional model with arbitrary $\alpha_1, \ldots, \alpha_k$, we test the null hypothesis $\alpha_1 = \cdots = \alpha_k$; restricted to the parametric submodel $\alpha_h = \alpha(\cdot; \theta_h)$, $h = 1, \ldots, k$, we test the null hypothesis $\theta_1 = \cdots = \theta_k$; and after the reparametrization to local parameters (8.2.9), we test the null hypothesis $\phi_1 = \cdots = \phi_k$.

In Section V.2.3, this approach only gave nice results in two special cases: the two-sample case $k = 2$ and the case of asymptotically proportional risk sets when the functions y_h (the limits in probability of the normalized "number at risk" processes $a_n^{-2}Y_h^{(n)}$) are proportional over time. At first instance, we will meet the same problem here. It will turn out to be necessary to extend the parametric model (8.2.8) to a model with a k-dimensional nuisance parameter, supposed to be the same for each of the k samples,

$$\alpha_h^{(n)}(t) = \alpha(t; \theta_h^{(n)}, \boldsymbol{\eta}^{(n)})$$

before conclusive general results are obtained, except in the case of asymptotically proportional risk sets. (Note that the nuisance parameter $\boldsymbol{\eta}$ is the *same* in each sample, so the null and alternative hypotheses remain the same, namely, the equality or nonequality of the k parameters θ_h.) This is a very reasonable extension since in the model (8.2.8), knowing the form of $\alpha(\cdot;\cdot)$ could mean that more statistical information is available for testing equality of the θ_h. Adding the k-dimensional nuisance parameter, with α depending locally on $\boldsymbol{\eta}$ in a special way, will serve to make testing in the resulting $2k$-dimensional parametric model as difficult as the original nonparametric one. The question is really why introduction of this parameter is *not* necessary in the second of the two special cases just mentioned.

To start with, we will work in the (too small) submodel (8.2.8) since the groundwork done there will immediately be of use in the final submodel with the nuisance parameter $\boldsymbol{\eta}$. In particular, we need to build up some general methods for computing the asymptotic power of χ^2-type tests with nuisance parameters (the common value of the θ_h under the null hypothesis is a one-dimensional nuisance parameter).

The approach is to try to verify the LAN property and then go on to calculate the asymptotic power of an asymptotically optimal test of the null hypothesis. If we are able, by a lucky choice of parametric model (i.e., by choice of γ), to get an optimal parametric test which is asymptotically equivalent to one of the nonparametric tests of Section V.2, then we have also established the same optimality property of that nonparametric test (see the discussion at the end of Section VIII.1.3 on notions of optimality for statistical tests).

Specifying regularity conditions for the LAN property to hold is now a trivial matter and could be tackled by writing down the form of Condition VI.1.1 for the parametric submodel at hand, or by verification of the conditions of the Jacod and Shiryayev theorem (Theorem VIII.2.1), or from first principles. We will leave this matter to the interested reader and here just make the assumption plausible, identifying the asymptotic score $\mathbf{U}$ for the local parameter $\boldsymbol{\phi} = (\phi_1, \ldots, \phi_k)$, which is a_n^{-1} times the score for $\boldsymbol{\theta}$. This leads to determining the crucial information matrix $\mathscr{I}$ for this problem via (8.1.9)–(8.1.11).

So informally, note that the score for θ_h (at $\theta_h = \theta_0$) in the model (8.2.8) is just

$$\int_0^\tau \frac{\partial}{\partial\theta} \log \alpha(t;\theta_0)\, \mathrm{d}M_h(t), \quad h = 1, \ldots, k,$$

where $M_h = N_h - \int Y_h \alpha$ is (under the null hypothesis) the hth component of the basic counting process martingale. The score for ϕ_h is, therefore, by (8.2.9) and (8.2.10),

$$a_n^{-1} \int_0^\tau \gamma\, \mathrm{d}M_h. \tag{8.2.11}$$

Under the assumption (5.2.14) of Proposition V.2.2, we may therefore expect (and it is easily established under the usual uniformity and boundedness conditions) that we have LAN with the vector of scores for the vector parameter $\boldsymbol{\phi} = (\phi_1, \ldots, \phi_k)^\mathsf{T}$ being asymptotically $\mathcal{N}(\mathbf{0}, \mathcal{I})$ distributed with asymptotic information matrix $\mathcal{I}$ with diagonal elements

$$\mathcal{I}_{hh} = \int_0^\tau \gamma^2 y_h \alpha, \quad h = 1, \ldots, k$$

and zero off-diagonal elements. It is, of course, not surprising that the estimation of the different components of $\boldsymbol{\theta}$, which each enter into a separate component of the compensator of $\mathbf{N}$, can be done (asymptotically) independently.

The asymptotic power of the optimal χ^2-type test (on $k-1$ degrees of freedom) of the hypothesis $\phi_1 = \cdots = \phi_k$ can be computed as follows. First, we reparametrize again to replace $\boldsymbol{\phi}$ by a vector, say $\mathbf{h}$, partitioned into $k-1$ and 1 components, respectively, the first $k-1$ being any set of linearly independent contrasts, for instance, the differences $\phi_h - \phi_{h+1}$. The last component could, for instance, be ϕ_k. This leads to an asymptotic problem of the following form (cf. the beginning of Section VIII.1.3): Based on

$$\mathbf{U} \sim \mathcal{N}(\mathcal{I}\mathbf{h}, \mathcal{I})$$

with $\mathbf{h}$ partitioned as

$$\mathbf{h} = \begin{pmatrix} \mathbf{h}_1 \\ \mathbf{h}_2 \end{pmatrix},$$

test the null hypothesis $\mathbf{h}_1 = \mathbf{0}_1$.

Put $\boldsymbol{\Sigma} = \mathcal{I}^{-1}$ and $\mathbf{X} = \mathcal{I}^{-1}\mathbf{U} \sim \mathcal{N}(\mathbf{h}, \boldsymbol{\Sigma})$. The (usual) optimal test is then based on $\mathbf{X}_1^\mathsf{T}\boldsymbol{\Sigma}_{11}^{-1}\mathbf{X}_1$, with noncentrality parameter $\mathbf{h}_1^\mathsf{T}\boldsymbol{\Sigma}_{11}^{-1}\mathbf{h}_1$.

This result has several alternative formulations which we will use in the rest of the section. By the usual results for the inverse of a partitioned matrix, we have the relationships $\boldsymbol{\Sigma}_{11}^{-1} = \mathcal{I}_{11} - \mathcal{I}_{12}\mathcal{I}_{22}^{-1}\mathcal{I}_{21}$ and $\boldsymbol{\Sigma}_{12} = -\boldsymbol{\Sigma}_{11}\mathcal{I}_{12}\mathcal{I}_{22}^{-1}$. Note that $\mathbf{U}_1 - \mathcal{I}_{12}\mathcal{I}_{22}^{-1}\mathbf{U}_2$ is the vector of residuals from the regressions of the components of $\mathbf{U}_1$ on $\mathbf{U}_2$, and $\mathcal{I}_{11} - \mathcal{I}_{12}\mathcal{I}_{22}^{-1}\mathcal{I}_{21}$ is the covariance matrix of this vector of residuals. We introduce

$$\mathbf{U}_{1|2} = \mathbf{U}_1 - \mathcal{I}_{12}\mathcal{I}_{22}^{-1}\mathbf{U}_2, \tag{8.2.12}$$

called the *effective score* for $\mathbf{h}_1$, and call

$$\mathcal{I}_{11|2} = \mathcal{I}_{11} - \mathcal{I}_{12}\mathcal{I}_{22}^{1}\mathcal{I}_{21} \tag{8.2.13}$$

the *effective information*. Then writing $\mathbf{X}_{1|2} = \mathcal{I}_{11|2}^{-1}\mathbf{U}_{1|2}$ with covariance matrix $\boldsymbol{\Sigma}_{11|2} = \mathcal{I}_{11|2}^{-1}$, we have $\mathbf{X}_{1|2} \sim \mathcal{N}(\mathbf{h}_1, \boldsymbol{\Sigma}_{11|2})$ and the optimal test statistic for testing $\mathbf{h}_1 = \mathbf{0}$ can be written both as $\mathbf{X}_{1|2}^\mathsf{T}\boldsymbol{\Sigma}_{11|2}^{-1}\mathbf{X}_{1|2}$ and as $\mathbf{U}_{1|2}^\mathsf{T}\mathcal{I}_{11|2}^{-1}\mathbf{U}_{1|2}$, with noncentrality parameter $\mathbf{h}_1^\mathsf{T}\boldsymbol{\Sigma}_{11|2}^{-1}\mathbf{h}_1$; exactly the same relationship as holds (for testing $\mathbf{h} = \mathbf{0}$) between $\mathcal{I}$, $\boldsymbol{\Sigma}$, $\mathbf{U}$, and $\mathbf{X}$.

Another important *coordinate-free* interpretation is arrived at after some straightforward algebra, which we give in a moment. Consider testing the hypothesis $\mathbf{A}\boldsymbol{\phi} = \mathbf{0}$ for some $l \times k$ matrix $\mathbf{A}$ of full rank $l < k$. As explained above, we can handle this situation by reparametrizing (augment $\mathbf{A}$ to a $k \times k$ nonsingular matrix). The resulting noncentrality parameter turns out to be

$$\boldsymbol{\phi}^{\mathsf{T}}(\mathscr{I} - \mathscr{I}\mathbf{C}(\mathbf{C}^{\mathsf{T}}\mathscr{I}\mathbf{C})^{-1}\mathbf{C}^{\mathsf{T}}\mathscr{I})\boldsymbol{\phi}, \tag{8.2.14}$$

where $\mathbf{C}$ is a $k \times (k - l)$ matrix whose columns span the null space of $\mathbf{A}$; it can be obtained, for instance, by augmenting $\mathbf{A}$ to a nonsingular matrix $(\mathbf{A}^{\mathsf{T}}\mathbf{B}^{\mathsf{T}})^{\mathsf{T}}$ with the inverse partitioned as $(\mathbf{D}\,\mathbf{C})$. This noncentrality parameter can be described in a coordinate-free way as the *squared length of the projection of* $\boldsymbol{\phi}$ *onto the orthogonal complement of the null space of* $\mathbf{A}$, *with respect to the inner product* $(\boldsymbol{\phi}, \boldsymbol{\phi}')_{\mathrm{H}} = \boldsymbol{\phi}^{\mathsf{T}}\mathscr{I}\boldsymbol{\phi}'$.

To prove (8.2.14), note that the information for $(\mathbf{A}^{\mathsf{T}}\mathbf{B}^{\mathsf{T}})^{\mathsf{T}}\boldsymbol{\phi}$ is $(\mathbf{D}\,\mathbf{C})^{\mathsf{T}}\mathscr{I}(\mathbf{D}\,\mathbf{C})$; the effective information (8.2.13) for $\mathbf{A}\boldsymbol{\phi}$ is, therefore $\mathbf{D}^{\mathsf{T}}\mathscr{I}\mathbf{D} - \mathbf{D}^{\mathsf{T}}\mathscr{I}\mathbf{C}(\mathbf{C}^{\mathsf{T}}\mathscr{I}\mathbf{C})^{-1}\mathbf{C}^{\mathsf{T}}\mathscr{I}\mathbf{D}$. The noncentrality parameter is, therefore,

$$\boldsymbol{\phi}^{\mathsf{T}}\mathbf{A}^{\mathsf{T}}(\mathbf{D}^{\mathsf{T}}\mathscr{I}\mathbf{D} - \mathbf{D}^{\mathsf{T}}\mathscr{I}\mathbf{C}(\mathbf{C}^{\mathsf{T}}\mathscr{I}\mathbf{C})^{-1}\mathbf{C}^{\mathsf{T}}\mathscr{I}\mathbf{D})\mathbf{A}\boldsymbol{\phi}$$

which, since

$$(\mathbf{D}\,\mathbf{C})\begin{pmatrix}\mathbf{A}\\ \mathbf{B}\end{pmatrix} = \mathbf{D}\mathbf{A} + \mathbf{C}\mathbf{B} = \mathbf{I},$$

equals

$$\begin{aligned}&\boldsymbol{\phi}^{\mathsf{T}}(\mathbf{I} - \mathbf{B}^{\mathsf{T}}\mathbf{C}^{\mathsf{T}})(\mathscr{I} - \mathscr{I}\mathbf{C}(\mathbf{C}^{\mathsf{T}}\mathscr{I}\mathbf{C})^{-1}\mathbf{C}^{\mathsf{T}}\mathscr{I})(\mathbf{I} - \mathbf{C}\mathbf{B})\boldsymbol{\phi}\\ &\quad = \boldsymbol{\phi}^{\mathsf{T}}(\mathscr{I} - \mathscr{I}\mathbf{C}(\mathbf{C}^{\mathsf{T}}\mathscr{I}\mathbf{C})^{-1}\mathbf{C}^{\mathsf{T}}\mathscr{I})\boldsymbol{\phi}.\end{aligned}$$

The projection of $\boldsymbol{\phi}$ onto the orthogonal complement of $\mathbf{C}$ is $\boldsymbol{\phi} - \mathbf{C}(\mathbf{C}^{\mathsf{T}}\mathscr{I}\mathbf{C})^{-1}\mathbf{C}^{\mathsf{T}}\mathscr{I}\boldsymbol{\phi}$ (its inner product with $\mathbf{C}$ is zero) and the squared length of this vector is the noncentrality parameter. Finally, since $(\mathbf{D}\,\mathbf{C})(\mathbf{A}^{\mathsf{T}}\mathbf{B}^{\mathsf{T}})^{\mathsf{T}} = \mathbf{I}$, we also have

$$\begin{pmatrix}\mathbf{A}\\ \mathbf{B}\end{pmatrix}(\mathbf{D}\,\mathbf{C}) = \begin{pmatrix}\mathbf{AD} & \mathbf{AC}\\ \mathbf{BD} & \mathbf{BC}\end{pmatrix} = \mathbf{I},$$

hence $\mathbf{AC} = \mathbf{0}$ or the columns of $\mathbf{C}$ span the null space of $\mathbf{A}$.

In our problem, we want to test $\mathbf{h}_1 = \mathbf{A}\boldsymbol{\phi} = \mathbf{0}$, where $\mathbf{A}$ can be taken as the $(k - 1) \times k$ matrix

$$\mathbf{A} = \begin{bmatrix} 1 & -1 & & & \\ & 1 & -1 & & \\ & & \cdots & & \\ & & & 1 & -1 \end{bmatrix}$$

(the rest of the elements are zeros). The null space of $\mathbf{A}$ is one-dimensional and is spanned by the vector $\mathbf{C} = \mathbf{1} = (1, \ldots, 1)^{\mathsf{T}}$.

The noncentrality parameter we are looking for is, therefore, by (8.2.14)

$$\zeta = \boldsymbol{\phi}^{\mathsf{T}}\left(\mathscr{I} - \frac{\mathscr{I}\mathbf{1}\mathbf{1}^{\mathsf{T}}\mathscr{I})}{\mathbf{1}^{\mathsf{T}}\mathscr{I}\mathbf{1}}\right)\boldsymbol{\phi}$$

or, since $\mathscr{I}$ is diagonal with

$$\mathscr{I}_{hh} = \sigma_h^2 = \int_0^\tau \gamma^2 y_h \alpha$$

[not to be confused with σ_{hj} of (5.2.12)],

$$\zeta = \sum_h \sigma_h^2 \left(\phi_h - \frac{\sum \sigma_j^2 \phi_j}{\sum \sigma_j^2}\right)^2 \tag{8.2.15}$$

($k - 1$ degrees of freedom). However, this noncentrality parameter seems not to bear any resemblance to the that resulting from the tests of Chapter V, except in the special case of asymptotically proportional risk sets which turned up as a tractable special case at the end of Section V.2.3. The case $k = 2$, which also was tractable in Section V.2.3, gives a different result.

When $k = 2$, (8.2.15) simplifies to

$$\zeta = \frac{\sigma_1^2 \sigma_2^2 (\phi_1 - \phi_2)^2}{\sigma_1^2 + \sigma_2^2},$$

which is larger than the noncentrality parameter of the test of the general class described in Theorem V.2.1 with $\kappa = \gamma$ and $k = 2$ (unless y_1 and y_2 are proportional). However, when we have asymptotically proportional risk sets with $y_h/y_{\bullet} = p_h$ (constants) and arbitrary k, then

$$\sigma_h^2 = p_h \int_0^\tau \gamma^2 y_{\bullet} \alpha$$

and

$$\zeta = \int_0^\tau \gamma^2 y_{\bullet} \alpha \sum_h p_h (\phi_h - \bar{\phi})^2$$

with

$$\bar{\phi} = \sum_h p_h \phi_h;$$

cf. (5.2.34).

In general though, even the best test of the class described in Section V.2 (which we will return to shortly) does not attain the power of the best parametric test for the hypothesis $\phi_1 = \cdots = \phi_k$ in the parametric model given by (8.2.8) and (8.2.9). Before one starts to doubt the effectiveness of our nonparametric tests however, one should realize that the parametric model (8.2.8) only partly catches the important quality of our nonparametric model ($\alpha_1, \ldots, \alpha_k$ arbitrary; test $\alpha_1 = \cdots = \alpha_k$). What we have missed is the fact that the α_h may have arbitrary shapes and our nonparametric tests are applicable whatever the shapes may be.

A first attempt to incorporate this would be to generalize (8.2.8) to the model

$$\alpha_h^{(n)}(t) = \alpha(t; \theta_h^{(n)}, \eta^{(n)})$$

for some one-dimensional nuisance parameter $\eta^{(n)}$ (taking the same value for each component $h = 1, \ldots, k$). Actually, this is still not quite enough freedom of shape: It turns out that we need to introduce a k-dimensional nuisance parameter $\boldsymbol{\eta}^{(n)}$: It has as many components as the parameter of interest $\boldsymbol{\theta}^{(n)}$.

So we consider the model

$$\alpha_h^{(n)}(t) = \alpha(t; \theta_h^{(n)}, \boldsymbol{\eta}^{(n)}) \tag{8.2.16}$$

with the local reparametrization

$$\boldsymbol{\theta}^{(n)} = \theta_0 \mathbf{1} + a_n^{-1} \boldsymbol{\phi},$$

$$\boldsymbol{\eta}^{(n)} = \boldsymbol{\eta}_0 + a_n^{-1} \boldsymbol{\zeta},$$

where both $\boldsymbol{\phi}$ and $\boldsymbol{\zeta}$ are k-dimensional. Let

$$\gamma = \frac{\partial}{\partial \theta} \log \alpha(\cdot; \theta_0, \boldsymbol{\eta}_0),$$

$$\gamma_h = \frac{\partial}{\partial \eta_h} \log \alpha(\cdot; \theta_0, \boldsymbol{\eta}_0),$$

We find, cf. (8.2.11), that the score for ϕ_h is $a_n^{-1} \int_0^\tau \gamma \, \mathrm{d}M_h$ and that the score for ζ_h is $a_n^{-1} \int_0^\tau \gamma_h \, \mathrm{d}M$. (These scores are a_n^{-1} times the scores for θ_h and η_h, respectively: We evaluate all scores at the "fixed point" in the null hypothesis $\boldsymbol{\theta} = \theta_0 \mathbf{1}$; $\boldsymbol{\eta} = \boldsymbol{\eta}_0$; $\boldsymbol{\phi} = \boldsymbol{\zeta} = \mathbf{0}$.) Hence, we may expect to have LAN with $2k \times 2k$ information matrix given schematically as

$$\begin{pmatrix} \operatorname{diag}(\int \gamma^2 y_h \alpha) & (\int \gamma \gamma_h y_l \alpha) \\ (\int \gamma \gamma_l y_h \alpha) & (\int \gamma_h \gamma_l y_. \alpha) \end{pmatrix} = \begin{pmatrix} \mathscr{I}^{\phi\phi} & \mathscr{I}^{\phi\zeta} \\ \mathscr{I}^{\zeta\phi} & \mathscr{I}^{\zeta\zeta} \end{pmatrix}. \tag{8.2.17}$$

The procedure for computing the noncentrality parameter of the asymptotically optimal parametric test of $\phi_1 = \cdots = \phi_k$ has been explained above, in terms of effective scores and effective information for (e.g.) $\phi_1 - \phi_2, \ldots, \phi_k - \phi_{k-1}$, or in terms of the projection of $\binom{\boldsymbol{\phi}}{\boldsymbol{\zeta}}$ into the orthogonal complement of the space spanned by $\binom{\mathbf{1}}{\mathbf{0}}$ and $\binom{\mathbf{0}}{\boldsymbol{\zeta}}$ ($\boldsymbol{\zeta}$ arbitrary) with respect to the information metric. We will do this calculation with a particular choice of $\gamma_1, \ldots, \gamma_k$. In fact, this choice *minimizes* the noncentrality parameter, and hence determines (locally at least) the *hardest* parametric submodel for testing $\alpha_1 = \cdots = \alpha_k$. This will be established in Section VIII.4 after we have developed some more powerful (and abstract) theory precisely aimed at this kind of application, which enables us to find the right $\gamma_1, \ldots, \gamma_k$ without guessing. However, for the time being, it is enough to point out that since our special choice of $\gamma_1, \ldots, \gamma_k$ will give an optimal "parametric" power equal to the power of our *nonparametric* test of Section V.2.3 (with $\kappa = \gamma$), this power must correspond to the *hardest* parametric model (the optimal power in any parametric model must be greater than the power of any available non-parametric test).

Consider then the choice

$$\gamma_h = \gamma y_h / y_., \tag{8.2.18}$$

where $y_. = \sum_h y_h$. This definition is legitimate since $y_h = 0$ where $y_. = 0$. Then, from (8.2.17), we see that

$$\mathscr{I}_{hl}^{\zeta\zeta} = \int_0^\tau \frac{\gamma y_h}{y_.} \frac{\gamma y_l}{y_.} y_. \alpha = \int_0^\tau \gamma^2 \frac{y_h y_l}{y_.} \alpha,$$

$$\mathscr{I}_{hl}^{\phi\zeta} = \int_0^\tau \gamma y_h \frac{\gamma y_l}{y_.} \alpha = \int_0^\tau \gamma^2 \frac{y_h y_l}{y_.} \alpha,$$

$$\mathscr{I}_{hl}^{\phi\phi} = \delta_{hl} \int_0^\tau \gamma^2 y_h \alpha.$$

Thus, $\mathscr{I}^{\phi\zeta} = \mathscr{I}^{\zeta\phi} = \mathscr{I}^{\zeta\zeta}$ and the effective information (8.2.13) for $\boldsymbol{\phi}$ is

$$\mathscr{I}^{\phi\phi|\zeta} = \mathscr{I}^{\phi\phi} - \mathscr{I}^{\phi\zeta}(\mathscr{I}^{\zeta\zeta})^{-1}\mathscr{I}^{\zeta\phi} = \mathscr{I}^{\phi\phi} - \mathscr{I}^{\zeta\zeta}$$

with

$$(\mathscr{I}^{\phi\phi|\zeta})_{hl} = \int_0^\tau \gamma^2 \frac{y_h}{y_.}\left(\delta_{hl} - \frac{y_l}{y_.}\right) y_. \alpha. \tag{8.2.19}$$

We note that $\sum_l (\mathscr{I}^{\phi\phi|\zeta})_{hl} = 0$, so the effective information matrix for $\boldsymbol{\phi}$ is singular. This means that formula (8.2.14) cannot be applied to get the desired noncentrality parameter (with $\mathbf{C} = \mathbf{1}$ and $\mathscr{I} = \mathscr{I}^{\phi\phi|\zeta}$, $\mathbf{C}^\mathsf{T}\mathscr{I}\mathbf{C} = 0$). However, the "coordinate-free" description [see the discussion below (8.2.14)] of the noncentrality parameter is still applicable: Since $\boldsymbol{\phi}^\mathsf{T}\mathscr{I}^{\phi\phi|\zeta}\mathbf{1} = 0$ for all $\boldsymbol{\phi}$, the projection of $\boldsymbol{\phi}$ into the orthogonal complement of the space spanned by $\mathbf{1}$ is $\boldsymbol{\phi}$ itself (it is already orthogonal to $\mathbf{1}$) and its squared length is, therefore, $\boldsymbol{\phi}^\mathsf{T}\mathscr{I}^{\phi\phi|\zeta}\boldsymbol{\phi}$ or

$$\int_0^\tau \gamma^2 \sum (\phi_h - \bar{\phi})^2 y_h \alpha, \tag{8.2.20}$$

where

$$\bar{\phi}(t) = \sum \frac{\phi_h y_h(t)}{y_.(t)},$$

under asymptotically proportional risk sets (proportional y_h) is independent of t and equal to the constant $\bar{\phi}$ introduced earlier and also appearing in Section V.2.3.

We compare this with the noncentrality parameter of the tests of Section V.2.3,

$$\zeta(\tau) = \boldsymbol{\xi}(\tau)^\mathsf{T}\boldsymbol{\Sigma}(\tau)^-\boldsymbol{\xi}(\tau),$$

where

$$(\boldsymbol{\Sigma}(\tau))_{hl} = \int_0^\tau \kappa^2 \frac{y_h}{y_\cdot}\left(\delta_{hl} - \frac{y_l}{y_\cdot}\right) y_\cdot \alpha,$$

$$(\boldsymbol{\xi}(\tau))_h = \int_0^\tau \kappa\gamma \frac{y_h}{y_\cdot}(\phi_h - \bar{\phi}) y_\cdot \alpha;$$

see (5.2.12), (5.2.26), and (5.2.29)–(5.2.32).

Suppose we have a test with weight function κ equal to (or just proportional to) γ. Then we see that

$$\boldsymbol{\Sigma}(\tau) = \mathscr{I}^{\phi\phi|\zeta}, \qquad \boldsymbol{\xi}(\tau) = \mathscr{I}^{\phi\phi|\zeta}\boldsymbol{\phi},$$

so that

$$\begin{aligned}\zeta(\tau) &= \boldsymbol{\phi}^\mathsf{T}\mathscr{I}^{\phi\phi|\zeta}(\mathscr{I}^{\phi\phi|\zeta})^- \mathscr{I}^{\phi\phi|\zeta}\boldsymbol{\phi} \\ &= \boldsymbol{\phi}^\mathsf{T}\mathscr{I}^{\phi\phi|\zeta}\boldsymbol{\phi}\end{aligned}$$

is exactly the same as (8.2.20). Thus, if the weight process $K^{(n)}$ of the counting process based test is chosen so that its asymptotic form κ is proportional to γ, its asymptotic power in the parametric model (8.2.16) coincides with that of the best parametric test for the corresponding parametric testing problem. Indeed, the two test statistics can be shown to be asymptotically equivalent.

This seems paradoxical because in Section V.2.3 we were unable to exhibit the best test of the family considered against these alternatives. In fact, a little further analysis there would have shown that maximizing $\zeta(\tau)$ by choice of κ has as *implicit* solution (since $\boldsymbol{\Sigma}$ and $\boldsymbol{\xi}$ are defined in terms of κ as well as vice versa)

$$\kappa = \left[\boldsymbol{\xi}^\mathsf{T}\boldsymbol{\Sigma}^- \begin{pmatrix} \frac{y_1}{y_\cdot}(\phi_1 - \bar{\phi}) \\ \vdots \\ \frac{y_k}{y_\cdot}(\phi_k - \bar{\phi}) \end{pmatrix} \Bigg/ \boldsymbol{\xi}^\mathsf{T}\boldsymbol{\Sigma}^- \left(\frac{y_h}{y_\cdot}\left(\delta_{hl} - \frac{y_l}{y_\cdot}\right)\right)\boldsymbol{\Sigma}^-\boldsymbol{\xi} \right]\gamma; \tag{8.2.21}$$

this is obtained by a standard variational argument. [Differentiate $\zeta(\tau)$ with respect to θ after κ has been replaced by $\kappa_{\text{opt}} + \theta\delta$, where δ is an arbitrary function; write the derivative at $\theta = 0$, which must be zero, as an integral of some function times δ. Since δ is arbitrary, this other term must be identically zero.] Now when the functions $y_h/y_\cdot$ are not constant (*non*proportional risk sets), we see that the optimal κ depends on $\boldsymbol{\phi}$. So the paradox disappears because the optimal parametric test we have found is not asymptotically *uniformly* most powerful, but only most powerful in a certain restricted class of tests: tests of $\phi_1 = \cdots = \phi_k$ whose power is invariant under transformations of the parameters of interest $(\phi_2 - \phi_1, \ldots, \phi_k - \phi_{k-1})$ which preserve the effective information for these parameters; equivalently, tests whose power is constant on the ellipsoids

$$\boldsymbol{\phi}^{\mathsf{T}} \mathscr{I}^{\phi\phi|\zeta} \boldsymbol{\phi} = c.$$

Now the test (8.2.21) achieves its maximum power at one particular point on this ellipsoid (and its mirror image on the other side) and has power at some points certainly smaller than that of the test with $\kappa = \gamma$ (except in the two special cases mentioned earlier). From the general theory of optimal testing, one may also say that the choice $\kappa = \gamma$ yields the best *average* power over this ellipsoid and maximizes the *minimum* power over the ellipsoid. It has best *constant* power on the ellipsoid. The reason to be interested in the ellipsoid is because on it the envelope power is constant: It contains alternatives equally far from the null in a statistical sense. Finally (without special reference to this ellipsoid any more), the choice $\kappa = \gamma$ minimizes the maximum shortcoming from maximal (envelope) power (most stringent) since it does this on each ellipsoid separately.

The two special cases deserve some further attention. Why do we not need to introduce the nuisance parameters ζ in the case of asymptotically proportional risk sets; in other words, why is the hardest parametric submodel only k-dimensional instead of $2k$-dimensional? In fact, one can always reduce to a k-dimensional model; suppose we set $\zeta = -\boldsymbol{\phi}$, then the score for $\boldsymbol{\phi}$ becomes its old score minus the old score for ζ and the information for $\boldsymbol{\phi}$ is just $\mathscr{I}^{\phi\phi|\zeta}$. So the difference between $2k$ and k parameters is not essential. What happens in the special case of proportional risk sets is that the γ_h defined just above (8.2.17) are all proportional to γ so that all the nuisance parameters ζ are confounded with the "general level" of the parameters $\boldsymbol{\phi}$ (the contrasts based on $\boldsymbol{\phi}$ are not affected).

What happens in (8.2.21) is that when we have asymptotically proportional risk sets, the functions $y_h/y_{.}$ and also $\bar{\phi}$ are constants so that the numerator and denominator of (8.2.21) (except for γ itself) are constants, giving a solution κ proportional to γ. When $k = 2$, what happens is more subtle: We have

$$\begin{bmatrix} \dfrac{y_1}{y_{.}} & (\phi_1 - \bar{\phi}) \\[2ex] \dfrac{y_2}{y_{.}} & (\phi_2 - \bar{\phi}) \end{bmatrix} = \frac{y_1 y_2}{y_{.}^2}(\phi_1 - \phi_2)\begin{pmatrix} +1 \\ -1 \end{pmatrix}$$

and

$$\begin{bmatrix} \dfrac{y_1}{y_{.}}\left(1 - \dfrac{y_1}{y_{.}}\right) & -\dfrac{y_1 y_2}{y_{.} y_{.}} \\[2ex] -\dfrac{y_1 y_2}{y_{.} y_{.}} & \dfrac{y_2}{y_{.}}\left(1 - \dfrac{y_2}{y_{.}}\right) \end{bmatrix} = \frac{y_1 y_2}{y_{.}^2}\begin{pmatrix} +1 & -1 \\ -1 & +1 \end{pmatrix}$$

so that after cancellation of the time varying but common factor $y_1 y_2 / y_{.}^2$, numerator and denominator are constants and the result is again: κ is proportional to γ.

We close this section with a remark on technical detail: When actually proving the optimality theorem corresponding to our informal discussion here, the extra conditions in Section V.2.3 concerning behavior of the test statistic under the alternatives may be omitted since joint weak convergence *under the null hypothesis* is easily obtained; contiguity theory then gives the behavior under the local alternatives [see Gill (1980a, Section 5.2)]. Thus, (5.2.14) to (5.2.16) need only hold under the null hypothesis, and (5.2.28) is superfluous, though we do need *noninformative censoring*.

Another remark is that, in the case $k = 2$, the optimality property of the best test of our class is much more easily stated: Since the parameter of interest is the scalar $\phi_1 - \phi_2$, the transformations preserving its effective information consist of just the sign change (invariance reduces to unbiasedness). The best test of the class is asymptotically most powerful *unbiased*. The one-sided version of the test is even asymptotically uniformly most powerful, without further qualification.

The one-sample case is easy to treat along the lines set out above, no nuisance parameters need to be introduced, and the best test of the class discussed in Section V.1.3 [see (5.1.10), (5.1.13), and Example V.1.7] is asymptotically most powerful unbiased (two-sided version) and asymptotically uniformly most powerful (one-sided).

VIII.2.4. Efficiency of Nonparametric Estimators

We now go further back, to Chapter IV, and consider efficiency of the estimators of Sections IV.1, IV.3, and IV.4 (Nelson–Aalen, Kaplan–Meier, and Aalen–Johansen).

Start by considering the asymptotic setup of Theorem IV.1.2. We note that the estimator $\hat{A}_{h_0}(t_0)$ (for a given fixed t_0 and fixed h_0) is *asymptotically linear* in the basic counting process martingales $M_l^{(n)} = N_l^{(n)} - \int \alpha_l Y_l$, $l = 1, \ldots, k$; thus, from the proof of the theorem, we see that

$$a_n(\hat{A}_{h_0}(t_0) - A_{h_0}(t_0)) - \int_0^\tau \frac{I(u \leq t_0)}{y_{h_0}(u)} a_n^{-1} \, \mathrm{d}M_{h_0}^{(n)}(u) \xrightarrow{\mathrm{P}} 0.$$

Asymptotically linear means asymptotically equivalent to a linear combination of stochastic integrals of deterministic integrands; the linear equivalent is called the influence function.

Define $\gamma(u) = I(u \leq t_0)/y_{h_0}(u)$ for certain fixed h_0 and t_0. Then we recognise

$$a_n^{-1} \int_0^\tau \gamma(u) \, \mathrm{d}M_{h_0}^{(n)}(u)$$

as the score for the local parameter ϕ in the parametric model

$$\alpha_{h_0}^{(n)}(t) = \alpha_{h_0}(t; \theta^{(n)}), \tag{8.2.22}$$

$$\theta^{(n)} = \theta_0 + a_n^{-1}\phi, \tag{8.2.23}$$

$$\frac{\partial}{\partial\theta}\log\alpha_{h_0}(t;\theta_0)=\gamma(t) \tag{8.2.24}$$

(other components of $\boldsymbol{\alpha}$ are supposed to be fixed); cf. (8.1.11) and (8.2.8)–(8.2.11).

Now, since $A_{h_0}(t_0)=\int_0^{t_0}\alpha_{h_0}(u)\,du$, we can consider $A_{h_0}(t_0)$ as a function of θ. Note that

$$\frac{\partial}{\partial\theta}A_{h_0}(t_0;\theta_0)=\int_0^{t_0}\gamma(u)\alpha_{h_0}(u)\,du.$$

Thus, if we replace γ by its rescaled version

$$\frac{I(u\leq t_0)/y_{h_0}(u)}{\int_0^{t_0}(\alpha_{h_0}(u)/y_{h_0}(u))\,du)}$$

the resulting model (8.2.22)–(8.2.24) has

$$\frac{\partial}{\partial\theta}A_{h_0}(t_0;\theta_0)=1. \tag{8.2.25}$$

This means that we can actually identify θ and $A_{h_0}(t_0;\theta)$, putting $\theta_0=A_{h_0}(t_0;\theta_0)$. For instance, one could specify

$$\alpha_{h_0}(t;\theta)=\alpha_{h_0}(t)\left(1+(\theta-\theta_0)\frac{I(t\leq t_0)/y_{h_0}(t_0)}{\int_0^{t_0}\alpha_{h_0}(u)/y_{h_0}(u)\,du}\right),$$

(for θ close enough to θ_0 that the resulting expression is non-negative) giving *exactly* on integration

$$A_{h_0}(t_0;\theta)=A_{h_0}(t_0)+(\theta-\theta_0)=\theta.$$

But even without this exact identification, (8.2.25) guarantees that the asymptotic Fisher information for estimating θ and that for estimating $A_{h_0}(t_0;\theta)$ are the same at $\theta=\theta_0$ [where $A_{h_0}(t_0;\theta_0)=A_{h_0}(t_0)$]; we have still locally identified θ and $A_{h_0}(t_0,\theta)$.

What is this asymptotic information? Since the score for θ at $\theta=\theta_0$ is $\int_0^\tau\gamma\,dM_{h_0}^{(n)}$, under the standard condition (4.1.16) the local score is asymptotically $\mathcal{N}(0,\int_0^\tau\gamma^2 y_{h_0}\alpha_{h_0})$ distributed and, therefore, $\int_0^\tau\gamma^2 y_{h_0}\alpha_{h_0}$ is the asymptotic information.

With our choice of (rescaled) γ, the information for ϕ is

$$\frac{\int_0^{t_0}\alpha_{h_0}/y_{h_0}}{(\int_0^{t_0}\alpha_{h_0}/y_{h_0})^2}$$

and, hence, the inverse information is just $\int_0^{t_0}\alpha_{h_0}/y_{h_0}$, exactly the same as the asymptotic variance of $a_n(\hat{A}_{h_0}^{(n)}(t_0)-A_{h_0}(t_0))$. Therefore, we have found a *parametric submodel* of our original infinite-dimensional or nonparametric model, such that the information bound for estimating $A_{h_0}(t_0,\theta)$ in the parametric model is *achieved* by the nonparametric estimator $\hat{A}_{h_0}(t_0)$. Regularity of the estimator $\hat{A}_{h_0}(t_0)$ is easily verified (for *any* parametric submodel), so $\hat{A}_{h_0}(t_0)$ is an estimator covered by the Hájek convolution theorem, Theorem VIII.1.2. Since the asymptotic variance of $\hat{A}_{h_0}(t_0)$ must exceed the information bound in *any* parametric submodel (since it is a candidate regular estimator for all submodels), we have apparently identified the hardest parametric submodel for the estimation of $A_{h_0}(t_0)$ and, moreover, shown that $\hat{A}_{h_0}(t_0)$ is an optimal regular estimator in the larger nonparametric model.

This scheme of proof extends to any finite collection of the estimators $\hat{A}_h(t_i)$, $h = 1, \ldots, k$, $i = 1, \ldots, j$, and also to any *compactly differentiable functions* (cf. Section II.8) of the $\hat{A}_h$. For the first step, we note that the estimators are asymptotically linear in the counting process martingales, giving a collection of scores for some local finite-dimensional model. The rescaling becomes not division of the score by a derivative but premultiplication of the vector of scores by a matrix of partial derivatives, "coincidentally" equal to the covariance matrix of the influence functions of the estimators (the coincidence arises here precisely because we are considering what is ultimately an efficient estimator). After rescaling, we have found a parametric submodel for which the parameters can be identified (at least locally) with the collection of $A_h(t_i)$ being estimated and for which the inverse Fisher information exactly equals the asymptotic covariance matrix of the $\hat{A}_h(t_i)$.

So each finite collection of $\hat{A}_h(t_i)$ is asymptotically efficient for the corresponding $A_h(t_i)$, in the sense of achieving an information bound for arbitrary regular estimators; and the bound equals the information bound in a certain hardest parametric submodel.

When we look at differentiable functions of the A_h, for instance, Kaplan–Meier (Section IV.3) or Aalen–Johansen estimators (Section IV.4), quantiles, restricted or trimmed means (Section IV.3.4), and so on, we note that by compact differentiability, such estimators are asymptotically linear functionals of the $\hat{A}_h$, hence also asymptotically linear in *their* influence functions, $a_n^{-1} \int_0^{(\cdot)} \mathrm{d}M_h^{(n)}/y_h$. Here, we make use of the functional delta-method together with the easily proven result

$$\sup_{t \in [0,\tau]} \left| a_n(\hat{A}_h(t) - A_h(t)) - a_n^{-1} \int_0^t y_h^{-1} \, \mathrm{d}M_h^{(n)} \right| \xrightarrow{\mathrm{P}} 0$$

as $n \to \infty$. Combining these facts, any compactly differentiable real functional of the $\hat{A}_h$ on a finite interval $[0, \tau]$ is asymptotically equivalent to $\sum_h a_n^{-1} \int_0^\tau \gamma_h \, \mathrm{d}M_h^{(n)}$ for some functions γ_h. We can now consider the one-dimensional parametric submodel for which

$$\frac{\partial}{\partial \theta} \log \alpha_h(t; \theta_0) = \gamma_h(t)$$

and follow the argument outlined above. Regularity carries over to compactly differentiable functionals and need not be explicitly verified again.

Thus, for any of the just mentioned functionals, we can prove asymptotic efficiency, and even explicitly identify a "hardest parametric submodel." The informal discussion we have just given can be taken as an introduction to the formal and abstract treatment of infinite-dimensional parameter estimation which follows next (Section VIII.3).

VIII.3. Infinite-Dimensional Parameter Spaces: the General Theory

In the previous section, we dealt with optimality issues for infinite-dimensional parameters in an informal way, by explicitly constructing hardest finite-dimensional parametric submodels and showing that the statistic or test under study is optimal for that submodel. This typically carries over to an optimality property for the original (infinite-dimensional) model.

Here we present, following van der Vaart (1988), a more formal abstract theory in which "hardest parametric submodels" do not have to be explicitly constructed. However, they are implicitly present in the background and, in fact, the theory shows how they can be found in a systematic way. The theory allows a number of powerful characterizations of asymptotically efficient estimators which are effective tools in applications.

VIII.3.1. Abstract Theory

We will only consider the estimation of parameters which are real-valued or take values in $D[0, \tau]$ (or a finite collection of such parameters jointly) since these are the only types of parameters which we have met in this volume. However, it is rather convenient (just as in Section II.8) to endow $D[0, \tau]$ with a different metric from the usual Skorohod metric: namely, we give it the supremum norm and, more generally, $B = \mathbb{R}^p \times (D[0, \tau])^q$ is given the max-supremum norm. This change is very convenient but does lead to a further break with tradition—weak convergence in $D[0, \tau]$, supremum norm has to be understood in the generalized sense of Dudley (1966) instead of, as is more familiar, in the sense of Billingsley (1968). This is expounded by Gaenssler (1983) and Pollard (1984). Thus, we endow $D[0, \tau]$ with the σ-algebra generated by the open balls (smaller than that generated by the open sets in the new norm) and $X_n \xrightarrow{\mathscr{D}} X$ in $D[0, \tau]$ means $\mathrm{E}f(X_n) \to \mathrm{E}f(X)$ for all continuous, *measurable*, bounded real f. In fact, on $D[0, \tau]$ the open ball σ-algebra for the supremum norm turns out to be the same as the Borel σ-algebra of the Skorohod metric, so the theories are even closer than at first sight appears. All the usual weak convergence theorems remain valid in this setup provided at least X takes values in a separable subset of $D[0, \tau]$, e.g., the set of all continuous functions on $[0, \tau]$. Moreover, $X_n \xrightarrow{\mathscr{D}} X$ with X having continuous paths in the Skorohod or in the Dudley sense of convergence in distribution are equivalent. So the reader may safely ignore the distinction on a first reading. (An even further generalized notion of weak convergence due to Hoffmann–Jørgensen [see Pollard (1990)], has more advantages.)

The general theorem we now present includes as the main ingredient an assumption of local asymptotic normality for a collection of families of finite-dimensional parametric submodels. This could be verified in later applications by using a version of the Jacod and Shiryayev theorem (Theorem VIII.2.1), by working from first principles, or using Condition VI.1.1 from the theory of maximum likelihood estimation (Theorem VIII.2.2). Since the LAN assumption is close to the assumption that the scores for a local parametric model are asymptotically normal with asymptotic variance the information matrix, we will in examples typically just write out the scores and make plausible their asymptotic normality.

We suppose that for each $n = 1, 2, \ldots$ we are given a statistical model—i.e., a measurable space $(\Omega^{(n)}, \mathscr{F}^{(n)})$ together with a family $\mathscr{P}^{(n)}$ of probability measures on it. We consider estimation of a parameter κ_n of the model, taking values in $B = \mathbb{R}^p \times (D[0, \tau])^q$. This means that κ_n is a map from $\mathscr{P}^{(n)}$ to B: For each probability distribution $\mathrm{P}^{(n)} \in \mathscr{P}^{(n)}$, we can associate a value $\kappa_n(\mathrm{P}^{(n)}) \in B$ of the parameter. Typically, we will have $\mathscr{P}^{(n)} = \{\mathrm{P}^{(n)}_{\theta,\phi}: \theta \in \Theta, \phi \in \Phi\}$ and $\kappa_n(\mathrm{P}^{(n)}_{\theta,\phi}) = \theta \in \Theta \subseteq B$; this is legimate if θ is identifiable.

Assumption VIII.3.1. (Local Asymptotic Normality.) There exist a Hilbert space $\mathbb{H}$ and a convex cone $C \subseteq \mathbb{H}$ such that C can be mapped into $\mathscr{P}^{(n)}$ for each n: To each $\mathbf{h} \in C$, we can associate an element $\mathrm{P}^{(n)}_{\mathbf{h}} \in \mathscr{P}^{(n)}$. We then have the LAN property for all finite-dimensional submodels of the models parametrized by $\mathbf{h} \in C$: For all r, and for all $\mathbf{h}_1, \ldots, \mathbf{h}_r \in C$,

$$\left(\log \frac{d\mathrm{P}^{(n)}_{\mathbf{h}_1}}{d\mathrm{P}^{(n)}_{\mathbf{0}}}, \ldots, \log \frac{d\mathrm{P}^{(n)}_{\mathbf{h}_r}}{d\mathrm{P}^{(n)}_{\mathbf{0}}}\right) \xrightarrow{\mathscr{D}(\mathrm{P}^{(n)}_{\mathbf{0}})} \mathscr{N}(-\tfrac{1}{2}\operatorname{diag}\boldsymbol{\Sigma}, \boldsymbol{\Sigma}) \tag{8.3.1}$$

where $(\boldsymbol{\Sigma})_{ij} = (\mathbf{h}_i, \mathbf{h}_j)_{\mathbb{H}}$ and diag of a matrix is the vector containing its diagonal elements.

This generalizes the finite-dimensional LAN property (8.1.8) in a natural way. An infinite-dimensional generalization of (8.1.9) is a little harder to make since we would have to replace the approximate vector of scores by a Hilbert space valued score vector (see Ibragimov and Khas'minskii, 1991); this would actually demand more than we need for the general theorems we give later. However, in applications, it is nice to work with a form of (8.1.9)–(8.1.11): This can be done, for given $\mathbf{h}_1, \ldots, \mathbf{h}_r$, by working with the r-dimensional parametric submodels $\mathrm{P}^{(n)}_{\sum_i \eta_i \mathbf{h}_i}$, parametrized by the vector $(\eta_1, \ldots, \eta_r)$. The inner product on $\mathbb{H}$ can now be identified with the asymptotic covariance between the scores for the different coefficients η_i of the $\mathbf{h}_i$ in such submodels. To determine it in practice, we simply look at one-dimensional submodels $\mathrm{P}^{(n)}_{\eta\mathbf{h}}$ for each fixed $\mathbf{h}$, determine the score $U^{(n)}_{\mathbf{h}}$ for the real parameter η at $\eta = 0$, and check that all finite collections of these scores are jointly asymptotically normal. The norm on $\mathbb{H}$ is the asymptotic variance of $U^{(n)}_{\mathbf{h}}$; the inner product is the asymptotic covariance. Note that the parametrization by $\mathbf{h}$ is already a local parametrization; no factors a_n have been mentioned in Assumption VIII.3.1. Typically such a factor will be involved in the definition of $\mathrm{P}_{\mathbf{h}}^{(n)}$.

In many applications, we will actually have $C = \mathbb{H}$ or C is dense in $\mathbb{H}$; for instance, if $\mathbb{H}$ is a space of square integrable functions, C might just consist of the bounded functions. In this case, the main theorems stated below are somewhat simpler.

The sequence $\mathrm{P}_{\mathbf{0}}^{(n)} \in \mathscr{P}^{(n)}$ constitutes the "fixed point" in the model in the neighborhood of which we try to estimate κ_n. We suppose, therefore, that the embedding of C into $\mathscr{P}^{(n)}$ is arranged in such a way that $\kappa_n(\mathrm{P}_{\mathbf{0}}^{(n)}) = \kappa_0$ is fixed. As $n \to \infty$, $\mathrm{P}_{\mathbf{h}}^{(n)}$ approaches $\mathrm{P}_{\mathbf{0}}^{(n)}$, so $\{\mathrm{P}_{\mathbf{h}}^{(n)} : \mathbf{h} \in C\}$ is a "local" parametrization, or reparametrization, of (perhaps only part of) $\mathscr{P}^{(n)}$. The next assumption states that the parameter of interest $\kappa_n(\mathrm{P}_{\mathbf{h}}^{(n)})$ then also approaches κ_0 in a smooth way.

Assumption VIII.3.2. (Differentiability of Parameter of Interest.) There exists $a_n \to \infty$ such that κ_n is differentiable at $\mathrm{P}_{\mathbf{0}}^{(n)}$ with rate sequence a_n; i.e., there exists a continuous linear map $\kappa' : \mathbb{H} \to B$ such that for all $\mathbf{h} \in \mathbb{H}$

$$a_n(\kappa_n(\mathrm{P}_{\mathbf{h}}^{(n)}) - \kappa_n(\mathrm{P}_{\mathbf{0}}^{(n)})) \to \kappa'(\mathbf{h}). \tag{8.3.2}$$

This assumption says that $\kappa_n(\mathrm{P}_{\mathbf{h}}^{(n)})$ is approximately $\kappa_0 + a_n^{-1}\kappa'(\mathbf{h})$. When $\mathscr{P}^{(n)}$ can also be parametrized "globally" as $\{\mathrm{P}_{\theta,\phi}^{(n)} : \theta, \phi \in \Theta \times \Phi\}$ and $\kappa_n(\mathrm{P}_{\theta,\phi}^{(n)}) = \theta$, we will typically let $\mathrm{P}_{\mathbf{h}}^{(n)}$ (local parametrization) be the same as $\mathrm{P}_{\theta_n,\phi_n}^{(n)}$ (global), where $\theta_n = \theta_n(\mathbf{h})$ is of the form $\theta_0 + a_n^{-1}$ times a smooth function of $\mathbf{h}$, zero at $\mathbf{h} = \mathbf{0}$, and similarly for ϕ_n (both possibly plus a term of smaller order in a_n). We cannot usually simply take $\theta_n(\mathbf{h}) = \theta_0 + a_n^{-1}\mathbf{h}$ (or plus a term of smaller order) since this would force $\mathbf{h}$ to be an element of B, which, however, is not necessarily a Hilbert space. We definitely need local Hilbert space structure—the Hilbert space inner product is nothing else than the (infinite-dimensional) Fisher information for our asymptotic statistical problem.

Before stating the main theorems, we need to define regular estimators and we need just a little more notation. Suppose T_n is an estimator sequence for κ_n. We say T_n is *regular* if

$$a_n(T_n - \kappa_n(\mathrm{P}_{\mathbf{h}}^{(n)})) \xrightarrow{\mathscr{D}(\mathrm{P}_{\mathbf{h}}^{(n)})} Z \tag{8.3.3}$$

for each $\mathbf{h} \in C$ where the limiting distribution does not depend on $\mathbf{h}$; thus, T_n responds in the proper way to local changes in κ_n.

The space B has a dual B^*: all continuous, linear (and, therefore, measurable) maps from B to $\mathbb{R}$. So, for $b^* \in B^*$, $b^* \circ \kappa'$ is a continuous linear map from $\mathbb{H}$ to $\mathbb{R}$. By self-duality (in other words, by the Riesz representation theorem), it may be identified with an element of $\mathbb{H}$ itself, which we denote by κ'_{b^*}; this means that κ'_{b^*} is the unique element of $\mathbb{H}$ such that $(\kappa'_{b^*}, \mathbf{h})_{\mathbb{H}} = b^*(\kappa'(\mathbf{h}))$ for all $\mathbf{h} \in \mathbb{H}$. Let $\tilde{\kappa}'_{b^*}$ denote the projection of κ'_{b^*} into the closed linear span of C, $\mathrm{lin}(C)$. Another way to say this is that $\tilde{\kappa}'_{b^*}$ is the unique element of $\mathrm{lin}(C)$ such that $(\tilde{\kappa}'_{b^*}, \mathbf{h})_{\mathbb{H}} = b^*(\kappa'(\mathbf{h}))$ for all $\mathbf{h} \in \mathrm{lin}(C)$ since this space is also a Hilbert space under the same inner product and so the Riesz representation theorem applies here too. Being a projection means that $\kappa'_{b*} - \tilde{\kappa}'_{b*}$ is orthogonal to every element of lin(C) with respect to the inner product on $\mathbb{H}$.

We can now state an infinite-dimensional or abstract version of the Hájek convolution theorem (Theorem VIII.1.2 or VIII.1.3). Those theorems can be extracted from this one by taking as local reparametrization simply $\boldsymbol{\theta}_n(\mathbf{h}) = \boldsymbol{\theta}_0 + a_n^{-1}\mathbf{h}$, where $\boldsymbol{\theta}_0$ and $\mathbf{h}$ are both in $\mathbb{R}^p$. However, in the local reparametrization $\mathbb{R}^p$ is considered as a Hilbert space by giving it the inner product $(\mathbf{h}, \mathbf{h}')_{\mathbb{H}} = \mathbf{h}^\top \mathscr{I} \mathbf{h}$, rather than as a Banach space with the Euclidean norm.

Theorem VIII.3.1 (Convolution Theorem). *Suppose T_n is a regular estimator of a differentiable parameter κ_n in a LAN model. Then the limit in distribution Z of $a_n(T_n - \kappa_0)$ under $\mathrm{P}_{\mathbf{0}}^{(n)}$ is the convolution $Z = X + Y$ of two independent random elements of B, where the distribution of X is characterized by*

$$b^*(X) \sim \mathcal{N}(0, \|\tilde{\kappa}'_{b*}\|_{\mathbb{H}}^2) \tag{8.3.4}$$

for each $b^* \in B^*$.

If T_n is regular and achieves this bound, we say that T_n is (*asymptotically*) *efficient* for κ_n.

Optimal testing of *parametric* hypotheses can be described just as in Section VIII.1.3. The main result is that an optimal test of a hypothesis $\kappa_n(\mathrm{P}^{(n)}) = \kappa_0$, where κ_n is a differentiable sequence of Euclidean parameters, is asymptotically equivalent to the natural asymptotically χ^2-distributed quadratic form based on an optimal estimator of κ_n. Testing optimality has to be understood here just as in Section VIII.1.3 (e.g., as asymptotically minimax with respect to the now infinite-dimensional ellipsoids of local alternatives where the asymptotic envelope power is constant).

A theory of optimal testing of nonparametric hypotheses, however, is not available and the remarks we made in Section VIII.1.3 on this topic suggests that a meaningful theory cannot exist.

Note that if lin(C) is all of C, no projection is necessary and the asymptotic variance is the squared norm of κ'_{b*}. One can always avoid doing projections by determining C by consideration of parametric submodels, then finding out the inner product from Assumption VIII.3.1, and finally letting $\mathbb{H}$ be the closed linear span of C with respect to the corresponding norm. However, when considering a big nonparametric model and various parametric or semiparametric submodels of it at the same time, it is convenient to keep $\mathbb{H}$ fixed, corresponding to the big model, and let various choices of C correspond to the submodels of interest.

We can now state some valuable characterizations of efficiency.

Theorem VIII.3.2. *Suppose Assumptions* VIII.3.1 *and* VIII.3.2 *hold* (*LAN and differentiability of κ_n*) *and suppose T_n is regular and coordinatewise efficient, then T_n is efficient.*

Recall that κ_n takes values in B, equal to $D[0, \tau]$ or to $\mathbb{R}$ or to a product of several copies of each or either. By a coordinate of κ_n we mean the value of one of the real components of κ_n, or the evaluation of one of the $D[0, \tau]$ components at a given $t \in [0, \tau]$. Thus, coordinatewise efficiency means that each of these coordinates of κ_n is estimated efficiently by the corresponding coordinate of T_n.

The previous theorem (coordinatewise efficiency plus regularity implies full efficiency) is pretty but not much use on its own. The next theorem, which can be paraphrased as "regular and asymptotically linear in the score vector implies efficient," shows how efficiency of an estimator of a real-valued parameter can be easily verified. (In an i.i.d. setting, the same theorem is called "regular with influence function in the tangent space implies efficient.") We will typically apply it, in turn, to each of the coordinates of a regular estimator of an infinite-dimensional parameter of interest; the two theorems together then give efficiency of our estimator.

Theorem VIII.3.3. *Suppose T_n is regular for a real parameter κ_n and suppose Assumptions* VIII.3.1 *and* VIII.3.2 *(LAN and differentiability) hold. Suppose, moreover, an $\mathbf{h} \in C$ exists such that*

$$a_n(T_n - \kappa_0) - \left(\log \frac{\mathrm{d}\mathrm{P}_{\mathbf{h}}^{(n)}}{\mathrm{d}\mathrm{P}_{\mathbf{0}}^{(n)}} + \frac{1}{2}\|\mathbf{h}\|_{\mathbb{H}}^2\right) \xrightarrow{\mathrm{P}_{\mathbf{0}}^{(n)}} 0. \tag{8.3.5}$$

Then T_n is efficient.

The next theorem is in some sense a converse to Theorem VIII.3.2. It is the abstract version of the result "asymptotically linear with optimal influence function implies efficient and regular" which we have used in previous sections. The theorem, however, is still restricted to the estimation of a real (or possibly finite-dimensional) parameter; there is no truly infinite-dimensional version of this one.

Theorem VIII.3.4. *Suppose T_n is an estimator of a real parameter κ_n and Assumption* VIII.3.1 *and* VIII.3.2 *(LAN and differentiability of the parameter) hold. Let $\mathbf{h} = \tilde{\kappa}'_\iota$ with ι the identity map from $\mathbb{R}$ to $\mathbb{R}$. If* (8.3.5) *then holds, T_n is regular (and, hence, efficient).*

Finally, we give a valuable theorem which allows us to transfer efficiency from a given estimator, possibly infinite-dimensional, to any smooth function of it. Smoothness just means compactly differentiable; see Section II.8.

Theorem VIII.3.5. *Suppose Assumptions* VIII.3.1 *and* VIII.3.2 *(LAN and differentiability of the parameter) hold and T_n is an efficient (and regular) estimator for κ_n. Suppose $\phi: B \to B'$ is a compactly (Hadamard) differentiable map from B to another space (of the type considered here) B'. Then $\phi \circ \kappa_n$ is also a differentiable parameter and $\phi(T_n)$ is efficient (and regular) for $\phi(\kappa_n)$.*

The rest of this section is devoted to an example which exemplifies the previous theorems in a familiar context, a parametric counting process model.

VIII.3.2. The Parametric Case

In this subsection, we reconsider the parametric setup of Section VIII.2.2. This is very important for the nonparametric models which follow in Section VIII.4 because the central part of the new theory—Assumption VIII.3.1—is, as we said earlier, just the LAN property for a class of parametric submodels, and Theorem VIII.2.2 gives us conditions under which the LAN property is true. Also, Theorem VIII.2.2 gives us the interpretation for the centered log-likelihood ratio,

$$\log \frac{d\mathrm{P}_{\mathbf{h}}^{(n)}}{d\mathrm{P}_{\mathbf{0}}^{(n)}} + \frac{1}{2}\|\mathbf{h}\|_{\mathbb{H}}^{2}, \tag{8.3.6}$$

appearing in Theorems VIII.3.3 and VIII.3.4. This is just the score function for the one-dimensional submodel $\{\mathrm{P}_{\eta\mathbf{h}}^{(n)}: \eta \in \mathbb{R}\}$; more precisely, under the conditions of Theorem VIII.2.2 (i.e., under Condition VI.1.1) for this one-dimensional model, (8.3.6) is asymptotically equivalent to the score

$$\frac{\partial}{\partial \eta} \log \frac{d\mathrm{P}_{\eta\mathbf{h}}^{(n)}}{d\mathrm{P}_{\mathbf{0}}^{(n)}}\bigg|_{\eta=0}. \tag{8.3.7}$$

Suppose for simplicity that the counting process N is one-dimensional (otherwise replace stochastic integrals with respect to $M^{(n)}$ by sums of integrals with respect to $M_h^{(n)}$, $h = 1, \ldots, k_n$). We let the parameter $\boldsymbol{\theta} \in \Theta \subseteq \mathbb{R}^p$ be p-dimensional.

Consider the usual local reparametrization of our global model,

$$\boldsymbol{\theta}_n(\mathbf{h}) = \boldsymbol{\theta}_0 + a_n^{-1}\mathbf{h}.$$

By this reparametrization, we implicitly define $\mathrm{P}_{\mathbf{h}}^{(n)}$ for all or some $\mathbf{h} \in \mathbb{R}^p$. The Hilbert space $\mathbb{H}$ will be just $\mathbb{R}^p$, but the inner product in $\mathbb{H}$ will be determined by the asymptotic information matrix $\mathscr{I}$ instead of being the usual Euclidean inner product.

In Theorem VIII.2.2, we showed that (under the regularity conditions of that theorem), at $\boldsymbol{\theta} = \boldsymbol{\theta}_0$,

$$\log \frac{d\mathrm{P}_{\mathbf{h}}^{(n)}}{d\mathrm{P}_{\mathbf{0}}^{(n)}} + \frac{1}{2}\mathbf{h}^{\mathsf{T}}\mathscr{I}\mathbf{h} = \mathbf{h}^{\mathsf{T}} a_n^{-1} \int_0^{\tau} \frac{\partial}{\partial \boldsymbol{\theta}} \log \lambda_{\boldsymbol{\theta}_0}\, dM_{\boldsymbol{\theta}_0}^{(n)} + o_{\mathrm{P}}(1).$$

Moreover, the vector of stochastic integrals

$$a_n^{-1} \int_0^{\tau} \frac{\partial}{\partial \boldsymbol{\theta}} \log \lambda_{\boldsymbol{\theta}_0}\, dM_{\boldsymbol{\theta}_0}^{(n)},$$

equal to the scores for $\mathbf{h}$ at $\mathbf{h} = \mathbf{0}$, was shown to be asymptotically $\mathcal{N}(\mathbf{0}, \mathcal{I})$ distributed. Thus, for any finite collection $\mathbf{h}_1, \ldots, \mathbf{h}_r$, the vector

$$\left(\log \frac{\mathrm{dP}_{\mathbf{h}_1}^{(n)}}{\mathrm{dP}_{\mathbf{0}}^{(n)}}, \ldots, \log \frac{\mathrm{dP}_{\mathbf{h}_r}^{(n)}}{\mathrm{dP}_{\mathbf{0}}^{(n)}}\right)$$

is asymptotically $\mathcal{N}(-\frac{1}{2}\operatorname{diag}\boldsymbol{\Sigma}, \boldsymbol{\Sigma})$ distributed where $\boldsymbol{\Sigma}_{ij} = \mathbf{h}_i^{\mathsf{T}}\mathcal{I}\mathbf{h}_j$ and $\operatorname{diag}\boldsymbol{\Sigma}$ is the vector containing the diagonal elements of $\boldsymbol{\Sigma}$.

This statement is only true for $\mathbf{h}$ such that $\boldsymbol{\theta}_n(\mathbf{h})$ is eventually in Θ, so whether or not Assumption VIII.3.1 holds, and with what choice of C, depends on which $\mathbf{h}$ are feasible. If $\boldsymbol{\theta}_0$ is in the interior of Θ, then all $\mathbf{h}$ are feasible and $C = \mathbb{H}$. The inner product in $\mathbb{H}$ has to be

$$(\mathbf{h}, \mathbf{h}')_{\mathbb{H}} = \mathbf{h}^{\mathsf{T}}\mathcal{I}\mathbf{h}'.$$

Suppose we had restricted attention to a submodel, e.g., $\{\boldsymbol{\theta}: \boldsymbol{\gamma}(\boldsymbol{\theta}) = \mathbf{0}\}$ for some smooth map $\boldsymbol{\gamma}: \mathbb{R}^p \to \mathbb{R}^k$. We would then have only required $\boldsymbol{\theta}_n(\mathbf{h}) = \boldsymbol{\theta}_0 + a_n^{-1}\mathbf{h} + o(a_n^{-1})$ since we need to keep $\boldsymbol{\gamma}(\boldsymbol{\theta}_n(\mathbf{h})) = \mathbf{0}$. Then C would have been the subspace

$$\{\mathbf{h} \in \mathbb{R}^p: \mathrm{d}\boldsymbol{\gamma} \cdot \mathbf{h} = \mathbf{0}\},$$

where $\mathrm{d}\boldsymbol{\gamma}$ is the $k \times p$ matrix of partial derivatives of $\boldsymbol{\gamma}$ with respect to $\boldsymbol{\theta}$ at $\boldsymbol{\theta} = \boldsymbol{\theta}_0$.

Consider estimation of a function $\boldsymbol{\kappa}(\boldsymbol{\theta})$, where

$$\boldsymbol{\kappa}: \Theta \subseteq \mathbb{R}^p \to B = \mathbb{R}^l$$

is differentiable at $\boldsymbol{\theta} = \boldsymbol{\theta}_0$ with $l \times p$ matrix of partial derivatives $\mathrm{d}\boldsymbol{\kappa}$. We define naturally

$$\boldsymbol{\kappa}_n(\mathrm{P}_{\mathbf{h}}^{(n)}) = \boldsymbol{\kappa}(\boldsymbol{\theta}_n) = \boldsymbol{\kappa}(\boldsymbol{\theta}_0 + a_n^{-1}\mathbf{h})$$

and, therefore,

$$a_n(\boldsymbol{\kappa}_n(\mathrm{P}_{\mathbf{h}}^{(n)}) - \boldsymbol{\kappa}_n(\mathrm{P}_{\mathbf{0}}^{(n)})) \to \mathrm{d}\boldsymbol{\kappa} \cdot \mathbf{h} \quad \text{as } n \to \infty.$$

It seems, therefore, that Assumption VIII.3.2 (differentiability of $\boldsymbol{\kappa}_n$) is satisfied too, but first we must check that $\mathbf{h} \mapsto \kappa'(\mathbf{h}) = \mathrm{d}\boldsymbol{\kappa} \cdot \mathbf{h}$ is not only linear but also continuous as a map from $\mathbb{H}$ to B (in Assumption VIII.3.2, we required the derivative κ' to be continuous as well as linear).

Now we can write, if $\mathcal{I}$ is nonsingular,

$$\mathrm{d}\boldsymbol{\kappa} \cdot \mathbf{h} = \mathrm{d}\boldsymbol{\kappa}\mathcal{I}^{-1}\mathcal{I}\mathbf{h} = (\mathcal{I}^{-1}\,\mathrm{d}\boldsymbol{\kappa}^{\mathsf{T}}, \mathbf{h})_{\mathbb{H}},$$

where $\mathcal{I}^{-1}\,\mathrm{d}\boldsymbol{\kappa}^{\mathsf{T}}$ is thought of as a vector of l column vectors in $\mathbb{H}$. This shows that $\mathbf{h} \mapsto \mathrm{d}\boldsymbol{\kappa} \cdot \mathbf{h}$ is, indeed, a continuous linear map.

If $\mathcal{I}$ is singular, provided the rows of $\mathrm{d}\boldsymbol{\kappa}$ are in the null space of $\mathcal{I}$, we can still write

$$\mathrm{d}\boldsymbol{\kappa} \cdot \mathbf{h} = \mathrm{d}\boldsymbol{\kappa}\mathcal{I}^{-}\mathcal{I}\mathbf{h} = (\mathcal{I}^{-}\,\mathrm{d}\boldsymbol{\kappa}^{\mathsf{T}}, \mathbf{h})_{\mathbb{H}},$$

where $\mathcal{I}^{-}$ is the Moore–Penrose generalized inverse of $\mathcal{I}$. (This generalized inverse is the one which is obtained by writing $\mathscr{I}$ in terms of eigenvectors and eigenvalues, and then replacing the diagonal matrix of eigenvalues with the diagonal matrix containing the inverses of the nonzero eigenvalues, the zero eigenvalues left unaltered.) This shows that even with a singular information matrix, some functions of $\boldsymbol{\theta}$ can still be estimable.

By taking $\mathbb{H}$ as small as possible—it need not be larger than the closed linear span of C—it can become easier to check differentiability since the continuity of $\mathbf{h} \mapsto \mathbf{d}\boldsymbol{\kappa} \cdot \mathbf{h}$ need only be verified on a smaller space. So, by going over to the submodel $\boldsymbol{\gamma}(\boldsymbol{\theta}) = \mathbf{0}$, identifiability problems—manifesting themselves in singularity of $\mathscr{I}$—can sometimes be resolved.

We return to the case of a submodel later, but continue with the full model and assuming nonsingularity of $\mathscr{I}$. This means that there is no subspace to project: κ'_{b*} and $\tilde{\kappa}'_{b*}$ are the same, simplifying Theorem VIII.3.1, formula (8.3.4).

To determine the lower bound for estimation of $\boldsymbol{\kappa}(\boldsymbol{\theta})$ by Theorem VIII.3.1, we determine first the lower bound for $b^*(\boldsymbol{\kappa}(\boldsymbol{\theta}))$, where b^* is a continuous linear map from B to $\mathbb{R}$. This means that b^* is just a linear combination of components of $\boldsymbol{\kappa}$,

$$b^*(\boldsymbol{\kappa}(\boldsymbol{\theta})) = \mathbf{b}^\mathsf{T}\boldsymbol{\kappa}(\boldsymbol{\theta}), \quad \mathbf{b} \in \mathbb{R}^l.$$

We already found that $\boldsymbol{\kappa}'$ could be represented as $\mathscr{I}^{-1}\,\mathbf{d}\boldsymbol{\kappa}^\mathsf{T}$, so

$$\kappa'_{b*} = b^* \circ \boldsymbol{\kappa}' = (b^* \circ \boldsymbol{\kappa})' = \mathscr{I}^{-1}\,\mathbf{d}\boldsymbol{\kappa}^\mathsf{T}\mathbf{b}$$

thought of as a vector in $\mathbb{H}$ acting on $\mathbf{h} \in \mathbb{H}$ by taking their inner product

$$(\mathscr{I}^{-1}\,\mathbf{d}\boldsymbol{\kappa}^\mathsf{T}\mathbf{b}, \mathbf{h})_{\mathbb{H}} = \mathbf{b}^\mathsf{T}\,\mathbf{d}\boldsymbol{\kappa}\mathscr{I}^{-1}\mathscr{I}\mathbf{h} = \mathbf{b}^\mathsf{T}\,\mathbf{d}\boldsymbol{\kappa}\mathbf{h}.$$

As we just mentioned, we are working in the whole model with $\mathrm{lin}(C) = \mathbb{H}$ and $\kappa'_{b*} = b^* \circ \boldsymbol{\kappa}' = \mathscr{I}^{-1}\,\mathbf{d}\boldsymbol{\kappa}^\mathsf{T}\mathbf{b} \in \mathbb{H}$. So the projection $\tilde{\kappa}'_{b*}$ of this vector into $\mathrm{lin}(C)$ is itself, and its squared norm, the optimal asymptotic variance of $\mathbf{b}^\mathsf{T}\hat{\boldsymbol{\kappa}}_n$ where $\hat{\boldsymbol{\kappa}}_n$ is a regular estimator of $\boldsymbol{\kappa}$, is by (8.3.4)

$$\begin{aligned} \|\mathscr{I}^{-1}\,\mathbf{d}\boldsymbol{\kappa}^\mathsf{T}\mathbf{b}\|^2_{\mathbb{H}} &= (\mathscr{I}^{-1}\,\mathbf{d}\boldsymbol{\kappa}^\mathsf{T}\mathbf{b}, \mathscr{I}^{-1}\,\mathbf{d}\boldsymbol{\kappa}^\mathsf{T}\mathbf{b})_{\mathbb{H}} \\ &= \mathbf{b}^\mathsf{T}\,\mathbf{d}\boldsymbol{\kappa}\mathscr{I}^{-1}\mathscr{I}\mathscr{I}^{-1}\,\mathbf{d}\boldsymbol{\kappa}^\mathsf{T}\mathbf{b} = \mathbf{b}^\mathsf{T}(\mathbf{d}\boldsymbol{\kappa}\mathscr{I}^{-1}\,\mathbf{d}\boldsymbol{\kappa}^\mathsf{T})\mathbf{b}. \end{aligned}$$

The optimal covariance matrix of $\hat{\boldsymbol{\kappa}}_n$ is, therefore, $\mathbf{d}\boldsymbol{\kappa}\mathscr{I}^{-1}\,\mathbf{d}\boldsymbol{\kappa}^\mathsf{T}$, as we stated in Theorem VIII.1.3.

When we work in the submodel $\boldsymbol{\gamma}(\boldsymbol{\theta}) = \mathbf{0} \in \mathbb{R}^k$, C becomes the subspace of $\mathbb{R}^p$ orthogonal (with respect to the Euclidean norm) to the k columns of $\mathbf{d}\boldsymbol{\gamma}^\mathsf{T}$. In terms of $\mathbb{H}$ therefore, C is the subspace orthogonal to $\mathscr{I}^{-1}\,\mathbf{d}\boldsymbol{\gamma}^\mathsf{T}$. The projection of $\mathscr{I}^{-1}\,\mathbf{d}\boldsymbol{\kappa}^\mathsf{T}\mathbf{b}$ into the orthogonal complement (in $\mathbb{H}$) of C is

$$\begin{aligned} &\mathscr{I}^{-1}\,\mathbf{d}\boldsymbol{\kappa}^\mathsf{T}\mathbf{b} - (\mathscr{I}^{-1}\,\mathbf{d}\boldsymbol{\gamma}^\mathsf{T})(\mathscr{I}^{-1}\,\mathbf{d}\boldsymbol{\gamma}^\mathsf{T}, \mathscr{I}^{-1}\,\mathbf{d}\boldsymbol{\gamma}^\mathsf{T})^{-1}_{\mathbb{H}}(\mathscr{I}^{-1}\,\mathbf{d}\boldsymbol{\gamma}^\mathsf{T}, \mathscr{I}^{-1}\,\mathbf{d}\boldsymbol{\kappa}^\mathsf{T}\mathbf{b})_{\mathbb{H}} \\ &\quad = \mathscr{I}^{-1}\,\mathbf{d}\boldsymbol{\kappa}^\mathsf{T}\mathbf{b} - \mathscr{I}^{-1}\,\mathbf{d}\boldsymbol{\gamma}^\mathsf{T}(\mathbf{d}\boldsymbol{\gamma}\mathscr{I}^{-1}\,\mathbf{d}\boldsymbol{\gamma}^\mathsf{T})^{-1}(\mathbf{d}\boldsymbol{\gamma}\mathscr{I}^{-1}\,\mathbf{d}\boldsymbol{\kappa}^\mathsf{T})\mathbf{b} \end{aligned}$$

(this vector is orthogonal to the columns of $\mathscr{I}^{-1}\,\mathbf{d}\boldsymbol{\gamma}^\mathsf{T}$ and its difference with $\mathscr{I}^{-1}\,\mathbf{d}\boldsymbol{\kappa}^\mathsf{T}\mathbf{b}$ is a linear combination of the columns of $\mathscr{I}^{-1}\,\mathbf{d}\boldsymbol{\gamma}^\mathsf{T}$).

The squared length of this vector is

$$\mathbf{b}^\top \mathrm{d}\boldsymbol{\kappa}\mathscr{I}^{-1}\mathrm{d}\boldsymbol{\kappa}^\top\mathbf{b} - \mathbf{b}^\top \mathrm{d}\boldsymbol{\kappa}\mathscr{I}^{-1}\mathrm{d}\gamma^\top(\mathrm{d}\gamma\mathscr{I}^{-1}\mathrm{d}\gamma^\top)^{-1}\mathrm{d}\gamma\mathscr{I}^{-1}\mathrm{d}\boldsymbol{\kappa}^\top\mathbf{b},$$

so the optimal asymptotic covariance matrix of a regular estimator of $\boldsymbol{\kappa}(\boldsymbol{\theta})$ is

$$\mathrm{d}\boldsymbol{\kappa}\mathscr{I}^{-1}\mathrm{d}\boldsymbol{\kappa}^\top - \mathrm{d}\boldsymbol{\kappa}\mathscr{I}^{-1}\mathrm{d}\gamma^\top(\mathrm{d}\gamma\mathscr{I}^{-1}\mathrm{d}\gamma^\top)^{-1}\mathrm{d}\gamma\mathscr{I}^{-1}\mathrm{d}\boldsymbol{\kappa}^\top.$$

As we remarked, in the proof of Theorem VIII.2.2 we discovered that

$$\log\frac{\mathrm{dP}_{\mathbf{h}}^{(n)}}{\mathrm{dP}_{\mathbf{0}}^{(n)}} + \frac{1}{2}\|\mathbf{h}\|_{\mathbb{H}}^2$$

was asymptotically equivalent to $a_n^{-1}\mathbf{h}^\top\int_0^\tau(\partial/\partial\boldsymbol{\theta})\log\lambda_{\boldsymbol{\theta}_0}\,\mathrm{d}M_{\boldsymbol{\theta}_0}^{(n)}$. Theorems VIII.3.2–VIII.3.5 can now be applied to the maximum likelihood estimators of Chapter VI. Regularity of the estimators can be proved directly by going through the proof of their asymptotic normality, stated in Theorem VI.1.2, and this proof also yields their asymptotic equivalence with linear combinations of $a_n^{-1}\int_0^\tau(\partial/\partial\theta_j)\log\lambda_{\boldsymbol{\theta}_0}\,\mathrm{d}M_{\boldsymbol{\theta}_0}^{(n)}$. Theorem VIII.3.3, therefore, gives asymptotic efficiency of each component of the maximum likelihood estimator $\hat{\boldsymbol{\theta}}_n$, and Theorem VIII.3.2 extends this to the whole vector. Theorem VIII.3.5 gives efficiency of any differentiable function of $\hat{\boldsymbol{\theta}}_n$.

Alternatively, we can omit the regularity verification but use Theorem VIII.3.4 instead. To do this, we consider $\mathbf{b}^\top\hat{\boldsymbol{\theta}}_n$ as an estimator of $\mathbf{b}^\top\boldsymbol{\theta}$ for some vector $\mathbf{b}$. By the proof of Theorem VI.1.2, $\mathbf{b}^\top\hat{\boldsymbol{\theta}}_n$ is asymptotically equivalent to $a_n^{-1}\mathbf{b}^\top\mathscr{I}^{-1}\int_0^\tau(\partial/\partial\boldsymbol{\theta})\log\lambda_{\boldsymbol{\theta}_0}\,\mathrm{d}M_{\boldsymbol{\theta}_0}^{(n)}$. Thus, (8.3.5) holds with $\mathbf{h} = \mathscr{I}^{-1}\mathbf{b}$. Now taking $\boldsymbol{\kappa}(\boldsymbol{\theta}) = \boldsymbol{\theta}$, so $\boldsymbol{\kappa}$ is the identity mapping $\boldsymbol{\iota}$, we have already found that $(b^* \circ \boldsymbol{\iota})' = \mathscr{I}^{-1}\mathbf{b}$ and its projection into $\mathrm{lin}(C) = \mathbb{H}$ remains the same.

Therefore, Theorem VIII.3.4 gives that $\mathbf{b}^\top\hat{\boldsymbol{\theta}}_n$ is efficient and regular for $\mathbf{b}^\top\boldsymbol{\theta}$ for all $\mathbf{b} \in \mathbb{R}^p$, with asymptotic distribution $\mathscr{N}(0, \mathbf{b}^\top\mathscr{I}^{-1}\mathbf{b})$ under both fixed $\boldsymbol{\theta} = \boldsymbol{\theta}_0$ and under local sequences $\boldsymbol{\theta}_n$. By the Cramèr–Wold device, $\hat{\boldsymbol{\theta}}_n$ is $\mathscr{N}(\mathbf{0}, \mathscr{I}^{-1})$ distributed under the same sequences and, therefore, regular and efficient.

VIII.4. Semiparametric Counting Process Models

In this section, we look in turn at the Nelson–Aalen estimator, nonparametric k-sample tests, the Cox regression model, and the Aalen linear regression model, all from the abstract point of view of Section VIII.3.

The results of Section VIII.4.1 on the Nelson–Aalen estimator mildly improve those obtained in Section VIII.2.4. More importantly, the example gives a good illustration of the power of the abstract theory in the most simple setting. The optimality proof of the Nelson–Aalen estimator is extremely simple now: One just has to verify that the estimator is regular and is asymptotically linear in the space of asymptotic scores of the model (Theorems VIII.3.2 and VIII.3.3). Both points can be verified by inspection of the proof of asymptotic normality of the estimator, Theorem IV.1.2. The "hardest parametric submodels" of Section VIII.2.4 need not be explicitly exhibited though one can easily determine them with the abstract theory if one wishes. Extension to optimality of functionals such as the Kaplan–Meier and Aalen–Johansen estimators follows immediately from Theorem VIII.3.5, on the preservation of efficiency under smooth transformations (compactly differentiable transformations).

The subsection concludes with a summary of the main points of the derivation, showing how the abstract theory can be applied in more general examples.

Section VIII.4.2 similarly streamlines the analysis in Section VIII.2.3 of k-sample tests. It is shown how the tests arise as optimal tests for semiparametric transformation models and gives further insight into why the "direct approach" of Section V.2.3 only worked under the special cases of asymptotically proportional risk sets and the two-sample case. The hardest parametric submodel of Section VIII.2.3 comes out of a general information calculation. The best counting process based tests of Section V.2 are the asymptotically optimal tests among all possible tests for this semiparametric model; one can also say they are the best rank tests.

In Section VIII.4.3, we prove the semiparametric efficiency of the usual estimators in the Cox regression model, and in Section VIII.4.4 of the best "generalized least squares" type estimator in the Aalen linear regression model. Both results are obtained extremely easily following the same lines of argument as for the multiplicative intensity model. Section VIII.4.5 relates these results to nonparametric maximum likelihood properties of the various estimators and briefly discusses the topic of efficient estimation in partially specified models.

VIII.4.1. The Nelson–Aalen Estimator

Here, we take another look at the material of Section VIII.2.4, to exemplify the notions of Section VIII.3 and prepare for the main examples of subsequent subsections: the Cox and the Aalen semiparametric regression models.

To simplify notation, let us consider a univariate counting process satisfying the multiplicative intensity model with completely unknown baseline hazard α. We assume noninformative censoring (so the partial likelihood based on the counting process is the full likelihood) and work on an interval $\mathscr{T} = [0, \tau]$ such that $A(\tau) = \int_0^\tau \alpha < \infty$. For the asymptotics, we assume that the usual condition (4.1.16) holds,

$$\sup_{s \in [0,\tau]} |a_n^{-2} Y^{(n)}(s) - y(s)| \xrightarrow{\mathrm{P}} 0 \tag{8.4.1}$$

for some sequence $a_n \to \infty$ and some function y bounded away from zero on $[0, \tau]$.

This convergence in probability holds for a sequence of models indexed by $n = 1, 2, \ldots$ at a "fixed point" in the model with a given, fixed α. By contiguity, however, it extends to any local sequence of "alternatives."

Our first task is to exhibit a Hilbert space $\mathbb{H}$ and a cone $C \subseteq \mathbb{H}$ whose finite-dimensional subspaces parametrize a nice collection of local finite-dimensional submodels. The Hilbert space can be identified with the corresponding (asymptotic) score functions, asymptotically multivariate normally distributed under the sequence of probability measures belonging to the fixed α. The inner product is just the covariance of score functions. Since we just use $\mathbb{H}$ to parametrize local models, any convenient representation of $\mathbb{H}$ will do as well as any other. If the closed linear span of C is all of $\mathbb{H}$, there will be no need to carry out projections onto subspaces.

Since we know from Theorem VIII.2.2 [see also (8.2.22)–(8.2.24) of Section VIII.2.4] that the score for a local parametric model,

$$
\begin{aligned}
\alpha^{(n)}(t) &= \alpha(t; \theta^{(n)}),\\
\theta^{(n)} &= \theta_0 + a_n^{-1}\phi,\\
\alpha(t; \theta_0) &= \alpha(t),\\
\frac{\partial}{\partial\theta}\log\alpha(t; \theta_0) &= \gamma(t),
\end{aligned}
\tag{8.4.2}
$$

is $a_n^{-1}\int_0^\tau \gamma(t)\,\mathrm{d}M(t)$, under (8.4.1) asymptotically $\mathcal{N}(0, \int_0^\tau \gamma^2 y\alpha)$ distributed (under appropriate uniformity and boundedness conditions), it seems appropriate to aim at, for $\mathbb{H}$, the space of functions γ on $[0, \tau]$ such that $\|\gamma\|_{\mathbb{H}}^2 = \int_0^\tau \gamma^2 y\alpha < \infty$; the inner product is $(\gamma, \gamma')_{\mathbb{H}} = \int_0^\tau \gamma\gamma' y\alpha$.

For each $\gamma \in C \subseteq \mathbb{H}$, we must exhibit $\alpha(t; \theta)$ satisfying (8.4.2) and then we can take the $\mathrm{P}_\gamma^{(n)}$ of Assumption VIII.3.1 as the probability measure corresponding to $\alpha(\cdot; \theta_0 + a_n^{-1} 1)$. Possible choices (using the local parameter ϕ) would be

$$
\begin{aligned}
\alpha^{(n)}(t; \phi) &= \alpha(t)(1 + a_n^{-1}\phi\gamma(t)),\\
\alpha^{(n)}(t; \phi) &= \alpha(t)\exp(a_n^{-1}\phi\gamma(t)),\\
\alpha^{(n)}(t; \phi) &= \alpha(t)(1 + \tfrac{1}{2}a_n^{-1}\phi\gamma(t))^2.
\end{aligned}
$$

In each case, some further truncation or other modification may be necessary before the function $\alpha^{(n)}(t; \phi)$ really is a baseline hazard in the required class. For each $\gamma \in \mathbb{H}$, we have $\int_0^\tau \gamma^2 y\alpha < \infty$, but we also need that

$$
\int_0^\tau \alpha^{(n)}(t; \phi)\,\mathrm{d}t < \infty \quad \text{and} \quad \alpha^{(n)}(t; \phi) \geq 0 \quad \text{for all } t.
$$

The third choice for $\alpha^{(n)}$ is quite convenient since no modification at all turns out to be necessary and we can have $C = \mathbb{H}$; the square already assures us that $\int_0^\tau \gamma^2 y\alpha < \infty$ and y bounded from below makes $\int_0^\tau \gamma^2\alpha < \infty$; hence, $A^{(n)}(\tau, \varphi) < \infty$ for all ϕ. With the first choice, one would let C be all uniformly bounded elements of $\mathbb{H}$ which is a cone, dense in $\mathbb{H}$. For the second choice, the same specification of C would save messy integrability conditions.

Also, the LAN assumption is now easily verified. We can do this via the Jacod and Shiryayev theorem, Theorem VIII.2.1, taking the third choice for $\alpha^{(n)}$ but restricting C to the bounded functions of $\mathbb{H}$ as we may. Taking, first, a single $\mathbf{h} = \gamma$ (and $\phi = 1$), the left-hand side of (8.2.1) is

$$\int_0^\tau a_n^{-2}\gamma(t)^2\alpha(t)\,Y^{(n)}(t)\,\mathrm{d}t \tag{8.4.3}$$

and that of (8.2.2) is

$$\int_0^\tau a_n^{-2}\gamma(t)^2\alpha(t)\,Y^{(n)}(t)I(|\tfrac{1}{2}a_n^{-1}\gamma(t)| > \varepsilon)\,\mathrm{d}t. \tag{8.4.4}$$

By uniform convergence (in probability) of $a_n^{-2}Y^{(n)}$ to y and boundedness of the necessary integrals, (8.4.3) converges in probability to $\|\gamma\|^2_{\mathbb{H}}$. Since the integrand of (8.4.4) converges pointwise in probability to zero while being dominated by the integrand of (8.4.3), an application of Proposition II.5.3 [take $k_\delta(t) = \gamma(t)^2(y(t) + \delta)\alpha(t)$] gives us convergence in probability of (8.4.4) to zero. From (8.2.3), we get the (expected) bonus: $\log(\mathrm{dP}^{(n)}_\gamma/\mathrm{dP}^{(n)}) + \frac{1}{2}\|\gamma\|^2_{\mathbb{H}}$ is asymptotically equivalent to the score $a_n^{-1}\int_0^\tau \gamma\,\mathrm{d}M^{(n)}$ in the one-dimensional model $\mathrm{P}^{(n)}_{\theta\gamma}$ (where γ is fixed and the parameter θ is real) at $\theta = 0$. The extension of these arguments to the situation required for Assumption VIII.3.1 (a finite collection of γ considered simultaneously) is straightforward and is left to the reader to work through.

Alternatively, one can check LAN by looking at the parametric submodels obtained with γ replaced by $\sum_i \phi_i\gamma_i$ for some finite collection of γ_i, and use Theorem VIII.2.2 to get the same result.

We next need to check condition (8.3.2), differentiability of the parameter of interest (the function A considered as an element of $D[0, \tau]$ endowed with the supremum norm) as a function of $\gamma \in C \subseteq \mathbb{H}$. With the third specification

$$\alpha^{(n)}_\gamma(t) = \alpha(t)(1 + \tfrac{1}{2}a_n^{-1}\gamma(t))^2, \tag{8.4.5}$$

we find

$$A^{(n)}_\gamma(t) = A(t) + a_n^{-1}\int_0^t \gamma(s)\alpha(s)\,\mathrm{d}s + \tfrac{1}{4}a_n^{-2}\int_0^t \gamma(s)^2\alpha(s)\,\mathrm{d}s.$$

Thus [cf. (8.3.2)], as $n \to \infty$,

$$a_n(A^{(n)}_\gamma(t) - A(t)) \to \int_0^t \gamma(s)\alpha(s)\,\mathrm{d}s$$

uniformly in $t \in [0, \tau]$, so we take

$$A'(\gamma) = \int_0^{(\cdot)} \gamma\alpha;$$

this is, indeed, a continuous linear map from $\mathbb{H}$ to $D[0,\tau]$ since, for any two points γ, γ' in $\mathbb{H}$,

$$\left\| \int_0^{(\cdot)} \gamma\alpha - \int_0^{(\cdot)} \gamma'\alpha \right\|_\infty \leq \sqrt{\int_0^\tau (\gamma - \gamma')^2 \alpha A(\tau)}$$
$$\leq \|\gamma - \gamma'\|_{\mathbb{H}} \sqrt{\frac{A(\tau)}{\inf_{[0,\tau]} y}}.$$

Next we must check that $\hat{A}^{(n)}$ is a regular estimator of A in the sense of (8.3.3). Since $\mathrm{P}^{(n)}(Y^{(n)} > 0 \text{ on } [0,\tau]) \to 1$ as $n \to \infty$, by contiguity $\mathrm{P}_\gamma^{(n)}(Y^{(n)} > 0$ on $[0,\tau]) \to 1$ too. It suffices, therefore, to show that, under $\mathrm{P}_\gamma^{(n)}$, $a_n(\hat{A}^{(n)} - A_\gamma^{(n)}) = a_n \int (Y^{(n)})^{-1}\, \mathrm{d}M_\gamma^{(n)}$ converges in distribution to the same limit whatever γ, where $M_\gamma^{(n)}$ is the $\mathrm{P}_\gamma^{(n)}$ martingale $N^{(n)} - \int Y^{(n)}\alpha_\gamma^{(n)}$. Under condition (8.4.1) (which continues to hold under $\mathrm{P}_\gamma^{(n)}$, by contiguity again), together with definition (8.4.5), convergence of the predictable variation process in $\mathrm{P}_\gamma^{(n)}$ probability to $\int \alpha y$ is easy to check, and the Lindeberg condition also holds easily. So we have regularity by the martingale central limit theorem.

Now by the LAN condition, for each $\gamma \in C$ we have found a $\mathrm{P}_\gamma^{(n)}$ such that

$$\log \frac{\mathrm{dP}_\gamma^{(n)}}{\mathrm{dP}^{(n)}} - \frac{1}{2}\|\gamma\|_{\mathbb{H}}^2 - a_n^{-1} \int_0^\tau \gamma\, \mathrm{d}M^{(n)} \xrightarrow{\mathrm{P}^{(n)}} 0.$$

The proof of weak convergence of $a_n(\hat{A}_n - A)$, however, shows in particular that

$$a_n(\hat{A}_n(t) - A(t)) - a_n^{-1} \int_0^\tau \frac{I(s \leq t)}{y(s)}\, \mathrm{d}M^{(n)}(s) \xrightarrow{\mathrm{P}^{(n)}} 0.$$

So we have found $\gamma \in \mathbb{H}$ such that (8.3.5) holds and, therefore, by Theorem VIII.3.3, $\hat{A}_n(t)$ is efficient for $A(t)$ for each $t \in [0,\tau]$. By Theorem VIII.3.2, $\hat{A}_n$ is efficient for A. By Theorem VIII.3.5, every compactly differentiable functional of $\hat{A}_n$ is efficient for the same functional of A.

Exactly the same line of proof is available in the k-dimensional case. The Hilbert space $\mathbb{H}$ becomes the set of k-tuples of functions $\boldsymbol{\gamma} = (\gamma_1, \ldots, \gamma_k)$, with $\|\boldsymbol{\gamma}\|_{\mathbb{H}} = \sum_h \int \gamma_h^2 y_h \alpha_h$. Efficiency of $\hat{\mathbf{A}}^{(n)} = (\hat{A}_1, \ldots, \hat{A}_k)$ is proved as before and carries over to efficiency of, e.g., Kaplan–Meier- and Aalen–Johansen-type estimators.

VIII.4.1.1. *Technical Aspects*

The rest of the section is devoted to a collection of miscellaneous remarks on the proof just given and on the interpretation of its results. One point to be noted is that efficiency of the Nelson–Aalen estimator $\hat{A}^{(n)}$ as defined here is relative to the particular choice of $\mathbb{H}$ and its embedding into our nonparametric model indexed by the infinite-dimensional "parameter" α. There is no "canonical" way to do this unless one, a priori, gives the space of all α more structure. The choice reflects what one wishes to consider as "parametric submodels" of the nonparametric model. Two conflicting aspects are involved here: By allowing more submodels (making $\mathbb{H}$ large in some sense) one makes it easier for a given estimator to be efficient because it is easier to "find its influence function in the tangent space"; but, on the other hand, the requirement of *regularity* becomes more stringent. As we consider more submodels, we restrict our attention to a smaller class of estimators (paradoxically, the class of *estimands* becomes smaller too, even though the model is larger).

There are several ways out of this unsatisfactory situation. One way is to base efficiency concepts on asymptotic minimax theorems, avoiding all mention of regularity. From a mathematical point of view, this leads to a nice theory, but we feel such considerations are less close to statistical practice. Another way out is to only consider real or real vector parameters. By Theorem VIII.3.4, once we have shown efficiency, e.g., of $\hat{\mathbf{A}}(t_1), \ldots, \hat{\mathbf{A}}(t_j)$ for some fixed $t_1, \ldots, t_j$, either via Theorem VIII.3.3 or via direct calculation as indicated in Theorem VIII.3.4, we automatically have regularity of this finite-dimensional estimator.

So, regularity of an estimator of a real (or vector) parameter comes for free. We would also have regularity of the estimator $\hat{\mathbf{A}}^{(n)}$ considered as an element of $(D[0,\tau])^k$ by this route if we could prove tightness of $a_n(\hat{\mathbf{A}}^{(n)} - \mathbf{A}_\gamma)$ under any local sequence $\mathrm{P}_\gamma^{(n)}$.

This may be automatically true in our special situation and a proof might run as follows. There are, however, some details which are not yet clear.

By the martingale central limit theorem, we have convergence of $[a_n(\hat{\mathbf{A}}^{(n)} - \mathbf{A}^{*(n)})]$ under the "fixed point" $\mathrm{P}^{(n)}$ to the same (deterministic) limit as $\langle a_n(\hat{\mathbf{A}}^{(n)} - \mathbf{A}^{*(n)})\rangle$. Now $[a_n(\hat{\mathbf{A}}^{(n)} - \mathbf{A}^{*(n)})]$ consists of sums of squares and products of jumps of $a_n\hat{\mathbf{A}}^{(n)}$ and is, therefore, equal to $[a_n(\hat{\mathbf{A}}^{(n)} - \mathbf{A}_\gamma^{*(n)})]$ under any local sequence $\mathrm{P}_\gamma^{(n)}$. But convergence in probability of a nondecreasing process to a continuous, deterministic limit can be conjectured to imply convergence in probability of its compensator to the same limit [though perhaps some conditions are needed; see Aldous (1978b)]. So we would get convergence of the predictable variation process of $a_n(\hat{\mathbf{A}}^{*(n)} - \mathbf{A}_\gamma^{(n)})$ under each local sequence and, hence, by Jacod and Shiryayev (1987, Theorem VI.4.13) tightness of the process itself under every local sequence. By efficiency of the finite-dimensional coordinate projections, we have their regularity and, hence, their weak convergence under each local sequence, to the same limit. By tightness, we, therefore, have weak convergence of $a_n(\hat{\mathbf{A}}^{(n)} - \mathbf{A}_\gamma^{*(n)})$ to the same limit under each sequence $\mathrm{P}_\gamma^{(n)}$ and, hence, regularity of $\hat{\mathbf{A}}^{(n)}$.

To exemplify Theorems VIII.3.1 and VIII.3.4, we do an explicit calculation of $\tilde{\kappa}_t'$ in a particular case. Return to the one-dimensional case and consider the estimation of $A(t_1)$ for a given t_1. By the differentiability of A which we already verified, $A(t_1)$ is also differentiable with derivative

$$\gamma \mapsto \int_0^{t_1} \gamma(s)\alpha(s)\,\mathrm{d}s \tag{8.4.6}$$

(a continuous linear map from $\mathbb{H}$ to $\mathbb{R}$). By the self-duality of a Hilbert space, this map can be represented as the inner product of γ with some fixed element of $\mathbb{H}$, called $\tilde{\kappa}_\iota'$ in the theorem. By the definition of $\mathbb{H}$, we have

$$(\tilde{\kappa}_\iota', \gamma)_{\mathbb{H}} = \int_0^\tau \tilde{\kappa}_\iota'(s)\gamma(s)y(s)\alpha(s)\,\mathrm{d}s \tag{8.4.7}$$

which means

$$\tilde{\kappa}_\iota'(s) = \frac{I(s \leq t_1)}{y(s)}. \tag{8.4.8}$$

This element of $\mathbb{H}$ has squared norm

$$\|\tilde{\kappa}_\iota'\|_{\mathbb{H}}^2 = \int_0^{t_1} \frac{\alpha(s)}{y(s)}\,\mathrm{d}s,$$

which is, therefore, the optimal asymptotic variance of a regular estimator of $A(t_1)$.

By Theorem VIII.2.1 which we used in verifying Assumption VIII.3.1, and our specification of $\alpha_{\kappa_\iota}^{(n)}$,

$$\log \frac{\mathrm{d}P_{\kappa_\iota}^{(n)}}{\mathrm{d}P^{(n)}} + \frac{1}{2}\|\tilde{\kappa}_\iota'\|_{\mathbb{H}}^2$$

is asymptotically equivalent to

$$a_n^{-1}\int_0^\tau \tilde{\kappa}_\iota'(s)\,\mathrm{d}M^{(n)}(s) = a_n^{-1}\int_0^{t_1} \frac{\mathrm{d}M^{(n)}(s)}{y(s)}.$$

But from the proof of weak convergence of $a_n(\widehat{A}^{(n)} - A)$, $a_n(\widehat{A}^{(n)}(t_1) - A(t_1))$ is asymptotically equivalent to the same quantity. Theorem VIII.3.4, therefore, gives regularity and efficiency of $\widehat{A}^{(n)}(t_1)$.

VIII.4.1.2. *Summary of Approach*

We conclude by summarizing the main line of these proofs. There are two main parts, corresponding to analysis of the model and of a candidate efficient estimator:

— Find a rich collection of parametric submodels; the local parameters are given a Hilbert space structure by identifying each parameter value with its asymptotic (normally distributed) score and letting the inner product be the covariance of the corresponding scores.

— Show that the estimator of interest is asymptotically linear in the scores. If the estimator is regular, Theorem VIII.3.3 now establishes its efficiency.

Alternatively (especially for finite-dimensional estimands), calculate the derivative of the *estimand* by (8.3.2) and calculate the lower bound by (8.3.4). This is essentially the computation of the *hardest parametric submodel* as we did it in Section VIII.2. (In fact, we find the scaled score functions of the hardest submodel: inverse information times scores. The scores themselves are obtained by repeating the scaling operation.) Efficiency of an estimator can now be obtained by checking that it is regular and achieves the lower bound, or simply by checking that it is asymptotically equivalent to the scaled scores.

In the previous case, our Hilbert space was rather big, making calculations, especially the identification of $\tilde{\kappa}'_t$, quite easy. If our original model had been more restrictive, e.g., a parametric model, then $\mathbb{H}$ would have been smaller. One could still identify it with a class of functions γ with the same inner product via the scores $a_n^{-1}\int_0^\tau \gamma\, \mathrm{d}M^{(n)}$, thus with a *subspace* of the large space. The derivative of an estimand, for instance $A(t_1)$, would have been given by exactly the same map (8.4.6) or (8.4.7); however, the conclusion (8.4.8) is no longer valid since (8.4.7) only holds for γ in the subspace corresponding to the new, smaller model. One can proceed by solving (8.4.7) in the *big, original space*, and then *projecting* the solution (8.4.8) into the subspace of interest because, by definition, the derivative has to lie in this subspace but need only satisfy (8.4.7) for γ in the subspace.

This argument is another illustration of how the computation of hardest parametric submodels, leading in first instance to a variational problem, by local considerations typically reduces to a projection calculation in a Hilbert space.

VIII.4.2. k-Sample Tests Revisited

In this section, we review the developments of Section VIII.2.3 from the new "semiparametric" viewpoint, showing how the "hardest parametric submodels" of that section can be found automatically by determining the Hilbert space representation $\tilde{\kappa}'$ of the derivative of the parameter of interest. Also, we show how the parametric submodels there with a k-dimensional nuisance parameter can be embedded in a semiparametric transformation model, where the nuisance (shape) parameters are replaced by a completely unknown monotone transformation putting one into a one-dimensional parametric model. This corresponds to our interest in *nonparametric* k-sample tests, typically *rank tests.*

In Example V.2.14, we also considered the k-sample problem for censored survival data from the point of view of optimal nonparametric rank tests; that is, we suposed that the observations came from a time-transformed parametric model. The time-transformation was supposed completely unknown. Here, we do the same, in more generality; the parametric model need not be a location model for instance.

Consider the transformation model

$$\alpha(t;\theta,g) = \alpha_0(g(t);\theta)g'(t) \tag{8.4.9}$$

where $\alpha_0(u;\theta)$ is a parametric family of baseline hazard rates and g is a smooth transformation of the time axis with non-negative derivative g'. Suppose $\mathbf{N}$ is a k-variate counting process satisfying the multiplicative intensity model such that the deterministic parts of the intensities are of the form (8.4.9), each with the same transformation parameters g but with possibly differing "parametric" parameters θ_h, $h = 1, \ldots, k$, whose equality we want to test. Thus, $\mathbf{N}$ has intensity $(\alpha_1 Y_1, \ldots, \alpha_k Y_k)$ with

$$\alpha_h(t) = \alpha_0(g(t); \theta_h) g'(t). \tag{8.4.10}$$

Were g known, one could transform the counting process $\mathbf{N}$ to the new counting process $\mathbf{N} \circ g^{-1}$ with intensity $(\alpha_0(u; \theta_1) Y_1(g^{-1}(u)), \ldots, \alpha_0(u; \theta_k) Y_k(g^{-1}(u)))$, obtaining a truly parametric model.

We will test the null hypothesis $\theta_1 = \cdots = \theta_k$, deriving the optimal test at a "fixed point" where $\theta_1 = \cdots = \theta_k = \theta$, say, and at a given g. The optimal test is the same as the χ^2-type test based on an optimal estimator of $\boldsymbol{\theta} = (\theta_1, \ldots, \theta_k)$, rather on a set of $k - 1$ linearly independent contrasts based on $\boldsymbol{\theta}$. The "general level" of the θ_h is actually not identified since it can be absorbed into the unknown transformation g. This is not a problem (except that it will involve us again with generalized inverses instead of ordinary ones).

In a location model, this is obvious since one can subtract a constant from all the θ_h and absorb it into the transformation g. But also, in general, one can find a transformation turning the model, locally at least, into a location model, and then again the nonidentifiability becomes plausible.

Suppose now the transformation g actually belonged to a parametric family $g(t; \eta)$. Then at our fixed point we would have had the following score functions for $\theta_1, \ldots, \theta_k$ and η, respectively, where partial derivatives with respect to u refer to the first argument of α_0 [just after (8.4.9)]:

$$\int_0^\tau \frac{\partial}{\partial \theta} \log \alpha_0(g(t); \theta) \, \mathrm{d}M_h^{(n)}(t), \quad h = 1, \ldots, k,$$

$$\int_0^\tau \left(\frac{\partial}{\partial u} \log \alpha_0(g(t); \theta) \frac{\partial}{\partial \eta} g(t; \eta) + \frac{\partial}{\partial \eta} \log g'(t; \eta) \right) \mathrm{d}M_{\cdot}^{(n)}(t).$$

Let $\gamma(t) = (\partial/\partial\theta) \log \alpha_0(g(t); \theta)$ and let us suppose that for an *arbitrary* function $\xi(t)$ (and given θ and g), we can *find* a parametric model $g(t; \eta)$ with $g(t; 0) = g(t)$ and

$$\frac{\partial}{\partial u} \log \alpha_0(g(t); \theta) \frac{\partial}{\partial \eta} g(t; 0) + \frac{\partial}{\partial \eta} \log g'(t; 0) = \xi(t). \tag{8.4.11}$$

By writing g as an integral of g' and differentiating with respect to t, this equation can be transformed into a first-order differential equation for $z(t) = (\partial/\partial\eta) \log g'(t; 0)$. In fact, we only need the existence of a solution for k special ξ corresponding to hardest parametric submodels which we will derive in a moment, or even only for ξ approximating these arbitrarily well.

Then the score functions for $\theta_1, \ldots, \theta_k$ and η (at $\theta_1 = \cdots = \theta_k = \theta$, $\eta = 0$) are simply

$$\int_0^\tau \gamma(t)\,\mathrm{d}M_h^{(n)}(t), \quad h = 1, \ldots, k,$$

and

$$\int_0^\tau \xi(t)\,\mathrm{d}M_{\bullet}^{(n)}(t),$$

respectively. Next we transform to local parameters by taking

$$\theta_h = \theta + a_n^{-1}\phi_h,$$
$$\eta = a_n^{-1}\zeta.$$

Here, a_n is the rate sequence appearing in our usual conditions

$$\sup_{t \in [0,\tau]} |Y_h^{(n)}(t)/a_n^2 - y_h(t)| \xrightarrow{\mathrm{P}} 0$$

as $n \to \infty$. The score functions transform by the factor a_n^{-1}, and, provided all the following asymptotic variances are finite, they are asymptotically normally distributed with means zero and covariance matrix

$$\begin{bmatrix} \ddots & \vdots & 0 & \vdots \\ & \int_0^\tau \gamma^2 y_h \alpha & & \int_0^\tau \xi\gamma y_h \alpha \\ 0 & \vdots & \ddots & \vdots \\ \cdots & \int_0^\tau \xi\gamma y_h \alpha & \cdots & \int_0^\tau \xi^2 y_{\bullet}\alpha \end{bmatrix}.$$

Write $\mathbf{y}$ for the vector $(y_1, \ldots, y_k)^\top$ and $\operatorname{diag}(\mathbf{y})$ for the diagonal matrix with $\mathbf{y}$ on its diagonal. Define

$$\mathscr{I} = \int_0^\tau \gamma^2 \operatorname{diag} \mathbf{y}\alpha$$

and (for later use)

$$\mathbf{V} = \mathscr{I} - \int_0^\tau \gamma^2 \frac{\mathbf{y}\mathbf{y}^\top \alpha}{y_{\bullet}}; \tag{8.4.12}$$

let $\mathbf{1} = (1 \ldots 1)^\top$. We can write the above covariance matrix more compactly now as

$$\begin{pmatrix} \mathscr{I} & \int_0^\tau \xi\gamma\mathbf{y}\alpha \\ \int_0^\tau \xi\gamma\mathbf{y}^\top\alpha & \int_0^\tau \xi^2 y_{\bullet}\alpha \end{pmatrix}.$$

The joint asymptotic normality of the scores we have just derived means that Assumption VIII.3.1 is satisfied, where hopefully we may take $C = \mathbb{H}$ (so no projection is needed in Theorem VIII.3.1) and equal to be the space of pairs $(\boldsymbol{\phi}, \xi)$, with inner product

$$((\boldsymbol{\phi}, \xi), (\boldsymbol{\phi}', \xi'))_{\mathbb{H}} = \boldsymbol{\phi}^\top \mathscr{I} \boldsymbol{\phi}' + \boldsymbol{\phi}^\top\left(\int_0^\tau \xi'\gamma\mathbf{y}\alpha\right) + \left(\int_0^\tau \xi\gamma\mathbf{y}^\top\alpha\right)\boldsymbol{\phi}' + \int_0^\tau \xi\xi' y_{\bullet}\alpha. \tag{8.4.13}$$

Thus, $\mathbb{H} = \mathbb{R}^k \times L^2(y_{.}\alpha)$ with the inner product (8.4.13) [$L^2(y_{.}\alpha)$ denoting the space of functions such that the integral of their square times $y_{.}\alpha$ is bounded]. For $\mathbf{h} = (\boldsymbol{\phi}, \xi)$, $\mathrm{P}_{\mathbf{h}}^{(n)}$ is the model corresponding to $\boldsymbol{\theta} = \theta\mathbf{1} + a_n^{-1}\boldsymbol{\phi}$, $g = g(\cdot;\eta)$ with the family $g(\cdot;\eta)$ corresponding to the given ξ via (8.4.11), and finally $\eta = a_n^{-1}\zeta$ with $\zeta = 1$. In order not to have to compute a projection, we only need construct such models for $\mathbf{h} \in C$, a cone contained in $\mathbb{H}$ whose closed linear span is all of $\mathbb{H}$. For instance, it would be enough to show that $g(\cdot;\eta)$ can be constructed for *bounded* ξ. (At the final analysis, we in fact only need construct these models for ξ arbitrarily close to the "hardest" ξ coming out of the calculations at the end of this subsection.)

Consider estimation of $\kappa = \mathbf{b}^\mathsf{T}\boldsymbol{\theta}$ for some given fixed $\mathbf{b}$. The left-hand side of (8.3.2) is just

$$a_n(\mathbf{b}^\mathsf{T}(\theta\mathbf{1} + a_n^{-1}\boldsymbol{\phi}) - \mathbf{b}^\mathsf{T}(\theta\mathbf{1})) = \mathbf{b}^\mathsf{T}\boldsymbol{\phi},$$

so Assumption VIII.3.2 is satisfied for this parameter if the map

$$\kappa'(\boldsymbol{\phi}, \xi) \mapsto \mathbf{b}^\mathsf{T}\boldsymbol{\phi}$$

is a continuous, linear map from $\mathbb{H}$ to $\mathbb{R}$.

Linearity is obvious, but continuity is a little more subtle. The map is continuous if and only if it can be represented as an inner product with a fixed element of $\mathbb{H}$. So our problem is (for given $\mathbf{b}$): Determine $(\boldsymbol{\psi}, \beta) \in \mathbb{H}$ (if it exists) such that

$$((\boldsymbol{\phi}, \xi), (\boldsymbol{\psi}, \beta))_{\mathbb{H}} = \mathbf{b}^\mathsf{T}\boldsymbol{\phi} \quad \text{for all } (\boldsymbol{\phi}, \xi) \in \mathbb{H}.$$

The optimal limiting variance of a regular estimator of $\mathbf{b}^\mathsf{T}\boldsymbol{\theta}$ is, by Theorem VIII.3.1, given by the squared norm of $(\boldsymbol{\psi}, \beta)$ (no projection is necessary). From this expression (it will turn out to be a quadratic form in $\mathbf{b}$), we can determine the optimal limiting covariance matrix of regular estimators of $\boldsymbol{\theta}$ itself and, hence, the optimal form of a test of $\theta_1 = \cdots = \theta_k$. The function β in the solution will turn out to be a linear combination, depending on $\mathbf{b}$, of the k fixed functions (familiar from Sections V.2.3 and VIII.2.3) $\gamma y_h/y_{.}$. These functions therefore define a $k + 1$-dimensional subspace of $\mathbb{H}$ for which the projection problem has the same solution: Therefore, they determine the *hardest parametric submodel for this problem.*

From (8.4.13), the problem to be solved is therefore: For given $\mathbf{b}$, determine $(\boldsymbol{\psi}, \beta)$ such that

$$\boldsymbol{\phi}^\mathsf{T}\mathscr{I}\boldsymbol{\psi} + \boldsymbol{\phi}^\mathsf{T}\left(\int_0^\tau \beta\gamma\mathbf{y}\alpha\right) + \left(\int_0^\tau \xi\gamma\mathbf{y}^\mathsf{T}\alpha\right)\boldsymbol{\psi} + \int_0^\tau \beta\xi y_{.}\alpha = \boldsymbol{\phi}^\mathsf{T}\mathbf{b} \tag{8.4.14}$$

for all (ϕ, ξ). We rewrite this expression as a linear equation in ϕ:

$$\boldsymbol{\phi}^\mathsf{T}\left(\mathscr{I}\boldsymbol{\psi} + \int_0^\tau \beta\gamma\mathbf{y}\alpha - \mathbf{b}\right) + \left(\int_0^\tau \xi\gamma\mathbf{y}^\mathsf{T}\alpha\boldsymbol{\psi} + \int_0^\tau \beta\xi y_{.}\alpha\right) = 0$$

for all $(\boldsymbol{\phi}, \xi)$. The coefficients must therefore be zero:

$$\mathscr{I}\boldsymbol{\psi} + \int_0^\tau \beta\gamma\mathbf{y}\alpha - \mathbf{b} = \mathbf{0},$$

$$\int_0^\tau \xi\gamma\mathbf{y}^\top\alpha\boldsymbol{\psi} + \int_0^\tau \beta\xi y_.\alpha = 0$$

for all ξ. From the first equation, we get

$$\boldsymbol{\psi} = \mathscr{I}^{-1}\left(\mathbf{b} - \int_0^\tau \beta\gamma\mathbf{y}\alpha\right) \tag{8.4.15}$$

and from the second (since it is true for all ξ)

$$\beta = -\gamma\mathbf{y}^\top\boldsymbol{\psi}/y_.. \tag{8.4.16}$$

Substituting for β, we get a single equation for $\boldsymbol{\psi}$:

$$\mathscr{I}\boldsymbol{\psi} = \mathbf{b} + \left(\int_0^\tau \gamma^2\frac{\mathbf{y}\mathbf{y}^\top}{y_.}\alpha\right)\boldsymbol{\psi}$$

or

$$\mathbf{V}\boldsymbol{\psi} = \mathbf{b} \tag{8.4.17}$$

where $\mathbf{V}$ was defined in (8.4.12),

$$\mathbf{V} = \int_0^\tau \gamma^2(\operatorname{diag}\mathbf{y})\alpha - \int_0^\tau \gamma^2\frac{\mathbf{y}\mathbf{y}^\top}{y_.}\alpha.$$

Now $\mathbf{V}$ is singular: $\mathbf{V1} = \mathbf{0}$; but it has maximal rank $k - 1$ as long as one can find a chain of pairs covering $\{1, 2, \ldots, k\}$ such that $\int_0^\tau \gamma^2(y_h y_l/y_.)\alpha > 0$ for each (h, l) in the chain; see Section V.2.2. $\mathbf{V}$ is also symmetric, so it has a Moore– Penrose generalized inverse $\mathbf{V}^-$ such that $\mathbf{V}\mathbf{V}^- = \mathbf{V}^-\mathbf{V}$ is the projection onto the orthogonal complement (with respect to the Euclidean norm) of the null space of $\mathbf{V}$, the linear span of $\mathbf{1}$, so $\mathbf{V}\mathbf{V}^- = \mathbf{I} - \mathbf{1}\mathbf{1}^\top/\mathbf{1}^\top\mathbf{1}$. (The Moore–Penrose generalized inverse is obtained by inverting the nonzero eigenvalues in the eigenvector–eigenvalue decomposition of $\mathbf{V}$.)

This means that (8.4.17) can be uniquely solved to give

$$\boldsymbol{\psi} = \mathbf{V}^-\mathbf{b}$$

if and only if $\mathbf{b}$ is in the range of $\mathbf{V}$, i.e., if and only if $\mathbf{1}^\top\mathbf{b} = 0$ (for then, this equation does give $\mathbf{V}\boldsymbol{\psi} = \mathbf{V}\mathbf{V}^-\mathbf{b} = (\mathbf{I} - \mathbf{1}\mathbf{1}^\top/\mathbf{1}^\top\mathbf{1})\mathbf{b} = \mathbf{b}$).

Now (8.4.15) and (8.4.16) were necessary and sufficient conditions for (8.4.14), so we have shown that (8.4.14) has a solution $(\boldsymbol{\psi}, \beta)$ if and only if $\mathbf{1}^\top\mathbf{b} = 0$, and the solution is then

$$\begin{aligned} \boldsymbol{\psi} &= \mathbf{V}^-\mathbf{b}, \\ \beta &= -\gamma\mathbf{y}^\top\mathbf{V}^-\mathbf{b}/y_.. \end{aligned} \tag{8.4.18}$$

Note that from (8.4.16), we already see linear combinations of $\gamma y_h/y_.$ emerging for the "least favorable" nuisance parameters scores. When we look at all k components of $\boldsymbol{\theta}$ simultaneously, we will need each of these nuisance scores to determine the hardest parametric submodel, exactly as we found in Section VIII.2.3.

The next task is to determine $\|(\boldsymbol{\psi}, \beta)\|^2_{\mathbb{H}}$. By (8.4.18) and (8.4.13), this is equal to

$$\mathbf{b}^\mathsf{T}\mathbf{V}^-\mathscr{I}\mathbf{V}^-\mathbf{b} - 2\mathbf{b}^\mathsf{T}\mathbf{V}^- \int_0^\tau \frac{\mathbf{y}\mathbf{y}^\mathsf{T}}{y_\cdot}\alpha\mathbf{V}^-\mathbf{b} + \mathbf{b}^\mathsf{T}\mathbf{V}^- \int_0^\tau \frac{\gamma^2\mathbf{y}\mathbf{y}^\mathsf{T}}{y_\cdot}\alpha\mathbf{V}^-\mathbf{b}$$

$$= \mathbf{b}^\mathsf{T}\mathbf{V}^-\mathbf{V}\mathbf{V}^-\mathbf{b} = \mathbf{b}^\mathsf{T}\mathbf{V}^-\mathbf{b}.$$

But $\mathbf{V}$ is exactly the same matrix as $\mathscr{I}^{\phi\phi|\zeta}$ given by (8.2.19) in Section VIII.2.3, and the optimal asymptotic variance $\mathbf{b}^\mathsf{T}\mathbf{V}^-\mathbf{b}$ for estimating a contrast $\mathbf{b}^\mathsf{T}\boldsymbol{\theta}$ gives the same optimal noncentrality parameter $\boldsymbol{\phi}^\mathsf{T}\mathbf{V}\boldsymbol{\phi}$ of the best test of $\mathbf{b}^\mathsf{T}\boldsymbol{\theta} = 0$ for all contrasts $\mathbf{b}$.

The special cases $k = 2$ and proportional y_h are worth looking at from this point of view too. For instance, when $k = 2$, one finds that $\mathbf{V}$ is proportional to the 2×2 matrix with ones on the diagonal and minus ones outside, whose generalized inverse is of the same form. Under asymptotically proportional risk sets, $\mathbf{V}$ is proportional to a multinomial covariance matrix which has a generalized inverse familiar from the quadratic form of the χ^2 goodness-of-fit test. These details are left to the reader.

VIII.4.3. The Cox Regression Model

Here we sketch a proof of asymptotically efficiency of the usual estimators (and asymptotic optimality of the usual tests) in the Cox regression model on now familiar lines: We calculate the score functions for a family of local parametric submodels, make plausible the asymptotic normality of the scores, and show that the usual estimators are asymptotically linear in the scores. A verification of regularity plus Theorems VIII.3.2 and VIII.3.3 then do the rest. Alternatively, one may explicitly calculate the "efficient scores" and use Theorems VIII.3.4 and VIII.3.2; regularity need not be verified (for finite-dimensional parameters of interest, at least; for infinite-dimensional parameters, one must check that tightness is preserved under local sequences).

LAN itself may be verified by direct methods or by appeal to Theorems VIII.2.2 or VIII.2.1. Since we only need the LAN property for a dense subset of our infinite-dimensional local parametrization, we may freely impose boundedness restrictions or other convenient restrictions on these parameters.

Recall that our model assumes that the intensity of N_{hi}, $h = 1, \ldots, k$, $i = 1, \ldots, n$, is

$$\lambda_{hi}(t) = \alpha_{h0}(t)\exp(\boldsymbol{\beta}^\mathsf{T}\mathbf{Z}_{hi}(t))\,Y_{hi}(t).$$

Consider local sequences with $a_n = n^{1/2}$ and (for instance)

$$\alpha_h^{(n)}(t) = \alpha_{h0}(t)(1 + \tfrac{1}{2}a_n^{-1}\gamma_h(t))^2,$$
$$\boldsymbol{\beta}^{(n)} = \boldsymbol{\beta}_0 + a_n^{-1}\boldsymbol{\phi},$$

where $\boldsymbol{\alpha}_0$ is fixed and the local parameter $(\boldsymbol{\gamma}, \boldsymbol{\phi}) = (\gamma_1, \ldots, \gamma_k, \boldsymbol{\phi})$ consists of k functions and a vector in $\mathbb{R}^p$.

If we take a fixed $(\boldsymbol{\gamma}, \boldsymbol{\phi})$, we can consider a one-dimensional model with real parameter θ obtained by replacing $(\boldsymbol{\gamma}, \boldsymbol{\phi})$ by $\theta(\boldsymbol{\gamma}, \boldsymbol{\phi}) = (\theta\gamma_1, \ldots, \theta\gamma_k, \theta\boldsymbol{\phi})$ in the definitions of $\alpha_h^{(n)}$ and $\boldsymbol{\beta}^{(n)}$. In this model, we have

$$\frac{\partial}{\partial\theta}\log\lambda_{hi}(t)\bigg|_{\theta=0} = a_n^{-1}(\gamma_h(t) + \boldsymbol{\phi}^{\mathsf{T}}\mathbf{Z}_{hi}(t)).$$

Thus, the score for θ is

$$\sum_h a_n^{-1}\int_0^\tau \gamma_h \,\mathrm{d}M_{h.} + \boldsymbol{\phi}^{\mathsf{T}}\sum_{h,i} a_n^{-1}\int_0^\tau \mathbf{Z}_{hi}\,\mathrm{d}M_{hi}. \tag{8.4.19}$$

This martingale has jumps of the order of size of a_n^{-1} and predictable variation process

$$a_n^{-2}\sum_h \int \gamma_h^2 \sum_i \lambda_{hi} + a_n^{-2}\boldsymbol{\phi}^{\mathsf{T}}\sum_h \int \sum_i \mathbf{Z}_{hi}^{\otimes 2}\lambda_{hi}\boldsymbol{\phi} + 2a_n^{-2}\sum_h \int \gamma_h \sum_i \mathbf{Z}_{hi}^{\mathsf{T}}\lambda_{hi}\boldsymbol{\phi}$$
$$= a_n^{-2}\sum_h \int \gamma_h^2 S_h^{(0)}\alpha_{h0} + a_n^{-2}\boldsymbol{\phi}^{\mathsf{T}}\sum_h \int \mathbf{S}_h^{(2)}\alpha_{h0}\boldsymbol{\phi} + 2a_n^{-2}\boldsymbol{\phi}^{\mathsf{T}}\int \sum_h \gamma_h \mathbf{S}_h^{(1)}\alpha_{h0}.$$

By Condition VII.2.1, assuming $\int_0^\tau \gamma_h^2 s_h^{(0)}\alpha_{h0} < 0$ for each h, the predictable variation process converges to the deterministic function

$$\sum_h \left(\int \gamma_h^2 s_h^{(0)}\alpha_{h0} + 2\boldsymbol{\phi}^{\mathsf{T}}\int \gamma_h \mathbf{s}_h^{(1)}\alpha_{h0} + \boldsymbol{\phi}^{\mathsf{T}}\int \mathbf{s}_h^{(2)}\alpha_{h0}\boldsymbol{\phi}\right).$$

We may, therefore, expect the LAN property of Assumption VIII.3.1 to hold, with $\mathbb{H} = \{(\boldsymbol{\gamma}, \boldsymbol{\phi})\}$ and

$$\|(\boldsymbol{\gamma}, \boldsymbol{\phi})\|_{\mathbb{H}}^2 = \sum_h \left(\int_0^\tau \gamma_h^2 s_h^{(0)}\alpha_{h0} + 2\boldsymbol{\phi}^{\mathsf{T}}\int_0^\tau \gamma_h \mathbf{s}_h^{(1)}\alpha_{h0} + \boldsymbol{\phi}^{\mathsf{T}}\int_0^\tau \mathbf{s}_h^{(2)}\alpha_{h0}\boldsymbol{\phi}\right).$$

From this,

$$((\boldsymbol{\gamma}, \boldsymbol{\phi}), (\boldsymbol{\gamma}', \boldsymbol{\phi}'))_{\mathbb{H}}$$
$$= \sum_h \left(\int_0^\tau \gamma_h\gamma_h' s_h^{(0)}\alpha_{h0} + \boldsymbol{\phi}^{\mathsf{T}}\int_0^\tau \gamma_h' \mathbf{s}_h^{(1)}\alpha_{h0} + \int_0^\tau \gamma_h \mathbf{s}_h^{(1)^{\mathsf{T}}}\boldsymbol{\phi}' + \boldsymbol{\phi}^{\mathsf{T}}\int_0^\tau \mathbf{s}_h^{(2)}\alpha_{h0}\boldsymbol{\phi}'\right).$$

We also need to check the differentiability Assumption VIII.3.2 for estimating $\boldsymbol{\beta}$ and for estimating $A_{h0} = \int \alpha_{h0}$, $h = 1, \ldots, k$. It is easy to check that

(with a prime denoting the derivative of the functional)

$$\boldsymbol{\beta}'(\boldsymbol{\gamma}, \boldsymbol{\phi}) = \boldsymbol{\phi}$$

and

$$A'_{h0}(\boldsymbol{\gamma}, \boldsymbol{\phi}) = \int \gamma_h \alpha_{h0}$$

are the limits in (8.3.2) (the first is trivial, the second was done in Section VIII.4.1), but not so easy to check that these are *continuous* functions of $(\boldsymbol{\gamma}, \boldsymbol{\phi})$ with respect to the $\mathbb{H}$ norm. One way to check this is to set about finding the representations of $\beta'_h(\boldsymbol{\gamma}, \boldsymbol{\phi})$ and $A'_{h0}(\boldsymbol{\gamma}, \boldsymbol{\phi})(t)$ as inner products of $(\boldsymbol{\gamma}, \boldsymbol{\phi})$ with fixed elements of $\mathbb{H}$ as in Section VIII.4.2; if we can do this and, moreover, the norm of the representative of $A'_{h0}(t)$ is bounded in t, we have proved differentiability.

There is another way to proceed here since we already have analyzed estimators which we expect to be efficient. We see if (8.3.5) is true and then use Theorem VIII.3.4 to guess the representations. If these work, we have efficiency and regularity of $\hat{\boldsymbol{\beta}}$ and of the finite-dimensional coordinate projections of the $\hat{A}_{h0}$.

Now Theorem VII.2.2 essentially showed that $a_n(\hat{\boldsymbol{\beta}} - \boldsymbol{\beta})$ is asymptotically equivalent to

$$\sum_h \mathscr{I}^{-1} a_n^{-1} \int_0^\tau \left(\sum_i \mathbf{Z}_{hi} \, \mathrm{d}M_{hi} - \mathbf{e}_h \, \mathrm{d}M_{h.} \right), \tag{8.4.20}$$

where [see part (e) of Condition VII.2.1]

$$\mathscr{I} = \sum_h \int_0^\tau \mathbf{v}_h s_h^{(0)} \alpha_{h0},$$

$$\mathbf{e}_h = \frac{\mathbf{s}_h^{(1)}}{s_h^{(0)}} \quad \text{and} \quad \mathbf{v}_h = \frac{\mathbf{s}_h^{(2)}}{s_h^{(0)}} - \frac{\mathbf{s}_h^{(1)} \mathbf{s}_h^{(1)\mathsf{T}}}{s_h^{(0)} s_h^{(0)}}.$$

Thus, $a_n(\mathbf{b}^\mathsf{T}\hat{\boldsymbol{\beta}} - \mathbf{b}^\mathsf{T}\boldsymbol{\beta})$, by comparison of $\mathbf{b}^\mathsf{T}$ times (8.4.20) and (8.4.19), is equivalent to the score (8.4.19) with

$$\gamma_h = -\mathbf{e}_h^\mathsf{T} \mathscr{I}^{-1} \mathbf{b},$$
$$\boldsymbol{\phi} = \mathscr{I}^{-1} \mathbf{b}.$$

Now we check that for *any* $(\boldsymbol{\gamma}, \boldsymbol{\phi})$,

$$((-\mathbf{e}_h^\mathsf{T} \mathscr{I}^{-1} \mathbf{b}, \mathscr{I}^{-1} \mathbf{b}), (\boldsymbol{\gamma}, \boldsymbol{\phi}))_{\mathbb{H}} = \mathbf{b}^\mathsf{T} \boldsymbol{\phi} \tag{8.4.21}$$

and

$$\|(-\mathbf{e}_h^\mathsf{T} \mathscr{I}^{-1} \mathbf{b}, \mathscr{I}^{-1} \mathbf{b})\|_{\mathbb{H}}^2 < \infty. \tag{8.4.22}$$

This will show in one blow (see Theorem VIII.3.4) that $\boldsymbol{\beta}$ is a differentiable parameter and that $\hat{\boldsymbol{\beta}}$ is asymptotically efficient and regular for $\boldsymbol{\beta}$. Moreover, (8.4.22) will give the optimal asymptotic variance.

The calculation of (8.4.21) is as follows:

$$
\begin{aligned}
&((-\mathbf{e}_h^{\mathsf{T}}\mathscr{I}^{-1}\mathbf{b}, \mathscr{I}^{-1}\mathbf{b}), (\boldsymbol{\gamma}, \boldsymbol{\phi}))_{\mathbb{H}} \\
&= \sum_h \mathbf{b}^{\mathsf{T}}\mathscr{I}^{-1}\left(\int_0^\tau \mathbf{s}_h^{(2)}\alpha_{h0}\boldsymbol{\phi} + \int_0^\tau \mathbf{s}_h^{(1)}\gamma_h\alpha_{h0} - \int_0^\tau \mathbf{e}_h\mathbf{s}_h^{(1)\mathsf{T}}\alpha_{h0}\boldsymbol{\phi}\right. \\
&\qquad \left. - \int_0^\tau \mathbf{e}_h\gamma_h s_h^{(0)}\alpha_{h0}\right) \\
&= \mathbf{b}^{\mathsf{T}}\mathscr{I}^{-1}\mathscr{I}\boldsymbol{\phi} = \mathbf{b}^{\mathsf{T}}\phi.
\end{aligned}
$$

Applied to $(\boldsymbol{\gamma}, \boldsymbol{\phi}) = (-\mathbf{e}_h\mathscr{I}^{-1}\mathbf{b}, \mathscr{I}^{-1}\mathbf{b})$, we find

$$
\|(-\mathbf{e}_h\mathscr{I}^{-1}\mathbf{b}, \mathscr{I}^{-1}\mathbf{b})\|_{\mathbb{H}}^2 = \mathbf{b}^{\mathsf{T}}\mathscr{I}^{-1}\mathbf{b} < \infty.
$$

Thus, the optimal asymptotic covariance matrix is $\mathscr{I}^{-1}$, giving us the complete desired results for $\hat{\boldsymbol{\beta}}$.

Theorem VII.2.3 showed that

$$
a_n(\hat{A}_{h0} - A_{h0}) + \int \mathbf{e}_h^{\mathsf{T}}\alpha_{h0}\cdot a_n(\hat{\boldsymbol{\beta}} - \boldsymbol{\beta})
$$

is asymptotically equivalent to

$$
a_n^{-1}\int (s_h^{(0)})^{-1}\,\mathrm{d}M_{h\bullet},
$$

and, therefore,

$$
a_n(\hat{A}_{h0} - A_{h0})
$$

is asymptotically equivalent to

$$
\begin{aligned}
&a_n^{-1}\int (s_h^{(0)})^{-1}\,\mathrm{d}M_{h\bullet} + a_n^{-1}\sum_{h'}\int \mathbf{e}_h^{\mathsf{T}}\alpha_{h0}\mathscr{I}^{-1}\int_0^\tau \sum_i \mathbf{Z}_{h'i}\,\mathrm{d}M_{h'i} \\
&\qquad - a_n^{-1}\sum_{h'}\int \mathbf{e}_h^{\mathsf{T}}\alpha_{h0}\mathscr{I}^{-1}\int_0^\tau \mathbf{e}_{h'}\,\mathrm{d}M_{h'\bullet}.
\end{aligned}
$$

From this in the point t, we can recognize, for instance, the score (8.4.19) with

$$
\gamma_{h'} = \delta_{hh'}(s_h^{(0)})^{-1}I_{[0,t]} - \left(\int_0^t \mathbf{e}_h^{\mathsf{T}}\alpha_{h0}\right)\mathscr{I}^{-1}\mathbf{e}_{h'}, \quad h' = 1, \ldots, k,
$$

and

$$
\boldsymbol{\phi} = \mathscr{I}^{-1}\left(\left(\int_0^t \mathbf{e}_h\alpha_{h0}\right)\right).
$$

The inner product of this vector with an arbitrary $(\boldsymbol{\gamma}, \boldsymbol{\phi}) \in \mathbb{H}$ is

$$\int_0^t \gamma_h \alpha_{h0} - \int_0^t \mathbf{e}_h^\mathsf{T} \alpha_{h0} \mathscr{I}^{-1} \sum_{h'} \int_0^\tau \mathbf{e}_{h'} \gamma_h s_{h'}^{(0)} \alpha_{h'0} + \boldsymbol{\phi}^\mathsf{T} \int_0^t (s_h^{(0)})^{-1} \mathbf{s}_h^{(1)} \alpha_{h0}$$

$$- \boldsymbol{\phi}^\mathsf{T} \sum_{h'} \int_0^\tau \mathbf{s}_{h'}^{(1)} \mathbf{e}_{h'}^\mathsf{T} \alpha_{h'0} \mathscr{I}^{-1} \int_0^t \mathbf{e}_h \alpha_{h0} + \sum_{h'} \int_0^\tau \gamma_{h'} (\mathbf{s}_{h'}^{(1)})^\mathsf{T} \alpha_{h'0} \mathscr{I}^{-1} \int_0^t \mathbf{e}_h \alpha_{h0}$$

$$+ \boldsymbol{\phi}^\mathsf{T} \sum_{h'} \int_0^\tau \mathbf{s}_{h'}^{(2)} \alpha_{h'0} \mathscr{I}^{-1} \int_0^t \mathbf{e}_h \alpha_{h0}$$

$$= \int_0^t \gamma_h \alpha_{h0}.$$

Its own square norm—its inner product with itself—is, therefore,

$$\int_0^t (s_h^{(0)})^{-1} \alpha_{h0} - \left(\int_0^t \mathbf{e}_h^\mathsf{T} \alpha_{h0} \right) \mathscr{I}^{-1} \left(\int_0^t \mathbf{e}_h \alpha_{h0} \right) < \infty. \tag{8.4.23}$$

These results give us asymptotic efficiency of $\hat{A}_{h0}(t)$ for each h and t separately, with optimal asymptotic variance (8.4.23). Clearly, the calculations extend to any finite collection of these estimators jointly, or even to a compactly differentiable real vector functional of $\hat{\boldsymbol{\beta}}$ and $\hat{A}_{10}, \ldots, \hat{A}_{k0}$. To get a truly infinite-dimensional optimality result, we must verify tightness of $a_n(\hat{A}_{h0} - A_{h0})$ under local sequences.

VIII.4.4. Aalen's Linear Regression Model

Arguments on exactly the same lines as those in the preceding section (the Cox regression model) show that the *weighted least squares* estimators (7.4.1) and (7.4.11)–(7.4.13), are asymptotically efficient in the nonparametric linear regression model of Section VII.4, under suitable asymptotic stability conditions.

Recall that the model (in the univariate case) is specified by

$$\lambda_i(t) = \sum_{j=1}^p Y_{ij}(t)\beta_j(t)$$

for certain observable processes Y_{ij} and functions β_j whose integrals are to be estimated. We have by the way replaced p by $p - 1$ and reindexed.

Consider the local sequences obtained by replacing β_j by $\beta_j(1 + \frac{1}{2}a_n^{-1}\gamma_j)^2$ for functions γ_j; $a_n^{-2} = n$. On further replacing γ_j by $\theta\gamma_j$, $\theta \in \mathbb{R}$, we obtain a one-dimensional parametric model $\lambda_i(t;\theta)$ with

$$\frac{\partial}{\partial\theta} \log \lambda_i(t;0) = a_n^{-1} \frac{\sum_j Y_{ij}\gamma_j\beta_j}{\sum_j Y_{ij}\beta_j}$$

and, therefore, with score

$$a_n^{-1} \sum_i \int_0^\tau \frac{\sum_j Y_{ij}\gamma_j\beta_j}{\sum_j Y_{ij}\beta_j} \mathrm{d}M_i = a_n^{-1} \sum_i \int_0^\tau \sum_j \frac{Y_{ij}\beta_j\gamma_j}{\lambda_i} \mathrm{d}M_i.$$

We will take as $\mathbb{H}$ the space of functions $\boldsymbol{\gamma} = (\gamma_1, \ldots, \gamma_p)$ with norm as yet to be specified.

In matrix notation, the score can be written as

$$a_n^{-1} \int_0^\tau \boldsymbol{\gamma}^\mathsf{T}(\operatorname{diag} \boldsymbol{\beta})\mathbf{Y}^\mathsf{T}(\operatorname{diag} \boldsymbol{\lambda})^{-1}\, d\mathbf{M} \tag{8.4.24}$$

(diag of a vector is a diagonal matrix). The corresponding martingale has jumps of the order of a_n^{-1} and predictable variance process

$$a_n^{-2} \int \boldsymbol{\gamma}^\mathsf{T}(\operatorname{diag} \boldsymbol{\beta})(\mathbf{Y}^\mathsf{T}(\operatorname{diag} \boldsymbol{\lambda})^{-1}\mathbf{Y})(\operatorname{diag} \boldsymbol{\beta})\boldsymbol{\gamma}. \tag{8.4.25}$$

The in probability limit of this as $n \to \infty$, at $t = \tau$, will give us the norm (and inner product) in $\mathbb{H}$.

The weighted least squares estimator $\hat{\mathbf{B}}$ can be written, when rank $\mathbf{Y}(t) = p$ for all $t \in [0, \tau]$, as

$$\hat{\mathbf{B}} = \mathbf{B} + \int \mathbf{Y}^-\, d\mathbf{M}$$

with

$$\mathbf{Y}^- = (\mathbf{Y}^\mathsf{T}(\operatorname{diag} \hat{\boldsymbol{\lambda}})^{-1}\mathbf{Y})^{-1}\mathbf{Y}^\mathsf{T}(\operatorname{diag} \hat{\boldsymbol{\lambda}})^{-1},$$

where $\hat{\boldsymbol{\lambda}}$ is based on a preliminary estimate of $\boldsymbol{\beta}$. A suitable form of uniform consistency of this estimate will ensure the asymptotic equivalence of $a_n(\hat{\mathbf{B}} - \mathbf{B})$ and

$$a_n \int (\mathbf{Y}^\mathsf{T}(\operatorname{diag} \boldsymbol{\lambda})^{-1}\mathbf{Y})^{-1}\mathbf{Y}^\mathsf{T}(\operatorname{diag} \boldsymbol{\lambda})^{-1}\, d\mathbf{M}. \tag{8.4.26}$$

Suppose that $a_n^{-2}(\mathbf{Y}^\mathsf{T}(\operatorname{diag} \boldsymbol{\lambda})^{-1}\mathbf{Y}) \to \mathbf{V}$ uniformly on $[0, \tau]$, in probability, where $\mathbf{V}$ is a nonsingular $p \times p$ matrix-valued function. This is a modification of the second part of Assumption A in Theorem VII.4.1 [convergence of $a_n^{-2}\mathbf{Y}^\mathsf{T}\mathbf{Y}$, without the weight $(\operatorname{diag} \boldsymbol{\lambda})^{-1}$].

We may now expect, from (8.4.24) and (8.4.25), that the LAN property will hold with $\mathbb{H}$ the set of all $\boldsymbol{\gamma}$ such that

$$\|\boldsymbol{\gamma}\|_{\mathbb{H}} = \int_0^\tau \boldsymbol{\gamma}^\mathsf{T} \operatorname{diag} \boldsymbol{\beta} \mathbf{V} \operatorname{diag} \boldsymbol{\beta} \boldsymbol{\gamma}$$

is finite. C will be at least dense in $\mathbb{H}$, if not all of $\mathbb{H}$. The inner product can be deduced from the norm:

$$(\boldsymbol{\gamma}, \boldsymbol{\gamma}')_{\mathbb{H}} = \int_0^\tau \boldsymbol{\gamma}^\mathsf{T} \operatorname{diag} \boldsymbol{\beta} \mathbf{V} \operatorname{diag} \boldsymbol{\beta} \boldsymbol{\gamma}'.$$

For given j and t, the derivative of $B_j(t)$ as a function of $\boldsymbol{\gamma}$ (see Assumption VIII.3.2) is easily checked just as in Section VIII.4.1 to be

$$B'_{j,t}(\gamma) = \int_0^t \gamma_j(s)\beta_j(s)\,\mathrm{d}s$$

(a linear and hopefully also continuous mapping from $\mathbb{H}$ to $\mathbb{R}$, as we will check in a moment by exhibiting a representative of $\mathbb{H}$ whose inner product with γ gives the same mapping).

Now, from (8.4.26), we may expect that $a_n(\hat{\mathbf{B}}_j(t) - \mathbf{B}_j(t))$ is asymptotically equivalent to

$$a_n^{-1} \int_0^\tau (I_{[0,t]}\mathbf{e}_j^\top \mathbf{V}^{-1}(\operatorname{diag}\boldsymbol{\beta})^{-1}) \operatorname{diag}\boldsymbol{\beta}\mathbf{Y}^\top(\operatorname{diag}\boldsymbol{\lambda})^{-1}\,\mathrm{d}\mathbf{M},$$

where $\mathbf{e}_j$ is the jth unit vector.

This suggests trying

$$\gamma = I_{[0,t]}(\operatorname{diag}\boldsymbol{\beta})^{-1}\mathbf{V}^{-1}\mathbf{e}_j$$

as representative in $\mathbb{H}$ of $B'_{j,t}$. For any $\gamma \in \mathbb{H}$, we have

$$\begin{aligned}(I_{[0,t]}(\operatorname{diag}\boldsymbol{\beta})^{-1}\mathbf{V}^{-1}\mathbf{e}_j, \gamma)_{\mathbb{H}} &= \int_0^\tau I_{[0,t]}\mathbf{e}_j^\top\mathbf{V}^{-1}(\operatorname{diag}\boldsymbol{\beta})^{-1}\operatorname{diag}\boldsymbol{\beta}\mathbf{V}\operatorname{diag}\boldsymbol{\beta}\gamma\\ &= \int_0^t \mathbf{e}_j^\top(\operatorname{diag}\boldsymbol{\beta})\gamma\\ &= \int_0^t \beta_j\gamma_j\end{aligned}$$

as required. In particular, putting $\gamma = I_{[0,t]}(\operatorname{diag}\boldsymbol{\beta})^{-1}\mathbf{V}^{-1}\mathbf{e}_j$ gives

$$\|I_{[0,t]}(\operatorname{diag}\boldsymbol{\beta})^{-1}\mathbf{V}^{-1}\mathbf{e}_j\|_{\mathbb{H}}^2 = \int_0^t \beta_j\beta_j^{-1}(\mathbf{V}^{-1})_{jj} = \int_0^t (\mathbf{V}^{-1})_{jj}.$$

In general, we will find, for any finite linear combination of $B_j(t_i)$, as derivative the corresponding linear combination of efficient scores, and, hence, as the (i, j), (i', j') element of the optimal covariance matrix

$$\int_0^{t_i \wedge t_{i'}} (\mathbf{V}^{-1})_{jj'}.$$

Theorem VIII.3.4 and the Cramér–Wold device, therefore, give the efficiency of the finite-dimensional coordinate projections of $\hat{\mathbf{B}}$. When tightness has been verified under local sequences, this extends to the efficiency of $\hat{\mathbf{B}}$ itself (regularity follows from tightness and finite-dimensional regularity) by Theorem VIII.3.2. Finally, Theorem VIII.3.5 gives the efficiency of all compactly differentiable functionals of $\hat{\mathbf{B}}$. Real vector functionals can alternatively be handled by more direct calculations.

VIII.4.5. Nonparametric Maximum Likelihood. Partially Specified Models.

VIII.4.5.1. Optimality of the NPMLE

In Section IV.1.5, we derived the Nelson–Aalen estimator as the nonparametric maximum likelihood estimator. This is extended and generalized to the Kaplan–Meier estimator in Section IV.3, to the Aalen–Johansen estimator in IV.4, and to the usual estimators in the Cox regression model in Section VII.2. Now, in the preceding subsections, we have seen that these estimators are also asymptotically efficient. Is there a connection? Is this related to the efficiency properties of linear nonparametric tests and to their interpretation as score tests in Cox regression models with suitably chosen covariates?

The answer is: Yes. The bare bones of an explanation could run as follows. We saw toward the end of Section IV.1.5 that an NPMLE could be often thought of as a solution of a set of score equations, for a large class of parametric submodels within the nonparametric model of interest. Now in the parametric case, a solution of estimating equations is typically asymptotically equivalent to a linear combination of the same equations by a Taylor expansion argument (see Section VI.2 on M-estimators). Possibly by using an infinite-dimensional version of this method, one could hope to show that the same holds for the NPMLE: It is asymptotically equivalent to a linear combination of the score equations used to define it.

But this means that the estimator is asymptotically linear "with influence function in the tangent space" in the sense of Theorem VIII.3.3. Now, provided that the estimator is *regular* (which just means a small amount of uniformity and continuity in its approach to its limiting distribution), Theorem VIII.3.3 will give us *efficiency* too.

Consequently, we should not be surprised when an NPMLE turns out to be asymptotically efficient. Moreover, the fact that efficiency is connected to asymptotic linearity in score functions of parametric submodels goes a long way to justify the last discussed approach to the definition of an NPMLE, based on solving score equations, in Section IV.1.5. However, the argument so far does hang on a silken thread: To justify a first-order Taylor expansion in the first place, we must already assume *consistency* of the estimator.

Now even in the parametric case, consistency of the maximum likelihood estimator is a delicate matter, usually needing compactness assumptions or special structural properties (for instance, strict concavity of the log-likelihood, implying uniqueness of the solution of the likelihood equations, makes life a lot more pleasant). The fact that the expected score equals zero, implies "Fisher consistency" of the NPMLE (the population version of the likelihood equations has the right solution), but one needs a lot more (continuity of the solution under perturbations of the data) to get consistency. In the infinite-dimensional case, there are a lot of ways the NPMLE can fail to work and, indeed, examples abound of nice models for which the NPMLE is very badly behaved.

To summarize: There are no guarantees, but the easier it is to compute the NPMLE, the more likely it is (and the easier it will also be to prove) that it has all the nice properties one could want, including asymptotic efficiency. We, therefore, use the principle in the next chapter to propose new statistical methods in semiparametric frailty models and though little has been proved yet, so far all the signs are positive that these methods have excellent asymptotic properties.

Some first steps in making a formal theory of the above ideas were taken by Gill (1989, 1991b); see also Gill (1991a).

VIII.4.5.2. Partially Specified Models

So far, all our efficiency results are in models satisfying the property of *non-informative censoring*: Even if there are unknown parameters involved in censoring, covariates, other components of the basic counting process, the partial likelihood for the components of interest is equal to the full likelihood for the parameters of interest. There have been various attempts to generalize these efficiency results by dropping the noninformative censoring assumption and, therefore, to work in a set up with *partially specified models*: Only some aspects of the phenomenon of interest are statistically modelled, the rest is left in some sense free; see Section III.5 for some concrete examples.

One attempt due to Dzhaparidze (1985) and Dzhaparidze and Spreij (1990) is to restrict the class of allowed estimators. If the model specifies a martingale structure (by defining the compensator of a counting process), it makes sense to only consider estimators which are *asymptotically linear* in the same martingales; or in other words, to suppose that nice estimators are at least asymptotically equivalent to M-estimators.

All reasonable estimators which only use the information specified by the model should have this form. Now, a regularity assumption and standard Cauchy–Schwarz-based arguments (as in the Cramér–Rao inequality) leads to a characterization of "asymptotically best estimator," with our old favorites of this chapter topping the bill for their respective models.

Another approach due to Greenwood and Wefelmeyer (1989, 1990, 1991c) is not to restrict the class of estimators, but to extend the class of probability distributions in the model: One says that the model is precisely the class of *all* probability distributions allowing the counting process to have a compensator of such and such a form. This means an explicit introduction of a very highly dimensional nuisance parameter, and a critical comment is that it is unlikely the "user" really does mean "the rest of the model" to be *completely* arbitrary. However, now we are back in a standard situation dealt with in this chapter. We do have to check that this large model does include, as score functions of parametric submodels, the score functions coming from the *partial likelihood* based on the counting process of interest.

All works out nicely again, and our old favorites gain yet another optimality property.

VIII.5. Bibliographic Remarks

Section VIII.1

The theory of contiguity and local asymptotic normality was developed by LeCam (1960, 1969, 1986) over a number of papers and books. The convolution theorem is due to Hájek (1970). The famous three lemmas of LeCam were actually first so named by Hájek and Šidák (1967). The theory is nicely described in LeCam and Yang (1990) who also gave excellent historical and bibliographical discussions.

Asymptotic optimality in testing theory, since the fundamental contributions of Wald (1943), has been somewhat neglected in modern accounts of the theory though Hájek (1962) was especially interested in asymptotic optimality of nonparametric testing procedures. The treatment here is largely based on Choi (1989). We have emphasized Wald's (1943) notion of most stringent tests, which seems to us the only way to give the usual testing procedures of applied statistics convincing large sample optimality properties.

Section VIII.2

The general theorem on LAN for counting processes given here is due to Jacod and Shiryayev (1987, Theorem X.1.12), whose book contains many further results putting the present theorem in a much broader context. Results on LAN for various counting process models have been given by, among others, Hjort (1986) (general parametric models), Höpfner (1991) (Markov processes observed on an increasing time interval), Janssen (1989) (the simple random censorship model), and van Pul (1992a) (software reliability models). The asymptotic efficiency of an appropriate nonparametric linear counting process based test against certain parametric or semiparametric alternatives has been known in various degrees of generality for a long time. Already, Aalen (1975) showed that the two-sample log-rank test, in its one-sided form, was an asymptotically optimal *similar* test against one-sided proportional hazards alternatives, in a counting process setup. His result is, however, rather difficult to appreciate and has been later neglected. In our opinion, the optimality property he derives is not "asymptotically most powerful similar" but more strongly and simply "asymptotically most powerful" in the semiparametric model where the baseline hazard is unknown. He interpreted his result as a statement that the log-rank test is the best rank test; this is a correct interpretation as having unknown baseline hazard puts one in a transformation model (see Section VIII.4.2) for which by invariance considerations one would restrict attention to rank tests anyway.

Gill (1980a) showed optimality of the log-rank test as well as of other linear rank tests, in the two-sample case with asymptotically proportional risk sets, against the appropriate parametric alternatives. As we have seen, one needs to go semiparametric, or at least incorporate fairly arbitrary nuisance parameters, before a general optimality result (as Aalen's was) can be established. However, Gill did suggest (following Aalen) that it should be possible to show that such a test as the log-rank test was the asymptotically optimal rank test, whether or not one had proportional risk sets.

Earlier work on optimality of rank tests was done by Peto and Peto (1972) and Prentice (1978). These papers make the right proposals but do not give a rigorous optimality theory.

Tarone and Ware (1977) used an analogy with contingency table analysis to propose "best" rank tests getting a rather different answer, however. Incidentally, these authors also mentioned that the Dvoretsky (1972) martingale central limit theorem could be used to get the asymptotic results, though they avoided using the word martingale in the journal publication of their work. Schoenfeld (1980b, 1981) gave an informal approach to asymptotic optimality obtaining correct expressions for asymptotic efficacies against parametric alternatives and deriving the best tests of the class; unfortunately, much of this work was omitted in the final journal publication.

Other authors such as Pons (1981), Hjort (1984), Cuzick (1985), and, later, Janssen (1989, 1991) made further connections between the various classes of statistics and established further efficiency results. Their work only showed asymptotic efficiency in the case of asymptotically proportional risk sets. The present book seems to be the first place where Aalen's (1975) original idea—that the log-rank test was in complete generality the asymptotically efficient rank test against proportional hazards alternatives—has been borne out and extended to other semiparametric alternatives and to the k-sample case, though Sasieni (private communication) derived results similar to ours of Section VIII.4.2 on transformation models. Such models were also studied from the point of view of estimation theory in Klaassen (1988) and Bickel, Klaassen, Ritov, and Wellner (1993).

Hjort (1985a) discussed projected work (which unfortunately never materialized) by himself and Helgeland on asymptotic efficiency (in the sense of the convolution theorem) of the Nelson–Aalen estimator; and, indeed, Aalen (1977; unpublished manuscript) proved the result too, by a novel approach based on approximating finite-dimensional models of larger and larger dimension.

Asymptotic optimality of the Kaplan–Meier estimator (in the simple random censorship model) was first established by Wellner (1982). Miller (1983) discusses the *inefficiency* of the estimator when one could have assumed a parametric model, which is, of course, especially severe when estimating tail probabilities under heavy censoring. Van der Vaart (1988, 1991) proved optimality of the Kaplan–Meier estimator, making use of the special feature of the simple random censorship model (with censoring distribution unknown) that the distribution of the data is completely arbitrary so that the model is, in fact, completely nonparametric. This implies that the empirical distribution of the data is efficient as an estimator thereof, and by preservation of efficiency under smooth transformations, the Kaplan–Meier estimator is efficient for the underlying survival function. (This is the kind of approach we have described in Sections VIII.3 and VIII.4.) Keiding and Gill (1990) sketched the parallel arguments for the simple (i.i.d.) random truncation model. Weits (1991) shows second-order optimality of the Kaplan–Meier estimator, and also that if smoothness assumptions can be made, the Kaplan–Meier estimator can be improved to second order by smoothing. For small sample sizes, the possible improvement can be striking. Wefelmeyer (1991) showed efficiency of the Nelson–Aalen and Kaplan–Meier estimators by similar techniques to ours.

This volume is the first place where optimality of the Aalen–Johansen estimator is studied.

Section VIII.3

Infinite-dimensional optimality theory, or as one could say optimality in semiparametric models, has precursors in the work of Stein (1956), Levit (1973, 1975, 1978), and Koshevnik and Levit (1976). More recent contributors were Pfanzagl (1982), Ibragimov and Khas'minskii (1991), van der Vaart (1988), and Bickel, Klaassen, Ritov, and Wellner (1993), among others. Here, we have chosen van der Vaart's (1988) approach, reformulating his results and definitions for the i.i.d. case to the general LAN case, the LAN assumption being all that is needed to make the proofs work. Wefelmeyer (1991) and Greenwood and Wefelmeyer (1990, 1991a, 1991b) also gave a general theory in the context of semimartingale models, with many of the same examples as we have taken in Section VIII.4, as well as others [for instance, the nonparametric regression model of McKeague and Utikal (1990a) discussed in Section VII.5].

Van der Vaart (1988) gives a nice and unified treatment of the convolution and local asymptotic minimax theorems. His theory of preservation of efficiency under smooth transformations [see also van der Vaart (1991)] was developed at the request of one of us (Gill) and serves us very well here. He makes it clear that as far as deriving lower bounds is concerned, there really does not exist an "infinite-dimensional theory"—the finite-dimensional case is already rich enough and all the hard work has to be done there. Stronger forms of LAN involving uniformity over the parameter space are needed for showing that the lower bounds are, in general, sharp; see, e.g., Ibragimov and Khas'minskii (1991). Since we do not go into this side of things, we can make do with the weakest form of the LAN assumption.

The general theory of van der Vaart (1988, 1991) is available for parameters taking values in metrizable vector spaces, in particular, normed vector spaces, under some regularity conditions; our restriction allows us to avoid these technicalities.

Section VIII.4

We have already discussed the literature on the efficiency of the Nelson–Aalen estimator and of nonparametric k-sample tests. The Cox regression model was treated informally in different ways by Efron (1977) and Oakes (1977). In particular, the former showed optimality of the Cox partial likelihood estimators if the baseline hazard is arbitrary enough. More formal results were obtained by Begun, Hall, Huang, and Wellner (1983), Begun and Wellner (1983), Sasieni (1989), Klaassen (1989), Greenwood and Wefelmeyer (1990, 1991a), and Slud (1986).

Just as with the k-sample tests, there is a coincidental efficiency *without* introducing nuisance parameters at $\boldsymbol{\beta} = \mathbf{0}$, and this has led some authors to state that the Cox maximum partial likelihood estimators are only efficient at $\boldsymbol{\beta} = \mathbf{0}$. This is true when one considers the baseline hazard as known, but not true when it is unknown. The phenomenon is equivalent to the equally coincidental parametric efficiency of censored data linear rank tests at asymptotically proportional risk sets, which can be seen by formulating the k-sample problem as a (Cox) regression problem.

The optimality of the appropriate choice of generalized least squares estimator in Aalen's linear regression model was shown in Greenwood and Wefelmeyer (1990, 1991a).

CHAPTER IX

Frailty Models

IX.1. Introduction

In this chapter, we consider models in which the intensity process for the counting process depends partly on an unobservable random variable, supposed to act multiplicatively on the intensity, so that a large value of the variable increases the intensity throughout the whole time interval. If the counting process is registering failures or deaths, then this underlying variable can be thought of as a *frailty* or accident-proneness which increases the susceptibility to failure. Moreover, we may let several components of a multivariate counting process share the same frailty variable and in this way model positive statistical dependence between the individual counting processes. There exists a fairly large literature on frailty models in survival analysis (e.g., Vaupel et al., 1979; Clayton and Cuzick, 1985a; Hougaard, 1987; Oakes, 1989). Our presentation based on counting processes in this chapter is new but follows to a large extent Nielsen et al. (1992).

Let us first consider some examples of models obtained by introducing a frailty variable into some of the models of previous chapters. In the following examples, $\mathbf{N}$ is, in each case, a multivariate counting process with components N_i, N_{il}, or N_{ihl} as the case may be, where components with the same value of the first index i share the same frailty variable Z_i. Typically, i refers to family or litter, h to stratum (e.g., treatment group), and l to individuals within family and/or stratum. The intensity process of $\mathbf{N}$, denoted by $\boldsymbol{\lambda}$, has corresponding components λ_i, λ_{il}, or λ_{ihl}. There will be no necessity for the l or h, l indices to be balanced over the i's; for some i, some combinations may be missing altogether.

EXAMPLE IX.1.1. A k-Sample Frailty Model

Define the counting process

$$\mathbf{N} = (N_{ihl}: i = 1, \ldots, n; h = 1, \ldots, k; l = 1, \ldots, n_{ih})$$

with intensity process

$$\lambda_{ihl}(t) = Z_i Y_{ihl}(t)\alpha_h(t). \tag{9.1.1}$$

An example of this k-sample model with dependence between some of the individuals would be a study of the mortality in a number (n) of families ($i = 1, \ldots, n$) with $k = 4$ corresponding to mothers ($h = 1, n_{i1} = 1$), fathers ($h = 2, n_{i2} = 1$), daughters ($h = 3, n_{i3} = 0, 1, 2, \ldots$), and sons ($h = 4, n_{i4} = 0, 1, 2, \ldots$). Another example is a litter-matched study where individual (h, l) in litter (i) is treated with treatment h ($h = 1, \ldots, k$) and where individuals $l = 1, \ldots, n_{ih}$ from the same litter may be dependent. A further special case is then a study of n matched pairs ($i = 1, \ldots, n$) with $k = 2$ treatments and $n_{i1} = n_{i2} = 1$ member from each pair in each treatment group; see also Examples V.3.4 and VII.2.13. □

EXAMPLE IX.1.2. A Multiplicative Regression Model with Frailty

In the model (9.1.1) of the previous example, one may specialize to a proportional hazards model (for given values of the frailty variables) where

$$\lambda_{ihl}(t) = Z_i Y_{ihl}(t)e^{\beta_h}\alpha(t). \tag{9.1.2}$$

In the litter-matched example mentioned above, this corresponds to an assumption of the treatment effects being modelled by regression parameters $\beta_h, h = 1, \ldots, k$ (e.g., with $\beta_1 = 0$).

More generally, there may be observable and possibly time-dependent covariates $\mathbf{X}_{ihl}(t)$, $\exp(\boldsymbol{\beta}^\mathsf{T}\mathbf{x}_{ihl}(t))$ acting multiplicatively on baseline hazards $\alpha_{h0}(t)$:

$$\lambda_{ihl}(t) = Z_i Y_{ihl}(t)e^{\boldsymbol{\beta}^\mathsf{T}\mathbf{X}_{ihl}(t)}\alpha_{h0}(t). \tag{9.1.3}$$

A possible interpretation of a univariate version of (9.1.3) (i.e., $k = 1$, $n_1 = 1$) would be a Cox regression model where some time-fixed covariates (say $X_{1i}, \ldots, X_{mi}$ with regression coefficients $\gamma_1, \ldots, \gamma_m$) were left out or not observed and where

$$Z_i = \exp(\gamma_1 X_{1i} + \cdots + \gamma_m X_{mi}).$$

Alternatively, as in Example IX.1.1, (9.1.3) may be used to model dependence between individuals. □

All models considered in Examples IX.1.1 and IX.1.2 may be used to test goodness-of-fit of the standard models without frailty; we shall return to this in Examples IX.4.2 and IX.4.3.

In what follows, we study both semiparametric models—in which α_h (or

α_{h0} in Example IX.1.2), $h = 1, \ldots, k$, are completely unknown baseline hazards—and parametric models for α_h. As seen in the examples, there are some limitations to our approach. First, the frailty variable is supposed to act multiplicatively on the intensity process with a value which does not depend on time. Second, for each (litter or family) i, only one frailty variable Z_i is allowed though there may be several types of events ($h = 1, \ldots, k$) that an individual l may experience. Finally, the Z_i are everywhere modelled parametrically; in fact, we shall assume them to be gamma distributed with unknown (inverse) scale parameter η and unknown shape parameter ν. Obviously, this is not the only possibility. It has been shown in the regression context (Elbers and Ridder, 1982) that the model with arbitrary frailty distribution *with finite mean* is identifiable, so one could, in principle, allow both the frailty distribution and the underlying hazard to vary freely. Some authors have worked with parametric baseline hazard (e.g., piecewise constant) and nonparametric frailty distribution. Finally, many other parametric options than the convenient and conventional gamma are possible (Heckman and Singer, 1982, 1984a, 1984b; Hougaard, 1984, 1986a, 1986b, 1987; Vaupel, 1990a, 1990b; Aalen, 1990).

Our approach to these models will be through parametric and nonparametric maximum likelihood since this works so well for their "frailty-less" analogues and leads to a basically very straightforward analysis, generalizing the familiar analyses discussed in Chapter VI and Sections IV.1.5, VII.2.1 and VII.2.2 in a very direct way. However, so far it has not been possible to work out a large sample theory for this approach (nor for any competing general approach), but positive experience with both simulation experiments and practical data analyses exist (Nielsen et al., 1992). Recently, Murphy (1991) has constructed a nice consistency proof which can potentially be generalized.

IX.2. Model Construction

In the following, we will work with the simplest model (9.1.1) with $k = 1$ since all the main ideas for the other more complex models are already present here. We briefly return to these more complex models in Section IX.4. Moreover, we simplify further by replacing the double index il by the single one i (this simplification corresponds to aggregating over l), implicitly assuming that each N_i can have more than one jump. Thus, our initial model is: $\mathbf{N} = (N_i: i = 1, \ldots, n)$ is a multivariate counting process with intensity process $\boldsymbol{\lambda}$ satisfying

$$\lambda_i(t) = Z_i Y_i(t) \alpha(t) \tag{9.2.1}$$

for some observable predictable process Y_i, an unknown baseline hazard function α, and unobservable random variables Z_i, independently drawn from the gamma (ν, η) distribution. We consider maximum likelihood estimation of α or $A = \int \alpha$ and (v, η) and collect the Z_i, $(N_i(t))$ and $(Y_i(t))$ into vector random variables and processes, denoted by $\mathbf{Z}$, $\mathbf{N}$, and $\mathbf{Y}$.

It will turn out to be very fruitful to consider the statistical problem as an incomplete or missing data problem: The complete but unobservable data would consist of the just mentioned triple; the incomplete, actually available, data just consist of the last two components. The complete data problem is, of course, a very easy one: We can absorb the frailty variables into the random part $Y_i(t)$ of the intensity process, and the models reduce to the standard ones which the frailty models are intended to generalize. Inference about the parameters of the frailty distribution is also immediate and in a sense orthogonal to inference about the other parameters. So we will emphasize the interplay between the complete and incomplete data viewpoints, in particular complete and incomplete data likelihoods and intensities. The fact that the complete data problem is so simple makes the EM algorithm especially appropriate for solving the incomplete data problem at hand.

An important point is that when the parameters of the gamma distribution are unknown, an identifiability problem arises—we can divide and multiply Z_i and $\alpha(t)$ by the same constant, thereby staying in the gamma model, but with a new scale and the same shape parameter. Therefore, when we consider the parameters of the gamma distribution as unknown, we make the restriction $v = \eta$ and only refer to the second parameter η. This means that the gamma distribution is scaled to have mean $v/\eta = 1$; the variance of the Z_i is $\xi = v/\eta^2 = \eta^{-1}$.

A further very important point is that from this we can see that we also have an interesting model in the case $\eta \to \infty$; namely, the variance of the (unit mean) frailty variable tends to zero, so the model becomes the frailty-less starting point with all Z_i identically equal to one. We will often want to statistically test this submodel. At first hand, one would say that this puts us in a nonstandard problem with regard to the use of generalized likelihood ratio tests or score tests: We are testing a parameter value at the boundary of the parameter space. However, we will see later that when we parametrize by (to begin with, positive) $\xi = \eta^{-1}$, not only is the value $\xi = 0$ a legitimate parameter value, but also negative values close enough to zero are allowed. We cannot extend the joint distribution of $\mathbf{Z}$, $\mathbf{N}$, $\mathbf{Y}$ to allow negative ξ, so a frailty interpretation of the model is no longer possible [see, however, Oakes (1989), for a discussion of the classical survival data situation], but we can extend in a natural way the marginal distribution of the observables $\mathbf{N}$ and $\mathbf{Y}$. So the score test based on the usual asymptotic theory of the important hypothesis $\xi = 0$ ($\eta = \infty$) is valid, as well as both two-sided and one-sided generalized likelihood ratio tests. This phenomenon is analogous to the well-known situation with negative estimated variances in variance component models.

Now the specification (9.2.1) implies that the unobservable random vector $\mathbf{Z}$ is $\mathscr{G}_0$-measurable in the filtration $(\mathscr{G}_t)$ with respect to which the intensity of $\mathbf{N}$ is defined. When calculating likelihood functions, we have to consider intensities with respect to the smaller filtration generated by the data, for instance, the filtration $(\mathscr{F}_t) = (\sigma\{\mathbf{N}(s), \mathbf{Y}(s), s \leq t\})$; see Section III.2. To make the step from the larger to the smaller filtration easier (it will involve taking conditional expectations), we shall impose a product structure on the underlying complete data sample space, thereby slightly strengthening the assumption (9.2.1). Since $\mathbf{Z}$ is implicitly assumed to be $\mathscr{G}_0$-measurable, one would like to interpret (9.2.1) as stating that, conditional on $\mathbf{Z} = \mathbf{z}$, $\mathbf{N}$ has the intensity process $\boldsymbol{\lambda}$ satisfying

$$\lambda_i(t) = z_i Y_i(t)\alpha(t). \tag{9.2.2}$$

We need to make this statement precise, and it will be helpful to make it as strong as possible too. Without assuming any more structure on our sample space, we would become involved in a careful treatment of regular conditional distributions; to avoid this, we build the existence of such conditional distributions explicitly into the model, as follows. Let $(\Omega', \mathscr{G}', (\mathrm{P}'_\eta))$ be the natural probability space on which $\mathbf{Z}$ is defined: simply n-dimensional Euclidean space with its Borel σ-algebra and the family of probability measures corresponding to the possible (independent and identical) gamma distributions of the coordinates of $\mathbf{Z} = \boldsymbol{\omega}'$. Let $(\Omega'', (\mathscr{G}''_t), (\mathrm{P}''_{\alpha\mathbf{z}}))$ be a filtered probability space, with a family of probability measures and a multivariate counting process $\mathbf{N}$ defined on it such that, under the probability measure $\mathrm{P}''_{\alpha\mathbf{z}}$, $\mathbf{N}$ has intensity process $\boldsymbol{\lambda}$ satisfying (9.2.2) for some predictable process $\mathbf{Y}$. Let $\boldsymbol{\omega}'' = (\mathbf{N}, \mathbf{Y})$. Now form the product space $(\Omega, \mathscr{G}, (\mathrm{P}_{\eta\alpha}))$ endowed with the probability measures $\mathrm{P}_{\eta\alpha}$ obtained by taking the mixture over $\mathbf{z}$, according to P'_η, of $\mathrm{P}''_{\alpha\mathbf{z}}$. On the new probability space, $\mathbf{Z}$ and $(\mathbf{N}, \mathbf{Y})$ are both defined as the same functions of the respective coordinates of $\boldsymbol{\omega} = (\boldsymbol{\omega}', \boldsymbol{\omega}'')$ as on the old spaces. (This was why we omitted the prime and double prime when first introducing $\mathbf{Z}$, $\mathbf{N}$, $\mathbf{Y}$.) Moreover, with respect to the filtration $(\mathscr{G}_t)$ defined by taking $\mathscr{G}_t = \mathscr{G}' \otimes \mathscr{G}''_t$, $\mathbf{N}$ has intensity (9.2.1); while conditional on $\mathbf{Z}$, in other words conditional on the product of $\mathscr{G}'$ and the trivial σ-algebra on Ω'', $\mathbf{N}$ really does have intensity (9.2.2).

To these obvious parts of the product construction, we add a smaller filtration than $(\mathscr{G}_t)$, namely, $(\mathscr{F}_t)$ defined by taking $\mathscr{F}_t$ to be the product of the trivial σ-algebra on Ω' with $\mathscr{G}''_t$. One can, therefore, safely identify $\mathscr{F}_t$ with $\mathscr{G}''_t$, but it now has the interpretation as being the filtration generated by the observable part of the model. The larger filtration corresponds to adding to the data, the hypothetical observation of $\mathbf{Z}$ at time zero.

IX.3. Likelihoods and Intensities

We can now start preparing for the statistical analysis of the model, in which α will either be completely unknown or be parametrized by some Euclidean parameter $\boldsymbol{\theta}$. In Sections III.2 and III.4, we discussed that informative censoring could reduce the efficiency of the inference procedures in models without frailty. An important difference is that now it is *essential* to assume *noninformative censoring* for the mere *validity* of the inference, not just for its possible efficiency, and *with regard to the frailty variable* $\mathbf{z}$ in fact, not with respect to α. The reason for this is that we will need to integrate over $\mathbf{z}$ to discover the observable consequences of the model [(9.2.1) and (9.2.2) cannot be used directly for statistical inference since $\mathbf{Z}$ is unobservable]. Since, in the non-self-exciting case, specifying the intensity process of a counting process only specifies a partial likelihood, our model so far is only partially specified (Section III.5). If the nonspecified parts also involve $\mathbf{z}$, then we cannot tell what the effect of integrating $\mathbf{z}$ out of the *full* likelihood will be; see also the discussion of censoring depending on covariates in Example III.2.9.

So, we suppose to begin with that the filtration $(\mathscr{G}_t'')$ is of the type discussed in Sections III.2, III.4, and III.5, i.e., generated by a "large" marked point process with marks of innovative and noninnovative types, corresponding to jumps of $\mathbf{N}$ and to the other events (e.g., censorings) determining $\mathbf{Y}$, respectively. It is important that the large marked point process only carries marks which are actually observed, i.e., we have already made the reduction from an even larger marked point process including events which actually got censored. The fact that the conditional intensity of $\mathbf{N}$ is, indeed, given by (9.2.2) will generally mean that we have already made a hidden assumption of *independent censorship* (Section III.2.2):

Assumption IX.3.1. *Conditional on* $\mathbf{Z} = \mathbf{z}$, *censoring is independent.*

We then add to this the main assumption for the subsequent analysis (see also Section III.2.3):

Assumption IX.3.2. *Conditional on* $\mathbf{Z} = \mathbf{z}$, *censoring is noninformative of* $\mathbf{z}$.

Now by this assumption, the partial conditional likelihood based on $\mathbf{N}$ is

$$\mathrm{dP}''_{\mathbf{z}\alpha} = \prod_t \left\{ \prod_i (\lambda_{i\alpha\mathbf{z}}(t))^{\Delta N_i(t)} (1 - \lambda_{\cdot\alpha\mathbf{z}}(t)\,\mathrm{d}t)^{1-\Delta N_\cdot(t)} \right\}. \tag{9.3.1}$$

Here, $\lambda_{i\alpha\mathbf{z}}(t) = z_i Y_i(t)\alpha(t)$ and a dot, as usual, denotes summation over the relevant index, here over i. Considered as a function of $\mathbf{z}$, (9.3.1) is proportional to the conditional density of the data $(\mathbf{N}, \mathbf{Y})$ given $\mathbf{Z} = \mathbf{z}$. Hence, the conditional density of $\mathbf{Z}$ given the data, by Bayes's rule, is proportional to the "complete data (partial) likelihood" $L^{\mathscr{G}}(\alpha, \eta)$: the product of (9.3.1) and the marginal density of $\mathbf{Z}$, itself the product of the n components' gamma densities. Substituting the specification of $\lambda_{i\alpha\mathbf{z}}(t)$ and evaluating the product integral using (2.6.2), we find that this equals

$$L^{\mathscr{G}}(\alpha, \eta) = \prod_i \left\{ p(z_i; \nu, \eta) \prod_t (z_i Y_i(t)\,\mathrm{d}A(t))^{\Delta N_i(t)} \exp\left(-z_i \int_0^\tau Y_i(s)\,\mathrm{d}A(s) \right) \right\}, \tag{9.3.2}$$

where p is the gamma density

$$p(z; \nu, \eta) = \frac{z^{\nu-1}\eta^{\nu}e^{-\eta z}}{\Gamma(\nu)}, \tag{9.3.3}$$

τ denotes the end of the observation period (e.g., $\tau = \infty$) and A is the integrated baseline hazard $\int \alpha$. But now we see that, as a function of $\mathbf{z}$, (9.3.2) is proportional to

$$\prod_i \left\{ z_i^{\nu + N_i(\tau) - 1} \exp\left(-z_i\eta - z_i \int_0^\tau Y_i(t)\,\mathrm{d}A(t) \right) \right\};$$

in other words, conditional on the data, the Z_i are still independent and gamma distributed but now with parameters $\nu + N_i(\tau)$, $\eta + \int_0^\tau Y_i(t)\,\mathrm{d}A(t)$. Now, as we already stated, for given (ν, η), (9.3.2) is either a full or a partial likelihood for α based on observation of $\mathbf{Z}, \mathbf{N}, \mathbf{Y}$; a full likelihood in the case that *conditional on* $\mathbf{Z} = \mathbf{z}$, *censoring is noninformative on* α. If we integrate $\mathbf{z}$ out of this expression, we obtain correspondingly either the full likelihood for α based on observation of $\mathbf{N}$ and $\mathbf{Y}$ or else something rather more complicated which we will call a *marginal partial likelihood.* In fact, under the key Assumption IX.3.2, it does hold that this *marginal partial likelihood* equals the desired *partial marginal likelihood*, i.e., the partial likelihood for α based on $\mathbf{N}$ in the "marginal" or "incomplete data experiment" when $\mathbf{Z}$ is not observed (i.e., the experiment of interest). It was proved by Gill (1992) that "the marginal partial likelihood equals the partial marginal likelihood under noninformative censoring for the variable being margined out." We will, nevertheless, also sketch a direct verification of the equality since doing so provides us with further valuable insight into our model. Before this however, we also mention the analogous statements for the frailty parameter η. Again, if *conditional on* $\mathbf{Z} = \mathbf{z}$, *censoring is noninformative for the frailty parameter*, then (9.3.2) is the full likelihood for this parameter based on the "complete data" $\mathbf{Z}, \mathbf{N}, \mathbf{Y}$; and, on integrating out $\mathbf{z}$, we get the full likelihood based on the actually available "incomplete data" $\mathbf{N}, \mathbf{Y}$. Otherwise, (9.3.2) is a partial likelihood and we claim by the same theorem that the integration provides a partial (marginal) likelihood, just as for the survival parameter.

Now, as promised, we give a direct verification of the marginal/partial interchange. Using the fact

$$\int_0^\infty z^{\kappa-1}e^{-\lambda z}\,\mathrm{d}z = \frac{\Gamma(\kappa)}{\lambda^\kappa}, \tag{9.3.4}$$

we find by direct integration over $\mathbf{z}$ in (9.3.2) that the "marginal partial" likelihood equals

$$L^{\mathscr{F}}(\alpha, \eta) = \prod_i \left\{ \frac{\eta^\nu}{\Gamma(\nu)} \frac{\Gamma(\nu + N_i(\tau))}{(\eta + \int_0^\tau Y_i(t)\,\mathrm{d}A(t))^{\nu + N_i(\tau)}} \prod_t (Y_i(t)\,\mathrm{d}A(t))^{\Delta N_i(t)} \right\}. \tag{9.3.5}$$

To compute the "partial marginal" likelihood, we must first compute the intensity of $\mathbf{N}$ with respect to the smaller filtration $(\mathscr{F}_t)$. Now, by exactly the same calculation as that which led us from (9.3.1) to (9.3.2), if we start with the data available at time $s-$ instead of at the time τ, the product over all t in (9.3.1) is replaced by a product over $t < s$, and, in (9.3.2), the integral is taken from 0 to $s-$ instead of from 0 to τ. Thus, we find that conditional on the data from the time interval $[0, s)$, i.e., conditional on $\mathscr{F}_{s-}$, the Z_i are independent and gamma $(\nu + N_i(s-), \eta + \int_0^{s-} Y_i(t)\,\mathrm{d}A(t))$ distributed.

The expectation of a gamma (κ, λ) variable is κ/λ. So conditional on $\mathscr{F}_{s-}$, the expectation of Z_i is $(\nu + N_i(s-))/(\eta + \int_0^{s-} Y_i(t)\,\mathrm{d}A(t))$. Therefore, by the innovation theorem (Section II.4.2), the $(\mathscr{F}_t)$-compensator of $\mathbf{N}$ is $\boldsymbol{\lambda}^{\mathscr{F}}$ with components

$$\lambda_{i\eta\alpha}^{\mathscr{F}}(t) = \frac{\nu + N_i(t-)}{\eta + \int_0^{t-} Y_i(s)\,\mathrm{d}A(s)} Y_i(t)\alpha(t). \tag{9.3.6}$$

Now with this specification one may show, following Section II.7, that the partial likelihood for η and α with respect to the observed data filtration $(\mathscr{F}_t)$ is, indeed, given by (9.3.5).

A further important point is that our whole frailty construction, as model for the observable data, is operationally identical with the specification (9.3.6) of the intensity of $\mathbf{N}$ with respect to the observed data filtration. We could have started with (9.3.6) as model in its own right, and only for the purposes of motivation derived its formal equivalence with a frailty model later. The statistical analyses we present also are really analyses of the model (9.3.6) which use this formal identity for computational purposes only. This point makes the following observation meaningful. In (9.3.6), divide the numerator and denominator of the first term on the right-hand side by $\eta = \nu$ and replace the resulting occurrences of η^{-1} by the equivalent parameter ξ. We see that the model now makes sense as an intensity model also in the case $\xi = 0$ and even in the case $\xi < 0$ ($\eta > +\infty$!), provided this first term stays non-negative, which is the case if ξ is larger than the maximum of the $2n$ negative quantities $-(N_i(\tau))^{-1}$ and $-(\int_0^\tau Y_i(s)\,\mathrm{d}A(s))^{-1}$. This lower bound is random and, in particular, could become arbitrarily close to zero as $n \to \infty$. However if the model is such that N_i and Y_i are bounded and τ is such that $A(\tau)$ is finite, a strictly negative lower bound can be given uniformly in n. In an actual data analysis, this suggests artificially censoring $\mathbf{N}$ at a suitable finite τ smaller than the real upper time limit of observation; we sacrifice some possible asymptotic efficiency but probably improve the approximation suggested by the asymptotic theory to the statistical behavior of the resulting estimators. That this is, indeed, the case was demonstrated in the simulations conducted by Nielsen et al. (1992) and we shall return to it in Example IX.4.1.

IX.4. Parametric and Nonparametric Maximum Likelihood Estimation with the EM-Algorithm

With these preparations done, we can finally turn to a discussion of the computation of parametric and nonparametric maximum likelihood estimators.

IX.4.1. Parametric Models

In the parametric case, α depends on some Euclidean parameter, $\boldsymbol{\theta}$ say. One possibility is then to make direct use of the observed data (partial) likelihood (9.3.5) [see, e.g., Aalen (1987a, 1987b, 1988b, 1990)]. This expression is not so complicated and a direct numerical maximization of it over $\boldsymbol{\theta}$ and η is often quite straightforward. This situation is within the framework of Section VI.1 with the intensity process given by (9.3.6). In particular, conditions for the usual large sample properties of the maximum likelihood estimators $\hat{\boldsymbol{\theta}}$ and $\hat{\eta}$ and the corresponding likelihood-based test statistics are given there. Also, by our earlier discussion of the legitimacy of the model extended past $\xi = 0$, the usual likelihood-based tests of $\xi = 0$ should, indeed, be asymptotically valid.

However, another natural possibility (Gill, 1985) is to use the EM-algorithm (Dempster, Laird and Rubin, 1977); this is especially attractive if the maximum likelihood estimator with complete data (observation of $\mathbf{Z}$) is easy to calculate. Let us first consider the case when the parameters of the frailty distribution are fixed. Even though they will generally be fixed at the same value, we keep the different symbols for the inverse scale and shape parameters, as this makes the formulas easier to understand. Note that in (9.3.2), if we take logarithms and consider the resulting expression as a function of $\alpha_{\boldsymbol{\theta}}$ to be maximized over $\boldsymbol{\theta}$, $\mathbf{Z}$ only appears linearly, in fact in the term $-Z_i \int_0^\tau Y_i(t)\,\mathrm{d}A_{\boldsymbol{\theta}}(t)$. So the E-step of the EM-algorithm consists of replacing Z_i in this part of the complete data log-likelihood by its conditional expectation (under current parameter values) given the data $\mathbf{N}$, $\mathbf{Y}$; this is

$$\text{E-step:} \qquad \hat{Z}_i = \frac{\nu + N_i(\tau)}{\eta + \int_0^\tau Y_i(t)\,\mathrm{d}A_{\boldsymbol{\theta}}(t)}. \tag{9.4.1}$$

The M-step is then: Compute $\hat{\boldsymbol{\theta}}$, the maximum likelihood estimator for $\boldsymbol{\theta}$, based on the multiplicative intensity model for $\mathbf{N}$ with intensity process given by

$$\text{M-step:} \qquad \hat{\lambda}_{i\boldsymbol{\theta}}(t)\,\mathrm{d}t = (\hat{Z}_i Y_i(t))\alpha_{\boldsymbol{\theta}}(t)\,\mathrm{d}t; \tag{9.4.2}$$

cf. (9.2.1).

By the general theory of the EM-algorithm, if this algorithm converges, it converges to a stationary point of $\log L(\boldsymbol{\theta}, \eta)$, i.e., to a solution of

$$\frac{\partial}{\partial \boldsymbol{\theta}} \log L(\boldsymbol{\theta}, \eta) = \mathbf{0}.$$

Under further conditions, it actually maximizes the log-likelihood. Note that it is irrelevant to the EM-algorithm, seen as a computational method for finding a stationary point of $\log L$, whether we are dealing with a proper or only a partial likelihood; what makes the EM-algorithm work is the formal stucture relating the complete and incomplete data likelihoods (integration over $\mathbf{z}$) which is preserved here by the marginal/partial result. Also note that in the frailty-less model ($\eta = +\infty$), the E-step (9.4.1) always gives $\hat{Z}_i = 1$ and the M-step (9.4.2), then corresponds to ordinary maximum likelihood estimation.

To estimate $\eta = \xi^{-1}$, we suggest running the EM algorithm to get an estimate of $\boldsymbol{\theta}$ for given η at each of a series of values of η. Then we plot the log incomplete data profile likelihood (maximized in $\boldsymbol{\theta}$) as a function of η (better still, as a function of $\xi = \eta^{-1}$) and numerically or graphically choose the global MLE $\hat{\eta}$, after which $\hat{\boldsymbol{\theta}}$ is explicitly calculated at this value if necessary. Standard errors and generalized likelihood ratio test statistics can now also be calculated directly from (9.3.5) and as discussed earlier the usual likelihood-based tests of $\xi = 0$ should be asymptotically valid.

IX.4.2. Nonparametric Models

The estimation of the nonparametric form of the model in which α is completely unknown is just as simple. Following the approach of Section IV.1.5, we take the likelihood function (9.3.5), obtained by integrating $\mathbf{z}$ out of (9.3.2), as the form of the likelihood function for the extended model (or rather, the extended parameter space) in which the cumulative baseline hazard measure A is now not necessarily absolutely continuous. The maximum likelihood estimator for A will be discrete with jumps at the jump times of $\mathbf{N}$ only. So, its computation comes down to maximizing (9.3.5) over such A, replacing $\mathrm{d}A(t)$ by the jump of A at that time point.

We still have the same relation between the corresponding versions of (9.3.2) and (9.3.5): The latter is obtained from the former by integrating out $\mathbf{z}$, and, as a function of $\mathbf{z}$, (9.3.2) is still proportional to (9.3.3), a product of gamma densities with the parameters described earlier. So we can still use the EM-algorithm to maximize (9.3.5): The E-step is unchanged as in (9.4.1) and the M-step corresponds to the generalization of (9.4.2) obtained on replacing $\alpha(t)\,\mathrm{d}t$ by $\mathrm{d}A(t)$ and $\lambda_i(t)\,\mathrm{d}t$ by $\mathrm{d}\Lambda_i(t)$. Therefore, the M-step is to calculate the Nelson–Aalen estimator as if $\mathbf{Z}$ had been observed (and was equal to $\hat{\mathbf{Z}}$):

$$\text{M-step:} \qquad \hat{A}(t) = \int_0^t \frac{\mathrm{d}N_{\cdot}(s)}{\sum_i \hat{Z}_i Y_i(s)}. \tag{9.4.3}$$

To estimate $\eta = \xi^{-1}$, we propose taking the partially maximized discrete analogue of (9.3.5) as a profile likelihood for ξ.

We conjecture that the resulting estimator of A has properties similar to the Nelson–Aalen estimator in a frailty-less model; at least, under similar regularity conditions and perhaps on censoring the data at a time τ satisfying $A(\tau) < \infty$ [otherwise, we can get arbitrarily large values of the denominator of (9.4.1), even if Y_i is uniformly bounded]. That this is the case is strongly suggested by the simulation results of Nielsen et al. (1992). However, so far, no complete proof has been constructed of this though Murphy (1991) has some promising initial results based on methods from sieve estimation. Also, we conjecture *efficiency* of the estimators though the hardest parametric submodel (see Section VIII.2) does not have an explicit solution in this case.

Several authors have constructed very similar estimators (Clayton and Cuzick, 1985a; Self and Prentice, 1986), in the former case using completely different principles, but only ad hoc approximations and heuristics are available to justify them.

By an ordinary delta-method calculation, one can, in principle, hope to estimate the asymptotic variance of the estimator $\hat{A}$ using the inverse of the matrix of second derivatives of the logarithm of the discrete version of (9.3.5) taking as parameters $\Delta A(t)$; cf. Section IV.1.5. However, more work needs to be done to find less computationally demanding procedures.

Finally, we remark on other versions of the model, in particular those mentioned in Examples IX.1.1 and IX.1.2 as well as on the identifiability of the various models. The model (9.1.1) is almost as easily dealt with as (9.2.1); we have much of the same structure and the only difference in the M-step is that not one but several cumulative baseline hazards have to be estimated, simply with Nelson–Aalen estimators with the unobservable Z_i replaced by their current predictions. In the numerator and denominator of the E-step (9.4.1), we sum over the strata (h) in obvious fashion.

The proportional hazards models (9.1.2) and (9.1.3) can also be dealt with by noting that, if the frailty $\mathbf{Z}$ were observed, then the usual Cox maximum partial likelihood estimator of the regression coefficient based on (7.2.7) together with the estimator (7.2.5) of the baseline hazard are the nonparametric maximum likelihood estimators when extending the likelihood functions (9.3.2) and (9.3.5) to allow discrete A in exactly the same way as we have done here (Sections VII.2.1 and VII.2.2). The log-likelihood for baseline hazard and regression coefficient still involves the Z_i multiplicatively so the M-step is simply a Cox analysis with Z_i replaced by its current prediction. In the E-step, each $Y_i(t)$ is simply multiplied by the corresponding multiplicative factor depending on the regression coefficient and the observable covariates. As above, variance estimates may, in principle, be obtained from the matrix of second-order derivatives of minus the logarithm of (9.3.5), taking as parameters the frailty parameter η, the regression coefficients $\boldsymbol{\beta}$, and the jumps $\Delta A_0(t)$ in the cumulative baseline hazard. However, because of the computational complexity of this approach, we have not pursued this possibility any further in the following examples. Instead, where possible, we quote likelihood ratio test statistics to indicate the uncertainty of the parameter estimates.

The model (9.2.1) is unidentifiable if each N_i makes at most one jump; whatever gamma distribution is used leads to the same value of the incomplete data log-likelihood. In general, we either need covariates or replicates for meaningful estimation in frailty models.

IX.4.3. Examples

Let us finally consider some practical examples of frailty models returning to the models studied in Examples IX.1.1 and IX.1.2. In the examples, we apply the large sample results conjectured above; ξ was estimated by maximizing the profile likelihood using a simple numerical procedure and the standard deviation of $\hat{\xi}$ in Example IX.4.1 was estimated from the curvature of a parabola approximating the log profile likelihood in a neighborhood of $\hat{\xi}$. The null hypothesis H_0: $\xi = 0$ of no frailty was tested using the (two-sided) likelihood ratio test statistic $-2\log Q$ based on the log profile likelihood.

EXAMPLE IX.4.1. Heritability of Life Length. Association Between the Life Times of Adoptees and Their Parents

The data were introduced in Example I.3.15. Our main interest focuses on the correlation between the life time of the adoptee (AD) and that of the adoptive mother (AM), the adoptive father (AF), the biological mother (BM), or the biological father (BF). We consider nonparametric models of the form (9.1.1) with $k = 3$ strata corresponding to male adoptees ($h = 1$), female adoptees ($h = 2$), and the relevant parent ($h = 3$), i.e., n_{i1} and n_{i2} are 0 or 1 depending on the sex of the adoptee, whereas $n_{i3} = 1$.

The results are shown in Table IX.1. It is seen that we everywhere have weak and insignificant associations between the life times with small values of $\hat{\xi}$. If all life times are censored at $t = 70$ years, however, the estimates of ξ get considerably larger. This artificial censoring as discussed above is moti-

Table IX.1. Estimated Frailty Parameters $\hat{\xi}$ with 95% Confidence Intervals and Associated Likelihood Ratio Test Statistics $-2\log Q$

Relation	$\hat{\xi}$	Confidence Interval	$-2\log Q$	P
		Lifetimes Unrestricted		
AD–BF	0.01	(−0.16, 0.18)	0.01	0.94
AD–BM	0.11	(−0.09, 0.30)	1.22	0.27
AD–AF	0.09	(−0.07, 0.26)	1.31	0.25
AD–AM	−0.03	(−0.18, 0.11)	0.20	0.65
		Lifetimes Censored at $t = 70$		
AD–BF	0.14	(−0.12, 0.40)	1.19	0.27
AD–BM	0.37	(0.05, 0.69)	6.46	0.01
AD–AF	0.17	(−0.12, 0.45)	1.50	0.22
AD–AM	−0.24	(−0.46, −0.01)	3.57	0.06

Note: AD: adoptee; BF: biological father; BM: biological mother; AF: adoptive father; AM: adoptive mother.

vated both from the simulation results of Nielsen et al. (1992) and from the fact that the study by Sørensen et al. (1988) concerned *early mortality*. We now see a significant positive correlation ($P = 0.01$) between the life time of the adoptee and that of his or her biological mother. For adoptive mothers, the estimate is negative in which case the model does not possess a frailty interpretation; see Section IX.1. □

EXAMPLE IX.4.2. Survival in Matched Pairs. A Frailty Model for Remission Lengths in Leukemia

The data were introduced in Example I.3.6 and analyzed in a matched-pairs model in Example VII.2.13. In the present example, we consider a frailty model of the form (9.1.2) with $n = 21$ pairs ($i = 1, \ldots, 21, h = 1, 2, n_{ih} \equiv 1$) and proportional hazards between patients treated with placebo and patients treated with 6-MP. Here the interest does not concern the frailty parameter ξ but rather the estimation of the treatment effect β or $\exp(\beta)$ in the presence of a possible within-pair correlation. The estimates become $\hat{\xi} = -0.21$ and $\exp(\hat{\beta}) = 0.26$. The frailty parameter is not significantly different from 0 (likelihood ratio test statistic $-2 \log Q = 0.83$ corresponding to $P = 0.36$ when referred to the χ_1^2 distribution). Under H_0: $\xi = 0$, the calculations, as explained above, reduce to those of Example VII.2.13, giving $\exp(\hat{\beta}) = 0.22$ with an approximate 95% confidence interval (0.10, 0.49). Thus, the association within pairs seems to be very small and the treatment effect is the same. □

EXAMPLE IX.4.3. A Frailty Model for Survival with Malignant Melanoma

In this last example we once more consider the melanoma data introduced in Example I.3.1.

The purpose of this example is to use frailty models to study goodness-of-fit of some of the Cox regression models analyzed in Sections VII.2 and VII.3. The results are shown in Table IX.2. The "final" model obtained in Example VII.3.4 was stratified by ulceration and included log (tumor thickness) as the only covariate with $\hat{\beta} = 0.589$ [$\widehat{\text{s.d.}}(\hat{\beta}) = 0.175$]. If we add a frailty variable to that model we obtain a model of the form (9.1.3) with $k = 2$ strata (ulceration versus no ulceration) and including the time-fixed covariate log (tumor thickness). The estimated regression coefficient for this covariate becomes $\hat{\beta} = 0.957$ somewhat larger than in the frailty-less model. The estimate of the frailty parameter is $\hat{\xi} = 1.780$, but the likelihood ratio test statistic for H_0: $\xi = 0$ takes the insignificant value 1.50 ($P = 0.22$ when referred to the χ_1^2 distribution). This shows that with a frailty model as an alternative we cannot reject the "final" model for the melanoma data.

Next, as an illustration we instead consider the model where ulceration enters as a regression variable rather than as a stratification variable. The estimated effects are shown in Table IX.2 (unstratified models). In this case, an added frailty variable is highly significant: $\hat{\xi} = 4.22$ with a likelihood ratio

Table IX.2. Frailty Models for the Melanoma Data

	Models Stratified by Ulceration			
	No Frailty		Frailty	
Covariate	$\hat{\beta}$	$-2\log Q$	$\hat{\beta}$	$-2\log Q$
log(thickness)	0.589	11.53	0.957	(*)
Frailty ($\hat{\xi}$)			1.780	1.50
$-2\max\log L$	575.25		573.75	
	Unstratified Models			
	No Frailty		Frailty	
Covariate	$\hat{\beta}$	$-2\log Q$	$\hat{\beta}$	$-2\log Q$
log(thickness)	0.610	12.24	1.370	16.89
Ulceration	0.971	10.12	1.690	8.55
Frailty ($\hat{\xi}$)			4.22	7.20
$-2\max\log L$	639.72		632.52	

Note: (*): not meaningful because the frailty model without covariates is not identifiable.

test statistic of 7.20. Recalling from Examples VII.3.1 and VII.3.4 that the ratio between the hazard functions for patients with and patients without ulceration *decreased* with time, we see that the frailty model is able to catch this particular kind of deviation from proportionality.

It should be emphasized, however, that an insignificant frailty variable added to a Cox regression model cannot be interpreted as an overall acceptance criterion for the model. To illustrate this, consider the model with sex as the only covariate included. Here, $\hat{\beta} = 0.662$ [$\widehat{\text{s.d.}}(\hat{\beta}) = 0.265$]; cf. Example VII.2.5. In this model, the estimated parameter for an added frailty variable is $\hat{\xi} = 1.48$, whereas the estimated sex effect becomes $\hat{\beta} = 0.892$. The likelihood ratio test statistic for the frailty parameter is $-2\log Q = 674.25 - 674.08 = 0.17$. This is because the proportional hazards when only sex is included fits the data perfectly well (Examples VII.3.1, VII.3.4, and VII.3.5) although important covariates like tumor thickness and ulceration are left out. □

IX.5. Bibliographic Remarks

Gamma mixture models have quite a long tradition in actuarial mathematics, see, e.g., Norberg (1990). An early contribution to the literature about random effects models for *survival data* was the paper by Vaupel, Manton, and Stallard (1979). They discussed various consequences of using the gamma distribution to describe unobserved heterogeneity between individuals and introduced the term "frailty" for the unobserved covariate. Later, both a demographic literature [e.g., Manton et al. (1981, 1986); Vaupel and Yashin (1985a, 1985b); Vaupel (1988)], an econometric literature [e.g., Lancaster and Nickell (1980); Heckman and Singer (1982, 1984a, 1984b)], and a statistical literature has developed. In the latter, there are obvious links to the discussion of misspecified regression models (see the Bibliographic Remarks to Chapter VII) and multivariate survival time distributions (see Section X.3 and below). Also related is the discussion by Aalen (1988a) of a Markov process model with random time change.

In a series of papers, Hougaard (1984, 1986a, 1986b, 1987) discussed (among other things) the impact of using different frailty distributions. One of his major points was to emphasize that in a regression context with dependence among some individuals, a frailty distribution with a finite mean is, in principle, identifiable from the margins alone as shown by Elbers and Ridder (1982). Therefore, the frailty parameter measures more than dependence. He then argued that, for instance, the *positive stable distributions* might be a better choice—a family of distributions which in addition to having an infinite mean preserves proportional hazards. Hougaard et al. (1992) compared results from analyses of the correlation between the life times of Danish twins using different frailty distributions. Other frailty distributions have been discussed by Heckman and Singer (1982, 1984a, 1984b) and Aalen (1990), whereas Clayton and Cuzick (1985a, 1985b), Self and Prentice (1986), and Klein (1992) used the gamma distribution as we have done above. Finally, a Bayesian approach using *Gibbs sampling* has been applied to the gamma frailty model by Clayton (1991).

For bivariate survival data, the frailty approach gives an explicit formulation of the joint survival function for the two margins [e.g., Clayton, (1978); Oakes (1982b); Clayton and Cuzick (1985a); Oakes (1986a, 1986b, 1989)]. Other approaches to the analysis of bivariate survival data focus on *tests for independence* [e.g., Weier and Basu (1980); Cuzick (1982); Oakes (1982a); Pons (1986); Dabrowska (1986); Pons and Turckheim (1991)], estimation of the *marginal distributions* in the presence of a possible, unspecified dependence [e.g., Huster et al. (1989); Wei et al. (1989); Liang et al. (1990)] or nonparametric estimation of the *joint distribution.* We return to the latter topic in Section X.3.

CHAPTER X

Multivariate Time Scales

A recurrent theme of this volume has been the importance of keeping detailed track of the various time dimensions relevant to each specific problem. So far, either our problems have had only one relevant time dimension or we have focused on one time dimension as "basic" or "underlying," regarding others as (methodologically) secondary.

In this chapter, we first (in Section X.1) briefly survey how we have handled *several time scales* so far in several of the empirical examples. We also add comments on the range of our methodology in analyzing the Lexis diagram and semi-Markov processes, and we introduce the idea of Arjas (1985, 1986) of always letting calendar time ("real time") be the basic time variable.

In Section X.2, we study sequential analysis of clinical trials with *staggered entry* where the interest is on modelling intensities as a function of duration on study, but the sequential analysis is to be performed in calendar time. Section X.3 discusses the generalization of the Kaplan–Meier estimator to nonparametric estimation of a *multivariate distribution function* under censoring, together with the concepts of multivariate hazard and odds measures and their estimates.

In this final chapter, our main tools, martingales and stochastic integrals, do not always suffice for the analysis of the models, and alternative approaches are often necessary. Asymptotic normality will often be provided by empirical process central limit theory rather than martingale central limit theory. In Section X.3, a prominent methodological role is played by the multivariate product-integral.

X.1. Examples of Several Time Scales

EXAMPLE X.1.1. Survival with Malignant Melanoma

The problem and the data were presented in Example I.3.1. In Examples IV.1.11 and VII.4.3, mortality from all causes was studied *either* as function of time since operation ("duration") *or* as function of age of the patient. In Example IV.1.11, it was concluded that a reasonable model for mortality from all causes at duration t since operation is to write the hazard as $\gamma_0 + \mu_i(t)$, where $\mu_i(t)$ is the mortality rate for a person in the general population of the same sex as patient i, evaluated at the age that patient i has at duration t. □

EXAMPLE X.1.2. Survival of Diabetics in the County of Fyn

The problem and the data were presented in Example I.3.2. In Example VII.2.14, we summarized analyses of absolute and relative mortality, using age as the basic time variable, calendar time and diabetes duration as covariates in the regression analysis. The somewhat subtle distinction between using age at disease onset as time-fixed covariate and diabetes duration as time-dependent covariate was discussed. □

EXAMPLE X.1.3. Survival with Liver Cirrhosis

The problem and the data were presented in Example I.3.4. Example VII.2.15 summarized the analysis of a clinical trial of treatments of liver cirrhosis. The basic time was duration on trial, while age was also of interest; this was entered as a covariate in the regression analysis. It was noted that if age enters linearly into the regression, the same model results whether age at entry or current age is used. □

EXAMPLE X.1.4. Psychiatric Admissions for Women Who Have Just Given Birth

The problem and the data were presented in Example I.3.10. The analysis is summarized in Example VII.2.16. The basic time variable is time since birth, but it is discussed at some length that duration since last discharge from the psychiatric ward is also important. In the particular example, there are problems with the complete observation of the duration. A third time variable, the age of the woman, also plays a role, although this is regarded as constant over the brief time interval (1 year) here considered.

The particular framework of semi-Markov processes for modelling duration dependence is discussed further in Example X.1.7. □

Example X.1.5. Nephropathy and Mortality among Insulin-Dependent Diabetics

The problem and the data were presented in Example I.3.11. In this example, the transitions between the three states, 0: alive with diabetes but without diabetic nephropathy (DN), 1: alive with DN, and 2: dead, are studied; see Example VII.2.11. The relevant time scales are age, calendar time, duration of diabetes, and (for the $1 \to 2$ transition) duration of DN, the latter two entering into the model through inclusion of the three covariates calendar time, age at onset of diabetes, and age at onset of DN. The incidence of DN ($0 \to 1$), the relative mortality before nephropathy ($0 \to 2$), and the absolute mortality for patients with nephropathy ($1 \to 2$) depended on these covariates as shown in Table VII.2.1. □

It is a common feature of these examples that, for each individual, the time scales run simultaneously: An increase in calendar time equals an increase in age equals an increase in duration, etc. This is the reason that we may often handle the estimation problem by introducing time-fixed covariates, as mentioned.

A general class of situations of this type is covered by the following example.

Example X.1.6. The Lexis Diagram

The Lexis diagram is a (time, age)-coordinate system, representing individual lives by line segments of unit slope, joining (time, age) of the birth and death; cf. Figure X.1.1.

Although calendar time and age certainly run simultaneously, with one-dimensional increments of slope 1 in the diagram, interest often centers on the death intensity $\mu(t, a)$ considered as function of the two independently varying arguments t (for time) and a (for age). Keiding (1990) surveyed possible approaches for non- and semiparametric inference on $\mu(t, a)$.

Most approaches use one time variable (usually most fruitfully age) as underlying, entering calendar time as a covariate. This may be done in a proportional hazards semiparametric (Cox) model or in a less restrictive regression model such as those surveyed in Section VII.5. Possibilities are the penalized likelihood or local likelihood modifications of the Cox model or the kernel-smoother approach of Beran (1981), Dabrowska (1987), and McKeague and Utikal (1990a). As noted by Keiding (1990), the latter essentially involves a bivariate smoothing of the point pattern of deaths in the Lexis diagram, each point being weighted inversely to (a smoothed version of) the density of persons at risk.

The proportional hazard models and their generalizations separate the effects of time and age in a similar fashion as the (age-period, age-cohort, age-period-cohort) log-linear intensity models with piecewise constant inten-

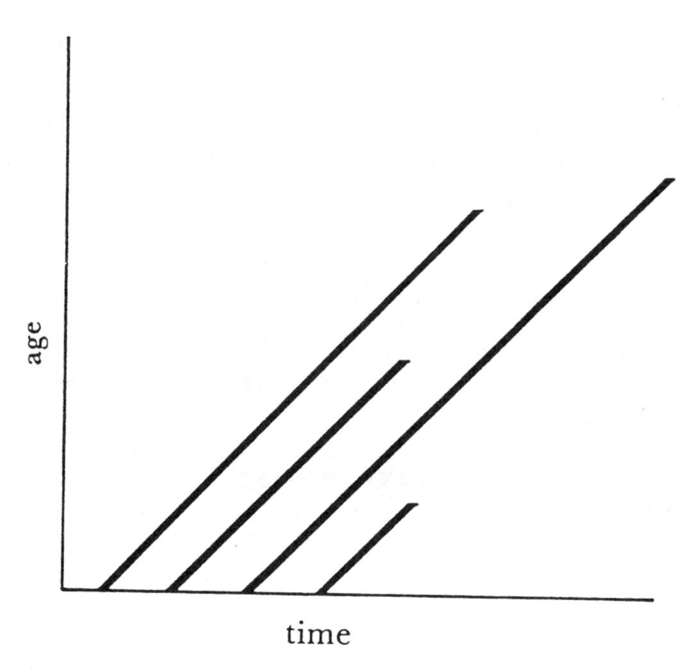

Figure X.1.1. A Lexis diagram.

sities commonly used in epidemiology (cf. Example VI.1.3). See Keiding (1990) for further references.

The Lexis diagram is a very useful tool for analysing clinical trials with staggered entry [cf. Example I.3.5 and Section X.2]; in this case, "age" is interpreted as duration on study. □

Semi-Markov processes are at the borderline of what may be handled by the present methodology, as illustrated by the next two examples.

EXAMPLE X.1.7. The Illness-Death Process

The illness-death process was discussed in Examples III.1.11 and III.3.5. The states 0, 1, and 2 denote "healthy," "diseased," and "dead" and the intensities $\alpha_{01}(t)$, $\alpha_{02}(t)$, and $\alpha_{12}(t, d)$ depend on time and (for α_{12}) duration in the diseased state. The trivariate counting process

$$\mathbf{N}(t) = (N_{01}(t), N_{02}(t), N_{12}(t))$$

counts the number of relevant transitions in $[0, t]$. Here, $N_{0h}(t)$, $h = 1, 2$, has intensity process $\alpha_{0h}(t) Y_0(t)$ with $Y_0(t) = 1 - N_{01}(t-) - N_{02}(t-)$, whereas $N_{12}(t)$ has intensity process $\alpha_{12}(t, t - T) Y_1(t)$ with $Y_1(t) = N_{01}(t-) - N_{02}(t-)$ and $T = \inf\{t\colon N_{01}(t) = 1\}$ the time of transition from 0 to 1, if this ever occurs.

We shall focus attention on the estimation of the intensity α_{12}, which depends on the two simultaneously running time scales: time t and duration in state 1, $d = t - T$.

In the situation where $\alpha_{12}(t, d)$ depends only on t and not on d, the illness–death process is a Markov process, and the estimation problem is within the multiplicative intensity model framework as specified in Examples III.1.4 and IV.1.9 for Markov processes.

On the other hand, assume that $\alpha_{12}(t, d)$ depends only on duration d; since time and duration in state 0 coincide for individuals still in 0, the process is then a standard *semi-Markov process*. The intensity process of $N_{12}(t)$ is then $\alpha_{12}(t - T)Y_1(t)$ which does not factor into a product of a deterministic function and an observable process; hence, the model is not a multiplicative intensity model. However, using duration $d = t - T$ as the basic time variable, that is, defining $K(d) = N_{12}(d + T)$, $U(d) = Y_1(d + T)$, K is a counting process with intensity process $\alpha_{12}(d)U(d)$ with respect to the filtration $(\mathscr{F}_d) = (\mathscr{G}_\tau \vee \mathscr{K}_d)$, where $\mathscr{G}_\tau = \sigma\{(N_{01}(t), N_{02}(t)): 0 < t < \tau\}$ and $\mathscr{K}_d = \sigma\{K(d): 0 < d < \infty\}$. It follows that the multiplicative intensity model *is* satisfied in this time scale, and our standard machinery applies. For n i.i.d. replications, take

$$K(d) = \sum_{i=1}^{n} N_{12}^{(i)}(d + T_i), \qquad U(d) = \sum_{i=1}^{n} Y_1^{(i)}(d + T_i)$$

with obvious notation.

This time transformation was noted by Aalen (1975, Section 5E) and developed by Voelkel and Crowley (1984). □

Example X.1.8. General Semi-Markov Processes

The time transformation trick in the above example will not work in semi-Markov processes allowing transitions back and forth, that is, for at least one state, it is possible to return once it has been left (Gill, 1980b).

Consider a semi-Markov process with k states, for our purpose most conveniently specified by transition intensities $\alpha_{hj}(d)$, $h, j = 1, \ldots, k$, $h \neq j$, depending on duration in state h, $h = 1, \ldots, k$. Define $J_0, J_1, \ldots$ as that state which is occupied before jump 1, 2, ... and let $N_{hj}(t)$ be the number of (direct) transitions from h to j in $[0, t]$. For $N(t) = \sum_h \sum_j N_{hj}(t)$, the total number of transitions in $[0, t]$, $Z(t) = J_{N(t)}$ is the state currently occupied and is called a semi-Markov process. If X_i is the ith sojourn time (duration), $S_0 = 0$, $S_i = X_1 + \cdots + X_i$, $L(t) = t - S_{N(t-)}$ is a left-continuous version of the backward recurrence time. The counting pocess $N_{hj}(t)$ has intensity process given by $Y_h(t)\alpha_{hj}(L(t))$ [where $Y_h(t) = I(Z(t-) = h)$] with respect to the self-exciting filtration of $(N_{hj}(t))$, but (as in the previous example) this is not a multiplicative intensity model.

As in the previous example, let us attempt to exploit the intrinsic time scale of this process: duration rather than "calendar" time. Define

$K_{hj}(d)$ = the number of durations in h observed to take on a value $\leq d$ and to be followed by a jump to j,
$U_h(d)$ = the number of durations in h observed to take on a value $\geq d$.

We define

$$H_{hj}(d) = K_{hj}(d) - \int_0^d U_h(s)\alpha_{hj}(s)\,\mathrm{d}s,$$

but there is no filtration making H_{hj} a martingale: We can no longer "put the past behind us definitively" via a time transformation trick as in the previous example.

However, Gill [1980b, (27) and (28)] showed that H_{hj} has the following properties which are the same as for the counting process martingales M_{hj} generated by Markov processes (cf. Theorem II.6.8).

$$\mathrm{E}H_{hj}(d) = 0,$$

that is,

$$\mathrm{E}K_{hj}(d) = \int_0^d \mathrm{E}U_h(s)\alpha_{hj}(s)\,\mathrm{d}s,$$

$$\operatorname{cov}(H_{hj}(d), H_{h'j'}(d')) = \begin{cases} 0 & \text{if } (h,j) \neq (h',j') \\ \mathrm{E}K_{hj}(d \wedge d') & \text{if } (h,j) = (h',j'), \end{cases}$$

implying, in particular, that $H_{hj}(\cdot)$ has uncorrelated increments. These results are enough to derive identical asymptotic results as for Markov processes (see Example IV.1.9), although alternative techniques to the martingale central limit theory are needed (Gill 1980b). Transition *probabilities* may also be estimated.

Gill (1983b) noted that the likelihood representations of the Markov and semi-Markov processes "look the same" in a sense to be specified in a moment. In light of the development of generalized maximum likelihood tools, this fact might help explain the similarity of the asymptotic results; cf. Section IV.1.5 (in particular the page 229) and Section VIII.4.5 (pages 654–655).

Consider (possibly censored) observation of a Markov process as in Section IV.4. This is a multiplicative intensity model (N_{hj} has intensity process $\lambda_{hj} = Y_h\alpha_{hj}$) and the likelihood, under noninformative censoring, can be written as

$$\prod_t \prod_h \left(\prod_{j \neq h} \mathrm{d}A_{hj}^{\mathrm{d}N_{hj}} (1 - \mathrm{d}A_{h.})^{Y_h - \mathrm{d}N_{h.}} \right).$$

This expression has been written in a form allowing generalization to discrete-time processes and has a "product of independent multinomials" interpretation: Given $\mathscr{F}_{t-}$, $(\mathrm{d}N_{hj}, j \neq h)$ has a multinomial $(Y_h; \mathrm{d}A_{hj}, j \neq h)$ distribution for $h = 1, \ldots, k$, independent over h. Now for a semi-Markov process, again writing $N_{hj}(t)$ for the number of jumps observed from state h to state j in $[0, t]$, with intensity process $\lambda_{hj}(t) = Y_h(t)\alpha_{hj}(L(t))$ and compensator $\Lambda_{hj} = \int \lambda_{hj}$, the likelihood under noninformative censoring can be written

$$\prod_t \prod_h \left(\prod_{j \neq h} \mathrm{d}\Lambda_{hj}^{\mathrm{d}N_{hj}} (1 - \mathrm{d}\Lambda_{h\bullet})^{Y_h - \mathrm{d}N_{h.}} \right).$$

However, introducing $u = L(t)$ and reordering the product-integral according to values of u, this likelihood is proportional to

$$\prod_u \prod_h \left(\prod_{j \neq h} \mathrm{d}A_{hj}^{\mathrm{d}K_{hj}} (1 - \mathrm{d}A_{h\bullet})^{U_h - \mathrm{d}K_{h.}} \right).$$

So the likelihood in the semi-Markov case has exactly the same form as function of the data (K_{hj}, U_h) as it had in the Markov case with respect to (N_{hj}, Y_h). This means that the NPMLE of (A_{hj}) is the same function of the data in each case, and, moreover, its large sample distribution has the same form (the likelihoods are the same, so the information is the same, hence the inverse information too); see Section IV.1.5.

More generally, one could envisage a Cox regression model for semi-Markov processes. By the interpretation of the Cox estimators as the NPMLE, one may expect the same estimators to have the same asymptotic properties (in particular, to have identical asymptotic covariance structure) as in the usual case, even though the martingale approach of Section VII.2 is not available. The same remarks can be made on the periodic Cox regression model of Pons and Turckheim (1988a); see Example VII.2.15.

A special case of semi-Markov processes occurs in *testing with replacement*, where we, in connection with Example III.2.11, remarked that the martingale techniques do not work. While Gill (1980b, 1981) obtained asymptotic results by generalizing the approach of Breslow and Crowley (1974), Sellke (1988) proved weak convergence of the Nelson–Aalen estimator of the component lifetime distribution in a similar fashion as in Sellke and Siegmund's (1983) work on sequential analysis of the proportional hazards model. Sellke studied the counting process in the (time, duration)-plane (which we would call a Lexis diagram) and identified an approximating two-dimensional process with a certain orthogonal martingale structure; this made the usual martingale central limit theorems applicable. □

EXAMPLE X.1.9. Arjas's Real Time Approach

In the previous example, we noted that for semi-Markov models the "counting process approach" could not be used to study the large sample properties of *nonparametric* estimators of the transition intensities. The reason for this was that there exists no filtration relative to "duration time" d which makes $H_{hj} = K_{hj} - \int U_h \alpha_{hj}$ a martingale (notation as in Example X.1.8). But for *parametric* models, where the transition intensities $\alpha_{hj}(d, \boldsymbol{\theta})$ depend on a q-dimensional Euclidean parameter $\boldsymbol{\theta}$, there is no need to define the $K_{hj}(d)$ counting the number of jumps (from h to j) with durations (in h) less than d. These are only essential in the nonparametric setup.

Instead, we can work directly with the counting processes $N_{hj}(t)$, which have intensity processes $\lambda_{hj}(t) = Y_h(t)\alpha_{hj}(L(t); \boldsymbol{\theta})$ [still with $Y_h(t) = I(Z(t-) = h)$], relative to the self-exciting filtration in "real time" t.

For n i.i.d. replications of the semi-Markov process, the log-likelihood function becomes (with obvious notation)

$$C_\tau(\boldsymbol{\theta}) = \int_0^\tau \sum_{i=1}^n \sum_{h \neq j} \log\{Y_h^{(i)}(t)\alpha_{hj}(L^{(i)}(t), \boldsymbol{\theta})\} \, \mathrm{d}N_{hj}^{(i)}(t)$$
$$- \int_0^\tau \sum_{i=1}^n \sum_{h \neq j} Y_h^{(i)}(t)\alpha_{hj}(L^{(i)}(t), \boldsymbol{\theta}) \, \mathrm{d}t$$

[cf. (6.1.2)], so that the score statistics are given by

$$U_\tau^l(\boldsymbol{\theta}) = \int_0^\tau \sum_{i=1}^n \sum_{h \neq j} \frac{\partial}{\partial \theta_l} \log \alpha_{hj}(L^{(i)}(t), \boldsymbol{\theta}) \, \mathrm{d}N_{hj}^{(i)}(t)$$
$$- \int_0^\tau \sum_{i=1}^n \sum_{h \neq j} Y_h^{(i)}(t) \frac{\partial}{\partial \theta_l} \alpha_{hj}(L^{(i)}(t), \boldsymbol{\theta}) \, \mathrm{d}t$$

for $l = 1, \ldots, q$. Since

$$M_{hj}^{(i)}(t) = N_{hj}^{(i)}(t) - \int_0^\tau Y_{hj}^{(i)}(s)\alpha_{hj}(L^{(i)}(s), \boldsymbol{\theta}) \, \mathrm{d}s$$

are orthogonal local square integrable martingales when evaluated at the true parameter value, the score statistics are still martingales (in "real time" t) when integrated up to t instead of τ for this parameter value. From this, the usual large sample properties of the maximum likelihood estimator $\hat{\boldsymbol{\theta}}$ [solving $U_\tau(\boldsymbol{\theta}) = \mathbf{0}$] follow. In fact, the results of Section VI.1.2 are general enough to cover the present situation as a special case, and, therefore, the regularity conditions needed here can be found from Condition VI.1.1.

Independent right-censoring can easily be introduced into the above setup by letting $Y_h^{(i)}(t) = I(Z^{(i)}(t-) = h, U_i \leq t)$ for some censoring times $U_1, \ldots, U_n$. Independent left-truncation and filtering are, in theory, feasible as well, but we then have to assume that the backward recurrence times $L^{(i)}(t)$ are available for all i and t for which $Y_h^{(i)}(t) = 1$ for some h, and this may quite often not be the case in practical applications (cf. the discussion just above Example III.5.6).

The idea of letting "real time" (i.e., calendar time) be the basic time variable for parametric models is due to Arjas (1985, 1986). As illustrated by him, this approach works for other situations as well where problems occur for nonparametric methods, e.g., for the situation with staggered entry discussed in the following section. □

X.2. Sequential Analysis of Censored Survival Data with Staggered Entry

In most of the practical examples of clinical trials used in this volume [such as presented in Examples I.3.1 (malignant melanoma), I.3.4 (liver cirrhosis), I.3.5 (breast cancer), and I.3.6 (leukemia)], the patients entered the trial sequentially in calendar time, whereas the time dimension of main interest was duration on trial. During most of our discussion, we have (more or less tacitly) assumed that, in the statistical analysis, these duration variables have been realigned to all start at time (= duration) 0, and the counting process martingales, etc., were then defined in the duration time scale, after alignment.

This device is not completely satisfactory for all purposes, especially when calendar time monitoring is desired. Already, in Chapter III (cf. Examples III.2.9–III.2.12), we mentioned that the censoring process may not always be adapted with respect to a natural filtration leaving the intensities for the separate individuals as they "should be." Thus, problems arise if a decision to stop the trial is based on the survival experience of patients who had much longer duration than some of those still at risk. However, even if the random censoring model is assumed, it is not directly possible to develop an approximating Gaussian martingale in continuous calendar time in the manner described in Section V.4.2.

Most of the literature (see Bibliograhic Remarks in Section X.4) has concentrated on two-sample tests, and we present here aspects of the so far most definitive results for this case, due to Gu and Lai (1991). According to the notation of Example V.2.1, let the survival times X_{hi} for $i = 1, \dots, n_h$, $h = 1, 2$, be independent non-negative random variables with distribution function F_h and hazard function α_h, $h = 1, 2$. Let $n = n_1 + n_2$. For each individual, there are also specified censoring times U_{hi} and entry times T_{hi}. Here, T_{hi} are on the calendar time scale, whereas X_{hi} and U_{hi} are on a duration (on trial) scale. Individual i in group h is followed from calendar time T_{hi} until death (at calendar time $T_{hi} + X_{hi}$) or censoring (at calendar time $T_{hi} + U_{hi}$) whichever comes first. Define the counting process (with bivariate time)

$$N_{hi}(t, a) = I(T_{hi} + X_{hi} \leq t, X_{hi} \leq U_{hi}, X_{hi} \leq a)$$

which is the indicator function that individual i in group h was observed to die before (calendar) time t and duration a. In a Lexis diagram (cf. Figure X.1.1), with abscissa calendar time, henceforth to be denoted time, and ordinate duration, $N_{hi}(t, a)$ counts an event happening in the lower left quadrant at (t, a). Similarly, define

$$Y_{hi}(t, a) = I(t - T_{hi} \geq a, X_{hi} \geq a, U_{hi} \geq a)$$

as the indicator that (h, i) is at risk of dying at duration a and before time t:

This requires arrival earlier than time $t - a$ and no death or censoring before duration a. Let

$$N_h = \sum_{i=1}^{n_h} N_{hi} \quad \text{and} \quad Y_h = \sum_{i=1}^{n_h} Y_{hi} \tag{10.2.1}$$

and let $N_{\bullet} = N_1 + N_2$, $Y_{\bullet} = Y_1 + Y_2$. Define the σ-algebra $\mathscr{F}_{t,a}$ generated by events happening before time t and duration a; formally:

$$\begin{aligned}\mathscr{F}_{t,a} = \sigma\{&I(T_{hi} \leq t), T_{hi}I(T_{hi} \leq t), I(X_{hi} \leq a \wedge U_{hi} \wedge (t - T_{hi})^+),\\ &X_{hi}I(X_{hi} \leq a \wedge U_{hi} \wedge (t - T_{hi})^+),\\ &I(U_{hi} \leq a \wedge X_{hi} \wedge (t - T_{hi})^+),\\ &U_{hi}I(U_{hi} \leq a \wedge X_{hi} \wedge (t - T_{hi})^+), i = 1, \ldots, n_h, h = 1, 2\}.\end{aligned} \tag{10.2.2}$$

Then, for fixed t, under certain independence assumptions

$$N_h(t, a) - \int_0^a \alpha_h(s) Y_h(t, s)\,\mathrm{d}s$$

is a martingale with respect to the filtration $(\mathscr{F}_{t,a})_{a \geq 0}$, and standard Nelson–Aalen estimators are given by

$$\hat{A}_h(t, a) = \int_0^a \frac{J_h(t, s)}{Y_h(t, s)} N_h(t, \mathrm{d}s) \tag{10.2.3}$$

with $J_h(t, s) = I(Y_h(t, s) > 0)$. From these Nelson–Aalen estimators defined (for each fixed time t) on the vertical lines in the Lexis diagram (cf. Example X.1.6), we want to consider the test statistic process with time t as the time scale:

$$Z(t) = \int_0^\infty L(t, s)[\hat{A}_1(t, \mathrm{d}s) - \hat{A}_2(t, \mathrm{d}s)],$$

where $L(t, a)$ is a weight process to be chosen [cf. (5.2.7)]. At the outset, we only assume $L(t, \cdot)$ to be predictable with respect to $(\mathscr{F}_{t,a})_{a \geq 0}$ for each t.

Gu and Lai (1991) derived the following asymptotic distribution results under the null hypothesis $F_1 = F_2$.

Theorem X.2.1. *Let (X_{hi}, U_{hi}, T_{hi}), $h = 1, 2$, $i = 1, 2, \ldots$ be independent random vectors with non-negative components such that (U_{hi}, T_{hi}) is independent of X_{hi} for all h and i. Let F_h be the distribution function of X_{hi}. Define $N_{hi}^{(n)}$, $N_h^{(n)}$, $Y_{hi}^{(n)}$, $Y_h^{(n)}$ as above; cf.* (10.2.1). *Assume that as $n \to \infty$ under the hypothesis $F_1 = F_2$,*

$$y_h(t, a) = \lim_{n \to \infty} \frac{1}{n} \mathrm{E}(Y_h^{(n)}(t, a))$$

exists and is positive and continuous for all $a \geq 0$ and all t. Define

$$Z_n(t) = \int_0^\infty L_n(t, s)[\hat{A}_1^{(n)}(t, \mathrm{d}s) - \hat{A}_2^{(n)}(t, \mathrm{d}s)],$$

where for every t, $L_n(t,\cdot)$ is predictable with respect to $(\mathscr{F}_{t,a}^{(n)})_{a\geq 0}$, $\mathscr{F}_{t,a}^{(n)}$ being defined in (10.2.2) and $\hat{A}_h^{(n)}(t,a)$ in (10.2.3), for each n. Let $t_0 > 0$, $K_n(t,a) = L_n(t,a)Y_{\bullet}^{(n)}(t,a)/\{Y_1^{(n)}(t,a)Y_2^{(n)}(t,a)\}$ and assume that a (nonrandom) function κ exists such that

$$\sup_{0\leq a\leq t\leq t_0} |K_n(t,a) - \kappa(t,a)|\, I(Y_{\bullet}^{(n)}(t,a)/n > \varepsilon) \xrightarrow{\mathrm{P}} 0$$

for every $\varepsilon > 0$ and that there exists $\alpha \in [0, \frac{1}{2})$ such that as $n \to \infty$

$$\sup_{0\leq t\leq t_0}\left\{\operatorname*{Tot.\,var.}_{0\leq a\leq t}[(Y_{\bullet}^{(n)}(t,a)/n)^{\alpha}K_n(t,a)] + \operatorname*{Tot.\,var.}_{0\leq a\leq t}[\{y_{\bullet}(t,a)\}^{\alpha}\kappa(t,a)]\right\}$$

is of order 1 in probability.

Then under the null hypothesis $F_1 = F_2$, as $n\to\infty$, $n^{-1/2}Z_n$ converges in $D[0,t_0]$ to a zero-mean Gaussian process $\{Z(t), 0\leq t\leq t_0\}$ with covariance function

$$\begin{aligned} c(t,s) &= \operatorname{cov}(Z(t), Z(s)) \\ &= \int_0^{t\wedge s} \kappa(t,a)\kappa(s,a)y_1(t\wedge s,a)y_2(t\wedge s,a)(y_{\bullet}(t\wedge s,a))^{-1}\alpha(a)\,\mathrm{d}a. \end{aligned}$$

Note that most of these conditions and results are direct generalizations of Theorem V.2.1; cf. Proposition V.2.2, (5.2.9), and Example V.2.1.

As a specific example, define a Tarone–Ware class of weight processes

$$K_n(t,a) = g(Y_{\bullet}^{(n)}(t,a)/n),$$

where g is a continuous function on $[0,1]$ with $\text{Tot.var}_{0\leq x\leq 1}[x^{\alpha}g(x)] < \infty$ for some $\alpha\in[0,\frac{1}{2})$. As usual, $g\equiv 1$ yields a log-rank test and $g(x) = x$, the Gehan generalization of the Wilcoxon test. In this situation,

$$\kappa(t,a) = g(y_{\bullet}(t,a)).$$

The limiting Gaussian process Z has independent increments (that is, is a martingale) only if the covariance function $c(t,s)$ is a function of $t\wedge s$. For this to happen for the Tarone–Ware class, either g should be constant (as for the log-rank statistic) or $y_{\bullet}$ should be independent of t, as would be the case for simultaneous entry, $T_{hi} = 0$ for all h, i. It is noted, in particular, that Z is not a martingale for the Gehan statistic.

However, a modification in the spirit of Prentice (see Example V.2.1) is useful. Define, for each t, the Kaplan–Meier estimator of $S = 1 - F$ in the combined sample

$$\hat{S}_n(t,a) = \prod_0^a \left(1 - \frac{N_{\bullet}^{(n)}(t,\mathrm{d}s)}{Y_{\bullet}^{(n)}(t,s)}\right)$$

and consider the class of weight processes given by

$$K_n(t,a) = g(\hat{S}_n(t,a-)). \tag{10.2.4}$$

Then $\kappa(t, a) = g(S(a))$ and we may write $c(t, s) = \sigma^2(t \wedge s)$ with

$$\sigma^2(t) = \int_0^t g^2(S(a)) y_1(t, a) y_2(t, a) (y_.(t, a))^{-1} \alpha(a)\, da.$$

Thus, for this class, $Z(t)$ *is* a martingale. Again, $g = 1$ corresponds to the log-rank test, whereas $g(x) \equiv x$ gives the Prentice/Peto and Peto generalization of the Wilcoxon test. It may, however, be shown that Efron's two-sample test does not lead to a limiting martingale.

Gu and Lai (1991) went on to study asymptotic properties under local alternatives, including asymptotically optimal choice of weight processes against given classes of local alternatives, etc. A vital part of Gu and Lai's proofs consisted of establishing a weak convergence result in $D([0, t_0] \times [0, t_0])$, as $n \to \infty$, of the random fields

$$W_h^{(n)}(t, a) = n^{-1/2} \int_0^a \frac{J_h^{(n)}(t, u)}{(Y_h^{(n)}(t, u)/n)^\alpha} M_h^{(n)}(t, du)$$

for all $\alpha \in [0, \frac{1}{2})$, where

$$M_h^{(n)}(t, a) = N_h^{(n)}(t, a) - \int_0^a \alpha_h(u) Y_h^{(n)}(t, u)\, du.$$

Thus, we note that the intrinsic bivariate time scale has to be taken directly into account in the proof.

When using the above results for sequential testing, an estimate of the variance of $Z_n(t)$ is required. It is a corollary of Gu and Lai's results that for weight processes of the form (10.2.4) the obvious generalization of our usual variance estimator (5.2.8) is valid, indeed, as $n \to \infty$:

$$\frac{1}{n} \hat{\sigma}_n^2(t) = \frac{1}{n} \int_0^t L_n^2(t, a) \{Y_1^{(n)}(t, a) Y_2^{(n)}(t, a)\}^{-1} N_.^{(n)}(t, da) \xrightarrow{\text{P}} \sigma^2(t) = \operatorname{var}(Z(t)).$$

For the particular case of the log-rank test, we have

$$\hat{\sigma}_n^2(t) = \int_0^t Y_1^{(n)}(t, a) Y_2^{(n)}(t, a) \{Y_.^{(n)}(t, a)\}^{-2} N_.^{(n)}(t, da).$$

Following a remark by Siegmund (1985, Section V.6), we may note that in a clinical trial with staggered entry, where each arriving patient is allocated to group h with probability $\frac{1}{2}$ and censoring in the two treatment groups follows the same distribution, it is intuitively clear that under the null hypothesis

$$Y_1^{(n)}(t, a) Y_2^{(n)}(t, a) \{Y_.^{(n)}(t, a)\}^{-2} \xrightarrow{\text{P}} \tfrac{1}{4},$$

so that

$$\hat{\sigma}_n^2(t) \approx \tfrac{1}{4} N_.^{(n)}(t, t).$$

A repeated significance test may now be set up by stopping sampling with rejection of H_0 as soon as $|Z_n(t)|/\hat{\sigma}_n(t) > b$, where the fractile b may be determined from the approximating Brownian motion theory, as mentioned in Section V.4.2. One would usually specify a maximum time t_0 at which a definitive decision (possibly accepting H_0) has to be reached. A modification of this procedure is to specify in advance a finite number of time points $t_1, \ldots, t_n$ at which tests [based on $|Z_n(t)|/\hat{\sigma}_n(t)$] are to be performed, so that periodic reviews of the trial may be performed while controlling the total significance level, the total probability of rejecting H_0 if it is true. For specific procedures, see Tsiatis (1982), Slud (1984), and Siegmund (1985, Section V.6).

Alternatively, one might use a Wald-type sequential test with stopping as soon as $|Z_n(t)|$ exceeds some constant a; see Jones and Whitehead (1979) and the above references.

The methodology of sequential tests becomes rather more complicated for choices of weight process $K_n(t, a)$ that do not lead to an approximating martingale $Z(t)$. Slud and Wei (1982) carried through a construction of a repeated significance test based on the Gehan statistic, for which $K_n(t, a) = Y_{\bullet}(t, a)/n$, thus improving on previous suggestions by Jones and Whitehead (1979, cf. correction note) who had assumed the asymptotic independent increment property to be true also for Gehan's statistic.

The simplification in the case of the log-rank test and the Prentice test correspond to the interpretation of these tests as score tests in a semiparametric model; cf. Example VII.2.4 and Sections VIII.2.3 and VIII.4.2. Even with nuisance parameters present (here infinite-dimensional), martingale properties of likelihood processes ensure that the score, with nuisance parameters estimated by maximum likelihood, has approximatey independent increments with variance equal to the increment of the effective information for the parameter of interest. (The Gehan test also has a likelihood interpretation; however, the model for which it has this property depends on the average censoring distribution. In a sequential setting, the test is actually testing against *different* alternatives as time proceeds.)

Similar comments apply to the sequential behavior of a MLE or even NPMLE. Let the martingale U be a score process for a real parameter θ. The corresponding maximum likelihood estimator is asymptotically equivalent to $\theta + \langle U \rangle^{-1} U$. Now if $\langle U \rangle$ is deterministic,

$$\begin{aligned}\operatorname{cov}(\langle U \rangle^{-1}(s) U(s), \langle U \rangle^{-1}(t) U(t)) &= \frac{1}{\langle U \rangle^{-1}(s) \langle U \rangle^{-1}(t)} \langle U \rangle (s \wedge t) \\ &= \langle U \rangle^{-1}(s \vee t).\end{aligned}$$

Thus, $\langle U \rangle^{-1} U$ has approximately the covariance structure of a *reverse* martingale. Sellke (personal communication) has shown that the Kaplan–Meier estimator, considered as a process in *calendar time* for data arising in a sequential setting, has asymptotically the distribution of a *reverse* Gaussian martingale. This is a consequence of the NPMLE interpretation of the estimator and the remarks we have just made (generalized to the multivariate case).

X.3. Nonparametric Estimation of the Multivariate Survival Function

In this section, we discuss multivariate generalization of the Kaplan–Meier estimator. Possible applications are to the modelling of subsequent stages of an illness, or of survival times of family members, or of the joint distribution of a survival time and a biochemical measurement, censored by a varying upper bound to its measurement. The aim is to estimate the multivariate survival function on the basis of such multivariate censored data without assuming any special structure relating the different time coordinates (as was the case with the frailty models of Chapter IX or the models of the previous sections of this chapter). The result could be used as a nonparametric baseline for testing goodness-of-fit of semiparametric frailty models, Markov models, or semi-Markov models for dependent survival times.

In Section X.3.1, we generalize the notion of hazard function or hazard measure to multivariate time. We show how the multivariate survival function can be reconstructed by product-integration of various kinds from its hazard measures, generalizing Theorem II.6.6 to higher dimensions. In Section X.3.2, this is used to define plug-in estimators based on the natural multivariate generalization of the Nelson–Aalen estimator. Large sample theory is available based on the functional delta-method.

A big difference with the one-dimensional case is that in higher dimensions, there are actually many nonequivalent representations of the survival function in terms of the hazard, leading to a great variety of different estimators. Here, we concentrate on the representation by Dabrowska (1988, 1989b), but first discuss a proposal by Bickel (personal communication) and also mention later another by Prentice and Cai (1991, 1992).

The results of Sections X.3.1 and X.3.2 are developed in the context of the natural multivariate generalization of the classical random censorship model. In Section X.3.3, we will evaluate these results and see whether a role can be played by the theory of multivariate time martingales and whether our results allow extension to multidimensional independent censoring, truncation, and filtering in some form or other. The conclusion is that though some use can be made of the theory of weak martingales (Pons, 1986; Pons and Turckheim, 1991), in general multivariate time does not seem to allow a rich stochastic process based generalization as does one-dimensional survival analysis. Notions of independent censoring in multivariate time are so far very restrictive and possess many unsatisfactory features.

The three subsections can be read quite independently of one another.

X.3.1. Multivariate Hazard and Survival

The idea of Section IV.3 was to see the survival function as a functional of the hazard function or hazard measure. Since a natural estimator of the hazard was readily available (the Nelson–Aalen estimator), substitution provided an estimator of the survival function. We, therefore, try to follow the same route, considering first the definition of *multivariate hazard measures* for a k-dimensional survival time $\mathbf{X} = (X_1, \ldots, X_k)$, and their relation with the survival function. Generalizing the Nelson–Aalen estimator turns out to be easy (Section X.3.2). However, generalizing the relation between hazard and survival is rather tricky. It will turn out that it is not enough to consider just the full multivariate hazard measure on its own; we need also to consider so-called *marginal* and *conditional hazard measures* too. Also, there will be many possible representations, not just one as in the one-dimensional case.

More details and heuristics on this subject are contained in Gill (1990), though the approach here is slightly different.

X.3.1.1. *Preliminaries*

First some notation. For vectors $\mathbf{s}, \mathbf{t} \in \mathbb{R}_+^k = [0, \infty)^k$, we write

$$\mathbf{s} \le \mathbf{t} \Leftrightarrow s_i \le t_i, \quad i = 1, \ldots, k,$$

$$\mathbf{s} \ll \mathbf{t} \Leftrightarrow s_i < t_i, \quad i = 1, \ldots, k,$$

$$(\mathbf{s}, \mathbf{t}] = \{\mathbf{x} \in \mathbb{R}_+^k : \mathbf{s} \ll \mathbf{x} \le \mathbf{t}\} = \bigotimes_{i=1}^{k} (s_i, t_i].$$

To emphasize its special nature, we will, to begin with, refer to the generalized interval $(\mathbf{s}, \mathbf{t}]$ as a *hyperrectangle*; one could also have spoken of a

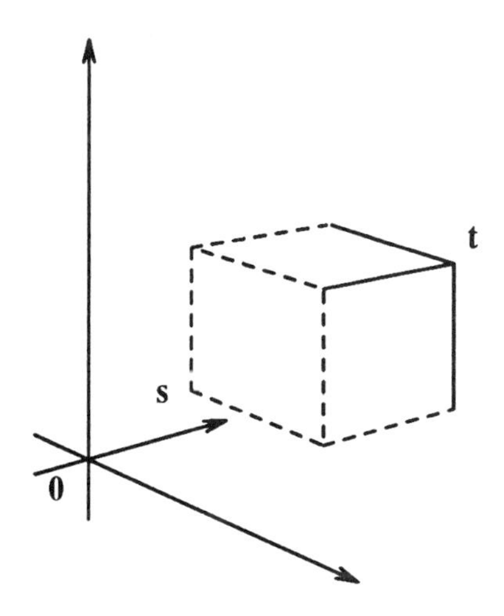

Figure X.3.1. The hyperrectangle $(\mathbf{s}, \mathbf{t}]$.

hyperinterval or k-dimensional interval. (Later we will just call it an interval.) This hyperrectangle has 2^k corners joined by ribs; we call $\mathbf{t}$ the top corner and $\mathbf{s}$ the bottom corner.

A function of bounded variation on $\mathbb{R}^k$ generates a measure which assigns to $(\mathbf{s}, \mathbf{t}]$ what we call the *generalized difference* of the function over the hyperrectangle: This is the value of the function at the "top" corner $\mathbf{t}$ minus its values at all corners one step down from the top, plus its values one step further down, and so on, until one reaches the bottom corner $\mathbf{s}$. [Think in two dimensions of the formula $F(t_1, t_2) - F(t_1, s_2) - F(s_1, t_2) + F(s_1, s_2) = (F(t_1, t_2) - F(s_1, t_2)) - (F(t_1, s_2) - F(s_1, s_2))$ for the probability of the rectangle $((s_1, s_2), (t_1, t_2)]$ in terms of a distribution function F.]

Let E be the set of indices (to be thought of as the names of the components of the multivariate survival time) $\{1, \ldots, k\}$. For $\mathbf{t} \in \mathbb{R}_+^k$ and $\varnothing \subset C \subseteq E$, we write $\mathbf{t}_C$ for the lower-dimensional vector $(t_i\colon i \in C)$. For $\mathbf{s}, \mathbf{t} \in \mathbb{R}_+^k$ and $C, D \subset E$, $C \cap D = \varnothing$, we write $(\mathbf{s}_C, \mathbf{t}_D) \in \mathbb{R}_+^{|C \cup D|}$ for the vector with components s_i for $i \in C$ and t_i for $i \in D$, where $|C \cup D|$ is the number of elements in $C \cup D$.

Let $\mathbf{X}$ be a multivariate survival time in $\mathbb{R}_+^k$, strictly positive so $\mathrm{P}(\mathbf{X} \gg \mathbf{0}) = 1$. Its survival function is

$$S(\mathbf{t}) = \mathrm{P}(\mathbf{X} \gg \mathbf{t}), \quad \mathbf{t} \in \mathbb{R}_+^k.$$

A subset of the components of $\mathbf{X}$, say $\mathbf{X}_C$ for $C \subseteq E$, has marginal (multivariate) survival function

$$S_C(\mathbf{t}_C) = \mathrm{P}(\mathbf{X}_C \gg \mathbf{t}_C) = \mathrm{P}(\mathbf{X}_C \gg \mathbf{t}_C, \mathbf{X}_{E \setminus C} \gg \mathbf{0}_{E \setminus C}) = S(\mathbf{t}_C, \mathbf{0}_{E \setminus C}).$$

For $C, D \subseteq E$, $C \cap D = \varnothing$, we can look at the survival function of $\mathbf{X}_C$ conditional on $\mathbf{X}_D \geq \mathbf{s}_D$ say, this is

$$S_{C|D}(\mathbf{t}_C | \mathbf{s}_D) = \mathrm{P}(\mathbf{X}_C \gg \mathbf{t}_C | \mathbf{X}_D \geq \mathbf{s}_D) = \frac{S_{C \cup D}(\mathbf{t}_C, \mathbf{s}_D -)}{S_D(\mathbf{s}_D -)}.$$

The natural definition of the multivariate hazard of $\mathbf{X}$ is as the measure

$$A(\mathrm{d}\mathbf{t}) = \frac{(-1)^{|E|} S(\mathrm{d}\mathbf{t})}{S(\mathbf{t}-)}, \tag{10.3.1}$$

to be interpreted as $\mathrm{P}(\mathbf{X} \in [\mathbf{t}, \mathbf{t} + \mathrm{d}\mathbf{t}) | \mathbf{X} \geq \mathbf{t})$ [cf. (2.1.1)] with the corresponding integrated or cumulative hazard function

$$A(\mathbf{t}) = \int_{(\mathbf{0}, \mathbf{t}]} \frac{(-1)^{|E|} \mathrm{d}S}{S_-}.$$

The alternating sign here ensures A is a positive measure whatever the number of elements of E.

One might have expected from (10.3.1) that S could be reconstructed from A alone. This is, however, not true. Boundary conditions are needed which in the one-dimensional case are trivial but in the multivariate case involve the $S(\mathbf{t}_C, \mathbf{0}_{E \setminus C})$ for $C \subset E$, in other words, the marginal multivariate survival functions.

Corresponding to these marginal and (for later use) also the conditional survival functions, we, therefore, also define *marginal* and *conditional* hazard measures

$$A_C(\mathrm{d}\mathbf{t}_C) = \frac{(-1)^{|C|}S_C(\mathrm{d}\mathbf{t}_C)}{S_C(\mathbf{t}_C-)},$$
$$A_{C|D}(\mathrm{d}\mathbf{t}_C|\mathbf{s}_D) = \frac{(-1)^{|C|}S_{C\cup D}(\mathrm{d}\mathbf{t}_C, \mathbf{s}_D-)}{S_{C\cup D}(\mathbf{t}_C-, \mathbf{s}_D-)}. \tag{10.3.2}$$

One may interpret an element of this last measure by the heuristic

$$A_{C|D}(\mathrm{d}\mathbf{t}_C|\mathbf{s}_D) = \mathrm{P}(\mathbf{X}_C \in [\mathbf{t}_C, \mathbf{t}_C + \mathrm{d}\mathbf{t}_C)|\mathbf{X}_C \geq \mathbf{t}_C, \mathbf{X}_D \geq \mathbf{s}_D).$$

For example, in the two-dimensional case one must distinguish between the two-dimensional hazard $A(\mathrm{d}t_1, \mathrm{d}t_2) = S(\mathrm{d}t_1, \mathrm{d}t_2)/S(t_1-, t_2-)$, two marginal hazards $A_1(\mathrm{d}t_1) = -S_1(\mathrm{d}t_1)/S_1(t_1-)$ and $A_2(\mathrm{d}t_2) = -S_2(\mathrm{d}t_2)/S_2(t_2-)$, and finally two conditional hazards $A_{1|2}(\mathrm{d}t_1|t_2) = -S(\mathrm{d}t_1, t_2-)/S(t_1-, t_2-)$ and $A_{2|1}(\mathrm{d}t_2|t_1) = -S(t_1-, \mathrm{d}t_2)/S(t_1-, t_2-)$.

X.3.1.2. *Survival in Terms of Hazard:* I, *the Volterra Representation*

We can now derive a representation for S in terms of its full and marginal hazard measures. For $\mathbf{t}$ such that $S(\mathbf{t}-) > 0$, we first write using (10.3.1)

$$\begin{aligned}\mathrm{P}(\mathbf{X} \leq \mathbf{t}) &= \int_{(\mathbf{0},\mathbf{t}]} S_- \,\mathrm{d}A = (-1)^{|E|}\int_{(\mathbf{0},\mathbf{t}]} \mathrm{d}S \\ &= \sum_{\varnothing \subseteq C \subseteq E} (-1)^{|C|}S(\mathbf{t}_C, \mathbf{0}_{E\setminus C}) \\ &= (-1)^{|E|}S(\mathbf{t}) + 1 + \sum_{\varnothing \subset C \subset E} (-1)^{|C|}S_C(\mathbf{t}_C).\end{aligned}$$

The middle line here is the generalized difference of the function $(-1)^{|E|}S$ over the hyperrectangle $(\mathbf{0}, \mathbf{t}]$. The whole equation is actually just a version of the inclusion–exclusion formula for expressing the probability of the union of the events $\{X_i > t_i\}$ in terms of the probabilities of the events themselves and of their intersections. For instance, when $k = 2$, it specializes to $\mathrm{P}(X_1 \leq t_1, X_2 \leq t_2) = \mathrm{P}(X_1 > t_1, X_2 > t_2) + 1 - \mathrm{P}(X_1 > t_1) - \mathrm{P}(X_2 > t_2)$ which can be rearranged as $1 - \mathrm{P}(X_1 \leq t_1$ and $X_2 \leq t_2) = \mathrm{P}(X_1 > t_1$ or $X_2 > t_2) = \mathrm{P}(X_1 > t_1) + \mathrm{P}(X_2 > t_2) - \mathrm{P}(X_1 > t_1$ and $X_2 > t_2)$.

Next, multiplying the equation by $(-1)^{|E|}$ and rearranging terms, we obtain

$$\begin{aligned}S(\mathbf{t}) &= -(-1)^{|E|}\left(1 + \sum_{\varnothing \subset C \subset E} (-1)^{|C|}S_C(\mathbf{t}_C)\right) \\ &\quad + \int_{\mathbf{s}\in(\mathbf{0},\mathbf{t}]} S(\mathbf{s}-)(-1)^{|E|}A(\mathrm{d}\mathbf{s}) \\ &= W(\mathbf{t}) + \int_{\mathbf{s}\in(\mathbf{0},\mathbf{t}]} S(\mathbf{s}-)A^{\pm}(\mathrm{d}\mathbf{s}),\end{aligned} \tag{10.3.3}$$

where

$$W(\mathbf{t}) = -(-1)^{|E|}\left(1 + \sum_{\varnothing \subset C \subset E} (-1)^{|C|} S_C(\mathbf{t}_C)\right) \tag{10.3.4}$$

and

$$A^{\pm} = (-1)^{|E|} A. \tag{10.3.5}$$

For instance, in the two-dimensional case, $W(t_1, t_2) = -1 + S_1(t_1) + S_2(t_2)$ and $A^{\pm}(t_1, t_2) = A(t_1, t_2)$. Now, if the full cumulative hazard A and all the lower-dimensional survival functions S_C, $C \subset E$, are known, so that W and $A^{\pm}$ are fixed, (10.3.3) can be thought of as an integral equation for the unknown survival function S, very similar to the Volterra equation (2.6.6). Indeed, the equation does have a similar and unique solution, not in terms of the product-integral of $A^{\pm}$ but in terms of its *Péano series* [cf. (2.6.8)]:

$$\mathscr{P}(\mathbf{s}, \mathbf{t}; A^{\pm}) = 1 + \sum_{j=1}^{\infty} \int \cdots \int_{\mathbf{s} \ll \mathbf{u}_1 \ll \cdots \ll \mathbf{u}_j \leq \mathbf{t}} A^{\pm}(\mathrm{d}\mathbf{u}_1) \cdots A^{\pm}(\mathrm{d}\mathbf{u}_j). \tag{10.3.6}$$

The solution of (10.3.3) is [cf. (2.6.7)]

$$S(\mathbf{t}) = W(\mathbf{t}) + \int_{\mathbf{s} \in (\mathbf{0}, \mathbf{t}]} W(\mathbf{s}-) A^{\pm}(\mathrm{d}\mathbf{s}) \mathscr{P}(\mathbf{s}, \mathbf{t}; A^{\pm}). \tag{10.3.7}$$

In the one-dimensional case (but only then), $\mathscr{P}(s, t; A^{\pm}) = \prod_{(s,t]}(1 + \mathrm{d}A^{\pm}) = \prod_{(s,t]}(1 - \mathrm{d}A)$.

Now we have, in fact, shown by the above how S can be built up recursively from all the multivariate hazards A_C, $C \subseteq E$; apply (10.3.7) first to get the one-dimensional marginals from the one-dimensional hazards, then to get the two-dimensional marginals from the one-dimensional marginals and the two-dimensional hazard, and so on. This representation is due to Bickel (personal communication). The corresponding plug-in estimator, to be discussed in Section X.3.2, is called the *Volterra* estimator.

X.3.1.3. *Survival in Terms of Hazard:* II, *the Dabrowska Representation*

Though the result above is of theoretical significance, its practical importance is not so great since there exists another way of rebuilding S from its hazard measures which has more implications for model-building as well as leading to statistically more efficient estimators. This method [due to Dabrowska (1988)] involves the *conditional* hazard measures as well as the marginal measures. Here, we shall first discuss it in terms of some related quantities, the so-called *iterated odds ratio meaures* or *cumulant measures.* Instead of the Péano series, true product-integrals occur now. The method gives a generalization of the relation between hazard and survival which is much closer in spirit to the one-dimensional result than the method based on the Péano series.

The previous approach leading to the Volterra representation (10.3.7) and building on the Péano series (10.3.6) generalized the fact that the one-dimensional survival function is related to the one-dimensional hazard measure through a Volterra equation $S(t) = 1 - \int_0^t S(s-)\,dA(s)$. One may also see the relation between S and A in a different and more fundamental light, and this leads to an alternative multivariate formulation. This new relation is a relation between *multiplicative* and *additive* interval functions, used by Gill and Johansen (1990) in their treatment of product-integration.

Let X be a (one-dimensional positive) survival time and consider the conditional probabilities

$$S(s,t) = \mathrm{P}(X > t | X > s) = S(t)/S(s), \quad s < t.$$

We consider $S(s,t)$ as a function of the intervals $(s,t]$ contained in some fixed interval $(0,\tau]$, where τ satisfies $S(\tau-) > 0$. Note that as an interval function, S satisfies the following four properties:

(i) $S(t,t) = 1$ for all t;
(ii) $\lim_{t \downarrow s} S(s,t) = 1$ for all s;
(iii) $S(s,t)S(t,u) = S(s,u)$ for all $s \le t \le u$;
(iv) $S - 1$ is dominated by a finite measure; in fact,

$$|S(s,t) - 1| = \left|\frac{S(t)}{S(s)} - 1\right| = \left|\frac{S(t) - S(s)}{S(s)}\right| \le \frac{F(t) - F(s)}{S(\tau-)}$$

for all $s < t \le \tau$, so $S - 1$ is dominated by $(S(\tau-))^{-1}F$.

[We can write $S(\tau-)$ instead of $S(\tau)$ in the last inequality because of the implied restriction $s < \tau$.] An interval function (denoted by μ) satisfying these four properties, with $S(\tau-)^{-1}F$ replaced by any other finite measure in (iv), is called a *dominated multiplicative interval function.*

By Theorem 2 of Gill and Johansen (1990), which we cite in a moment, *any dominated* multiplicative *interval function is the product-integral of a dominated* additive *interval function. Conversely, any dominated* additive *interval function is the "sum-integral" of a dominated* multiplicative *interval function*; the term sum-integral is explained below. In the particular case of a one-dimensional survival time, the additive interval function corresponding to the multiplicative interval function S is the negative hazard measure $A^{\pm} = -A$, considered as an interval function by the natural definition

$$A^{\pm}(s,t) = A^{\pm}(t) - A^{\pm}(s) = -(A(t) - A(s)).$$

It satisfies the four parallel properties:

(i) $A^{\pm}(t,t) = 0$ for all t;
(ii) $\lim_{t \downarrow s} A^{\pm}(s,t) = 0$ for all s;
(iii) $A^{\pm}(s,t) + A^{\pm}(t,u) = A^{\pm}(s,u)$ for all $s \le t \le u$;
(iv) $A^{\pm}$ is dominated by a finite measure; in fact,

$$|A^{\pm}(s,t)| \leq A(t) - A(s)$$

for all $s < t \leq \tau$, so $A^{\pm}$ is dominated by A.

An interval function (denoted by α) satisfying these four properties, with A replaced by any other finite measure in (iv), is called a *dominated additive interval function.* (The symbol α, therefore, has a completey different meaning in this section from the rest of the book).

Here is the general result connecting S to $A^{\pm}$, including the needed definition of *sum-integral* parallel to the product-integral we are already familiar with:

Theorem X.3.1. *Let μ be a dominated multiplicative interval function. Then there exists a unique dominated, additive interval function α such that*

$$\mu(s,t) = \prod_{(s,t]} (1 + \mathrm{d}\alpha) \quad \textit{for all } s \leq t \tag{10.3.8}$$

and α is given by

$$\alpha(s,t) = \int_{(s,t]} \mathrm{d}(\mu - 1). \tag{10.3.9}$$

Conversely, if α is a dominated additive interval function, then there exists a unique dominated, multiplicative interval function μ such that (10.3.9) *holds, and μ is given by* (10.3.8). *The product- and sum-integrals are both defined as limits, as the mesh of the partition tends to zero, of approximating finite products and sums over subintervals partitioning the interval* $(s,t]$.

We illustrate the meaning of the sum-integral by considering again the application to a one-dimensional survival function. Take μ to be the multiplicative interval function S, and α to be the additive interval function $A^{\pm} = -A$, the negative hazard. Relation (10.3.8) is then just the formula

$$\frac{S(t)}{S(s)} = \prod_{(s,t]} (1 - \mathrm{d}A),$$

whereas (10.3.9) is the less familiar

$$-A(s,t) = \int_{(s,t]} \mathrm{d}(S - 1)$$

or, taking $s = t_0 < t_1 < \cdots < t_n = t$ to be a partition of $(s,t]$, by definition of the sum-integral,

$$-A(s,t) = \lim \sum_{i=1}^{n} (S - 1)(t_{i-1}, t_i) = \lim \sum_{i=1}^{n} \left(\frac{S(t_i)}{S(t_{i-1})} - 1\right)$$
$$= \lim \sum_{i=1}^{n} \frac{S(t_i) - S(t_{i-1})}{S(t_{i-1})}$$

$$= -\lim \sum_{i=1}^{n} \frac{F(t_i) - F(t_{i-1})}{S(t_{i-1})}$$

$$= -\int_s^t \frac{dF(u)}{S(u-)}.$$

Here, the limit is taken as $\max_{i=1}^n |t_i - t_{i-1}| \to 0$ and the last step follows by dominated convergence; the final integral is an ordinary measure-theoretic Lebesgue–Stieltjes integral.

We are concerned with the multivariate generalization of Theorem X.3.1. In fact, there is also a generalisation to allow *matrix-valued* α and μ, which Gill and Johansen (1990) show is the key to understanding matrix product-integration and its application to Markov processes (see Sections II.6 and IV.4). The correspondence between multiplicative and additive interval functions becomes the correspondence between transition probability matrices (multiplicative, by the Markov property) and intensity measures (additive, being matrices of ordinary additive measures).

The generalization we need now keeps α and μ scalar, but lets the time variable t be replaced by the vector $\mathbf{t}$. The interval $(s, t]$ is replaced by the hyperrectangle $(\mathbf{s}, \mathbf{t}]$. Partitions of intervals by subintervals become partitions of hyperrectangles by subhyperrectangles (one needs only consider "rectangular partitions" obtained as a product of partitions of each time axis separately). We consider additive and multiplicative rectangle functions $\alpha(\mathbf{s}, \mathbf{t})$, $\mu(\mathbf{s}, \mathbf{t})$, where additivity and multiplicativity refer to partitions of rectangles. In this setup, the definitions of dominated multiplicative and additive interval functions remain completely unaltered, and Theorem X.3.1 remains exactly the same.

To apply Theorem X.3.1 to multivariate survival functions, we need to find a multivariate analogue of the multiplicative interval function $S(s, t) = \mathrm{P}(X > t | X > s) = S(t)/S(s)$. Note we meet a ratio rather than a difference of the underlying survival function S. This suggests looking at generalized ratios (i.e., ratios of ratios of ratios ...) just as one looks at generalized differences (differences of differences of differences ...) when generating an *additive* interval function, i.e., a measure, from a function. So the natural definition is the following:

$$S(\mathbf{s}, \mathbf{t}) = \prod_{\varnothing \subseteq C \subseteq E} S(\mathbf{s}_C, \mathbf{t}_{E \setminus C})^{(-1)^{|C|}}. \tag{10.3.10}$$

The right-hand side is the value of the survival function S at the top corner $\mathbf{t}$ [written as $(\mathbf{s}_C, \mathbf{t}_{E \setminus C})$ with $C = \varnothing$] of the rectangle $(\mathbf{s}, \mathbf{t}]$, divided by its values at all corners one step down from the top (C has one element), multiplied by its values one step further down (C has two elements), and so on, until one reaches the bottom corner $\mathbf{s}$ (where $C = E$).

We call the interval function S defined in this way the *iterated odds ratio* for the interval (hyperrectangle) $(\mathbf{s}, \mathbf{t}]$. Its multiplicativity [condition (iii)] is easy to see: Consider two adjacent hyperrectangles whose union is another bigger hyperrectangle. The two have a common face with 2^{k-1} corners. But the survival function at all these corners always appears in the denominator of one of the interval functions and in the numerator of the other. On multiplication, the terms coming from the common face disappear and one is left with the terms at the corners of the big hyperrectangle, in numerator or denominator as required.

It is also easy to check that the iterated odds ratios are "equal to one on the diagonal" [condition (i)] and are "right-continuous" [condition (ii)]. To apply Theorem X.3.1, it remains to check the domination property (iv). This is delicate (and amusing once the idea is understood), but we will leave it until the end of this section, and first try to better understand the interpretation of the iterated odds ratios and the modelling consequences of the result of Theorem X.3.1. To start with, we must explain the name.

In fact, $S(\mathbf{s}, \mathbf{t})$ has an interpretation in terms of the $2 \times 2 \times \cdots \times 2$ contingency table for the events $s_i < X_i \le t_i$ versus $X_i > t_i$, with respect to the conditional distribution of $\mathbf{X}$ given $\mathbf{X} \gg \mathbf{s}$. [The same contingency table was used by Clayton (1978), inspired by Irwin (1949) and Lancaster (1949, 1965), to motivate the frailty model for dependent survival times we studied in Chapter IX.] Consider first the two-dimensional case: We have by definition

$$S((s_1, s_2), (t_1, t_2)) = \frac{S(t_1, t_2)S(s_1, s_2)}{S(s_1, t_2)S(t_1, s_2)}$$

since there are four subsets C to consider, two of them ($\varnothing$ and E) having an even number of elements, and two ($\{1\}$ and $\{2\}$) having an odd number. We can now rewrite $S(\mathbf{s}, \mathbf{t})$ as

$$S((s_1, s_2), (t_1, t_2)) = \frac{S(t_1, t_2)/S(s_1, t_2)}{S(t_1, s_2)/S(s_1, s_2)}$$

$$= \frac{\mathrm{P}(X_1 > t_1 | X_2 > t_2)/\mathrm{P}(X_1 > s | X_2 > t_2)}{\mathrm{P}(X_1 > t_1 | X_2 > s_2)/\mathrm{P}(X_1 > s_1 | X_2 > s_2)}.$$

So, $S(\mathbf{s}, \mathbf{t})$ is the ratio of the conditional odds for $X_1 > t_1$ against $X_1 > s_1$, under the conditions $X_2 > t_2$ and $X_2 > s_2$, respectively. If X_1 and X_2 are independent, this *odds ratio* will equal 1. "Positive dependence" between X_1 and X_2 will express itself in an odds ratio larger than 1 since "increasing X_2 leads to a higher odds on X_1 being large." Negative dependence corresponds to an odds ratio smaller than 1. In fact, we will see in a moment that if the odds ratio equals 1 for all $\mathbf{s} \le \mathbf{t}$, then X_1 and X_2 are independent. So, in two dimensions, $S - 1$ is a measure of dependence indexed by all intervals $(\mathbf{s}, \mathbf{t}]$.

In one dimension, the odds "ratio" is just the odds itself, $\mathrm{P}(X_1 > t_1 | X_1 > s_1)$. In higher dimensions, the k-dimensional iterated odds ratio is the ratio of two $k-1$-dimensional iterated odds ratios; i.e., the ratio of the iterated odds ratios for $(X_1, \ldots, X_{k-1})$ and the interval $((s_1, \ldots, s_{k-1}), (t_1, \ldots, t_{k-1})]$, conditional on $X_k > t_k$ and conditional on $X_k > s_k$. Now it measures multidimensional dependence or interaction: If the dependence between $X_1, \ldots, X_{k-1}$ increases as X_k increases, one has a positive interaction (increasing interdependence) between $X_1, \ldots, X_k$ and the iterated odds ratio is larger than 1.

The result of Theorem X.3.1 [when we have verified the domination property (iv)] is that a dominated *additive* interval function (therefore, an ordinary signed measure) exists, let us call it L, such that

$$S(\mathbf{s},\mathbf{t}) = \prod_{(\mathbf{s},\mathbf{t}]} (1 + \mathrm{d}L), \tag{10.3.11}$$

where the product-integral is understood as the limit of approximating finite products over rectangular partitions of the hyperrectangle $(\mathbf{s},\mathbf{t}]$ into small subhyperrectangles. We call L the *iterated odds ratio measure* or *cumulant measure* and consider it as a measure of k-dimensional interaction (a measure of dependence when $k = 2$ and just a description of the marginal distribution when $k = 1$). Note that there is an L-measure, denoted L_C, for each subset of components $\mathbf{X}_C$ of $\mathbf{X}$. Since the odds ratio for a small rectangle $(\mathbf{t},\mathbf{t}+\mathrm{d}\mathbf{t}]$ is

$$S(\mathbf{t},\mathbf{t}+\mathrm{d}\mathbf{t}) = 1 + L(\mathrm{d}\mathbf{t})$$

and a ratio of 1 corresponds to independence, we may interpret an L of zero as corresponding to zero-interaction or independence; a positive L corresponds to positive interaction or dependence, and similarly for a negative L. Of course, things may be more complicated: L may take different signs in different regions of space.

By Theorem X.3.1, we may calculate L by the sum-integral

$$L(\mathbf{s},\mathbf{t}) = \int_{(\mathbf{s},\mathbf{t}]} L(\mathrm{d}\mathbf{u}) = \int_{(\mathbf{s},\mathbf{t}]} \mathrm{d}(S - 1); \tag{10.3.12}$$

in other words, the L-measure of a rectangle $(\mathbf{s},\mathbf{t}]$ is approximated by just adding together the deviations from independence (or interactions) $S(\mathbf{u},\mathbf{u}+\mathrm{d}\mathbf{u}) - 1$ of small rectangles $(\mathbf{u},\mathbf{u}+\mathrm{d}\mathbf{u}]$ forming a partition of $(\mathbf{s},\mathbf{t}]$.

Though one can build the *interval function* S from its L-measure, this does not quite imply that one can reconstruct the *survival function* S from L also. There are edge effects just as when building the multivariate survival function from its hazards. However, edge effects just mean that lower-dimensional or marginal survival functions are also involved, and one can hope to recursively build up S not just from L but from all L_C, $C \subseteq E$.

Taking in (10.3.10) and (10.3.11) the special choice $\mathbf{s} = \mathbf{0}$ and interchanging C and $E \backslash C$, we find

$$\prod_{(\mathbf{0},\mathbf{t}]} (1 + \mathrm{d}L) = S(\mathbf{0},\mathbf{t}) = \prod_{\varnothing \subseteq C \subseteq E} S(\mathbf{0}_C, \mathbf{t}_{E\backslash C})^{(-1)^{|C|}} = \prod_{\varnothing \subseteq C \subseteq E} S_C(\mathbf{t}_C)^{(-1)^{|E\backslash C|}}. \tag{10.3.13}$$

For instance, in two dimensions,

$$S(\mathbf{0}, \mathbf{t}) = \frac{S(t_1, t_2)}{S_1(t_1)S_2(t_2)}.$$

To recover the original survival function $S(\mathbf{t})$ from (10.3.13), we must first multiply by all the $k - 1$-dimensional marginals. This can be achieved by multiplying by all product-integrals of $k - 1$-dimensional L measures, but this also has an effect on the $k - 2$-dimensional margins, and so on. It turns out that the whole process can simply be continued, *multiplying* by *all* lower-dimensional product-inegrals of L-measures results exactly in the desired survival function (no *division* is required):

$$S(\mathbf{t}) = \prod_{\varnothing \subset C \subseteq E} \prod_{(\mathbf{0}_C, \mathbf{t}_C]} (1 + \mathrm{d}L_C). \tag{10.3.14}$$

If all X_i are independent, then all iterated odds ratio measures from dimension 2 onward are zero, as can be seen from (10.3.12). Equation (10.3.14) then expresses the joint survival function as a product of its one-dimensional marginals since the one-dimensional L_i are just the one-dimensional negative hazards $-A_i$, and (10.3.14) becmes $S(\mathbf{t}) = \prod_i \prod_{(0, t_i]}(1 - \mathrm{d}A_i)$.

The iterated odds ratio measures can be expressed in terms of the conditional hazard measures. The idea of this representation is to use the inclusion–exclusion principle to rewrite conditional probabilities (given $\mathbf{X} \gg \mathbf{t}$) of events such as $X_i > t_i + \mathrm{d}t_i$ in terms of the complementary events $X_i \in (t_i, t_i + \mathrm{d}t_i]$. The result is the pretty formula

$$1 + L(\mathrm{d}\mathbf{t}) = \prod_{C \subseteq E} \left(1 + \sum_{\varnothing \subset B \subseteq C} (-1)^{|B|} A_{B|E \setminus B}(\mathrm{d}\mathbf{t}_B | \mathbf{t}_{E \setminus B})\right)^{(-1)^{|E \setminus C|}}. \tag{10.3.15}$$

Though this formula is actually very easy to interpret heuristically (it is the centerpiece in the approach of Gill (1990) to the Dabrowska representation), its mathematical interpretation and justification are quite intricate since it involves several nonstandard operations on measures which have to be properly defined and shown to have the expected properties. The main use of the formula is in proving asymptotic results by the functional delta-method, and we mention it again briefly in the next section.

Combination of (10.3.14) and (10.3.15) gives an expression for the multivariate survival function in terms of *all* the conditional hazard measures $A_{B|C}$ for $B, C \subseteq E$, $B \cap C = \varnothing$. In the discrete case, the representation can be evaluated by taking finite products over the atoms of the distribution of $\mathbf{X}$. In the continuous case, when S, A, and L all have densities, simple explicit formulas are available. Here we give the formulas for dimensions $k = 1, 2, 3$, and 4, writing $l_{12\ldots k}$ for the density of $L_{\{1,2,\ldots,k\}}$, $\alpha_{12\ldots j}$ for $\alpha_{\{1,2\ldots j\}|\{j+1,\ldots,k\}}$, etc.

$$l_1 = -\alpha_1 \quad (k = 1),$$

$$l_{12} = \alpha_{12} - \alpha_1\alpha_2 \quad (k = 2),$$

$$l_{123} = -\alpha_{123} + \alpha_{12}\alpha_3 + \alpha_{13}\alpha_2 + \alpha_{23}\alpha_1 - 2\alpha_1\alpha_2\alpha_3 \quad (k = 3),$$

$$l_{1234} = \alpha_{1234} - \alpha_{123}\alpha_4 - \alpha_{124}\alpha_3 - \alpha_{134}\alpha_2 - \alpha_{234}\alpha_1 - \alpha_{12}\alpha_{34} - \alpha_{13}\alpha_{24}$$
$$- \alpha_{24}\alpha_{23} + 2\alpha_{12}\alpha_3\alpha_4 + 2\alpha_{13}\alpha_2\alpha_4 + 2\alpha_{14}\alpha_2\alpha_3 + 2\alpha_{34}\alpha_1\alpha_2$$
$$+ 2\alpha_{24}\alpha_1\alpha_3 + 2\alpha_{23}\alpha_1\alpha_4 - 6\alpha_1\alpha_2\alpha_3\alpha_4 \quad (k = 4),$$

and so on. Each line can be obtained from the preceding one by "differentiating with respect to the new variable" and using the rule

$$\mathrm{d}_k\alpha_{1\ldots j} = -\alpha_{1\ldots j,k} + \alpha_{1\ldots j}\alpha_k.$$

Note that the sum of all the coefficients in the kth row for $k = 2$ and onward is zero. This reflects the following fact: If the k variables are independent of one another, then the conditional hazard measure of one group of variables given another is equal to the product of marginal (one-dimensional) hazards taken over the first group; and at independence, all the odds measures (from the second onward) are zero.

In the general formula, one sums over all partitions of $\{1,\ldots,k\}$, with coefficients equal to minus one to the power of the size of the partition minus one, times the factorial of the same number. For instance, the '6' in the last row of this formula is $(4 - 1)!$, since we have a partition of the set $\{1, 2, 3, 4\}$ into the four elements $\{1\}$, $\{2\}$, $\{3\}$, $\{4\}$. The same formulas occur in the theory of cumulants, used to describe multivariate dependence in both continuous distributions and discrete ones (contingency tables); see Streitberg (1990). Dabrowska (1992) used the theory of iterated odds ratio measures for proposing interesting semiparametric models of dependence.

Given a complete set of marginal and conditional hazard measures, we now have two quite different ways of rebuilding the survival functuion: one via the Péano series (10.3.7) with (10.3.4) and (10.3.5) and the other via the product-integrals of the cumulant measures (10.3.14) and (10.3.15). In fact, from dimension 3 onward, many hybrid forms are possible since one can mix the rules for getting the higher marginal survival functions from the lower. We return to this choice when considering estimation in the next section.

X.3.1.4. *Checking the Domination Property*

It remains to verify the dominaton property of the iterated odds ratios $S(\mathbf{s}, \mathbf{t})$. This rather technical material is not needed elsewhere.

Let us first look at the two-dimensional case which will give the required insight for the general case. For $\mathbf{s} \leq \mathbf{t} \leq \tau$, $\mathbf{s} \neq \mathbf{t}$, we have

$$|S(\mathbf{s}, \mathbf{t}) - 1| = \left| \frac{S(t_1, t_2)S(s_1, s_2)}{S(s_1, t_2)S(t_1, s_2)} - 1 \right|$$
$$\leq S^*(\tau)^{-2} |S(t_1, t_2)S(s_1, s_2) - S(s_1, t_2)S(t_1, s_2)|,$$

where we write $S^*(\tau)$ as shorthand for $\mathrm{P}(\mathbf{X} \gg \tau \text{ or } \mathbf{X} = \tau)$; this may be different from $\mathrm{P}(\mathbf{X} \geq \tau)$; taking account of the difference allows us to get a slightly stronger result. Now let a, b, c, d be the probabilities in the 2×2 table:

	$X_1 \in (s_1, t_1]$	$X_1 > t_1$
$X_2 \in (s_2, t_2]$	a	b
$X_2 > t_2$	c	d

Then the last inequality can be rewritten as

$$\begin{aligned} |S(\mathbf{s},\mathbf{t}) - 1| &\le S^*(\boldsymbol{\tau})^{-2}|d(a+b+c+d) - (c+d)(b+d)| \\ &= S^*(\boldsymbol{\tau})^{-2}|ad - bc| \\ &\le S^*(\boldsymbol{\tau})^{-2}(a + bc) \\ &= S^*(\boldsymbol{\tau})^{-2}(\mathrm{P}(\mathbf{X} \in (\mathbf{s},\mathbf{t}]) + \mathrm{P}(X_1 \in (s_1,t_1])\mathrm{P}(X_2 \in (s_2,t_2])). \end{aligned}$$

The right-hand side, a constant time the joint probability measure of X_1 and X_2 plus the product of their marginals, is a finite measure on $(\mathbf{0}, \boldsymbol{\tau}]$; hence, the domination property holds.

Exactly the same kind of bound holds, in general, by taking account of the same magic cancellation of unwanted terms. Let $\boldsymbol{\tau}$ be fixed and satisfy $S^*(\boldsymbol{\tau}) = \mathrm{P}(\mathbf{X} \gg \boldsymbol{\tau} \text{ or } \mathbf{X} = \boldsymbol{\tau}) > 0$. From (10.3.10), we can write, for $\mathbf{s} \le \mathbf{t} \le \boldsymbol{\tau}$, $\mathbf{s} \neq \mathbf{t}$,

$$S(\mathbf{s},\mathbf{t}) - 1 = \frac{\prod_{\text{even } C} S(\mathbf{s}_C, \mathbf{t}_{E\setminus C}) - \prod_{\text{odd } C} S(\mathbf{s}_C, \mathbf{t}_{E\setminus C})}{\prod_{\text{odd } C} S(\mathbf{s}_C, \mathbf{t}_{E\setminus C})},$$

where $\emptyset \subseteq C \subseteq E$. Now, by the inclusion–exclusion principle,

$$\begin{aligned} &\mathrm{P}(\mathbf{X} \gg \mathbf{s},\, X_i > t_i \text{ for all } i \in E\setminus C) \\ &\quad = \mathrm{P}(\mathbf{X} \gg \mathbf{s}) - \mathrm{P}(\mathbf{X} \gg \mathbf{s},\, X_i \le t_i \text{ for some } i \in E\setminus C) \\ &\quad = \mathrm{P}(\mathbf{X} \gg \mathbf{s}) - \sum_{i \in E\setminus C} \mathrm{P}(\mathbf{X} \gg \mathbf{s}, X_i \le t_i) \\ &\qquad + \sum_{i \neq j \in E\setminus C} \mathrm{P}(\mathbf{X} \gg \mathbf{s}, X_i \le t_i, X_j \le t_j) - \cdots \end{aligned}$$

or, in other words,

$$\begin{aligned} S(\mathbf{s}_C, \mathbf{t}_{E\setminus C}) &= S(\mathbf{s}) + \sum_{\emptyset \subset B \subseteq E\setminus C} (-1)^{|B|}\mathrm{P}(\mathbf{X} \in (\mathbf{s}_B, \mathbf{t}_B] \times (\mathbf{s}_{E\setminus B}, \boldsymbol{\infty}_{E\setminus B})) \\ &= \sum_{\emptyset \subseteq B \subseteq E\setminus C} (-1)^{|B|}\mathrm{P}(\mathbf{X} \in (\mathbf{s}_B, \mathbf{t}_B] \times (\mathbf{s}_{E\setminus B}, \boldsymbol{\infty}_{E\setminus B})). \end{aligned}$$

Interchanging the roles of C and $E\setminus C$ in the numerator and neglecting a possible sign change (if $|E|$ is odd), we get

$$S(\mathbf{s},\mathbf{t}) - 1 = \pm\frac{\prod_{\text{even } C} \sum_{\emptyset \subseteq B \subseteq C} \phi_B - \prod_{\text{odd } C} \sum_{\emptyset \subseteq B \subseteq C} \phi_B}{\prod_{\text{odd } C} S(\mathbf{s}_C, \mathbf{t}_{E\setminus C})}, \tag{10.3.16}$$

where

$$\phi_B = (-1)^{|B|}\mathbf{P}(\mathbf{X} \in (\mathbf{s}_B, \mathbf{t}_B] \times (\mathbf{s}_{E\setminus B}, \boldsymbol{\infty}_{E\setminus B})).$$

Now when we expand the numerator of (10.3.16), an amazing cancellation occurs: Products of ϕ_B where the sets B do not cover E cancel out, leaving just products of sets which do cover E. Before proving this, we illustrate it when $E = \{1, 2\}$:

$$\begin{aligned}\prod_{\text{even } C}\sum_{B \subseteq C}\phi_B - \prod_{\text{odd } C}\sum_{B \subseteq C}\phi_B &= (\phi_{12} + \phi_1 + \phi_2 + \phi_\varnothing)\phi_\varnothing \\ &\quad - (\phi_1 + \phi_\varnothing)(\phi_2 + \phi_\varnothing) \\ &= (\phi_{12}\phi_\varnothing - \phi_1\phi_2),\end{aligned} \tag{10.3.17}$$

where $\{1,2\} \cup \varnothing = E$, $\{1\} \cup \{2\} = E$.

In general, consider one element $i \in E$ and split the sums and products in (10.3.17) according to whether or not i is included in B, C: We get

$$\left(\prod_{\text{even } C, i \notin C}\sum_{B \subseteq C}\phi_B\right)\prod_{\text{odd } C, i \notin C}\left(\sum_{B \subseteq C}\phi_B + \sum_{B \subseteq C}\phi_{B \cup \{i\}}\right)$$
$$-\left(\prod_{\text{odd } C, i \notin C}\sum_{B \subseteq C}\phi_B\right)\prod_{\text{even } C, i \notin C}\left(\sum_{B \subseteq C}\phi_B + \sum_{B \subseteq C}\phi_{B \cup \{i\}}\right).$$

The terms which nowhere include i are then

$$\left(\prod_{\text{even } C, i \notin C}\sum_{B \subseteq C}\phi_B\right)\prod_{\text{odd } C, i \notin C}\sum_{B \subseteq C}\phi_B - \left(\prod_{\text{odd } C, i \notin C}\sum_{B \subseteq C}\phi_B\right)\left(\prod_{\text{even } C, i \notin C}\sum_{B \subseteq C}\phi_B\right)$$
$$= 0;$$

thus, each term in the expansion of (10.3.17)—a sum of products of ϕ_B—includes a B containing i.

The result is that (10.3.16) can be bounded in absolute value by $S^*(\tau)^{-2^{|E|-1}}$ times a sum of products of ϕ_B, where each term has $\bigcup B = E$. Consider such a term $\prod \phi_{B_i}$. For each B_i, choose $C_i \subseteq B_i$ such that $\bigcup C_i = E$ and the C_i are disjoint. Now bound $\prod \phi_{B_i}$ by $\prod \mathbf{P}(\mathbf{X}_{C_i} \in (\mathbf{s}_C, \mathbf{t}_{C_i}])$. These are finite measures, so we have obtained the required result.

X.3.2. Nonparametric Estimation

Now we consider the statistical problem of estimating the survival function given a randomly censored sample from it. Let $\mathbf{X}_1, \ldots, \mathbf{X}_n$ be independent survival vectors from the survival function S and, independently of them, let $\mathbf{U}_1, \ldots, \mathbf{U}_n$ be independent censoring vectors from the censoring survival function $\mathbf{S}_{\mathbf{U}}$. Consider estimation of S ($= S_{\mathbf{X}}$) on a rectangle $(\mathbf{0}, \boldsymbol{\tau}]$, where $\boldsymbol{\tau}$ satisfies $S(\boldsymbol{\tau})S_{\mathbf{U}}(\boldsymbol{\tau}-) > 0$. The observations are $\tilde{\mathbf{X}}_i$, $\mathbf{D}_i$, $i = 1, \ldots, n$, where

$$\tilde{X}_{ij} = X_{ij} \wedge U_{ij}, \qquad D_{ij} = I(X_{ij} \leq U_{ij}).$$

Recall definition (10.3.2) of the conditional hazard $A_{C|D}$. To motivate estimators, think of the discrete case and write $\tilde{\mathbf{X}}$, $\mathbf{D}$ for a generic observation and $\mathbf{X}$, $\mathbf{U}$ for the unobservable random vectors which generated it. Let $\mathbf{t} \in \mathbb{R}_+^k$ be fixed and let $C, D \subseteq E$ be disjoint. On the event $\tilde{\mathbf{X}}_{C\cup D} \geq \mathbf{t}_{C\cup D}$, the events

$$\mathbf{X}_C = \mathbf{t}_C \quad \text{and} \quad \tilde{\mathbf{X}}_C = \mathbf{t}_C, \mathbf{D}_C = \mathbf{1}_C$$

coincide. Conditional on $\tilde{\mathbf{X}}_{C\cup D} \geq \mathbf{t}_{C\cup D}$, $\mathbf{X}_{C\cup D}$ and $\mathbf{U}_{C\cup D}$ remain independent. Therefore,

$$\mathrm{P}(\tilde{\mathbf{X}}_C = \mathbf{t}_C, \mathbf{D}_C = \mathbf{1}_C | \tilde{\mathbf{X}}_{C\cup D} \geq \mathbf{t}_{C\cup D}) = A_{C|D}(\{\mathbf{t}_C\}|\mathbf{t}_D).$$

This suggests (whether the distributions are continuous or not) estimating $A_{C|D}(\cdot|\mathbf{t}_D)$ (where D may be empty) by the discrete measure with atoms

$$\frac{\#\{i: \tilde{\mathbf{X}}_{iC} = \mathbf{t}_C, \mathbf{D}_{iC} = \mathbf{1}_C, \tilde{\mathbf{X}}_{iD} \geq \mathbf{t}_D\}}{\#\{i: \tilde{\mathbf{X}}_{iC\cup D} \geq \mathbf{t}_{C\cup D}\}}$$

at each $\mathbf{t}_C$ such that the numerator is nonzero.

This is simply a multivariate Nelson–Aalen estimator. Define

$$N_{C|D}(\mathbf{t}_C|\mathbf{t}_D) = \#\{i: \tilde{\mathbf{X}}_{iC} \leq \mathbf{t}_C, \mathbf{D}_{iC} = \mathbf{1}_C, \tilde{\mathbf{X}}_{iD} \geq \mathbf{t}_D\},$$
$$Y_B(\mathbf{t}_B) = \#\{i: \tilde{\mathbf{X}}_{iB} \geq \mathbf{t}_B\}.$$

Then we have set

$$\hat{A}_{C|D}(\mathbf{t}_C|\mathbf{t}_D) = \int_{(\mathbf{0}_C, \mathbf{t}_C]} \frac{N_{C|D}(\mathrm{d}\mathbf{s}_C|\mathbf{t}_D)}{Y_{C\cup D}(\mathbf{s}_C, \mathbf{t}_D)}. \tag{10.3.18}$$

In Section X.3.3, we will discuss the possibility of using multivariate time martingale theory to derive properties of these estimators. For the time being, we note that the processes $N_{C|D}$ and Y_C, divided by n, are (in the i.i.d. case) simple multivariate empirical processes to which the Glivenko–Cantelli theorem and the Donsker theorem can be applied (Pollard, 1984). Multidimensional integration and product-integration and basic algebraic operations are all that are needed to build from these empirical processes the estimators $\hat{A}_{C|D}$ and corresponding $\hat{S}$. Indeed, consistency and asymptotic normality can be derived as an application of the functional delta-method; see Gill (1990) and van der Laan (1990) for more details.

As we have implied earlier, the Volterra estimator based on (10.3.4)–(10.3.7) and (10.3.18) and the Dabrowska estimator based on (10.3.14), (10.3.15), and (10.3.18) are generally different, only coinciding in some degenerate cases (e.g., when there is no censoring at all, *both* simplify to the empirical distribution survival function). Neither is generally a proper survival function: Negative mass can be assigned to some points (in fact, the Dabrowska estimator assigns a nonvanishing proportion of negative mass, to a nonvanishing fraction of points). Neither estimator is generally asymptotically efficient. A curious exception is that at complete independence of all censoring and all survival times, the Dabrowska estimator is efficient. Generally, the Dabrowska estimator does better (and better than the many competing estimators in the literature). The Volterra estimator is especially poor at large survival times where the risk sets are small.

Both estimators are easy to compute: Since everything becomes discrete, one can build up solutions recursively by always moving outward on the lattice of possible mass points. The Volterra equation (10.3.4) can be used directly in this recursive computation (the Péano series does not have to be separately calculated). Similarly, the Dabrowska estimator can be calculated recursively from (10.3.14) and the *empirical odds measures*, where the odds ratio for a quadrant $[\mathbf{t}_C, \infty_C)$, needed in building up L_C, is calculated in the obvious way from the observations with $\tilde{\mathbf{X}}_{iC} \geq \mathbf{t}_C$, for each of which it is known to which cell of the $2^{|C|}$ table formed by the categories "$= t_j$" versus "$> t_j$," $j \in C$, the underlying $\mathbf{X}_{iC}$ belongs. To be specific, if for the ith observation $\tilde{\mathbf{X}}_{iC} \geq \mathbf{t}_C$, then we know for each $j \in C$ that $X_{ij} = t_j$ if $\tilde{X}_{ij} = t_j$ and $D_{ij} = 1$ and otherwise that $X_{ij} > t_j$. Now we can build the $2 \times 2 \times \cdots \times 2$ table based on these observations, categorized according to whether the jth coordinate, for $j \in C$, is known to be $\geq t_j$ or also known to be $> t_j$. This is actually a table where one category of each variable j, $j \in C$, is contained in the other, but that is not important. For each $\mathbf{t}_C$, we compute the iterated odds ratio for the corresponding table. For most $\mathbf{t}_C$, the empirical odds ratio equals one and the L_C measure is zero; the only points which can supply a nonzero atom to L_C are those $\mathbf{t}_C$ for which, for each $j \in C$, there is an observation i with $\tilde{X}_{ij} = t_j$, $D_{ij} = 1$. Thus, one looks only at the lattice of points formed by the uncensored observations in each dimension and takes a product in (10.3.14) over the points which lie in the hyperrectangle of interest.

Another estimator with many similar features has recently been described by Prentice and Cai (1991, 1992). They showed that the covariance function between the empirical counting processes for the marginal failure times enters into another Volterra-type equation for the survival function, so that a natural estimator of this covariance can be used to build up an estimator of the survival function. The value of this approach is that just like the L-measures, Prentice and Cai's covariance function gives a natural description of dependence structure and hence a way to come up with interesting semiparametric models of dependence. For instance, the frailty models of Chapter IX could be alternatively specified through L-measures and covariance structure of special form. It seems as though the multivariate hazards themselves are not so useful in this respect.

The inefficiency and the nonequivalence of the estimators is troublesome and related to the fact that neither is a nonparametric maximum likelihood estimator. The NPMLE for this problem is not well understood at all. It is severely nonunique. (For further discussion, see the Bibliographic Remarks to this section at the end of the present chapter.) The Volterra estimator is based on a nice parametrization of the model by the unconditional hazard measures A_C, $C \subseteq E$, but the estimator does not make use of all the data. The Dabrowska estimator uses more of the data, but takes no account of the many constraints which relate all the conditional hazard functions $A_{C|D}$ (by the Volterra representation, we can express the conditional ones in terms of the unconditional).

An optimal estimator of S is still unknown. It will presumably be based on a variant of parametric maximum likelihood applied to a suitably smoothed or discretized version of the originally nonparametric model.

X.3.3. Multivariate Time Martingales

The estimators of the previous section will be reasonable not just under the specific random censorship model considered there, but also under a weaker "independent censoring" assumption. For the multivariate (and conditional) Nelson–Aalen estimators to be sensible, it is just necessary that observations for which $\mathbf{X}_{iC}$ is known to lie in $[\mathbf{t}_C, \mathbf{\infty}_C)$ (in other words, observations for which $\tilde{\mathbf{X}}_{iC} \geq \mathbf{t}_C$), are a random sample from the conditional distribution of $\mathbf{X}_C$ given $\mathbf{X}_C \geq \mathbf{t}_C$.

Obviously, also truncation and filtering is allowed under the same hypothesis: We just need that the "risk set" at $\mathbf{t}_C$ is typical for the uncensored population conditional on $\mathbf{X}_C \geq \mathbf{t}_C$. In this sense, there is a complete analogy with univariate survival analysis and one can go on to make this informal analogy mathematically exact through the notion of *weak martingale*, and through the theory of stochastic integration of predictable processes with respect to weak martingales in multivariate time.

Consider a stochastic process $M = (M(\mathbf{t}), \mathbf{0} \leq \mathbf{t} \leq \mathbf{\tau})$, adapted to a filtration $(\mathscr{F}_\mathbf{t})$. A filtration in multivariate time has to be increasing with respect to the natural *partial order* on the time variable, $\mathbf{s} \leq \mathbf{t}$ if and only if $s_i \leq t_i$ for all i. For pairs of time points which are not ordered, no restriction is put on the two σ-algebras. The process M is called a *weak martingale* if the conditional expectation of the generalized difference of M over a generalized interval $(\mathbf{s}, \mathbf{t}]$ given the past at the lower end-point, $\mathscr{F}_\mathbf{s}$, is zero for all $\mathbf{s} \leq \mathbf{t}$; cf. (2.1.7) and (2.1.8), the latter rewritten as $\mathrm{E}(M(\mathbf{t}) - M(\mathbf{s}) | \mathscr{F}_\mathbf{s}) = 0$.

We can define a predictable process in multivariate time in the obvious way as a member of the class of processes generated by the left-continuous, adapted processes. It is now possible also to develop a theory of stochastic integration. Under integrability conditions, the integral of a predictable process with respect to a weak martingale is again a weak martingale. The intuitive background to this is, of course, the fact that since $\mathrm{E}(H(\mathbf{u}) M(\mathrm{d}\mathbf{u}) | \mathscr{F}_{\mathbf{u}-}) = 0$ for all $\mathbf{u} \in (\mathbf{s}, \mathbf{t}]$, addition over small hyperrectangles $[\mathbf{u}, \mathbf{u} + \mathrm{d}\mathbf{u})$ partitioning $[\mathbf{s} + \mathrm{d}\mathbf{s}, \mathbf{t} + \mathrm{d}\mathbf{t}) = (\mathbf{s}, \mathbf{t}]$ should give the weak martingale property $\mathrm{E}(\int_{(\mathbf{s}, \mathbf{t}]} H \, \mathrm{d}M | \mathscr{F}_\mathbf{s}) = 0$; cf. Section II.1.

This theory is described in detail by Pons (1986), who went on to show that some properties of the multivariate Nelson–Aalen estimators can be obtained exactly as in the univariate case by considering stochastic integrals of the predictable process Y^{-1} with respect to the *counting process weak martingale* $N - \int Y \, \mathrm{d}A$. This leads to proposals for tests for independence in the bivariate case since independence can then be characterized by saying that the joint hazard measure is the product of the marginal measures. We also see that *independent censoring and filtering*, in the sense of having a predictable observation process J, preserves the weak martingale property of these *compensated counting processes.*

However, there are two drawbacks to this at first sight promising theory. The first is that there is no weak martingale central limit theory. The reason for this is that though weak martingales can be seen to have uncorrelated increments across ordered time points, in the sense that $\mathrm{E}(M(d\mathbf{s})M(d\mathbf{t})) = 0$ for $\mathbf{s} \leq \mathbf{t}$, no special structure is given to correlations across unordered time points, and multivariate time is only partially and not totally ordered. Thus, even if predictable variation processes could be introduced, many very different weak martingales (even if continuous and with deterministic predictable variation) would share the same predictable variation. So there is no characterization of an essentially unique Gaussian weak martingale through predictable variation as is the case in one-dimensional time, and hence no corresponding convergence theorem. For instance, even though the multivariate Nelson–Aalen estimators can be shown to be asymptotically Gaussian (and even asymptotic weak martingales) under appropriate conditions (and using empirical process central limit theory), the asymptotic covariance structure is complicated and not essentially unique as was the case in one-dimensional time. This means that to build asymptotic tests, confidence bands, etc., one has to estimate a whole covariance structure and do specific calculations for that structure in each separate application, rather than there being one basic process (Brownian bridge or motion) which appears in all limiting situations. In practice, this means the use of bootstrap procedures and reliance on empirical process theory rather than on martingale theory. Many useful results have been obtained in this direction; see especially Pons and Turckheim (1991) who among other things continue and improve Pons's (1986) work on testing independence.

Some stronger theory of stochastic integration in multivariate time is available; for instance, the square-integrability based theory of Norberg (1989). However, this demands more structure on the filtration, corresponding to some kind of independence across different time coordinates, and is, therefore, not useful for our applications.

The other big drawback is that the predictability of an observation process in multivariate time is actually a rather strong property, so that many apparently well-behaved situations with multivariate censored survival times do not satisfy the independent censoring assumption.

We illustrate this with some simple examples. Consider two dependent survival times X_1, X_2. Suppose these times are actually embedded in the same real one-dimensional time, for instance both survival periods begin at time zero in one-dimensional time. Suppose the observation of the two time points X_1, X_2 is censored in real time by a single censoring time U. This results in a censored pair $X_1 \wedge U$, $X_2 \wedge U$, together with the corresponding indicators. If U is independent of $\mathbf{X} = (X_1, X_2)$, then $\mathbf{X}$ is independent of $\mathbf{U} = (U, U)$ and we have predictable censoring in the bivariate time scale. However, modest (predictable) dependence in real time can easily destroy the two-dimensional predictability and, consequently, destroy the possibility of using the statistical procedures of the previous section. For instance, if $U = X_1 + 1$, then censoring of X_2 gives information on X_1 and it is easy to see the bivariate and conditional Nelson–Aalen estimators will fail.

A similar example is when X_1 and X_2 are consecutive and adjacent periods of time, censored in real time by a single, independent censoring time U. In bivariate time, $\mathbf{X}$ is now censored by $(U, U - X_1)$ and independence is lost.

Consequently, one must be very wary: Predictable censoring in real time has nothing to do with predictable censoring in multivariate time, and one cannot apply multivariate Nelson–Aalen estimators to data which is superficially described as multivariate censored data without first carefully investigating the censoring mechanism.

One should note that the multivariate notion of independent censoring remains invariant under *independent* monotone transformations of *each* time coordinate. This is a very strong demand. Time variables which are linked together in some way in real time will most likely have rather special relationships which are disturbed by different monotone transformations (which can easily reverse the forward sequence of cause and effect in real time).

X.4. Bibliographic Remarks

Section X.1

Modern statistical approaches to inference from the Lexis diagram were surveyed by Keiding (1990). The Lexis diagram is also crucial in a statistical approach to age-specific incidence and prevalence (Keiding, 1991). Analysis of *prevalent cohort* data (with focus on the interplay between the three time scales: calendar time, age, and duration) was studied by Brookmeyer and Gail (1987), Brookmeyer and Liao (1990), Wang (1991), and Keiding (1992).

The role of semi-Markov processes in statistical inference was discussed by several contributors to the conference volume edited by Janssen (1986), in particular Commenges (1986), Cox (1986), Keiding (1986), and McClean (1986).

Section X.2

We refer to Section V.5 for bibliographic remarks concerning sequential methods for the case of simultaneous entry.

Sequential methods for clinical trials with staggered entry have been developed by the authors referenced in Section X.2. See also Sellke and Siegmund (1983) for a clever construction of an approximating martingale enabling the derivation of the asymptotic sequential log-rank statistics, the practical surveys by Olschewski and Schumacher (1986) and Whitehead (1992), and the useful summary of the various approaches to scoring and weighting for fixed-sample and sequential two-sample rank tests for survival data of Gu et al. (1991). An interesting discussion of the role of sequential analysis in clinical trials followed the papers of Bather (1985) and Armitage (1985).

In a series of papers almost completely isolated from the above development, Majumdar and Sen (1978) and Sinha and Sen (1982) developed sequential methods for clinical trials with staggered entry and random withdrawals; see the survey by Sen (1985, Section 2.6). As was the case for Gu and Lai (1991), Sen and co-workers based their development on weak convergence results in the (time, duration)-plane (a Lexis diagram); see Sen (1976). A detailed reconciliation of these two lines of work does not seem to exist.

Section X.3

Many proposals for estimation of the bivariate survival function in the presence of bivariate censored data have been made. The usual NPMLE and self-consistency principle do not lead to a consistent estimator: The NPMLE does not have to be consistent (Tsai, Leurgans, and Crowley, 1986). Therefore, most proposals are explicit estimators based on various representations of the bivariate survival function in terms of distribution functions of the data, the earliest being the so-called pathwise Kaplan–Meier estimator: One writes the joint survival function as the marginal survival function of one of the two variables times the conditional survival function of the other; see Muñoz (1980), Campbell (1981, 1982), Campbell and Földes (1982), Langberg and Shaked (1982), Korwar and Dahihya (1982), Hanley and Parnes (1983), and Burke (1988). The result depends on which component one takes first. A more sophisticated proposal was made by Tsai, Leurgans, and Crowley (1986) involving density estimation. However, the behavior of these estimators, and also that of the Volterra estimator, in practice seems rather poor (Bakker, 1990, Pruitt, 1991c) compared to those we discuss next.

We have concentrated on the proposal by Dabrowska (1988, 1989b). More recently, Prentice and Cai (1991, 1992) proposed a very interesting estimator, closely related to Dabrowska's though it has some similarity with the Volterra estimator which we also discussed. Dabrowska's multivariate product-limit estimator, based on an elegant representation of a multivariate survival function in terms of its conditional multivariate hazard measure, had the best practical performance of the then proposed estimators in a simulation study by Bakker (1990), whereas the Prentice and Cai estimator seems to have a similar performance (Prentice and Cai, 1991, 1992). These estimators are smooth functionals of the empirical distributions of the data so that such results as consistency, asymptotic normality, correctness of the bootstrap, consistent estimation of the asymptotic variance, all hold by application of the functional delta-method: Details were given by Gill (1990). The analysis of the Prentice and Cai estimator is similar [see Gill, van der Laan, and Wellner (1993)]. None of these ad hoc estimators are asymptotically efficient and are not even strictly speaking distribution functions (Pruitt 1991a, Bakker, 1990).

Pruitt (1991b) proposed an implicitly defined estimator which is the solution of an ad hoc modification of the self-consistency equation (generalized score equation). He derived and illustrated intuitively nice properties of his estimator. The usual kind of asymptotic result for this estimator, again inefficient, were proved by van der Laan (1991). Its empirical behavior is again very good; see Pruitt (1991c).

The main problem with the NPMLE for this problem is that each singly-censored observation requires probability mass $1/n$ to be distributed over a certain half-line in the plane on which the underlying survival times are known to lie. However, since no other observations lie on this line, there is no way to sensibly distribute the mass. Pruitt's (1991b) proposal went some way to solving this: Replace the line by a thin strip. Van der Laan (1992) showed that the idea can be carried through much further to get an efficient estimator. Lines are replaced by strips and the EM-algorithm can be used to calculate the NPMLE for an approximation to the original model, which becomes better as the sample size gets larger and the strips narrower.

Very little work has been done on more general multivariate modelling, one exception being Pons (1989) who introduced a form of the Cox regression model for bivariate hazards.

Appendix: The Melanoma Survival Data and Standard Mortality Tables for the Danish Population 1971–75

Table A.1. The Melanoma Survival Data (Collected at Odense University Hospital, Denmark, by K.T. Drzewiecki, cf. Example I.3.1)

Number (i)	Survival Time ($\tilde{X}_i$)	Indicator (D_i)	Sex	Age	Year of Operation	Tumor Thickness	Ulceration
1	10	3	1	76	1972	6.76	1
2	30	3	1	56	1968	0.65	0
3	35	2	1	41	1977	1.34	0
4	99	3	0	71	1968	2.90	0
5	185	1	1	52	1965	12.08	1
6	204	1	1	28	1971	4.84	1
7	210	1	1	77	1972	5.16	1
8	232	3	0	60	1974	3.22	1
9	232	1	1	49	1968	12.88	1
10	279	1	0	68	1971	7.41	1
11	295	1	0	53	1969	4.19	1
12	355	3	0	64	1972	0.16	1
13	386	1	0	68	1965	3.87	1
14	426	1	1	63	1970	4.84	1
15	469	1	0	14	1969	2.42	1
16	493	3	1	72	1971	12.56	1
17	529	1	1	46	1971	5.80	1
18	621	1	1	72	1972	7.06	1
19	629	1	1	95	1968	5.48	1
20	659	1	1	54	1972	7.73	1
21	667	1	0	89	1968	13.85	1
22	718	1	1	25	1967	2.34	1
23	752	1	1	37	1973	4.19	1
24	779	1	1	43	1967	4.04	1

Table A.1 (*continued*)

Number (i)	Survival Time ($\tilde{X}_i$)	Indicator (D_i)	Sex	Age	Year of Operation	Tumor Thickness	Ulceration
25	793	1	1	68	1970	4.84	1
26	817	1	0	67	1966	0.32	0
27	826	3	0	86	1965	8.54	1
28	833	1	0	56	1971	2.58	1
29	858	1	0	16	1967	3.56	0
30	869	1	0	42	1965	3.54	0
31	872	1	0	65	1968	0.97	0
32	967	1	1	52	1970	4.83	1
33	977	1	1	58	1967	1.62	1
34	982	1	0	60	1970	6.44	1
35	1041	1	1	68	1967	14.66	0
36	1055	1	0	75	1967	2.58	1
37	1062	1	1	19	1966	3.87	1
38	1075	1	1	66	1971	3.54	1
39	1156	1	0	56	1970	1.34	1
40	1228	1	1	46	1973	2.24	1
41	1252	1	0	58	1971	3.87	1
42	1271	1	0	74	1971	3.54	1
43	1312	1	0	65	1970	17.42	1
44	1427	3	1	64	1972	1.29	0
45	1435	1	1	27	1969	3.22	0
46	1499	2	1	73	1973	1.29	0
47	1506	1	1	56	1970	4.51	1
48	1508	2	1	63	1973	8.38	1
49	1510	2	0	69	1973	1.94	0
50	1512	2	0	77	1973	0.16	0
51	1516	1	1	80	1968	2.58	1
52	1525	3	0	76	1970	1.29	1
53	1542	2	0	65	1973	0.16	0
54	1548	1	0	61	1972	1.62	0
55	1557	2	0	26	1973	1.29	0
56	1560	1	0	57	1973	2.10	0
57	1563	2	0	45	1973	0.32	0
58	1584	1	1	31	1970	0.81	0
59	1605	2	0	36	1973	1.13	0
60	1621	1	0	46	1972	5.16	1
61	1627	2	0	43	1973	1.62	0
62	1634	2	0	68	1973	1.37	0
63	1641	2	1	57	1973	0.24	0
64	1641	2	0	57	1973	0.81	0
65	1648	2	0	55	1973	1.29	0
66	1652	2	0	58	1973	1.29	0
67	1654	2	1	20	1973	0.97	0
68	1654	2	0	67	1973	1.13	0
69	1667	1	0	44	1971	5.80	1

Table A.1 (*continued*)

Number (i)	Survival Time ($\tilde{X}_i$)	Indicator (D_i)	Sex	Age	Year of Operation	Tumor Thickness	Ulceration
70	1678	2	0	59	1973	1.29	0
71	1685	2	0	32	1973	0.48	0
72	1690	1	1	83	1971	1.62	0
73	1710	2	0	55	1973	2.26	0
74	1710	2	1	15	1973	0.58	0
75	1726	1	0	58	1970	0.97	1
76	1745	2	0	47	1973	2.58	1
77	1762	2	0	54	1973	0.81	0
78	1779	2	1	55	1973	3.54	1
79	1787	2	1	38	1973	0.97	0
80	1787	2	0	41	1973	1.78	1
81	1793	2	0	56	1973	1.94	0
82	1804	2	0	48	1973	1.29	0
83	1812	2	1	44	1973	3.22	1
84	1836	2	0	70	1972	1.53	0
85	1839	2	0	40	1972	1.29	0
86	1839	2	1	53	1972	1.62	1
87	1854	2	0	65	1972	1.62	1
88	1856	2	1	54	1972	0.32	0
89	1860	3	1	71	1969	4.84	1
90	1864	2	0	49	1972	1.29	0
91	1899	2	0	55	1972	0.97	0
92	1914	2	0	69	1972	3.06	0
93	1919	2	1	83	1972	3.54	0
94	1920	2	1	60	1972	1.62	1
95	1927	2	1	40	1972	2.58	1
96	1933	1	0	77	1972	1.94	0
97	1942	2	0	35	1972	0.81	0
98	1955	2	0	46	1972	7.73	1
99	1956	2	0	34	1972	0.97	0
100	1958	2	0	69	1972	12.88	0
101	1963	2	0	60	1972	2.58	0
102	1970	2	1	84	1972	4.09	1
103	2005	2	0	66	1972	0.64	0
104	2007	2	1	56	1972	0.97	0
105	2011	2	0	75	1972	3.22	1
106	2024	2	0	36	1972	1.62	0
107	2028	2	1	52	1972	3.87	1
108	2038	2	0	58	1972	0.32	1
109	2056	2	0	39	1972	0.32	0
110	2059	2	1	68	1972	3.22	1
111	2061	1	1	71	1968	2.26	0
112	2062	1	0	52	1965	3.06	0
113	2075	2	1	55	1972	2.58	1
114	2085	3	0	66	1970	0.65	0

Table A.1 (*continued*)

Number (i)	Survival Time ($\tilde{X}_i$)	Indicator (D_i)	Sex	Age	Year of Operation	Tumor Thickness	Ulceration
115	2102	2	1	35	1972	1.13	0
116	2103	1	1	44	1966	0.81	0
117	2104	2	0	72	1972	0.97	0
118	2108	1	0	58	1969	1.76	1
119	2112	2	0	54	1972	1.94	1
120	2150	2	0	33	1972	0.65	0
121	2156	2	0	45	1972	0.97	0
122	2165	2	1	62	1972	5.64	0
123	2209	2	0	72	1971	9.66	0
124	2227	2	0	51	1971	0.10	0
125	2227	2	1	77	1971	5.48	1
126	2256	1	0	43	1971	2.26	1
127	2264	2	0	65	1971	4.83	1
128	2339	2	0	63	1971	0.97	0
129	2361	2	1	60	1971	0.97	0
130	2387	2	0	50	1971	5.16	1
131	2388	1	1	40	1966	0.81	0
132	2403	2	0	67	1971	2.90	1
133	2426	2	0	69	1971	3.87	0
134	2426	2	0	74	1971	1.94	1
135	2431	2	0	49	1971	0.16	0
136	2460	2	0	47	1971	0.64	0
137	2467	1	0	42	1965	2.26	1
138	2492	2	0	54	1971	1.45	0
139	2493	2	1	72	1971	4.82	1
140	2521	2	0	45	1971	1.29	1
141	2542	2	1	67	1971	7.89	1
142	2559	2	0	48	1970	0.81	1
143	2565	1	1	34	1970	3.54	1
144	2570	2	0	44	1970	1.29	0
145	2660	2	0	31	1970	0.64	0
146	2666	2	0	42	1970	3.22	1
147	2676	2	0	24	1970	1.45	1
148	2738	2	0	58	1970	0.48	0
149	2782	1	1	78	1969	1.94	0
150	2787	2	1	62	1970	0.16	0
151	2984	2	1	70	1969	0.16	0
152	3032	2	0	35	1969	1.29	0
153	3040	2	0	61	1969	1.94	0
154	3042	1	0	54	1967	3.54	1
155	3067	2	0	29	1969	0.81	0
156	3079	2	1	64	1969	0.65	0
157	3101	2	1	47	1969	7.09	0
158	3144	2	1	62	1969	0.16	0

Table A.1 (*continued*)

Number (i)	Survival Time ($\tilde{X}_i$)	Indicator (D_i)	Sex	Age	Year of Operation	Tumor Thickness	Ulceration
159	3152	2	0	32	1969	1.62	0
160	3154	3	1	49	1969	1.62	0
161	3180	2	0	25	1969	1.29	0
162	3182	3	1	49	1966	6.12	0
163	3185	2	0	64	1969	0.48	0
164	3199	2	0	36	1969	0.64	0
165	3228	2	0	58	1969	3.22	1
166	3229	2	0	37	1969	1.94	0
167	3278	2	1	54	1969	2.58	0
168	3297	2	0	61	1968	2.58	1
169	3328	2	1	31	1968	0.81	0
170	3330	2	1	61	1968	0.81	1
171	3338	1	0	60	1967	3.22	1
172	3383	2	0	43	1968	0.32	0
173	3384	2	0	68	1968	3.22	1
174	3385	2	0	4	1968	2.74	0
175	3388	2	1	60	1968	4.84	1
176	3402	2	1	50	1968	1.62	0
177	3441	2	0	20	1968	0.65	0
178	3458	3	0	54	1967	1.45	0
179	3459	2	0	29	1968	0.65	0
180	3459	2	1	56	1968	1.29	1
181	3476	2	0	60	1968	1.62	0
182	3523	2	0	46	1968	3.54	0
183	3667	2	0	42	1967	3.22	0
184	3695	2	0	34	1967	0.65	0
185	3695	2	0	56	1967	1.03	0
186	3776	2	1	12	1967	7.09	1
187	3776	2	0	21	1967	1.29	1
188	3830	2	1	46	1967	0.65	0
189	3856	2	0	49	1967	1.78	0
190	3872	2	0	35	1967	12.24	1
191	3909	2	1	42	1967	8.06	1
192	3968	2	0	47	1967	0.81	0
193	4001	2	0	69	1967	2.10	0
194	4103	2	0	52	1966	3.87	0
195	4119	2	1	52	1966	0.65	0
196	4124	2	0	30	1966	1.94	1
197	4207	2	1	22	1966	0.65	0
198	4310	2	1	55	1966	2.10	0
199	4390	2	0	26	1965	1.94	1
200	4479	2	0	19	1965	1.13	1
201	4492	2	1	29	1965	7.06	1
202	4668	2	0	40	1965	6.12	0

Table A.1 (*continued*)

Number (i)	Survival Time ($\tilde{X}_i$)	Indicator (D_i)	Sex	Age	Year of Operation	Tumor Thickness	Ulceration
203	4688	2	0	42	1965	0.48	0
204	4926	2	0	50	1964	2.26	0
205	5565	2	0	41	1962	2.90	0

Note: Survival time ($\tilde{X}_i$) in days, death indicator (D_i) (1—dead from malignant melanoma, 2—alive 1 January 1978, 3—dead from other cause), sex (1—man, 0—woman), age in years, year of operation, tumor thickness in mm, and ulceration (1—present, 0—absent) for 205 patients with malignant melanoma.

Table A.2. Mortality Rates per 100,000 Years for Danish Males 1971–75; See Figure IV.1.8 in Example IV.1.11

Age (years)	0	1	2	3	4	5	6	7	8	9
0– 9	1442	101	72	62	50	50	53	53	49	45
10–19	39	37	37	39	53	67	82	101	118	127
20–29	119	114	119	110	96	96	100	103	104	100
30–39	111	122	124	126	134	153	169	189	212	221
40–49	247	271	305	349	357	401	448	478	523	589
50–59	693	763	821	880	950	1058	1157	1281	1423	1561
60–69	1726	1921	2148	2328	2565	2910	3170	3417	3648	4164
70–79	4503	4871	5358	5886	6427	6853	7341	8084	8830	9574
80–89	10388	11436	12331	13284	14718	16216	18189	19463	20675	22908
90–99	24936	27196	29660	32347	35278	38474	41960	45762	49908	54430

Table A.3. Mortality Rates per 100,000 Years for Danish Females 1971–75; See Figure IV.1.8 in Example IV.1.11

Age (years)	0	1	2	3	4	5	6	7	8	9
0– 9	1030	71	45	48	44	36	39	30	25	27
10–19	25	23	24	24	31	39	40	36	35	38
20–29	40	43	40	41	44	43	42	44	50	54
30–39	59	65	72	84	94	101	118	122	130	160
40–49	175	202	225	236	275	291	308	347	398	423
50–59	449	488	522	557	586	623	687	756	802	850
60–69	931	1038	1121	1209	1356	1476	1579	1757	1978	2179
70–79	2417	2689	2975	3327	3725	4205	4754	5325	5949	6620
80–89	7251	8145	9362	10319	11292	12462	14338	16289	17521	19603
90–99	21477	23835	26452	29357	32580	36158	40128	44534	49425	54852

References

Aalen, O.O. (1972). Estimering av risikorater for prevensjonsmidlet 'spiralen' (in Norwegian). Graduate thesis in statistics, Institute of Mathematics, University of Oslo.

Aalen, O.O. (1975). Statistical inference for a family of counting processes. PhD thesis, University of California, Berkeley.

Aalen, O.O. (1976). Nonparametric inference in connection with multiple decrement models. *Scand. J. Statist.* **3**, 15–27.

Aalen, O.O. (1977). Weak convergence of stochastic integrals related to counting processes. *Z. Wahrsch. verw. Geb.* **38**, 261–277. Correction: **48**, 347 (1979).

Aalen, O.O. (1978a). Nonparametric estimation of partial transition probabilities in multiple decrement models. *Ann. Statist.* **6**, 534–545.

Aalen, O.O. (1978b). Nonparametric inference for a family of counting processes. *Ann. Statist.* **6**, 701–726.

Aalen, O.O. (1980). A model for non-parametric regression analysis of counting processes. *Springer Lect. Notes Statist.* **2**, 1–25. Mathematical Statistics and Probability Theory. W. Klonecki, A. Kozek, and J. Rosiński, editors.

Aalen, O.O. (1982a). Discussion of the paper by P.K. Andersen, Ø. Borgan, R.D. Gill and N. Keiding. *Internat. Statist. Rev.* **50**, 244–246.

Aalen, O.O. (1982b). Practical applications of the non-parametric statistical theory for counting processes. Statistical Research Report 82/2, Institute of Mathematics, University of Oslo.

Aalen, O.O. (1987a). Two examples of modelling heterogeneity in survival analysis. *Scand. J. Statist.* **14**, 19–25.

Aalen, O.O. (1987b). Mixing distributions on a Markov chain. *Scand. J. Statist.* **14**, 281–289.

Aalen, O.O. (1988a). Dynamic description of a Markov chain with random time scale. *Math. Scientist* **13**, 90–103.

Aalen, O.O. (1988b). Heterogeneity in survival analysis. *Statist. Med.* **7**, 1121–1137.

Aalen, O.O. (1989). A linear regression model for the analysis of life times. *Statist. Med.* **8**, 907–925.

Aalen, O.O. (1990). Modelling heterogeneity in survival analysis by the compound Poisson distribution. Technical report, Section of Medical Statistics, University of Oslo. Has appeared in *Ann. Appl. Probab.* **2**, 951–972, 1992.

Aalen, O.O., Borgan, Ø., Keiding, N., and Thormann, J. (1980). Interaction between life history events: nonparametric analysis of prospective and retrospective data in the presence of censoring. *Scand. J. Statist.* **7**, 161–171.

Aalen, O.O. and Hoem, J.M. (1978). Random time changes for multivariate counting processes. *Scand. Actuar. J.* 81–101.

Aalen, O.O. and Johansen, S. (1978). An empirical transition matrix for nonhomogeneous Markov chains based on censored observations. *Scand. J. Statist.* **5**, 141–150.

Aitkin, J., Anderson, D., Francis, B., and Hinde, J. (1989). *Statistical Modelling in GLIM*. Clarendon Press, Oxford.

Aitkin, M. and Clayton, D.G. (1980). The fitting of exponential, Weibull and extreme value distributions to complex censored survival data using GLIM. *Appl. Statist.* **29**, 156–163.

Akritas, M.G. (1986). Bootstrapping the Kaplan–Meier estimator. *J. Amer. Statist. Assoc.* **81**, 1032–1038.

Akritas, M.G. (1988). Pearson-type goodness-of-fit tests: the univariate case. *J. Amer. Statist. Assoc.* **83**, 222–230.

Albers, W. (1988a). Combined rank tests for randomly censored paired data. *J. Amer. Statist. Assoc.* **83**, 1159–1162.

Albers, W. (1988b). Rank tests for regression and k-sample rank tests under random censorship. *Statist. Probab. Lett.* **6**, 315–319.

Albers, W. and Akritas, M.G. (1987). Combined rank tests for the two-sample problem with randomly censored data. *J. Amer. Statist. Assoc.* **82**, 648–655.

Aldous, D.J. (1978a). Stopping times and tightness. *Ann. Probab.* **6**, 335–340.

Aldous, D.J. (1978b). On weak convergence of stochastic processes viewed in the Strasbourg manner. Unpublished manuscript.

Altman, D.G. and Andersen, P.K. (1986). A note on the uncertainty of a survival probability estimated from Cox's regression model. *Biometrika* **73**, 722–724.

Altman, D.G. and Andersen, P.K. (1989). Bootstrap investigation of the stability of a Cox regression model. *Statist. Med.* **8**, 771–783.

Altmann, S.A. and Altmann, J. (1970) *Baboon Ecology: African Field Research*, Bibliotheca Primatologica **12**. S. Karger, Basel.

Altshuler, B. (1970). Theory for the measurement of competing risks in animal experiments. *Math. Biosci.* **6**, 1–11.

Aly, E.-E. A.A., Csörgő, M., and Horváth, L. (1985). Strong approximations of the quantile process of the product limit estimator. *J. Multivar. Anal.* **16**, 185–210.

Andersen, A.R., Christiansen, J.S., Andersen, J.K., Kreiner, S., and Deckert, T. (1983a). Diabetic nephropathy in Type 1 (insulin-dependent) diabetes: an epidemiological study. *Diabetologia* **25**, 496–501.

Andersen, O. (1985). *Dødelighed og erhverv 1970–*80, Statistiske undersøgelser **41**. Danmarks Statistik, Copenhagen.

Andersen, P.K. (1982). Testing goodness-of-fit of Cox's regression and life model. *Biometrics* **38**, 67–77. Correction: **40**, 1217 (1984).

Andersen, P.K. (1983a). Comparing survival distributions via hazard ratio estimates. *Scand. J. Statist.* **10**, 77–85.

Andersen, P.K. (1983b). Testing for proportional hazards. Research Report 83/1, Statistical Research Unit, University of Copenhagen.

Andersen, P.K. (1986). Time-dependent covariates and Markov processes. In Moolgavkar, S.H. and Prentice, R.L., editors, *Modern Statistical Methods in Chronic Disease Epidemiology*, pp. 82–103. Wiley, New York.

Andersen, P.K. (1988). Multistate models in survival analysis: a study of nephropathy and mortality in diabetes. *Statist. Med.* **7**, 661–670.

Andersen, P.K., Borch-Johnsen, K., Deckert, T., Green, A., Hougaard, P., Keiding, N., and Kreiner, S. (1985). A Cox regression model for the relative mortality and its application to diabetes mellitus survival data. *Biometrics* **41**, 921–932.

Andersen, P.K. and Borgan, Ø. (1985). Counting process models for life history data: A review (with discussion). *Scand. J. Statist.* **12**, 97–158.
Andersen, P.K., Borgan, Ø., Gill, R.D., and Keiding, N. (1982). Linear non-parametric tests for comparison of counting processes, with application to censored survival data (with discussion). *Internat. Statist. Rev.* **50**, 219–258. Amendment: **52**, 225 (1984).
Andersen, P.K., Borgan, Ø., Gill, R.D., and Keiding, N. (1988). Censoring, truncation and filtering in statistical models based on counting processes. *Contemp. Math.* **80**, 19–60.
Andersen, P.K., Christensen, E., Fauerholdt, L., and Schlichting, P. (1983b). Measuring prognosis using the proportional hazards model. *Scand. J. Statist.* **10**, 49–52.
Andersen, P.K., Christensen, E., Fauerholdt, L., and Schlichting, P. (1983c). Evaluating prognoses based on the proportional hazards model. *Scand. J. Statist.* **10**, 141–144.
Andersen, P.K. and Gill R.D. (1982). Cox's regression model for counting processes: A large sample study. *Ann. Statist.* **10**, 1100–1120.
Andersen, P.K., Hansen, L.S., and Keiding, N. (1991a). Non- and semi-parametric estimation of transition probabilities from censored observations of a non-homogeneous Markov process. *Scand. J. Statist.* **18**, 153–167.
Andersen, P.K., Hansen, L.S., and Keiding, N. (1991b). Assessing the influence of reversible disease indicators on survival. *Statist. Med.* **10**, 1061–1067.
Andersen, P.K. and Rasmussen, N.K. (1986). Psychiatric admissions and choice of abortion. *Statist. Med.* **5**, 243–253.
Andersen, P.K. and Væth, M. (1984). *Statistisk analyse af overlevelsesdata ved lægevidenskabelige undersøgelser*. FaDL, Copenhagen.
Andersen, P.K. and Væth, M. (1989). Simple parametric and nonparametric models for excess and relative mortality. *Biometrics* **45**, 523–535.
Anderson, J.A. and Senthilselvan, A. (1980). Smooth estimates for the hazard function. *J. Roy. Statist. Soc. B* **42**, 322–327.
Anderson, T.W. (1955). The integral of a symmetric unimodal function over a symmetric convex set and some probability inequalities. *Proc. Amer. Math. Soc.* **6**, 170–176.
Antoniadis, A. (1989). A penalty method for nonparametric estimation of the intensity function of a counting process. *Ann. Inst. Statist. Math.* **41**, 781–807.
Antoniadis, A. and Grégoire, G. (1990). Penalized likelihood estimation for rates with censored survival data. *Scand. J. Statist.* **17**, 43–63.
Aranda-Ordaz, F.J. (1983). An extension of the proportional-hazards model for grouped data. *Biometrics* **39**, 109–117.
Arjas, E. (1985). Discussion of the paper by P.K. Andersen and Ø. Borgan. *Scand. J. Statist.* **12**, 150–153.
Arjas, E. (1986). Stanford heart tansplantation data revisited: a real time approach. In Moolgavkar, S.H. and Prentice, R.L., editors, *Modern Statistical Methods in Chronic Disease Epidemiology*, pp. 65–81. Wiley, New York.
Arjas, E. (1988). A graphical method for assessing goodness of fit in Cox's proportional hazards model. *J. Amer. Statist. Assoc.* **83**, 204–212.
Arjas, E. (1989). Survival models and martingale dynamics (with discussion). *Scand. J. Statist.* **16**, 177–225.
Arjas, E. and Haara, P. (1984). A marked point process approach to censored failure data with complicated covariates. *Scand. J. Statist.* **11**, 193–209.
Arjas, E. and Haara, P. (1988a). A note on the asymptotic normality in the Cox regression model. *Ann. Statist.* **16**, 1133–1140.
Arjas, E. and Haara, P. (1988b). A note on the exponentiality of total hazards before failure. *J. Multivar. Anal.* **26**, 207–218.
Arjas, E. and Haara, P. (1992a). Observation scheme and likelihood. *Scand. J. Statist.* **19**, 111–132.
Arjas, E. and Haara, P. (1992b). Periodic inspections in a longitudinal study: Viewing

occult tumors through a filter. In Klein, J.P. and Goel, P.K., editors, *Survival Analysis: State of the Art*, pp. 329–344. Kluwer, Dordrecht.
Arjas, E., Haara, P., and Norros, I. (1992). Filtering the histories of a partially observed marked point process. *Stoch. Proc. Appl.* **40**, 225–250.
Arley, N. (1943). *On the Theory of Stochastic Processes and their Application to the Theory of Cosmic Radiation.* G. E. C. Gads Forlag, Copenhagen.
Armitage, P. (1985). The search for optimality in clinical trials (with discussion). *Internat. Statist. Rev.* **53**, 15–24.
Aven, T. (1985). A theorem for determining the compensator of a counting process. *Scand. J. Statist.* **12**, 69–72.
Aven, T. (1986). Bayesian inference in parametric counting process models. *Scand. J. Statist.* **13**, 87–97.
Ayer, M., Brunk, H.D., Ewing, G.M., Reid, W.T., and Silverman, E. (1955). An empirical distribution function for sampling with incomplete information. *Ann. Math. Statist.* **26**, 641–647.
Baddeley, A.J. and Gill, R.D. (1992). Kaplan-Meier estimators for interpoint distance distributions of spatial point processes. Preprint 718, Dept. of Mathematics, University of Utrecht.
Bahadur, R.R. (1967). Rates of convergence of estimates and some test statistics. *Ann. Math. Statist.* **38**, 303–324.
Bailey, K.R. (1983). The asymptotic joint distribution of regression and survival parameter estimates in the Cox regression model. *Ann. Statist.* **11**, 39–58.
Bakker, D.M. (1990). Two nonparametric estimators of the survival function of bivariate right censored observations. Report BS-R9035, Centre for Mathematics and Computer Science, Amsterdam.
Barlow, R.E., Bartholomew, D.J., Bremner, J.M., and Brunk, H.D. (1972). *Statistical Inference under Order Restrictions.* Wiley, New York.
Barlow, R.E. and Campo, R. (1975). Total time on test processes and application to failure data analysis. In Barlow, R.E., Fussell, J., and Singpurwalla, N.D., editors, *Reliability and Fault Tree Analysis*, pp. 451–481. SIAM, Philadelphia.
Barlow, W.E. and Prentice, R.L. (1988). Residuals for relative risk regression. *Biometrika* **75**, 65–74.
Barndorff-Nielsen, O.E. and Sørensen, M. (1992). Asymptotic likelihood theory for stochastic processes. A review. Has appeared in *Internat. Statist. Rev.* **62**, 133–165, 1994.
Bartoszyński, R., Brown, B.W., McBride, C.M., and Thompson, J.R. (1981). Some nonparametric techniques for estimating the intensity function of a cancer related nonstationary Poisson process. *Ann. Statist.* **9**, 1050–1060.
Bather, J.A. (1985). On the allocation of treatments in sequential medical trials (with discussion). *Internat. Statist. Rev.* **53**, 1–13.
Becker, N.G. (1989). *Analysis of infectious disease data.* Chapman and Hall, London.
Becker, R.A., Chambers, J.M., and Wilks, A.R. (1988). *The new S language. A programming environment for data analysis and graphics.* Wadsworth, Belmont.
Bednarski, T. (1989). On sensitivity of Cox's estimator. *Statist. Decisions* **7**, 215–228.
Bednarski, T. (1991). Robust estimation in Cox regression model. Technical report, Institute of Mathematics, Polish Academy of Sciences.
Begun, J.M., Hall, W.J., Huang, W.-M., and Wellner, J.A. (1983). Information and asymptotic efficiency in parametric-nonparametric models. *Ann. Statist.* **11**, 432–452.
Begun, J.M. and Reid, N. (1983). Estimating the relative risk with censored data. *J. Amer. Statist. Assoc.* **78**, 337–341.
Begun, J.M. and Wellner, J.A. (1983). Asymptotic efficiency of relative risk estimates. In Sen, P.K., editor *Contributions to Statistics, Essays in Honour of Norman L. Johnson*, pp. 47–62. North-Holland, Amsterdam.
Belsley, D.A., Kuh, E., and Welsch, R.E. (1980). *Regression Diagnostics.* Wiley, New

York.
Beran, R.J. (1977). Estimating a distribution function. *Ann. Statist.* **5**, 400–404.
Beran, R.J. (1981). Nonparametric regression with randomly censored survival data. Technical report, University of California, Berkeley.
Bergman, B. (1985). On reliability theory and its applications (with discussion). *Scand. J. Statist.* **12**, 1–41.
Bernstein, S. (1926). Sur l'extension du théorème limite du calcul des probabilités aux sommes de quantités dépendantes. *Math. Ann.* **97**, 1–59.
Bichteler, K. (1979). Stochastic integrators. *Bull. Amer. Math. Soc.* **1**, 761–765.
Bickel, P.J., Klaassen, C.A., Ritov, Y., and Wellner, J.A. (1993). *Efficient and adaptive inference in semiparametric models.* Johns Hopkins University Press, Baltimore.
Bie, O., Borgan, Ø., and Liestøl, K. (1987). Confidence intervals and confidence bands for the cumulative hazard rate function and their small sample properties. *Scand. J. Statist.* **14**, 221–233.
Billingsley, P. (1961). *Statistical Inference for Markov Processes.* University of Chicago Press, Chicago.
Billingsley, P. (1968). *Convergence of Probability Measures.* Wiley, New York.
Bishop, Y.M.M., Fienberg, S.E., and Holland, P.W. (1975). *Discrete multivariate analysis: Theory and practice.* MIT Press, Cambridge, MA.
Blossfeld, H.-P., Hamerle, A., and Mayer, K.U. (1989). *Event History Analysis. Statistical Theory and Application in the Social Sciences.* Lawrence Erlbaum, Hillsdale, NJ.
BMDP (1988). (Dixon, W.J., editor), *BMDP Statistical Software.* University of California Press, Berkeley.
Boel, R., Varaiya, P., and Wong, E. (1975a). Martingales on jump processes I: Representation results. *SIAM J. Control* **13**, 999–1021.
Boel, R., Varaiya, P., and Wong, E. (1975b). Martingales on jump processes II: Applications. *SIAM J. Control* **13**, 1022–1061.
Böhmer, P.E. (1912). Theorie der unabhängigen Warscheinlichkeiten. *Rapports, Mém. et Procés—verbaux 7e Congrès Internat. Act., Amsterdam* **2**, 327–343.
Borch-Johnsen, K., Andersen, P.K., and Deckert, T. (1985). The effect of proteinuria on relative mortality in Type 1 (insulin-dependent) diabetes mellitus. *Diabetologia* **28**, 590–596.
Borch-Johnsen, K., Kreiner, S., and Deckert, T. (1986). Mortality of Type 1 (insulin-dependent) diabetes mellitus in Denmark. *Diabetologia* **29**, 767–772.
Borgan, Ø. (1980). Applications of non-homogeneous Markov chains to medical studies. Nonparametric analysis for prospective and retrospective data. In Victor, N., Lehmacher, W., and van Eimeren, W., editors, *Explorative Datenanalyse, Frühjahrstagung München* 1980, *Proceedings*, Medizinische Informatik und Statistik 26, pp. 102–115. Springer-Verlag, Heidelberg.
Borgan, Ø. (1984). Maximum likelihood estimation in parametric counting process models, with applications to censored failure time data. *Scand. J. Statist.* **11**, 1–16. Correction: **11**, 275 (1984).
Borgan, Ø. and Gill, R.D. (1982). Case-control studies in a Markov chain setting. Preprint SW 89/82, Mathematical Centre, Amsterdam.
Borgan, Ø., Goldstein, L. and Langholz, B. (1992). Methods for the analysis of sampled cohort data in the Cox proportional hazards model. Statist. Res. Rep. 7/92, Institute of Mathematics, University of Oslo.
Borgan, Ø. and Liestøl, K. (1990). A note on confidence intervals and bands for the survival curve based on transformations. *Scand. J. Statist.* **17**, 35–41.
Borgan, Ø., Liestøl, K., and Ebbesen, P. (1984). Efficiencies of experimental designs for an illness–death model. *Biometrics* **40**, 627–638.
Bowman, A.W. (1984). An alternative method of cross-validation for the smoothing of density estimates. *Biometrika* **71**, 353–360.
Breiman, L. (1968). *Probability.* Addison-Wesley, Reading, MA.

Brémaud, P. (1972). *A martingale approach to point processes.* PhD thesis, Electrical Research Laboratory, Berkeley.
Brémaud, P. (1981). *Point Processes and Queues: Martingale Dynamics.* Springer-Verlag, New York.
Brémaud, P. and Jacod, J. (1977). Processus ponctuels et martingales: Résultats récents sur la modélisation et le filtrage. *Adv. Appl. Probab.* **9**, 362–416.
Breslow, N.E. (1970). A generalized Kruskal–Wallis test for comparing K samples subject to unequal patterns of censorship. *Biometrika* **57**, 579–594.
Breslow, N.E. (1972). Discussion of the paper by D.R. Cox. *J. Roy. Statist. Soc. B* **34**, 216–217.
Breslow, N.E. (1974). Covariance analysis of censored survival data. *Biometrics* **30**, 89–99.
Breslow, N.E. (1975). Analysis of survival data under the proportional hazards model. *Internat. Statist. Rev.* **43**, 45–58.
Breslow, N.E. and Crowley, J.J. (1974). A large sample study of the life table and product limit estimates under random censorship. *Ann. Statist.* **2**, 437–453.
Breslow, N.E. and Day, N.E. (1987). *Statistical methods in cancer research.* Volume II—*The design and analysis of cohort studies*, IARC Scientific Publications 82. International Agency for Research on Cancer, Lyon.
Breslow, N.E., Lubin, J.H., Marek, P., and Langholz, B. (1983). Multiplicative models and cohort analysis. *J. Amer. Statist. Assoc.* **78**, 1–12.
Bretagnolle, J. and Huber-Carol, C. (1988). Effects of omitting covariates in Cox's model for survival data. *Scand. J. Statist.* **15**, 125–138.
Brookmeyer, R. and Crowley, J.J. (1982a). A confidence interval for the median survival time. *Biometrics* **38**, 29–41.
Brookmeyer, R. and Crowley, J.J. (1982b). A k-sample median test for censored data. *J. Amer. Statist. Assoc.* **77**, 433–440.
Brookmeyer, R. and Gail, M.H. (1987). Biases in prevalent cohorts. *Biometrics* **43**, 739–749.
Brookmeyer, R. and Liao, J. (1990). The analysis of delays in disease reporting: methods and results for the acquired immunodeficiency syndrome. *Amer. J. Epidem.* **132**, 355–365.
Brown, B.M. (1971). Martingale central limit theorems. *Ann. Math. Statist.* **42**, 59–66.
Brown, T.C. (1988). The Doob–Meyer decomposition and stochastic calculus. Technical report, Department of Mathematics, University of Western Australia.
Buckley, J.D. (1984). Additive and multiplicative models for relative survival rates. *Biometrics* **40**, 51–62.
Buckley, J.D. and James, I.R. (1979). Linear regression with censored data. *Biometrika* **66**, 429–436.
Burke, M.D. (1988). An almost sure representation of a multivariate product-limit estimator under random censorship. *Statist. Decisions* **6**, 89–108.
Burke, M.D., Csörgő, S., and Horváth, L. (1981). Strong approximations of some biometric estimates under random censorship. *Z. Wahrsch. verw. Geb.* **56**, 87–112. Correction: **79**, 51–57 (1988).
Cain, K.C. and Lange, N.T. (1984). Approximate case influence for the proportional hazards regression model with censored data. *Biometrics* **40**, 493–499.
Campbell, G. (1981). Nonparametric bivariate estimation with randomly censored data. *Biometrika* **68**, 417–422.
Campbell, G. (1982). Asymptotic properties of several nonparametrical multivariate distribution function estimators under random censoring. In Crowley, J.J. and Johnson, R.A., editors, *Lecture notes—Monograph series* 2, *Survival Analysis*, pp. 243–256. Institute of Mathematical Statistics. Hayward, California.
Campbell, G. and Földes, A. (1982). Large-sample properties of nonparametric bivariate estimators with censored data. In Gnedenko, B.V., Puri, M.L., and Vincze,

I., editors, *Nonparametric Statistical Inference*, pp. 103–122. North-Holland, Amsterdam.
Chang, M.N. (1990). Weak convergence of a self-consistent estimator of the survival function with doubly censored data. *Ann. Statist.* **18**, 391–404.
Chang, M.N. and Yang, G.L. (1987). Strong consistency of a nonparametric estimator of the survival function with doubly censored data. *Ann. Statist.* **15**, 1536–1547.
Chen, C.-H. and George, S.L. (1985). The bootstrap and identification of prognostic factors via Cox's proportional hazards regression model. *Statist. Med.* **4**, 39–46.
Chen, Y.Y., Hollander, M., and Langberg, N.A. (1982). Small sample results for the Kaplan–Meier estimator. *J. Amer. Statist. Assoc.* **77**, 141–144.
Chiang, C.L. (1968). *Introduction to Stochastic Processes in Biostatistics.* Wiley, New York.
Choi, S. (1989). On asymptotically optimal tests. PhD thesis, University of Rochester.
Christensen, E., Schlichting, P., Andersen, P.K., Fauerholdt, L., Juhl, E., Poulsen, H., and Tygstrup, N., for The Copenhagen Study Group for Liver Diseases. (1985). A therapeutic index that predicts the individual effects of prednisone in patients with cirrhosis. *Gastroenterology* **88**, 156–165.
Christensen, E., Schlichting, P., Andersen, P.K., Fauerholdt, L., Schou, G., Pedersen, B.V., Juhl, E., Poulsen, H., and Tygstrup, N., for The Copenhagen Study Group for Liver Diseases. (1986). Updating prognosis and therapeutic effect evaluation in cirrhosis using Cox's multiple regression model for time dependent variables. *Scand. J. Gastroenterol.* **21**, 163–174.
Chung, C.F. (1986). Formulae for probabilities associated with Wiener and Brownian bridge processes. Technical Report 79, Laboratory for Research in Statistics and Probability, Carleton University. Ottawa, Canada.
Chung, C.F. (1987). Wiener pack: a subroutine package for computing probabilities associated with Wiener and Brownian bridge processes. Paper 87-12, Geological Survey of Canada.
Chung, K.-L. (1974). *A Course on Probability Theory*, second edition. Academic Press, New York.
Clayton, D.G. (1978). A model for association in bivariate life tables and its application in epidemiological studies of familial tendency in chronic disease incidence. *Biometrika* **65**, 141–151.
Clayton, D.G. (1991). A Monte Carlo method for Bayesian inference in frailty models. *Biometrics* **47**, 467–485.
Clayton, D.G. and Cuzick, J. (1985a). Multivariate generalizations of the proportional hazards model (with discussion). *J. Roy. Statist. Soc. A* **148**, 82–117.
Clayton, D.G. and Cuzick, J. (1985b). The semi-parametric Pareto model for regression analysis of survival times. *Bull. Internat. Statist. Inst.* **51**(4), 23.3: 1–18.
Cohen, J.E. (1969). Natural primate troops and a stochastic population model. *Amer. Natur.* **103**, 455–477.
Commenges, D. (1986). Semi-Markov and non-homogeneous Markov models in medicine. In Janssen, J., editor, *Semi-Markov models. Theory and Application*, pp. 423–436. Plenum Press, New York and London.
Cook, R.D. and Weisberg, S. (1982). *Residuals and Influence in Regression.* Chapman and Hall, New York.
Copenhagen Study Group for Liver Diseases. (1969). Effect of prednisone on the survival of patients with cirrhosis of the liver. *Lancet* i 119–121.
Copenhagen Study Group for Liver Diseases. (1974). Sex, ascites and alcoholism in survival of patients with cirrhosis. Effect of prednisone. *New Engl. J. Med.* **291**, 271–273.
Courrège, P. and Priouret, P. (1965). Temps d'arrêt d'un fonction aléatoire. *Publ. Inst. Stat. Univ. Paris* **14**, 245–274.

Cox, D.R. (1959). The analysis of exponentially distributed life-times with two types of failure. *J. Roy. Statist. Soc. B* **21**, 411–421.
Cox, D.R. (1972). Regression models and life-tables (with discussion). *J. Roy. Statist. Soc. B* **34**, 187–220.
Cox, D.R. (1975). Partial likelihood. *Biometrika* **62**, 269–276.
Cox, D.R. (1979). A note on the graphical analysis of survival data. *Biometrika* **66**, 188–190.
Cox, D.R. (1986). Some remarks on semi-Markov processes in medical statistics. In Janssen, J., editor, *Semi-Markov models. Theory and Application*, pp. 411–421. Plenum Press, New York and London.
Cox, D.R. and Oakes, D. (1984). *Analysis of Survival Data.* Chapman and Hall, London.
Cox, D.R. and Snell, E.J. (1968). A general definition of residuals (with discussion). *J. Roy. Statist. Soc. B* **30**, 248–275.
Cramér, H. (1945). *Mathematical methods of statistics.* Almqvist and Wiksell, Uppsala.
Crowley, J.J. and Breslow, N.E. (1975). Remarks on the conservatism of $\sum (O - E)^2/E$ in survival data. *Biometrics* **31**, 957–961.
Crowley, J.J. and Hu, M. (1977). Covariance analysis of heart transplant survival data. *J. Amer. Statist. Assoc.* **72**, 27–36.
Crowley, J.J., Liu, P.Y., and Voelkel, J.G. (1982). Estimation of the ratio of hazard functions. In Crowley, J.J. and Johnson, R.A., editors, *Lecture notes—Monograph series* 2, *Survival Analysis*, pp. 56–73. Institute of Mathematical Statistics. Hayward, California.
Crowley, J.J. and Storer, B.E. (1983). Comment on 'A Reanalysis of the Stanford Heart Transplant Data', by M. Aitkin, N. Laird, and B. Francis. *J. Amer. Statist. Assoc.* **78**, 277–281.
Csörgő, S. and Horváth, L. (1982a). On cumulative hazard processes under random censorship. *Scand. J. Statist.* **9**, 13–21.
Csörgő, S. and Horváth, L. (1982b). On random censorship from the right. *Acta Sci. Math. (Szeged)* **44**, 23–34.
Csörgő, S. and Horváth, L. (1983). The rate of strong uniform consistency of the product-limit estimator. *Z. Wahrsch. verw. Geb.* **62**, 411–426.
Csörgő, S. and Horváth, L. (1985). The baboons come down from the trees quite normally. In Grossmann, W., Pflug, G., Vincze, I., and Wertz, W., editors, *Proc. of the 4th Pannonian Symp. on Math. Stat., Bad Tatzmannsdorf, Austria 1983*, pp. 95–106. Reidel, Dordrecht.
Csörgő, S. and Horváth, L. (1986). Confidence bands from censored samples. *Canad. J. Statist.* **14**, 131–144.
Cuzick, J. (1982). Rank tests for association with right censored data. *Biometrika* **69**, 351–364.
Cuzick, J. (1985). Asymptotic properties of censored linear rank tests. *Ann. Statist.* **13**, 133–141.
Cuzick, J. (1988). Rank regression. *Ann. Statist.* **16**, 1369–1389.
Dabrowska, D.M. (1986). Rank tests for independence for bivariate censored data. *Ann. Statist.* **14**, 250–264.
Dabrowska, D.M. (1987). Non-parametric regression with censored survival time data. *Scand. J. Statist.* **14**, 181–198.
Dabrowska, D.M. (1988). Kaplan–Meier estimate on the plane. *Ann. Statist.* **16**, 1475–1489.
Dabrowska, D.M. (1989a). Rank tests for matched pair experiments with censored data. *J. Multivar. Anal.* **28**, 88–114.
Dabrowska, D.M. (1989b). Kaplan-Meier estimate on the plane: Weak convergence, LIL, and the bootstrap. *J. Multivar. Anal.* **29**, 308–325.

Dabrowska, D.M. (1990). Signed-rank tests for censored matched pairs. *J. Amer. Statist. Assoc.* **85**, 478–485.
Dabrowska, D.M. (1992). Product integrals and measures of dependence. Unpublished manuscript, University of California, Los Angeles.
Dabrowska, D.M. and Doksum, K.A. (1988). Partial likelihood in transformation models with censored data. *Scand. J. Statist.* **15**, 1–24.
Dabrowska, D.M., Doksum, K.A., Feduska, N.J., Husing, R., and Neville, P. (1992). Methods for comparing cumulative hazard functions in a semi-proportional hazard model. *Statist-Med.* **11**, 1465–1476.
Dabrowska, D.M., Doksum, K.A., and Song, J.-K. (1989). Graphical comparison of cumulative hazards for two populations. *Biometrika* **76**, 763–773.
Daley, D.J. and Vere-Jones, D. (1988). *An Introduction to the Theory of Point Processes.* Springer-Verlag, New York.
Daniels, H.E. (1945). The statistical theory of the strength of bundles of threads, I. *Proc. Roy. Soc. A* **183**, 405–435.
David, H. and Moeschberger, M.L. (1978). *The theory of competing risks.* Charles Griffin, London.
Davidsen, M. and Jacobsen, M. (1991). Weak convergence of two-sided stochastic integrals, with an application to models for left truncated survival data. In Prabhu, N.U. and Basawa, I.V., editors, *Statistical Inference in Stochastic Processes*, pp. 167–182, Marcel Dekker, New York.
Dellacherie, C. (1972). *Capacités et Processus Stochastiques.* Springer-Verlag, Berlin.
Dellacherie, C. (1980). Un survol de la théorie de l'integrale stochastique. *Stoch. Proc. Appl.* **10**, 115–144.
Dellacherie, C. and Meyer, P.-A. (1982). *Probability and Potentials (B).* North-Holland, Amsterdam.
Dempster, A.P., Laird, N.M., and Rubin, D.B. (1977). Maximum likelihood estimation from incomplete data via the EM algorithm (with discussion). *J. Roy. Statist. Soc. B* **39**, 1–38.
Devroye, L. (1987). *A Course in Density Estimation.* Birkhäuser, Boston.
Devroye, L. and Györfi, L. (1985). *Nonparametric Density Estimation: The L_1 View.* Wiley, New York.
Dewanji, A. and Kalbfleisch, J.D. (1986). Nonparametric methods for survival/sacrifice experiments. *Biometrics* **42**, 325–341.
Dobrushin, R.L. (1953). Generalization of Kolmogorov's equations for a Markov process with a finite number of possible states. *Mat. Sb. (N.S.)* **33**, 567–596 (in Russian).
Doksum, K.A. (1974). Tailfree and neutral random probabilities and their posteriori distributions. *Ann. Probab.* **2**, 183–201.
Doksum, K.A. (1987). An extension of partial likelihood methods for proportional hazards models to general transformation models. *Ann. Statist.* **15**, 325–345.
Doksum, K.A. and Yandell, B.S. (1984). Tests for exponentiality. In Krishnaiah P.R. and Sen, P.K., editors. *Handbook of Statistics* 4, pp. 579–611. North-Holland, Amsterdam.
Doléans-Dade, C. (1970). Quelques applications de la formula de changement de variable pour les semimartingales. *Z. Wahrsch. verw. Geb.* **16**, 181–194.
Doléans-Dade, C. and Meyer, P.-A. (1970). Intégrales stochastiques par rapport aux martingales locales. In *Séminaire Probab. IV*, Lecture Notes in Mathematics **124**, pp. 77–107. Springer-Verlag, New York.
Dolivo, F. (1974). Counting processes and integrated conditional rates: a martingale approach with application to detection. PhD thesis, University of Michigan.
Dollard, J.D. and Friedman, C.N. (1979). *Product Integration with Applications to Differential Equations* (with an appendix by P.R. Masani). Addison-Wesley, Reading, MA.

Doob, J.L. (1940). Regularity properties of certain families of chance variables. *Trans. Amer. Math. Soc.* **47**, 455–486.

Doob, J.L. (1953). *Stochastic Processes.* Wiley, New York.

Doss, H. and Gill, R.D. (1992). A method for obtaining weak convergence results for quantile processes, with applications to censored survival data. *J. Amer. Statist. Assoc.* **87**, 869–877.

Downton, F. (1972). Discussion of the paper by D.R. Cox *J. Roy. Statist. Soc. B.* **34**, 202–205.

Drzewiecki, K.T. and Andersen, P.K. (1982). Survival with malignant melanoma. A regression analysis of prognostic factors. *Cancer* **49**, 2414–2419.

Drzewiecki, K.T., Ladefoged, C., and Christensen, H.E. (1980a). Biopsy and prognosis for cutaneous malignant melanomas in clinical stage 1. *Scand. J. Plast. Reconstr. Surg.* **14**, 141–144.

Drzewiecki, K.T., Christensen, H.E., Ladefoged, C., and Poulsen, H. (1980b). Clinical course of cutaneous malignant melanoma related to histopathological criteria of primary tumour. *Scand. J. Plast. Reconstr. Surg.* **14**, 229–234.

Dudley, R.M. (1966). Weak convergence of probabilities on nonseparable metric spaces and empirical measures on Euclidean spaces. *Illinois J. Math.* **10**, 109–126.

Durbin, J. (1973). *Distribution Theory for Tests Based on the Sample Distribution Function.* SIAM, Philadelphia.

Dvoretsky, A. (1972). Asymptotic normality for sums of dependent random variables. In *Proc. Sixth Berkeley Symp. Math. Statist. Probab.* 2, pp. 513–535. University of California Press, Berkeley.

Dzhaparidze, K. (1985). On asymptotic inference about intensity parameters of a counting process. *Bull. Internat. Statist. Inst.* **51**(4), 23.2:1–15.

Dzhaparidze, K. and Spreij, P.J.C. (1990). On second order optimality of regular projective estimators: Part I. Report BS-R9029, Centre for Mathematics and Computer Science, Amsterdam. Has appeared in *Stochastics* **42**, 53–65, 1993.

Ederer, F., Axtell, L.M., and Cutler, S.J. (1961). The relative survival rate: A statistical methodology. *Nat. Cancer Inst. Monographs* **6**, 101–121.

Efron, B. (1967). The two sample problem with censored data. In *Proceedings of the Fifth Berkeley Symposium on Mathematical Statistics and Probability* **4**, 831–853. Prentice-Hall New York.

Efron, B. (1977). The efficiency of Cox's likelihood function for censored data. *J. Amer. Statist. Assoc.* **72**, 557–565.

Efron, B. (1979). Bootstrap methods: Another look at the jackknife. *Ann. Statist.* **7**, 1–26.

Efron, B. (1981). Censored data and the bootstrap. *J. Amer. Statist. Assoc.* **76**, 312–319.

Efron, B. and Johnstone, I.M. (1990). Fisher's information in terms of the hazard rate. *Ann. Statist.* **18**, 38–62.

Elbers, C. and Ridder, G. (1982). True and spurious duration dependence: the identifiability of the proportional hazard model. *Rev. Economic Stud.* **49**, 403–409.

Epstein, B. and Sobel, M. (1953). Life testing. *J. Amer. Statist. Assoc.* **48**, 486–502.

Fabian, V. and Hannan, J. (1982). On estimation and adaptive estimation for locally asymptotically normal families. *Z. Wahrsch. verw. Geb.* **59**, 459–478.

Fabian, V. and Hannan, J. (1985). *Introduction to Probability and Mathematical Statistics.* Wiley, New York.

Ferguson, T.S. (1973). A Bayesian analysis of some nonparametric problems. *Ann. Statist.* **1**, 209–230.

Ferguson, T.S. and Phadia, E.G. (1979). Bayesian nonparametric estimation based on censored data. *Ann. Statist.* **7**, 163–186.

Fix, E. and Neyman, J. (1951). A simple stochastic model of recovery, relapse, death and loss of patients. *Human Biol.* **23**, 205–241.

Fleiss, J.L. (1981). *Statistical Methods for Rates and Proportions*, second edition. Wiley, New York.

Fleming, T.R. (1978a). Nonparametric estimation for nonhomogeneous Markov processes in the problem of competing risks. *Ann. Statist.* **6**, 1057–1070.

Fleming, T.R. (1978b). Asymptotic distribution results in competing risks estimation. *Ann. Statist.* **6**, 1071–1079.

Fleming, T.R. and Harrington, D.P. (1981). A class of hypothesis tests for one and two samples of censored survival data. *Commun. Statist.* **10**, 763–794.

Fleming, T.R. and Harrington, D.P. (1991). *Counting Processes and Survival Analysis.* Wiley, New York.

Fleming, T.R., Harrington, D.P., and O'Sullivan, M. (1987). Supremum versions of the log-rank and generalized Wilcoxon statistics. *J. Amer. Statist. Assoc.* **82**, 312–320.

Fleming, T.R., O'Fallon, J.R., O'Brien, P.C., and Harrington, D.P. (1980). Modified Kolmogorov–Smirnov test procedures with application to arbitrarily right censored data. *Biometrics* **36**, 607–626.

Forsén, L. (1979). The efficiency of selected moment methods in Gompertz–Makeham graduation of mortality. *Scand. Actuar. J.* 167–178.

Fréchet, M. (1937). Sur la notion de différentielle dans l'analyse générale. *J. Math. Pures Appl.* **16**, 233–250.

Freireich, E.J., Gehan, E., Frei, E., Schroeder, L.R., Wolman, I.J., Anbari, R., Burgert, E.O., Mills, S.D., Pinkel, D., Selawry, O.S., Moon, J.H., Gendel, B.R., Spurr, C.L., Storrs, R., Haurani, F., Hoogstraten, B., and Lee, S. (1963). The effect of 6-Mercaptopurine on the duration of steroid-induced remissions in acute leukemia: a model for evaluation of other potentially useful therapy. *Blood* **21**, 699–716.

Frydman, H. (1991). Nonparametric maximum likelihood estimation for a periodically observed Markov "illness–death" process, with application to diabetes survival data. Research Report 91/3, Statistical Research Unit, University of Copenhagen.

Frydman, H. (1992). A nonparametric estimation procedure for a periodically observed three state Markov process, with application to AIDS. *J. Roy. Statist. Soc. B* **54**, 853–866.

Gaenssler, P. (1983). *Empirical Processes.* IMS Lecture Notes—monograph series 3. Hayward, California.

Gail, M.H. (1975). A review and critique of some models used in competing risk analysis. *Biometrics* **31**, 209–222.

Gail, M.H. (1986). Adjusting for covariates that have the same distribution in exposed and unexposed cohorts. In Moolgavkar, S.H. and Prentice, R.L., editors, *Modern Statistical Methods in Chronic Disease Epidemiology*, pp. 1–18. Wiley, New York.

Gail, M.H., Tan, W.Y., and Piantadosi, S. (1988). Tests for no treatment effect in randomized clinical trials. *Biometrika* **75**, 57–64.

Gail, M.H., Wieand, S., and Piantadosi, S. (1984). Biased estimates of treatment effect in randomized experiments with nonlinear regressions and omitted covariates. *Biometrika* **71**, 431–444.

Gart, J.J., Krewski, D., Lee, P.N., Tarone, R.E., and Wahrendorf, J. (1986). *Statistical methods in cancer research.* Volume *III*. *The design and analysis of long-term animal experiments.* IARC Scientific Publications. 79. International Agency for Research in Cancer. Lyon.

Gasser, T. and Müller, H.G. (1979). Kernel estimation of regression functions. In *Smoothing Techniques for Curve Estimation*, Lecture Notes in Mathematics 757, pp. 23–68. Springer-Verlag, Berlin.

Gasser, T., Müller, H.-G., and Mammitzsch, V. (1985). Kernels for nonparametric curve estimation. *J. Roy. Statist. Soc. B* **47**, 238–252.

Gatsonis, C., Hsieh, H.K., and Korwar, R. (1985). Simple nonparametric tests for a known standard survival based on censored data. *Commun. Statist.—Teor. Meth.* **14**, 2137–2162.

Gehan, E.A. (1965). A generalized Wilcoxon test for comparing arbitrarily singly censored samples. *Biometrika* **52**, 203–223.

Geurts, J.H.J. (1985). Some small-sample non-proportional hazards results for the Kaplan–Meier estimator. *Statist. Neerland.* **39**, 1–13.

Geurts, J.H.J. (1987). On the small-sample performance of Efron's and Gill's version of the product limit estimator under nonproportional censoring. *Biometrics* **43**, 683–692.

Geurts, W.A.J., Hasselaar, M.M.A., and Verhagen, J.H. (1988). Large sample theory for statistical inference in several software reliability models. Technical Report MS-8807, Centre for Mathematics and Computer Science, Amsterdam.

Gill, R.D. (1980a). *Censoring and Stochastic Integrals*, Mathematical Centre Tracts 124. Mathematisch Centrum, Amsterdam.

Gill, R.D. (1980b). Nonparametric estimation based on censored observations of a Markov renewal process. *Z. Wahrsch. verw. Geb.* **53**, 97–116.

Gill, R.D. (1981). Testing with replacement and the product limit estimator. *Ann. Statist.* **9**, 853–860.

Gill, R.D. (1983a). Large sample behavior of the product-limit estimator on the whole line. *Ann. Statist.* **11**, 49–58.

Gill, R.D. (1983b). Discussion of the papers by Helland and Kurtz. *Bull. Internat. Statist. Inst.* **50**(3), 239–243.

Gill, R.D. (1984). Understanding Cox's regression model: A martingale approach. *J. Amer. Statist. Assoc.* **79**, 441–447.

Gill, R.D. (1985). Discussion of the paper by D. Clayton and J. Cuzick, *J. Roy. Statist. Soc. A* **148**, 108–109.

Gill, R.D. (1986a). On estimating transition intensities of a Markov process with aggregated data of a certain type: 'Occurrences but no exposures.' *Scand. J. Statist.* **13**, 113–134.

Gill, R.D. (1986b). The total time on test plot and the cumulative total time on test statistic for a counting process. *Ann. Statist.* **14**, 1234–1239.

Gill, R.D. (1989). Non- and semi-parametric maximum likelihood estimators and the von Mises method (Part 1). *Scand. J. Statist.* **16**, 97–128.

Gill, R.D. (1990). Multivariate survival analysis. Preprint 621, Department of Mathematics, University of Utrecht. Has appeared in *Theor. Prob. Appl.* **37**, 18–31 and 284–301, 1992.

Gill, R.D. (1991a). Multistate life-tables and regression models. Preprint 659, Department of Mathematics, University Utrecht. Has appeared in *Math. Pop. Studies* **3**, 259–276, 1992.

Gill, R.D. (1991b). Non- and semi-parametric maximum likelihood estimators and the von Mises Method, Part II. Preprint 664, Department of Mathematics, University Utrecht. Has appeared joint with A.W. van der Vaart in *Scand. J. Statist.* **20**, 271–288, 1993.

Gill, R.D. (1992). Marginal partial likelihood. *Scand. J. Statist.* **79**, 133–137.

Gill, R.D. and Johansen, S. (1990). A survey of product-integration with a view towards application in survival analysis. *Ann. Statist.* **18**, 1501–1555.

Gill, R.D. and Schumacher, M. (1987). A simple test of the proportional hazards assumption. *Biometrika* **74**, 289–300.

Gill, R.D., van der Laan, M.J., and Wellner, J.A. (1993). Inefficient estimators for three multivariate models. Preprint 767, Department of Mathematics, University of Utrecht.

Gillespie, M.J. and Fisher, L. (1979). Confidence bands for the Kaplan–Meier survival curve estimates. *Ann. Statist.* **7**, 920–924.

Goldstein, L. and Langholz, B. (1992). Asymptotic theory for nested case-control sampling in the Cox regression model. *Ann. Statist.* **20**, 1903–1928.

Gray, R.J. (1988). A class of k-sample tests for comparing the cumulative incidence of a competing risk. *Ann. Statist.* **16**, 1141–1154.
Gray, R.J. (1990). Some diagnostic methods of Cox regression models through hazard smoothing. *Biometrics* **46**, 93–102.
Gray, R.J. and Pierce, D.A. (1985). Goodness-of-fit tests for censored data. *Ann. Statist.* **13**, 552–563.
Green, A., Borch-Johnsen, K., Andersen, P.K., Hougaard, P., Keiding, N., Kreiner, S., and Deckert, T. (1985). Relative mortality of Type 1 (insulin-dependent) diabetes in Denmark: 1933–1981. *Diabetologia* **28**, 339–342.
Green, A., Hauge, M., Holm, N.V., and Rasch, L.L. (1981). Epidemiological studies of diabetes mellitus in Denmark. II. A prevalence study based on insulin prescriptions. *Diabetologia* **20**, 468–470.
Green, A. and Hougaard, P. (1984). Epidemiological studies of diabetes mellitus in Denmark: 5. Mortality and causes of death among insulin-treated diabetic patients. *Diabetologia* **26**, 190–194.
Greenwood, M. (1926). The natural duration of cancer. In *Reports on Public Health and Medical Subjects* 33, pp. 1–26. His Majesty's Stationery Office, London.
Greenwood, P.E. (1988). Partially specified semimartingale experiments. *Contemp. Math.* **80**, 1–17.
Greenwood, P.E. and Wefelmeyer, W. (1989). Partially and fully specified semiparametric models and efficiency. Preprints in Statistics 124, Dept. Math., University of Köln.
Greenwood, P.E. and Wefelmeyer, W. (1990). Efficiency of estimators for partially specified filtered models. *Stoch. Proc. Appl.* **36**, 353–370.
Greenwood, P.E. and Wefelmeyer, W. (1991a). Efficient estimating equations for nonparametric filtered models. In Prabhu, N.U. and Basawa, I.V., editors, *Statistical Inference in Stochastic Processes*, pp. 107–141. Marcel Dekker, New York.
Greenwood, P.E. and Wefelmeyer, W. (1991b). Efficient estimation in a nonlinear counting process regression model. *Canad. J. Statist.* **19**, 165–178.
Greenwood, P.E. and Wefelmeyer, W. (1991c). On optimal estimating functions for partially specified models. In Godambe, V.P., editor, *Estimating Functions*, pp. 147–160. Oxford University Press, Oxford.
Grimmett, G.R. and Stirzaker, D.R. (1982). *Probability and Random Processes.* Oxford University Press, Oxford.
Groeneboom, P. and Wellner, J.A. (1992). Information Bounds and Nonparametric Maximum Likelihood Estimation. Birkhäuser Verlag, Basel.
Gross, A.J. and Clark, V.A. (1975). *Survival Distributions: Reliability Applications in the Biomedical Sciences.* Wiley, New York.
Gross, S.T. and Huber, C. (1987). Matched pair experiments: Cox and maximum likelihood estimation. *Scand. J. Statist.* **14**, 27–41.
Gross, S.T. and Huber-Carol, C. (1992). Regression models for truncated survival data. *Scand. J. Statist.* **19**, 193–213.
Grüger, J. (1986). Nichtparametrische Analyse sporadisch beobachtbarer Krankheits-Verlaufsdaten. Dissertation, Universität Dortmund.
Grüger, J., Kay, R., and Schumacher, M. (1991). The validity of inferences based on incomplete observations in disease state models. *Biometrics* **47**, 595–605.
Gu, M.G., Lai, T.L., and Lan, K.K.G. (1991). Rank tests based on censored data and their sequential analogues. *Amer. J. Math. Management Sci.* **11**, 147–176.
Gu, M.G. and Lai, T.L. (1991). Weak convergence of time-sequential censored rank statistics with applications to sequential testing in clinical trials. *Ann. Statist.* **19**, 1403–1433.
Guilbaud, O. (1988). Exact Kolmogorov-type tests for left-truncated and/or right-censored data. *J. Amer. Statist. Assoc.* **83**, 213–221.
Haara, P. (1987). A note on the asymptotic behaviour of the empirical score in Cox's

regression model for counting processes. In *Proceedings of the 1st World Congress of the Bernoulli Society*, pp. 139–142. VNU Science Press, Tashkent, Soviet Union.
Habib, M.G. and Thomas, D.R. (1986). Chi-square goodness-of-fit tests for randomly censored data. *Ann. Statist.* **14**, 759–765.
Hájek, J. (1962). Asymptotically most powerful rank order tests. *Ann. Math. Statist.* **33**, 1124–1147.
Hájek, J. (1970). A characterization of limiting distributions of regular estimators. *Z. Wahrsch. verw. Geb.* **14**, 323–330.
Hájek, J. (1972). Local asymptotic minimax and admissibility in estimation. *Proc. Sixth Berkeley Symp. Math. Statist. Probab.* **1**, 245–261. University of California Press, Berkeley, California.
Hájek, J. and Sĭdák, Z. (1967). *Theory of Rank Tests.* Academic Press, New York.
Hald, A. (1949). Maximum likelihood estimation of the parameters of a normal distribution which is truncated at a known point. *Skand. Aktuarietidskr.* **32**, 119–134.
Hald, A. (1952). *Statistical Theory with Engineering Applications.* Wiley, New York.
Hall, P. (1990). Using the bootstrap to estimate mean squared error and select the smoothing parameter in nonparametric problems. *J. Multivar. Anal.* **32**, 177–203.
Hall, P. and Marron, J.S. (1987). Estimation of integrated squared density derivatives. *Statist. Probab. Lett.* **6**, 109–115.
Hall, W.J. and Wellner J.A. (1980). Confidence bands for a survival curve from censored data. *Biometrika* **67**, 133–143.
Halley, E. (1693). An estimate of the degrees of the mortality of mankind drawn from curious tables of the births and funerals at the city of Breslaw. *Philos. Trans. Roy. Soc. London* **17**, 596–610. [Reprinted in *J. Inst. Act.* **112**, 278–301 (1985).]
Hanley, J.A. and Parnes, M.N. (1983). Nonparametric estimation of a multivariate distribution in the presence of censoring. *Biometrics* **39**, 129–139.
Härdle, W. (1990). *Applied nonparametric regression.* Cambridge University Press, Cambridge.
Harrington, D.P. and Fleming, T.R. (1982). A class of rank test procedures for censored survival data. *Biometrika* **69**, 133–143.
Hastie, T. and Tibshirani, R. (1986). Generalized additive models (with discussion). *Statist. Sci.* **1**, 297–319.
Heckman, J.J. and Singer, B. (1982). Population heterogeneity in demographic models. In Land, K.C. and Rogers, A., editors, *Multidimensional Mathematical Demography*, Chap. 12, pp. 567–599. Academic Press, New York.
Heckman, J.J. and Singer, B. (1984a). Econometric duration analysis. *J. Econometrics* **24**, 63–132.
Heckmann, J.J. and Singer, B. (1984b). A method for minimizing the impact of distributional assumptions in econometric models for duration data. *Econometrica* **52**, 271–320.
Heesterman, C.C. and Gill, R.D. (1992). A central limit theorem for M-estimators by the von Mises method. *Statist. Neerland.* **46**, 165–177.
Heitjan, D.F. (1993). Ignorability and coarse data: some biomedical examples. *Biometrics* **49**, 1099–1109.
Heitjan, D.F. and Rubin, D.B. (1991). Ignorability and coarse data. *Ann. Statist.* **19**, 2244–2253.
Helland, I.S. (1982). Central limit theorems for martingales with discrete or continuous time. *Scand. J. Statist.* **9**, 79–94.
Helland, I.S. (1983). Applications of central limit theorems for martingales with continuous time. *Bull. Int. Statist. Inst.* **50**(1), 346–360.
Hjort, N.L. (1984). Local asymptotic power of linear nonparametric tests for comparison of counting processes. Report 763, Norwegian Computing Center.

Hjort, N.L. (1985a). Discussion of the paper by P.K. Andersen and Ø. Borgan, *Scand. J. Statist.* **12**, 141–150.
Hjort, N.L. (1985b). Bootstrapping Cox's regression model. Technical Report 241, Department of Statistics, Stanford University, CA.
Hjort, N.L. (1986). Bayes estimators and asymptotic efficiency in parametric counting process models. *Scand. J. Statist.* **13**, 63–85.
Hjort, N.L. (1990a). Goodness of fit tests in models for life history data based on cumulative hazard rates. *Ann. Statist.* **18**, 1221–1258.
Hjort, N.L. (1990b). Nonparametric Bayes estimators based on beta processes in models for life history data. *Ann. Statist.* **18**, 1259–1294.
Hjort, N.L. (1992a). On inference in parametric survival data models. *Internat. Statist. Rev.* **60**, 355–387.
Hjort, N.L. (1992b). Semiparametric estimation of parametric hazard rates. In Klein, J.P. and Goel, P.K., editors, *Survival Analysis: State of the Art*, pp. 211–236. Kluwer, Dordrecht.
Hoel, D.G. (1972). A representation of mortality data by competing risks. *Biometrics* **28**, 475–488.
Hoel, D.G. and Walburg, H.E. (1972). Statistical analysis of survival experiments. *J. Nat. Cancer Inst.* **49**, 361–372.
Hoem, J.M. (1969a). Fertility rates and reproduction rates in a probabilistic setting. *Biométrie-Praximétrie* **10**, 38–66. Correction: **11**, 20 (1970).
Hoem, J.M. (1969b). Purged and partial Markov chains. *Skand. Aktuarietidskr.* **52**, 147–155.
Hoem, J.M. (1971). Point estimation of forces of transition in demographic models. *J. Roy. Statist. Soc. B* **33**, 275–289.
Hoem, J.M. (1972). On the statistical theory of analytic graduation. In *Proceedings of the Sixth Berkeley Symposium on Mathematical Statistics and Probability* **1**, 569–600. University of California Press, Berkeley, California.
Hoem, J.M. (1976). The statistical theory of demographic rates. A review of current developments (with discussion). *Scand. J. Statist.* **3**, 169–185.
Hoem, J.M. (1987). Statistical analysis of a multiplicative model and its application to the standardization of vital rates. A review. *Internat. Statist. Rev.* **55**, 119–152.
Hoem, J.M and Aalen, O.O. (1978). Actuarial values of payment streams. *Scand. Actuar. J.* 38–47.
Hollander, M. and Peña, E. (1989). Families of confidence bands for the survival function under the general random censorship model and the Koziol–Green model. *Canad. J. Statist.* **17**, 59–74.
Hollander, M. and Proschan, F. (1979). Testing to determine the underlying distribution using randomly censored data. *Biometrics* **35**, 393–401.
Hollander, M. and Proschan, F. (1984). Nonparametric concepts and methods in reliability. In Krishnaiah, P.R. and Sen, P.K., editors, *Handbook of Statistics* 4, 613–655. North-Holland, Amsterdam.
Holt, J.D. and Prentice, R.L. (1974). Survival analysis in twin studies and matched-pair experiments. *Biometrika* **61**, 17–30.
Höpfner, R. (1991). On statistics of Markov step processes: representation of log-likelihood ratio processes in filtered local models. Preprint, Institute of Mathematical Stochastics, University of Freiburg.
Horváth, L. and Yandell, B.S. (1987). Convergence rates for the bootstrapped product limit process. *Ann. Statist.* **15**, 1155–1173.
Hougaard, P. (1984). Life table methods for heterogeneous populations: Distributions describing the heterogeneity. *Biometrika* **71**, 75–83.
Hougaard, P. (1986a). Survival models for heterogeneous popualtions derived from stable distributions. *Biometrika* **73**, 387–396.

Hougaard, P. (1986b). A class of multivariate failure time distributions. *Biometrika* **73**, 671–678.
Hougaard, P. (1987). Modelling multivariate survival. *Scand. J. Statist.* **14**, 291–304.
Hougaard, P., Harvald, B., and Holm, N.V. (1992). Measuring the similarities between the life times of adult Danish twins born 1881–1930. *J. Amer. Statist. Assoc.* **87**, 17–24.
Huber, P.J. (1967). The behavior of maximum likelihood estimates under nonstandard conditions. In *Proceedings of the Fifth Berkeley Symp. Math. Statist. Probab.* **1**, 221–233. Prentice-Hall, New York.
Huber, P.J. (1981). *Robust Statistics.* Wiley, New York.
Huffer, F.W. and McKeague, I.W. (1991). Weighted least squares estimation for Aalen's additive risk model. *J. Amer. Statist. Assoc.* **86**, 114–129.
Huster, W.J., Brookmeyer, R., and Self, S.G. (1989). Modelling paired survival data with covariates. *Biometrics* **45**, 145–156.
Hyde, J. (1977). Testing survival under right censoring and left truncation. *Biometrika* **64**, 225–230.
Hyde, J. (1980). Survival analysis with incomplete observations. In Miller, R.G., Efron, B., Brown, B.W., and Moses, L.E., editors, *Biostatistics Casebook*, pp. 31–46. Wiley, New York.
Ibragimov, I.A. and Khas'minskii, R.Z. (1981). *Statistical Estimation; Asymptotic Theory.* Springer-Verlag, New York.
Ibragimov, I.A. and Khas'minskii, R.Z. (1991). Asymptotically normal families of distributions and effective estimation. *Ann. Statist.* **19**, 1681–1724.
Irwin, J.O. (1949). A note on the subdivision of χ^2 into components. *Biometrika* **36**, 130–134.
Itô, K. (1944). Stochastic integral. *Proc. Imperial Acad. Tokyo* **20**, 519–524.
Itô, K. (1951). On a formula concerning stochastic differentials. *Nagoya J. Math.* **3**, 55–65.
Itô, K. and Watanabe, S. (1965). Transformation of Markov processes by multiplicative functionals. *Ann. Inst. Fourier* **15**, 15–30.
Izenman, A.J. (1991). Recent developments in nonparametric density estimation. *J. Amer. Statist. Assoc.* **86**, 205–224.
Jacobsen, M. (1982). *Statistical Analysis of Counting Processes*, Lecture Notes in Statistics 12. Springer-Verlag, New York.
Jacobsen, M. (1984). Maximum likelihood estimation in the multiplicative intensity model: A survey. *Internat. Statist. Rev.* **52**, 193–207.
Jacobsen, M. (1989a). Existence and unicity of MLEs in discrete exponential family distributions. *Scand. J. Statist.* **16**, 335–349.
Jacobsen, M. (1989b). Right censoring and martingale methods for failure time data. *Ann. Statist.* **17**, 1133–1156.
Jacobsen, M. and Keiding, N. (1991). Random censoring and coarsening at random. Research Report 91/6, Statistical Research Unit. University of Copenhagen.
Jacobsen, N., Badsberg, J.H., Lönnqvist, B., Ringdén, O., Volin, L., Rajantie, J., Nikoskelainen, J., and Keiding, N., for The Nordic Bone Marrow Transplantation Group. (1990). Graft-versus-leukemia activity associated with CMV-seropositive donor, posttransplant CMV infection, young donor age and chronic graft-versus-host disease in bone marrow allograft recipients. *Bone Marrow Transplantation* **5**, 413–418.
Jacobsen, N., Badsberg, J.H., Lönnqvist, B., Ringdén, O., Volin, L., Ruutu, T., Rajantie, J., Siimes, M.A., Nikoskelainen, J., Toivanen, A., Andersen, H.K., Keiding, N., and Gahrton, G., for The Nordic Bone Marrow Transplantation Group. (1987). Predictive factors for chronic graft-versus-host disease and leukaemic relapse after allogeneic bone marrow transplantation. In Baum, S.J., Santon, G.W., ad Takaku,

F., editors, *Recent Advances and Future Directions in Bone Marrow Transplantation*, pp. 161–164. Springer-Verlag, New York.
Jacod, J. (1973). On the stochastic intensity of a random point process over the half-line. Technical Report 15, Department of Statistics, Princeton University.
Jacod, J. (1975). Multivariate point processes: Predictable projection, Radon–Nikodym derivatives, representation of martingales. *Z. Wahrsch. verw. Geb.* **31**, 235–253.
Jacod, J. (1979). *Calcul stochastique et problèmes de martingales*, Lecture Notes in Mathematics 714. Springer-Verlag, Berlin.
Jacod, J. (1987). Partial likelihood process and asymptotic normality. *Stoch. Proc. Appl.* **26**, 47–71.
Jacod, J. (1990a). Regularity, partial regularity, partial information process. *Probability Theory Related Fields* **86**, 305–335.
Jacod, J. (1990b). Sur le processus de vraisemblance partielle. *Ann. Inst. Henri Poincaré* **26**, 299–329.
Jacod, J. and Mémin, J. (1980). Sur la convergence des semimartingales vers un processus à accroissements indépendantes. In *Séminaire de Probabilite XIV*, Lecture Notes in Mathematics 784, 227–248. Springer-Verlag, Berlin.
Jacod, J. and Shiryayev, A.N. (1987). *Limit Theorems for Stochastic Processes.* Springer-Verlag, Berlin.
James, I.R. and Smith, P.J. (1984). Consistency results for linear regression with censored data. *Ann. Statist.* **12**, 590–600.
Janssen, A. (1989). Local asymptotic normality for randomly censored models with applications to rank tests. *Statist. Neerland.* **43**, 109–125.
Janssen, A. (1991). Optimal k-sample tests for randomly censored data. *Scand. J. Statist.* **18**, 135–152.
Janssen, A. and Milbrodt, M. (1991). Rényi type goodness of fit tests with adjusted principle direction of alternatives. Technical report, Department of Mathematics, University of Siegen.
Janssen, J., editor (1986). *Semi-Markov Models. Theory and Application.* Plenum Press, New York and London.
Jelinski, Z. and Moranda, P. (1972). Software reliability research. *Statistical Computer Performance Evaluation*, pp. 466–484. Academic Press. New York.
Jennison, C. and Turnbull, B.W. (1985). Repeated confidence intervals for the median survival time. *Biometrika* **72**, 619–625.
Jespersen, N.C.B (1986). Dichotomizing a continuous covariate in the Cox regression model. Research Report 86/2, Statistical Research Unit, University of Copenhagen.
Johansen, S. (1978). The product limit estimator as maximum likelihood estimator. *Scand. J. Statist.* **5**, 195–199.
Johansen, S. (1983). An extension of Cox's regression model. *Internat. Statist. Review* **51**, 258–262.
Johansen, S. (1986). Product integrals and Markov processes. *CWI Newsletter* (12), 3–13. Originally appeared (1977) as Preprint 3, Institute of Mathematical Statistics, University of Copenhagen.
Johnson, M.E., Tolley, H.D., Bryson, M.C., and Goldman, A.S. (1982). Covariate analysis of survival data: A small-sample study of Cox's model. *Biometrics* **38**, 685–698.
Jolivet, E., Reyne, Y., and Teyssier, J. (1983). Approche méthodologique de la répartition nycthémérale des prises d'aliment chez le lapin domestique en croissance. *Reproduction Nutrition Développement* **23**, 13–24.
Jones, D. and Whitehead, J. (1979). Sequential forms of the logrank and modified Wilcoxon tests for censored data. *Biometrika* **66**, 105–113. Amendment: **68**, 576 (1981).

Jones, M.P. and Crowley, J.J. (1989). A general class of nonparametric tests for survival analysis. *Biometrics* **45**, 157–170.
Jones, M.P. and Crowley, J.J. (1990). Asymptotic properties of a general class of nonparametric tests for survival analysis. *Ann. Statist.* **18**, 1203–1220.
Kabanov, Y., Liptser, R.S., and Shiryayev, A.N. (1976). Criteria of absolute continuity of measures corresponding to multivariate point processes. In *Proc. Third Japan–USSR Symposium*, Lecture Notes in Mathematics 550, pp. 232–252. Springer-Verlag, Berlin.
Kalbfleisch, J.D. (1978a). Likelihood methods and nonparametric tests. *J. Amer. Statist. Assoc.* **73**, 167–170.
Kalbfleisch, J.D. (1978b). Nonparametric Bayesian analysis of survival time data. *J. Roy. Statist. Soc. B* **40**, 214–221.
Kalbfleisch, J.D. and Lawless, J.F. (1985). The analysis of panel data under a Markov assumption. *J. Amer. Statist. Assoc.* **80**, 863–871.
Kalbfleisch, J.D. and Lawless, J.F. (1989). Inference based on retrospective ascertainment: An analysis of the data on transfusion-related AIDS. *J. Amer. Statist. Assoc.* **84**, 360–372.
Kalbfleisch, J.D. and MacKay, R.J. (1979). On constant-sum models for censored survival data. *Biometrika* **66**, 87–90.
Kalbfleisch, J.D. and Prentice, R.L. (1973). Marginal likelihoods based on Cox's regression and life model. *Biometrika* **60**, 267–278.
Kalbfleisch, J.D. and Prentice, R.L. (1980). *The Statistical Analysis of Failure Time Data.* Wiley, New York.
Kaplan, E.L. and Meier, P. (1958). Non-parametric estimation from incomplete observations. *J. Amer. Statist. Assoc.* **53**, 457–481, 562–563.
Kardaun, O. (1983). Statistical analysis of male larynx-cancer patients—a case study. *Statist. Neerland.* **37**, 103–126.
Karr, A.F. (1986). *Point Processes and their Statistical Inference.* Marcel Dekker, New York.
Karr, A.F. (1987). Maximum likelihood estimation in the multiplicative intensity model via sieves. *Ann. Statist.* **15**, 473–490.
Kay, R. (1977). Proportional hazards regression models and the analysis of censored survival data. *Appl. Statist.* **26**, 227–237.
Kay, R. (1984). Goodness-of-fit methods for the proportional hazards model: A review. *Rev. Épidem. Santé Publ.* **32**, 185–198.
Kay, R. (1986). A Markov model for analyzing cancer markers and disease states in survival studies. *Biometrics* **42**, 855–865.
Keiding, N. (1975). Maximum likelihood estimation in the birth-and-death process. *Ann. Statist.* **3**, 363–372. Correction: **6**, 472 (1978).
Keiding, N. (1977). Statistical comments on Cohen's application of a simple stochastic population model to natural primate troops. *Amer. Natur.* **111**, 1211–1219.
Keiding, N. (1986). Statistical analysis of semi-Markov models based on the theory of counting processes. In Janssen, J., editor, *Semi-Markov Models. Theory and Applications*, pp. 301–315. Plenum Press, New York and London.
Keiding, N. (1987). The method of expected number of deaths 1786–1886–1986. *Internat. Statist. Rev.* **55**, 1–20.
Keiding, N. (1990). Statistical inference in the Lexis diagram. *Phil. Trans. Roy. Soc. London A* **332**, 487–509.
Keiding, N. (1991). Age-specific incidence and prevalence: a statistical perspective (with discussion). *J. Roy. Statist. Soc. A* **154**, 371–412.
Keiding, N. (1992). Independent delayed entry. In Klein, J.P. and Goel, P.K., editors, *Survival Analysis: State of the Art*, pp. 309–326. Kluwer, Dordrecht.
Keiding, N. and Andersen, P.K. (1989). Nonparametric estimation of transition inten-

sities and transition probabilities: a case study of a two-state Markov process. *Appl. Statist.* **38**, 319–329.

Keiding, N., Bayer, T., and Watt-Boolsen, S. (1987). Confirmatory analysis of survival data using left truncation of the life times of primary survivors. *Statist. Med.* **6**, 939–944.

Keiding, N. and Gill, R.D. (1988). Random truncation models and Markov processes. Report MS-R8817, Centre for Mathematics and Computer Science, Amsterdam.

Keiding, N. and Gill, R.D. (1990). Random truncation models and Markov processes. *Ann. Statist.* **18**, 582–602.

Keiding, N., Hansen, B.E., and Holst, C. (1990). Nonparametric estimation of disease incidence from a cross-sectional sample of a stationary population. In Gabriel, J.P., Lefèvre, C., and Picard, P., editors, *Stochastic Processes in Epidemic Theory*, Lecture Notes in Biomathematics 86, pp. 36–45. Springer-Verlag, Berlin.

Keiding, N., Holst, C., and Green, A. (1989). Retrospective estimation of diabetes incidence from information in a current prevalent population and historical mortality. *Amer. J. Epidemiol.* **130**, 588–600.

Kellerer, A.M. and Chmelevsky, D. (1983). Small-sample properties of censored-data rank tests. *Biometrics* **39**, 675–682.

Khmaladze, E.V. (1981). Martingale approach to the goodness of fit tests. *Theory Probab. Appl.* **26**, 246–265.

Khmaladze, E.V. (1988). An innovation approach to goodness-of-fit tests in R^m. *Ann. Statist.* **16**, 1503–1516.

Kiefer, J. and Wolfowitz, J. (1956). Consistency of the maximum likelihood estimator in the presence of infinitely many nuisance parameters. *Ann. Math. Statist.* **27**, 887–906.

Kim, J. (1990). Conditional bootstrap methods for censored data. PhD thesis, Florida State University.

King, G. and Hardy, G.F. (1880). Notes on the practical application of Mr. Makeham's formula to the graduation of mortality tables (with discussion). *J. Inst. Actuar.* **22**, 191–231.

Klaassen, C.A.J. (1988). Efficient estimation in the Clayton–Cuzick model for survival data. Report TW-88-06, Department of Mathematics, University of Leiden.

Klaassen, C.A.J. (1989). Efficient estimation in the Cox model for survival data. In Mandl, P. and Husková, M., editors, *Proc. Fourth Prague Symp. Asympt. Statist.*, pp. 313–319. Charles University, Prague.

Klein, J.P. (1988). Small sample properties of censored data estimators of the cumulative hazard rate, survivor function, and estimators of their variance. Research Report 88/7, Statistical Research Unit, University of Copenhagen.

Klein, J.P. (1991). Small sample moments of some estimators of the variance of the Kaplan–Meier and Nelson-Aalen estimators. *Scand. J. Statist.* **18**, 333–340.

Klein, J.P. (1992). Semiparametric estimation of random effects using the Cox model based on the EM algorithm. *Biometrics* **48**, 795–806.

Kłopotowski, A. (1980). Mixtures of infinitely divisible distributions as limit laws for sums of dependent random variables. *Z. Wahrsch. verw. Geb.* **51**, 101–115.

Kofoed-Enevoldsen, A., Borch-Johnsen, K., Kreiner, S., Nerup, J., and Deckert, T. (1987). Declining incidence of persistent proteinuria in Type 1 (insulin-dependent) diabetic patients in Denmark. *Diabetes* **36**, 205–209.

Korwar, R. and Dahihya, R. (1982). Estimation of a bivariate distribution function from incomplete observations. *Commun. Statist. (A)* **11**, 887–897.

Koshevnik, Yu. A. and Levit, B. Ya. (1976). On a nonparametric analogue of the information matrix. *Theory Prob. Appl.* **21**, 738–753.

Koul, H., Susarla, V., and Van Ryzin, J. (1981). Regression analysis with randomly right censored data. *Ann. Statist.* **9**, 1276–1288.

Koziol, J.A. (1978). A two-sample Cramér–von Mises test for randomly censored data. *Biometrical. J.* **20**, 603–608.

Koziol, J.A. and Byar, D.P. (1975). Percentage points of the asymptotic distributions of one and two sample K-S statistics for truncated or censored data. *Technometrics* **17**, 507–510.

Koziol, J.A. and Green, S.B. (1976). A Cramér–von Mises statistic for randomly censored data. *Biometrika* **63**, 465–474.

Koziol, J.A. and Yuh, Y.S. (1982). Omnibus two-sample test procedures with randomly censored data. *Biometrical J.* **24**, 743–750.

Kumazawa, Y. (1987). A note on an estimator of life expectancy with random censorship. *Biometrika* **74**, 655–658.

Kunita, H. and Watanabe, S. (1967). On square integrable martingales. *Nagoya Math. J.* **30**, 209–245.

Kurtz, T.G. (1983). Gaussian approximations for Markov chains and counting processes. *Bull. Internat. Statist. Inst.* **50**(1), 361–375.

Lagakos, S.W. (1979). General right censoring and its impact on the analysis of survival data. *Biometrics* **35**, 139–156.

Lagakos, S.W. (1981). The graphical evaluation of explanatory variables in proportional hazards regression models. *Biometrika* **68**, 93–98.

Lagakos, S.W. (1988). The loss in efficiency from misspecifying covariates in proportional hazards regression models. *Biometrika* **75**, 156–160.

Lagakos, S.W., Barraj, L.M., and DeGruttola, V. (1988). Nonparametric analysis of truncated survival data, with application to AIDS. *Biometrika* **75**, 515–523.

Lagakos, S.W. and Schoenfeld, D. (1984). Properties of proportional hazards score tests under misspecified regression models. *Biometrics* **40**, 1037–1048.

Lagakos, S.W. and Williams, J.S. (1978). Models for censored survival analysis: A cone class of variable-sum models. *Biometrika* **65**, 181–189.

Lai, T.L. and Ying, Z. (1991a). Estimating a distribution function with truncated and censored data. *Ann. Statist.* **19**, 417–442.

Lai, T.L. and Ying, Z. (1991b). Large sample theory of a modified Buckley–James estimator for regression analysis with censored data. *Ann. Statist.* **19**, 1370–1402.

Lancaster, H.O. (1949). The derivation and partition of χ^2 in certain discrete distributions. *Biometrika* **36**, 117–129. Correction: **37**, 452 (1950).

Lancaster, H.O. (1965). The Helmert matrices. *Amer. Math. Monthly* **72**, 1–12.

Lancaster, T. and Nickell, S. (1980). The analysis of reemployment probabilities for the unemployed. *J. Roy. Statist. Soc. A* **143**, 141–165.

Langberg, N.A. and Shaked, M. (1982). On the identifiability of multivariate life distribution functions. *Ann. Probab.* **10**, 773–779.

Latta, R.B. (1981). A Monte Carlo study of some two-sample rank tests with censored data. *J. Amer. Statist. Assoc.* **76**, 713–719.

Lawless, J.F. (1982). *Statistical Models and Methods for Lifetime Data.* Wiley, New York.

LeCam, L. (1953). On some asymptotic properties of maximum likelihood estimators and related Bayes' estimates. *Univ. Calif. Publ. Statist.* **1**(11), 277–330.

LeCam, L. (1960). Locally asymptotically normal families of distributions. *Univ. Calif. Publ. Statist.* **3**, 738–753.

LeCam, L. (1969). Théorie Asymptotique de la Décision Statistique. University of Montréal.

LeCam, L. (1970). On the assumptions used to prove asymptotic normality of maximum likelihood estimators. *Ann. Math. Statist.* **41**, 802–828.

LeCam, L. (1972). Limits of experiments. *Proc. Sixth Berkeley Symp. Math. Statist. Probab.* **1**, 245–261. University of California Press, Berkeley, California.

LeCam, L. (1986). *Asymptotic Methods in Statistical Decision Theory.* Springer-Verlag, New York.

LeCam, L. and Yang, G.L. (1990). *Asymptotics in Statistics. Some Basic Concepts.* Springer-Verlag, New York.
Lee, E.T. (1980). *Statistical Models for Survival Data Analysis.* Lifetime Learning Publications, Belmont, California.
Lee, E.T. and Desu, M.M. (1972). A computer program for comparing k samples with right-censored data. *Computer Programs Biomed.* **2**, 315–321.
Lee, E.T., Desu, M.M., and Gehan, E.A. (1975). A Monte Carlo study of the power of some two-sample tests. *Biometrika* **62**, 425–432.
Lee, K.L., Harrell, F.E., Tolley, H.D., and Rosati, R.A. (1983). A comparison of test statistics for assessing the effects of concomitant variables in survival analysis. *Biometrics* **39**, 341–350.
Lenglart, E. (1977). Relation de domination entre deux processus. *Ann. Inst. Henri Poincaré* **13**, 171–179.
Leurgans, S. (1983). Three classes of censored data rank tests: Strengths and weaknesses under censoring. *Biometrika* **70**, 651–658.
Leurgans, S. (1984). Asymptotic behavior of two-sample rank tests in the presence of random censoring. *Ann. Statist.* **12**, 572–589.
Levit, B. Ya. (1973). On optimality of some statistical estimates. In: Hájek, J., editor, *Proc. Prague Symp. Asymptotic Statistics* **2**, 215–238. Charles University, Prague.
Levit, B. Ya. (1975). Efficiency of a class of nonparametric estimates. *Theor. Probab. Appl.* **20**, 738–754.
Levit, B. Ya. (1978). Infinite dimensional information inequalities. *Theory Prob. Appl.* **23**, 371–377.
Lévy, P. (1935). Propriétées asymptotiques des sommes de variables aléatoires enchaînées. *Bull. Sci. Math.* **59**, 84–96, 108–128.
Li, G. and Doss, H. (1991). A chi-square goodness-of-fit test, with applications to models with life history data. Tech. Report D-6, Department of Statistics, Florida State University.
Liang, K.-Y., Self, S.G. and Chang, Y.-C. (1990). Modelling marginal hazards in multivariate failure-time data. Technical report, The Johns Hopkins University and Fred Hutchinson Cancer Research Center.
Lin, D.Y. (1991). Goodness of fit for the Cox regression model based on a class of parameter estimators. *J. Amer. Statist. Assoc.* **86**, 725–728.
Lininger, L., Gail, M.H., Green, S.B., and Byar, D.P. (1979). Comparison of four tests for equality of survival curves in the presence of stratification and censoring. *Biometrika* **66**, 419–428.
Link, C.L. (1984). Confidence intervals for the survival function using Cox's proportional-hazard model with covariates. *Biometrics* **40**, 601–610.
Liptser, R.S. and Shiryayev, A.N. (1980). A functional central limit theorem for semimartingales. *Theory Probab. Appl.* **25**, 667–688.
Liptser, R.S. and Shiryayev, A.N. (1989). *Theory of Martingales.* Kluwer, Dordrecht.
Little, R.J.A. and Rubin, D.B. (1987). *Statistical Analysis with Missing Data.* Wiley, New York.
Littlewood, B. (1980). Theories of software reliability: How good are they and how can they be improved? *IEEE Trans. Software Eng.* **6**, 489–500.
Lo, S.H. and Singh, K. (1986). The product-limit estimator and the bootstrap: some asymptotic representations. *Probab. Theory Related Fields* **71**, 455–465.
Louis, T.A. (1981). Nonparametric analysis of an accelerated failure time model. *Biometrika* **68**, 381–390.
Lustbader, E.D. (1980). Time dependent covariates in survival analysis. *Biometrika* **67**, 697–698.
Lustbader, E.D. (1986). Relative risk regression diagnostics. In Moolgavkar, S.H. and Prentice, R.L., editors, *Modern Statistical Methods in Chronic Disease Epidemiology*, pp. 121–139. Wiley, New York.

Major, P. and Rejtő, L. (1988). Strong embedding of the estimator of the distribution function under random censorship. *Ann. Statist.* **16**, 1113–1132.

Majumdar, H. and Sen, P.K. (1978). Nonparametric testing for simple regression under progressive censoring with staggering entry and random withdrawal. *Commun. Statist.—Theor. Meth.* **A7**, 349–371.

Mantel, N. (1966). Evaluation of survival data and two new rank order statistics arising in its consideration. *Cancer Chemother. Rep.* **50**, 163–170.

Manton, K.G. and Stallard, E. (1988). *Chronic Disease Modelling*. Griffin, London.

Manton, K.G., Stallard, E., and Vaupel, J.W. (1981). Methods for comparing the mortality experience of heterogeneous populations. *Demography* **18**, 389–410.

Manton, K.G., Stallard, E., and Vaupel, J.W. (1986). Alternative models for the heterogeneity of mortality risks among the aged. *J. Amer. Statist. Assoc.* **81**, 635–644.

Mardia, K.V., Kent, J.T., and Bibby, J.M. (1979). *Multivariate analysis*. Academic Press, London.

Marron, J.S. and Padgett, W.J. (1987). Asymptotically optimal bandwidth selection for kernel density estimators for randomly censored data. *Ann. Statist.* **15**, 1520–1535.

Matthews, D.E. (1988). Likelihood-based confidence intervals for functions of many parameters. *Biometrika* **75**, 139–144.

Mau, J. (1985). Statistical modelling via partitioned counting processes. *J. Statist. Planning Inf.* **12**, 171–176.

Mau, J. (1986a). Nonparametric estimation of the integrated intensity of an unobservable transition in a Markov illness–death process. *Stoch. Proc. Appl.* **21**, 275–289.

Mau, J. (1986b). On a graphical method for the detection of time-dependent effects of covariates in survival data. *Appl. Statist.* **35**, 245–255.

Mau, J. (1986c). Sequential and repeated significance tests for counting processes. Research Report Series of the Statistics Project 4/87, SFB175 Implantology, University of Tübingen.

Mau, J. (1987). Monitoring of therapeutical studies via partitioned counting processes. Research Report Series of the Statistics Project 1/87, SFB175 Implantology, University of Tübingen.

Mauro, D. (1985). A combinatoric approach to the Kaplan–Meier estimator. *Ann. Statist.* **13**, 142–149.

McClean, S. (1986). Semi-Markov models for manpower planning. In Janssen, J., editor, *Semi-Markov Models. Theory and Applications*, pp. 283–300. Plenum Press, New York and London.

McCullagh, P. and Nelder, J.A. (1989). *Generalized Linear Models*, 2nd edition. Chapman and Hall, London.

McKeague, I.W. (1986). Estimation for a semimartingale regression model using the method of sieves. *Ann. Statist.* **14**, 579–589.

McKeague, I.W. (1988a). Asymptotic theory for weighted least squares estimators in Aalen's additive risk model. *Contemp. Math.* **80**, 139–152.

McKeague, I.W. (1988b). A counting process approach to the regression analysis of grouped survival data. *Stoch. Proc. Appl.* **28**, 221–239.

McKeague, I.W. and Utikal, K.J. (1990a). Inference for a nonlinear counting process regression model. *Ann. Statist.* **18**, 1172–1187.

McKeague, I.W. and Utikal, K.J. (1990b). Identifying nonlinear covariate effects in semimartingale regression models. *Probab. Theory Related Fields* **87**, 1–25.

McKeague, I.W. and Utikal, K.J. (1991). Goodness-of-fit tests for additive hazards and proportional hazards models. *Scand. J. Statist.* **18**, 177–195.

McKnight, B. (1985). Discussion of session on statistical tests for carcinogenic effects. In *Proceedings of the Symposium on Long-Term Animal Carcinogenicity Studies: A*

Statistical Perspective, pp. 107–111. American Statistical Association, Washington DC.
McKnight, B. and Crowley, J.J. (1984). Tests for differences in tumor incidence based on animal carcinogenesis experiments. *J. Amer. Statist. Assoc.* **79**, 639–648.
McLeish, D.L. (1974). Dependent central limit theorems and invariance principles. *Ann. Probab.* **2**, 620–628.
Medley, G.F., Billard, L., Cox, D.R., and Anderson, R.A (1988). The distribution of the incubation period for the acquired immunodeficiency syndrome (AIDS). *Proc. Roy. Soc. London B* **233**, 367–377.
Meier, P. (1975). Estimation of a distribution function from incomplete observations. In Gani, J., editor, *Perspectives in Probability and Statistics*, pp. 67–87. Applied Probability Trust, Sheffield.
Métivier, M. and Pellaumail, J. (1980). *Stochastic Integration.* Academic Press, New York.
Meyer, P.-A. (1962). A decomposition theorem for supermartingales. *Illinois J. Math.* **6**, 193–205.
Meyer, P.-A. (1966). *Probability and Potentials.* Blaisdell, Waltham, MA.
Meyer, P.-A. (1967). Intégrales stochastiques. In *Séminaire de Probabilités I.*, Lecture Notes in Mathematics 39, 72–94. Springer-Verlag, Berlin.
Meyer, P.-A. (1976). Un cours sur les intégrales stochastiques. In *Séminaire de Probabilités X.*, Lecture Notes in Mathematics 511, pp. 245–400. Springer-Verlag, Berlin.
Millar, P.W. (1979). Asymptotic minimax theorems for the sample distribution function. *Z. Wahrsch. verw. Geb.* **48**, 233–252.
Miller, R.G. (1976). Least squares regression with censored data. *Biometrika* **63**, 449–464.
Miller, R.G. (1983). What price Kaplan–Meier? *Biometrics* **39**, 1077–1082.
Miller, R.G. and Siegmund, D. (1982). Maximally selected chi-square statistics. *Biometrics* **38**, 1011–1016.
Moek, G. (1984). Comparison of some software reliability models for simulated and real failure data. *Internat. J. Model Simul.* **4**, 29–41.
Moreau, T., O'Quigley, J., and Lellouch, J. (1986). On D. Schoenfeld's approach for testing the proportional hazards assumption. *Biometrika* **73**, 513–515.
Moreau, T., O'Quigley, J., and Mesbah, M. (1985). A global goodness-of-fit statistic for the proportional hazards model. *Appl. Statist.* **34**, 212–218.
Morgan, T.M. (1986). Reader Reaction: Omitting covariates from the proportional hazards model. *Biometrics* **42**, 993–995.
Müller, H.-G. (1984). Smooth optimum kernel estimators of densities, regression curves and modes. *Ann. Statist.* **12**, 766–774.
Müller, H.-G. (1988). *Nonparametric Regression Analysis of Longitudinal Data*, Lecture Notes in Statistics 46. Springer-Verlag, New York.
Müller, H.-G. and Wang, J.-L. (1990). Locally adaptive hazard smoothing. *Probab. Theory Related Fields* **85**, 523–538.
Muñoz, A. (1980). Nonparametric estimation from censored bivariate observations. Technical Report 60, Division of Biostatistics, Stanford University.
Murphy, S.A. (1991). Consistency in a proportional hazards model incorporating a random effect. Preprint 98, Department of Statistics, Pennsylvania State University. Has appeared in *Ann. Statist.* **22**, 712–731, 1994.
Musa, J.D. (1975). A theory of software reliability and its applications. *Trans. Software Eng.* **1**, 312–327.
Musa, J.D., Iannino, A., and Okumoto, K. (1987). *Software Reliability: Measurement, Prediction, Application.* McGraw-Hill, New York.
Næs, T. (1982). The asymptotic distribution of the estimator for the regression parameter in Cox's regression model. *Scand. J. Statist.* **9**, 107–115.

Nagelkerke, N.J.D., Oosting, J., and Hart, A.A.M. (1984). A simple test for goodness of fit of Cox's proportional hazards model. *Biometrics* **40**, 483–486.

Nair, V.N. (1981). Plots and tests for goodness of fit with randomly censored data. *Biometrika* **68**, 99–103.

Nair, V.N. (1984). Confidence bands for survival functions with censored data: A comparative study. *Technometrics* **26**, 265–275.

Nelson, W. (1969). Hazard plotting for incomplete failure data. *J. Qual. Technol.* **1**, 27–52.

Nelson, W. (1972). Theory and applications of hazard plotting for censored failure data. *Technometrics* **14**, 945–965.

Nelson, W. (1982). *Applied Life Data Analysis*. Wiley, New York.

Neuhaus, G. (1988). Asymptotically optimal rank tests for the two-sample problem with randomly censored data. *Commun. Statist. Theory Meth.* **17**, 2037–2058.

Nielsen, G.G., Gill, R.D., Andersen, P.K., and Sørensen, T.I.A. (1992). A counting process approach to maximum likelihood estimation in frailty models. *Scand. J. Statist.* **19**, 25–43.

Norberg, R. (1990). Risk theory and its statistics environment. *Statistics* **21**, 273–299.

Norberg, T. (1989). Stochastic integration on lattices. Preprint 1989-08, Department of Mathematics, Chalmers University, Göteborg.

Oakes, D. (1977). The asymptotic information in censored survival data. *Biometrika* **59**, 472–474.

Oakes, D. (1981). Survival times: Aspects of partial likelihood (with discussion). *Internat. Statist. Rev.* **49**, 235–264.

Oakes, D. (1982a). A concordance test for independence in the presence of censoring. *Biometrics* **38**, 451–455.

Oakes, D. (1982b). A model for association in bivariate survival data. *J. Roy. Statist. Soc. B* **44**, 414–422.

Oakes, D. (1986a). A model for bivariate survival data. In Moolgavkar, S.H. and Prentice, R.L., editors, *Modern Statistical Methods in Chronic Disease Epidemiology*, pp. 151–166. Wiley, New York.

Oakes, D. (1986b). Semiparametric inference in a model for association in bivariate survival data. *Biometrika* **73**, 353–361.

Oakes, D. (1989). Bivariate survival models induced by frailties. *J. Amer. Statist. Assoc.* **84**, 487–493.

O'Brien, P.C. and Fleming, T.R. (1987). A paired Prentice–Wilcoxon test for censored paired data. *Biometrics* **43**, 169–180.

Öhman, M.L. (1990). A Monte Carlo study of some censored data Wilcoxon rank tests. *Biometrical J.* **32**, 721–735.

Olschewski, M. and Schumacher, M. (1986). Sequential analysis of survival times in clinical trials. *Biometrical J.* **28**, 273–293.

O'Quigley, J. and Moreau, T. (1984). Testing the proportional hazards regression model against some general alternatives. *Rev. Epidem. Santé Publ.* **34**, 199–205.

O'Quigley, J. and Moreau, T. (1986). Cox's regression model: computing a goodness of fit statistic. *Computer Methods Programs Biomed.* **22**, 253–256.

O'Sullivan, F. (1988). Fast computation of fully automated log-density and log-hazard estimators. *SIAM J. Sci. Statist. Comput.* **9**, 363–379.

O'Sullivan, F. (1989). Nonparametric estimation in the Cox proportional hazards model. Technical report, University of California, Berkeley.

Peace, K.E. and Flora, R.E. (1978). Size and power assessments of tests of hypotheses on survival parameters. *J. Amer. Statist. Assoc.* **73**, 129–132.

Péano, G. (1888). Intégration par séries des équations différentielles linéaires. *Math. Ann.* **32**, 450–456.

Pepe, M.S. and Fleming, T.R. (1989). Weighted Kaplan–Meier statistics: A class of distance tests for censored survival data. *Biometrics* **45**, 497–507.

Pepe, M.S. and Fleming, T.R. (1991). Weighted Kaplan–Meier statistics: Large sample and optimality considerations. *J. Roy. Statist. Soc. B* **53**, 341–352.
Pepe, M.S., Longton, G., and Thornquist, M. (1991). A qualifier Q for the survival function to describe the prevalence of a transient condition. *Statist. Med.* **10**, 413–421.
Peterson, A.V. (1977). Expressing the Kaplan–Meier estimator as a function of empirical subsurvival functions. *J. Amer. Statist. Assoc.* **72**, 854–858.
Peto, R. and Peto, J. (1972). Asymptotically efficient rank invariant test procedures (with discussion). *J. Roy. Statist. Soc. A* **135**, 185–206.
Peto, R. and Pike, M.C. (1973). Conservatism of the approximation $\sum(O - E)^2/E$ in the log rank test for survival data or tumor incidence data. *Biometrics* **29**, 579–584.
Peto, R., Pike, M.C., Armitage, P., Breslow, N.E., Cox, D.R., Howard, S.V., Mantel, N., McPherson, K., Peto, J., and Smith, P.G. (1977). Design and analysis of randomized clinical trials requiring prolonged observation of each patient, II. *Br. J. Cancer* **35**, 1–39.
Pettitt, A.N. (1982). Inference for the linear model using a likelihood based on ranks. *J. Roy. Statist. Soc. B* **44**, 234–243.
Pettitt, A.N. (1983). Approximate methods using ranks for regression with censored data. *Biometrika* **70**, 121–132.
Pettitt, A.N. (1984). Proportional odds model for survival data and estimates using ranks. *Appl. Statist.* **33**, 169–175.
Pettitt, A.N. and Stephens, M.A. (1976). Modified Cramér–von Mises statistics for censored data. *Biometrika* **63**, 291–298.
Pfanzagl, J. (with W. Wefelmeyer). (1982). *Contributions to a General Asymptotic Statistical Theory*, Lecture Notes in Statistics 13, Springer-Verlag, Berlin.
Phelan, M.J. (1990). Bayes estimation from a Markov renewal process. *Ann. Statist.* **18**, 603–616.
Pierce, D.A., Stewart, W.H., and Kopecky, K.J. (1979). Distribution-free regression analysis of grouped survival data. *Biometrics* **35**, 785–793.
Pocock, S.J. (1983). *Clinical Trials. A Practical Approach.* Wiley, Chichester.
Pocock, S.J., Gore, S.M., and Kerr, G.R. (1982). Long-term survival analysis: The curability of breast cancer. *Statist. Med.* **1**, 93–104.
Pollard, D. (1984). *Convergence of Stochastic Processes.* Springer-Verlag, New York.
Pollard, D. (1990). *Empirical Processes: Theory and Applications*, Regional Conference Series in Probability and Statistics 2. Institute of Mathematical Statistics, Hayward, CA.
Pons, O. (1981). Test sur la loi d'un processus ponctuel. *C.R. Acad. Sci. Paris A* **292**, 91–94.
Pons, O. (1986). A test of independence for two censored survival times. *Scand. J. Statist.* **13**, 173–185.
Pons, O. (1989). Nonparametric model and Cox model for bivariate survival data. Rapport technique de Biométrie 89-02, INRA, Laboratoire de Biométrie, Jouy-en-Josas, France.
Pons, O. and de Turckheim, E. (1987). Estimation in Cox's periodic model with a histogram-type estimator for the underlying intensity. *Scand. J. Statist.* **14**, 329–345.
Pons, O. and de Turckheim, E. (1988a). Cox's periodic regression model. *Ann. Statist.* **16**, 678–693.
Pons, O. and de Turckheim, E. (1988b). *Modèle de régression de Cox périodique et étude d'un comportement alimentaire*, Cahier de Biométrie **1**. INRA, Jouy-en-Josas, France.
Pons, O. and de Turckheim, E. (1991). Tests of independence for bivariate censored data based on the empirical joint hazard function. *Scand. J. Statist.* **18**, 21–37.

Præstgaard, J. (1991). Nonparametric estimators of actuarial values. *Scand. Actuar. J.*, 129–143.

Prentice, R.L. (1978). Linear rank tests with right censored data. *Biometrika* **65**, 167–179. Correction: **70**, 304 (1983).

Prentice, R.L. (1982). Covariate measurement errors and parameter estimation in a failure time regression model. *Biometrika* **69**, 331–342.

Prentice, R.L. (1986a). A case-cohort design for epidemiologic cohort studies and disease prevention trials. *Biometrika* **73**, 1–11.

Prentice, R.L. (1986b). On the design of synthetic case-control studies. *Biometrics* **42**, 301–310.

Prentice, R.L. and Breslow, N.E. (1978). Retrospective studies and failure time models. *Biometrika* **65**, 153–158.

Prentice, R.L. and Cai, J. (1991). Covariance and survivor function estimation using censored multivariate failure time data. *Biometrika* **79**, 495–512 (1992).

Prentice, R.L. and Cai, J. (1992). A covariance function for bivariate survival data and a bivariate survivor function. In Klein, J.P. and Goel, P.K., editors, *Survival Analysis: State of the Art*, pp. 393–406. Kluwer, Dordrecht.

Prentice, R.L. and Gloeckler, L.A. (1978). Regression analysis of grouped survival data with application to breast cancer data. *Biometrics* **34**, 57–67.

Prentice, R.L., Kalbfleisch, J.D., Peterson, A.V., Flournoy, N., Farewell, V.T., and Breslow, N.E. (1978). The analysis of failure time data in the presence of competing risks. *Biometrics* **34**, 541–554.

Prentice, R.L. and Marek, P. (1979). A qualitative discrepancy between censored data rank tests. *Biometrics* **35**, 861–867.

Prentice, R.L. and Self, S.G. (1983). Asymptotic distribution theory for Cox-type regressions models with general relative risk form. *Ann. Statist.* **11**, 804–813.

Prentice, R.L., Williams, B.J., and Peterson, A.V. (1981). On the regression analysis of multivariate failure time data. *Biometrika* **68**, 373–379.

Protter, P. (1990). *Stochastic Integration and Differential Equations (a new approach)*. Springer-Verlag, New York.

Pruitt, R.C. (1991a). On negative mass assigned by the bivariate Kaplan–Meier estimator. *Ann. Statist.* **19**, 443–453.

Pruitt, R.C. (1991b). Strong consistency of self-consistent estimators: general theory and an application to bivariate survival analysis. Technical Report No. 543, University of Minnesota.

Pruitt, R.C. (1991c). Small sample comparison of five bivariate survival curve estimators. Technical Report No. 559, University of Minnesota.

Pyke, R. and Shorack, G.R. (1968). Weak convergence of a two-sample empirical process and a new approach to the Chernoff–Savage theorems. *Ann. Math. Statist.* **39**, 755–771.

Rabelais, J. (1542). *Gargantua I, book 20.*

Ramlau-Hansen, H. (1981). Udglatning med kernefunktioner i forbindelse med tælleprocesser. Working paper 41, Laboratory of Actuarial Mathematics, University of Copenhagen.

Ramlau-Hansen, H. (1983a). Smoothing counting process intensities by means of kernel functions. *Ann. Statist.* **11**, 453–466.

Ramlau-Hansen, H. (1983b). The choice of a kernel function in the graduation of counting process intensities. *Scand. Actuar. J.* 165–182.

Ramlau-Hansen, H., Jespersen, N.C.B., Andersen, P.K., Borch-Johnsen, K., and Deckert, T. (1987). Life insurance for insulin-dependent diabetics. *Scand. Actuar. J.* 19–36.

Rao, C.R. (1973). *Linear Statistical Inference and Its Applications*, second edition. Wiley, New York.

Rasmussen, N.K. (1983). *Abort—et valg? Fødsler, fødselsbegrænsning og svangerskabsafbrydelse. Baggrund og årsager til udviklingen i aborttallet.* FaDL, Copenhagen.
Rebolledo, R. (1978). Sur les applications de la théorie des martingales à l'étude statistique d'une famille de processus ponctuels. In Springer Lecture Notes in Mathematics 636, pp. 27–70. Springer-Verlag, Berlin.
Rebolledo, R. (1979). La méthode des martingales appliquée a l'étude de la convergence en loi de processus. *Mem. Soc. Math. France* **62**.
Rebolledo, R. (1980a). Central limit theorems for local martingales. *Z. Wahrsch. verw. Geb.* **51**, 269–286.
Rebolledo, R. (1980b). The central limit theorem for semimartingales: necessary and sufficient conditions. Unpublished manuscript.
Reeds, J.A. (1976). On the definition of von Mises functionals. PhD thesis, Department of Statistics, Harvard University. Research Report S-44.
Reid, N. (1981). Influence functions for censored data. *Ann. Statist.* **9**, 78–92.
Reid, N. and Crépeau, H. (1985). Influence functions for proportional hazards regression. *Biometrika* **72**, 1–9.
Rice, J. and Rosenblatt, M. (1976). Estimation of the log survivor function and hazard function. *Sankhyā A* **38**, 60–78.
Ritov, Y. (1990). Estimation in a linear regression model with censored data. *Ann. Statist.* **18**, 303–328.
Ritov, Y. and Wellner, J.A. (1988). Censoring, martingales and the Cox model. *Contemp. Math.* **80**, 191–220.
Rubin, D.B. (1976). Inference and missing data. *Biometrika* **63**, 581–592.
Rudemo, M. (1982). Empirical choice of histogram and kernel density estimators. *Scand. J. Statist.* **9**, 65–78.
Rudin, W. (1976). *Principles of Mathematical Analysis*, third edition. McGraw-Hill, New York.
Samuelsen, S.O. (1985). Ikke-parametrisk estimering av fordelingsfunksjoner når datamaterialet er venstre- eller dobbelsensurert. Asymptotisk teori. Cand. real. thesis, Department of Mathematics, University of Oslo.
Samuelsen, S.O. (1989). Asymptotic theory for non-parametric estimators from doubly censored data. *Scand. J. Statist.* **16**, 1–21.
Sander, J.M. (1975). The weak convergence of quantiles of the product limit estimator. Technical Report 5, Department of Statistics, Stanford University.
SAS Institute Inc. (1985). *SAS User's Guide: Statistics, Version 5 Edition.* Cary, NC.
Sasieni, P. (1989). Beyond the Cox model: extensions of the model and alternative estimators. PhD thesis, Department of Statistics, University of Washington.
Schäfer, H. (1985). A note on data-adaptive kernel estimation of the hazard and density function in the random censorship situation. *Ann. Statist.* **13**, 818–820.
Schlichting, P., Christensen, E., Andersen, P.K., Fauerholdt, L., Juhl, E., Poulsen, H., and Tygstrup, N., for The Copenhagen Study Group for Liver Diseases. (1983). Prognostic factors in cirrhosis identified by Cox's regression model. *Hepatology* **3**, 889–895.
Schlichting, P., Christensen, E., Fauerholdt, L., Poulsen, H., Juhl, E., and Tygstrup, N., for The Copenhagen Study Group for Liver Diseases. (1982a). Prednisone and chronic liver disease. II. Clinical versus morphological criteria for selection of patients for prednisone treatment. *Liver* **2**, 113–118.
Schlichting, P., Fauerholdt, L., Christensen, E., Poulsen, H., Juhl, E., and Tygstrup, N., for The Copenhagen Study Group for Liver Diseases. (1982b). Clinical relevance of restrictive morphological criteria for the diagnosis of cirrhosis in liver biopsies. *Liver* **1**, 56–61.
Schlichting, P., Fauerholdt, L., Christensen, E., Poulsen, H., Juhl, E., and Tygstrup, N., for The Copenhagen Study Group for Liver Diseases. (1982c). Prednisone

treatment of chronic liver disease. I. Chronic aggressive hepatitis as a therapeutic marker. *Liver* **2**, 104–112.
Schoenfeld, D. (1980a). Chi-squared goodness-of-fit tests for the proportional hazards regression model. *Biometrika* **67**, 145–153.
Schoenfeld, D. (1980b). The asymptotic properties of rank tests for the censored two-sample problem. Preprint, Harvard School of Public Health.
Schoenfeld, D. (1981). The asymptotic properties of nonparametric tests for comparing survival distributions. *Biometrika* **68**, 316–319.
Schoenfeld, D. (1982). Partial residuals for the proportional hazards regression model. *Biometrika* **69**, 239–241.
Scholz, F.W. (1980). Towards a unified definition of maximum likelihod. *Canad. J. Statist.* **8**, 193–203.
Schou, G. and Væth, M. (1980). A small sample study of occurrence/exposure rates for rare events. *Scand. Actuar. J.* 209–225.
Schumacher, M. (1984). Two-sample tests of Cramér–von Mises and Kolmogorov–Smirnov type for randomly censored data. *Internat. Statist. Rev.* **52**, 263–281.
Self, S.G. and Prentice, R.L. (1982). Commentary on Andersen and Gill's Cox's regression model for counting processes: A large sample study. *Ann. Statist.* **10**, 1121–1124.
Self, S.G. and Prentice, R.L. (1986). Incorporating random effects into multivariate relative risk regression models. In Moolgavkar, S.H. and Prentice, R.L., editors, *Modern Statistical Methods in Chronic Disease Epidemiology*, pp. 167–177. Wiley, New York.
Self, S.G. and Prentice, R.L. (1988). Asymptotic distribution theory and efficiency results for case-cohort studies. *Ann. Statist.* **16**, 64–81.
Sellke, T. (1988). Weak convergence of the Aalen estimator for a censored renewal process. In Gupta, S.S. and Berger, J.O., editors, *Statistical Decision Theory and Related Topics IV*, Vol. 2, pp. 183–194. Springer-Verlag, New York.
Sellke, T. and Siegmund, D. (1983). Sequential analysis of the proportional hazards model. *Biometrika* **70**, 315–326.
Sen, P.K. (1976). A two-dimensional functional permutational central limit theorem for linear rank statistics. *Ann. Probab.* **4**, 13–26.
Sen, P.K. (1981). *Sequential Nonparametrics.* Wiley, New York
Sen, P.K. (1985). *Theory and Applications of Sequential Nonparametrics*, Regional Conference Series in Applied Mathematics 49, SIAM, Philadelphia.
Senthilselvan, A. (1987). Penalized likelihood estimation of hazard and intensity functions. *J. Roy. Statist. Soc. B* **49**, 170–174.
Serfling, R.J. (1980). *Approximation Theorems of Mathematical Statistics.* Wiley, New York.
Sheehy, A. and Wellner, J.A. (1990a). Uniform Donsker classes of functions. Report 189, Department of Statistics, University of Washington. Has appeared in *Ann. Probab.* **20**, 1983–2030, 1992.
Sheehy, A. and Wellner, J.A. (1990b). The delta method and the bootstrap for nonlinear functions of empirical distribution functions via Hadamard derivatives. Report, Department of Statistics, University of Washington.
Shiryayev, A.N. (1981). Martingales: Recent developments, results and applications. *Internat. Statist. Rev.* **49**, 199–233.
Shorack, G.R. and Wellner, J.A. (1986). *Empirical Processes.* Wiley, New York. Corrections and changes: Technical Report 167, Department of Statistics, University of Washington (1989).
Siegmund, D. (1985). *Sequential Analysis. Tests and Confidence Intervals.* Springer-Verlag, New York.
Silverman, B. (1986). *Density Estimation for Statistics and Data Analysis.* Chapman and Hall, London.

Sinha, A.N. and Sen, P.K. (1982). Tests based on empirical processes for progressive censoring schemes with staggering entry and random withdrawal. *Sankhyā B* **44**, 1–18.

Slud, E.V. (1984). Sequential linear rank tests for two-sample censored survival data. *Ann. Statist.* **12**, 551–571.

Slud, E.V. (1986). Inefficiency of inferences with the partial likelihood. *Commun. Statist. Theory Meth.* **15**, 3333–3351.

Slud, E.V. (1991). Partial likelihood for continuous-time stochastic processes. Preprint, Mathematics Department, University of Maryland.

Slud, E.V., Byar, D.P., and Green, S.B. (1984). A comparison of reflected versus test-based confidence intervals for the median survival time based on censored data. *Biometrics* **40**, 587–600.

Slud, E.V. and Wei, L.J. (1982). Two-sample repeated significance tests based on the modified Wilcoxon statistic. *J. Amer. Statist. Assoc.* **77**, 862–868.

Snyder, D.L. (1972). Filtering and detection for doubly stochastic Poisson processes. *IEEE Trans. Inf. Theory* **18**, 97–102.

Snyder, D.L. (1975). *Random Point Processes.* Wiley, New York.

Solomon, P.J. (1984). Effect of misspecification of regression models in the analysis of survival data. *Biometrika* **71**, 291–298. Amendment: **73**, 245 (1986).

Sørensen, T.I.A., Nielsen, G.G., Andersen, P.K., and Teasdale, T.W. (1988). Genetic and environmental influences on premature death in adult adoptees. *New Engl. J. Med.* **318**, 727–732.

SPSS (1981). *Update 7–9. New Procedures and Facilities for Releases 7–9.* Hull, C.H. and Nie, N.H. editors. McGraw-Hill, New York.

Stein, C. (1956). Efficient nonparametric testing and estimation. *Proc. Third Berkeley Symp. Math. Stat. Probab.* **1**, 187–195. University of California Press, Berkeley, California.

Storer, B.E. and Crowley, J.J. (1985). A diagnostic for Cox regression and general conditional likelihoods. *J. Amer. Statist. Assoc.* **80**, 139–147.

Strasser, H. (1985). *Mathematical Theory of Statistics.* de Gruyter, Berlin.

Streitberg, B. (1990). Lancaster interactions revisited. *Ann. Statist.* **18**, 1878–1885.

Struthers, C.A. (1984). Asymptotic properties of linear rank tests with censored data. PhD thesis, Statistics Department, University of Waterloo.

Struthers, C.A. and Kalbfleisch, J.D. (1986). Misspecified proportional hazard models. *Biometrika* **73**, 363–369.

Stute, W. and Wang, J.-L. (1993). The strong law under random censorship. *Ann. Statist.* **21**, 1591–1607.

Susarla, V. and Van Ryzin, J. (1976). Nonparametric Bayesian estimation of survival curves from incomplete observations. *J. Amer. Statist. Assoc.* **61**, 897–902.

Susarla, V. and Van Ryzin, J. (1980). Large sample theory for an estimator of the mean survival time from censored samples. *Ann. Statist.* **8**, 1002–1016.

Svensson, Å. (1990). Asymptotic estimation in counting processes with parametric intensities based on one realization. *Scand. J. Statist.* **17**, 23–33.

Sverdrup, E. (1965). Estimates and test procedures in connection with stochastic models for deaths, recoveries and transfers between different states of health. *Skand. Aktuarietidskr.* **48**, 184–211.

Tanner, M.A. (1983). A note on the variable kernel estimator of the hazard function for randomly censored data. *Ann. Statist.* **11**, 994–998.

Tanner, M.A. and Wong, W.H. (1983). The estimation of the hazard function from randomly censored data by the kernel method. *Ann. Statist.* **11**, 989–993.

Tanner, M.A. and Wong, W.H. (1984). Data-based nonparametric estimation of the hazard function with applications to model diagnostics and exploratory analysis. *J. Amer. Statist. Assoc.* **79**, 174–182.

Tarone, R.E. (1975). Tests for trend in life table analysis. *Biometrika* **62**, 679–682.

Tarone, R.E. and Ware, J.H. (1977). On distribution-free tests for equality for survival distributions. *Biometrika* **64**, 156–160.

Temkin, N.R. (1978). An analysis for transient states with application to tumor shrinkage. *Biometrics* **34**, 571–580.

Therneau, T.M., Grambsch, P.M., and Fleming, T.R. (1990). Martingale-based residuals for survival models. *Biometrika* **77**, 147–160.

Thomas, D.C. (1977). Addendum to: Methods of cohort analysis: Appraisal by application to asbestos mining. By F.D.K. Liddell, J.C. McDonald and D.C. Thomas. *J. Roy. Statist. Soc. A* **140**, 469–491.

Thomas, D.C. (1983). Non-parametric estimation and tests of fit for dose-response relations. *Biometrics* **39**, 263–268.

Thomas, D.R. and Grunkemeier, G.L. (1975). Confidence interval estimation of survival probabilities for censored data. *J. Amer. Statist. Assoc.* **70**, 865–871.

Thomsen, B.L. (1988). A note on the modelling of continuous covariates in Cox's regression model. Research Report 88/5, Statistical Research Unit, University of Copenhagen.

Tsai, W.-Y. (1990). Testing the assumption of independence of truncation time and failure time. *Biometrika* **77**, 169–177.

Tsai, W.-Y., Leurgans, S., and Crowley, J.J. (1986). Nonparametric estimation of a bivariate survival function in presence of censoring. *Ann. Statist.* **14**, 1351–1365.

Tsiatis, A.A. (1975). A nonidentifiability aspect of the problem of competing risks. *Proc. Nat. Acad. Sci. USA* **72**, 20–22.

Tsiatis, A.A. (1981). A large sample study of Cox's regression model. *Ann. Statist.* **9**, 93–108.

Tsiatis, A.A. (1982). Repeated significance testing for a general class of statistics used in censored surivival analysis. *J. Amer. Statist. Assoc.* **77**, 855–861.

Tsiatis, A.A. (1990). Estimating regression parameters using linear rank tests for censored data. *Ann. Statist.* **18**, 354–372.

Tuma, N.B. and Hannan, M.T. (1984). *Social Dynamics: Models and Methods*. Academic Press, New York.

Tuma, N.B., Hannan, M.T., and Groeneveld, L.P. (1979). Dynamic analysis of event histories. *Amer. J. Sociol.* **84**, 820–854.

Turnbull, B.W. (1974). Nonparametric estimation of a survivorship function with doubly censored data. *J. Amer. Statist. Assoc.* **69**, 169–173.

Turnbull, B.W. (1976). The empirical distribution function with arbitrarily grouped, censored and truncated data. *J. Roy. Statist. Soc. B* **38**, 290–295.

van der Laan, M.J. (1990). Dabrowska's multivariate product limit estimator and the delta-method. Master's thesis, Department of Mathematics, University of Utrecht.

van der Laan, M.J. (1991). Analysis of Pruitt's estimator of the bivariate survival function. Preprint 648, Department of Mathematics, University Utrecht.

van der Laan, M.J. (1992). Efficient estimator of the bivariate survival function for right censored data. Technical report 377, Department of Statistics, University of California, Berkeley.

van der Vaart, A.W. (1987). Maximum likelihod estimation in general parameter spaces. Unpublished manuscript.

van der Vaart, A.W. (1988). Statistical estimation for large parameter spaces. CWI tracts 44, Centre for Mathematics and Computer Science, Amsterdam.

van der Vaart, A.W. (1991). Efficiency and Hadamard differentiability. *Scand. J. Statist.* **18**, 63–75.

van der Vaart, A.W. and Wellner, J.A. (1990). Prohorov and continuous mapping theorems in the Hoffmann–Jørgensen weak convergence theory, with applications to convolution and asymptotic minimax theorems. Technical Report 157, University of Washington.

van Lambalgen, M. (1987a). Random sequences. PhD thesis, University of Amsterdam.

van Lambalgen, M. (1987b). Von Mises definition of random sequences reconsidered. *J. Symbol. Logic* **55**, 1143–1167.

van Pul, M.C.J. (1992a). Asymptotic properties of a class of statistical models in software reliability. *Scand. J. Statist.* **19**, 1–23.

van Pul, M.C.J. (1992b). Simulations on the Jelinski–Moranda model of software reliability; application of some parametric bootstrap methods. *Statist. Comp.* **2**, 121–136.

van Schuppen, J. and Wong, E. (1974). Translation of local martingales under a change of law. *Ann. Probab.* **2**, 879–888.

Vaupel, J.W. (1988). Inherited frailty and longevity. *Demography* **25**, 227–287.

Vaupel, J.W. (1990a). Kindred lifetimes: frailty models in population genetics. In Adams, J., Hermalin, A., Lam, D., and Smouse, P., editors, *Convergent Issues in Genetics and Demography*, Chap. 10, pp. 157–173. Oxford University Press, Oxford.

Vaupel, J.W. (1990b). Relatives' risks: frailty models of life history data. *Theoret. Pop. Biol.* **37**, 220–234.

Vaupel, J.W., Manton, K.G., and Stallard, E. (1979). The impact of heterogeneity in individual frailty on the dynamics of mortality. *Demography* **16**, 439–454.

Vaupel, J.W. and Yashin, A. (1985a). The deviant dynamics of death in heterogeneous populations. In Tuma, N.B., editor, *Sociological Methodology*, pp. 179–211. Jossey-Bass, London.

Vaupel, J.W. and Yashin, A. (1985b). Heterogeneity's ruses: Some surprising effects of selection on population dynamics. *Amer. Statist.* **39**, 176–185.

Ville, J. (1939). *Étude Critique de la Notion de Collectif.* Gauthiers-Villars, Paris.

Voelkel, J.G. and Crowley, J.J. (1984). Nonparametric inference for a class of semi-Markov processes with censored observations. *Ann. Statist.* **12**, 142–160.

Volterra, V. (1887). Sulle equazioni differenziali lineari. *Rend.Accad. Lincei* (*Series 4*) **3**, 393–396.

von Mises, R. (1947). On the asymptotic distribution of differentiable statistical functions. *Ann. Math. Statist.* **18**, 309–348.

von Weizsäcker, H. and Winkler, G. (1990). *Stochastic Integrals. An Introduction.* Vieweg, Braunschweig.

Wagner, S.S. and Altmann, S.A. (1973). What time do the baboons come down from the trees? (An estimation problem). *Biometrics* **29**, 623–635.

Walburg, H.E. and Cosgrove, G.E. (1969). Reticular neoplasms in irradiated and unirradiated germfree mice. In Mirand, E.A. and Back, N., editors, *Germfree Biology, Experimental and Clinical Aspects*, Advances in Experimental Medicine and Biology 3, pp. 135–141. Plenum Press, New York.

Wald, A. (1943). Tests of statistical hypotheses concerning several parameters when the number of observations is large. *Trans. Amer. Math. Soc.* **54**, 426–482.

Wang, J.G. (1987). A note on the uniform consistency of the Kaplan–Meier estimator. *Ann. Statist.* **15**, 1313–1316.

Wang, M.-C. (1991). Nonparametric estimation from cross-sectional survival data. *J. Amer. Statist. Assoc.* **86**, 130–143.

Wang, M.-C., Jewell, N.P., and Tsai, W.-Y. (1986). Asymptotic properties of the product limit estimate under random truncation. *Ann. Statist.* **14**, 1597–1605.

Ware, J.H. and DeMets, D.L. (1976). Reanalysis of some baboon descent data. *Biometrics* **32**, 459–463.

Watson, G.S. and Leadbetter, M.R. (1964a). Hazard analysis I. *Biometrika* **51**, 175–184.

Watson, G.S. and Leadbetter, M.R. (1964b). Hazard analysis II. *Sankhyā Ser. A* **26**, 101–116.

Watt-Boolsen, S., Ottesen, G., Andersen, J.A., Bayer, T., Jespersen, N.C.B., Keiding, N., Mouridsen, H.T., Dombernowsky, P., and Blichert-Toft, M. (1989). Significance of incisional biopsy in breast carcinoma: results from a clinical trial with intended excisional biopsy. *Eur. J. Surg. Oncol.* **15**, 33–37.

Wefelmeyer, W. (1991). Efficient estimation in multiplicative counting process models. *Statist. Decisions* **9**, 301–317.

Wei, L.J. (1984). Testing goodness-of-fit for the proportional hazards model with censored observations. *J. Amer. Statist. Assoc.* **79**, 649–652.

Wei, L.J. and Gail, M.H. (1983). Nonparametric estimation for a scale-change with censored observations. *J. Amer. Statist. Assoc.* **78**, 382–388.

Wei, L.J., Lin, D.Y., and Weissfeld, L. (1989). Regression analysis of multivariate incomplete failure time data by modeling marginal distributions. *J. Amer. Statist. Assoc.* **84**, 1065–1073.

Wei, L.J., Ying, Z., and Lin, D.Y. (1990). Linear regression analysis of censored survival data based on rank tests. *Biometrika* **77**, 845–852.

Weier, D.R. and Basu, A.P. (1980). An investigation of Kendall's τ modified for censored data with applications. *J. Statist. Plan. Inf.* **4**, 381–390.

Weits, E. (1991). Integrability of statistical models and the second order optimality of a smoothed Kaplan–Meier estimator. Preprint 673, Department of Mathematics, University of Utrecht. Has appeared in *Scand. J. Statist.* **20**, 111–132, 1993.

Wellek, S. (1990). A nonparametric model for product-limit estimation under right censoring and left truncation. *Commun. Statist.—Stoch. Models* **6**, 561–592.

Wellner, J.A. (1982). Asymptotic optimality of the product limit estimator. *Ann. Statist.* **10**, 595–602.

Wellner, J.A. (1985). A heavy censoring limit theorem for the product limit estimator. *Ann. Statist.* **13**, 150–162.

Whitehead, J. (1992). *The design and analysis of sequential clinical trials*, second edition. E. Horwood Ltd., Chichester.

Wiener, N. (1923). Differential-space. *J. Math. Phys.* **2**, 131–174.

Wijers, B.J. (1991). Consistent non-parametric estimation for a one-dimensional line segment process observed in an interval. Preprint 683, Department of Mathematics, University of Utrecht.

Williams, J.A. and Lagakos, S.W. (1977). Models for censored survival analysis: Constant sum and variable sum models. *Biometrika* **64**, 215–224.

Winter, B.B., Földes, A., and Rejtő, L. (1978). Glivenko–Cantelli theorems for the product limit estimate. *Problems Contr. Inf. Theory* **7**, 213–225.

Wong, W.H. (1986). Theory of partial likelihood. *Ann. Statist.* **14**, 88–123.

Woodroofe, M. (1985). Estimating a distribution function with truncated data. *Ann. Statist.* **13**, 163–177. Correction: **15**, 883 (1987).

Woolson, R.F. (1981). Rank tests and a one-sample logrank test for comparing observed survival data to a standard population. *Biometrics* **37**, 687–696.

Yandell, B.S. (1983). Nonparametric inference for rates with censored survival data. *Ann. Statist.* **11**, 1119–1135.

Yang, G. (1977). Life expectancy under random censorship. *Stoch. Proc. Appl.* **6**, 33–39.

Ying, Z. (1989). A note on the asymptotic properties of the product-limit estimator on the whole line. *Statist. Probab. Lett.* **7**, 311–314.

Zucker, D.M. and Karr, A.F. (1990). Nonparametric survival analysis with time-dependent covariate effects: A penalized partial likelihood approach. *Ann. Statist.* **18**, 329–353.

Author Index

Subject Index

Springer Series in Statistics

(continued from p. ii)

Pollard: Convergence of Stochastic Processes.
Pratt/Gibbons: Concepts of Nonparametric Theory.
Read/Cressie: Goodness-of-Fit Statistics for Discrete Multivariate Data.
Reinsel: Elements of Multivariate Time Series Analysis.
Reiss: A Course on Point Processes.
Reiss: Approximate Distributions of Order Statistics: With Applications to Non-parametric Statistics.
Rieder: Robust Asymptotic Statistics.
Rosenbaum: Observational Studies.
Ross: Nonlinear Estimation.
Sachs: Applied Statistics: A Handbook of Techniques, 2nd edition.
Särndal/Swensson/Wretman: Model Assisted Survey Sampling.
Schervish: Theory of Statistics.
Seneta: Non-Negative Matrices and Markov Chains, 2nd edition.
Shao/Tu: The Jackknife and Bootstrap.
Siegmund: Sequential Analysis: Tests and Confidence Intervals.
Simonoff: Smoothing Methods in Statistics.
Small: The Statistical Theory of Shape.
Tanner: Tools for Statistical Inference: Methods for the Exploration of Posterior Distributions and Likelihood Functions, 3rd edition.
Tong: The Multivariate Normal Distribution.
van der Vaart/Wellner: Weak Convergence and Empirical Processes: With Applications to Statistics.
Vapnik: Estimation of Dependences Based on Empirical Data.
Weerahandi: Exact Statistical Methods for Data Analysis.
West/Harrison: Bayesian Forecasting and Dynamic Models.
Wolter: Introduction to Variance Estimation.
Yaglom: Correlation Theory of Stationary and Related Random Functions I: Basic Results.
Yaglom: Correlation Theory of Stationary and Related Random Functions II: Supplementary Notes and References.